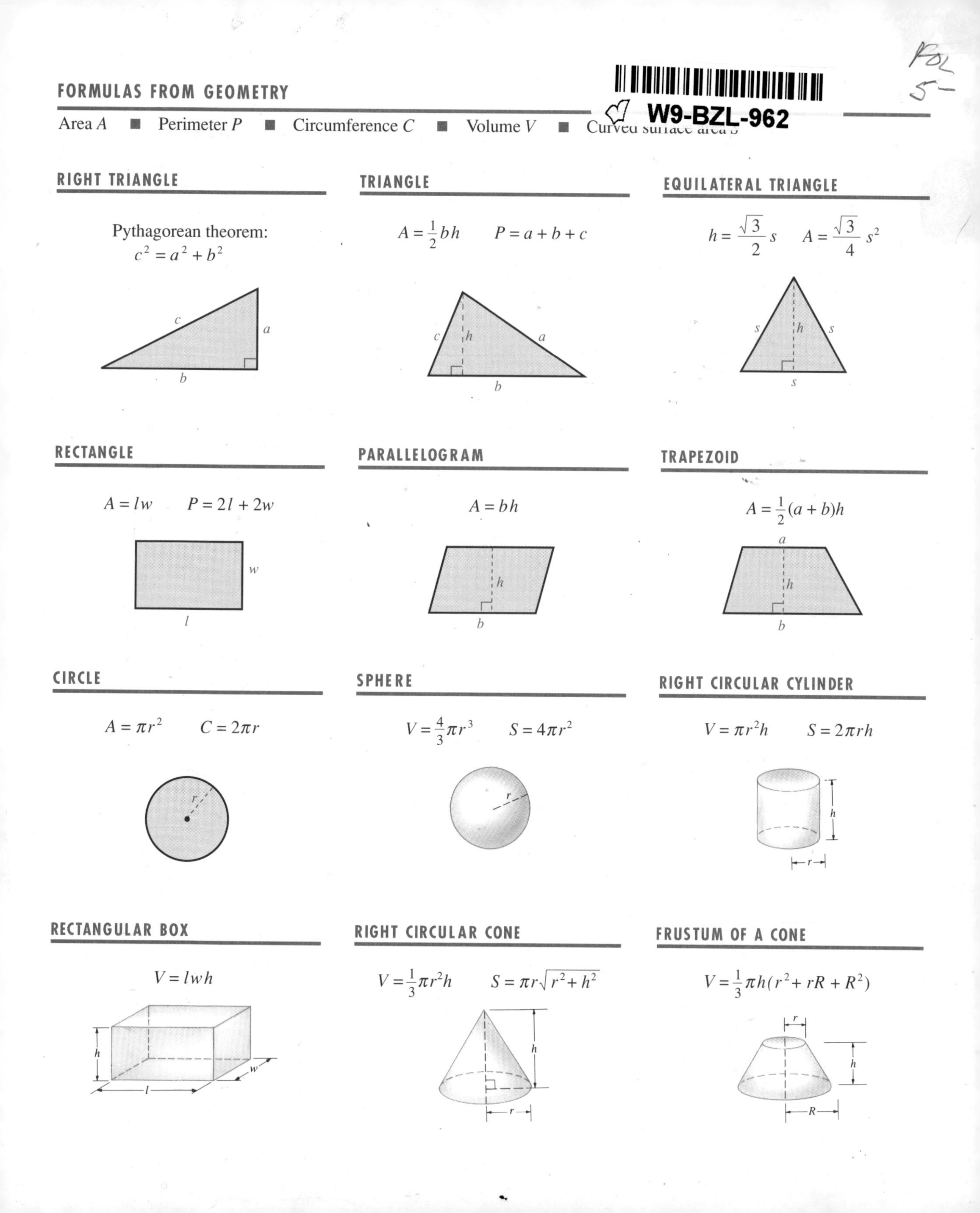

FORMULAS FROM GEOMETRY

Area A ■ Perimeter P ■ Circumference C ■ Volume V ■ Curved surface area S

RIGHT TRIANGLE

Pythagorean theorem:
$$c^2 = a^2 + b^2$$

TRIANGLE

$$A = \frac{1}{2}bh \qquad P = a + b + c$$

EQUILATERAL TRIANGLE

$$h = \frac{\sqrt{3}}{2}s \qquad A = \frac{\sqrt{3}}{4}s^2$$

RECTANGLE

$$A = lw \qquad P = 2l + 2w$$

PARALLELOGRAM

$$A = bh$$

TRAPEZOID

$$A = \frac{1}{2}(a + b)h$$

CIRCLE

$$A = \pi r^2 \qquad C = 2\pi r$$

SPHERE

$$V = \frac{4}{3}\pi r^3 \qquad S = 4\pi r^2$$

RIGHT CIRCULAR CYLINDER

$$V = \pi r^2 h \qquad S = 2\pi r h$$

RECTANGULAR BOX

$$V = lwh$$

RIGHT CIRCULAR CONE

$$V = \frac{1}{3}\pi r^2 h \qquad S = \pi r\sqrt{r^2 + h^2}$$

FRUSTUM OF A CONE

$$V = \frac{1}{3}\pi h(r^2 + rR + R^2)$$

ALGEBRA AND TRIGONOMETRY WITH ANALYTIC GEOMETRY

THE PRINDLE, WEBER, AND SCHMIDT SERIES IN MATHEMATICS

Althoen and Bumcrot, *Introduction to Discrete Mathematics*
Brown and Sherbert, *Introductory Linear Algebra with Applications*
Buchthal and Cameron, *Modern Abstract Algebra*
Burden and Faires, *Numerical Analysis*, Fourth Edition
Cullen, *Linear Algebra and Differential Equations*
Cullen, *Mathematics for the Biosciences*
Eves, *In Mathematical Circles*
Eves, *Mathematical Circles Adieu*
Eves, *Mathematical Circles Revisited*
Eves, *Mathematical Circles Squared*
Eves, *Return to Mathematical Circles*
Fletcher and Patty, *Foundations of Higher Mathematics*
Geltner and Peterson, *Geometry for College Students*
Gilbert and Gilbert, *Elements of Modern Algebra*, Second Edition
Gobran, *Beginning Algebra*, Fourth Edition
Gobran, *College Algebra*
Gobran, *Intermediate Algebra*, Fourth Edition
Gordon, *Calculus and the Computer*
Hall, *Algebra for College Students*
Hall and Bennett, *College Algebra with Applications*, Second Edition
Hartfiel and Hobbs, *Elementary Linear Algebra*
Hunkins and Mugridge, *Applied Finite Mathematics*, Second Edition
Kaufmann, *Algebra for College Students*, Third Edition
Kaufmann, *Algebra with Trigonometry for College Students*, Second Edition
Kaufmann, *College Algebra*
Kaufmann, *College Algebra and Trigonometry*
Kaufmann, *Elementary Algebra for College Students*, Third Edition
Kaufmann, *Intermediate Algebra for College Students*, Third Edition
Kaufmann, *Precalculus*
Kaufmann, *Trigonometry*
Keisler, *Elementary Calculus: An Infinitesimal Approach*, Second Edition
Kirkwood, *Introduction to Real Analysis*
Laufer, *Discrete Mathematics and Applied Modern Algebra*
Nicholson, *Linear Algebra with Applications*
Pasahow, *Mathematics for Electronics*

Powers, *Elementary Differential Equations*

Powers, *Elementary Differential Equations with Boundary-Value Problems*

Powers, *Elementary Differential Equations with Linear Algebra*

Proga, *Arithmetic and Algebra*, Second Edition

Proga, *Basic Mathematics*, Second Edition

Radford, Vavra, and Rychlicki, *Introduction to Technical Mathematics*

Radford, Vavra, and Rychlicki, *Technical Mathematics with Calculus*

Rice and Strange, *Calculus and Analytic Geometry for Engineering Technology*

Rice and Strange, *Plane Trigonometry*, Fifth Edition

Rice and Strange, *Technical Mathematics*

Rice and Strange, *Technical Mathematics and Calculus*

Schelin and Bange, *Mathematical Analysis for Business and Economics*, Second Edition

Strnad, *Introductory Algebra*

Swokowski, *Algebra and Trigonometry with Analytic Geometry*, Seventh Edition

Swokowski, *Calculus with Analytic Geometry*, Second Alternate Edition

Swokowski, *Calculus with Analytic Geometry*, Fourth Edition

Swokowski, *Fundamentals of Algebra and Trigonometry*, Seventh Edition

Swokowski, *Fundamentals of College Algebra*, Seventh Edition

Swokowski, *Fundamentals of Trigonometry*, Seventh Edition

Swokowski, *Precalculus: Functions and Graphs*, Fifth Edition

Tan, *Applied Calculus*

Tan, *Applied Finite Mathematics*, Second Edition

Tan, *Calculus for the Managerial, Life, and Social Sciences*

Tan, *College Mathematics*, Second Edition

Venit and Bishop, *Elementary Linear Algebra*, Third Edition

Venit and Bishop, *Elementary Linear Algebra*, Alternate Second Edition

Willard, *Calculus and Its Applications*, Second Edition

Wood and Capell, *Arithmetic*

Wood, Capell, and Hall, *Developmental Mathematics*, Third Edition

Wood, Capell, and Hall, *Intermediate Algebra*

Wood, Capell, and Hall, *Introductory Algebra*

Zill, *A First Course in Differential Equations with Applications*, Fourth Edition

Zill, *Calculus with Analytic Geometry*, Second Edition

Zill, *Differential Equations with Boundary-Value Problems*, Second Edition

SEVENTH EDITION

ALGEBRA AND TRIGONOMETRY WITH ANALYTIC GEOMETRY

EARL W. SWOKOWSKI
Marquette University

PWS • KENT PUBLISHING COMPANY • Boston

PWS-KENT
Publishing Company

20 Park Plaza
Boston, Massachusetts 02116

PWS-KENT Publishing Company is a division of Wadsworth, Inc.

93 92 91 90 89 / 10 9 8 7 6 5 4 3 2 1

LIBRARY OF CONGRESS CATALOGING-IN-PUBLICATION DATA

Swokowski, Earl W.
Algebra and trigonometry with analytic geometry / Earl W. Swokowski. —7th ed.
p. cm.
Includes index.
ISBN 0-534-91712-7
1. Algebra. 2. Trigonometry. 3. Geometry, Analytic. I. Title.
QA152.2.S965 1989 88-31244
512'.1—dc 19 CIP

SPONSORING EDITOR ■ David Geggis
PRODUCTION COORDINATOR ■ Elise Kaiser
MANUFACTURING COORDINATOR ■ Marcia A. Locke
PRODUCTION AND DESIGN ■ Kathi Townes
TEXT COMPOSITION ■ Syntax International Pte. Ltd.
TECHNICAL ARTWORK ■ Hayden Graphics, J & R Services
TEXT MANUFACTURING ■ Arcata Graphics / Halliday Lithograph
COVER MANUFACTURING ■ New England Book Components, Inc.
COVER IMAGE ■ *Hommage à Apollinaire* (1982–1984) by Jan Dibbets, Collection Exxon Corporation

Printed in the United States of America.

PREFACE

In this seventh edition of *Algebra and Trigonometry with Analytic Geometry*, one of my main goals was to make discussions of concepts easier to understand without sacrificing the mathematical soundness of the previous edition. Solutions to many examples have been rewritten so that important steps stand out clearly. More applied problems have been added to stress the usefulness of the subject matter. Exercise sets have also been improved and expanded. The comments that follow highlight some of the changes and features of this edition.

CHANGES FOR THE SEVENTH EDITION

- The binomial theorem is discussed at the end of Chapter 1 so that instructors who wish to postpone this topic may do so without disrupting the review of basic algebraic techniques.
- Lines are considered early in Chapter 3 to provide more experience in sketching graphs prior to the function concept.
- The graph of a function is introduced in the same section as the definition of function and is applied to graphical interpretations of domain and range.
- A definition of inverse function is stated without using the formulas $f^{-1}(f(x)) = x$ and $f(f^{-1}(y)) = y$.
- In Section 4.1 the maximum or minimum value of a quadratic function is obtained by employing a formula for the vertex of a parabola.
- In Chapter 5 the definition of the number e is stated using the notation $(1 + \frac{1}{n})^n \to e$ as $n \to \infty$.
- Section 5.3 on logarithms has been revised extensively and the change of base formula is given more emphasis.
- The discussion of radian measure in Section 6.1 has been simplified, and greater emphasis is placed on the length of arc formula $s = r\theta$.
- The fundamental identities are introduced in Section 6.2 after the definition of the trigonometric functions; however, their use is limited to simple illustrations.
- In Section 6.6, amplitudes, periods, and phase shifts are emphasized for sketching graphs that involve the sine and cosine functions.
- Section 6.7 extends the discussion of trigonometric graphs to the tangent, cotangent, secant, and cosecant functions.
- At the beginning of Section 7.2, methods of obtaining solutions of three simple trigonometric equations are considered before investigating more complicated equations.
- The graphical significance of cofunctions of complementary angles is illustrated by means of a right triangle in Section 7.3.
- Section 7.6 on inverse trigonometric functions contains more examples and a brief discussion of the inverse secant function.
- The *cis* notation for the trigonometric form of a complex number is introduced in Section 8.3.
- The dot product of vectors is discussed in Section 8.6.
- Section 9.3 contains an improved notation for row operations on matrices, and the reduced echelon form is emphasized for finding solutions of systems of linear equations.

- In Section 10.1 the discussion of the axiom of mathematical induction has been deleted so that proofs by induction can be treated earlier.
- Chapter 11 includes applications involving the eccentricity and reflective properties of conics, orbits of planets, and LORAN navigation.

FEATURES

EXAMPLES Each section contains carefully chosen examples to help students understand and assimilate new concepts. Whenever feasible, applications are included. An innovation in this edition are the labeled *illustrations*, which are brief demonstrations of the use of definitions, laws, and theorems.

EXERCISES Approximately half the exercises differ from those in the previous edition. Considerable effort has gone into carefully grading these new problems. Students can obtain practice on all topics in a section by working either the odd-numbered or even-numbered exercises.

APPLIED PROBLEMS The previous edition contained a great variety of applied problems from many fields. Approximately 140 new exercises and examples involving applications have been added to further strengthen this important aspect of the text.

CALCULATORS It is possible to work most of the exercises without a calculator; however, for some problems its use is advisable in order to shorten numerical computations. Certain exercises that definitely require a calculator include the instruction *use a calculator*.

TEXT DESIGN The text has been redesigned so that concepts and examples stand out. Many figures have been added to help visualize important aspects of problems.

FLEXIBILITY Syllabi from schools that have used previous editions attest to the flexibility of the text. Sections and chapters can be rearranged in different ways, depending on the objectives and the length of the course.

SUPPLEMENTS

Users of this text may obtain the following supplements from the publisher:

Study Guide and Solutions Manual, by Jeffery A. Cole of Anoka-Ramsey Community College, contains selected solutions and strategies for solving typical exercises.

Instructor's Solutions Manual, by Jeffery A. Cole, provides solutions and answers for all the exercises.

PWStest is an algorithm-driven test generator for the Apple II or IBM-PC and compatibles, which can accommodate testing by either chapter or learning objective.

Printed Test Bank has three alternate forms of tests for each chapter along with instructor answer keys.

Transparency Masters contain figures adapted from the text and statements of key definitions and theorems.

True BASIC™ Pre-Calculus software, by John Kemeny and Thomas Kurtz, is ideal for classroom demonstrations, individual study, and problem-solving.

ACKNOWLEDGMENTS

I wish to thank Michael Cullen of Loyola Marymount University and Jeffery Cole of Anoka-Ramsey Community College for their significant input on the exercise sets. Michael's ideas are the basis of the new applied problems that, together with those he supplied for the previous edition, provide a strong motivation for the mathematical concepts introduced in the text. Jeffery contributed a fine assortment of well-graded new drill exercises and also worked every problem in the text, checking and rechecking answers for accuracy.

This revision has benefited from the comments of many instructors who have used my texts. In particular, I wish to thank the following individuals, who reviewed all or portions of the manuscript for this edition:

Kathryn L. Ainsworth, *University of Louisville*
Daniel D. Anderson, *University of Iowa*
Paul R. Boltz, *Harrisburg Area Community College*
William E. Coppage, *Wright State University*
Carol A. Edwards, *St. Louis Community College at Florissant Valley*
James H. Gaunt, *Florida State University*
Albert A. Grasser, *Oakland Community College*

James E. Hall, *Westminster College*
John Kubicek, *Southwest Missouri State University*
J. Walter Lynch, *Georgia Southern College*
Gordon D. Mock, *Western Illinois University*
Frances O. McDonald, *Delgado Community College*
E. James Peake, *Iowa State University*
Martin Peres, *Broward Community College, North Campus*
Elizabeth A. Sirjani, *Washington State University*
George L. Szoke, *University of Akron*
Marvel D. Townsend, *University of Florida*
Jan Vandever, *South Dakota State University*
Mary Voxman, *University of Idaho*
Bostwick F. Wyman, *Ohio State University*

I am grateful for the excellent cooperation of the staff of PWS-KENT. Two people deserve special mention. Managing Editor David Geggis supervised the project and offered many helpful suggestions. His expertise, acquired through many years of publishing, made a difficult task much easier. My production editor, Kathi Townes, did an outstanding job of keeping her author's writing style consistent, honest, and to the point. The present form of the book was greatly influenced by their efforts, and I owe them both a debt of gratitude.

In addition to all the persons named here, I express my sincere appreciation to the many unnamed students and teachers who have helped shape my views on mathematics education.

EARL W. SWOKOWSKI

CONTENTS

CHAPTER 7

CHAPTER 8

CHAPTER 9

CHAPTER 10

CHAPTER 11

APPENDICES

CHAPTER 1

FUNDAMENTAL CONCEPTS OF ALGEBRA

The material in this chapter is basic to the study of algebra. We begin by discussing properties of real numbers. Next we turn our attention to exponents and radicals, and how they may be used to simplify complicated algebraic expressions. In the last section we consider the *binomial theorem*.

1.1 WHAT IS ALGEBRA?

The word *algebra* comes from the title of the book *ilm al-jabr w'al muqabala*, which was written in the ninth century by the Arabian mathematician al-Khworizimi. The title has been translated as *the science of restoration and reduction*, which means transposing and combining similar terms (of an equation). The Latin translation of *al-jabr* led to the name of the branch of mathematics we now call *algebra*.

Algebra has evolved from the operations and rules of arithmetic. The study of arithmetic begins with addition, multiplication, subtraction, and division of numbers:

$$4 + 7, \qquad (37)(681), \qquad 79 - 22, \qquad 40 \div 8.$$

In *algebra* we introduce symbols or letters—such as a, b, c, d, x, y—to denote *arbitrary* numbers and, instead of special cases, we often consider *general* statements:

$$a + b, \qquad cd, \qquad x - y, \qquad x \div a.$$

This *language of algebra* serves a twofold purpose. First, we may use it as a shorthand to abbreviate and simplify long or complicated statements. Second, it provides a convenient means of generalizing many specific statements. To illustrate this second use, we know that

$$2 + 3 = 3 + 2, \qquad 4 + 7 = 7 + 4, \qquad 1 + 8 = 8 + 1,$$

and so on. This property may be phrased as: *If two numbers are added, then the order of addition is immaterial; that is, the same result is obtained whether the second number is added to the first, or the first number is added to the second.* This lengthy description can be shortened and simplified by means of the following algebraic statement, using a and b to represent arbitrary numbers:

$$a + b = b + a.$$

The generality of algebra is illustrated by many formulas used in science and industry. For example, if an airplane flies at a constant rate of 300 mi/hr (miles per hour) for 2 hours, then the distance it travels is (300)(2) = 600 miles. If the airplane's rate is 250 mi/hr and the elapsed time is 3 hours, then the distance traveled is (250)(3) = 750 miles. If we introduce symbols and let r denote the constant rate, t the elapsed time, and d the distance traveled, then the two situations we have described are special cases of the general algebraic formula

$$d = rt.$$

If specific numerical values for r and t are known, we may find the distance d by substitution. We may also use the formula to solve related problems. For example, suppose the distance between two cities is 645 miles, and we wish to find the constant rate that would enable an airplane to cover that distance in 2 hours 30 minutes. Thus, we are given

$$d = 645 \text{ miles} \quad \text{and} \quad t = 2.5 \text{ hours},$$

and we must find r. Since $d = rt$, it follows that

$$r = \frac{d}{t} = \frac{645}{2.5} = 258 \text{ mi/hr}.$$

Thus, if an airplane flies at a constant rate of 258 mi/hr, then it will travel 645 miles in 2 hours 30 minutes. This indicates how the introduction of an algebraic formula allows us not only to solve special cases conveniently but also to enlarge the scope of our knowledge by suggesting new problems that can be considered.

We have given several elementary illustrations of the value of algebraic methods. There are an unlimited number of situations where a symbolic approach may lead to insights and solutions that would be impossible to obtain using only numerical processes. As you proceed through this text and go on either to more advanced courses in mathematics or to fields that employ mathematics, you will become even more aware of the importance and the power of algebraic techniques.

1.2 REAL NUMBERS

Real numbers are employed throughout mathematics, and you should be acquainted with symbols that represent them, such as

$$1, \quad 73, \quad -5, \quad \tfrac{49}{12}, \quad \sqrt{2}, \quad 0, \quad \sqrt[3]{-85}, \quad 0.33333\ldots, \quad \text{and} \quad 596.25.$$

Throughout this chapter lowercase letters $a, b, c, x, y, \ldots$ represent arbitrary real numbers.

The real numbers are **closed** relative to operations of addition (denoted by $+$) and multiplication (denoted by $\cdot$); that is, to every pair a, b of real numbers there corresponds a unique real number $a + b$ (the **sum** of a and b) and a unique real number $a \cdot b$ (the **product** of a and b). We also use ab for the product $a \cdot b$. If a and b denote the same real number, we write $a = b$. An expression of this type is an **equality**.

The special real numbers 0 (**zero**) and 1 (**one**) have the properties $a + 0 = a$ and $a \cdot 1 = a$ for every real number a. These numbers are sometimes referred to as the **additive identity** and **multiplicative identity**, respectively. Each real number a has a **negative**, $-a$, such that $a + (-a) = 0$, and each *nonzero* real number a has a **reciprocal**, $\frac{1}{a}$, such that $a\left(\frac{1}{a}\right) = 1$. We also call $-a$ the **additive inverse** of a, and $\frac{1}{a}$ the **multiplicative inverse** of a. These and other important properties of real numbers are listed in the following box.

PROPERTIES OF ADDITION AND MULTIPLICATION

Property	Addition	Multiplication
Commutative	$a + b = b + a$	$ab = ba$
Associative	$a + (b + c) = (a + b) + c$	$a(bc) = (ab)c$
Identities	$a + 0 = 0$	$a \cdot 1 = a$
	$0 + a = a$	$1 \cdot a = a$
Inverses	$a + (-a) = 0$	$a\left(\frac{1}{a}\right) = 1 \quad \text{if } a \neq 0$
	$(-a) + a = 0$	$\left(\frac{1}{a}\right)a = 1 \quad \text{if } a \neq 0$
Distributive	$a(b + c) = ab + ac$ $(a + b)c = ac + bc$	

The symbol a^{-1} may be used for $\frac{1}{a}$, as in the following definition.

DEFINITION OF a^{-1}

$$a^{-1} = \frac{1}{a}; \qquad a \neq 0$$

Note that if $a \neq 0$, then

$$a \cdot a^{-1} = 1 = a^{-1} \cdot a.$$

Since $a + (b + c)$ and $(a + b) + c$ are always equal, we may use $a + b + c$ to denote this real number. We use abc for either $a(bc)$ or $(ab)c$. Similarly, if four or more real numbers a, b, c, d are added or multiplied, we may write $a + b + c + d$ for their sum and $abcd$ for their product, regardless of how the numbers are grouped or interchanged.

The two distributive properties are useful for finding products of the type illustrated in the next example.

EXAMPLE ▪ 1

If a, b, c, and d denote real numbers, show that

$$(a + b)(c + d) = ac + bc + ad + bd.$$

SOLUTION We shall use the two distributive properties. Let us write $(a + b)$ in color to emphasize that we regard this expression as a single real number in the first multiplication.

$$\begin{aligned}(a + b)(c + d) &= (a + b)c + (a + b)d \\ &= (ac + bc) + (ad + bd) \\ &= ac + bc + ad + bd.\end{aligned}$$

The following is a basic property of equality.

PROPERTY OF EQUALITY

If $a = b$ and $c = d$, then $a + c = b + d$ and $ac = bd$.

This property states that the same number may be added to both sides of an equality, and both sides of an equality may be multiplied by the same number. We will use this property and the next theorem extensively in Chapter 2 to help find solutions of equations.

THEOREM

$ab = 0$ if and only if $a = 0$ or $b = 0$.

In the preceding theorem the phrase *if and only if* has a twofold character. It means that *if* $ab = 0$, then $a = 0$ or $b = 0$ and, *conversely*, if $a = 0$ or $b = 0$, then $ab = 0$. Consequently, if both $a \neq 0$ and $b \neq 0$, then $ab \neq 0$; that is, *the product of two nonzero real numbers is not zero.*

The following properties of negatives are true.

PROPERTIES OF NEGATIVES

$$-(-a) = a$$
$$(-a)b = -(ab) = a(-b)$$
$$(-a)(-b) = ab$$
$$(-1)a = -a$$

The operations of **subtraction** $(-)$ and **division** $(\div)$ are defined as follows.

SUBTRACTION AND DIVISION

$$a - b = a + (-b)$$
$$a \div b = a\left(\frac{1}{b}\right) = ab^{-1}; \qquad b \neq 0$$

We use either a/b or $\frac{a}{b}$ for $a \div b$ and refer to a/b as the **quotient of *a* and *b*** or the **fraction *a* over *b***. The numbers a and b are the **numerator** and **denominator**, respectively, of the fraction a/b. Since 0 has no multiplicative inverse, a/b is not defined if $b = 0$; that is, *division by zero is not defined.* Note that

$$1 \div b = \frac{1}{b} = b^{-1}; \quad b \neq 0.$$

The following properties of quotients are true, provided all denominators are nonzero real numbers.

PROPERTIES OF QUOTIENTS

$$\frac{a}{b} = \frac{c}{d} \quad \text{if and only if} \quad ad = bc$$
$$\frac{ad}{bd} = \frac{a}{b}, \qquad \frac{a}{-b} = \frac{-a}{b} = -\frac{a}{b}$$
$$\frac{a}{b} + \frac{c}{b} = \frac{a+c}{b}, \qquad \frac{a}{b} + \frac{c}{d} = \frac{ad+bc}{bd}$$
$$\frac{a}{b} \cdot \frac{c}{d} = \frac{ac}{bd}, \qquad \frac{a}{b} \div \frac{c}{d} = \frac{a}{b} \cdot \frac{d}{c} = \frac{ad}{bc}$$

The **positive integers**, $1, 2, 3, 4, \ldots$, result from adding the real number 1 successively to itself. The **negative integers** are $-1, -2, -3, -4, \ldots$. The **integers** consist of all positive and negative integers together with the real number 0. We sometimes list the integers as follows:

$$\ldots, \quad -4, \quad -3, \quad -2, \quad -1, \quad 0, \quad 1, \quad 2, \quad 3, \quad 4, \quad \ldots$$

If a, b and c are integers and $c = ab$, then a and b are **factors**, or **divisors**, of c. For example, since

$$6 = 2 \cdot 3 = (-2)(-3) = 1 \cdot 6 = (-1)(-6),$$

we see that $1, -1, 2, -2, 3, -3, 6$, and -6 are factors of 6.

A positive integer p different from 1 is **prime** if its only positive factors are 1 and p. The first few primes are 2, 3, 5, 7, 11, 13, 17, and 19. The **Fundamental Theorem of Arithmetic** states that every positive integer different from 1 can be expressed as a product of primes in one and only one way (except for order of factors). Some examples are:

$$12 = 2 \cdot 2 \cdot 3, \qquad 126 = 2 \cdot 3 \cdot 3 \cdot 7, \qquad 540 = 2 \cdot 2 \cdot 3 \cdot 3 \cdot 3 \cdot 5.$$

A **rational number** is a real number of the form a/b, where a and b are integers and $b \neq 0$. Real numbers that are not rational are **irrational numbers**. One common irrational number is the ratio of the circumference of a circle to its diameter, denoted by π. We often approximate π by the decimal 3.1416 or by the rational number $\frac{22}{7}$. We use the notation $\pi \approx 3.1416$ to indicate that π *is approximately equal to* 3.1416.

There is no rational number b such that $b^2 = 2$, where b^2 denotes $b \cdot b$. However, there is an *irrational* number, denoted by $\sqrt{2}$ (the **square root of 2**), such that $(\sqrt{2})^2 = 2$.

Real numbers may be expressed as decimals. Decimal representations for rational numbers either are terminating or are nonterminating and repeating. For example, we can show by division that

$$\tfrac{5}{4} = 1.25 \quad \text{and} \quad \tfrac{177}{55} = 3.2181818\ldots,$$

where the digits 1 and 8 in the representation of $\frac{177}{55}$ repeat indefinitely. Decimal representations for irrational numbers are always nonterminating and nonrepeating.

Real numbers may be represented by points on a line l such that for each real number a there corresponds one and only one point on l, and conversely, to each point P on l there corresponds precisely one real number. This is called a **one-to-one correspondence**. We first choose an arbitrary point O, called the **origin**, and associate with it the real number 0. Points associated with the integers are then determined by laying off successive line segments of equal length on either side of O, as illustrated in Figure 1. The points corresponding to rational numbers, such as $\frac{23}{5}$ and $-\frac{1}{2}$, are obtained by subdividing the equal line segments. Points associated with certain irrational numbers, such as $\sqrt{2}$, can be found by construction. (See Exercise 45.) To every irrational number there corresponds a unique point on l and, conversely, every point that is not associated with a rational number corresponds to an irrational number.

FIGURE 1

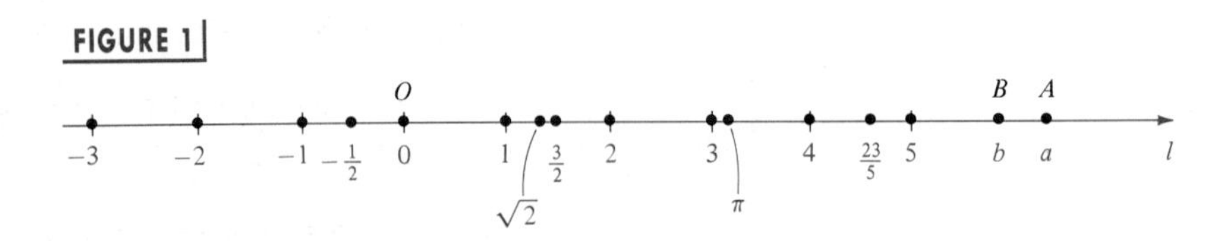

The number a that is associated with a point A on l is the **coordinate** of A. An assignment of coordinates to points on l is a **coordinate system** for l, and l is a **coordinate line**, or a **real line**. A direction can be assigned to l by taking the **positive direction** along l to the right and the **negative direction** to the left. The positive direction is noted by placing an arrowhead on l as shown in Figure 1.

The numbers that correspond to points to the right of O in Figure 1 are **positive real numbers**. Numbers that correspond to points to the left of O are **negative real numbers**. *The real number* 0 *is neither positive nor negative.*

If a and b are real numbers and $a - b$ is positive, we say that ***a* is greater than *b***, written $a > b$. An equivalent statement is that ***b* is less than *a***, written $b < a$. The symbols $>$ and $<$ are **inequality signs**. The expressions $a > b$ and $b < a$ are **inequalities**. From the manner in which we constructed the coordinate line l in Figure 1, we see that if A and B are points with coordinates a and b, respectively, then $a > b$ (or $b < a$) *if and only if* A *lies to the right of* B. The following definitions are stated for reference, for real numbers a and b.

DEFINITIONS OF > AND <

$a > b$ means $a - b$ is positive.

$b < a$ means $a - b$ is positive.

Note that the expressions $a > b$ and $b < a$ have the same meaning.

EXAMPLE ▪ 2

Replace the symbol □ with either $<$, $>$, or $=$ so that the resulting statement is true:

(a) $5 \square 3$ (b) $-6 \square -2$ (c) $\frac{1}{3} \square 0.33$ (d) $0 \square -4$

SOLUTION From the definitions of $>$ and $<$ we see that

(a) $5 > 3$, since $5 - 3 = 2$ is positive.

(b) $-6 < -2$, since $-2 - (-6) = -2 + 6 = 4$ is positive.

(c) $\frac{1}{3} > 0.33$, since $\frac{1}{3} - 0.33 = \frac{1}{3} - \frac{33}{100} = \frac{1}{300}$ is positive.

(d) $0 > -4$, since $0 - (-4) = 0 + 4 = 4$ is positive.

The next law enables us to compare, or *order*, any two real numbers.

TRICHOTOMY LAW

> For any real numbers a and b, one and only one of the following is true:
>
> $$a = b, \quad a > b, \quad \text{or} \quad a < b$$

We refer to the **sign** of a real number as positive or negative if the number is positive or negative, respectively. Two real numbers *have the same sign* if both are positive or both are negative. The numbers have *opposite signs* if one is positive and the other is negative. The next result about the signs of products and quotients of two real numbers a and b can be proved using properties of negatives and quotients.

LAWS OF SIGNS

> (i) If a and b have the same sign, then ab and $\frac{a}{b}$ are positive.
>
> (ii) If a and b have opposite signs, then ab and $\frac{a}{b}$ are negative.

The converses of the laws of signs are also true. For example, if a quotient is negative, then the numerator and denominator have opposite signs.

The notation $a \geq b$ (**a is greater than or equal to b**) means that either $a > b$ or $a = b$ (but not both). For example, $a^2 \geq 0$ for every real number a. The symbol $a \leq b$ (**a is less than or equal to b**) means that either $a < b$ or $a = b$. The expression $a < b < c$ means that both $a < b$ and $b < c$ and we say that **b is between a and c**. Similarly, the expression $c > b > a$ means that both $c > b$ and $b > a$. Thus,

$$1 < 5 < \tfrac{11}{2}, \qquad -4 < \tfrac{2}{3} < \sqrt{2}, \qquad 3 > -6 > -10.$$

There are other types of inequalities. For example, $a < b \le c$ means both $a < b$ and $b \le c$. Similarly, $a \le b < c$ means both $a \le b$ and $b < c$. Finally, $a \le b \le c$ means both $a \le b$ and $b \le c$.

If a is a real number, then it is the coordinate of some point A on a coordinate line, and the symbol $|a|$ denotes the number of units (or distance) between A and the origin, without regard to direction. The nonnegative number $|a|$ is the *absolute value* of a. Referring to Figure 2, we see that for the point with coordinate -4, we have $|-4| = 4$. Similarly, $|4| = 4$. In general, *if a is negative we change its sign to find $|a|$. If a is nonnegative, then $|a| = a$.* The next definition summarizes this discussion.

FIGURE 2

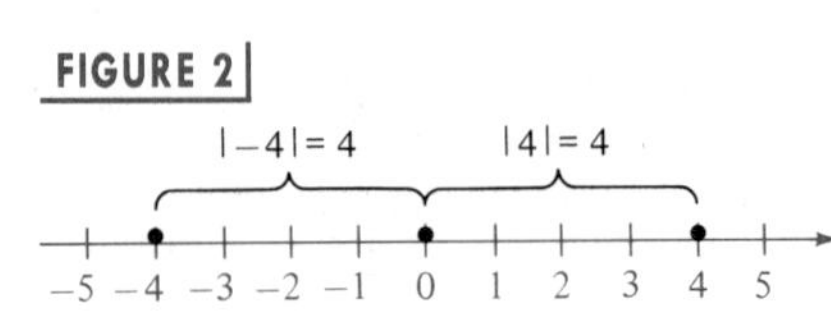

DEFINITION

The **absolute value** of a real number a, denoted by $|a|$, is

$$|a| = \begin{cases} a & \text{if } a \ge 0. \\ -a & \text{if } a < 0. \end{cases}$$

EXAMPLE 3

Find $|3|$, $|-3|$, $|0|$, $|\sqrt{2} - 2|$, and $|2 - \sqrt{2}|$.

SOLUTION Since 3, $2 - \sqrt{2}$, and 0 are nonnegative,

$$|3| = 3, \quad |2 - \sqrt{2}| = 2 - \sqrt{2}, \quad \text{and} \quad |0| = 0.$$

Since -3 and $\sqrt{2} - 2$ are negative, we use the formula $|a| = -a$ to obtain

$$|-3| = -(-3) = 3 \quad \text{and} \quad |\sqrt{2} - 2| = -(\sqrt{2} - 2) = 2 - \sqrt{2}.$$

In the preceding example, $|-3| = |3|$ and $|2 - \sqrt{2}| = |\sqrt{2} - 2|$. In general, we have the following.

THEOREM

$|a| = |-a|$ for every real number a.

We shall use the concept of absolute value to define the distance between any two points on a coordinate line. First note that the distance

FIGURE 3

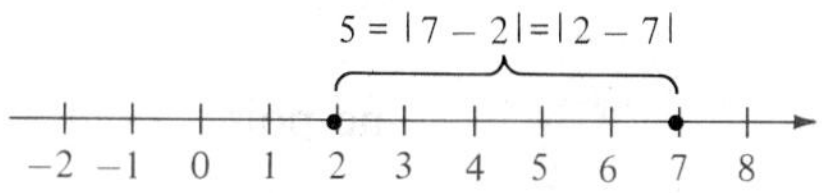

between the points with coordinates 2 and 7, shown in Figure 3, equals 5 units. This distance is the difference, $7 - 2$, obtained by subtracting the smaller coordinate from the larger. If we employ absolute values, then since $|7 - 2| = |2 - 7|$, it is unnecessary to be concerned about the order of subtraction. This motivates the next definition.

DEFINITION

Let a and b be the coordinates of two points A and B, respectively, on a coordinate line. The **distance between A and B**, denoted by $d(A, B)$, is

$$d(A, B) = |b - a|.$$

The number $d(A, B)$ is the **length of the line segment AB**.

Since $d(B, A) = |a - b|$ and $|b - a| = |a - b|$, we see that

$$d(A, B) = d(B, A).$$

Note that the distance between the origin O and the point A is

$$d(O, A) = |a - 0| = |a|.$$

This agrees with the geometric interpretation of absolute value illustrated in Figure 2. The formula $d(A, B) = |b - a|$ is true regardless of the signs of a and b, as illustrated in the next example.

EXAMPLE ▪ 4

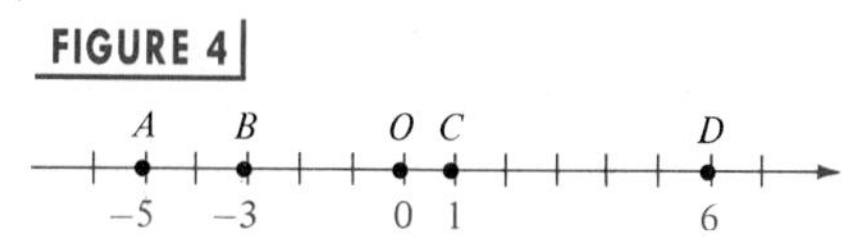

FIGURE 4

Let A, B, C, and D have coordinates -5, -3, 1, and 6, respectively, on a coordinate line. (See Figure 4.) Find $d(A, B)$, $d(C, B)$, $d(O, A)$, and $d(C, D)$.

SOLUTION Using the definition of the distance between points, we obtain the distances:

$$d(A, B) = |-3 - (-5)| = |-3 + 5| = |2| = 2$$
$$d(C, B) = |-3 - 1| = |-4| = 4$$
$$d(O, A) = |-5 - 0| = |-5| = 5$$
$$d(C, D) = |6 - 1| = |5| = 5$$

The concept of absolute value has uses other than finding distances between points: it is employed whenever we are interested in the "magnitude" or "numerical value" of a real number without regard to its sign.

1.2 EXERCISES

1 If $x < 0$ and $y > 0$, determine the sign of the real number:

(a) x^2y (b) xy^2 (c) $\dfrac{x}{y} + x$

(d) $y - x$ (e) $\dfrac{x - y}{xy}$ (f) x^4y^3

2 The rational numbers $\frac{22}{7}$ and $\frac{355}{113}$ are two useful approximations to π. Calculate the distances $d(\frac{22}{7}, \pi)$ and $d(\frac{335}{113}, \pi)$ to determine which is the best approximation.

Exer. 3–6: Replace the symbol □ with either <, >, or = so that the resulting statement is true.

3 (a) $-7 \square -4$ (b) $\dfrac{\pi}{2} \square 1.57$ (c) $\sqrt{225} \square 15$

4 (a) $-3 \square -5$ (b) $\dfrac{\pi}{4} \square 0.8$ (c) $\sqrt{289} \square 17$

5 (a) $\frac{1}{11} \square 0.09$ (b) $\frac{2}{3} \square 0.6666$ (c) $\frac{22}{7} \square \pi$

6 (a) $\frac{1}{7} \square 0.143$ (b) $\frac{5}{6} \square 0.833$ (c) $\sqrt{2} \square 1.4$

Exer. 7–18: Express the statement as an inequality.

7 x is negative.

8 y is nonnegative.

9 The reciprocal of z is nonpositive.

10 w is greater than or equal to -4.

11 q is less than or equal to π.

12 c is between $\frac{1}{5}$ and $\frac{1}{3}$.

13 The absolute value of x is greater than 7.

14 b is positive.

15 t is not less than 5.

16 p is not greater than -2.

17 The reciprocal of f is at most 14.

18 The absolute value of x is less than or equal to 4.

Exer. 19–24: Rewrite the number without using the absolute value symbol.

19 (a) $|-3 - 2|$ (b) $|-5| - |2|$ (c) $|7| + |-4|$

20 (a) $|-11 + 1|$ (b) $|6| - |-3|$ (c) $|8| + |-9|$

21 (a) $(-5)|3 - 6|$ (b) $|-6|/(-2)$ (c) $|-7| + |4|$

22 (a) $(4)|6 - 7|$ (b) $5/|-2|$ (c) $|-1| + |-9|$

23 (a) $|4 - \pi|$ (b) $|\pi - 4|$ (c) $|\sqrt{2} - 1.5|$

24 (a) $|\sqrt{3} - 1.7|$ (b) $|1.7 - \sqrt{3}|$ (c) $|\frac{1}{5} - \frac{1}{3}|$

Exer. 25–28: The given numbers are coordinates of points A, B, and C, respectively, on a coordinate line. Find the distance:

(a) $d(A, B)$ (b) $d(B, C)$ (c) $d(C, B)$ (d) $d(A, C)$

25 3, 7, -5

26 -6, -2, 4

27 -9, 1, 10

28 8, -4, -1

Exer. 29–36: Rewrite the expression without using the absolute value symbol.

29 $|3 + x|$ if $x < -3$

30 $|5 - x|$ if $x > 5$

31 $|2 - x|$ if $x < 2$

32 $|7 + x|$ if $x \geq -7$

33 $|a - b|$ if $a < b$

34 $|a - b|$ if $a > b$

35 $|x^2 + 4|$

36 $|-x^2 - 1|$

Exer. 37–44: Replace the symbol □ with either = or ≠ so that the resulting statement is true for all real numbers a, b, c, and d, whenever the expressions are defined.

37 $\dfrac{ab + ac}{a} \square b + ac$

38 $\dfrac{ab + ac}{a} \square b + c$

39 $\dfrac{b + c}{a} \square \dfrac{b}{a} + \dfrac{c}{a}$

40 $\dfrac{a + c}{b + d} \square \dfrac{a}{b} + \dfrac{c}{d}$

41 $(a \div b) \div c \square a \div (b \div c)$

42 $(a - b) - c \square a - (b - c)$

43 $\dfrac{a - b}{b - a} \square -1$

44 $-(a + b) \square -a + b$

45 The point on a coordinate line corresponding to $\sqrt{2}$ may be determined by constructing a right triangle with sides of length 1, as shown in the figure. Determine the points

EXERCISE 45

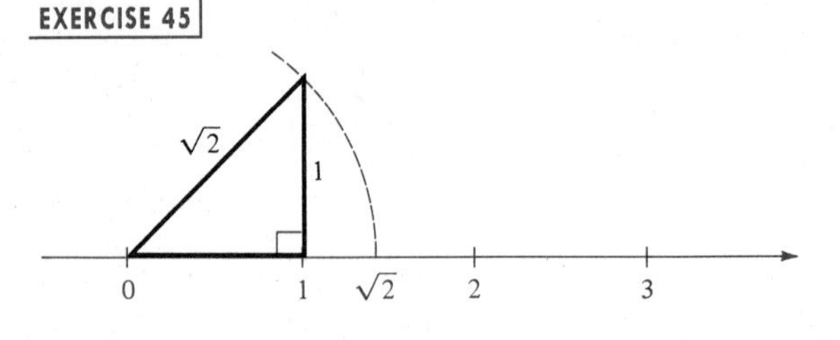

that correspond to $\sqrt{3}$ and $\sqrt{5}$, respectively. (*Hint:* Use the Pythagorean theorem.)

46 A circle of radius 1 rolls along a coordinate line in the positive direction, as shown in the figure. If point P is initially at the origin, find the coordinate of P:

(a) after one complete revolution.

(b) after two complete revolutions.

(c) after ten complete revolutions.

EXERCISE 46

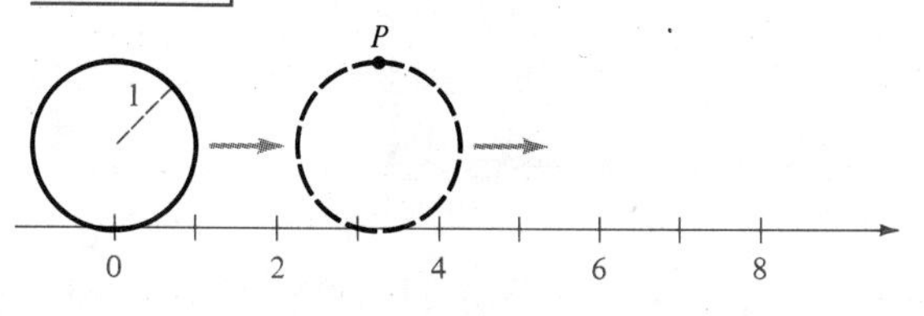

47 Geometric proofs of properties of real numbers were first given by the ancient Greeks. Compute the area of the rectangle shown in the figure in two ways to establish the distributive property $a(b + c) = ab + ac$ for positive real numbers a, b, and c.

EXERCISE 47

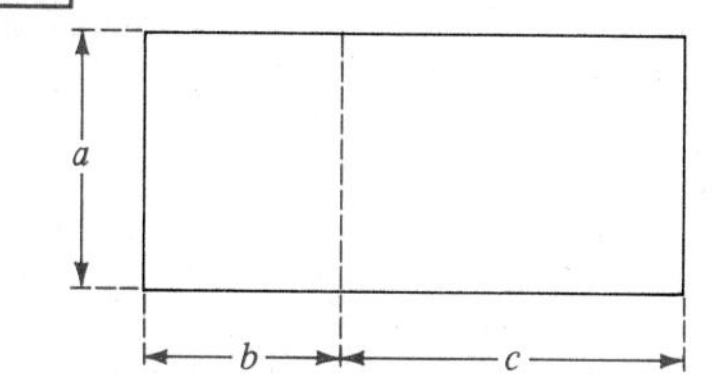

48 Rational approximations to square roots can be found using a formula discovered by the ancient Babylonians. Let x_1 be the first rational approximation for $\sqrt{n}$. If we let

$$x_2 = \frac{1}{2}\left(x_1 + \frac{n}{x_1}\right),$$

then x_2 will be a better approximation for $\sqrt{n}$, and we can repeat the computation with x_2 replacing x_1. Starting with $x_1 = \frac{17}{10}$, find the next two rational approximations for $\sqrt{3}$.

1.3 EXPONENTS AND RADICALS

The symbol a^2 denotes the real number $a \cdot a$, and a^3 is used in place of $a \cdot a \cdot a$. In general, we have the following, for every positive integer n.

DEFINITION OF a^n

$$a^n = a \cdot a \cdot a \cdots a \qquad (n \text{ factors of } a)$$

We call a^n a **power** of a, and refer to a^n as ***a* to the *n*th power** or, simply, *a to the n*. The positive integer n is the **exponent** of a. Note that $a^1 = a$.

ILLUSTRATION a^n

- $(\frac{1}{2})^5 = \frac{1}{2} \cdot \frac{1}{2} \cdot \frac{1}{2} \cdot \frac{1}{2} \cdot \frac{1}{2} = \frac{1}{32}$
- $(-3)^3 = (-3)(-3)(-3) = -27$
- $(\sqrt{2})^4 = \sqrt{2}\sqrt{2}\sqrt{2}\sqrt{2} = (\sqrt{2})^2(\sqrt{2})^2 = (2)(2) = 4$

It is important to note that if n is a positive integer, then an expression such as $3a^n$ means $3(a^n)$, not $(3a)^n$. The real number 3 is the **coefficient** of a^n in the expression $3a^n$. Similarly, $-3a^n$ means $(-3)a^n$, not $(-3a)^n$.

ILLUSTRATION ca^n

- $5 \cdot 2^3 = 5 \cdot 8 = 40$
- $-5 \cdot 2^3 = -5 \cdot 8 = -40$
- $(5 \cdot 2)^3 = 10^3 = 1000$

We extend the definition of a^n to nonpositive exponents as follows.

DEFINITION

$$a^0 = 1 \quad \text{and} \quad a^{-n} = \frac{1}{a^n}; \qquad a \neq 0$$

The following laws are indispensable when we work with exponents.

LAWS OF EXPONENTS

If a and b are real numbers and m and n are integers, then

(i) $a^m a^n = a^{m+n}$; (ii) $(a^m)^n = a^{mn}$; (iii) $(ab)^n = a^n b^n$;

(iv) $\left(\frac{a}{b}\right)^n = \frac{a^n}{b^n}$ if $b \neq 0$; (v) $\frac{a^m}{a^n} = a^{m-n} = \frac{1}{a^{n-m}}$ if $a \neq 0$.

PROOF To prove (i) for positive integers m and n, we write

$$a^m a^n = \underbrace{a \cdot a \cdot a \cdots a}_{m \text{ factors of } a} \cdot \underbrace{a \cdot a \cdot a \cdots a}_{n \text{ factors of } a}.$$

Since the total number of factors a on the right is $m + n$, this expression is equal to a^{m+n}.

Similarly, for (ii) we write

$$(a^m)^n = \underbrace{a^m \cdot a^m \cdot a^m \cdots a^m}_{n \text{ factors of } a^m}$$

and count the number of times a appears as a factor on the right-hand side. Since $a^m = a \cdot a \cdot a \cdots a$, with a occurring as a factor m times, and since the number of such groups of m factors is n, the total number of factors is $m \cdot n$. The cases $m \leq 0$ or $n \leq 0$ can be proved using the definition of nonpositive exponents.

Laws (iii) and (iv) may be proved in similar fashion. Law (v) is clear if $m = n$. For the case $m > n$, the integer $m - n$ is positive, and we may write

$$\frac{a^m}{a^n} = \frac{a^n a^{m-n}}{a^n} = \frac{a^n}{a^n} \cdot a^{m-n} = 1 \cdot a^{m-n} = a^{m-n}.$$

A similar argument may be used if $n > m$. ❑

We can extend laws of exponents to rules such as $(abc)^n = a^n b^n c^n$ and $a^m a^n a^p = a^{m+n+p}$.

ILLUSTRATION | LAWS OF EXPONENTS

- $x^5 x^6 = x^{5+6} = x^{11}$
- $(y^5)^7 = y^{5 \cdot 7} = y^{35}$
- $(rs)^7 = r^7 s^7$
- $\left(\dfrac{p}{q}\right)^{10} = \dfrac{p^{10}}{q^{10}}$
- $\dfrac{c^8}{c^3} = c^{8-3} = c^5$
- $\dfrac{u^3}{u^8} = u^{3-8} = u^{-5} = \dfrac{1}{u^5}$

To **simplify** *an expression involving powers of variables* means to change it to an expression in which each variable appears only once, and all exponents are positive. *We shall assume that variables in denominators represent nonzero real numbers.*

EXAMPLE ▪ 1

Simplify the expression:

(a) $(3x^3y^4)(4xy^5)$ **(b)** $(2a^2b^3c)^4$

(c) $\left(\dfrac{2r^3}{s}\right)^2\left(\dfrac{s}{r^3}\right)^3$ **(d)** $(u^{-2}v^3)^{-3}$

SOLUTION

(a)
$$\begin{aligned}(3x^3y^4)(4xy^5) &= (3)(4)x^3xy^4y^5 \\ &= 12x^4y^9\end{aligned}$$

(b)
$$\begin{aligned}(2a^2b^3c)^4 &= 2^4(a^2)^4(b^3)^4c^4 \\ &= 16a^8b^{12}c^4\end{aligned}$$

(c)
$$\begin{aligned}\left(\frac{2r^3}{s}\right)^2\left(\frac{s}{r^3}\right)^3 &= \left(\frac{2^2r^6}{s^2}\right)\left(\frac{s^3}{r^9}\right) \\ &= 2^2\left(\frac{r^6}{r^9}\right)\left(\frac{s^3}{s^2}\right) \\ &= 4\left(\frac{1}{r^3}\right)(s) = \frac{4s}{r^3}\end{aligned}$$

(d) $$(u^{-2}v^3)^{-3} = u^6v^{-9} = \frac{u^6}{v^9}$$

The following theorem is useful for problems that involve negative exponents.

THEOREM

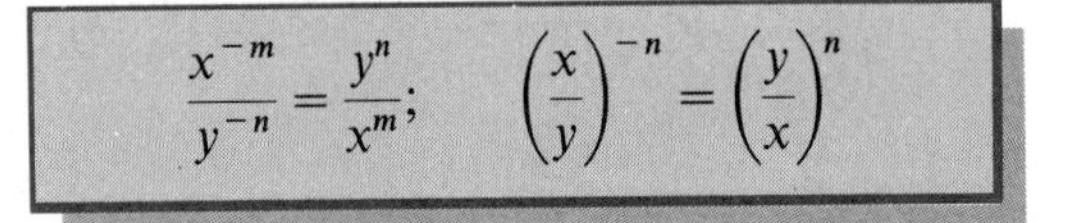

$$\frac{x^{-m}}{y^{-n}} = \frac{y^n}{x^m}; \quad \left(\frac{x}{y}\right)^{-n} = \left(\frac{y}{x}\right)^n$$

PROOF Using properties of negative exponents and quotients,

$$\frac{x^{-m}}{y^{-n}} = \frac{1/x^m}{1/y^n} = \frac{1}{x^m} \cdot \frac{y^n}{1} = \frac{y^n}{x^m};$$

$$\left(\frac{x}{y}\right)^{-n} = \frac{x^{-n}}{y^{-n}} = \frac{y^n}{x^n} = \left(\frac{y}{x}\right)^n. \quad \square$$

EXAMPLE ▪ 2

Simplify:

(a) $\dfrac{8x^3y^{-5}}{4x^{-1}y^2}$ (b) $\left(\dfrac{u^2}{2v}\right)^{-3}$

SOLUTION Applying the preceding theorem and laws of exponents,

(a) $$\frac{8x^3y^{-5}}{4x^{-1}y^2} = \frac{8x^3}{4y^2} \cdot \frac{y^{-5}}{x^{-1}} = \frac{8x^3}{4y^2} \cdot \frac{x}{y^5} = \frac{2x^4}{y^7}$$

(b) $$\left(\frac{u^2}{2v}\right)^{-3} = \left(\frac{2v}{u^2}\right)^3 = \frac{8v^3}{u^6}$$

We next define the **principal *n*th root** $\sqrt[n]{a}$ of a real number a.

DEFINITION OF $\sqrt[n]{a}$

Let n be a positive integer greater than 1 and let a be a real number.

(i) If $a > 0$, then $\sqrt[n]{a}$ is the *positive* real number b such that $b^n = a$.

(ii) If $a < 0$ and n is odd, then $\sqrt[n]{a}$ is the *negative* real number b such that $b^n = a$.

(iii) $\sqrt[n]{0} = 0$.

Complex numbers (see Section 2.4) are required to define $\sqrt[n]{a}$ if $a < 0$ and n is an *even* positive integer, since for all real numbers b, $b^n \geq 0$ whenever n is even.

If $n = 2$, we write $\sqrt{a}$ instead of $\sqrt[2]{a}$ and call $\sqrt{a}$ the **principal square root of** a or, simply, the **square root** of a. The number $\sqrt[3]{a}$ is the (principal) **cube root** of a.

ILLUSTRATION $\sqrt[n]{a}$

- $\sqrt{16} = 4$ since $4^2 = 16$.
- $\sqrt[3]{-8} = -2$ since $(-2)^3 = -8$.
- $\sqrt[4]{81} = 3$ since $3^4 = 81$.
- $\sqrt[5]{\frac{1}{32}} = \frac{1}{2}$ since $(\frac{1}{2})^5 = \frac{1}{32}$.

Note that $\sqrt{16} \neq \pm 4$, since by definition, roots of positive real numbers are positive.

To complete our terminology, the expression $\sqrt[n]{a}$ is a **radical**, the number a is the **radicand**, and n is the **index** of the radical. The symbol $\sqrt{}$ is a **radical sign**.

If $\sqrt{a} = b$, then $b^2 = a$; that is, $(\sqrt{a})^2 = a$. If $\sqrt[3]{a} = b$, then $b^3 = a$, or $(\sqrt[3]{a})^3 = a$. In general, (i) of the next theorem is true.

THEOREM

Let n be a positive integer.

(i) $(\sqrt[n]{a})^n = a$ if $\sqrt[n]{a}$ exists.

(ii) $\sqrt[n]{a^n} = a$ if $a > 0$, or if $a < 0$ and n is odd.

(iii) $\sqrt[n]{a^n} = |a|$ if n is even.

If $a > 0$, then (iii) of the theorem reduces to (ii). We also see from (iii) that $\sqrt{x^2} = |x|$ for every real number x. In particular, if $x \geq 0$, then $\sqrt{x^2} = x$; however, if $x < 0$, then $\sqrt{x^2} = -x$, which is positive.

ILLUSTRATION $\sqrt[n]{a^n}$

- $\sqrt{5^2} = 5$
- $\sqrt[3]{(-2)^3} = -2$
- $\sqrt[4]{3^4} = 3$
- $\sqrt{(-3)^2} = \sqrt{9} = 3 = |-3|$

The following laws are true for positive integers m and n, provided the indicated roots exist.

LAWS OF RADICALS

(i) $\sqrt[n]{ab} = \sqrt[n]{a}\sqrt[n]{b}$; (ii) $\sqrt[n]{\dfrac{a}{b}} = \dfrac{\sqrt[n]{a}}{\sqrt[n]{b}}$; (iii) $\sqrt[m]{\sqrt[n]{a}} = \sqrt[mn]{a}$.

EXAMPLE ▪ 3

Verify each statement:

(a) $\sqrt{50} = 5\sqrt{2}$ **(b)** $\sqrt[3]{-108} = -3\sqrt[3]{4}$ **(c)** $\sqrt[3]{\sqrt{64}} = 2$

SOLUTION

(a) $$\sqrt{50} = \sqrt{25 \cdot 2} = \sqrt{25}\sqrt{2} = 5\sqrt{2}$$

(b) $$\sqrt[3]{-108} = \sqrt[3]{(-27)(4)} = \sqrt[3]{-27}\sqrt[3]{4} = -3\sqrt[3]{4}$$

(c) We apply (iii) of the laws of radicals with $m = 3$ and $n = 2$:

$$\sqrt[3]{\sqrt{64}} = \sqrt[6]{64} = \sqrt[6]{2^6} = 2$$

As a check, we have $\sqrt[3]{\sqrt{64}} = \sqrt[3]{8} = 2$.

If c is a real number and c^n occurs as a factor in a radical of index n, then we can remove c from the radicand if the sign of c is taken into account. For example, if $c > 0$, or if $c < 0$ and n is *odd*, then

$$\sqrt[n]{c^n d} = \sqrt[n]{c^n}\,\sqrt[n]{d} = c\sqrt[n]{d}.$$

provided $\sqrt[n]{d}$ exists. If $c < 0$ and n is *even*, then

$$\sqrt[n]{c^n d} = \sqrt[n]{c^n}\,\sqrt[n]{d} = |c|\sqrt[n]{d}.$$

provided $\sqrt[n]{d}$ exists.

To avoid considering positive and negative cases separately in examples and exercises involving radicals, *we shall assume that all letters a, b, c, d, x, y, and so on represent positive real numbers.*

To *simplify a radical* means to remove factors from the radicand until no factor in the radicand has an exponent equal to or greater than the index of the radical, and the index is as low as possible.

EXAMPLE ▪ 4

Simplify the radical (all letters denote positive real numbers):

(a) $\sqrt[3]{320}$ (b) $\sqrt[3]{16x^3y^8z^4}$ (c) $\sqrt{3a^2b^3}\sqrt{6a^5b}$

SOLUTION

(a) $$\sqrt[3]{320} = \sqrt[3]{64 \cdot 5} = \sqrt[3]{4^3 \cdot 5} = \sqrt[3]{4^3}\sqrt[3]{5} = 4\sqrt[3]{5}$$

(b) $$\begin{aligned}\sqrt[3]{16x^3y^8z^4} &= \sqrt[3]{(2^3x^3y^6z^3)(2y^2z)}\\ &= \sqrt[3]{(2xy^2z)^3(2y^2z)}\\ &= \sqrt[3]{(2xy^2z)^3}\sqrt[3]{2y^2z}\\ &= 2xy^2z\sqrt[3]{2y^2z}\end{aligned}$$

(c) $$\begin{aligned}\sqrt{3a^2b^3}\sqrt{6a^5b} &= \sqrt{18a^7b^4}\\ &= \sqrt{(9a^6b^4)(2a)}\\ &= \sqrt{(3a^3b^2)^2(2a)}\\ &= \sqrt{(3a^3b^2)^2}\sqrt{2a}\\ &= 3a^3b^2\sqrt{2a}\end{aligned}$$

If the denominator of a quotient contains a radical, we sometimes multiply the numerator and denominator by some expression so that the resulting denominator does not contain a radical, as illustrated in Example 5(a) and (b). This process is referred to as **rationalizing a denominator**.

When the radicand is a quotient, we often use the technique illustrated in Example 5(c) and (d).

EXAMPLE ▪ 5

Rationalize the denominator:

(a) $\dfrac{1}{\sqrt{5}}$ (b) $\dfrac{1}{\sqrt[3]{x}}$ (c) $\sqrt{\dfrac{2}{3}}$ (d) $\sqrt[5]{\dfrac{x}{y^2}}$

SOLUTION

(a) $$\frac{1}{\sqrt{5}} = \frac{1}{\sqrt{5}} \cdot \frac{\sqrt{5}}{\sqrt{5}} = \frac{\sqrt{5}}{(\sqrt{5})^2} = \frac{\sqrt{5}}{5}$$

(b) $$\frac{1}{\sqrt[3]{x}} = \frac{1}{\sqrt[3]{x}}\,\frac{\sqrt[3]{x^2}}{\sqrt[3]{x^2}} = \frac{\sqrt[3]{x^2}}{\sqrt[3]{x^3}} = \frac{\sqrt[3]{x^2}}{x}$$

(c) $$\sqrt{\frac{2}{3}}=\sqrt{\frac{2}{3}\cdot\frac{3}{3}}=\sqrt{\frac{6}{3^2}}=\frac{\sqrt{6}}{\sqrt{3^2}}=\frac{\sqrt{6}}{3}$$

(d) $$\sqrt[5]{\frac{x}{y^2}}=\sqrt[5]{\frac{x}{y^2}\cdot\frac{y^3}{y^3}}=\frac{\sqrt[5]{xy^3}}{\sqrt[5]{y^5}}=\frac{\sqrt[5]{xy^3}}{y}$$

If we use a calculator to find decimal approximations of radicals, there is no advantage in rationalizing denominators, such as $1/\sqrt{5}=\sqrt{5}/5$ or $\sqrt{2/3}=\sqrt{6}/3$, as we did in Example 5. However, for *algebraic* simplifications, changing expressions to such forms is sometimes desirable. Similarly, in advanced mathematics courses such as calculus, changing $1/\sqrt[3]{x}$ to $\sqrt[3]{x^2}/x$ (as in part (b) of the example) would make a problem *more* complicated. In such courses it is simpler to work with the expression $1/\sqrt[3]{x}$ than with its rationalized form.

We next use radicals to define *rational exponents.*

DEFINITION OF RATIONAL EXPONENTS

Let m/n be a rational number and n a positive integer. If a is a real number such that $\sqrt[n]{a}$ exists, then

$$a^{1/n}=\sqrt[n]{a},\qquad a^{m/n}=(\sqrt[n]{a})^m=\sqrt[n]{a^m},\qquad a^{m/n}=(a^{1/n})^m=(a^m)^{1/n}.$$

ILLUSTRATION $a^{m/n}$

- $x^{1/3}=\sqrt[3]{x}$
- $x^{3/5}=(\sqrt[5]{x})^3=\sqrt[5]{x^3}$

The laws of exponents are true for rational exponents. In Chapter 5 we shall consider *irrational* exponents, such as $3^{\sqrt{2}}$ or 5^{π}.

To *simplify an expression involving rational powers of variables* we change it to an expression in which each variable appears only once, and all exponents are positive.

EXAMPLE ▪ 6

Simplify:

(a) $(-27)^{2/3}(4)^{-5/2}$ (b) $(r^2s^6)^{1/3}$ (c) $\left(\frac{2x^{2/3}}{y^{1/2}}\right)^2\left(\frac{3x^{-5/6}}{y^{1/3}}\right)$

SOLUTION

(a) $$(-27)^{2/3}(4)^{-5/2}=(\sqrt[3]{-27})^2(\sqrt{4})^{-5}=(-3)^2(2)^{-5}=\tfrac{9}{32}$$

(b)
$$(r^2s^6)^{1/3} = (r^2)^{1/3}(s^6)^{1/3} = r^{2/3}s^2$$

(c)
$$\left(\frac{2x^{2/3}}{y^{1/2}}\right)^2\left(\frac{3x^{-5/6}}{y^{1/3}}\right) = \left(\frac{4x^{4/3}}{y}\right)\left(\frac{3x^{-5/6}}{y^{1/3}}\right) = \frac{12x^{1/2}}{y^{4/3}}$$

Rational exponents are useful for problems that involve radicals that do not have the same index, as illustrated in the next example.

EXAMPLE ▪ 7

Change to an expression containing one radical $\sqrt[n]{a^m}$:

(a) $\sqrt[3]{a}\sqrt{a}$ (b) $\dfrac{\sqrt[4]{a}}{\sqrt[3]{a^2}}$

SOLUTION Introducing rational exponents, we obtain:

(a)
$$\sqrt[3]{a}\sqrt{a} = a^{1/3}a^{1/2} = a^{(1/3)+(1/2)} = a^{5/6} = \sqrt[6]{a^5}$$

(b)
$$\frac{\sqrt[4]{a}}{\sqrt[3]{a^2}} = \frac{a^{1/4}}{a^{2/3}} = a^{(1/4)-(2/3)} = a^{-5/12} = \frac{1}{a^{5/12}} = \frac{1}{\sqrt[12]{a^5}}$$

In the sciences it is often necessary to work with very large or very small numbers and to compare the relative magnitudes of very large or small quantities. We usually represent a large or small number a in *scientific form*, using $\times$ to denote multiplication.

SCIENTIFIC FORM

$$a = c \times 10^n \quad \text{for} \quad 1 \le c < 10 \text{ and } n \text{ an integer}$$

The distance a ray of light travels in one year is approximately 5,900,000,000,000 miles. This number may be written in scientific form as 5.9×10^{12}. The positive exponent 12 indicates that the decimal point should be moved 12 places to the *right*. The notation works equally well for small numbers. To illustrate, the weight of an oxygen molecule is estimated to be 0.000000000000000000000053 grams, or in scientific form, 5.3×10^{-23} grams. The negative exponent indicates that the decimal point should be moved 23 places to the *left*.

ILLUSTRATION **SCIENTIFIC FORM**

- $513 = 5.13 \times 10^2$
- $20{,}700 = 2.07 \times 10^4$
- $92{,}000{,}000 = 9.2 \times 10^7$
- $0.000648 = 6.48 \times 10^{-4}$
- $0.00000000043 = 4.3 \times 10^{-10}$

Many calculators employ scientific form in their display panels: for the number $c \times 10^n$, the 10 is suppressed and only the exponent is shown. For example, to find $(4{,}500{,}000)^2$ on a scientific calculator, we could enter the integer 4,500,000 and press the x^2 (or squaring) key. The display panel would show

2.025 13 or 2.025E 13

We would translate this as 2.025×10^{13}. Thus,

$$(4{,}500{,}000)^2 = 20{,}250{,}000{,}000{,}000.$$

Calculators may also use scientific form in the entry of numbers. The user's manual of a calculator will give specific details.

As a final remark, applied problems often include numbers that are obtained by various types of measurements and, hence, are approximations to exact values. Such answers should be rounded off, since the final result of a calculation cannot be more accurate than the data that has been used. For example, if the length and width of a rectangle are measured to two-decimal-place accuracy, we cannot expect more than two-decimal-place accuracy in the calculated value of the area of the rectangle. If a number a is written in scientific form as $a = c \times 10^k$ for $1 \le c < 10$, and if c is rounded off to n decimal places, then we say that a is accurate (or has been rounded off) to $n + 1$ **significant figures**.

ILLUSTRATION **SIGNIFICANT FIGURES**

Number of significant figures in $a = 37.2638$	Approximation to a
5	37.264
4	37.26
3	37.3
2	37
1	40

In Exercises 1.3, whenever an index of a radical is even (or a rational exponent m/n with n even is employed), we will assume that the letters that appear in the radicand denote positive real numbers.

1.3 EXERCISES

Exer. 1–10: Express the number in the form a/b for integers a and b.

1 $(-\frac{2}{3})^4$

2 $(-3)^3$

3 $\dfrac{2^{-3}}{3^{-2}}$

4 $\dfrac{2^0 + 0^2}{2 + 0}$

5 $(-2)^3 + 3^{-2}$

6 $(-\frac{3}{2})^4 - 2^{-4}$

7 $16^{-3/4}$

8 $9^{5/2}$

9 $(-0.008)^{2/3}$

10 $(0.008)^{-2/3}$

Exer. 11–46: Simplify.

11 $(\frac{1}{2}x^4)(16x^5)$

12 $(-3x^{-2})(4x^4)$

13 $\dfrac{(2x^3)(3x^2)}{(x^2)^3}$

14 $\dfrac{(2x^2)^3}{4x^4}$

15 $(\frac{1}{6}a^5)(-3a^2)(4a^7)$

16 $(-4b^3)(\frac{1}{6}b^2)(-9b^4)$

17 $\dfrac{(6x^3)^2}{(2x^2)^3}$

18 $\dfrac{(3y^3)(2y^2)^2}{(y^4)^3}$

19 $(3u^7v^3)(4u^4v^{-5})$

20 $(x^2yz^3)(-2xz^2)(x^3y^{-2})$

21 $(8x^4y^{-3})(\frac{1}{2}x^{-5}y^2)$

22 $\left(\dfrac{4a^2b}{a^3b^2}\right)\left(\dfrac{5a^2b}{2b^4}\right)$

23 $(\frac{1}{3}x^4y^{-3})^{-2}$

24 $(-2xy^2)^5\left(\dfrac{x^7}{8y^3}\right)$

25 $(3y^3)^4(4y^2)^{-3}$

26 $(-3a^2b^{-5})^3$

27 $(-2r^4s^{-3})^{-2}$

28 $(2x^2y^{-5})(6x^{-3}y)(\frac{1}{3}x^{-1}y^3)$

29 $(5x^2y^{-3})(4x^{-5}y^4)$

30 $(-2r^2s)^5(3r^{-1}s^3)^2$

31 $\left(\dfrac{3x^5y^4}{x^0y^{-3}}\right)^2$

32 $(4a^2b)^4\left(\dfrac{-a^3}{2b}\right)^2$

33 $(4a^{3/2})(2a^{1/2})$

34 $(-6x^{7/5})(2x^{8/5})$

35 $(3x^{5/6})(8x^{2/3})$

36 $(8r)^{1/3}(2r^{1/2})$

37 $(27a^6)^{-2/3}$

38 $(25z^4)^{-3/2}$

39 $(8x^{-2/3})x^{1/6}$

40 $(3x^{1/2})(-2x^{5/2})$

41 $\left(\dfrac{-8x^3}{y^{-6}}\right)^{2/3}$

42 $\left(\dfrac{-y^{3/2}}{y^{-1/3}}\right)^3$

43 $\left(\dfrac{x^6}{9y^{-4}}\right)^{-1/2}$

44 $\left(\dfrac{c^{-4}}{16d^8}\right)^{3/4}$

45 $\dfrac{(x^6y^3)^{-1/3}}{(x^4y^2)^{-1/2}}$

46 $a^{4/3}a^{-3/2}a^{1/6}$

Exer. 47–52: Rewrite the expression using rational exponents.

47 $\sqrt[4]{x^3}$

48 $\sqrt[3]{x^5}$

49 $\sqrt[3]{(a+b)^2}$

50 $\sqrt{a + \sqrt{b}}$

51 $\sqrt{x^2 + y^2}$

52 $\sqrt[3]{r^3 - s^3}$

Exer. 53–56: Rewrite the expression using a radical.

53 (a) $4x^{3/2}$ (b) $(4x)^{3/2}$

54 (a) $4 + x^{3/2}$ (b) $(4 + x)^{3/2}$

55 (a) $8 - y^{1/3}$ (b) $(8 - y)^{1/3}$

56 (a) $8y^{1/3}$ (b) $(8y)^{1/3}$

Exer. 57–80: Simplify the expression and rationalize the denominator when appropriate.

57 $\sqrt{81}$

58 $\sqrt[3]{-125}$

59 $\sqrt[5]{-64}$

60 $\sqrt[4]{256}$

61 $\dfrac{1}{\sqrt[3]{2}}$

62 $\sqrt{\dfrac{1}{7}}$

63 $\sqrt{9x^{-4}y^6}$

64 $\sqrt{16a^8b^{-2}}$

65 $\sqrt[3]{8a^6b^{-3}}$

66 $\sqrt[4]{81r^5s^8}$

67 $\sqrt{\dfrac{3x}{2y^3}}$

68 $\sqrt{\dfrac{1}{3x^3y}}$

69 $\sqrt[3]{\dfrac{2x^4y^4}{9x}}$

70 $\sqrt[3]{\dfrac{3x^2y^5}{4x}}$

71 $\sqrt[4]{\dfrac{5x^8y^3}{27x^2}}$

72 $\sqrt[4]{\dfrac{x^7y^{12}}{125x}}$

73 $\sqrt[5]{\dfrac{5x^7y^2}{8x^3}}$

74 $\sqrt[5]{\dfrac{3x^{11}y^3}{9x^2}}$

75 $\sqrt[4]{(3x^5y^{-2})^4}$

76 $\sqrt[6]{(2u^{-3}v^4)^6}$

77 $\sqrt[5]{\dfrac{8x^3}{y^4}}\sqrt[5]{\dfrac{4x^4}{y^2}}$

78 $\sqrt{5xy^7}\sqrt{10x^3y^3}$

79 $\sqrt[3]{3t^4v^2}\sqrt[3]{-9t^{-1}v^4}$

80 $\sqrt[3]{(2r-s)^3}$

Exer. 81–86: Replace the symbol □ with either = or ≠ so that the resulting statement is true for all values of the variables whenever the expressions are defined. Give a reason for your answer.

81 $(a^r)^2 \,\square\, a^{(r^2)}$

82 $(a^2+1)^{1/2} \,\square\, a+1$

83 $a^xb^y \,\square\, (ab)^{xy}$

84 $\sqrt{a^r} \,\square\, (\sqrt{a})^r$

85 $\sqrt[n]{\dfrac{1}{c}} \,\square\, \dfrac{1}{\sqrt[n]{c}}$

86 $a^{1/k} \,\square\, \dfrac{1}{a^k}$

87 The mass of a hydrogen atom is approximately 0.0000000000000000000000017 grams. Express this number in scientific form.

88 The mass of an electron is approximately 9.1×10^{-31} kilograms. Express this number in decimal form.

Exer. 89–90: Express the number in scientific form.

89 (a) 427,000 (b) 0.000000098 (c) 810,000,000

90 (a) 85,200 (b) 0.0000055 (c) 24,900,000

Exer. 91–92: Express the number in decimal form.

91 (a) 8.3×10^5 (b) 2.9×10^{-12} (c) 5.63×10^8

92 (a) 2.3×10^7 (b) 7.01×10^{-9} (c) 1.23×10^{10}

93 In astronomy, distances to stars are measured in light years. One light year is the distance a ray of light travels in one year. If the speed of light is 186,000 miles per second, approximate the number of miles in one light year.

94 (a) Astronomers have estimated that the Milky Way galaxy contains 100 billion stars. Express this number in scientific form.

(b) The diameter d of the Milky Way galaxy is estimated as 100,000 light years. Express d in miles. (Refer to Exercise 93.)

95 The number of hydrogen atoms in a mole is Avogadro's number, 6.02×10^{23}. If one mole of the gas has a mass of 1.01 grams, estimate the mass of a hydrogen atom.

96 The population dynamics of many fish are characterized by extremely high fertility rates among adults and very low survival rates among the young. A mature halibut may lay as many as 2.5 million eggs, but only 0.00035% of the offspring survive to the age of 3 years. Use scientific form to calculate the number of halibut that live to age 3.

97 The longest movie ever made is a 1970 British film that runs for 48 hours. Assuming that the film speed is 24 frames per second, approximate the total number of frames in this film. Express your answer in scientific form.

98 One of the largest known prime numbers is $2^{44497}-1$. At the time that this number was computed, it took one of the world's fastest computers about 60 days to verify that the number is prime. This computer is capable of performing 2×10^{11} calculations per second. Use scientific form to estimate the number of calculations needed to perform this computation.

Exer. 99–104: Use a calculator.

99 One of the oldest banks in the U.S. is the Bank of America, founded in 1812. If \$200 had been deposited at that time into an account that paid 4% annual interest, then 180 years later the amount would have grown to $200(1.04)^{180}$ dollars. Calculate this sum.

100 The distance d (in miles) that can be seen from the top of a tall building of height h feet can be approximated by $d = 1.2\sqrt{h}$. Approximate the distance that can be seen from the top of the Chicago Sears Tower, which is 1454 feet tall.

101 The length-weight relationship for the Pacific halibut is given by the formula $L = 0.46\sqrt[3]{W}$ for W in kg and L in meters. The largest documented catch is a halibut that weighed 230 kg. Estimate its length.

102 The length-weight relationship for the sei whale is given by the formula $W = 0.0016L^{2.43}$ for W in tons and L in feet. Estimate the weight of a whale that is 25 feet long.

103 O'Carroll's formula is used to handicap weight lifters. If a lifter who weighs b kg lifts w kg of weight, then the handicapped weight W is given by

$$W = \frac{w}{\sqrt[3]{b-35}}.$$

Suppose two lifters weighing 75 kg and 120 kg lift weights of 180 kg and 250 kg, respectively. Use O'Carroll's formula to determine the superior weight lifter.

104 The surface area S of the human body (in square feet) can be estimated from height h (in inches) and weight w (in pounds) using the formula

$$S = (0.1091)w^{0.425}h^{0.725}.$$

This formula is used to estimate total body fat.

(a) Estimate the body surface area of an individual 6 feet tall, weighing 175 pounds.

(b) If a person is 5 feet 6 inches tall, what is the effect of a 10% increase in weight on body surface area S?

1.4 ALGEBRAIC EXPRESSIONS

We sometimes use the notation and terminology of sets. A **set** is a collection of objects of some type, and the objects are **elements** of the set. Capital letters $A, B, C, R, S, \ldots$ are often used to denote sets, and lowercase letters $a, b, x, y, \ldots$ usually represent elements of sets. Throughout this book $\mathbb{R}$ denotes the set of real numbers and $\mathbb{Z}$, the set of integers. If S is a set, then $a \in S$ means that a is an element of S, and $a \notin S$ signifies that a is not an element of S. If every element of S is also an element of a set T, then S is a **subset** of T. For example, $\mathbb{Z}$ is a subset of $\mathbb{R}$. Two sets S and T are **equal**, denoted by $S = T$, if S and T contain precisely the same elements. The notation $S \neq T$ means that S and T are not equal.

We frequently use letters to represent arbitrary elements of a set. For example, we may use x to denote a real number when we do not wish to specify a *particular* real number. A letter that is used to represent *any* element of a set is a **variable**. A letter that represents a *specific* element is a **constant**. Letters near the end of the alphabet, such as x, y, and z, are often used for variables. Letters at the beginning of the alphabet, such as a, b, and c, often denote constants. Throughout this text, unless otherwise specified, variables represent real numbers.

The **domain of a variable** is the set of real numbers represented by the variable. To illustrate, $\sqrt{x}$ is a real number if and only if $x \geq 0$, so the domain of x is the set of nonnegative real numbers. Similarly, for the expression $1/(x - 2)$, we must exclude $x = 2$ to avoid division by zero; consequently, the domain of x is the set of all real numbers different from 2.

If the elements of a set S have a certain property, we sometimes write $S = \{x: \quad \}$ and state the property describing the variable x in the space after the colon. For example, $\{x : x > 3\}$ denotes the set of all real numbers greater than 3.

For finite sets we sometimes list all the elements of the set within braces. Thus, if the set T consists of the first five positive integers, we may write $T = \{1, 2, 3, 4, 5\}$. When we describe sets in this way, the order used in listing the elements is irrelevant, so we could also write $T = \{1, 3, 2, 4, 5\}$, $T = \{4, 3, 2, 5, 1\}$, and so on.

If we begin with any collection of variables and real numbers, then an **algebraic expression** is the result obtained by applying additions, subtractions, multiplications, divisions, powers, or the taking of roots.

ILLUSTRATION | ALGEBRAIC EXPRESSIONS

- $x^3 - 5x + \dfrac{6}{\sqrt{x}}$
- $\dfrac{2xy + 3x}{y - 1}$
- $\dfrac{4yz^{-2} + \left(\dfrac{-7}{x + w}\right)^5}{\sqrt[3]{y^2 + 5z}}$

If specific numbers are substituted for the variables in an algebraic expression, the resulting real number is the **value** of the expression for these numbers. In the preceding illustration, the value of the first expression for $x = 4$ is

$$4^3 - 5(4) + \frac{6}{\sqrt{4}} = 64 - 20 + 3 = 47.$$

When working with algebraic expressions, we assume that domains are chosen so that variables do not represent numbers which make the expressions meaningless. Thus, we assume that denominators are not zero and roots always exist.

Certain algebraic expressions are given special names. If x is a variable, then a **monomial** in x is an expression of the form ax^n for a real number a and a nonnegative integer n. A *polynomial in x* is a sum of monomials in x. Another way of stating this is as follows.

DEFINITION

A **polynomial in *x*** is a sum of the form

$$a_nx^n + a_{n-1}x^{n-1} + \cdots + a_1x + a_0,$$

where n is a nonnegative integer and each coefficient a_k is a real number.

Each expression a_kx^k in the sum is a **term** of the polynomial. If a coefficient a_k is zero, we usually delete the term a_kx^k. The coefficient a_n of the highest power of x is the **leading coefficient** of the polynomial and, if $a_n \neq 0$, the polynomial has **degree *n***.

By definition, two polynomials are **equal** if and only if they have the same degree and corresponding coefficients are equal. If all the coefficients of a polynomial are zero, it is called the **zero polynomial** and is denoted by 0. The degree of the zero polynomial is undefined. If c is a nonzero real number, then c is a polynomial of degree 0. Such polynomials (together with the zero polynomial) are **constant polynomials**.

ILLUSTRATION | POLYNOMIALS

- $3x^4 + 5x^3 + (-7)x + 4$ (degree 4)
- $x^8 + 9x^2 + (-2)x$ (degree 8)
- $5x^2 + 1$ (degree 2)
- $7x + 2$ (degree 1)
- 5 (degree 0)

If some coefficients of a polynomial are negative, we often use minus signs between appropriate terms. To illustrate,

$$3x^2 + (-5)x + (-7) = 3x^2 - 5x - 7.$$

We may also consider polynomials in other variables. For example, $\frac{2}{5}z^2 - 3z^7 + 8 - \sqrt{5}\,z^4$ is a polynomial in z of degree 7. We often arrange the terms in order of decreasing powers of the variable; thus, we write

$$\tfrac{2}{5}z^2 - 3z^7 + 8 - \sqrt{5}\,z^4 = -3z^7 - \sqrt{5}\,z^4 + \tfrac{2}{5}z^2 + 8.$$

We may think of a polynomial in x as an algebraic expression obtained by employing a finite number of additions, subtractions, and multiplications involving x. If an algebraic expression contains divisions or roots involving a variable x, then it is not a polynomial in x.

ILLUSTRATION | NONPOLYNOMIALS

- $\dfrac{1}{x} + 3x$
- $\dfrac{x - 5}{x^2 + 2}$
- $3x^2 + \sqrt{x} - 2$

Since polynomials, and the monomials that make up polynomials, represent real numbers, they have all of the properties described in Section 1.2. Thus, if additions, multiplications, and subtractions are carried out with polynomials, we may simplify the results by using properties of real numbers.

EXAMPLE ▪ 1

(a) Find the sum $(x^3 + 2x^2 - 5x + 7) + (4x^3 - 5x^2 + 3)$.

(b) Find the difference $(x^3 + 2x^2 - 5x + 7) - (4x^3 - 5x^2 + 3)$.

SOLUTION We rearrange terms and use properties of real numbers:

(a)
$$\begin{aligned}(x^3 + 2x^2 - 5x + 7) + (4x^3 - 5x^2 + 3) &= x^3 + 4x^3 + 2x^2 - 5x^2 - 5x + 7 + 3\\ &= (1 + 4)x^3 + (2 - 5)x^2 - 5x + (7 + 3)\\ &= 5x^3 - 3x^2 - 5x + 10\end{aligned}$$

(b) $(x^3 + 2x^2 - 5x + 7) - (4x^3 - 5x^2 + 3)$

$$\begin{aligned} &= x^3 + 2x^2 - 5x + 7 - 4x^3 + 5x^2 - 3 \\ &= x^3 - 4x^3 + 2x^2 + 5x^2 - 5x + 7 - 3 \\ &= (1 - 4)x^3 + (2 + 5)x^2 - 5x + (7 - 3) \\ &= -3x^3 + 7x^2 - 5x + 4 \end{aligned}$$

Example 1 illustrates that we can obtain the sum of any two polynomials in x by adding coefficients of like powers of x. We find the difference of two polynomials by subtracting coefficients of like powers. The intermediate grouping steps in the solutions to Example 1 were shown for completeness. You may omit these steps after you become proficient with such manipulations. In order to multiply two polynomials, we use the distributive properties together with the laws of exponents, and combine like terms, as illustrated in the next example.

EXAMPLE ▪ 2

Find the product $(x^2 + 5x - 4)(2x^3 + 3x - 1)$.

SOLUTION We begin by using a distributive property and treating $2x^3 + 3x - 1$ as a single real number:

$$\begin{aligned} &(x^2 + 5x - 4)(2x^3 + 3x - 1) \\ &\quad = x^2(2x^3 + 3x - 1) + 5x(2x^3 + 3x - 1) - 4(2x^3 + 3x - 1) \\ &\quad = 2x^5 + 3x^3 - x^2 + 10x^4 + 15x^2 - 5x - 8x^3 - 12x + 4 \\ &\quad = 2x^5 + 10x^4 + (3 - 8)x^3 + (-1 + 15)x^2 + (-5 - 12)x + 4 \\ &\quad = 2x^5 + 10x^4 - 5x^3 + 14x^2 - 17x + 4 \end{aligned}$$

We can also find the product by treating $x^2 + 5x - 4$ as a single number in the first step.

We may consider polynomials in more than one variable. For example, a polynomial in *two* variables, x and y, is a finite sum of terms, each of the form ax^my^k for some real number a and nonnegative integers m and k. An example is

$$3x^4y + 2x^3y^5 + 7x^2 - 4xy + 8y - 5.$$

We may also have polynomials in three variables, such as x, y, z, or for that matter, in *any* number of variables. Addition, subtraction, and multiplication are performed using properties of real numbers, as illustrated in the following example.

EXAMPLE ▪ 3

Find the product $(x^2 + xy + y^2)(x - y)$.

SOLUTION We begin by treating $x^2 + xy + y^2$ as a single real number:

$$\begin{aligned}(x^2 + xy + y^2)(x - y) &= (x^2 + xy + y^2)x - (x^2 + xy + y^2)y \\ &= x^3 + x^2y + xy^2 - x^2y - xy^2 - y^3 \\ &= x^3 - y^3\end{aligned}$$

The next example illustrates division of a polynomial by a monomial.

EXAMPLE ▪ 4

Express as a polynomial in x and y:

$$\frac{6x^2y^3 + 4x^3y^2 - 10xy}{2xy}$$

SOLUTION

$$\begin{aligned}\frac{6x^2y^3 + 4x^3y^2 - 10xy}{2xy} &= \frac{6x^2y^3}{2xy} + \frac{4x^3y^2}{2xy} - \frac{10xy}{2xy} \\ &= 3xy^2 + 2x^2y - 5\end{aligned}$$

Certain products occur so frequently in algebra that they deserve special attention. We list some of these next, using letters to represent real numbers. You can check the validity of each formula by performing the multiplication.

PRODUCT FORMULAS

(i) $(x + y)(x - y) = x^2 - y^2$

(ii) $(ax + b)(cx + d) = acx^2 + (ad + bc)x + bd$

(iii) $(x + y)^2 = x^2 + 2xy + y^2$

(iv) $(x - y)^2 = x^2 - 2xy + y^2$

(v) $(x + y)^3 = x^3 + 3x^2y + 3xy^2 + y^3$

(vi) $(x - y)^3 = x^3 - 3x^2y + 3xy^2 - y^3$

Since the variables x and y in these formulas represent real numbers, they may be replaced by algebraic expressions, as illustrated in the next example.

EXAMPLE ▪ 5

Find the product:

(a) $(2r^2 - \sqrt{s})(2r^2 + \sqrt{s})$ **(b)** $\left(\sqrt{c} + \frac{1}{\sqrt{c}}\right)^2$ **(c)** $(2a - 5b)^3$

SOLUTION

(a) We use product formula (i) with $x = 2r^2$ and $y = \sqrt{s}$:

$$\begin{aligned}(2r^2 - \sqrt{s})(2r^2 + \sqrt{s}) &= (2r^2)^2 - (\sqrt{s})^2 \\ &= 4r^4 - s\end{aligned}$$

(b) We use product formula (iii) with $x = \sqrt{c}$ and $y = \frac{1}{\sqrt{c}}$:

$$\begin{aligned}\left(\sqrt{c} + \frac{1}{\sqrt{c}}\right)^2 &= (\sqrt{c})^2 + 2\sqrt{c} \cdot \frac{1}{\sqrt{c}} + \left(\frac{1}{\sqrt{c}}\right)^2 \\ &= c + 2 + \frac{1}{c}\end{aligned}$$

(c) We apply product formula (vi) with $x = 2a$ and $y = 5b$:

$$\begin{aligned}(2a - 5b)^3 &= (2a)^3 - 3(2a)^2(5b) + 3(2a)(5b)^2 - (5b)^3 \\ &= 8a^3 - 60a^2b + 150ab^2 - 125b^3\end{aligned}$$

If a polynomial is a product of other polynomials, then each polynomial in the product is a **factor** of the original polynomial. **Factoring** is the process of expressing a polynomial as a product. For example, since $x^2 - 9 = (x + 3)(x - 3)$, the polynomials $x + 3$ and $x - 3$ are factors of $x^2 - 9$.

Factoring is an important process in mathematics, since it may be used to reduce the study of a complicated expression to the study of several simpler expressions. For example, properties of the polynomial $x^2 - 9$ can be determined by examining the factors $x + 3$ and $x - 3$.

In this book we shall be interested primarily in **nontrivial factors** of polynomials; that is, factors that contain polynomials of degree greater than zero. An exception to this rule is that if the coefficients are restricted to *integers*, then we usually remove a common integral factor from each term of the polynomial. For example,

$$4x^2y + 8z^3 = 4(x^2y + 2z^3).$$

An integer $a > 1$ is *prime* if it cannot be written as a product of two positive integers greater than 1. Similarly, a polynomial with coefficients in some set S of numbers is **prime**, or **irreducible** over S, if it cannot be written as a product of two polynomials of positive degree with coefficients in S. A polynomial may be irreducible over one set S but not over another. For example, $x^2 - 2$ is irreducible over the rational numbers, since it cannot be expressed as a product of two polynomials of positive degree that have *rational* coefficients. However, $x^2 - 2$ is *not* irreducible over the real numbers, since we can write

$$x^2 - 2 = (x + \sqrt{2})(x - \sqrt{2}).$$

Similarly, $x^2 + 1$ is irreducible over the real numbers but, as we shall see in Section 2.4, not over the complex numbers.

Every polynomial $ax + b$ of degree 1 is irreducible.

Before we factor a polynomial, we must specify the number system (or set) from which the coefficients of the factors are to be chosen. In this chapter we shall use the rule that *if a polynomial has integral coefficients, then the factors should be polynomials with integral coefficients.* To **factor a polynomial** means to express it as a product of irreducible polynomials.

ILLUSTRATION | FACTORED POLYNOMIALS

- $x^2 + x - 6 = (x + 3)(x - 2)$
- $4x^2 - 9y^2 = (2x - 3y)(2x + 3y)$

It is usually difficult to factor polynomials of degree greater than 2. In simple cases the following formulas may be useful. Each formula can be verified by multiplication.

FACTORING FORMULAS

$x^2 - y^2 = (x + y)(x - y)$	(difference of two squares)
$x^3 - y^3 = (x - y)(x^2 + xy + y^2)$	(difference of two cubes)
$x^3 + y^3 = (x + y)(x^2 - xy + y^2)$	(sum of two cubes)

EXAMPLE ▪ 6

Factor the polynomial:

(a) $25r^2 - 49s^2$ **(b)** $81x^4 - y^4$ **(c)** $16x^4 - (y - 2z)^2$

SOLUTION

(a) We apply the difference of two squares formula with $x = 5r$ and $y = 7s$:

$$25r^2 - 49s^2 = (5r)^2 - (7s)^2 = (5r + 7s)(5r - 7s).$$

(b) We write $81x^4 = (9x^2)^2$ and $y^4 = (y^2)^2$ and apply the difference of two squares formula twice:

$$\begin{aligned} 81x^4 - y^4 &= (9x^2)^2 - (y^2)^2 \\ &= (9x^2 + y^2)(9x^2 - y^2) \\ &= (9x^2 + y^2)(3x + y)(3x - y) \end{aligned}$$

(c) We write $16x^4 = (4x^2)^2$ and apply the difference of two squares formula:

$$\begin{aligned} 16x^4 - (y - 2z)^2 &= (4x^2)^2 - (y - 2z)^2 \\ &= [(4x^2) + (y - 2z)][(4x^2) - (y - 2z)] \\ &= (4x^2 + y - 2z)(4x^2 - y + 2z) \end{aligned}$$

EXAMPLE ▪ 7

Factor the polynomial:

(a) $a^3 + 64b^3$ **(b)** $8x^6 - 27y^9$

SOLUTION

(a) We apply the sum of two cubes formula with $x = a$ and $y = 4b$:

$$\begin{aligned} a^3 + 64b^3 &= a^3 + (4b)^3 \\ &= (a + 4b)[a^2 - a(4b) + (4b)^2] \\ &= (a + 4b)(a^2 - 4ab + 16b^2) \end{aligned}$$

(b) We use the difference of two cubes formula:

$$\begin{aligned} 8x^6 - 27y^9 &= (2x^2)^3 - (3y^3)^3 \\ &= (2x^2 - 3y^3)[(2x^2)^2 + (2x^2)(3y^3) + (3y^3)^2] \\ &= (2x^2 - 3y^3)(4x^4 + 6x^2y^3 + 9y^6) \end{aligned}$$

A factorization of the polynomial $px^2 + qx + r$, where p, q, and r are integers, must be of the form $(ax + b)(cx + d)$ for integers a, b, c, d. It follows that $ac = p$, $bd = r$, and $ad + bc = q$. Evidently, only a limited number of choices for a, b, c, and d satisfy these conditions. If none of the choices work, then $px^2 + qx + r$ is prime. This method is also applicable to polynomials of the form $px^2 + qxy + ry^2$—the factorization must be of the form $(ax + by)(cx + dy)$.

EXAMPLE ▪ 8

Factor the polynomial:

(a) $6x^2 - 7x - 3$ **(b)** $12x^2 - 36xy + 27y^2$

(c) $4x^4y - 11x^3y^2 + 6x^2y^3$

SOLUTION

(a) If we write

$$6x^2 - 7x - 3 = (ax + b)(cx + d),$$

then $ac = 6$, $bd = -3$, and $ad + bc = -7$. Trying various possibilities, we arrive at the factorization:

$$6x^2 - 7x - 3 = (2x - 3)(3x + 1)$$

(b) Since each term has 3 as a factor, we begin by writing

$$12x^2 - 36xy + 27y^2 = 3(4x^2 - 12xy + 9y^2).$$

If a factorization of $4x^2 - 12xy + 9y^2$ as a product of two first-degree polynomials exists, then it must be of the form

$$4x^2 - 12xy + 9y^2 = (ax + by)(cx + dy)$$

with $ac = 4$, $bd = 9$, and $ad + bc = -12$. By trial we obtain

$$4x^2 - 12xy + 9y^2 = (2x - 3y)(2x - 3y) = (2x - 3y)^2.$$

Thus,

$$12x^2 - 36xy + 27y^2 = 3(4x^2 - 12xy + 9y^2) = 3(2x - 3y)^2.$$

(c) Since each term has x^2y as a factor, we begin by writing

$$4x^4y - 11x^3y^2 + 6x^2y^3 = x^2y(4x^2 - 11xy + 6y^2).$$

By trial we obtain the factorization:

$$4x^4y - 11x^3y^2 + 6x^2y^3 = x^2y(4x - 3y)(x - 2y)$$

If a sum contains four or more terms, it may be possible to group the terms in a suitable manner, and then find a factorization by using distributive properties. This technique is illustrated in the next example.

EXAMPLE ▪ 9

Factor the polynomial:

(a) $4ac + 2bc - 2ad - bd$ (b) $3x^3 + 2x^2 - 12x - 8$

SOLUTION

(a) We group the first two terms and the last two terms and then proceed as follows:

$$\begin{aligned} 4ac + 2bc - 2ad - bd &= (4ac + 2bc) - (2ad + bd) \\ &= 2c(2a + b) - d(2a + b) \end{aligned}$$

At this stage we have not factored the given expression because the right-hand side has the form

$$2ck - dk \quad \text{with} \quad k = 2a + b.$$

However, we can write

$$2ck - dk = (2c - d)k = (2c - d)(2a + b)$$

Hence

$$\begin{aligned} 4ac + 2bc - 2ad - bd &= 2c(2a + b) - d(2a + b) \\ &= (2c - d)(2a + b) \end{aligned}$$

(b) We group the first two terms and the last two terms and then proceed as follows:

$$\begin{aligned} 3x^3 + 2x^2 - 12x - 8 &= (3x^3 + 2x^2) - (12x + 8) \\ &= x^2(3x + 2) - 4(3x + 2) \\ &= (x^2 - 4)(3x + 2) \end{aligned}$$

Finally, using the difference of two squares formula, we obtain the factorization:

$$3x^3 + 2x^2 - 12x - 8 = (x + 2)(x - 2)(3x + 2)$$

1.4 EXERCISES

Exer. 1–42: Express as a polynomial.

1 $(3x^3 + 4x^2 - 7x + 1) + (9x^3 - 4x^2 - 6x)$

2 $(7x^3 + 2x^2 - 11x) + (-3x^3 - 2x^2 + 5x - 3)$

3 $(4x^3 + 5x - 3) - (3x^3 + 2x^2 + 5x - 7)$

4 $(6x^3 - 2x^2 + x - 2) - (8x^2 - x - 2)$

5 $(2u + 3)(u - 4) + 4u(u - 2)$

6 $(3u - 1)(u + 2) + 7u(u + 1)$

7 $(3x + 5)(2x^2 + 9x - 5)$

8 $(7x - 4)(x^3 - x^2 + 6)$

9 $(t^2 + 2t - 5)(3t^2 - t + 2)$

10 $(r^2 - 8r - 2)(-r^2 + 3r - 1)$

11 $(x + 1)(2x^2 - 2)(x^3 + 5)$

12 $(2x - 1)(x^2 - 5)(x^3 - 1)$

13 $\dfrac{8x^2y^3 - 10x^3y}{2x^2y}$

14 $\dfrac{6a^3b^3 - 9a^2b^2 + 3ab^4}{3ab^2}$

15 $\dfrac{3u^3v^4 - 2u^5v^2 + (u^2v^2)^2}{u^3v^2}$

16 $\dfrac{6x^2yz^3 - xy^2z}{xyz}$

17 $(2x + 5)(3x - 7)$

18 $(3x - 4)(2x + 9)$

19 $(5x + 7y)(3x + 2y)$

20 $(4x - 3y)(x - 5y)$

21 $(2x + 3y)(2x - 3y)$

22 $(5x + 4y)(5x - 4y)$

23 $(x + 2)^2(x - 2)^2$

24 $(x + y)^2(x - y)^2$

25 $(\sqrt{x} + \sqrt{y})(\sqrt{x} - \sqrt{y})$

26 $(\sqrt{x} + \sqrt{y})^2(\sqrt{x} - \sqrt{y})^2$

27 $(x^2 + 9)(x^2 - 4)$

28 $(x^2 + 1)(x^2 - 16)$

29 $(3x + 2y)^2$

30 $(5x - 4y)^2$

31 $(x^2 - 3y^2)^2$

32 $(2x^2 + 5y^2)^2$

33 $(x^{1/3} - y^{1/3})(x^{2/3} + x^{1/3}y^{1/3} + y^{2/3})$

34 $(x^{1/3} + y^{1/3})(x^{2/3} - x^{1/3}y^{1/3} + y^{2/3})$

35 $(x - 2y)^3$

36 $(x + 3y)^3$

37 $(2x + 3y)^3$

38 $(3x - 4y)^3$

39 $(a + b - c)^2$

40 $(x^2 + x + 1)^2$

41 $(2x + y - 3z)^2$

42 $(x - 2y + 3z)^2$

Exer. 43–96: Factor the polynomial.

43 $rs + 4st$

44 $4u^2 - 2uv$

45 $3a^2b^2 - 6a^2b$

46 $10xy + 15xy^2$

47 $3x^2y^3 - 9x^3y^2$

48 $16x^5y^2 + 8x^3y^3$

49 $15x^3y^5 - 25x^4y^2 + 10x^6y^4$

50 $121r^3s^4 + 77r^2s^4 - 55r^4s^3$

51 $8x^2 - 53x - 21$

52 $7x^2 + 10x - 8$

53 $x^2 + 3x + 4$

54 $3x^2 - 4x + 2$

55 $6x^2 + 7x - 20$

56 $12x^2 - x - 6$

57 $12x^2 - 29x + 15$

58 $21x^2 + 41x + 10$

59 $4x^2 - 20x + 25$

60 $9x^2 + 24x + 16$

61 $25z^2 + 30z + 9$

62 $16z^2 - 56z + 49$

63 $45x^2 + 38xy + 8y^2$

64 $50x^2 + 45xy - 18y^2$

65 $36r^2 - 25t^2$

66 $81r^2 - 16t^2$

67 $z^4 - 64w^2$

68 $9y^4 - 121x^2$

69 $x^4 - 4x^2$

70 $x^3 - 25x$

71 $x^2 + 25$

72 $4x^2 + 9$

73 $75x^2 - 48y^2$

74 $64x^2 - 36y^2$

75 $64x^3 + 27$

76 $125x^3 - 8$

77 $64x^3 - y^6$

78 $216x^9 + 125y^3$

79 $343x^3 + y^9$

80 $x^6 - 27y^3$

81 $2ax - 6bx + ay - 3by$

82 $2ay^2 - axy + 6xy - 3x^2$

83 $3x^3 + 3x^2 - 27x - 27$

84 $5x^3 + 10x^2 - 20x - 40$

85 $x^4 + 2x^3 - x - 2$

86 $x^4 - 3x^3 + 8x - 24$

87 $a^3 - a^2b + ab^2 - b^3$

88 $6w^8 + 17w^4 + 12$

89 $a^6 - b^6$

90 $x^8 - 16$

91 $x^2 + 4x + 4 - 9y^2$

92 $x^2 - 4y^2 - 6x + 9$

93 $y^6 + 7y^3 - 8$

94 $8c^6 + 19c^3 - 27$

95 $x^{16} - 1$

96 $4x^3 + 4x^2 + x$

Exer. 97–98: The ancient Greeks gave geometric proofs of the factoring formulas for the difference of two squares and the difference of two cubes. Establish the formula for the special case described.

97 Compute the areas of regions I and II in the figure to establish the difference of two squares formula for the special case $x > y$.

EXERCISE 97

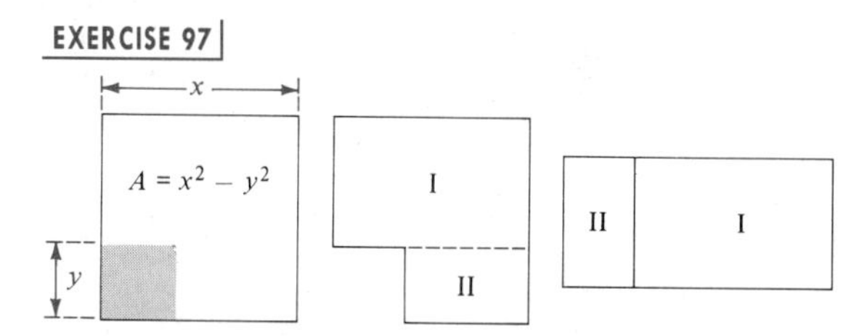

98 Compute the volumes of boxes I, II, and III in the figure to establish the difference of two cubes formula for the special case $x > y$.

EXERCISE 98

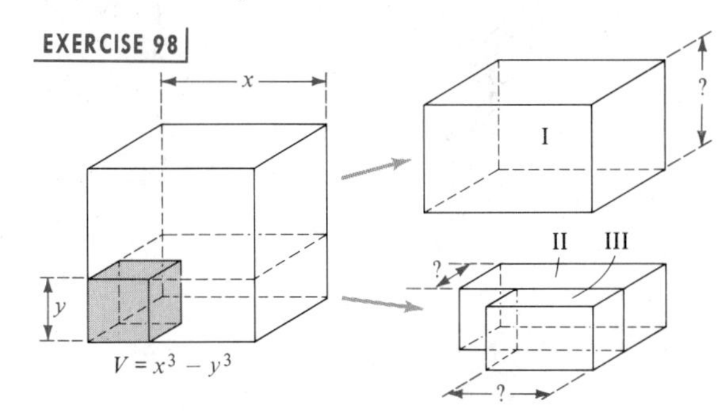

1.5 FRACTIONAL EXPRESSIONS

A **fractional expression** is a quotient p/q of two algebraic expressions p and q. Substituting real numbers for the variables gives us a quotient of two real numbers. Since division by zero is not allowed, we shall always assume that *when considering a fractional expression p/q, the variables are restricted to numbers that produce a nonzero value for the denominator q.*

A **rational expression** is a fractional expression p/q in which both the numerator p and the denominator q are polynomials.

ILLUSTRATION | RATIONAL EXPRESSIONS

- $\dfrac{x^2 - 5x + 1}{x^3 + 7}$
- $\dfrac{z^2x^4 - 3yz}{5z}$
- $\dfrac{x^3 - 3x^2y + 3xy^2 - y^3}{x^2 - y^2}$

Since the variables in a fractional expression represent real numbers, we may use the properties of quotients in Section 1.2, replacing the letters a, b, c, and d with algebraic expressions. The following property is of particular importance:

$$\frac{ad}{bd} = \frac{a}{b} \cdot \frac{d}{d} = \frac{a}{b} \cdot 1 = \frac{a}{b}$$

This simplification process is sometimes phrased *a common factor in the numerator and denominator of a quotient may be canceled*. In practice, we usually show this cancellation by drawing a slash through the common factor.

ILLUSTRATION | CANCELED COMMON FACTORS

- $\dfrac{a\not{d}}{b\not{d}} = \dfrac{a}{b}$
- $\dfrac{m\not{n}}{\not{n}pq} = \dfrac{m}{pq}$
- $\dfrac{\not{p}q\not{r}}{\not{r}\not{p}v} = \dfrac{q}{v}$

A rational expression is *simplified*, or reduced to lowest terms, if the numerator and denominator have no common factors other than 1 and -1. To *simplify a rational expression*, we factor both numerator and denominator into prime factors and then cancel common factors, as in the following illustration.

ILLUSTRATION | SIMPLIFIED RATIONAL EXPRESSIONS

- $\dfrac{3x^2 - 5x - 2}{x^2 - 4} = \dfrac{(3x + 1)\cancel{(x - 2)}}{(x + 2)\cancel{(x - 2)}} = \dfrac{3x + 1}{x + 2}$
- $\dfrac{2 - x - 3x^2}{6x^2 - x - 2} = \dfrac{-(3x^2 + x - 2)}{6x^2 - x - 2} = -\dfrac{\cancel{(3x - 2)}(x + 1)}{\cancel{(3x - 2)}(2x + 1)} = -\dfrac{x + 1}{2x + 1}$
- $\dfrac{(x^2 + 8x + 16)(x - 5)}{(x^2 - 5x)(x^2 - 16)} = \dfrac{(x + 4)^{\cancel{2}\,1}\cancel{(x - 5)}}{x\cancel{(x - 5)}\cancel{(x + 4)}(x - 4)} = \dfrac{x + 4}{x(x - 4)}$

EXAMPLE ▪ 1

Perform the indicated operation and simplify:

(a) $\dfrac{x^2 - 6x + 9}{x^2 - 1} \cdot \dfrac{2x - 2}{x - 3}$ **(b)** $\dfrac{x + 2}{2x - 3} \div \dfrac{x^2 - 4}{2x^2 - 3x}$

SOLUTION

(a)

$$\frac{x^2-6x+9}{x^2-1}\cdot\frac{2x-2}{x-3}=\frac{(x-3)^{\not 2}\,2\not(x-1)}{(x+1)\not(x-1)\not(x-3)}$$

$$=\frac{2(x-3)}{x+1}$$

(b)

$$\frac{x+2}{2x-3}\div\frac{x^2-4}{2x^2-3x}=\frac{x+2}{2x-3}\cdot\frac{2x^2-3x}{x^2-4}$$

$$=\frac{\not(x+2)x\not(2x-3)}{\not(2x-3)\not(x+2)(x-2)}=\frac{x}{x-2}$$

To add or subtract two fractional expressions, we usually find a common denominator and use the following properties of quotients:

$$\frac{a}{d}+\frac{c}{d}=\frac{a+c}{d};\qquad \frac{a}{d}-\frac{c}{d}=\frac{a-c}{d}.$$

If the denominators of the expressions are not the same, we may obtain a common denominator by multiplying the numerator and denominator of each fraction by suitable expressions. We often use the **least common denominator (lcd)** of the two fractions. To find the lcd we obtain the prime factorization for each denominator and then form the product of the different prime factors by using the *greatest* exponent that appears with each prime factor. Let us begin with a numerical example of this technique.

EXAMPLE ▪ 2

Express as a simplified rational number:

$$\frac{7}{24}+\frac{5}{18}$$

SOLUTION The prime factorizations of the denominators of the fractions are $24 = 2^3 \cdot 3$ and $18 = 2 \cdot 3^2$. To find the lcd we form the product of the different prime factors by using the highest exponent associated with each factor. This gives us $2^3 \cdot 3^2$. We now change each fraction to an equivalent fraction with denominator $2^3 \cdot 3^2$ and add:

$$\frac{7}{24}+\frac{5}{18}=\frac{7}{2^3\cdot 3}+\frac{5}{2\cdot 3^2}=\frac{7\cdot 3}{2^3\cdot 3^2}+\frac{5\cdot 2^2}{2^3\cdot 3^2}$$

$$=\frac{21}{72}+\frac{20}{72}=\frac{41}{72}$$

The method for finding the lcd for rational expressions is analogous to the process illustrated in Example 2. The only difference is that we use factorizations of polynomials instead of integers.

EXAMPLE ▪ 3

Perform the operations and simplify:

$$\frac{6}{x(3x-2)}+\frac{5}{3x-2}-\frac{2}{x^2}$$

SOLUTION The denominators are already in factored form. Evidently, the lcd is $x^2(3x-2)$. To obtain three fractions having a denominator of $x^2(3x-2)$, we multiply numerator and denominator of the first fraction by x, of the second by x^2, and of the third by $3x-2$. This gives us

$$\begin{aligned}\frac{6}{x(3x-2)}+\frac{5}{3x-2}-\frac{2}{x^2}&=\frac{6x}{x^2(3x-2)}+\frac{5x^2}{x^2(3x-2)}-\frac{2(3x-2)}{x^2(3x-2)}\\&=\frac{6x+5x^2-6x+4}{x^2(3x-2)}=\frac{5x^2+4}{x^2(3x-2)}.\end{aligned}$$

EXAMPLE ▪ 4

Perform the operations and simplify:

$$\frac{2x+5}{x^2+6x+9}+\frac{x}{x^2-9}+\frac{1}{x-3}$$

SOLUTION We begin by factoring denominators:

$$\frac{2x+5}{x^2+6x+9}+\frac{x}{x^2-9}+\frac{1}{x-3}=\frac{2x+5}{(x+3)^2}+\frac{x}{(x+3)(x-3)}+\frac{1}{x-3}$$

Since the lcd is $(x+3)^2(x-3)$, we multiply numerator and denominator of the first fraction by $x-3$, of the second by $x+3$, and of the third by $(x+3)^2$ and then add:

$$\begin{aligned}&\frac{(2x+5)(x-3)}{(x+3)^2(x-3)}+\frac{x(x+3)}{(x+3)^2(x-3)}+\frac{(x+3)^2}{(x+3)^2(x-3)}\\&\qquad=\frac{(2x^2-x-15)+(x^2+3x)+(x^2+6x+9)}{(x+3)^2(x-3)}\\&\qquad=\frac{4x^2+8x-6}{(x+3)^2(x-3)}=\frac{2(2x^2+4x-3)}{(x+3)^2(x-3)}\end{aligned}$$

It is sometimes necessary to simplify quotients in which the numerator and denominator are not polynomials, as illustrated in the next example.

EXAMPLE ▪ 5

Simplify:

$$\frac{1 - \dfrac{2}{x+1}}{x - \dfrac{1}{x}}$$

SOLUTION We change the numerator and denominator into single fractions and then use the rule for dividing quotients:

$$\frac{1 - \dfrac{2}{x+1}}{x - \dfrac{1}{x}} = \frac{\dfrac{(x+1) - 2}{x+1}}{\dfrac{x^2 - 1}{x}} = \frac{\dfrac{x-1}{x+1}}{\dfrac{x^2-1}{x}}$$

$$= \frac{x-1}{x+1} \cdot \frac{x}{x^2-1} = \frac{(x-1)x}{(x+1)(x+1)(x-1)}$$

$$= \frac{x}{(x+1)^2}$$

The denominators of fractional expressions may contain sums or differences that involve radicals. Some denominators can be rationalized, as shown in the next example.

EXAMPLE ▪ 6

Rationalize the denominator of the fractional expression:

$$\frac{1}{\sqrt{x} + \sqrt{y}}$$

SOLUTION By the difference of two squares formula,

$$(\sqrt{x})^2 - (\sqrt{y})^2 = (\sqrt{x} + \sqrt{y})(\sqrt{x} - \sqrt{y});$$

that is,

$$x - y = (\sqrt{x} + \sqrt{y})(\sqrt{x} - \sqrt{y}).$$

Thus we can rationalize the denominator of the given fraction by multiplying both numerator and denominator by $\sqrt{x} - \sqrt{y}$ as follows:

$$\begin{aligned}\frac{1}{\sqrt{x}+\sqrt{y}} &= \frac{1}{\sqrt{x}+\sqrt{y}} \cdot \frac{\sqrt{x}-\sqrt{y}}{\sqrt{x}-\sqrt{y}} \\ &= \frac{\sqrt{x}-\sqrt{y}}{(\sqrt{x})^2-(\sqrt{y})^2} \\ &= \frac{\sqrt{x}-\sqrt{y}}{x-y}\end{aligned}$$

In the following example we rationalize a *numerator*.

EXAMPLE ▪ 7

Rationalize the numerator of

$$\frac{\sqrt{x+h}-\sqrt{x}}{h}$$

if x and h are positive.

SOLUTION We multiply numerator and denominator by the expression $\sqrt{x+h}+\sqrt{x}$ and simplify:

$$\begin{aligned}\frac{\sqrt{x+h}-\sqrt{x}}{h} &= \frac{\sqrt{x+h}-\sqrt{x}}{h} \cdot \frac{\sqrt{x+h}+\sqrt{x}}{\sqrt{x+h}+\sqrt{x}} \\ &= \frac{(\sqrt{x+h})^2-(\sqrt{x})^2}{h(\sqrt{x+h}+\sqrt{x})} \\ &= \frac{(x+h)-x}{h(\sqrt{x+h}+\sqrt{x})} \\ &= \frac{h}{h(\sqrt{x+h}+\sqrt{x})} \\ &= \frac{1}{\sqrt{x+h}+\sqrt{x}}\end{aligned}$$

For certain types of problems in calculus it is necessary to simplify expressions of the types given in the next two examples. (Also see Exercises 63–68.)

EXAMPLE ▪ 8

Simplify:

$$\frac{\dfrac{1}{(x+h)^2} - \dfrac{1}{x^2}}{h}.$$

SOLUTION

$$\begin{aligned}
\frac{\dfrac{1}{(x+h)^2} - \dfrac{1}{x^2}}{h} &= \frac{\dfrac{x^2 - (x+h)^2}{(x+h)^2x^2}}{h} \\
&= \frac{x^2 - (x^2 + 2xh + h^2)}{(x+h)^2x^2h} \\
&= \frac{x^2 - x^2 - 2xh - h^2}{(x+h)^2x^2h} \\
&= \frac{-\cancel{h}(2x+h)}{(x+h)^2x^2\cancel{h}} \\
&= -\frac{2x+h}{(x+h)^2x^2}
\end{aligned}$$

EXAMPLE ▪ 9

Simplify:

$$\frac{3x^2(2x+5)^{1/2} - x^3(\frac{1}{2})(2x+5)^{-1/2}(2)}{[(2x+5)^{1/2}]^2}$$

SOLUTION

$$\begin{aligned}
\frac{3x^2(2x+5)^{1/2} - x^3(\frac{1}{2})(2x+5)^{-1/2}(2)}{[(2x+5)^{1/2}]^2} &= \frac{3x^2(2x+5)^{1/2} - \dfrac{x^3}{(2x+5)^{1/2}}}{2x+5} \\
&= \frac{\dfrac{3x^2(2x+5) - x^3}{(2x+5)^{1/2}}}{2x+5} \\
&= \frac{6x^3 + 15x^2 - x^3}{(2x+5)^{1/2}(2x+5)} \\
&= \frac{5x^3 + 15x^2}{(2x+5)^{3/2}} = \frac{5x^2(x+3)}{(2x+5)^{3/2}}
\end{aligned}$$

An alternative solution is to eliminate the negative power $(2x+5)^{-1/2}$ in the given expression first, by multiplying numerator and denominator by $(2x+5)^{1/2}$ as follows:

$$\frac{3x^2(2x+5)^{1/2} - x^3(\frac{1}{2})(2x+5)^{-1/2}(2)}{[(2x+5)^{1/2}]^2} \cdot \frac{(2x+5)^{1/2}}{(2x+5)^{1/2}}$$

$$= \frac{3x^2(2x+5) - x^3(\frac{1}{2})(2)}{(2x+5)(2x+5)^{1/2}}$$

$$= \frac{6x^3 + 15x^2 - x^3}{(2x+5)^{3/2}}$$

The remainder of the simplification is the same.

1.5 EXERCISES

Exer. 1–4: Write the expression as a simplified rational number.

1 $\frac{3}{50} + \frac{7}{30}$

2 $\frac{4}{63} + \frac{5}{42}$

3 $\frac{5}{24} - \frac{3}{20}$

4 $\frac{11}{54} - \frac{7}{72}$

Exer. 5–56: Simplify the expression.

5 $\frac{2x^2 + 7x + 3}{2x^2 - 7x - 4}$

6 $\frac{2x^2 + 9x - 5}{3x^2 + 17x + 10}$

7 $\frac{y^2 - 25}{y^3 - 125}$

8 $\frac{y^2 - 9}{y^3 + 27}$

9 $\frac{12 + r - r^2}{r^3 + 3r^2}$

10 $\frac{10 + 3r - r^2}{r^4 + 2r^3}$

11 $\frac{9x^2 - 4}{3x^2 - 5x + 2} \cdot \frac{9x^4 - 6x^3 + 4x^2}{27x^4 + 8x}$

12 $\frac{4x^2 - 9}{2x^2 + 7x + 6} \cdot \frac{4x^4 + 6x^3 + 9x^2}{8x^7 - 27x^4}$

13 $\frac{5a^2 + 12a + 4}{a^4 - 16} \div \frac{25a^2 + 20a + 4}{a^2 - 2a}$

14 $\frac{a^3 - 8}{a^2 - 4} \div \frac{a}{a^3 + 8}$

15 $\frac{6}{x^2 - 4} - \frac{3x}{x^2 - 4}$

16 $\frac{15}{x^2 - 9} - \frac{5x}{x^2 - 9}$

17 $\frac{2}{3s + 1} - \frac{9}{(3s + 1)^2}$

18 $\frac{4}{(5s - 2)^2} + \frac{s}{5s - 2}$

19 $\frac{2}{x} + \frac{3x + 1}{x^2} - \frac{x - 2}{x^3}$

20 $\frac{5}{x} - \frac{2x - 1}{x^2} + \frac{x + 5}{x^3}$

21 $\frac{3t}{t + 2} + \frac{5t}{t - 2} - \frac{40}{t^2 - 4}$

22 $\frac{t}{t + 3} + \frac{4t}{t - 3} - \frac{18}{t^2 - 9}$

23 $\frac{4x}{3x - 4} + \frac{8}{3x^2 - 4x} + \frac{2}{x}$

24 $\frac{12x}{2x + 1} - \frac{3}{2x^2 + x} + \frac{5}{x}$

25 $\frac{2x}{x + 2} - \frac{8}{x^2 + 2x} + \frac{3}{x}$

26 $\frac{5x}{2x + 3} - \frac{6}{2x^2 + 3x} + \frac{2}{x}$

27 $\frac{p^4 + 3p^3 - 8p - 24}{p^3 - 2p^2 - 9p + 18}$

28 $\frac{2ac + bc - 6ad - 3bd}{6ac + 2ad + 3bc + bd}$

29 $3 + \frac{5}{u} + \frac{2u}{3u + 1}$

30 $4 + \frac{2}{u} - \frac{3u}{u + 5}$

31 $\frac{2x + 1}{x^2 + 4x + 4} - \frac{6x}{x^2 - 4} + \frac{3}{x - 2}$

32 $\frac{2x + 6}{x^2 + 6x + 9} + \frac{5x}{x^2 - 9} + \frac{7}{x - 3}$

33 $\dfrac{\frac{b}{a} - \frac{a}{b}}{\frac{1}{a} - \frac{1}{b}}$

34 $\dfrac{\frac{1}{x + 2} - 3}{\frac{4}{x} - x}$

35 $\dfrac{\dfrac{x}{y^2}-\dfrac{y}{x^2}}{\dfrac{1}{y^2}-\dfrac{1}{x^2}}$

36 $\dfrac{\dfrac{r}{s}+\dfrac{s}{r}}{\dfrac{r^2}{s^2}-\dfrac{s^2}{r^2}}$

37 $\dfrac{\dfrac{5}{x+1}+\dfrac{2x}{x+3}}{\dfrac{x}{x+1}+\dfrac{7}{x+3}}$

38 $\dfrac{\dfrac{3}{w}-\dfrac{6}{2w+1}}{\dfrac{5}{w}+\dfrac{8}{2w+1}}$

39 $\dfrac{(x+h)^2-3(x+h)-(x^2-3x)}{h}$

40 $\dfrac{(x+h)^3+5(x+h)-(x^3+5x)}{h}$

41 $\dfrac{\dfrac{1}{(x+h)^3}-\dfrac{1}{x^3}}{h}$

42 $\dfrac{\dfrac{1}{x+h}-\dfrac{1}{x}}{h}$

43 $\dfrac{\dfrac{4}{3x+3h-1}-\dfrac{4}{3x-1}}{h}$

44 $\dfrac{\dfrac{5}{2x+2h+3}-\dfrac{5}{2x+3}}{h}$

45 $(2x^2-3x+1)(4)(3x+2)^3(3)+(3x+2)^4(4x-3)$

46 $(6x-5)^3(2)(x^2+4)(2x)+(x^2+4)^2(3)(6x-5)^2(6)$

47 $(x^2-4)^{1/2}(3)(2x+1)^2(2)+(2x+1)^3(\frac{1}{2})(x^2-4)^{-1/2}(2x)$

48 $(3x+2)^{1/3}(2)(4x-5)(4)+(4x-5)^2(\frac{1}{3})(3x+2)^{-2/3}(3)$

49 $(3x+1)^6(\frac{1}{2})(2x-5)^{-1/2}(2)+(2x-5)^{1/2}(6)(3x+1)^5(3)$

50 $(x^2+9)^4(-\frac{1}{3})(x+6)^{-4/3}+(x+6)^{-1/3}(4)(x^2+9)^3(2x)$

51 $\dfrac{(6x+1)^3(27x^2+2)-(9x^3+2x)(3)(6x+1)^2(6)}{(6x+1)^6}$

52 $\dfrac{(x^2-1)^4(2x)-x^2(4)(x^2-1)^3(2x)}{(x^2-1)^8}$

53 $\dfrac{(x^2+4)^{1/3}(3)-(3x)(\frac{1}{3})(x^2+4)^{-2/3}(2x)}{[(x^2+4)^{1/3}]^2}$

54 $\dfrac{(1-x^2)^{1/2}(2x)-x^2(\frac{1}{2})(1-x^2)^{-1/2}(-2x)}{[(1-x^2)^{1/2}]^2}$

55 $\dfrac{(4x^2+9)^{1/2}(2)-(2x+3)(\frac{1}{2})(4x^2+9)^{-1/2}(8x)}{[(4x^2+9)^{1/2}]^2}$

56 $\dfrac{(3x+2)^{1/2}(\frac{1}{3})(2x+3)^{-2/3}(2)-(2x+3)^{1/3}(\frac{1}{2})(3x+2)^{-1/2}(3)}{[(3x+2)^{1/2}]^2}$

Exer. 57–62: Rationalize the denominator.

57 $\dfrac{\sqrt{t}+5}{\sqrt{t}-5}$

58 $\dfrac{\sqrt{t}-4}{\sqrt{t}+4}$

59 $\dfrac{81x^2-16y^2}{3\sqrt{x}-2\sqrt{y}}$

60 $\dfrac{16x^2-y^2}{2\sqrt{x}-\sqrt{y}}$

61 $\dfrac{1}{\sqrt[3]{a}-\sqrt[3]{b}}$ (*Hint:* Multiply numerator and denominator by $\sqrt[3]{a^2}+\sqrt[3]{a}\sqrt[3]{b}+\sqrt[3]{b^2}$.)

62 $\dfrac{1}{\sqrt[3]{x}+\sqrt[3]{y}}$

Exer. 63–68: Rationalize the numerator.

63 $\dfrac{\sqrt{a}-\sqrt{b}}{a^2-b^2}$

64 $\dfrac{\sqrt{b}+\sqrt{c}}{b^2-c^2}$

65 $\dfrac{\sqrt{2(x+h)+1}-\sqrt{2x+1}}{h}$

66 $\dfrac{\sqrt{x}-\sqrt{x+h}}{h\sqrt{x}\sqrt{x+h}}$

67 $\dfrac{\sqrt{1-x-h}-\sqrt{1-x}}{h}$

68 $\dfrac{\sqrt[3]{x+h}-\sqrt[3]{x}}{h}$ (*Hint:* Compare with Exercise 61.)

1.6 THE BINOMIAL THEOREM

A sum $a+b$ is a **binomial**. Let us consider the expression $(a+b)^n$ for any positive integer n. A general formula for *expanding* $(a+b)^n$, that is, for expressing it as a sum, is given by the **binomial theorem**. We shall begin by considering some special cases. If we actually perform the multiplications, we obtain

$$(a + b)^2 = a^2 + 2ab + b^2$$
$$(a + b)^3 = a^3 + 3a^2b + 3ab^2 + b^3$$
$$(a + b)^4 = a^4 + 4a^3b + 6a^2b^2 + 4ab^3 + b^4$$
$$(a + b)^5 = a^5 + 5a^4b + 10a^3b^2 + 10a^2b^3 + 5ab^4 + b^5.$$

The expansions of $(a + b)^n$ for $n = 2, 3, 4,$ and 5 have the following properties:

(i) There are $n + 1$ terms, the first being a^n and the last b^n.

(ii) Proceeding from any term to the next, the power of a decreases by 1 and the power of b increases by 1. For each term the sum of the exponents of a and b is n.

(iii) Each term has the form $(c)a^{n-k}b^k$, where the coefficient c is a real number and $k = 0, 1, 2, \ldots, n$.

(iv) The following formula is true for each of the first n terms of the expansion:

$$\frac{(\text{coefficient of term}) \cdot (\text{exponent of } a)}{\text{number of term}} = \text{coefficient of next term}.$$

The following table illustrates property (iv) for the expansion of $(a + b)^5$.

Number of term	Coefficient of term	Exponent of a	Coefficient of next term
1	1	5	$\frac{1 \cdot 5}{1} = 5$
2	5	4	$\frac{5 \cdot 4}{2} = 10$
3	10	3	$\frac{10 \cdot 3}{3} = 10$
4	10	2	$\frac{10 \cdot 2}{4} = 5$
5	5	1	$\frac{5 \cdot 1}{5} = 1$

By using the information in the table (and remembering that the sum of the exponents is 5 for each term), we get the expansion

$$(a + b)^5 = a^5 + 5a^4b + 10a^3b^2 + 10a^2b^3 + 5ab^4 + b^5,$$

which agrees with the product we obtained at the beginning of this discussion.

Let us next consider $(a + b)^n$ for an arbitrary positive integer n. The first term is a^n, which has coefficient 1. If we assume that property (iv) is true, we obtain the successive coefficients listed in the following table.

Number of term	Coefficient of term	Exponent of a	Coefficient of next term
1	1	n	$\frac{1 \cdot n}{1} = \frac{n}{1}$
2	$\frac{n}{1}$	$n - 1$	$\frac{n(n-1)}{1 \cdot 2}$
3	$\frac{n(n-1)}{1 \cdot 2}$	$n - 2$	$\frac{n(n-1)(n-2)}{1 \cdot 2 \cdot 3}$
4	$\frac{n(n-1)(n-2)}{1 \cdot 2 \cdot 3}$	$n - 3$	$\frac{n(n-1)(n-2)(n-3)}{1 \cdot 2 \cdot 3 \cdot 4}$

The last coefficient belongs to the fifth term. Similarly, the coefficient of the sixth term is

$$\frac{n(n-1)(n-2)(n-3)(n-4)}{1 \cdot 2 \cdot 3 \cdot 4 \cdot 5}.$$

The coefficient of the $(k + 1)$st term in the expansion of $(a + b)^n$ is the following.

$(k + 1)$st COEFFICIENT OF $(a + b)^n$

$$\frac{n(n-1)(n-2)(n-3)\cdots(n-k+1)}{1 \cdot 2 \cdot 3 \cdot 4 \cdots k}; \qquad k = 0, 1, 2, \ldots, n$$

The $(k + 1)$st coefficient can be written in a compact form by using **factorial notation**. If n is any positive integer, the symbol $n!$ (n *factorial*) is defined as follows.

DEFINITION OF $n!$ $(n > 0)$

$$n! = n(n-1)(n-2)\cdots 1$$

Thus, $n!$ is the product of the first n positive integers.

ILLUSTRATION $n!$

- $1! = 1$
- $2! = 2 \cdot 1$
- $3! = 3 \cdot 2 \cdot 1 = 6$
- $4! = 4 \cdot 3 \cdot 2 \cdot 1 = 24$
- $5! = 5 \cdot 4 \cdot 3 \cdot 2 \cdot 1 = 120$
- $6! = 6 \cdot 5 \cdot 4 \cdot 3 \cdot 2 \cdot 1 = 720$
- $7! = 7 \cdot 6 \cdot 5 \cdot 4 \cdot 3 \cdot 2 \cdot 1 = 5040$

Notice the rapid growth of $n!$ as n increases. We can show with the aid of a calculator that

$$20! \approx 2.4329 \times 10^{18} \qquad \text{and} \qquad 50! \approx 3.0414 \times 10^{64}.$$

To ensure that certain formulas will be true for all *nonnegative* integers, we define 0! as follows.

DEFINITION OF 0!

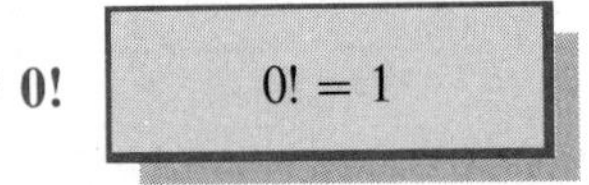

$$0! = 1$$

Note that for $n \geq 1$,

$$n! = n[(n-1)!] = n(n-1)!$$

For larger values of n we can write

$$n! = n(n-1)(n-2)! = n(n-1)(n-2)(n-3)!$$

and so on. The formula for the $(k+1)$st coefficient in the expansion of $(a+b)^n$ can be written as follows:

$$\frac{n(n-1)(n-2)\cdots(n-k+1)}{1 \cdot 2 \cdot 3 \cdots k} = \frac{\dfrac{n(n-1)(n-2)\cdots(n-k+1)(n-k)!}{(n-k)!}}{k!}$$

$$= \frac{\dfrac{n!}{(n-k)!}}{k!} = \frac{n!}{k!(n-k)!}$$

These numbers are **binomial coefficients** and are often denoted by either of the symbols $\binom{n}{k}$ or $C(n, k)$. Thus we have the following.

$(k+1)$st COEFFICIENT OF $(a+b)^n$ (Alternative form)

$$\binom{n}{k} = C(n,k) = \frac{n!}{k!(n-k)!}, \qquad k = 0, 1, 2, \ldots, n$$

The symbols $\binom{n}{k}$ and $C(n, k)$ are sometimes read *n choose k*.

EXAMPLE ▪ 1

Find $\binom{5}{0}$, $\binom{5}{1}$, $\binom{5}{2}$, $\binom{5}{3}$, $\binom{5}{4}$, and $\binom{5}{5}$.

SOLUTION These six numbers are the coefficients in the expansion of $(a+b)^5$, which we tabulated earlier in this section. By definition,

$$\binom{5}{0} = \frac{5!}{0!(5-0)!} = \frac{5!}{1 \cdot 5!} = 1$$

$$\binom{5}{1} = \frac{5!}{1!(5-1)!} = \frac{5!}{1 \cdot 4!} = \frac{5 \cdot 4!}{4!} = 5$$

$$\binom{5}{2} = \frac{5!}{2!(5-2)!} = \frac{5!}{2!3!} = \frac{5 \cdot 4 \cdot 3!}{2!3!} = \frac{20}{2} = 10$$

$$\binom{5}{3} = \frac{5!}{3!(5-3)!} = \frac{5!}{3!2!} = \frac{5 \cdot 4 \cdot 3!}{3!2!} = \frac{20}{2} = 10$$

$$\binom{5}{4} = \frac{5!}{4!(5-4)!} = \frac{5!}{4!1!} = \frac{5 \cdot 4!}{4!} = 5$$

$$\binom{5}{5} = \frac{5!}{5!(5-5)!} = \frac{1}{0!} = \frac{1}{1} = 1$$

The binomial theorem may be stated as follows. (A proof is given in Chapter 7.)

THE BINOMIAL THEOREM

$$(a+b)^n = a^n + \binom{n}{1}a^{n-1}b + \binom{n}{2}a^{n-2}b^2 + \cdots + \binom{n}{k}a^{n-k}b^k + \cdots + \binom{n}{n-1}ab^{n-1} + b^n$$

An alternative statement of the binomial theorem is as follows.

THE BINOMIAL THEOREM (Alternative form)

$$(a+b)^n = a^n + na^{n-1}b + \frac{n(n-1)}{2!}a^{n-2}b^2 + \cdots$$
$$+ \frac{n(n-1)(n-2)\cdots(n-k+1)}{k!}a^{n-k}b^k$$
$$+ \cdots + nab^{n-1} + b^n$$

The following examples may be solved by using either the general formulas for the binomial theorem or by repeated use of property (iv) stated at the beginning of this section.

EXAMPLE ▪ 2

Find the binomial expansion of $(2x + 3y^2)^4$.

SOLUTION We use the binomial theorem with $a = 2x$, $b = 3y^2$, and $n = 4$:

$$(2x+3y^2)^4 = (2x)^4 + \binom{4}{1}(2x)^3(3y^2) + \binom{4}{2}(2x)^2(3y^2)^2$$
$$+ \binom{4}{3}(2x)(3y^2)^3 + (3y^2)^4.$$

This can be written

$$(2x+3y^2)^4 = 16x^4 + 4(8x^3)(3y^2) + 6(4x^2)(9y^4) + 4(2x)(27y^6) + 81y^8$$
$$= 16x^4 + 96x^3y^2 + 216x^2y^4 + 216xy^6 + 81y^8.$$

The next example illustrates that if either a or b is negative, then the terms of the expansion are alternately positive and negative.

EXAMPLE ▪ 3

Expand $\left(\frac{1}{x} - 2\sqrt{x}\right)^5$.

SOLUTION The binomial coefficients for $(a + b)^5$ were calculated in Example 1. Thus, if we let $a = 1/x$, $b = -2\sqrt{x}$, and $n = 5$ in the binomial theorem, then we obtain

$$\left(\frac{1}{x} - 2\sqrt{x}\right)^5 = \left(\frac{1}{x}\right)^5 + 5\left(\frac{1}{x}\right)^4(-2\sqrt{x}) + 10\left(\frac{1}{x}\right)^3(-2\sqrt{x})^2$$
$$+ 10\left(\frac{1}{x}\right)^2(-2\sqrt{x})^3 + 5\left(\frac{1}{x}\right)(-2\sqrt{x})^4 + (-2\sqrt{x})^5,$$

which can be written as

$$\left(\frac{1}{x} - 2\sqrt{x}\right)^5 = \frac{1}{x^5} - \frac{10}{x^{7/2}} + \frac{40}{x^2} - \frac{80}{x^{1/2}} + 80x - 32x^{5/2}.$$

To find a specific term in the expansion of $(a + b)^n$, it is convenient to first find the exponent k that is to be assigned to b. Notice that, by the binomial theorem, *the exponent of b is always one less than the number of the term.* Once k is found, the exponent of a is $n - k$ and the coefficient is $\binom{n}{k}$.

EXAMPLE ▪ 4

Find the fifth term in the expansion of $(x^3 + \sqrt{y})^{13}$.

SOLUTION Let $a = x^3$ and $b = \sqrt{y}$. The exponent of b in the fifth term is 4 and hence the exponent of a is 9. From the discussion of the preceding paragraph we obtain

$$\binom{13}{4}(x^3)^9(\sqrt{y})^4 = \frac{13!}{4!(13-4)!}x^{27}y^2 = \frac{13 \cdot 12 \cdot 11 \cdot 10}{4!}x^{27}y^2$$
$$= 715x^{27}y^2.$$

There is an interesting triangular array of numbers called **Pascal's triangle**, which can be used to obtain binomial coefficients. The numbers are arranged as follows.

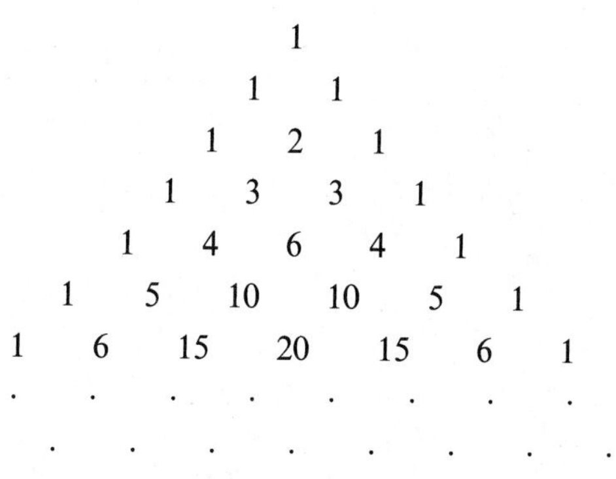

The numbers in the second row are the coefficients in the expansion of $(a + b)^1$; those in the third row are the coefficients determined by $(a + b)^2$; those in the fourth row are obtained from $(a + b)^3$, and so on. Each number in the array that is different from 1 can be found by adding the two numbers in the previous row that appear above and immediately to the left and right of the number, as illustrated in the solution of the next example.

EXAMPLE ▪ 5

Find the eighth row of Pascal's triangle and use it to expand $(a + b)^7$.

SOLUTION Let us rewrite the seventh row and then use the process described previously. In the following display the arrows indicate which two numbers in row seven are added to obtain the numbers in row eight.

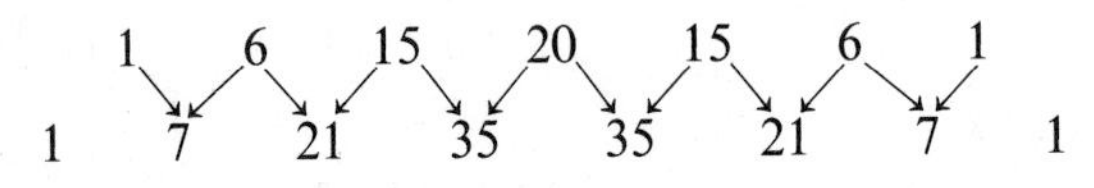

The eighth row gives us the coefficients in the expansion of $(a + b)^7$:

$$(a + b)^7 = a^7 + 7a^6b + 21a^5b^2 + 35a^4b^3 + 35a^3b^4 + 21a^2b^5 + 7ab^6 + b^7$$

Pascal's triangle is useful for expanding small powers of $a + b$; however, to expand large powers or to find a specific term, as in Example 4, the general formula given by the binomial theorem is more useful.

1.6 EXERCISES

Exer. 1–12: Evaluate the expression.

1 $7!$

2 $8!$

3 $3!4!$

4 $5!0!$

5 $\dfrac{2!}{4!}$

6 $\dfrac{6!}{3!}$

7 $\dbinom{5}{5}$

8 $\dbinom{7}{0}$

9 $\dbinom{7}{5}$

10 $\dbinom{8}{4}$

11 $\dbinom{13}{4}$

12 $\dbinom{52}{2}$

Exer. 13–24: Expand and simplify the expression.

13 $(a + b)^6$

14 $(a + b)^7$

15 $(a + b)^4$

16 $(a - b)^5$

17 $(3x - 5y)^4$

18 $(2t - s)^5$

19 $(\frac{1}{3}x + y^2)^5$

20 $(\frac{1}{2}c + d^3)^4$

21 $\left(\dfrac{1}{x^2} + 3x\right)^6$

22 $\left(\dfrac{1}{x^3} - 2x\right)^5$

23 $\left(\sqrt{x} + \dfrac{1}{\sqrt{x}}\right)^5$

24 $\left(\sqrt{x} - \dfrac{1}{\sqrt{x}}\right)^5$

Exer. 25–40: Without expanding completely, find the indicated term(s) in the expansion of the expression.

25 $(3c^{2/5} + c^{4/5})^{25}$; first three terms

26 $(x^3 + 5x^{-2})^{20}$; first three terms

27 $(4b^{-1} - 3b)^{15}$; last three terms

28 $(s - 2t^3)^{12}$; last three terms

29 $\left(\dfrac{3}{c} + \dfrac{c^2}{4}\right)^7$; sixth term

30 $(3a^2 - \sqrt{b})^9$; fifth term

31 $(\frac{1}{3}u + 4v)^8$; seventh term

32 $(3x^2 - y^3)^{10}$; fourth term

33 $(x^{1/2} + y^{1/2})^8$; middle term

34 $(rs^2 + t)^7$; two middle terms

35 $\left(3x - \dfrac{1}{4x}\right)^6$; term that does not contain x

36 $(xy - 3y^{-3})^8$; term that does not contain y

37 $(2y + x^2)^8$; term that contains x^{10}

38 $(x^2 - 2y^3)^5$; term that contains y^6

39 $(3b^3 - 2a^2)^4$; term that contains b^9

40 $(\sqrt{c} + \sqrt{d})^8$; term that contains c^2

41 Approximate $(1.2)^{10}$ by using the first three terms in the expansion of $(1 + 0.2)^{10}$, and compare your answer with that obtained using a calculator.

42 Approximate $(0.9)^4$ by using the first three terms in the expansion of $(1 - 0.1)^4$, and compare your answer with that obtained using a calculator.

Exer. 43–44: Simplify the expression using the binomial theorem.

43 $\dfrac{(x + h)^4 - x^4}{h}$

44 $\dfrac{(x + h)^5 - x^5}{h}$

1.7 REVIEW

Define or discuss each of the following.

Commutative properties of real numbers ■ Associative properties ■ Distributive properties ■ Integers ■ Prime number ■ Rational and irrational numbers ■ Coordinate line ■ A number a is greater than a number b ■ A number a is less than a number b ■ Absolute value of a real number ■ The distance between points on a coordinate line ■ Exponent ■ Laws of exponents ■ Principal nth root ■ Radical notation ■ Rationalizing a denominator ■ Rational exponents ■ Scientific form for real numbers ■ Variable ■ Domain of a variable ■ Algebraic expression ■ Polynomial ■ Degree of a polynomial ■ Prime polynomial ■ Irreducible polynomial ■ Rational expression ■ Factorial notation ■ The binomial theorem ■ Binomial coefficients ■ Pascal's triangle

1.7 EXERCISES

1 Express as a simplified rational number:

(a) $(\frac{2}{3})(-\frac{5}{8})$ (b) $\frac{3}{4} + \frac{6}{5}$

(c) $\frac{5}{8} - \frac{6}{7}$ (d) $\frac{3}{4} \div \frac{6}{5}$

2 Replace the symbol □ with either $<$, $>$, or $=$ so that the resulting statement is true:

(a) $-0.1 \;\square\; -0.01$ (b) $\sqrt{9} \;\square\; -3$

(c) $\frac{1}{6} \;\square\; 0.166$

3 Express as an inequality:

(a) x is negative.

(b) a is between $\frac{1}{2}$ and $\frac{1}{3}$.

(c) The absolute value of x is not greater than 4.

4 Rewrite without using the absolute value symbol:

(a) $|-7|$ (b) $\dfrac{|-5|}{-5}$ (c) $|3^{-1} - 2^{-1}|$

5 If points A, B, and C on a coordinate line have coordinates -8, 4, and -3, respectively, find the following:

(a) $d(A, C)$ (b) $d(C, A)$ (c) $d(B, C)$

6 Determine if the expression is true for all values of the variables, whenever the expressions are defined.

(a) $(x + y)^2 = x^2 + y^2$ (b) $\dfrac{1}{\sqrt{x + y}} = \dfrac{1}{\sqrt{x}} + \dfrac{1}{\sqrt{y}}$

(c) $\dfrac{1}{\sqrt{c} - \sqrt{d}} = \dfrac{\sqrt{c} + \sqrt{d}}{c - d}$

Exer. 7–28: Simplify the expression and rationalize the denominator when appropriate.

7 $(3a^2b)^2(2ab^3)$

8 $\dfrac{6r^3y^2}{2r^5y}$

9 $\dfrac{(3x^2y^{-3})^{-2}}{x^{-5}y}$

10 $\left(\dfrac{a^{2/3}b^{3/2}}{a^2b}\right)^6$

11 $(-2p^2q)^3\left(\dfrac{p}{4q^2}\right)^2$

12 $c^{-4/3}c^{3/2}c^{1/6}$

13 $\left(\dfrac{xy^{-1}}{\sqrt{z}}\right)^4 \div \left(\dfrac{x^{1/3}y^2}{z}\right)^3$

14 $\left(\dfrac{-64x^3}{z^6y^9}\right)^{2/3}$

15 $[(a^{2/3}b^{-2})^3]^{-1}$

16 $\dfrac{(3u^2v^5w^{-4})^3}{(2uv^{-3}w^2)^4}$

17 $\dfrac{r^{-1}+s^{-1}}{(rs)^{-1}}$

18 $(u+v)^3(u+v)^{-2}$

19 $\sqrt[3]{(x^4y^{-1})^6}$

20 $\sqrt[3]{8x^5y^3z^4}$

21 $\dfrac{1}{\sqrt[3]{4}}$

22 $\sqrt{\dfrac{a^2b^3}{c}}$

23 $\sqrt[3]{4x^2y}\,\sqrt[3]{2x^5y^2}$

24 $\sqrt[4]{(-4a^3b^2c)^2}$

25 $\dfrac{1}{\sqrt{t}}\left(\dfrac{1}{\sqrt{t}}-1\right)$

26 $\sqrt{\sqrt[3]{(c^3d^6)^4}}$

27 $\dfrac{\sqrt{12x^4y}}{\sqrt{3x^2y^5}}$

28 $\sqrt[3]{(a+2b)^3}$

Exer. 29–32: Rationalize the denominator.

29 $\dfrac{1-\sqrt{x}}{1+\sqrt{x}}$

30 $\dfrac{1}{\sqrt{a}+\sqrt{a-2}}$

31 $\dfrac{81x^2-y^2}{3\sqrt{x}+\sqrt{y}}$

32 $\dfrac{3+\sqrt{x}}{3-\sqrt{x}}$

33 Express in scientific form:

(a) 93,700,000,000 (b) 0.00000402

34 Express as a decimal:

(a) 6.8×10^7 (b) 7.3×10^{-4}

35 The body of an average person contains 5.5 liters of blood and about 5 million red blood cells per cubic millimeter of blood. Remembering that 1 liter $= 10^6$ mm^3, estimate the number of red blood cells in an average person's circulatory system. Express the answer in scientific form.

36 A healthy heart beats 70 to 90 times per minute. If an individual lives to age 80, estimate the number of heartbeats in his or her lifetime. Express the answer in scientific form.

Exer. 37–52: Express as a polynomial.

37 $(3x^3-4x^2+x-7)+(x^4-2x^3+3x^2+5)$

38 $(4z^4-3z^2+1)-z(z^3+4z^2-4)$

39 $(x+4)(x+3)-(2x-1)(x-5)$

40 $(4x-5)(2x^2+3x-7)$

41 $(3y^3-2y^2+y+4)(y^2-3)$

42 $(3x+2)(x-5)(5x+4)$

43 $(a-b)(a^3+a^2b+ab^2+b^3)$

44 $\dfrac{9p^4q^3-6p^2q^4+5p^3q^2}{3p^2q^2}$

45 $(3a-5b)(2a+7b)$

46 $(4r^2-3s)^2$

47 $(13a^2+4b)(13a^2-4b)$

48 $(a^3-a^{-3})^2$

49 $(2a+b)^3$

50 $(c^2-d^2)^3$

51 $(3x+2y)^2(3x-2y)^2$

52 $(a+b+c+d)^2$

Exer. 53–64: Factor the polynomial.

53 $60xw+70w$

54 $2r^4s^3-8r^2s^5$

55 $28x^2+4x-9$

56 $16a^4+24a^2b^2+9b^4$

57 $2wy+3yx-8wz-12zx$

58 $2c^3-12c^2+3c-18$

59 $8x^3+64y^3$

60 $u^3v^4-u^6v$

61 p^8-q^8

62 $x^4-8x^3+16x^2$

63 w^6+1

64 $3x+6$

Exer. 65–76: Simplify the expression.

65 $\dfrac{6x^2-7x-5}{4x^2+4x+1}$

66 $\dfrac{r^3-t^3}{r^2-t^2}$

67 $\dfrac{6x^2-5x-6}{x^2-4} \div \dfrac{2x^2-3x}{x+2}$

68 $\dfrac{2}{4x-5}-\dfrac{5}{10x+1}$

69 $\dfrac{7}{x+2}+\dfrac{3x}{(x+2)^2}-\dfrac{5}{x}$

70 $\dfrac{x+x^{-2}}{1+x^{-2}}$

71 $\dfrac{1}{x}-\dfrac{2}{x^2+x}-\dfrac{3}{x+3}$

72 $(a^{-1}+b^{-1})^{-1}$

73 $\dfrac{x + 2 - \dfrac{3}{x + 4}}{\dfrac{x}{x + 4} + \dfrac{1}{x + 4}}$

74 $\dfrac{\dfrac{x}{x + 2} - \dfrac{4}{x + 2}}{x - 3 - \dfrac{6}{x + 2}}$

75 $(x^2 + 1)^{3/2}(4)(x + 5)^3 + (x + 5)^4(\frac{3}{2})(x^2 + 1)^{1/2}(2x)$

76 $\dfrac{(4 - x^2)(\frac{1}{3})(6x + 1)^{-2/3}(6) - (6x + 1)^{1/3}(-2x)}{(4 - x^2)^2}$

77 Expand and simplify $(x^2 - 3y)^6$.

Exer. 78–80: Without expanding completely, find the indicated term(s) in the expansion of the expression.

78 $(a^{2/5} + 2a^{-3/5})^{20}$; first three terms

79 $(b^3 - \frac{1}{2}c^2)^9$; sixth term

80 $(2c^3 + 5c^{-2})^{10}$; term that does not contain c

CHAPTER 2

For hundreds of years one of the main concerns in algebra has been the study of equations. More recently, inequalities have achieved the same degree of importance. Both of these topics are used extensively in applications of mathematics. In this chapter we shall discuss several methods for solving equations and inequalities.

EQUATIONS AND INEQUALITIES

2.1 LINEAR EQUATIONS

If x is a variable, the expressions

$$x + 3 = 0, \qquad x^2 - 5 = 4x, \qquad (x^2 - 9)\sqrt[3]{x + 1} = 0$$

are **equations** in x. A number a is a **solution**, or **root**, of an equation if a true statement is obtained when a is substituted for x. We say that a **satisfies** the equation. The equation $x^2 - 5 = 4x$ has 5 as a solution, since substitution gives us

$$(5)^2 - 5 = 4(5), \quad \text{or} \quad 20 = 20,$$

which is a true statement. To **solve an equation** means to find all the solutions.

An equation is an **identity** if every number in the domain of the variable is a solution of the equation.

ILLUSTRATION | IDENTITY

■ $\dfrac{1}{x^2 - 4} = \dfrac{1}{(x + 2)(x - 2)}$

An equation in x is a **conditional equation** if there are numbers in the domain of x that are *not* solutions.

The solutions of an equation depend on the type of numbers that are allowed. For example, if only rational numbers are allowed, then the equation $x^2 = 2$ has no solution, since there is no rational number whose square is 2. However, if we allow *real* numbers, then the solutions are $-\sqrt{2}$ and $\sqrt{2}$. Similarly, the equation $x^2 = -1$ has no real solution; however, we shall see later that this equation has solutions if *complex* numbers are allowed.

Two equations are **equivalent** if they have exactly the same solutions.

ILLUSTRATION | EQUIVALENT EQUATIONS

■ $2x + 1 = 7, \quad 2x = 7 - 1, \quad 2x = 6, \quad x = 3$

■ $3x^2 = 12, \quad x^2 = 4, \quad x^2 - 4 = 0, \quad (x + 2)(x - 2) = 0$

One method of solving an equation is to replace it with a list of equivalent equations, each in some sense simpler than the preceding one, ending the list with an equation for which the solutions are obvious. This is accomplished by using properties of real numbers. For example, we may add the same expression to both sides of an equation. Similarly, we may subtract the same expression from both sides. We can also multiply or divide both sides of an equation by an expression that represents a *non-zero* real number.

EXAMPLE ▪ 1

Solve the equation $6x - 7 = 2x + 5$.

SOLUTION The equations in the following list are equivalent:

$$6x - 7 = 2x + 5$$
$$(6x - 7) + 7 = (2x + 5) + 7$$
$$6x = 2x + 12$$
$$6x - 2x = (2x + 12) - 2x$$
$$4x = 12$$
$$\tfrac{1}{4}(4x) = \tfrac{1}{4}(12)$$
$$x = 3$$

Since the last equation has exactly one solution, 3, it follows that 3 is the only solution of the original equation, $6x - 7 = 2x + 5$.

In the process of finding solutions of an equation, we may introduce errors because of incorrect manipulations or mistakes in arithmetic. Thus, it is advisable to check answers by substitution in the original equation. To check the solution in Example 1, we substitute 3 for x in the equation $6x - 7 = 2x + 5$, obtaining

$$6(3) - 7 = 2(3) + 5$$
$$18 - 7 = 6 + 5$$
$$11 = 11,$$

which is a true statement.

We sometimes say that an equation such as that given in Example 1 *has the solution* $x = 3$. As another illustration, the equation $x^2 = 4$ has solutions $x = 2$ and $x = -2$.

If we multiply both sides of an equation by an expression that equals zero for some value of x, then the equation obtained may not be equivalent to the original one, as illustrated in the following example.

EXAMPLE ▪ 2

Solve $\dfrac{3x}{x-2} = 1 + \dfrac{6}{x-2}$.

SOLUTION We multiply both sides by $x - 2$ and simplify:

$$\left(\frac{3x}{x-2}\right)(x-2) = (1)(x-2) + \left(\frac{6}{x-2}\right)(x-2)$$

$$3x = (x-2) + 6$$

$$3x = x + 4$$

$$2x = 4$$

$$x = 2$$

We then check to verify that 2 is a solution of the original equation. Substituting 2 for x, we obtain

$$\frac{3(2)}{2-2} = 1 + \frac{6}{2-2}$$

$$\frac{6}{0} = 1 + \frac{6}{0}.$$

Since division by 0 is not defined, 2 is not a solution. Actually, the given equation has no solution, for we have shown that if the equation is true for some value of x, then that value must be 2. However, as we have seen, 2 is not a solution.

The preceding example indicates that *we must check answers that are obtained after we multiply both sides of an equation by an expression that contains a variable.*

A **linear equation** is an equation that can be written

$$ax + b = 0$$

for real numbers a and b with $a \neq 0$. To solve this equation we first subtract b from both sides and then multiply by $1/a$:

$$ax + b = 0$$

$$ax = -b$$

$$x = \frac{-b}{a}$$

Thus *the linear equation $ax + b = 0$ has precisely one solution, $x = -b/a$.*

EXAMPLE ▪ 3

Solve $(8x - 2)(3x + 4) = (4x + 3)(6x - 1)$.

SOLUTION The following equations are equivalent:

$$\begin{aligned}(8x - 2)(3x + 4) &= (4x + 3)(6x - 1)\\ 24x^2 + 26x - 8 &= 24x^2 + 14x - 3\\ 26x - 8 &= 14x - 3\\ 26x - 14x &= -3 + 8\\ 12x &= 5\\ x &= \tfrac{5}{12}\end{aligned}$$

Hence, the solution of the given equation is $\frac{5}{12}$.

EXAMPLE ▪ 4

Solve $\dfrac{3}{2x - 4} - \dfrac{5}{x + 3} = \dfrac{2}{x - 2}$.

SOLUTION If we rewrite the equation as

$$\frac{3}{2(x - 2)} - \frac{5}{x + 3} = \frac{2}{x - 2},$$

we see that the lcd of the three fractions is $2(x - 2)(x + 3)$. Multiplying both sides by this lcd gives us

$$\frac{3}{2(x - 2)}\,2(x - 2)(x + 3) - \frac{5}{x + 3}\,2(x - 2)(x + 3) = \frac{2}{x - 2}\,2(x - 2)(x + 3)$$

which reduces to

$$3(x + 3) - 10(x - 2) = 4(x + 3).$$

We simplify:

$$\begin{aligned}3x + 9 - 10x + 20 &= 4x + 12\\ 3x - 10x - 4x &= 12 - 9 - 20\\ -11x &= -17\\ x &= \tfrac{17}{11}.\end{aligned}$$

Since we multiplied by an expression involving x, we must check this result in the given equation. Substituting for x, we have

$$\frac{3}{2(\frac{17}{11}) - 4} - \frac{5}{\frac{17}{11} + 3} = \frac{2}{\frac{17}{11} - 2}.$$

This reduces to $-\frac{22}{5} = -\frac{22}{5}$. Thus, the given equation has the solution $\frac{17}{11}$.

Sometimes it is difficult to determine if an equation is conditional or is an identity. An identity will often be indicated when, after applying properties of real numbers, we obtain an equation of the form $p = p$ for some expression p. To illustrate, if we multiply both sides of the equation

$$\frac{x}{x^2 - 4} = \frac{x}{(x + 2)(x - 2)}$$

by $x^2 - 4$, we obtain $x = x$. This result alerts us that the equation may be an identity; however, it does not prove anything. A standard method for verifying that an equation is an identity is to show, using properties of real numbers, that the expression that appears on one side of the equation can be transformed into the expression that appears on the other side. This transformation is easy to do in the above illustration, since we know that $x^2 - 4 = (x + 2)(x - 2)$. To show that an equation is *not* an identity, we need only find one real number in the domain of the variable that fails to satisfy the equation.

2.1 EXERCISES

Exer. 1–42: Solve the equation.

1 $-3x + 4 = -1$

2 $2x - 2 = -9$

3 $4(2y + 5) = 3(5y - 2)$

4 $6(2y + 3) - 3(y - 5) = 0$

5 $\frac{1}{5}x + 2 = 3 - \frac{2}{7}x$

6 $\frac{5}{3}x - 1 = 4 + \frac{2}{3}x$

7 $0.3(3 + 2x) + 1.2x = 3.2$

8 $1.5x - 0.7 = 0.4(3 - 5x)$

9 $\dfrac{3 + 5x}{5} = \dfrac{4 - x}{7}$

10 $\dfrac{2x - 9}{4} = 2 + \dfrac{x}{12}$

11 $\dfrac{13 + 2x}{4x + 1} = \dfrac{3}{4}$

12 $\dfrac{3}{7x - 2} = \dfrac{9}{3x + 1}$

13 $8 - \dfrac{5}{x} = 2 + \dfrac{3}{x}$

14 $\dfrac{3}{y} + \dfrac{6}{y} - \dfrac{1}{y} = 11$

15 $(3x - 2)^2 = (x - 5)(9x + 4)$

16 $(x + 5)^2 + 3 = (x - 2)^2$

17 $(5x - 7)(2x + 1) - 10x(x - 4) = 0$

18 $(2x + 9)(4x - 3) = 8x^2 - 12$

19 $\dfrac{3x + 1}{6x - 2} = \dfrac{2x + 5}{4x - 13}$

20 $\dfrac{5x + 2}{10x - 3} = \dfrac{x - 8}{2x + 3}$

21 $\dfrac{2}{5} + \dfrac{4}{10x + 5} = \dfrac{7}{2x + 1}$

22 $\dfrac{-5}{3x - 9} + \dfrac{4}{x - 3} = \dfrac{5}{6}$

23 $\dfrac{3}{2x - 4} - \dfrac{5}{3x - 6} = \dfrac{3}{5}$

24 $\dfrac{9}{2x + 6} - \dfrac{7}{5x + 15} = \dfrac{2}{3}$

25 $2 - \dfrac{5}{3x - 7} = 2$

26 $\dfrac{6}{2x + 11} + 5 = 5$

27 $\dfrac{1}{2x - 1} = \dfrac{4}{8x - 4}$

28 $\dfrac{4}{5x + 2} - \dfrac{12}{15x + 6} = 0$

29 $\dfrac{7}{y^2 - 4} - \dfrac{4}{y + 2} = \dfrac{5}{y - 2}$

30 $\dfrac{4}{2u-3}+\dfrac{10}{4u^2-9}=\dfrac{1}{2u+3}$

31 $(x+3)^3-(3x-1)^2=x^3+4$

32 $(x-1)^3=(x+1)^3-6x^2$

33 $\dfrac{9x}{3x-1}=2+\dfrac{3}{3x-1}$

34 $\dfrac{2x}{2x+3}+\dfrac{6}{4x+6}=5$

35 $\dfrac{1}{x+4}+\dfrac{3}{x-4}=\dfrac{3x+8}{x^2-16}$

36 $\dfrac{2}{2x+3}+\dfrac{4}{2x-3}=\dfrac{5x+6}{4x^2-9}$

37 $\dfrac{4}{x+2}+\dfrac{1}{x-2}=\dfrac{5x-6}{x^2-4}$

38 $\dfrac{2}{2x+5}+\dfrac{3}{2x-5}=\dfrac{10x+5}{4x^2-25}$

39 $\dfrac{2}{2x+1}-\dfrac{3}{2x-1}=\dfrac{-2x+7}{4x^2-1}$

40 $\dfrac{3}{2x+5}+\dfrac{4}{2x-5}=\dfrac{14x+3}{4x^2-25}$

41 $\dfrac{5}{2x+3}+\dfrac{4}{2x-3}=\dfrac{14x+3}{4x^2-9}$

42 $\dfrac{-3}{x+4}+\dfrac{7}{x-4}=\dfrac{-5x+4}{x^2-16}$

Exer. 43–48: Show that the equation is an identity.

43 $(4x-3)^2-16x^2=9-24x$

44 $(3x-4)(2x+1)+5x=6x^2-4$

45 $\dfrac{x^2-9}{x+3}=x-3$

46 $\dfrac{x^3+8}{x+2}=x^2-2x+4$

47 $\dfrac{3x^2+8}{x}=\dfrac{8}{x}+3x$

48 $\dfrac{49x^2-25}{7x-5}=7x+5$

Exer. 49–50: For what value of c is the number a a solution of the equation?

49 $4x+1+2c=5c-3x+6;\quad a=-2$

50 $3x-2+6c=2c-5x+1;\quad a=4$

51 Determine values for a and b such that $\frac{5}{3}$ is a solution of the equation $ax+b=0$. Are these the only possible values for a and b? Explain.

Exer. 52–55: Determine if the two equations are equivalent.

52 (a) $x^2=16,\quad x=4$ (b) $x=\sqrt{9},\quad x=3$

53 (a) $x^2=25,\quad x=5$ (b) $x=\sqrt{64},\quad x=8$

54 (a) $\dfrac{7x}{x-5}=\dfrac{42}{x-5},\quad x=6$ (b) $\dfrac{7x}{x-5}=\dfrac{35}{x-5},\quad x=5$

55 (a) $\dfrac{8x}{x-7}=\dfrac{72}{x-7},\quad x=9$ (b) $\dfrac{8x}{x-7}=\dfrac{56}{x-7},\quad x=7$

Exer. 56–57: Find an equation that is not equivalent to the preceding equation.

56
$$\begin{aligned} x^2-x-2&=x^2-4\\ (x+1)(x-2)&=(x+2)(x-2)\\ x+1&=x+2\\ 1&=2 \end{aligned}$$

57
$$\begin{aligned} x+3&=0\\ 5x-4x&=-3\\ 5x+6&=4x+3\\ x^2+5x+6&=x^2+4x+3\\ (x+2)(x+3)&=(x+1)(x+3)\\ x+2&=x+1\\ 2&=1 \end{aligned}$$

2.2 APPLIED PROBLEMS

Formulas involving several variables occur in many applications of mathematics. Sometimes it is necessary to solve for a specific variable in terms of the remaining variables that appear in the formula, as the next three examples illustrate.

FIGURE 1

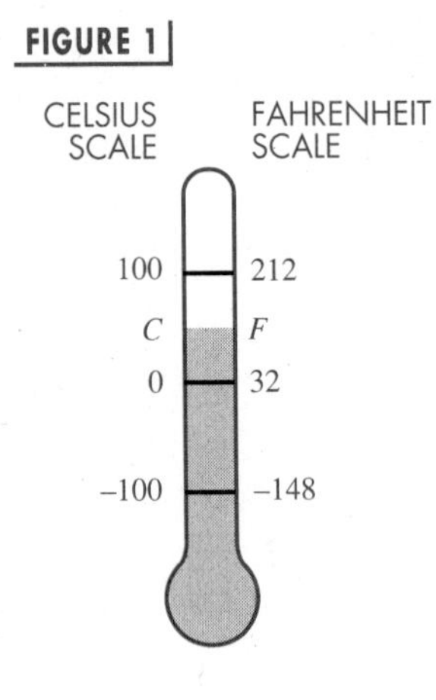

FIGURE 2

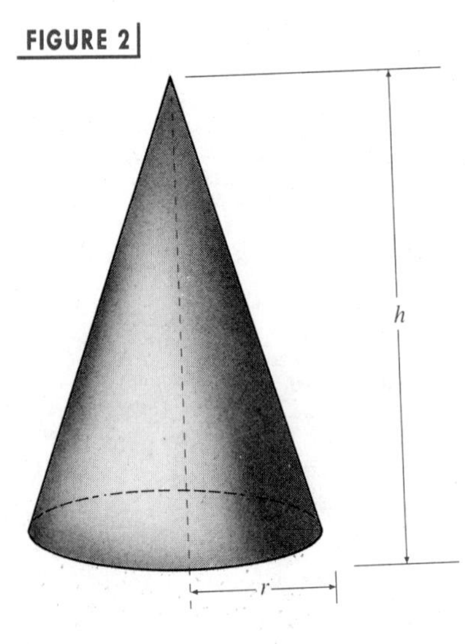

EXAMPLE ▪ 1

The Celsius and Fahrenheit temperature scales are shown on the thermometer in Figure 1. The relationship between the temperature readings C and F is given by $C = \frac{5}{9}(F - 32)$. Solve for F.

SOLUTION

$$C = \tfrac{5}{9}(F - 32)$$
$$\tfrac{9}{5}C = F - 32$$
$$\tfrac{9}{5}C + 32 = F$$
$$F = \tfrac{9}{5}C + 32$$

EXAMPLE ▪ 2

A right circular cone of altitude h and base radius r is shown in Figure 2. Its volume V is given by $V = \frac{1}{3}\pi r^2 h$. Solve this equation

(a) for h (b) for r.

SOLUTION Multiplying both sides of $V = \frac{1}{3}\pi r^2 h$ by 3 gives us

$$3V = \pi r^2 h, \quad \text{or} \quad \pi r^2 h = 3V.$$

(a) To solve for h, we divide both sides by πr^2 and simplify:

$$\frac{\pi r^2 h}{\pi r^2} = \frac{3V}{\pi r^2}$$
$$h = \frac{3V}{\pi r^2}$$

(b) To solve for r, we first divide both sides of $\pi r^2 h = 3V$ by πh and simplify:

$$\frac{\pi r^2 h}{\pi h} = \frac{3V}{\pi h}$$
$$r^2 = \frac{3V}{\pi h}.$$

Taking the square root of both sides and using the fact that $r > 0$ gives us

$$r = \sqrt{\frac{3V}{\pi h}}.$$

EXAMPLE ▪ 3

FIGURE 3

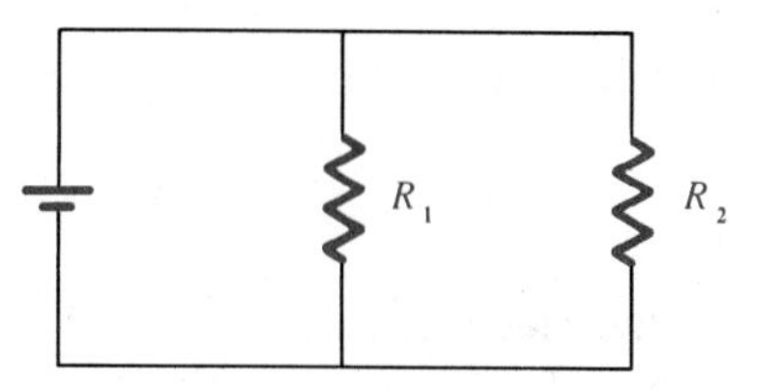

In electrical theory the formula

$$\frac{1}{R} = \frac{1}{R_1} + \frac{1}{R_2}$$

is used to find the total resistance R when two resistors R_1 and R_2 are connected in parallel, as illustrated in Figure 3. Solve for R_1.

SOLUTION The following equations are equivalent to the given equation:

$$\frac{1}{R_1} + \frac{1}{R_2} = \frac{1}{R}$$

$$\frac{1}{R_1} = \frac{1}{R} - \frac{1}{R_2}$$

$$\frac{1}{R_1} = \frac{R_2 - R}{RR_2}$$

If two nonzero numbers are equal, then so are their reciprocals. Hence

$$R_1 = \frac{RR_2}{R_2 - R}.$$

Equations are used to solve practical problems. We shall use the terminology *applied problem* for any problem that involves an application of mathematics to some other field. Due to the unlimited variety of applied problems, it is difficult to state specific rules for finding solutions. However, we may develop a general strategy for attacking such problems. The following guidelines may be helpful, provided the problem can be formulated in terms of an equation in one variable.

GUIDELINES SOLVING APPLIED PROBLEMS

1 If the problem is stated in written words, read it carefully several times and think about the given facts, together with the unknown quantity that is to be found.

2 Introduce a letter to denote the unknown quantity. This is one of the most crucial steps in the solution. Phrases containing words such as *what, find, how much, how far,* or *when* should alert you to the unknown quantity.

3 If feasible, draw a picture and label it appropriately.

4 Make a list of known facts, together with any relationships that involve the unknown quantity. A relationship may often be described by an equation in which written statements, instead of letters or numbers, appear on one or both sides of the equal sign.

5 After analyzing the list in step 4, formulate an equation that describes precisely what is stated in words.

6 Solve the equation formulated in step 5.

7 Check the solutions obtained in step 6 by referring to the original statement of the problem. Verify that the solution agrees with the stated conditions. ❑

EXAMPLE ▪ 4

A student has test scores of 64 and 78. What score on a third test will give the student an average of 80?

SOLUTION We shall follow the guidelines. Reading the problem carefully (step 1), we note that the unknown quantity is the score on the third test. Accordingly, we introduce a letter (step 2) for this quantity:

$$x = \text{score on the third test.}$$

Drawing a picture (step 3) is inappropriate for this problem, so we go on to step 4, and look for relationships involving x. Since the average of the three scores is found by adding them and dividing by 3, we may write

$$\frac{64 + 78 + x}{3} = \text{average of the three scores 64, 78, and } x.$$

From the statement of the problem we see that

$$80 = \text{average desired.}$$

Thus, x must satisfy the equation

$$\frac{64 + 78 + x}{3} = 80.$$

This equation is the one referred to in step 5 of the guidelines. We next solve the equation (step 6):

$$64 + 78 + x = 240$$

$$142 + x = 240$$

$$x = 98$$

We check the solution by referring to the original statement (step 7). If the three test scores are 64, 78, and 98, then the average is

$$\frac{64 + 78 + 98}{3} = \frac{240}{3} = 80.$$

Hence, a score of 98 on the third test will give the student an average of 80.

In the remaining examples try to identify the explicit guidelines that are used to arrive at solutions.

EXAMPLE ▪ 5

A store holding a clearance sale advertises that all prices have been discounted 20%. If a certain article is on sale for \$28, what was its price before the sale?

SOLUTION We begin by noting that the unknown quantity is the presale price. It is convenient to arrange our work as follows, with the quantities measured in dollars:

$$x = \text{presale price}$$

$$0.20x = \text{discount}$$

$$28 = \text{sale price}$$

The sale price is determined as follows:

$$(\text{presale price}) - (\text{discount}) = (\text{sale price}),$$

that is,

$$x - 0.20x = 28.$$

We solve this equation as follows:

$$0.8x = 28$$

$$x = \frac{28}{0.8} = 35.$$

Thus, the presale price was \$35.

To check this answer we note that if a \$35 article is discounted 20%, then the discount (in dollars) is $(0.20)(35) = 7$, and the selling price is $35 - 7$, or \$28.

The following formula is often used to determine the interest on money that is invested or loaned for a short period of time.

SIMPLE INTEREST FORMULA

> If a sum of money P (the **principal**) is invested at a simple interest rate r (expressed as a decimal), then the **simple interest** I at the end of t years is $I = Prt$.

For example, if $P = \$1000$ and the rate is $r = 8\% = 0.08$, then the interest after one year ($t = 1$) is

$$I = 1000(0.08)(1) = 80.$$

If the principal \$1000 is invested for two years ($t = 2$), the interest is

$$I = 1000(0.08)(2) = 160.$$

We shall use the simple interest formula in the next example.

EXAMPLE ■ 6

An investment firm has \$100,000 to invest for a client and decides to invest it in two stocks A and B. The expected annual rate of return, or interest, for stock A is 15%, but there is some risk involved, and the client does not wish to invest more than \$50,000 in this stock. The annual rate of return on the more stable stock B is anticipated to be 10%. Determine if there is a way of investing the money so that the annual interest is:

(a) \$12,000 (b) \$13,000

SOLUTION Using the simple interest formula $I = Prt$ with $t = 1$, the annual interest is given by $I = Pr$. If we let x denote the amount invested in stock A, then $100{,}000 - x$ will be invested in stock B. This leads to the following equalities:

$$\begin{aligned} x &= \text{amount invested in stock A at 15\%} \\ 100{,}000 - x &= \text{amount invested in stock B at 10\%} \\ 0.15x &= \text{annual interest from stock A} \\ 0.10(100{,}000 - x) &= \text{annual interest from stock B} \end{aligned}$$

Adding the interest from both stocks we obtain the following:

$$\begin{aligned} \text{total annual interest} &= 0.15x + 0.10(100{,}000 - x) \\ &= 10{,}000 + 0.05x \end{aligned}$$

(a) The total annual interest is \$12,000 if:

$$\begin{aligned} 10{,}000 + 0.05x &= 12{,}000 \\ 0.05x &= 2{,}000 \\ x &= 40{,}000 \end{aligned}$$

Thus \$40,000 should be invested in stock A and the remaining \$60,000 should be invested in stock B. Since the amount invested in stock A is less than \$50,000, this manner of investing the money meets the requirement of the client.

(b) The total annual interest is \$13,000 if:

$$\begin{aligned} 10{,}000 + 0.05x &= 13{,}000 \\ 0.05x &= 3{,}000 \\ x &= 60{,}000 \end{aligned}$$

Thus \$60,000 should be invested in Stock A and the remaining \$40,000 in Stock B. This does *not* meet the requirement that not more than \$50,000 is to be invested in stock A. Hence the firm cannot invest the client's money in stocks A and B such that the total annual interest is \$13,000.

In certain applications it is necessary to mix two substances to obtain a prescribed mixture. For such problems it is helpful to draw a picture, as illustrated in the next two examples.

EXAMPLE ▪ 7

A chemist has 10 ml of a solution that contains a 30% concentration of acid. How many ml of pure acid must be added in order to increase the concentration to 50%?

SOLUTION Since we wish to find the amount of pure acid to add, we let

$$x = \text{ml of pure acid to be added.}$$

A method for visualizing the problem is illustrated in Figure 4.

FIGURE 4

Original 30% mixture **Pure acid** **New 50% mixture**

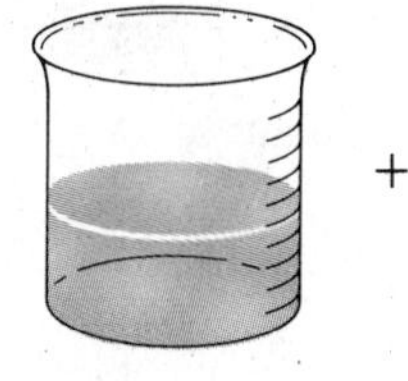
+

=
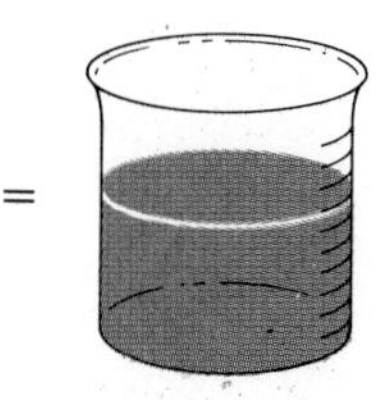

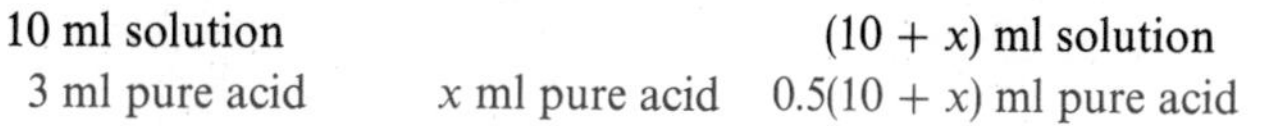

We can express the amount of pure acid in the final solution as either $3 + x$ or $0.5(10 + x)$. This leads to the following equivalent equations:

$$3 + x = 0.5(10 + x)$$

$$3 + x = 5 + 0.5x$$

$$0.5x = 2$$

$$x = \frac{2}{0.5} = 4.$$

Hence, 4 ml of the acid should be added to the original solution.

To check, we note that if 4 ml of acid are added to the given solution, then the new solution contains 14 ml, 7 of which are acid. This is the desired 50% concentration.

EXAMPLE ▪ 8

A radiator contains 8 quarts of a mixture of water and antifreeze. If 40% of the mixture is antifreeze, how much of the mixture should be drained and replaced by pure antifreeze so that the resultant mixture will contain 60% antifreeze?

SOLUTION Let

x = the number of quarts of mixture to be drained.

Since there were 8 quarts in the original 40% mixture, we may picture the problem as shown in Figure 5.

FIGURE 5

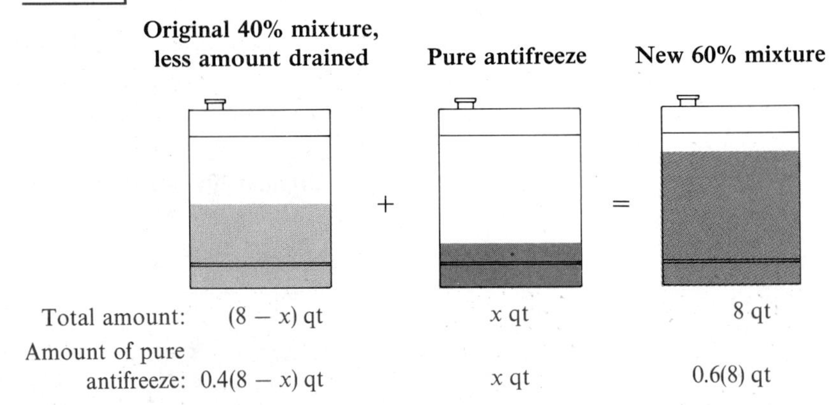

The number of quarts of pure antifreeze in the final mixture can be expressed as either $0.4(8 - x) + x$ or $0.6(8)$, which leads to the equation

$$0.4(8 - x) + x = 0.6(8).$$

We now solve for x:

$$3.2 - 0.4x + x = 4.8$$
$$-0.4x + x = 4.8 - 3.2$$
$$0.6x = 1.6$$
$$x = \frac{1.6}{0.6} = \frac{16}{6} = \frac{8}{3}.$$

Thus, $\frac{8}{3}$ quarts should be drained from the original mixture.

To check, let us first note that the amount of antifreeze in the original 8-quart mixture was 0.4(8), or 3.2 quarts. In draining $\frac{8}{3}$ quarts of the original 40% mixture, we lose $0.4(\frac{8}{3})$ quarts of antifreeze, and hence, $3.2 - 0.4(\frac{8}{3})$ quarts of antifreeze remain after draining. If we then add $\frac{8}{3}$ quarts of pure antifreeze, the amount of antifreeze in the final mixture is

$$3.2 - 0.4(\tfrac{8}{3}) + \tfrac{8}{3} = 4.8 \text{ quarts.}$$

This number, 4.8, is 60% of 8.

Many applied problems involve objects that move at a constant, or uniform, rate. If an object travels at a constant (or average) rate r, then the distance d covered in time t is given by $d = rt$. Of course, we assume that the units are properly chosen; that is, if r is in feet per second, then t is in seconds and d is in feet.

EXAMPLE ▪ 9

Two cities are connected by means of a highway. A car leaves city A at 1:00 P.M. and travels at a constant rate of 40 mi/hr toward city B. Thirty minutes later, another car leaves A and travels toward B at a constant rate of 55 mi/hr. At what time will the second car overtake the first car?

SOLUTION Let t denote the number of hours after 1:00 P.M. for the second car to overtake the first. Since the second car leaves A at 1:30 P.M., it has traveled $\frac{1}{2}$ hour less than the first. This leads to the following table.

Car	Rate (mi/hr)	Hours traveled	Miles traveled
First car	40	t	$40t$
Second car	55	$t - \frac{1}{2}$	$55(t - \frac{1}{2})$

The second car overtakes the first car when the number of miles traveled by the two cars is equal; that is,

$$55(t - \tfrac{1}{2}) = 40t.$$

We now solve for t:

$$55t - \frac{55}{2} = 40t$$

$$15t = \frac{55}{2}$$

$$t = \frac{55}{30} = \frac{11}{6}.$$

Thus, $t = 1\frac{5}{6}$ hours, or equivalently, 1 hour 50 minutes after 1:00 P.M. Consequently, the second car overtakes the first at 2:50 P.M.

To check our answer we note that at 2:50 P.M. the first car has traveled for $1\frac{5}{6}$ hours and its distance from A is $40(\frac{11}{6}) = \frac{220}{3}$ miles. At 2:50 P.M. the second car has traveled for $1\frac{1}{3}$ hours and is $55(\frac{4}{3}) = \frac{220}{3}$ miles from A. Hence, they are together at 2:50 P.M.

EXAMPLE ▪ 10

Tom can do a certain job in 3 hours, and Bob can do the same job in 4 hours. If they work together, how long will it take them to do the job?

SOLUTION Let

t = number of hours for Tom and Bob to do the job together.

It is convenient to introduce the *part* of the job done in 1 hour as follows:

$\frac{1}{3}$ = part of job done by Tom in 1 hour

$\frac{1}{4}$ = part of job done by Bob in 1 hour

$\frac{1}{t}$ = part of job done by Tom and Bob together in 1 hour.

We have assumed that both Tom and Bob work at a steady pace throughout the job, and that there is no gain or loss in efficiency when they work together. Since

$$\begin{pmatrix}\text{part done by}\\ \text{Tom in 1 hour}\end{pmatrix} + \begin{pmatrix}\text{part done by}\\ \text{Bob in 1 hour}\end{pmatrix} = \begin{pmatrix}\text{part done by Tom and}\\ \text{Bob together in 1 hour}\end{pmatrix}$$

we obtain
$$\frac{1}{3}+\frac{1}{4}=\frac{1}{t}, \quad \text{or} \quad \frac{7}{12}=\frac{1}{t}.$$

Solving for t gives us
$$7t=12, \quad \text{or} \quad t=\frac{12}{7}.$$

Thus, by working together, Tom and Bob can do the job in $1\frac{5}{7}$ hours, or approximately 1 hour 43 minutes.

2.2 EXERCISES

Exer. 1–14: The formula occurs in the indicated application. Solve for the specified variable.

1 $I = Prt$ for P (simple interest)

2 $d = rt$ for t (distance traveled)

3 $A = \frac{1}{2}bh$ for h (area of a triangle)

4 $C = 2\pi r$ for r (circumference of a circle)

5 $P = 2l + 2w$ for w (perimeter of a rectangle)

6 $A = P + Prt$ for r (principal plus interest)

7 $R = \dfrac{V}{I}$ for I (Ohm's law in electrical theory)

8 $K = \frac{1}{2}mv^2$ for m (kinetic energy)

9 $F = g\dfrac{mM}{d^2}$ for m (Newton's law of gravitation)

10 $\dfrac{1}{R} = \dfrac{1}{R_1} + \dfrac{1}{R_2} + \dfrac{1}{R_3}$ for R_2
(three resistors connected in parallel)

11 $S = \pi r\sqrt{r^2 + h^2}$ for h (surface area of a cone)

12 $s = \frac{1}{2}gt^2 + v_0t$ for v_0 (distance an object falls)

13 $A = \frac{1}{2}(b_1 + b_2)h$ for b_1 (area of a trapezoid)

14 $S = 2(lw + hw + hl)$ for h
(surface area of a rectangular box)

15 A student in an algebra course has test scores of 75, 82, 71, and 84. What score on the next test will raise the student's average to 80?

16 Before the final exam, a student has test scores of 72, 80, 65, 78, and 60. If the final exam counts as one-third of the final grade, what score must the student receive in order to have a final average of 76?

17 Refer to Example 1. The relationship between the temperature reading F on the Fahrenheit scale and the temperature reading C on the Celsius scale is given by the formula $C = \frac{5}{9}(F - 32)$. Find the temperature at which the reading is the same on both scales.

18 Shown in the figure is a cross section of a design for a two-story home. The center height h of the second story has not yet been determined. Find h such that the second story will have the same cross-sectional area as the first story.

EXERCISE 18

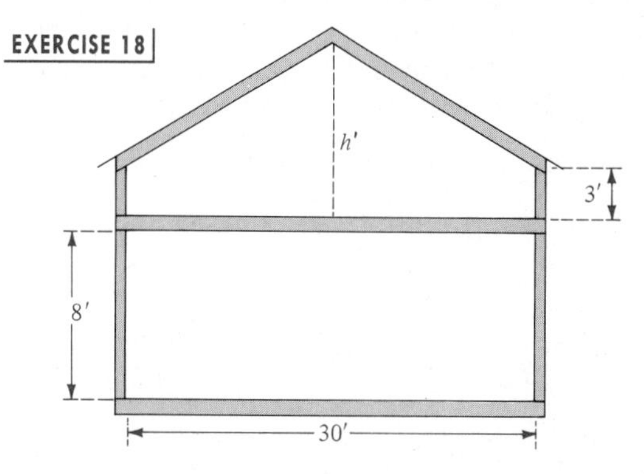

19 The prenatal growth of a fetus more than 12 weeks old can be approximated by the formula $L = 1.53t - 6.7$ for length L in cm and the age t in weeks. Prenatal length can be determined by ultrasound. Estimate the age of a fetus whose length is 28 cm.

20 Based on discus records from the Olympics, the winning distance can be approximated by $d = 175 + 1.75t$ for d in feet and $t = 0$ corresponding to the year 1948.

(a) Predict the winning distance for the 1992 Summer Olympics in Barcelona, Spain.

(b) Estimate the year in which the winning toss will be about 260 feet.

21 A stained glass window is being designed in the shape of a rectangle surmounted by a semicircle, as shown in the figure. The width of the window is to be 3 feet, but the height h is yet to be determined. If 24 ft^2 of glass is to be used, find the height h.

EXERCISE 21

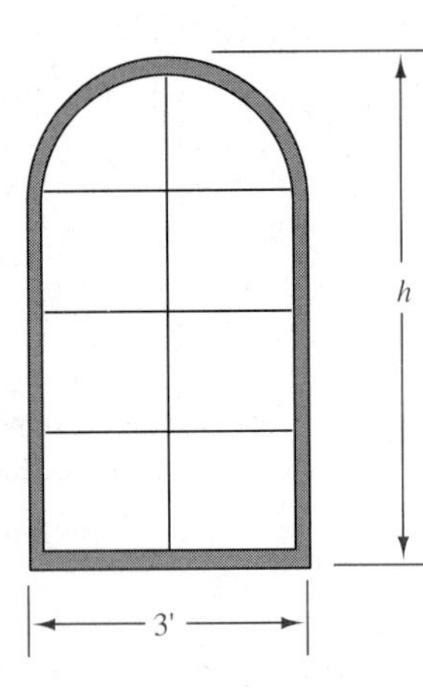

22 The wafer cone shown in the figure is to hold 8 in.^{3} of ice cream when filled to the bottom. The diameter of the cone is 2 inches and the top of the ice cream has the shape of a hemisphere. Find the height h of the cone.

EXERCISE 22

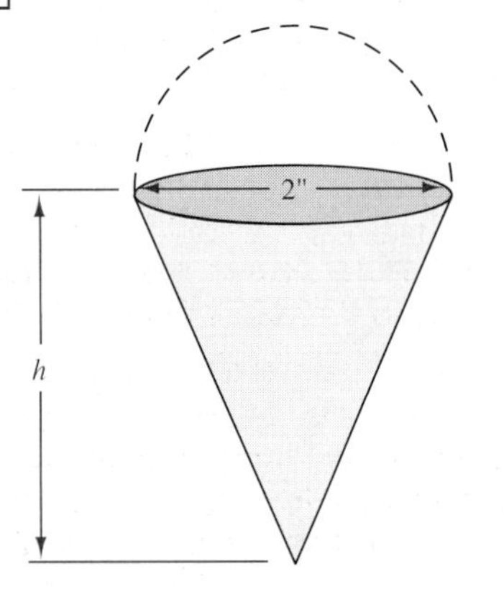

23 Two boys who are 224 meters apart start walking toward each other at the same instant at rates of 1.5 m/sec and 2 m/sec, respectively (see figure).

(a) When will they meet?

(b) How far will each have walked?

EXERCISE 23

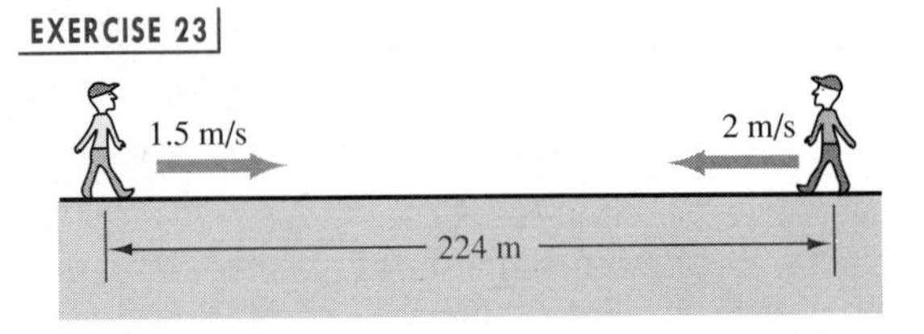

24 A runner starts at the beginning of a runner's path and runs at a constant rate of 6 mi/hr. Five minutes later a second runner begins at the same point, running at a rate of 8 mi/hr and following the same course. How long will it take the second runner to catch up with the first?

25 A ring that weighs 80 grams is made of silver and gold. By measuring the displacement of the ring in water, it has been determined that the ring has a volume of 5 cm^3. Gold weighs 19.3 g/cm^3 and silver weighs 10.5 g/cm^3. How many grams of gold does the ring contain?

26 A pharmacist is to prepare 15 ml of special eye drops for a glaucoma patient. The eye-drop solution must have a 2% active ingredient, but the pharmacist only has 10% solution and 1% solution in stock. How much of each type of solution should be used to fill the prescription?

27 In a certain medical test designed to measure carbohydrate tolerance, an adult drinks 7 ounces of a 30% glucose solution. When the test is administered to a child, the glucose concentration must be decreased to 20%. How much 30% glucose solution and how much water are used to prepare 7 ounces of 20% glucose solution?

28 Theophyline, an asthma medicine, is to be prepared from an elixir with a drug concentration of 5 mg/ml and a cherry-flavored syrup that is to be added to hide the taste of the drug. How much of each must be used to prepare 100 ml of solution with a drug concentration of 2 mg/ml?

29 British sterling silver is a copper-silver alloy that is 7.5% copper by weight. How many grams of pure copper and how many grams of British sterling silver should be used to prepare 200 grams of a copper-silver alloy that is 10% copper by weight?

30 A city government has approved the construction of a \$50 million sports arena. Up to \$30 million will be raised by selling bonds that pay simple interest at a rate of 12% annually. The remaining amount (up to \$40 million) will be obtained by borrowing money from an insurance company at a simple interest rate of 10%. Determine if the arena can be financed so that the annual interest is \$5.2 million.

31 An algebra student has won \$100,000 in a lottery and wishes to deposit it in saving accounts in two financial institutions. One account pays 8% simple interest, but deposits are insured only to \$50,000. The second account pays 6.4% simple interest and deposits are insured up to \$100,000. Determine if the money can be invested so that it is fully insured and earns annual interest of \$7,500.

32 Six hundred people attended the premiere of the latest horror film. Adult tickets cost \$5, and children were admitted for \$2. If box office receipts totaled \$2400, how many children attended the premiere?

33 A large grain silo is to be constructed in the shape of a circular cylinder with a hemisphere attached to the top (see figure). The diameter of the silo is to be 30 feet, but the height is yet to be determined. Find the total height of the silo that will result in a capacity of $11{,}250\pi \text{ ft}^3$.

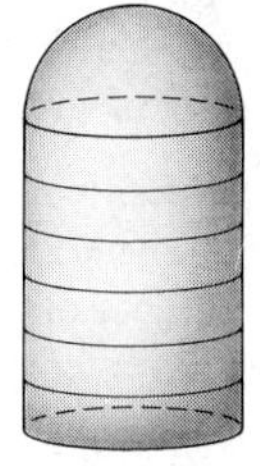

34 Every cross section of a drainage ditch is an isosceles trapezoid with a small base of 3 feet and a height of 1 foot (see figure). Determine the width of the larger base that would give the ditch a cross-sectional area of 5 ft^2.

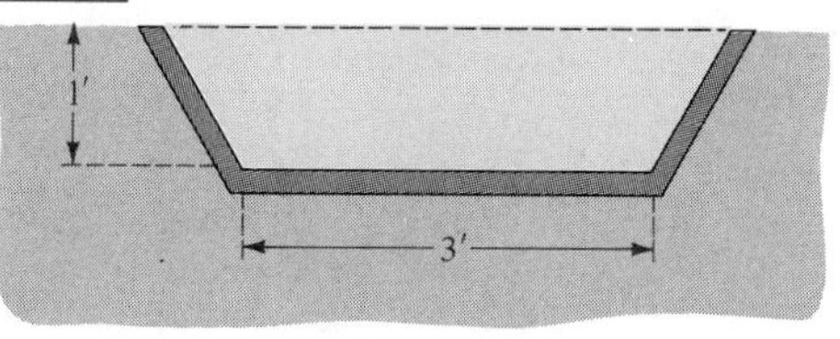

35 At 6 A.M. a snowplow, traveling at a steady rate, begins to clear a highway leading out of town. At 8 A.M. an automobile starts down the highway at a speed of 30 mi/hr and meets the plow 30 minutes later. Approximately how fast is the snowplow traveling?

36 An automobile 20 feet long overtakes a truck that is 40 feet long and traveling at 50 mi/hr (see figure). At what constant speed must the automobile travel in order to pass the truck in 5 seconds?

EXERCISE 36

37 It takes a boy 90 minutes to mow his father's yard, but his sister can mow it in 60 minutes. How long would it take them to mow the lawn if they worked together, using two lawnmowers?

38 Using water from one hose, a swimming pool can be filled in 8 hours. A second, larger hose used alone can fill the pool in 5 hours. How long would it take to fill the pool if both hoses were used simultaneously?

39 A salesperson purchased an automobile that was advertised as averaging 25 mi/gal in the city and 40 mi/gal on the highway. A recent sales trip that covered 1800 miles required 51 gallons of gasoline. Assuming that the advertised mileage estimates were correct, how many miles were driven in the city?

40 Two boys own two-way radios that have a maximum range of 2 miles. One of the boys leaves a certain point at 1:00 P.M., walking due north at a rate of 4 mi/hr. The second boy leaves the same point at 1:15 P.M., traveling due south at 6 mi/hr. When will they be unable to communicate with one another?

41 A farmer plans to use 180 feet of fencing to enclose a rectangular region, using part of a straight river bank instead of fencing as one side of the rectangle. Find the area of the region if the length of the side parallel to the river

bank is

(a) twice the length of an adjacent side.

(b) one-half the length of an adjacent side.

(c) the same as the length of an adjacent side.

42 A consulting engineer's time is billed at \$60 per hour, and his assistant's is billed at \$20 per hour. For a certain job a customer received a bill for \$580. If the assistant worked 5 hours less than the engineer, how much time did each bill on the job?

43 It takes a boy 45 minutes to deliver the newspapers on his route; however, if his sister helps, it takes them only 20 minutes. How long would it take his sister to deliver the newspapers by herself?

44 A water tank can be emptied by using one pump for 5 hours. A second, smaller pump can empty the tank in 8 hours. If the larger pump is started at 1:00 P.M., at what time should the smaller pump be started so that the tank will be emptied at 5:00 P.M.?

45 A boy can row a boat at a constant rate of 5 mi/hr in still water, as indicated in the figure. He rows upstream for 15 minutes, and then rows downstream, returning to his starting point in another 12 minutes.

(a) Find the rate of the current.

(b) Find the total distance traveled.

EXERCISE 45

Upstream net speed = $5 - x$ mi/hr

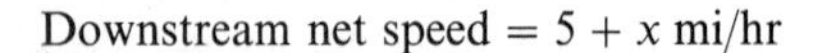
Downstream net speed = $5 + x$ mi/hr

46 A woman begins jogging at 3:00 P.M., running due north at a 6-minute-mile pace. Later she reverses direction and runs due south at a 7-minute-mile pace. If she returns to her starting point at 3:45 P.M., find the total number of miles she ran.

47 The intelligence quotient (IQ) is determined by multiplying the quotient of a person's mental age and chronological age by 100.

(a) Find the IQ of a 12-year-old child whose mental age is 15.

(b) Find the mental age of a person 15 years old whose IQ is 140.

48 In electrical theory Ohm's law states that $I = V/R$ for the current I in amperes, electromotive force V in volts, and the resistance R in ohms. In a certain circuit $V = 110$ and $R = 50$. If V and R are to be changed by the same numerical amount, what change in them will cause I to double?

49 The cost of installing insulation in a particular two-bedroom home is \$1080. Present monthly heating costs average \$60, but the insulation is expected to reduce heating costs by 10%. How many months will it take to recover the cost of the insulation?

50 A workman's basic hourly wage is \$10, but he receives one and a half times his hourly rate for any hours worked in excess of 40 per week. If his paycheck for the week is \$595, how many hours of overtime did he work?

51 After playing 100 games, a major-league baseball team has a record of 0.650. The team then wins only 50% of its games for the remainder of the season. After how many additional games will the team record be 0.600?

52 A bullet is fired horizontally at a target, and the sound of its impact is heard 1.5 seconds later. If the speed of the bullet is 3300 ft/sec and the speed of sound is 1100 ft/sec, how far away is the target?

2.3 QUADRATIC EQUATIONS

DEFINITION

A **quadratic equation** in x is an equation of the form

$$ax^2 + bx + c = 0$$

for real numbers a, b, and c with $a \neq 0$.

One method for solving quadratic equations uses the fact that if p and q represent real numbers, then $pq = 0$ if and only if $p = 0$ or $q = 0$. It follows that if $ax^2 + bx + c$ can be expressed as a product of two first-degree polynomials, then solutions can be found by setting each factor equal to 0, as illustrated in the next two examples. This is the **method of factoring**.

EXAMPLE ▪ 1

Solve $3x^2 = 10 - x$.

SOLUTION First we write the equation in the form $ax^2 + bx + c = 0$:

$$3x^2 + x - 10 = 0$$

Next we factor the left-hand side of this equation:

$$(3x - 5)(x + 2) = 0$$

Setting each factor equal to 0 gives us

$$3x - 5 = 0, \qquad x + 2 = 0.$$

The solutions of these linear equations are

$$x = \tfrac{5}{3}, \qquad x = -2.$$

The fact that $\frac{5}{3}$ and -2 are roots of $3x^2 = 10 - x$ may be checked by substitution.

EXAMPLE ▪ 2

Solve $x^2 + 16 = 8x$.

SOLUTION We begin with an equivalent equation that has all nonzero terms on one side:

$$x^2 - 8x + 16 = 0$$

Factoring gives us $$(x - 4)^2 = 0.$$

Setting each factor $x - 4$ equal to zero, we obtain $x - 4 = 0$. Hence, the given equation has one solution, $x = 4$.

Since $x - 4$ appears as a factor twice in the previous solution, we call 4 a **double root**, or **root of multiplicity 2**, of the equation $x^2 + 16 = 8x$.

If a quadratic equation has the form $x^2 = d$ for some $d > 0$, then $x^2 - d = 0$, or equivalently,

$$(x + \sqrt{d})(x - \sqrt{d}) = 0.$$

Setting each factor equal to zero gives us the solutions $-\sqrt{d}$ and $\sqrt{d}$. We frequently use $\pm\sqrt{d}$ (*plus or minus* $\sqrt{d}$) to represent both solutions. Thus, for $d > 0$, we have proved the following.

THEOREM

If $x^2 = d$, then $x = \pm\sqrt{d}$.

The process of solving $x^2 = d$ as indicated in the box is referred to as *taking the square root of both sides of the equation*. Note that this procedure gives us a positive and a negative square root, not just the principal square root (which was defined in Section 1.3).

EXAMPLE ▪ 3

Solve the equation $x^2 = 5$.

SOLUTION Taking the square root of both sides gives us $x = \pm\sqrt{5}$. Thus, the solutions are $\sqrt{5}$ and $-\sqrt{5}$.

In the work to follow it will be necessary to replace an expression of the form $x^2 + kx$ by $(x + d)^2$, where k and d are real numbers. This procedure, **completing the square** for $x^2 + kx$, is accomplished by adding $(k/2)^2$ as follows:

$$x^2 + kx + \left(\frac{k}{2}\right)^2 = \left(x + \frac{k}{2}\right)^2.$$

In words, *add the square of half the coefficient of* x to $x^2 + kx$. Let us restate this fact for reference.

COMPLETING THE SQUARE

> To complete the square for $x^2 + kx$, add $\left(\frac{k}{2}\right)^2$.

EXAMPLE ▪ 4

Complete the square for $x^2 + 5x$.

SOLUTION The square of half the coefficient of x is $(\frac{5}{2})^2$. Thus,

$$x^2 + 5x + (\tfrac{5}{2})^2 = (x + \tfrac{5}{2})^2.$$

Consider any quadratic equation

$$ax^2 + bx + c = 0, \quad a \neq 0.$$

Dividing both sides by a gives us

$$x^2 + \frac{b}{a}x + \frac{c}{a} = 0,$$

or

$$x^2 + \frac{b}{a}x = -\frac{c}{a}.$$

We next complete the square for $x^2 + (b/a)x$ by adding the square of half the coefficient of x. Of course, *to maintain equality we must add* $\left(\frac{b}{2a}\right)^2$ *to both sides of the equation* as follows:

$$x^2 + \frac{b}{a}x + \left(\frac{b}{2a}\right)^2 = \left(\frac{b}{2a}\right)^2 - \frac{c}{a},$$

which can be written

$$\left(x + \frac{b}{2a}\right)^2 = \frac{b^2 - 4ac}{4a^2}.$$

If $b^2 - 4ac \geq 0$, then

$$x + \frac{b}{2a} = \pm\sqrt{\frac{b^2 - 4ac}{4a^2}}$$

and

$$x = -\frac{b}{2a} \pm \sqrt{\frac{b^2 - 4ac}{4a^2}}.$$

We may write the radical in the last equation as

$$\pm\sqrt{\frac{b^2 - 4ac}{4a^2}} = \pm\frac{\sqrt{b^2 - 4ac}}{\sqrt{(2a)^2}} = \pm\frac{\sqrt{b^2 - 4ac}}{|2a|}.$$

Since $|2a| = 2a$ if $a > 0$, or $|2a| = -2a$ if $a < 0$, we see that in all cases

$$x = -\frac{b}{2a} \pm \frac{\sqrt{b^2 - 4ac}}{2a}.$$

We have shown that if the quadratic equation $ax^2 + bx + c = 0$ has real roots, then they are given by the numbers in the last formula. Moreover, it can be shown by direct substitution that the numbers do satisfy the equation. This gives us the *quadratic formula*.

QUADRATIC FORMULA

If $a \neq 0$, the roots of $ax^2 + bx + c = 0$ are given by

$$x = \frac{-b \pm \sqrt{b^2 - 4ac}}{2a}.$$

The number $b^2 - 4ac$ under the radical sign in the quadratic formula is the **discriminant** of the quadratic equation. The discriminant can be used to determine the nature of the roots of the equation, as in the following theorem.

THEOREM

Let $ax^2 + bx + c = 0$ be a quadratic equation.

(i) If $b^2 - 4ac > 0$, the equation has two real and unequal roots.

(ii) If $b^2 - 4ac = 0$, the equation has one root of multiplicity 2.

(iii) If $b^2 - 4ac < 0$, the equation has no real root.

EXAMPLE ▪ 5

Solve $4x^2 + x - 3 = 0$.

SOLUTION Let $a = 4$, $b = 1$, and $c = -3$ in the quadratic formula:

$$x = \frac{-1 \pm \sqrt{1 - 4(4)(-3)}}{2(4)} = \frac{-1 \pm \sqrt{49}}{8} = \frac{-1 \pm 7}{8}.$$

Hence, the solutions are

$$x = \frac{-1 + 7}{8} = \frac{3}{4} \quad \text{and} \quad x = \frac{-1 - 7}{8} = -1.$$

Example 5 can also be solved by factoring. Writing $(4x - 3)(x + 1) = 0$ and setting each factor equal to zero gives us $x = \frac{3}{4}$ and $x = -1$.

EXAMPLE ▪ 6

Solve $2x(3 - x) = 3$.

SOLUTION To use the quadratic formula we must write the equation in the form $ax^2 + bx + c = 0$. The following equations are equivalent:

$$\begin{aligned} 2x(3 - x) &= 3 \\ 6x - 2x^2 &= 3 \\ -2x^2 + 6x - 3 &= 0 \\ 2x^2 - 6x + 3 &= 0 \end{aligned}$$

We now let $a = 2$, $b = -6$, and $c = 3$ in the quadratic formula, obtaining

$$x = \frac{6 \pm \sqrt{(-6)^2 - 4(2)(3)}}{2(2)} = \frac{6 \pm \sqrt{12}}{4} = \frac{6 \pm 2\sqrt{3}}{4}$$

Since 2 is a factor of the numerator and denominator, we can simplify the last fraction as follows:

$$\frac{2(3 \pm \sqrt{3})}{2 \cdot 2} = \frac{3 \pm \sqrt{3}}{2}$$

Hence, the solutions are $\dfrac{3 + \sqrt{3}}{2}$ and $\dfrac{3 - \sqrt{3}}{2}$.

The following example illustrates the case of a double root.

EXAMPLE ▪ 7

Solve $9x^2 - 30x + 25 = 0$.

SOLUTION Let $a = 9$, $b = -30$, $c = 25$ in the quadratic formula:

$$\begin{aligned} x &= \frac{30 \pm \sqrt{(-30)^2 - 4(9)(25)}}{2(9)} \\ &= \frac{30 \pm \sqrt{900 - 900}}{18} \\ &= \frac{30 \pm 0}{18} = \frac{5}{3}. \end{aligned}$$

Consequently, the equation has one (double) root $\frac{5}{3}$. (Note that the discriminant is zero).

The next example illustrates the use of a quadratic equation in solving an equation that contains rational expressions.

EXAMPLE ▪ 8

Solve $\dfrac{2x}{x-3} + \dfrac{5}{x+3} = \dfrac{36}{x^2 - 9}$.

SOLUTION We may assume that $x \neq \pm 3$, since substitution of 3 or -3 for x produces a zero denominator. We multiply both sides of the equation by the lcd, $(x + 3)(x - 3)$, and simplify:

$$\begin{aligned} 2x(x + 3) + 5(x - 3) &= 36 \\ 2x^2 + 6x + 5x - 15 - 36 &= 0 \\ 2x^2 + 11x - 51 &= 0 \\ (x - 3)(2x + 17) &= 0 \end{aligned}$$

The number 3 cannot be a solution; however, $-\frac{17}{2}$ checks in the given equation and hence $-\frac{17}{2}$ is the only solution.

Many applied problems lead to quadratic equations. One is illustrated in the following example.

EXAMPLE ▪ 9

A box with a square base and no top is to be made from a square piece of tin by cutting out 3-inch squares from each corner and folding up the sides. If the box is to hold 48 in.^{3}, what size piece of tin should be used?

FIGURE 6

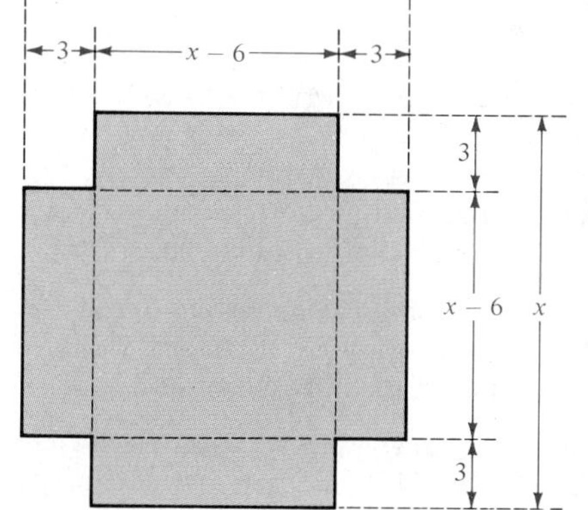

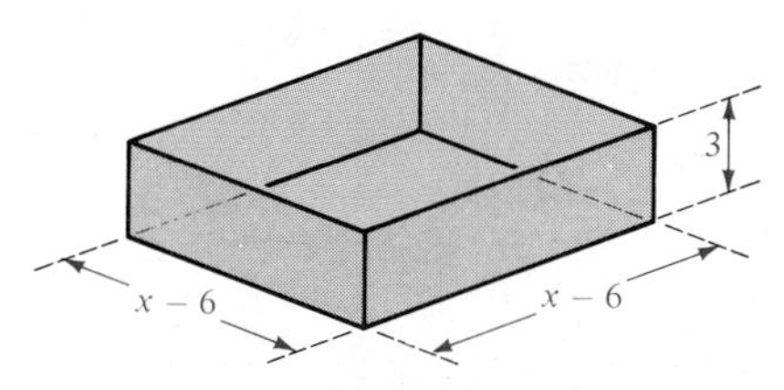

SOLUTION We begin by drawing the picture in Figure 6, letting x denote the length of the side of the piece of tin.

Since the area of the base of the box is $(x-6)^2$ and the height is 3, we obtain

$$\text{volume of box} = 3(x-6)^2.$$

Since the box is to hold 48 in.^{3},

$$3(x-6)^2 = 48.$$

We now solve for x:

$$(x-6)^2 = 16$$
$$x - 6 = \pm 4$$
$$x = 6 \pm 4$$

Consequently, $x = 10$ or $x = 2$.

Let us now check each of these numbers. Referring to Figure 6, we see that 2 is unacceptable, since no box is possible in this case. However, if we begin with a 10-inch square of tin, cut out 3-inch corners, and fold, we obtain a box having dimensions 4 inches, 4 inches, and 3 inches. The box has the desired volume of 48 in.^{3}. Thus, 10 inches is the answer to the problem.

As illustrated in Example 9, even though an equation is formulated correctly, it is possible to arrive at meaningless solutions due to the physical nature of a given problem. Such solutions should be discarded. For example, we would not accept the answer -7 years for the age of an individual, nor $\sqrt{50}$ for the number of automobiles in a parking lot.

2.3 EXERCISES

Exer. 1–4: Solve the equation.

1 $x^2 = 169$

2 $x^2 = 361$

3 $25x^2 = 9$

4 $16x^2 = 49$

Exer. 5–6: Determine the values of c that complete the square for the expression.

5 (a) $x^2 + 9x + c$ (b) $x^2 - 8x + c$

(c) $x^2 + cx + 36$ (d) $x^2 + cx + \frac{49}{4}$

6 (a) $x^2 + 13x + c$ (b) $x^2 - 6x + c$

(c) $x^2 + cx + 25$ (d) $x^2 + cx + \frac{81}{4}$

Exer. 7–16: Solve the equation by factoring.

7 $6x^2 + x - 12 = 0$

8 $4x^2 + x - 14 = 0$

9 $15x^2 - 12 = -8x$

10 $15x^2 - 14 = 29x$

11 $2x(4x + 15) = 27$

12 $x(3x + 10) = 77$

13 $75x^2 + 35x - 10 = 0$

14 $48x^2 + 12x - 90 = 0$

15 $12x^2 + 60x + 75 = 0$

16 $4x^2 - 72x + 324 = 0$

17 $\dfrac{2x}{x+3} + \dfrac{5}{x} - 4 = \dfrac{18}{x^2 + 3x}$

18 $\dfrac{5x}{x-2} + \dfrac{3}{x} + 2 = \dfrac{-6}{x^2 - 2x}$

19 $\dfrac{5x}{x-3} + \dfrac{4}{x+3} = \dfrac{90}{x^2 - 9}$

20 $\dfrac{3x}{x-2} + \dfrac{1}{x+2} = \dfrac{-4}{x^2 - 4}$

Exer. 21–30: Solve the equation by using the quadratic formula.

21 $6x^2 - x = 2$

22 $5x^2 + 13x - 6 = 0$

23 $x^2 + 4x + 2 = 0$

24 $x^2 - 6x - 3 = 0$

25 $2x^2 - 3x - 4 = 0$

26 $3x^2 + 5x + 1 = 0$

27 $\frac{3}{2}z^2 - 4z - 1 = 0$

28 $\frac{5}{3}s^2 + 3s + 1 = 0$

29 $\dfrac{5}{w^2} - \dfrac{10}{w} + 2 = 0$

30 $\dfrac{x+1}{3x+2} = \dfrac{x-2}{2x-3}$

31 The boundary of a city is a circle of diameter 5 miles. As shown in the figure, a straight highway runs through the center of the city from A to B. The highway department is planning to build a 6-mile-long freeway from A to a point P on the outskirts and then to B. Find the distance from A to P. (*Hint:* APB is a right triangle.)

EXERCISE 31

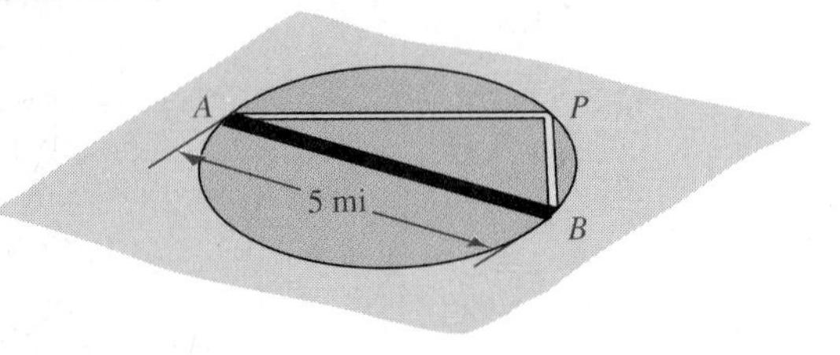

32 The boundary of a city is a circle of diameter 10 miles. Within the last decade, the city has grown in area by 16π square miles (or about $50\ \text{mi}^2$). Assuming the city was always circular in shape, find the corresponding change in distance from the center of town to the boundary.

33 A rectangular plot of ground having dimensions 26 feet by 30 feet is surrounded by a walk of uniform width. If the area of the walk is $240\ \text{ft}^2$, what is its width?

34 A manufacturer of tin cans wishes to construct a right circular cylindrical can of height 20 cm and of capacity $3000\ \text{cm}^3$ (see figure). Find the inner radius r of the can.

EXERCISE 34

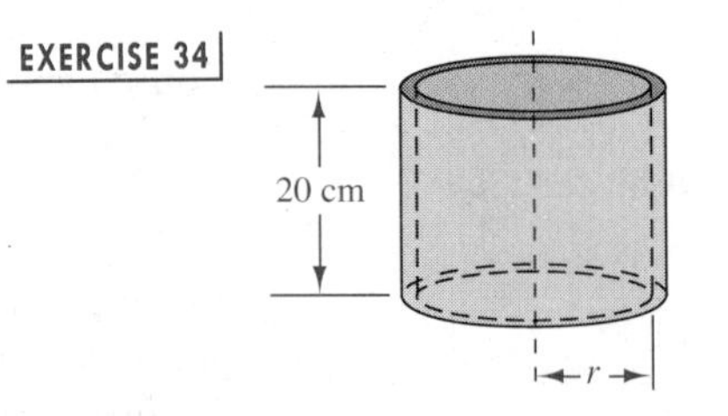

35 An airplane flying north at 200 mi/hr passed over a point on the ground at 2:00 P.M. Another airplane at the same altitude passed over the point at 2:30 P.M., flying east at 400 mi/hr (see figure on page 83).

(a) If t denotes the time in hours after 2:30 P.M., express the distance d between the airplanes in terms of t.

(b) At what time after 2:30 P.M. were the airplanes 500 miles apart?

EXERCISE 35

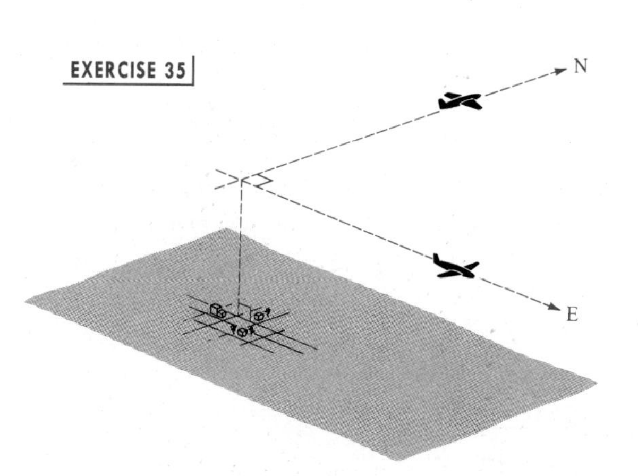

36 Refer to Example 9. A box with an open top is to be constructed by cutting out 3-inch squares from a rectangular sheet of tin whose length is twice its width. What size sheet will produce a box having a volume of 60 in.^{3}?

37 Two surveyors with two-way radios leave the same point at 9:00 A.M., one walking due south at 4 mi/hr and the other due west at 3 mi/hr. How long can they communicate with one another if each radio has a maximum range of 2 miles?

38 A farmer plans to enclose a rectangular region, using part of his barn for one side and fencing for the other three sides. If he wants the side parallel to the barn to be twice the length of an adjacent side, and the area of the region to be 128 ft^2, how many feet of fencing should he purchase?

39 A baseball is thrown straight upward with an initial speed of 64 ft/sec. The number of feet s above the ground after t seconds is given by $s = -16t^2 + 64t$.

(a) When will the baseball be 48 feet above the ground?

(b) When will it hit the ground?

(c) What is its maximum height?

40 A particle of charge -1 is placed on a coordinate line at $x = -2$ and a particle of charge -2 is placed at $x = 2$, as shown in the figure. If a particle of charge $+1$ is placed at a position x between -2 and 2, Coulomb's law in electrical theory asserts that the *net force* F acting on this particle is given by

$$F = \frac{-k}{(x+2)^2} + \frac{2k}{(2-x)^2}$$

for some constant $k > 0$. Determine the position at which the net force is zero.

EXERCISE 40

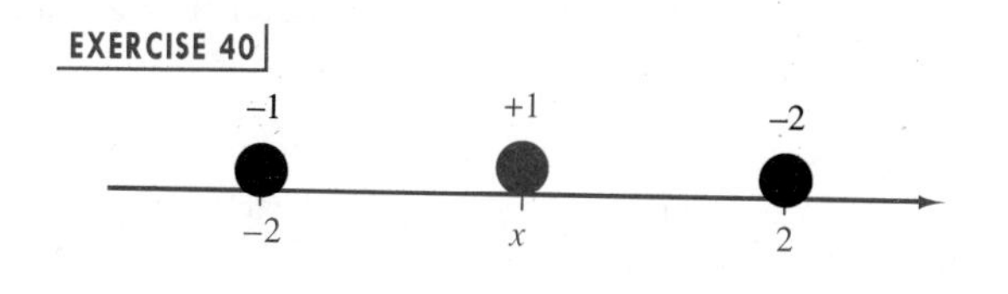

41 The distance that a car travels between the time the driver makes the decision to hit the brakes and the time the car actually stops is the braking distance. If a car is traveling v mi/hr, the braking distance d (in feet) is approximated by $d = v + (v^2/20)$.

(a) Find the braking distance when v is 55 mi/hr.

(b) If a driver decides to brake 120 feet from a stop sign, how fast can the car be going and still stop by the time it reaches the sign?

42 A boy drops a rock off a cliff and into the ocean below. If he hears the splash 4 seconds later, how high is the cliff? (*Hint:* Assume that the rock travels $16t^2$ feet after t seconds and that the speed of sound is 1100 ft/sec.)

43 The temperature T (in °C) at which water boils is related to the elevation h (in meters above sea level) by the formula

$$h = 1000(100 - T) + 580(100 - T)^2$$

for $95 \le T \le 100$.

(a) At what elevation does water boil at a temperature of 98 °C?

(b) The elevation of Mt. Everest is 29,000 feet (or 8840 meters). At what temperature will water boil at the top of this mountain? (*Hint:* Let $x = 100 - T$ and use the quadratic formula.)

44 A square vegetable garden is to be enclosed with a fence. If the fence costs \$1 per foot, and if the cost of preparing the soil is 50 cents per ft^2, determine the size of the garden that can be set up for \$120.

45 A 24 inch × 36 inch sheet of paper is to be used for a poster, with the shorter side at the bottom. The margins at the sides and top are to have the same width, and the bottom margin is to be twice as wide as the other margins. Find the width of the margins if the printed area is to be 661.5 in.^{2}.

46 A company sells running shoes to dealers at a rate of \$20 per pair if less than 50 pairs are ordered. If 50 or more pairs are ordered (up to 600), the price per pair is reduced at a rate of 2 cents times the number ordered. How many pairs can a dealer purchase for \$4200?

47 Refer to Example 9. A pizza box with square base is to be made from a rectangular sheet of cardboard by cutting out six 1-inch squares from the corners and the middle sections, and folding up the sides (see figure). If the area of the base is to be 144 in.^{2}, what size piece of cardboard should be used?

EXERCISE 47

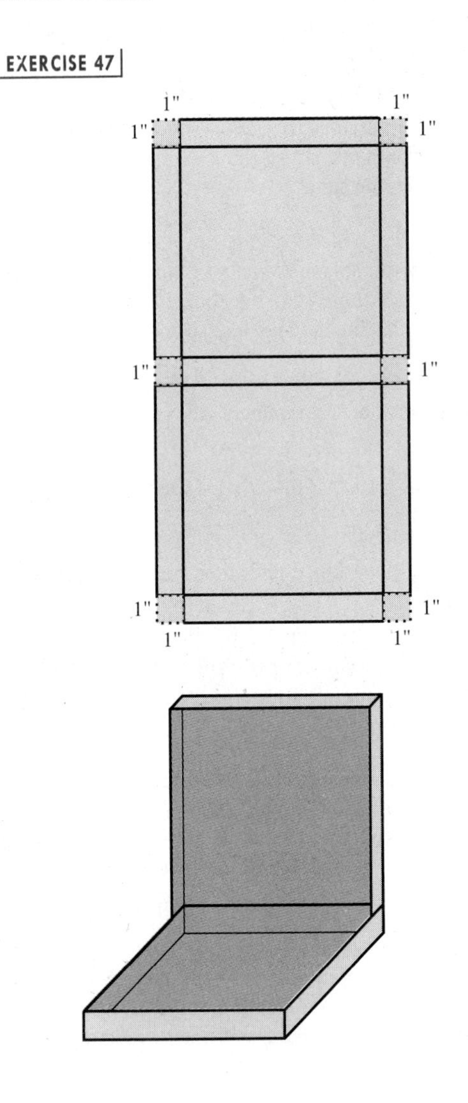

48 The rate at which a tablet of vitamin C begins to dissolve depends on the surface area of the tablet. One brand of tablet is 2 cm long and is in the shape of a cylinder with hemispheres of diameter 0.5 cm attached to both ends (see figure). A second brand of tablet is to be manufactured in the shape of a right circular cylinder of altitude 0.5 cm.

(a) Find the diameter of the second tablet so that its surface area is equal to that of the first tablet.

(b) Find the volume of each tablet.

EXERCISE 48

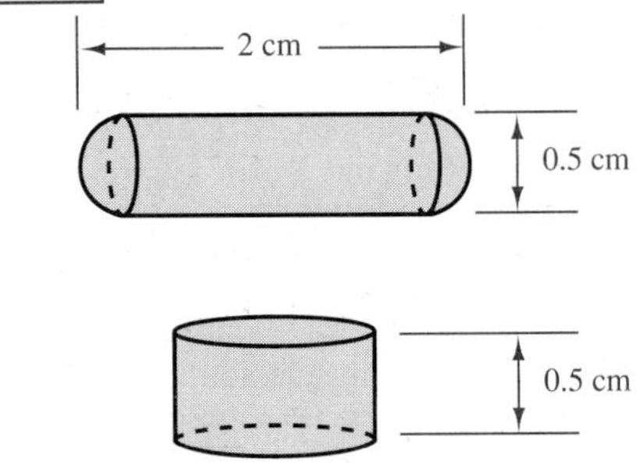

49 A closed cylindrical oil drum of height 4 feet is to be constructed so that the total surface area is 10π ft^2. Find the diameter of the drum.

50 When a popular brand of CD player is priced at \$300 per unit, a stereo store sells 15 units per week. Each time the price is reduced by \$10, however, the sales increase by 2 per week. What selling price will result in weekly revenues of \$7000?

51 The speed of the current in a stream is 5 mi/hr. It takes a girl 30 minutes longer to paddle a canoe 1.2 miles upstream than the same distance downstream. What is her rate in still water?

52 Two square wire frames are to be constructed from a piece of wire 100 inches long. If the area enclosed by one frame is to be one-half the area enclosed by the other, find the dimensions of each frame. Neglect the thickness of the wire.

53 In a round-robin softball tournament, each pair of teams meets once. Four games can be played each day, and the tournament organizers have rented the field for one week.

The number of pairings of n teams is $\frac{1}{2}n(n-1)$. How many teams should be invited?

54 A *diagonal* of a polygon is a line segment joining any two nonadjacent vertices. The 20 diagonals of an octagon are shown in the figure. The number of diagonals in a polygon with n sides is $\frac{1}{2}n(n-3)$. How many sides must a polygon have if the number of diagonals is 35? Is there a polygon with 100 diagonals?

EXERCISE 54

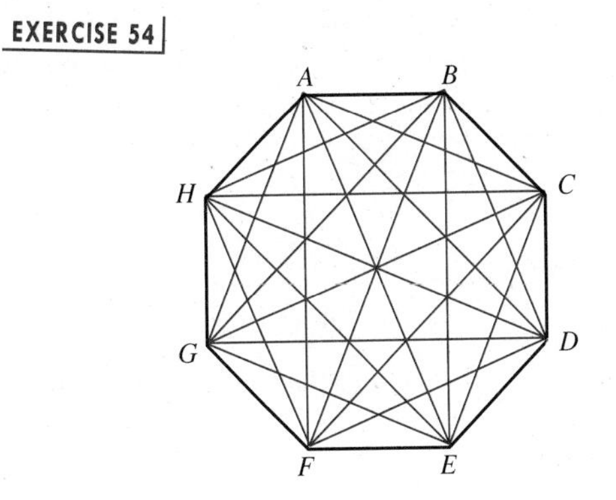

2.4 COMPLEX NUMBERS

Real numbers are needed to find solutions of equations. To illustrate, the nonnegative integers $0, 1, 2, 3, \ldots$ do not contain solutions of equations such as $x + 5 = 0$ or $x + 24 = 0$. To solve such equations we need the system of integers, which contains negatives $-1, -2, -3, \ldots$. In this expanded number system we find the solution -5 for the equation $x + 5 = 0$, and -24 for $x + 24 = 0$.

Similarly, to solve the equation $3x + 5 = 0$, we must enlarge the set of integers to the rational numbers, thereby obtaining the solution $x = -\frac{5}{3}$.

The set of rational numbers is still not large enough to solve every equation; for example, $x^2 = 5$ has no rational solution. Thus, we must again expand our number system to include irrational numbers, such as $\sqrt{5}$ and $-\sqrt{5}$. This leads us to the real number system $\mathbb{R}$, which contains all rational and irrational numbers.

Since squares of real numbers are never negative, $\mathbb{R}$ does not contain solutions of equations of the form $x^2 = -a$ for $a > 0$. To solve such equations we need a larger number system $\mathbb{C}$, the **complex number system**, which contains $\mathbb{R}$ and also contains numbers whose squares are negative.

We begin by introducing the **imaginary unit**, denoted by i, which has the following properties.

PROPERTIES OF i

$$i = \sqrt{-1}; \qquad i^2 = -1.$$

The letter i does not represent a real number. It is a new mathematical entity that will enable us to obtain the number system $\mathbb{C}$, which contains solutions of every algebraic equation.

Since i, together with $\mathbb{R}$, is to be contained in $\mathbb{C}$, we must consider products of the form bi for a real number b and also expressions of the form $a + bi$ for real numbers a and b. This motivates the next definition.

DEFINITION

A **complex number** is an expression of the form $a + bi$, where a and b are real numbers and $i^2 = -1$.

We next define equality of two complex numbers $a + bi$ and $c + di$.

DEFINITION OF EQUALITY

$a + bi = c + di$ if and only if $a = c$ and $b = d$.

The operations of addition and multiplication of complex numbers are defined as follows.

ADDITION AND MULTIPLICATION

$$(a + bi) + (c + di) = (a + c) + (b + d)i$$

$$(a + bi)(c + di) = (ac - bd) + (ad + bc)i$$

It is unnecessary to memorize the definitions of addition and multiplication of complex numbers. Instead, *we may treat all symbols as though they represent real numbers, with exactly one exception: we replace i^2 by -1.* Note that if we use this technique for multiplication, we obtain

$$\begin{aligned}(a + bi)(c + di) &= (a + bi)c + (a + bi)(di)\\ &= ac + (bi)c + a(di) + (bi)(di)\\ &= ac + (bc)i + (ad)i + (bd)(i^2)\\ &= ac + (bc)i + (ad)i + (bd)(-1)\\ &= ac + (bd)(-1) + (ad)i + (bc)i\\ &= (ac - bd) + (ad + bc)i,\end{aligned}$$

which agrees with the definition of multiplication of complex numbers.

EXAMPLE ▪ 1

Express in the form $a + bi$, where a and b are real numbers:

(a) $(3 + 4i) + (2 + 5i)$ (b) $(3 + 4i)(2 + 5i)$

SOLUTION

(a)
$$(3 + 4i) + (2 + 5i) = (3 + 2) + (4 + 5)i = 5 + 9i$$

(b)
$$\begin{aligned}(3 + 4i)(2 + 5i) &= (3 + 4i)2 + (3 + 4i)(5i) \\ &= 6 + 8i + 15i + 20i^2 \\ &= 6 + 20(-1) + 23i \\ &= -14 + 23i\end{aligned}$$

The set $\mathbb{R}$ of real numbers may be identified with the set of complex numbers of the form $a + 0i$. It is also convenient to denote the complex number $0 + bi$ by bi. Thus

$$(a + 0i) + (0 + bi) = (a + 0) + (0 + b)i = a + bi.$$

Hence we may regard $a + bi$ as the sum of two complex numbers a and bi (that is, $a + 0i$ and $0 + bi$).

The **identity element** relative to addition is 0 (or, equivalently, $0 + 0i$); that is,

$$(a + bi) + 0 = a + bi$$

for every complex number $a + bi$.

If $(-a) + (-b)i$ is added to $a + bi$, we obtain 0. This implies that $(-a) + (-b)i$ is the **additive inverse** of $a + bi$; that is,

$$-(a + bi) = (-a) + (-b)i.$$

Subtraction of complex numbers is defined using additive inverses:

$$(a + bi) - (c + di) = (a + bi) + [-(c + di)].$$

Since $-(c + di) = (-c) + (-d)i$, it follows that

$$(a + bi) - (c + di) = (a - c) + (b - d)i.$$

ILLUSTRATION | SUBTRACTION

■ $(3 + 4i) - (7 + 2i) = (-4) + 2i$

If c, d, and k are real numbers, then

$$k(c + di) = (k + 0i)(c + di) = (kc - 0d) + (kd + 0c)i,$$

that is,
$$k(c + di) = kc + (kd)i.$$

If we are asked to write an expression in the form $a + bi$, we shall also accept the form $a - di$, since $a - di = a + (-d)i$. This is illustrated in the next example.

EXAMPLE ▪ 2

Express in the form $a + bi$, where a and b are real numbers:

(a) $4(2 + 5i) - (3 - 4i)$ (b) $(4 - 3i)(2 + i)$

(c) $i(3 - 2i)^2$ (d) i^{51}

SOLUTION

(a) $$4(2 + 5i) - (3 - 4i) = 8 + 20i - 3 + 4i = 5 + 24i$$

(b) $$(4 - 3i)(2 + i) = 8 - 6i + 4i - 3i^2 = 11 - 2i$$

(c) $$i(3 - 2i)^2 = i(9 - 12i + 4i^2) = i(5 - 12i) = 5i - 12i^2 = 12 + 5i$$

(d) Taking successive powers of i, we obtain

$$i^1 = i, \quad i^2 = -1, \quad i^3 = -i, \quad i^4 = 1,$$

and then the cycle starts over:

$$i^5 = i, \quad i^6 = i^2 = -1, \quad \text{and so on.}$$

In particular, $$i^{51} = i^{48}i^3 = (i^4)^{12}i^3 = (1)^{12}i^3 = i^3 = -i.$$

DEFINITION

The **conjugate** of a complex number $a + bi$ is $a - bi$.

The two properties of conjugates in the next box follow from the definitions of addition and multiplication of complex numbers.

PROPERTIES OF CONJUGATES

$$(a + bi) + (a - bi) = 2a$$
$$(a + bi)(a - bi) = a^2 + b^2$$

The preceding formulas show that *the sum and product of a complex number and its conjugate are real numbers*. Conjugates are useful for finding the **multiplicative inverse** $\dfrac{1}{a + bi}$ of $a + bi$, or for simplifying the quotient $\dfrac{a + bi}{c + di}$ of two complex numbers, as illustrated in the next example.

EXAMPLE ▪ 3

Express the fraction in the form $a + bi$ for real numbers a and b:

(a) $\dfrac{1}{9 + 2i}$ (b) $\dfrac{7 - i}{3 - 5i}$

SOLUTION We multiply numerator and denominator by the conjugate of the denominator:

(a)
$$\frac{1}{9 + 2i} = \frac{1}{9 + 2i} \cdot \frac{9 - 2i}{9 - 2i} = \frac{9 - 2i}{81 + 4} = \frac{9}{85} - \frac{2}{85}i$$

(b)
$$\frac{7 - i}{3 - 5i} = \frac{7 - i}{3 - 5i} \cdot \frac{3 + 5i}{3 + 5i} = \frac{21 - 3i + 35i - 5i^2}{9 + 25}$$
$$= \frac{26 + 32i}{34} = \frac{13}{17} + \frac{16}{17}i$$

If p is any positive real number, then the equation $x^2 = -p$ has solutions in $\mathbb{C}$. One solution is $i\sqrt{p}$, since

$$(i\sqrt{p})^2 = i^2(\sqrt{p})^2 = (-1)p = -p.$$

Similarly, $-i\sqrt{p}$ is also a solution.

The next definition is motivated by $(i\sqrt{r})^2 = -r$ for $r > 0$.

DEFINITION

If r is a positive real number, then the **principal square root** of $-r$ is denoted by $\sqrt{-r}$ and is defined by

$$\sqrt{-r} = i\sqrt{r} = \sqrt{r}\,i.$$

Care must be taken *not* to write $\sqrt{ri}$ when $\sqrt{r}\,i$ is intended.

ILLUSTRATION $\sqrt{-r}$

- $\sqrt{-9} = i\sqrt{9} = i(3) = 3i$
- $\sqrt{-5} = i\sqrt{5} = \sqrt{5}\, i$
- $\sqrt{-1} = i\sqrt{1} = i$

The radical sign must be used with caution when the radicand is negative. For example, the formula $\sqrt{a}\sqrt{b} = \sqrt{ab}$, which holds for positive real numbers, is not true when a and b are both negative, as shown below:

$$\sqrt{-3}\sqrt{-3} = (i\sqrt{3})(i\sqrt{3}) = i^2(\sqrt{3})^2 = (-1)3 = -3,$$

and
$$\sqrt{(-3)(-3)} = \sqrt{9} = 3.$$

Hence,
$$\sqrt{-3}\sqrt{-3} \neq \sqrt{(-3)(-3)}.$$

If only *one* of a or b is negative, then $\sqrt{a}\sqrt{b} = \sqrt{ab}$. In general, we shall not apply laws of radicals if radicands are negative. Instead, we shall change the form of radicals before performing any operations, as illustrated in the next example.

EXAMPLE ▪ 4

Express $(5 - \sqrt{-9})(-1 + \sqrt{-4})$ in the form $a + bi$ for real numbers a and b.

SOLUTION

$$\begin{aligned}(5 - \sqrt{-9})(-1 + \sqrt{-4}) &= (5 - i\sqrt{9})(-1 + i\sqrt{4})\\ &= (5 - 3i)(-1 + 2i)\\ &= -5 + 3i + 10i - 6i^2\\ &= -5 + 13i + 6 = 1 + 13i\end{aligned}$$

In the previous section we proved that if a, b, and c are real numbers such that $b^2 - 4ac \geq 0$, and if $a \neq 0$, then the solutions of the quadratic equation $ax^2 + bx + c = 0$ are

$$\frac{-b + \sqrt{b^2 - 4ac}}{2a} \quad \text{and} \quad \frac{-b - \sqrt{b^2 - 4ac}}{2a}.$$

We may now extend this fact to include $b^2 - 4ac < 0$. The same manipulations used to obtain the quadratic formula, together with the developments in this section, show that if $b^2 - 4ac < 0$, then the solutions of $ax^2 + bx + c = 0$ are the two *complex* numbers given above. Notice that the solutions are conjugates of one another.

EXAMPLE ▪ 5

Solve $5x^2 + 2x + 1 = 0$.

SOLUTION By the quadratic formula,

$$x = \frac{-2 \pm \sqrt{4 - 20}}{10} = \frac{-2 \pm \sqrt{-16}}{10} = \frac{-2 \pm 4i}{10} = \frac{-1 \pm 2i}{5}.$$

Thus the solutions of the equation are $-\frac{1}{5} + \frac{2}{5}i$ and $-\frac{1}{5} - \frac{2}{5}i$.

EXAMPLE ▪ 6

Solve $x^3 - 1 = 0$.

SOLUTION Using the difference of two cubes factoring formula (see page 31), we may write the equation $x^3 - 1 = 0$ as

$$(x - 1)(x^2 + x + 1) = 0.$$

Setting each factor equal to zero and solving the resulting equations, we obtain the solutions

$$1, \qquad \frac{-1 \pm \sqrt{1 - 4}}{2} = \frac{-1 \pm \sqrt{3}i}{2}$$

or, equivalently,

$$1, \qquad -\frac{1}{2} + \frac{\sqrt{3}}{2}i, \qquad -\frac{1}{2} - \frac{\sqrt{3}}{2}i.$$

The three solutions of $x^3 - 1 = 0$ are the **cube roots of unity**.

2.4 EXERCISES

Exer. 1–34: Write the expression in the form $a + bi$, where a and b are real numbers.

1 $(5 - 2i) + (-3 + 6i)$

2 $(-5 + 7i) + (4 + 9i)$

3 $(7 - 6i) - (-11 - 3i)$

4 $(-3 + 8i) - (2 + 3i)$

5 $(3 + 5i)(2 - 7i)$

6 $(-2 + 6i)(8 - i)$

7 $(1 - 3i)(2 + 5i)$

8 $(8 + 2i)(7 - 3i)$

9 $(5 - 2i)^2$

10 $(6 + 7i)^2$

11 $i(3 + 4i)^2$

12 $i(2 - 7i)^2$

13 $(3 + 4i)(3 - 4i)$

14 $(4 + 9i)(4 - 9i)$

15 i^{43}

16 i^{92}

17 i^{73}

18 i^{66}

19 $\dfrac{3}{2 + 4i}$

20 $\dfrac{5}{2 - 7i}$

21 $\dfrac{1 - 7i}{6 - 2i}$

22 $\dfrac{2 + 9i}{-3 - i}$

23 $\dfrac{-4+6i}{2+7i}$

24 $\dfrac{-3-2i}{5+2i}$

25 $\dfrac{4-2i}{-5i}$

26 $\dfrac{-2+6i}{3i}$

27 $(2+5i)^3$

28 $(3-2i)^3$

29 $(2-\sqrt{-4})(3-\sqrt{-16})$

30 $(-3+\sqrt{-25})(8-\sqrt{-36})$

31 $\dfrac{4+\sqrt{-81}}{7-\sqrt{-64}}$

32 $\dfrac{5-\sqrt{-121}}{1+\sqrt{-25}}$

33 $\dfrac{\sqrt{-36}\sqrt{-49}}{\sqrt{-16}}$

34 $\dfrac{\sqrt{-25}}{\sqrt{-16}\sqrt{-81}}$

Exer. 35–38: Find the values of x and y.

35 $8+(3x+y)i=2x-4i$

36 $(x-y)+3i=7+yi$

37 $x^3+(2x-y)i=-8-3i$

38 $(3x+2y)-y^3i=9-27i$

Exer. 39–54: Find the solutions of the equation.

39 $x^2-6x+13=0$

40 $x^2-2x+26=0$

41 $x^2+4x+13=0$

42 $x^2+8x+17=0$

43 $x^2-5x+20=0$

44 $x^2+3x+6=0$

45 $4x^2+x+3=0$

46 $-3x^2+x-5=0$

47 $x^3+125=0$

48 $x^3-27=0$

49 $x^4=256$

50 $x^4=81$

51 $4x^4+25x^2+36=0$

52 $27x^4+21x^2+4=0$

53 $x^3+3x^2+4x=0$

54 $8x^3-12x^2+2x-3=0$

Exer. 55–60: If $z=a+bi$ is a complex number, its conjugate is often denoted by $\bar{z}$, that is, $\bar{z}=a-bi$. Verify the property.

55 $\overline{z+w}=\bar{z}+\bar{w}$

56 $\overline{z-w}=\bar{z}-\bar{w}$

57 $\overline{z^2}=(\bar{z})^2$

58 $\overline{z\cdot w}=\bar{z}\cdot\bar{w}$

59 $\bar{\bar{z}}=z$

60 $\bar{z}=z$ if and only if z is real.

2.5 MISCELLANEOUS EQUATIONS

If we can express an equation in factored form *with zero on one side*, then we may often obtain solutions by setting each factor equal to zero. For example, if p, q, r are expressions in x and if $pqr=0$, then either $p=0$, $q=0$, or $r=0$.

EXAMPLE ▪ 1

Solve $x^3+2x^2-x-2=0$.

SOLUTION The left side may be factored by grouping:

$$x^3+2x^2-x-2=0$$
$$x^2(x+2)-(x+2)=0$$
$$(x^2-1)(x+2)=0$$
$$(x+1)(x-1)(x+2)=0$$

Setting each factor equal to 0 gives us

$$x+1=0, \qquad x-1=0, \qquad x+2=0,$$

or, equivalently,

$$x = -1, \qquad x = 1, \qquad x = -2.$$

We may check that these three numbers are solutions by substitution in the original equation.

EXAMPLE ▪ 2

Solve $x^{3/2} = x^{1/2}$.

SOLUTION

$$x^{3/2} = x^{1/2}$$
$$x^{3/2} - x^{1/2} = 0$$
$$x^{1/2}(x - 1) = 0$$

Setting each factor equal to 0, we obtain

$$x^{1/2} = 0, \qquad x - 1 = 0.$$

This gives us the solutions $x = 0$ and $x = 1$.

In Example 2 it would have been *incorrect* to divide both sides of the equation $x^{3/2} = x^{1/2}$ by $x^{1/2}$, obtaining $x = 1$, since the solution $x = 0$ would be lost. In general, *avoid dividing both sides of an equation by an expression that contains variables*; always *factor* instead.

If a given equation involves radicals or fractional exponents, we often raise both sides to a positive power. The solutions of the new equation always contain the solutions of the given equation. For example, the solutions of

$$2x - 3 = \sqrt{x + 6}$$

are also solutions of

$$(2x - 3)^2 = (\sqrt{x + 6})^2.$$

In some cases the new equation has *more* solutions than the given equation. To illustrate, if we are given the equation $x = 3$ and we square both sides, we obtain $x^2 = 9$. Note that the given equation has only one solution, 3, but the new equation has two solutions, 3 and -3. Any solution of the new equation that is not a solution of the given equation is an **extraneous solution**. Since extraneous solutions may occur, it is *absolutely essential* to check all solutions obtained after raising both sides of an equation to an even power.

EXAMPLE ▪ 3

Solve $\sqrt[3]{x^2 - 1} = 2$.

SOLUTION We cube both sides to obtain a new equation:

$$(\sqrt[3]{x^2 - 1})^3 = 2^3$$
$$x^2 - 1 = 8$$
$$x^2 = 9$$
$$x = \pm 3$$

Hence, the only possible solutions of $\sqrt[3]{x^2 - 1} = 2$ are 3 or -3.

Let us check each of these numbers by substitution in $\sqrt[3]{x^2 - 1} = 2$. Substituting 3 for x in the equation, we obtain $\sqrt[3]{3^2 - 1} = 2$, or $\sqrt[3]{8} = 2$, which is a true statement. Thus, 3 is a solution. Similarly, -3 is a solution. Hence, the solutions of the given equation are 3 and -3.

EXAMPLE ▪ 4

Solve $3 + \sqrt{3x + 1} = x$.

SOLUTION We first isolate the radical on one side:

$$\sqrt{3x + 1} = x - 3.$$

Next we square both sides and simplify:

$$(\sqrt{3x + 1})^2 = (x - 3)^2$$
$$3x + 1 = x^2 - 6x + 9$$
$$x^2 - 9x + 8 = 0$$
$$(x - 1)(x - 8) = 0.$$

Since the last equation has solutions 1 and 8, it follows that 1 and 8 are the only possible solutions of the original equation.

Check each of these by substitution in $3 + \sqrt{3x + 1} = x$. Letting $x = 1$ gives us

$$3 + \sqrt{4} = 1, \quad \text{or} \quad 5 = 1,$$

which is false. Consequently, 1 is not a solution. Letting $x = 8$ in the given equation we obtain

$$3 + \sqrt{25} = 8, \quad \text{or} \quad 3 + 5 = 8,$$

which is true. Hence, the equation $3 + \sqrt{3x + 1} = x$ has only one solution, $x = 8$.

For an equation involving several radicals, it may be necessary to raise sides to powers several times, as in the next example.

EXAMPLE ▪ 5

Solve $\sqrt{2x-3} - \sqrt{x+7} + 2 = 0$.

SOLUTION Let us begin by writing

$$\sqrt{2x-3} = \sqrt{x+7} - 2.$$

Squaring both sides, we obtain

$$2x - 3 = (x+7) - 4\sqrt{x+7} + 4,$$

which simplifies to $\qquad x - 14 = -4\sqrt{x+7}.$

Square both sides of the last equation and simplify:

$$\begin{aligned} x^2 - 28x + 196 &= 16(x+7) \\ x^2 - 28x + 196 &= 16x + 112 \\ x^2 - 44x + 84 &= 0 \\ (x-42)(x-2) &= 0. \end{aligned}$$

Thus the only possible solutions are 42 and 2.

Check each of these by substitution in the given equation. Substituting $x = 42$ gives us

$$\begin{aligned} \sqrt{84-3} - \sqrt{42+7} + 2 &= 0 \\ 9 - 7 + 2 &= 0, \end{aligned}$$

which is false. Hence, 42 is not a solution. If we substitute $x = 2$, we obtain

$$\begin{aligned} \sqrt{4-3} - \sqrt{2+7} + 2 &= 0 \\ 1 - 3 + 2 &= 0, \end{aligned}$$

which is true. Hence, the given equation has one solution, $x = 2$.

An equation is of **quadratic type** if it can be written in the form

$$au^2 + bu + c = 0$$

with $a \neq 0$ and u an expression in some variable. If we find the solutions in terms of u, then the solutions of the given equation can be obtained by referring to the specific form of u.

EXAMPLE ▪ 6

Solve $x^{2/3} + x^{1/3} - 6 = 0$.

SOLUTION If we let $u = x^{1/3}$, then the equation can be written

$$u^2 + u - 6 = 0.$$

Factoring, we obtain

$$(u + 3)(u - 2) = 0.$$

This equation has solutions $\quad u = -3, \quad u = 2.$

Since $u = x^{1/3}$, $\quad x^{1/3} = -3, \quad x^{1/3} = 2.$

Cubing gives us $\quad x = -27, \quad x = 8.$

Check each of these by substitution. Letting $x = -27$ in the given equation, we obtain

$$(-27)^{2/3} + (-27)^{1/3} - 6 = 9 - 3 - 6 = 0.$$

Thus, -27 is a solution. Similarly, 8 is a solution. Hence, the solutions of the given equation are -27 and 8.

EXAMPLE ▪ 7

Solve $x^4 - 3x^2 + 1 = 0$.

SOLUTION Letting $u = x^2$ gives us

$$u^2 - 3u + 1 = 0.$$

By the quadratic formula,

$$u = \frac{3 \pm \sqrt{9 - 4}}{2} = \frac{3 \pm \sqrt{5}}{2}.$$

Since $u = x^2$,

$$x^2 = \frac{3 \pm \sqrt{5}}{2}, \qquad x = \pm\sqrt{\frac{3 \pm \sqrt{5}}{2}}.$$

Thus, there are four solutions:

$$\sqrt{\frac{3 + \sqrt{5}}{2}}, \quad -\sqrt{\frac{3 + \sqrt{5}}{2}}, \quad \sqrt{\frac{3 - \sqrt{5}}{2}}, \quad -\sqrt{\frac{3 - \sqrt{5}}{2}}.$$

Using a calculator, we obtain the approximations ± 1.62 and ± 0.62. No check is required since we did not raise both sides of an equation to a power.

FIGURE 7

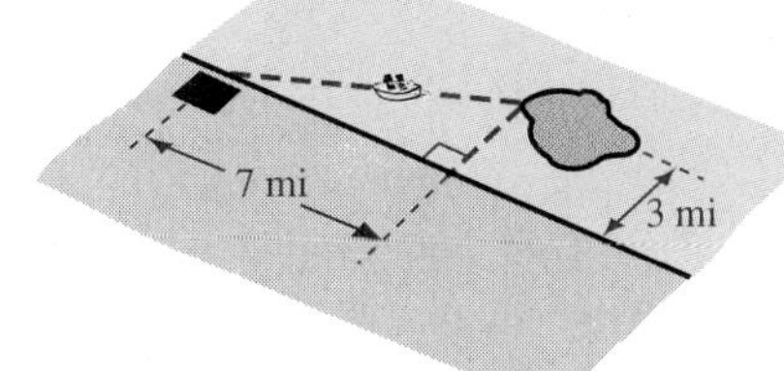

FIGURE 8

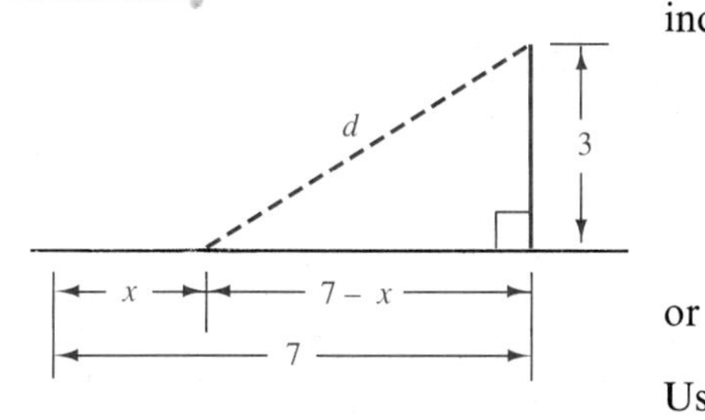

EXAMPLE ▪ 8

A passenger ferry makes trips from a town to an island community that is 7 miles downshore from the town and 3 miles off a straight shoreline. As shown in Figure 7, the ferry travels along the shoreline to some point and then proceeds directly to the island. If the ferry travels 12 mi/hr along the shoreline and 10 mi/hr as it moves out to sea, determine the routes that have a travel time of 45 minutes.

SOLUTION Let x denote the distance traveled along the shoreline. This leads to the sketch in Figure 8, where d is the distance from a point on the shoreline to the island. Applying the Pythagorean theorem to the indicated right triangle,

$$\begin{aligned} d^2 &= (7-x)^2 + 3^2 \\ &= 49 - 14x + x^2 + 9 \\ &= x^2 - 14x + 58 \end{aligned}$$

or

$$d = \sqrt{x^2 - 14x + 58}.$$

Using distance = (rate)(time), or, equivalently, time = (distance)/(rate) gives us the following table.

	along the shoreline	away from shore
Distance (mi)	x	$\sqrt{x^2 - 14x + 58}$
Rate (mi/hr)	12	10
Time (hr)	$\frac{x}{12}$	$\frac{\sqrt{x^2 - 14x + 58}}{10}$

The time for the complete trip is the sum of the two expressions in the last row of the table. Since the rate is in mi/hr we must, for consistency, express this time (45 minutes) as $\frac{3}{4}$ hour. Thus we have the following equation:

$$\frac{x}{12} + \frac{\sqrt{x^2 - 14x + 58}}{10} = \frac{3}{4}.$$

Equivalently,

$$\frac{\sqrt{x^2 - 14x + 58}}{10} = \frac{3}{4} - \frac{x}{12}$$

$$\frac{\sqrt{x^2 - 14x + 58}}{10} = \frac{9 - x}{12}.$$

Multiplying both sides by the lcd, 60, gives us

$$6\sqrt{x^2 - 14x + 58} = 5(9 - x).$$

We next square both sides:

$$36(x^2 - 14x + 58) = 25(9 - x)^2.$$

Multiplying the factors and simplifying, we can show that the last equation may be written in the form

$$11x^2 - 54x + 63 = 0.$$

Factoring, we obtain

$$(x - 3)(11x - 21) = 0.$$

Setting each factor equal to 0 leads to

$$x = 3, \qquad x = \frac{21}{11}.$$

These numbers are also solutions of the original equation. Hence there are two possible routes with a travel time of 45 minutes: the ferry may travel along the shoreline either 3 miles or $\frac{21}{11} \approx 1.9$ miles before proceeding to the island.

2.5 EXERCISES

Exer. 1–38: Solve the equation.

1 $9x^3 - 18x^2 - 4x + 8 = 0$

2 $3x^3 - 4x^2 - 27x + 36 = 0$

3 $4x^4 + 10x^3 = 6x^2 + 15x$

4 $15x^5 - 20x^4 = 6x^3 - 8x^2$

5 $y^{3/2} = 5y$

6 $y^{4/3} = -3y$

7 $\sqrt{7 - 5x} = 8$

8 $\sqrt{2x - 9} = \frac{1}{3}$

9 $2 + \sqrt[3]{1 - 5t} = 0$

10 $\sqrt[3]{6 - s^2} + 5 = 0$

11 $\sqrt[5]{2x^2 + 1} - 2 = 0$

12 $\sqrt[4]{2x^2 - 1} = x$

13 $\sqrt{7 - x} = x - 5$

14 $\sqrt{3 - x} - x = 3$

15 $3\sqrt{2x - 3} + 2\sqrt{7 - x} = 11$

16 $\sqrt{2x + 15} - 2 = \sqrt{6x + 1}$

17 $x = 4 + \sqrt{4x - 19}$

18 $x = 3 + \sqrt{5x - 9}$

19 $x + \sqrt{5x + 19} = -1$

20 $x - \sqrt{-7x - 24} = -2$

21 $\sqrt{7 - 2x} - \sqrt{5 + x} = \sqrt{4 + 3x}$

22 $4\sqrt{1 + 3x} + \sqrt{6x + 3} = \sqrt{-6x - 1}$

23 $\sqrt{11 + 8x} + 1 = \sqrt{9 + 4x}$

24 $2\sqrt{x} - \sqrt{x - 3} = \sqrt{5 + x}$

25 $\sqrt{2\sqrt{x + 1}} = \sqrt{3x - 5}$

26 $\sqrt{5\sqrt{x}} = \sqrt{2x - 3}$

27 $\sqrt{1 + 4\sqrt{x}} = \sqrt{x} + 1$

28 $\sqrt{x + 1} = \sqrt{x - 1}$

29 $x^4 - 25x^2 + 144 = 0$

30 $2x^4 - 10x^2 + 8 = 0$

31 $5y^4 - 7y^2 + 1 = 0$

32 $3y^4 - 5y^2 + 1 = 0$

33 $3x^{2/3} + 4x^{1/3} - 4 = 0$

34 $2y^{1/3} - 3y^{1/6} + 1 = 0$

35 $6w - 23w^{1/2} + 20 = 0$

36 $2x^{-2/3} - 7x^{-1/3} - 15 = 0$

37 $\left(\dfrac{t}{t+1}\right)^2 - \dfrac{2t}{t+1} - 8 = 0$

38 $6u^{-1/2} - 13u^{-1/4} + 6 = 0$

39 As sand leaks out of a certain container it forms a pile that has the shape of a right circular cone whose height is always one-half the diameter of the base. What is the diameter of the base at the instant that $144\,\text{cm}^3$ of sand has leaked out?

40 The volume of a spherical weather balloon is $10\frac{2}{3}\,\text{ft}^3$. In order to lift a transmitter and meteorological equipment, the balloon is inflated with an additional $25\frac{1}{3}\,\text{ft}^3$ of helium. How much does its diameter increase?

41 The cube rule in political science is an empirical formula that is said to predict the percentage y of seats in the U.S. House of Representatives that will be won by a political party from the popular vote for the party's presidential candidate. If x denotes the percentage vote for a party's presidential candidate, then the cube rule states that

$$y = \frac{x^3}{x^3 + (1-x)^3}.$$

What percentage of the vote will the presidential candidate need to capture in order for the candidate's party to win 60% of the House seats?

42 A conical paper cup is to have a height of 3 inches. Find the radius of the cone that will result in a surface area of $6\pi\,\text{in.}^2$. (*Hint:* The lateral surface area of a cone is given by $S = \pi r\sqrt{r^2 + h^2}$.)

43 A power line is to be installed across a river that is 1 mile wide to a town that is 5 miles downstream (see figure). It costs \$7500 per mile to lay the cable underwater and \$6000 per mile overland. Determine how the cable should be installed if \$35,000 has been allocated for this project.

EXERCISE 43

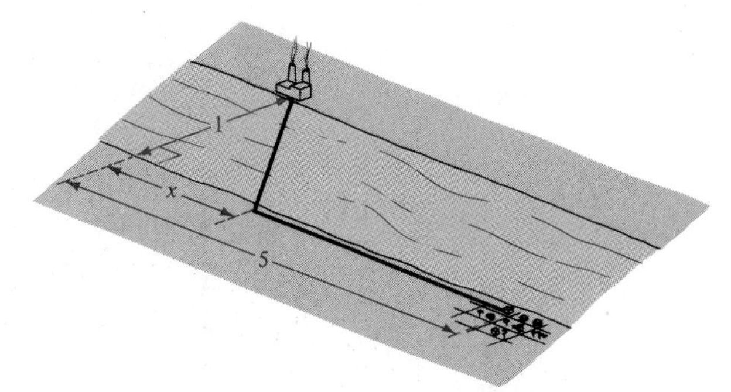

44 Adolphe Quetelet (1796–1874), the director of the Brussels Observatory from 1832 to 1874, was the first person to attempt to fit a mathematical expression to human growth data. Given that h denotes height in meters and t age in years, Quetelet's formula for males in Brussels can be expressed as

$$h + \frac{h}{h_M - h} = at + \frac{h_0 + t}{1 + \frac{4}{3}t}$$

with $h_0 = 0.5$ the height at birth, $h_M = 1.684$ the final adult male height, and $a = 0.545$.

(a) Find the expected height of a 12 year old male.

(b) At what age should 50% of the adult height be reached?

2.6 INEQUALITIES

Consider the inequality

$$x^2 - 3 < 2x + 4$$

containing the variable x. If numbers such as 4 or 5 are substituted for x, we obtain the false statements $13 < 12$ or $22 < 14$, respectively. Other numbers, such as 1 or 2, produce the true statements $-2 < 6$ or $1 < 8$. If a true statement is obtained when x is replaced by a real number a, then a is a **solution** of the inequality. Thus 1 and 2 are solutions of the inequality $x^2 - 3 < 2x + 4$, but 4 and 5 are not solutions. To **solve** an inequality means to find all solutions. Two inequalities are **equivalent** if they have exactly the same solutions.

As with equations, to solve an inequality we replace it with a list of equivalent inequalities, ending with an inequality for which the solutions are obvious. The properties in the following theorem are often useful.

THEOREM ON INEQUALITIES

Let a, b, and c be real numbers.

(i) If $a > b$ and $b > c$, then $a > c$.

(ii) If $a > b$, then $a + c > b + c$.

(iii) If $a > b$, then $a - c > b - c$.

(iv) If $a > b$ and $c > 0$, then $ac > bc$.

(v) If $a > b$ and $c < 0$, then $ac < bc$.

PROOF We will use the fact that both the sum and product of any two positive real numbers are positive. To prove (i) we first note that if $a > b$ and $b > c$, then $a - b$ and $b - c$ are both positive. Consequently, the sum $(a - b) + (b - c)$ is positive. Since the sum reduces to $a - c$, we see that $a - c$ is positive, which means that $a > c$.

To establish (ii) we again note that if $a > b$, then $a - b$ is positive. Since $(a + c) - (b + c) = a - b$, it follows that $(a + c) - (b + c)$ is positive, that is, $a + c > b + c$.

If $a > b$, then by (ii), $a + (-c) > b + (-c)$, or equivalently, $a - c > b - c$. This proves (iii).

To prove (iv) observe that if $a > b$ and $c > 0$, then $a - b$ and c are both positive and hence, so is the product $(a - b)c$. Consequently, $ac - bc$ is positive, that is, $ac > bc$.

Finally, to prove (v) we first note that if $c < 0$, then $0 - c$, or $-c$, is positive. In addition, if $a > b$, then $a - b$ is positive and hence, the product $(a - b)(-c)$ is positive. However, $(a - b)(-c) = -ac + bc$ and therefore $bc - ac$ is positive. This means that $bc > ac$, or $ac < bc$. ❑

The converses of (ii)–(v) are true:

If $a + c > b + c$, then $a > b$.

If $a - c > b - c$, then $a > b$.

If $ac > bc$ and $c > 0$, then $a > b$.

If $ac < bc$ and $c < 0$, then $a > b$.

Similar results are true for other inequalities and the symbols $\leq$ and $\geq$. Thus, if $a < b$, then $a + c < b + c$; if $a \leq b$ and $c < 0$, then $ac \geq bc$, and so on.

If x represents a real number, then by properties (ii) or (iii) of the theorem, adding or subtracting the same expression in x on both sides of an inequality leads to an equivalent inequality. By (iv) we may multiply both sides of an inequality by an expression containing x if we are certain that the expression is positive for all values of x under consideration. To illustrate, multiplication by $x^4 + 3x^2 + 5$ would be permissible, since this expression is always positive. If we multiply both sides of an inequality by an expression that is always negative, such as $-7 - x^2$, then by (v) the inequality sign is reversed.

EXAMPLE ▪ 1

Solve the inequality $-3x + 4 > 11$.

SOLUTION The following inequalities are equivalent:

$$-3x + 4 > 11$$
$$(-3x + 4) - 4 > 11 - 4$$
$$-3x > 7$$
$$(-\tfrac{1}{3})(-3x) < (-\tfrac{1}{3})(7)$$
$$x < -\tfrac{7}{3}$$

Thus, the solutions of $-3x + 4 > 11$ consist of all real numbers x such that $x < -\frac{7}{3}$.

EXAMPLE ▪ 2

Solve $4x - 3 < 2x + 5$.

SOLUTION The following are equivalent inequalities:

$$4x - 3 < 2x + 5$$
$$(4x - 3) + 3 < (2x + 5) + 3$$
$$4x < 2x + 8$$
$$4x - 2x < (2x + 8) - 2x$$
$$2x < 8$$
$$\tfrac{1}{2}(2x) < \tfrac{1}{2}(8)$$
$$x < 4$$

Hence, the solutions of the given inequality consist of all real numbers x such that $x < 4$.

We can represent solutions of inequalities graphically. The **graph** of a set of real numbers is the collection of points on a coordinate line that correspond to the numbers. To **sketch a graph** we darken, or color, an appropriate portion of the line. The graph corresponding to the solutions of $x < 4$ in Example 2 consists of all points to the left of the point with coordinate 4 and is sketched in Figure 9, where it is understood that the black portion extends indefinitely to the left. The parenthesis in the figure indicates that the point corresponding to 4 is not part of the graph.

FIGURE 9

0 4

If $a < b$, the symbol (a, b) is sometimes used for all real numbers between a and b; this set is an **open interval**.

OPEN INTERVAL

$$(a, b) = \{x : a < x < b\}$$

The expression to the right of the equal sign is translated *the set of all x such that $a < x < b$*. The numbers a and b are the **endpoints** of the interval. The graph of (a, b) consists of all points on a coordinate line that lie between the points corresponding to a and b. In Figure 10 we have sketched the graph of a general open interval (a, b) and also the special open intervals $(-1, 3)$ and $(2, 4)$. *The parentheses on the graphs indicate that the endpoints of the intervals are not included.* For convenience, we use the terms *open interval* and *graph of an open interval* interchangeably.

FIGURE 10

Open intervals (a, b), $(-1, 3)$, and $(2, 4)$

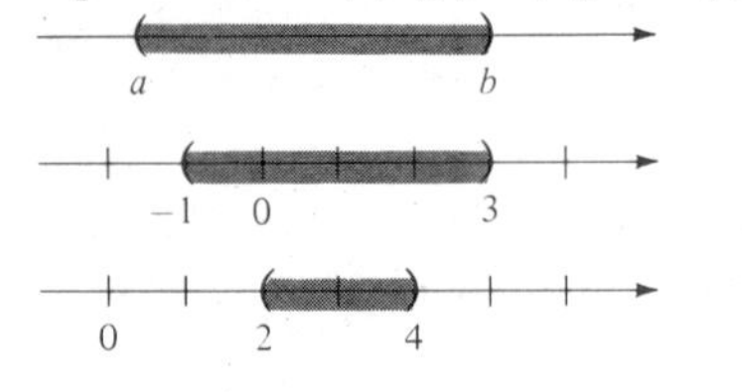

EXAMPLE ▪ 3

Solve the inequality $-6 < 2x - 4 < 2$ and represent the solutions graphically.

SOLUTION A real number x is a solution of the given inequality if and only if it is a solution of *both* of the inequalities

$$-6 < 2x - 4 \quad \text{and} \quad 2x - 4 < 2.$$

The first inequality is equivalent to each of the following:

$$-6 < 2x - 4$$
$$-6 + 4 < (2x - 4) + 4$$
$$-2 < 2x$$
$$\tfrac{1}{2}(-2) < \tfrac{1}{2}(2x)$$
$$-1 < x$$
$$x > -1$$

The second inequality is equivalent to each of the following:

$$2x - 4 < 2$$
$$2x < 6$$
$$x < 3$$

Thus, x is a solution of the given inequality if and only if *both*

$$x > -1 \quad \text{and} \quad x < 3,$$

that is, $$-1 < x < 3.$$

FIGURE 11

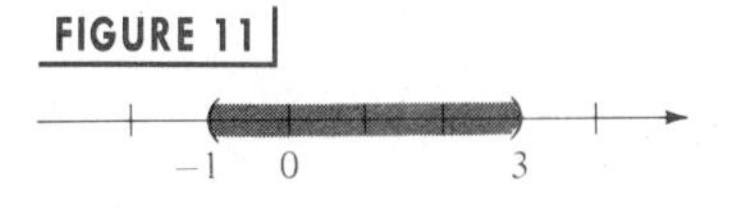

Hence, the solutions are all numbers in the open interval $(-1, 3)$. The graph is sketched in Figure 11.

An alternative (and shorter) method is to solve both inequalities simultaneously:

$$-6 < 2x - 4 < 2$$
$$-6 + 4 < 2x < 2 + 4$$
$$-2 < 2x < 6$$
$$-1 < x < 3$$

If we wish to include an endpoint of an interval, a bracket is used instead of a parenthesis. If $a < b$, then **closed intervals**, denoted by $[a, b]$, and **half-open intervals**, denoted by $[a, b)$ or $(a, b]$, are defined as follows.

CLOSED AND HALF-OPEN INTERVALS

$$[a, b] = \{x : a \leq x \leq b\}$$
$$[a, b) = \{x : a \leq x < b\}$$
$$(a, b] = \{x : a < x \leq b\}$$

FIGURE 12

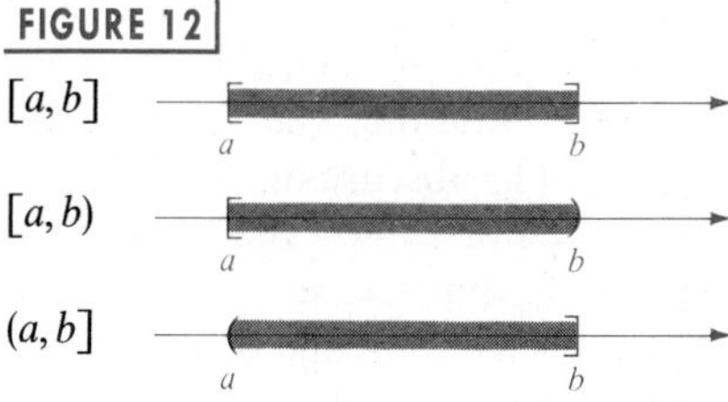

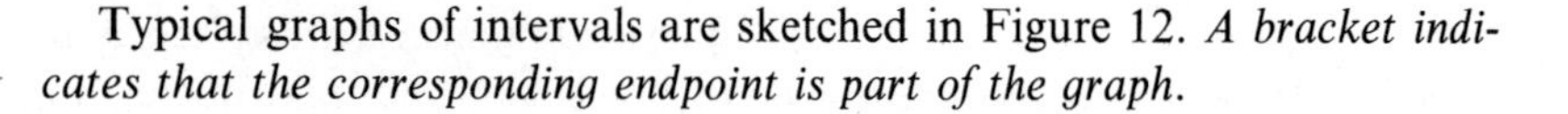

Typical graphs of intervals are sketched in Figure 12. *A bracket indicates that the corresponding endpoint is part of the graph.*

EXAMPLE ▪ 4

Solve $-5 \leq \dfrac{4 - 3x}{2} < 1$ and represent the solutions graphically.

SOLUTION A number x is a solution of the given inequality if and only if it satisfies both of the inequalities

$$-5 \leq \frac{4 - 3x}{2} \quad \text{and} \quad \frac{4 - 3x}{2} < 1.$$

We can either work with each inequality separately or solve both simultaneously, as follows:

$$-5 \le \frac{4-3x}{2} < 1$$

$$-10 \le 4 - 3x < 2$$

$$-10 - 4 \le -3x < 2 - 4$$

$$-14 \le -3x < -2$$

$$(-\tfrac{1}{3})(-14) \ge (-\tfrac{1}{3})(-3x) > (-\tfrac{1}{3})(-2)$$

$$\tfrac{14}{3} \ge x > \tfrac{2}{3}$$

$$\tfrac{2}{3} < x \le \tfrac{14}{3}$$

FIGURE 13

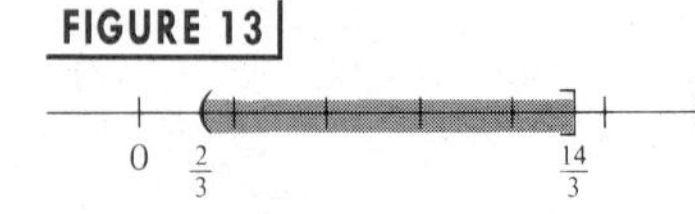

Thus, the solutions of the inequality are all numbers in the half-open interval $(\frac{2}{3}, \frac{14}{3}]$. The graph is sketched in Figure 13.

To describe solutions of inequalities such as $x < 4$, $x > -2$, $x \le 7$, or $x \ge 3$, it is convenient to use **infinite intervals**, defined as follows.

INFINITE INTERVALS

$$(a, \infty) = \{x : x > a\}$$
$$[a, \infty) = \{x : x \ge a\}$$
$$(-\infty, a) = \{x : x < a\}$$
$$(-\infty, a] = \{x : x \le a\}$$

FIGURE 14

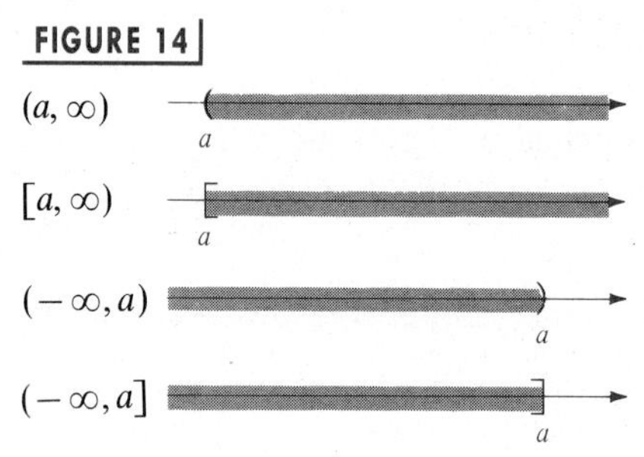

For example, $(1, \infty)$ represents all real numbers greater than 1. The symbol ∞ is read *infinity* and is merely a notational device. It does not represent a real number. Typical graphs of infinite intervals for an arbitrary real number a are sketched in Figure 14. The absence of a parenthesis or bracket on the right of the graph for (a, ∞) or $[a, \infty)$ and on the left for $(-\infty, a)$ or $(-\infty, a]$ indicates that the graph, shown as the black portion, extends indefinitely. The set $\mathbb{R}$ of real numbers is sometimes denoted by $(-\infty, \infty)$. Note that the solutions in Example 1 are the numbers in the infinite interval $(-\infty, -\frac{7}{3})$, and those in Example 2 are in $(-\infty, 4)$. The graph of $(-\infty, 4)$ is sketched in Figure 9.

EXAMPLE ▪ 5

Solve $\dfrac{1}{x-2} > 0$ and represent the solutions graphically.

FIGURE 15

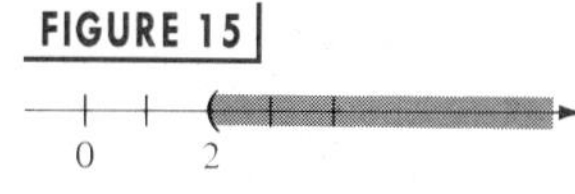

SOLUTION Since the numerator is positive, the fraction is positive if and only if $x - 2 > 0$, or equivalently, $x > 2$. Thus, the solutions are all numbers in the infinite interval $(2, \infty)$. See Figure 15.

EXAMPLE ▪ 6

FIGURE 16

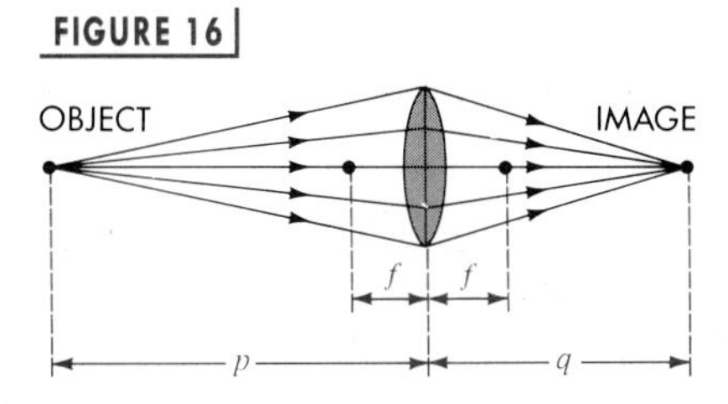

As illustrated in Figure 16, if a convex lens has focal length f cm and if an object is placed a distance p cm from the lens with $p > f$, then the distance q from the lens to the image is related to p and f by the formula

$$\frac{1}{p} + \frac{1}{q} = \frac{1}{f}.$$

If $f = 5$ cm, how close must the object be to the lens for the image to be more than 12 cm from the lens?

SOLUTION We are given that $f = 5$ and $p > 5$. The following equations are equivalent:

$$\frac{1}{p} + \frac{1}{q} = \frac{1}{5}$$

$$\frac{1}{q} = \frac{1}{5} - \frac{1}{p} = \frac{p-5}{5p}$$

$$q = \frac{5p}{p-5}$$

The image is more than 12 cm from the lens if $12 < q$; that is, if

$$12 < \frac{5p}{p-5}$$

Since $p > 5$, the denominator $p - 5$ is positive, and we may multiply both sides of the preceding inequality by $p - 5$, obtaining the following equivalent inequalities:

$$12(p-5) < 5p$$

$$12p - 60 < 5p$$

$$7p < 60$$

$$p < \tfrac{60}{7}$$

Since $p > 5$, we obtain $\quad 5 < p < \tfrac{60}{7}$.

2.6 EXERCISES

1 Given $-7 < -3$, determine what inequality is obtained if
(a) 5 is added to both sides.
(b) 4 is subtracted from both sides.
(c) both sides are multiplied by $\frac{1}{3}$.
(d) both sides are multiplied by $-\frac{1}{3}$.

2 Given $4 > -5$, determine what inequality is obtained if
(a) 7 is added to both sides.
(b) -5 is subtracted from both sides.
(c) both sides are divided by 6.
(d) both sides are divided by -6.

Exer. 3–12: Express the inequality as an interval, and sketch its graph.

3 $x < -2$

4 $x \leq 5$

5 $x \geq 4$

6 $x > -3$

7 $-2 < x \leq 4$

8 $-3 \leq x < 5$

9 $3 \leq x \leq 7$

10 $-3 < x < -1$

11 $5 > x \geq -2$

12 $-3 \geq x > -5$

Exer. 13–20: Express the interval as an inequality in the variable x.

13 $(-5, 8]$

14 $[0, 4)$

15 $[-4, -1]$

16 $(3, 7)$

17 $[4, \infty)$

18 $(-3, \infty)$

19 $(-\infty, -5)$

20 $(-\infty, 2]$

Exer. 21–46: Solve the inequality and express the solutions as an interval.

21 $3x - 2 > 14$

22 $2x + 5 \leq 7$

23 $-2 - 3x \geq 2$

24 $3 - 5x < 11$

25 $2x + 5 < 3x - 7$

26 $x - 8 > 5x + 3$

27 $9 + \frac{1}{3}x \geq 4 - \frac{1}{2}x$

28 $\frac{1}{4}x + 7 \leq \frac{1}{3}x - 2$

29 $-3 < 2x - 5 < 7$

30 $4 \geq 3x + 5 > -1$

31 $3 \leq \dfrac{2x - 3}{5} < 7$

32 $-2 < \dfrac{4x + 1}{3} \leq 0$

33 $4 > \dfrac{2 - 3x}{7} \geq -2$

34 $5 \geq \dfrac{6 - 5x}{3} > 2$

35 $0 \leq 4 - \frac{1}{3}x < 2$

36 $-2 < 3 + \frac{1}{4}x \leq 5$

37 $(2x - 3)(4x + 5) \leq (8x + 1)(x - 7)$

38 $(x - 3)(x + 3) \geq (x + 5)^2$

39 $(x - 4)^2 > x(x + 12)$

40 $2x(6x + 5) < (3x - 2)(4x + 1)$

41 $\dfrac{4}{3x + 2} \geq 0$

42 $\dfrac{3}{2x + 5} \leq 0$

43 $\dfrac{-2}{4 - 3x} > 0$

44 $\dfrac{-3}{2 - x} < 0$

45 $\dfrac{2}{(1 - x)^2} > 0$

46 $\dfrac{4}{x^2 + 4} < 0$

47 Temperature readings on the Fahrenheit and Celsius scales are related by the equation $C = \frac{5}{9}(F - 32)$. What values of F correspond to $30 \leq C \leq 40$?

48 Ohm's law in electrical theory states that if R denotes the resistance of an object (in ohms), V the potential difference across the object (in volts), and I the current that flows through it (in amperes), then $R = V/I$ (see figure). If the voltage is 110, what values of the resistance will result in a current that does not exceed 10 amperes?

EXERCISE 48

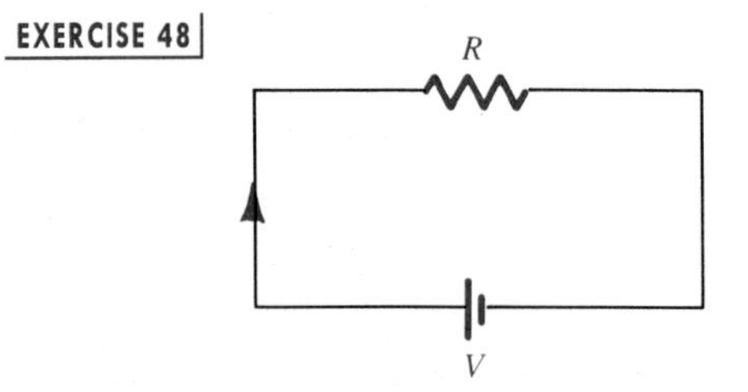

49 According to Hooke's law, the force F (in pounds) required to stretch a certain spring x inches beyond its natural length is given by the formula $F = (4.5)x$ (see figure). If $10 \leq F \leq 18$, what are the corresponding values for x?

EXERCISE 49

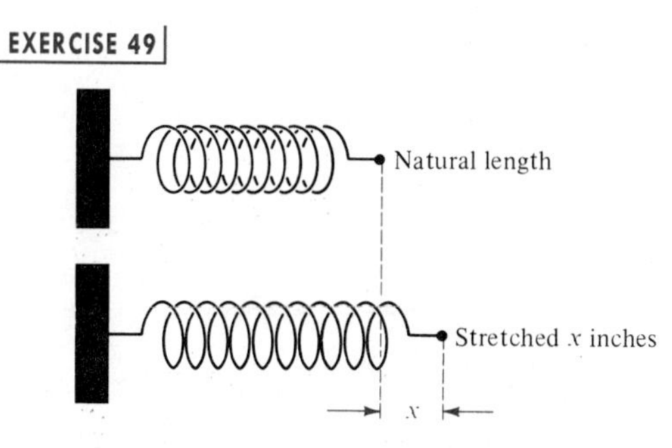

50 If two resistors R_1 and R_2 are connected in parallel in an electrical circuit, the net resistance R is given by

$$1/R = (1/R_1) + (1/R_2).$$

If $R_1 = 10$ ohms, what values of R_2 will result in a net resistance of less than 5 ohms?

51 Shown in the figure is a simple magnifier consisting of a convex lens. The object to be magnified is positioned so that its distance p from the lens is less than the focal length f. The linear magnification M is the ratio of the image size to the object size. It is shown in physics that $M = f/(f - p)$. If $f = 6$ cm, how far should the object be placed from the lens so that its image appears at least three times as large? (Compare with Example 6.)

EXERCISE 51

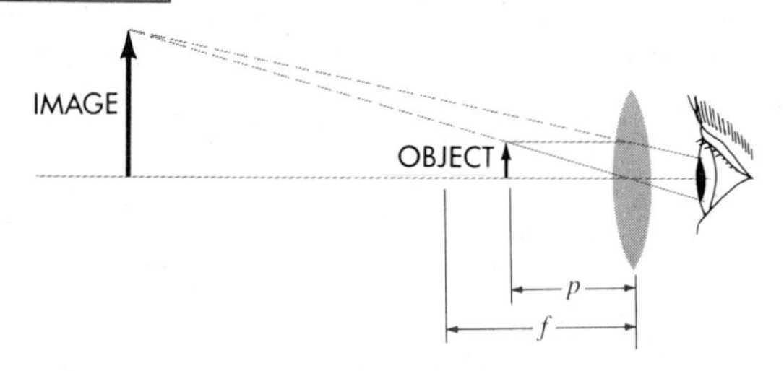

52 An astronaut's weight decreases after being launched into space until a state of weightlessness is achieved. The weight of a 125-pound astronaut at an altitude of x km above sea level is given by

$$W = 125\left(\frac{6400}{6400 + x}\right)^2.$$

At what altitudes is the astronaut's weight less than 5 pounds?

53 A construction firm is trying to decide which of two models of a crane to purchase. Model A costs \$50,000 and requires \$4000 per year to maintain. Model B has an initial cost of \$40,000 and maintenance cost of \$5500 per year. For how many years must model A be used before it becomes more economical than B?

54 **(a)** If $0 < a < b$, prove that $(1/a) > (1/b)$. Why is the restriction $0 < a$ necessary?

(b) If $0 < a < b$, prove that $a^2 < b^2$. Why is the restriction $0 < a$ necessary?

2.7 MORE ON INEQUALITIES

If a is a real number, the inequality $|a| < 1$ is equivalent to $-1 < a < 1$. Thus a is in the open interval $(-1, 1)$. In general, if b is any positive real number, we have the following.

PROPERTIES OF ABSOLUTE VALUES ($b > 0$)

(i) $|a| < b$ if and only if $-b < a < b$.

(ii) $|a| > b$ if and only if $a < -b$ or $a > b$.

(iii) $|a| = b$ if and only if $a = b$ or $a = -b$.

Properties (ii) and (iii) are also true if $b = 0$. If $b \geq 0$, then:

$$|a| \leq b \quad \text{if and only if} \quad -b \leq a \leq b$$

$$|a| \geq b \quad \text{if and only if} \quad a \geq b \quad \text{or} \quad a \leq -b.$$

EXAMPLE ▪ 1

Solve and represent the solutions graphically:

(a) $|x| < 4$ **(b)** $|x| > 4$ **(c)** $|x| = 4$

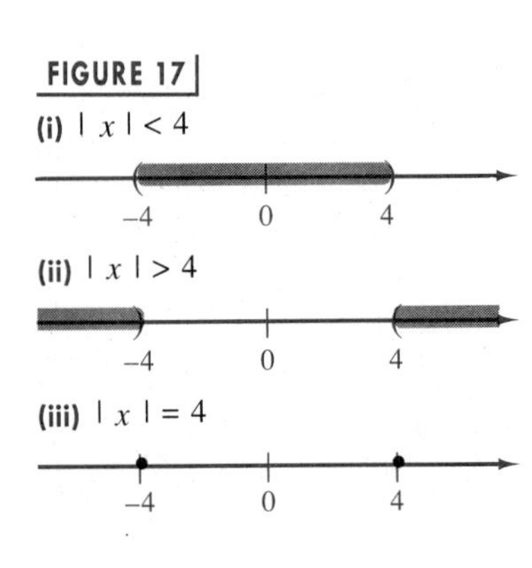

FIGURE 17

SOLUTION

(a) Using property (i) with $a = x$ and $b = 4$, we see that

$$|x| < 4 \quad \text{if and only if} \quad -4 < x < 4.$$

Hence, the solutions are all real numbers in the open interval $(-4, 4)$. The graph is shown in Figure 17(i).

(b) By property (ii), $|x| > 4$ means that either $x > 4$ or $x < -4$. Thus, the solutions consist of all real numbers in the two infinite intervals $(-\infty, -4)$ and $(4, \infty)$, as illustrated in Figure 17(ii).

(c) By property (iii)

$$|x| = 4 \quad \text{if and only if} \quad x = 4 \quad \text{or} \quad x = -4.$$

These solutions are illustrated in Figure 17(iii).

For the solutions obtained in Example 1(b), we may use the **union symbol**, $\cup$, and write

$$(-\infty, -4) \cup (4, \infty)$$

to denote all real numbers that are in either $(-\infty, -4)$ or $(4, \infty)$.

EXAMPLE ▪ 2

Solve $|x - 3| < 0.5$ and represent the solutions graphically.

SOLUTION By property (i) of absolute values with $a = x - 3$ and $b = 0.5$, the inequality is equivalent to the following:

$$-0.5 < x - 3 < 0.5$$

$$-0.5 + 3 < (x - 3) + 3 < 0.5 + 3$$

$$2.5 < x < 3.5$$

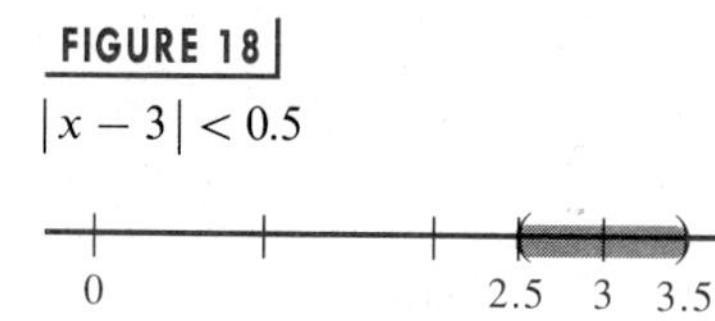

FIGURE 18 $|x - 3| < 0.5$

Thus the solutions are the real numbers in the open interval (2.5, 3.5). The graph is sketched in Figure 18.

EXAMPLE ▪ 3

If a and δ denote real numbers and $\delta > 0$, solve the inequality $|x - a| < \delta$ and represent the solutions graphically.

SOLUTION We may proceed as in Example 2. Thus,

$$-\delta < x - a < \delta$$

$$a - \delta < (x - a) + a < a + \delta$$

$$a - \delta < x < a + \delta.$$

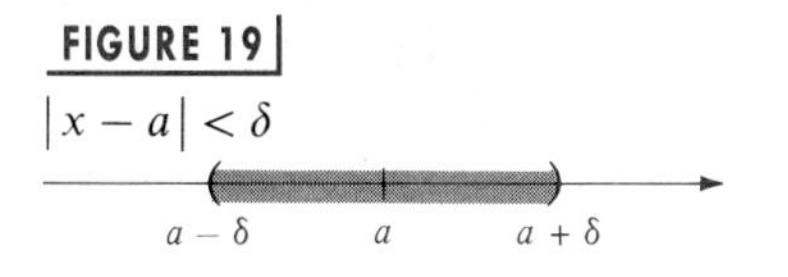

FIGURE 19 $|x - a| < \delta$

The solutions of the inequality consist of all real numbers in the open interval $(a - \delta, a + \delta)$. A typical graph is sketched in Figure 19.

Note that Example 2 is the special case of Example 3 with $a = 3$ and $\delta = 0.5$. Inequalities of this type occur in many branches of mathematics, including calculus.

EXAMPLE ▪ 4

Solve $|2x + 3| > 9$ and illustrate the solutions graphically.

SOLUTION By property (ii) of absolute values with $a = 2x + 3$ and $b = 9$, the solutions of the inequality are the solutions of the following:

$$2x + 3 < -9 \quad \text{or} \quad 2x + 3 > 9$$

$$2x < -12 \quad \text{or} \quad 2x > 6$$

$$x < -6 \quad \text{or} \quad x > 3$$

FIGURE 20

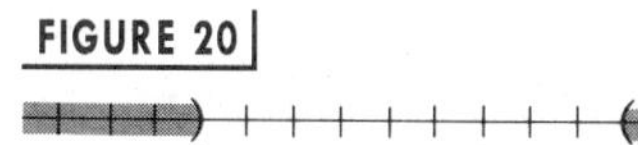

Consequently, the solutions of the inequality $|2x + 3| > 9$ consist of the numbers in $(-\infty, -6) \cup (3, \infty)$. The graph is sketched in Figure 20.

To solve inequalities involving polynomials of degree greater than 1, we may use the following theorem, which will be discussed further in Section 4.2. In the statement of the theorem, the phrase *successive solutions* c and d means that there are no other solutions between c and d.

THEOREM

Let $a_n x^n + \cdots + a_1 x + a_0$ be a polynomial. If the real numbers c and d are successive solutions of the equation

$$a_n x^n + \cdots + a_1 x + a_0 = 0,$$

then when x is in the open interval (c, d) either all values of the polynomial are positive or all values are negative.

This theorem implies that given successive solutions c and d, if we choose *any* number k, such that $c < k < d$, and if the value of the polynomial is positive for $x = k$, then the polynomial is positive for *every* x in (c, d). Similarly, if the polynomial is negative for $x = k$, then it is negative throughout (c, d). We call the value of the polynomial at $x = k$ a **test value** of the polynomial at k. Test values may also be used on infinite intervals of the form $(-\infty, a)$ or (a, ∞), provided the polynomial equation has no solutions in these intervals. The use of test values is demonstrated in the following examples.

EXAMPLE ■ 5

Solve $2x^2 - x < 3$ and represent the solutions graphically.

SOLUTION To use test values *it is essential to have all nonzero terms on one side* of the inequality sign. Thus, we begin by writing

$$2x^2 - x - 3 < 0.$$

Factoring gives us $\qquad (x + 1)(2x - 3) < 0.$

We see from the factored form that the equation $2x^2 - x - 3 = 0$ has solutions -1 and $\frac{3}{2}$. For reference, let us plot the corresponding points on a real axis, as in Figure 21(i). These points divide the axis into three parts and determine the following intervals:

FIGURE 21

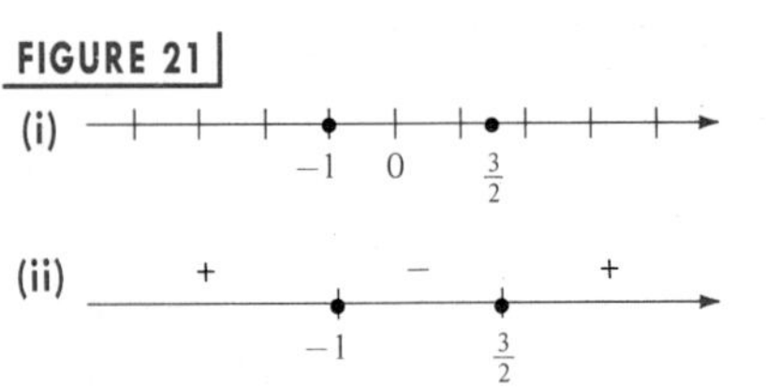

$$(-\infty, -1), \qquad (-1, \tfrac{3}{2}), \qquad (\tfrac{3}{2}, \infty).$$

We now determine the sign of the polynomial $2x^2 - x - 3$ in each interval by using a suitable test value.

If we choose -2 in $(-\infty, -1)$, then the polynomial $2x^2 - x - 3$ has the value

$$2(-2)^2 - (-2) - 3 = 8 + 2 - 3 = 7.$$

Since 7 is positive, it follows from the preceding theorem that the polynomial $2x^2 - x - 3$ is positive for every x in $(-\infty, -1)$, as indicated by the $+$ to the left of -1 in Figure 21(ii).

If we choose 0 in $(-1, \frac{3}{2})$, then the polynomial has the value

$$2(0)^2 - (0) - 3 = -3.$$

Since -3 is negative, $2x^2 - x - 3 < 0$ for every x in $(-1, \frac{3}{2})$, as indicated by the $-$ in Figure 21(ii).

Finally, choosing 2 in $(\frac{3}{2}, \infty)$, we obtain

$$2(2)^2 - 2 - 3 = 8 - 2 - 3 = 3,$$

and since 3 is positive, $2x^2 - x - 3 > 0$ throughout $(\frac{3}{2}, \infty)$, as indicated by the $+$ to the right of $\frac{3}{2}$ in Figure 21(ii).

We summarize our results in the following table.

Interval	$(-\infty, -1)$	$(-1, \frac{3}{2})$	$(\frac{3}{2}, \infty)$
k	-2	0	2
Test value of $2x^2 - x - 3$ at k	7	-3	3
Sign of $2x^2 - x - 3$ in interval	$+$	$-$	$+$

FIGURE 22

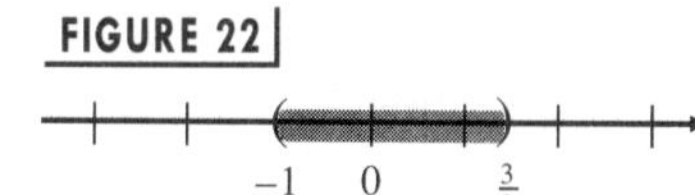

Thus, the solutions of $2x^2 - x - 3 < 0$, or equivalently, $2x^2 - x < 3$, are the real numbers in the open interval $(-1, \frac{3}{2})$. This interval is sketched in Figure 22.

EXAMPLE ▪ 6

Solve $x^2 > 7x - 10$ and represent the solutions graphically.

SOLUTION As in Example 5, we take all terms to one side of the inequality sign and factor:

$$x^2 - 7x + 10 > 0$$

$$(x - 2)(x - 5) > 0$$

FIGURE 23

Points corresponding to the solutions 2 and 5 of $x^2 - 7x + 10 = 0$ are plotted in Figure 23. Referring to the figure, we obtain the following intervals:

$$(-\infty, 2), \qquad (2, 5), \qquad (5, \infty)$$

We next use test values to determine the sign of $x^2 - 7x + 10$ in each interval. The following table summarizes results. (*Check each entry.*)

Interval	$(-\infty, 2)$	$(2, 5)$	$(5, \infty)$
k	0	3	6
Test value of $x^2 - 7x + 10$ at k	$0^2 - 7(0) + 10 = 10$	$3^2 - 7(3) + 10 = -2$	$6^2 - 7(6) + 10 = 4$
Sign of $x^2 - 7x + 10$ in interval	+	−	+

FIGURE 24

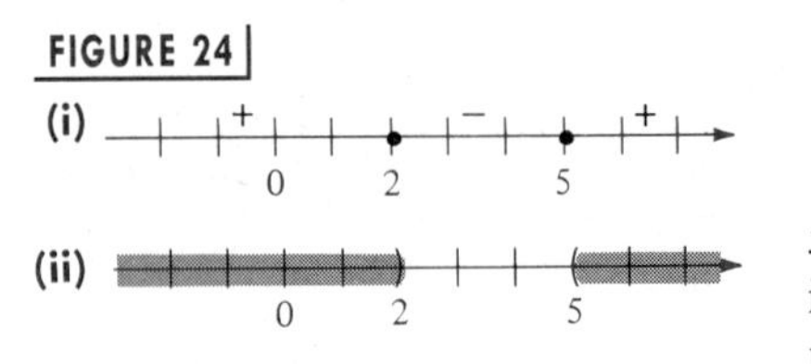

Figure 24(i) illustrates where $x^2 - 7x + 10$ is positive or negative. Thus, $x^2 - 7x + 10 > 0$ if x is in either $(-\infty, 2)$ or $(5, \infty)$. The solutions of the inequality are given by $(-\infty, 2) \cup (5, \infty)$. The graph is sketched in Figure 24(ii).

EXAMPLE ▪ 7

Solve the inequality $\dfrac{x+1}{x+3} \leq 2$ and represent the solutions graphically.

SOLUTION We cannot multiply both sides of the inequality by $x + 3$, since this expression may be positive or negative. Hence we take all nonzero terms to one side of the inequality symbol and simplify:

$$\frac{x+1}{x+3} \leq 2$$

$$\frac{x+1}{x+3} - 2 \leq 0$$

$$\frac{x+1-2(x+3)}{x+3} \leq 0$$

$$\frac{-x-5}{x+3} \leq 0$$

Multiplying both sides of the last inequality by -1, we obtain

$$\frac{x+5}{x+3} \geq 0.$$

FIGURE 25

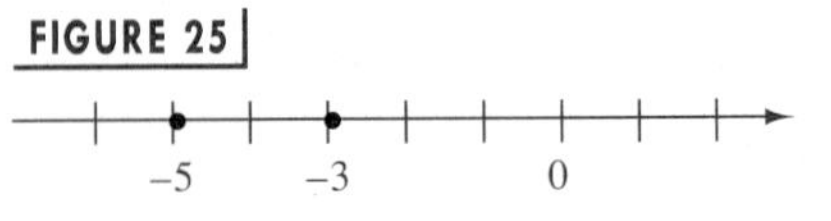

The numerator and denominator of $(x + 5)/(x + 3)$ equal zero at $x = -5$ and $x = -3$, respectively. For reference, we plot these points in Figure 25. Note that -5 is a solution of $(x + 5)/(x + 3) \geq 0$, but -3 is *not* a solution since a zero denominator occurs if -3 is substituted for x. The points in the figure determine the following intervals:

$$(-\infty, -5), \qquad (-5, -3), \qquad (-3, \infty)$$

Since $(x + 5)/(x + 3)$ is a quotient of two polynomials, it is always positive or always negative throughout each interval. (The sign of this quotient is the same as the sign of the product $(x + 5)(x + 3)$.) As in preceding examples, we may use test values to determine the sign in each interval. The following table summarizes our results.

Interval	$(-\infty, -5)$	$(-5, -3)$	$(-3, \infty)$
k	-6	-4	0
Test value of $(x + 5)/(x + 3)$ at k	$\frac{1}{3}$	-1	$\frac{5}{3}$
Sign of $(x + 5)/(x + 3)$ in interval	+	−	+

FIGURE 26

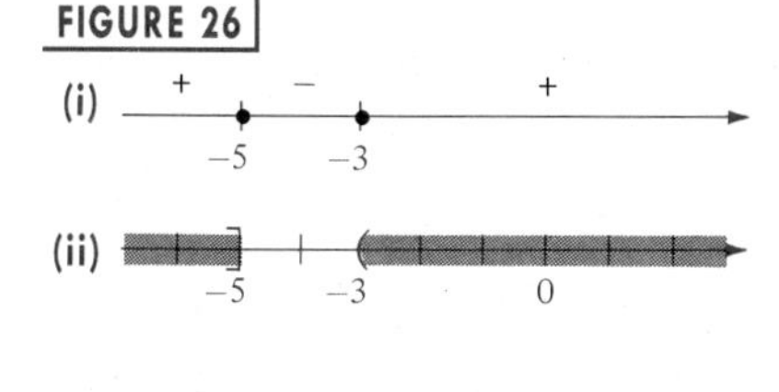

Figure 26(i) indicates where $(x + 5)/(x + 3)$ is positive or negative. Thus, the solutions of $(x + 5)/(x + 3) > 0$ are given by $(-\infty, -5) \cup (-3, \infty)$. The solutions of $(x + 5)/(x + 3) \geq 0$ are given by $(-\infty, -5] \cup (-3, \infty)$. (Remember that the bracket on $(-\infty, -5]$ means that we include the endpoint -5 of the interval.) The graph is sketched in Figure 26.

EXAMPLE ▪ 8

Solve $(x + 2)(x - 1)(x - 5) > 0$ and represent the solutions graphically.

FIGURE 27

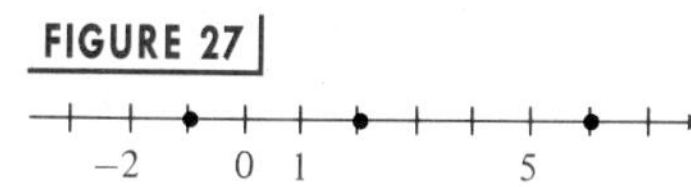

SOLUTION The expression $(x + 2)(x - 1)(x - 5)$ is zero at -2, 1, and 5. The corresponding points are plotted on a real axis in Figure 27. These points determine four intervals:

$$(-\infty, -2), \qquad (-2, 1), \qquad (1, 5), \qquad (5, \infty)$$

We next use test values, as shown in the following table.

Interval	$(-\infty, -2)$	$(-2, 1)$	$(1, 5)$	$(5, \infty)$
k	-3	0	2	6
Test value of $(x + 2)(x - 1)(x - 5)$ at k	$(-1)(-4)(-8) = -32$	$(2)(-1)(-5) = 10$	$(4)(1)(-3) = -12$	$(8)(5)(1) = 40$
Sign of $(x + 2)(x - 1)(x - 5)$ in interval	−	+	−	+

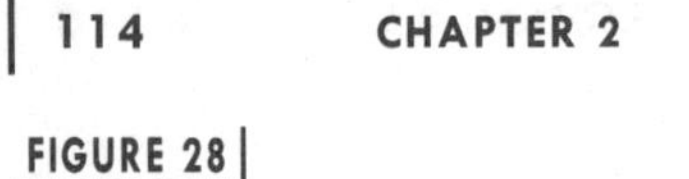

FIGURE 28

(i)

(ii)

Figure 28(i) shows where $(x+2)(x-1)(x-5)$ is positive or negative. Thus, the solutions of $(x+2)(x-1)(x-5) > 0$ are given by $(-2, 1) \cup (5, \infty)$. The graph is sketched in Figure 28(ii).

EXAMPLE ▪ 9

For a drug to have a beneficial effect, its concentration in the bloodstream must exceed a certain value, the *minimum therapeutic level*. Suppose that the concentration c of a drug t hours after it is taken orally is given by $c = 20t/(t^2 + 4)$ mg/liter. If the minimum therapeutic level is 4 mg/liter, determine when this level is exceeded.

SOLUTION The minimum therapeutic level, 4 mg/liter, is exceeded if $c > 4$, or equivalently, if $4 < c$. Thus we must solve the inequality

$$4 < \frac{20t}{t^2 + 4}.$$

Since $t^2 + 4 > 0$ for every t, we may multiply both sides by $t^2 + 4$ and simplify:

$$4t^2 + 16 < 20t$$
$$4t^2 - 20t + 16 < 0$$
$$t^2 - 5t + 4 < 0$$
$$(t-1)(t-4) < 0$$

The factors in the last inequality determine the time values $t = 1$ hr and $t = 4$ hr. (These are times at which $c = 4$.) As in previous examples, we may use test values—say $t = 0$, $t = 2$, and $t = 5$—to show that $(t-1)(t-4) < 0$ for every t in the interval $(1, 4)$. Hence the minimum therapeutic level is exceeded if $1 < t < 4$.

2.7 EXERCISES

Exer. 1–56: Solve the inequality and express the solutions in terms of intervals when possible.

1 $|x| < 3$

2 $|x| \le 7$

3 $|x| \ge 5$

4 $|-x| > 2$

5 $|x + 3| < 0.01$

6 $|x - 4| \le 0.03$

7 $|x + 2| \ge 0.001$

8 $|x - 3| > 0.002$

9 $|2x + 5| < 4$

10 $|3x - 7| \ge 5$

11 $|6 - 5x| \le 3$

12 $|-11 - 7x| > 6$

13 $|7x + 2| > -2$

14 $|6x - 5| \le -2$

15 $|3x - 9| > 0$

16 $|5x + 2| \le 0$

17 $\left|\dfrac{2 - 3x}{5}\right| \ge 2$

18 $\left|\dfrac{2x + 5}{3}\right| < 1$

19 $\dfrac{3}{|5-2x|} < 2$

20 $\dfrac{2}{|2x+3|} \ge 5$

21 $(3x+1)(5-10x) > 0$

22 $(2-3x)(4x-7) \ge 0$

23 $(x+2)(x-1)(4-x) \le 0$

24 $(x-5)(x+3)(-2-x) < 0$

25 $x^2 - x - 6 < 0$

26 $x^2 + 4x + 3 \ge 0$

27 $x^2 - 2x - 5 > 3$

28 $x^2 - 4x - 17 \le 4$

29 $x(2x+3) \ge 5$

30 $x(3x-1) \le 4$

31 $6x - 8 > x^2$

32 $x + 12 \le x^2$

33 $x^2 < 16$

34 $x^2 > 9$

35 $25x^2 - 9 < 0$

36 $25x^2 - 9x < 0$

37 $16x^2 \ge 9x$

38 $16x^2 > 9$

39 $x^4 + 5x^2 \ge 36$

40 $x^4 + 15x^2 < 16$

41 $x^3 + 2x^2 - 4x - 8 \ge 0$

42 $2x^3 - 3x^2 - 2x + 3 \le 0$

43 $\dfrac{x-2}{x^2-3x-10} \ge 0$

44 $\dfrac{x+5}{x^2-7x+12} \le 0$

45 $\dfrac{-3x}{x^2-9} > 0$

46 $\dfrac{2x}{16-x^2} < 0$

47 $\dfrac{x+1}{2x-3} > 2$

48 $\dfrac{x-2}{3x+5} \le 4$

49 $\dfrac{1}{x-2} \ge \dfrac{3}{x+1}$

50 $\dfrac{2}{2x+3} \le \dfrac{2}{x-5}$

51 $\dfrac{4}{3x-2} \le \dfrac{2}{x+1}$

52 $\dfrac{3}{5x+1} \ge \dfrac{1}{x-3}$

53 $\dfrac{x}{3x-5} \le \dfrac{2}{x-1}$

54 $\dfrac{x}{2x-1} \ge \dfrac{3}{x+2}$

55 $x^3 > x$

56 $x^4 \ge x^2$

57 *Guinness Book of World Records* reports instances of German shepherds making vertical leaps of over 10 feet when scaling walls. If a dog's distance s (in feet) off the ground after t seconds is given by $s = -16t^2 + 24t + 1$, for how many seconds is the dog more than 9 feet off the ground?

58 If the length of the pendulum in a grandfather clock is l cm, then its period T (in seconds) is given by $T = 2\pi\sqrt{l/g}$ for a gravitational constant g. If, under certain conditions, $g = 980$ and $98 \le l \le 100$, what is the corresponding range for T?

59 The braking distance d (in feet) of a car traveling v mi/hr is approximated by $d = v + (v^2/20)$. Determine velocities that result in braking distances of less than 75 feet.

60 For a satellite to maintain an orbit of altitude h km, its velocity (in km/sec) must equal $626.4/\sqrt{h+R}$, where $R = 6372$ km is the radius of the earth. What velocities will result in orbits of altitude more than 100 km from the earth's surface?

61 For a particular salmon population, the relationship between the number S of spawners and the number R of offspring that survive to maturity is given by the formula $R = 4500S/(S + 500)$. Under what conditions is $R > S$?

62 The speed of sound in air at 0 °C (or 273 °K) is 1087 ft/sec but this speed increases as the temperature rises. If T is temperature in °K, the speed of sound v at this temperature is given by $v = 1087\sqrt{T/273}$. At what temperatures does the speed of sound exceed 1100 ft/sec?

63 Refer to Example 9. To treat arrhythmia (irregular heartbeat) a drug is fed intravenously into the bloodstream. Suppose that the concentration c of the drug after t hours is given by $c = 3.5t/(t+1)$ mg/liter. If the minimum therapeutic level is 1.5 mg/liter, determine when this level is exceeded.

64 The population density D (in people/mi^2) in a large city is related to the distance x from the center of the city by $D = 5000x/(x^2 + 36)$. In what areas of the city does the population density exceed 400 people/mi^2?

2.8 REVIEW

Define or discuss each of the following.

Solution of an equation ■ Root of an equation ■ Identity ■ Conditional equation ■ Equivalent equation ■ Linear equation ■ Quadratic equation ■ Completing the square ■

The quadratic formula ■ Discriminant of a quadratic equation ■ Complex number ■ Conjugate of a complex number ■ Extraneous solution ■ Properties of inequalities ■ Solution of an inequality ■ Equivalent inequalities ■ The graph of a set of real numbers ■ Intervals: closed, open, half-open, infinite

2.8 EXERCISES

Exer. 1–34: Solve the equation or inequality.

1 $\frac{3x+1}{5x+7} = \frac{6x+11}{10x-3}$

2 $2 - \frac{1}{x} = 1 + \frac{4}{x}$

3 $\frac{2}{x+5} - \frac{3}{2x+1} = \frac{5}{6x+3}$

4 $\frac{7}{x-2} - \frac{6}{x^2-4} = \frac{3}{2x+4}$

5 $\frac{1}{\sqrt{x}} - 2 = \frac{1-2\sqrt{x}}{\sqrt{x}}$

6 $2x^2 + 5x - 12 = 0$

7 $x(3x+4) = 5$

8 $\frac{x}{3x+1} = \frac{x-1}{2x+3}$

9 $(x-2)(x+1) = 3$

10 $4x^4 - 33x^2 + 50 = 0$

11 $x^{2/3} - 2x^{1/3} - 15 = 0$

12 $20x^3 + 8x^2 - 35x - 14 = 0$

13 $5x^2 = 2x - 3$

14 $x^2 + \frac{1}{3}x + 2 = 0$

15 $6x^4 + 29x^2 + 28 = 0$

16 $x^4 - 3x^2 + 1 = 0$

17 $\frac{1}{x} + 6 = \frac{5}{\sqrt{x}}$

18 $\sqrt[3]{4x-5} - 2 = 0$

19 $\sqrt{7x+2} + x = 6$

20 $\sqrt{x+4} = \sqrt[4]{6x+19}$

21 $\sqrt{3x+1} - \sqrt{x+4} = 1$

22 $10 - 7x < 4 + 2x$

23 $-\frac{1}{2} < \frac{2x+3}{5} < \frac{3}{2}$

24 $(3x-1)(10x+4) \geq (6x-5)(5x-7)$

25 $\frac{6}{10x+3} < 0$

26 $|4x+7| < 21$

27 $|16-3x| \geq 5$

28 $2 < |x-6| < 4$

29 $10x^2 + 11x > 6$

30 $x(x-3) \leq 10$

31 $\frac{3}{2x+3} < \frac{1}{x-2}$

32 $\frac{x+1}{x^2-25} \leq 0$

33 $x^3 > x^2$

34 $(x^2-x)(x^2-5x+6) < 0$

Exer. 35–40: Solve for the specified variable.

35 $V = \frac{4}{3}\pi r^3$ for r (volume of a sphere)

36 $F = \frac{\pi P R^4}{8VL}$ for R (blood flow in an arteriole)

37 $A = 2\pi r(r+h)$ for r (surface area of a closed cylinder)

38 $F = g\frac{mM}{d^2}$ for d (Newton's law of gravitation)

39 $s = \frac{1}{2}gt^2 + v_0 t$ for t (distance an object falls)

40 $V = \frac{1}{3}\pi h(r^2 + R^2 + rR)$ for r (volume of a frustum of a cone)

Exer. 41–46: Write the expression in the form $a + bi$, where a and b are real numbers.

41 $(7+5i) + (-8+3i)$

42 $(4+2i)(-5+4i)$

43 $(3+8i)^2$

44 $\frac{1}{9-2i}$

45 $\frac{6-3i}{2+7i}$

46 $\frac{20-8i}{4i}$

47 The longest drive to the center of a square city from the outskirts is 10 miles. Within the last decade, the city has expanded in area by 50 mi^2. Assuming the city has always been square in shape, find the corresponding change in the longest drive to the center of the city.

48 The surface area S of the membrane of a spherical cell of radius r is given by $S = 4\pi r^2$. If $r = 6$ microns, what change in radius will increase the surface area of the cell membrane by 25%?

49 An airplane flew with the wind for 30 minutes and returned the same distance in 45 minutes. If the cruising speed of the airplane was 320 mi/hr, what was the speed of the wind?

50 A hospital dietician wishes to prepare a 10-ounce meat-vegetable dish that will provide 7 grams of protein. Each ounce of vegetable supplies $\frac{1}{2}$ gram of protein, and each ounce of meat supplies 1 gram of protein. How much of each should be used?

51 A solution of ethyl alcohol that is 75% alcohol by weight is to be used as a bactericide. The solution is to be made by adding water to a 95% ethyl alcohol solution. How many grams of each should be used to prepare 400 grams of the bactericide?

52 A large solar heating panel requires 120 gallons of a fluid that is 30% antifreeze. The fluid comes in either a 50% solution or a 20% solution. How many gallons of each should be used to prepare the 120-gallon solution?

53 When two resistors R_1 and R_2 are connected in parallel, the net resistance R is given by $1/R = (1/R_1) + (1/R_2)$. If $R_1 = 5$ ohms, what is the value of R_2 such that the net resistance is 2 ohms?

54 A tugboat can bring a large ship into port in 2 hours, while a smaller tug can do the job in 3 hours. Predict the number of hours needed for both tugs to tow the ship to port together, as shown in the figure.

EXERCISE 54

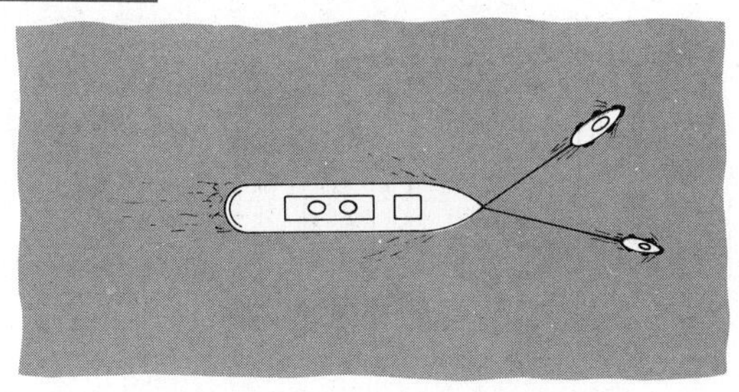

55 A boat that carries 10 gallons of gasoline travels at 20 mi/hr with fuel consumption of 16 mi/gal when operated at full throttle in still water. The boat is moving upstream into a 5-mi/hr current. How far upstream can the boat travel and return on the 10 gallons of gasoline if it is operated at full throttle during the entire trip?

56 A high-speed train makes a 400-mile nonstop run between two major cities in $5\frac{1}{2}$ hours. The train travels 100 mi/hr in the country, but safety regulations require that it travel only 25 mi/hr when passing through a city. How many hours are spent traveling through cities?

57 A contractor wishes to design a rectangular sunken bath with 40 ft^2 of bathing area. A one-foot-wide tile strip is to surround the bathing area. The total length of the tiled area is to be twice the width. Find the dimensions of the bathing area.

58 A kennel owner has 270 feet of fencing material to be used to divide a rectangular area into 10 equal pens, as shown in the figure. Find dimensions that would allow 100 ft^2 for each pen.

EXERCISE 58

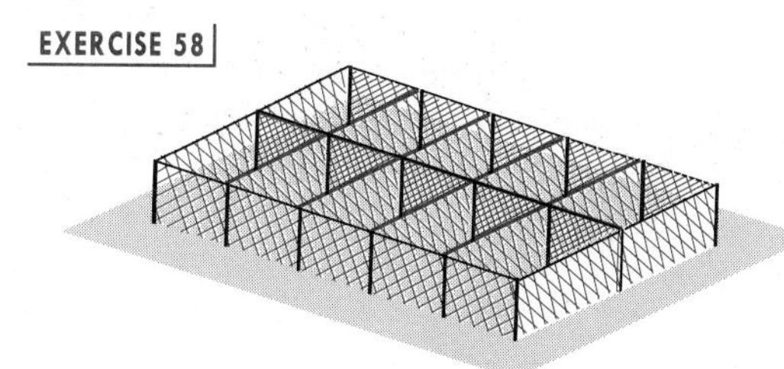

59 A North-South highway intersects an East-West highway at a point P. An automobile crosses P at 10:00 A.M., traveling east at a constant rate of 20 mi/hr. At that same instant another automobile is 2 miles north of P, traveling south at 50 mi/hr.

(a) Find a formula for the distance d between the automobiles at time t (hours) after 10:00 A.M.

(b) At what time will the automobiles be 104 miles apart?

60 A sales representative for a company estimates that gasoline consumption for her automobile averages 28 miles per gallon on the highway and 22 miles per gallon in the city. On a recent trip she covered 627 miles and used 24 gallons of gasoline. How much of the trip was spent driving in the city?

61 An open-topped aquarium is to be constructed with sides 6 feet long and square ends, as shown in the figure.

EXERCISE 61

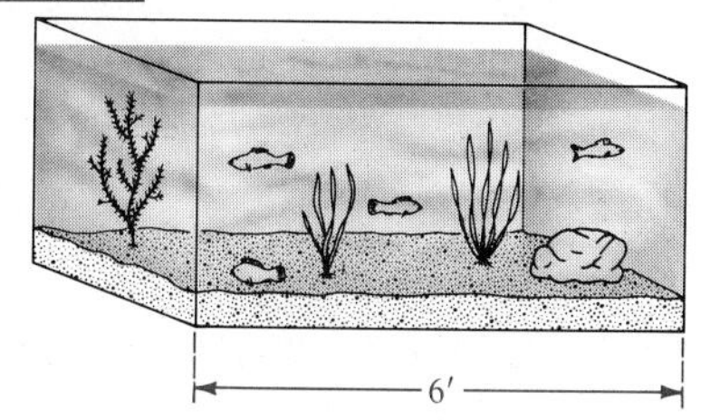

(a) Find the height if its volume is to be 48 ft^3.

(b) Find the height if 44 ft^2 of glass is to be used.

62 The length of a rectangular reflecting pool is to be four times its width, and a sidewalk of width 6 feet will surround the pool. If a total area of 1440 ft^2 has been set aside for construction, what are the dimensions of the pool?

63 The population P (in thousands) of a small town is expected to increase according to the formula

$$P = 15 + \sqrt{3t + 2}$$

for time t in years. When will the population be 20,000?

64 A recent college graduate has job offers for a sales position in two computer firms. Job A pays $25,000 per year plus a 5% commission. Job B pays only $20,000 per year but the commission rate is 10%. How much yearly business must the salesman do for the second job to be more lucrative?

65 Boyle's law for a certain gas states that $pv = 200$ for pressure p (in lb/in.^{2}) and volume v (in in.^{3}). If $25 \leq v \leq 50$, what is the corresponding range for p?

66 The Lorentz contraction formula in relativity theory relates the length L of an object moving at a velocity of V mi/sec with respect to an observer to its length L_0 at rest. If c is the speed of light, then

$$L^2 = L_0^2\left(1 - \frac{V^2}{c^2}\right).$$

For what velocities will L be less than $\frac{1}{2}L_0$? State your answer in terms of c.

CHAPTER 3

FUNCTIONS AND GRAPHS

In the first three sections we consider coordinate systems and graphs in two dimensions. The remainder of the chapter contains a discussion of one of the most important concepts in mathematics—*function*.

3.1 COORDINATE SYSTEMS IN TWO DIMENSIONS

FIGURE 1

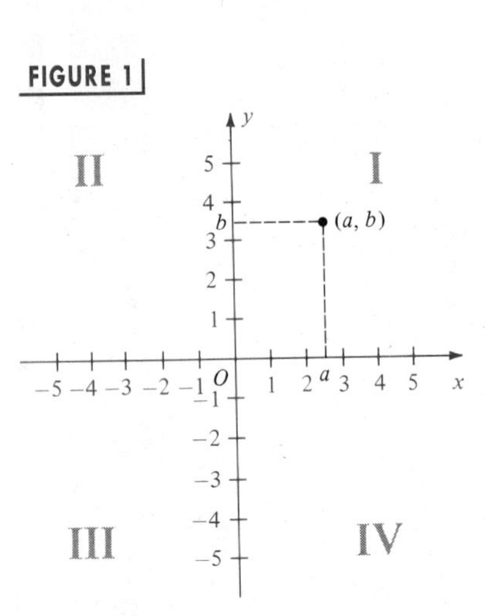

In Section 1.2 we discussed a method of assigning coordinates to points on a line. A coordinate system can also be introduced in a plane by means of *ordered pairs*. The term **ordered pair** refers to two real numbers, with one designated as the first number and the other as the second. The symbol (a, b) denotes the ordered pair consisting of the real numbers a and b with a first and b second.

Ordered pairs have many uses. We used them in Section 2.6 to denote open intervals. In this section they will represent points in a plane. Although ordered pairs are employed in different situations, there is little chance that we will confuse their uses, since it should always be clear from our discussion whether the symbol (a, b) represents an interval, a point, or some other mathematical concept.

Two ordered pairs (a, b) and (c, d) are **equal**, and we write

$$(a, b) = (c, d) \quad \text{if and only if} \quad a = c \quad \text{and} \quad b = d.$$

This statement implies, in particular, that $(a, b) \neq (b, a)$ if $a \neq b$.

A **rectangular**, or **Cartesian* coordinate system** may be introduced in a plane by considering two perpendicular coordinate lines in the plane that intersect in the origin O on each line. Unless specified otherwise, we choose the same unit of length on each line. Usually one of the lines is horizontal with positive direction to the right, and the other line is vertical with positive direction upward, as indicated by the arrowheads in Figure 1. The two lines are **coordinate axes**, and the point O is the **origin**. We usually refer to the horizontal line as the **x-axis** and the vertical line as the **y-axis**, and label them x and y, respectively. The plane is then a **coordinate plane**, or an ***xy*-plane**. In certain applications different labels, such as s or t, are used for the coordinate axes, and we refer to the system by its labels, such as an *st*-plane. The coordinate axes divide the plane into four parts called the **first, second, third**, and **fourth quadrants** and labeled I, II, III, and IV, respectively (see Figure 1). Points on the axes do not belong to any quadrant.

*The term *Cartesian* is used in honor of the French mathematician and philosopher René Descartes (1596–1650), who was one of the first to employ such coordinate systems.

FIGURE 2

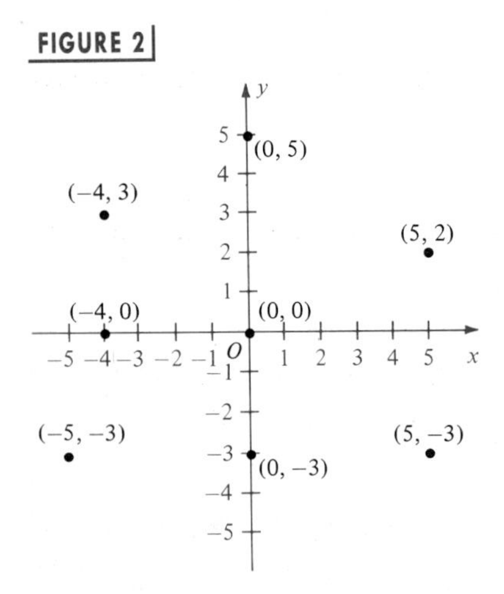

Each point P in an xy-plane may be assigned a unique ordered pair (a, b) as shown in Figure 1. The number a is the **x-coordinate** (or **abscissa**) of P, and b is the **y-coordinate** (or **ordinate**). We say that P *has coordinates* (a, b). Conversely, every ordered pair (a, b) determines a point P in the xy-plane with coordinates a and b. We often refer to the *point* (a, b), or $P(a, b)$, meaning the point P with x-coordinate a and y-coordinate b. To **plot a point** $P(a, b)$, we locate P in a coordinate plane and represent it by a dot, as illustrated by some points plotted in Figure 2.

We may use the following formula to find the distance between two points in a coordinate plane.

DISTANCE FORMULA

The distance $d(P_1, P_2)$ between any two points $P_1(x_1, y_1)$ and $P_2(x_2, y_2)$ in a coordinate plane is

$$d(P_1, P_2) = \sqrt{(x_2 - x_1)^2 + (y_2 - y_1)^2}.$$

FIGURE 3

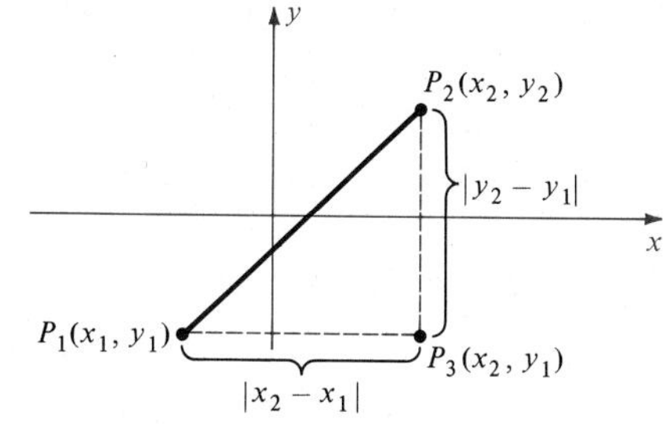

PROOF If $x_1 \neq x_2$ and $y_1 \neq y_2$, then, as illustrated in Figure 3, the points P_1, P_2, and $P_3(x_2, y_1)$ are vertices of a right triangle. By the Pythagorean theorem,

$$[d(P_1, P_2)]^2 = [d(P_1, P_3)]^2 + [d(P_3, P_2)]^2.$$

From the figure we see that

$$d(P_1, P_3) = |x_2 - x_1| \quad \text{and} \quad d(P_3, P_2) = |y_2 - y_1|.$$

Since $|a|^2 = a^2$ for every real number a, we may write

$$[d(P_1, P_2)]^2 = (x_2 - x_1)^2 + (y_2 - y_1)^2.$$

Taking the square root of each side of the last equation gives us the distance formula.

If $y_1 = y_2$, the points P_1 and P_2 lie on the same horizontal line, and

$$d(P_1, P_2) = |x_2 - x_1| = \sqrt{(x_2 - x_1)^2}.$$

Similarly, if $x_1 = x_2$ the points are on the same vertical line, and

$$d(P_1, P_2) = |y_2 - y_1| = \sqrt{(y_2 - y_1)^2}.$$

These are special cases of the distance formula.

Although we referred to the points shown in Figure 3, our proof is independent of the positions of P_1 and P_2. ❑

When applying the distance formula, note that $d(P_1, P_2) = d(P_2, P_1)$ and, hence, the order in which we subtract the x-coordinates and the y-coordinates of the points is immaterial.

EXAMPLE ▪ 1

Show that the triangle with vertices $A(-1, -3)$, $B(6, 1)$, and $C(2, -5)$ is a right triangle, and find its area.

SOLUTION From plane geometry, a triangle is a right triangle if the sum of the squares of two of its sides is equal to the square of the remaining

FIGURE 4

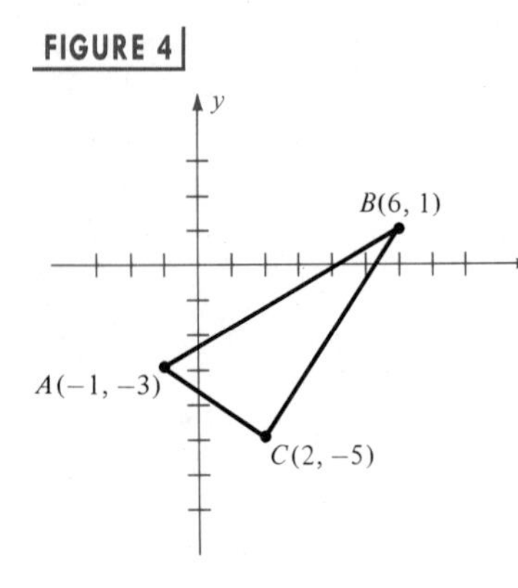

side. The triangle with vertices A, B, C is shown in Figure 4. Using the distance formula,

$$d(A, B) = \sqrt{(-1-6)^2 + (-3-1)^2} = \sqrt{49+16} = \sqrt{65}$$
$$d(B, C) = \sqrt{(6-2)^2 + (1+5)^2} = \sqrt{16+36} = \sqrt{52}$$
$$d(A, C) = \sqrt{(-1-2)^2 + (-3+5)^2} = \sqrt{9+4} = \sqrt{13}.$$

Since $[d(A, B)]^2 = [d(B, C)]^2 + [d(A, C)]^2$, the triangle is a right triangle with hypotenuse AB.

The area of a triangle with base b and altitude h is $\frac{1}{2}bh$. Referring to Figure 4, we let

$$b = d(B, C) = \sqrt{52} \quad \text{and} \quad h = d(A, C) = \sqrt{13}.$$

Hence the area of the triangle is

$$\tfrac{1}{2}bh = \tfrac{1}{2}\sqrt{52}\sqrt{13} = \tfrac{1}{2}\cdot 2\sqrt{13}\sqrt{13} = 13.$$

We can find the midpoint of a line segment by using the following formula.

MIDPOINT FORMULA

> The midpoint M of the line segment from $P_1(x_1, y_1)$ to $P_2(x_2, y_2)$ is
>
> $$\left(\frac{x_1 + x_2}{2}, \frac{y_1 + y_2}{2}\right).$$

FIGURE 5

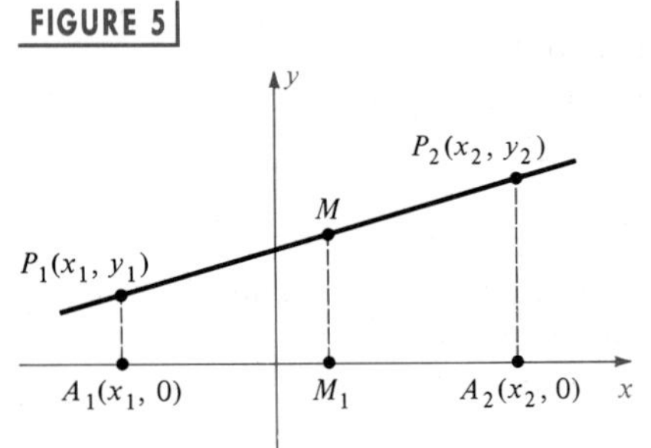

PROOF The lines through P_1 and P_2 parallel to the y-axis intersect the x-axis at $A_1(x_1, 0)$ and $A_2(x_2, 0)$. From plane geometry, the line through M parallel to the y-axis bisects the segment A_1A_2 at point M_1 (see Figure 5). If $x_1 < x_2$, then $x_2 - x_1 > 0$, and hence $d(A_1, A_2) = x_2 - x_1$. Since M_1 is halfway from A_1 to A_2, the x-coordinate of M_1 is

$$\begin{aligned} x_1 + \tfrac{1}{2}(x_2 - x_1) &= x_1 + \tfrac{1}{2}x_2 - \tfrac{1}{2}x_1 \\ &= \tfrac{1}{2}x_1 + \tfrac{1}{2}x_2 \\ &= \frac{x_1 + x_2}{2}. \end{aligned}$$

This is the *average* of the numbers x_1 and x_2. It follows that the x-coordinate of M is also $(x_1 + x_2)/2$. Similarly, the y-coordinate of M is $(y_1 + y_2)/2$. These formulas hold for all positions of P_1 and P_2. ❑

EXAMPLE ▪ 2

Find the midpoint M of the line segment from $P_1(-2, 3)$ to $P_2(4, -2)$, and verify that $d(P_1, M) = d(P_2, M)$.

SOLUTION By the midpoint formula, the coordinates of M are

$$\left(\frac{-2+4}{2}, \frac{3+(-2)}{2}\right), \quad \text{or} \quad \left(1, \frac{1}{2}\right).$$

FIGURE 6

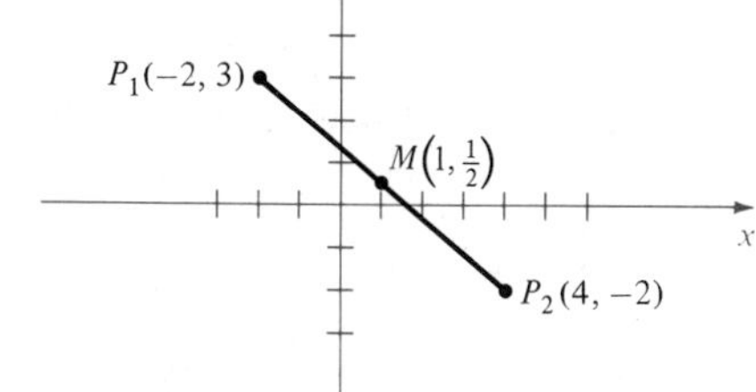

The three points P_1, P_2, and M are plotted in Figure 6. Using the distance formula,

$$d(P_1, M) = \sqrt{(-2-1)^2 + (3-\tfrac{1}{2})^2} = \sqrt{9 + \tfrac{25}{4}}$$

$$d(P_2, M) = \sqrt{(4-1)^2 + (-2-\tfrac{1}{2})^2} = \sqrt{9 + \tfrac{25}{4}}.$$

Hence, $d(P_1, M) = d(P_2, M)$.

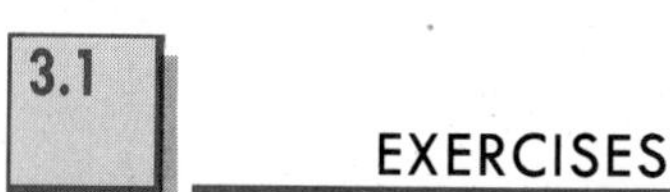

3.1 EXERCISES

1 Plot the points $A(5, -2)$, $B(-5, -2)$, $C(5, 2)$, $D(-5, 2)$, $E(3, 0)$, and $F(0, 3)$ on a coordinate plane.

2 Plot the points $A(-3, 1)$, $B(3, 1)$, $C(-2, -3)$, $D(0, 3)$, and $E(2, -3)$ on a coordinate plane. Draw the line segments AB, BC, CD, DE, and EA.

3 Plot the points $A(0, 0)$, $B(1, 1)$, $C(3, 3)$, $D(-1, -1)$, and $E(-2, -2)$. Describe the set of all points of the form (x, x) for a real number x.

4 Plot the points $A(0, 0)$, $B(1, -1)$, $C(2, -2)$, $D(-1, 1)$, and $E(-3, 3)$. Describe the set of all points of the form $(a, -a)$ for a real number a.

5 Describe the set of all points $P(x, y)$ in a coordinate plane such that:

(a) $x = -2$ (b) $y = 3$

(c) $x \geq 0$ (d) $xy > 0$

(e) $y < 0$ (f) $x = 0$

6 Describe the set of all points $P(x, y)$ in a coordinate plane such that:

(a) $y = -2$ (b) $x = -4$

(c) $x/y < 0$ (d) $xy = 0$

(e) $y > 1$ (f) $y = 0$

Exer. 7–12:

(a) Find the distance $d(A, B)$ between A and B.

(b) Find the midpoint of the segment AB.

7 $A(4, -3)$, $B(6, 2)$

8 $A(-2, -5)$, $B(4, 6)$

9 $A(-5, 0)$, $B(-2, -2)$

10 $A(6, 2)$, $B(6, -2)$

11 $A(7, -3)$, $B(3, -3)$

12 $A(-4, 7)$, $B(0, -8)$

Exer. 13–14: Show that the triangle with vertices A, B, and C is a right triangle, and find its area.

13 $A(8, 5)$, $B(1, -2)$, $C(-3, 2)$

14 $A(-6, 3)$, $B(3, -5)$, $C(-1, 5)$

15 Show that the points $A(-4, 2)$, $B(1, 4)$, $C(3, -1)$, and $D(-2, -3)$ are vertices of a square.

16 Show that the points $A(-4, -1)$, $B(0, -2)$, $C(6, 1)$, and $D(2, 2)$ are vertices of a parallelogram.

17 Given $A(-3, 8)$, find the coordinates of the point B such that $M(5, -10)$ is the midpoint of AB.

18 Given $A(5, -8)$ and $B(-6, 2)$, find the point on AB that is three-fourths of the way from A to B.

19 Given $A(-4, -3)$ and $B(6, 1)$, prove that $P(5, -11)$ is on the perpendicular bisector of AB.

20 Given $A(-4, -3)$, and $B(6, 1)$, find a formula that expresses the fact that $P(x, y)$ is on the perpendicular bisector of AB.

21 Find a formula that expresses the fact that $P(x, y)$ is a distance 5 from the origin. Describe the set of all such points.

22 If r is a positive real number, find a formula that states that $P(x, y)$ is a distance r from a fixed point $C(h, k)$. Describe the set of all such points.

23 Find all points on the y-axis that are 6 units from the point $(5, 3)$.

24 Find all points on the x-axis that are 5 units from $(-2, 4)$.

25 Let S denote the set of points of the form $(2x, x)$ for a real number x. Find the point in S that is in the third quadrant and is a distance 5 from the point $(1, 3)$.

26 Let S denote the set of points of the form (x, x) for a real number x. Find all points in S that are a distance 3 from the point $(-2, 1)$.

27 For what values of a is the distance between $(a, 3)$ and $(5, 2a)$ greater than $\sqrt{26}$?

28 Given the points $A(-2, 0)$ and $B(2, 0)$, find a formula not containing radicals that expresses the fact that the sum of the distances from $P(x, y)$ to A and to B, respectively, is 5.

29 Prove that the midpoint of the hypotenuse of any right triangle is equidistant from the vertices. (*Hint:* Label the vertices of the triangle $O(0, 0)$, $A(a, 0)$, and $B(0, b)$.)

30 Prove that the diagonals of any parallelogram bisect each other. (*Hint:* Label three of the vertices of the parallelogram $O(0, 0)$, $A(a, b)$, and $C(0, c)$.)

3.2 GRAPHS

If W is a set of ordered pairs, we may consider the point $P(x, y)$ in a coordinate plane that corresponds to the ordered pair (x, y) in W. The **graph** of W is the set of all such points. To *sketch the graph of W*, we illustrate the significant features of the graph on a coordinate plane.

EXAMPLE ▪ 1

Sketch the graph of $W = \{(x, y): |x| \le 2, |y| \le 1\}$.

SOLUTION The set W consists of all ordered pairs (x, y) such that $|x| \le 2$ and $|y| \le 1$. These inequalities are equivalent to $-2 \le x \le 2$ and $-1 \le y \le 1$. Thus, the graph of W consists of all points within and on the boundary of the rectangular region shown in Figure 7.

FIGURE 7

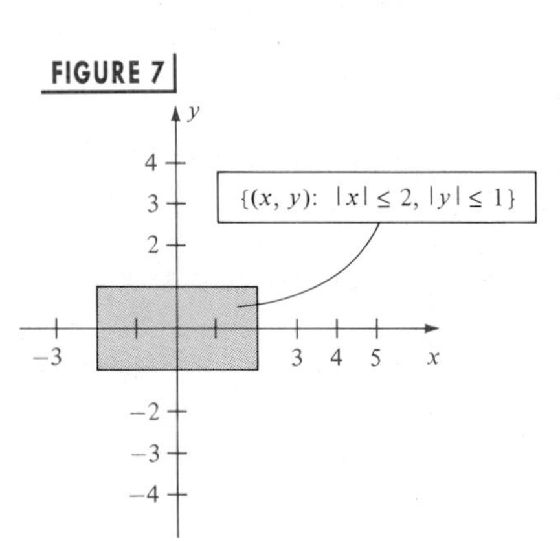

EXAMPLE ▪ 2

Sketch the graph of $W = \{(x, y): y = 2x - 1\}$.

SOLUTION We wish to find the points (x, y) that correspond to the ordered pairs (x, y) in W. It is convenient to list coordinates of several such

FIGURE 8

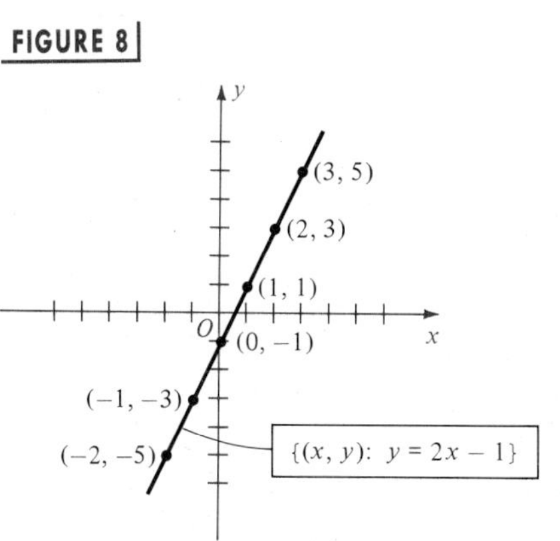

points in tabular form, where for each x, we obtain the value for y from $y = 2x - 1$:

x	-2	-1	0	1	2	3
y	-5	-3	-1	1	3	5

The points with these coordinates appear to lie on a line and we sketch the graph in Figure 8. Ordinarily, the few points we have plotted would not be enough to illustrate the graph; however, in this elementary case we can be reasonably sure that the graph is a line. In the next section we will prove this fact.

The **x-intercepts** of a graph are the x-coordinates of points at which the graph intersects the x-axis. The **y-intercepts** are the y-coordinates of points at which the graph intersects the y-axis. The graph in Figure 8 has one x-intercept, $\frac{1}{2}$, and one y-intercept, -1.

It is impossible to sketch the entire graph in Example 2, since we may assign values to x that are numerically as large as desired. Nevertheless, we call the drawing in Figure 8 *the graph of W* or *a sketch of the graph.* It is understood that the drawing is only a device for visualizing the actual graph and the line does not terminate as shown in the figure. In general, the sketch of a graph should illustrate enough of the graph so that the remaining parts are evident.

Given an equation in x and y, an ordered pair (a, b) is a **solution** of the equation if equality is obtained when we substitute a for x and b for y. For example, $(2, 3)$ is a solution of $y = 2x - 1$, since substitution of 2 for x and 3 for y leads to $3 = 4 - 1$, or $3 = 3$. Two equations in x and y are **equivalent** if they have the same solutions. The solutions of an equation in x and y determine a set W of ordered pairs, and we define the **graph of an equation in x and y** as the graph of W. Note that the graph of the equation $y = 2x - 1$ is the same as the graph of the set W in Example 2 (see Figure 8).

For some of the equations we shall consider in this chapter, the technique we will use for sketching the graph consists of plotting a sufficient number of points until some pattern emerges, and then sketching the graph accordingly. This is an elementary and often inaccurate) way to arrive at the graph; however, it is a method often employed in introductory courses. As we progress through this text, methods will be introduced that will enable us to sketch more accurate graphs without plotting many points. Still more efficient techniques are developed in advanced courses in mathematics.

EXAMPLE ■ 3

Sketch the graph of the equation $y = x^2$.

FIGURE 9

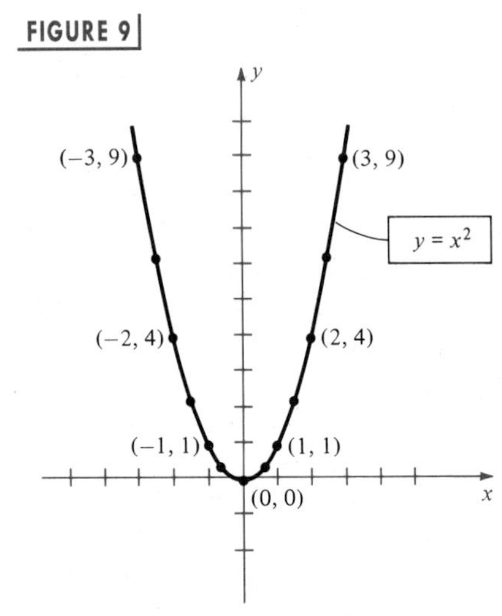

SOLUTION To sketch the graph, we must plot more points than in the previous example. Increasing successive x-coordinates by $\frac{1}{2}$, we obtain a table of coordinates:

x	-3	$-\frac{5}{2}$	-2	$-\frac{3}{2}$	-1	$-\frac{1}{2}$	0	$\frac{1}{2}$	1	$\frac{3}{2}$	2	$\frac{5}{2}$	3
y	9	$\frac{25}{4}$	4	$\frac{9}{4}$	1	$\frac{1}{4}$	0	$\frac{1}{4}$	1	$\frac{9}{4}$	4	$\frac{25}{4}$	9

Larger values of $|x|$ produce larger values of y. For example, the points $(4, 16)$, $(5, 25)$, and $(6, 36)$ are on the graph, as are $(-4, 16)$, $(-5, 25)$, and $(-6, 36)$. Plotting the points given by the table and drawing a smooth curve through these points gives us the sketch in Figure 9, in which several points are labeled.

The graph in Figure 9 is a **parabola**. In this case, the y-axis is the **axis of the parabola**. The lowest point $(0, 0)$ is the **vertex** of the parabola, and we say that the parabola **opens upward**. If we invert the graph, as would be the case for $y = -x^2$, then the parabola **opens downward**, and the vertex $(0, 0)$ is the highest point on the graph. In general, the graph of *any* equation of the form $y = ax^2$ for $a \neq 0$ is a parabola with vertex $(0, 0)$. Parabolas may also open to the right or to the left (see Example 4). In Chapter 4 we shall consider parabolas having axes *parallel* to either the x- or y-axis.

If the coordinate plane in Figure 9 is folded along the y-axis, then the graph that lies in the left half of the plane coincides with that in the right half. We say that the **graph is symmetric with respect to the y-axis**. As in Figure 10(i), a graph is symmetric with respect to the y-axis provided that

FIGURE 10 Symmetries

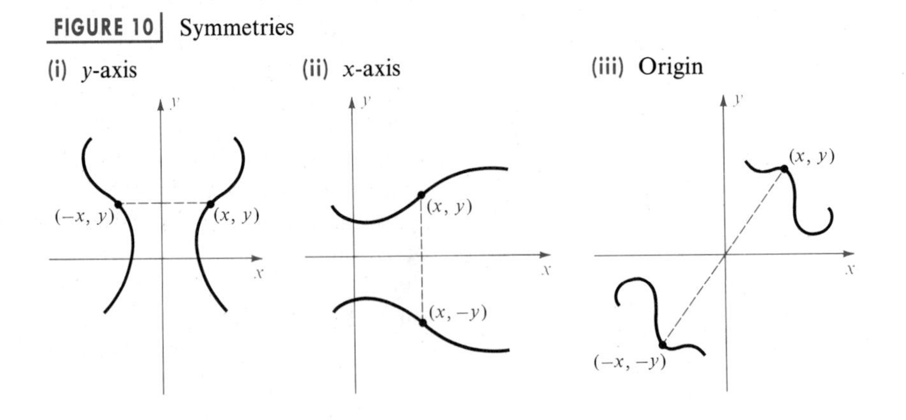

the point $(-x, y)$ is on the graph whenever (x, y) is on the graph. As in Figure 10(ii), a **graph is symmetric with respect to the *x*-axis** if whenever a point (x, y) is on the graph, then $(x, -y)$ is also on the graph. Certain graphs possess a **symmetry with respect to the origin**, meaning that if a point (x, y) is on the graph, then $(-x, -y)$ is also on the graph, as illustrated in Figure 10(iii).

The following tests are useful for investigating the three types of symmetry for graphs of equations in x and y.

TESTS FOR SYMMETRY

(i) The graph of an equation is symmetric with respect to the y-axis if substitution of $-x$ for x leads to an equivalent equation.

(ii) The graph of an equation is symmetric with respect to the x-axis if substitution of $-y$ for y leads to an equivalent equation.

(iii) The graph of an equation is symmetric with respect to the origin if the simultaneous substitution of $-x$ for x and $-y$ for y leads to an equivalent equation.

If, in the equation of Example 3, we substitute $-x$ for x, we obtain $y = (-x)^2$, which is equivalent to $y = x^2$. Hence, by symmetry test (i), the graph is symmetric with respect to the y-axis.

If a graph is symmetric with respect to an axis, it is sufficient to determine the graph in half of the coordinate plane, since we may sketch the remainder by taking a mirror image, or reflection, through the axis of symmetry.

EXAMPLE ▪ 4

Sketch the graph of the equation $y^2 = x$.

SOLUTION Since substitution of $-y$ for y does not change the equation, the graph is symmetric with respect to the x-axis (see symmetry test (ii)). Thus, it is sufficient to plot points with nonnegative y-coordinates and then reflect through the x-axis. Since $y^2 = x$, the y-coordinates of points above the x-axis are given by $y = \sqrt{x}$. Coordinates of some points on the graph are listed below. The graph is sketched in Figure 11.

x	0	1	2	3	4	9
y	0	1	$\sqrt{2} \approx 1.4$	$\sqrt{3} \approx 1.7$	2	3

FIGURE 11

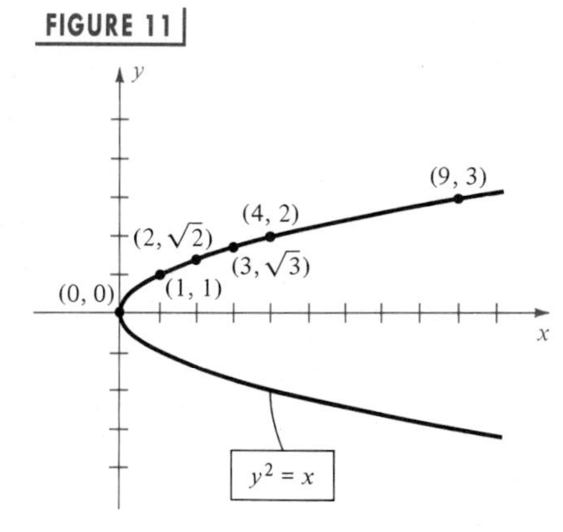

The graph is a parabola that opens to the right, with its vertex at the origin. In this case the x-axis is the axis of the parabola.

EXAMPLE ■ 5

Sketch the graph of the equation $4y = x^3$.

SOLUTION If we substitute $-x$ for x and $-y$ for y, then

$$4(-y) = (-x)^3, \quad \text{or} \quad -4y = -x^3.$$

FIGURE 12

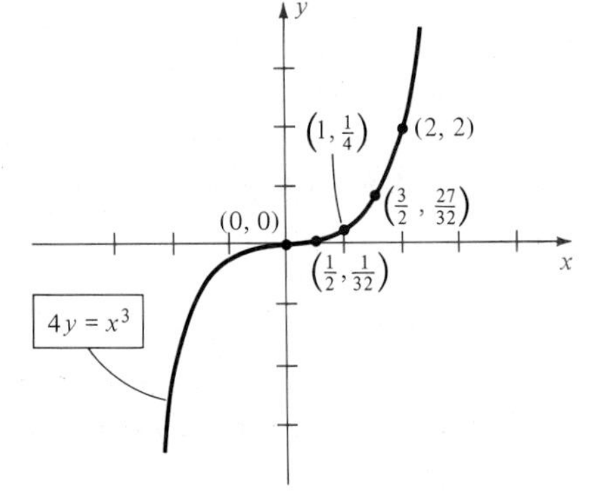

Multiplying both sides by -1, we see that the last equation has the same solutions as the equation $4y = x^3$. Hence, from symmetry test (iii), the graph is symmetric with respect to the origin. The following table lists coordinates of some points on the graph.

x	0	$\frac{1}{2}$	1	$\frac{3}{2}$	2	$\frac{5}{2}$
y	0	$\frac{1}{32}$	$\frac{1}{4}$	$\frac{27}{32}$	2	$\frac{125}{32}$

By symmetry we see that the points $(-1, -\frac{1}{4})$, $(-2, -2)$, and so on, are also on the graph. Plotting points leads to the sketch in Figure 12.

FIGURE 13

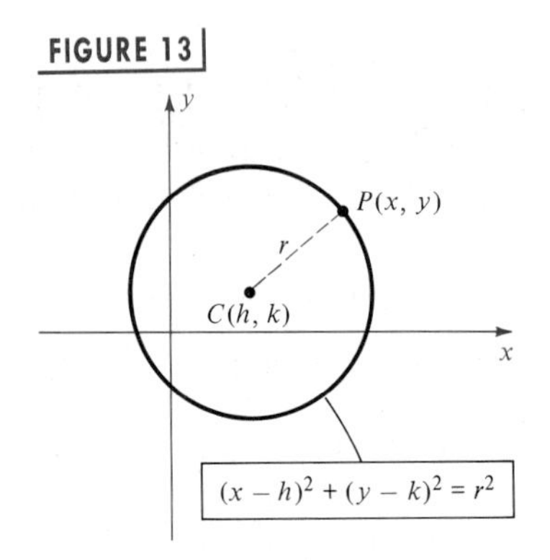

If $C(h, k)$ is a point in a coordinate plane, then a circle with center C and radius $r > 0$ consists of all points in the plane that are r units from C. As shown in Figure 13, a point $P(x, y)$ is on the circle if and only if $d(C, P) = r$ or, by the distance formula, if and only if

$$\sqrt{(x - h)^2 + (y - k)^2} = r.$$

This equation is equivalent to the following equation, which is called the **equation of a circle of radius *r* and center (*h*, *k*)**.

EQUATION OF A CIRCLE

$$(x - h)^2 + (y - k)^2 = r^2; \qquad r > 0$$

If $h = 0$ and $k = 0$, this equation reduces to $x^2 + y^2 = r^2$, which is an equation of a circle of radius r with center at the origin (see Figure 14). If $r = 1$, we call the graph a **unit circle**.

FIGURE 14

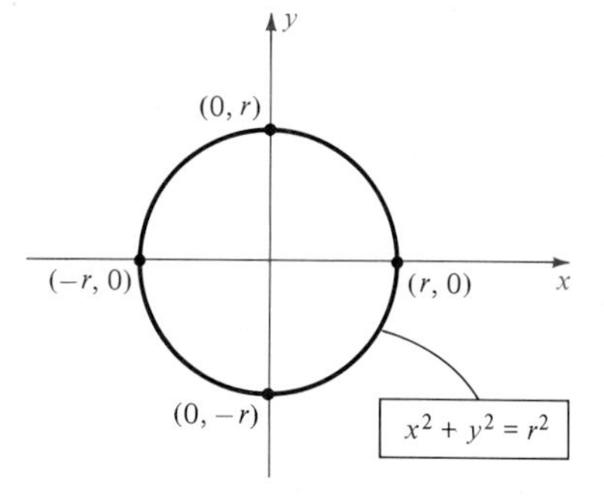

FIGURE 15

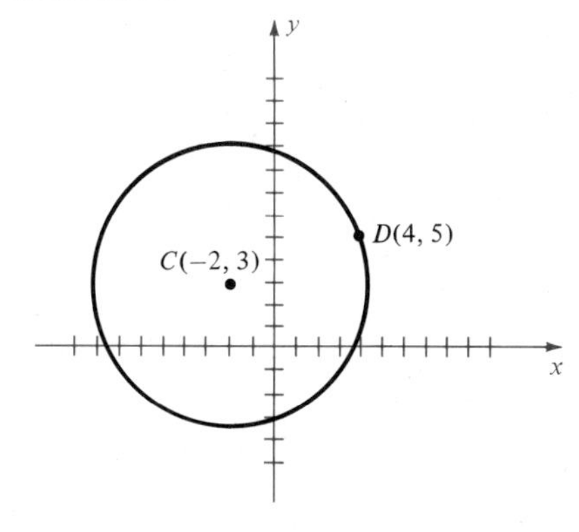

EXAMPLE ▪ 6

Find an equation of the circle that has center $C(-2, 3)$ and contains the point $D(4, 5)$.

SOLUTION The circle is illustrated in Figure 15. Since D is on the circle, the radius r is $d(C, D)$. By the distance formula,

$$r = \sqrt{(-2-4)^2 + (3-5)^2} = \sqrt{36+4} = \sqrt{40}.$$

Using the equation of a circle with $h = -2$, $k = 3$, and $r = \sqrt{40}$, we obtain

$$(x+2)^2 + (y-3)^2 = 40$$

$$x^2 + y^2 + 4x - 6y - 27 = 0.$$

Squaring terms of $(x-h)^2 + (y-k)^2 = r^2$ and simplifying leads to an equation of the form

$$x^2 + y^2 + ax + by + c = 0$$

for some real numbers a, b, and c. Conversely, if we begin with the last equation, it is always possible, by *completing squares* (see page 77), to obtain an equation of the form

$$(x-h)^2 + (y-k)^2 = d.$$

The method will be illustrated in Example 7. If $d > 0$ the graph is a circle with center (h, k) and radius $r = \sqrt{d}$. If $d = 0$ the graph consists of only one point (h, k). Finally, if $d < 0$ the equation has no real solutions and hence there is no graph.

EXAMPLE ▪ 7

Find the center and radius of the circle with equation

$$x^2 + y^2 - 4x + 6y - 3 = 0.$$

SOLUTION We begin by arranging the equation as follows:

$$(x^2 - 4x) + (y^2 + 6y) = 3$$

Next we complete the squares for the expressions within parentheses. Of course, to obtain an equivalent equation, we must add the numbers to *both* sides of the equation. To complete the square for an expression of the form $x^2 + ax$, we add the square of half the coefficient of x, that is, $(a/2)^2$, to both sides of the equation (see page 77). Similarly, for

$y^2 + by$, we add $(b/2)^2$ to both sides. In this example $a = -4$, $b = 6$, $(a/2)^2 = (-2)^2 = 4$, and $(b/2)^2 = 3^2 = 9$. These additions lead to:

$$(x^2 - 4x + 4) + (y^2 + 6y + 9) = 3 + 4 + 9$$
$$(x - 2)^2 + (y + 3)^2 = 16.$$

Hence, the center is $(2, -3)$ and the radius is $\sqrt{16} = 4$.

3.2 EXERCISES

Exer. 1–8: Sketch the graph of the set W.

1 $W = \{(x, y): |x| \geq 2,\ |y| < 3\}$

2 $W = \{(x, y): |x| < 5,\ |y| \geq 2\}$

3 $W = \{(x, y): |x| \geq 4,\ |y| \geq 3\}$

4 $W = \{(x, y): |x| < 1,\ |y| < 5\}$

5 $W = \{(x, y): |x + 2| \leq 1,\ |y - 3| < 5\}$

6 $W = \{(x, y): |x - 1| \geq 4,\ |y + 2| > 3\}$

7 $W = \{(x, y): |x - 2| \leq 5,\ |y - 4| > 2\}$

8 $W = \{(x, y): |x + 3| > 2,\ |y + 1| \leq 3\}$

Exer. 9–28: Sketch the graph of the equation.

9 $y = 2x - 3$

10 $y = 3x + 2$

11 $y = -x + 1$

12 $y = -2x - 3$

13 $y = -4x^2$

14 $y = \frac{1}{3}x^2$

15 $y = 2x^2 - 1$

16 $y = -x^2 + 2$

17 $y = -\frac{1}{2}x^3$

18 $y = \frac{1}{2}x^3$

19 $y = x^3 - 8$

20 $y = -x^3 + 1$

21 $y = \sqrt{x}$

22 $y = \sqrt{-x}$

23 $y = \sqrt{x} - 4$

24 $y = \sqrt{x - 4}$

25 $y = -\sqrt{16 - x^2}$

26 $y = \sqrt{4 - x^2}$

27 $x = \sqrt{9 - y^2}$

28 $x = -\sqrt{25 - y^2}$

Exer. 29–36: Describe the graph of the equation.

29 $x^2 + y^2 = 11$

30 $x^2 + y^2 = 7$

31 $4x^2 + 4y^2 = 25$

32 $9x^2 + 9y^2 = 1$

33 $(x + 3)^2 + (y - 2)^2 = 9$

34 $(x - 4)^2 + (y + 2)^2 = 4$

35 $(x + 3)^2 + y^2 = 16$

36 $x^2 + (y - 2)^2 = 25$

Exer. 37–48: Find an equation of the circle that satisfies the stated conditions.

37 Center $C(2, -3)$, radius 5

38 Center $C(-4, 1)$, radius 3

39 Center $C(\frac{1}{4}, 0)$, radius $\sqrt{5}$

40 Center $C(\frac{3}{4}, -\frac{2}{3})$, radius $3\sqrt{2}$

41 Center $C(-4, 6)$, passing through $P(1, 2)$

42 Center at the origin, passing through $P(4, -7)$

43 Center $C(-3, 6)$, tangent to the y-axis

44 Center $C(4, -1)$, tangent to the x-axis

45 Tangent to both axes, center in the second quadrant, radius 4

46 Tangent to both axes, center in the fourth quadrant, radius 3

47 Endpoints of a diameter $A(4, -3)$ and $B(-2, 7)$

48 Endpoints of a diameter $A(-5, 2)$ and $B(3, 6)$

Exer. 49–62: Find the center and radius of the circle with the given equation.

49 $x^2 + y^2 - 4x + 6y - 36 = 0$

50 $x^2 + y^2 + 8x - 10y + 37 = 0$

51 $x^2 + y^2 + 4y - 117 = 0$

52 $x^2 + y^2 + 6x - 72 = 0$

53 $x^2 + y^2 - 10x + 18 = 0$

54 $x^2 + y^2 - 12y + 31 = 0$

55 $2x^2 + 2y^2 - 12x + 4y - 15 = 0$

56 $4x^2 + 4y^2 + 24x - 16y + 39 = 0$

57 $9x^2 + 9y^2 + 12x - 6y + 4 = 0$

58 $3x^2 + 3y^2 - 3x + 2y + 1 = 0$

59 $x^2 + y^2 + 4x - 2y + 5 = 0$

60 $x^2 + y^2 - 6x + 4y + 13 = 0$

61 $x^2 + y^2 - 2x - 8y + 19 = 0$

62 $x^2 + y^2 + 4x + 6y + 16 = 0$

3.3 LINES

The following concept is fundamental to the study of lines. All lines referred to are considered to be in a coordinate plane.

DEFINITION

Let l be a line that is not parallel to the y-axis, and let $P_1(x_1, y_1)$ and $P_2(x_2, y_2)$ be distinct points on l. The **slope m** of l is

$$m = \frac{y_2 - y_1}{x_2 - x_1}.$$

If l is parallel to the y-axis, then the slope is not defined.

FIGURE 16
Positive slope

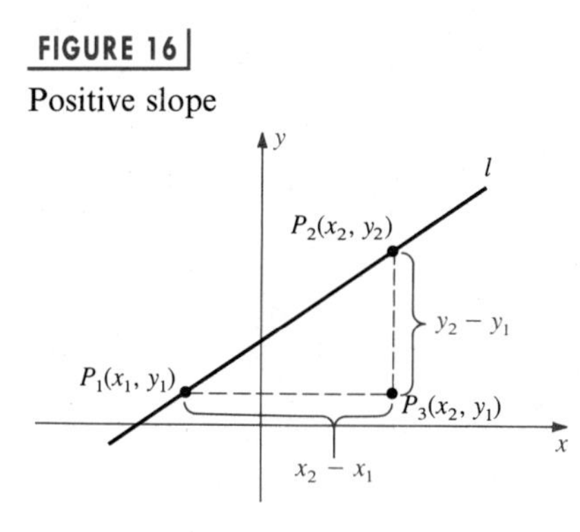

FIGURE 17
Negative slope

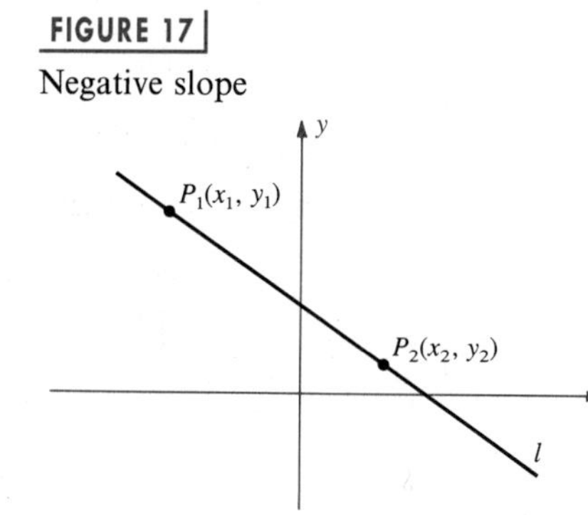

Typical points P_1 and P_2 on a line l are shown in Figures 16 and 17. The numerator $y_2 - y_1$ in the formula for m is the vertical change in direction from P_1 to P_2 and may be positive, negative, or zero. The denominator $x_2 - x_1$ is the horizontal change from P_1 to P_2, and it may be positive or negative, but never zero, because l is not parallel to the y-axis if the slope exists.

When finding the slope of a line it is immaterial which point we label as P_1 or as P_2, since

$$\frac{y_2 - y_1}{x_2 - x_1} = \frac{y_2 - y_1}{x_2 - x_1} \cdot \frac{(-1)}{(-1)} = \frac{y_1 - y_2}{x_1 - x_2}.$$

Consequently, we may assume that the points are labeled so that $x_1 < x_2$, as in Figures 16 and 17. In this situation, $x_2 - x_1 > 0$, and hence the slope is positive, negative, or zero, depending on whether $y_2 > y_1$, $y_2 < y_1$, or $y_2 = y_1$. The slope of the line shown in Figure 16 is positive. The slope of the line shown in Figure 17 is negative.

FIGURE 18

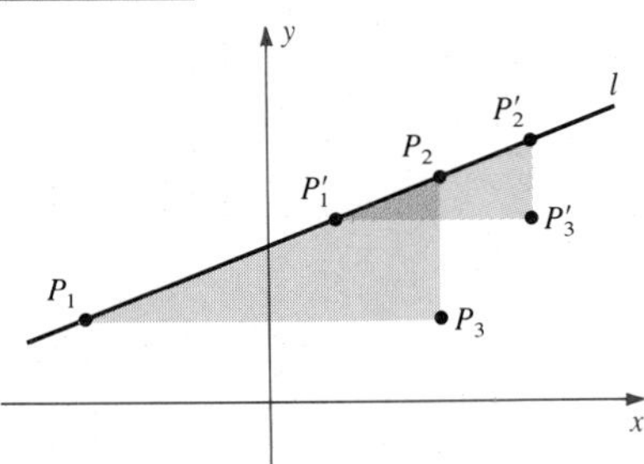

A **horizontal line** is a line parallel to the x-axis. Note that *a line is horizontal if and only if its slope is* 0. A **vertical line** is a line parallel to the y-axis. The slope of a vertical line is undefined.

The definition of slope is independent of the two points that are chosen on l. If other points $P'_1(x'_1, y'_1)$ and $P'_2(x'_2, y'_2)$ are used, then as in Figure 18, the triangle with vertices P'_1, P'_2, and $P'_3(x'_2, y'_1)$ is similar to the triangle with vertices P_1, P_2, and $P_3(x_2, y_1)$. Since the ratios of corresponding sides of similar triangles are equal,

$$\frac{y_2 - y_1}{x_2 - x_1} = \frac{y'_2 - y'_1}{x'_2 - x'_1}.$$

EXAMPLE ▪ 1

Sketch the line through each pair of points, and find its slope:

(a) $A(-1, 4)$ and $B(3, 2)$ **(b)** $A(2, 5)$ and $B(-2, -1)$

(c) $A(4, 3)$ and $B(-2, 3)$ **(d)** $A(4, -1)$ and $B(4, 4)$

SOLUTION The lines are sketched in Figure 19. Using the definition of slope,

(a) $m = \dfrac{2 - 4}{3 - (-1)} = \dfrac{-2}{4} = -\dfrac{1}{2}.$

(b) $m = \dfrac{5 - (-1)}{2 - (-2)} = \dfrac{6}{4} = \dfrac{3}{2}.$

(c) $m = \dfrac{3 - 3}{-2 - 4} = \dfrac{0}{-6} = 0.$

(d) the slope is undefined because the line is vertical. Note that if the formula for m is used, the denominator is zero.

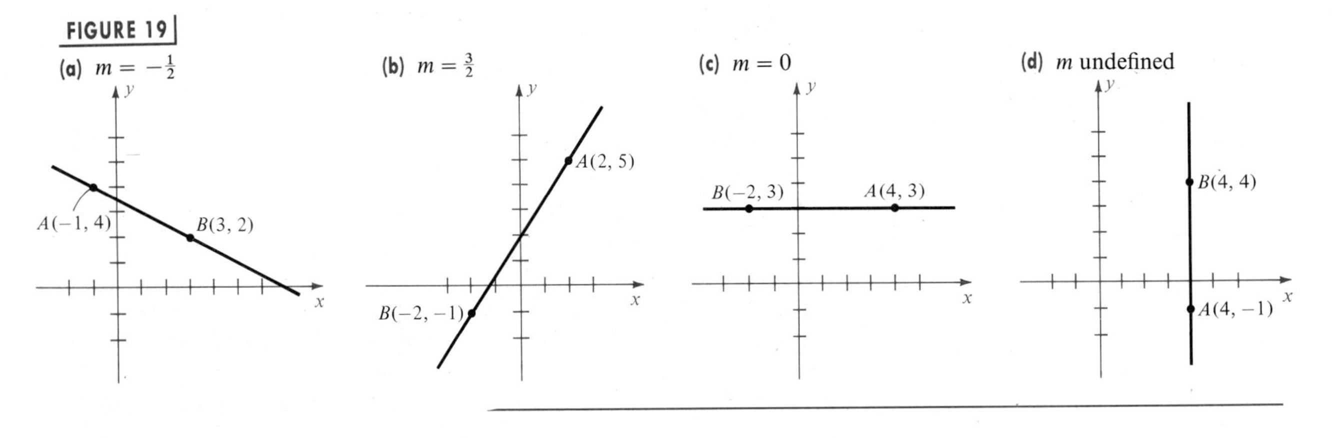

EXAMPLE ▪ 2

Construct a line through $P(2, 1)$ that has (a) slope $\frac{5}{3}$; (b) slope $-\frac{5}{3}$.

SOLUTION If the slope of a line is a/b and b is positive, then for every change of b units in the horizontal direction, the line rises or falls $|a|$ units, depending on whether a is positive or negative.

(a) If $P(2, 1)$ is on the line and $m = \frac{5}{3}$, we can obtain another point on the line by starting at P and moving 3 units to the right and 5 units upward. This gives us the point $Q(5, 6)$, and the line is determined (see Figure 20).

(b) If $P(2, 1)$ is on the line and $m = -\frac{5}{3}$, we move 3 units to the right and 5 units downward, obtaining the line through $Q(5, -4)$, as in Figure 21.

FIGURE 20 $m = \frac{5}{3}$ FIGURE 21 $m = -\frac{5}{3}$

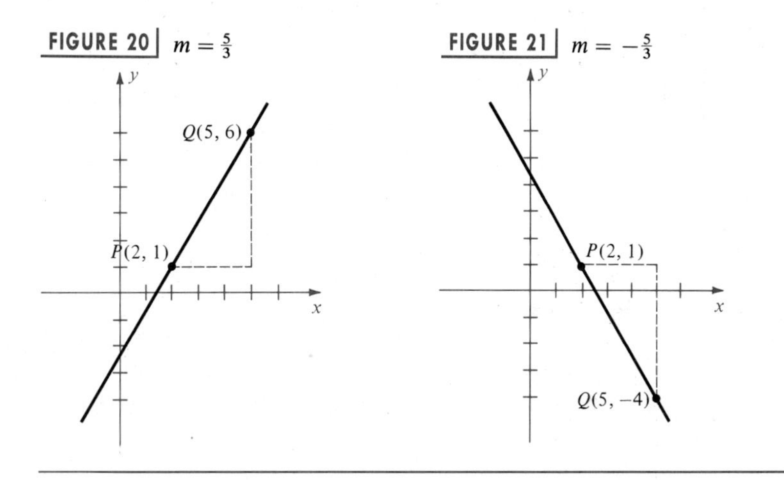

It is not difficult to obtain an equation whose graph is a given line. We shall begin with the simplest case, in which the line is either vertical or horizontal.

THEOREM

(i) The graph of the equation $x = a$ is a vertical line with x-intercept a.

(ii) The graph of the equation $y = b$ is a horizontal line with y-intercept b.

PROOF The equation $x = a$ may be written in the form $x + (0)y = a$. Some typical solutions of this equation are $(a, -2)$, $(a, 1)$, and $(a, 0)$. Evidently, every solution has the form (a, y), where y may have any value and

FIGURE 22

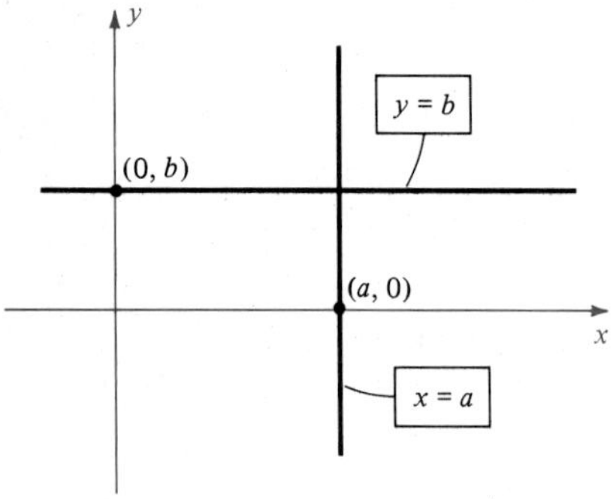

a is fixed. It follows that the graph of $x = a$ is a line with x-intercept a and parallel to the y-axis, as illustrated in Figure 22. This proves (i). Part (ii) is proved in similar fashion. ❑

Let us next find an equation of a line l through a point $P_1(x_1, y_1)$ with slope m (only one such line exists). If $P(x, y)$ is any point with $x \neq x_1$ (see Figure 23), then P is on l if and only if the slope of the line through P_1 and P is m, that is,

$$\frac{y - y_1}{x - x_1} = m.$$

FIGURE 23

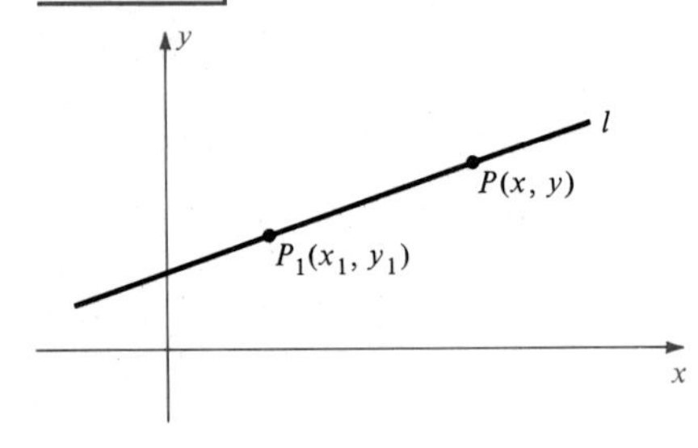

This equation may be written in the form

$$y - y_1 = m(x - x_1).$$

Note that (x_1, y_1) is a solution of the last equation and hence the points on l are precisely the points that correspond to the solutions. This equation for l is referred to as the **point-slope form**.

POINT-SLOPE FORM

> An equation for the line through the point (x_1, y_1) with slope m is
>
> $$y - y_1 = m(x - x_1).$$

EXAMPLE ▪ 3

Find an equation of the line through the points $A(1, 7)$ and $B(-3, 2)$.

SOLUTION The slope m of the line is

$$m = \frac{7 - 2}{1 - (-3)} = \frac{5}{4}.$$

We may use the coordinates of either A or B for (x_1, y_1) in the point-slope form. Using $A(1, 7)$ gives us the following equivalent equations for the line:

$$y - 7 = \tfrac{5}{4}(x - 1)$$
$$4y - 28 = 5x - 5$$
$$5x - 4y + 23 = 0$$

The point-slope form may be rewritten as $y = mx - mx_1 + y_1$, which is of the form

$$y = mx + b$$

with $b = -mx_1 + y_1$. The real number b is the y-intercept of the graph, as we may see by setting $x = 0$. Since the equation $y = mx + b$ displays the slope m and y-intercept b of l, it is called the **slope-intercept form** for the equation of a line. Conversely, if we start with $y = mx + b$, we may write

$$y - b = m(x - 0).$$

Comparing this equation with the point-slope form, we see that the graph is a line with slope m and passing through the point $(0, b)$. This gives us the next result.

SLOPE-INTERCEPT FORM

> The graph of the equation $y = mx + b$ is a line having slope m and y-intercept b.

We have shown that every line is the graph of an equation of the form

$$ax + by + c = 0$$

for real numbers a, b, and c such that a and b are not both zero. We call such an equation a **linear equation** in x and y. Let us show, conversely, that the graph of $ax + by + c = 0$, with a and b not both zero, is always a line. If $b \neq 0$, we may solve for y, obtaining

$$y = \left(-\frac{a}{b}\right)x + \left(-\frac{c}{b}\right)$$

which, by the slope-intercept form, is an equation of a line with slope $-a/b$ and y-intercept $-c/b$. If $b = 0$ but $a \neq 0$, we may solve for x, obtaining $x = -c/a$, which is the equation of a vertical line with x-intercept $-c/a$. This discussion establishes the following result.

GENERAL FORM

> The graph of a linear equation $ax + by + c = 0$ is a line and, conversely, every line is the graph of a linear equation.

For simplicity, we use the terminology *the line* $ax + by + c = 0$, rather than *the line with equation* $ax + by + c = 0$.

EXAMPLE ■ 4

Sketch the graph of $2x - 5y = 8$.

SOLUTION From the general form for a linear equation, the graph is a line, and hence it is sufficient to find two points on the graph. Let us find the x- and y-intercepts. Substituting $y = 0$ in the given equation, we obtain the x-intercept 4. Substituting $x = 0$, we see that the y-intercept is $-\frac{8}{5}$. Plotting the points $(4, 0)$ and $(0, -\frac{8}{5})$ leads to the graph in Figure 24.

FIGURE 24

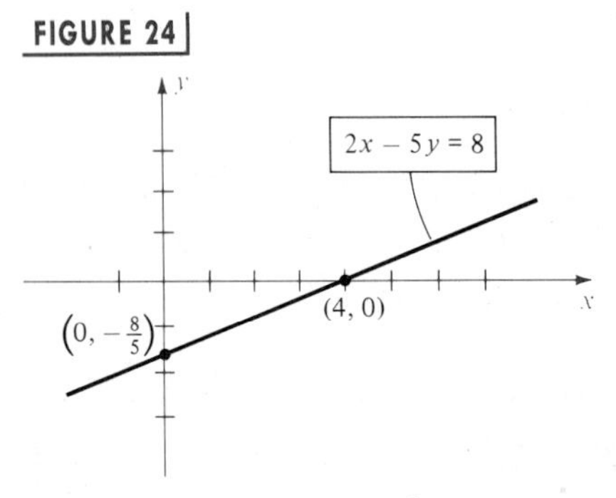

Another method of solution is to express the given equation in slope-intercept form, as follows:

$$2x - 5y = 8$$
$$5y = 2x - 8$$
$$y = \tfrac{2}{5}x - \tfrac{8}{5}$$

The last equation is of the form $y = mx + b$ with $m = \frac{2}{5}$ and $b = -\frac{8}{5}$. We may then sketch a line through $(0, -\frac{8}{5})$ with slope $\frac{2}{5}$.

The next theorem specifies the relationship between parallel lines and slope.

THEOREM

> Two nonvertical lines are parallel if and only if they have the same slope.

PROOF Let l_1 and l_2 be distinct lines of slopes m_1 and m_2. If the y-intercepts are b_1 and b_2, then by the slope-intercept form the lines have equations

$$y = m_1x + b_1 \quad \text{and} \quad y = m_2x + b_2.$$

The lines intersect at some point (x, y) if and only if

$$m_1x + b_1 = m_2x + b_2$$

and

$$(m_1 - m_2)x = b_2 - b_1.$$

Since $l_1 \neq l_2$, the last equation can be solved for x if and only if $m_1 - m_2 \neq 0$. We have shown that the lines l_1 and l_2 intersect if and only if $m_1 \neq m_2$. Hence they do *not* intersect (are parallel) if and only if $m_1 = m_2$. ❑

EXAMPLE ▪ 5

Find an equation of the line through the point $(5, -7)$ that is parallel to the line $6x + 3y - 4 = 0$.

SOLUTION Let us express the given equation in slope-intercept form. We begin by solving for y as follows:

$$3y = -6x + 4$$
$$y = -2x + \tfrac{4}{3}.$$

The last equation is in slope-intercept form with slope $m = -2$. Since parallel lines have the same slope, the required line also has slope -2. Using the point-slope form gives us the following:

$$y + 7 = -2(x - 5)$$
$$y + 7 = -2x + 10$$
$$2x + y - 3 = 0$$

The next result gives us conditions for perpendicular lines.

THEOREM

> Two lines with slope m_1 and m_2 are perpendicular if and only if
> $$m_1 m_2 = -1.$$

FIGURE 25

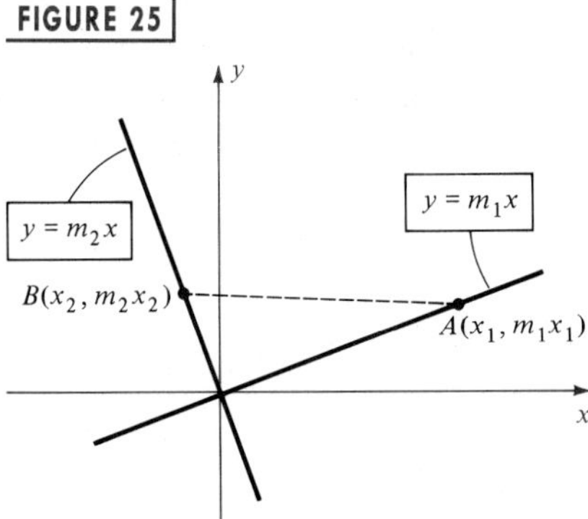

PROOF For simplicity, let us consider the special case of two lines that intersect at the origin O, as illustrated in Figure 25. Equations of these lines are $y = m_1x$ and $y = m_2x$. If, as in the figure, we choose points $A(x_1, m_1x_1)$ and $B(x_2, m_2x_2)$ different from O on the lines, then the lines are perpendicular if and only if angle AOB is a right angle. Applying the Pythagorean theorem, angle AOB is a right angle if and only if

$$[d(A, B)]^2 = [d(O, B)]^2 + [d(O, A)]^2$$

or, by the distance formula,

$$(m_2x_2 - m_1x_1)^2 + (x_2 - x_1)^2 = (m_2x_2)^2 + x_2^2 + (m_1x_1)^2 + x_1^2.$$

Squaring terms and simplifying gives us

$$-2m_1m_2x_1x_2 - 2x_1x_2 = 0.$$

Dividing both sides by $-2x_1x_2$, we see that $m_1m_2 + 1 = 0$. Thus, the lines are perpendicular if and only if $m_1m_2 = -1$.

The same type of proof may be given if the lines intersect at *any* point (a, b). ❑

A convenient way to remember the conditions for perpendicularity is to note that m_1 and m_2 must be *negative reciprocals* of one another, that is, $m_1 = -1/m_2$ and $m_2 = -1/m_1$.

EXAMPLE ▪ 6

Find an equation of the line that passes through the point $(5, -7)$ and is perpendicular to the line $6x + 3y - 4 = 0$.

SOLUTION We considered the line $6x + 3y - 4 = 0$ in Example 5 and found that its slope is -2. Hence, the slope of the required line is the negative reciprocal $-[1/(-2)]$, or $\frac{1}{2}$. Applying the point-slope form gives us the following:

$$y + 7 = \tfrac{1}{2}(x - 5)$$

$$2y + 14 = x - 5$$

$$x - 2y - 19 = 0$$

EXAMPLE ▪ 7

Find an equation for the perpendicular bisector of the line segment from $A(1, 7)$ to $B(-3, 2)$.

SOLUTION By the midpoint formula, the midpoint M of the segment AB is $(-1, \frac{9}{2})$. Since the slope of AB is $\frac{5}{4}$ (see Example 3), it follows from the preceding theorem that the slope of the perpendicular bisector is $-\frac{4}{5}$. Applying the point-slope form, we obtain

$$y - \tfrac{9}{2} = -\tfrac{4}{5}(x + 1).$$

Multiplying both sides by 10 and simplifying leads to $8x + 10y - 37 = 0$.

Two variables x and y are **linearly related** if $y = ax + b$ for constants a and b with $a \neq 0$. Linear relationships between variables occur frequently in applied problems. The following example gives one illustration. For other applications, see Exercises 33–44.

EXAMPLE ▪ 8

The relationship between the air temperature T (in °F) and the altitude h (in feet above sea level) is approximately linear. When the temperature at sea level is 60°, an increase of 5000 feet in altitude lowers the air temperature about 18°.

(a) Express T in terms of h, and sketch the graph on an hT-coordinate system.

(b) Approximate the air temperature at an altitude of 15,000 feet.

(c) Approximate the altitude at which the temperature is 0°.

SOLUTION

(a) If T is linearly related to h, then

$$T = ah + b$$

for some constants a and b. Since $T = 60$ when $h = 0$,

$$60 = a(0) + b, \quad \text{or} \quad b = 60.$$

Thus,

$$T = ah + 60.$$

In addition, if $h = 5000$, then $T = 60 - 18 = 42$. Substituting these values into the formula $T = ah + 60$, we obtain:

$$42 = a(5000) + 60$$
$$5000a = -18$$
$$a = -\frac{18}{5000} = -\frac{9}{2500}$$

Hence the (approximate) formula for T is

$$T = -\frac{9}{2500}h + 60.$$

FIGURE 26

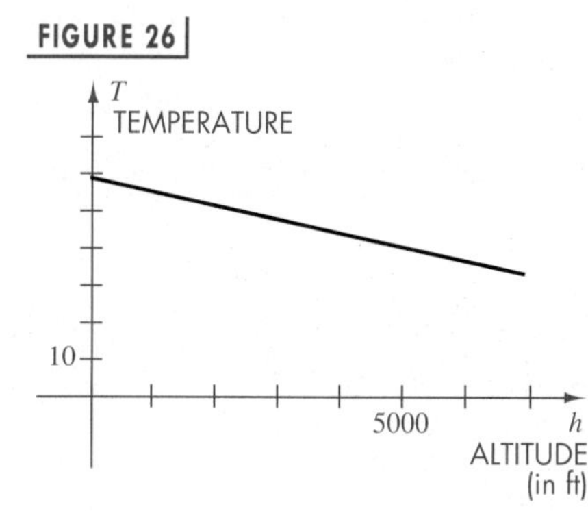

The graph is sketched in Figure 26 with different scales on the axes.

(b) Using the formula for T obtained in part (a), the (approximate) temperature (in °F) when $h = 15,000$ is

$$T = -\frac{9}{2500}(15,000) + 60 = -54 + 60 = 6.$$

(c) To find the altitude h that corresponds to $T = 0°$, we substitute 0 for T in the formula for T from part (a):

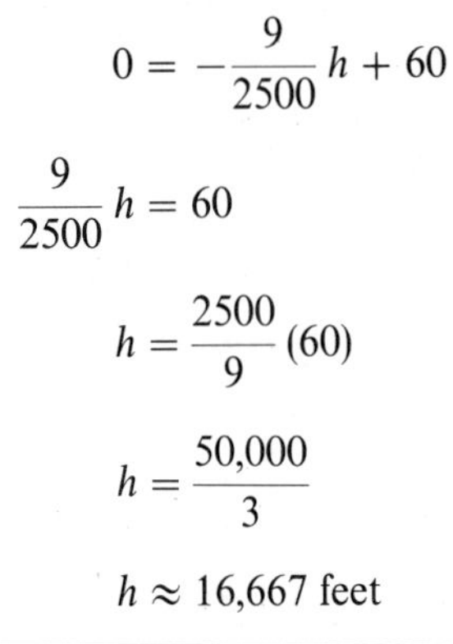

$$0 = -\frac{9}{2500}h + 60$$

$$\frac{9}{2500}h = 60$$

$$h = \frac{2500}{9}(60)$$

$$h = \frac{50{,}000}{3}$$

$$h \approx 16{,}667 \text{ feet}$$

3.3 EXERCISES

Exer. 1–4: Plot the points A and B, and find the slope of the line through A and B.

1 $A(-3, 2)$, $B(5, -4)$ **2** $A(4, -1)$, $B(-6, 7)$

3 $A(2, 5)$, $B(-7, 5)$ **4** $A(5, -1)$, $B(5, 6)$

Exer. 5–8: Use slopes to show that the points are vertices of the specified polygon.

5 $A(-3, 1)$, $B(5, 3)$, $C(3, 0)$, $D(-5, -2)$; parallelogram

6 $A(2, 3)$, $B(5, -1)$, $C(0, -6)$, $D(-6, 2)$; trapezoid

7 $A(6, 15)$, $B(11, 12)$, $C(-1, -8)$, $D(-6, -5)$; rectangle

8 $A(1, 4)$, $B(6, -4)$, $C(-15, -6)$; right triangle

9 If three consecutive vertices of a parallelogram are $A(-1, -3)$, $B(4, 2)$, and $C(-7, 5)$, find the fourth vertex.

10 Let $A(x_1, y_1)$, $B(x_2, y_2)$, $C(x_3, y_3)$, and $D(x_4, y_4)$ denote the vertices of an arbitrary quadrilateral. Show that the line segments joining midpoints of adjacent sides form a parallelogram.

Exer. 11–22: Find an equation of the line through the point A that satisfies the given condition.

11 $A(5, -2)$; **(a)** parallel to the y-axis; **(b)** perpendicular to the y-axis

12 $A(-4, 2)$; **(a)** parallel to the x-axis **(b)** perpendicular to the x-axis

13 $A(5, -3)$; slope -4

14 $A(-1, 4)$; slope $\frac{2}{3}$

15 $A(4, 0)$; slope -3

16 $A(0, -2)$; slope 5

17 $A(4, -5)$; through $B(-3, 6)$

18 $A(-1, 6)$; x-intercept 5

19 $A(2, -4)$; parallel to the line $5x - 2y = 4$

20 $A(-3, 5)$; parallel to the line $x + 3y = 1$

21 $A(7, -3)$; perpendicular to the line $2x - 5y = 8$

22 $A(4, 5)$; perpendicular to the line $3x + 2y = 7$

Exer. 23–24: Find an equation for the perpendicular bisector of the line segment AB.

23 $A(3, -1)$, $B(-2, 6)$ **24** $A(4, 2)$, $B(-2, 10)$

Exer. 25–26: Find an equation for the line that bisects the given quadrants.

25 Quadrants II, IV **26** Quadrants I, III

Exer. 27–30: Use the slope-intercept form to find the slope and y-intercept of the given line, and sketch its graph.

27 $2x = 15 - 3y$ **28** $7x = -4y - 8$

29 $4x - 3y = 9$ **30** $x - 5y = -15$

Exer. 31–32: If a line l has nonzero x- and y-intercepts a and b, respectively, then its *intercept form* is

$$\frac{x}{a} + \frac{y}{b} = 1.$$

Find the intercept form for the given line.

31 $4x - 2y = 6$

32 $x - 3y = -2$

33 Six years ago a house was purchased for $89,000. This year it was appraised at $125,000.

(a) Assuming that the value of the house is linearly related to time, find a formula that specifies the value at any time after the purchase date.

(b) When was the house worth $103,000?

34 Charles' law for gases states that if the pressure remains constant, then the relationship between the volume V that a gas occupies and its temperature T (in °C) is given by $V = V_0(1 + \frac{1}{273}T)$.

(a) What is the significance of V_0?

(b) What increase in temperature is needed to increase the volume from V_0 to $2V_0$?

(c) Sketch the graph of the equation on a TV-plane for the case $V_0 = 100$ and $T \geq -273$.

35 The electrical resistance R (in ohms) for a pure metal wire is linearly related to its temperature T (in °C) by the formula $R = R_0(1 + aT)$ for some constants a and $R_0 > 0$.

(a) What is the significance of R_0?

(b) At absolute zero ($T = -273$ °C), $R = 0$. Find a.

(c) At 0 °C, silver wire has a resistance of 1.25 ohms. At what temperature is the resistance doubled?

36 The freezing point of water is 0 °C, or 32 °F. The boiling point is 100 °C, or 212 °F.

(a) Find a linear relationship between the temperature in °F and the temperature in °C.

(b) What temperature increase in °F corresponds to an increase in temperature of 1 °C?

37 The expected weight W (in tons) of an adult humpback whale is related to its length L (in feet) by the linear equation $W = 1.70L - 42.8$.

(a) Estimate the weight of a 30-foot humpback whale.

(b) If the error in estimating the length could be as large as 2 feet, what is the corresponding error for the weight estimate?

38 Newborn blue whales measure approximately 24 feet and weigh 3 tons. Young whales are nursed for 7 months and, after weaning, measure an amazing 53 feet and weigh 23 tons. Let L and W denote the length (in feet) and the weight (in tons), respectively, of a whale that is t months of age.

(a) If L and t are linearly related, what is the daily increase in a young whale's length? (Use 1 month = 30 days.)

(b) If W and t are linearly related, what is the daily increase in a young whale's weight?

39 The amount of heat H (in joules) required to convert one gram of water into vapor is linearly related to the temperature T (in °C) of the atmosphere. At 10 °C this conversion requires 2480 joules, and each increase in temperature of 15 °C lowers the amount of heat needed by 40 joules. Express H in terms of T.

40 In exercise physiology, aerobic power P is defined in terms of maximum oxygen intake. For altitudes up to 1800 meters, aerobic power is optimal, that is, 100%. Beyond 1800 meters, P decreases linearly from the maximum of 100% to a value near 40% at 5000 meters.

(a) Express aerobic power P in terms of altitude h (in meters) for $1800 \leq h \leq 5000$.

(b) Estimate aerobic power at Mexico City (altitude 2400 meters), the site of the 1968 Summer Olympic Games.

41 The owner of an ice-cream franchise must pay the parent company $1000 per month plus 10% of the profit. The rest of the operating cost of the franchise is a fixed cost of $2600 per month for items such as utilities and labor. Suppose that 50% of the monthly revenue is profit.

(a) If the monthly take-home pay M of the owner and the monthly revenue R are linearly related, express M in terms of R.

(b) How much revenue must be collected to break even?

42 Pharmacological products must specify recommended dosages for adults and children. Two formulas for modification of adult dosage levels for young children are:

Cowling's Rule: $y = \frac{1}{24}(t + 1)a$

Friend's Rule: $y = \frac{2}{25}ta$

where a denotes adult dose (in mg) and t denotes the age of the child (in years).

(a) If $a = 100$, graph the two linear equations on the same axes for $0 \leq t \leq 12$.

(b) For what age do the two formulas specify the same dosage?

43 In the video game shown in the figure, airplanes fly from left to right along the path $y = 1 + (1/x)$ and can shoot their bullets in the tangent direction at creatures placed along the x-axis at $x = 1, 2, 3, 4, 5$. From calculus, the slope of the tangent line to the path at $P(1, 2)$ and $Q(\frac{3}{2}, \frac{5}{3})$ is $m = -1$ and $m = -\frac{4}{9}$, respectively. Determine whether a creature will be hit if the player shoots when the plane is at (a) P; (b) Q.

EXERCISE 43

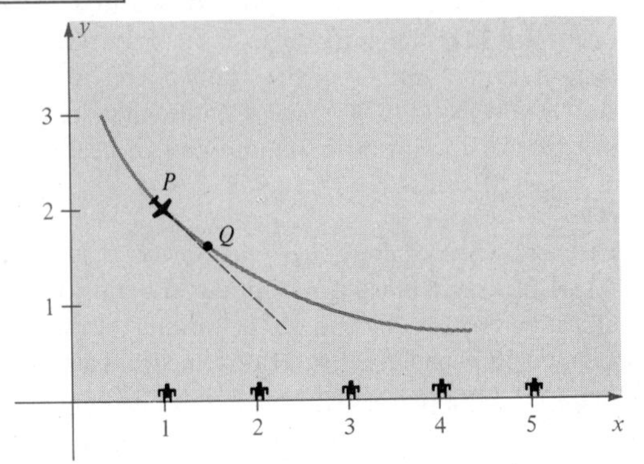

44 A hammer thrower is working on his form in a small practice area. The hammer spins, generating a circle with a radius of 5 feet, and when released, hits a tall screen that is 50 feet from the center of the throwing area. Let coordinate axes be introduced as shown in the figure (not to scale).

(a) If the hammer is released at $(-4, -3)$ and travels in the tangent direction, where will it hit the screen?

(b) If the hammer is to hit at $(0, -50)$, where on the circle should it be released?

EXERCISE 44

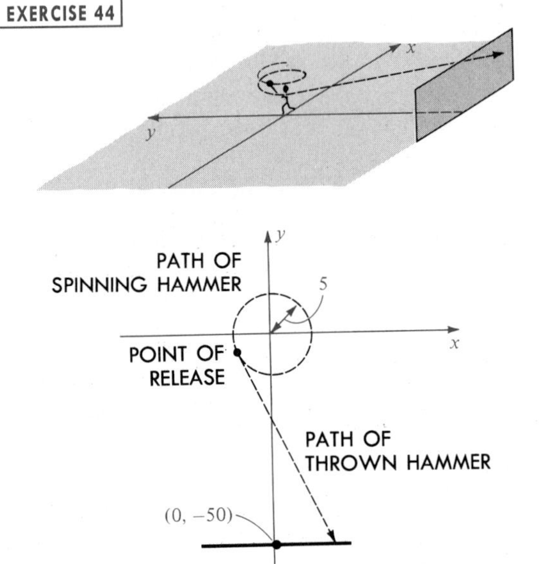

3.4 DEFINITION OF FUNCTION

The notion of **correspondence** occurs frequently in everyday life. As examples,

to each book in a library there corresponds the number of pages in the book;

to each human being there corresponds a birth date;

if the temperature of the air is recorded throughout a day, then at each instant of time there corresponds a temperature.

These examples of correspondence involve two sets, D and E. In our first example D denotes the set of books in a library and E the set of positive integers. For each book x in D there corresponds a positive integer y in E—namely, the number of pages in the book.

We sometimes depict correspondences by diagrams of the type shown in Figure 27, where the sets D and E are represented by points within

FIGURE 27

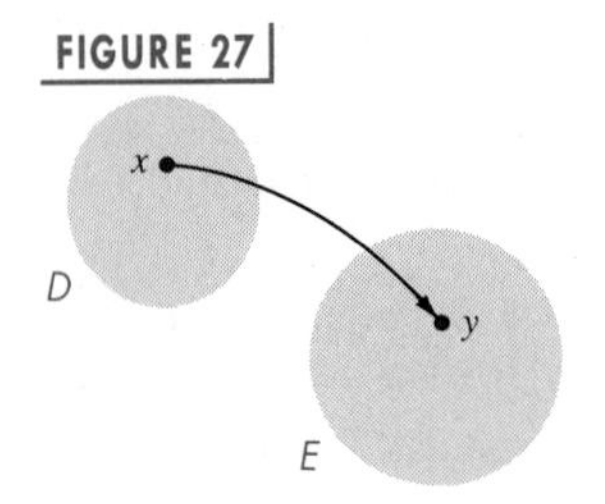

regions in a plane. The curved arrow indicates that the element y of E corresponds to the element x of D. The two sets may have elements in common. As a matter of fact, we often have $D = E$.

Our examples indicate that *to each x in D there corresponds one and only one y in E*; that is, *y is unique* for a given x. However, the same element of E may correspond to different elements of D. For example, two books may have the same number of pages, two people may have the same birthday, and so on.

In most of our work D and E will be sets of numbers. To illustrate, let both D and E denote the set $\mathbb{R}$ of real numbers, and to each real number x let us assign its square x^2. Thus, to 3 we assign 9, to -5 we assign 25, and to $\sqrt{2}$ we assign 2. This gives us a correspondence from $\mathbb{R}$ to $\mathbb{R}$.

Each of the preceding examples of a correspondence is a *function*, which we define as follows.

DEFINITION

> A **function** f from a set D to a set E is a correspondence that assigns to each element x of D a unique element y of E.

The element y of E is the **value** of f at x and is denoted by $f(x)$, read as f *of* x). The set D is the **domain** of the function. The **range** of f is the subset of E consisting of all possible values $f(x)$ for x in D.

FIGURE 28

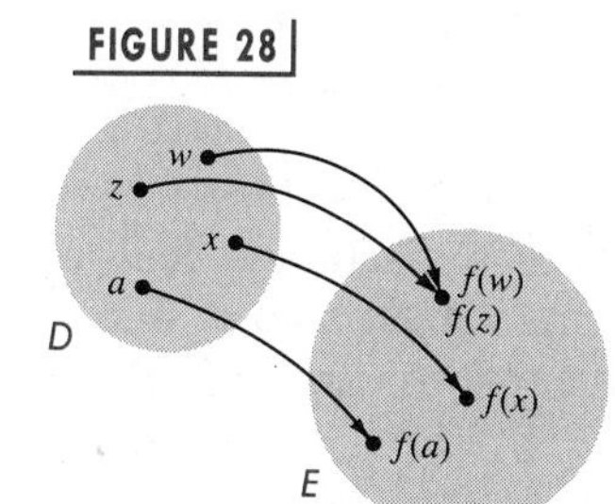

Consider the diagram in Figure 28. The curved arrows indicate that the elements $f(x)$, $f(w)$, $f(z)$, and $f(a)$ of E correspond to the elements x, w, z, and a of D. It is important to remember that *to each x in D there is assigned precisely one value $f(x)$ in E*; however, different elements of D, such as w and z in Figure 28, may have the same value in E.

The symbols

$$D \xrightarrow{f} E, \qquad f: D \to E, \qquad \text{and}$$

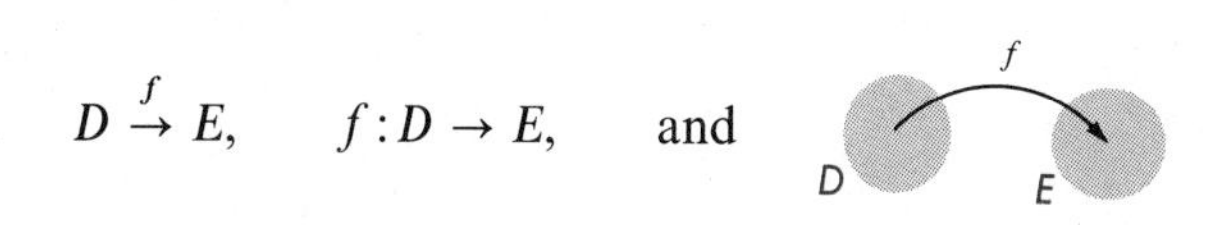

signify that f is a function from D to E. Initially the notations f and $f(x)$ may be confusing. Remember that f is used to represent the function. It is neither in D nor in E. However, $f(x)$ is an element of E—the element that f assigns to x.

Two functions f and g from D to E are **equal**, and we write

$$f = g \quad \text{provided} \quad f(x) = g(x) \quad \text{for every } x \text{ in } D.$$

For example, if $g(x) = \frac{1}{2}(2x^2 - 6) + 3$ and $f(x) = x^2$ for every x in $\mathbb{R}$, then $g = f$.

EXAMPLE ▪ 1

Let f be the function with domain $\mathbb{R}$ such that $f(x) = x^2$ for every x in $\mathbb{R}$.

(a) Find $f(-6)$, $f(\sqrt{3})$, and $f(a + b)$ for real numbers a and b.

(b) What is the range of f?

SOLUTION

(a) We find values of f by substituting for x in the equation $f(x) = x^2$:

$$f(-6) = (-6)^2 = 36$$

$$f(\sqrt{3}) = (\sqrt{3})^2 = 3$$

$$f(a + b) = (a + b)^2 = a^2 + 2ab + b^2$$

(b) By definition, the range of f consists of all numbers of the form $f(x) = x^2$ for x in $\mathbb{R}$. Since the square of every real number is nonnegative, the range is contained in the set of all nonnegative real numbers. Moreover, every nonnegative real number c is a value of f, since $f(\sqrt{c}) = (\sqrt{c})^2 = c$. Hence, the range of f is the set of all nonnegative real numbers.

If a function is defined as in Example 1, the symbols used for the function and variable are immaterial; that is, expressions such as $f(x) = x^2$, $f(s) = s^2$, $g(t) = t^2$, and $k(r) = r^2$ all define the same function. This is true because if a is any number in the domain, then the same value a^2 is obtained regardless of which expression is employed.

In the remainder of our work the phrase *f is a function* will mean that the domain and range are sets of real numbers. If a function is defined by means of an expression, as in Example 1, and the domain D is not stated explicitly, then we will consider D to be the totality of real numbers x such that $f(x)$ is real. To illustrate, if $f(x) = \sqrt{x - 2}$, then the domain is assumed to be the set of real numbers x such that $\sqrt{x - 2}$ is real, that is, $x - 2 \geq 0$, or $x \geq 2$. Thus, the domain is the infinite interval $[2, \infty)$. If x is in the domain, we say that f **is defined at** x, or that $f(x)$ **exists**. If a set S is contained in the domain, f **is defined on** S. The terminology f **is undefined at** x means that x is not in the domain of f.

EXAMPLE ▪ 2

Let $g(x) = \dfrac{\sqrt{4 + x}}{1 - x}$.

(a) Find the domain of g. **(b)** Find $g(5)$, $g(-2)$, $g(-a)$, $-g(a)$.

SOLUTION

(a) The expression $g(x)$ is a real number if and only if the radicand $4 + x$ is nonnegative and the denominator $1 - x$ is not equal to 0. Thus, $g(x)$ exists if and only if

$$4 + x \geq 0 \quad \text{and} \quad 1 - x \neq 0,$$

or, equivalently,

$$x \geq -4 \quad \text{and} \quad x \neq 1.$$

We may express the domain in terms of intervals as $[-4, 1) \cup (1, \infty)$.

(b) To find values of g, we substitute for x:

$$g(5) = \frac{\sqrt{4+5}}{1-5} = \frac{\sqrt{9}}{-4} = -\frac{3}{4}$$

$$g(-2) = \frac{\sqrt{4+(-2)}}{1-(-2)} = \frac{\sqrt{2}}{3}$$

$$g(-a) = \frac{\sqrt{4+(-a)}}{1-(-a)} = \frac{\sqrt{4-a}}{1+a}$$

$$-g(a) = -\frac{\sqrt{4+a}}{1-a} = \frac{\sqrt{4+a}}{a-1}$$

FIGURE 29

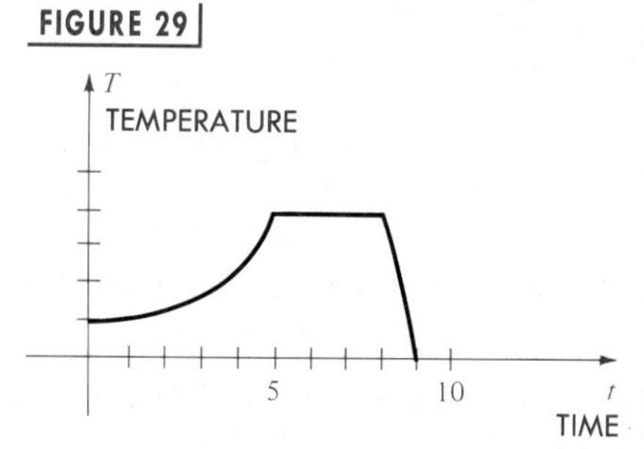

We often use graphs to describe the variation of physical quantities. For example, a scientist may use the graph in Figure 29 to indicate the temperature T of a certain solution at various times t during an experiment. The sketch shows that the temperature increased gradually from time $t = 0$ to time $t = 5$, did not change between $t = 5$ and $t = 8$, and then decreased rapidly from $t = 8$ to $t = 9$. This visual aid reveals the variation of T more clearly than a long table of numerical values would.

Similarly, if f is a function, we may use a graph to indicate the change in $f(x)$ as x varies through the domain of f. Specifically, we have the following definition.

DEFINITION

> The **graph of a function** f is the graph of the equation $y = f(x)$ for x in the domain of f.

We often attach the label $y = f(x)$ to a sketch of the graph. If $P(a, b)$ is a point on the graph, then the y-coordinate b is the function value $f(a)$, as

FIGURE 30

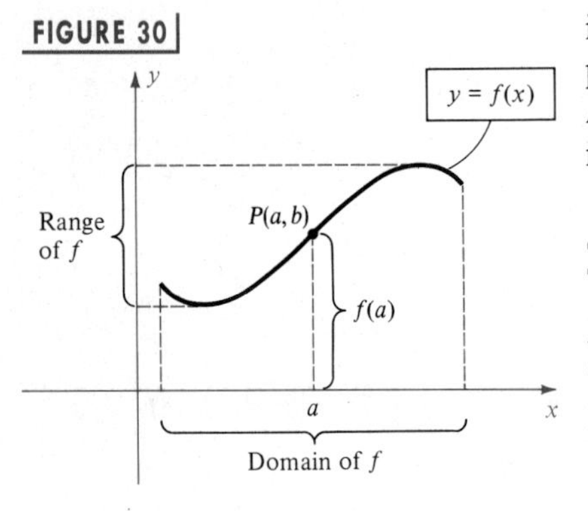

illustrated in Figure 30. The figure exhibits the domain of f (the set of possible values of x) and the range of f (the corresponding values of y). Although we have pictured the domain and range as closed intervals, they may be infinite intervals or other sets of real numbers.

It is important to note that since there is a unique value $f(a)$ for each a in the domain, only *one* point on the graph of f has x-coordinate a. Thus, *every vertical line intersects the graph of a function in at most one point*. Consequently, the graph of a function cannot be a figure such as a circle, in which a vertical line may intersect the graph in two or more points.

The x-intercepts of the graph of a function f are the solutions of the equation $f(x) = 0$. These numbers are the **zeros** of the function. The y-intercept of the graph is $f(0)$, if it exists.

EXAMPLE ▪ 3

Let $f(x) = \sqrt{x - 1}$.

(a) Sketch the graph of f. **(b)** Find the domain and range of f.

SOLUTION

(a) By definition, the graph of f is the graph of the equation $y = \sqrt{x - 1}$. The following table lists coordinates of several points on the graph.

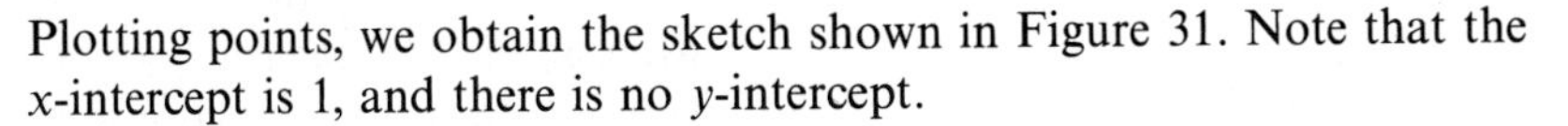

x	1	2	3	4	5	6
y	0	1	$\sqrt{2} \approx 1.4$	$\sqrt{3} \approx 1.7$	2	$\sqrt{5} \approx 2.2$

FIGURE 31

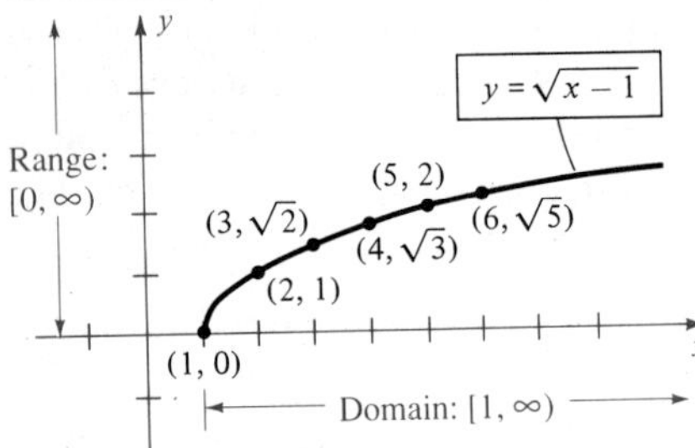

Plotting points, we obtain the sketch shown in Figure 31. Note that the x-intercept is 1, and there is no y-intercept.

(b) Referring to Figure 31, the domain of f consists of all real numbers x such that $x \geq 1$; or, equivalently, the interval $[1, \infty)$. The range of f is the set of all real numbers y such that $y \geq 0$ or, equivalently, $[0, \infty)$.

In Example 3, as x increases, the function value $f(x)$ also increases, and the graph of f *rises* (see Figure 31). A function of this type is said to be *increasing*. For certain functions $f(x)$ decreases as x increases. In this case the graph *falls* and f is a *decreasing* function. In general, we shall consider functions that increase or decrease on certain intervals, as in the following definition.

FIGURE 32

(i) Increasing function

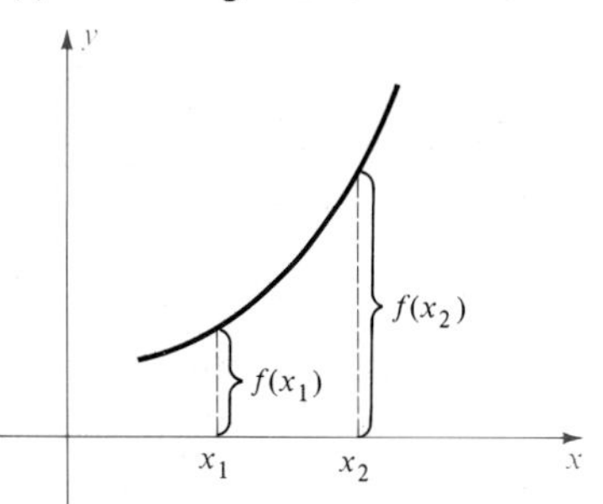

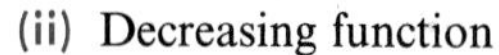

(ii) Decreasing function

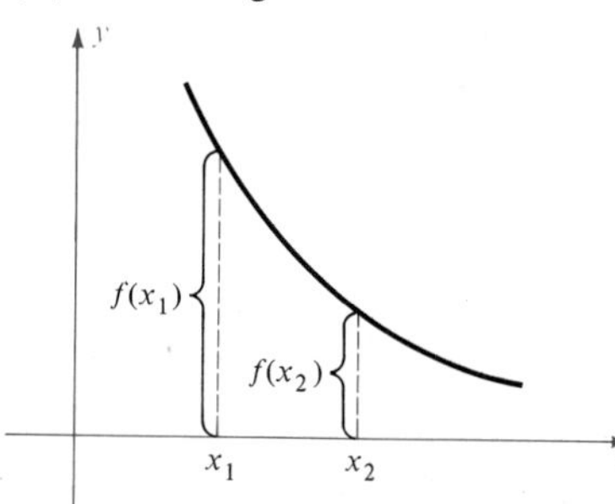

(iii) Constant function

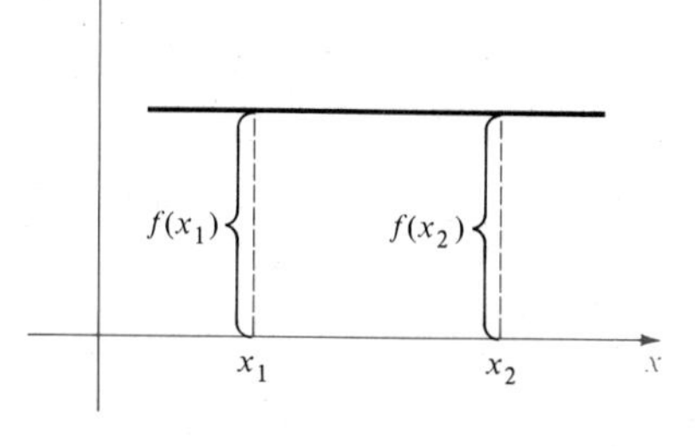

DEFINITION

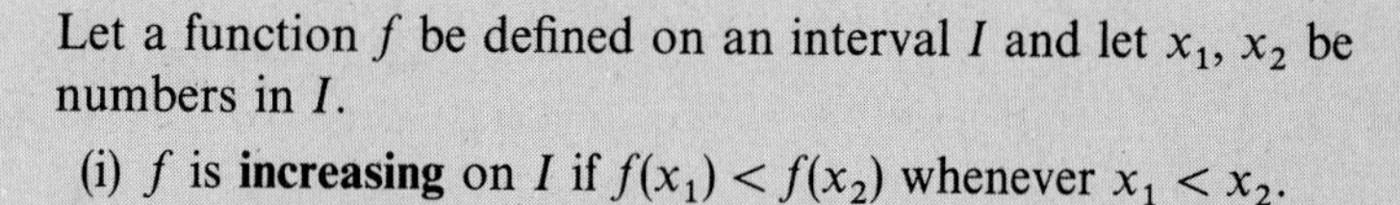

Let a function f be defined on an interval I and let x_1, x_2 be numbers in I.

(i) f is **increasing** on I if $f(x_1) < f(x_2)$ whenever $x_1 < x_2$.

(ii) f is **decreasing** on I if $f(x_1) > f(x_2)$ whenever $x_1 < x_2$.

(iii) f is **constant** on I if $f(x_1) = f(x_2)$ for every x_1 and x_2.

Illustrations of this definition are shown in Figure 32, where the interval I is not indicated. Note that if f is constant, then the graph is part of a horizontal line. If $f(x) = c$ for every x, then f is called a **constant function**.

We shall use the phrases *f is increasing* and *$f(x)$ is increasing* interchangeably. This will also be done for the terms *decreasing* and *constant*.

EXAMPLE ▪ 4

Let $f(x) = \sqrt{9 - x^2}$.

(a) Sketch the graph of f.

(b) Find the domain and range of f.

(c) Find the intervals on which f is increasing or decreasing.

SOLUTION

(a) By definition, the graph of f is the graph of the equation $y = \sqrt{9 - x^2}$. We know from our work with circles in Section 3.2 that the graph of $x^2 + y^2 = 9$ is a circle of radius 3 with center at the origin. Solving the equation $x^2 + y^2 = 9$ for y gives us $y = \pm\sqrt{9 - x^2}$. It follows that the graph of f is the *upper half* of the circle, as illustrated in Figure 33.

FIGURE 33

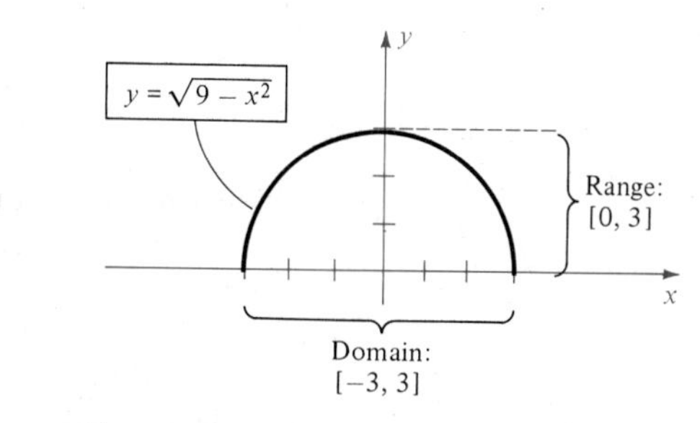

(b) Referring to Figure 33, we see that the domain of f is the closed interval $[-3, 3]$, and the range of f is the interval $[0, 3]$.

(c) The graph rises as x increases from -3 to 0, and hence f is increasing on the interval $[-3, 0]$. The graph falls as x increases from 0 to 3, and hence f is decreasing on $[0, 3]$.

EXAMPLE ▪ 5

Let $f(x) = 2x + 3$.

(a) Sketch the graph of f.

(b) Find the domain and range of f.

(c) Determine where f is increasing or decreasing.

SOLUTION

(a) The graph of f is the graph of the equation $y = 2x + 3$ and hence, from Section 3.3, is a line having slope 2 and y-intercept 3, as illustrated in Figure 34.

FIGURE 34

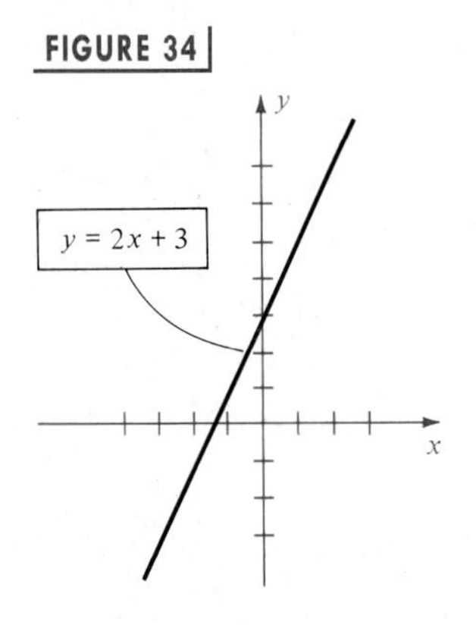

(b) Since the values of x and y may be any real numbers, both the domain and range of f are $\mathbb{R}$.

(c) As x increases, the graph rises; that is, $f(x_1) < f(x_2)$ *whenever* $x_1 < x_2$. Hence f is increasing throughout its domain.

If, as illustrated in Example 5, $f(x) = ax + b$ for some constants a and b, then the graph of f is a line, and f is called a **linear function**.

Many formulas that occur in mathematics and the sciences determine functions. For instance, the formula $A = \pi r^2$ for the area A of a circle of radius r assigns to each positive real number r a unique value of A. This determines a function f such that $f(r) = \pi r^2$, and we may write $A = f(r)$. The letter r, which represents an arbitrary number from the domain of f, is an **independent variable**. The letter A, which represents a number from the range of f, is a **dependent variable**, since its value depends on the number assigned to r. If two variables r and A are related in this manner, we say that *A is a function of r*. As another example, if an automobile travels at a uniform rate of 50 mi/hr, then the distance d (miles) traveled in time t (hours) is given by $d = 50t$, and hence the distance *d is a function of time t.*

EXAMPLE ▪ 6

A steel storage tank for propane gas is to be constructed in the shape of a right circular cylinder of altitude 10 feet with a hemisphere attached to each end. The radius r is yet to be determined. Express the volume V of the tank as a function of r.

FIGURE 35

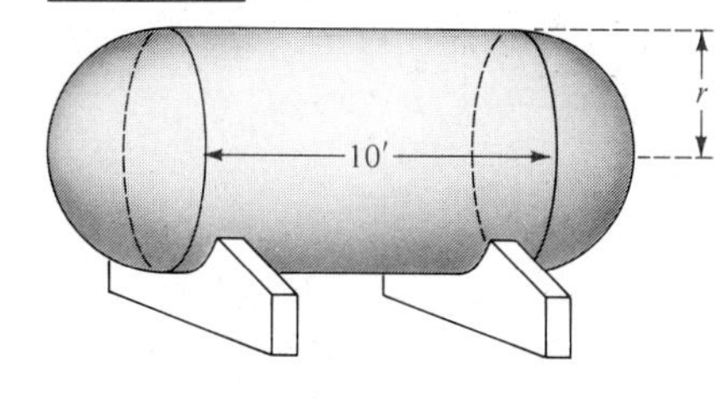

SOLUTION The tank is sketched in Figure 35. We may find the volume of the cylindrical part of the tank by multiplying the altitude 10 by the area πr^2 of the base of the cylinder. This gives us

$$\text{volume of cylinder} = 10(\pi r^2) = 10\pi r^2.$$

The two hemispherical ends, taken together, form a sphere of radius r. Using the formula for the volume of a sphere, we obtain

$$\text{volume of the two ends} = \tfrac{4}{3}\pi r^3.$$

Thus, the volume V of the tank is

$$V = \tfrac{4}{3}\pi r^3 + 10\pi r^2.$$

This formula expresses V as a function of r. In factored form,

$$V = \tfrac{1}{3}\pi r^2(4r + 30) = \tfrac{2}{3}\pi r^2(2r + 15).$$

EXAMPLE ▪ 7

Two ships leave port at the same time, one sailing west at a rate of 17 mi/hr and the other sailing south at 12 mi/hr. If t is the time (in hours) after their departure, express the distance d between the ships as a function of t.

FIGURE 36

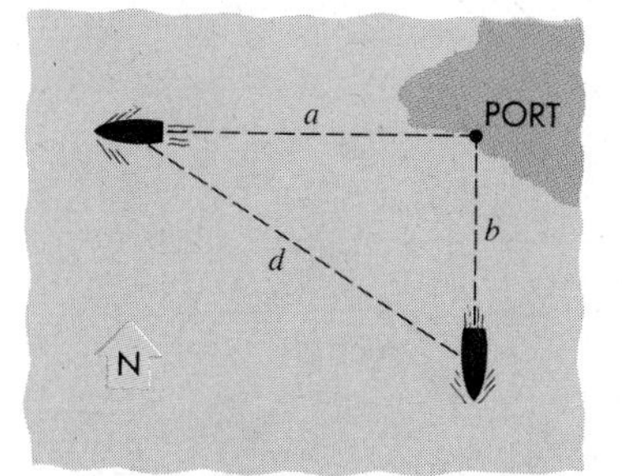

SOLUTION To help visualize the problem we begin by drawing a picture and labeling it, as in Figure 36. Using the Pythagorean theorem,

$$d^2 = a^2 + b^2, \quad \text{or} \quad d = \sqrt{a^2 + b^2}.$$

Since distance = (rate)(time) and the rates are 17 and 12, respectively,

$$a = 17t \quad \text{and} \quad b = 12t.$$

Substitution in $d = \sqrt{a^2 + b^2}$ gives us

$$d = \sqrt{(17t)^2 + (12t)^2} = \sqrt{289t^2 + 144t^2} = \sqrt{433t^2} = \sqrt{433}\,t$$

and

$$d \approx (20.8)t.$$

3.4 EXERCISES

1 If $f(x) = -x^2 - x - 4$, find $f(-2)$, $f(0)$, and $f(4)$.

2 If $f(x) = -x^3 - x^2 + 3$, find $f(-3)$, $f(0)$, and $f(2)$.

3 If $f(x) = \sqrt{x-4} - 3x$, find $f(4)$, $f(8)$, and $f(13)$.

4 If $f(x) = \dfrac{x}{x-3}$, find $f(-2)$, $f(0)$, and $f(3)$.

Exer. 5–8: If a and h are real numbers, find:

(a) $f(a)$ (b) $f(-a)$

(c) $-f(a)$ (d) $f(a+h)$

(e) $f(a) + f(h)$ (f) $\dfrac{f(a+h) - f(a)}{h}$, provided $h \neq 0$

5 $f(x) = 5x - 2$ 6 $f(x) = 3 - 4x$

7 $f(x) = x^2 - x + 3$ 8 $f(x) = 2x^2 + 3x - 7$

Exer. 9–12: If a is a positive real number, find:

(a) $g\left(\dfrac{1}{a}\right)$ (b) $\dfrac{1}{g(a)}$ (c) $g(\sqrt{a})$ (d) $\sqrt{g(a)}$

9 $g(x) = 4x^2$ 10 $g(x) = 2x - 5$

11 $g(x) = \dfrac{2x}{x^2 + 1}$ 12 $g(x) = \dfrac{x^2}{x+1}$

Exer. 13–20: Find the domain of f.

13 $f(x) = \sqrt{2x+7}$ 14 $f(x) = \sqrt{8-3x}$

15 $f(x) = \sqrt{9 - x^2}$ 16 $f(x) = \sqrt{x^2 - 25}$

17 $f(x) = \dfrac{x+1}{x^3 - 4x}$ 18 $f(x) = \dfrac{4x}{6x^2 + 13x - 5}$

19 $f(x) = \dfrac{\sqrt{2x-3}}{x^2 - 5x + 4}$ 20 $f(x) = \dfrac{\sqrt{4x-3}}{x^2 - 4}$

Exer. 21–30:

(a) Sketch the graph of f.

(b) Find the domain D and range R of f.

(c) Find the intervals on which f is increasing, decreasing, or constant.

21 $f(x) = 3x - 2$ 22 $f(x) = -2x + 3$

23 $f(x) = 4 - x^2$ 24 $f(x) = x^2 - 1$

25 $f(x) = \sqrt{x+4}$ 26 $f(x) = \sqrt{4-x}$

27 $f(x) = -2$ 28 $f(x) = 3$

29 $f(x) = 4/x$ 30 $f(x) = 1/x^2$

31 An open box is to be made from a rectangular piece of cardboard having dimensions 20 inches × 30 inches by cutting out identical squares of area x^2 from each corner and turning up the sides (see figure). Express the volume V of the box as a function of x.

EXERCISE 31

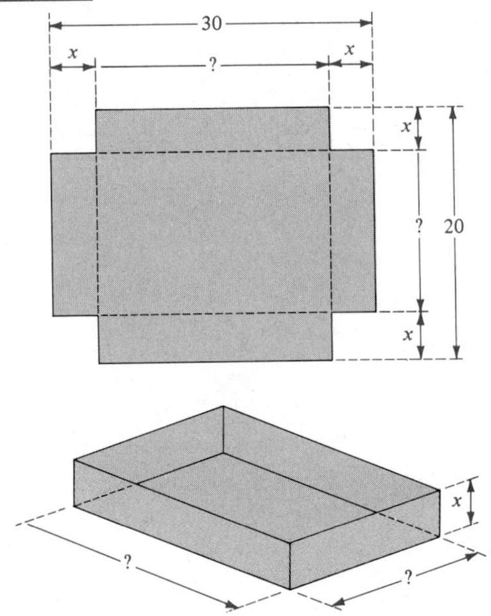

32 An aquarium of height 1.5 feet is to have a volume of 6 ft³. Let x denote the length of the base, and let y denote the width (see figure).

(a) Express y as a function of x.

EXERCISE 32

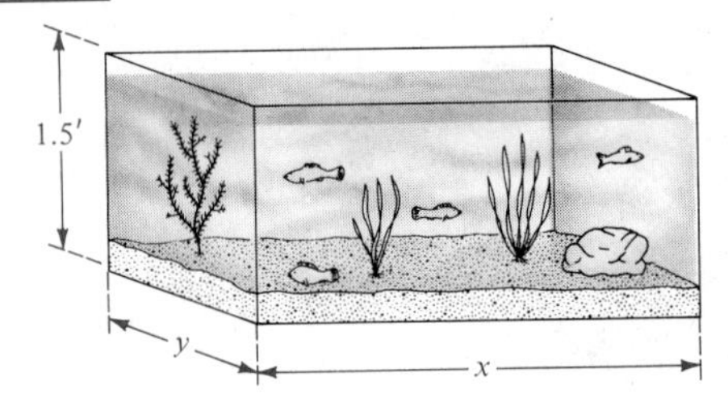

(b) Express the total number of square feet of glass needed as a function of x.

33 A small office building is to contain 500 ft^2 of floor space. The simple floor plans are shown in the figure.

(a) Express the length y of the building as a function of the width x.

(b) If the walls cost \$100 per running foot, express the cost C of the walls as a function of the width x. (Disregard the wall space above the doors.)

EXERCISE 33

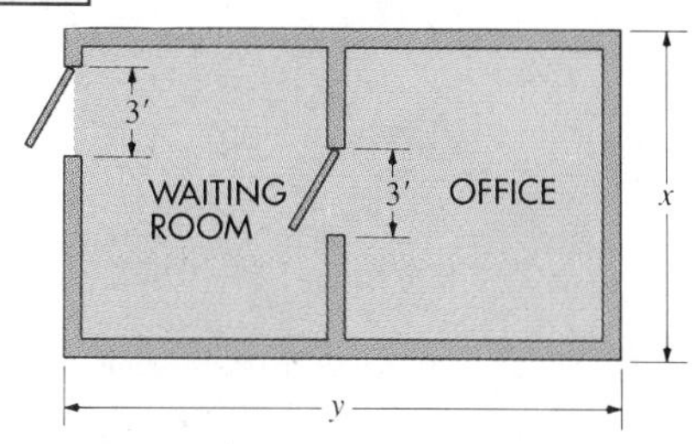

34 Refer to Example 6. A steel storage tank for propane gas is to be constructed in the shape of a right circular cylinder of altitude 10 feet with a hemisphere attached to each end. The radius r is yet to be determined. Express the surface area S of the tank as a function of r.

35 A hot-air balloon is released at 1:00 P.M. and rises vertically at a rate of 2 m/sec. An observation point is situated 100 meters from a point on the ground directly below the balloon (see figure). If t denotes the time (in seconds) after 1:00 P.M., express the distance d between the balloon and the observation point as a function of t.

EXERCISE 35

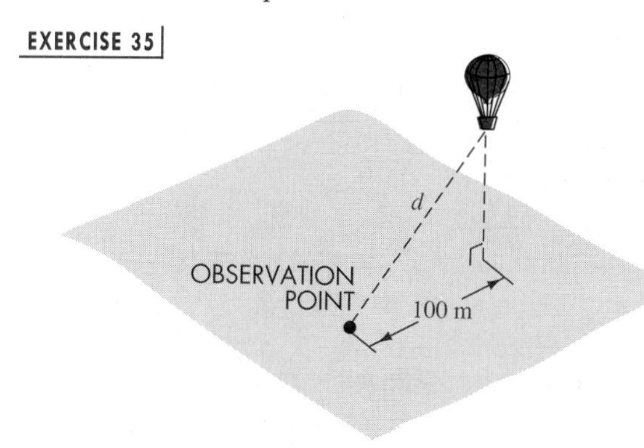

36 Triangle ABC is inscribed in a semicircle of diameter 15 (see figure).

(a) If x denotes the length of side AC, express the length y of side BC as a function of x. (*Hint:* Angle ACB is a right angle.)

(b) Express the area of triangle ABC as a function of x, and find the domain of this function.

EXERCISE 36

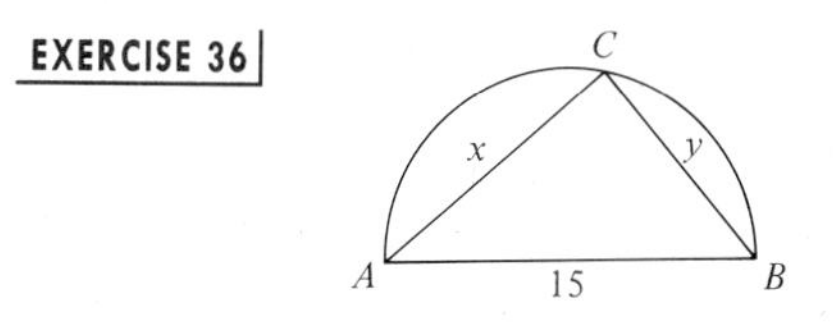

37 From an exterior point P that is h units from a circle of radius r, a tangent line is drawn to the circle (see figure). Let y denote the distance from the point P to the point of tangency T.

(a) Express y as a function of h. (*Hint:* If C is the center of the circle, then PT is perpendicular to CT.)

(b) If r is the radius of the earth, and h is the altitude of a space shuttle, then we can derive a formula for the maximum distance (to the earth) that an astronaut can see from the shuttle. In particular, if $h = 200$ miles and $r \approx 4000$ miles, approximate y.

EXERCISE 37

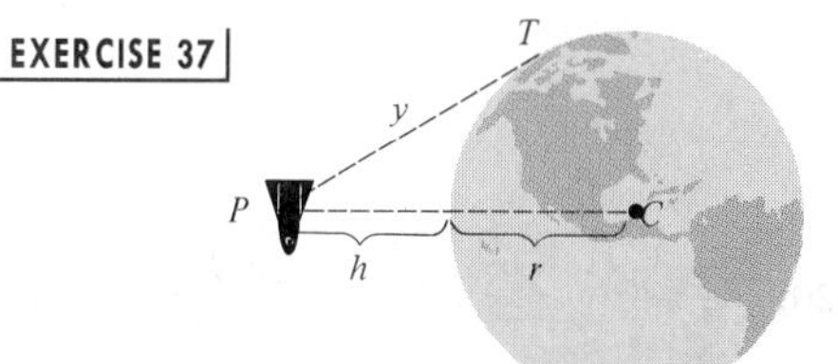

38 The figure illustrates the apparatus for a tightrope walker. Two poles are set 50 feet apart, but the point of attachment P for the rope is yet to be determined.

(a) Express the length L of the rope as a function of the distance x from point P to the ground.

(b) If the total walk is to be 75 feet, determine the height of the point of attachment P.

EXERCISE 38

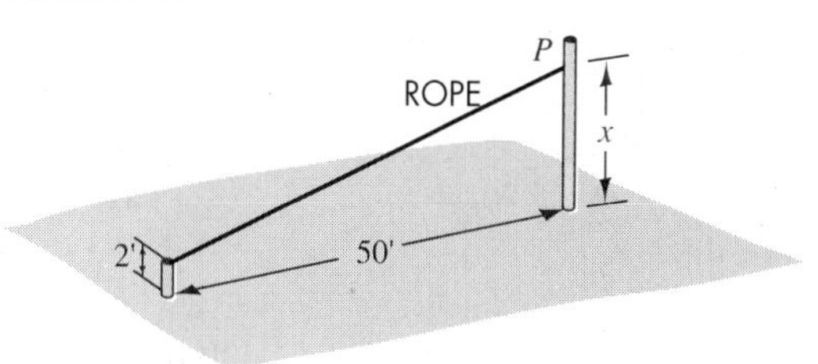

39 The relative positions of an aircraft runway and a 20-foot-tall control tower are shown in the figure. The beginning of the runway is at a perpendicular distance of 300 feet from the base of the tower. If x denotes the distance an airplane has moved down the runway, express the distance d between the airplane and the control booth as a function of x.

EXERCISE 39

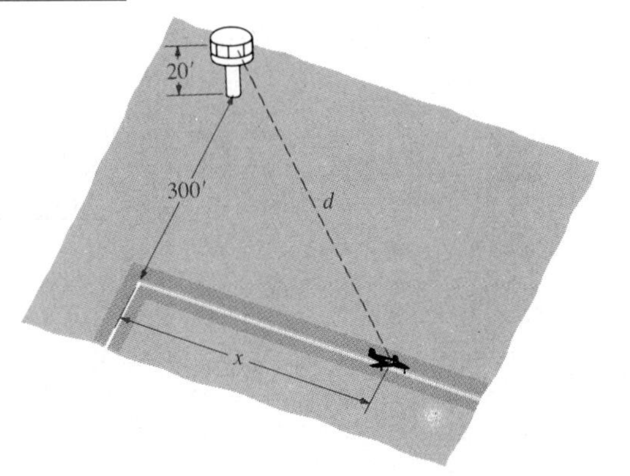

40 A man in a rowboat that is 2 miles from the nearest point A on a straight shoreline wishes to reach a house located at a point B that is 6 miles further downshore (see figure). He plans to row to a point P that is between A and B and is x miles from the house, and then walk the remainder of the distance. Suppose he can row at a rate of 3 mi/hr and can walk at a rate of 5 mi/hr. If T is the total time required to reach the house, express T as a function of x.

EXERCISE 40

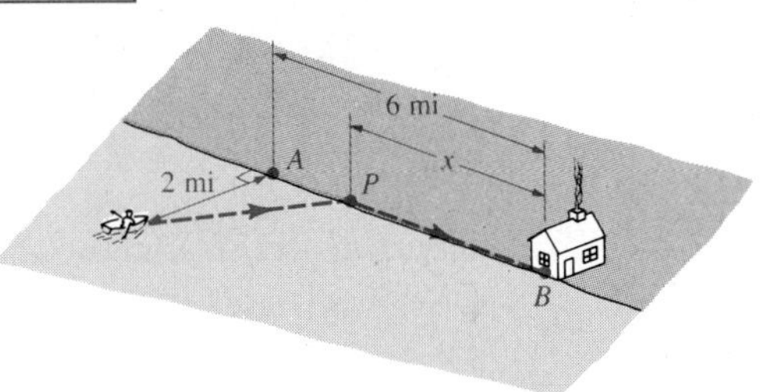

3.5 GRAPHS OF FUNCTIONS

In this section we discuss aids for sketching the graphs of certain types of functions. Let us begin with the following definition.

DEFINITION

Let f be a function such that $-x$ is in the domain D whenever x is in D.

(i) f is **even** if $f(-x) = f(x)$ for every x in D.

(ii) f is **odd** if $f(-x) = -f(x)$ for every x in D.

EXAMPLE ▪ 1

Determine if f is even, odd, or neither even nor odd:

(a) $f(x) = 3x^4 - 2x^2 + 5$ (b) $f(x) = 2x^5 - 7x^3 + 4x$

(c) $f(x) = x^3 + x^2$

SOLUTION Let x be any real number.

(a)
$$f(-x) = 3(-x)^4 - 2(-x)^2 + 5$$
$$= 3x^4 - 2x^2 + 5 = f(x)$$

Hence, f is even.

(b)
$$f(-x) = 2(-x)^5 - 7(-x)^3 + 4(-x)$$
$$= -2x^5 + 7x^3 - 4x$$
$$= -(2x^5 - 7x^3 + 4x) = -f(x)$$

Thus, f is odd.

(c)
$$f(-x) = (-x)^3 + (-x)^2 = -x^3 + x^2$$

Since $f(-x) \neq f(x)$ and $f(-x) \neq -f(x)$, the function f is neither even nor odd.

The next theorem gives useful information about the graphs of even and odd functions.

THEOREM ON SYMMETRY

(i) The graph of an even function is symmetric with respect to the y-axis.

(ii) The graph of an odd function is symmetric with respect to the origin.

PROOF If f is even, then $f(-x) = f(x)$, and hence the equation $y = f(x)$ is not changed if $-x$ is substituted for x. Statement (i) now follows from symmetry test (i) of Section 3.2. Statement (ii) is proved in similar fashion using symmetry test (ii). ❑

EXAMPLE ▪ 2

Let $f(x) = |x|$.

(a) Sketch the graph of f.

(b) Find the domain and range of f.

(c) Find the intervals on which f is increasing or decreasing.

SOLUTION

(a) Since $|-x| = |x|$, we see that $f(-x) = f(x)$. Hence, f is an even function and by the preceding theorem, the graph is symmetric with respect to the y-axis. If $x \geq 0$, then $f(x) = x$ and hence the points (x, x) in the first quadrant are on the graph of f. Some special cases are $(0, 0)$, $(1, 1)$,

FIGURE 37

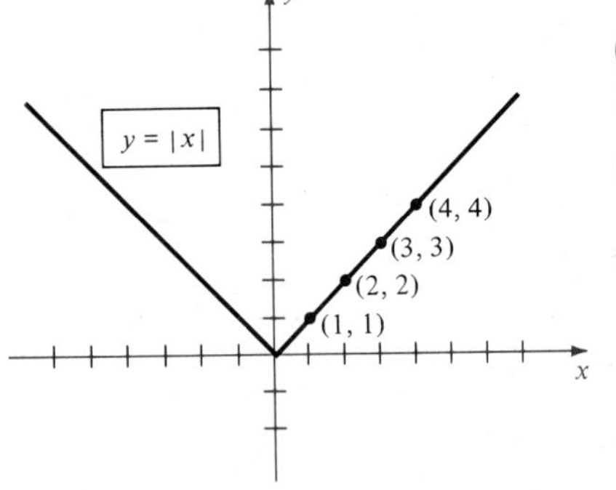

(2, 2), (3, 3), and (4, 4). Plotting points and using symmetry leads to the sketch in Figure 37.

(b) Referring to the graph, we see that the domain of f is $\mathbb{R}$, and the range is $[0, \infty)$.

(c) The function is decreasing on $(-\infty, 0]$ and increasing on $[0, \infty)$.

EXAMPLE ▪ 3

If $f(x) = \dfrac{1}{x}$, sketch the graph of f.

SOLUTION The domain of f is the set of all nonzero real numbers. The function is odd, since

$$f(-x) = \frac{1}{-x} = -\frac{1}{x} = -f(x).$$

Hence the graph is symmetric with respect to the origin. If x is positive, so is $f(x)$, and thus no part of the graph lies in quadrant IV. Coordinates of several points on the graph of $y = 1/x$ for $x > 0$ are listed in the following table.

FIGURE 38

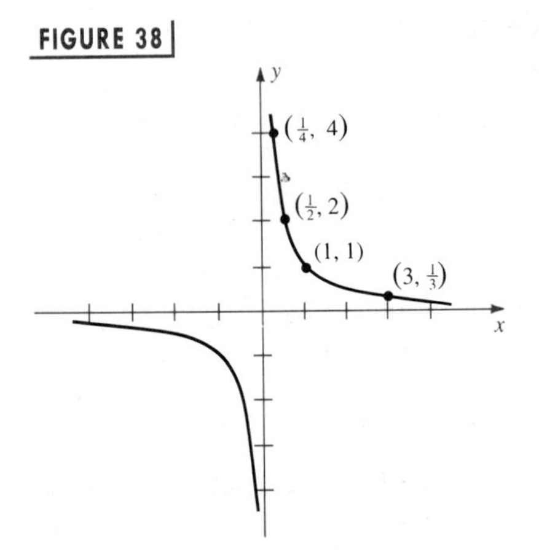

x	$\frac{1}{4}$	$\frac{1}{2}$	1	2	3	4
y	4	2	1	$\frac{1}{2}$	$\frac{1}{3}$	$\frac{1}{4}$

Plotting points leads to part of the graph in quadrant I (see Figure 38). Using symmetry with respect to the origin, we obtain the points $(-\frac{1}{4}, -4)$, $(-\frac{1}{2}, -2)$, $(-1, -1)$, . . . , which give us part of the graph in quadrant III.

Note that if x is close to zero and $x > 0$, then $1/x$ is large. As x increases through positive values, $1/x$ decreases and is close to zero when x is large. Functions of this type (*rational functions*) will be discussed in detail in Section 4.6.

If f is a function and we know the graph of $y = f(x)$, it is easy to obtain the graphs of

$$y = f(x) + c \quad \text{and} \quad y = f(x) - c$$

for any positive real number c. As illustrated in Figure 39, for $y = f(x) + c$ we merely add c to the y-coordinate of each point on the graph of $y = f(x)$. This *shifts* the graph of f *upward* a distance c. For $y = f(x) - c$ with

$c > 0$, we subtract c from each y-coordinate, thereby shifting the graph of f a distance c *downward*, as shown in Figure 40.

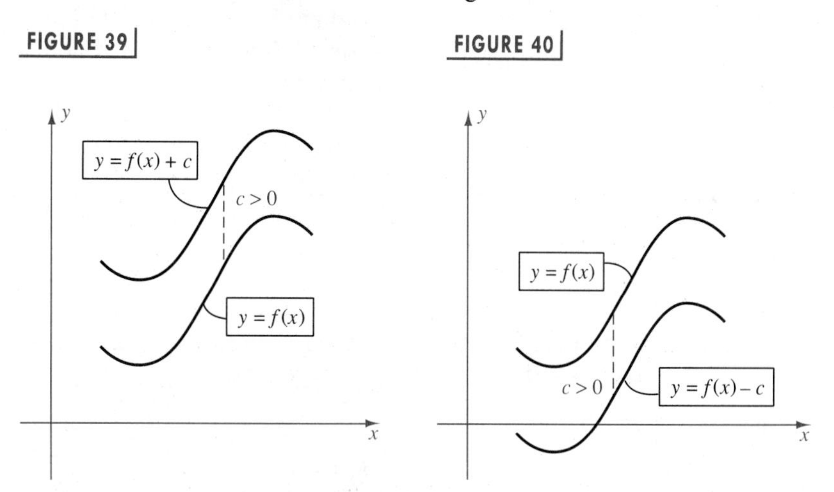

These **vertical shifts** of graphs of functions are summarized in the next box.

VERTICAL SHIFTS OF GRAPHS ($c > 0$)

To obtain the graph of:	shift the graph of $y = f(x)$:
$y = f(x) + c$	c units upward
$y = f(x) - c$	c units downward

EXAMPLE ▪ 4

Let $f(x) = x^2 + c$. Sketch the graph of f if:

(a) $c = 4$ **(b)** $c = -2$

FIGURE 41

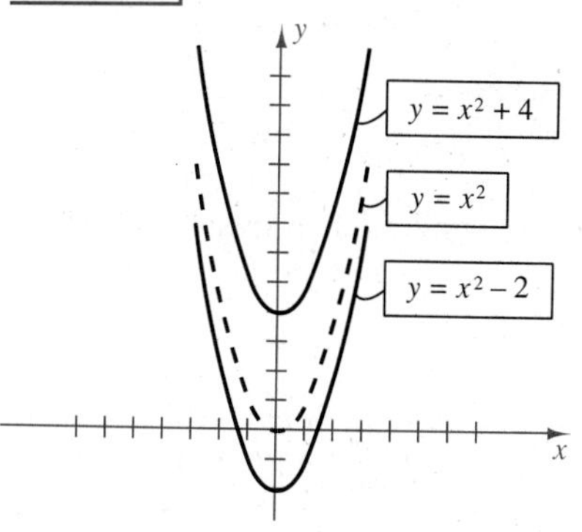

SOLUTION We shall sketch both graphs on the same coordinate plane. The graph of $y = x^2$ was sketched in Figure 9, and, for reference, is represented by dashes in Figure 41.

(a) To find the graph of $y = x^2 + 4$ we add 4 to the y-coordinate of each point on the graph of $y = x^2$. This amounts to shifting the graph of $y = x^2$ upward 4 units, as shown in the figure.

(b) For $c = -2$ we decrease y-coordinates by 2 and, hence, the graph of $y = x^2 - 2$ may be obtained by shifting the graph of $y = x^2$ downward 2 units.

Each graph is a parabola symmetric with respect to the y-axis.

FIGURE 42

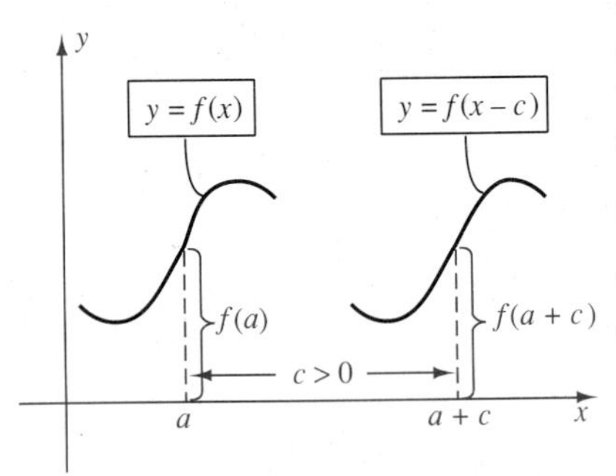

We can also consider **horizontal shifts** of graphs. Specifically, if $c > 0$, consider the graphs of $y = f(x)$ and $y = f(x - c)$ sketched on the same coordinate plane, as illustrated in Figure 42. Since $f(a) = f([a + c] - c)$, we see that the point with x-coordinate a on the graph of $y = f(x)$ has the same y-coordinate as the point with x-coordinate $a + c$ on the graph of $y = f(x - c)$. This implies that the graph of $y = f(x - c)$ can be obtained by shifting the graph of $y = f(x)$ *to the right* c units. Similarly, the graph of $y = f(x + c)$ can be obtained by shifting the graph of f *to the left* c units, as shown in Figure 43. These rules are listed for reference in the next box.

HORIZONTAL SHIFTS OF GRAPHS ($c > 0$)

To obtain the graph of:	shift the graph of $y = f(x)$:
$y = f(x - c)$	c units to the right
$y = f(x + c)$	c units to the left

FIGURE 43

Horizontal and vertical shifts are also referred to as *translations*.

EXAMPLE ▪ 5

Sketch the graph of f: **(a)** $f(x) = (x - 4)^2$ **(b)** $f(x) = (x + 2)^2$

SOLUTION The graph of $y = x^2$ is sketched with dashes in Figure 44.

(a) According to the rules for horizontal shifts, shifting the graph of $y = x^2$ to the right 4 units gives us the graph of $y = (x - 4)^2$ shown in the figure.

(b) Shifting the graph of $y = x^2$ to the left 2 units leads to the graph of $y = (x + 2)^2$ shown in the figure.

FIGURE 44

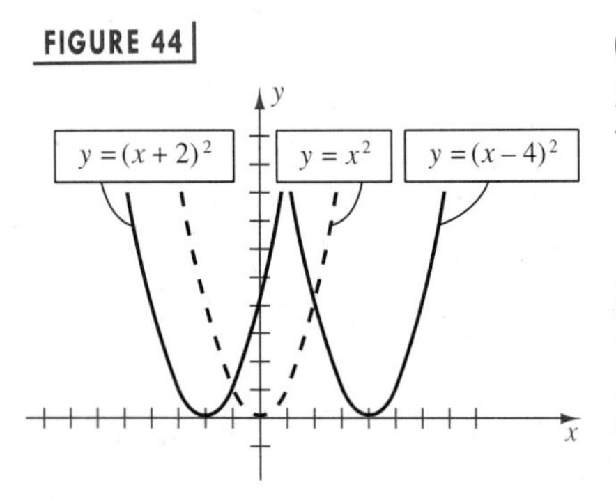

To obtain the graph of $y = cf(x)$ for some real number c, we may *multiply* the y-coordinates of points on the graph of $y = f(x)$ by c. For example, if $y = 2f(x)$, we double y-coordinates, or if $y = \frac{1}{2}f(x)$, we multiply each y-coordinate by $\frac{1}{2}$. If $c > 0$ (and $c \neq 1$), we shall refer to this procedure as **stretching** the graph of $y = f(x)$.

EXAMPLE ▪ 6

Sketch the graph of the equation: **(a)** $y = 4x^2$ **(b)** $y = \frac{1}{4}x^2$

FIGURE 45

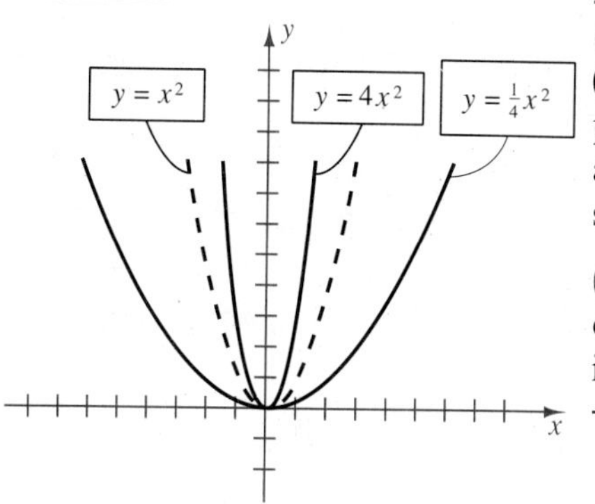

SOLUTION

(a) To sketch the graph of $y = 4x^2$ we may refer to the graph of $y = x^2$ (shown with dashes in Figure 45), and multiply the y-coordinate of each point by 4. This gives us a narrower parabola that is sharper at the vertex, as illustrated in the figure. To obtain the correct shape, we should plot several points, such as $(0, 0)$, $(\frac{1}{2}, 1)$, and $(1, 4)$.

(b) The graph of $y = \frac{1}{4}x^2$ may be sketched by multiplying y-coordinates of points on the graph of $y = x^2$ by $\frac{1}{4}$. The graph is a wider parabola that is flatter at the vertex, as shown in Figure 45.

We may obtain the graph of $y = -f(x)$ by multiplying the y-coordinate of each point on the graph of $y = f(x)$ by -1. Thus, every point (a, b) on the graph of $y = f(x)$ that lies above the x-axis determines a point $(a, -b)$ on the graph of $y = -f(x)$ that lies below the x-axis. Similarly, if (c, d) lies below the x-axis (that is, $d < 0$), then $(c, -d)$ lies above the x-axis. The graph of $y = -f(x)$ is a **reflection** of the graph of $y = f(x)$ through the x-axis.

FIGURE 46

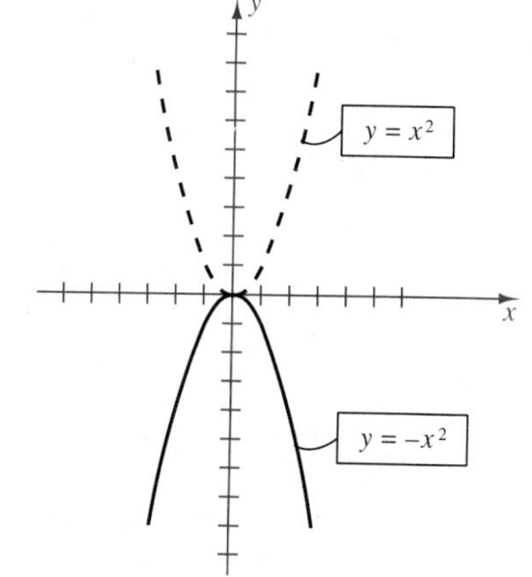

EXAMPLE ▪ 7

Sketch the graph of $y = -x^2$.

SOLUTION The graph may be found by plotting points; however, since the graph of $y = x^2$ is well known, we sketch it with dashes, as in Figure 46, and then multiply y-coordinates of points by -1. This procedure gives us the reflection through the x-axis indicated in the figure.

Functions are sometimes described by more than one expression, as in the next examples. We call such functions **piecewise-defined functions**.

FIGURE 47

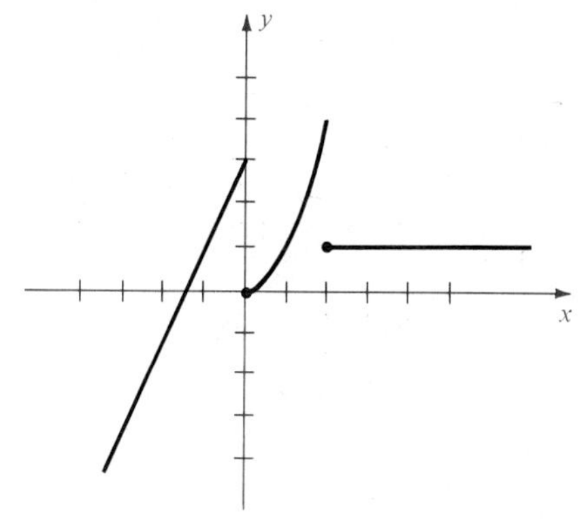

EXAMPLE ▪ 8

Sketch the graph of the function f if

$$f(x) = \begin{cases} 2x + 3 & \text{if } x < 0 \\ x^2 & \text{if } 0 \le x < 2 \\ 1 & \text{if } x \ge 2 \end{cases}$$

SOLUTION If $x < 0$, then $f(x) = 2x + 3$. Thus, if x is negative, the expression $2x + 3$ should be used to find function values. Consequently, if $x < 0$, then the graph of f coincides with the line $y = 2x + 3$, and we sketch that portion of the graph to the left of the y-axis, as indicated in Figure 47.

If $0 \le x < 2$, we use x^2 to find values of f, and therefore this part of the graph of f coincides with the graph of the parabola $y = x^2$. We then sketch the part of the graph of f between $x = 0$ and $x = 2$, as indicated in the figure.

Finally, if $x \geq 2$, the values of f are always 1. The graph of f for $x \geq 2$ is the horizontal half-line illustrated in Figure 47.

EXAMPLE ▪ 9

If x is any real number, then there exist consecutive integers n and $n + 1$ such that $n \leq x < n + 1$. Let f be the function defined as follows: If $n \leq x < n + 1$, then $f(x) = n$. Sketch the graph of f.

SOLUTION The x- and y-coordinates of some points on the graph may be listed as follows:

Values of x	$f(x)$
...	...
$-2 \leq x < -1$	-2
$-1 \leq x < 0$	-1
$0 \leq x < 1$	0
$1 \leq x < 2$	1
$2 \leq x < 3$	2
...	...

FIGURE 48

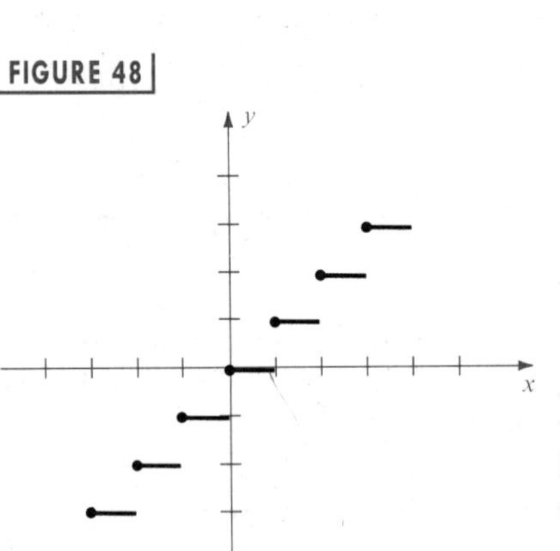

Since f is a constant function whenever x is between successive integers, the corresponding part of the graph is a segment of a horizontal line. Part of the graph is sketched in Figure 48. The graph continues indefinitely to the right and to the left.

The symbol $[\![x]\!]$ is often used to denote the largest integer n such that $n \leq x$. For example $[\![1.6]\!] = 1$, $[\![\sqrt{5}]\!] = 2$, $[\![\pi]\!] = 3$, and $[\![-3.5]\!] = -4$. Using this notation, the function f of Example 9 may be defined by $f(x) = [\![x]\!]$. We refer to f as the **greatest integer function**.

3.5 EXERCISES

Exer. 1–10: Determine if f is even, odd, or neither even nor odd.

1 $f(x) = 5x^3 + 2x$

2 $f(x) = |x| - 3$

3 $f(x) = 3x^4 + 2x^2 - 5$

4 $f(x) = 7x^5 - 4x^3$

5 $f(x) = 8x^3 - 3x^2$

6 $f(x) = 12$

7 $f(x) = \sqrt{x^2 + 4}$

8 $f(x) = 3x^2 - 5x + 1$

9 $f(x) = \sqrt[3]{x^3 - x}$

10 $f(x) = x^3 - \dfrac{1}{x}$

Exer. 11–22: Sketch, on the same coordinate plane, the graphs of f for the given values of c. (Make use of symmetry, vertical shifts, horizontal shifts, stretching, or reflecting.)

11 $f(x) = |x| + c,\quad c = -3, 1, 3$

12 $f(x) = |x - c|,\quad c = -3, 1, 3$

13 $f(x) = -x^2 + c,\quad c = -4, 2, 4$

14 $f(x) = 2x^2 - c,\quad c = -4, 2, 4$

15 $f(x) = x^3 + c,\quad c = -4, 0, 1$

16 $f(x) = -x^3 + c,\quad c = -1, 0, 4$

17 $f(x) = 2\sqrt{x} + c,\quad c = -2, 0, 2$

18 $f(x) = \sqrt{9 - x^2} + c,\quad c = -3, 0, 3$

19 $f(x) = 2\sqrt{x - c}\quad c = -2, 0, 2$

20 $f(x) = -2(x - c)^2,\quad c = -2, 0, 3$

21 $f(x) = c\sqrt{4 - x^2},\quad c = -2, 1, 3$

22 $f(x) = (x + c)^3,\quad c = -2, 1, 2$

Exer. 23–24: The graph of a function f with domain $0 \le x \le 4$ is shown in the figure. Sketch the graph of the given equation.

23

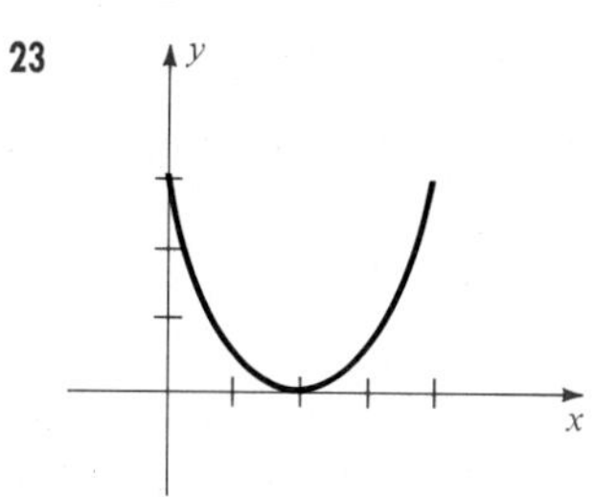

(a) $y = f(x + 3)$
(b) $y = f(x - 3)$
(c) $y = f(x) + 3$
(d) $y = f(x) - 3$
(e) $y = -3f(x)$
(f) $y = -\frac{1}{3}f(x)$
(g) $y = -f(x + 2) - 3$
(h) $y = f(x - 2) + 3$

24

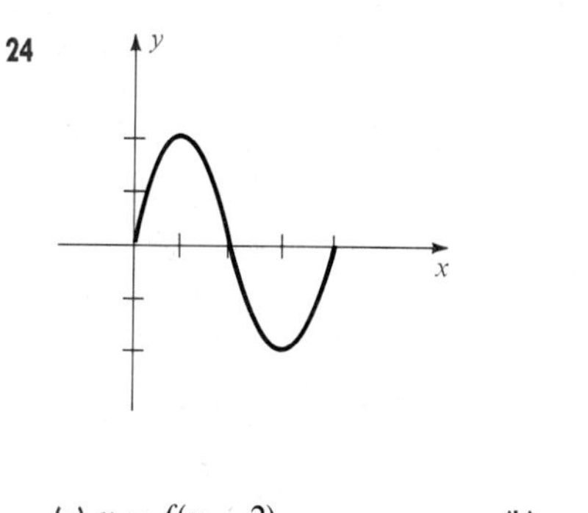

(a) $y = f(x - 2)$
(b) $y = f(x + 2)$
(c) $y = f(x) - 2$
(d) $y = f(x) + 2$
(e) $y = -2f(x)$
(f) $y = -\frac{1}{2}f(x)$
(g) $y = -f(x + 4) - 2$
(h) $y = f(x - 4) + 2$

Exer. 25–28: The graph of a function f is shown, together with graphs of three other functions (a)–(c). Use properties of symmetry, shifts, and reflecting to find equations for graphs (a)–(c) in terms of f.

25

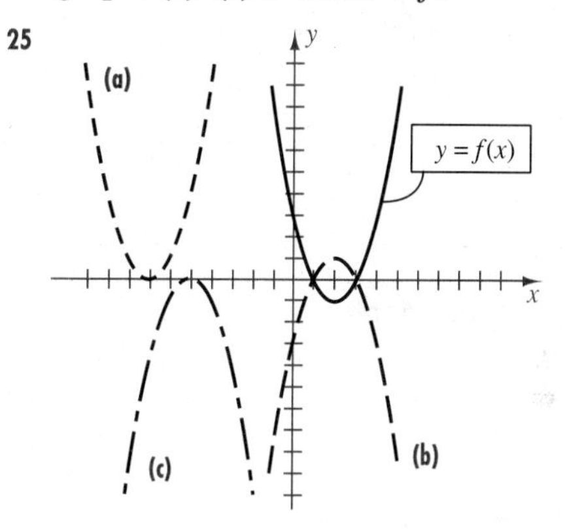

26

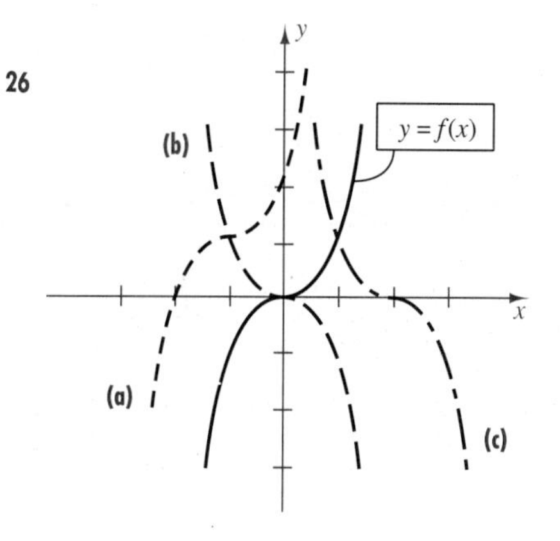

27

28

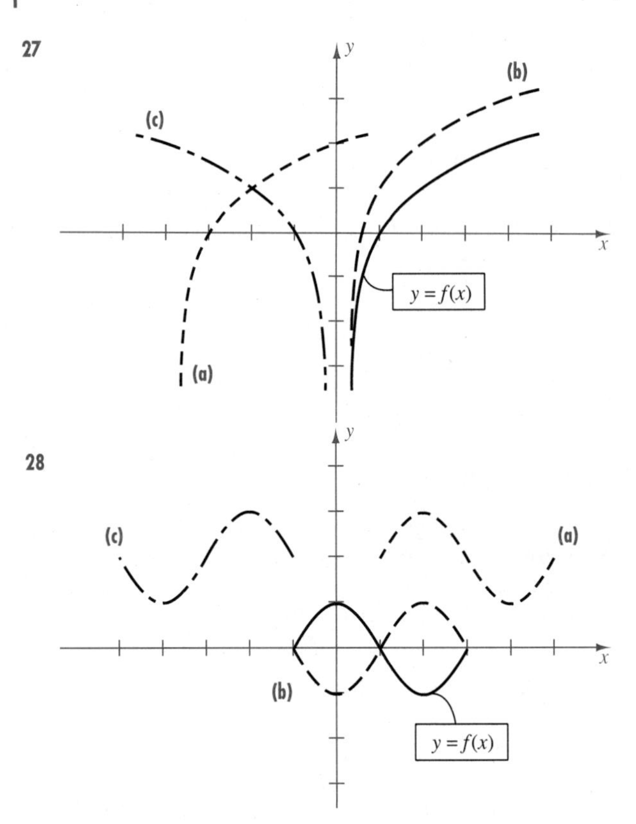

Exer. 29–36: Sketch the graph of f.

29 $f(x) = \begin{cases} 3 & \text{if } x \le -1 \\ -2 & \text{if } x > -1 \end{cases}$

30 $f(x) = \begin{cases} -1 & \text{if } x \text{ is an integer} \\ -2 & \text{if } x \text{ is not an integer} \end{cases}$

31 $f(x) = \begin{cases} 3 & \text{if } x < -2 \\ -x + 1 & \text{if } |x| \le 2 \\ -3 & \text{if } x > 2 \end{cases}$

32 $f(x) = \begin{cases} -2x & \text{if } x < -1 \\ x^2 & \text{if } -1 \le x < 1 \\ -2 & \text{if } x \ge 1 \end{cases}$

33 $f(x) = \begin{cases} x + 2 & \text{if } x \le -1 \\ x^3 & \text{if } |x| < 1 \\ -x + 3 & \text{if } x \ge 1 \end{cases}$

34 $f(x) = \begin{cases} x - 3 & \text{if } x \le -2 \\ -x^2 & \text{if } -2 < x < 1 \\ -x + 4 & \text{if } x \ge 1 \end{cases}$

35 $f(x) = \begin{cases} \dfrac{x^2 - 1}{x + 1} & \text{if } x \ne -1 \\ 2 & \text{if } x = -1 \end{cases}$

36 $f(x) = \begin{cases} \dfrac{x^2 - 4}{2 - x} & \text{if } x \ne 2 \\ 1 & \text{if } x = 2 \end{cases}$

Exer. 37–38: The symbol $[\![x]\!]$ denotes values of the greatest integer function. Sketch the graph of f.

37 (a) $f(x) = [\![x - 3]\!]$ (b) $f(x) = [\![x]\!] - 3$
(c) $f(x) = 2[\![x]\!]$ (d) $f(x) = [\![2x]\!]$

38 (a) $f(x) = [\![x + 2]\!]$ (b) $f(x) = [\![x]\!] + 2$
(c) $f(x) = \frac{1}{2}[\![x]\!]$ (d) $f(x) = [\![\frac{1}{2}x]\!]$

Explain why the graph of the equation is not the graph of a function.

39 $x = y^2$ 40 $x = -|y|$

3.6 OPERATIONS ON FUNCTIONS

Functions are often defined in terms of sums, differences, products, and quotients of various expressions. For example, if

$$h(x) = x^2 + \sqrt{5x + 1},$$

we may regard $h(x)$ as a sum of values of the simpler functions f and g

defined by

$$f(x) = x^2 \quad \text{and} \quad g(x) = \sqrt{5x + 1}.$$

We refer to the function h as the *sum* of f and g.

In general, suppose f and g are *any* functions. Let I be *the intersection of their domains*, that is, the numbers *common* to both domains. The **sum** of f and g is the function h defined by

$$h(x) = f(x) + g(x)$$

for every x in I.

It is convenient to denote h by the symbol $f + g$. Thus the value of $f + g$ at x is $f(x) + g(x)$; that is,

$$(f + g)(x) = f(x) + g(x).$$

The **difference** $f - g$ and the **product** fg of f and g are defined by

$$(f - g)(x) = f(x) - g(x) \quad \text{and} \quad (fg)(x) = f(x)g(x)$$

for x in I. The **quotient** f/g of f by g is given by

$$\left(\frac{f}{g}\right)(x) = \frac{f(x)}{g(x)}$$

for x in I and $g(x) \neq 0$.

EXAMPLE ▪ 1

If $f(x) = \sqrt{4 - x^2}$ and $g(x) = 3x + 1$, find the sum, difference, and product of f and g, and the quotient of f by g, and specify the domains.

SOLUTION The domain of f is the closed interval $[-2, 2]$ and the domain of g is $\mathbb{R}$. Consequently, the intersection of their domains is $[-2, 2]$, and we obtain:

$$(f + g)(x) = \sqrt{4 - x^2} + (3x + 1), \qquad -2 \leq x \leq 2$$

$$(f - g)(x) = \sqrt{4 - x^2} - (3x + 1), \qquad -2 \leq x \leq 2$$

$$(fg)(x) = \sqrt{4 - x^2}(3x + 1), \qquad -2 \leq x \leq 2$$

$$\left(\frac{f}{g}\right)(x) = \frac{\sqrt{4 - x^2}}{3x + 1}, \qquad -2 \leq x \leq 2 \text{ and } x \neq -\tfrac{1}{3}$$

A function f is a **polynomial function** if $f(x)$ is a polynomial (see page 26). A polynomial function may be regarded as a sum of functions whose values are of the form cx^k for a real number c and a nonnegative integer k.

An **algebraic function** is a function that can be expressed in terms of finite sums, differences, products, quotients, or roots of polynomial functions.

ILLUSTRATION ALGEBRAIC FUNCTION

- $f(x) = 5x^4 - 2\sqrt[3]{x} + \dfrac{x(x^2 + 5)}{\sqrt{x^3 + \sqrt{x}}}$

Functions that are not algebraic are **transcendental**. The exponential and logarithmic functions considered in Chapter 5 are examples of transcendental functions.

We conclude this section by describing an important method of using two functions f and g to obtain a third function. Suppose D, E, and K are sets of real numbers. Let f be a function from D to E, and let g be a function from E to K. Using arrow notation we may write

$$D \xrightarrow{f} E \xrightarrow{g} K.$$

FIGURE 49

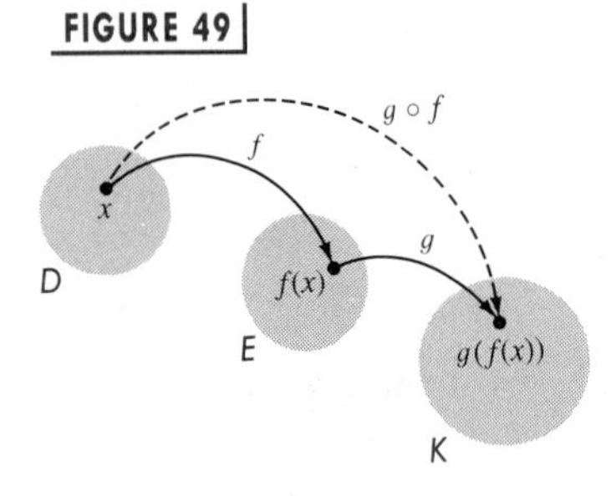

We shall use f and g to define a function from D to K.

For every x in D, the number $f(x)$ is in E. Since the domain of g is E, we may then find the number $g(f(x))$ in K. By associating $g(f(x))$ with x, we obtain a function from D to K called the *composite function* of g by f. This is illustrated in Figure 49, where the dashed arrow indicates the correspondence we have defined from D to K.

We sometimes use an operation symbol $\circ$ and denote a composite function as $g \circ f$ (read *g circle f*). The following definition summarizes our discussion.

DEFINITION

Let f be a function from D to E and let g be a function from E to K. The **composite function $g \circ f$** is the function from D to K defined by

$$(g \circ f)(x) = g(f(x))$$

for every x in D.

If f is a function from D to E, and if the domain of g is a *subset* E' of E, then the domain of $g \circ f$ consists of all numbers x in D such that $f(x)$ is in E'.

EXAMPLE ▪ 2

Let $f(x) = x^3$ and $g(x) = 5x^2 + 2x + 1$.

(a) Find $(g \circ f)(x)$. **(b)** Find the domain of $g \circ f$.

SOLUTION

(a) By definition,

$$(g \circ f)(x) = g(f(x)) = g(x^3).$$

Since $g(x^3)$ means that x^3 should be substituted for x in the expression for $g(x)$, we have

$$g(x^3) = 5(x^3)^2 + 2(x^3) + 1.$$

Consequently,
$$(g \circ f)(x) = 5x^6 + 2x^3 + 1.$$

(b) The domain of g is $\mathbb{R}$. Since the function value $f(x)$ is in $\mathbb{R}$ for every x, the domain of $g \circ f$ is $\mathbb{R}$.

EXAMPLE ▪ 3

Let $f(x) = x - 2$ and $g(x) = 5x + \sqrt{x}$.

(a) Find $(g \circ f)(x)$. **(b)** Find the domain of $g \circ f$.

SOLUTION

(a) Formal substitutions give us the following:

$$\begin{aligned}
(g \circ f)(x) &= g(f(x)) && \text{(definition of } g \circ f\text{)} \\
&= g(x - 2) && \text{(definition of } f\text{)} \\
&= 5(x - 2) + \sqrt{x - 2} && \text{(definition of } g\text{)} \\
&= 5x - 10 + \sqrt{x - 2} && \text{(simplifying)}
\end{aligned}$$

(b) The domain of g is the set of nonnegative real numbers. The function values $f(x) = x - 2$ are nonnegative if $x - 2 \geq 0$, that is, if $x \geq 2$. Hence the domain of $g \circ f$ is the interval $[2, \infty)$.

Given f and g, it may also be possible to find $(f \circ g)(x) = f(g(x))$, as illustrated in the next example.

EXAMPLE ▪ 4

Let $f(x) = x^2 - 1$ and $g(x) = 3x + 5$.

(a) Find $(f \circ g)(x)$. **(b)** Find $(g \circ f)(x)$.

SOLUTION

(a)
$$\begin{aligned} (f \circ g)(x) &= f(g(x)) && \text{(definition of } f \circ g) \\ &= f(3x + 5) && \text{(definition of } g) \\ &= (3x + 5)^2 - 1 && \text{(definition of } f) \\ &= 9x^2 + 30x + 24 && \text{(simplifying)} \end{aligned}$$

(b)
$$\begin{aligned} (g \circ f)(x) &= g(f(x)) && \text{(definition of } g \circ f) \\ &= g(x^2 - 1) && \text{(definition of } f) \\ &= 3(x^2 - 1) + 5 && \text{(definition of } g) \\ &= 3x^2 + 2 && \text{(simplifying)} \end{aligned}$$

We see from Example 4 that $f(g(x))$ and $g(f(x))$ are not always the same, that is, $f \circ g \neq g \circ f$.

In some applied problems it is necessary to express a quantity y as a function of time t. The following example illustrates that it is often easier to introduce a third variable x, then express x as a function of t—that is, $x = g(t)$. Next express y as a function of x—that is, $y = f(x)$—and finally form the composite function given by $y = f(x) = f(g(t))$.

EXAMPLE ▪ 5

A meteorologist is inflating a spherical balloon with helium gas. If the radius of the balloon is changing at a rate of 1.5 cm/sec, express the volume V of the balloon as a function of time t (in seconds).

SOLUTION Let x denote the radius of the balloon. If we assume that the radius is 0 initially, then after t seconds

$$x = 1.5t \qquad \text{(radius of balloon after } t \text{ seconds).}$$

To illustrate, after 1 second the radius is 1.5 cm; after 2 seconds, it is 3.0 cm; after 3 seconds, it is 4.5 cm; and so on.

Next we write

$$V = \tfrac{4}{3}\pi x^3 \qquad \text{(volume of a sphere of radius } x).$$

This gives us a composite function relationship in which V is a function of x, and x is a function of t. By substitution, we obtain

$$V = \tfrac{4}{3}\pi x^3 = \tfrac{4}{3}\pi(1.5t)^3 = \tfrac{4}{3}\pi(\tfrac{3}{2}t)^3 = \tfrac{4}{3}\pi(\tfrac{27}{8}t^3).$$

Simplifying, we obtain the following formula for V as a function of t:

$$V = \tfrac{9}{2}\pi t^3$$

3.6 EXERCISES

Exer. 1–6: Find the sum, difference, and product of f and g, and the quotient of f by g, and specify the domain of each.

1 $f(x) = x^2 + 2, \quad g(x) = 2x^2 - 1$

2 $f(x) = x^2 + x, \quad g(x) = x^2 - 3$

3 $f(x) = \sqrt{x + 5}, \quad g(x) = \sqrt{x + 5}$

4 $f(x) = \sqrt{3 - 2x}, \quad g(x) = \sqrt{x + 4}$

5 $f(x) = \dfrac{2x}{x - 4}, \quad g(x) = \dfrac{x}{x + 5}$

6 $f(x) = \dfrac{x}{x - 2}, \quad g(x) = \dfrac{3x}{x + 4}$

Exer. 7–24: Find (a) $(f \circ g)(x)$ and (b) $(g \circ f)(x)$.

7 $f(x) = 2x - 5, \quad g(x) = 3x + 7$

8 $f(x) = 5x + 2, \quad g(x) = 6x - 1$

9 $f(x) = 3x^2 + 4, \quad g(x) = 5x$

10 $f(x) = 3x - 1, \quad g(x) = 4x^2$

11 $f(x) = 2x^2 + 3x - 4, \quad g(x) = 2x - 1$

12 $f(x) = 5x - 7, \quad g(x) = 3x^2 - x + 2$

13 $f(x) = 4x, \quad g(x) = 2x^3 - 5x$

14 $f(x) = x^3 + 2x^2, \quad g(x) = 3x$

15 $f(x) = x^2 - 3x, \quad g(x) = \sqrt{x + 2}$

16 $f(x) = \sqrt[3]{x + 2}, \quad g(x) = x^3 - 3$

17 $f(x) = \dfrac{x}{3x + 2}, \quad g(x) = \dfrac{2}{x}$

18 $f(x) = \dfrac{x}{x - 2}, \quad g(x) = \dfrac{3}{x}$

19 $f(x) = |x|, \quad g(x) = -7$

20 $f(x) = 5, \quad g(x) = x^2$

21 $f(x) = x^2, \quad g(x) = \dfrac{1}{x^3}$

22 $f(x) = \dfrac{1}{x - 1}, \quad g(x) = x - 1$

23 $f(x) = \dfrac{3x + 5}{2}, \quad g(x) = \dfrac{2x - 5}{3}$

24 $f(x) = x^3 + 5, \quad g(x) = \sqrt[3]{x - 5}$

Exer. 25–30: Use the method of Example 5 to solve.

25 A fire has started in a dry open field and spreads in the form of a circle. If the radius of this circle increases at the rate of 6 ft/min, express the total fire area as a function of time t (in minutes).

26 A 100-foot-long cable of diameter 4 inches is submerged in seawater. Due to corrosion, the surface area of the cable decreases at the rate of 750 in.^{2} per year. Express the diameter of the cable as a function of time. (Ignore corrosion at the ends of the cable.)

27 A hot-air balloon rises vertically as a rope attached to the base of the balloon is released at the rate of 5 ft/sec (see figure). The pulley that releases the rope is 20 feet from a platform where passengers board the balloon. Express the height of the balloon as a function of time.

EXERCISE 27

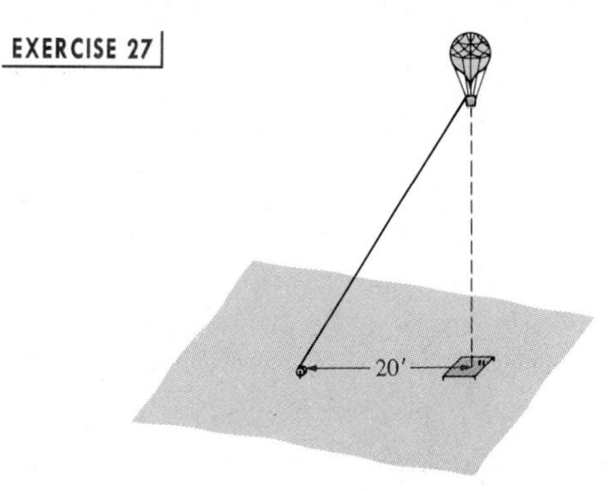

28 The diameter d of a cube is the distance between two opposite vertices. Express d as a function of the edge x of the cube. (*Hint:* First express the diagonal y of a face as a function of x.)

29 Refer to Exercise 38 of Section 3.4. The tightrope walker moves up the rope at a steady rate of 1 ft/sec. If the rope is attached 30 feet up the pole, express the height h of the walker above the ground as a function of time t. (*Hint:* Let d denote the total distance traveled along the wire. First express d as a function of t, and then h as a function of d.)

30 Refer to Exercise 39 of Section 3.4. When the airplane is 500 feet down the runway, it has reached and will maintain a speed of 150 ft/sec (or about 100 mi/hr) until takeoff. Express the distance of the plane from the control tower as a function of time t (in seconds). (*Hint:* In the figure, first write x as a function of t.)

3.7 INVERSE FUNCTIONS

A function f may have the same value for different numbers in its domain. For example, if $f(x) = x^2$, then $f(2) = 4$ and $f(-2) = 4$, but $2 \neq -2$. To define *the inverse of a function,* it is essential that different numbers in the domain *always* give different values of f. Such functions are called *one-to-one*.

DEFINITION

A function f with domain D and range R is a **one-to-one function** if whenever $a \neq b$ in D, then $f(a) \neq f(b)$ in R.

FIGURE 50

The arrow diagram in Figure 50 illustrates a one-to-one function, since each function value in the range R corresponds to *exactly one* element in the domain D. An equivalent statement is that *if $f(w) = f(z)$ for w and z in D, then $w = z$.* The function illustrated in Figure 28 is not one-to-one, since $f(w) = f(z)$ but $w \neq z$.

EXAMPLE ▪ 1

(a) If $f(x) = 3x + 2$, prove that f is one-to-one.

(b) If $g(x) = x^4 + 2x^2$, prove that g is not one-to-one.

SOLUTION

(a) Suppose that $f(w) = f(z)$ for some numbers w and z in the domain. This gives us

$$3w + 2 = 3z + 2$$
$$3w = 3z$$
$$w = z.$$

Hence f is one-to-one.

(b) To show that a function *is* one-to-one requires a *general* proof, as in part (a). To show that g is *not* one-to-one we need only find two distinct real numbers in the domain that produce the same function value. For example, $-1 \neq 1$, but $g(-1) = g(1)$. Since g is an even function, there are many other pairs a and $-a$ with this property.

As stated in the next theorem, a function that either increases throughout its domain or decreases throughout its domain is one-to-one.

THEOREM

(i) If f is an increasing function throughout its domain, then f is one-to-one.

(ii) If f is a decreasing function throughout its domain, then f is one-to-one.

PROOF If f is an increasing function, then as illustrated in Figure 51, the function value $f(x)$ increases as x increases. Thus if $a < b$, then $f(a) < f(b)$, and if $b < a$, then $f(b) < f(a)$. Hence if $a \neq b$, then $f(a) \neq f(b)$; that is, f is one-to-one.

A similar proof may be given for a decreasing function f (see Figure 52). ❑

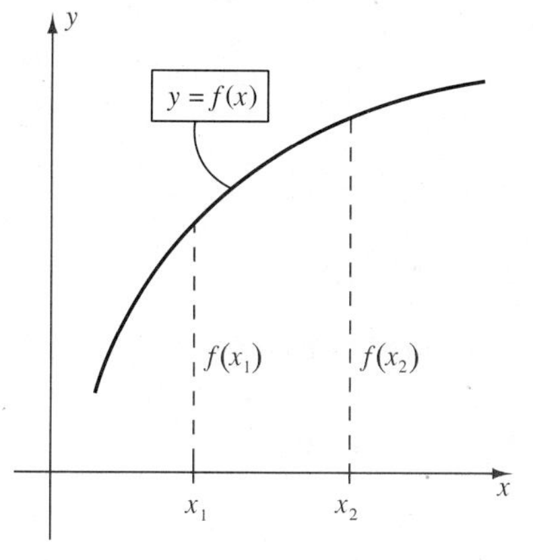

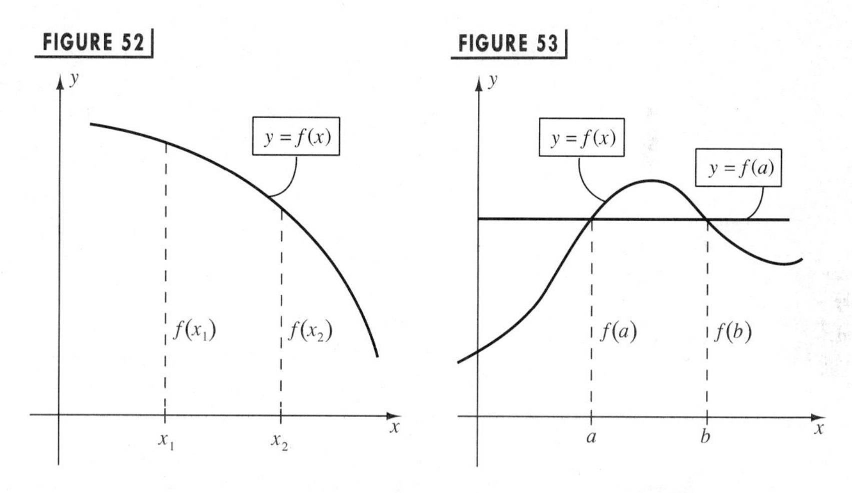

The function whose graph is illustrated in Figure 53 is not one-to-one, since $a \neq b$, but $f(a) = f(b)$. Note that the horizontal line $y = f(a)$ (or $y = f(b)$) intersects the graph in more than one point. Thus *if some horizontal line intersects the graph of a function f in more than one point, then f is not one-to-one.*

Let f be a one-to-one function with domain D and range R. Thus, for each number y in R, there is *exactly one* number x in D such that $y = f(x)$,

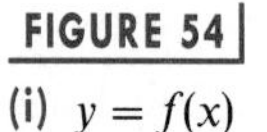

FIGURE 54

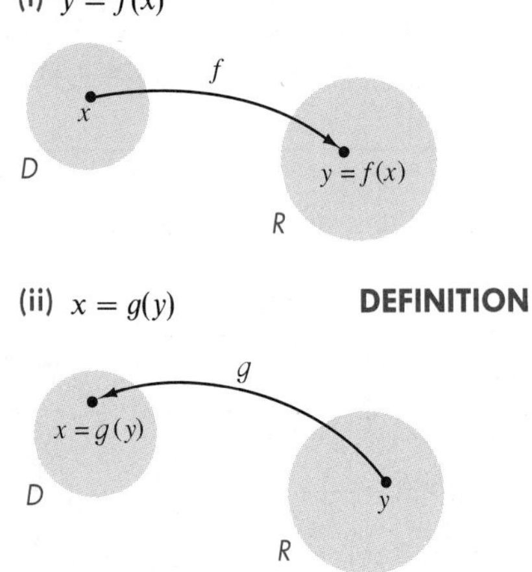

as illustrated by the arrow in Figure 54(i). Since x is *unique*, we may define a function g from R to D by means of the following rule:

$$x = g(y).$$

As in Figure 54(ii), *g reverses the correspondence given by f*. We call g the *inverse function* of f, as in the following definition.

DEFINITION

> Let f be a one-to-one function with domain D and range R. A function g with domain R and range D is the **inverse function** of f, provided the following condition is true for every x in D and every y in R:
>
> $$y = f(x) \quad \text{if and only if} \quad x = g(y).$$

Remember that to define the inverse of a function f, *it is absolutely essential that f be one-to-one*. The most common examples of one-to-one functions are those that are increasing or decreasing throughout their domains. The following theorem is useful when verifying that a function g is the inverse of f.

THEOREM

> Let f be a one-to-one function with domain D and range R. If g is a function with domain R and range D, then g is the inverse function of f if and only if both of the following conditions are true:
>
> (i) $g(f(x)) = x$ for every x in D.
>
> (ii) $f(g(y)) = y$ for every y in R.

PROOF Conditions (i) and (ii) of the theorem are illustrated in Figure 55(i) and (ii), respectively, where the black arrow indicates that f is a function from D to R, and the colored arrow indicates that g is a function from R to D.

Note that in (i) we first apply f to the number x in D, obtaining the function value $f(x)$ in R, and then we apply g to $f(x)$, obtaining the number $g(f(x))$ in D. Condition (i) of the theorem states that $g(f(x)) = x$ for every x.

In (ii) we use the reverse order for the functions. We first apply g to the number y in R, obtaining the function value $g(y)$ in D, and then we apply f to $g(y)$, obtaining the number $f(g(y))$ in R. Condition (ii) of the theorem states that $f(g(y)) = y$ for every y.

FIGURE 55

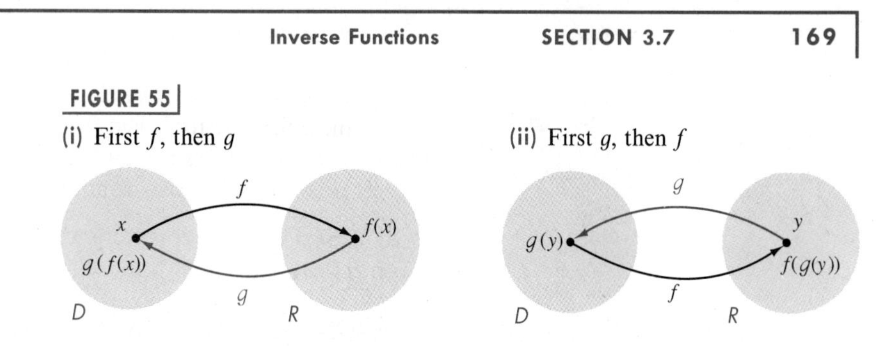

To prove the theorem we must show that if g is the inverse function of f, then (i) and (ii) are true and, conversely, if (i) and (ii) are true, then g is the inverse function of f.

First suppose that g is the inverse function of f. By the definition of inverse function,

$$y = f(x) \quad \text{if and only if} \quad x = g(y)$$

for every x in D and every y in R. An equivalent statement is

$$f(x) = y \quad \text{if and only if} \quad g(y) = x.$$

Substituting $f(x)$ for y in the equation $g(y) = x$ we obtain condition (i):

$$g(f(x)) = x \quad \text{for every } x \text{ in } D.$$

Similarly, substituting $g(y)$ for x in the equation $f(x) = y$, we obtain condition (ii):

$$f(g(y)) = y \quad \text{for every } y \text{ in } R.$$

We have now proved that if g is the inverse function of f, then (i) and (ii) are true.

To prove the converse of the theorem, let g be a function with domain R and range D, and assume that conditions (i) and (ii) are true. To show that g is the inverse function of f, we must prove that

$$y = f(x) \quad \text{if and only if} \quad x = g(y)$$

for every x in D and every y in R.

First suppose that $y = f(x)$. Since (i) is true, $g(f(x)) = x$; that is, $g(y) = x$. This shows that if $y = f(x)$, then $x = g(y)$.

Next suppose that $x = g(y)$. Since (ii) is true, $f(g(y)) = y$; that is, $f(x) = y$. This shows that if $x = g(y)$, then $y = f(x)$, which completes the proof. ❑

If a function f has an inverse g, we often denote g by f^{-1}. The -1 used in this notation should not be mistaken for an exponent; that is, $f^{-1}(y)$ *does not mean* $1/[f(y)]$. The reciprocal $1/[f(y)]$ may be denoted by $[f(y)]^{-1}$.

When we considered functions in previous sections we usually let x denote an arbitrary number in the domain. Similarly, for the inverse function f^{-1}, we may wish to consider $f^{-1}(x)$, *where x is in the domain R of f^{-1}*. In this event, the two conditions in the theorem are written

(i) $f^{-1}(f(x)) = x$ for every x in D.

(ii) $f(f^{-1}(x)) = x$ for every x in R.

Figure 54 contains a hint for finding the inverse of a one-to-one function in certain cases: If possible, we *solve the equation $y = f(x)$ for x in terms of y*, obtaining an equation of the form $x = g(y)$. If the two conditions $g(f(x)) = x$ and $f(g(x)) = x$ are true for every x in the domains of f and g, respectively, then g is the required inverse function f^{-1}. The following guidelines summarize this procedure; in guideline 2, in anticipation of finding f^{-1}, we write $x = f^{-1}(y)$ instead of $x = g(y)$.

GUIDELINES FINDING f^{-1} IN SIMPLE CASES

1 Verify that f is a one-to-one function (or that f is increasing or decreasing) throughout its domain.

2 Solve the equation $y = f(x)$ for x in terms of y, obtaining an equation of the form $x = f^{-1}(y)$.

3 Verify the two conditions

$$f^{-1}(f(x)) = x \quad \text{and} \quad f(f^{-1}(x)) = x$$

for every x in the domains of f and f^{-1}, respectively. ❑

The success of this method depends on the nature of the equation $y = f(x)$, since we must be able to solve for x in terms of y. For this reason, we include *simple cases* in the title of the guidelines.

EXAMPLE ▪ 2

Let $f(x) = 3x - 5$. Find the inverse function of f.

SOLUTION We shall follow the three guidelines. First, we note that the graph of the linear function f is a line of slope 3, and hence f is increasing throughout $\mathbb{R}$. Thus, the inverse function f^{-1} exists. Moreover, since the domain and range of f is $\mathbb{R}$, the same is true for f^{-1}.

As in guideline 2, we consider the equation

$$y = 3x - 5$$

and then solve for x in terms of y, obtaining

$$x = \frac{y + 5}{3}.$$

We now formally let

$$f^{-1}(y) = \frac{y+5}{3}.$$

Since the symbol used for the variable is immaterial, we may also write

$$f^{-1}(x) = \frac{x+5}{3}.$$

Finally, we verify that the two conditions

$$f^{-1}(f(x)) = x \quad \text{and} \quad f(f^{-1}(x)) = x$$

are satisfied. Thus,

$$\begin{aligned} f^{-1}(f(x)) &= f^{-1}(3x-5) && \text{(definition of } f\text{)} \\ &= \frac{(3x-5)+5}{3} && \text{(definition of } f^{-1}\text{)} \\ &= x && \text{(simplifying)} \end{aligned}$$

Also,

$$\begin{aligned} f(f^{-1}(x)) &= f\left(\frac{x+5}{3}\right) && \text{(definition of } f^{-1}\text{)} \\ &= 3\left(\frac{x+5}{3}\right) - 5 && \text{(definition of } f\text{)} \\ &= x && \text{(simplifying)} \end{aligned}$$

These verifications prove that the inverse function of f is given by

$$f^{-1}(x) = \frac{x+5}{3}.$$

FIGURE 56

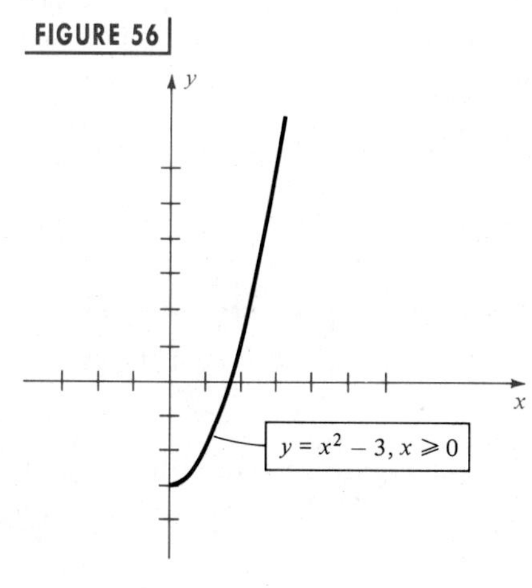

EXAMPLE ▪ 3

Let $f(x) = x^2 - 3$ for $x \geq 0$. Find the inverse function of f.

SOLUTION The graph of f is sketched in Figure 56. The domain D is $[0, \infty)$, and the range R is $[-3, \infty)$. Since f is increasing on D, it has an inverse function f^{-1} that has domain R and range D.

As in guideline 2, we consider the equation

$$y = x^2 - 3$$

and solve for x, obtaining

$$x = \pm\sqrt{y+3}.$$

Since x is nonnegative, we reject $x = -\sqrt{y+3}$ and let

$$f^{-1}(y) = \sqrt{y+3} \quad \text{or, equivalently,} \quad f^{-1}(x) = \sqrt{x+3}.$$

Finally, we verify that $f^{-1}(f(x)) = x$ for x in $D = [0, \infty)$ and that $f(f^{-1}(x)) = x$ for x in $R = [-3, \infty)$. Thus,

$$f^{-1}(f(x)) = f^{-1}(x^2 - 3) = \sqrt{(x^2-3)+3} = \sqrt{x^2} = x \quad \text{if } x \geq 0,$$

and

$$f(f^{-1}(x)) = f(\sqrt{x+3}) = (\sqrt{x+3})^2 - 3 = (x+3) - 3 = x \quad \text{if } x \geq -3.$$

Thus the inverse function is given by

$$f^{-1}(x) = \sqrt{x+3} \quad \text{for } x \geq -3.$$

FIGURE 57

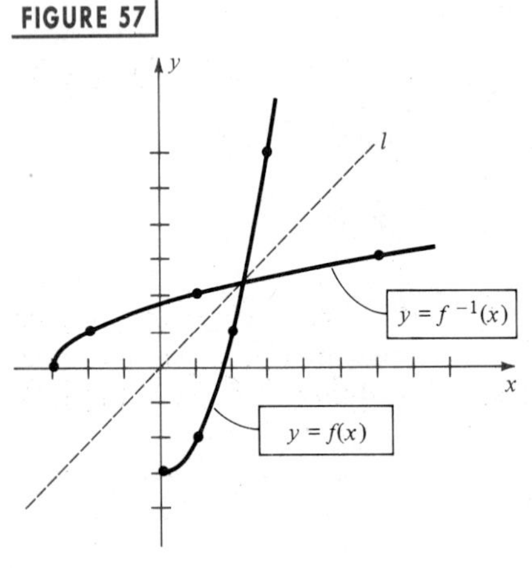

There is an interesting relationship between the graphs of a function f and its inverse function f^{-1}. We first note that $b = f(a)$ means the same thing as $a = f^{-1}(b)$. These equations imply that the point (a, b) is on the graph of f if and only if the point (b, a) is on the graph of f^{-1}.

As an illustration, in Example 3 we found that the functions f and f^{-1} given by

$$f(x) = x^2 - 3 \quad \text{and} \quad f^{-1}(x) = \sqrt{x+3}$$

are inverse functions of one another, provided that x is suitably restricted. Some points on the graph of f are $(0, -3)$, $(1, -2)$, $(2, 1)$, and $(3, 6)$. Corresponding points on the graph of f^{-1} are $(-3, 0)$, $(-2, 1)$, $(1, 2)$, and $(6, 3)$. The graphs of f and f^{-1} are sketched on the same coordinate plane in Figure 57. If the page is folded along the line l that bisects quadrants I and III (as indicated by the dashes in the figure), then the graphs of f and f^{-1} coincide. Note that an equation for l is $y = x$. The two graphs are *reflections* of one another through the line l (or *symmetric* with respect to l). This is typical of the graph of every function f that has an inverse function f^{-1} (see Exercise 34).

FIGURE 58

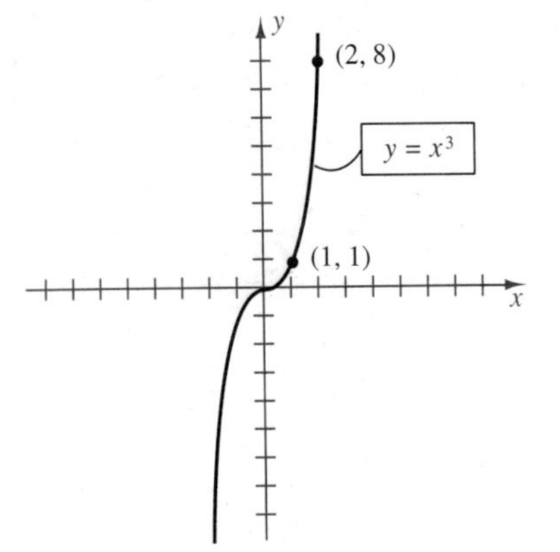

EXAMPLE ▪ 4

Let $f(x) = x^3$. Find the inverse function f^{-1} of f, and sketch the graphs of f and f^{-1} on the same coordinate plane.

SOLUTION The graph of f is sketched in Figure 58. Note that f is an odd function and hence the graph is symmetric with respect to the origin (see page 153).

Since f is increasing throughout its domain $\mathbb{R}$, it has an inverse function f^{-1}. As in guideline 2, we consider the equation

$$y = x^3$$

and solve for x by taking the cube root of each side, obtaining

$$x = y^{1/3} = \sqrt[3]{y}.$$

We now let

$$f^{-1}(y) = \sqrt[3]{y}, \quad \text{or, equivalently,} \quad f^{-1}(x) = \sqrt[3]{x}.$$

Next we verify the two conditions of guideline 3. Thus, for every x in $\mathbb{R}$,

$$f^{-1}(f(x)) = f^{-1}(x^3) = \sqrt[3]{x^3} = x$$

and

$$f(f^{-1}(x)) = f(\sqrt[3]{x}) = (\sqrt[3]{x})^3 = x.$$

FIGURE 59

The graph of f^{-1}, that is, the graph of the equation $y = \sqrt[3]{x}$ may be obtained by reflecting the graph in Figure 58 through the line $y = x$, as shown in Figure 59. Several points on the graph of f^{-1} are $(0, 0)$, $(1, 1)$, and $(8, 2)$.

3.7 EXERCISES

Exer. 1–12: Is the function f one-to-one?

1 $f(x) = 3x - 7$

2 $f(x) = \dfrac{1}{x - 2}$

3 $f(x) = x^2 - 9$

4 $f(x) = x^2 + 4$

5 $f(x) = \sqrt{x}$

6 $f(x) = \sqrt[3]{x}$

7 $f(x) = |x|$

8 $f(x) = 3$

9 $f(x) = \sqrt{4 - x^2}$

10 $f(x) = 2x^3 - 4$

11 $f(x) = 1/x$

12 $f(x) = 1/x^2$

Exer. 13–16: Prove that f and g are inverse functions of one another, and sketch the graphs of f and g on the same coordinate plane.

13 $f(x) = 3x - 2; \quad g(x) = \dfrac{x + 2}{3}$

14 $f(x) = x^2 + 5, \ x \le 0; \quad g(x) = -\sqrt{x - 5}, \ x \ge 5$

15 $f(x) = -x^2 + 3, \ x \ge 0; \quad g(x) = \sqrt{3 - x}, \ x \le 3$

16 $f(x) = x^3 - 4; \quad g(x) = \sqrt[3]{x + 4}$

Exer. 17–32: Find the inverse function of f.

17 $f(x) = 3x + 5$

18 $f(x) = 7 - 2x$

19 $f(x) = \dfrac{1}{3x - 2}$

20 $f(x) = \dfrac{1}{x + 3}$

21 $f(x) = \dfrac{3x + 2}{2x - 5}$

22 $f(x) = \dfrac{4x}{x - 2}$

23 $f(x) = 2 - 3x^2, \quad x \le 0$

24 $f(x) = 5x^2 + 2, \quad x \ge 0$

25 $f(x) = 2x^3 - 5$

26 $f(x) = -x^3 + 2$

27 $f(x) = \sqrt{3 - x}$

28 $f(x) = \sqrt{4 - x^2}$, $0 \le x \le 2$

29 $f(x) = \sqrt[3]{x} + 1$

30 $f(x) = (x^3 + 1)^5$

31 $f(x) = x$

32 $f(x) = -x$

33 (a) Prove that the linear function defined by $f(x) = ax + b$ for $a \ne 0$ has an inverse function, and find $f^{-1}(x)$.

(b) Does a constant function have an inverse? Explain.

34 Show that the graph of f^{-1} is the reflection of the graph of f through the line $y = x$ by verifying the following conditions:

(i) If $P(a, b)$ is on the graph of f, then $Q(b, a)$ is on the graph of f^{-1}.

(ii) The midpoint of line segment PQ is on the line $y = x$.

(iii) The line PQ is perpendicular to the line $y = x$.

Exer. 35–38: The graph of a one-to-one function f is shown.

(a) Use the reflection property to sketch the graph of f^{-1}.

(b) Find the domain D and range R of the function f.

(c) Find the domain R and range D of the inverse function f^{-1}.

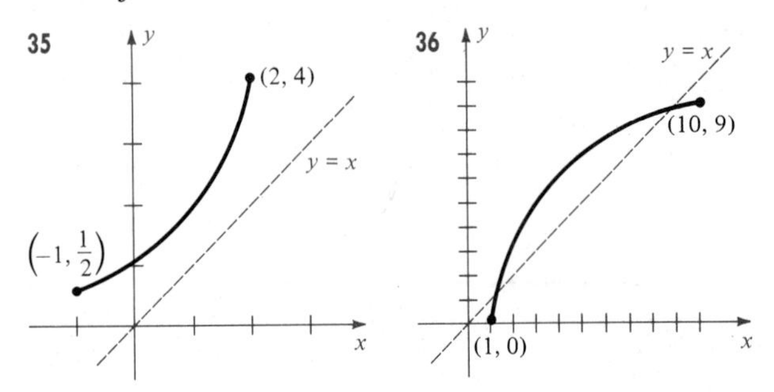

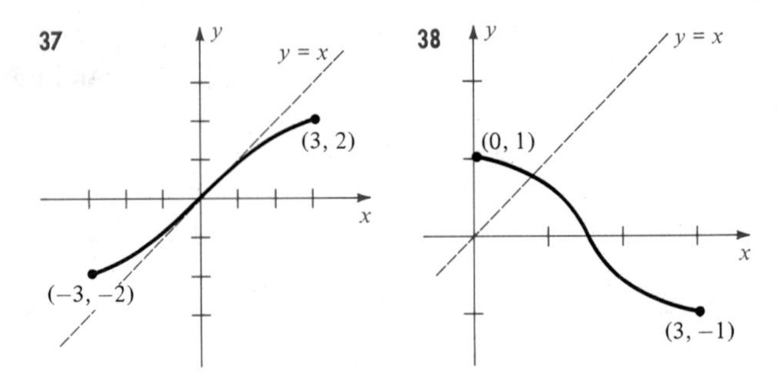

39 Verify that $f(x) = f^{-1}(x)$ if:

(a) $f(x) = -x + b$

(b) $f(x) = \dfrac{ax + b}{cx - a}$ for $c \ne 0$

(c) $f(x)$ has this graph:

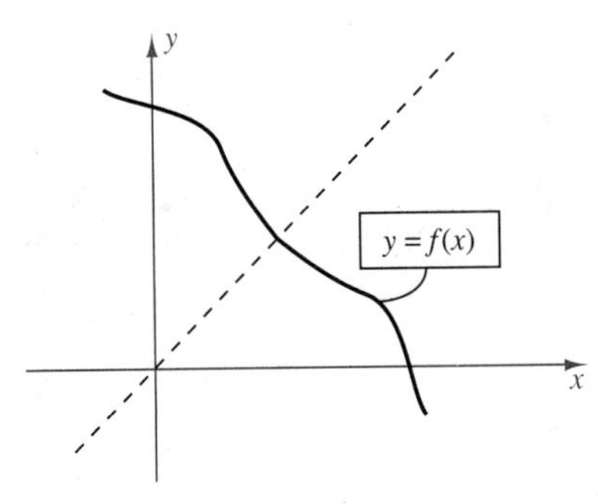

40 Let n be any positive integer. Find the inverse function of f if:

(a) $f(x) = x^n$ for $x \ge 0$.

(b) $f(x) = x^{m/n}$ for $x \ge 0$ and m any positive integer.

3.8 VARIATION

In some scientific investigations, the terminology of *variation*, or *proportion*, is used to describe relationships among variable quantities. In the following definition we shall not specify domains of variables. In any particular problem the domains should be evident.

DEFINITION

(i) The phrase **y varies directly as x**, or **y is directly proportional to x**, means that $y = kx$ for some real number $k \neq 0$.

(ii) The phrase **y varies inversely as x**, or **y is inversely proportional to x**, means that $y = \dfrac{k}{x}$ for some real number $k \neq 0$.

The number k is the **constant of variation**, or the **constant of proportionality**. For example, if an automobile is moving at a rate of 50 mi/hr, then the distance d it travels in t hours is given by $d = 50t$. Hence, as in (i), the distance d is directly proportional to the time t, and the constant of proportionality is 50.

The formula $A = \pi r^2$ for the area of a circle states that the area A varies directly as the square of the radius r. The constant of proportionality is π. The formula $V = \frac{4}{3}\pi r^3$ for the volume of a sphere of radius r states that the volume V is directly proportional to the cube of the radius. The constant of proportionality in this case is $\frac{4}{3}\pi$.

There are many other types of variation. If x, y, and z are variables and $y = kxz$ for some real number k, then *y varies directly as the product of x and z*, or **y varies jointly as x and z**. If $y = k(x/z)$, then *y varies directly as x and inversely as z*. As a final illustration, if a variable s varies directly as the product of u and the cube of v and inversely as the square of w, then

$$s = k\frac{uv^3}{w^2}$$

for some real number k.

In many applied problems the constant of proportionality can be determined by examining experimental facts, as illustrated in the following example.

EXAMPLE ▪ 1

If the temperature remains constant, then the pressure of an enclosed gas is inversely proportional to the volume. The pressure of a certain gas within a spherical balloon of radius 9 inches is 20 lb/in.^{2}. If the radius of the balloon increases to 12 inches, approximate the new pressure of the gas.

SOLUTION The original volume is $\frac{4}{3}\pi(9)^3 = 972\pi$ in.^{3}. If we denote the pressure by P and the volume by V, then by (ii) of the definition,

$$P = \frac{k}{V}$$

for some real number k. Since $P = 20$ when $V = 972\pi$,

$$20 = \frac{k}{972\pi}$$

and hence $k = 20(972\pi) = 19{,}440\pi$. Consequently, a formula for P is

$$P = \frac{19{,}440\pi}{V}.$$

If the radius is 12 inches, then $V = \frac{4}{3}\pi(12)^3 = 2304\pi$ in.^{3}. Substituting this number for V in the previous equation gives us

$$P = \frac{19{,}440\pi}{2304\pi} = \frac{135}{16} = 8.4375.$$

Thus, the pressure is approximately 8.4 lb/in.^{2} when the radius is 12 inches.

EXAMPLE ▪ 2

The weight that can be safely supported by a beam with a rectangular cross section varies directly as the product of the width and square of the depth of the cross section, and inversely as the length of the beam. If a 2-inch by 4-inch beam that is 8 feet long safely supports a load of 500 pounds, what weight can be safely supported by a 2-inch by 8-inch beam that is 10 feet long? (Assume that the width is the *shorter* dimension of the cross section.)

SOLUTION If the width, depth, length, and weight are denoted by w, d, l, and W, respectively, then

$$W = k\frac{wd^2}{l}.$$

According to the given data,

$$500 = k\frac{2(4^2)}{8}.$$

Solving for k we obtain $k = 125$, and hence the formula for W is

$$W = 125\left(\frac{wd^2}{l}\right).$$

To answer the question we substitute $w = 2$, $d = 8$, and $l = 10$, obtaining

$$W = 125\left(\frac{2 \cdot 8^2}{10}\right) = 1600 \text{ pounds.}$$

3.8 EXERCISES

Exer. 1–12: Express each statement as a formula, and determine the constant of proportionality from the given conditions.

1 a is directly proportional to v. If $v = 30$, then $a = 12$.

2 s varies directly as t. If $t = 10$, then $s = 18$.

3 r varies directly as s and inversely as t. If $s = -2$ and $t = 4$, then $r = 7$.

4 w varies directly as z and inversely as the square root of u. If $z = 2$ and $u = 9$, then $w = 6$.

5 y is directly proportional to the square of x and inversely proportional to the cube of z. If $x = 5$ and $z = 3$, then $y = 25$.

6 q is inversely proportional to the sum of x and y. If $x = 0.5$ and $y = 0.7$, then $q = 1.4$.

7 c is directly proportional to the product of the square of a and the cube of b. If $a = 7$ and $b = -2$, then $c = 16$.

8 r is directly proportional to the product of s and v and inversely proportional to the cube of p. If $s = 2$, $v = 3$, and $p = 5$, then $r = 40$.

9 y is directly proportional to x and inversely proportional to the square of z. If $x = 4$ and $z = 3$, then $y = 16$.

10 y is directly proportional to x and inversely proportional to the sum of r and s. If $x = 3$, $r = 5$, and $s = 7$, then $y = 2$.

11 y is directly proportional to the square root of x and inversely proportional to the cube of z. If $x = 9$ and $z = 2$, then $y = 5$.

12 y is directly proportional to the square of x and inversely proportional to the square root of z. If $x = 5$ and $z = 16$, then $y = 10$.

13 The pressure acting at a point in a liquid is directly proportional to the distance from the surface of the liquid to the point. In a certain oil tank the pressure at a depth of 2 feet is 118 lb/ft^2. Find the pressure at a depth of 5 feet.

14 Hooke's law states that the force required to stretch a spring x units beyond its natural length is directly proportional to x. If a weight of 4 pounds stretches a spring from its natural length of 10 inches to a length of 10.3 inches, what weight will stretch it to a length of 11.5 inches?

15 The electrical resistance of a wire varies directly as its length and inversely as the square of its diameter. If a wire 100 feet long of diameter 0.01 inch has a resistance of 25 ohms, find the resistance in a wire made of the same material that has a diameter of 0.015 inch and is 50 feet long.

16 The intensity of illumination I from a source of light varies inversely as the square of the distance d from the source. If a searchlight has an intensity of 1,000,000 candlepower at 50 feet, what is the intensity at a distance of 1 mile?

17 The period of a simple pendulum—that is, the time required for one complete oscillation—varies directly as the square root of its length. If a pendulum 2 feet long has a period of 1.5 seconds, find the period of a pendulum 6 feet long.

18 A circular cylinder is often used in physiology as a simple representation of a human limb.

(a) Show that the volume V of a cylinder varies directly as the product of the length L and the square of the circumference C.

(b) The formula obtained in part (a) can be used to approximate the volume of a limb from length and circumference measurements. Suppose the (average) circumference of a man's forearm is 22 cm and the average length is 27 cm. Approximate the volume of the forearm.

19 Kepler's third law states that the period T of a planet (the time needed to make one complete revolution about the sun) is directly proportional to the $\frac{3}{2}$ power of the average distance d from the sun. For the planet earth, $T = 365$ days and $d = 93$ million miles. Venus is 67 million miles from the sun. Estimate the period of Venus.

20 A motorcycle daredevil has made a jump of 150 feet. His speed coming off the ramp was 70 mi/hr. It is known from physics that the range of a projectile is directly proportional to the square of the velocity. If he can attain 80 mi/hr coming off the ramp and maintain proper balance, estimate the possible length of such a jump.

21 Police can sometimes estimate the speed V at which an automobile was traveling before the brakes were applied from the length L of the skid marks. Suppose that on a dry surface $L = 50$ feet when $V = 35$ mi/hr. Assuming that the speed is directly proportional to the square root of the length of the skid marks, estimate the initial speed if the skid measures 150 feet.

22 Coulomb's law in electrical theory asserts that the force F of attraction between two oppositely charged particles varies directly as the product of the magnitudes Q_1 and Q_2 of the charges, and inversely as the square of the distance d between the particles.

(a) Find a formula for F.

(b) What is the effect of reducing the distance between the particles by a factor of one-fourth?

23 Threshold weight W is defined to be that weight beyond which risk of death increases significantly. For middle-aged males, this weight is directly proportional to the third power of the height h. For a 6-foot male, W is about 200 pounds. Estimate the threshold weight for an individual who is 5 feet 6 inches tall.

24 The ideal gas law asserts that the volume V that a gas occupies is directly proportional to the product of the number n of moles of gas and the temperature T (in °K) and is inversely proportional to the pressure P (in atmospheres). What is the effect on the volume if the number of moles is doubled but the temperature and pressure are reduced by a factor of one-half?

25 Poiseuille's law asserts that the blood flow rate F (in liters/minute) through major arteries is directly proportional to the product of the fourth power of the radius r and the blood pressure P. During heavy exercise, normal flow rates sometimes triple. If the radius increases by 10%, approximately how much harder must the heart pump?

26 Suppose 200 trout are caught, tagged, and released in a lake's general population. Let y denote the number of tagged fish that are recovered when a sample of n trout are caught later. The validity of the mark-recapture method for estimating the lake's total trout population is based on the assumption that y is directly proportional to n. If 10 tagged trout are recovered from a sample of 300, estimate the total trout population of the lake.

3.9 REVIEW

Define or discuss each of the following.

Ordered pair ▪ Rectangular coordinate system in a plane ▪ Coordinate axes ▪ Quadrants ▪ Coordinates of a point ▪ Distance formula ▪ Midpoint formula ▪ Graph of an equation in x and y ▪ Tests for symmetry ▪ Equation of a circle ▪ Unit circle ▪ Slope of a line ▪ Point-slope form ▪ Slope-intercept form ▪ Linear equation in x and y ▪ Function ▪ Domain and range of a function ▪ Graph of a function ▪ Increasing function ▪ Decreasing function ▪ Constant function ▪ Linear function ▪ Even function ▪ Odd function ▪ Vertical shifts of graphs ▪ Horizontal shifts of graphs ▪ Stretching of graphs ▪ Reflections of graphs ▪ Piecewise-defined function ▪ Greatest integer function ▪ Operations on functions ▪ Polynomial function ▪ Algebraic function ▪ Transcendental function ▪ Composite function of two functions ▪ One-to-one function ▪ Inverse function ▪ Variation

3.9 EXERCISES

1 Show that the triangle with vertices $A(3, 1)$, $B(-5, -3)$, and $C(4, -1)$ is a right triangle, and find its area.

2 Given points $P(-5, 9)$ and $Q(-8, -7)$, find

(a) the midpoint of the segment PQ.

(b) a point T such that Q is the midpoint of PT.

3 Describe the set of all points (x, y) in a coordinate plane such that $y/x < 0$.

4 Find the slope of the line through $C(11, -5)$ and $D(-8, 6)$.

5 Show that the points $A(-3, 1)$, $B(1, -1)$, $C(4, 1)$, and $D(3, 5)$ are vertices of a trapezoid.

6 Find an equation of the circle that has center $C(7, -4)$ and passes through the point $Q(-3, 3)$.

7 Find an equation of the circle that has center $C(-5, -1)$ and is tangent to the line $x = 4$.

8 Express the equation $8x + 3y - 24 = 0$ in slope-intercept form.

9 Find an equation of the line through $A(\frac{1}{2}, -\frac{1}{3})$ that is

(a) parallel to the line $6x + 2y + 5 = 0$.

(b) perpendicular to the line $6x + 2y + 5 = 0$.

10 Find an equation of the line that has x-intercept -3 and passes through the center of the circle with equation $x^2 + y^2 - 4x + 10y + 26 = 0$.

Exer. 11–21: Sketch the graph of the equation.

11 $2y + 5x - 8 = 0$

12 $x = 3y + 4$

13 $x + 5 = 0$

14 $2y - 7 = 0$

15 $y = \sqrt{1 - x}$

16 $3x - 7y^2 = 0$

17 $9y^2 + 2x = 0$

18 $y^2 = 16 - x^2$

19 $x^2 + y^2 + 4x - 16y + 64 = 0$

20 $x^2 + y^2 - 8x = 0$

21 $y - x^2 = 1$

22 Find the domain and range of f if:

(a) $f(x) = \sqrt{3x - 4}$

(b) $f(x) = 1/(x + 3)^2$

23 If $f(x) = x/\sqrt{x + 3}$ find (a)–(g):

(a) $f(1)$

(b) $f(-1)$

(c) $f(0)$

(d) $f(-x)$

(e) $-f(x)$

(f) $f(x^2)$

(g) $(f(x))^2$

Exer. 24–30:

(a) Sketch the graph of f.

(b) Find the domain D and range R of f.

(c) Find the intervals on which f is increasing, decreasing, or constant.

24 $f(x) = |x + 3|$

25 $f(x) = \dfrac{1 - 3x}{2}$

26 $f(x) = \sqrt{2 - x}$

27 $f(x) = 1 - \sqrt{x + 1}$

28 $f(x) = 9 - x^2$

29 $f(x) = 1000$

30 $f(x) = \begin{cases} x^2 & \text{if } x < 0 \\ 3x & \text{if } 0 \le x < 2 \\ 6 & \text{if } x \ge 2 \end{cases}$

31 Sketch the graphs of the following equations, making use of shifting, stretching, or reflecting.

(a) $y = \sqrt{x}$

(b) $y = \sqrt{x + 4}$

(c) $y = \sqrt{x} + 4$

(d) $y = 4\sqrt{x}$

(e) $y = \frac{1}{4}\sqrt{x}$

(f) $y = -\sqrt{x}$

32 Determine if f is even, odd, or neither even nor odd:

(a) $f(x) = \sqrt[3]{x^3 + 4x}$

(b) $f(x) = \sqrt[3]{3x^2 - x^3}$

(c) $f(x) = \sqrt[3]{x^4 + 3x^2 + 5}$

Exer. 33–34: Find (a) $(f \circ g)(x)$ and (b) $(g \circ f)(x)$.

33 $f(x) = 2x^2 - 5x + 1$, $g(x) = 3x + 2$

34 $f(x) = \sqrt{3x + 2}$, $g(x) = 1/x^2$

Exer. 35–36: (a) Find $f^{-1}(x)$ and (b) sketch the graphs of f and f^{-1} on the same coordinate plane.

35 $f(x) = 10 - 15x$

36 $f(x) = 9 - 2x^2$, $x \le 0$

37 For children between ages 6 and 10, the height y (in inches) and age t (in years) are frequently linearly related. The height of a certain boy is 48 inches at age 6 and 50.5 inches at age 7.

(a) Express y in terms of t.

(b) Sketch the graph of the equation in part (a), and interpret the slope.

(c) Predict the boy's height at age 10.

38 An automobile presently gets 20 mi/gal but is in need of a tune-up that will cost \$50. The tune-up will improve gasoline mileage by 10%.

(a) If gasoline costs \$1.25 per gallon, find a linear function that gives the cost C of driving x miles without the tune-up.

(b) Find a linear function that gives the cost C of driving x miles with the tune-up.

(c) How many miles must the automobile be driven before the tune-up saves money?

39 An open rectangular storage shelter consisting of two vertical sides, 4 feet wide, and a flat roof is to be attached to an existing structure as illustrated in the figure. The flat roof is made of tin and costs \$5 per square foot, and the other two sides are made of plywood costing \$2 per square foot.

(a) If \$400 is available for construction, express the length y as a function of the height x.

(b) Express the volume V inside the shelter as a function of x.

EXERCISE 39

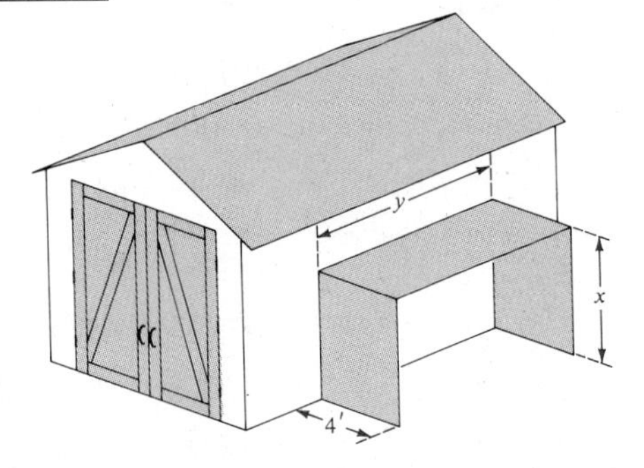

40 A company plans to manufacture a container having the shape of a right circular cylinder, open at the top, and having a capacity of 24π in.^{3}. If the cost of the material for the bottom is 30 cents per in.^{2} and that for the curved sides is 10 cents per in.^{2}, express the total cost C of the material as a function of the radius r of the base of the container.

41 A cross section of a rectangular pool of dimensions 80 feet by 40 feet is shown in the figure. The pool is being filled with water at the rate of 10 ft^3 per minute.

(a) Express the volume V as a function of depth h at the deep end for $0 \le h \le 6$ and then for $6 < h \le 9$.

(b) Express the volume V as a function of time t.

(c) Express h as a function of t.

EXERCISE 41

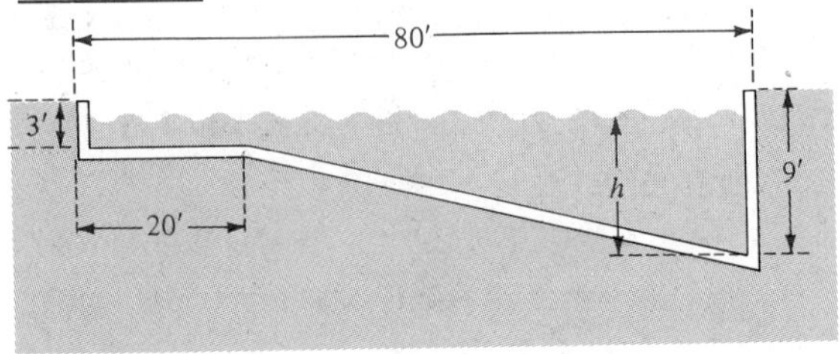

42 Water in a paper conical filter drips into a cup as shown in the figure. Suppose 5 in.^{3} of water is poured into the filter. Let x denote the height of the water in the filter and let y denote the height of the water in the cup.

(a) Express the radius r shown in the figure as a function of x. (*Hint:* Use similar triangles.)

(b) Express the height y of the water in the cup as a function of x. (*Hint:* What is the sum of the two volumes shown in the figure?)

EXERCISE 42

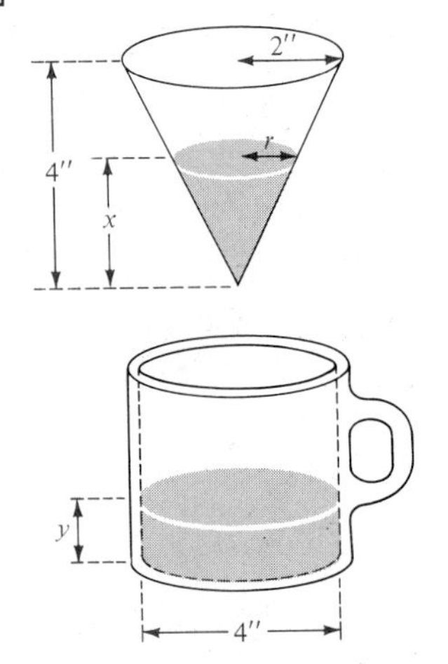

43 The shape of the first spacecraft in the Apollo program was a frustum of a right circular cone, a solid formed by truncating a cone by a plane parallel to its base. For the frustum shown in the figure, the radii a and b have already been determined.

(a) Use similar triangles to express y as a function of h.

(b) Express the volume of the frustum as a function of h.

(c) If $a = 6$ feet and $b = 3$ feet, for what value of h is the volume of the frustum $600\ \text{ft}^3$?

EXERCISE 43

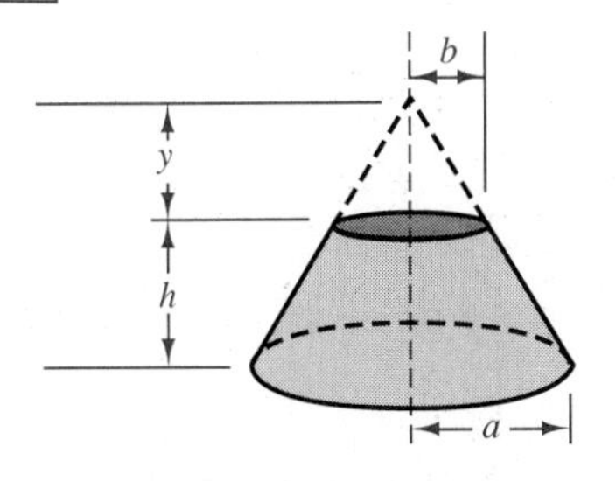

44 In a certain county, the average number of telephone calls per day between any two cities is directly proportional to the product of their populations and inversely proportional to the square of the distance between them. Cities A and B are 25 miles apart and have populations of 10,000 and 5000, respectively. Telephone records indicate an average of 2000 calls per day between the two cities. Estimate the average number of calls per day between city A and another city of 15,000 people that is 100 miles from A.

45 The power P generated by a wind rotor is directly proportional to the product of the area A swept out by the blades and the third power of the wind velocity v. Suppose the diameter of the circular area swept out by the blades is 10 ft, and $P = 3000$ watts when $v = 20$ mi/hr. Find the power generated when the wind velocity is 30 mi/hr.

CHAPTER 4

Polynomial functions are the most basic functions in algebra. We discuss techniques for sketching their graphs in the first two sections. We then turn our attention to division, and study methods for finding zeros of polynomial functions. In the last section we consider quotients of polynomial functions, that is, *rational functions*.

POLYNOMIAL FUNCTIONS AND RATIONAL FUNCTIONS

4.1 QUADRATIC FUNCTIONS

Among the most important functions in mathematics are those defined as follows.

DEFINITION

A function f is a **polynomial function** if

$$f(x) = a_n x^n + a_{n-1}x^{n-1} + \cdots + a_1 x + a_0$$

where the coefficients $a_0, a_1, \ldots, a_n$ are real numbers and the exponents are nonnegative integers.

If $a_n \neq 0$ in the preceding definition, then a_n is the **leading coefficient** of $f(x)$ and f has **degree n**. A polynomial function of degree 1 is a linear function. If the degree is 2, then, as in the next definition, f is a *quadratic function*.

DEFINITION

A function f is a **quadratic function** if

$$f(x) = ax^2 + bx + c$$

for real numbers a, b, and c with $a \neq 0$.

If $b = c = 0$ in the preceding definition, then $f(x) = ax^2$, and the graph is a parabola with vertex at the origin, opening upward if $a > 0$ or downward if $a < 0$ (see Section 3.5, Figures 45–46). If $b = 0$ and $c \neq 0$, then

$$f(x) = ax^2 + c,$$

and, from our discussion of vertical shifts in Section 3.5, the graph is a parabola with vertex at the point $(0, c)$ on the y-axis. Some typical graphs resulting from vertical shifts are sketched in Section 3.5, Figure 41. The following example contains additional illustrations.

EXAMPLE ▪ 1

Sketch the graph of f if:

(a) $f(x) = -\frac{1}{2}x^2$ (b) $f(x) = -\frac{1}{2}x^2 + 4$

FIGURE 1

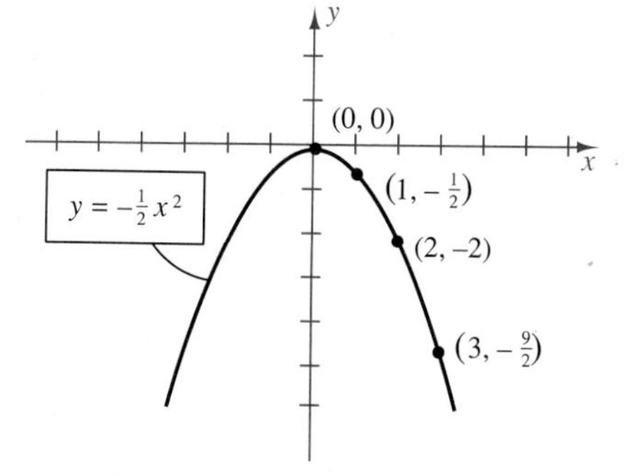

FIGURE 2

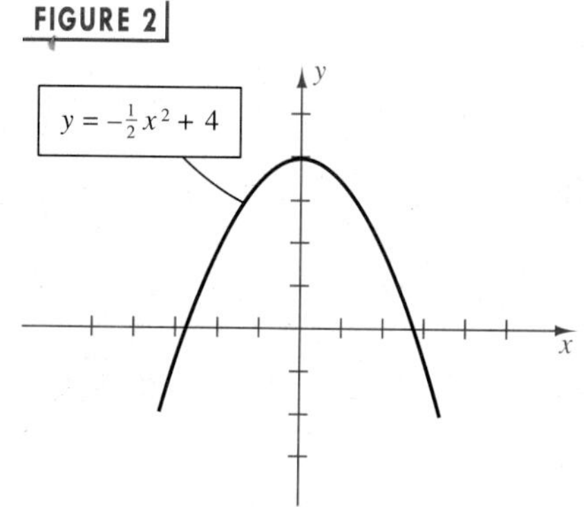

SOLUTION

(a) By a symmetry test the graph of $y = -\frac{1}{2}x^2$ is symmetric with respect to the y-axis. It is similar in shape, but wider than the parabola $y = -x^2$, sketched in Figure 46 of Section 3.5. Several points on the graph are $(0, 0)$, $(1, -\frac{1}{2})$, $(2, -2)$ and $(3, -\frac{9}{2})$. Plotting and using symmetry, we obtain the sketch in Figure 1.

(b) To find the graph of $y = -\frac{1}{2}x^2 + 4$, we shift the graph of $y = -\frac{1}{2}x^2$ upward a distance 4, obtaining the sketch in Figure 2.

If $f(x) = ax^2 + bx + c$ and $b \neq 0$, then by completing the square (see page 77), we can change the form of $f(x)$ to

$$f(x) = a(x - h)^2 + k$$

for some real numbers h and k. This technique is illustrated in the next example.

EXAMPLE ▪ 2

If $f(x) = 3x^2 + 24x + 50$, express $f(x)$ in the form $a(x - h)^2 + k$.

SOLUTION Before we complete the square, *it is essential that we factor out the coefficient of* x^2 *from the first two terms of* $f(x)$, as follows:

$$\begin{aligned} f(x) &= 3x^2 + 24x + 50 \\ &= 3(x^2 + 8x \qquad) + 50. \end{aligned}$$

We may now complete the square for the expression $x^2 + 8x$ by adding the square of one-half the coefficient of x, that is, $(\frac{8}{2})^2$, or 16. However, if we add 16 to the expression within parentheses, then, because of the factor 3, we are actually adding 48 to $f(x)$. Hence, we must compensate by subtracting 48:

$$\begin{aligned} f(x) &= 3(x^2 + 8x \qquad) + 50 \\ &= 3(x^2 + 8x + 16) + 50 - 48 \\ &= 3(x + 4)^2 + 2, \end{aligned}$$

which has the desired form with $a = 3$, $h = -4$, and $k = 2$.

If $f(x) = ax^2 + bx + c$, then by completing the square as in Example 2, we see that the graph of f is the same as the graph of an equation of the form

$$y = a(x - h)^2 + k.$$

FIGURE 3

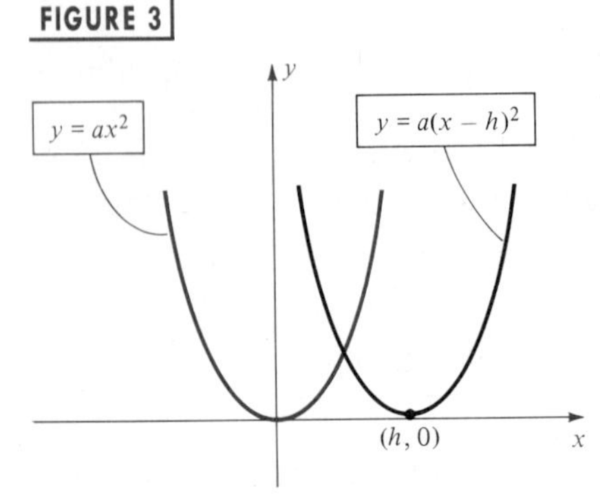

From the discussion of horizontal shifts in Section 3.5, we can find the graph of $y = a(x - h)^2$ by shifting the graph of $y = ax^2$ to either the left or right, depending on the sign of h. Thus, $y = a(x - h)^2$ is an equation of a parabola that has vertex $(h, 0)$ and a vertical axis. A typical graph is sketched in Figure 3 for the case in which both a and h are positive. Since the graph of $y = a(x - h)^2 + k$ can be obtained from that of $y = a(x - h)^2$ by a *vertical* shift of $|k|$ units, it follows that *the graph of a quadratic function f is a parabola that has a vertical axis.* The sketch in Figure 4 illustrates one possible graph. If $a > 0$, the point (h, k) is the lowest point on the parabola, and the function f has a **minimum value** $f(h) = k$. If $a < 0$, the point (h, k) is the highest point on the parabola, and the function f has a **maximum value** $f(h) = k$.

We have obtained the following result.

STANDARD EQUATION: PARABOLA WITH VERTICAL AXIS

The graph of the equation

$$y = a(x - h)^2 + k$$

for $a \neq 0$ is a parabola that has vertex (h, k) and a vertical axis. The parabola opens upward if $a > 0$ or downward if $a < 0$.

FIGURE 4

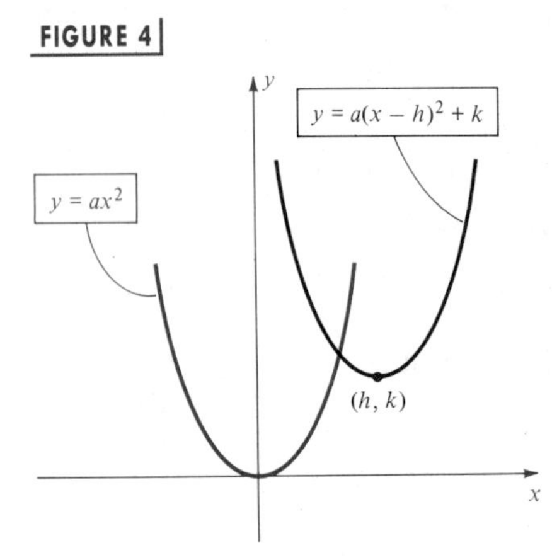

For convenience, we often refer to the *parabola* $y = ax^2 + bx + c$ when considering the graph of this equation.

EXAMPLE 3

Let $f(x) = 2x^2 - 6x + 4$.

(a) Sketch the graph of f. **(b)** Find the minimum value of $f(x)$.

SOLUTION

(a) The graph of f, a parabola, is the same as the graph of the equation $y = 2x^2 - 6x + 4$. Completing the square, we obtain

$$\begin{aligned} y &= 2x^2 - 6x + 4 \\ &= 2(x^2 - 3x \qquad) + 4 \\ &= 2(x^2 - 3x + \tfrac{9}{4}) + (4 - \tfrac{9}{2}) \\ &= 2(x - \tfrac{3}{2})^2 - \tfrac{1}{2} \end{aligned}$$

which has the form of the standard equation with $a = 2$, $h = \frac{3}{2}$, and $k = -\frac{1}{2}$. Hence the vertex (h, k) of the parabola is $(\frac{3}{2}, -\frac{1}{2})$. Since $a = 2 > 0$, the parabola opens upward.

FIGURE 5

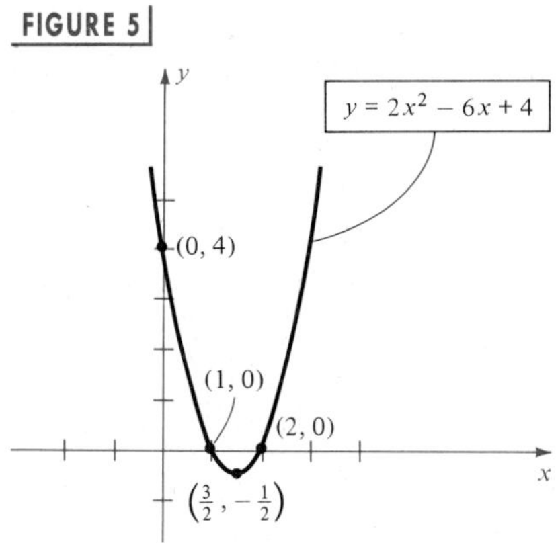

The y-intercept of the graph is $f(0) = 4$. To find the x-intercepts we solve $2x^2 - 6x + 4 = 0$, or the equivalent equation $2(x - 1)(x - 2) = 0$, obtaining $x = 1$ and $x = 2$. Plotting the vertex together with the x- and y-intercepts provides enough points for a reasonably accurate sketch (see Figure 5).

(b) The minimum value of f occurs at the vertex, and hence is $f(\frac{3}{2}) = -\frac{1}{2}$.

EXAMPLE ▪ 4

Let $f(x) = 8 - 2x - x^2$.

(a) Sketch the graph of f. **(b)** Find the maximum value of $f(x)$.

SOLUTION

(a) The graph is a parabola with a vertical axis. Let us consider the equation $y = f(x)$, and find the vertex by completing the square:

$$\begin{aligned} y &= -x^2 - 2x + 8 \\ &= -(x^2 + 2x \qquad) + 8 \\ &= -(x^2 + 2x + 1) + 8 + 1 \\ &= -(x + 1)^2 + 9 \end{aligned}$$

FIGURE 6

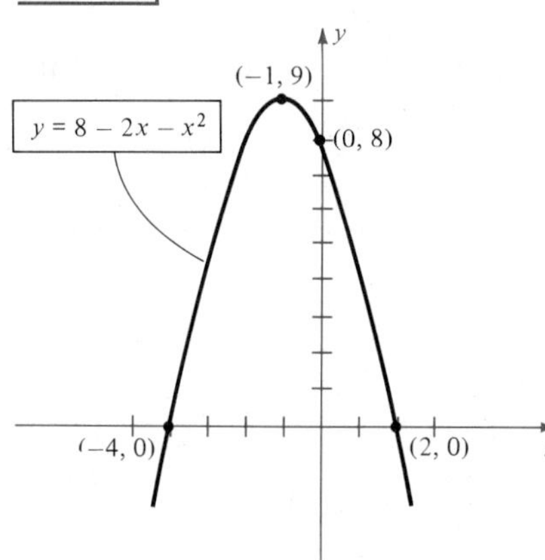

Comparing this with the standard equation of a parabola, we see that $h = -1$, $k = 9$, and hence the vertex is $(-1, 9)$. Since $a = -1 < 0$, the parabola opens downward.

To find the x-intercepts we solve the equation $8 - 2x - x^2 = 0$, or, equivalently, $x^2 + 2x - 8 = 0$. Factoring gives us $(x + 4)(x - 2) = 0$, and hence the intercepts are $x = -4$ and $x = 2$. The y-intercept is 8. Using this information gives us the sketch in Figure 6.

(b) The maximum value of f occurs at the vertex: $f(-1) = 9$.

The next result gives us a simple formula for locating the vertex of a parabola.

VERTEX OF A PARABOLA

> The vertex of the parabola $y = ax^2 + bx + c$ has x-coordinate
>
> $$\frac{-b}{2a}.$$

PROOF Let us begin by writing $y = ax^2 + bx + c$ as

$$y = a\left(x^2 + \frac{b}{a}x \qquad\right) + c.$$

Next we complete the square by adding $\left(\dfrac{1}{2}\dfrac{b}{a}\right)^2$ to the expression within parentheses:

$$y = a\left(x^2 + \frac{b}{a}x + \frac{b^2}{4a^2}\right) + \left(c - \frac{b^2}{4a}\right).$$

Note that if $b^2/(4a^2)$ is added *inside* the parentheses, then because of the factor a on the *outside* we have actually added $b^2/(4a)$ to y. Therefore we must compensate by subtracting $b^2/(4a)$. The preceding equality may be written

$$y = a\left(x + \frac{b}{2a}\right)^2 + \left(c - \frac{b^2}{4a}\right).$$

This is the equation of a parabola that has vertex (h, k) with $h = -b/(2a)$ and $k = c - b^2/(4a)$. ❑

It is unnecessary to remember the formula for the y-coordinate of the vertex of the parabola in the preceding result. Once the x-coordinate is found, we can calculate the y-coordinate by substituting $-b/(2a)$ for x in the equation of the parabola.

EXAMPLE ▪ 5

Find the vertex of the parabola $y = 2x^2 - 6x + 4$.

SOLUTION We considered this parabola in Example 3 and found the vertex by completing the square. We shall now use the vertex formula with $a = 2$ and $b = -6$, obtaining the x-coordinate

$$\frac{-b}{2a} = \frac{-(-6)}{2(2)} = \frac{6}{4} = \frac{3}{2}.$$

We next find the y-coordinate by substituting $\frac{3}{2}$ for x in the given equation:

$$y = 2(\tfrac{3}{2})^2 - 6(\tfrac{3}{2}) + 4 = -\tfrac{1}{2}.$$

Thus the vertex is $(\frac{3}{2}, -\frac{1}{2})$ (see Figure 5).

Since the graph of $f(x) = ax^2 + bx + c$ for $a \neq 0$ is a parabola, we can use the vertex formula to help find the maximum or minimum value of a quadratic function. Specifically, since the x-coordinate of the vertex is $-b/(2a)$, the maximum or minimum value is the y-coordinate $f(-b/(2a))$. In the next example, we solve an applied problem involving a quadratic function.

EXAMPLE ▪ 6

A long rectangular sheet of metal, 12 inches wide, is to be made into a rain gutter by turning up two sides so that they are perpendicular to the sheet. How many inches should be turned up to give the gutter its greatest capacity?

FIGURE 7

SOLUTION The gutter is illustrated in Figure 7. If x denotes the number of inches turned up on each side, the width of the base of the gutter is $12 - 2x$ inches. The capacity will be greatest when the cross-sectional area of the rectangle with sides of lengths x and $12 - 2x$, has its greatest value. Letting $f(x)$ denote this area,

$$\begin{aligned} f(x) &= x(12 - 2x) \\ &= 12x - 2x^2 \\ &= -2x^2 + 12x. \end{aligned}$$

This equation has the form $f(x) = ax^2 + bx + c$ with $a = -2$, $b = 12$, and $c = 0$. The graph of f is a parabola that has a vertical axis and opens downward. The maximum value of f occurs at $x = -b/(2a) = -12/(-4)$, or $x = 3$. Thus 3 inches should be turned up on each side to achieve maximum capacity.

Parabolas (and hence quadratic functions) occur in applications of mathematics to the physical world. For example, if a projectile is fired, and we assume that it is acted upon only by the force of gravity (that is, air resistance and other outside factors are ignored), then the path of the projectile is parabolic. Properties of parabolas are used in the design of mirrors for telescopes and searchlights and in the construction of radar antennas.

4.1 EXERCISES

Exer. 1–4: Sketch the graph of f for the indicated value of c or a.

1 $f(x) = 4x^2 + c$; (a) $c = 3$, (b) $c = -3$

2 $f(x) = -3x^2 + c$; (a) $c = -1$, (b) $c = 1$

3 $f(x) = ax^2 + 2$; (a) $a = 3$, (b) $a = -\frac{1}{3}$

4 $f(x) = ax^2 - 3$; (a) $a = -4$, (b) $a = \frac{1}{4}$

Exer. 5–10: Sketch the graph of f, and label the vertex.

5 $f(x) = \frac{1}{3}x^2 - 2$

6 $f(x) = 5 - 2x^2$

7 $f(x) = -3x^2 - 4$

8 $f(x) = \frac{1}{2}x^2 + 2$

9 $f(x) = x^2 - 4x$

10 $f(x) = -x^2 - 6x$

Exer. 11–14: Use the quadratic formula to find the zeros of f.

11 $f(x) = -12x^2 + 11x + 15$

12 $f(x) = -4x^2 + 4x - 1$

13 $f(x) = 9x^2 + 24x + 16$

14 $f(x) = 6x^2 + 7x - 24$

Exer. 15–18: Express $f(x)$ in the form $a(x - h)^2 + k$.

15 $f(x) = 2x^2 - 12x + 22$

16 $f(x) = 5x^2 + 20x + 17$

17 $f(x) = -3x^2 - 6x - 5$ **18** $f(x) = -4x^2 + 16x - 13$

Exer. 19–28: Sketch the graph of f, and find the maximum or minimum value of $f(x)$.

19 $f(x) = 4x^2 - 4x + \frac{1}{4}$ **20** $f(x) = 6x^2 + 4x + \frac{4}{3}$

21 $f(x) = -2x^2 + 20x - 43$ **22** $f(x) = -3x^2 - 6x - 6$

23 $f(x) = x^2 + 4x + 9$ **24** $f(x) = 2x^2 - 4x - 11$

25 $f(x) = \frac{1}{3}x^2 + \frac{8}{3}x - \frac{2}{3}$ **26** $f(x) = \frac{2}{5}x^2 - \frac{12}{5}x + \frac{23}{5}$

27 $f(x) = -\frac{3}{4}x^2 + 9x - 34$ **28** $f(x) = -\frac{1}{2}x^2 - 2x + 3$

29 Flights of leaping animals typically have parabolic paths. The figure illustrates a frog jump superimposed on a coordinate system. The length of the leap is 9 feet and the maximum height off the ground is 3 feet. Find a quadratic function f that specifies the path of the frog.

EXERCISE 29

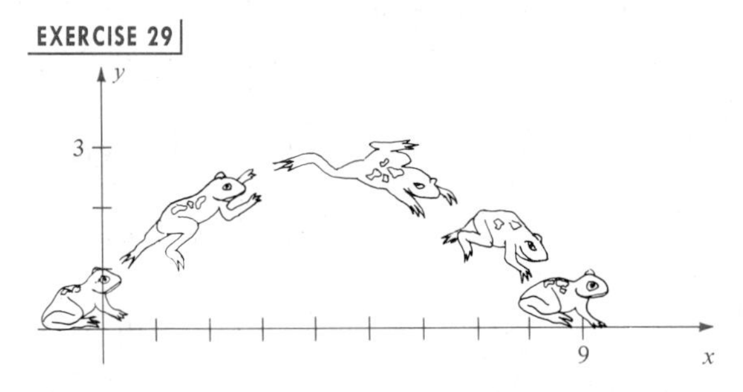

30 In the 1940s, the human cannonball stunt was performed regularly by Emmanuel Zacchini for The Ringling Brothers and Barnum & Bailey Circus. The tip of the cannon rose 15 feet off the ground, and the total horizontal distance traveled was 175 feet. When the cannon is aimed at a 45-degree angle, an equation of the parabolic flight has the form $y = ax^2 + x + c$ (see figure).

(a) Find values for a and c that correspond to the given information.

(b) Find the maximum height attained by the human cannonball for the given information.

EXERCISE 30

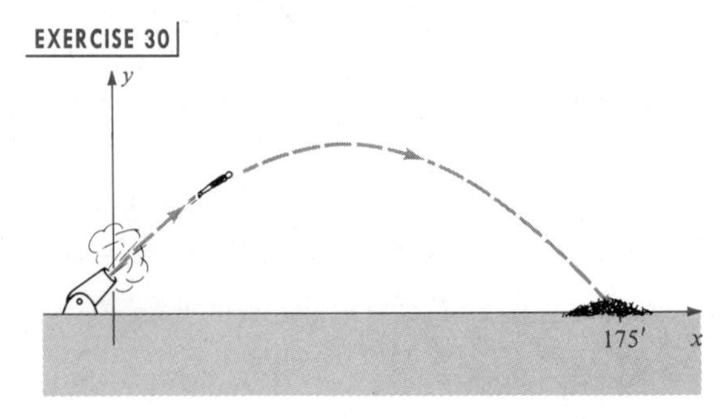

31 One section of a suspension bridge has its weight uniformly distributed between twin towers that are 400 feet apart and rise 90 feet above the horizontal roadway (see figure). A cable strung between the tops of the towers has the shape of a parabola, and its center point is 10 feet above the roadway. Suppose coordinate axes are introduced, as shown in the figure.

(a) Find an equation for the parabola.

(b) Nine equally spaced vertical cables are used to support the bridge (see figure). Find the total length of these supports.

EXERCISE 31

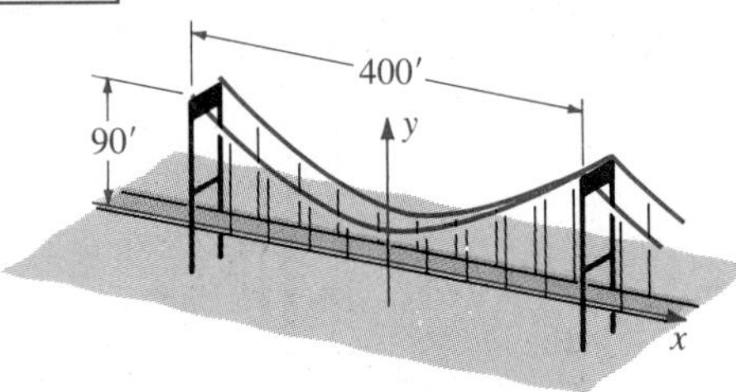

32 Traffic engineers are designing a stretch of highway that will connect a horizontal highway with one having a 20% grade (i.e., slope $\frac{1}{5}$), as illustrated in the figure. The smooth transition is to take place over a horizontal distance of 800 feet using a parabolic piece of highway to connect points A and B. If the equation of the parabolic segment is $y = ax^2 + bx + c$, it can be shown that the slope of the tangent line at the point $P(x, y)$ on the parabola is given by $m = 2ax + b$.

(a) Find an equation of the parabola that has a tangent line of slope 0 at A and $\frac{1}{5}$ at B.

(b) Find the coordinates of B.

EXERCISE 32

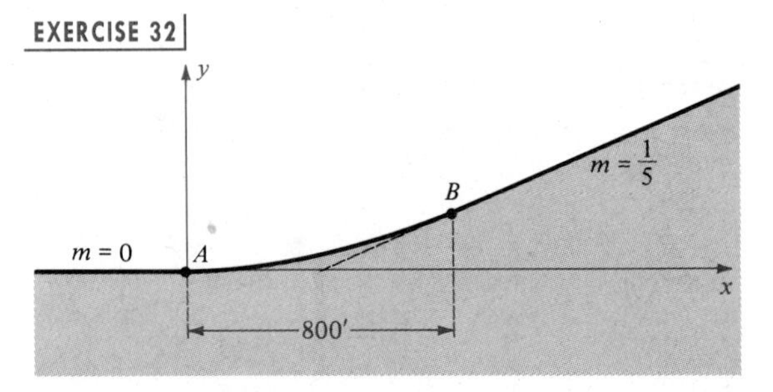

33 Find an equation of a parabola that has a vertical axis and passes through the points $A(2, 3)$, $B(-1, 6)$, and $C(1, 0)$.

34 Prove that there is exactly one line of a given slope m that intersects the parabola $x^2 = 4py$ in exactly one point, and that its equation is $y = mx - pm^2$.

35 A person standing on the top of a building projects an object directly upward with a velocity of 144 ft/sec. Its height $s(t)$ in feet above the ground after t seconds is given by $s(t) = -16t^2 + 144t + 100$.

(a) What is its maximum height?

(b) What is the height of the building?

36 A rocket is shot straight up into the air with an initial velocity of v_0 ft/sec, and its height $s(t)$ in feet above the ground after t seconds is given by $s(t) = -16t^2 + v_0 t$.

(a) The rocket hits the ground after 4 seconds. What is its initial velocity v_0?

(b) What is the maximum height attained by the rocket?

37 The growth rate y (in pounds per month) of infants is related to their present weight x (in pounds) by the formula $y = cx(21 - x)$ for some constant $c > 0$. At what weight is the rate of growth a maximum?

38 The number of miles M that an automobile can travel on one gallon of gasoline is a function of its speed v (in mi/hr). If

$$M = -\tfrac{1}{30}v^2 + \tfrac{5}{2}v \quad \text{for } 0 < v < 70,$$

find the most economical speed for a trip.

39 One thousand feet of chain-link fence is to be used to construct six cages for a zoo exhibit. The design is shown in the figure.

(a) Express the width y as a function of the length x.

(b) Express the total enclosed area A of the exhibit as a function of x.

(c) Find the dimensions that maximize the enclosed area.

EXERCISE 39

40 A man wishes to put a fence around a rectangular field and then subdivide the field into three smaller rectangular plots by placing two fences parallel to one of the sides. If he can afford only 1000 yards of fencing, what dimensions will give the maximum rectangular area?

41 A piece of wire 24 inches long is bent into the shape of a rectangle having width x and length y.

(a) Express y as a function of x.

(b) Express the area A of the rectangle as a function of x.

(c) Show that the area A is greatest if the rectangle is a square.

42 A company sells running shoes to dealers at a rate of \$20 per pair if less than 50 pairs are ordered. If a dealer orders 50 or more pairs (up to 600), the price per pair is reduced at a rate of 2 cents times the number ordered. What size order will produce the maximum amount of money for the company?

43 A boy tosses a baseball from the edge of a plateau down a hill as illustrated in the figure. The ball, thrown at an angle of 45 degrees, lands 50 feet down the hill, which is defined by the line $4y + 3x = 0$. Using calculus, it can be shown that the path of the baseball is given by the equation $y = ax^2 + x + c$ for some constants a and c.

(a) Ignoring the height of the boy, find an equation for the path.

(b) What is the maximum height of the ball *off the ground*?

EXERCISE 43

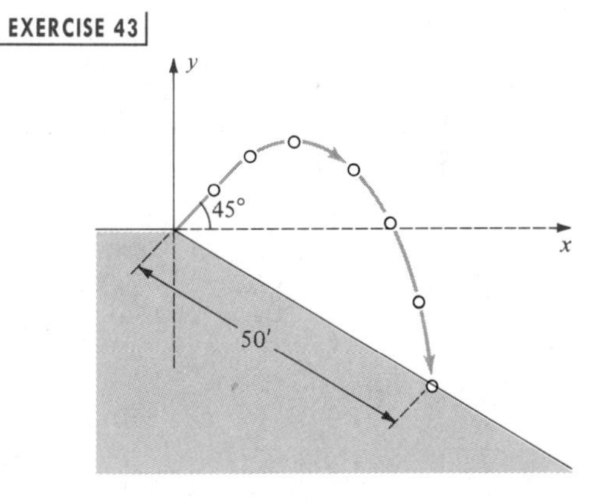

44 A cable television firm presently serves 5000 households and charges \$20 per month. A marketing survey indicates that each decrease of \$1 in the monthly charge will result in 500 new customers. Let $R(x)$ denote the total monthly revenue when the monthly charge is x dollars.

(a) Determine the revenue function R.

(b) Sketch the graph of R and find the value of x that results in maximum monthly revenue.

4.2 GRAPHS OF POLYNOMIAL FUNCTIONS OF DEGREE GREATER THAN 2

Let f be a polynomial function of degree n; that is,

$$f(x) = a_n x^n + a_{n-1}x^{n-1} + \cdots + a_1 x + a_0$$

with $a_n \neq 0$. The domain of f is $\mathbb{R}$. If the degree is odd, then the range of f is also $\mathbb{R}$; however, if the degree is even, then the range is an infinite interval of the form $(-\infty, a]$ or $[a, \infty)$. These general facts are illustrated by the graphs in this section.

Recall that if $f(c) = 0$, then c is a **zero** of f, or of $f(x)$. We also call c a **solution**, or **root**, of the equation $f(x) = 0$. The real zeros of f are the x-intercepts of the graph of f.

If a polynomial function has degree 0, then $f(x) = a$ for some nonzero real number a, and the graph is a horizontal line. Graphs of polynomial functions of degree 1 (linear functions) are lines. Polynomial functions of degree 2 (quadratic functions) have parabolas for their graphs. In this section we shall study graphs of polynomial functions of degree greater than 2.

If f has degree n, and all the coefficients except a_n are zero, then

$$f(x) = ax^n \quad \text{for some } a = a_n \neq 0.$$

In this case, if $n = 1$, the graph of f is a line that passes through the origin. If $n = 2$, the graph is a parabola with vertex at the origin. Two examples for $n = 3$ were discussed in Chapter 3. (See Figures 12 and 58 in Chapter 3.) Additional illustrations are given in the next example.

EXAMPLE 1

Sketch the graph of f if:

(a) $f(x) = \frac{1}{2}x^3$ (b) $f(x) = -\frac{1}{2}x^3$

SOLUTION

(a) The following table lists several points on the graph of $y = \frac{1}{2}x^3$.

x	0	$\frac{1}{2}$	1	$\frac{3}{2}$	2	$\frac{5}{2}$
y	0	$\frac{1}{16} \approx 0.06$	$\frac{1}{2}$	$\frac{27}{16} \approx 1.7$	4	$\frac{125}{16} \approx 7.8$

Since f is an odd function, the graph of f is symmetric with respect to the origin, and hence points such as $(-\frac{1}{2}, -\frac{1}{16})$ and $(-1, -\frac{1}{2})$ are also on the graph. The graph is sketched in Figure 8.

FIGURE 8

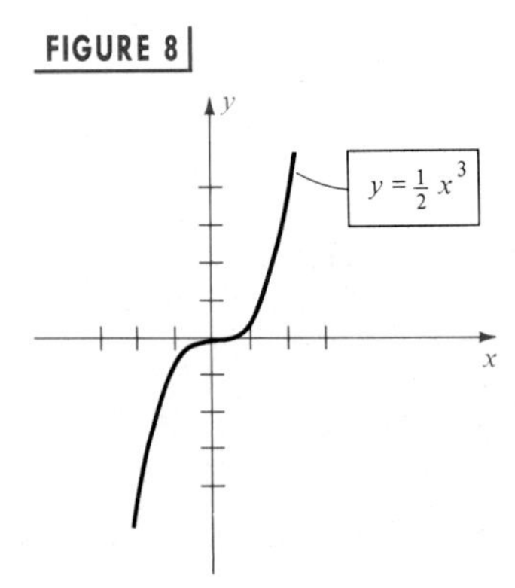

FIGURE 9

(b) If $y = -\frac{1}{2}x^3$, the graph can be obtained from that in part (a) by multiplying all y-coordinates by -1; that is, by reflecting the graph in part (a) through the x-axis. This gives us the sketch in Figure 9.

If $f(x) = ax^n$ and n is an *odd* positive integer, then f is an odd function and the graph of f for $a > 0$ is similar in shape to that in Figure 8; however, as n increases, the graph rises more sharply for $x > 1$. If $a < 0$ we reflect the graph through the x-axis, as in Figure 9.

If $f(x) = ax^n$ and n is an *even* integer, then f is an even function, and the graph of f is symmetric with respect to the y-axis, as illustrated in Figure 10 for the case $a = 1$. Note that as the exponent increases, the graph becomes flatter at the origin. It also rises more rapidly for $x > 1$. If $a < 0$, the graph lies below the x-axis.

FIGURE 10

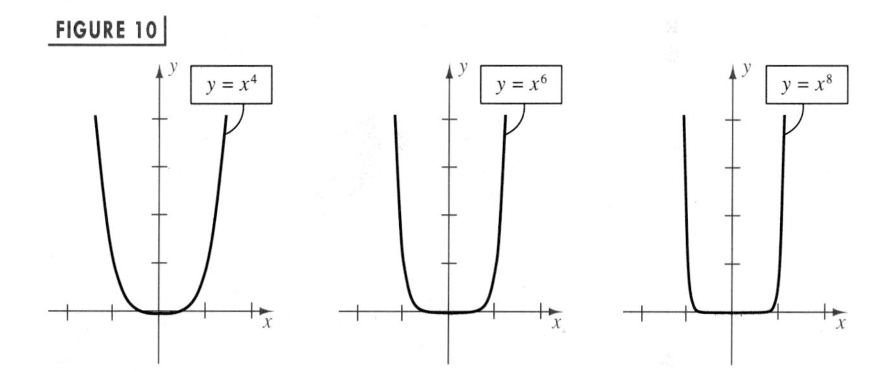

FIGURE 11

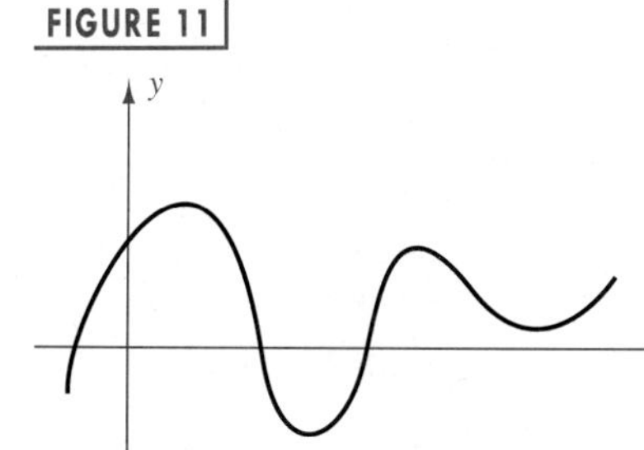

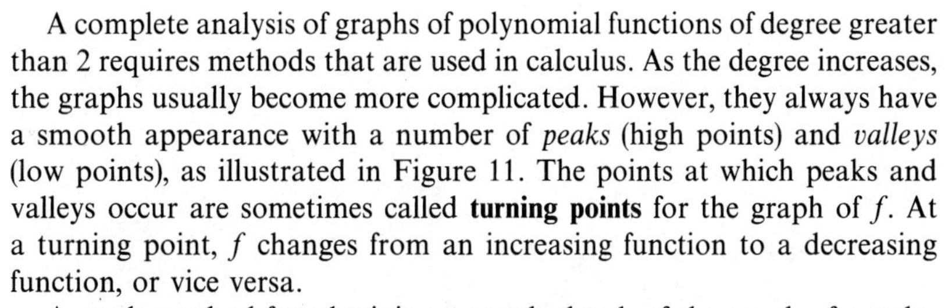

A complete analysis of graphs of polynomial functions of degree greater than 2 requires methods that are used in calculus. As the degree increases, the graphs usually become more complicated. However, they always have a smooth appearance with a number of *peaks* (high points) and *valleys* (low points), as illustrated in Figure 11. The points at which peaks and valleys occur are sometimes called **turning points** for the graph of f. At a turning point, f changes from an increasing function to a decreasing function, or vice versa.

A crude method for obtaining a rough sketch of the graph of a polynomial function is to plot many points and then fit a curve to the resulting configuration; however, this is usually an extremely tedious procedure. This method is based on a property of polynomial functions called **continuity**. Continuity, which is studied extensively in calculus, implies that a small change in x produces a small change in $f(x)$. The next theorem specifies another important property of polynomial functions. The proof is omitted, since it requires advanced mathematical methods.

INTERMEDIATE VALUE THEOREM FOR POLYNOMIAL FUNCTIONS

> If f is a polynomial function and $f(a) \neq f(b)$ for $a < b$, then f takes on every value between $f(a)$ and $f(b)$ in the interval $[a, b]$.

FIGURE 12

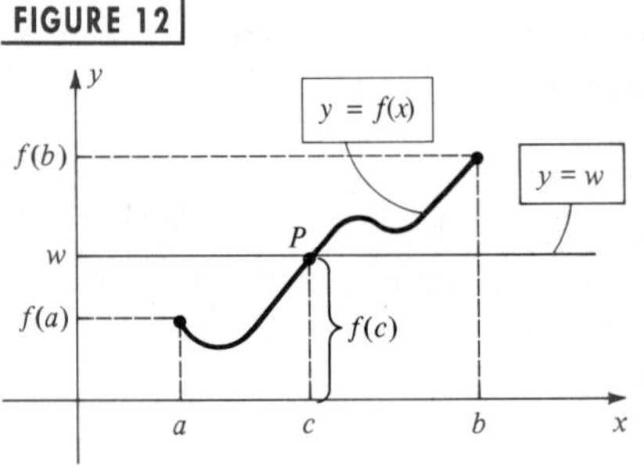

The intermediate value theorem states that if w is any number between $f(a)$ and $f(b)$, then there is a number c between a and b such that $f(c) = w$. If we regard the graph of the polynomial function f as extending continuously from the point $(a, f(a))$ to the point $(b, f(b))$, as illustrated in Figure 12, then for any number w between $f(a)$ and $f(b)$, it appears that a horizontal line with y-intercept w should intersect the graph in at least one point P. The x-coordinate c of P is a number such that $f(c) = w$.

A consequence of the intermediate value theorem is that if $f(a)$ and $f(b)$ have opposite signs, then there is at least one number c between a and b such that $f(c) = 0$; that is, f has a zero at c. Thus, if the point $(a, f(a))$ on the graph of a polynomial function lies below the x-axis, and the point $(b, f(b))$ lies above the x-axis, or vice versa, then the graph crosses the x-axis at least once between $x = a$ and $x = b$, as illustrated in Figure 13.

FIGURE 13

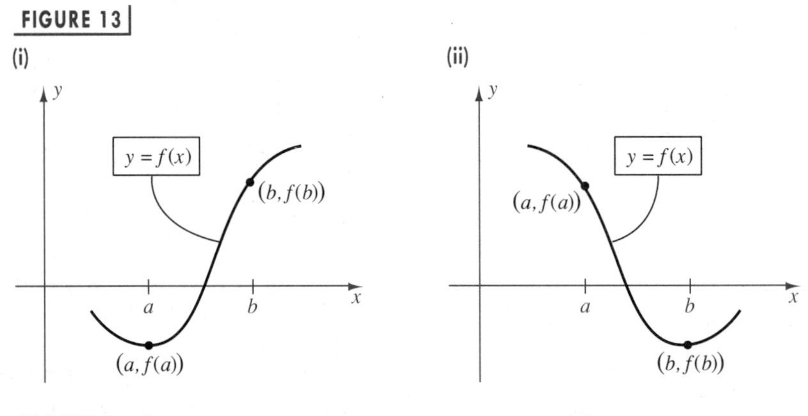

EXAMPLE ▪ 2

Show that $f(x) = x^5 + 2x^4 - 6x^3 + 2x - 3$ has a zero between 1 and 2.

SOLUTION Substituting 1 and 2 for x gives us the function values:

$$f(1) = 1 + 2 - 6 + 2 - 3 = -4$$

$$f(2) = 32 + 32 - 48 + 4 - 3 = 17$$

Since $f(1)$ and $f(2)$ have opposite signs, we see that $f(c) = 0$ for at least one real number c between 1 and 2.

Example 2 illustrates a scheme for locating real zeros of polynomials. By using a method of *successive approximation*, each zero can be approximated to any degree of accuracy by locating it in smaller and smaller intervals (see Exercises 39–44).

A by-product of the preceding discussion is that if c and d are *successive* real zeros of $f(x)$, that is, there are no other zeros between c and d, then *$f(x)$ does not change sign on the interval (c, d)*. Thus, if we choose any number k such that $c < k < d$, and if $f(k)$ is positive, then $f(x)$ is positive throughout (c, d). Similarly, if $f(k)$ is negative, then $f(x)$ is negative throughout (c, d). We shall call $f(k)$ a **test value** for $f(x)$ on the interval (c, d). Test values may also be used on infinite intervals of the form $(-\infty, a)$ or (a, ∞), provided that $f(x)$ has no zero on these intervals. The use of test values in graphing is similar to the technique used for inequalities in Section 2.7.

EXAMPLE ▪ 3

Let $f(x) = x^3 + x^2 - 4x - 4$.

(a) Find all values of x such that $f(x) > 0$, and all values of x such that $f(x) < 0$.
(b) Sketch the graph of f.

SOLUTION

(a) The graph of f lies above the x-axis for values of x such that $f(x) > 0$, and it lies below the x-axis if $f(x) < 0$. We may factor $f(x)$ by grouping terms:

$$\begin{aligned} f(x) &= (x^3 + x^2) - (4x + 4) \\ &= x^2(x + 1) - 4(x + 1) \\ &= (x^2 - 4)(x + 1) \\ &= (x + 2)(x - 2)(x + 1) \end{aligned}$$

FIGURE 14

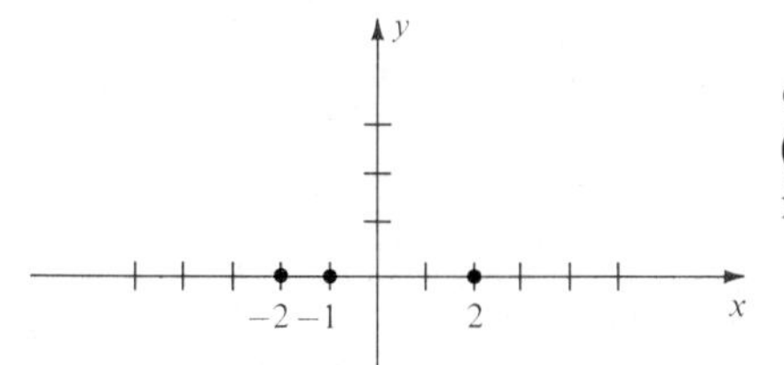

We see from the last equation that the zeros of $f(x)$ (the x-intercepts of the graph) are -2, -1, and 2. The corresponding points on the graph (see Figure 14) divide the x-axis into four parts, and we consider the open intervals

$$(-\infty, -2), \quad (-2, -1), \quad (-1, 2), \quad (2, \infty).$$

The sign of $f(x)$ in each of these intervals can be determined by finding a suitable test value. Thus, if we choose -3 in $(-\infty, -2)$, then

$$\begin{aligned} f(-3) &= (-3)^3 + (-3)^2 - 4(-3) - 4 \\ &= -27 + 9 + 12 - 4 = -10. \end{aligned}$$

Since the test value $f(-3) = -10$ is negative, $f(x)$ is negative throughout the interval $(-\infty, -2)$.

FIGURE 15

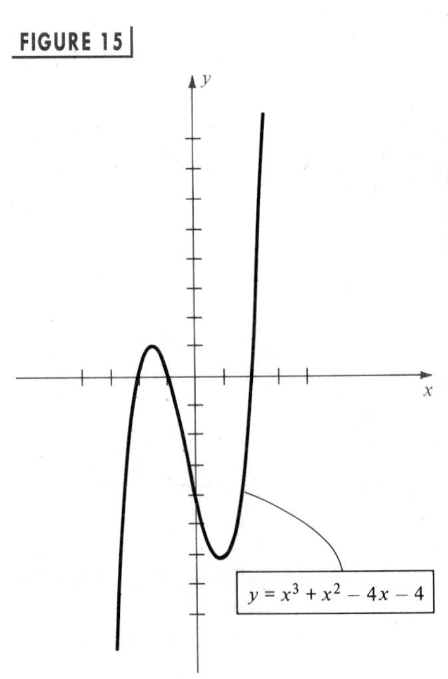

If we choose $-\frac{3}{2}$ in the interval $(-2, -1)$, then the corresponding test value is

$$f(-\tfrac{3}{2}) = (-\tfrac{3}{2})^3 + (-\tfrac{3}{2})^2 - 4(-\tfrac{3}{2}) - 4$$
$$= -\tfrac{27}{8} + \tfrac{9}{4} + 6 - 4 = \tfrac{7}{8}.$$

Since $f(-\frac{3}{2}) = \frac{7}{8}$ is positive, $f(x)$ is positive throughout $(-2, -1)$. The following table summarizes these facts and lists suitable test values and results for the remaining two intervals.

Interval	$(-\infty, -2)$	$(-2, -1)$	$(-1, 2)$	$(2, \infty)$
Test value	$f(-3) = -10$	$f(-\frac{3}{2}) = \frac{7}{8}$	$f(0) = -4$	$f(3) = 20$
Sign of $f(x)$	$-$	$+$	$-$	$+$
Position of graph	Below x-axis	Above x-axis	Below x-axis	Above x-axis

(b) Using the information from the table and plotting several points gives us the graph in Figure 15. To find the turning points of the graph, it would be necessary to use methods developed in calculus.

The graph of every polynomial function of degree 3 has an S-shaped appearance similar to that shown in Figure 15, or it has an inverted version of that graph if the coefficient of x^3 is negative. However, sometimes the graph may have only one x-intercept or the S shape may be elongated, as in Figures 8 and 9.

EXAMPLE ▪ 4

Let $f(x) = x^4 - 4x^3 + 3x^2$.

(a) Find all values of x such that $f(x) > 0$, and all x such that $f(x) < 0$.

(b) Sketch the graph of f.

SOLUTION

(a) We shall follow the same steps used in the solution of Example 2. Thus, we begin by factoring $f(x)$:

$$f(x) = x^2(x^2 - 4x + 3)$$
$$= x^2(x - 1)(x - 3)$$

The zeros of $f(x)$, that is, the x-intercepts of the graph are, in *increasing* order,

$$0, \quad 1, \quad \text{and} \quad 3.$$

The corresponding points on the graph divide the x-axis into four parts, and we consider the open intervals

$$(-\infty, 0), \quad (0, 1), \quad (1, 3), \quad (3, \infty).$$

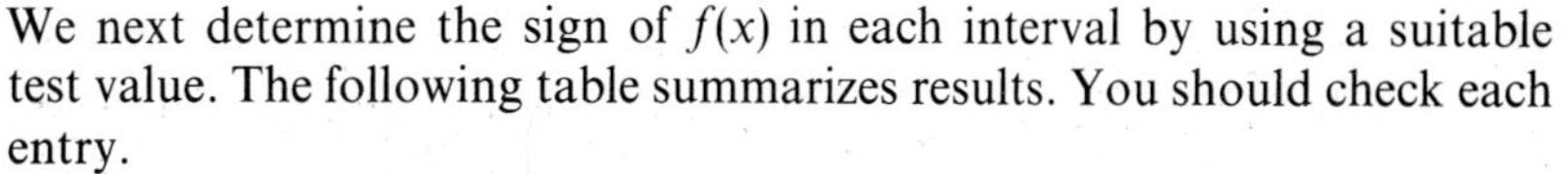
We next determine the sign of $f(x)$ in each interval by using a suitable test value. The following table summarizes results. You should check each entry.

FIGURE 16

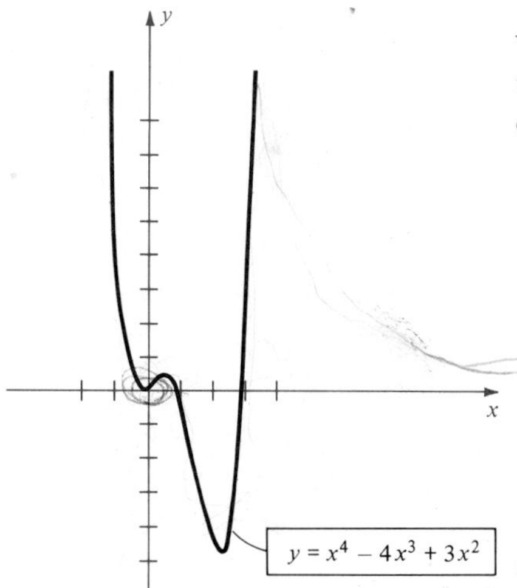

Interval	$(-\infty, 0)$	$(0, 1)$	$(1, 3)$	$(3, \infty)$
Test value	$f(-1) = 8$	$f(\frac{1}{2}) = \frac{5}{16}$	$f(2) = -4$	$f(4) = 48$
Sign of $f(x)$	$+$	$+$	$-$	$+$
Position of graph	Above x-axis	Above x-axis	Below x-axis	Above x-axis

(b) Making use of the information in the table and plotting several points gives us the sketch in Figure 16.

4.2 EXERCISES

Exer. 1–4: Sketch the graph of f for the indicated value of c or a.

1 $f(x) = 2x^3 + c$; **(a)** $c = 3$; **(b)** $c = -3$

2 $f(x) = -2x^3 + c$; **(a)** $c = -2$; **(b)** $c = 2$

3 $f(x) = ax^3 + 2$; **(a)** $a = 2$; **(b)** $a = -\frac{1}{3}$

4 $f(x) = ax^3 - 3$; **(a)** $a = -2$; **(b)** $a = \frac{1}{4}$

Exer. 5–20:

(a) Find all values of x such that $f(x) > 0$, and all x such that $f(x) < 0$.

(b) Sketch the graph of f.

5 $f(x) = \frac{1}{4}x^3 - 2$

6 $f(x) = -\frac{1}{9}x^3 - 3$

7 $f(x) = -\frac{1}{16}x^4 + 1$

8 $f(x) = x^5 + 1$

9 $f(x) = x^4 - 4x^2$

10 $f(x) = 9x - x^3$

11 $f(x) = -x^3 + 3x^2 + 10x$

12 $f(x) = x^4 + 3x^3 - 4x^2$

13 $f(x) = \frac{1}{6}(x + 2)(x - 3)(x - 4)$

14 $f(x) = -\frac{1}{8}(x + 4)(x - 2)(x - 6)$

15 $f(x) = x^3 + 2x^2 - 4x - 8$

16 $f(x) = x^3 - 3x^2 - 9x + 27$

17 $f(x) = x^4 - 6x^2 + 8$

18 $f(x) = -x^4 + 12x^2 - 27$

19 $f(x) = x^2(x + 2)(x - 1)^2(x - 2)$

20 $f(x) = x^3(x + 1)^2(x - 2)(x - 4)$

21 Let $f(x)$ be a polynomial such that the coefficient of every odd power of x is 0. Show that f is an even function.

22 Let $f(x)$ be a polynomial such that the coefficient of every even power of x is 0. Show that f is an odd function.

23 If $f(x) = 3x^3 - kx^2 + x - 5k$, find a number k such that the graph of f contains the point $(-1, 4)$.

24 If one zero of $f(x) = x^3 - 2x^2 - 16x + 16k$ is 2, find two other zeros.

Exer. 25–30: Show that f has a zero between a and b.

25 $f(x) = x^3 - 4x^2 + 3x - 2;\quad a = 3,\quad b = 4$

26 $f(x) = 2x^3 + 5x^2 - 3;\quad a = -3,\quad b = -2$

27 $f(x) = -x^4 + 3x^3 - 2x + 1;\quad a = 2,\quad b = 3$

28 $f(x) = 2x^4 + 3x - 2;\quad a = \frac{1}{2},\quad b = \frac{3}{4}$

29 $f(x) = x^5 + x^3 + x^2 + x + 1;\quad a = -\frac{1}{2},\quad b = -1$

30 $f(x) = x^5 - 3x^4 - 2x^3 + 3x^2 - 9x - 6;\quad a = 3,\quad b = 4$

31 The third-degree Legendre polynomial

$$P(x) = \tfrac{1}{2}(5x^3 - 3x)$$

occurs in the solution of heat transfer problems in physics and engineering.

(a) Find all values of x such that $P(x) > 0$, and all x such that $P(x) < 0$.

(b) Sketch the graph of P.

32 The fourth-degree Chebyshev polynomial

$$f(x) = 8x^4 - 8x^2 + 1$$

occurs in statistical studies. Find all values of x such that $f(x) > 0$. (*Hint:* Let $z = x^2$ and use the quadratic formula.)

33 An open box is to be made from a rectangular piece of cardboard having dimensions 20 inches × 30 inches, by cutting out identical squares of area x^2 from each corner and turning up the sides (see Exercise 31 of Section 3.4). Show that the volume of the box is given by the third-degree polynomial

$$V(x) = x(20 - 2x)(30 - 2x).$$

(a) Find all positive values of x such that $V(x) > 0$.

(b) Sketch the graph of V for $x > 0$.

34 The frame for a shipping crate is to be constructed from 24 feet of 2 × 2 lumber (see figure).

(a) If the crate is to have square ends of side x feet, express the outer volume V of the crate as a function of x. (Disregard the thickness of the lumber.)

(b) Sketch the graph of V for $x > 0$.

EXERCISE 34

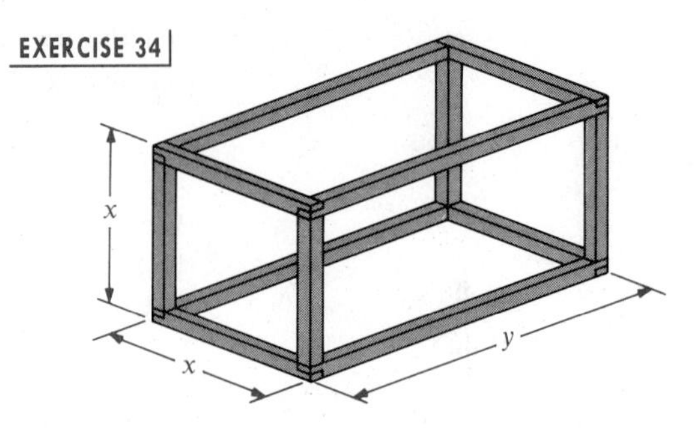

35 A meteorologist determines that the temperature T (in °F) on a certain cold winter day was given by

$$T = \tfrac{1}{20}t(t - 12)(t - 24)$$

for time t (in hours) and $t = 0$ corresponding to 6 A.M.

(a) When was $T > 0$ and when was $T < 0$?

(b) Sketch a graph that depicts the temperature T for $0 \le t \le 24$.

(c) Show that the temperature was 32 °F sometime between 12 noon and 1 P.M. (*Hint:* Use the intermediate value theorem.)

36 A diver stands on the very end of a diving board before beginning a dive. The deflection d of the board at a position s feet from the stationary end is given by the third-degree polynomial

$$d = cs^2(3L - s) \quad \text{for } 0 \le s \le L$$

for the length L of the board and a positive constant c that depends on the weight of the diver and on the physical properties of the board (see figure). Suppose the board is 10 feet long.

(a) If the deflection at the end of the board is 1 foot, find c.

(b) Show that the deflection is $\frac{1}{2}$ foot somewhere between $s = 6.5$ and $s = 6.6$.

EXERCISE 36

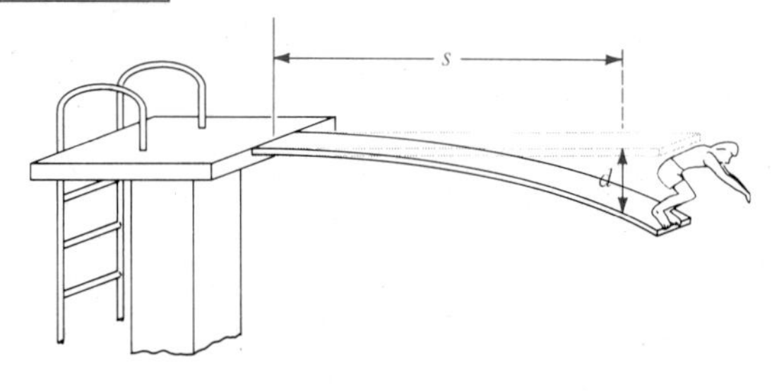

37 A herd of 100 deer is introduced to a small island. The herd at first increases rapidly, but eventually the food resources of the island dwindle and the population declines. Suppose that the number $N(t)$ of deer after t years is given by

$$N(t) = -t^4 + 21t^2 + 100.$$

(a) Determine the positive values of t for which $N(t) > 0$. Does the population become extinct? If so, when?

(b) Sketch the graph of N for $t > 0$.

38 Refer to Exercise 37. It can be shown by means of calculus that the rate R at which the deer population grows (or declines) at time t is given by

$$R = -4t^3 + 42t \text{ (deer per year).}$$

(a) When does the population cease to grow?

(b) Determine the positive values of t for which $R > 0$.

39 If a zero of a polynomial lies between two integers, then a method of *successive approximation* may be used to approximate the zero to any degree of accuracy. Let $f(x) = x^3 - 3x + 1$.

(a) Show that $f(x)$ has a zero between 1 and 2.

(b) By increasing values of x by tenths, show that $f(x)$ has a zero between 1.5 and 1.6.

(c) By increasing values of x by hundredths, show that $f(x)$ has a zero between 1.53 and 1.54. (This zero is said to be *isolated between successive hundredths* on the interval $[1, 2]$.)

Exer. 40–44: Refer to Exercise 39(c). Isolate a zero of $f(x)$ between successive hundredths on the interval $[a, b]$.

40 $f(x) = x^3 + 5x - 3;\quad a = 0,\quad b = 1$

41 $f(x) = 2x^3 - 4x^2 - 3x + 1;\quad a = 2,\quad b = 3$

42 $f(x) = x^4 - 4x^3 + 3x^2 - 8x + 2;\quad a = 3,\quad b = 4$

43 $f(x) = x^5 - 2x^2 + 4;\quad a = -2,\quad b = -1$

44 $f(x) = x^4 - 2x^3 + 10x - 25;\quad a = 2,\quad b = 3$

4.3 PROPERTIES OF DIVISION

In this section symbols such as $f(x)$ and $g(x)$ will be used to denote polynomials in x. If a polynomial $g(x)$ is a factor of a polynomial $f(x)$, then $f(x)$ is **divisible** by $g(x)$. For example, the polynomial $x^4 - 16$ is divisible by $x^2 - 4$, by $x^2 + 4$, by $x + 2$, and by $x - 2$.

The polynomial $x^4 - 16$ is not divisible by $x^2 + 3x + 1$; however, we can use *long division* to find a *quotient* and *remainder*, as in the following illustration.

ILLUSTRATION | LONG DIVISION

$$\begin{array}{r|l}
 & x^2 - 3x + 8 \quad \textit{(quotient)} \\
\hline
x^2 + 3x + 1 & x^4 \qquad\qquad\qquad\qquad - 16 \\
 & x^4 + 3x^3 + x^2 \\
 & -3x^3 - x^2 \\
 & -3x^3 - 9x^2 - 3x \\
 & 8x^2 + 3x - 16 \\
 & 8x^2 + 24x + 8 \\
 & -21x - 24 \quad \textit{(remainder)}
\end{array}$$

The long division process ends when we arrive at a polynomial (the remainder) that either is 0 or has smaller degree than the divisor. The result of the long division in the preceding illustration can be written

$$\frac{x^4 - 16}{x^2 + 3x + 1} = (x^2 - 3x + 8) + \left(\frac{-21x - 24}{x^2 + 3x + 1}\right).$$

Multiplying both sides of this equation by $x^2 + 3x + 1$, we obtain

$$x^4 - 16 = (x^2 + 3x + 1)(x^2 - 3x + 8) + (-21x - 24).$$

This example illustrates the following theorem, which we state without proof.

DIVISION ALGORITHM FOR POLYNOMIALS

If $f(x)$ and $p(x)$ are polynomials and if $p(x) \neq 0$, then there exist unique polynomials $q(x)$ and $r(x)$ such that

$$f(x) = p(x)q(x) + r(x)$$

where either $r(x) = 0$ or the degree of $r(x)$ is less than the degree of $p(x)$. The polynomial $q(x)$ is the **quotient** and $r(x)$ is the **remainder** in the division of $f(x)$ by $p(x)$.

An interesting special case occurs if $f(x)$ is divided by a first-degree polynomial of the form $x - c$ where c is a real number. If $x - c$ is a factor of $f(x)$, then

$$f(x) = (x - c)q(x)$$

for some quotient $q(x)$; that is, the remainder $r(x)$ is 0. If $x - c$ is not a factor of $f(x)$, then the degree of the remainder $r(x)$ is less than the degree of $x - c$, and hence $r(x)$ must have degree 0. This means that the remainder is a nonzero number. Consequently, for every $x - c$ we have

$$f(x) = (x - c)q(x) + d$$

where the remainder d is some real number (possibly $d = 0$). If c is substituted for x in the equation $f(x) = (x - c)q(x) + d$, we obtain

$$f(c) = (c - c)q(c) + d,$$

which reduces to $f(c) = d$. This proves the following theorem.

REMAINDER THEOREM

> If a polynomial $f(x)$ is divided by $x - c$, then the remainder is $f(c)$.

EXAMPLE ■ 1

If $f(x) = x^3 - 3x^2 + x + 5$, use the remainder theorem to find $f(2)$.

SOLUTION According to the remainder theorem, $f(2)$ is the remainder when $f(x)$ is divided by $x - 2$. By long division,

$$
\begin{array}{r|l}
 & x^2 - x - 1 \\
\hline
x - 2 & x^3 - 3x^2 + x + 5 \\
 & x^3 - 2x^2 \\
\hline
 & -x^2 + x \\
 & -x^2 + 2x \\
\hline
 & -x + 5 \\
 & -x + 2 \\
\hline
 & 3
\end{array}
$$

Hence, $f(2) = 3$. We may check this fact by direct substitution:

$$f(2) = 2^3 - 3(2)^2 + 2 + 5 = 3.$$

FACTOR THEOREM

> A polynomial $f(x)$ has a factor $x - c$ if and only if $f(c) = 0$.

PROOF By the remainder theorem, $f(x) = (x - c)q(x) + f(c)$ for some quotient $q(x)$. If $f(c) = 0$, then $f(x) = (x - c)q(x)$; that is, $x - c$ is a factor of $f(x)$. Conversely, if $x - c$ is a factor, then the remainder upon division of $f(x)$ by $x - c$ must be 0, and hence, by the remainder theorem, $f(c) = 0$. ❑

The factor theorem is useful for finding factors of polynomials, as illustrated in the next example.

EXAMPLE ■ 2

Show that $x - 2$ is a factor of the polynomial $f(x) = x^3 - 4x^2 + 3x + 2$.

SOLUTION Since $f(2) = 8 - 16 + 6 + 2 = 0$, we see from the factor theorem that $x - 2$ is a factor of $f(x)$. Another method of solution would be to divide $f(x)$ by $x - 2$ and show that the remainder is 0. The quotient in the division would be another factor of $f(x)$.

EXAMPLE ▪ 3

Find a polynomial $f(x)$ of degree 3 that has zeros 2, -1, and 3.

SOLUTION By the factor theorem, $f(x)$ has factors $x - 2$, $x + 1$, and $x - 3$. Thus

$$f(x) = a(x - 2)(x + 1)(x - 3),$$

where any nonzero value may be assigned to a. If we let $a = 1$ and multiply, we obtain

$$f(x) = x^3 - 4x^2 + x + 6.$$

To apply the remainder theorem it is necessary to divide a polynomial $f(x)$ by $x - c$. The method of **synthetic division** may be used to simplify this work. The following rules state how to proceed. The method can be justified by a careful (and lengthy) comparison with the method of long division.

GUIDELINES SYNTHETIC DIVISION OF $a_nx^n + a_{n-1}x^{n-1} + \cdots + a_1x + a_0$ BY $x - c$

1 Begin with the following display, supplying zeros for any missing coefficients in the given polynomial.

$$\begin{array}{r|llllll} c & a_n & a_{n-1} & a_{n-2} & \cdots & a_1 & a_0 \\ & & & & & & \\ \hline & a_n & & & & & \end{array}$$

2 Multiply a_n by c and place the product ca_n underneath a_{n-1}, as indicated by the arrow in the following display. (This arrow, and others, is used only to help clarify these rules, and will not appear in *specific* synthetic divisions.) Next find the sum $b_1 = a_{n-1} + ca_n$ and place it below the line as shown.

$$\begin{array}{r|llllllll} c & a_n & a_{n-1} & a_{n-2} & \cdots & & a_1 & a_0 \\ & & ca_n & cb_1 & cb_2 & \cdots & cb_{n-2} & cb_{n-1} \\ \hline & a_n & b_1 & b_2 & \cdots & b_{n-2} & b_{n-1} & r \end{array}$$

3 Multiply b_1 by c and place the product cb_1 underneath a_{n-2}, as indicated by another arrow. Next find the sum $b_2 = a_{n-2} + cb_1$ and place it below the line as shown.

4 Continue this process, as indicated by the arrows, until the final sum $r = a_0 + cb_{n-1}$ is obtained. The numbers

$$a_n, \quad b_1, \quad b_2, \quad \ldots, \quad b_{n-2}, \quad b_{n-1}$$

are the coefficients of the quotient $q(x)$; that is,

$$q(x) = a_n x^{n-1} + b_1 x^{n-2} + \cdots + b_{n-2}x + b_{n-1},$$

and r is the remainder. ❑

The following examples illustrate synthetic division for some special cases.

EXAMPLE ▪ 4

Use synthetic division to find the quotient and remainder if the polynomial $2x^4 + 5x^3 - 2x - 8$ is divided by $x + 3$.

SOLUTION Since the divisor is $x + 3$, the c in the expression $x - c$ is -3. Hence, the synthetic division takes this form:

-3	2	5	0	-2	-8
		-6	3	-9	33
	2	-1	3	-11	25

The first four numbers in the third row are the coefficients of the quotient $q(x)$ and the last number is the remainder r. Thus,

$$q(x) = 2x^3 - x^2 + 3x - 11 \quad \text{and} \quad r = 25.$$

Synthetic division can be used to find values of polynomial functions, as illustrated in the next example.

EXAMPLE ▪ 5

If $f(x) = 3x^5 - 38x^3 + 5x^2 - 1$, use synthetic division to find $f(4)$.

SOLUTION By the remainder theorem, $f(4)$ is the remainder when $f(x)$ is divided by $x - 4$. Dividing synthetically, we obtain

4	3	0	-38	5	0	-1
		12	48	40	180	720
	3	12	10	45	180	719

Consequently, $f(4) = 719$.

Synthetic division may be employed to help find zeros of polynomials. By the method illustrated in the preceding example, $f(c) = 0$ if and only if the remainder in the synthetic division by $x - c$ is 0.

EXAMPLE ▪ 6

Show that -11 is a zero of the polynomial

$$f(x) = x^3 + 8x^2 - 29x + 44.$$

SOLUTION Dividing synthetically by $x - (-11) = x + 11$ gives us

$$\begin{array}{r|rrrr} -11 & 1 & 8 & -29 & 44 \\ & & -11 & 33 & -44 \\ \hline & 1 & -3 & 4 & 0 \end{array}$$

Thus, $f(-11) = 0$.

Example 6 shows that the number -11 is a solution of the equation $x^3 + 8x^2 - 29x + 44 = 0$. In Section 4.5 we shall use synthetic division to find rational solutions of equations.

4.3 EXERCISES

Exer. 1–6: Find the quotient and remainder if $f(x)$ is divided by $p(x)$.

1 $f(x) = 2x^4 - x^3 - 3x^2 + 7x - 12;\quad p(x) = x^2 - 3$

2 $f(x) = 3x^4 + 2x^3 - x^2 - x - 6;\quad p(x) = x^2 + 1$

3 $f(x) = 3x^3 + 2x - 4;\quad p(x) = 2x^2 + 1$

4 $f(x) = 3x^3 - 5x^2 - 4x - 8;\quad p(x) = 2x^2 + x$

5 $f(x) = 7x + 2;\quad p(x) = 2x^2 - x - 4$

6 $f(x) = -5x^2 + 3;\quad p(x) = x^3 - 3x + 9$

Exer. 7–10: Use the remainder theorem to find $f(c)$.

7 $f(x) = 3x^3 - x^2 + 5x - 4;\quad c = 2$

8 $f(x) = 2x^3 + 4x^2 - 3x - 1;\quad c = 3$

9 $f(x) = x^4 - 6x^2 + 4x - 8;\quad c = -3$

10 $f(x) = x^4 + 3x^2 - 12;\quad c = -2$

Exer. 11–14: Use the factor theorem to show that $x - c$ is a factor of $f(x)$.

11 $f(x) = x^3 + x^2 - 2x + 12;\quad c = -3$

12 $f(x) = x^3 + x^2 - 11x + 10;\quad c = 2$

13 $f(x) = x^{12} - 4096;\quad c = -2$

14 $f(x) = x^4 - 3x^3 - 2x^2 + 5x + 6;\quad c = 2$

Exer. 15–22: Use synthetic division to find the quotient and remainder if the first polynomial is divided by the second.

15 $2x^3 - 3x^2 + 4x - 5;\quad x - 2$

16 $3x^3 - 4x^2 - x + 8;\quad x + 4$

17 $x^3 - 8x - 5;\quad x + 3$

18 $5x^3 - 6x^2 + 15;\quad x - 4$

19 $3x^5 + 6x^2 + 7;\quad x + 2$

20 $-2x^4 + 10x - 3;\quad x - 3$

21 $4x^4 - 5x^2 + 1;\quad x - \frac{1}{2}$

22 $9x^3 - 6x^2 + 3x - 4;\quad x - \frac{1}{3}$

Exer. 23–28: Use synthetic division to find $f(c)$.

23 $f(x) = 2x^3 + 3x^2 - 4x + 4;\quad c = 3$

24 $f(x) = -x^3 + 4x^2 + x;\quad c = -2$

25 $f(x) = 0.3x^3 + 0.04x - 0.034;\quad c = -0.2$

26 $f(x) = 8x^5 - 3x^2 + 7; \quad c = \frac{1}{2}$

27 $f(x) = x^2 + 3x - 5; \quad c = 2 + \sqrt{3}$

28 $f(x) = x^3 - 3x^2 - 8; \quad c = 1 + \sqrt{2}$

Exer. 29–32: Use synthetic division to show that c is a zero of $f(x)$.

29 $f(x) = 3x^4 + 8x^3 - 2x^2 - 10x + 4; \quad c = -2$

30 $f(x) = 4x^3 - 9x^2 - 8x - 3; \quad c = 3$

31 $f(x) = 4x^3 - 6x^2 + 8x - 3; \quad c = \frac{1}{2}$

32 $f(x) = 27x^4 - 9x^3 + 3x^2 + 6x + 1; \quad c = -\frac{1}{3}$

Exer. 33–34: Find all values of k such that $f(x)$ is divisible by the given linear polynomial.

33 $f(x) = kx^3 + x^2 + k^2x + 3k^2 + 11; \quad x + 2$

34 $f(x) = k^2x^3 - 4kx + 3; \quad x - 1$

Exer. 35–36: Show that $x - c$ is not a factor of $f(x)$ for any real number c.

35 $f(x) = 3x^4 + x^2 + 5$

36 $f(x) = -x^4 - 3x^2 - 2$

37 Find the remainder if the polynomial

$$3x^{100} + 5x^{85} - 4x^{38} + 2x^{17} - 6$$

is divided by $x + 1$.

Exer. 38–40: Use the factor theorem to verify the statement.

38 $x - y$ is a factor of $x^n - y^n$ for every positive integer n.

39 $x + y$ is a factor of $x^n - y^n$ for every positive even integer n.

40 $x + y$ is a factor of $x^n + y^n$ for every positive odd integer n.

41 Let $P(x, y)$ be a first-quadrant point on the line $y = 6 - x$ and consider the vertical line segment PQ shown in the figure.

EXERCISE 41

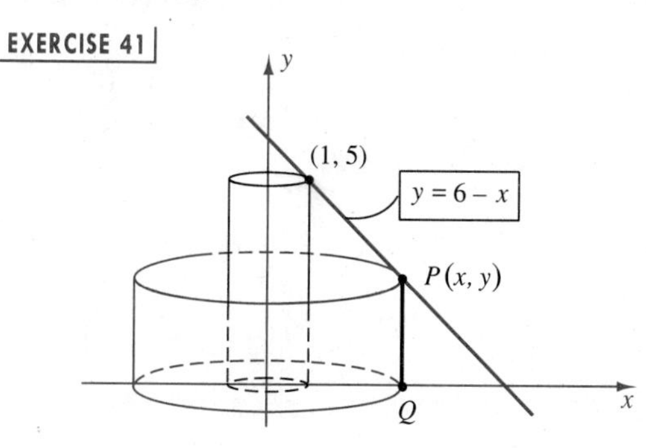

(a) If PQ is rotated about the y-axis, determine the volume V of the resulting cylinder.

(b) For what point $P(x, y)$ with $x \neq 1$ is the volume V in part (a) the same as the volume of the cylinder of radius 1 and altitude 5 shown in the figure?

42 The strength of a rectangular beam is directly proportional to the product of its width and the square of the depth of a cross section (see figure). A beam of width 1.5 feet has been cut from a cylindrical log of radius 1 foot. Find the width of a second rectangular beam of equal strength that could have been cut from the log.

EXERCISE 42

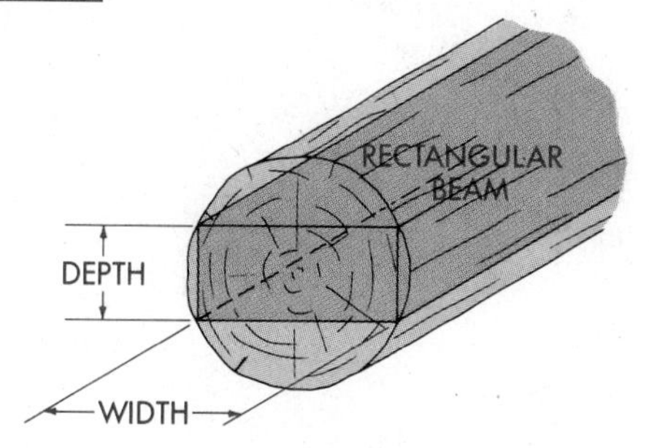

43 An arch has the shape of the parabola $y = 4 - x^2$. A rectangle is fit under the arch by selecting a point (x, y) on the parabola (see figure).

(a) Express the area A of the rectangle in terms of x.

(b) If $x = 1$, the rectangle has base 2 and height 3. Find the base of a second rectangle that has the same area.

EXERCISE 43

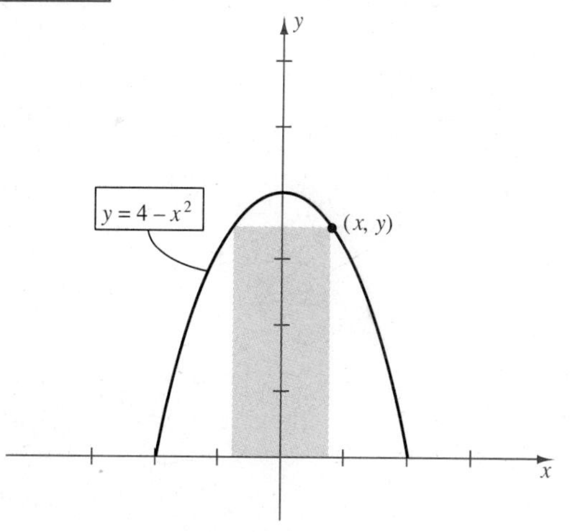

4.4 ZEROS OF POLYNOMIALS

The zeros of a polynomial $f(x)$ are the solutions of the equation $f(x) = 0$. Each zero is an x-intercept of the graph of f. In applied fields, calculators and computers are usually employed to find or approximate zeros. However, before you use a calculator, it is worth knowing what type of zeros to expect. Some questions we could ask are: How many zeros are real or complex? How many real zeros are positive or negative? How many are rational or irrational? Are the zeros large or small in value? In this and the following section we shall discuss results that help answer some of these questions. These results form the basis of the *theory of equations*.

The factor and remainder theorems can be extended to the system of complex numbers. Thus, a complex number $c = a + bi$ is a zero of a polynomial $f(x)$ if and only if $x - c$ is a factor of $f(x)$. Except in special cases, zeros of polynomials are very difficult to find. For example, $f(x) = x^5 - 3x^4 + 4x^3 - 4x - 10$ has no obvious zeros. Although we have no formula that can be used to find the zeros, the next theorem states that there is at *least* one zero.

FUNDAMENTAL THEOREM OF ALGEBRA

> If a polynomial $f(x)$ has positive degree and complex coefficients, then $f(x)$ has at least one complex zero.

The usual proof of this theorem requires results from the field of mathematics called *functions of a complex variable*. A prerequisite for studying this field is a strong background in calculus. The first proof of the fundamental theorem of algebra was given by the German mathematician Carl Friedrich Gauss (1777–1855), who is considered by many to be the greatest mathematician of all time.

As a special case of the fundamental theorem, if all the coefficients of $f(x)$ are real, then $f(x)$ has at least one complex zero. If $a + bi$ is a complex zero, it may happen that $b = 0$, in which case we refer to the number a as a **real zero**. We can use the factor theorem to obtain the following useful corollary of the fundamental theorem.

COROLLARY

> Every polynomial of positive degree has a factor of the form $x - c$ for some complex number c.

The corollary enables us, at least in theory, to express every polynomial $f(x)$ of positive degree as a product of polynomials of degree 1, as in the next theorem.

THEOREM

> If $f(x)$ is a polynomial of degree $n > 0$, then there exist n complex numbers $c_1, c_2, \ldots, c_n$ such that
>
> $$f(x) = a(x - c_1)(x - c_2) \cdots (x - c_n)$$
>
> where a is the leading coefficient of $f(x)$. Each number c_k is a zero of $f(x)$.

PROOF If $f(x)$ has degree $n > 0$, then applying the preceding corollary gives us

$$f(x) = (x - c_1)f_1(x)$$

for a complex number c_1 and a polynomial $f_1(x)$ of degree $n - 1$. If $n - 1 > 0$, we may apply the corollary again, obtaining

$$f_1(x) = (x - c_2)f_2(x)$$

for a complex number c_2 and a polynomial $f_2(x)$ of degree $n - 2$. Hence,

$$f(x) = (x - c_1)(x - c_2)f_2(x).$$

Continuing this process, after n steps we arrive at a polynomial $f_n(x)$ of degree 0. Thus, $f_n(x) = a$ for some nonzero number a, and we may write

$$f(x) = a(x - c_1)(x - c_2) \cdots (x - c_n)$$

such that each complex number c_k is a zero of $f(x)$. The leading coefficient of the polynomial on the right-hand side in the last equation is a and hence a is the leading coefficient of $f(x)$. This completes the proof. ❑

COROLLARY

> A polynomial of degree $n > 0$ has at most n different complex zeros.

PROOF We shall give an indirect proof. Suppose $f(x)$ has *more* than n different complex zeros. Let us choose $n + 1$ of these zeros and label them

$c_1, c_2, \ldots, c_n$, and c. We may use the c_k to obtain the factorization indicated in the statement of the theorem. Substituting c for x and using the fact that $f(c) = 0$, we obtain

$$0 = a(c - c_1)(c - c_2) \cdots (c - c_n).$$

However, each factor on the right-hand side is different from zero because $c \neq c_k$ for every k. Since the product of nonzero numbers cannot equal zero, we have a contradiction. ❑

EXAMPLE ▪ 1

Find a polynomial $f(x)$ of degree 3 with zeros 2, -1, and 3 such that $f(1) = 5$.

SOLUTION By the factor theorem, $f(x)$ has factors $x - 2$, $x + 1$, and $x - 3$. No other factors of degree 1 exist, since, by the factor theorem, another linear factor $x - c$ would produce a fourth zero of $f(x)$, contrary to the preceding corollary. Hence, $f(x)$ has the form

$$f(x) = a(x - 2)(x + 1)(x - 3)$$

for some number a. Since $f(1) = 5$, we see that

$$a(1 - 2)(1 + 1)(1 - 3) = 5$$
$$4a = 5.$$

Consequently, $a = \frac{5}{4}$ and

$$f(x) = \tfrac{5}{4}(x - 2)(x + 1)(x - 3).$$

We multiply the four factors to obtain the polynomial

$$f(x) = \tfrac{5}{4}x^3 - 5x^2 + \tfrac{5}{4}x + \tfrac{15}{2}.$$

The numbers $c_1, c_2, \ldots, c_n$ in the preceding theorem are not necessarily all different. To illustrate, the polynomial $f(x) = x^3 + x^2 - 5x + 3$ has the factorization

$$f(x) = (x + 3)(x - 1)(x - 1).$$

If a factor $x - c$ occurs m times in the factorization, then c is a **zero of multiplicity m** of $f(x)$, or a **root of multiplicity m** of the equation $f(x) = 0$. In the preceding illustration, 1 is a zero of multiplicity 2 and -3 is a zero of multiplicity 1.

EXAMPLE ▪ 2

Find the zeros of the polynomial $f(x) = (x - 2)(x - 4)^3(x + 1)^2$, and state the multiplicity of each.

SOLUTION We see from the factored form that $f(x)$ has three distinct zeros, 2, 4, and -1. The zero 2 has multiplicity 1, the zero 4 has multiplicity 3, and the zero -1 has multiplicity 2. Note that $f(x)$ has degree 6.

If $f(x)$ is a polynomial of degree n and $f(x) = a(x - c_1)(x - c_2) \cdots (x - c_n)$, then the n complex numbers $c_1, c_2, \ldots, c_n$ are zeros of $f(x)$. If a zero of multiplicity m is counted as m zeros, this tells us that $f(x)$ has at least n zeros (not necessarily all different). Combining this with the fact that $f(x)$ has at most n zeros gives us the next result.

THEOREM

> If $f(x)$ is a polynomial of degree $n > 0$ and if a zero of multiplicity m is counted m times, then $f(x)$ has precisely n zeros.

EXAMPLE ▪ 3

Express $f(x) = x^5 - 4x^4 + 13x^3$ as a product of linear factors, and find the five zeros of $f(x)$.

SOLUTION We begin by writing

$$f(x) = x^3(x^2 - 4x + 13).$$

By the quadratic formula, the zeros of the polynomial $x^2 - 4x + 13$ are

$$\frac{4 \pm \sqrt{16 - 52}}{2} = \frac{4 \pm \sqrt{-36}}{2} = \frac{4 \pm 6i}{2} = 2 \pm 3i.$$

Hence, by the factor theorem, $x^2 - 4x + 13$ has factors $x - (2 + 3i)$ and $x - (2 - 3i)$, and we obtain the factorization

$$f(x) = x \cdot x \cdot x \cdot (x - 2 - 3i)(x - 2 + 3i).$$

Since $x - 0$ occurs as a factor three times, the number 0 is zero of multiplicity 3, and the five zeros of $f(x)$ are 0, 0, 0, $2 + 3i$, and $2 - 3i$.

We may use the next theorem to obtain information about the zeros of a polynomial $f(x)$ with real coefficients. In the statement of the theorem we assume that the terms of $f(x)$ are arranged in order of decreasing powers of x, and that terms with zero coefficients are deleted. We also assume that the **constant term**, that is, the term that does not contain x, is different from 0. We say there is a **variation of sign** in $f(x)$ if two consecutive coefficients have opposite signs. To illustrate, the polynomial

$$f(x) = 2x^5 - 7x^4 + 3x^2 + 6x - 5$$

has three variations in sign—one variation from $2x^5$ to $-7x^4$, a second from $-7x^4$ to $3x^2$, and a third from $6x$ to -5.

The theorem also refers to the variations of sign in $f(-x)$. Using the previous illustration, note that

$$\begin{aligned} f(-x) &= 2(-x)^5 - 7(-x)^4 + 3(-x)^2 + 6(-x) - 5 \\ &= -2x^5 - 7x^4 + 3x^2 - 6x - 5. \end{aligned}$$

Hence, there are two variations of sign in $f(-x)$—one from $-7x^4$ to $3x^2$, and a second from $3x^2$ to $-6x$.

DESCARTES' RULE OF SIGNS

Let $f(x)$ be a polynomial with real coefficients and nonzero constant term.

(i) The number of positive real solutions of the equation $f(x) = 0$ either is equal to the number of variations of sign in $f(x)$ or is less than that number by an even integer.

(ii) The number of negative real solutions of the equation $f(x) = 0$ either is equal to the number of variations of sign in $f(-x)$ or is less than that number by an even integer.

The proof of Descartes' rule will not be given.

EXAMPLE ▪ 4

Discuss the number of possible positive and negative real solutions and nonreal complex solutions of the equation

$$2x^5 - 7x^4 + 3x^2 + 6x - 5 = 0.$$

SOLUTION The polynomial $f(x)$ on the left-hand side of the equation is the same as the one given in the illustration preceding the statement of Descartes' rule. Since there are three variations of sign in $f(x)$, the equation has either three positive real solutions or one positive real solution.

Since $f(-x) = -2x^5 - 7x^4 + 3x^2 - 6x - 5$ has two variations of sign, the given equation has either two negative solutions or no negative solution. The solutions that are not real numbers are complex numbers of the form $a + bi$ for real numbers a and b with $b \neq 0$. The following table summarizes the various possibilities that can occur for solutions of the equation.

Total number of solutions	5	5	5	5
Number of positive real solutions	1	1	3	3
Number of negative real solutions	0	2	0	2
Number of nonreal, complex solutions	4	2	2	0

Descartes' rule stipulates that the constant term of the polynomial is different from 0. If the constant term is 0, as in the equation

$$x^4 - 3x^3 + 2x^2 - 5x = 0,$$

we factor out the lowest power of x, obtaining

$$x(x^3 - 3x^2 + 2x - 5) = 0.$$

Thus one solution is $x = 0$, and we apply Descartes' rule to the polynomial $x^3 - 3x^2 + 2x - 5$ to determine the nature of the remaining three solutions.

EXAMPLE ▪ 5

Discuss the nature of the roots of the equation $3x^5 + 4x^3 + 2x - 5 = 0$.

SOLUTION The polynomial $f(x)$ on the left-hand side of the equation has one variation of sign and hence, by (i) of Descartes' rule, the equation has precisely one positive real root. Since

$$\begin{aligned} f(-x) &= 3(-x)^5 + 4(-x)^3 + 2(-x) - 5 \\ &= -3x^5 - 4x^3 - 2x - 5, \end{aligned}$$

$f(-x)$ has no variation of sign and hence, by (ii) of Descartes' rule, there is no negative real root. Thus, the equation has one real root and four nonreal, complex roots.

When applying Descartes' rule, we count roots of multiplicity k as k roots. For example, given $x^2 - 2x + 1 = 0$, the polynomial $x^2 - 2x + 1$ has two variations of sign and hence the equation has either two positive real roots or none. The factored form of the equation is $(x - 1)^2 = 0$, and hence 1 is a root of multiplicity 2.

We conclude this section with a discussion of *bounds* for the real solutions of an equation $f(x) = 0$ if $f(x)$ is a polynomial with real coefficients. By definition, a real number b is an **upper bound** for the solutions if no solution is greater than b. A real number a is a **lower bound** for the solutions if no solution is less than a. Thus, if r is a real solution of $f(x) = 0$, then $a \le r \le b$; that is, r is in the closed interval $[a, b]$. Note that upper and lower bounds are not unique, since any number greater than b is also an upper bound, and any number less than a is a lower bound.

We may use synthetic division to find upper and lower bounds for the solutions of $f(x) = 0$. Recall that if we divide $f(x)$ synthetically by $x - c$, then the third row in the division process contains the coefficients of the quotient $q(x)$ together with the remainder $f(c)$. The following theorem indicates how this third row may be used to find upper and lower bounds for the real solutions.

BOUNDS FOR REAL ZEROS OF POLYNOMIALS

Suppose that $f(x)$ is a polynomial with real coefficients and positive leading coefficient, and that $f(x)$ is divided synthetically by $x - c$.

(i) If $c > 0$ and if all numbers in the third row of the division process are either positive or zero, then c is an upper bound for the real solutions of the equation $f(x) = 0$.

(ii) If $c < 0$ and if the numbers in the third row of the division process are alternately positive and negative (and a 0 in the third row is considered to be either positive or negative), then c is a lower bound for the real solutions of the equation $f(x) = 0$.

A general proof of this result can be patterned after the solution given in the next example.

EXAMPLE ▪ 6

Find upper and lower bounds for the real solutions of the equation $2x^3 + 5x^2 - 8x - 7 = 0$.

SOLUTION We divide the polynomial $2x^3 + 5x^2 - 8x - 7$ synthetically by $x - 1$ and $x - 2$:

$$\begin{array}{r|rrrr} 1 & 2 & 5 & -8 & -7 \\ & & 2 & 7 & -1 \\ \hline & 2 & 7 & -1 & -8 \end{array} \qquad \begin{array}{r|rrrr} 2 & 2 & 5 & -8 & -7 \\ & & 4 & 18 & 20 \\ \hline & 2 & 9 & 10 & 13 \end{array}$$

The third row of the synthetic division by $x - 1$ contains negative numbers, and hence (i) of the result on bounds for real zeros does not apply. However, since all numbers in the third row of the synthetic division by $x - 2$ are positive, it follows from (i) that 2 is an upper bound for the real solutions of the equation. This fact is also evident if we express the division by $x - 2$ in the division algorithm form

$$2x^3 + 5x^2 - 8x - 7 = (x - 2)(2x^2 + 9x + 10) + 13,$$

for if $x > 2$, then the right-hand side of the equation is positive, and hence is not zero. Consequently, $2x^3 + 5x^2 - 8x - 7$ is not zero if $x > 2$.

We now find the lower bound. After some trial-and-error attempts using $x - (-1)$, $x - (-2)$, and $x - (-3)$, we find that synthetic division by $x - (-4)$ gives us

$$\begin{array}{r|rrrr} -4 & 2 & 5 & -8 & -7 \\ & & -8 & 12 & -16 \\ \hline & 2 & -3 & 4 & -23 \end{array}$$

Since the numbers in the third row are alternately positive and negative, it follows from (ii) of the preceding theorem that -4 is a lower bound for the real solutions. This can also be proved by expressing the division by $x + 4$ in the form

$$2x^3 + 5x^2 - 8x - 7 = (x + 4)(2x^2 - 3x + 4) - 23,$$

for if $x < -4$, then the right-hand side of this equation is negative, and therefore is not zero. Hence, $2x^3 + 5x^2 - 8x - 7$ is not zero if $x < -4$.

Since lower and upper bounds for the real solutions are -4 and 2, respectively, it follows that all real solutions are in the closed interval $[-4, 2]$.

FIGURE 17

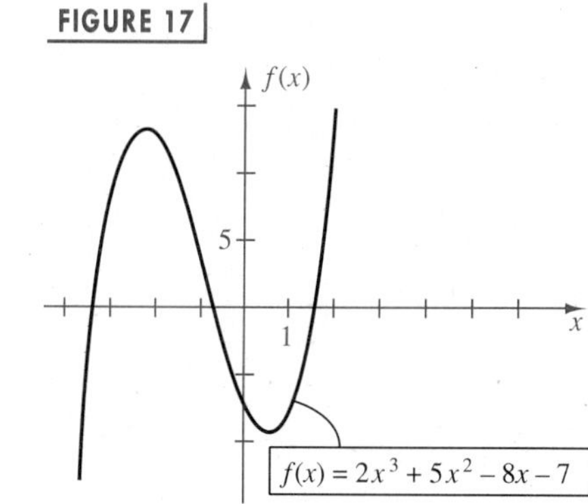

Many calculators can be programmed to approximate zeros of polynomials. Some calculators also have a graphics capability. Figure 17 is a computer-generated graph of the equation $f(x) = 2x^3 + 5x^2 - 8x - 7$ discussed in the preceding example. The graph shows that the three zeros of f are in the intervals $[-4, -3]$, $[-1, 0]$, and $[1, 2]$, respectively.

4.4 EXERCISES

Exer. 1–6: Find a polynomial $f(x)$ of degree 3 that has the indicated zeros and satisfies the given condition.

1 $-1, 2, 3$; $f(-2) = 80$

2 $-5, 2, 4$; $f(3) = -24$

3 $-4, 3, 0$; $f(2) = -36$

4 $-3, -2, 0$; $f(-4) = 16$

5 $-2i, 2i, 3$; $f(1) = 20$

6 $-3i, 3i, 4$; $f(-1) = 50$

7 Find a polynomial of degree 4 such that both -4 and 3 are zeros of multiplicity 2.

8 Find a polynomial of degree 4 such that both -5 and 2 are zeros of multiplicity 2.

9 Find a polynomial $f(x)$ of degree 6 such that 0 and 3 are both zeros of multiplicity 3 and $f(2) = -24$.

10 Find a polynomial $f(x)$ of degree 7 such that -2 and 2 are both zeros of multiplicity 2, 0 is a zero of multiplicity 3, and $f(-1) = 27$.

11 Find the third-degree polynomial function whose graph is shown in the figure.

EXERCISE 11

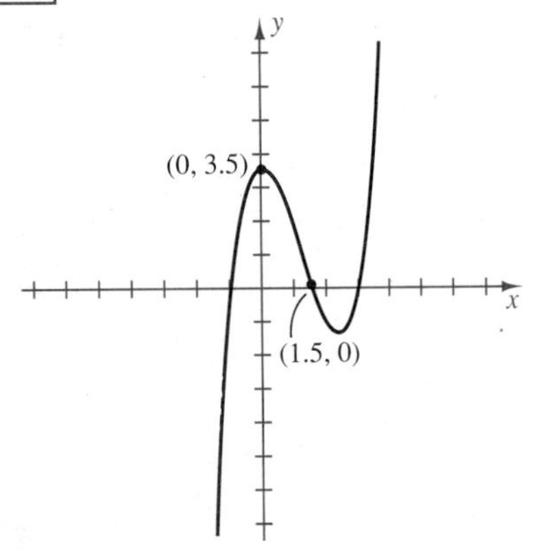

12 Find the fourth-degree polynomial function whose graph is shown in the figure.

EXERCISE 12

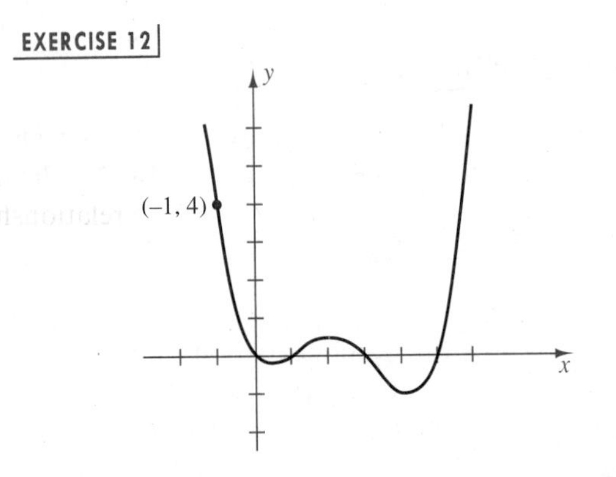

Exer. 13–14: If the graph of a polynomial function is tangent to the x-axis at $x = c$, then c is a zero of multiplicity 2 or greater. Find the polynomial function of degree 3 whose graph is shown in the figure.

13

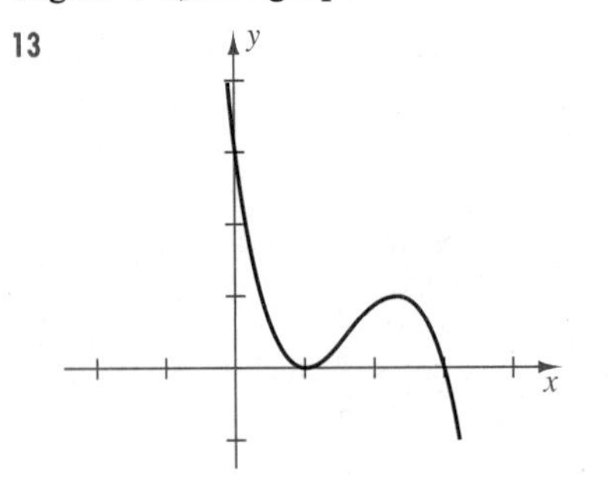

14

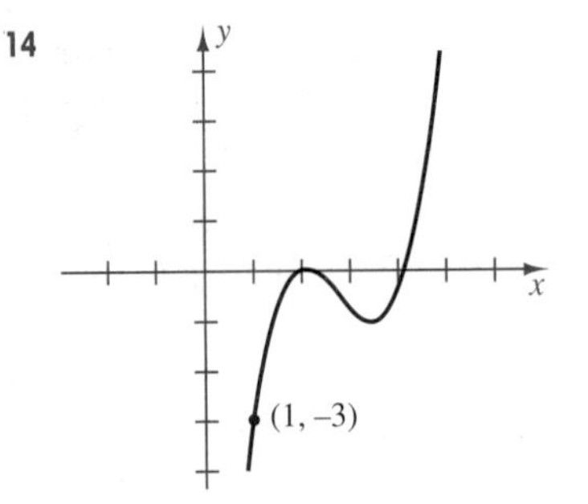

15 A scientist has limited data on the temperature T (in °C) during a 24-hour period. If t denotes time in hours and $t = 0$ corresponds to midnight, find the fourth-degree polynomial that fits the information in the following table.

t (hours)	0	5	12	19	24
T (°C)	0	0	10	0	0

16 A polynomial $f(x)$ of degree 3 with zeros c_1, c_2, and c_3 and with $f(c) = 1$ for $c_2 < c < c_3$ is a third-degree Lagrange interpolation polynomial. Find an explicit formula for $f(x)$ in terms of c_1, c_2, c_3, and c.

Exer. 17–24: Find the zeros of $f(x)$ and state the multiplicity of each zero.

17 $f(x) = x^2(3x + 2)(2x - 5)^3$

18 $f(x) = x(x + 1)^4(3x - 7)^2$

19 $f(x) = 4x^5 + 12x^4 + 9x^3$

20 $f(x) = (4x^2 - 5)^2$

21 $f(x) = (x^2 + x - 12)^3(x^2 - 9)^2$

22 $f(x) = (6x^2 + 7x - 5)^4(4x^2 - 1)^2$

23 $f(x) = x^4 + 7x^2 - 144$

24 $f(x) = x^4 + 21x^2 - 100$

Exer. 25–28: Show that the number is a zero of $f(x)$ of the given multiplicity, and express $f(x)$ as a product of linear factors.

25 $f(x) = x^4 + 7x^3 + 13x^2 - 3x - 18$; -3 (multiplicity 2)

26 $f(x) = x^4 - 9x^3 + 22x^2 - 32$; 4 (multiplicity 2)

27 $f(x) = x^6 - 4x^5 + 5x^4 - 5x^2 + 4x - 1$;
1 (multiplicity 5)

28 $f(x) = x^5 + x^4 - 6x^3 - 14x^2 - 11x - 3$;
-1 (multiplicity 4)

Exer. 29–36: Use Descartes' rule of signs to determine the number of possible positive, negative, and nonreal complex solutions of the equation.

29 $4x^3 - 6x^2 + x - 3 = 0$

30 $5x^3 - 6x - 4 = 0$

31 $4x^3 + 2x^2 + 1 = 0$

32 $3x^3 - 4x^2 + 3x + 7 = 0$

33 $3x^4 + 2x^3 - 4x + 2 = 0$

34 $2x^4 - x^3 + x^2 - 3x + 4 = 0$

35 $x^5 + 4x^4 + 3x^3 - 4x + 2 = 0$

36 $2x^6 + 5x^5 + 2x^2 - 3x + 4 = 0$

Exer. 37–42: Find the smallest and largest integers that are upper and lower bounds, respectively, for the real solutions of the equation.

37 $x^3 - 4x^2 - 5x + 7 = 0$

38 $2x^3 - 5x^2 + 4x - 8 = 0$

39 $x^4 - x^3 - 2x^2 + 3x + 6 = 0$

40 $2x^4 - 9x^3 - 8x - 10 = 0$

41 $2x^5 - 13x^3 + 2x - 5 = 0$

42 $3x^5 + 2x^4 - x^3 - 8x^2 - 7 = 0$

43 Let $f(x)$ and $g(x)$ be polynomials of degree not greater than n, where n is a positive integer. Show that if $f(x)$ and $g(x)$ are equal in value for more than n distinct values of x, then $f(x)$ and $g(x)$ are identical; that is, coefficients of like powers are the same. (*Hint:* Write

$$f(x) = a_nx^n + a_{n-1}x^{n-1} + \cdots + a_1x + a_0$$

$$g(x) = b_nx^n + b_{n-1}x^{n-1} + \cdots + b_1x + b_0$$

and consider

$$h(x) = f(x) - g(x) = (a_n - b_n)x^n + \cdots + (a_0 - b_0).$$

Show that $h(x)$ has more than n distinct zeros and conclude that $a_k = b_k$ for every k.)

4.5 COMPLEX AND RATIONAL ZEROS OF POLYNOMIALS

Example 3 of the preceding section illustrates an interesting fact about polynomials with real coefficients: The two complex zeros $2 + 3i$ and $2 - 3i$ of $x^5 - 4x^4 + 13x^3$ are conjugates of one another. The relationship is not accidental, since the following general result is true.

THEOREM

> If a polynomial $f(x)$ of degree $n > 0$ has real coefficients, and if z is a complex zero of $f(x)$, then the conjugate $\bar{z}$ of z is also a zero of $f(x)$.

PROOF We may write

$$f(x) = a_nx^n + a_{n-1}x^{n-1} + \cdots + a_1x + a_0,$$

where each coefficient a_k is a real number and $a_n \neq 0$. If $f(z) = 0$, then

$$a_nz^n + a_{n-1}z^{n-1} + \cdots + a_1z + a_0 = 0.$$

If two complex numbers are equal, then so are their conjugates. Hence, the conjugate of the left-hand side of the last equation equals the conjugate of the right-hand side; that is,

$$\overline{a_nz^n + a_{n-1}z^{n-1} + \cdots + a_1z + a_0} = \bar{0} = 0$$

The fact that $\bar{0} = 0$ follows from $\bar{0} = \overline{0 + 0i} = 0 - 0i = 0$.

It is not difficult to prove that if z and w are complex numbers, then $\overline{z+w} = \bar{z} + \bar{w}$. More generally, the conjugate of *any* sum of complex numbers is the sum of the conjugates. Consequently,

$$\overline{a_n z^n} + \overline{a_{n-1} z^{n-1}} + \cdots + \overline{a_1 z} + \overline{a_0} = 0.$$

Using Exercises 56–58 of Section 2.4, if z and w are complex numbers, then $\overline{z \cdot w} = \bar{z} \cdot \bar{w}$, $\overline{z^n} = \bar{z}^n$ for every positive integer n, and $\bar{z} = z$ if and only if z is real. Thus, for every k,

$$\overline{a_k z^k} = \overline{a_k} \cdot \overline{z^k} = \overline{a_k} \cdot \bar{z}^k = a_k \bar{z}^k$$

and therefore

$$a_n \bar{z}^n + a_{n-1} \bar{z}^{n-1} + \cdots + a_1 \bar{z} + a_0 = 0.$$

The last equation states that $f(\bar{z}) = 0$, which completes the proof. ❑

EXAMPLE ▪ 1

Find a polynomial $f(x)$ of degree 4 that has real coefficients and zeros $2 + i$ and $-3i$.

SOLUTION By the preceding theorem, $f(x)$ must also have zeros $2 - i$ and $3i$. Applying the factor theorem, $f(x)$ has the following factors:

$$x - (2 + i), \qquad x - (2 - i), \qquad x - (-3i), \qquad x - (3i).$$

Multiplying these four factors gives us:

$$\begin{aligned} f(x) &= [x - (2 + i)][x - (2 - i)](x + 3i)(x - 3i) \\ &= (x^2 - 4x + 5)(x^2 + 9) \\ &= x^4 - 4x^3 + 14x^2 - 36x + 45 \end{aligned}$$

If a polynomial with real coefficients is factored as on page 207, some of the factors $x - c_k$ may have a complex coefficient c_k. However, it is always possible to obtain a factorization into polynomials with real coefficients, as stated in the next theorem.

THEOREM

> Every polynomial with real coefficients and positive degree n can be expressed as a product of linear and quadratic polynomials with real coefficients such that the quadratic factors have no real zeros.

PROOF Since $f(x)$ has precisely n complex zeros $c_1, c_2, \ldots, c_n$, we may write

$$f(x) = a(x - c_1)(x - c_2) \cdots (x - c_n)$$

where a is the leading coefficient of $f(x)$. Of course, some of the zeros may be real. In such cases we obtain the linear factors referred to in the statement of the theorem.

If a zero c_k is not real, then by the preceding theorem the conjugate $\overline{c_k}$ is also a zero of $f(x)$ and hence must be one of the numbers $c_1, c_2, \ldots, c_n$. This implies that both $x - c_k$ and $x - \overline{c_k}$ appear in the factorization of $f(x)$. If those factors are multiplied, we obtain

$$(x - c_k)(x - \overline{c_k}) = x^2 - (c_k + \overline{c_k})x + c_k\overline{c_k},$$

which has *real* coefficients, since $c_k + \overline{c_k}$ and $c_k\overline{c_k}$ are real numbers (see page 88). Thus, if c_k is a complex zero, then the product $(x - c_k)(x - \overline{c_k})$ is a quadratic polynomial that is irreducible over $\mathbb{R}$. This completes the proof. ❑

EXAMPLE ▪ 2

Express $x^4 - 2x^2 - 3$ as a product (a) of linear polynomials and (b) of linear and quadratic polynomials with real coefficients that are irreducible over $\mathbb{R}$.

SOLUTION

(a) We can find the zeros of the given polynomial by solving the equation $x^4 - 2x^2 - 3 = 0$, which may be regarded as quadratic in x^2. Solving for x^2 by means of the quadratic formula, we obtain

$$x^2 = \frac{2 \pm \sqrt{4 + 12}}{2} = \frac{2 \pm 4}{2}.$$

Thus, $x^2 = 3$ or $x^2 = -1$. Hence, the zeros of $x^4 - 2x^2 - 3$ are $\sqrt{3}$, $-\sqrt{3}$, i, and $-i$, and we obtain the factorization

$$x^4 - 2x^2 - 3 = (x - \sqrt{3})(x + \sqrt{3})(x - i)(x + i).$$

(b) Multiplying the last two factors in the factorization from part (a) gives us

$$x^4 - 2x^2 - 3 = (x - \sqrt{3})(x + \sqrt{3})(x^2 + 1),$$

which is of the form stated in the preceding theorem.

The solution of this example could also have been obtained by factoring the original expression without first finding the zeros. Thus,

$$\begin{aligned} x^4 - 2x^2 - 3 &= (x^2 - 3)(x^2 + 1) \\ &= (x + \sqrt{3})(x - \sqrt{3})(x + i)(x - i). \end{aligned}$$

We have already pointed out that it is generally very difficult to find the zeros of a polynomial of high degree. However, if all the coefficients are integers or rational numbers, there is a method for finding the *rational* zeros, if they exist. The method is a consequence of the following theorem.

THEOREM ON RATIONAL ZEROS

Suppose that $f(x) = a_n x^n + a_{n-1}x^{n-1} + \cdots + a_1 x + a_0$ is a polynomial with integral coefficients. If c/d is a rational zero of $f(x)$ such that c and d have no common prime factors, then c is a factor of a_0 and d is a factor of a_n.

PROOF Assume that $c > 0$. (The proof for $c < 0$ is similar.) Let us show that c is a factor of a_0. The case $c = 1$ is trivial, since 1 is a factor of *any* number. Thus, suppose $c \neq 1$. In this case $c/d \neq 1$, for if $c/d = 1$, we obtain $c = d$, and since c and d have no prime factor in common, this implies that $c = d = 1$, a contradiction. Hence, in the following discussion we have $c \neq 1$ and $c \neq d$.

Since $f(c/d) = 0$,

$$a_n(c^n/d^n) + a_{n-1}(c^{n-1}/d^{n-1}) + \cdots + a_1(c/d) + a_0 = 0.$$

We multiply by d^n and then add $-a_0 d^n$ to both sides:

$$\begin{aligned} a_n c^n + a_{n-1}c^{n-1}d + \cdots + a_1 c d^{n-1} &= -a_0 d^n \\ c(a_n c^{n-1} + a_{n-1}c^{n-2}d + \cdots + a_1 d^{n-1}) &= -a_0 d^n \end{aligned}$$

The last equation shows that c is a factor of the integer $a_0 d^n$. Since c and d have no common factor, c is a factor of a_0. A similar argument may be used to prove that d is a factor of a_n. ❑

The technique of using the preceding theorem for finding rational solutions of equations with integral coefficients is illustrated in the following example.

EXAMPLE ▪ 3

Find all rational solutions of the equation

$$3x^4 + 14x^3 + 14x^2 - 8x - 8 = 0.$$

SOLUTION The problem is equivalent to finding the rational zeros of the polynomial on the left-hand side of the equation. According to the preceding theorem, if c/d is a rational zero and $c > 0$, then c is a divisor of -8 and d is a divisor of 3. Hence, the possible choices for c are 1, 2, 4, and 8, and the choices for d are ± 1 and ± 3. Consequently, any rational roots are included among the numbers ± 1, ± 2, ± 4, ± 8, $\pm \frac{1}{3}$, $\pm \frac{2}{3}$, $\pm \frac{4}{3}$, and $\pm \frac{8}{3}$. We can reduce the number of possibilities by finding upper and lower bounds for the real solutions; however, we shall not do so here. It is necessary to determine which of the numbers, if any, are zeros. Synthetic division is an appropriate method for this task. After perhaps many trial-and-error attempts, we obtain:

$$\begin{array}{r|rrrrr} -2 & 3 & 14 & 14 & -8 & -8 \\ & & -6 & -16 & 4 & 8 \\ \hline & 3 & 8 & -2 & -4 & 0 \end{array}$$

This result shows that -2 is a zero. Moreover, the synthetic division provides the coefficients of the quotient in the division of the polynomial by $x + 2$. Hence, we have the following factorization of the given polynomial:

$$(x + 2)(3x^3 + 8x^2 - 2x - 4)$$

The remaining solutions of the equation must be zeros of the second factor, and therefore we use that polynomial to check for solutions. Again proceeding by trial and error, we ultimately find that synthetic division by $x + \frac{2}{3}$ gives us the following result:

$$\begin{array}{r|rrrr} -\frac{2}{3} & 3 & 8 & -2 & -4 \\ & & -2 & -4 & 4 \\ \hline & 3 & 6 & -6 & 0 \end{array}$$

Therefore, $-\frac{2}{3}$ is also a zero.

The remaining zeros are solutions of the equation $3x^2 + 6x - 6 = 0$. Dividing both sides by 3 gives us the equivalent equation $x^2 + 2x - 2 = 0$. By the quadratic formula, this equation has solutions

$$\frac{-2 \pm \sqrt{4 - 4(-2)}}{2} = \frac{-2 \pm \sqrt{12}}{2} = \frac{-2 \pm 2\sqrt{3}}{2}.$$

Hence, the given polynomial has two rational roots, -2 and $-\frac{2}{3}$, and two irrational roots, $-1 + \sqrt{3} \approx 0.732$ and $-1 - \sqrt{3} \approx -2.732$.

The graph in Figure 18 indicates that the four zeros of the polynomial $3x^4 + 14x^3 + 14x^2 - 8x - 8$ are in the intervals $[-3, -2]$, $[-1, 0]$ and $[0, 1]$. A graph of this type can be used to discover rational zeros more quickly than the trial-and-error method we have used.

FIGURE 18

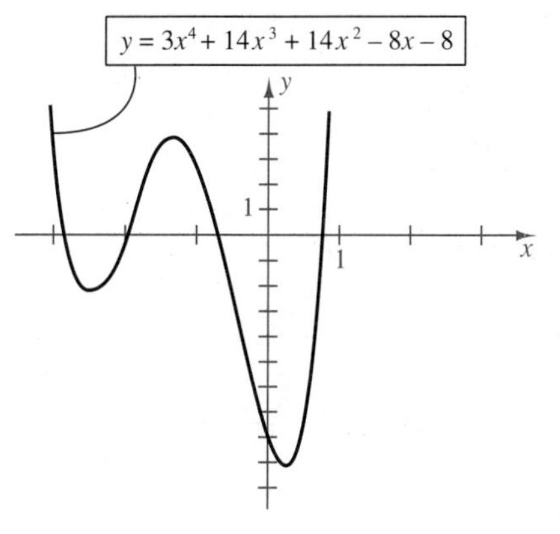

The theorem on rational zeros may be applied to equations with rational coefficients. We merely multiply both sides of the equation by the lcd of all the coefficients to obtain an equation with integral coefficients, and then proceed as in Example 3.

EXAMPLE ▪ 4

A grain silo has the shape of a right circular cylinder with a hemisphere attached to the top. If the total height of the structure is 30 feet, find the radius of the cylinder that results in a total volume of $1008\pi\ \text{ft}^3$.

FIGURE 19

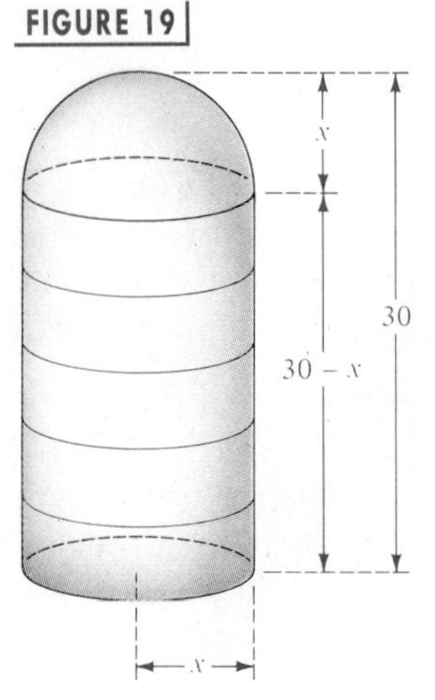

SOLUTION Let x denote the radius of the cylinder. A sketch of the silo, labeled appropriately, is shown in Figure 19. Since the volume of the cylinder is $\pi x^2(30 - x)$, and the volume of the hemisphere is $\frac{2}{3}\pi x^3$, we must solve equation

$$\pi x^2(30 - x) + \tfrac{2}{3}\pi x^3 = 1008\pi.$$

Each of the following is equivalent to the preceding equation:

$$3x^2(30 - x) + 2x^3 = 3024$$
$$90x^2 - 3x^3 + 2x^3 = 3024$$
$$90x^2 - x^3 = 3024$$
$$x^3 - 90x^2 + 3024 = 0$$

To look for rational roots, we first factor 3024 into primes, obtaining $3024 = 2^4 \cdot 3^3 \cdot 7$. It follows that some of the positive factors of 3024 are

$$1,\ 2,\ 3,\ 4,\ 6,\ 8,\ 9,\ 12,\ \ldots$$

Dividing synthetically, we eventually arrive at

$$\begin{array}{r|rrrr} 6 & 1 & -90 & 0 & 3024 \\ & & 6 & -504 & -3024 \\ \hline & 1 & -84 & -504 & 0 \end{array}$$

FIGURE 20

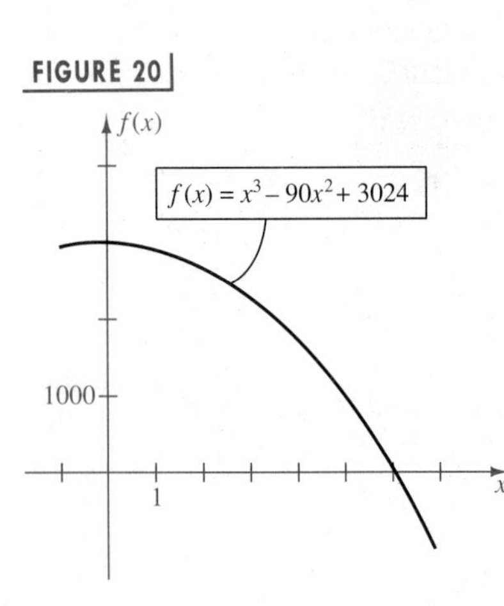

Thus, 6 is a solution of the equation $x^3 - 90x^2 + 3024 = 0$.

The remaining two solutions of the equation can be found by solving $x^2 - 84x - 504 = 0$, which we obtain from the quotient in the synthetic division. It is not difficult to show that neither of the solutions of this quadratic equation satisfy the conditions of the problem. Thus the desired radius is 6 feet.

The graph of $f(x) = x^3 - 90x^2 + 3024$ in Figure 20 clearly shows the zero $x = 6$. (The *complete* graph would also indicate a negative zero and a large positive zero.)

4.5

EXERCISES

Exer. 1–10: Find a polynomial with real coefficients that has the given zero(s) and degree.

1 $3 + 2i$; degree 2

2 $-4 + 3i$; degree 2

3 $2, -2 - 5i$; degree 3

4 $-3, 1 - 7i$; degree 3

5 $-1, 0, 3 + i$; degree 4

6 $0, 2, -2 - i$; degree 4

7 $4 + 3i, -2 + i$; degree 4

8 $3 + 5i, -1 - i$; degree 4

9 $0, -2i, 1 - i$; degree 5

10 $0, 3i, 4 + i$; degree 5

11 Does there exist a polynomial of degree 3 with real coefficients that has zeros 1, -1, and i? Justify your answer.

12 The complex number i is a zero of the polynomial $f(x) = x^3 - ix^2 + 2ix + 2$; however, the conjugate $-i$ of i is not a zero. Why doesn't this contradict the first theorem of this section?

Exer. 13–30: Find all solutions of the equation.

13 $x^3 - x^2 - 10x - 8 = 0$

14 $x^3 + x^2 - 14x - 24 = 0$

15 $2x^3 - 3x^2 - 17x + 30 = 0$

16 $12x^3 + 8x^2 - 3x - 2 = 0$

17 $x^4 + 3x^3 - 30x^2 - 6x + 56 = 0$

18 $3x^5 - 10x^4 - 6x^3 + 24x^2 + 11x - 6 = 0$

19 $2x^3 - 7x^2 - 10x + 24 = 0$

20 $2x^3 - 3x^2 - 8x - 3 = 0$

21 $6x^3 + 19x^2 + x - 6 = 0$

22 $6x^3 + 5x^2 - 17x - 6 = 0$

23 $8x^3 + 18x^2 + 45x + 27 = 0$

24 $3x^3 - x^2 + 11x - 20 = 0$

25 $x^4 - x^3 - 9x^2 - 3x - 36 = 0$

26 $3x^4 + 16x^3 + 28x^2 + 31x + 30 = 0$

27 $9x^4 + 15x^3 - 20x^2 - 20x + 16 = 0$

28 $15x^4 + 4x^3 + 11x^2 + 4x - 4 = 0$

29 $4x^5 + 12x^4 - 41x^3 - 99x^2 + 10x + 24 = 0$

30 $4x^5 + 24x^4 - 13x^3 - 174x^2 + 9x + 270 = 0$

Exer. 31–38: Show that the equation has no rational root.

31 $x^3 + 3x^2 - 4x + 6 = 0$

32 $3x^3 - 4x^2 + 7x + 5 = 0$

33 $x^5 - 3x^3 + 4x^2 + x - 2 = 0$

34 $2x^5 + 3x^3 + 7 = 0$

35 $3x^4 + 7x^3 + 3x^2 - 8x - 10 = 0$

36 $2x^4 - 3x^3 + 6x^2 - 24x + 5 = 0$

37 $8x^4 + 16x^3 - 26x^2 - 12x + 15 = 0$

38 $5x^5 + 2x^4 + x^3 - 10x^2 - 4x - 2 = 0$

39 If n is an odd positive integer, prove that a polynomial of degree n with real coefficients has at least one real zero.

40 Show that the theorem on page 215 is not necessarily true if $f(x)$ has complex coefficients.

41 Complete the proof of the theorem on rational zeros by showing that d is a factor of a_n.

42 If a polynomial of the form

$$x^n + a_{n-1}x^{n-1} + \cdots + a_1x + a_0,$$

where each a_k is an integer, has a rational root r, show that r is an integer and is a factor of a_0.

43 An open box is to be made from a rectangular piece of cardboard having dimensions 20 inches $\times$ 30 inches, by removing squares of area x^2 from each corner and turning up the sides.

(a) Show that there are two boxes that have a volume of 1000 in.^{3}.

(b) Which box has the smaller surface area?

44 The frame for a shipping crate is to be constructed from 24 feet of 2×2 lumber. Assuming the crate is to have square ends of length x feet, determine the value(s) of x that result in a volume of 4 ft^3. (Compare Exercise 34 of Section 4.2.)

45 A meteorologist determines that the temperature T (in °F) on a certain cold winter day was given by

$T = \frac{1}{20}t(t - 12)(t - 24)$ for $0 \le t \le 24$, where t is time in hours and $t = 0$ corresponds to 6 A.M. At what time(s) of the day was the temperature 32 °F? (Compare Exercise 35 of Section 4.2.)

46 A herd of 100 deer is introduced to a small island. Assuming the number $N(t)$ of deer after t years is given by $N(t) = -t^4 + 21t^2 + 100$ (for $t > 0$), determine when the herd size exceeds 180. (Compare Exercise 37 of Section 4.2.)

47 A right triangle has area 30 ft^2 and a hypotenuse that is 1 foot longer than one of its sides.

(a) If x denotes the length of this side, show that $2x^3 + x^2 - 3600 = 0$.

(b) Show that there is a single positive root of the equation in part (a) and that this root is less than 13.

(c) Find the lengths of the sides of the triangle.

48 A storage tank for propane gas is to be constructed in the shape of a right circular cylinder of altitude 10 feet with a hemisphere attached to each end. Determine the radius x so that the resulting volume is 27π ft^3. (Compare Example 6 of Section 3.4.)

49 A storage shelter is to be constructed in the shape of a cube with a triangular prism forming the roof (see figure). The length x of a side of the cube is yet to be determined.

(a) If the total height of the structure is 6 feet, show that its volume V is given by $V = x^3 + \frac{1}{2}x^2(6 - x)$.

(b) Determine x so that the volume is 80 ft^3.

EXERCISE 49

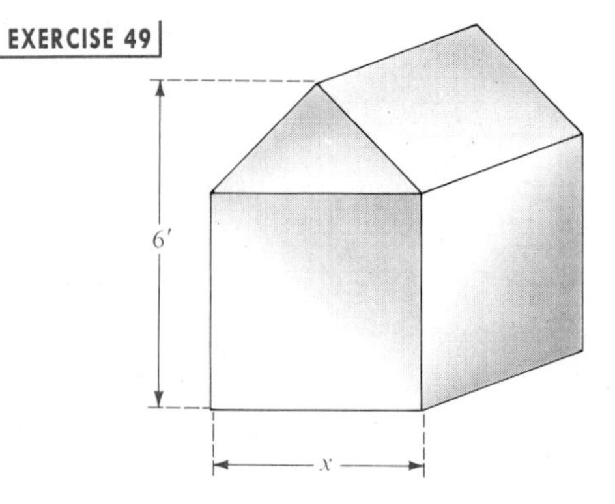

50 A canvas camping tent has the shape of a pyramid with a square base. An 8-foot pole will form the center support, as illustrated in the figure. Find the length x of a side of the base so that the total canvas needed for the sides and bottom is 384 ft^2.

EXERCISE 50

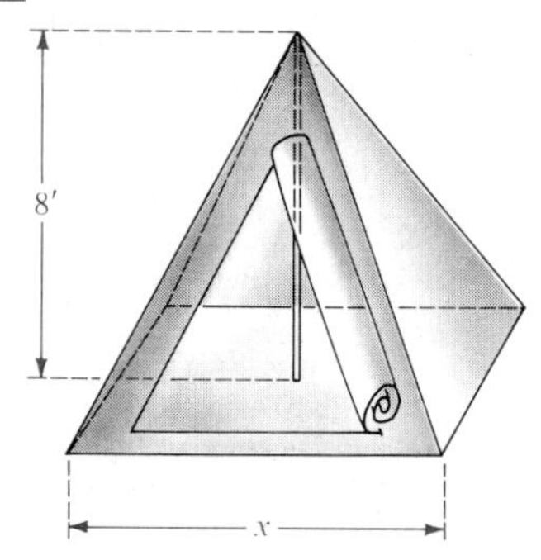

4.6 RATIONAL FUNCTIONS

A function f is a **rational function** if

$$f(x) = \frac{g(x)}{h(x)}$$

for polynomials $g(x)$ and $h(x)$. The zeros of the numerator and denominator are of major importance. *Throughout this section we shall assume that $g(x)$ and $h(x)$ have no common factors* and hence no common zeros. If $g(c) = 0$, then $f(c) = g(c)/h(c) = 0$. However, if $h(c) = 0$, then $f(c)$ is undefined. As indicated in the next example, the investigation of $f(x)$ requires special attention if x is near a zero of the denominator $h(x)$.

EXAMPLE ▪ 1

Sketch the graph of f if $f(x) = \dfrac{1}{x-2}$.

SOLUTION The numerator 1 is never zero and hence $f(x)$ has no zero. This means that the graph has no x-intercept.

The denominator $x - 2$ is zero at $x = 2$. If x is close to 2 and $x > 2$, then $f(x)$ is large, as shown in the following table of function values.

x	2.1	2.01	2.001	2.0001
$f(x)$	10	100	1000	10,000

If x is close to 2 and $x < 2$, then $|f(x)|$ is large, but $f(x)$ is negative, as in the next table.

x	1.9	1.99	1.999	1.9999
$f(x)$	-10	-100	-1000	$-10{,}000$

Several other values of $f(x)$ are displayed below.

x	-8	-1	0	1	3	4	12
$f(x)$	$-\frac{1}{10}$	$-\frac{1}{3}$	$-\frac{1}{2}$	-1	1	$\frac{1}{2}$	$\frac{1}{10}$

FIGURE 21 $y = \dfrac{1}{x-2}$

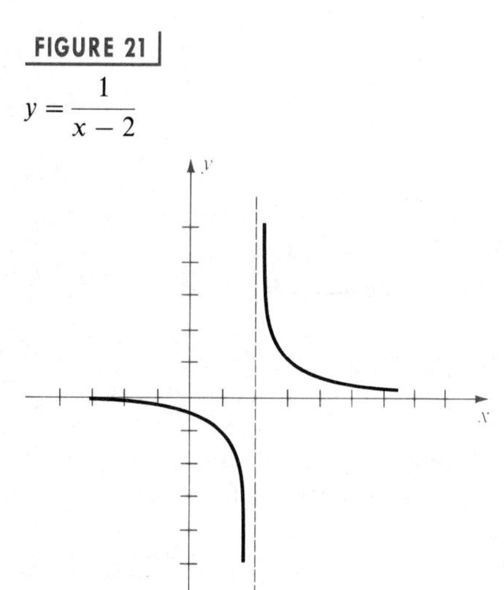

Observe that as $|x|$ increases, $f(x)$ approaches zero. Plotting points and noting what happens near $x = 2$ gives us the sketch in Figure 21. The dashed vertical line indicates where the denominator $x - 2$ of $f(x)$ is zero.

In Example 1, we can make $f(x)$ as large as desired by choosing x close to 2 (and $x > 2$). We denote this fact by writing

$$f(x) \to \infty \quad \text{as} \quad x \to 2^{+}.$$

We say that $f(x)$ *increases without bound as x approaches* 2 *from the right*. The symbol ∞ (read *infinity*) does not represent a real number, but is used to denote the variation of $f(x)$ that we have described.

For the case $x < 2$ in Example 1 we write

$$f(x) \to -\infty \quad \text{as} \quad x \to 2^-$$

and say that $f(x)$ *decreases without bound as x approaches* 2 *from the left.*

In general, the notation $x \to a^+$ will signify that x approaches a from the *right*, that is, through values *greater* than a. The symbol $x \to a^-$ will mean that x approaches a from the *left*, through values *less* than a. Some illustrations of the manner in which a function f may increase or decrease without bound, with the corresponding notation, are shown in Figure 22. In the figure, a is positive, but we can also have $a \leq 0$.

FIGURE 22

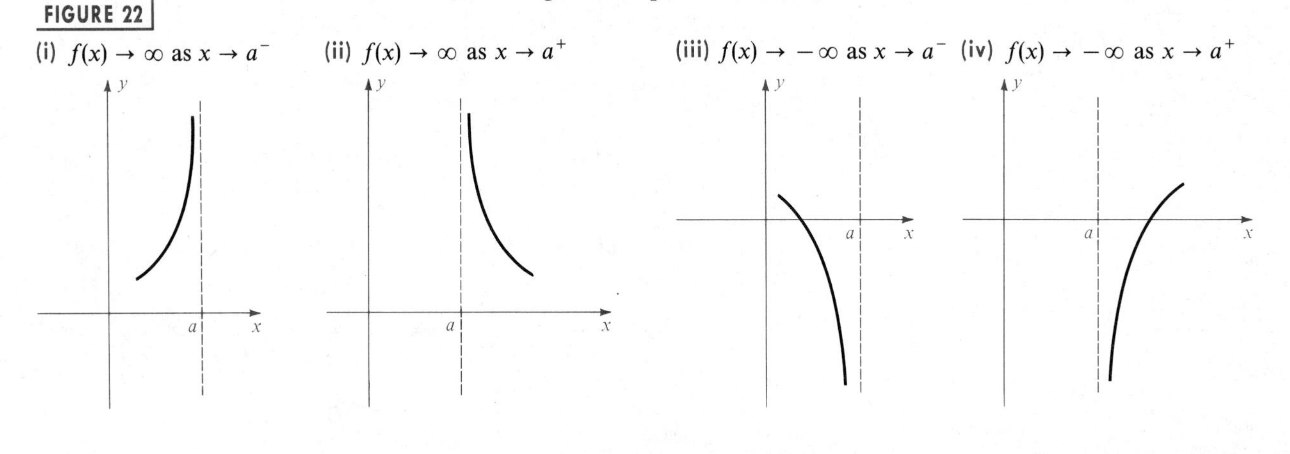

DEFINITION

> The line $x = a$ is a **vertical asymptote** for the graph of a function f if $f(x) \to \infty$ or $f(x) \to -\infty$ as x approaches a from either the left or right.

In Figure 22 the dashed lines represent vertical asymptotes. In Figure 21 the dashed line $x = 2$ is a vertical asymptote for the graph of $y = 1/(x - 2)$.

Vertical asymptotes are common for graphs of rational functions: If $f(x) = g(x)/h(x)$ and if $g(x)$ and $h(x)$ have no common factor, then the graph of f has a vertical asymptote $x = a$ for every real zero a of the denominator $h(x)$.

We are also interested in values $f(x)$ of a function f when $|x|$ is large. The following table lists some approximate values for $f(x) = 1/(x - 2)$.

x	1000	10,000	100,000	1,000,000
$f(x)$ (approx.)	0.001	0.0001	0.00001	0.000001

We can make $f(x) = 1/(x - 2)$ as close to 0 as we desire by choosing x sufficiently large. We express this symbolically by

$$f(x) \to 0 \quad \text{as} \quad x \to \infty,$$

which is read $f(x)$ *approaches* 0 *as* x *increases without bound.*

Several approximate values of $f(x)$ for $|x|$ large and $x < 0$ are listed in the next table.

x	$-1{,}000$	$-10{,}000$	$-100{,}000$	$-1{,}000{,}000$
$f(x)$ (approx.)	-0.001	-0.0001	-0.00001	-0.000001

We can make $f(x) = 1/(x - 2)$ arbitrarily close to zero by choosing $|x|$ sufficiently large and x negative. We express this symbolically by

$$f(x) \to 0 \quad \text{as} \quad x \to -\infty,$$

which is read $f(x)$ *approaches* 0 *as* x *decreases without bound.*

In Example 1 the line $y = 0$, that is, the x-axis, is a *horizontal asymptote* for the graph. In general, we have the following definition. The notation should be self-evident.

DEFINITION

The line $y = c$ is a **horizontal asymptote** for the graph of a function f if $f(x) \to c$ as $x \to \infty$ or as $x \to -\infty$.

Some typical horizontal asymptotes (for $x \to \infty$) are illustrated in Figure 23. The manner in which the graph approaches the line $y = c$ may vary, depending on the nature of the function. Similar sketches may be made for the case $x \to -\infty$. As in the third sketch, the graph of f may cross a horizontal asymptote.

FIGURE 23 $f(x) \to c$ as $x \to \infty$

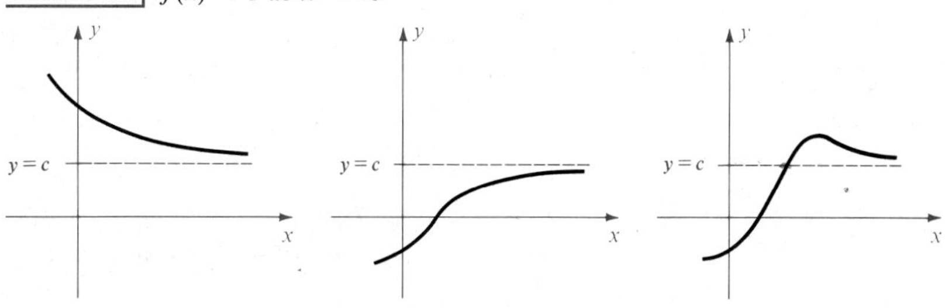

The following theorem is useful for finding horizontal asymptotes for the graph of a rational function.

THEOREM ON HORIZONTAL ASYMPTOTES

Let $f(x) = \dfrac{a_n x^n + a_{n-1}x^{n-1} + \cdots + a_1 x + a_0}{b_k x^k + b_{k-1}x^{k-1} + \cdots + b_1 x + b_0}$.

(i) If $n < k$, then the x-axis (the line $y = 0$) is a horizontal asymptote for the graph of f.

(ii) If $n = k$, then the line $y = a_n/b_k$ is a horizontal asymptote.

(iii) If $n > k$, the graph of f has no horizontal asymptote.

Proofs of (i) and (ii) of this theorem may be patterned after the solution to the following example. A similar argument can be given in case (iii).

EXAMPLE ▪ 2

Find the horizontal asymptotes for the graph of f if:

(a) $f(x) = \dfrac{3x - 1}{x^2 - x - 6}$ (b) $f(x) = \dfrac{5x^2 + 1}{3x^2 - 4}$

SOLUTION

(a) The degree of the numerator $3x - 1$ is less than the degree of the denominator $x^2 - x - 6$, and hence by (i) of the theorem, the x-axis is a horizontal asymptote. To verify this directly, we divide numerator and denominator of the quotient by x^2, obtaining

$$f(x) = \frac{\dfrac{3x - 1}{x^2}}{\dfrac{x^2 - x - 6}{x^2}} = \frac{\dfrac{3}{x} - \dfrac{1}{x^2}}{1 - \dfrac{1}{x} - \dfrac{6}{x^2}} \quad \text{for } x \neq 0.$$

If x is very large, then both $1/x$ and $1/x^2$ are close to 0, and hence,

$$f(x) \approx \frac{0 - 0}{1 - 0 - 0} = \frac{0}{1} = 0.$$

Thus, $$f(x) \to 0 \quad \text{as} \quad x \to \infty.$$

Since $f(x)$ is the y-coordinate of a point on the graph, the last statement means that the x-axis is a horizontal asymptote.

(b) If $f(x) = (5x^2 + 1)/(3x^2 - 4)$, then the numerator and denominator have the same degree, and hence by (ii) of the theorem, the line $y = \frac{5}{3}$ is a horizontal asymptote. We can also show that $y = \frac{5}{3}$ is a horizontal asymptote by dividing numerator and denominator of $f(x)$ by x^2, obtaining

$$f(x) = \frac{5 + \dfrac{1}{x^2}}{3 - \dfrac{4}{x^2}} \quad \text{for } x \neq 0.$$

Since $1/x^2 \to 0$ as $x \to \infty$, we see that

$$f(x) \to \frac{5 + 0}{3 - 0} = \frac{5}{3} \quad \text{as} \quad x \to \infty.$$

We next list some guidelines for sketching the graph of a rational function. Their use will be illustrated in Examples 3, 4, and 5.

GUIDELINES SKETCHING THE GRAPH OF $f(x) = \dfrac{g(x)}{h(x)}$ IF $g(x)$ AND $h(x)$ ARE POLYNOMIALS THAT HAVE NO COMMON FACTOR

STEP 1 Find the real zeros of the numerator $g(x)$ and use them to plot the points corresponding to the x-intercepts.

STEP 2 Find the real zeros of the denominator $h(x)$. For each zero a, the line $x = a$ is a vertical asymptote. Represent $x = a$ with dashes.

STEP 3 Find the intervals determined by the zeros of $g(x)$ and $h(x)$ and specify the sign of $f(x)$ in each interval. Use these signs to decide if the graph lies above or below the x-axis in each interval.

STEP 4 If $x = a$ is a vertical asymptote, use the information in Step 3 to determine if $f(x) \to \infty$ or if $f(x) \to -\infty$ for each case (i) $x \to a^-$ and (ii) $x \to a^+$. Note this by sketching a portion of the graph on each side of $x = a$.

STEP 5 Use the information in Step 3 to determine the manner in which the graph intersects the x-axis.

STEP 6 Apply the theorem on horizontal asymptotes. If there is a horizontal asymptote, represent it with dashes.

STEP 7 Sketch the graph, using the information found in the preceding steps and plotting points wherever necessary. ❑

EXAMPLE ▪ 3

Sketch the graph of f if $f(x) = \dfrac{x-1}{x^2 - x - 6}$.

SOLUTION We begin by factoring the denominator:

$$f(x) = \frac{x-1}{(x+2)(x-3)}$$

We shall sketch the graph by following the steps listed in the guidelines.

FIGURE 24

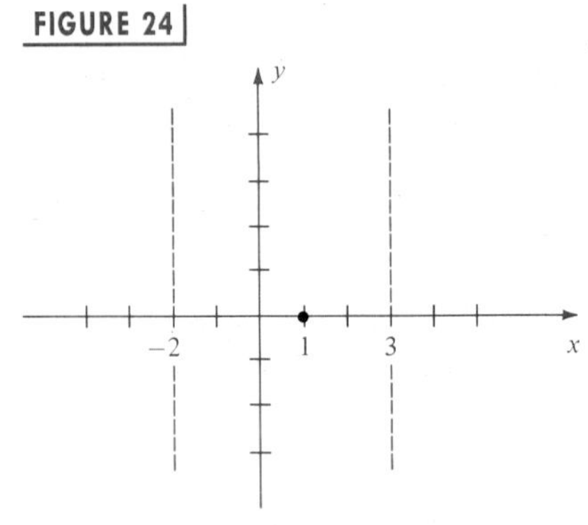

STEP 1 The numerator $x - 1$ has the zero 1, and we plot the point $(1, 0)$ on the graph, as shown in Figure 24.

STEP 2 The denominator has zeros -2 and 3. Hence, the lines $x = -2$ and $x = 3$ are vertical asymptotes, and we represent them with dashes, as in Figure 24.

STEP 3 The zeros -2, 1, and 3 of the numerator and denominator of $f(x)$ determine the intervals

$$(-\infty, -2), \quad (-2, 1), \quad (1, 3), \quad (3, \infty).$$

Since $f(x)$ is a quotient of two polynomials, we see from our work in Section 4.2 that $f(x)$ is always positive or always negative throughout each interval. Using test values to determine the sign of $f(x)$ gives us the following table.

Interval	$(-\infty, -2)$	$(-2, 1)$	$(1, 3)$	$(3, \infty)$
Test value	$f(-3) = -\frac{2}{3}$	$f(0) = \frac{1}{6}$	$f(2) = -\frac{1}{4}$	$f(4) = \frac{1}{2}$
Sign of $f(x)$	$-$	$+$	$-$	$+$
Position of graph	Below x-axis	Above x-axis	Below x-axis	Above x-axis

STEP 4 We shall use the fourth row of the preceding table to investigate the function value $f(x)$ near each vertical asymptote.

(i) Consider the vertical asymptote $x = -2$. Since the graph lies *below* the x-axis throughout the interval $(-\infty, -2)$, we see that

$$f(x) \to -\infty \quad \text{as} \quad x \to -2^-.$$

Since the graph is *above* the x-axis throughout the interval $(-2, 1)$,

$$f(x) \to \infty \quad \text{as} \quad x \to -2^+.$$

FIGURE 25

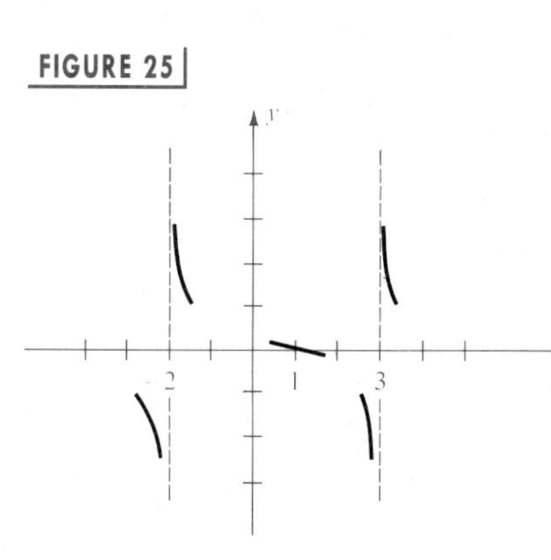

We note these facts in Figure 25 by sketching portions of the graph on each side of the line $x = -2$.

(ii) Consider the vertical asymptote $x = 3$. The graph lies *below* the x-axis throughout the interval $(1, 3)$, and hence

$$f(x) \to -\infty \quad \text{as} \quad x \to 3^-.$$

The graph lies *above* the x-axis throughout $(3, \infty)$, and hence

$$f(x) \to \infty \quad \text{as} \quad x \to 3^+.$$

We note these facts in Figure 25 by sketching portions of the graph on each side of $x = 3$.

STEP 5 Referring to the fourth row of the table in Step 3, we see that the graph crosses the x-axis at $(1, 0)$, as illustrated in Figure 25.

FIGURE 26

$y = \dfrac{x - 1}{x^2 - x - 6}$

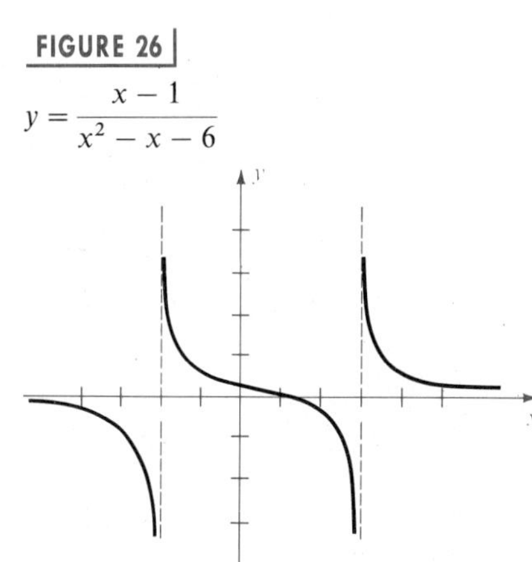

STEP 6 The degree of the numerator $x - 1$ is less than the degree of the denominator $x^2 - x - 6$. Hence, by (i) of the theorem on horizontal asymptotes, the x-axis is a horizontal asymptote.

STEP 7 We use the information found in Steps 4, 5, and 6, and plot several points to obtain the sketch in Figure 26.

EXAMPLE ▪ 4

Sketch the graph of f if $f(x) = \dfrac{x^2}{x^2 - x - 2}$.

SOLUTION Factoring the denominator gives us

$$f(x) = \frac{x^2}{(x + 1)(x - 2)}.$$

FIGURE 27

We again follow the guidelines listed earlier.

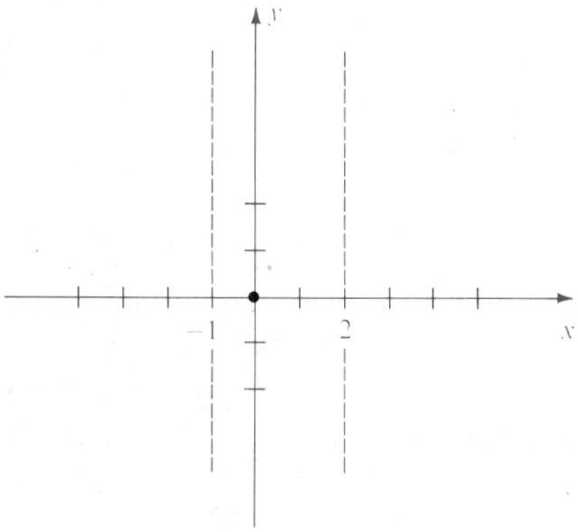

STEP 1 The numerator x^2 has 0 as a zero, and hence the graph intersects the x-axis at $(0, 0)$, as shown in Figure 27.

STEP 2 Since the denominator has zeros -1 and 2, the lines $x = -1$ and $x = 2$ are vertical asymptotes, and we represent them with dashes, as in Figure 27.

STEP 3 The intervals determined by the zeros 0, -1, and 2 are

$$(-\infty, -1), \quad (-1, 0), \quad (0, 2), \quad (2, \infty).$$

Following the procedure in Step 3 of the guidelines, we arrive at the following table.

Interval	$(-\infty, -1)$	$(-1, 0)$	$(0, 2)$	$(2, \infty)$
Test value	$f(-2) = 1$	$f(-\frac{1}{2}) = -\frac{1}{5}$	$f(1) = -\frac{1}{2}$	$f(3) = \frac{9}{4}$
Sign of $f(x)$	$+$	$-$	$-$	$+$
Position of graph	Above x-axis	Below x-axis	Below x-axis	Above x-axis

STEP 4 We refer to the fourth row of the preceding table and proceed as follows:

(i) Consider the vertical asymptote $x = -1$. Since the graph is above the x-axis in $(-\infty, -1)$, we see that

$$f(x) \to \infty \quad \text{as} \quad x \to -1^-.$$

FIGURE 28

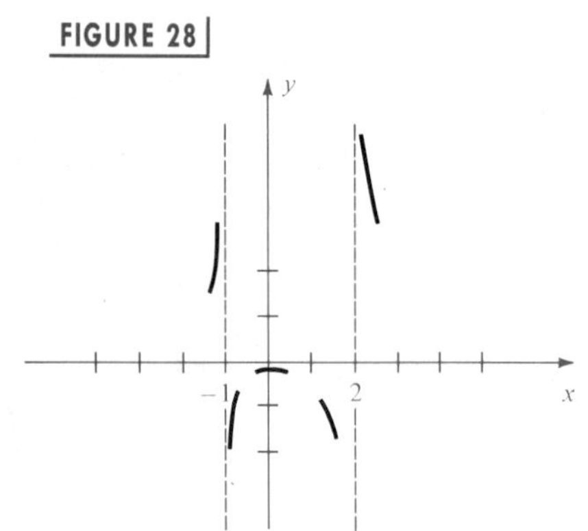

Since the graph is below the x-axis in $(-1, 0)$, we have

$$f(x) \to -\infty \quad \text{as} \quad x \to -1^+.$$

We note these facts in Figure 28 by sketching portions of the graph on each side of $x = -1$.

(ii) Consider the vertical asymptote $x = 2$. The graph is below the x-axis in the interval $(0, 2)$, and hence

$$f(x) \to -\infty \quad \text{as} \quad x \to 2^-.$$

The graph is above the x-axis in $(2, \infty)$, and hence

$$f(x) \to \infty \quad \text{as} \quad x \to 2^+.$$

We note these facts in Figure 28.

FIGURE 29

$$y = \frac{x^2}{x^2 - x - 2}$$

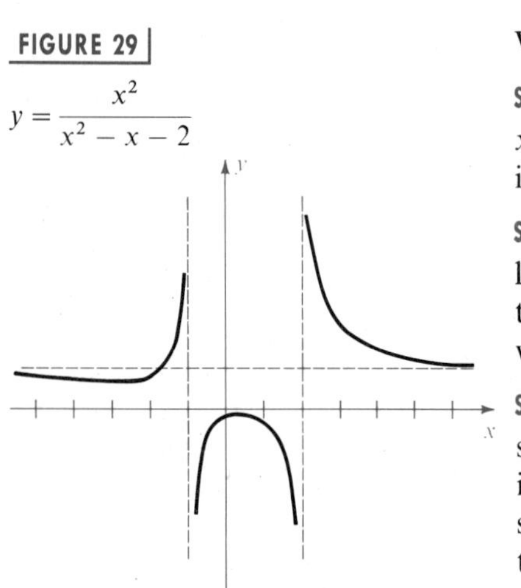

STEP 5 Referring to the table in Step 3, we see that the graph lies below the x-axis in both of the intervals $(-1, 0)$ and $(0, 2)$. Consequently, the graph intersects, but does not cross, the x-axis at $(0, 0)$.

STEP 6 The numerator and denominator of $f(x)$ have the same degree and leading coefficient 1. Hence, by (ii) of the theorem on horizontal asymptotes, the line $y = \frac{1}{1} = 1$ is a horizontal asymptote. We sketch this line with dashes in Figure 29.

STEP 7 Using the information found in Steps 4, 5, and 6, and plotting several points, we obtain the graph sketched in Figure 29. The graph intersects the horizontal asymptote at $x = -2$; we may verify this by solving the equation $x^2/(x^2 - x - 2) = 1$. The fact that the graph lies below the horizontal asymptote if $x < -2$ and above it if $-2 < x < -1$ may be verified by plotting points.

EXAMPLE ▪ 5

Sketch the graph of f if $f(x) = \dfrac{2x^4}{x^4 + 1}$.

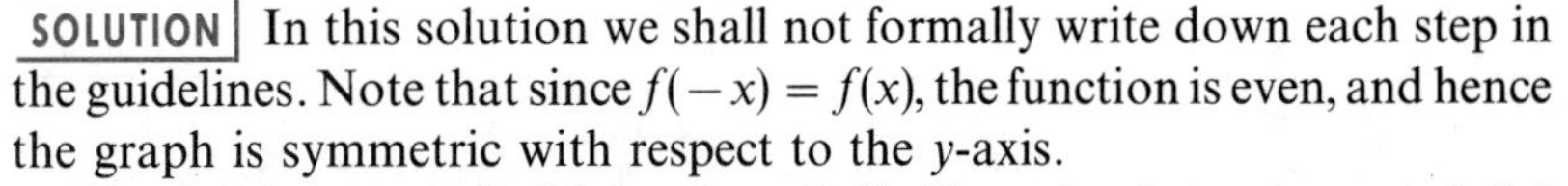

SOLUTION In this solution we shall not formally write down each step in the guidelines. Note that since $f(-x) = f(x)$, the function is even, and hence the graph is symmetric with respect to the y-axis.

FIGURE 30

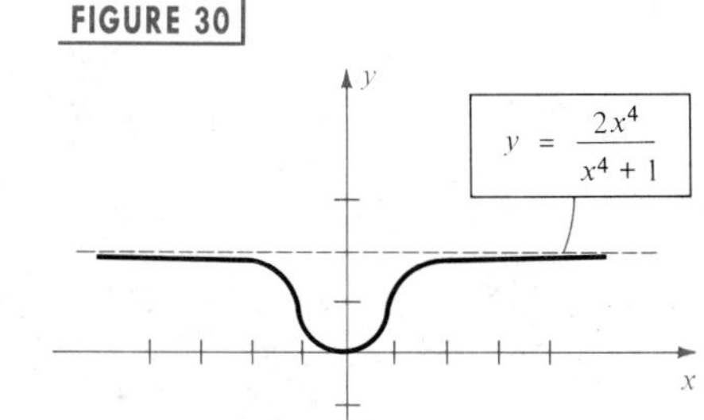

The graph intersects the x-axis at $(0, 0)$. Since the denominator of $f(x)$ has no real zero, the graph has no vertical asymptote.

The numerator and denominator of $f(x)$ have the same degree. Since the leading coefficients are 2 and 1, respectively, we see from (ii) of the theorem on horizontal asymptotes that the line $y = \frac{2}{1} = 2$ is a horizontal asymptote. We represent this line with dashes in Figure 30.

Plotting several points and making use of the symmetry with respect to the y-axis leads to the sketch in Figure 30.

If $f(x) = g(x)/h(x)$ for polynomials $g(x)$ and $h(x)$, and *if the degree of $g(x)$ is one greater than the degree of $h(x)$*, then the graph of f has an **oblique asymptote** $y = ax + b$; that is, the graph approaches this line as $x \to \infty$ or as $x \to -\infty$. To find the oblique asymptote we may use long division to express $f(x)$ in the form

$$f(x) = \frac{g(x)}{h(x)} = (ax + b) + \frac{r(x)}{h(x)}$$

where either $r(x) = 0$ or the degree of $r(x)$ is less than the degree of $h(x)$. From (i) of the theorem on horizontal asymptotes,

$$\frac{r(x)}{h(x)} \to 0 \quad \text{as} \quad x \to \infty \quad \text{or as} \quad x \to -\infty.$$

Consequently, $f(x)$ gets closer and closer to $ax + b$ as $|x|$ increases without bound.

EXAMPLE ▪ 6

Find all the asymptotes and sketch the graph of f if $f(x) = \dfrac{x^2 - 9}{2x - 4}$.

SOLUTION A vertical asymptote occurs if $2x - 4 = 0$, that is, if $x = 2$.

The degree of the numerator of $f(x)$ is greater than the degree of the denominator. Hence by (iii) of the theorem on horizontal asymptotes,

FIGURE 31

there is no horizontal asymptote. However, since the degree of the numerator $x^2 - 9$ is *one* greater than the degree of the denominator $2x - 4$, the graph has an oblique asymptote. By long division, we obtain

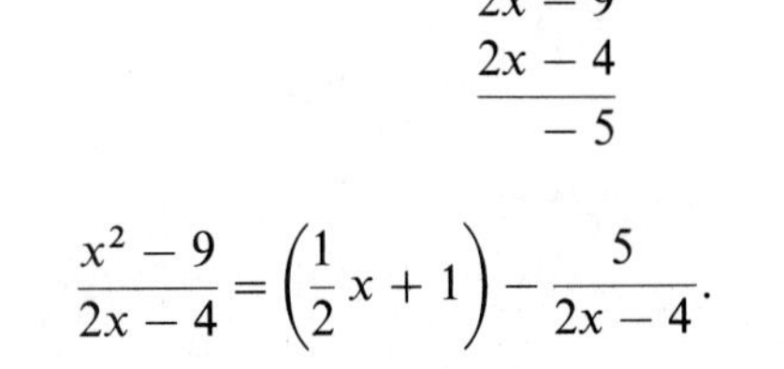

Therefore,
$$\frac{x^2 - 9}{2x - 4} = \left(\frac{1}{2}x + 1\right) - \frac{5}{2x - 4}.$$

FIGURE 32

As we indicated in the discussion preceding this example, the line $y = \frac{1}{2}x + 1$ is an oblique asymptote. This line and the vertical asymptote $x = 2$ are sketched with dashes in Figure 31.

The x-intercepts of the graph are the solutions of the equation $x^2 - 9 = 0$, and hence are 3 and -3. The y-intercept is $f(0) = \frac{9}{4}$. The corresponding points are plotted in Figure 31. We may now show that the graph has the shape indicated in Figure 32.

Graphs of rational functions may become increasingly complicated as the degrees of the polynomials in the numerator and denominator increase. Techniques developed in calculus must be employed for a thorough treatment of such graphs.

4.6 EXERCISES

Exer. 1–22: Sketch the graph of f.

1 $f(x) = \dfrac{3}{x - 4}$

2 $f(x) = \dfrac{-3}{x + 3}$

3 $f(x) = \dfrac{-3x}{x + 2}$

4 $f(x) = \dfrac{4x}{2x - 5}$

5 $f(x) = \dfrac{x - 2}{x^2 - x - 6}$

6 $f(x) = \dfrac{x + 1}{x^2 + 2x - 3}$

7 $f(x) = \dfrac{-4}{(x - 2)^2}$

8 $f(x) = \dfrac{2}{(x + 1)^2}$

9 $f(x) = \dfrac{x - 3}{x^2 - 1}$

10 $f(x) = \dfrac{x + 4}{x^2 - 4}$

11 $f(x) = \dfrac{2x^2 - 2x - 4}{x^2 + x - 12}$

12 $f(x) = \dfrac{-3x^2 - 3x + 6}{x^2 - 9}$

13 $f(x) = \dfrac{-x^2 - x + 6}{x^2 + 3x - 4}$

14 $f(x) = \dfrac{x^2 - 3x - 4}{x^2 + x - 6}$

15 $f(x) = \dfrac{3x^2 - 3x - 36}{x^2 + x - 2}$

16 $f(x) = \dfrac{2x^2 + 4x - 48}{x^2 + 3x - 10}$

17 $f(x) = \dfrac{-2x^2 + 10x - 12}{x^2 + x}$

18 $f(x) = \dfrac{2x^2 + 8x + 6}{x^2 - 2x}$

19 $f(x) = \dfrac{x - 1}{x^3 - 4x}$

20 $f(x) = \dfrac{x^2 - 2x + 1}{x^3 - 9x}$

21 $f(x) = \dfrac{-3x^2}{x^2 + 1}$

22 $f(x) = \dfrac{x^2 - 4}{x^2 + 1}$

Exer. 23–26: Find the oblique asymptote, and sketch the graph of f.

23 $f(x) = \dfrac{x^2 - x - 6}{x + 1}$

24 $f(x) = \dfrac{2x^2 - x - 3}{x - 2}$

25 $f(x) = \dfrac{8 - x^3}{2x^2}$

26 $f(x) = \dfrac{x^3 + 1}{x^2 - 9}$

Exer. 27–32: Simplify $f(x)$ and sketch the graph of f.

27 $f(x) = \dfrac{2x^2 + x - 6}{x^2 + 3x + 2}$

28 $f(x) = \dfrac{x^2 - x - 6}{x^2 - 2x - 3}$

29 $f(x) = \dfrac{x - 1}{1 - x^2}$

30 $f(x) = \dfrac{x + 2}{x^2 - 4}$

31 $f(x) = \dfrac{x^2 + x - 2}{x + 2}$

32 $f(x) = \dfrac{x^3 - 2x^2 - 4x + 8}{x - 2}$

33 A cylindrical container for storing radioactive waste is to be constructed from lead. This container must be 6 inches thick (see figure). The volume of the outside cylinder shown in the figure is to be $16\pi \text{ ft}^3$.

(a) Express the height h of the inside cylinder as a function of the inside radius r.

(b) Show that the inside volume is given by a rational function V such that

$$V(r) = \pi r^2 \left[\frac{16}{(r + 0.5)^2} - 1 \right].$$

(c) What values of r must be excluded in part (b)?

EXERCISE 33

34 Young's rule is a formula that is used for modifying adult drug dosage levels for young children. If a denotes the adult dose (in mg), and if t is the age of the child (in years), then the child's dose y is given by $y = ta/(t + 12)$. Sketch the graph of this equation for $t > 0$ and $a = 100$.

35 Salt water of concentration 0.1 pounds of salt per gallon flows into a large tank that initially contains 50 gallons of pure water.

(a) If the flow rate of salt water into the tank is 5 gallons per minute, what is the volume $V(t)$ of water and the amount $A(t)$ of salt in the tank at time t?

(b) Show that the salt concentration at time t is given by $c(t) = t/(10t + 100)$.

(c) Discuss the behavior of $c(t)$ as $t \to \infty$.

36 The total number of inches of rain during a storm of length t hours can be approximated by a rational function R such that

$$R(t) = \frac{at}{t + b}$$

for positive constants a and b that depend on the geographical locale.

(a) Discuss the variation of $R(t)$ as $t \to \infty$.

(b) The intensity I of the rainfall is defined to be $I = R(t)/t$ inches/hour. If $a = 2$ and $b = 8$, sketch the graph of R and I on the same coordinate plane for $t > 0$.

37 Coulomb's law in electrical theory states that the force of attraction F between two charged particles is inversely proportional to the square of the distance between the particles and directly proportional to the product of the charges. Suppose a particle of charge $+1$ is placed on a coordinate line between two particles of charge -1, as shown in the figure.

(a) Show that the net force acting on the particle of charge $+1$ is given by

$$F(x) = -\frac{k}{x^2} + \frac{k}{(x - 2)^2}$$

for some $k > 0$.

(b) Let $k = 1$ and sketch the graph of F for $0 < x < 2$.

EXERCISE 37

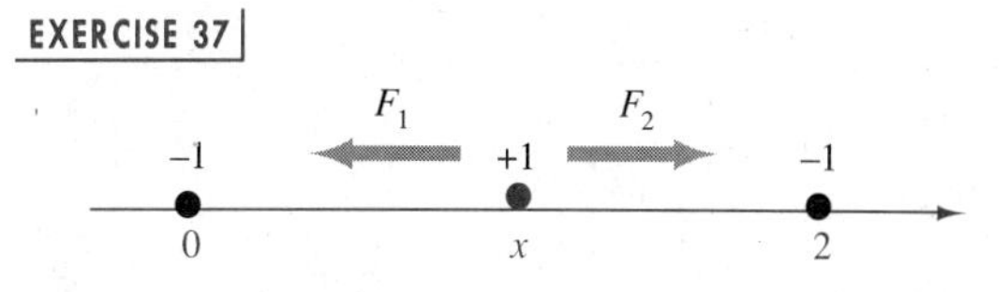

38 Biomathematicians have proposed many different functions for describing the effect of light on the rate at which photosynthesis can take place. If the function is to be realistic, then it must exhibit the photoinhibition effect; that is, the rate of production P of photosynthesis must decrease to 0 as the light intensity I reaches high levels (see figure). Which of the following functions might be used and which may not be used? Why?

(a) $P = \dfrac{aI}{b + I}$ (b) $P = \dfrac{aI}{b + I^2}$

EXERCISE 38

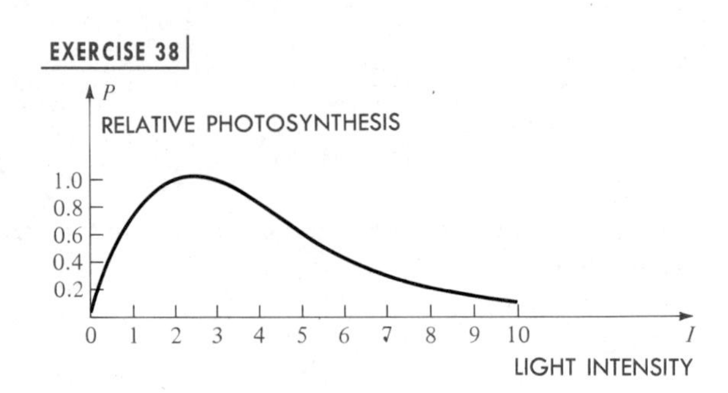

4.7 REVIEW

Define or discuss each of the following.

Polynomial function ■ Quadratic function ■ Graph of a quadratic function ■ Maximum or minimum values of a quadratic function ■ Intermediate value theorem ■ Graphs of polynomial functions of degree greater than 2 ■ Division algorithm for polynomials ■ Remainder theorem ■ Factor theorem ■ Synthetic division ■ Fundamental theorem of algebra ■ Multiplicity of a zero of a polynomial ■ The number of zeros of a polynomial ■ Descartes' rule of signs ■ Bounds for solutions of an equation ■ Rational zeros of a polynomial function ■ Rational function ■ Graph of a rational function ■ Vertical and horizontal asymptotes ■ Oblique asymptotes

4.7 EXERCISES

Exer. 1–4: Sketch the graph of the equation.

1 $x = y^2 + 6y + 16$

2 $4y^2 + x = 10$

3 $4y = (x + 2)(x - 1)^2(3 - x)$

4 $y = \frac{1}{15}(x^5 - 20x^3 + 64x)$

Exer. 5–16: Sketch the graph of f.

5 $f(x) = (x - 4)^2$

6 $f(x) = 12x^2 + 5x - 3$

7 $f(x) = (x + 2)^3$

8 $f(x) = 2x^2 + x^3 - x^4$

9 $f(x) = x^3 + 2x^2 - 8x$

10 $f(x) = x^6 - 32$

11 $f(x) = \dfrac{-2}{(x + 1)^2}$

12 $f(x) = \dfrac{1}{(x - 1)^3}$

13 $f(x) = \dfrac{3x^2}{16 - x^2}$

14 $f(x) = \dfrac{x}{(x + 5)(x^2 - 5x + 4)}$

15 $f(x) = \dfrac{x^2 + 2x - 8}{x + 3}$

16 $f(x) = \dfrac{x^4 - 16}{x^3}$

Exer. 17–18: Find the maximum or minimum value of $f(x)$.

17 $f(x) = 5x^2 + 30x + 49$

18 $f(x) = -3x^2 + 30x - 82$

19 The interior of a half-mile race track consists of a rectangle with semicircles at two opposite ends. Find the dimensions that will maximize the area of the rectangle.

20 At 1:00 P.M. ship A is 30 miles due south of ship B and is sailing north at a rate of 15 mi/hr. If ship B is sailing west at a rate of 10 mi/hr, find the time at which the distance between the ships is minimal (see figure).

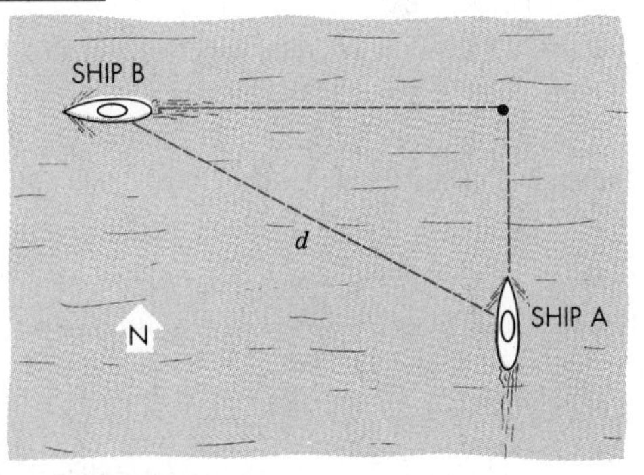

Exer. 21–24: Find the quotient and remainder if $f(x)$ is divided by $p(x)$.

21 $f(x) = 3x^5 - 4x^3 + x + 5,\quad p(x) = x^3 - 2x + 7$

22 $f(x) = 7x^2 + 3x - 10,\quad p(x) = x^3 - x^2 + 10$

23 $f(x) = 9x + 4,\quad p(x) = 2x - 5$

24 $f(x) = 4x^3 - x^2 + 2x - 1,\quad p(x) = x^2$

25 If $f(x) = -4x^4 + 3x^3 - 5x^2 + 7x - 10$, use the remainder theorem to find $f(-2)$.

26 Use the remainder theorem to show that $x - 3$ is a factor of $2x^4 - 5x^3 - 4x^2 + 9$.

Exer. 27–28: Use synthetic division to find the quotient and remainder if $f(x)$ is divided by $p(x)$.

27 $f(x) = 6x^5 - 4x^2 + 8,\quad p(x) = x + 2$

28 $f(x) = 2x^3 + 5x^2 - 2x + 1,\quad p(x) = x - \sqrt{2}$

Exer. 29–30: Find a polynomial with real coefficients that has the indicated zeros and degree and satisfies the given condition.

29 $-3 + 5i, -1;\quad$ degree 3; $\quad f(1) = 4$

30 $1 - i, 3, 0;\quad$ degree 4; $\quad f(2) = -1$

31 Find a polynomial of degree 7 such that -3 is a zero of multiplicity 2 and 0 is a zero of multiplicity 5.

32 Show that 2 is a zero of multiplicity 3 of the polynomial $x^5 - 4x^4 - 3x^3 + 34x^2 - 52x + 24$ and express this polynomial as a product of linear factors.

Exer. 33–34: Find the zeros of the polynomial and state the multiplicity of each zero.

33 $(x^2 - 2x + 1)^2(x^2 + 2x - 3)$

34 $x^6 + 2x^4 + x^2$

Exer. 35–36:

(a) Use Descartes' rule of signs to determine the number of positive, negative, and nonreal complex solutions.

(b) Find the smallest and largest integers that are upper and lower bounds, respectively, of the real solutions.

35 $2x^4 - 4x^3 + 2x^2 - 5x - 7 = 0$

36 $x^5 - 4x^3 + 6x^2 + x + 4 = 0$

37 Show that $7x^6 + 2x^4 + 3x^2 + 10$ has no real zero.

Exer. 38–40: Find all solutions of the equation.

38 $x^4 + 9x^3 + 31x^2 + 49x + 30 = 0$

39 $16x^3 - 20x^2 - 8x + 3 = 0$

40 $x^4 - 7x^2 + 6 = 0$

41 An aspirin tablet in the shape of a right circular cylinder has height $\frac{1}{3}$ cm and radius $\frac{1}{2}$ cm. The manufacturer also wishes to market the aspirin in capsule form. The capsule is to be $\frac{3}{2}$ cm long, in the shape of a right circular cylinder with hemispheres attached at both ends (see figure).

(a) If r denotes the radius of a hemisphere, find a formula for the volume of the capsule.

(b) Find the radius of the capsule so that its volume is equal to that of the tablet.

EXERCISE 41

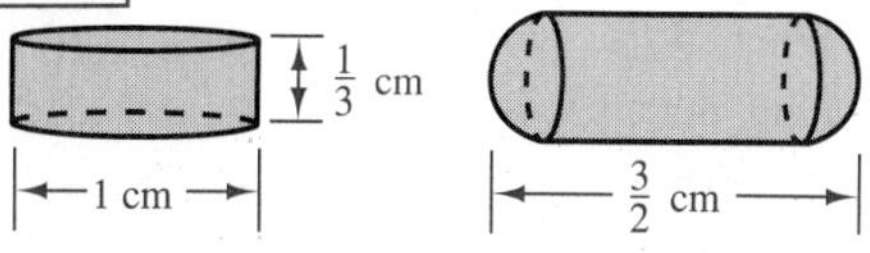

42 A rectangle made of elastic material is to be made into a cylinder by joining edge AD to edge BC, as shown in the figure. A wire of fixed length l is placed along the diagonal of the rectangle to support the structure. Let x denote the height of the cylinder.

(a) Express the volume V of the cylinder in terms of x.

(b) For what positive values of x is $V > 0$?

EXERCISE 42

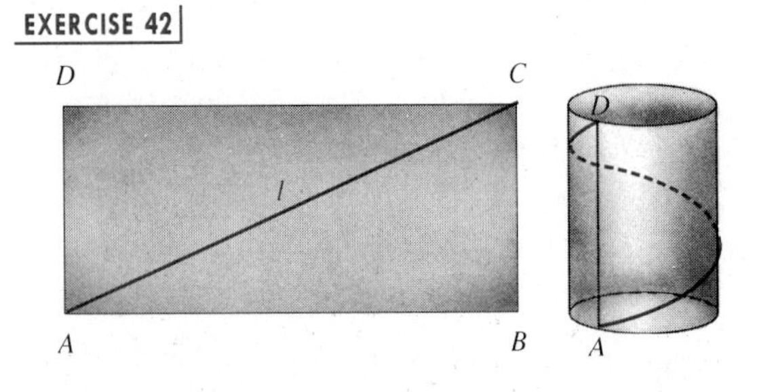

43 A horizontal beam l feet long is supported at one end and unsupported at the other end (see figure). If the beam is subjected to a uniform load, and if y denotes the deflection of the beam at a position x feet from the supported end, then it can be shown that $y = cx^2(x^2 - 4lx + 6l^2)$ for a positive constant c that depends on the weight of the load and the physical properties of the beam.

(a) If the beam is 10 feet long and the deflection at the unsupported end of the board is 2 feet, find c.

(b) Show that the deflection is 1 foot somewhere between $x = 6.1$ and $x = 6.2$.

EXERCISE 43

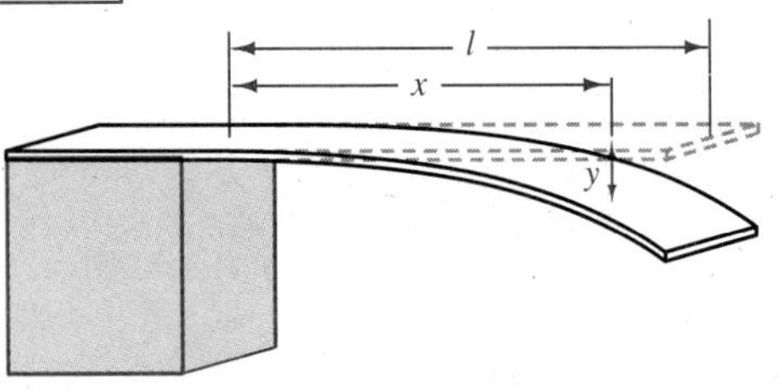

44 A rocket is fired up a hillside, following a path given by $y = -0.016x^2 + 1.6x$. The hillside has slope $\frac{1}{5}$ as illustrated in the figure.

(a) Where does the rocket land?

(b) Find the maximum height of the rocket *above the ground.*

EXERCISE 44

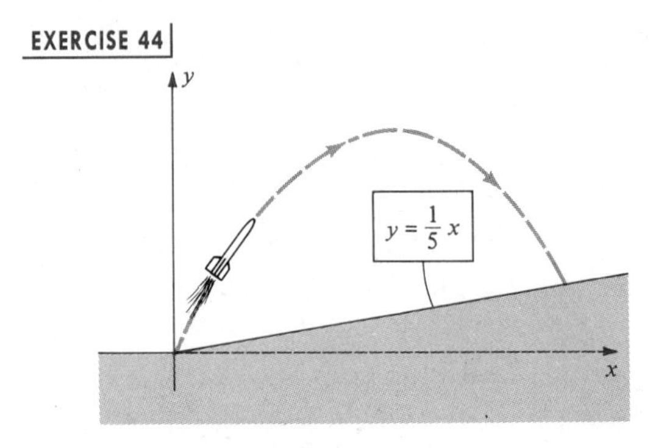

45 When a particular basketball player leaps straight up for a dunk, his distance $f(t)$ in feet off the ground after t seconds is given by $f(t) = -\frac{1}{2}gt^2 + 16t$.

(a) If $g = 32$, find the player's hang time; that is, the total number of seconds that the player is in the air.

(b) Find the player's vertical leap; that is, the maximum distance of his feet from the floor.

(c) On the moon, $g = \frac{32}{6}$. Rework parts (a) and (b) for a player on the moon.

46 The cost $C(x)$ of cleaning up x percent of an oil spill that has washed ashore increases greatly as x approaches 100. Suppose that

$$C(x) = \frac{20x}{101 - x} \quad \text{(thousand dollars).}$$

(a) Compare $C(100)$ to $C(90)$.

(b) Sketch the graph of C for $0 < x < 100$.

47 In biochemistry, the general threshold-response curve is the graph of an equation

$$R = \frac{kS^n}{S^n + a^n},$$

for the chemical response R when the level of the substance being acted on is S and for positive constants a, k, and n. An example is the removal rate R of alcohol from the bloodstream by the liver when the blood alcohol concentration is S.

(a) Find an equation of the horizontal asymptote for the graph.

(b) In the case of alcohol removal, $n = 1$ and a typical value of k is 0.22 gram/liter/minute. What is the interpretation of k in this setting?

CHAPTER 5

EXPONENTIAL AND LOGARITHMIC FUNCTIONS

Exponential and logarithmic functions have applications in almost every field of human endeavor. They are especially useful in chemistry, biology, physics, and engineering to describe the manner in which quantities vary. In this chapter we shall examine properties of these functions and consider many of their applications.

5.1 EXPONENTIAL FUNCTIONS

Throughout this chapter the letter a will denote a positive real number. Sometimes we will impose additional restrictions, such as $a>1$ or $0<a<1$. In Chapter 1 we defined a^r for every rational number r as follows: if m and n are integers with $n>0$, then $a^{m/n}=\sqrt[n]{a^m}$. Using methods developed in calculus, we can define a^x for every *real* number x. To illustrate, for a^π we could use the nonterminating decimal representation 3.1415926 . . . for π and consider the following *rational* powers of a:

$$a^3, \quad a^{3.1}, \quad a^{3.14}, \quad a^{3.141}, \quad a^{3.1415}, \quad a^{3.14159}, \quad \ldots$$

If a^x is properly defined, then each successive power gets closer to a^π. In this chapter we shall assume that a positive real number a^x can be obtained in similar fashion for every real number x and that the laws of exponents are valid in this more general setting.

Since to each real number x there corresponds a unique real number a^x, we can define a function as follows.

DEFINITION

Let a be a positive real number different from 1. The **exponential function f with base a** is given by

$$f(x)=a^x$$

for every real number x.

The following theorem can be proved.

THEOREM

(i) If $a>1$, then $f(x)=a^x$ is increasing on $\mathbb{R}$.
(ii) If $0<a<1$, then $f(x)=a^x$ is decreasing on $\mathbb{R}$.

FIGURE 1

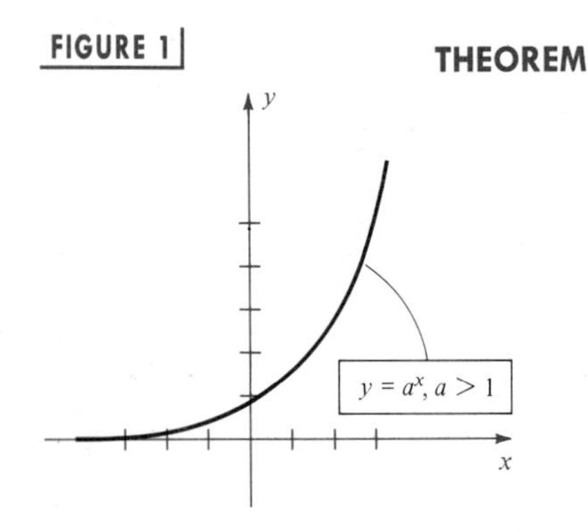

Part (i) of the theorem states that for $a>1$, the graph of f rises as x increases. The graph has the general appearance of the graph in Figure 1; however, the exact shape depends on the value of a. The domain of f is $\mathbb{R}$ and the range is the set of positive real numbers. Since $a^0=1$, the y-intercept is 1. As x decreases through negative values, the graph approaches the x-axis but never intersects it, since $a^x>0$ for every x. Thus

FIGURE 2

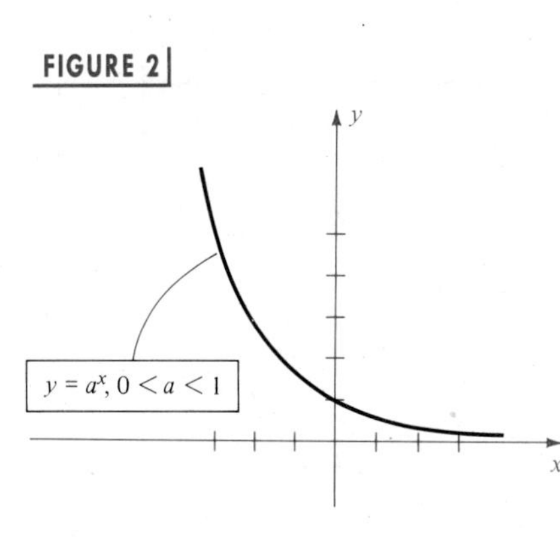

the x-axis is a *horizontal asymptote*. As x increases through positive values, the graph rises rapidly. This type of variation is characteristic of the **exponential law of growth**, and f is sometimes called a **growth function**.

If $0 < a < 1$, as in (ii) of the theorem, the graph of $f(x) = a^x$ has the general shape illustrated in Figure 2. As x increases through positive values the graph approaches the x-axis asymptotically. This type of variation of an exponential function is known as **exponential decay**.

The next three examples illustrate special cases of exponential functions.

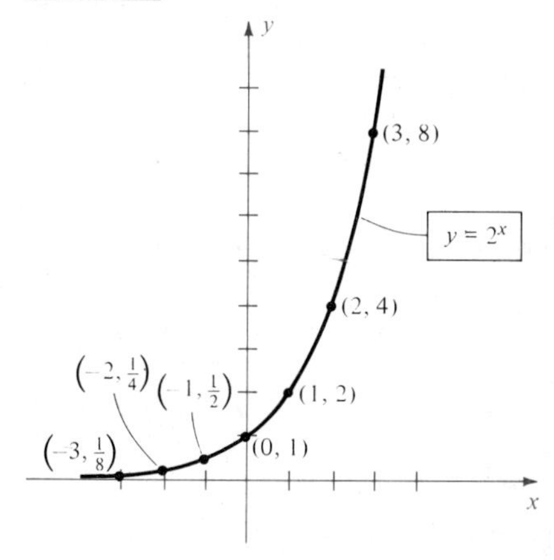

EXAMPLE ▪ 1

If $f(x) = 2^x$, sketch the graph of f.

SOLUTION Coordinates of some points on the graph of $y = 2^x$ are listed in the following table.

x	-3	-2	-1	0	1	2	3	4
y	$\frac{1}{8}$	$\frac{1}{4}$	$\frac{1}{2}$	1	2	4	8	16

Plotting points and using the fact that f is increasing gives us the sketch in Figure 3.

EXAMPLE ▪ 2

If $f(x) = (\frac{3}{2})^x$ and $g(x) = 3^x$, sketch the graphs of f and g on the same coordinate plane.

SOLUTION The following table displays coordinates for several points on the graphs.

x	-2	-1	0	1	2	3	4
$(\frac{3}{2})^x$	$\frac{4}{9} \approx 0.4$	$\frac{2}{3} \approx 0.7$	1	$\frac{3}{2}$	$\frac{9}{4} \approx 2.3$	$\frac{27}{8} \approx 3.4$	$\frac{81}{16} \approx 5.1$
3^x	$\frac{1}{9} \approx 0.1$	$\frac{1}{3} \approx 0.3$	1	3	9	27	81

Plotting points, we obtain the graphs in Figure 4.

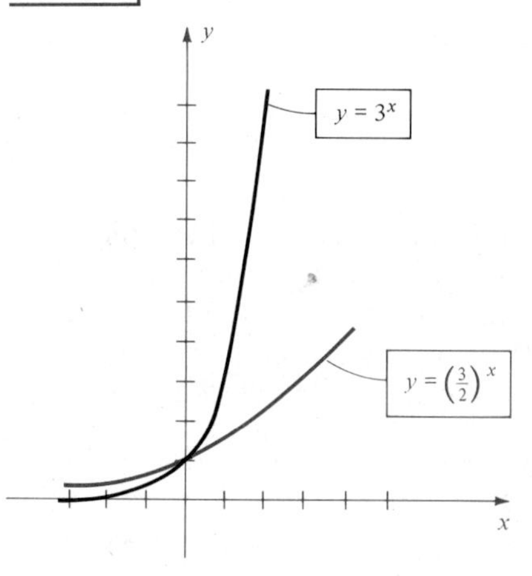

Example 2 illustrates the fact that if $1 < a < b$, then $a^x < b^x$ for positive values of x and $b^x < a^x$ for negative values of x. In particular, since $\frac{3}{2} < 2 < 3$, the graph of $y = 2^x$ in Example 1 lies between the graphs of f and g in Example 2.

EXAMPLE ▪ 3

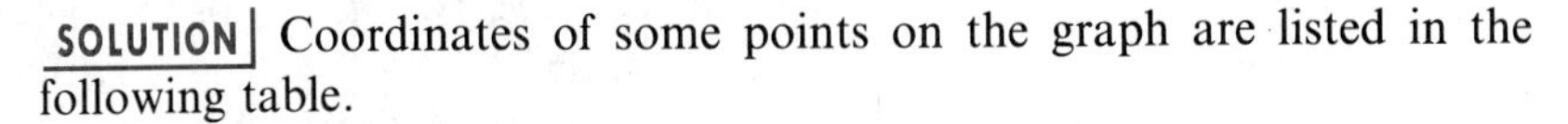

Sketch the graph of the equation $y = (\frac{1}{2})^x$.

FIGURE 5

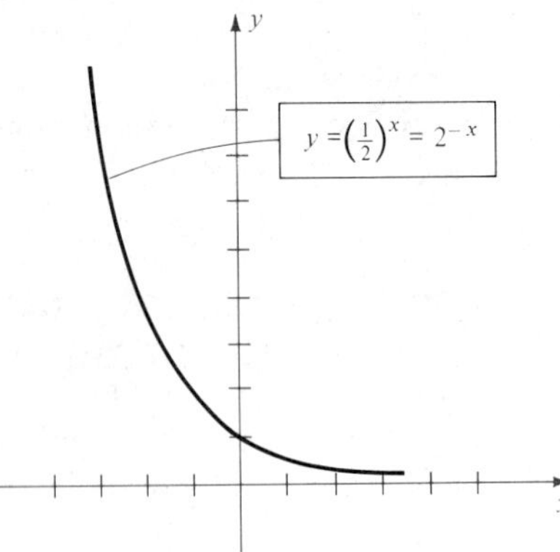

SOLUTION Coordinates of some points on the graph are listed in the following table.

x	-3	-2	-1	0	1	2	3
$(\frac{1}{2})^x$	8	4	2	1	$\frac{1}{2}$	$\frac{1}{4}$	$\frac{1}{8}$

The graph is sketched in Figure 5. Since $(\frac{1}{2})^x = 2^{-x}$, the graph is the same as the graph of the equation $y = 2^{-x}$. Note that the graph is a reflection, through the y-axis, of the graph of $y = 2^x$ in Figure 3.

In advanced mathematics and applications we consider functions such that $f(x) = a^p$ for some expression p in x. The next example illustrates the case $p = -x^2$.

EXAMPLE ▪ 4

If $f(x) = 2^{-x^2}$, sketch the graph of f.

FIGURE 6

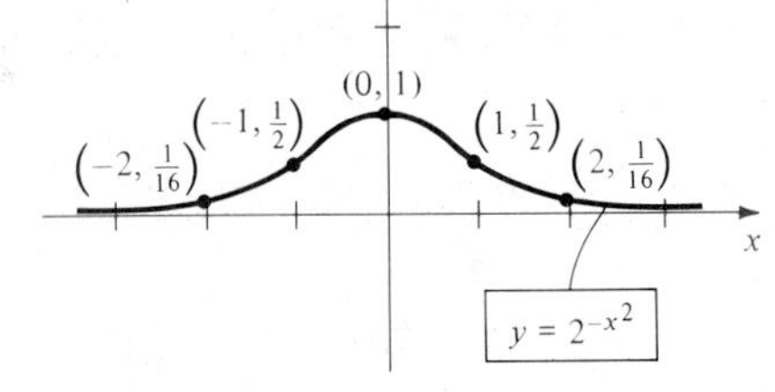

SOLUTION If we rewrite $f(x)$ as

$$f(x) = \frac{1}{2^{(x^2)}}$$

we see that as x increases through positive values, the point $(x, f(x))$ approaches the x-axis. Thus, the x-axis is a horizontal asymptote for the graph. The maximum value of f is $f(0) = 1$. Since f is an even function, the graph is symmetric with respect to the y-axis. Some points on the graph are $(0, 1)$, $(1, \frac{1}{2})$, and $(2, \frac{1}{16})$. Plotting and using symmetry gives us the sketch in Figure 6.

FIGURE 7

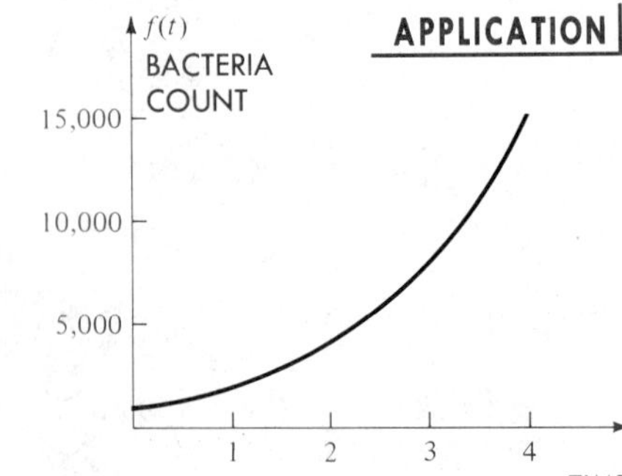

APPLICATION BACTERIAL GROWTH

Exponential functions occur in the study of the growth of certain populations. As an illustration, it might be observed experimentally that the number of bacteria in a culture doubles every day. If 1000 bacteria are present at the start, then the experimenter would obtain the readings listed below, where t is the time in days and $f(t)$ is the bacteria count at time t.

t (time)	0	1	2	3	4
$f(t)$ (bacteria count)	1000	2000	4000	8000	16,000

It appears that $f(t) = (1000)2^t$. With this formula we can predict the number of bacteria present at any time t. For example, at $t = 1.5 = \frac{3}{2}$,

$$f(t) = (1000)2^{3/2} \approx 2828.$$

The graph of f is sketched in Figure 7.

APPLICATION

RADIOACTIVE DECAY

Certain physical quantities *decrease* exponentially. In such cases if a is the base of the exponential function, then $0 < a < 1$. One of the most common examples is the decay of a radioactive substance. As an illustration, the polonium isotope ^{210}Po has a half-life of approximately 140 days; that is, given any amount, one-half of it will disintegrate in 140 days. If 20 mg of ^{210}Po is present initially, then the following table indicates the amount remaining after various intervals of time.

FIGURE 8

Decay of polonium 210

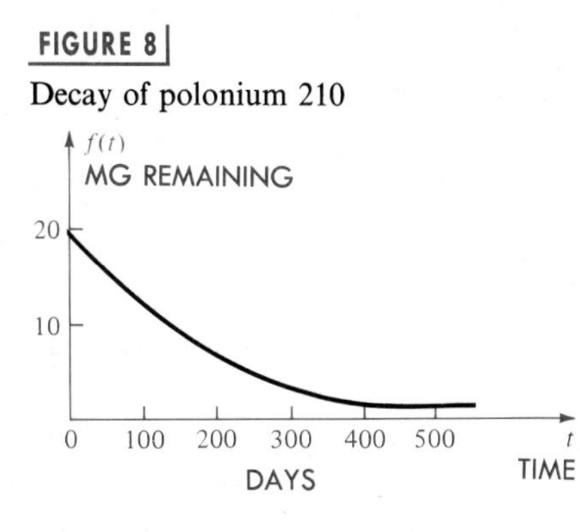

t (days)	0	140	280	420	560
Amount remaining (mg)	20	10	5	2.5	1.25

The sketch in Figure 8 illustrate the exponential nature of the disintegration.

APPLICATION

COMPOUND INTEREST

Compound interest provides a good illustration of exponential growth. If a sum of money P, the *principal*, is invested at a *simple* interest rate r, then the interest at the end of one interest period is the product Pr when r is expressed as a decimal (see page 66). For example, if $P = \$1000$ and the interest rate is 9% per year, then $r = 0.09$, and the interest at the end of one year is \$1000(0.09), or \$90.

If the interest is reinvested with the principal at the end of the interest period, then the new principal is

$$P + Pr, \quad \text{or} \quad P(1 + r).$$

Note that to find the new principal we may multiply the original principal by $(1 + r)$. In the preceding illustration the new principal is \$1000(1.09), or \$1090.

After another interest period has elapsed, the new principal may be found by multiplying $P(1 + r)$ by $(1 + r)$. Thus, the principal after two interest periods is $P(1 + r)^2$. If we continue to reinvest, the principal after three periods is $P(1 + r)^3$; after four it is $P(1 + r)^4$; and in general, the amount A invested after k interest periods is

$$A = P(1 + r)^k.$$

Interest accumulated by means of this formula is **compound interest**. Note that A is expressed in terms of an exponential function with base $1 + r$.

The interest period may be measured in years, months, weeks, days, or any other suitable unit of time. When applying the formula for A, remember that *r is the interest rate per interest period expressed as a decimal.* For example, if the rate is stated as 6% *per year compounded monthly*, then the rate per month is $\frac{6}{12}\%$, or equivalently, 0.5%. Thus, $r = 0.005$ and k is the number of months. If \$100 is invested at this rate, then the formula for A is

$$A = 100(1 + 0.005)^k = 100(1.005)^k$$

Generally, suppose that r is the yearly interest rate (expressed as a decimal) and that interest is compounded n times per year. The interest rate per interest period is r/n. If the principal P is invested for t years, then the number of interest periods is nt, and the amount A after t years is given by the following formula.

COMPOUND INTEREST FORMULA

$$A = P\left(1 + \frac{r}{n}\right)^{nt}$$

EXAMPLE ■ 5

Suppose that \$1000 is invested at an interest rate of 9% compounded monthly. Find the new amount of principal after 5 years; after 10 years; after 15 years. Illustrate graphically the growth of the investment.

FIGURE 9

Compound interest:
$A = 1000(1.0075)^{12t}$

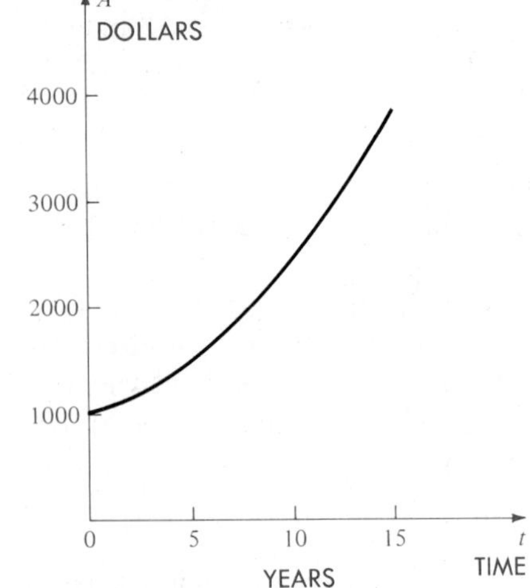

SOLUTION Applying the compound interest formula with $r = 0.09$, $n = 12$, and $P = \$1000$, the amount after t years is

$$A = 1000\left(1 + \frac{0.09}{12}\right)^{12t} = 1000(1.0075)^{12t}.$$

Substituting $t = 5$, 10, and 15, we obtain the following amounts:

after 5 years: $A = 1000(1.0075)^{60} = \$1565.68$

after 10 years: $A = 1000(1.0075)^{120} = \$2451.36$

after 15 years: $A = 1000(1.0075)^{180} = \$3838.04$

The exponential nature of the increase is indicated by the fact that during the first five years, the growth in the investment is \$565.68; during the second five-year period, the growth is \$885.68; and during the last five-year period, it is \$1368.68.

The sketch in Figure 9 illustrates the growth of \$1000 invested over a period of 15 years.

5.1 EXERCISES

Exer. 1–24: Sketch the graph of f.

1 $f(x) = 4^x$

2 $f(x) = 4^{-x}$

3 $f(x) = 5^{-x}$

4 $f(x) = 5^x$

5 $f(x) = (\frac{1}{4})^x$

6 $f(x) = (\frac{1}{4})^{-x}$

7 $f(x) = (\frac{2}{5})^{-x}$

8 $f(x) = (\frac{2}{5})^x$

9 $f(x) = -3^x$

10 $f(x) = -(\frac{1}{3})^x$

11 $f(x) = 2^x - 4$

12 $f(x) = (\frac{1}{3})^x - 3$

13 $f(x) = -(\frac{1}{2})^x + 4$

14 $f(x) = -3^x + 9$

15 $f(x) = 2^{|x|}$

16 $f(x) = 2^{-|x|}$

17 $f(x) = 2^{x+3}$

18 $f(x) = 3^{x+2}$

19 $f(x) = 2^{3-x}$

20 $f(x) = 3^{-2-x}$

21 $f(x) = 3^{1-x^2}$

22 $f(x) = 2^{-(x+1)^2}$

23 $f(x) = 3^x + 3^{-x}$

24 $f(x) = 3^x - 3^{-x}$

Exer. 25–26: Why are the following not allowed when considering a^x?

25 $a \leq 0$

26 $a = 1$

Exer. 27–28: Compare the graph of the equation to the graph of $y = a^x$.

27 $y = -a^x$

28 $y = a^{-x}$

29 One hundred elk, each one year old, are introduced into a game preserve. The number $N(t)$ alive after t years is predicted to be $N(t) = 100(0.9)^t$. Estimate the number of elk alive after

(a) 1 year. (b) 5 years. (c) 10 years.

30 A drug is eliminated from the body through urine. The initial dose is 10 mg and the amount $A(t)$ in the body t hours later is given by $A(t) = 10(0.8)^t$.

(a) Estimate the amount of the drug in the body 8 hours after the initial dose.

(b) What percentage of the drug still in the body is eliminated each hour?

31 The number of bacteria in a certain culture increased from 600 to 1800 between 7:00 A.M. and 9:00 A.M. Assuming exponential growth, the number $f(t)$ of bacteria t hours after 7:00 A.M. is given by $f(t) = 600(3)^{t/2}$.

(a) Estimate the number of bacteria in the culture at 8:00 A.M.; 10:00 A.M.; 11:00 A.M.

(b) Sketch the graph of f from $t = 0$ to $t = 4$.

32 According to Newton's law of cooling, the rate at which an object cools is directly proportional to the difference in temperature between the object and the surrounding medium. The face of a household iron cools from 125° to 100° in 30 minutes in a room that has been at a constant temperature of 75°. Using calculus, the temperature $f(t)$ of the face after t hours of cooling is given by $f(t) = 50(2)^{-2t} + 75$.

(a) Assuming $t = 0$ corresponds to 1:00 P.M., approximate to the nearest tenth of a degree the temperature at 2:00 P.M., 3:30 P.M., and 4:00 P.M.

(b) Sketch the graph of f from $t = 0$ to $t = 4$.

33 The radioactive bismuth isotope ^{210}Bi has a half-life of 5 days; that is, the number of radioactive particles will decrease to one-half the number in 5 days. If there are 100 mg of ^{210}Bi present at $t = 0$, then the amount $f(t)$ remaining after t days is given by $f(t) = 100(2)^{-t/5}$.

(a) How much ^{210}Bi remains after 5 days? 10 days? 12.5 days?

(b) Sketch the graph of f from $t = 0$ to $t = 30$.

34 An important problem in oceanography is to determine the amount of light that can penetrate to various ocean depths. The Beer-Lambert law asserts that an exponential function I such that $I(x) = I_0 a^x$ should be used to model this phenomenon (see figure). For a certain

EXERCISE 34

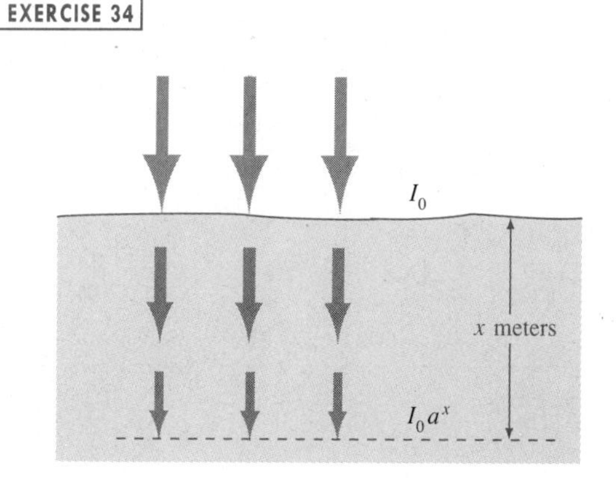

location, $I(x) = 10(0.4)^x$ is the amount of light (in calories/cm^2/second) reaching a depth of x meters.

(a) Find the amount of light at a depth of 2 meters.

(b) Sketch the graph of I from $x = 0$ to $x = 5$.

35 The half-life of radium is 1600 years; that is, given any quantity, one-half of it will disintegrate in 1600 years. If the initial amount is q_0 milligrams, then the quantity $q(t)$ remaining after t years is given by $q(t) = q_0 2^{kt}$. Find k.

36 If 10 grams of salt are added to a quantity of water, then the amount $q(t)$ that is undissolved after t minutes is given by $q(t) = 10(\frac{4}{5})^t$. Sketch a graph that shows the value $q(t)$ at any time from $t = 0$ to $t = 10$.

Exer. 37–44: Use a calculator.

37 If \$1000 is invested at a rate of 12% per year compounded monthly, find the principal after

(a) 1 month. **(b)** 2 months.

(c) 6 months. **(d)** 1 year.

38 If a savings fund pays interest at a rate of 10% compounded semiannually, how much money invested now will amount to \$5000 after one year?

39 If a certain make of automobile is purchased for C dollars, then its trade-in value $v(t)$ at the end of t years is given by $v(t) = 0.78C(0.85)^{t-1}$. If the original cost is \$10,000, calculate to the nearest dollar the value after

(a) 1 year. **(b)** 4 years. **(c)** 7 years.

40 If the value of real estate increases at a rate of 10% per year, then after t years the value V of a house purchased for P dollars is given by $V = P(1.1)^t$. A graph for the value of a house purchased for \$80,000 in 1986 is shown in the figure. Find the value of the house in 1990.

EXERCISE 40

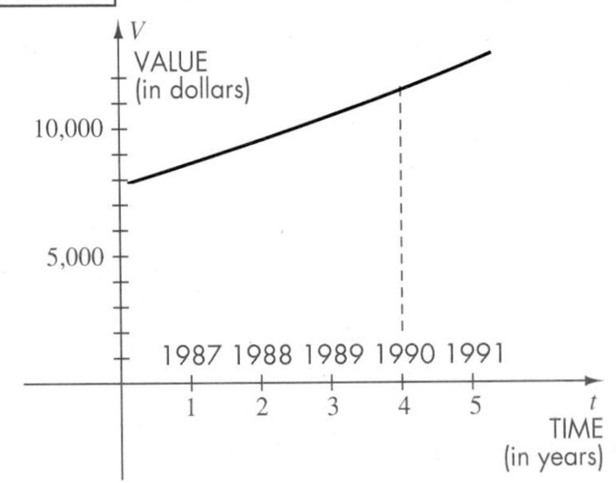

41 If \$1000 is invested at an interest rate of 6% per year compounded quarterly, find the principal at the end of

(a) one year. **(b)** two years.

(c) five years. **(d)** ten years.

42 A certain department store requires its credit card customers to pay interest on any unpaid bill at the rate of 18% per year compounded monthly. If a customer buys a television set for \$500 on credit and then makes no payments for one year, how much is owed at the end of the year?

43 The declining balance method is an accounting method of depreciation in which the amount of depreciation taken each year is a fixed percentage of the present value of the item. If y is the value of the item after x years, the depreciation taken is ay for some a with $0 < a < 1$, and the new value is $(1 - a)y$.

(a) If the initial value of the item is y_0, show that $y = y_0(1 - a)^x$.

(b) At the end of N years, the item has a salvage value of s dollars. A taxpayer wishes to choose a depreciation rate so that the value of the item after N years will equal the salvage value (see figure). Show that $a = 1 - \sqrt[N]{s/y_0}$.

EXERCISE 43

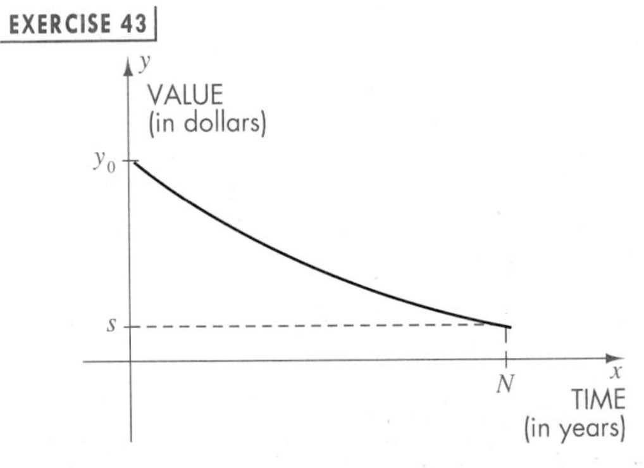

44 Glottochronology is a method of dating a language at a particular stage, based on the theory that over a long period of time, linguistic changes take place at a fairly constant rate. Suppose a language originally had N_0 basic words, and that at time t, measured in millennia (1 millennium = 1000 years), the number $N(t)$ of basic words that remain in common use is given by $N(t) = N_0(0.805)^t$.

(a) What percentage of basic words is lost every 100 years?

(b) If $N_0 = 200$, sketch the graph of N for $0 \leq t \leq 5$.

Exer. 45–48: Some lending institutions calculate the monthly payment M on a loan of L dollars at an interest rate r (expressed as a decimal) by using the formula

$$M = \frac{Lrk}{12(k-1)}$$

where $k = \left(1 + \frac{r}{12}\right)^{12t}$ and t is the number of years that the loan is in effect.

45 **(a)** Find the monthly payment on a 30-year $90,000 mortgage for a home if the interest rate is 12%.

(b) Find the total interest paid on the loan in part (a).

46 Find the largest 25-year home mortgage that can be obtained at an interest rate of 10% if the monthly payment is to be $800.

47 An automobile dealer offers customers no-down-payment 3-year loans at an interest rate of 15%. If a customer can afford to pay $220 per month, find the price of the most expensive automobile that can be purchased.

48 **(a)** The owner of a small business wishes to finance a new computer by borrowing $3000 for two years at an interest rate of 12.5%. Find the monthly payment.

(b) Find the total interest paid on the loan in part (a).

5.2 THE NATURAL EXPONENTIAL FUNCTION

The *compound interest formula* discussed in the preceding section is

$$A = P\left(1 + \frac{r}{n}\right)^{nt}$$

where P is the principal invested, r is the interest rate (expressed as a decimal), n is the number of interest periods per year, and t is the number of years that the principal is invested. The next example illustrates what happens if the rate and total time invested are fixed, but the *interest period* is varied.

EXAMPLE ▪ 1

Suppose $1000 is invested at a compound interest rate of 9%. Find the new amount of principal after one year if the interest is compounded quarterly; monthly; weekly; daily; hourly; each minute.

SOLUTION If we let $P = \$1000$, $t = 1$, and $r = 0.09$ in the compound interest formula, then

$$A = 1000\left(1 + \frac{0.09}{n}\right)^{n}$$

for n interest periods per year. The values of n we wish to consider are listed in the following table, where we have assumed that there are 365 days in a year, and hence (365)(24) = 8760 hours and (8760)(60) = 525,600 minutes. (In actual business transactions an investment year is considered to be 360 days.)

Interest period	Quarter	Month	Week	Day	Hour	Minute
n	4	12	52	365	8760	525,600

Using the compound interest formula (and a calculator) we obtain the amounts given in the following table.

Interest period	Principal after one year
Quarter	$1000\left(1+\frac{0.09}{4}\right)^{4} = \$1093.08$
Month	$1000\left(1+\frac{0.09}{12}\right)^{12} = \$1093.81$
Week	$1000\left(1+\frac{0.09}{52}\right)^{52} = \$1094.09$
Day	$1000\left(1+\frac{0.09}{365}\right)^{365} = \$1094.16$
Hour	$1000\left(1+\frac{0.09}{8760}\right)^{8760} = \$1094.17$
Minute	$1000\left(1+\frac{0.09}{525{,}600}\right)^{525{,}600} = \$1094.17$

Note that, in the preceding example, after we reach an interest period of one hour, the number of interest periods per year has no effect on the final amount. If interest had been compounded each *second*, the result would still be \$1094.17, since we round off A to the nearest cent. (Some decimal places *beyond* the first two *do* change.) Thus, the amount approaches a fixed value as n increases. Interest is said to be **compounded continuously** if the number n of time periods per year increases without bound.

If we let $P = 1$, $r = 1$, and $t = 1$ in the compound interest formula, we obtain

$$A = \left(1 + \frac{1}{n}\right)^n.$$

The expression on the right-hand side of the equation occurs in calculus. In Example 1 we considered a similar situation: as n increased, A approached a limiting value. The same phenomenon occurs for this formula, as illustrated by the following table, which was obtained using a calculator.

n	Approximation to $\left(1+\frac{1}{n}\right)^n$
1	2.00000000
10	2.59374246
100	2.70481383
1000	2.71692393
10,000	2.71814593
100,000	2.71826824
1,000,000	2.71828047
10,000,000	2.71828169
100,000,000	2.71828181
1,000,000,000	2.71828183

It can be proved that as n increases without bound, the value of $[1 + (1/n)]^n$ approaches a certain irrational number, denoted by e. The number e arises in the investigation of many physical phenomena. An approximation is $e \approx 2.71828$. Using the notation we developed for rational functions in Section 4.6, we denote this fact as follows.

THE NUMBER e

If n is a positive integer, then

$$\left(1+\frac{1}{n}\right)^n \to e \approx 2.71828 \quad \text{as} \quad n \to \infty.$$

FIGURE 10

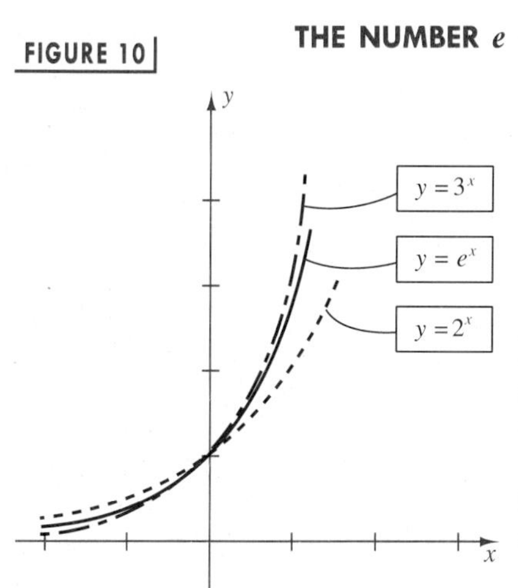

The function f defined by $f(x) = e^x$ is the **natural exponential function**. It is one of the most important functions in advanced mathematics and applications. Since $2 < e < 3$, the graph of $y = e^x$ lies between the graphs of $y = 2^x$ and $y = 3^x$, as shown in Figure 10.

A brief table of values of e^x and e^{-x} is given in Table 2 of Appendix II. Some calculators have an $\boxed{e^x}$ key for approximating values of the natural exponential function and a $\boxed{y^x}$ key that can be used for any positive base y. Approximations of e^x can then be found by calculating $(2.71828)^x$.

EXAMPLE ▪ 2

Sketch the graph of f if

$$f(x) = \frac{e^x + e^{-x}}{2}.$$

SOLUTION Note that f is an even function, because

$$f(-x) = \frac{e^{-x} + e^{-(-x)}}{2} = \frac{e^{-x} + e^x}{2} = f(x).$$

FIGURE 11

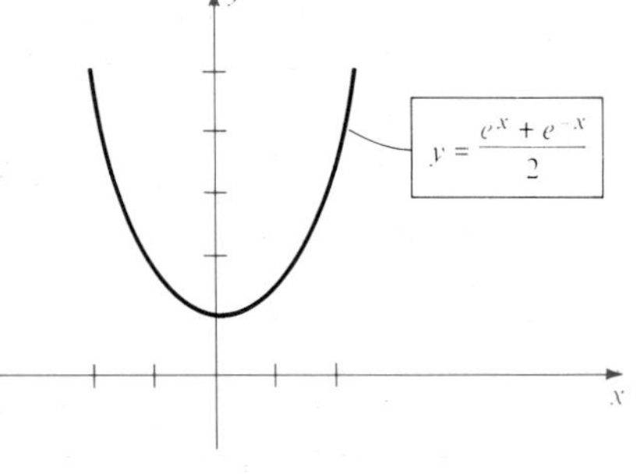

Thus, the graph is symmetric with respect to the y-axis. Using a calculator or Table 2 of Appendix II, we obtain the following approximations of $f(x)$.

x	0	0.5	1.0	1.5	2.0
$f(x)$ (approx.)	1	1.13	1.54	2.35	3.76

Plotting points and using symmetry with respect to the y-axis gives us the sketch in Figure 11.

APPLICATION FLEXIBLE CABLES

FIGURE 12

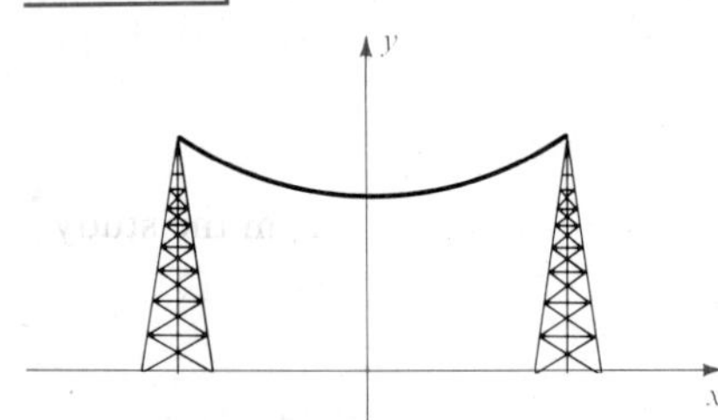

The function f of Example 2 occurs in applied mathematics and engineering, where it is called the **hyperbolic cosine function**. This function can be used to describe the shape of a uniform flexible cable, or chain, whose ends are supported from the same height, such as for telephone or power lines (see Figure 12). If we introduce a coordinate system, as indicated in the figure, then it can be shown that an equation that corresponds to the shape of the cable is $y = (a/2)(e^{x/a} + e^{-x/a})$ for a real number a. The graph is called a **catenary**, after the Latin word for *chain*. The function in Example 2 is the special case in which $a = 1$.

APPLICATION RADIOTHERAPY

One of the many fields in which exponential functions with base e play an important role is *radiotherapy*, the treatment of tumors by X rays or radioactive substances. Of major interest is the fraction of cells in a tumor that survives a treatment. This *surviving fraction* depends not only on the energy and nature of the radiation, but also on the depth, size, and characteristics of the tumor itself. The exposure to radiation may be thought of as a number of potentially damaging events, where only one *hit* is required to kill a tumor cell. Suppose that each cell has exactly one *target* that must be hit. If k denotes the average target size of a tumor cell, and if x is the number of damaging events (the *dose*), then the surviving fraction $f(x)$ is given by

$$f(x) = e^{-kx}.$$

This is called the *one-target–one-hit surviving fraction.*

Next, suppose that each cell has n targets and that hitting any one of the targets results in the death of a cell. In this case, the *n-target–one-hit surviving fraction* is

$$f(x) = 1 - (1 - e^{-kx})^n.$$

The graph of f may be analyzed to determine what effect increasing the dosage x will have on decreasing the surviving fraction of tumor cells. Note that $f(0) = 1$; that is, if there is no dose, then all cells survive. As an example, if $k = 1$ and $n = 2$, then

$$\begin{aligned} f(x) &= 1 - (1 - e^{-x})^2 \\ &= 1 - (1 - 2e^{-x} + e^{-2x}) \\ &= 2e^{-x} - e^{-2x}. \end{aligned}$$

FIGURE 13

Surviving fraction of tumor cells after a radiation treatment

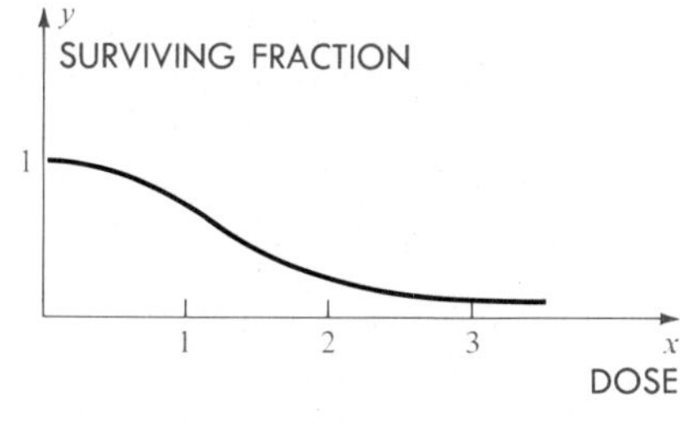

A complete analysis of the graph of f requires calculus. The graph is sketched in Figure 13. The *shoulder* on the curve near the point $(0, 1)$ represents the threshold nature of the treatment; that is, a small dose results in very little tumor cell elimination. Note that for a large x, an increase in dosage has little effect on the surviving fraction. To determine the ideal dose that should be administered to a patient, specialists in radiation therapy must also take into account the number of healthy cells that are killed during a treatment.

Problems of the type illustrated in the next example occur in the study of calculus (see also Exercises 9–12).

EXAMPLE ▪ 3

If $f(x) = x^2(-2e^{-2x}) + 2xe^{-2x}$, find the zeros of f.

SOLUTION We may factor $f(x)$ as follows:

$$\begin{aligned} f(x) &= 2xe^{-2x} - 2x^2e^{-2x} \\ &= 2xe^{-2x}(1 - x) \end{aligned}$$

To find the zeros of f, we solve the equation $f(x) = 0$. Since $e^{-2x} > 0$ for every x, we see that $f(x) = 0$ if and only if $x = 0$ or $1 - x = 0$. Thus, the zeros of f are 0 and 1.

5.2 EXERCISES

Exer. 1–7: Use the graph of $y = e^x$ to help sketch the graph of f.

1 (a) $f(x) = e^{-x}$ (b) $f(x) = -e^x$

2 (a) $f(x) = e^{2x}$ (b) $f(x) = 2e^x$

3 (a) $f(x) = e^{x+4}$ (b) $f(x) = e^x + 4$

4 (a) $f(x) = e^{-2x}$ (b) $f(x) = -2e^x$

5 $f(x) = \dfrac{e^{2x} + e^{-2x}}{2}$

6 $f(x) = \dfrac{e^x - e^{-x}}{2}$

7 $f(x) = \dfrac{2}{e^x + e^{-x}}$

(*Hint:* Take reciprocals of y-coordinates in Example 2.)

8 In statistics the probability density function for the normal distribution is defined by

$$f(x) = \frac{1}{\sigma\sqrt{2\pi}} e^{-z^2/2} \quad \text{with } z = \frac{x - \mu}{\sigma}$$

for real numbers μ and $\sigma > 0$. (μ is the *mean* and σ^2 is the *variance* of the distribution.) Sketch the graph of f for the case $\sigma = 1$ and $\mu = 0$.

Exer. 9–12: Find the zero(s) of f.

9 $f(x) = xe^x + e^x$

10 $f(x) = -x^2e^{-x} + 2xe^{-x}$

11 $f(x) = x^3(4e^{4x}) + 3x^2e^{4x}$

12 $f(x) = x^2(2e^{2x}) + 2xe^{2x} + e^{2x} + 2xe^{2x}$

Exer. 13–14: Simplify the expression.

13 $\dfrac{(e^x + e^{-x})(e^x + e^{-x}) - (e^x - e^{-x})(e^x - e^{-x})}{(e^x + e^{-x})^2}$

14 $\dfrac{(e^x - e^{-x})^2 - (e^x + e^{-x})^2}{(e^x + e^{-x})^2}$

Exer. 15–26: Use a calculator.

15 An exponential function W such that $W(t) = W_0e^{kt}$ for $k > 0$ describes the first month of growth for crops such as maize, cotton, and soybeans. The function value $W(t)$ is the total weight in mg, W_0 is the weight on the day of emergence, and t is the time in days. If, for a species of soybean, $k = 0.2$ and $W_0 = 68$ mg, predict the weight at the end of the month ($t = 30$).

16 Refer to Exercise 15. It is often difficult to measure the weight W_0 of a plant when it first emerges from the soil. If, for a species of cotton, $k = 0.21$ and the weight after 10 days is 575 mg, estimate W_0.

17 The 1980 population of the United States was approximately 227 million, and the population has been growing at a rate of 0.7% per year. The population $N(t)$, t years after 1980, may be approximated by $N(t) = 227e^{0.007t}$. A graph of N is shown in the figure. If this growth trend continues, predict the population in the year 2000.

EXERCISE 17

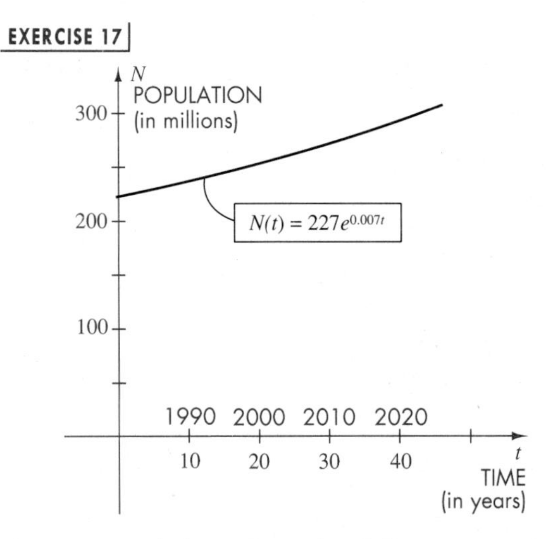

18 The 1985 population estimate for India was 762 million, and the population has been growing at a rate of about 2.2% per year. The population $N(t)$, t years later, may be approximated by $N(t) = 762e^{0.022t}$. Assuming that this rapid growth rate continues, estimate the population of India in the year 2000.

19 In fishery science, a cohort is the collection of fish that results from one annual reproduction. It is usually assumed that the number of fish $N(t)$ still alive after t years is given by an exponential function. For Pacific halibut, $N(t) = N_0e^{-0.2t}$, where N_0 is the initial size of the cohort. What percentage of the original number is still alive after 10 years?

20 The radioactive tracer ^{51}Cr can be used to locate the position of the placenta in a pregnant woman. Often the tracer must be ordered from a medical lab. If A_0 units (microcuries) are shipped, then because of radioactive decay, the number of units $A(t)$ present after t days is given by $A(t) = A_0e^{-0.0249t}$. If 35 units are shipped and it takes 2 days for the tracer to arrive, how many units are then available for the test? If 35 units are needed for the test, how many units should be shipped?

21 In 1978, the population of blue whales in the southern hemisphere was thought to number 5000. Since whaling has been outlawed and an abundant food supply is available, the population $N(t)$ is expected to grow exponentially according to the formula $N(t) = 5000e^{0.0036t}$ with t in years and $t = 0$ corresponding to 1978. Predict the population in the year

(a) 1990. **(b)** 2000.

22 The length (in cm) of many common commercial fish t years old can be approximated by a von-Bertalanffy growth function $f(t) = a(1 - be^{-kt})$ for constants a, b, and k.

(a) For Pacific halibut, $a = 200$, $b = 0.956$, and $k = 0.18$. Estimate the length of a typical 10-year-old halibut.

(b) What is the interpretation of the constant a in the formula?

23 Under certain conditions the atmospheric pressure p (in inches) at altitude h feet is given by $p = 29e^{-0.000034h}$. What is the pressure at an altitude of 40,000 feet?

24 Starting with c milligrams of the polonium isotope ^{210}Po, the amount remaining after t days may be approximated by $A = ce^{-0.00495t}$. If the initial amount is 50 milligrams, approximate, to the nearest hundredth, the amount remaining after

(a) 30 days. **(b)** 180 days. **(c)** 365 days.

25 The Jenss model is generally regarded as the most accurate formula for predicting the height of preschool children. If y is the height (in cm) and x is age (in years), then

$$y = 79.041 + 6.39x - e^{3.261 - 0.993x}$$

for $\frac{1}{4} \le x \le 6$. From calculus, the rate of growth R (in cm/year) is given by $R = 6.39 + 0.993e^{3.261 - 0.993x}$. Find the height and rate of growth of a typical one year old.

26 A very small spherical particle (on the order of 5 microns in diameter) is projected into still air with an initial velocity of v_0 m/sec, but its velocity decreases because of drag forces. Its velocity t seconds later is given by $v(t) = v_0e^{-at}$ for some $a > 0$, and the total distance $s(t)$ the particle travels is given by

$$s(t) = \frac{v_0}{a}(1 - e^{-at}).$$

(a) The stopping distance is the total distance traveled by the particle before it comes to rest. Approximate the stopping distance in terms of v_0 and a.

(b) Use the formula in part (a) to estimate the stopping distance if $v_0 = 10$ m/sec and $a = 8 \times 10^5$.

5.3 LOGARITHMIC FUNCTIONS

If $f(x) = a^x$ and $a > 1$, then f is increasing throughout $\mathbb{R}$; however, if $0 < a < 1$, then f is decreasing (see Figures 1 and 2). Thus, if $a > 0$ and $a \neq 1$, then f is a one-to-one function and hence has an inverse function f^{-1} that is also one-to-one (see Section 3.7). The inverse of the exponential function with base a is the **logarithmic function with base *a*** and is denoted by $\mathbf{log}_a$. Its values are denoted by $\log_a (x)$ or $\log_a x$, read *the logarithm of x with base a* (or *to* the base *a*). Since

$$y = f^{-1}(x) \quad \text{if and only if} \quad x = f(y),$$

the definition of $\log_a$ may be expressed as follows.

DEFINITION OF $\log_a x$

Let a be a positive real number different from 1. The logarithmic function with base a is defined by

$$y = \log_a x \quad \text{if and only if} \quad x = a^y$$

for every $x > 0$ and every real number y.

Note that

$$\text{if} \quad y = \log_a x, \quad \text{then} \quad x = a^y = a^{\log_a x}.$$

Thus, $\log_a x$ *is the exponent to which* a *must be raised in order to obtain* x. This is statement (i) of the next theorem.

THEOREM

(i) $a^{\log_a x} = x$ for every $x > 0$
(ii) $\log_a a^x = x$ for every x
(iii) $\log_a a = 1$
(iv) $\log_a 1 = 0$

PROOF We have already proved (i).

To prove (ii) we use the definition of $\log_a x$ with x replaced by a^x. Thus

$$\text{if} \quad y = \log_a a^x, \quad \text{then} \quad a^x = a^y.$$

Since the exponential function with base a is one-to-one, it follows that $x = y$, that is, $x = \log_a a^x$.

To prove (iii) and (iv) we use (ii) with $x = 1$ and $x = 0$, respectively. ❑

EXAMPLE ▪ 1

Find (a) $\log_2 8$, (b) $\log_{10} 10{,}000$, and (c) $10^{\log_{10} 2}$.

SOLUTION We use the preceding theorem:

(a) By (ii), $\log_2 8 = \log_2 2^3 = 3$

(b) Using (ii), $\log_{10} 10{,}000 = \log_{10} 10^4 = 4$

(c) From (i), $10^{\log_{10} 2} = 2$

EXAMPLE ▪ 2

Find s if:

(a) $\log_4 2 = s$ (b) $\log_5 s = 2$ (c) $\log_s 8 = 3$

SOLUTION We use the definition of $\log_a x$.

(a) If $\log_4 2 = s$, then $4^s = 2$ and hence $s = \frac{1}{2}$.

(b) If $\log_5 s = 2$, then $5^2 = s$ and hence $s = 25$.

(c) If $\log_s 8 = 3$, then $s^3 = 8$ and hence $s = \sqrt[3]{8} = 2$.

EXAMPLE ▪ 3

Solve the equation $\log_4 (5 + x) = 3$.

SOLUTION If $\log_4 (5 + x) = 3$, then by the definition of logarithm:

$$5 + x = 4^3$$
$$5 + x = 64$$
$$x = 59$$

The following laws are fundamental for all work with logarithms of positive real numbers u and w.

LAWS OF LOGARITHMS

(i) $\log_a (uw) = \log_a u + \log_a w$

(ii) $\log_a \dfrac{u}{w} = \log_a u - \log_a w$

(iii) $\log_a (u^c) = c \log_a u$ for every real number c.

PROOF To prove (i) we begin by letting

$$r = \log_a u \qquad \text{and} \qquad s = \log_a w.$$

Applying the definition of logarithm, $a^r = u$ and $a^s = w$. Consequently,

$$a^r a^s = uw$$

and hence,
$$a^{r+s} = uw.$$

By the definition of logarithm, the last equation is equivalent to

$$r + s = \log_a (uw).$$

Since $r = \log_a u$ and $s = \log_a w$, we obtain

$$\log_a u + \log_a w = \log_a uw.$$

This completes the proof of law (i).

To prove (ii) we begin as in the proof of (i), but divide a^r by a^s, obtaining

$$\frac{a^r}{a^s} = \frac{u}{w}$$

$$a^{r-s} = \frac{u}{w}.$$

Using the definition of logarithm, we may write the last equation as

$$r - s = \log_a \frac{u}{w}.$$

Substituting for r and s gives us

$$\log_a u - \log_a w = \log_a \frac{u}{w}$$

This proves law (ii).

Finally, if c is any real number, then

$$(a^r)^c = u^c$$

$$a^{cr} = u^c.$$

By the definition of logarithm, the last equality implies that

$$cr = \log_a u^c.$$

Substituting for r, we obtain

$$c \log_a u = \log_a u^c.$$

This proves law (iii). ❑

There are no laws for expressing $\log_a (u + w)$ or $\log_a (u - w)$ in terms of simpler logarithms. It is evident that

$$\log_a (u + w) \neq \log_a u + \log_a w,$$

since the sum on the right-hand side equals $\log_a (uw)$. Similarly,

$$\log_a (u - w) \neq \log_a u - \log_a w,$$

since the difference on the right-hand side equals $\log_a (u/w)$.

The following examples illustrate uses of the laws of logarithms.

EXAMPLE ▪ 4

Express $\log_a \dfrac{x^3\sqrt{y}}{z^2}$ in terms of logarithms of x, y, and z.

SOLUTION We write $\sqrt{y}$ as $y^{1/2}$ and use the laws of logarithms:

$$\begin{aligned}\log_a \frac{x^3y^{1/2}}{z^2} &= \log_a (x^3y^{1/2}) - \log_a z^2 && \text{[law (ii)]}\\ &= \log_a x^3 + \log_a y^{1/2} - \log_a z^2 && \text{[law (i)]}\\ &= 3\log_a x + \tfrac{1}{2}\log_a y - 2\log_a z. && \text{[law (iii)]}\end{aligned}$$

EXAMPLE ▪ 5

Express $\frac{1}{3}\log_a (x^2 - 1) - \log_a y - 4\log_a z$ as one logarithm.

SOLUTION We use laws of logarithms:

$$\begin{aligned}&\tfrac{1}{3}\log_a (x^2 - 1) - \log_a y - 4\log_a z\\ &\qquad = \log_a (x^2 - 1)^{1/3} - \log_a y - \log_a z^4 && \text{[law (iii)]}\\ &\qquad = \log_a \sqrt[3]{x^2 - 1} - (\log_a y + \log_a z^4) && \text{[algebra]}\\ &\qquad = \log_a \sqrt[3]{x^2 - 1} - \log_a yz^4 && \text{[law (i)]}\\ &\qquad = \log_a \frac{\sqrt[3]{x^2 - 1}}{yz^4} && \text{[law (ii)]}\end{aligned}$$

EXAMPLE ▪ 6

Solve the equation $\log_2 (2x + 3) = \log_2 11 + \log_2 3$.

SOLUTION By law (i), the equation is equivalent to the following:

$$\log_2 (2x + 3) = \log_2 (11 \cdot 3)$$

$$\log_2 (2x + 3) = \log_2 33.$$

Since $\log_2$ is a one-to-one function,

$$2x + 3 = 33, \quad \text{or} \quad 2x = 30.$$

Hence, the solution is $x = 15$.

EXAMPLE ▪ 7

Solve the equation $\log_4 (x + 6) - \log_4 10 = \log_4 (x - 1) - \log_4 2$.

SOLUTION The equation is equivalent to

$$\log_4 (x + 6) - \log_4 (x - 1) = \log_4 10 - \log_4 2.$$

Applying law (ii),

$$\log_4 \left(\frac{x + 6}{x - 1}\right) = \log_4 \frac{10}{2}$$

Since $\log_4$ is a one-to-one function,

$$\frac{x + 6}{x - 1} = 5.$$

The last equation implies that

$$x + 6 = 5x - 5, \quad \text{or} \quad 4x = 11.$$

Thus, the solution is $x = \frac{11}{4}$.

Extraneous solutions sometimes occur in the process of solving equations that involve logarithms, as illustrated in the next example.

EXAMPLE ▪ 8

Solve $\log_2 x + \log_2 (x + 2) = 3$.

SOLUTION We apply law (i):

$$\log_2 x(x + 2) = 3$$

$$\log_2 (x^2 + 2x) = 3$$

Using the definition of logarithm gives us the following equivalent equations:

$$x^2 + 2x = 2^3$$

$$x^2 + 2x - 8 = 0$$

$$(x + 4)(x - 2) = 0$$

Hence $x = -4$ or $x = 2$. We can check that $x = 2$ is a solution of the original equation; however, $x = -4$ is not a solution, since x must be positive for $\log_2 x$ to exist.

Logarithmic functions occur frequently in applications. In particular, if two variables u and v are related such that u is an exponential function of v, then v is a logarithmic function of u.

EXAMPLE ▪ 9

The number N of bacteria in a certain culture after t hours is given by $N = (1000)2^t$. Express t as a logarithmic function of N with base 2.

SOLUTION If $N = (1000)2^t$, then $2^t = \dfrac{N}{1000}$. Changing to logarithmic form, we obtain

$$t = \log_2 \frac{N}{1000}.$$

5.3 EXERCISES

Exer. 1–6: Use the definition of logarithm to change the equation to logarithmic form.

1 $4^3 = 64$

2 $3^5 = 243$

3 $4^{-3} = \frac{1}{64}$

4 $3^{-4} = \frac{1}{81}$

5 $t^r = s$

6 $c^p = d$

Exer. 7–12: Use the definition of logarithm to change the equation to exponential form.

7 $\log_{10} 1000 = 3$

8 $\log_3 81 = 4$

9 $\log_3 \frac{1}{243} = -5$

10 $\log_4 \frac{1}{256} = -4$

11 $\log_t r = p$

12 $\log_v w = q$

Exer. 13–18: Evaluate.

13 $\log_5 125$

14 $\log_3 243$

15 $\log_4 \frac{1}{16}$

16 $\log_2 128$

17 $10^{\log_{10} 3}$

18 $e^{\log_e 2}$

Exer. 19–24: Express in terms of logarithms of x, y, and z.

19 $\log_a \dfrac{x^3 w}{y^2 z^4}$

20 $\log_a \dfrac{y^5 w^2}{x^4 z^3}$

21 $\log_a \dfrac{\sqrt[3]{z}}{x\sqrt{y}}$

22 $\log_a \dfrac{\sqrt{y}}{x^4 \sqrt[3]{z}}$

23 $\log_a \sqrt[4]{\dfrac{x^7}{y^5 z}}$

24 $\log_a \left(x \sqrt[3]{\dfrac{y^4}{z^5}} \right)$

Exer. 25–28: Express as one logarithm.

25 $2 \log_a x + \frac{1}{3} \log_a (x - 2) - 5 \log_a (2x + 3)$

26 $5 \log_a x - \frac{1}{2} \log_a (3x - 4) - 3 \log_a (5x + 1)$

27 $\log_a (x^3 y^2) - 2 \log_a x \sqrt[3]{y} - 3 \log_a \dfrac{x}{y}$

28 $2 \log_a \dfrac{y^3}{x} - 3 \log_a y + \frac{1}{2} \log_a x^4 y^2$

Exer. 29–34: Replace the symbol $\square$ with either $=$ or $\neq$ so that the resulting statement is true. Give a reason for your answer.

29 $\log_a x + \log_a y \;\square\; \log_a (x + y)$

30 $\log_a \dfrac{x}{y} \;\square\; \dfrac{\log_a x}{\log_a y}$

31 $\log_a x - \log_a y \;\square\; \log_a (x \div y)$

32 $\log_a \sqrt{x} \;\square\; \frac{1}{2} \log_a x$

33 $\log_{ab} xy \;\square\; \log_a x + \log_b y$

34 $\log_a x^2 \;\square\; (\log_a x)^2$

Exer. 35–52: Solve the equation.

35 $\log_3 (x - 4) = 2$

36 $\log_2 (x - 5) = 4$

37 $\log_9 x = \frac{3}{2}$

38 $\log_4 x = -\frac{3}{2}$

39 $\log_5 x^2 = -2$

40 $\log_{10} x^2 = -4$

41 $\log_6 (2x - 3) = \log_6 12 - \log_6 3$

42 $\log_4 (3x + 2) = \log_4 5 + \log_4 3$

43 $2 \log_3 x = 3 \log_3 5$

44 $3 \log_2 x = 2 \log_2 3$

45 $\log_2 x - \log_2 (x + 1) = 3 \log_2 4$

46 $\log_3 (x + 2) - \log_3 x = 2 \log_3 4$

47 $\log_5 (-4 - x) + \log_5 3 = \log_5 (2 - x)$

48 $\log_5 x + \log_5 (x + 6) = \frac{1}{2} \log_5 9$

49 $\log_2 (x + 7) + \log_2 x = 3$

50 $\log_6 (x + 5) + \log_6 x = 2$

51 $\log_3 (x + 3) + \log_3 (x + 5) = 1$

52 $\log_3 (x - 2) + \log_3 (x - 4) = 2$

53 Solve $A = B \cdot 2^{Ct} + D$ for t using logarithms with base 2.

54 Solve $L = M \cdot 3^{t/N} - P$ for t using logarithms with base 3.

55 Starting with q_0 milligrams of radium, the amount q remaining after t years is $q = q_0(2)^{-t/1600}$. Use logarithms with base 2 to solve for t in terms of q and q_0.

56 The radioactive bismuth isotope ^{210}Bi disintegrates according to the law $Q = k(2)^{-t/5}$ for t in days. Use logarithms with base 2 to solve for t in terms of Q and k.

57 The relationship between the demand D for a product and its selling price p is often given by

$$\log_a D = \log_a c - k \log_a p$$

for positive constants c and k. Use the laws of logarithms to solve this equation for D.

58 Pareto's law for capitalist countries states that the relationship between annual income x and the number y of individuals whose income exceeds x is

$$\log_{10} y = \log_{10} b - k \log_{10} x$$

with b and k positive. The value of k is approximately 1.5. Use the laws of logarithms to solve this equation for y.

59 The basic source of genetic diversity is mutation, or changes in the chemical structure of genes. If a gene mutates at a constant rate m, and if other evolutionary forces are negligible, then the frequency F of the original gene after t generations is given by $F = F_0(1 - m)^t$, where F_0 is the frequency at $t = 0$.

(a) Solve the equation for t using logarithms with base 10.

(b) If $m = 5 \times 10^{-5}$, after how many generations does $F = \frac{1}{2}F_0$?

60 Manufacturers sometimes use empirically based formulas to predict the time required to produce the nth item on an assembly line for an integer n. If $T(n)$ denotes the time required to assemble the nth item, and T_1 denotes the time required for the first item, or prototype, then typically $T(n) = T_1 n^{-k}$ for some positive constant k.

(a) For many airplanes the time required to assemble the second airplane, $T(2)$, is equal to $(0.80)T_1$. Find the value of k.

(b) Express the time required to assemble the fourth airplane in terms of T_1.

(c) Express the time required to assemble the $(2n)$th airplane in terms of $T(n)$.

5.4 GRAPHS OF LOGARITHMIC FUNCTIONS

Since the logarithmic function $\log_a$ is the inverse of the exponential function with base a, the graph of $y = \log_a x$ can be obtained by reflecting the graph of $y = a^x$ through the line $y = x$ that bisects quadrants I and III (see page 172). This is illustrated in Figure 14 for the case $a > 1$.

We can also sketch the graph by plotting points. Since

$$y = \log_a x \quad \text{if and only if} \quad x = a^y,$$

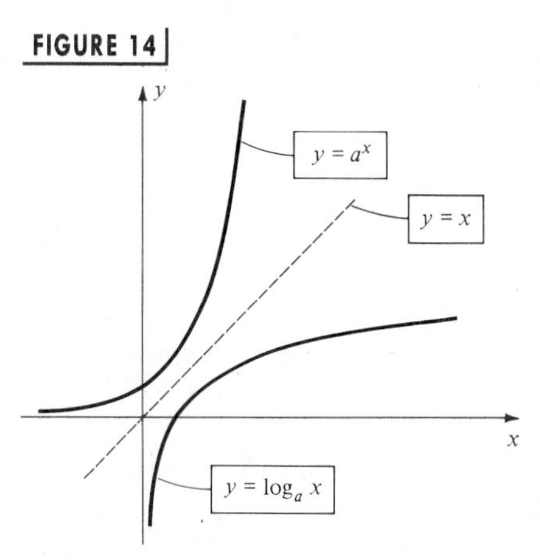

FIGURE 14

coordinates of points on the graph of $y = \log_a x$ may be found by using the equation $x = a^y$. Substituting for y and then finding x gives us the following table.

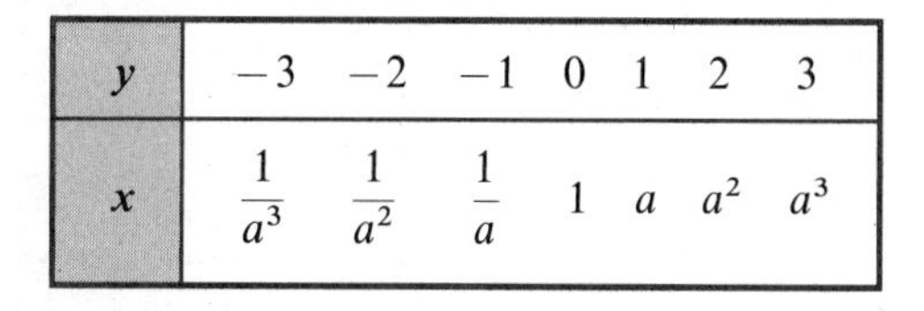

y	-3	-2	-1	0	1	2	3
x	$\frac{1}{a^3}$	$\frac{1}{a^2}$	$\frac{1}{a}$	1	a	a^2	a^3

If $a > 1$, we obtain the sketch in Figure 15(i). In this case f is an increasing function throughout its domain. If $0 < a < 1$, then the graph has the general shape shown in Figure 15(ii) and hence f is a decreasing function. (Logarithms with base less than 1 are seldom used in applications.)

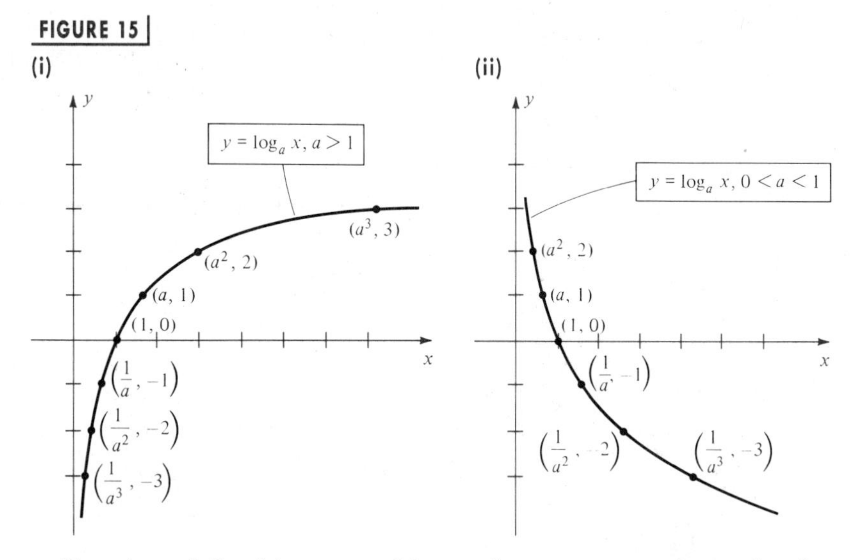

FIGURE 15

Functions defined in terms of $\log_a p$, for some expression p that involves x, are members of the logarithmic family; however, the graphs may differ from those sketched in Figures 14 and 15, as illustrated in the following examples.

EXAMPLE ▪ 1

Sketch the graph of f if $f(x) = \log_3(-x)$ for $x < 0$.

SOLUTION If $x < 0$, then $-x > 0$, and hence $\log_3(-x)$ exists. We wish to sketch the graph of the equation $y = \log_3(-x)$, which, by the definition of logarithm, is equivalent to $3^y = -x$. To find points on the graph of f, we may substitute for y in the equation $x = -3^y$ and then find x. The following table displays the coordinates of several such points.

FIGURE 16

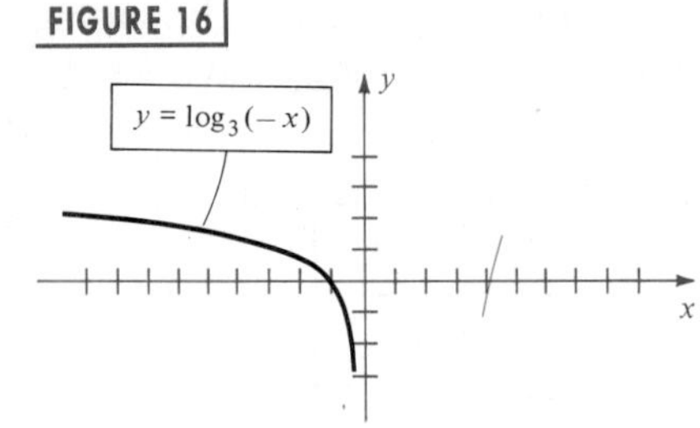

y	-2	-1	0	1	2
$x = -3^y$	$-\frac{1}{9}$	$-\frac{1}{3}$	-1	-3	-9

Plotting points given by the table leads to the sketch in Figure 16.

EXAMPLE ▪ 2

Sketch the graph of the equation $y = \log_3 |x|$ if $x \neq 0$.

SOLUTION Since $|x| > 0$ for every $x \neq 0$, the graph includes points corresponding to negative values of x as well as to positive values. If $x > 0$, then $|x| = x$ and hence, to the right of the y-axis, the graph coincides with the graph of $y = \log_3 x$ or, equivalently, $x = 3^y$. If $x < 0$, then $|x| = -x$ and the graph is the same as that of $y = \log_3 (-x)$ (see Example 1). The graph is sketched in Figure 17. Note that the graph is symmetric with respect to the y-axis.

FIGURE 17 $y = \log_3 |x|$

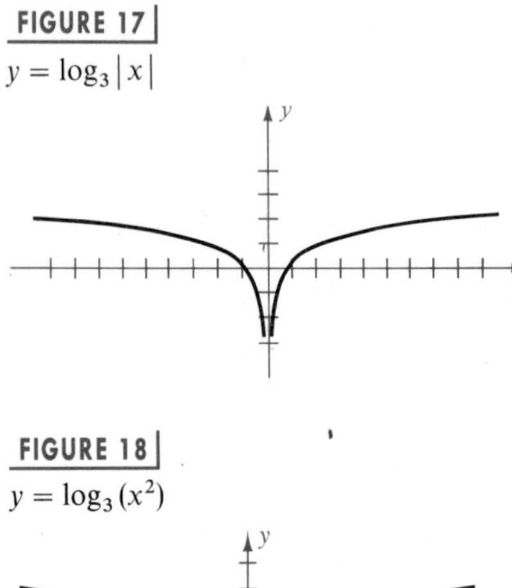

EXAMPLE ▪ 3

Sketch the graph of the equation:

(a) $y = \log_3 (x^2)$ **(b)** $y = 2 \log_3 x$

FIGURE 18 $y = \log_3 (x^2)$

SOLUTION

(a) Since $x^2 = |x|^2$, we may rewrite the equation as

$$y = \log_3 |x|^2.$$

Using a law of logarithms,

$$y = 2 \log_3 |x|.$$

We can obtain the graph of $y = 2 \log_3 |x|$ by multiplying the y-coordinates of points on the graph of $y = \log_3 |x|$ in Figure 17 by 2. (See the discussion of stretching in Section 3.5.) This gives us the graph in Figure 18.

(b) If $y = 2 \log_3 x$, then x must be positive. Hence the graph is identical to that part of the graph of $y = 2 \log_3 |x|$ in Figure 18 that lies to the right of the y-axis. This gives us Figure 19.

FIGURE 19

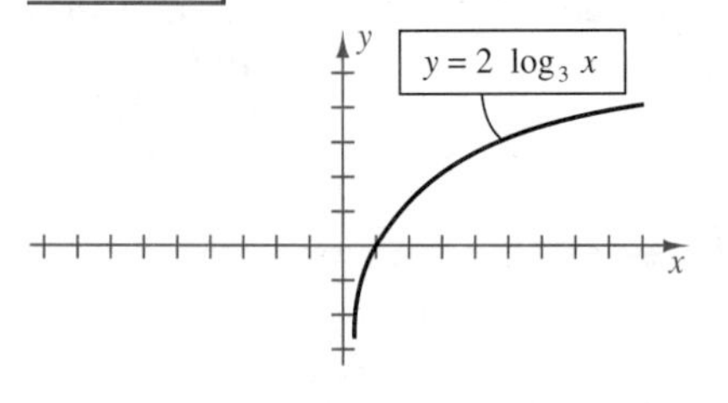

EXAMPLE ▪ 4

Sketch the graph of the equation:

(a) $y = \log_3 (x - 2)$ **(b)** $y = \log_3 x - 2$

SOLUTION

(a) The graph of $y = \log_3 x$ is the part of the graph in Figure 17 that lies to the right of the y-axis. It is sketched with dashes in Figure 20. From the discussion of horizontal shifts in Section 3.5, we can obtain the graph of $y = \log_3 (x - 2)$ by shifting the graph of $y = \log_3 x$ two units to the right, as shown in Figure 20.

(b) From the discussion of vertical shifts in Section 3.5, the graph of $y = \log_3 x - 2$ can be obtained by shifting the graph of $y = \log_3 x$ two units downward, as shown in Figure 21. Note that the x-intercept is given by $\log_3 x = 2$, or $x = 3^2 = 9$.

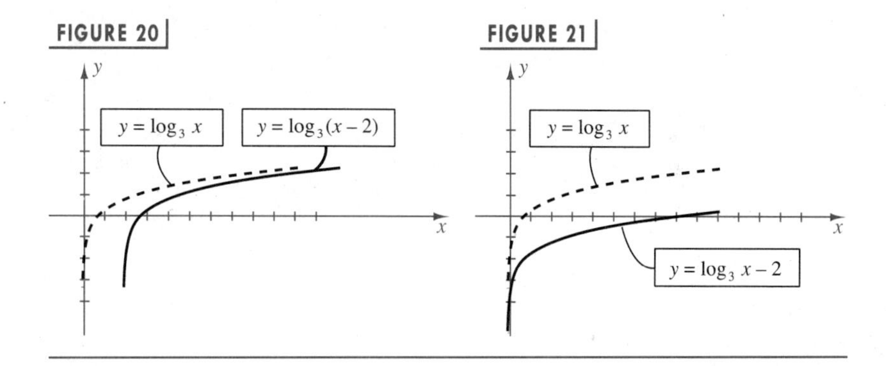

5.4 EXERCISES

1 Sketch the graph of f if $a = 4$:

(a) $f(x) = \log_a x$
(b) $f(x) = \log_a (-x)$
(c) $f(x) = -\log_a x$
(d) $f(x) = \log_a (x + 2)$
(e) $f(x) = \log_a x + 2$
(f) $f(x) = \log_a (x - 2)$
(g) $f(x) = \log_a x - 2$
(h) $f(x) = \log_a |x|$
(i) $f(x) = |\log_a x|$
(j) $f(x) = \log_{(1/a)} x$
(k) $f(x) = \log_{(1/a)} (x + 2)$
(l) $f(x) = \log_{(1/a)} x + 2$

2 Work Exercise 1 if $a = 5$.

Exer. 3–22: Sketch the graph of f.

3 $f(x) = \log_{10} x$

4 $f(x) = \log_e x$

5 $f(x) = \log_3 (3x)$

6 $f(x) = \log_4 (16x)$

7 $f(x) = 3 \log_3 x$

8 $f(x) = \frac{1}{3} \log_3 x$

9 $f(x) = \log_3 (x^2)$

10 $f(x) = \log_2 (x^2)$

11 $f(x) = \log_2 (x^3)$

12 $f(x) = \log_3 (x^3)$

13 $f(x) = \log_2 \sqrt{x}$

14 $f(x) = \log_2 \sqrt[3]{x}$

15 $f(x) = \log_3 \dfrac{1}{x}$

16 $f(x) = \log_2 \dfrac{1}{x}$

17 $f(x) = \log_3 (3 - x)$

18 $f(x) = \log_2 (2 - x)$

19 $f(x) = \dfrac{1}{\log_3 x}$

20 $f(x) = \dfrac{1}{(\log_3 x)^2}$

21 $f(x) = \log_2 |x - 5|$

22 $f(x) = \log_3 |x + 1|$

Exer. 23–30: Shown in the figure is the graph of a function f. Express $f(x)$ in terms of logarithms with base 2.

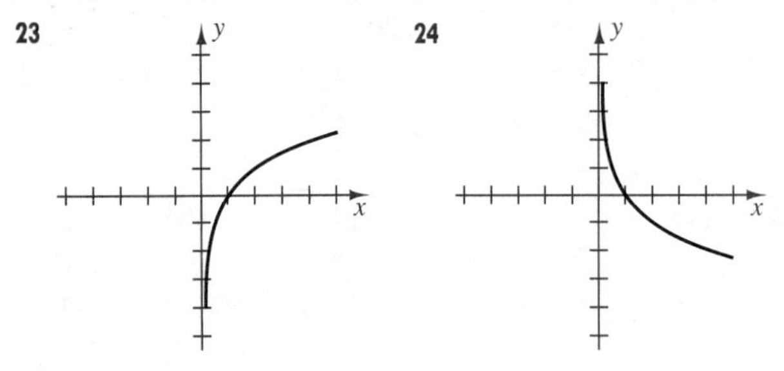

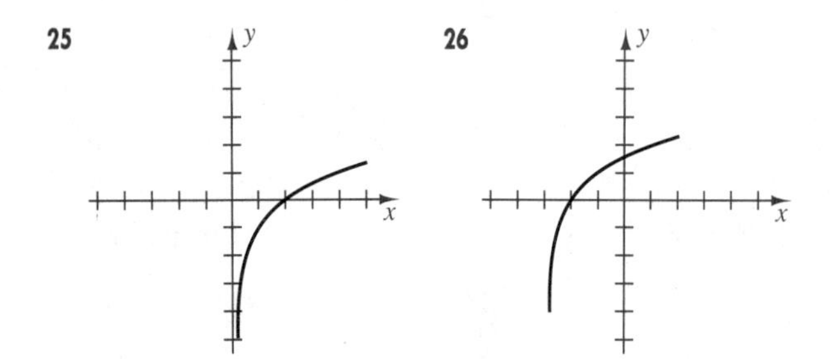

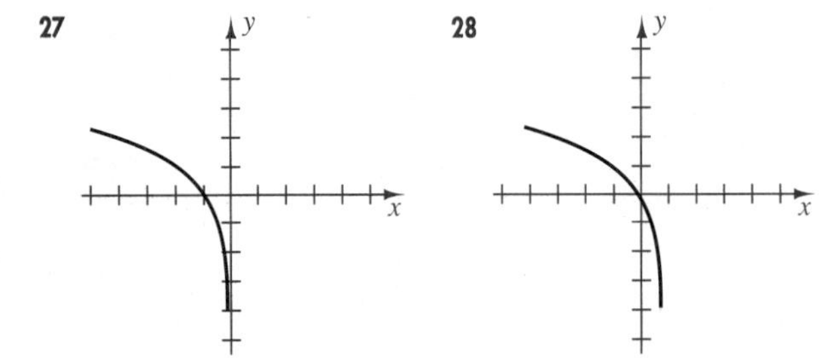

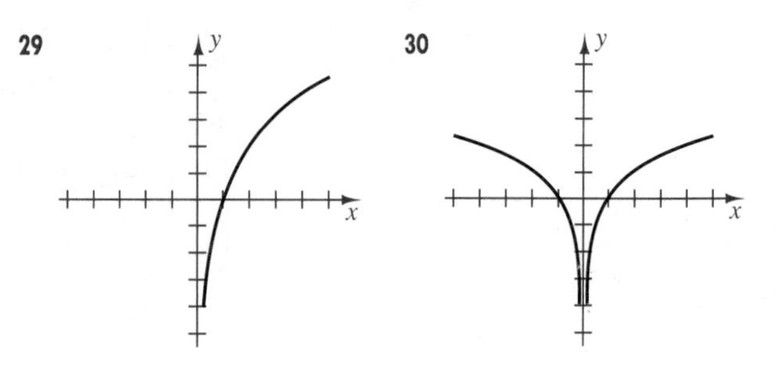

31 If v denotes the wind velocity (in m/sec) at a height of z meters above the ground, then under certain conditions $v = c \log_e (z/z_0)$ for a positive constant c, the base e of the natural exponential function, and the height z_0 at which the velocity is zero. Sketch the graph of this equation on a zv-plane for $c = 0.5$ and $z_0 = 0.1$ meter.

32 When a large number N of butterflies or moths has been collected from a certain locale, the expected number S of different species in the collection can be estimated from the formula

$$S = a \log_e \left(\frac{a + N}{a}\right)$$

for some $a > 0$ with e the base of the natural exponential function.

(a) Find S for $N = 0$ and $N = 1$.

(b) Sketch the graph of S for $N > 0$ if $a = 40$.

5.5 COMMON AND NATURAL LOGARITHMS

Before electronic calculators were invented, logarithms with base 10 were used for complicated numerical computations involving products, quotients, and powers of real numbers. Base 10 was employed because it is well suited for numbers that are expressed in decimal form. Logarithms with base 10 are **common logarithms**. The symbol $\log x$ is used as an abbreviation for $\log_{10} x$.

DEFINITION OF COMMON LOGARITHMS

$$\log x = \log_{10} x \quad \text{for every } x > 0.$$

Using the definition of $\log_a x$ with $a = 10$, we see that

$$y = \log x \quad \text{if and only if} \quad x = 10^y.$$

Since inexpensive calculators are now available, there is little need for common logarithms as a tool for computational work. However, base 10 does occur in applications, and hence many calculators have a LOG key that can be used to approximate common logarithms. Appendix II contains a table of common logarithms that may be used if a calculator is not available. The use of the table of common logarithms is explained in Appendix I.

EXAMPLE ▪ 1

Using the Richter scale, the magnitude R of an earthquake of intensity I is given by

$$R = \log \frac{I}{I_0}$$

for a certain minimum intensity I_0.

(a) Find R, assuming the intensity of an earthquake is $1000I_0$.

(b) Express I in terms of R and I_0.

SOLUTION

(a) If $I = 1000I_0$, then

$$R = \log \frac{I}{I_0} = \log \frac{1000I_0}{I_0} = \log 1000.$$

Using (ii) of the theorem in Section 5.3, we may write

$$\log 1000 = \log 10^3 = 3.$$

Hence $R = 3$.

(b) By the definition of logarithm with base 10,

$$\text{if} \quad R = \log \frac{I}{I_0}, \quad \text{then} \quad \frac{I}{I_0} = 10^R.$$

Thus,

$$I = I_0 \cdot 10^R.$$

The natural exponential function f is given by $f(x) = e^x$ (see Section 5.2). The logarithmic function with base e is the **natural logarithmic function**. The symbol **ln *x*** (read *ell-en of* x) is an abbreviation for $\log_e x$ and we refer to it as the **natural logarithm of *x***. Thus *the natural logarithmic and natural exponential functions are inverse functions of one another.*

DEFINITION OF NATURAL LOGARITHMS

$$\ln x = \log_e x \quad \text{for every} \quad x > 0.$$

Using the definition of $\log_a x$ with $a = e$, we see that

$$y = \ln x \quad \text{if and only if} \quad x = e^y.$$

Many calculators have a key labeled [LN] that can be used to approximate natural logarithms. A short table of values of $\ln x$ is given in Table 3 of Appendix II.

Since $e \approx 3$, the graph of $y = \ln x$ is similar in appearance to the graph of $y = \log_3 x$. The laws of logarithms for natural logarithms are as follows.

LAWS OF NATURAL LOGARITHMS

(i) $\ln (uv) = \ln u + \ln v$

(ii) $\ln \dfrac{u}{v} = \ln u - \ln v$

(iii) $\ln u^c = c \ln u$

If we substitute e for a in the theorem on page 264, we obtain the following.

THEOREM

(i) $e^{\ln x} = x$ for every $x > 0$

(ii) $\ln e^x = x$ for every x

(iii) $\ln e = 1$

(iv) $\ln 1 = 0$

EXAMPLE ▪ 2

Newton's law of cooling states that the rate at which an object cools is directly proportional to the difference in temperature between the object and its surrounding medium. Newton's law can be used to show that under certain conditions the temperature T of an object at time t is given by $T = 75e^{-2t}$. Express t as a function of T.

SOLUTION The equation $T = 75e^{-2t}$ may be written

$$e^{-2t} = \frac{T}{75}.$$

Using the definition of natural logarithm gives us

$$-2t = \log_e \frac{T}{75} = \ln \frac{T}{75}.$$

Consequently,

$$t = -\tfrac{1}{2} \ln \frac{T}{75}, \quad \text{or} \quad t = -\tfrac{1}{2}(\ln T - \ln 75).$$

An alternative solution is to start with the equation $e^{-2t} = T/75$ and take the natural logarithm of both sides, obtaining

$$\ln e^{-2t} = \ln \frac{T}{75}.$$

Since $\ln e^{-2t} = -2t \ln e = -2t(1) = -2t$, we see that

$$-2t = \ln \frac{T}{75}$$

The remainder of the solution is the same.

FIGURE 22

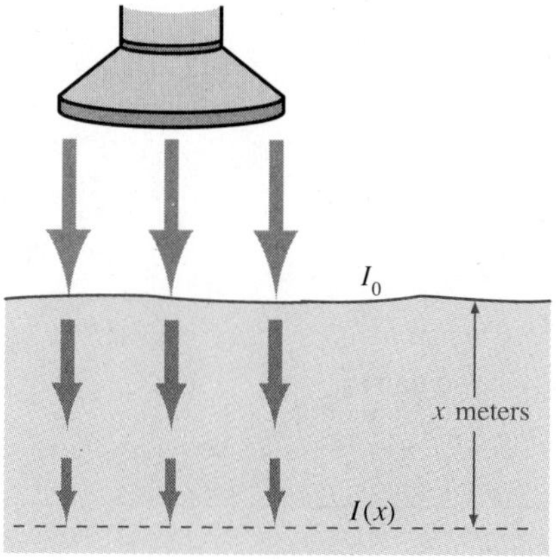

EXAMPLE ▪ 3

If a beam of light that has intensity I_0 is projected vertically downward into water, then its intensity $I(x)$ at a depth of x meters is $I(x) = I_0e^{-1.4x}$ (see Figure 22). At what depth is the intensity one-half its value at the surface?

SOLUTION At the surface $x = 0$, and the intensity is

$$I(0) = I_0e^0 = I_0.$$

We wish to find the value of x such that $I(x) = \frac{1}{2}I_0$, that is,

$$I_0e^{-1.4x} = \tfrac{1}{2}I_0$$

or

$$e^{-1.4x} = \tfrac{1}{2}.$$

Using the definition of natural logarithm (or taking the natural logarithm of both sides),

$$-1.4x = \ln \tfrac{1}{2}$$

or

$$-1.4x = \ln 1 - \ln 2 = 0 - \ln 2 = -\ln 2.$$

Hence,

$$x = \frac{-\ln 2}{-1.4} = \frac{\ln 2}{1.4}$$

Using a calculator or Table 3,

$$x \approx \frac{0.693}{1.4} \approx 0.5 \text{ meter.}$$

To solve certain problems, it is necessary to find x when given $\log x$. If a calculator has a key labeled [INV], we can accomplish this by entering $\log x$ and then pressing, successively, [INV] and [LOG]. Similarly, given $\ln x$, we can find x by entering $\ln x$ and pressing [INV] followed by [LN]. This procedure is illustrated in the next example. Some calculators have keys for [ALOG] and [EXP] that can be used in place of [INV] [LOG] and [INV] [LN], respectively. Since calculators may differ, the owner's manual should be consulted before solving problems of this type.

EXAMPLE ▪ 4

Use a calculator to approximate x to three decimal places if:

(a) $\log x = 1.7959$ **(b)** $\ln x = 4.7$

SOLUTION

(a) Given $\log x = 1.7959$,

Enter 1.7959 : 1.7959

Press [INV] [LOG] : 62.502876

Thus, $x \approx 62.503$.

(b) Given $\ln x = 4.7$,

Enter 4.7 : 4.7

Press [INV] [LN] : 109.94717

Hence, $x \approx 109.947$.

Alternatively, we could have solved Example 5(a) by noting that since common logarithms have base 10,

$$\text{if} \quad \log x = 1.7959, \quad \text{then} \quad x = 10^{1.7959}.$$

The number x can then be approximated using a $\boxed{10^x}$ key or a $\boxed{y^x}$ key.

Similarly, part (b) could be solved by using the fact that natural logarithms have base e. Specifically,

$$\text{if} \quad \ln x = 4.7, \quad \text{then} \quad x = e^{4.7}.$$

A calculator or Table 2 in Appendix II could then be used to approximate x.

It is sometimes necessary to *change the base* of a logarithm by expressing $\log_b u$ in terms of $\log_a u$, for some positive real number $b \neq 1$. We begin with the equivalent equations

$$v = \log_b u \quad \text{and} \quad b^v = u.$$

We next take the logarithm, base a, of both sides of the second equation:

$$\log_a b^v = \log_a u$$

$$v \log_a b = \log_a u.$$

Solving for v (that is, $\log_b u$), we obtain formula (i) in the next box.

CHANGE OF BASE FORMULAS

$$\text{(i)} \ \log_b u = \frac{\log_a u}{\log_a b} \qquad \text{(ii)} \ \log_b a = \frac{1}{\log_a b}$$

To obtain formula (ii) we let $u = a$ in (i) and use the fact that $\log_a a = 1$. If we let $a = e$ and $b = 10$ in the change of base formulas we obtain the following special cases:

$$\log u = \frac{\ln u}{\ln 10} \quad \text{and} \quad \log e = \frac{1}{\ln 10}.$$

Logarithms with base 2 are used in computer science; however, most scientific calculators have only $\boxed{\text{LN}}$ and $\boxed{\text{LOG}}$ keys. The next example indicates how to find logarithms with base 2 using a change of base formula.

EXAMPLE ▪ 5

Approximate $\log_2 5$ using:

(a) common logarithms **(b)** natural logarithms

SOLUTION We use change of base formula (i):

(a) Letting $b = 2$, $a = 10$, and $u = 5$,

$$\log_2 5 = \frac{\log 5}{\log 2} \approx \frac{0.699}{0.301} \approx 2.322$$

(b) Letting $b = 2$, $a = e$, and $u = 5$,

$$\log_2 5 = \frac{\ln 5}{\ln 2} \approx \frac{1.609}{0.693} \approx 2.322$$

5.5 EXERCISES

1 Use the Richter scale formula $R = \log (I/I_0)$ to find the magnitude of an earthquake that has intensity

(a) 100 times that of I_0.

(b) 10,000 times that of I_0.

(c) 100,000 times that of I_0.

2 Refer to Exercise 1. The largest recorded magnitudes of earthquakes have been between 8 and 9 on the Richter scale. Find the corresponding intensities in terms of I_0.

3 In the western U.S., the area A (in mi^2) affected by an earthquake is related to the magnitude R of the quake by the formula

$$R = 2.3 \log (A + 3000) - 5.1.$$

Solve for A in terms of R.

4 Refer to Exercise 3. For the eastern U.S., the area-magnitude formula takes the form

$$R = 2.3 \log (A + 34{,}000) - 7.5.$$

If A_1 is the area affected by an earthquake of magnitude R in the west and A_2 is the area affected by a similar quake in the east, find a formula for A_1/A_2 in terms of R.

5 The loudness of a sound, as experienced by the human ear, is based upon intensity levels. A formula used for finding the intensity level α, in decibels, that corresponds to a sound intensity I is $\alpha = 10 \log (I/I_0)$, where I_0 is a special value of I agreed to be the weakest sound that can be detected by the ear under certain conditions. Find α if

(a) I is 10 times as great as I_0.

(b) I is 1000 times as great as I_0.

(c) I is 10,000 times as great as I_0. (This is the intensity level of the average voice.)

6 Refer to Exercise 5. A sound intensity level of 140 decibels produces pain in the average human ear. Approximately how many times greater than I_0 must I be in order for α to reach this level?

Exer. 7–10: Chemists use a number denoted by pH to describe quantitatively the acidity or basicity of solutions. By definition, pH = $-\log [H^+]$ for the hydrogen ion concentration $[H^+]$ in moles per liter.

7 Approximate the pH of each substance:

(a) vinegar: $[H^+] \approx 6.3 \times 10^{-3}$

(b) carrots: $[H^+] \approx 1.0 \times 10^{-5}$

(c) sea water: $[H^+] \approx 5.0 \times 10^{-9}$

8 Approximate the hydrogen ion concentration $[H^+]$ in each substance:

(a) apples: pH ≈ 3.0 **(b)** beer: pH ≈ 4.2

(c) milk: pH ≈ 6.6

9 A solution is considered acidic if $[H^+] > 10^{-7}$ or basic if $[H^+] < 10^{-7}$. Find the corresponding inequalities involving pH.

10 Many solutions have a pH between 1 and 14. Find the corresponding range of $[H^+]$.

11 A simple electrical circuit consisting of a resistor and an inductor is shown in the figure. The current I at time t is given by $I = 20e^{-Rt/L}$ for resistance R and inductance L. Use natural logarithms to solve this equation for t.

EXERCISE 11

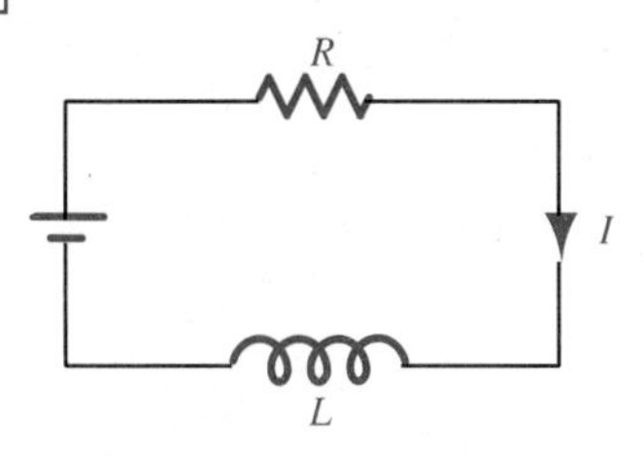

12 An electrical condenser with initial charge Q_0 is allowed to discharge. After t seconds the charge Q is $Q = Q_0e^{kt}$, where k is a constant. Use natural logarithms to solve this equation for t.

13 Under certain conditions the atmospheric pressure p at altitude h is given by $p = 29e^{-0.000034h}$. Use natural logarithms to solve for h as a function of p.

14 If p denotes the selling price (in dollars) of a commodity and x is the corresponding demand (in number sold per day), then frequently the relationship between p and x is $p = p_0e^{-ax}$ for positive constants p_0 and a. Express x as a function of p.

15 The Ehrenberg relation $\ln W = \ln 2.4 + (1.84)h$ is an empirically based formula relating the height h (in meters) to the weight W (in kg) for children aged 5 through 13 years old. The formula has been verified in many different countries. Express W as a function of h.

16 A rocket of mass m_1 is filled with fuel of initial mass m_2. Assuming frictional forces are neglected, the total mass m of the rocket at time t after ignition is related to its upward velocity v by the formula $v = -a \ln m + b$ for some constants a and b. At time $t = 0$, we have $v = 0$ and $m = m_1 + m_2$. At burnout, $m = m_1$. Use this information to find a formula in terms of one logarithm for the velocity of the rocket at burnout.

17 Let n be the average number of earthquakes per year that have magnitudes between R and $R + 1$ on the Richter scale. An approximate formula relating n and R is $\log n = 7.7 - (0.1)R$.

(a) Solve the equation for n in terms of R.

(b) Find n if $R = 4$, 5, and 6.

18 The energy E (in ergs) released during an earthquake of magnitude R may be approximated by the formula $\log E = 11.4 + (1.5)R$.

(a) Solve for E in terms of R.

(b) Find the energy released during the famous Alaskan quake of 1964, which measured 8.4 on the Richter scale.

19 A certain radioactive substance decays according to the formula $q(t) = q_0e^{-0.0063t}$ for the initial amount q_0 of the substance and time t in days. Approximate its half-life, that is, the number of days it takes for half of the substance to decay.

20 The air pressure $p(h)$ (lb/in.^{2}) at an altitude of h feet above sea level may be approximated by the formula $p(h) = 14.7e^{-0.0000385h}$. At approximately what altitude h is the air pressure

(a) 10 lb/in.^{2}?

(b) one-half its value at sea level?

21 Use the compound interest formula to determine how long it will take a sum of money to double if it is invested at a rate of 6% per year compounded monthly.

22 If interest is compounded continuously at the rate of 10% per year, the compound interest formula takes the form $A = Pe^{0.1t}$. Approximate the number of years it takes an initial deposit of \$6000 to grow to \$25,000.

23 The population $N(t)$ (in millions) of the United States t years after 1980 may be approximated by the formula $N(t) = 227e^{0.007t}$. When will the population be twice what it was in 1980?

24 The population $N(t)$ (in millions) of India t years after 1985 may be approximated by $N(t) = 762e^{0.022t}$. When will the population grow to one billion?

25 The Count model is a formula that can be used to predict the height of preschool children. If y is height (in cm) and x is age (in years), then

$$y = 70.228 + 5.104x + 9.222 \ln x$$

for $\frac{1}{4} \le x \le 6$. From calculus, the rate of growth R (in cm/year) is given by $R = 5.104 + (9.222/x)$. Predict the height and rate of growth of a typical two year old.

26 The vapor pressure P (in lb/in.^{2}), a measure of the volatility of a liquid, is related to its temperature T (in °F) by the Antoine equation

$$\log P = a + \frac{b}{c + T}$$

for constants a, b, and c. Vapor pressure increases rapidly with an increase in temperature. Express P as a function of T.

27 Let R denote the reaction of a subject to a stimulus of strength x. There are many possibilities for R and x. If the stimulus x is saltiness (in grams of salt per liter), R may be the subject's estimate of how salty the solution tastes, based on a scale from 0 to 10. One relationship between R and x is given by the Weber-Fechner formula $R(x) = a \log (x/x_0)$ for a positive constant a and threshold stimulus x_0.

(a) Find $R(x_0)$.

(b) Find a relationship between $R(x)$ and $R(2x)$.

28 Shown in the figure is a graph of $f(x) = (\ln x)/x$ for $x > 0$. The maximum value of $f(x)$ occurs at $x = e$.

EXERCISE 28

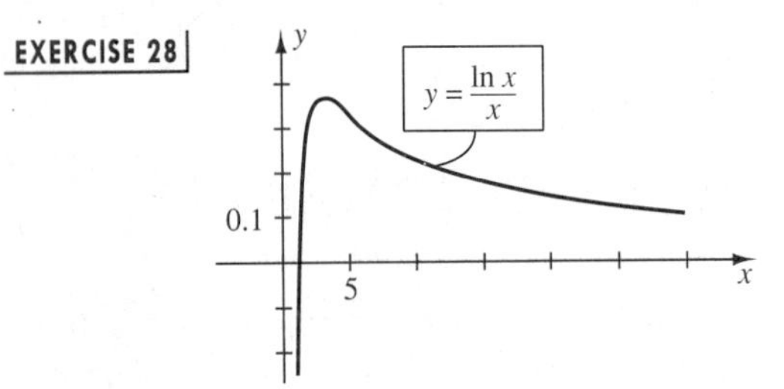

(a) The integers 2 and 4 have the unusual property that $2^4 = 4^2$. Show that if $x^y = y^x$, then $(\ln x)/x = (\ln y)/y$.

(b) Use the graph of f to explain why many pairs of real numbers satisfy the equation $x^y = y^x$.

Exer. 29–40: Approximate x to three significant figures.

29 $\log x = 3.6274$

30 $\log x = 1.8965$

31 $\log x = 0.9469$

32 $\log x = 4.9680$

33 $\log x = -1.6253$

34 $\log x = -2.2118$

35 $\ln x = 2.3$

36 $\ln x = 3.7$

37 $\ln x = 0.05$

38 $\ln x = 0.95$

39 $\ln x = -1.6$

40 $\ln x = -5$

Exer. 41–44: Evaluate using a change of base formula.

41 $\log_5 6$

42 $\log_2 20$

43 $\dfrac{\log_5 16}{\log_5 4}$

44 $\dfrac{\log_7 243}{\log_7 3}$

Exer. 45–46: Sketch the graph of f and use the change of base formulas to approximate the y-intercept.

45 $f(x) = \log_2 (x + 3)$

46 $f(x) = \log_3 (x + 5)$

Exer. 47–48: Sketch the graph of f and use the change of base formulas to approximate the x-intercept.

47 $f(x) = 4^x - 3$

48 $f(x) = 3^x - 6$

5.6 EXPONENTIAL AND LOGARITHMIC EQUATIONS

If variables in equations appear as exponents or logarithms, the equations are **exponential**, or **logarithmic equations**, respectively. The following examples illustrate techniques for solving such equations.

EXAMPLE ▪ 1

Solve the equation $3^x = 21$.

SOLUTION We take the common logarithm of both sides of $3^x = 21$ and use (iii) of the laws of logarithms:

$$\log(3^x) = \log 21$$
$$x \log 3 = \log 21$$
$$x = \frac{\log 21}{\log 3}$$

If an approximation is desired, we may use a calculator or table:

$$x \approx \frac{1.3222}{0.4771} \approx 2.77$$

A partial check on the solution is to note that since $3^2 = 9$ and $3^3 = 27$, the number x such that $3^x = 21$ should lie between 2 and 3, somewhat closer to 3 than to 2.

We could have solved for x by using natural logarithms instead of common logarithms. In this case,

$$x = \frac{\ln 21}{\ln 3} \approx \frac{3.0445}{1.0986} \approx 2.77.$$

EXAMPLE ▪ 2

Solve the equation $5^{2x+1} = 6^{x-2}$.

SOLUTION If we take the common logarithm of both sides of the equation and use (iii) of the laws of logarithms, we obtain

$$(2x + 1) \log 5 = (x - 2) \log 6.$$

We may now solve for x:

$$2x \log 5 + \log 5 = x \log 6 - 2 \log 6$$
$$2x \log 5 - x \log 6 = -\log 5 - 2 \log 6$$
$$x(2 \log 5 - \log 6) = -(\log 5 + \log 6^2)$$
$$x = \frac{-(\log 5 + \log 36)}{2 \log 5 - \log 6}$$
$$x = \frac{-\log(5 \cdot 36)}{\log 5^2 - \log 6}.$$
$$x = -\frac{\log 180}{\log \frac{25}{6}}.$$

An approximation to x is

$$x \approx -\frac{2.2553}{0.6198} \approx -3.64.$$

Natural logarithms could have been used instead:

$$x = -\frac{\ln 180}{\ln \frac{25}{6}} \approx -\frac{5.1929569}{1.4271164} \approx -3.64$$

EXAMPLE ▪ 3

Solve the equation $\log(5x - 1) - \log(x - 3) = 2$.

SOLUTION The equation may be written

$$\log \frac{5x - 1}{x - 3} = 2.$$

Using the definition of logarithm with $a = 10$ gives us

$$\frac{5x - 1}{x - 3} = 10^2.$$

Consequently,

$$\begin{aligned} 5x - 1 &= 10^2(x - 3) \\ 5x - 1 &= 100x - 300 \\ 299 &= 95x. \\ x &= \tfrac{299}{95}. \end{aligned}$$

You should check that this is the solution of the equation.

EXAMPLE ▪ 4

Solve the equation $\dfrac{5^x - 5^{-x}}{2} = 3$.

SOLUTION Multiplying both sides of the equation by 2 gives us

$$5^x - 5^{-x} = 6.$$

If we now multiply both sides by 5^x, we obtain

$$5^{2x} - 1 = 6(5^x),$$

which may be written

$$(5^x)^2 - 6(5^x) - 1 = 0.$$

Letting $u = 5^x$ leads to the quadratic equation

$$u^2 - 6u - 1 = 0$$

in the variable u. Applying the quadratic formula,

$$u = \frac{6 \pm \sqrt{36 + 4}}{2} = 3 \pm \sqrt{10},$$

that is, $5^x = 3 \pm \sqrt{10}$. Since 5^x is never negative, the number $3 - \sqrt{10}$ must be discarded; therefore,

$$5^x = 3 + \sqrt{10}.$$

We take the common logarithm of both sides and use (iii) of the laws of logarithms to solve for x:

$$x \log 5 = \log (3 + \sqrt{10})$$

$$x = \frac{\log (3 + \sqrt{10})}{\log 5}$$

To approximate x, we write $3 + \sqrt{10} \approx 6.16$, obtaining

$$x \approx \frac{\log 6.16}{\log 5} \approx \frac{0.7896}{0.6990} \approx 1.13.$$

Natural logarithms could also have been used to obtain

$$x = \frac{\ln (3 + \sqrt{10})}{\ln 5}.$$

EXAMPLE ▪ 5

The Beer-Lambert law states that the amount of light I that penetrates to a depth of x meters in an ocean is given by $I = I_0 a^x$, where $0 < a < 1$ and I_0 is the amount of light at the surface. (Compare Example 3 of Section 5.5.)

(a) Solve for x using common logarithms.

(b) Solve for x using natural logarithms.

(c) If $a = \frac{1}{4}$, what is the depth at which $I = 0.01 I_0$? (This depth determines the zone where photosynthesis can take place.)

SOLUTION

(a) We first take the common logarithm of both sides of the equation $I = I_0 a^x$ and apply laws of logarithms:

$$\begin{aligned} \log I &= \log (I_0 a^x) \\ &= \log I_0 + \log a^x \\ &= \log I_0 + x \log a \end{aligned}$$

Hence,
$$x \log a = \log I - \log I_0 = \log (I/I_0).$$

and
$$x = \frac{\log (I/I_0)}{\log a}.$$

(b) Replacing log with ln throughout part (a), we obtain

$$x = \frac{\ln (I/I_0)}{\ln a}$$

(c) Letting $I = 0.01I_0$ and $a = \frac{1}{4}$ in the formula for x obtained in part (a),

$$x = \frac{\log (0.01I_0/I_0)}{\log \frac{1}{4}} = \frac{\log (0.01)}{\log 1 - \log 4} = \frac{-2}{-\log 4}.$$

An approximation is

$$x \approx \frac{-2}{-0.6021} \approx 3.32 \text{ meters.}$$

5.6 EXERCISES

Exer. 1–18: Find the exact solution, using common logarithms, and a two-decimal-place approximation of each solution, when appropriate.

1 $5^x = 8$

2 $4^x = 3$

3 $3^{4-x} = 5$

4 $(\frac{1}{3})^x = 100$

5 $3^{x+4} = 2^{1-3x}$

6 $4^{2x+3} = 5^{x-2}$

7 $2^{2x-3} = 5^{x-2}$

8 $3^{2-3x} = 4^{2x+1}$

9 $2^{-x} = 8$

10 $2^{-x^2} = 5$

11 $\log x = 1 - \log (x - 3)$

12 $\log (5x + 1) = 2 + \log (2x - 3)$

13 $\log (x^2 + 4) - \log (x + 2) = 3 + \log (x - 2)$

14 $\log (x - 4) - \log (3x - 10) = \log (1/x)$

15 $5^x + 125(5^{-x}) = 30$

16 $3(3^x) + 9(3^{-x}) = 28$

17 $4^x - 3(4^{-x}) = 8$

18 $2^x - 6(2^{-x}) = 6$

Exer. 19–24: Solve the equation without using a calculator or table.

19 $\log (x^2) = (\log x)^2$

20 $\log \sqrt{x} = \sqrt{\log x}$

21 $\log (\log x) = 2$

22 $\log \sqrt{x^3 - 9} = 2$

23 $x^{\sqrt{\log x}} = 10^8$

24 $\log (x^3) = (\log x)^3$

Exer. 25–28: Use common logarithms to solve for x in terms of y.

25 $y = \dfrac{10^x + 10^{-x}}{2}$

26 $y = \dfrac{10^x - 10^{-x}}{2}$

27 $y = \dfrac{10^x - 10^{-x}}{10^x + 10^{-x}}$

28 $y = \dfrac{10^x + 10^{-x}}{10^x - 10^{-x}}$

Exer. 29–32: Use natural logarithms to solve for x in terms of y.

29 $y = \dfrac{e^x - e^{-x}}{2}$

30 $y = \dfrac{e^x + e^{-x}}{2}$

31 $y = \dfrac{e^x + e^{-x}}{e^x - e^{-x}}$

32 $y = \dfrac{e^x - e^{-x}}{e^x + e^{-x}}$

33 The current I in a certain electrical circuit at time t is given by

$$I = \frac{V}{R}(1 - e^{-Rt/L})$$

for the electromotive force V, resistance R, and inductance L (see figure). Use natural logarithms to solve the equation for t.

EXERCISE 33

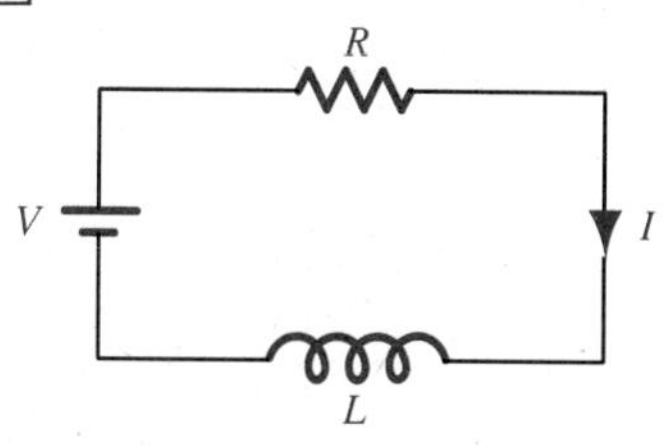

34 Solve the compound interest formula

$$A = P\left(1 + \frac{r}{n}\right)^{nt}$$

for t by using natural logarithms.

35 Refer to Example 5. The most important zone in the sea from the viewpoint of marine biology is the photic zone, in which photosynthesis takes place. That zone must end at the depth where about 1% of the surface light penetrates. In very clear waters in the Caribbean, 50% of the light at the surface reaches a depth of about 13 meters. Estimate the depth of the photic zone.

36 In contrast to the situation described in Exercise 35, in parts of New York harbor, 50% of the surface light does not reach a depth of 10 cm. Estimate the depth of the photic zone.

37 A drug is eliminated from the body through urine. The initial dose is 10 mg and the amount $A(t)$ in the body t hours later is given by $A(t) = 10(0.8)^t$. In order for the drug to be effective, at least 2 mg must be in the body.

(a) Determine when only 2 mg are left.

(b) What is the half-life of the drug?

38 Radioactive iodine ^{131}I is frequently used in tracer studies involving the thyroid gland. The substance decays according to the formula $A(t) = A_0 a^{-t}$ for the initial dose A_0 and time t in days. Find a assuming the half-life of ^{131}I is eight days.

39 The technique of carbon-14 (^{14}C) dating is used to determine the age of archaeological and geological specimens. The formula $T = -8310 \ln x$ is sometimes used to predict the age T (in years) of a bone fossil, where x is the percentage (expressed as a decimal) of ^{14}C still present in the fossil.

(a) Estimate the age of a bone fossil that contains 4% as much ^{14}C as an equal amount of carbon in present-day bone.

(b) Approximate the percentage of ^{14}C present in a fossil that is 10,000 years old.

40 A country presently has coal reserves of 50 million tons. Last year 6.5 million tons were consumed. Based on past years' data and population projections, the rate of consumption R (in million tons/year) is expected to increase according to the formula $R = 6.5e^{0.02t}$, and the total amount T (in million tons) of coal that will be used in t years is given by $T = 325(e^{0.02t} - 1)$. If the country uses only its own resources, when will the coal reserves be depleted?

41 Based on present birth and death rates, the population of Kenya is expected to increase according to the formula $N = 20.2e^{0.041t}$ with N in millions and $t = 0$ corresponding to 1985. How many years will it take for the population to double?

42 Refer to Exercise 44 of Section 5.1. If a language originally had N_0 basic words of which $N(t)$ are still in use, then $N(t) = N_0(0.805)^t$ for time t measured in millennia. After how many years are one-half the basic words still in use, that is, when does $N(t) = \frac{1}{2}N_0$?

43 An urban density model is a formula that relates the population density D (in thousands/mi^2) to the distance x (in miles) from the center of the city. The formula $D = ae^{-bx}$ for the central density a and coefficient of decay b has been found to be appropriate for many large U.S. cities. For the city of Atlanta in 1970, $a = 5.5$ and $b = 0.10$. At what distance was the population density less than 2000 per square mile?

44 The energy $E(x)$ of an electron after passing through material of thickness x is given by $E(x) = E_0 e^{-x/x_0}$ with E_0 the initial electron energy and x_0 the radiation length.

(a) Express the energy of an electron after it passes through material of thickness x_0 in terms of E_0.

(b) Express the thickness at which the electron loses 99% of its initial energy in terms of x_0.

5.7 REVIEW

Define or discuss each of the following.

The exponential function with base a ■ The natural exponential function ■ The logarithmic function with base a ■ Laws of logarithms ■ Common logarithms ■ The natural logarithmic function

EXERCISES

Exer. 1–10: Evaluate without using a calculator or table.

1 $\log_2 \frac{1}{16}$

2 $\log_5 \sqrt[3]{5}$

3 $6^{\log_6 4}$

4 $10^{3 \log 2}$

5 $\log 1{,}000{,}000$

6 $\ln e$

7 $\log_4 2$

8 $\log_\pi 1$

9 $e^{\ln 5}$

10 $\log \log 10^{10}$

Exer. 11–24: Sketch the graph of f.

11 $f(x) = 3^{x+2}$

12 $f(x) = (\frac{3}{5})^x$

13 $f(x) = (\frac{3}{2})^{-x}$

14 $f(x) = 3^{-2x}$

15 $f(x) = 3^{-x^2}$

16 $f(x) = 1 - 3^{-x}$

17 $f(x) = \log_6 x$

18 $f(x) = \log_4 (x^2)$

19 $f(x) = 2 \log_4 x$

20 $f(x) = \log_2 (x + 4)$

21 $f(x) = e^{x/2}$

22 $f(x) = \frac{1}{2}e^x$

23 $f(x) = e^{x-2}$

24 $f(x) = e^{2-x}$

Exer. 25–34: Solve the equation without using a calculator or table.

25 $\log_8 (x - 5) = \frac{2}{3}$

26 $\log_4 (x + 1) = 2 + \log_4 (3x - 2)$

27 $2 \log_3 (x + 3) - \log_3 (x + 1) = 3 \log_3 2$

28 $\log \sqrt[4]{x + 1} = \frac{1}{2}$

29 $2^{5-x} = 6$

30 $3^{(x^2)} = 7$

31 $2^{5x+3} = 3^{2x+1}$

32 $e^{\ln (x+1)} = 3$

33 $x^2(-2xe^{-x^2}) + 2xe^{-x^2} = 0$

34 $\ln x = 1 + \ln (x + 1)$

35 Express $\log x^4 \sqrt[3]{y^2/z}$ in terms of logarithms of x, y, and z.

36 Express $\log (x^2/y^3) + 4 \log y - 6 \log \sqrt{xy}$ as one logarithm.

Exer. 37–38: Solve the equation for x in terms of y.

37 $y = \dfrac{1}{10^x - 10^{-x}}$

38 $y = \dfrac{1}{10^x + 10^{-x}}$

Exer. 39–44: Approximate x using (a) a calculator; (b) tables.

39 $\log x = 1.8938$

40 $\log x = -2.4260$

41 $\ln x = 1.8$

42 $\ln x = -0.75$

43 $x = \ln 6.6$

44 $x = \log 8.4$

45 The number of bacteria in a certain culture at time t is given by $Q(t) = 2(3^t)$ for t measured in hours and $Q(t)$ in thousands.

(a) What is the initial number of bacteria?

(b) Find the number after 10 minutes; 30 minutes; 1 hour.

46 If \$1000 is invested at a rate of 12% compounded four times per year, what is the principal after one year?

47 Radioactive iodine ^{131}I is frequently used in tracer studies involving the thyroid gland. The substance decays according to the formula $N = N_0(0.5)^{t/8}$ for the initial dose N_0 and time t.

(a) Sketch the graph of the equation if $N_0 = 64$.

(b) Show that the half-life of ^{131}I is eight days.

48 One thousand young trout are put in a fishing pond. Three months later, the owner estimates that about 600

trout are left. Find an exponential formula $N = N_0 a^t$ that fits this information and use it to estimate the number of trout left after one year.

49 Ten thousand dollars is invested in a savings fund in which interest is compounded continuously at the rate of 11% per year. The amount A in the account t years later is given by $A = 10{,}000e^{0.11t}$.

(a) When will the account contain $35,000?

(b) How long does it take money to double in the account?

50 The current $I(t)$ at time t in a certain electrical circuit is given by $I(t) = I_0 e^{-Rt/L}$ for a resistance R, inductance L, and initial current I_0 at time $t = 0$. At what time is the current $\frac{1}{100}I_0$?

51 The sound intensity level formula is $\alpha = 10 \log (I/I_0)$.

(a) Solve for I in terms of α and I_0.

(b) Show that a one-decibel rise in the intensity level α corresponds to a 26% increase in the intensity I.

52 Stars are classified into categories of brightness called magnitudes. The faintest stars (with light flux L_0) are assigned a magnitude of 6. Brighter stars are assigned magnitudes according to the formula

$$m = 6 - (2.5) \log \frac{L}{L_0}$$

for the light flux L from the star.

(a) Find m if $L = 10^{0.4}L_0$.

(b) Solve the formula for L in terms of m and L_0.

53 Refer to Exercise 22 of Section 5.2. The length L of a fish is related to its age by means of the von-Bertalanffy growth formula

$$L = a(1 - be^{-kt})$$

for positive constants a, b, and k that depend on the type of fish. Solve this equation for t to obtain a formula that can be used to estimate the age of a fish from a length measurement.

54 The weight W (in kg) of a female African elephant at age t (in years) may be approximated by

$$W = 2600(1 - 0.51e^{-0.075t})^3.$$

(a) Approximate the weight at birth.

(b) The graph illustrates the case of an adult female weighing 1800 kg. Estimate her age (i) from the graph and (ii) by using the formula for W.

EXERCISE 54

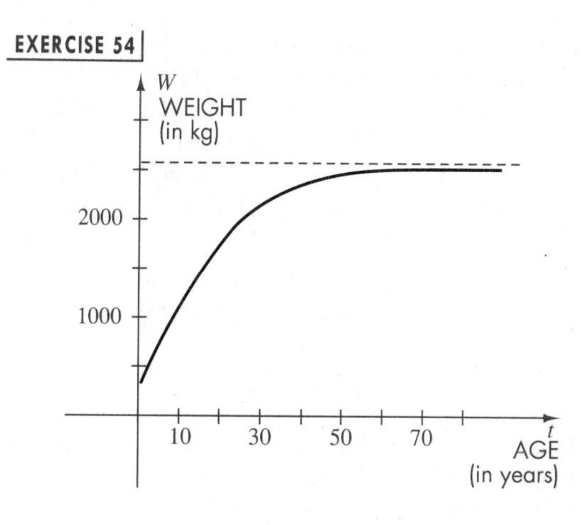

55 If the pollution of Lake Erie were stopped suddenly, it has been estimated that the level y of pollutants would decrease according to the formula $y = y_0 e^{-0.3821t}$ with time t in years and y_0 the pollutant level at which further pollution ceased. How many years would it take to clear 50% of the pollutants?

56 Radioactive strontium ^{90}Sr has been deposited into a large field by acid rain. If sufficient amounts make their way through the food chain to man, bone cancer can result. It has been determined that the radioactivity level in the field is 2.5 times the safe level S. ^{90}Sr decays according to the formula

$$A(t) = A_0 e^{-0.0239t}$$

for the present amount A_0 in the field and time t in years. For how many years will this field be contaminated?

57 If a 100-mg tablet of an asthma drug is taken orally, and if none of the drug is present in the body when the tablet is first taken, the total amount A in the bloodstream after t minutes is predicted to be

$$A = 100[1 - (0.9)^t] \quad \text{for } 0 \le t \le 10.$$

(a) Sketch the graph of the equation.

(b) Determine the number of minutes needed for 50 mg of the drug to have entered the bloodstream.

58 Refer to Exercise 3 of Section 5.5. For the Rocky Mountain and Central states, the area-magnitude formula has the form

$$R = 2.3 \log (A + 14{,}000) - 6.6.$$

An earthquake measures 4 on the Richter scale. Estimate the area A of the region that will feel the quake.

59 Under certain conditions, the logistic equation $y = K/(1 + ce^{-rt})$, for positive constants k and r, may be used to predict the number y in a population at time t. In a famous study of the growth of protozoa by Gause, a population of Paramecium caudata was found to be described by a logistic equation with $r = 1.1244$ and $K = 105$ and with t in days.

(a) What is the significance of the constant K?

(b) Find c if the initial population was 3 protozoa.

(c) In the study the fastest rate of growth occurred at $y = 52$. Determine t for this value of y.

60 Refer to Exercise 59. The growth in height of trees is frequently given by logistic equations. Suppose the height h (in feet) of a tree at age t (in years) is

$$h = \frac{120}{1 + 200e^{-0.2t}}.$$

A graph of this equation is shown in the figure.

(a) What is the height of the tree at age 10?

(b) At what age is the height 50 feet?

EXERCISE 60

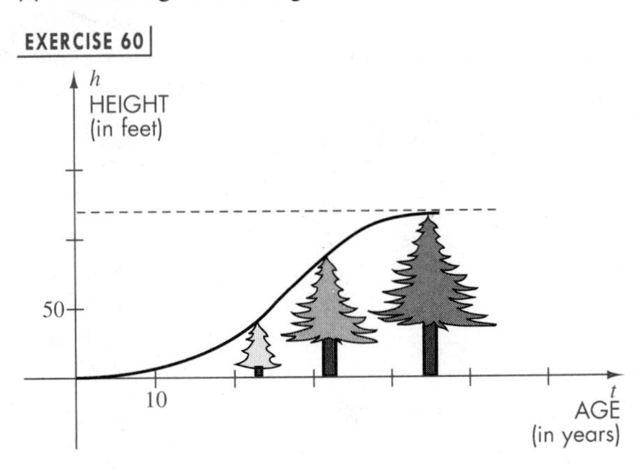

CHAPTER 6

THE TRIGONOMETRIC FUNCTIONS

We begin the chapter by discussing angles and how they are measured. We then introduce the trigonometric functions by the classical method of using ratios of sides of a right triangle. Applications involving right triangles are considered before trigonometric functions of any angle and of real numbers. The chapter concludes with graphing techniques that make use of amplitudes, periods, and phase shifts.

6.1

ANGLES

FIGURE 1

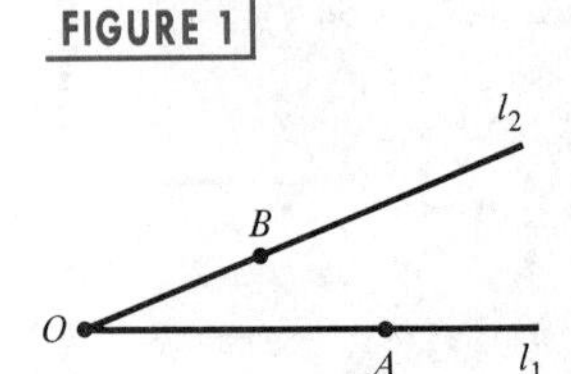

In geometry an **angle** is defined as the set of points determined by two rays, or half-lines, l_1 and l_2, having the same endpoint O (see Figure 1). If A and B are points on l_1 and l_2, respectively, we refer to **angle** $\boldsymbol{AOB}$. An angle may also be considered as two finite line segments with a common endpoint.

In trigonometry we often interpret angles as rotations of rays. Start with a fixed ray l_1 having endpoint O, and rotate it about O, in a plane, to a position specified by ray l_2. We call l_1 the **initial side**, l_2 the **terminal side**, and O the **vertex** of angle AOB. The amount or direction of rotation is not restricted in any way. We might let l_1 make several revolutions in either direction about O before coming to position l_2, as illustrated by the curved arrows in Figure 2. Thus many different angles have the same initial and terminal sides. Any two such angles are called **coterminal angles**.

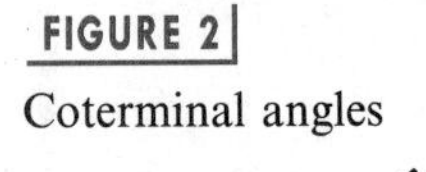

FIGURE 2
Coterminal angles

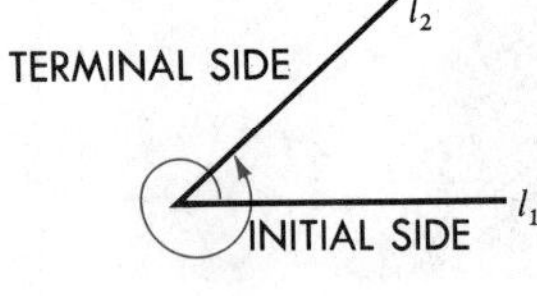

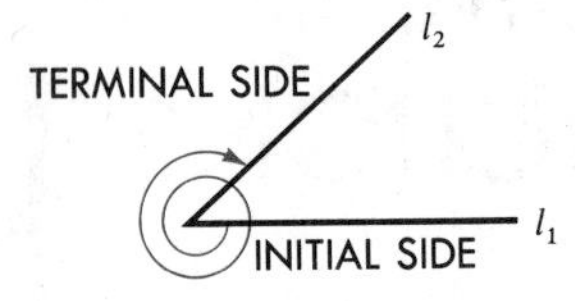

If we introduce a rectangular coordinate system, then the **standard position** of an angle is obtained by taking the vertex at the origin and letting the initial side l_1 coincide with the positive x-axis. If l_1 is rotated in a *counterclockwise* direction to the terminal position l_2, then the angle is considered **positive**. If l_1 is rotated in a *clockwise* direction, the angle is **negative**. We often denote angles by lowercase Greek letters such as α (*alpha*), β (*beta*), γ (*gamma*), θ (*theta*), φ (*phi*), and so on. Figure 3 contains sketches of two positive angles, α and β, and a negative angle γ. If the terminal side of an angle in standard position is in a certain quadrant, we say that the *angle* is in that quadrant. In Figure 3, α is in quadrant III, β is in quadrant I, and γ is in quadrant II.

FIGURE 3 Standard position of an angle

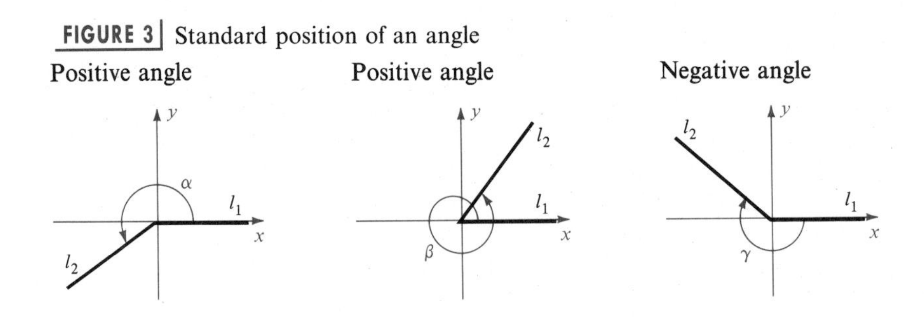

One unit of measurement for angles is the **degree**. The angle in standard position obtained by one complete revolution in the counterclockwise direction has measure 360 degrees, written 360°. Thus an angle of measure 1 degree (1°) is obtained by $\frac{1}{360}$ of one complete counterclockwise revolution. In Figure 4, several angles measured in degrees are shown in standard position on rectangular coordinate systems.

FIGURE 4

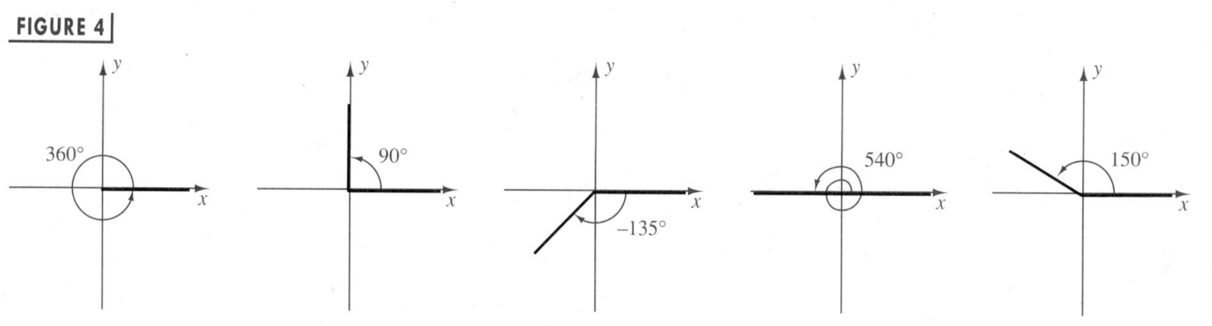

Throughout our work, a notation such as $\theta = 60°$ specifies an angle θ whose measure is 60°. We also refer to *an angle of* 60°, or *a* 60° *angle*, instead of the more precise (but cumbersome) phrase, *an angle having degree measure* 60°.

EXAMPLE ▪ 1

If $\theta = 60°$ is in standard position, find two positive angles and two negative angles that are coterminal with θ.

SOLUTION The angle θ is shown in standard position in the first sketch in Figure 5. To find positive coterminal angles, we may add 360° or 720° (or any other positive multiple of 360°) to θ, obtaining

$$60° + 360° = 420° \quad \text{and} \quad 60° + 720° = 780°.$$

These coterminal angles are also shown in Figure 5.

To find negative coterminal angles we may add $-360°$ or $-720°$ (or any negative multiple of 360°), obtaining

$$60° + (-360°) = -300° \quad \text{and} \quad 60° + (-720°) = -660°$$

as shown in Figure 5.

FIGURE 5

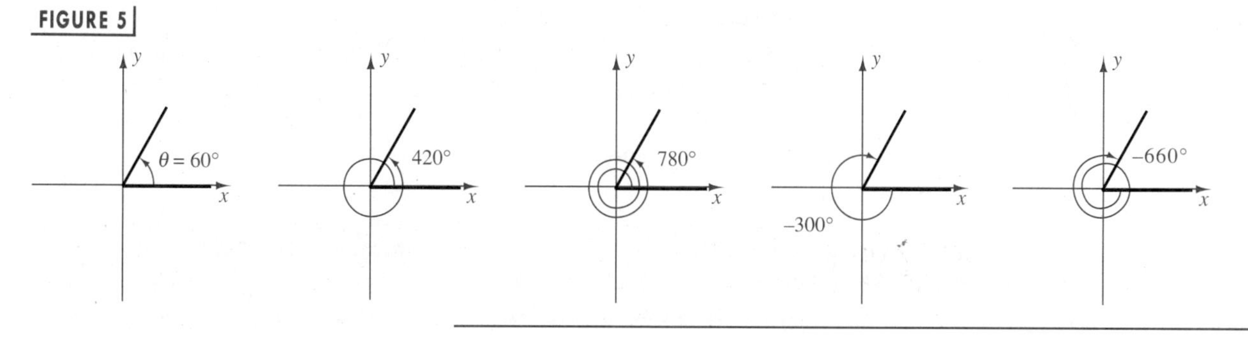

A **right angle** is a 90° angle. An angle θ is **acute** if $0° < \theta < 90°$ or **obtuse** if $90° < \theta < 180°$. Two acute angles are **complementary** if their sum is 90°. Two positive angles are **supplementary** if their sum is 180°.

ILLUSTRATION | COMPLEMENTARY ANGLES

- 20°, 70°
- 60°, 30°
- 7°, 83°

ILLUSTRATION | SUPPLEMENTARY ANGLES

- 45°, 135°
- 110°, 70°
- 17°, 163°

If smaller measurements than the degree are required, we can use tenths, hundredths, or thousandths of degrees. Alternatively, we can divide the degree into 60 equal parts, called **minutes** (denoted by ′), and each minute into 60 equal parts, called **seconds** (denoted by ″). Thus $1° = 60'$, and $1' = 60''$. The notation $\theta = 73°56'18''$ refers to an angle θ that has measure 73 degrees, 56 minutes, 18 seconds.

EXAMPLE ■ 2

Find the angle that is complementary to θ:

(a) $\theta = 25°43'37''$ **(b)** $\theta = 73.26°$

SOLUTION We wish to find $90° - \theta$. Let us arrange our work as follows:

(a)
$$\begin{array}{rl} 90° & = 89°59'60'' \\ \theta & = 25°43'37'' \\ \hline 90° - \theta & = 64°16'23'' \end{array}$$

(b)
$$\begin{array}{rl} 90° & = 90.00° \\ \theta & = 73.26° \\ \hline 90° - \theta & = 16.74° \end{array}$$

Degree measure for angles is used in applied areas such as surveying, navigation, and the design of mechanical equipment. In scientific applications that require calculus it is customary to employ *radian measure*. To define an angle of radian measure 1 we consider a circle of any radius r. A **central angle** of a circle is an angle whose vertex is at the center of the circle. If θ is the central angle shown in Figure 6, we say that the arc $\widehat{AP}$ of the circle **subtends** θ, or that θ **is subtended by** $\widehat{AP}$. If the length of $\widehat{AP}$ is equal to the radius r of the circle, then θ has a measure of one radian, as in the next definition.

FIGURE 6

DEFINITION OF RADIAN MEASURE

One radian is the measure of the central angle of a circle subtended by an arc equal in length to the radius of the circle.

If we consider a circle of radius r, then an angle α whose measure is 1 radian intercepts an arc $\widehat{AP}$ of length r, as illustrated in Figure 7(i). The angle β in Figure 7(ii) has radian measure 2, since it is subtended by an arc of length $2r$. Similarly, γ in (iii) of the figure has radian measure 3, since it is subtended by an arc of length $3r$.

FIGURE 7

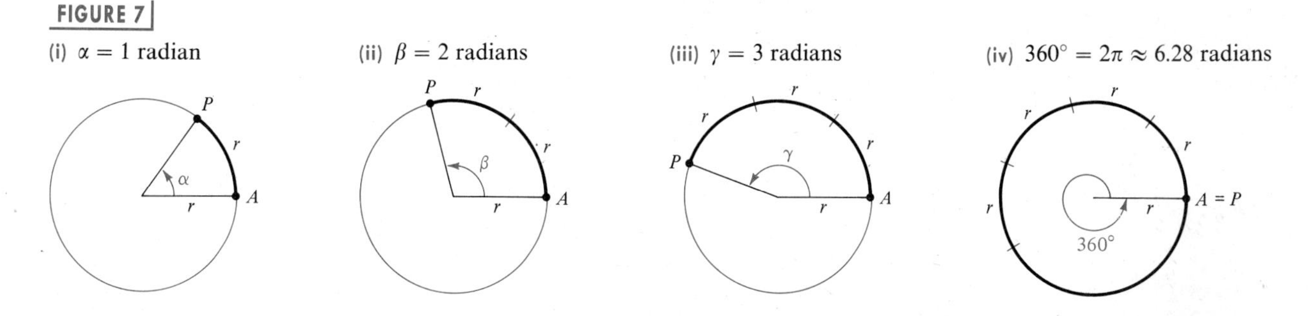

To find the radian measure corresponding to 360° we must find the number of times that a circular arc of length r can be laid off along the circumference (see Figure 7(iv)). This number is not an integer or even a rational number. Since the circumference of the circle is $2\pi r$, the number of times r units can be laid off is 2π. Thus an angle of measure 2π radians corresponds to the degree measure 360°, and we write $360° = 2\pi$ radians. This gives us the following relationships.

RELATIONSHIPS BETWEEN DEGREES AND RADIANS

$$180° = \pi \text{ radians}, \qquad 1° = \frac{\pi}{180} \text{ radian}, \qquad 1 \text{ radian} = \left(\frac{180}{\pi}\right)^{\circ}.$$

If we approximate $\pi/180$ and $180/\pi$, we obtain

$$1° \approx 0.01745 \text{ radian} \quad \text{and} \quad 1 \text{ radian} \approx 57.29578°.$$

The next theorem is a consequence of the preceding formulas.

THEOREM

(i) To change radian measure to degrees, multiply by $\dfrac{180}{\pi}$.

(ii) To change degree measure to radians, multiply by $\dfrac{\pi}{180}$.

When radian measure of an angle is used, no units will be indicated. Thus, if an angle has radian measure 5, we write $\theta = 5$ instead of $\theta = 5$ *radians*. There should be no confusion as to whether radian or degree measure is being used, since if θ has degree measure 5°, we write $\theta = 5°$, and *not* $\theta = 5$.

EXAMPLE ▪ 3

(a) Find the radian measure of θ if $\theta = 150°$ and if $\theta = 225°$.

(b) Find the degree measure of θ if $\theta = \frac{7\pi}{4}$ and if $\theta = \frac{\pi}{3}$.

SOLUTION

(a) By (ii) of the preceding theorem,

$$150° = 150\left(\frac{\pi}{180}\right) = \frac{5\pi}{6}$$

$$225° = 225\left(\frac{\pi}{180}\right) = \frac{5\pi}{4}$$

(b) By (i) of the theorem,

$$\frac{7\pi}{4} = \frac{7\pi}{4}\left(\frac{180}{\pi}\right)^{\circ} = 315°$$

$$\frac{\pi}{3} = \frac{\pi}{3}\left(\frac{180}{\pi}\right)^{\circ} = 60°$$

The following table displays the relationship between the radian and degree measures of several common angles. The entries may be checked by using the last theorem.

Radians	0	$\frac{\pi}{6}$	$\frac{\pi}{4}$	$\frac{\pi}{3}$	$\frac{\pi}{2}$	$\frac{2\pi}{3}$	$\frac{3\pi}{4}$	$\frac{5\pi}{6}$	π
Degrees	0°	30°	45°	60°	90°	120°	135°	150°	180°

Several angles in radian measure are shown in standard position in Figure 8.

FIGURE 8

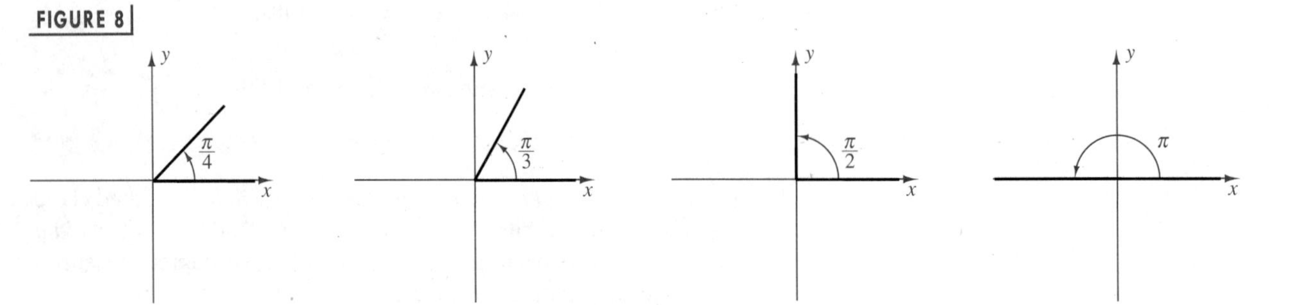

EXAMPLE ▪ 4

If the measure of an angle θ is 3 radians, approximate θ in terms of degrees, minutes, and seconds.

SOLUTION

$$3 \text{ radians} \approx 3\left(\frac{180}{\pi}\right)^\circ \approx 171.88734^\circ$$

$$3 \text{ radians} \approx 171^\circ + (0.88734)^\circ$$

Since $1^\circ = 60'$,

$$(0.88734)^\circ = (0.88734)(1^\circ) = (0.88734)(60') = 53.2404' = 53' + 0.2404'.$$

Since $1' = 60''$,

$$0.2404' = (0.2404)(1') = (0.2404)(60'') = 14.424'' \approx 14''.$$

Hence $$3 \text{ radians} \approx 171^\circ 53' 14''.$$

Some calculators have a key that can be used to convert the radian measure of an angle to degrees, and vice versa. Calculators may also have a [DMS] key for converting decimal degree measure to degrees, minutes, and seconds, and vice versa. Refer to the user's manual of a particular calculator for specific information. Since entries are made in terms of decimals, angles that are expressed in terms of degrees, minutes, and seconds must be changed to decimal form. If a [DMS] key is not available, we may follow the procedure given in the next example.

EXAMPLE ▪ 5

Express $19^\circ 47' 23''$ as a decimal, to the nearest ten-thousandth of a degree.

SOLUTION Since $1' = \frac{1}{60}^\circ$ and $1'' = \frac{1}{60}' = \frac{1}{3600}^\circ$,

$$19^\circ 47' 23'' = 19^\circ + \left(\frac{47}{60}\right)^\circ + \left(\frac{23}{3600}\right)^\circ$$

$$19^\circ 47' 23'' \approx 19^\circ + 0.7833^\circ + 0.0064^\circ$$

$$19^\circ 47' 23'' \approx 19.7897^\circ$$

The next result specifies the relationship between the length of a circular arc and the central angle that it subtends.

LENGTH OF A CIRCULAR ARC

> If an arc of length s on a circle of radius r subtends a central angle of radian measure θ, then
>
> $$s = r\theta.$$

FIGURE 9

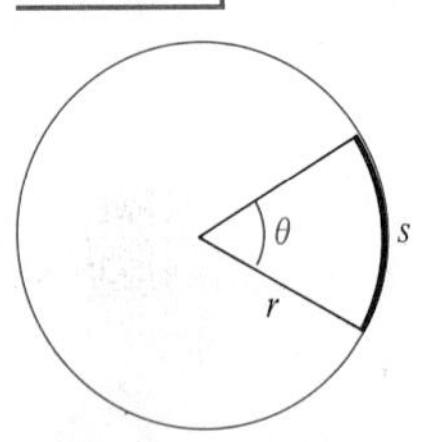

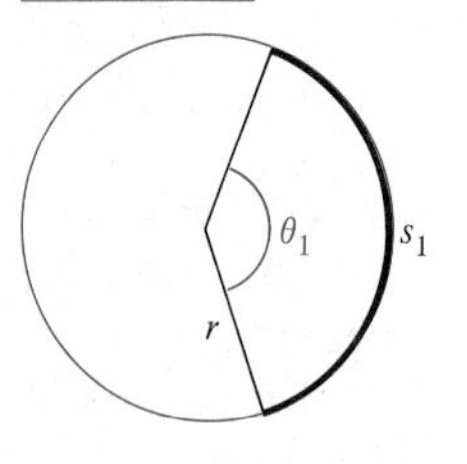

PROOF A typical arc of length s and the corresponding central angle θ are shown in Figure 9. Figure 10 shows an arc of length s_1 and central angle θ_1. If radian measure is used, then from plane geometry, the ratio of the angular measures is the same as the ratio of the lengths of arc, that is

$$\frac{\theta}{\theta_1} = \frac{s}{s_1}.$$

Thus, if θ_1 is twice as large as θ, then s_1 is twice as large as s, and so on. If we consider the special case in which θ_1 has radian measure 1 then, from the definition of radian, $s_1 = r$ and

$$\frac{\theta}{1} = \frac{s}{r}.$$

Multiplying both sides of this equation by r gives us $s = r\theta$. ❑

EXAMPLE ▪ 6

A central angle θ is subtended by an arc 10 cm long on a circle of radius 4 cm. Find the measure of θ in (a) radians and (b) degrees.

SOLUTION

(a) The formula $s = r\theta$ for the length of a circular arc may be rewritten as

$$\theta = \frac{s}{r}.$$

Substituting $s = 10$ and $r = 4$,

$$\theta = \frac{10}{4} = 2.5.$$

(b) To change the radian measure found in part (a) to degrees, we multiply by $180/\pi$. Thus

$$\theta = (2.5)\left(\frac{180}{\pi}\right)^\circ = \left(\frac{450}{\pi}\right)^\circ \approx 143.24^\circ.$$

FIGURE 11

Rotating wheel

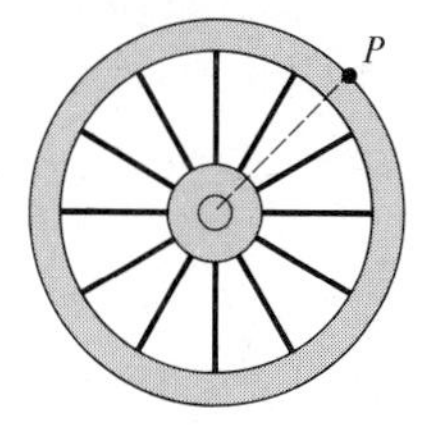

EXAMPLE ▪ 7

The **angular speed** of a wheel that is rotating at a constant rate is the angle generated in one unit of time by a line segment from the center of the wheel to a point P on the circumference (see Figure 11). Suppose that a machine contains a wheel of diameter 3 feet, rotating at a rate of 1600 rpm (revolutions per minute).

(a) Find the angular speed of the wheel.

(b) Find the speed at which a point P on the circumference of the wheel is moving.

SOLUTION

(a) Let O denote the center of the wheel and let P be a point on the circumference. Since the number of revolutions per minute is 1600, and since each revolution generates an angle of 2π radians, the angle generated by the line segment OP in one minute has radian measure $(1600)(2\pi)$; that is,

$$\text{angular speed} = (1600)(2\pi) = 3200\pi \text{ radians per minute.}$$

Note that the diameter of the wheel is irrelevant in finding the angular speed.

(b) The speed at which the point P moves is the distance it travels per minute. (This is sometimes called the **linear speed** of P.) We may find this distance by using the formula $s = r\theta$, with $r = \frac{3}{2}$ ft and $\theta = 3200\pi$ radians per minute. Thus

$$s = r\theta = \tfrac{3}{2}(3200\pi) = 4800\pi \text{ ft/min.}$$

To the nearest integer, this is approximately 15,080 ft/min, or 171.36 mi/hr. Unlike angular speed, the linear speed *is* dependent on the diameter of the wheel.

6.1 EXERCISES

Exer. 1–12: Place the angle with the indicated measure in standard position on a rectangular coordinate system, and find the measure of two positive angles and two negative angles that are coterminal with the given angle.

1 120° **2** 240° **3** 135°

4 315° **5** -30° **6** -150°

7 620° **8** 570° **9** $\dfrac{5\pi}{6}$

10 $\dfrac{2\pi}{3}$ **11** $-\dfrac{\pi}{4}$ **12** $-\dfrac{5\pi}{4}$

Exer. 13–24: Find the exact radian measure that corresponds to the degree measure.

13 150° **14** 120° **15** -60°

16 -135° **17** 225° **18** 210°

19 450° **20** 630° **21** 72°

22 54° **23** 100° **24** 95°

Exer. 25–36: Find the exact degree measure that corresponds to the radian measure.

25 $\frac{2\pi}{3}$ 26 $\frac{5\pi}{6}$ 27 $\frac{11\pi}{6}$

28 $\frac{4\pi}{3}$ 29 $\frac{3\pi}{4}$ 30 $\frac{11\pi}{4}$

31 $-\frac{7\pi}{2}$ 32 $-\frac{5\pi}{2}$ 33 7π

34 9π 35 $\frac{\pi}{9}$ 36 $\frac{\pi}{16}$

Exer. 37–40: Find an approximate measure of θ in terms of degrees, minutes, and seconds.

37 $\theta = 2$ 38 $\theta = 1.5$

39 $\theta = 5$ 40 $\theta = 4$

Exer. 41–44: Express the degree measure as a decimal, to the nearest ten-thousandth of a degree.

41 $37°41'$ 42 $83°17'$

43 $115°26'27''$ 44 $258°39'52''$

Exer. 45–48: Express the angle in terms of degrees, minutes, and seconds, to the nearest second.

45 $63.169°$ 46 $12.864°$

47 $310.6215°$ 48 $81.7238°$

49 Approximate the radian and degree measures of the central angle subtended by an arc of length 7 cm on a circle of radius 4 cm.

50 Approximate the radian and degree measures of the central angle subtended by an arc of length 3 feet on a circle of radius 20 inches.

51 Approximate the length of an arc that subtends a central angle of measure 50° on a circle of diameter 16 m.

52 Approximate the length of an arc that subtends a central angle of radian measure 2.2 on a circle of diameter 120 cm.

53 If a circular arc of length 10 cm subtends a central angle of radian measure 4, find the radius of the circle.

54 If a circular arc of length 3 km subtends a central angle of measure 20°, find the radius of the circle.

55 The distance between two points A and B on the earth is measured along a circle having center C at the center of the earth and radius equal to the distance from C to the surface (see figure). If the diameter of the earth is approximately 8000 miles, approximate the distance between A and B if angle ACB has the indicated measure:

(a) 60° (b) 45° (c) 30° (d) 10° (e) 1°

EXERCISE 55

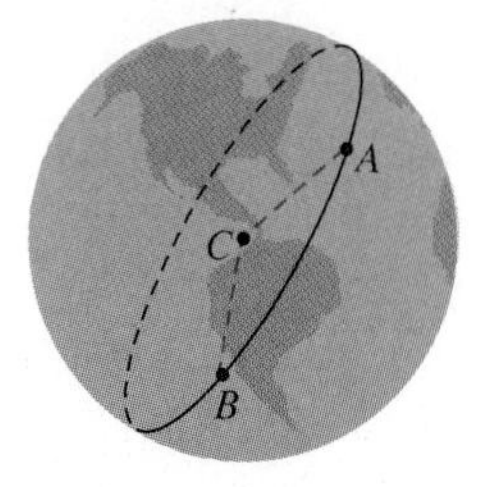

56 Refer to Exercise 55. If angle ACB has measure 1′, then the distance between A and B is a nautical mile. Approximate the number of land (statute) miles in a nautical mile.

57 Refer to Exercise 55. If two points A and B are 500 miles apart, find the measure of angle ACB in degrees and in radians.

Exer. 58–59: For a wheel of the given radius that is rotating at the indicated rate, find

(a) the angular speed of the wheel (in radians per minute).

(b) the speed of a point on the circumference of the wheel (in ft/min).

58 radius 9 in., 2400 rpm

59 radius 5 in., 40 rpm

60 A typical tire for a compact car is 22 inches in diameter. If the car is traveling at a rate of 60 mi/hr, find the number of revolutions the tire makes per minute.

61 For phonograph records, an LP album has a 12-inch diameter and a single has a 7-inch diameter. The album rotates at a rate of $33\frac{1}{3}$ rpm, and the single rotates at 45 rpm.

(a) Find the angular speed (in radians per minute) of the album and of the single.

(b) Find the linear speed (in ft/min) of a point on the circumference of the album and of the single.

62 A pendulum in a grandfather clock is 4 feet long and swings back and forth along a 6-inch arc. Approximate the angle (in degrees) through which the pendulum passes during one swing.

63 A large winch of diameter 3 feet is used to hoist cargo as shown in the figure.

(a) Find the distance the cargo is lifted if the winch rotates through an angle of radian measure $\frac{7\pi}{4}$.

(b) Find the angle (in radians) through which the winch must rotate in order to lift the cargo d feet.

EXERCISE 63

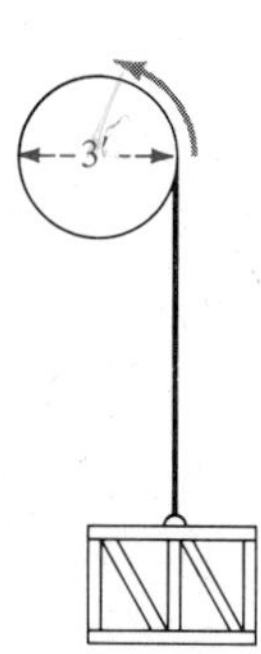

64 The sprocket assembly for a bicycle is shown in the figure. If the sprocket of radius r_1 rotates through an angle of θ_1 radians, find the corresponding angle of rotation for the sprocket of radius r_2.

EXERCISE 64

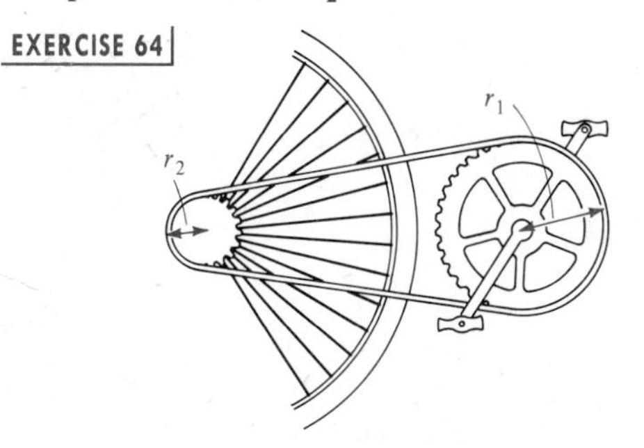

65 An expert cyclist can attain a speed of 40 mi/hr. If in Exercise 64 the sprocket assembly has $r_1 = 5$ inches, $r_2 = 2$ inches, and the wheel has a diameter of 28 inches, approximately how many revolutions per minute will produce a speed of 40 mi/hr? (*Hint:* First change 40 mi/hr to in./sec.)

6.2 TRIGONOMETRIC FUNCTIONS OF ACUTE ANGLES

Trigonometry was invented over 2000 years ago by the Greeks, who needed precise methods for measuring angles and sides of triangles. In fact, the word *trigonometry* was derived from the two Greek words *trigonon* (triangle) and *metria* (measurement).

We shall introduce the trigonometric functions in the manner that they originated historically—as ratios of sides of a right triangle. In Section 6.4 we extend these definitions to functions of arbitrary angles.

A triangle is a **right triangle** if one of its angles is a right angle. If θ is any acute angle, we may consider a right triangle having θ as one of its angles as in Figure 12, where the symbol ⌐ specifies the 90° angle. Six ratios can be obtained using the lengths a, b, c of the sides of the triangle:

FIGURE 12

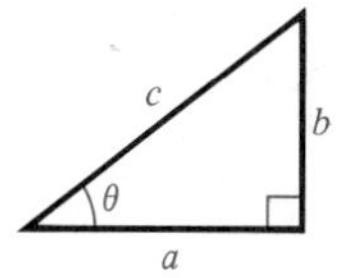

$$\frac{b}{c},\quad \frac{a}{c},\quad \frac{b}{a},\quad \frac{c}{b},\quad \frac{c}{a},\quad \frac{a}{b}$$

FIGURE 13

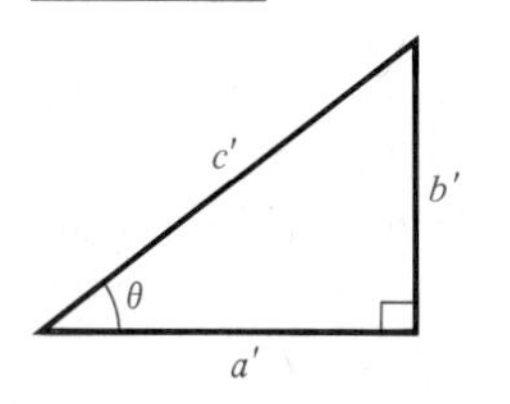

We can show that these ratios depend only on θ, and not on the size of the triangle, as follows. Suppose that θ is also an angle of a different right triangle, as shown in Figure 13. Since the two triangles have equal angles, they are similar, and therefore ratios of corresponding sides are proportional. For example,

$$\frac{b}{c} = \frac{b'}{c'}, \qquad \frac{a}{c} = \frac{a'}{c'}, \qquad \frac{b}{a} = \frac{b'}{a'}.$$

Thus, for each θ, the six ratios are uniquely determined and hence are functions of θ. They are called **trigonometric functions** and are designated as the **sine, cosine, tangent, cosecant, secant**, and **cotangent** functions, abbreviated **sin, cos, tan, csc, sec**, and **cot**, respectively. The symbol sin (θ), or sin θ, is used for the ratio b/c, which the sine function associates with θ. Values of the other five functions are denoted in similar fashion. To summarize, if θ is an acute angle of a right triangle, as in Figure 12, then by definition:

$$\sin\theta = \frac{b}{c} \qquad \csc\theta = \frac{c}{b}$$

$$\cos\theta = \frac{a}{c} \qquad \sec\theta = \frac{c}{a}$$

$$\tan\theta = \frac{b}{a} \qquad \cot\theta = \frac{a}{b}$$

The domain of each of the six trigonometric functions is the set of all acute angles. In Section 6.4 we will extend the domains to larger sets of angles and to real numbers.

If θ is the angle in Figure 12, we refer to the sides of the triangle of lengths a, b, and c as the **adjacent side, opposite side**, and **hypotenuse**, respectively. We shall use **adj, opp**, and **hyp** to denote the lengths of the sides. We may then represent the triangle as in Figure 14. Using this notation, the trigonometric functions may be expressed as follows.

FIGURE 14

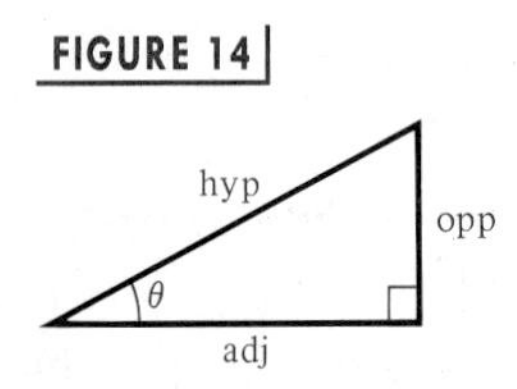

TRIGONOMETRIC FUNCTIONS OF ACUTE ANGLES

$$\sin\theta = \frac{\text{opp}}{\text{hyp}} \qquad \csc\theta = \frac{\text{hyp}}{\text{opp}}$$

$$\cos\theta = \frac{\text{adj}}{\text{hyp}} \qquad \sec\theta = \frac{\text{hyp}}{\text{adj}}$$

$$\tan\theta = \frac{\text{opp}}{\text{adj}} \qquad \cot\theta = \frac{\text{adj}}{\text{opp}}$$

These formulas can be applied to any right triangle without attaching the labels a, b, c to the sides. Since the lengths of the sides of a triangle are positive real numbers, *the values of the six trigonometric functions are positive for every acute angle* θ. Moreover, the hypotenuse is always greater than the adjacent or opposite side, and hence, $\sin\theta < 1$, $\cos\theta < 1$, $\csc\theta > 1$, and $\sec\theta > 1$ for every acute angle θ.

Note that since

$$\csc\theta = \frac{\text{hyp}}{\text{opp}} \quad \text{and} \quad \sin\theta = \frac{\text{opp}}{\text{hyp}},$$

$\sin\theta$ and $\csc\theta$ are reciprocals of one another, which gives us the two formulas in the first row of the next box. Similarly, $\cos\theta$ and $\sec\theta$ are reciprocals of one another, as are $\tan\theta$ and $\cot\theta$.

RECIPROCAL RELATIONSHIPS

$$\csc\theta = \frac{1}{\sin\theta} \qquad \sin\theta = \frac{1}{\csc\theta}$$

$$\sec\theta = \frac{1}{\cos\theta} \qquad \cos\theta = \frac{1}{\sec\theta}$$

$$\cot\theta = \frac{1}{\tan\theta} \qquad \tan\theta = \frac{1}{\cot\theta}$$

Several other important relationships among the trigonometric functions will be discussed at the end of this section.

EXAMPLE ▪ 1

If θ is an acute angle and $\cos\theta = \frac{3}{4}$, find the values of the trigonometric functions of θ.

SOLUTION We begin by sketching a right triangle having an acute angle θ with adj = 3 and hyp = 4, as shown in Figure 15. By the Pythagorean theorem,

FIGURE 15

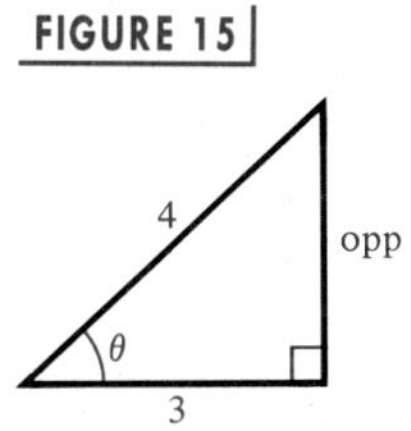

$$3^2 + (\text{opp})^2 = 4^2$$
$$(\text{opp})^2 = 16 - 9 = 7.$$

Taking square roots,

$$\text{opp} = \sqrt{7}.$$

Applying the definitions of trigonometric functions,

$$\sin\theta = \frac{\sqrt{7}}{4} \qquad \csc\theta = \frac{4}{\sqrt{7}} = \frac{4\sqrt{7}}{7}$$

$$\cos\theta = \frac{3}{4} \qquad \sec\theta = \frac{4}{3}$$

$$\tan\theta = \frac{\sqrt{7}}{3} \qquad \cot\theta = \frac{3}{\sqrt{7}} = \frac{3\sqrt{7}}{7}$$

EXAMPLE ▪ 2

Find the values of the trigonometric functions that correspond to θ:

(a) $\theta = 60^\circ$ **(b)** $\theta = 30^\circ$ **(c)** $\theta = 45^\circ$

SOLUTION Consider an equilateral triangle with sides of length 2. The median from one vertex to the opposite side bisects the angle at that vertex, as illustrated by the dashes in Figure 16. By the Pythagorean theorem, the side opposite 60° in the shaded right triangle has length $\sqrt{3}$. Using the preceding formulas, we obtain the values corresponding to 60° and 30° as follows:

FIGURE 16

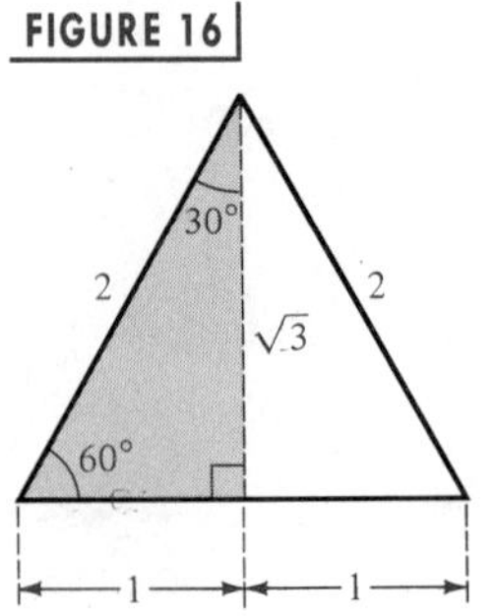

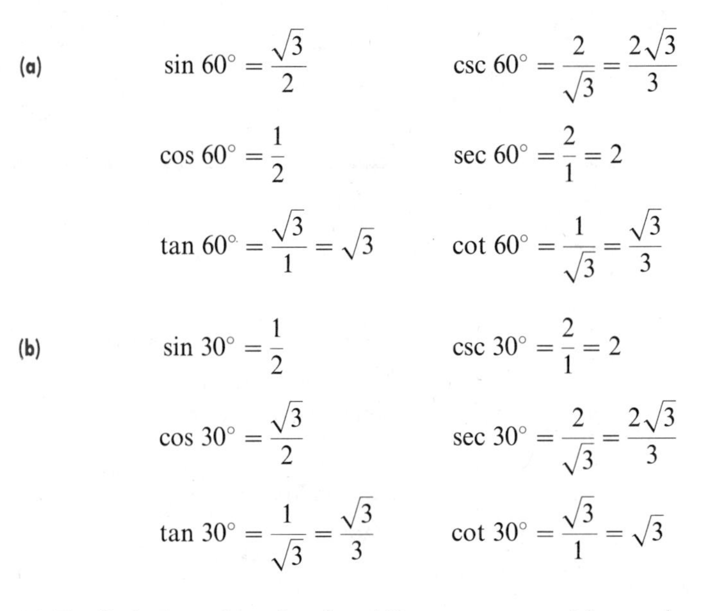

(a)

$$\sin 60^\circ = \frac{\sqrt{3}}{2} \qquad \csc 60^\circ = \frac{2}{\sqrt{3}} = \frac{2\sqrt{3}}{3}$$

$$\cos 60^\circ = \frac{1}{2} \qquad \sec 60^\circ = \frac{2}{1} = 2$$

$$\tan 60^\circ = \frac{\sqrt{3}}{1} = \sqrt{3} \qquad \cot 60^\circ = \frac{1}{\sqrt{3}} = \frac{\sqrt{3}}{3}$$

(b)

$$\sin 30^\circ = \frac{1}{2} \qquad \csc 30^\circ = \frac{2}{1} = 2$$

$$\cos 30^\circ = \frac{\sqrt{3}}{2} \qquad \sec 30^\circ = \frac{2}{\sqrt{3}} = \frac{2\sqrt{3}}{3}$$

$$\tan 30^\circ = \frac{1}{\sqrt{3}} = \frac{\sqrt{3}}{3} \qquad \cot 30^\circ = \frac{\sqrt{3}}{1} = \sqrt{3}$$

(c) To find the values for $\theta = 45^\circ$, we may consider an isosceles right triangle whose two equal sides have length 1, as illustrated in Figure 17. By the Pythagorean theorem, the length of the hypotenuse is $\sqrt{2}$. Hence, the values corresponding to 45° are:

FIGURE 17

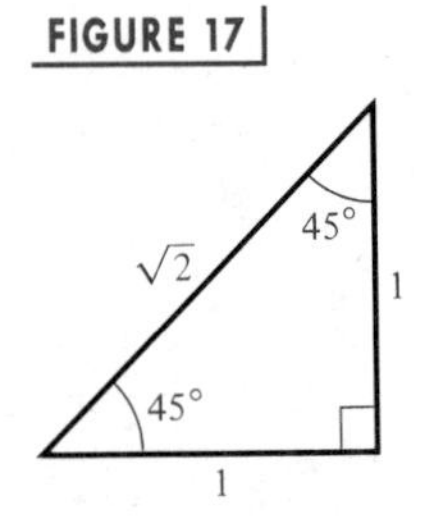

$$\sin 45^\circ = \frac{1}{\sqrt{2}} = \frac{\sqrt{2}}{2} = \cos 45^\circ \qquad \tan 45^\circ = \frac{1}{1} = 1$$

$$\csc 45^\circ = \frac{\sqrt{2}}{1} = \sqrt{2} = \sec 45^\circ \qquad \cot 45^\circ = \frac{1}{1} = 1$$

For reference, we list the values found in Example 2 in the following table, together with the radian measures of the angles. Two reasons for stressing these values are that they are exact and they occur frequently in work involving trigonometry. Because of the importance of these special values, it is a good idea either to memorize the table or to be able to find the values quickly by using triangles as in Example 2.

SPECIAL VALUES OF THE TRIGONOMETRIC FUNCTIONS

θ (radians)	θ (degrees)	$\sin\theta$	$\cos\theta$	$\tan\theta$	$\cot\theta$	$\sec\theta$	$\csc\theta$
$\frac{\pi}{6}$	30°	$\frac{1}{2}$	$\frac{\sqrt{3}}{2}$	$\frac{\sqrt{3}}{3}$	$\sqrt{3}$	$\frac{2\sqrt{3}}{3}$	2
$\frac{\pi}{4}$	45°	$\frac{\sqrt{2}}{2}$	$\frac{\sqrt{2}}{2}$	1	1	$\sqrt{2}$	$\sqrt{2}$
$\frac{\pi}{3}$	60°	$\frac{\sqrt{3}}{2}$	$\frac{1}{2}$	$\sqrt{3}$	$\frac{\sqrt{3}}{3}$	2	$\frac{2\sqrt{3}}{3}$

It is possible to approximate, to any degree of accuracy, the values of the trigonometric functions for any acute angle. Table 4 in Appendix II gives four-decimal-place approximations to many values. The use of Table 4 is explained in Appendix I. Table 5 can be used to find function values if θ is a two-decimal-place approximation of the radian measure of an angle and if $0 < \theta < 1.57$. (Note that $1.57 \approx \pi/2$.)

Scientific calculators have keys labeled SIN, COS, and TAN that can be used to approximate values of these functions. The values of csc, sec, and cot may then be found by means of the reciprocal key. *Before using a calculator to find function values that correspond to the radian measure of an acute angle, be sure that the calculator is in radian mode. For values corresponding to degree measure, select degree mode.*

As an illustration, to find $\sin 30^\circ$ on a typical calculator, place the calculator in degree mode, enter the number 30, and press the SIN key, obtaining $\sin 30^\circ = 0.5$, which is the exact value. Using the same procedure for 60° we obtain a decimal approximation to $\sqrt{3}/2$, such as

$$\sin 60^\circ \approx 0.8660254.$$

Similarly, to find a value such as $\cos 1.3$, where 1.3 is the radian measure of an acute angle, we place the calculator in radian mode, enter 1.3, and press COS, obtaining

$$\cos 1.3 \approx 0.2674988.$$

EXAMPLE ▪ 3

Approximate sec 54.8°.

SOLUTION Place a calculator in degree mode and use the reciprocal relationship $\sec\theta = 1/\cos\theta$:

Enter 54.8 :	54.8	(degree value of θ)
Press COS :	0.5764323	(value of $\cos\theta$)
Press 1/x :	1.7348091	(value of $1/\cos\theta$)

Hence $\sec 54.8° \approx 1.7348091$.

EXAMPLE ▪ 4

Approximate tan 67°29′.

SOLUTION For some calculators it is necessary to first express the angle 67°29′ as a decimal. In this case we calculate

$$29' = \left(\frac{29}{60}\right)^{\circ} \approx 0.4833333°$$

and then obtain

$$\tan 67°29' \approx \tan 67.4833333° \approx 2.4122286.$$

For other calculators the conversion of 67°29′ to 67.4833333° can be accomplished by using appropriate keys.

Inverse functions that correspond to trigonometric functions are discussed in Section 7.6, where we will define a function denoted by $\sin^{-1}$ such that

$$\sin^{-1}(\sin\theta) = \theta$$

provided

$$-\frac{\pi}{2} \le \theta \le \frac{\pi}{2} \quad \text{or} \quad -90° \le \theta \le 90°.$$

This fact can be used to find θ when given $\sin\theta$. For our present purpose, we need not be concerned about the restriction $-\pi/2 \le \theta \le \pi/2$ or $-90° \le \theta \le 90°$, since we are interested only in acute angles. The reason for this restriction is discussed in Section 7.6, where *inverse trigonometric functions* are treated in detail.

With some calculators, to use $\sin^{-1}$ we first press INV (for inverse) and then SIN. This technique is analogous to that used in Section 5.5, where given $\log x$, we pressed INV LOG to approximate x. Some calculators have a key labeled SIN^{-1} or ARCSIN that can be used to find values of $\sin^{-1}$ directly.

EXAMPLE ▪ 5

Let θ be an acute angle such that $\sin\theta = 0.5$. Use a calculator to approximate the measure of θ in **(a)** degrees and **(b)** radians.

SOLUTION

(a) Place the calculator in degree mode:

Enter 0.5 :	0.5	(the value of $\sin\theta$)
Press INV SIN :	30	(the degree measure of θ)

(b) Place the calculator in radian mode:

Enter 0.5 :	0.5	(the value of $\sin\theta$)
Press INV SIN :	0.5235988	(the radian measure of θ)

The last number is a decimal approximation for an angle of radian measure $\pi/6$.

In Section 7.6 we will also define functions denoted by $\cos^{-1}$ and $\tan^{-1}$ such that

$$\cos^{-1}(\cos\theta) = \theta \quad \text{if} \quad 0 \le \theta \le \pi \quad \text{or} \quad 0^\circ \le \theta \le 180^\circ$$

and

$$\tan^{-1}(\tan\theta) = \theta \quad \text{if} \quad -\frac{\pi}{2} < \theta < \frac{\pi}{2} \quad \text{or} \quad -90^\circ < \theta < 90^\circ.$$

These functions may be used in the same manner that $\sin^{-1}$ was used in Example 3.

EXAMPLE ▪ 6

Let θ be an acute angle such that $\cos\theta = 0.4271$. Approximate the degree measure of θ to the nearest **(a)** $(0.01)^\circ$; **(b)** $1'$.

SOLUTION

(a) Place a calculator in degree mode:

Enter 0.4271 :	0.4271	(the value of $\cos\theta$)
Press INV COS :	64.716341	(the degree measure of θ)

Thus $\theta \approx 64.72^\circ$.

(b) To find θ to the nearest $1'$ we use the method shown in Example 4 of the preceding section:

$$\begin{aligned}(0.716341)^\circ &= (0.716341)(1^\circ)\\ &= (0.716341)(60')\\ &= 42.98046' \approx 43'\end{aligned}$$

Hence $\theta \approx 64^\circ 43'$.

We can also use Table 4, together with the method of interpolation (see Appendix I), to approximate θ.

The formulas listed in the next box are, without doubt, the most important identities in trigonometry, because they may be used to simplify and unify many different aspects of the subject. Since the formulas are part of the foundation for work in trigonometry, they are called the *fundamental identities.*

Three of the fundamental identities involve squares, such as $(\sin\theta)^2$ and $(\cos\theta)^2$. In general, if n is an integer different from -1, then a power such as $(\cos\theta)^n$ is written $\cos^n\theta$. The symbols $\sin^{-1}\theta$ and $\cos^{-1}\theta$ are reserved for inverse trigonometric functions, which we will discuss in the next chapter. With this agreement on notation we have, for example,

$$\begin{aligned}\cos^2\theta &= (\cos\theta)^2 = (\cos\theta)(\cos\theta)\\ \tan^3\theta &= (\tan\theta)^3 = (\tan\theta)(\tan\theta)(\tan\theta)\\ \sec^4\theta &= (\sec\theta)^4 = (\sec\theta)(\sec\theta)(\sec\theta)(\sec\theta).\end{aligned}$$

Let us first list all the fundamental identities and then discuss the proofs. These formulas are true for every acute angle θ. We shall also see later that they are true for other angles and for real numbers.

FUNDAMENTAL IDENTITIES

$\csc\theta = \dfrac{1}{\sin\theta}$	$\tan\theta = \dfrac{\sin\theta}{\cos\theta}$	$\sin^2\theta + \cos^2\theta = 1$
$\sec\theta = \dfrac{1}{\cos\theta}$	$\cot\theta = \dfrac{\cos\theta}{\sin\theta}$	$1 + \tan^2\theta = \sec^2\theta$
$\cot\theta = \dfrac{1}{\tan\theta}$		$1 + \cot^2\theta = \csc^2\theta$

FIGURE 18

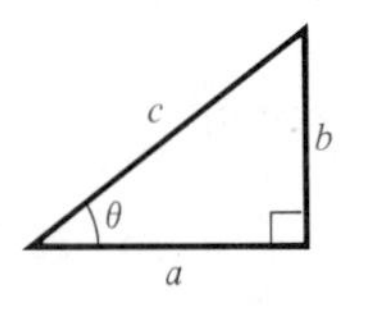

PROOF The three formulas in the left-hand column are reciprocal relationships, which we discussed earlier. The others can be proved by referring to the right triangle in Figure 18. Thus

$$\tan\theta = \frac{b}{a} = \frac{b/c}{a/c} = \frac{\sin\theta}{\cos\theta}.$$

and $$\cot\theta = \frac{1}{\tan\theta} = \frac{\cos\theta}{\sin\theta}.$$

If we apply the Pythagorean theorem to the triangle in Figure 18, we obtain

$$b^2 + a^2 = c^2.$$

Dividing both sides by c^2 leads to

$$\left(\frac{b}{c}\right)^2 + \left(\frac{a}{c}\right)^2 = 1$$

$$\sin^2\theta + \cos^2\theta = 1.$$

Dividing both sides of the last equation by $\cos^2\theta$ gives us

$$\frac{\sin^2\theta}{\cos^2\theta} + \frac{\cos^2\theta}{\cos^2\theta} = \frac{1}{\cos^2\theta}$$

and $$\left(\frac{\sin\theta}{\cos\theta}\right)^2 + 1 = \left(\frac{1}{\cos\theta}\right)^2.$$

Since $\sin\theta/\cos\theta = \tan\theta$ and $1/\cos\theta = \sec\theta$, we see that

$$\tan^2\theta + 1 = \sec^2\theta.$$

The proof that $1 + \cot^2\theta = \csc^2\theta$ is left as an exercise. ❑

We can use fundamental identities to express each trigonometric function in terms of any other trigonometric function. This is illustrated in the next two examples (see also Exercises 37–44).

EXAMPLE ▪ 7

If θ is an acute angle, express $\sin\theta$ in terms of $\cos\theta$.

SOLUTION We first rewrite the fundamental identity $\sin^2\theta + \cos^2\theta = 1$ as

$$\sin^2\theta = 1 - \cos^2\theta.$$

Taking the square root of both sides, we obtain

$$\sin\theta = \pm\sqrt{1 - \cos^2\theta}.$$

Since θ is an acute angle, $\sin\theta$ is positive, and hence the minus sign is extraneous. Thus

$$\sin\theta = \sqrt{1 - \cos^2\theta} \quad \text{for } 0 < \theta < \frac{\pi}{2}.$$

EXAMPLE ▪ 8

If θ is an acute angle, express $\tan\theta$ in terms of $\sin\theta$.

SOLUTION We begin with the fundamental identity

$$\tan\theta = \frac{\sin\theta}{\cos\theta}.$$

Next we express $\cos\theta$ in terms of $\sin\theta$. Since $\sin^2\theta + \cos^2\theta = 1$, we may write

$$\cos^2\theta = 1 - \sin^2\theta.$$

Taking square roots,

$$\cos\theta = \pm\sqrt{1 - \sin^2\theta}.$$

As in Example 7 we may discard the minus sign, since θ is acute. Hence

$$\tan\theta = \frac{\sin \partial}{\cos\theta} = \frac{\sin\theta}{\sqrt{1 - \sin^2\theta}} \quad \text{for } 0 < \theta < \frac{\pi}{2}.$$

In Chapter 7 we will use fundamental identities to simplify complicated trigonometric expressions.

6.2 EXERCISES

Exer. 1–10: Find the values of the trigonometric functions for the angle θ.

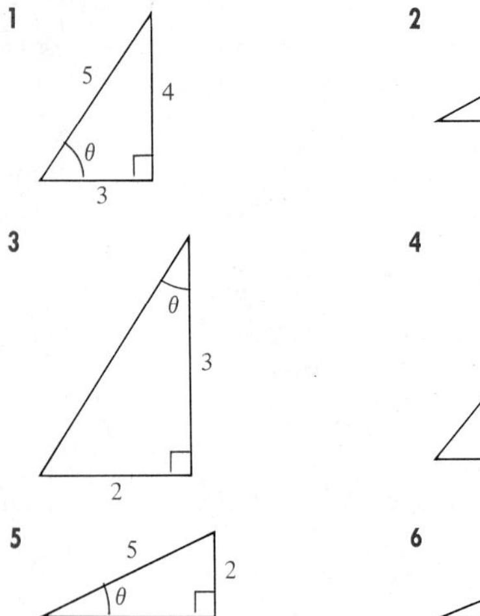

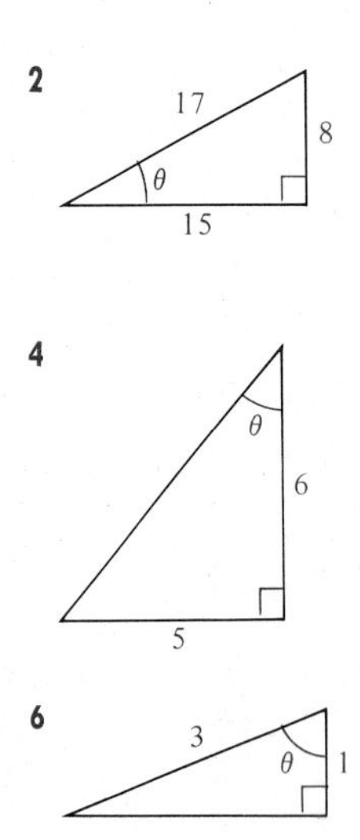

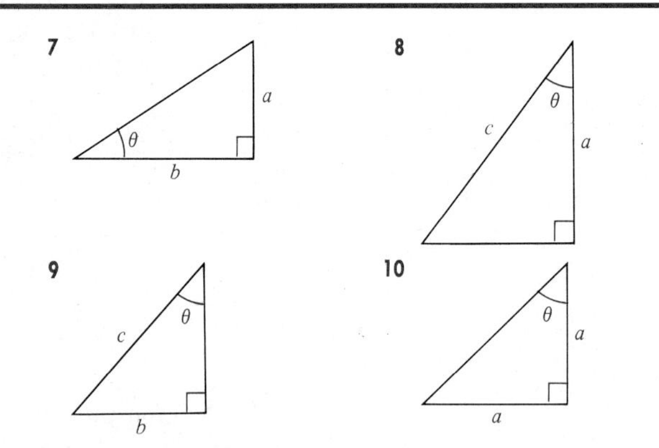

Exer. 11–16: Find the values of the trigonometric functions for the acute angle θ.

11 $\sin\theta = \frac{3}{5}$ **12** $\cos\theta = \frac{8}{17}$ **13** $\sec\theta = \frac{6}{5}$

14 $\csc\theta = 4$ **15** $\cos\theta = \frac{\sqrt{2}}{2}$ **16** $\sin\theta = \frac{\sqrt{3}}{2}$

Exer. 17–28: Approximate the function value to three decimal places.

17 $\cos 38°30'$
18 $\sin 73°20'$
19 $\cot 9°10'$
20 $\csc 43°40'$
21 $\sec 67°50'$
22 $\tan 21°10'$
23 $\cos 0.68$
24 $\sin 1.48$
25 $\cot 1.13$
26 $\csc 0.32$
27 $\sec 0.26$
28 $\tan 0.75$

Exer. 29–36: Use a calculator to approximate the acute angle θ to the nearest (a) $(0.01)°$ and (b) $1'$.

29 $\cos \theta = 0.8620$
30 $\sin \theta = 0.6612$
31 $\tan \theta = 3.7$
32 $\cos \theta = 0.8$
33 $\sin \theta = 0.4217$
34 $\tan \theta = 4.91$
35 $\sec \theta = 4.246$
36 $\csc \theta = 11$

Exer. 37–44: Use fundamental identities to write the first expression in terms of the second, for any acute angle θ.

37 $\cot \theta,\ \sin \theta$
38 $\tan \theta,\ \cos \theta$
39 $\sec \theta,\ \sin \theta$
40 $\csc \theta,\ \cos \theta$
41 $\tan \theta,\ \sec \theta$
42 $\cot \theta,\ \csc \theta$
43 $\sin \theta,\ \sec \theta$
44 $\cos \theta,\ \cot \theta$

Exer. 45–48: Prove that the formula is true for every acute angle θ.

45 (a) $\sin \theta \csc \theta = 1$
(b) $\cos \theta \sec \theta = 1$
(c) $\tan \theta \cot \theta = 1$

46 $1 + \cot^2 \theta = \csc^2 \theta$

47 $\log \csc \theta = -\log \sin \theta$

48 $\log \tan \theta = \log \sin \theta - \log \cos \theta$

49 Scientific calculators have [SIN], [COS], [TAN], [1/x] and [x²] keys.
(a) After entering θ, which key is pressed first to calculate $\cos^2 \theta$?
(b) How can $\csc \theta$ be calculated?
(c) How can $\cot^4 \theta$ be calculated?

50 If θ is the radian measure of an acute angle, the fifth-degree polynomial $P(\theta) = \theta - \frac{1}{6}\theta^3 + \frac{1}{120}\theta^5$ is an excellent approximation to $\sin \theta$ for $0 < \theta < \pi/4$. In fact, it gives four-decimal-place accuracy. Use $P(\theta)$ to estimate $\sin 0.5$ and compare the result to that obtained using the [SIN] key on a calculator.

6.3 APPLICATIONS INVOLVING RIGHT TRIANGLES

Trigonometry was developed to help solve problems involving angles and lengths of sides of triangles. Problems of that type are no longer the most important applications; however, questions about triangles still arise in physical situations. We shall restrict our discussion in this section to right triangles. Triangles that do not contain a right angle will be considered in Chapter 8.

We shall often use the following notation. The vertices of a triangle will be denoted by A, B, and C. The angles at A, B, and C will be denoted by α, β, and γ, respectively, and the lengths of the sides opposite these angles by a, b, and c, respectively. The triangle itself will be referred to as *triangle ABC*. If a triangle is a right triangle and if one of the acute angles and a side are known, or if two sides are given, then we may use the formulas in Section 6.2 that express the trigonometric functions as ratios of sides of a triangle to find the remaining parts.

In all examples *it is assumed that you know how to find trigonometric function values and angles by using either a calculator, tables, or results about special angles.*

EXAMPLE ▪ 1

In triangle ABC, $\gamma = 90°$, $\alpha = 34°$, and $b = 10.5$. Approximate the remaining parts.

SOLUTION Since the sum of the angles is 180° and $\alpha + \gamma = 124°$,

$$\beta = 180° - (\alpha + \gamma) = 180° - 124° = 56°.$$

FIGURE 19

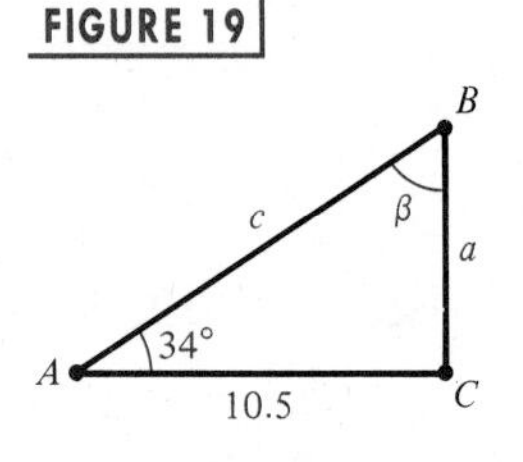

Referring to Figure 19 and using $\tan \alpha = \text{opp/adj}$, we obtain

$$\tan 34° = \frac{a}{10.5}.$$

Multiplying both sides by 10.5,

$$a = (10.5) \tan 34°$$

$$a \approx (10.5)(0.6745) \approx 7.1.$$

To find side c we can use either the cosine or secant functions. Using $\cos \alpha = \text{adj/hyp}$:

$$\cos 34° = \frac{10.5}{c}.$$

$$c = \frac{10.5}{\cos 34°}$$

$$c \approx \frac{10.5}{0.8290} \approx 12.7.$$

Using $\sec \alpha = \text{hyp/adj}$:

$$\sec 34° = \frac{c}{10.5}$$

$$c = (10.5) \sec 34°$$

$$c \approx (10.5)(1.2062) \approx 12.7$$

As illustrated in Example 1, when working with triangles, we usually round off answers. One reason for doing so is that in most applications, the lengths of sides of triangles and measures of angles are found by mechanical devices, and hence are only approximations to the exact values. Consequently, a number such as 10.5 in Example 1 is assumed to have been rounded off to the nearest tenth. We cannot expect more accuracy in the calculated values for the remaining sides and, therefore, they should also be rounded off to the nearest tenth.

In some problems a number with many digits, such as 36.4635, may be given for the side of a triangle. If a calculator is used, the number may be entered as usual; however, if Table 4 is employed, a number of this type should be rounded off before beginning any calculations. If a number a is written in the scientific form $a = c \times 10^k$ with $1 \le c < 10$, then before using Table 4, c should be rounded off to three decimal places. Another way of saying this is that x should be rounded off to four *significant figures*. Some examples will help to clarify the procedure. If $c = 36.4635$, we round off to 36.46. The number 684,279 should be rounded off to 684,300. For a decimal such as 0.096202 we use 0.09620. The reason for rounding off is that the values of the trigonometric functions in Table 4 have been rounded off to four significant figures, and hence we cannot expect more than four-figure accuracy in our computations.

When finding angles, answers should be rounded off as indicated in the following table.

Number of significant figures for sides	Round off degree measure of angles to the nearest
2	$1°$
3	$10'$ or $0.1°$
4	$1'$ or $0.01°$

Justification of this table requires a careful analysis of problems that involve approximate data.

EXAMPLE ▪ 2

In triangle ABC, $\gamma = 90°$, $a = 12.3$, and $b = 31.6$. Approximate the remaining parts.

FIGURE 20

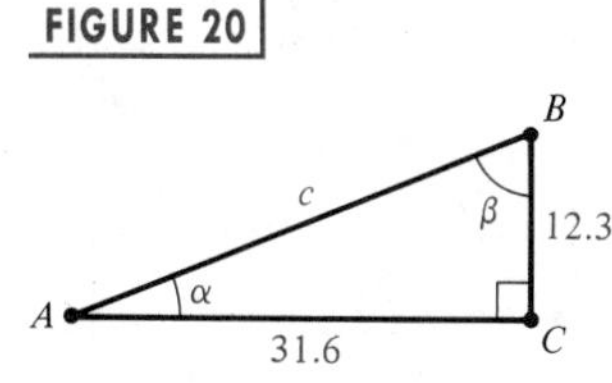

SOLUTION Referring to the triangle illustrated in Figure 20,

$$\tan \alpha = \frac{12.3}{31.6} \approx 0.3892.$$

Since the sides are given to three significant figures, the rule stated in the preceding table tells us that α should be rounded off to the nearest multiple of $10'$. If a calculator is used to approximate α, we use the degree mode and proceed as follows:

Enter 0.3892:	0.3892	(the value of $\tan \alpha$)
Press INV TAN:	$21.265988 \approx 21°16'$	
	$\approx 21°20'$	(the approximation to α)

Table 4 can also be used to find α.

Since $\alpha \approx 21°20'$,

$$\beta \approx 90° - 21°20', \quad \text{or} \quad \beta \approx 68°40'.$$

Again referring to Figure 20,

$$\sec \alpha = \frac{c}{31.6}, \quad \text{or} \quad c = (31.6) \sec \alpha$$

Thus $c \approx (31.6) \sec 21°20' \approx (31.6)(1.0736) \approx 33.9.$

Side c can also be found using other trigonometric functions. Thus from Figure 20 we see that

$$\cos \alpha = \frac{31.6}{c}$$

and hence $$c = \frac{31.6}{\cos 21°20'} \approx \frac{31.6}{0.9315} \approx 33.9.$$

Still another way to find c is to use the Pythagorean theorem:

$$c = \sqrt{(31.6)^2 + (12.3)^2} = \sqrt{1149.85} \approx 33.9$$

As illustrated in Figure 21, if an observer at point X sights an object, then the angle that the line of sight makes with the horizontal line l is the **angle of elevation** or the **angle of depression** of the object, if the object is above or below the horizontal line, respectively. We use this terminology in the next two examples.

FIGURE 21

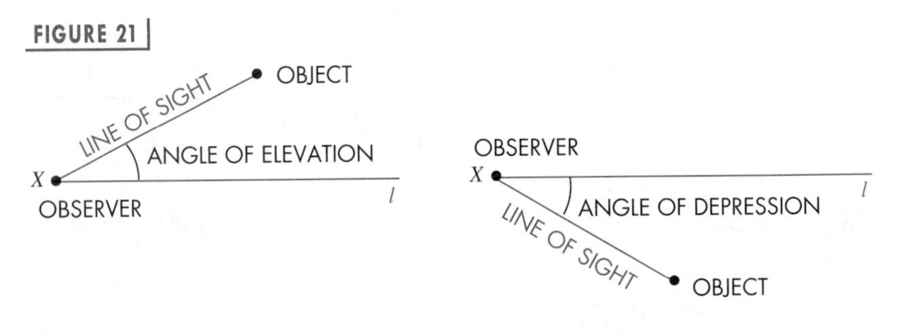

EXAMPLE ▪ 3

From a point on level ground 135 feet from the base of a tower, the angle of elevation of the top of the tower is 57°20′. Approximate the height of the tower.

FIGURE 22

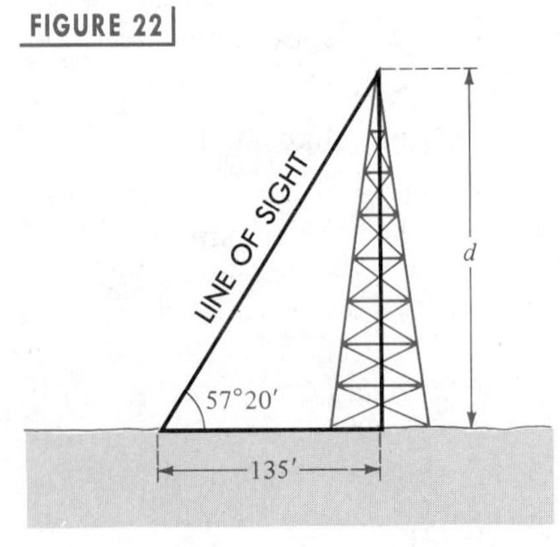

SOLUTION If we let d denote the height of the tower, then the given facts are represented by the triangle in Figure 22. Referring to the figure, we see that

$$\tan 57°20' = \frac{d}{135}, \quad \text{or} \quad d = (135) \tan 57°20'.$$

Thus, $d \approx (135)(1.560) \approx 211$ feet.

EXAMPLE ■ 4

From the top of a building that overlooks an ocean, an observer watches a boat sailing directly toward the building. If the observer is 100 feet above sea level and if the angle of depression of the boat changes from 25° to 40° during the period of observation, approximate the distance that the boat travels.

FIGURE 23

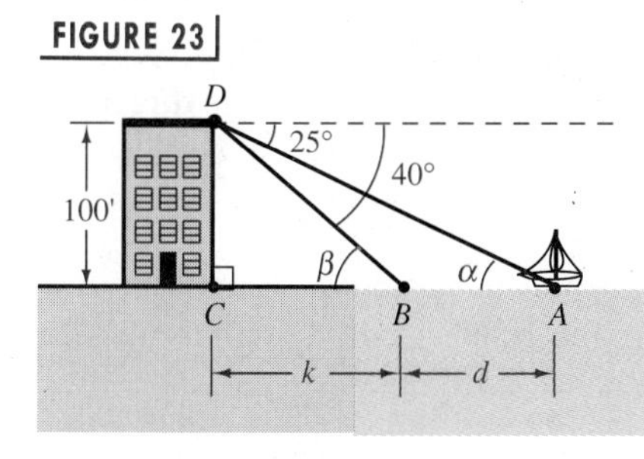

SOLUTION As in Figure 23, let A and B be the positions of the boat that correspond to the 25° and 40° angles, respectively. Suppose that the observer is at point D, and C is the point 100 feet directly below. Let d denote the distance the boat travels and let k denote the distance from B to C. If α and β denote angles DAC and DBC, respectively, then it follows from geometry that $\alpha = 25°$ and $\beta = 40°$.

From triangle BCD,

$$\cot \beta = \cot 40° = \frac{k}{100}$$

$$k = 100 \cot 40°.$$

From triangle DAC,

$$\cot \alpha = \cot 25° = \frac{d + k}{100}$$

$$d + k = 100 \cot 25°.$$

Consequently,

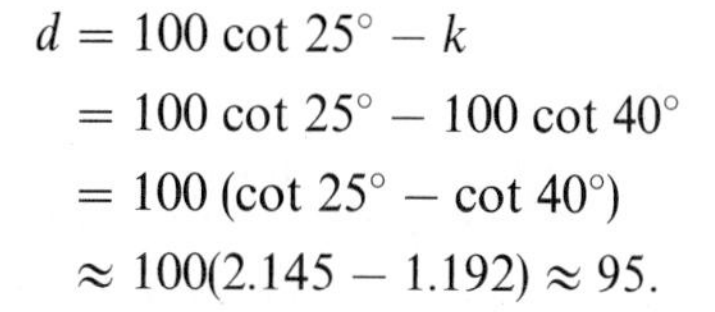

$$\begin{aligned} d &= 100 \cot 25° - k \\ &= 100 \cot 25° - 100 \cot 40° \\ &= 100 (\cot 25° - \cot 40°) \\ &\approx 100(2.145 - 1.192) \approx 95. \end{aligned}$$

Hence, $d \approx 95$ feet.

FIGURE 24

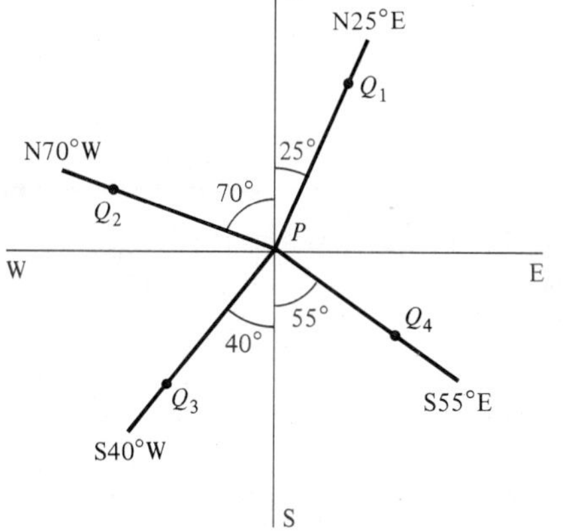

In certain navigation or surveying problems, the **direction**, or **bearing**, from a point P to a point Q is specified by stating the acute angle that segment PQ makes with the north-south line through P. We also state whether Q is north or south and east or west of P. Figure 24 illustrates four possibilities. The bearing from P to Q_1 is 25° east of north and is denoted by N25°E. We also refer to the **direction** N25°E, meaning the direction from P to Q_1. The bearings from P to Q_2, to Q_3, and to Q_4 are represented in a similar manner in the figure.

FIGURE 25

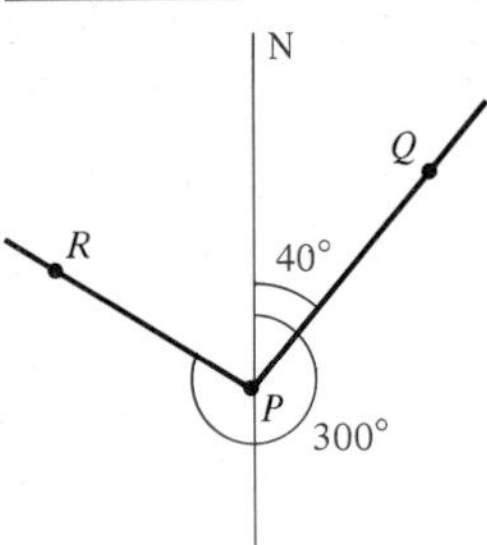

In air navigation, directions and bearings are specified by measuring from the north in a *clockwise* direction. In this case, a positive measure is assigned to the angle instead of the negative measure to which we are accustomed for clockwise rotations. Referring to Figure 25, we see that the direction of PQ is 40° and the direction of PR is 300°. These notations are used in Exercises 43–44.

EXAMPLE ▪ 5

Two ships leave port at the same time, one ship sailing in the direction N23°E at a speed of 11 mi/hr, and the second ship sailing in the direction S67°E at 15 mi/hr. Approximate the bearing from the second ship to the first, one hour later.

FIGURE 26

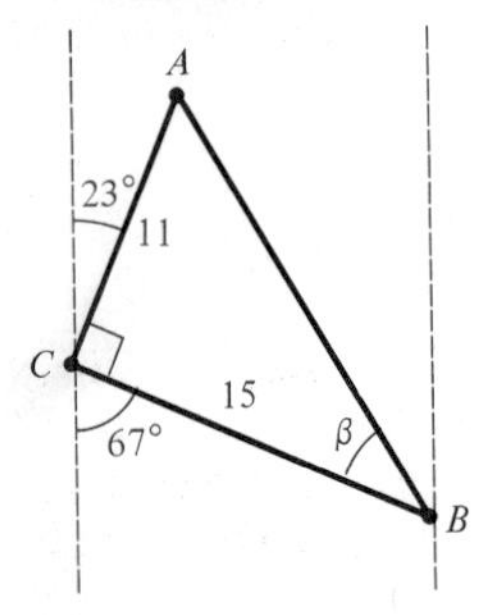

SOLUTION The sketch in Figure 26 indicates the positions of the first and second ships at points A and B after one hour. Point C represents the port. We wish to find the bearing from B to A. Note that

$$\angle ACB = 180° - (23° + 67°) = 90°$$

and hence triangle ABC is a right triangle. Thus

$$\tan \beta = \frac{11}{15} \approx 0.7333$$

and $$\beta \approx 36°.$$

FIGURE 27

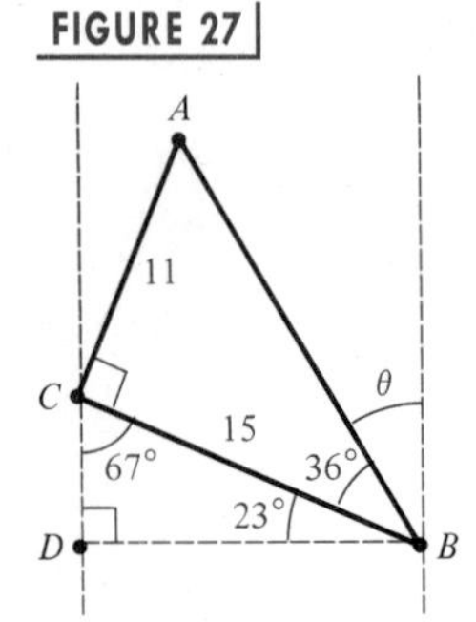

Referring to Figure 27 we see that

$$\angle CBD = 90° - 67° = 23°$$

and $$\angle ABD \approx 36° + 23°, \quad \text{or} \quad \angle ABD \approx 59°.$$

Hence $$\theta \approx 90° - 59°, \quad \text{or} \quad \theta \approx 31°,$$

which implies that the bearing from B to A is approximately N31°W.

6.3 EXERCISES

Exer. 1–8: Given the indicated parts of triangle ABC with $\gamma = 90°$, find the exact values of the remaining parts.

1 $\alpha = 30°, \quad b = 20$

2 $\beta = 45°, \quad b = 35$

3 $\beta = 45°, \quad c = 30$

4 $\alpha = 60°, \quad c = 6$

5 $a = 5, \quad b = 5$

6 $a = 4\sqrt{3}, \quad c = 8$

7 $b = 5\sqrt{3}, \quad c = 10\sqrt{3}$

8 $b = 7\sqrt{2}, \quad c = 14$

Exer. 9–16: Given the indicated parts of triangle ABC with $\gamma = 90°$, approximate the remaining parts.

9 $\alpha = 37°, \quad b = 24$

10 $\beta = 64°20', \quad a = 20.1$

11 $\beta = 71°51', \quad b = 240.0$

12 $\alpha = 31°10', \quad a = 510$

13 $a = 25, \quad b = 45$

14 $a = 31, \quad b = 9.0$

15 $c = 5.8, \quad b = 2.1$

16 $a = 0.42, \quad c = 0.68$

Exer. 17–24: Given the indicated parts of triangle ABC with $\gamma = 90°$, express the third part in terms of the first two.

17 $\alpha, c;\ \ b$

18 $\beta, c;\ \ b$

19 $\beta, b;\ \ a$

20 $\alpha, b;\ \ a$

21 $\alpha, a;\ \ c$

22 $\beta, a;\ \ c$

23 $a, c;\ \ b$

24 $a, b;\ \ c$

25 Approximate the angle of elevation of the sun if a boy 5.0 feet tall casts a shadow 4.0 feet long on level ground (see figure).

EXERCISE 25

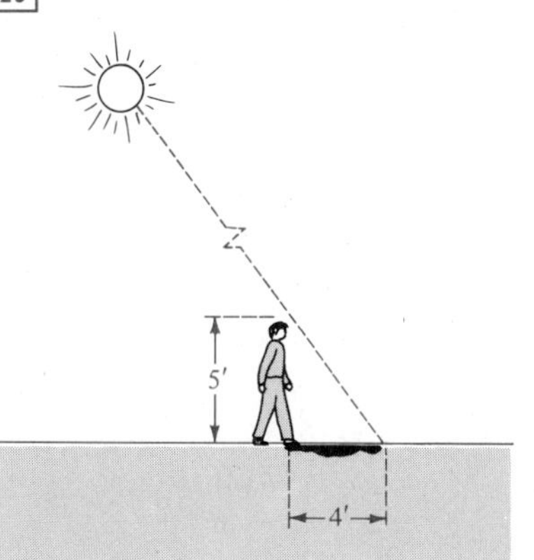

26 From a point 15 meters above level ground a surveyor measures the angle of depression of an object on the ground as 68°. Approximate the distance from the object to the point on the ground directly beneath the surveyor.

27 A girl flying a kite holds the string 4 feet above ground level. The string of the kite is taut and makes an angle of 60° with the horizontal (see figure). Approximate the height of the kite above level ground if 500 feet of string is payed out.

EXERCISE 27

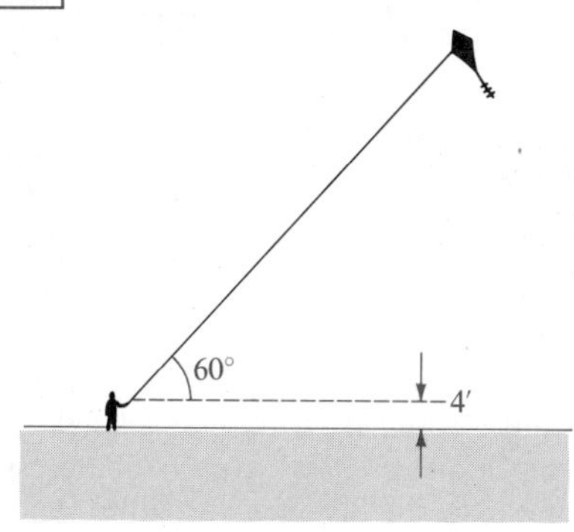

28 A regular octagon is inscribed in a circle of radius 12.0 cm. Approximate the perimeter of the octagon.

29 From a point P on level ground the angle of elevation of the top of a tower is 26°50′. From a point 25.0 meters closer to the tower and on the same line with P and the base of the tower, the angle of elevation of the top is 53°30′. Approximate the height of the tower.

30 A ladder 20 feet long leans against the side of a building and the angle between the ladder and the building is 22°.

(a) Approximate the distance from the bottom of the ladder to the building.

(b) If the distance from the bottom of the ladder to the building is increased by 3.0 feet, approximately how far does the top of the ladder move down the building?

31 As a hot-air balloon rises vertically, its angle of elevation from a point P on level ground 110 km from the point Q directly underneath the balloon changes from 19°20′ to 31°50′ (see figure). Approximately how far does the balloon rise during this period?

EXERCISE 31

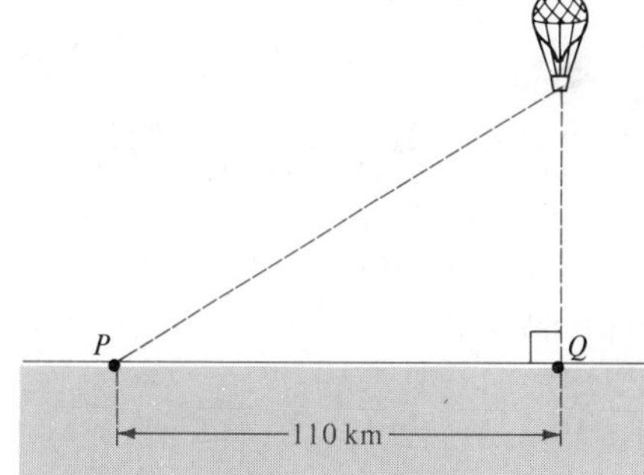

32 A guy wire is attached to the top of a radio antenna and to a point on horizontal ground that is 40.0 meters from the base of the antenna. If the wire makes an angle of 58°20′ with the ground, approximate the length of the wire.

33 The Pentagon is the largest office building in the world in terms of ground area. The perimeter of the building has the shape of a regular pentagon with each side 921 feet long. Find the area enclosed by the perimeter of the building.

34 From a point A that is 8.20 meters above level ground, the angle of elevation of the top of a building is $31°20'$ and the angle of depression of the base of the building is $12°50'$. Approximate the height of the building.

35 To find the distance d between two points P and Q on opposite shores of a lake, a surveyor locates a point R that is 50.0 meters from P such that RP is perpendicular to PQ as shown in the figure. Next, using a transit, the surveyor measures angle PRQ as $72°40'$. Find d.

EXERCISE 35

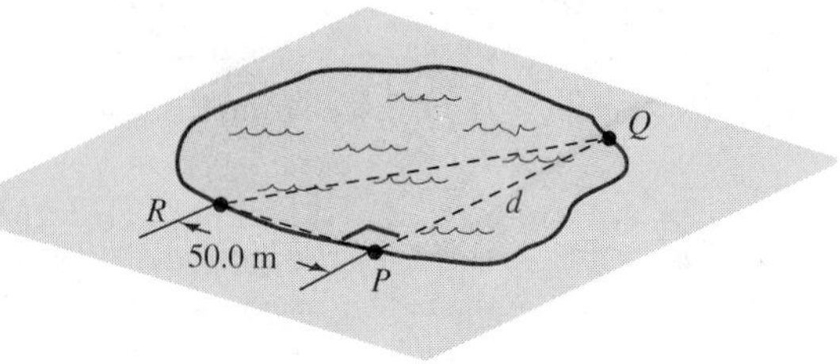

36 A builder wishes to construct a ramp 24 feet long that rises to a height of 5.0 feet above level ground. Approximate the angle that the ramp should make with the horizontal.

37 A rocket is fired at sea level and climbs at a constant angle of 75° through a distance of 10,000 feet. Approximate its altitude to the nearest foot.

38 A CB antenna is located on the top of a garage that is 16 feet tall. From a point on level ground that is 100 feet from a point directly below the antenna, the antenna subtends an angle of 12° as shown in the figure. Approximate the length of the antenna.

EXERCISE 38

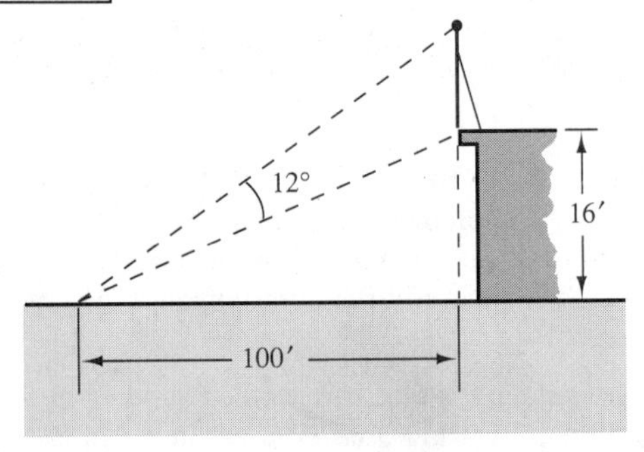

39 An airplane flying at an altitude of 10,000 feet passes directly over a fixed object on the ground. One minute later the angle of depression of the object is 42°. Approximate the speed of the airplane to the nearest mile per hour.

40 A motorist, traveling along a level highway at a speed of 60 km/hr directly toward a mountain, observes that between 1:00 P.M. and 1:10 P.M. the angle of elevation of the top of the mountain changes from 10° to 70°. Approximate the height of the mountain.

41 A pilot, flying at an altitude of 5000 feet, wishes to approach the numbers on a runway at an angle of 10°. Approximate, to the nearest 100 feet, the distance from the airplane to the numbers at the beginning of the descent.

42 To measure the height h of a cloud cover, a meteorology student directs a spotlight vertically upward from the ground. From a point on level ground that is d meters from the spotlight, the angle of elevation θ of the light image on the clouds is then measured (see figure).

(a) Express h in terms of d and θ.

(b) Approximate h if $d = 1000$ meters and $\theta = 59°$.

EXERCISE 42

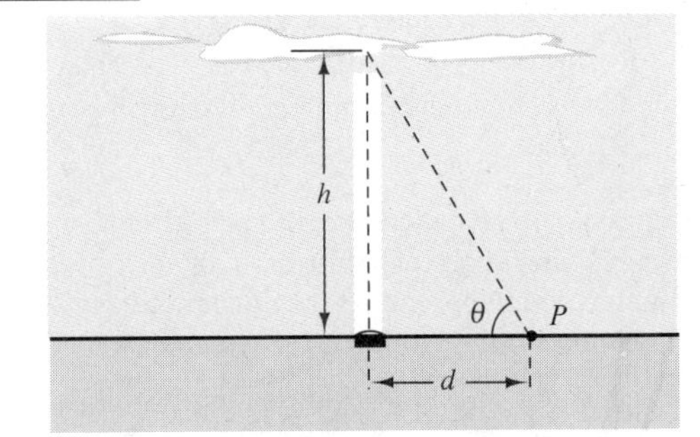

43 A ship leaves port at 1:00 P.M. and sails in the direction N34°W at a rate of 24 mi/hr. Another ship leaves port at 1:30 P.M. and sails in the direction N56°E at a rate of 18 mi/hr.

(a) Approximately how far apart are the ships at 3:00 P.M.?

(b) What is the bearing, to the nearest degree, from the first ship to the second?

44 From an observation point A a forest ranger sights a fire in the direction S35°50′W (see figure). From a point B, 5 miles due west of A, another ranger sights the same fire in the direction S54°10′E. Approximate, to the nearest tenth of a mile, the distance of the fire from A.

EXERCISE 44

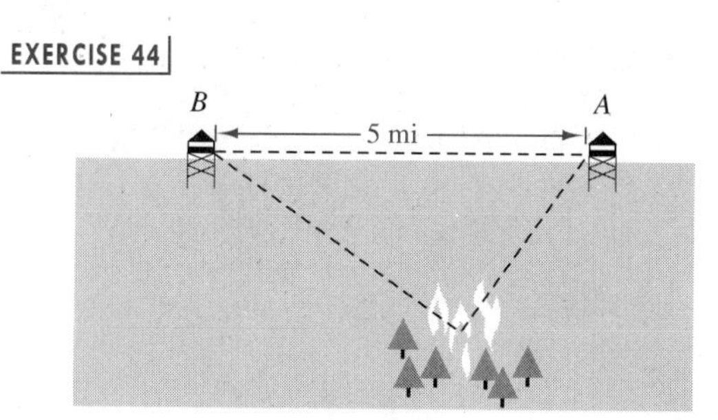

45 An airplane flying at a speed of 360 mi/hr flies from a point A in the direction 137° for 30 minutes and then flies in the direction 227° for 45 minutes. Approximate, to the nearest mile, the distance from the airplane to A.

46 An airplane takes off at a 10° angle and travels at the rate of 250 ft/sec. Approximately how long does the airplane take to reach an altitude of 15,000 feet?

47 A draw bridge is 150 feet long when stretched across a river. As shown in the figure, the two sections of the bridge can be rotated upward through an angle of 35°.

(a) If the water level is 15 feet below the closed bridge, find the distance d between the end of a section and the water level when the bridge is fully open.

(b) Approximately how far apart are the ends of the two sections when the bridge is fully opened, as shown in the figure?

EXERCISE 47

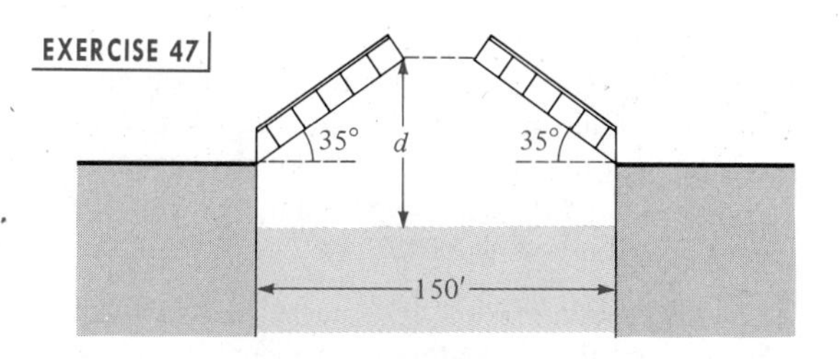

48 Shown in the figure is part of a design for a water slide. Find the total length of the slide to the nearest foot.

EXERCISE 48

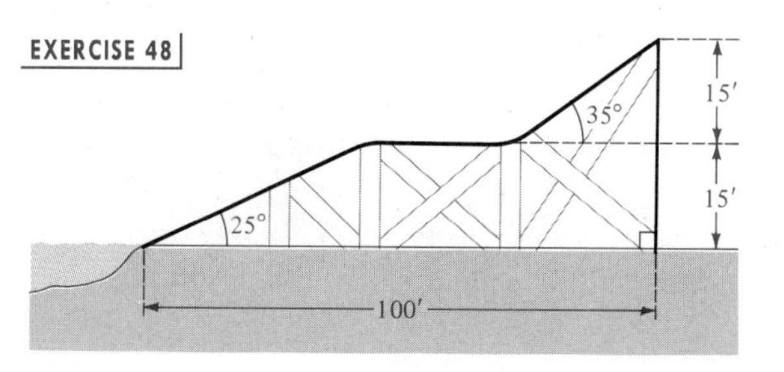

49 Shown in the figure is the screen for a simple video arcade game in which ducks move from A to B at the rate of 7 cm/sec. Bullets fired from point O travel 25 cm/sec. If a player shoots as soon as a duck appears at A, at what angle φ should the gun be aimed to score a direct hit?

EXERCISE 49

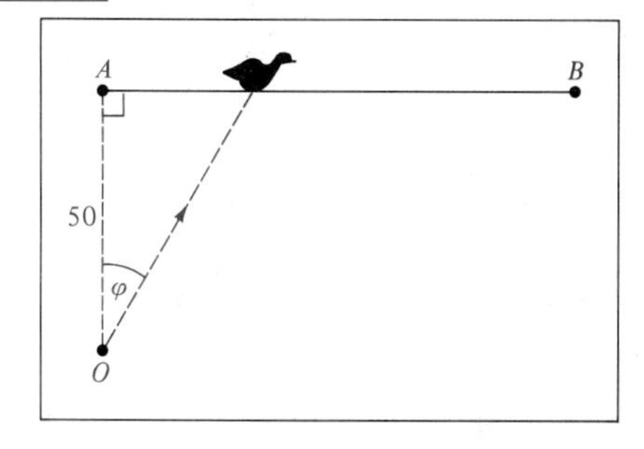

50 A conveyor belt 9 meters long can be hydraulically rotated up to an angle of 40° to unload cargo from airplanes (see figure).

(a) Find, to the nearest degree, the angle through which the conveyor belt should be rotated to reach a door that is 4 meters above the platform supporting the belt.

(b) Approximate the maximum height above the platform that the belt can reach.

EXERCISE 50

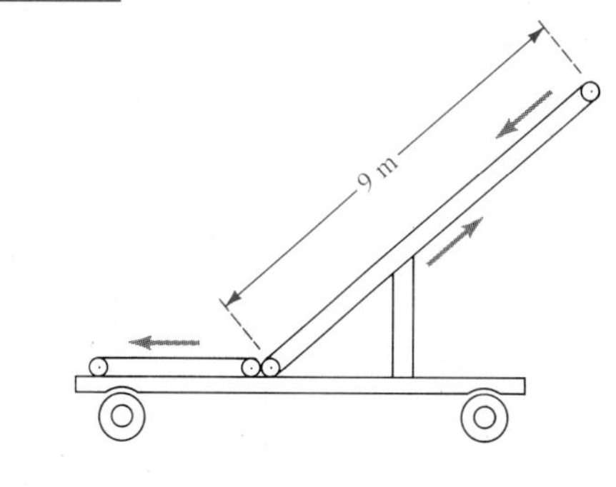

51 A rectangular box has dimensions $8 \times 6 \times 4$ in. Approximate, to the nearest tenth of a degree, the angle θ formed by a diagonal of the base and the diagonal of the box, as shown in the figure on page 308.

EXERCISE 51

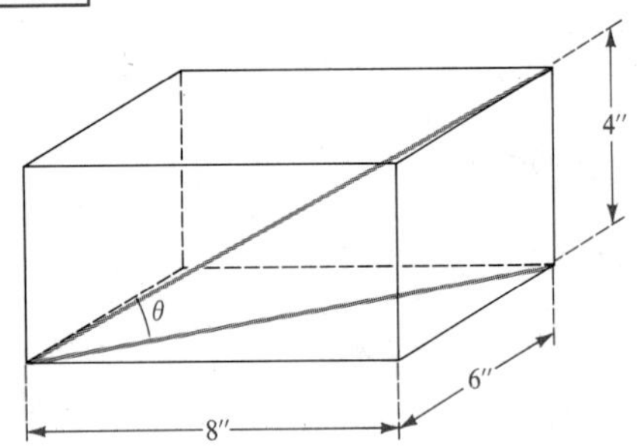

52 A conical paper cup has a radius of 2 inches. Approximate, to the nearest degree, the angle β (see figure) so that the cone will have a volume of 20 in.^{3}.

EXERCISE 52

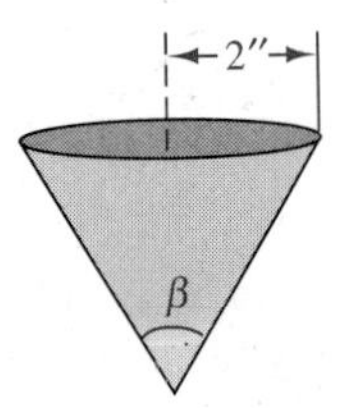

53 A spacelab circles the earth at an altitude of 380 miles. When an astronaut views the horizon of the earth, the angle θ shown in the figure is 65.8°. Use this information to estimate the radius r of the earth.

EXERCISE 53

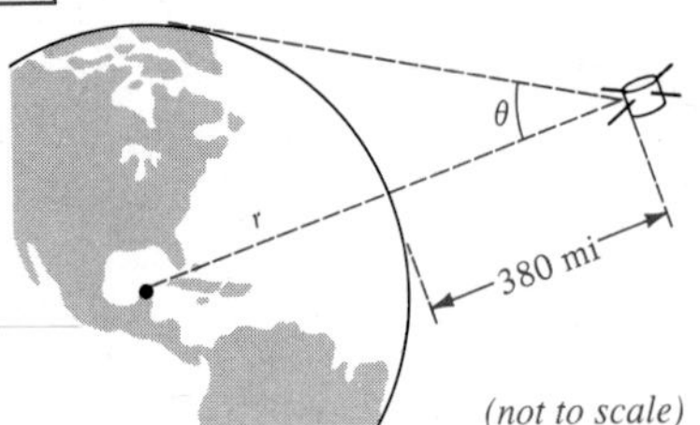

54 When viewed from a point P on the earth, the sun subtends an angle of 32′ as shown in the figure. Using 92,900,000 miles for the distance between P and the center of the sun, estimate the diameter of the sun.

EXERCISE 54

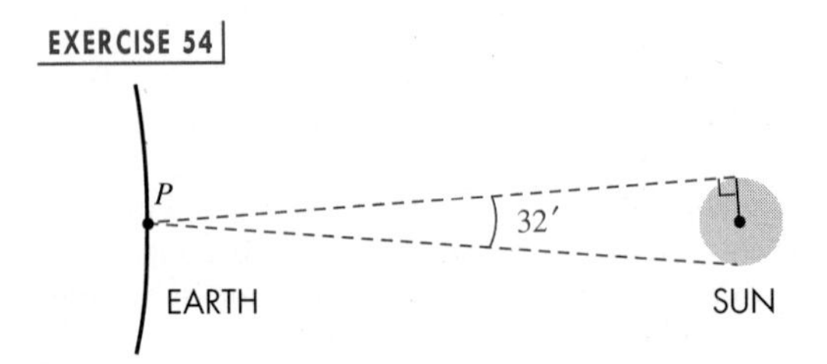

55 Shown in the first figure is a communications satellite with an equatorial orbit, that is, a nearly circular orbit in the plane determined by the earth's equator. If the satellite circles the earth at an altitude of $a = 22{,}300$ miles, its speed is the same as the rotational speed of earth, and the satellite appears to be stationary from an observer on the equator, that is, its orbit is synchronous.

(a) Use $R = 4000$ miles for the radius of the earth to determine the percentage of the equator that is within signal range of such a satellite.

(b) As shown in the second figure, three satellites are equally spaced in equatorial synchronous orbits. Use the value of θ obtained in part (a) to explain why all points on the equator are within signal range of at least one of the three satellites.

EXERCISE 55

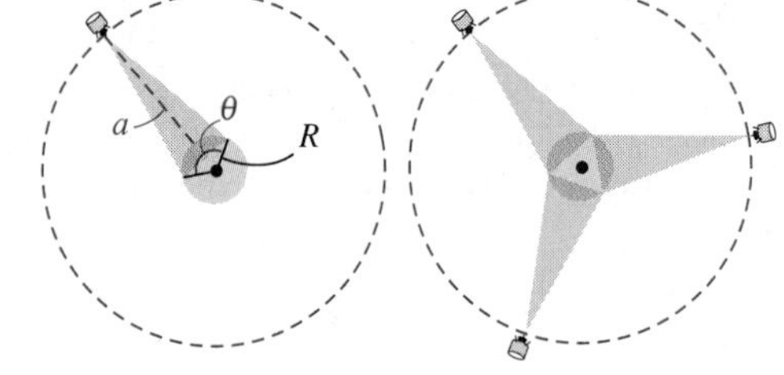

56 Refer to Exercise 55. Shown in the figure is the area served by a communications satellite circling a planet of radius R at an altitude a. The portion of the planet's surface within range of the satellite is a spherical cap of depth d and surface area $A = 2\pi Rd$.

(a) Express d in terms of R and θ.

(b) Estimate the percentage of the earth's surface that is within signal range of a single satellite in equatorial synchronous orbit.

EXERCISE 56

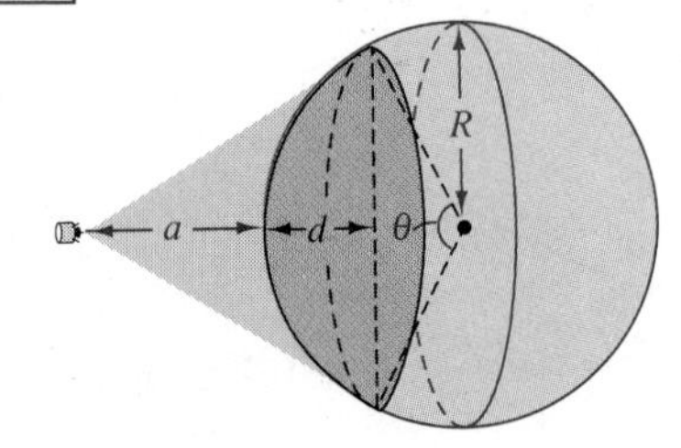

57 Generalize Exercise 26 to the case where the point is d meters above level ground and the angle of depression is α. Express the distance x in terms of d and α.

58 Generalize Exercise 27 to the case where the angle is α, the number of feet of string payed out is d, and the end of the string is held c feet above the ground. Express the height h of the kite in terms of α, d, and c.

59 Generalize Exercise 28 to the case of an n-sided polygon inscribed in a circle of radius r. Express the perimeter P in terms of n and r.

60 Generalize Exercise 29 to the case where the first angle is α, the second angle is β, and the distance between the two points is d. Express the height h of the tower in terms of d, α, and β.

61 Generalize Exercise 31 to the case where the distance from P to Q is d km and the angle of elevation changes from α to β.

62 Generalize Exercise 34 to the case where point A is d meters above ground and the angles of elevation and depression are α and β, respectively. Express the height h of the building in terms of d, α, and β.

6.4 TRIGONOMETRIC FUNCTIONS OF ANY ANGLE

Since many applied problems involve angles that are not acute, it is necessary to extend the definitions of the trigonometric functions. We make this extension by using the standard position of an angle θ on a rectangular coordinate system. If θ is acute, we have the situation illustrated in Figure 28, where we have chosen a point $P(a, b)$ on the terminal side of θ and where $c = \sqrt{a^2 + b^2}$. Referring to triangle OPQ,

$$\sin\theta = \frac{b}{c}, \qquad \cos\theta = \frac{a}{c}, \qquad \tan\theta = \frac{b}{a}.$$

FIGURE 28

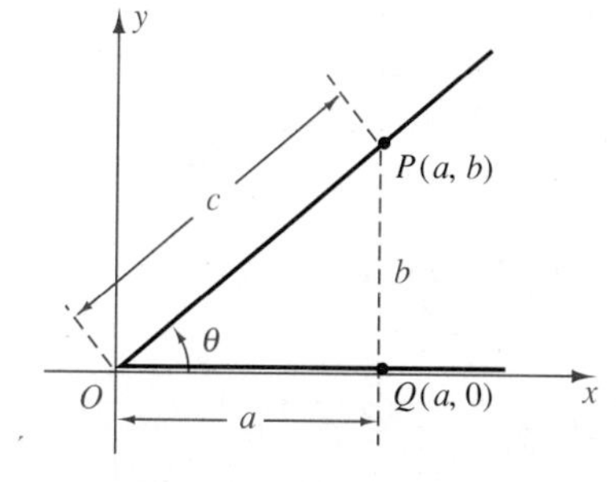

We now wish to consider angles of the type illustrated in Figure 29 (or any *other* angle, either positive, negative, or zero). We shall define the trigonometric functions so that their values agree with the definitions in Section 6.2 whenever the angle is acute.

FIGURE 29

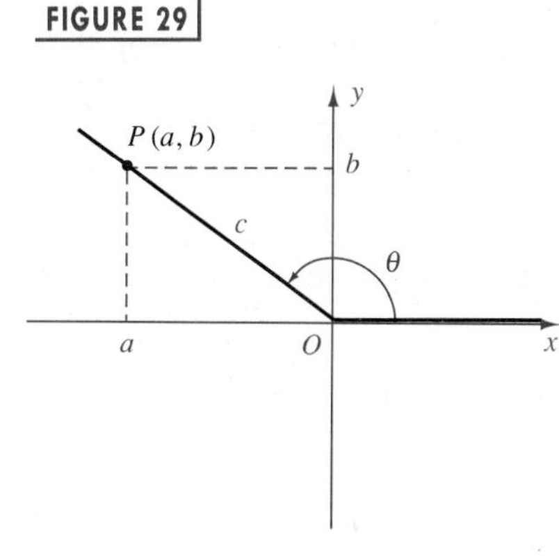

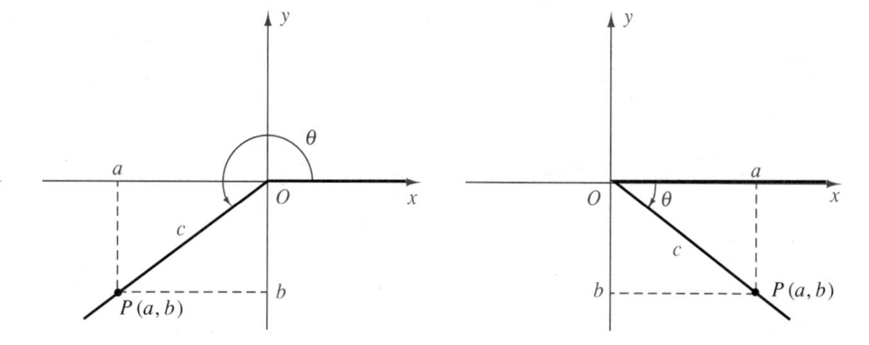

TRIGONOMETRIC FUNCTIONS OF ANY ANGLE

Let θ be an angle in standard position on a rectangular coordinate system and let $P(a, b)$ be any point other than O on the terminal side of θ. If $d(O, P) = c = \sqrt{a^2 + b^2}$, then

$$\sin\theta = \frac{b}{c} \qquad \csc\theta = \frac{c}{b} \quad (\text{if } b \neq 0)$$

$$\cos\theta = \frac{a}{c} \qquad \sec\theta = \frac{c}{a} \quad (\text{if } a \neq 0)$$

$$\tan\theta = \frac{b}{a} \quad (\text{if } a \neq 0) \qquad \cot\theta = \frac{a}{b} \quad (\text{if } b \neq 0)$$

We can show, using similar triangles, that the formulas in the previous definition do not depend on the point $P(a, b)$ that is chosen on the terminal side of θ.

The domains of the sine and cosine functions consist of all angles θ. However, $\tan\theta$ and $\sec\theta$ are undefined if $a = 0$, that is, if the terminal side of θ is on the y-axis. Thus the domain of both the tangent and secant functions consists of all angles *except* those having radian measure $(\pi/2) + \pi n$ for any integer n. Some special cases are $\pm\pi/2$, $\pm 3\pi/2$, and $\pm 5\pi/2$. The corresponding degree measures are $\pm 90°$, $\pm 270°$, and $\pm 450°$.

The domains of both the cotangent and cosecant functions consist of all angles except those that lead to $b = 0$, that is, all angles except those having terminal sides on the x-axis. These are the angles of radian measure πn (or degree measure $180° \cdot n$) for any integer n.

The following table summarizes this discussion.

Function	Domain
sin, cos	Every angle θ
tan, sec	Every angle θ except $\theta = \frac{\pi}{2} + \pi n$ for any integer n
cot, csc	Every angle θ except $\theta = \pi n$ for any integer n.

For all points $P(a, b)$ in the preceding definition, $|a| \leq c$ and $|b| \leq c$. Thus

$$|\sin\theta| \leq 1, \qquad |\cos\theta| \leq 1, \qquad |\csc\theta| \geq 1, \qquad |\sec\theta| \geq 1$$

for every θ in the domains of these functions.

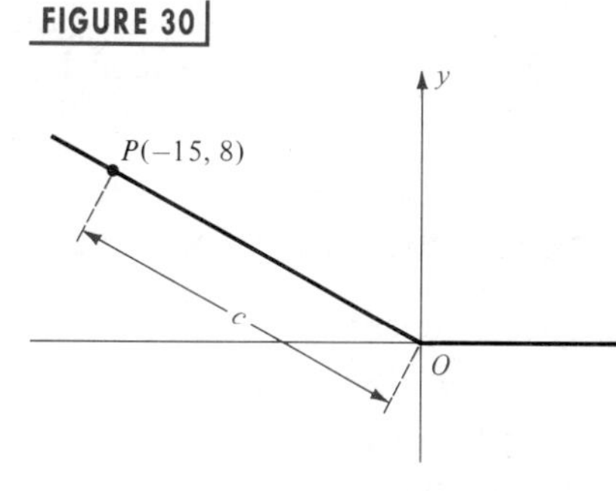

EXAMPLE ▪ 1

If θ is an angle in standard position on a rectangular coordinate system and if the point $P(-15, 8)$ is on the terminal side of θ, find the values of the trigonometric functions of θ.

SOLUTION The point $P(-15, 8)$ is shown in Figure 30. By the distance formula,

$$c = d(O, P) = \sqrt{(-15)^2 + 8^2} = \sqrt{225 + 64} = \sqrt{289} = 17.$$

We apply the definition of trigonometric functions of any angle with $a = -15$, $b = 8$, and $c = 17$, obtaining:

$$\sin\theta = \frac{8}{17} \qquad \csc\theta = \frac{17}{8}$$

$$\cos\theta = -\frac{15}{17} \qquad \sec\theta = -\frac{17}{15}$$

$$\tan\theta = -\frac{8}{15} \qquad \cot\theta = -\frac{15}{8}$$

EXAMPLE ▪ 2

An angle θ is in standard position on a rectangular coordinate system and its terminal side lies in quadrant III on the line $y = 3x$. Find the values of the trigonometric functions of θ.

FIGURE 31

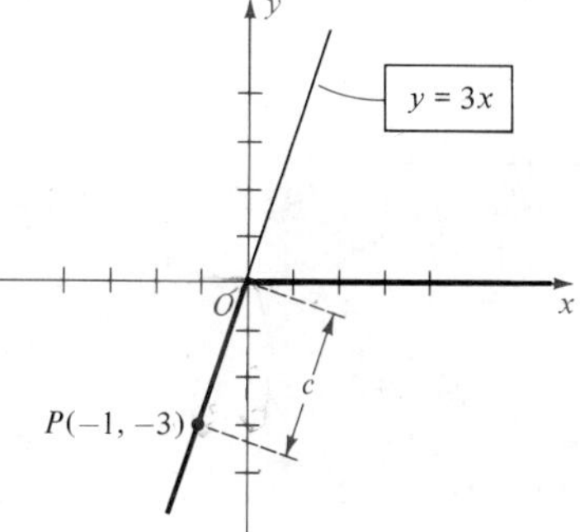

SOLUTION The graph of $y = 3x$ is sketched in Figure 31, together with the initial and terminal sides of θ. Since the terminal side of θ is in quadrant III, we begin by choosing a convenient third quadrant point, say $P(-1, -3)$ on the terminal side. The distance c from the origin O to P is

$$c = d(O, P) = \sqrt{(-1)^2 + (-3)^3} = \sqrt{10}.$$

Applying the definition with $a = -1$, $b = -3$, and $c = \sqrt{10}$ give us:

$$\sin\theta = \frac{-3}{\sqrt{10}} = -\frac{3\sqrt{10}}{10} \qquad \csc\theta = \frac{\sqrt{10}}{-3} = -\frac{\sqrt{10}}{3}$$

$$\cos\theta = \frac{-1}{\sqrt{10}} = -\frac{\sqrt{10}}{10} \qquad \sec\theta = \frac{\sqrt{10}}{-1} = -\sqrt{10}$$

$$\tan\theta = \frac{-3}{-1} = 3 \qquad \cot\theta = \frac{-1}{-3} = \frac{1}{3}$$

The definitions of the trigonometric functions may be applied if the terminal side of θ lies on either the x- or y-axis. This is illustrated by the next example.

EXAMPLE ▪ 3

If $\theta = \dfrac{3\pi}{2}$, find the values of the trigonometric functions of θ.

FIGURE 32

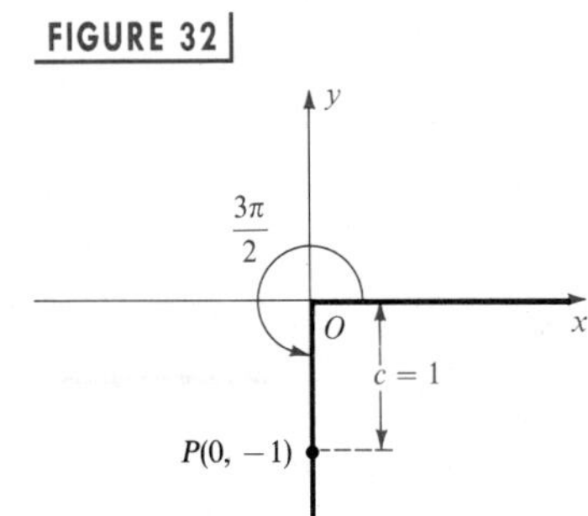

SOLUTION Note that $3\pi/2 = 270°$. Placing θ in standard position, the terminal side of θ coincides with the negative y-axis, as shown in Figure 32. To use the definition we may choose any point P on the terminal side of θ. For simplicity, we use $P(0, -1)$. In this case $c = 1$, $a = 0$, $b = -1$, and hence

$$\sin\frac{3\pi}{2} = \frac{-1}{1} = -1 \qquad \csc\frac{3\pi}{2} = \frac{1}{-1} = -1$$

$$\cos\frac{3\pi}{2} = \frac{0}{1} = 0 \qquad \cot\frac{3\pi}{2} = \frac{0}{-1} = 0$$

The tangent and secant functions are undefined, since the meaningless expressions $\tan\theta = (-1)/0$ and $\sec\theta = 1/0$ occur when we substitute in the appropriate formulas.

Let us determine the signs associated with values of the trigonometric functions. If θ is in quadrant II and $P(a, b)$ is a point on the terminal side, then a is negative, b is positive and hence, by definition, $\sin\theta$ and $\csc\theta$ are positive. The other four functions are negative. You should check the remaining quadrants. The following table indicates the signs in all four quadrants.

SIGNS OF THE TRIGONOMETRIC FUNCTIONS

Quadrant containing θ	Positive functions	Negative functions
I	All	None
II	sin, csc	cos, sec, tan, cot
III	tan, cot	sin, csc, cos, sec
IV	cos, sec	sin, csc, tan, cot

EXAMPLE ▪ 4

Find the quadrant containing θ if both $\sin\theta < 0$ and $\cos\theta > 0$.

SOLUTION Referring to the table of signs, we see that $\sin\theta < 0$ if θ is in quadrant III or IV, and $\cos\theta > 0$ if θ is in quadrant I or IV. Hence, for both conditions to be satisfied, θ' must be in quadrant IV.

The fundamental identities, which we have established for acute angles, are also true for trigonometric functions of any angle. The proofs are identical to those given in Section 6.2.

In calculus and other advanced courses it is necessary to consider trigonometric functions whose domains are sets of real numbers. The next definition indicates how to make the transition from angles to real numbers.

TRIGONOMETRIC FUNCTIONS OF REAL NUMBERS

The **value of a trigonometric function at a real number *t*** is its value at an angle of t radians, provided that value exists.

To find values of trigonometric functions of real numbers with a calculator, all that is necessary is to use the radian mode. When using Table 4 consult the column labeled t.

The following concept is needed when using tables to find values of trigonometric functions of any angle or real number. It is sometimes required when using a calculator to find all angles or real numbers that correspond to a given function value.

DEFINITION

Let θ be an angle in standard position such that the terminal side does not lie on a coordinate axis. The **reference angle** for θ is the acute angle θ' that the terminal side of θ makes with the positive or negative x-axis.

Figure 33 illustrates the reference angle θ' for an angle θ in each of the four quadrants.

FIGURE 33 Reference angles

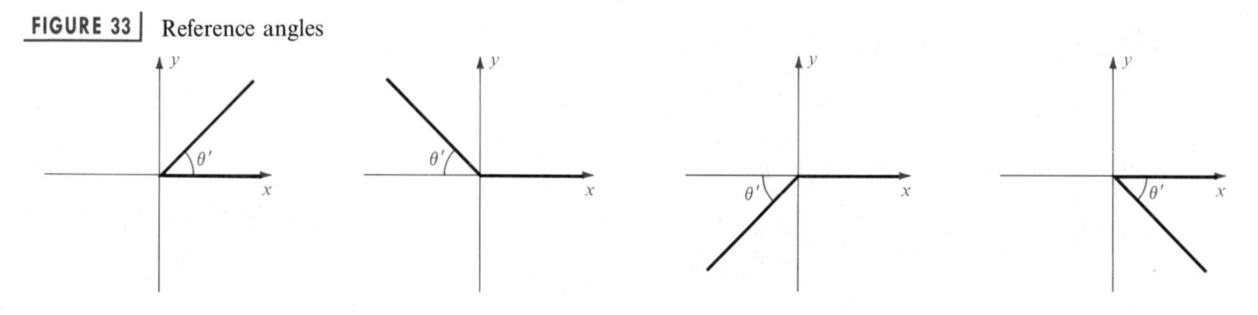

EXAMPLE ▪ 5

Sketch the reference angle θ' for θ and express the measure of θ' in both radians and degrees:

(a) $\theta = 315°$ **(b)** $\theta = -240°$ **(c)** $\theta = \dfrac{5\pi}{6}$ **(d)** $\theta = 4$

SOLUTION Each angle θ and its reference angle θ' is sketched in Figure 34.

FIGURE 34

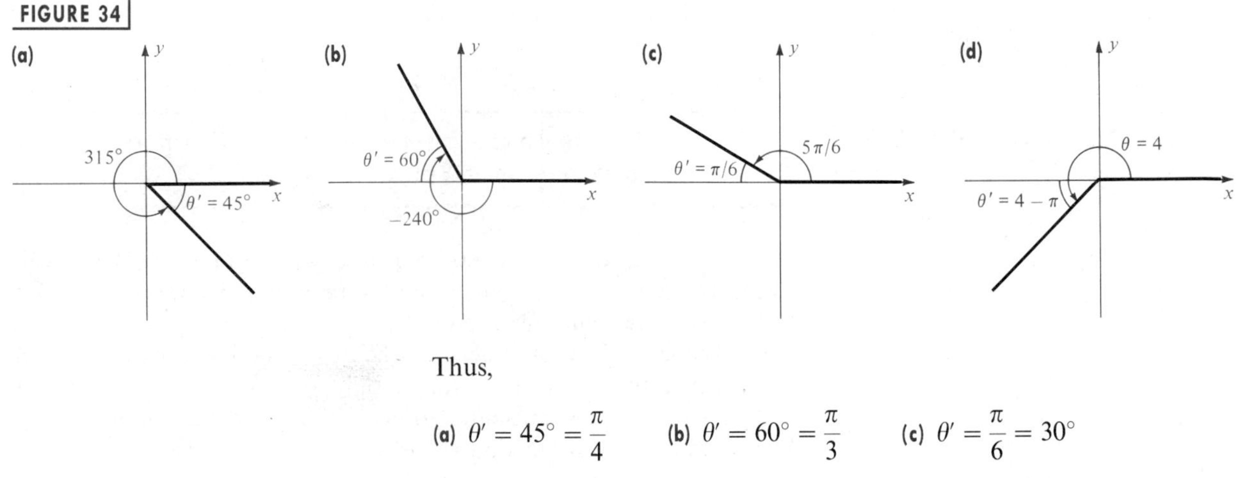

Thus,

(a) $\theta' = 45° = \dfrac{\pi}{4}$ **(b)** $\theta' = 60° = \dfrac{\pi}{3}$ **(c)** $\theta' = \dfrac{\pi}{6} = 30°$

(d) $\theta' = 4 - \pi = (4 - \pi)\left(\dfrac{180}{\pi}\right)^{\circ} \approx 49.2°$

If the terminal side of θ does not lie on a coordinate axis and if θ' is the reference angle for θ, then $0° < \theta' < 90°$ and $0 < \theta' < \pi/2$. Let $P(a, b)$ be a point on the terminal side of θ, and consider the point $Q(a, 0)$ on the x-axis. The sketches in Figure 35 illustrate a typical situation when θ is in each of the four quadrants.

FIGURE 35

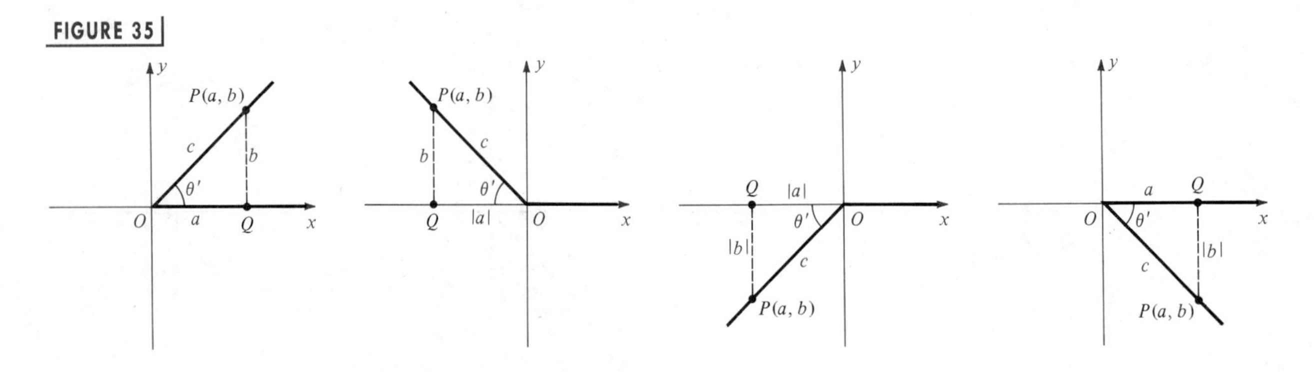

In all cases the lengths of the sides of triangle OPO are

$$d(O, Q) = |a|, \qquad d(Q, P) = |b|, \qquad d(O, P) = \sqrt{a^2 + b^2} = c.$$

We may apply the definitions of trigonometric functions of any angle and also use triangle OPQ to obtain the following formulas:

$$|\sin \theta| = \left|\frac{b}{c}\right| = \frac{|b|}{c} = \sin \theta'$$

$$|\cos \theta| = \left|\frac{a}{c}\right| = \frac{|a|}{c} = \cos \theta'$$

$$|\tan \theta| = \left|\frac{b}{a}\right| = \frac{|b|}{|a|} = \tan \theta'$$

These formulas lead to the rule stated in the next theorem. If the terminal side of θ is on a coordinate axis, then the definition of trigonometric functions should be used to find values.

THEOREM ON REFERENCE ANGLES

If an angle θ is in standard position and the terminal side is not on a coordinate axis, then to find the value of a trigonometric function at θ, determine its value for the reference angle θ' associated with θ and prefix the appropriate sign.

The "appropriate sign" mentioned in this theorem can be determined from the table of signs in this section.

EXAMPLE ▪ 5

Find $\sin \theta$, $\cos \theta$, and $\tan \theta$ if:

(a) $\theta = \dfrac{5\pi}{6}$ **(b)** $\theta = 315°$

SOLUTION The angles and their reference angles are sketched in Figure 34 (c) and (a). Using the theorem on reference angles and known results about special angles, we obtain

(a)

$$\sin \frac{5\pi}{6} = \sin \frac{\pi}{6} = \frac{1}{2}$$

$$\cos \frac{5\pi}{6} = -\cos \frac{\pi}{6} = -\frac{\sqrt{3}}{2}$$

$$\tan \frac{5\pi}{6} = -\tan \frac{\pi}{6} = -\frac{\sqrt{3}}{3}$$

(b)

$$\sin 315^\circ = -\sin 45^\circ = -\frac{\sqrt{2}}{2}$$

$$\cos 315^\circ = \cos 45^\circ = \frac{\sqrt{2}}{2}$$

$$\tan 315^\circ = -\tan 45^\circ = -1$$

If we use a calculator to approximate function values, reference angles are unnecessary. As an illustration, to find sin 210°, we place the calculator in degree mode, enter the number 210, and press the SIN key, obtaining $\sin 210^\circ = -0.5$, which is the exact value. Using the same procedure for 240° we obtain a decimal approximation:

$$\sin 240^\circ \approx -0.8660254$$

If we want to find the *exact* value of sin 240°, a calculator should not be used. In this case we find the reference angle 60° of 240° and use the theorem on reference angles, together with known results about special angles, to obtain

$$\sin 240^\circ = -\sin 60^\circ = -\frac{\sqrt{3}}{2}.$$

EXAMPLE ▪ 7

Approximate cot 837.4°.

SOLUTION Place a calculator in degree mode and use the reciprocal relationship $\cot \theta = 1/\tan \theta$ as follows:

Enter 837.4:	837.4	(degree value of θ)
Press TAN:	−1.9291956	(value of $\tan \theta$)
Press 1/x:	−0.5183508	(value of $1/\tan \theta$)

Thus $\cot 837.4^\circ \approx -0.5183508$.

In Section 6.2 we pointed out that the functions denoted by $\sin^{-1}$, $\cos^{-1}$, and $\tan^{-1}$ have the following properties.

$$\sin^{-1}(\sin \theta) = \theta \quad \text{for} \quad -\frac{\pi}{2} \le \theta \le \frac{\pi}{2} \quad \text{or} \quad -90^\circ \le \theta \le 90^\circ.$$

$$\cos^{-1}(\cos \theta) = \theta \quad \text{for} \quad 0 \le \theta \le \pi \quad \text{or} \quad 0^\circ \le \theta \le 180^\circ.$$

$$\tan^{-1}(\tan \theta) = \theta \quad \text{for} \quad -\frac{\pi}{2} < \theta < \frac{\pi}{2} \quad \text{or} \quad -90^\circ < \theta < 90^\circ.$$

When using a calculator to find θ, be aware of these restrictions on θ. For example, there are many values of θ such that $\tan\theta = -1$; however, a calculator gives only the value that is between $-\pi/2$ and 0 (or between $-90°$ and $0°$). If other values are desired, then reference angles may be employed, as illustrated in the next example.

EXAMPLE ▪ 8

If $\tan\theta = -0.4623$ and $0° \le \theta < 360°$, find θ to the nearest $0.1°$.

SOLUTION As pointed out in the preceding discussion, if we use a calculator (in degree mode) to find θ when $\tan\theta$ is negative, then the degree measure is in the interval $(-90°, 0°]$. In particular, we have the following:

Enter -0.4623:	-0.4623	(the value of $\tan\theta$)
Press [INV] [TAN]:	-24.811101	

Thus a degree approximation is $\theta \approx -24.8°$.

Since we wish to find values of θ between $0°$ and $360°$, we use the (approximate) reference angle $\theta' \approx 24.8°$. There are two possible values of θ such that $\tan\theta$ is negative—one in quadrant II, the other in quadrant IV. If θ is in quadrant II and $0° \le \theta < 360°$, we have the situation shown in Figure 36(i), and

$$\theta = 180° - \theta' \approx 180° - 24.8°, \quad \text{or} \quad \theta \approx 155.2°.$$

If θ is in quadrant IV and $0° \le \theta < 360°$, then as in Figure 36(ii),

$$\theta = 360° - \theta' \approx 360° - 24.8°, \quad \text{or} \quad \theta \approx 335.2°.$$

FIGURE 36

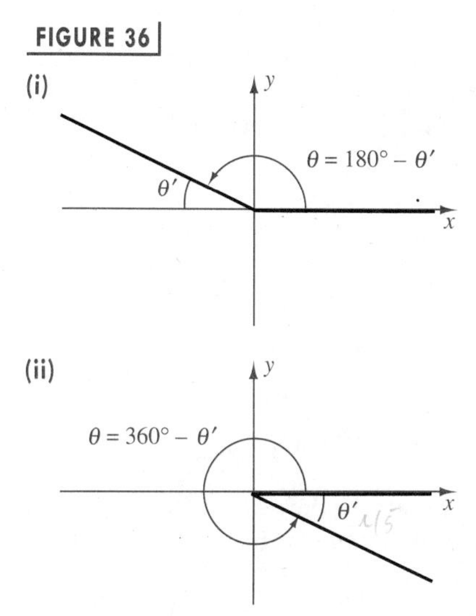

6.4 EXERCISES

Exer. 1–4: Find the values of the trigonometric functions of θ. Assume that θ is in standard position and point P is on the terminal side of θ.

1 $P(4, -3)$

2 $P(-8, -15)$

3 $P(-2, -5)$

4 $P(-1, 2)$

Exer. 5–12: Find the values of the trigonometric functions of θ. Assume that the terminal side of θ (in standard position) is in the specified quadrant and satisfies the given conditions.

5 II; on the line $y = -4x$

6 IV; on the line $3y + 5x = 0$

7 III; parallel to the line $2y - 7x + 2 = 0$

8 II; parallel to the line through the points $A(1, 4)$ and $B(3, -2)$

9 IV; perpendicular to the line through the points $A(5, 12)$ and $B(-3, -3)$

10 III; perpendicular to the line $5y + 12x + 5 = 0$

11 I; on a line having slope $\frac{4}{3}$

12 III; bisects the quadrant

Exer. 13–20: Find the values of the trigonometric functions of θ. Assume that θ is in standard position and has

the specified measure.

13 $90°$ 14 $0°$ 15 $180°$ 16 $-270°$

17 2π 18 $\frac{5\pi}{2}$ 19 $\frac{7\pi}{2}$ 20 3π

Exer. 21–30: Find the quadrant containing θ for the given conditions.

21 $\cos\theta > 0$ and $\sin\theta < 0$ 22 $\tan\theta < 0$ and $\cos\theta > 0$

23 $\sin\theta < 0$ and $\cot\theta > 0$ 24 $\sec\theta > 0$ and $\tan\theta < 0$

25 $\csc\theta > 0$ and $\sec\theta < 0$ 26 $\csc\theta > 0$ and $\cot\theta < 0$

27 $\sec\theta < 0$ and $\tan\theta > 0$ 28 $\sin\theta < 0$ and $\sec\theta$

29 $\cos\theta > 0$ and $\tan\theta > 0$ 30 $\cos\theta < 0$ and $\csc\theta < 0$

Exer. 31–38: Find the reference angle θ' for θ with the given measure.

31 (a) $240°$ (b) $340°$ (c) $-110°$

32 (a) $165°$ (b) $275°$ (c) $-202°$

33 (a) $130°40'$ (b) $-405°$ (c) $-260°35'$

34 (a) $335°20'$ (b) $-620°$ (c) $-185°40'$

35 (a) $\frac{3\pi}{4}$ (b) $\frac{4\pi}{3}$ (c) $-\frac{\pi}{6}$

36 (a) $\frac{7\pi}{6}$ (b) $\frac{2\pi}{3}$ (c) $-\frac{3\pi}{4}$

37 (a) $\frac{9\pi}{4}$ (b) $\frac{11\pi}{6}$ (c) $-\frac{2\pi}{3}$

38 (a) $\frac{8\pi}{3}$ (b) $\frac{7\pi}{4}$ (c) $-\frac{7\pi}{6}$

Exer. 39–44: Find the exact value.

39 (a) $\sin\frac{2\pi}{3}$ (b) $\sin\frac{4\pi}{3}$

40 (a) $\cos\frac{5\pi}{6}$ (b) $\cos\frac{7\pi}{6}$

41 (a) $\tan\left(-\frac{5\pi}{4}\right)$ (b) $\cot 315°$

42 (a) $\sin 210°$ (b) $\csc(-150°)$

43 (a) $\csc 300°$ (b) $\sec(-120°)$

44 (a) $\tan(-135°)$ (b) $\sec 225°$

Exer. 45–56: Approximate the number to three decimal places.

45 $\cos 213°20'$ 46 $\sin 98°10'$

47 $\tan 105°40'$ 48 $\cot 231°40'$

49 $\sec 294°10'$ 50 $\csc 320°50'$

51 $\sin 0.37$ 52 $\cos 0.65$

53 $\cos 1.46$ 54 $\sin 0.82$

55 $\tan 3$ 56 $\cot 6$

Exer. 57–64: Approximate, to the nearest 10°, the degree measures of all angles θ that are in the interval [0°, 360°).

57 $\sin\theta = -0.5640$ 58 $\cos\theta = 0.7490$

59 $\tan\theta = 2.798$ 60 $\cot\theta = -0.9601$

61 $\sec\theta = -1.116$ 62 $\csc\theta = 1.485$

63 $\cos\theta = 0.6604$ 64 $\sin\theta = -0.8225$

6.5 GRAPHS OF THE TRIGONOMETRIC FUNCTIONS

Let θ be an angle in standard position, let $P(a, b)$ be any point on the terminal side, and let $c = d(O, P) = \sqrt{a^2 + b^2}$ (see Figure 37). By the definitions of trigonometric functions of any angle,

$$\cos\theta = \frac{a}{c} \quad \text{and} \quad \sin\theta = \frac{b}{c}.$$

Let us study the variation of $\sin\theta$ and $\cos\theta$ as θ increases or decreases. Since P can be any point other than O on the terminal side, we may

FIGURE 37

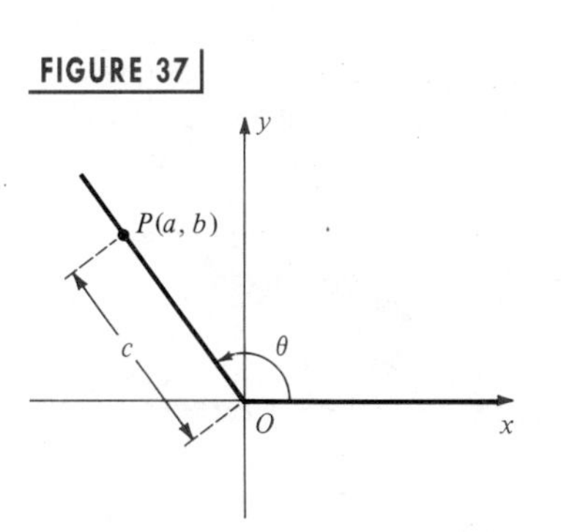

choose $P(a,b)$ such that $c = 1$ for every θ. Thus, *$P(a,b)$ is always on the unit circle U with center at the origin.* If $c = 1$, the formulas $\cos\theta = a/c$ and $\sin\theta = b/c$ take on the simple forms

$$\cos\theta = \frac{a}{1} = a \quad \text{and} \quad \sin\theta = \frac{b}{1} = b.$$

Hence the point $P(a,b)$ may be denoted as follows:

$$P(\cos\theta, \sin\theta)$$

FIGURE 38

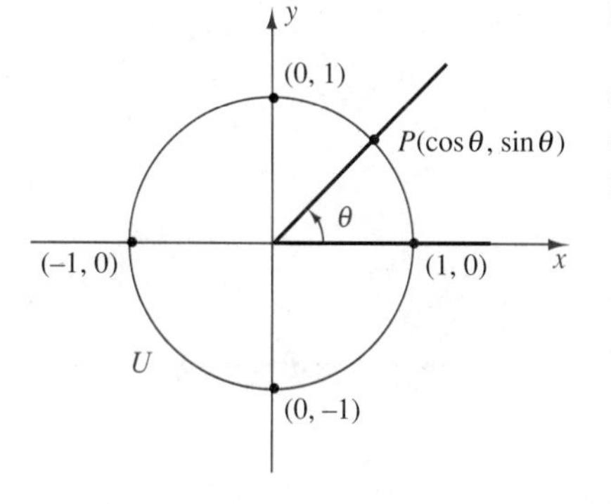

We now have the situation illustrated in Figure 38, where we have plotted points on U that are on the coordinate axes.

If we let θ increase from 0 to 2π radians, the point $P(\cos\theta, \sin\theta)$ travels around the unit circle U one time in the counterclockwise direction. By observing the variation of the x- and y-coordinates of P, we obtain the following table.

θ	$P(\cos\theta, \sin\theta)$	$\cos\theta$	$\sin\theta$
$0 \to \frac{\pi}{2}$	$(1,0) \to (0,1)$	$1 \to 0$	$0 \to 1$
$\frac{\pi}{2} \to \pi$	$(0,1) \to (-1,0)$	$0 \to -1$	$1 \to 0$
$\pi \to \frac{3\pi}{2}$	$(-1,0) \to (0,-1)$	$-1 \to 0$	$0 \to -1$
$\frac{3\pi}{2} \to 2\pi$	$(0,-1) \to (1,0)$	$0 \to 1$	$-1 \to 0$

The notation $0 \to \pi/2$ means that θ increases from 0 to $\pi/2$, and the notation $(1,0) \to (0,1)$ denotes the corresponding variation of $P(\cos\theta, \sin\theta)$ as it travels along U from $(1,0)$ to $(0,1)$. If θ increases from 0 to $\pi/2$, then $\sin\theta$ increases from 0 to 1, which is denoted by $0 \to 1$. Moreover, $\sin\theta$ takes on every value between 0 and 1. If θ increases from $\pi/2$ to π, then $\sin\theta$ decreases from 1 to 0, which is denoted by $1 \to 0$. Other entries in the table may be interpreted in similar fashion.

If θ increases from 2π to 4π, the point $P(\cos\theta, \sin\theta)$ in Figure 38 traces the unit circle U again, and the identical patterns for $\sin\theta$ and $\cos\theta$ are repeated; that is,

$$\sin(\theta + 2\pi) = \sin\theta \quad \text{and} \quad \cos(\theta + 2\pi) = \cos\theta$$

for every θ in the interval $[0, 2\pi]$. The same is true if θ increases from 4π to 6π, from 6π to 8π, and so on. In general, we have the following theorem.

THEOREM

$$\sin(\theta + 2\pi n) = \sin\theta \quad \text{and} \quad \cos(\theta + 2\pi n) = \cos\theta.$$

for every integer n.

The repetitive variation of the sine and cosine functions is *periodic* in the sense of the following direction.

DEFINITION

A function f is **periodic** if there exists a positive real number k such that

$$f(t + k) = f(t)$$

for every t in the domain of f. The least such positive real number k, if it exists, is the **period** of f.

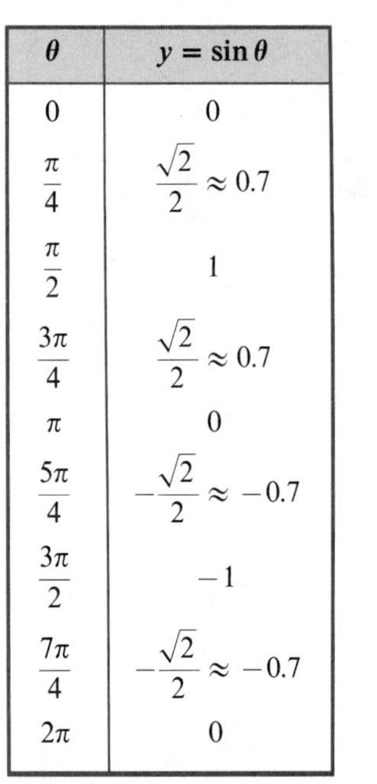

θ	$y = \sin\theta$
0	0
$\frac{\pi}{4}$	$\frac{\sqrt{2}}{2} \approx 0.7$
$\frac{\pi}{2}$	1
$\frac{3\pi}{4}$	$\frac{\sqrt{2}}{2} \approx 0.7$
π	0
$\frac{5\pi}{4}$	$-\frac{\sqrt{2}}{2} \approx -0.7$
$\frac{3\pi}{2}$	-1
$\frac{7\pi}{4}$	$-\frac{\sqrt{2}}{2} \approx -0.7$
2π	0

The period of the sine and cosine functions is 2π.

Let us sketch the graph of $y = \sin\theta$ on a θy-coordinate system, where θ is real number or the radian measure of an angle. The table in the margin lists coordinates of several points on the graph for $0 \le \theta \le 2\pi$. Additional points can be determined using results on special angles, such as $\sin(\pi/6) = \frac{1}{2}$ and $\sin(\pi/3) = \sqrt{3}/2 \approx 0.87$.

To sketch the graph for $0 \le \theta \le 2\pi$ we plot the points given by the table and remember that $\sin\theta$ increases in $[0, \pi/2]$, decreases in $[\pi/2, \pi]$ and $[\pi, 3\pi/2]$, and increases in $[3\pi/2, 2\pi]$. This gives us the sketch in Figure 39. Since the sine function is periodic, the pattern shown in Figure 39 is repeated to the right and left in intervals of length 2π. This gives us the sketch in Figure 40.

FIGURE 39

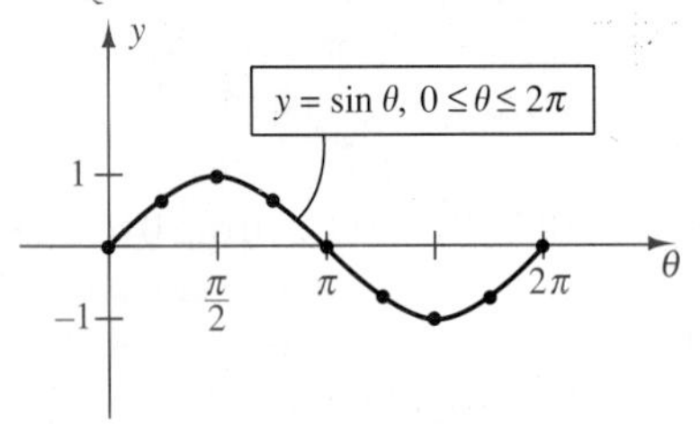

FIGURE 40

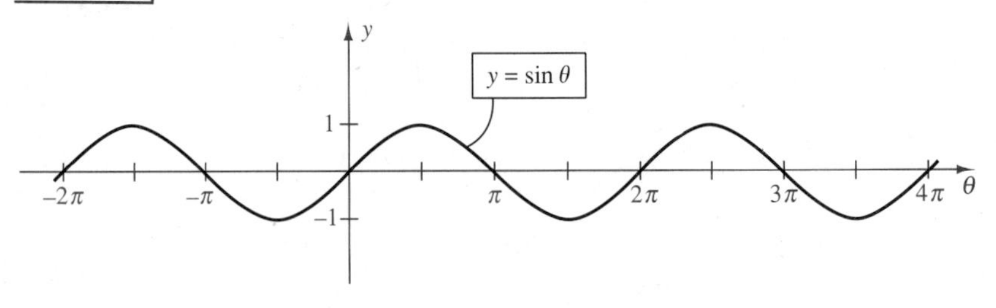

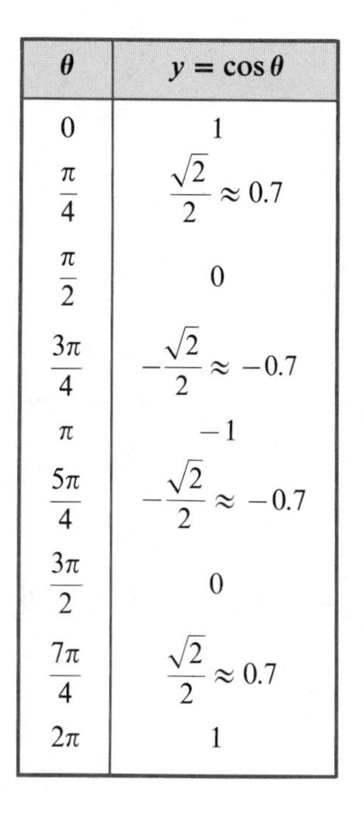

θ	$y = \cos\theta$
0	1
$\frac{\pi}{4}$	$\frac{\sqrt{2}}{2} \approx 0.7$
$\frac{\pi}{2}$	0
$\frac{3\pi}{4}$	$-\frac{\sqrt{2}}{2} \approx -0.7$
π	-1
$\frac{5\pi}{4}$	$-\frac{\sqrt{2}}{2} \approx -0.7$
$\frac{3\pi}{2}$	0
$\frac{7\pi}{4}$	$\frac{\sqrt{2}}{2} \approx 0.7$
2π	1

We can use the same procedure to sketch the graph of $y = \cos\theta$. The table in the margin lists coordinates of several points on the graph for $0 \le \theta \le 2\pi$. Plotting these points leads to the part of the graph shown in Figure 41. Repeating this pattern to the right and to the left in intervals of length 2π, we obtain the sketch in Figure 42.

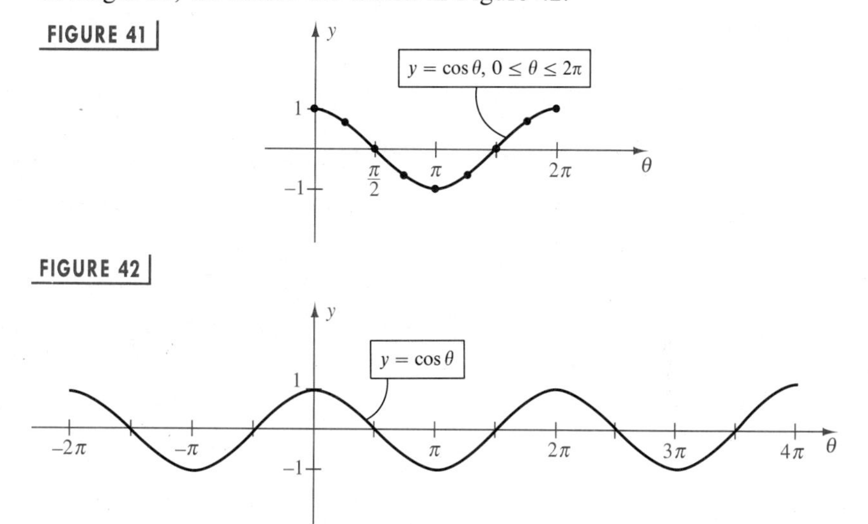

FIGURE 41

FIGURE 42

The part of the graph of the sine or cosine function corresponding to $0 \le \theta \le 2\pi$ is one **cycle**. For the sine function we also refer to a cycle as a **sine wave.**

The range of the sine and cosine functions consists of all real numbers in the closed interval $[-1, 1]$. Since $\csc\theta = 1/\sin\theta$ and $\sec\theta = 1/\cos\theta$, it follows that the range of the cosecant and secant functions consists of all real numbers having absolute value greater than or equal to 1.

As we shall see, the range of the tangent and cotangent functions consists of all real numbers.

Before discussing graphs of the other trigonometric functions, let us establish formulas that involve functions of $-\theta$ for any θ. Since a minus sign is involved, we call them *formulas for negatives*.

FORMULAS FOR NEGATIVES

$\sin(-\theta) = -\sin\theta$	$\csc(-\theta) = -\csc\theta$
$\cos(-\theta) = \cos\theta$	$\sec(-\theta) = \sec\theta$
$\tan(-\theta) = -\tan\theta$	$\cot(-\theta) = -\tan\theta$

PROOF Consider the unit circle U in Figure 43. As θ increases from 0 to 2π, the point $P(a, b)$ traces the unit circle U once in the counterclockwise direction and the point $Q(a, -b)$ corresponding to $-\theta$ traces U once in the clockwise direction. Applying the definitions of trigonometric functions

FIGURE 43

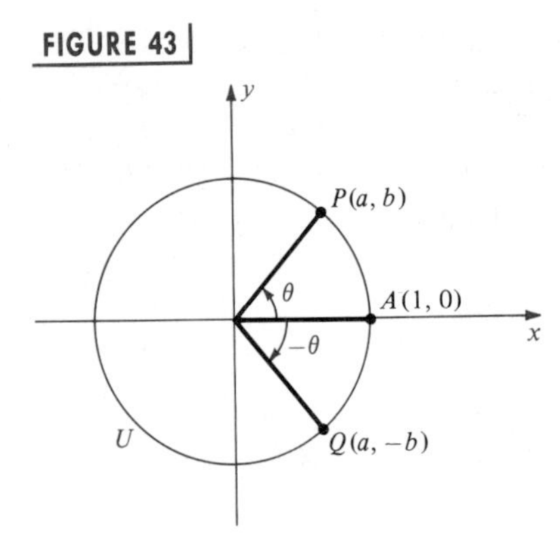

of any angle (with $c = 1$),

$$\sin(-\theta) = -b = -\sin\theta$$

$$\cos(-\theta) = a = \cos\theta$$

$$\tan(-\theta) = \frac{-b}{a} = -\frac{b}{a} = -\tan\theta.$$

The proofs of the remaining three formulas are similar. ❑

EXAMPLE

Use formulas for negatives to find the exact values of $\sin(-45°)$, $\cos(-30°)$, and $\tan\left(-\frac{\pi}{3}\right)$.

SOLUTION We use the appropriate formulas and results about special angles:

$$\sin(-45°) = -\sin 45° = -\frac{\sqrt{2}}{2}$$

$$\cos(-30°) = \cos 30° = \frac{\sqrt{3}}{2}$$

$$\tan\left(-\frac{\pi}{3}\right) = -\tan\frac{\pi}{3} = -\sqrt{3}$$

We may use the formulas for negatives to prove the following theorem.

THEOREM

(i) The cosine and secant functions are even.

(ii) The sine, tangent, cosecant, and cotangent functions are odd.

PROOF We shall prove the theorem for the cosine and sine functions.

If $f(\theta) = \cos\theta$, then

$$f(-\theta) = \cos(-\theta) = \cos\theta = f(\theta),$$

which means that the cosine function is even.

If $f(\theta) = \sin\theta$, then

$$f(-\theta) = \sin(-\theta) = -\sin\theta = -f(\theta).$$

Thus the sine function is odd. ❑

Since the sine function is odd, its graph is symmetric with respect to the origin (see Figure 40). Since the cosine function is even, its graph is symmetric with respect to the y-axis (see Figure 42).

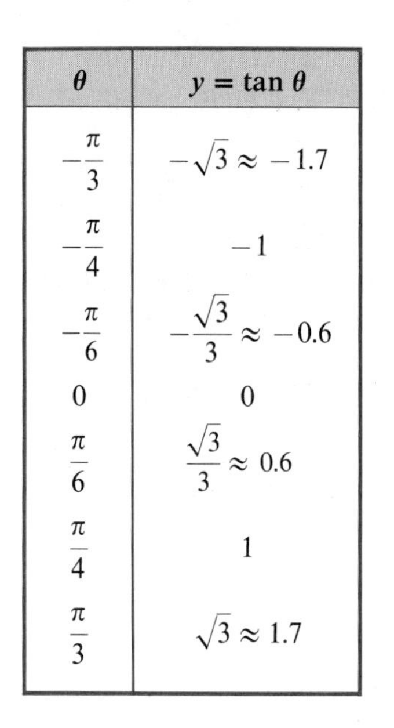

θ	$y = \tan\theta$
$-\frac{\pi}{3}$	$-\sqrt{3} \approx -1.7$
$-\frac{\pi}{4}$	-1
$-\frac{\pi}{6}$	$-\frac{\sqrt{3}}{3} \approx -0.6$
0	0
$\frac{\pi}{6}$	$\frac{\sqrt{3}}{3} \approx 0.6$
$\frac{\pi}{4}$	1
$\frac{\pi}{3}$	$\sqrt{3} \approx 1.7$

By the preceding theorem the tangent function is odd, and hence the graph of $y = \tan\theta$ is symmetric with respect to the origin. The table at the left lists some points on the graph if $-\pi/2 < \theta < \pi/2$. The corresponding points are plotted in Figure 44. The values of $\tan\theta$ near $\theta = \pi/2$ require special attention. If we consider $\tan\theta = \sin\theta/\cos\theta$, then as θ increases toward $\pi/2$, the numerator $\sin\theta$ approaches 1 and the denominator $\cos\theta$ approaches 0. Consequently, $\tan\theta$ takes on large positive values. Some approximations of $\tan\theta$ for θ close to $\pi/2 \approx 1.5708$ are:

$$\tan 1.5700 \approx 1{,}255.8$$
$$\tan 1.5703 \approx 2{,}014.8$$
$$\tan 1.5706 \approx 5{,}093.5$$
$$\tan 1.5707 \approx 10{,}381.3$$
$$\tan 1.57079 \approx 158{,}057.9$$

Notice how rapidly $\tan\theta$ increases as θ approaches $\pi/2$. We say that $\tan\theta$ *increases without bound as θ approaches $\pi/2$ through values less than $\pi/2$.* Similarly, if θ approaches $-\pi/2$ through values *greater* than $-\pi/2$, then $\tan\theta$ *decreases without bound.* We may denote this variation using the notation developed for rational functions in Section 4.6:

$$\tan\theta \to \infty \quad \text{as} \quad \theta \to \frac{\pi}{2}^{-}$$

$$\tan\theta \to -\infty \quad \text{as} \quad \theta \to -\frac{\pi}{2}^{+}$$

This variation of $\tan\theta$ in the open interval $(-\pi/2, \pi/2)$ is illustrated in Figure 45. The lines $\theta = \pi/2$ and $\theta = -\pi/2$ are vertical asymptotes for the graph. The same pattern is repeated in the open intervals $(-3\pi/2, -\pi/2)$, $(\pi/2, 3\pi/2)$, $(3\pi/2, 5\pi/2)$, and in similar intervals of length π, as shown in the figure. Thus, the *tangent function is periodic with period π.*

FIGURE 44

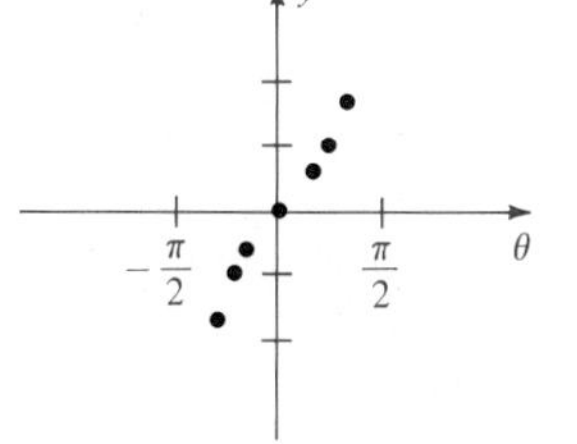

FIGURE 45 $y = \tan\theta$

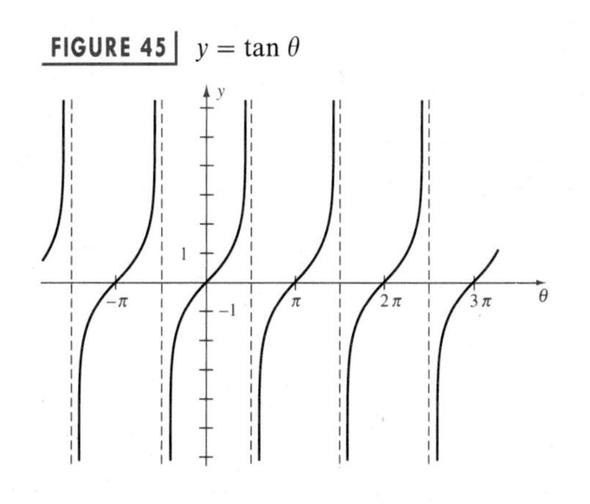

It is easy to sketch the graphs of the remaining three trigonometric functions. For example, since $\csc\theta = 1/\sin\theta$, we may find the y-coordinate of a point on the graph of the cosecant function by taking the reciprocal of the corresponding y-coordinate on the sine graph for every value of θ except $\theta = \pi n$ for any integer n. (If $\theta = \pi n$, $\sin\theta = 0$ and hence $1/\sin\theta$ is undefined.) As an aid to sketching the graph of the cosecant function, it is convenient to sketch the graph of the sine function (shown with dashes in Figure 46), and then take reciprocals to obtain points on the cosecant graph.

FIGURE 46 $y = \csc\theta$

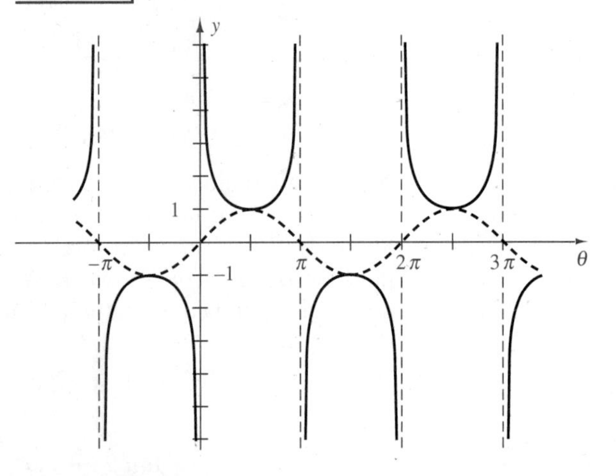

Notice the manner in which the cosecant function increases or decreases without bound as θ approaches πn, for any integer n. The graph has vertical asymptotes $x = \pi n$, as indicated in the figure.

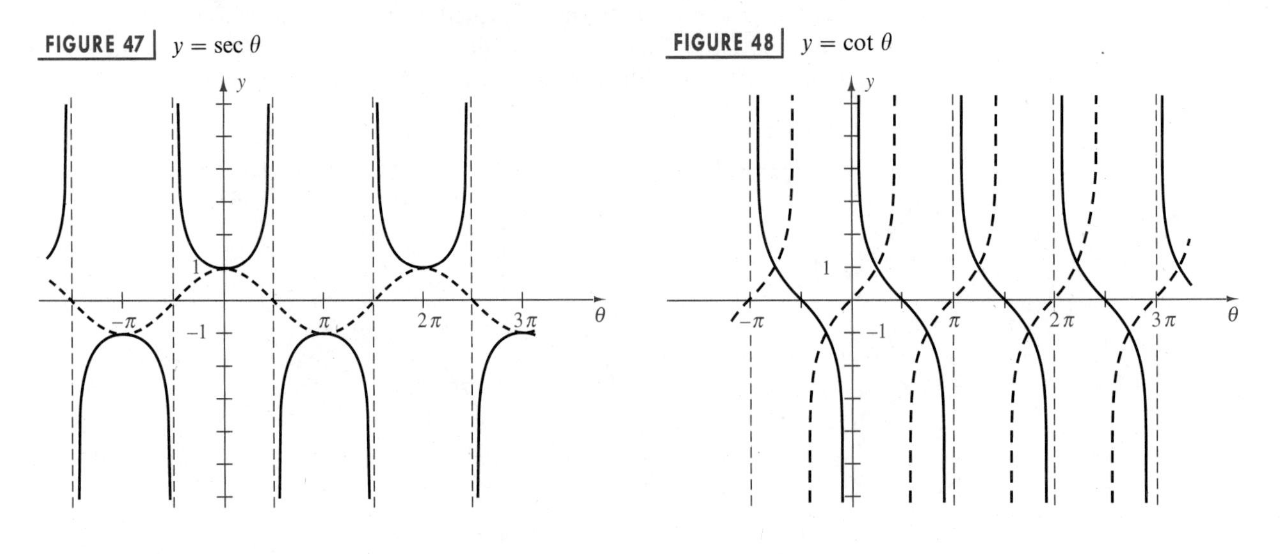

FIGURE 47 $y = \sec\theta$

FIGURE 48 $y = \cot\theta$

Since $\sec \theta = 1/\cos \theta$ and $\cot \theta = 1/\tan \theta$, we may obtain the graphs of the secant and cotangent functions by taking reciprocals of y-coordinates of points on the graphs of the cosine and tangent functions, as illustrated in Figures 47 and 48.

6.5 EXERCISES

Exer. 1–12: Use a formula for negatives to find the exact value.

1 $\sin(-30°)$

2 $\sin\left(-\frac{3\pi}{2}\right)$

3 $\cos\left(-\frac{3\pi}{4}\right)$

4 $\cos(-60°)$

5 $\tan(-45°)$

6 $\tan(-\pi)$

7 $\cot\left(-\frac{3\pi}{4}\right)$

8 $\cot(-225°)$

9 $\sec(-150°)$

10 $\sec\left(-\frac{\pi}{3}\right)$

11 $\csc\left(-\frac{7\pi}{6}\right)$

12 $\csc(-45°)$

Exer. 13–24: Complete the statement by referring to a graph of a trigonometric function (see Figures 40, 42, 45–48). (*Note:* $\theta \to a^+$ means θ approaches a through values *greater* than a; $\theta \to a^-$ means θ approaches a through values *less* than a; and, for a function f, $f(\theta) \to L$ means $f(\theta)$ approaches L.)

13 (a) As $\theta \to 0^+$, $\sin \theta \to$ __

(b) As $\theta \to -\frac{\pi}{2}^-$, $\sin \theta \to$ __

14 (a) As $\theta \to \pi^+$, $\sin \theta \to$ __

(b) As $\theta \to \frac{\pi}{6}^-$, $\sin \theta \to$ __

15 (a) As $\theta \to \frac{\pi}{4}^+$, $\cos \theta \to$ __

(b) As $\theta \to \pi^-$, $\cos \theta \to$ __

16 (a) As $\theta \to 0^+$, $\cos \theta \to$ __

(b) As $\theta \to -\frac{\pi}{3}^-$, $\cos \theta \to$ __

17 (a) As $\theta \to \frac{\pi}{4}^+$, $\tan \theta \to$ __

(b) As $\theta \to \frac{\pi}{2}^+$, $\tan \theta \to$ __

18 (a) As $\theta \to 0^+$, $\tan \theta \to$ __

(b) As $\theta \to -\frac{\pi}{2}^-$, $\tan \theta \to$ __

19 (a) As $\theta \to -\frac{\pi}{4}^-$, $\cot \theta \to$ __

(b) As $\theta \to 0^+$, $\cot \theta \to$ __

20 (a) As $\theta \to \frac{\pi}{6}^+$, $\cot \theta \to$ __

(b) As $\theta \to \pi^-$, $\cot \theta \to$ __

21 (a) As $\theta \to \frac{\pi}{2}^-$, $\sec \theta \to$ __

(b) As $\theta \to \frac{\pi}{4}^+$, $\sec \theta \to$ __

22 (a) As $\theta \to \frac{\pi}{2}^+$, $\sec \theta \to$ __

(b) As $\theta \to 0^-$, $\sec \theta \to$ __

23 (a) As $\theta \to 0^-$, $\csc \theta \to$ __

(b) As $\theta \to \frac{\pi}{2}^+$, $\csc \theta \to$ __

24 (a) As $\theta \to \pi^+$, $\csc \theta \to$ __

(b) As $\theta \to \frac{\pi}{4}^-$, $\csc \theta \to$ __

Exer. 25–48: Refer to the graph of the equation to find the values of θ in the specified interval that correspond to the given value of y.

25 $y = \sin \theta$, $[0, 4\pi]$; $y = -1$

26 $y = \sin \theta$, $[0, 4\pi]$; $y = 1$

27 $y = \cos \theta$, $[0, 4\pi]$; $y = 1$

28 $y = \cos \theta$, $[0, 4\pi]$; $y = -1$

29 $y = \cos\theta$, $[0, 4\pi]$; $y = \dfrac{\sqrt{2}}{2}$

30 $y = \cos\theta$, $[0, 4\pi]$; $y = -\frac{1}{2}$

31 $y = \sin\theta$, $[0, 4\pi]$; $y = \frac{1}{2}$

32 $y = \sin\theta$, $[0, 4\pi]$; $y = -\dfrac{\sqrt{2}}{2}$

33 $y = \tan\theta$, $\left(-\dfrac{\pi}{2}, \dfrac{3\pi}{2}\right)$; $y = 1$

34 $y = \tan\theta$, $\left(-\dfrac{\pi}{2}, \dfrac{3\pi}{2}\right)$; $y = \sqrt{3}$

35 $y = \cot\theta$, $(-\pi, \pi)$; $y = -1$

36 $y = \cot\theta$, $(-\pi, \pi)$; $y = 0$

37 $y = \cot\theta$, $(-\pi, \pi)$; $y = -\dfrac{1}{\sqrt{3}}$

38 $y = \cot\theta$, $(-\pi, \pi)$; $y = \sqrt{3}$

39 $y = \tan\theta$, $\left(-\dfrac{\pi}{2}, \dfrac{3\pi}{2}\right)$; $y = 0$

40 $y = \tan\theta$, $\left(-\dfrac{\pi}{2}, \dfrac{3\pi}{2}\right)$; $y = -\dfrac{1}{\sqrt{3}}$

41 $y = \sec\theta$, $\left(-\dfrac{\pi}{2}, \dfrac{3\pi}{2}\right)$; $y = 1$

42 $y = \sec\theta$, $\left(-\dfrac{\pi}{2}, \dfrac{3\pi}{2}\right)$; $y = 0$

43 $y = \csc\theta$, $(-\pi, \pi)$; $y = \sqrt{2}$

44 $y = \csc\theta$, $(-\pi, \pi)$; $y = -1$

45 $y = \csc\theta$, $(-\pi, \pi)$; $y = 0$

46 $y = \csc\theta$, $(-\pi, \pi)$; $y = 1$

47 $y = \sec\theta$, $\left(-\dfrac{\pi}{2}, \dfrac{3\pi}{2}\right)$; $y = -\sqrt{2}$

48 $y = \sec\theta$, $\left(-\dfrac{\pi}{2}, \dfrac{3\pi}{2}\right)$; $y = -1$

Exer. 49–52: Refer to the graph of the equation for the specified interval and real number a. Find all values of θ such that (a) $y = a$, (b) $y > a$, (c) $y < a$.

49 $y = \sin\theta$, $[-2\pi, 2\pi]$; $a = \frac{1}{2}$

50 $y = \cos\theta$, $[0, 4\pi]$; $a = \dfrac{\sqrt{3}}{2}$

51 $y = \cos\theta$, $[-2\pi, 2\pi]$; $a = -\frac{1}{2}$

52 $y = \sin\theta$, $[0, 4\pi]$; $a = -\dfrac{\sqrt{2}}{2}$

53 Determine the range of the secant function.

54 Determine the range of the cosecant function.

Exer. 55–58: Find the intervals between -2π and 2π on which the given function is (a) increasing. (b) decreasing.

55 secant

56 cosecant

57 tangent

58 cotangent

Exer. 59–72: Use the graph of a trigonometric function to sketch the graph of the equation without plotting points.

59 $y = 2 + \sin\theta$

60 $y = 3 + \cos\theta$

61 $y = \cos\theta - 2$

62 $y = \sin\theta - 1$

63 $y = 1 + \tan\theta$

64 $y = \cot\theta - 1$

65 $y = \sec\theta - 2$

66 $y = 1 + \csc\theta$

67 $y = |\sin\theta|$

68 $y = |\cos\theta|$

69 $y = |\sin\theta| + 2$

70 $y = -|\sin\theta| - 2$

71 $y = -|\cos\theta| + 1$

72 $y = |\cos\theta| - 3$

73 Practice sketching the graph of the sine function, taking different units of length on the horizontal and vertical axes. Practice sketching graphs of the cosine and tangent functions in the same manner. Continue this practice until you reach the stage at which, if you were awakened from a sound sleep in the middle of the night and asked to sketch one of these graphs, you could do so in less than thirty seconds.

74 Work Exercise 73 for the cosecant, secant, and cotangent functions.

75 Prove that the range of the tangent function is $\mathbb{R}$ by showing that if c is any real number, then there is an angle θ in standard position such that $\tan\theta = c$. (*Hint:* Let $P(x, y)$ be a point on the unit circle U, and consider the equation $\tan\theta = y/x = c$ with $x^2 + y^2 = 1$.)

76 Prove that the range of the cotangent function is $\mathbb{R}$.

6.6 TRIGONOMETRIC GRAPHS

In preceding sections we used $\sin \theta$ to denote values of the sine function at θ. Since we now wish to sketch graphs on an xy-coordinate system, we shall consider equations of the form $y = \sin x$. Instead of using a θy-coordinate system as in Section 6.5, we let x represent θ. We may think of x as the radian measure of an angle; however, in calculus, x is regarded as a real number. These are equivalent points of view, since the sine of an angle of x radians is the same as the sine of the real number x.

In this section we consider graphs of the equations

$$y = a \sin (bx + c) \quad \text{and} \quad y = a \cos (bx + c)$$

for real numbers a, b, and c. Our goal is to sketch such graphs without plotting many points. To do so we shall use facts about the graphs of the sine and cosine functions discussed in Section 6.5.

Let us begin by considering the special case $c = 0$ and $b = 1$, that is,

$$y = a \sin x \quad \text{and} \quad y = a \cos x.$$

FIGURE 49

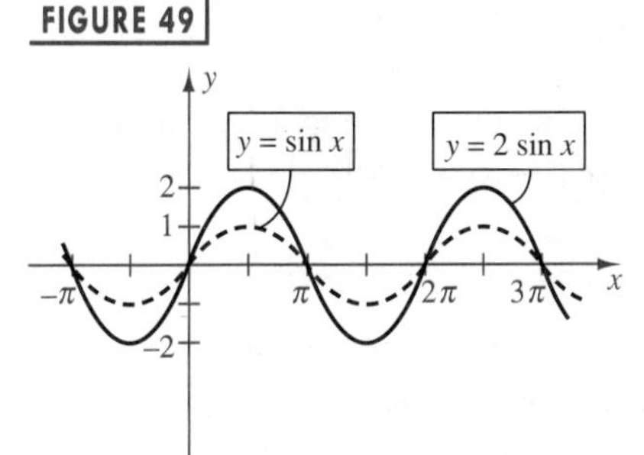

We can find y-coordinates of points on the graphs by multiplying y-coordinates of points on the graphs of $y = \sin x$ and $y = \cos x$ by a. To illustrate, if $y = 2 \sin x$, we multiply the y-coordinate of each point on the graph of $y = \sin x$ by 2. This gives us Figure 49, where for comparison we have shown the graph of $y = \sin x$ with dashes. This is the same procedure as that of stretching the graph of a function discussed in Section 3.5.

As another illustration, if $y = \frac{1}{2} \sin x$, we multiply y-coordinates of points on the graph of $y = \sin x$ by $\frac{1}{2}$. This compresses the graph of $y = \sin x$ by a factor of $\frac{1}{2}$, as illustrated in Figure 50.

FIGURE 50

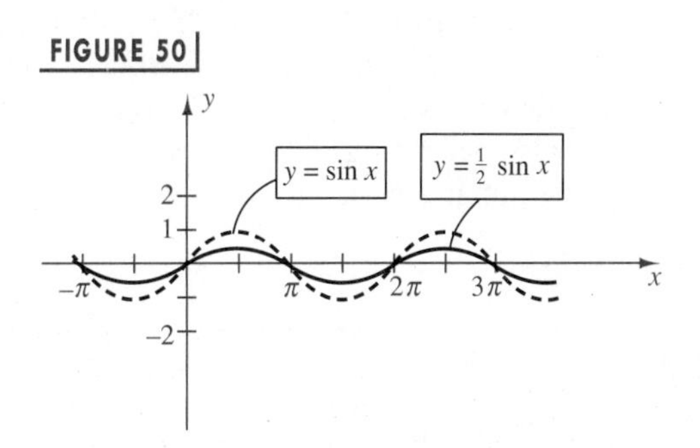

The following example illustrates a graph of $y = a \sin x$ with a negative.

EXAMPLE ▪ 1

Sketch the graph of the equation $y = -2 \sin x$.

FIGURE 51

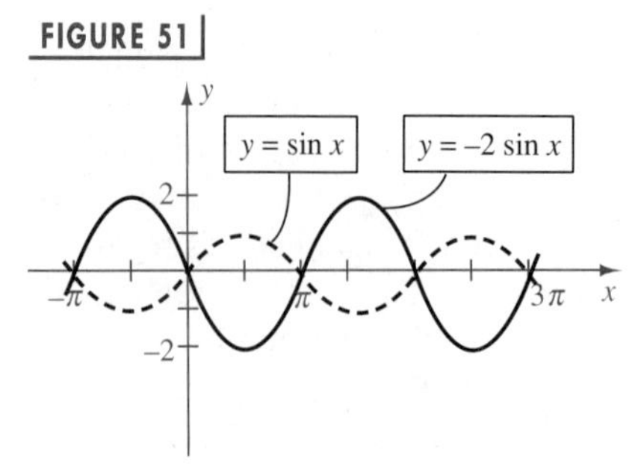

SOLUTION The graph of $y = -2 \sin x$ sketched in Figure 51 can be obtained by first sketching the graph of $y = \sin x$ (shown with dashes in the figure) and then multiplying y-coordinates by -2. An alternative method is to reflect the graph of $y = 2 \sin x$ (see Figure 49 through the x-axis.

For any $a \neq 0$, the graph of $y = a \sin x$ has the general appearance of one of the graphs illustrated in Figures 49–51. The amount of stretching of the graph of $y = \sin x$ and whether the graph is reflected is determined by the absolute value and sign of a. The largest y-coordinate $|a|$ is the **amplitude of the graph** or, equivalently, the **amplitude of the function *f*** given by $f(x) = a \sin x$. In Figures 49 and 51, the amplitude is 2. In Figure 50 the amplitude is $\frac{1}{2}$. Similar remarks and techniques apply if $y = a \cos x$.

EXAMPLE ▪ 2

Find the amplitude and sketch the graph of $y = 3 \cos x$.

FIGURE 52

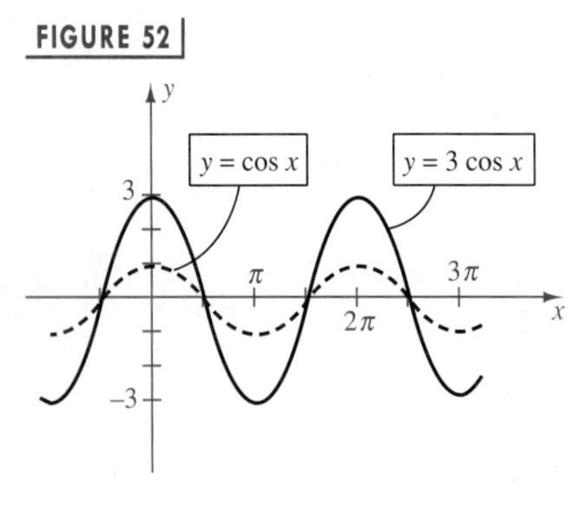

SOLUTION By the preceding discussion the amplitude is 3. As indicated in Figure 52, we first sketch the graph of $y = \cos x$ with dashes and then multiply y-coordinates by 3.

Let us next consider $y = a \sin bx$ and $y = a \cos bx$ for nonzero real numbers a and b. As before, the amplitude is $|a|$. If $b > 0$, then exactly one cycle occurs as bx increases from 0 to 2π or, equivalently, as x increases from 0 to $2\pi/b$. If $b < 0$, then $-b > 0$ and one cycle occurs if x increases from 0 to $2\pi/(-b)$. Thus the period of the function f given by $f(x) = a \sin bx$ or $f(x) = a \cos bx$ is $2\pi/|b|$. For convenience, we shall also refer to $2\pi/|b|$ as the period of the *graph* of f. The next theorem summarizes our discussion.

THEOREM

If $y = a \sin bx$ or $y = a \cos bx$ for nonzero real numbers a and b, then the graph has amplitude $|a|$ and period $\dfrac{2\pi}{|b|}$.

EXAMPLE ▪ 3

Find the amplitude and the period, and sketch the graph of $y = 3 \sin 2x$.

FIGURE 53

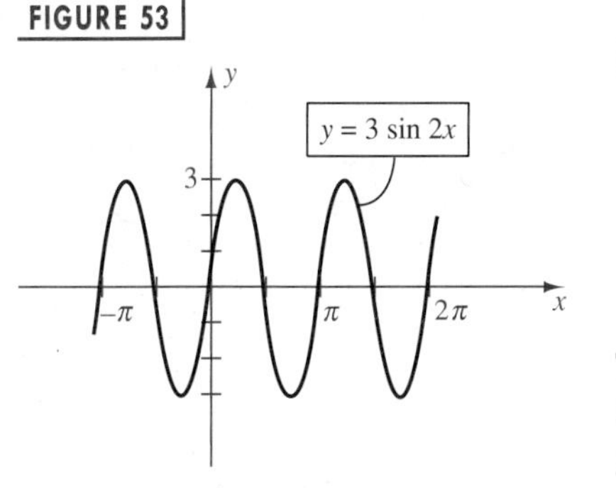

SOLUTION Using the preceding theorem with $a = 3$ and $b = 2$, we obtain the following:

$$\text{amplitude:} \quad |a| = |3| = 3$$

$$\text{period:} \quad \frac{2\pi}{|b|} = \frac{2\pi}{2} = \pi$$

Thus there is exactly one sine wave of amplitude 3 on the x-interval $[0, \pi]$. Sketching this wave and then extending the graph to the right and left gives us Figure 53.

EXAMPLE ▪ 4

Find the amplitude and the period, and sketch the graph of $y = 2 \sin \frac{1}{2}x$.

FIGURE 54

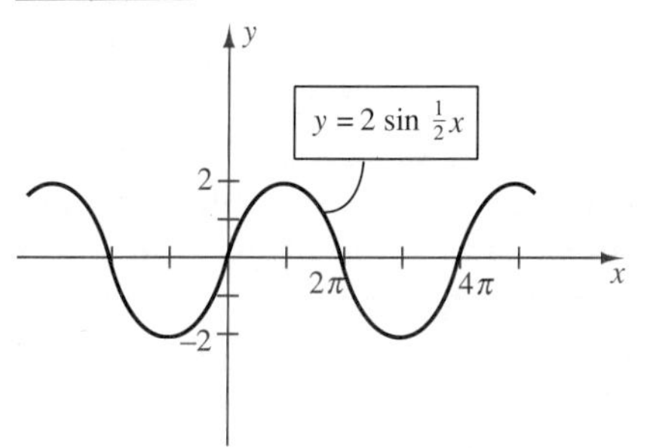

SOLUTION Using the preceding theorem with $a = 2$ and $b = \frac{1}{2}$, we obtain the following:

$$\text{amplitude:} \quad |a| = |2| = 2$$

$$\text{period:} \quad \frac{2\pi}{|b|} = \frac{2\pi}{\frac{1}{2}} = 4\pi$$

Thus there is one sine wave of amplitude 2 on the interval $[0, 4\pi]$. Sketching this wave and extending it left and right gives us the sketch in Figure 54.

If $y = a \sin bx$ and if b is a large positive number, then the period $2\pi/b$ is small and the sine waves are close together, with b sine waves on the interval $[0, 2\pi]$. For example, in Figure 53, $b = 2$ and we have two sine waves on $[0, 2\pi]$. If b is a small positive number, then the period $2\pi/b$ is large and the waves are far apart. To illustrate, if $y = \sin \frac{1}{10}x$, then $\frac{1}{10}$ of a sine wave occurs on $[0, 2\pi]$, and an interval 20π units long is required for one complete cycle. (See also Figure 54—for $y = 2 \sin \frac{1}{2}x$, one-half of a sine wave occurs in $[0, 2\pi]$.)

If $b < 0$ we can use the fact that $\sin(-x) = -\sin x$ to obtain the graph of $y = a \sin bx$. To illustrate, the graph of $y = \sin(-2x)$ is the same as the graph of $y = -\sin 2x$.

EXAMPLE ▪ 5

Find the amplitude and the period, and sketch the graph of $y = 2 \sin(-3x)$.

SOLUTION Since $\sin(-3x) = -\sin 3x$, we may write $y = -2 \sin 3x$. The amplitude is $|-2| = 2$ and the period is $2\pi/3$. Thus, there is one cycle on an interval of length $2\pi/3$. The minus sign indicates a reflection through the x-axis. If we consider the interval $[0, 2\pi/3]$ and sketch a sine wave of amplitude 2 (reflected through the x-axis), the shape of the graph is apparent. The part of the graph in the interval $[0, 2\pi/3]$ is repeated periodically, as illustrated in Figure 55.

FIGURE 55

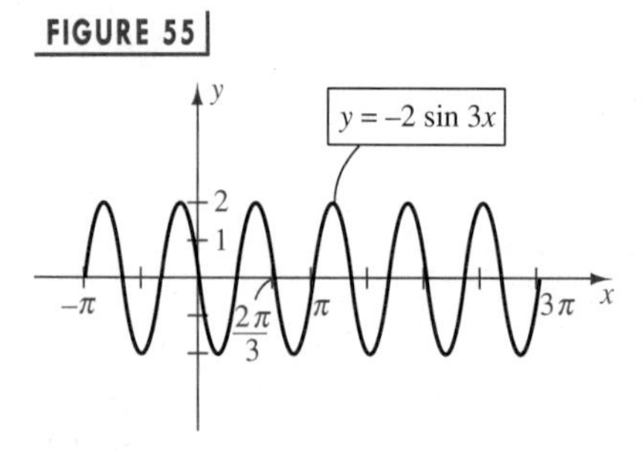

As in Section 3.5, if f is a function and c is a positive real number, then the graph of $y = f(x) + c$ can be obtained by shifting the graph of $y = f(x)$ vertically upward a distance c. For the graph of $y = f(x) - c$ we shift the graph of $y = f(x)$ vertically downward a distance c. In the next example we use this technique for a trigonometric graph.

EXAMPLE ▪ 6

Sketch the graph of $y = 2 \sin x + 3$.

SOLUTION It is important to note that $y \neq 2 \sin(x + 3)$. The graph of $y = 2 \sin x$ is sketched with dashes in Figure 56. If we shift this graph vertically upward a distance 3, we obtain the graph of $y = 2 \sin x + 3$.

FIGURE 56

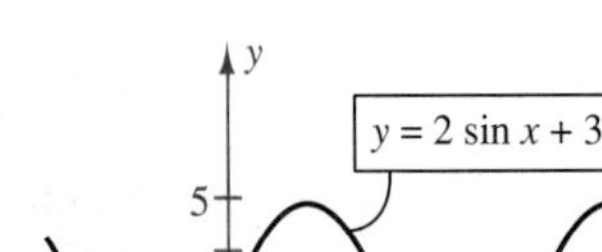

Let us next consider the graph of

$$y = a \sin(bx + c).$$

As before, the amplitude is $|a|$ and the period is $2\pi/|b|$. One cycle occurs if $bx + c$ increases from 0 to 2π. Hence we can find an interval containing exactly one sine wave by solving the equations

$$bx + c = 0 \quad \text{and} \quad bx + c = 2\pi.$$

The solutions are

$$x = -\frac{c}{b} \quad \text{and} \quad x = -\frac{c}{b} + \frac{2\pi}{b}.$$

The number $-c/b$ is the **phase shift** associated with the graph. The graph of $y = a \sin(bx + c)$ may be obtained by shifting the graph of $y = a \sin bx$ to the left if the phase shift is negative or to the right if the phase shift is positive.

Analogous results are true for $y = a \cos(bx + c)$. The next theorem summarizes our discussion.

THEOREM

> If $y = a \sin(bx + c)$ or $y = a \cos(bx + c)$ for nonzero real numbers a and b, then:
>
> (i) the amplitude is $|a|$ and the period is $\dfrac{2\pi}{|b|}$.
>
> (ii) the phase shift and an interval containing exactly one cycle can be found by solving the two equations
>
> $$bx + c = 0 \quad \text{and} \quad bx + c = 2\pi.$$

EXAMPLE ▪ 7

Find the amplitude, the period, and the phase shift, and sketch the graph of $y = 3 \sin\left(2x + \dfrac{\pi}{2}\right)$.

SOLUTION The equation is of the form $y = a \sin(bx + c)$ with $a = 3$, $b = 2$, and $c = \pi/2$. Thus the amplitude is $|a| = 3$ and the period is $2\pi/|b| = 2\pi/2 = \pi$.

By (ii) of the preceding theorem, the phase shift and an interval containing one sine wave can be found by solving the two equations

$$2x + \frac{\pi}{2} = 0 \quad \text{and} \quad 2x + \frac{\pi}{2} = 2\pi.$$

This gives us

$$x = -\frac{\pi}{4} \quad \text{and} \quad x = \frac{3\pi}{4}.$$

Thus the phase shift is $-\pi/4$, and one sine wave of amplitude 3 occurs on the interval $[-\pi/4, 3\pi/4]$. Sketching that wave and then repeating it to the right and left gives us the graph in Figure 57.

FIGURE 57

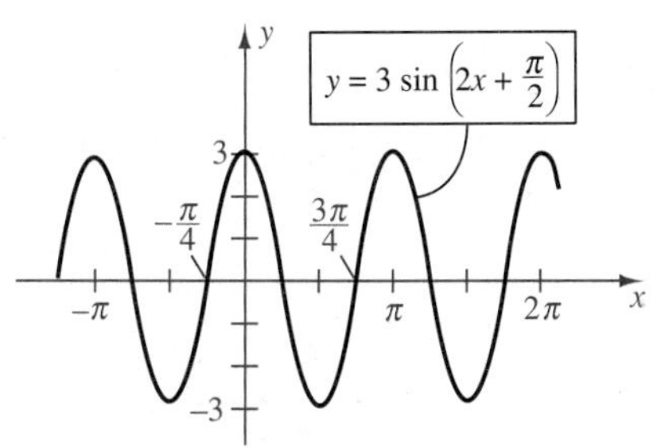

EXAMPLE ▪ 8

Find the amplitude, the period, and the phase shift, and sketch the graph of $y = 2\cos(3x - \pi)$.

FIGURE 58

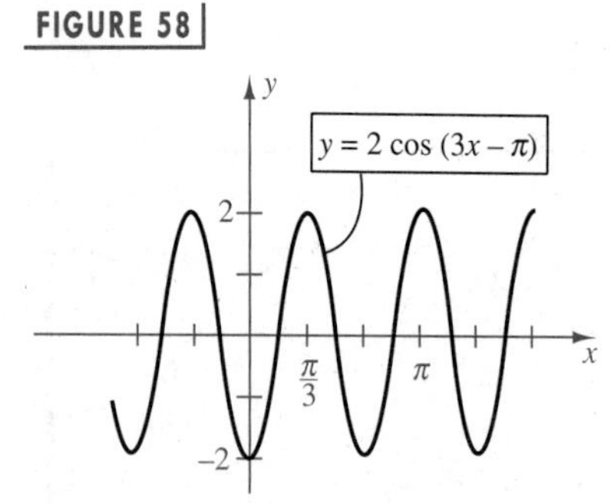

SOLUTION The equation has the form $y = a\cos(bx + c)$ with $a = 2$, $b = 3$, and $c = -\pi$. Thus the amplitude is $|a| = 2$ and the period is $2\pi/|b| = 2\pi/3$.

By (ii) of the previous theorem, the phase shift and an interval containing one cycle can be found by solving the two equations

$$3x - \pi = 0 \quad \text{and} \quad 3x - \pi = 2\pi.$$

This gives us $$x = \frac{\pi}{3} \quad \text{and} \quad x = \frac{3\pi}{3} = \pi.$$

Hence the phase shift is $\pi/3$ and one cosine-type cycle of amplitude 2 occurs on the interval $[\pi/3, \pi]$. Sketching that part of the graph and then repeating it to the right and left gives us the sketch in Figure 58.

Many phenomena that occur in nature vary in a cyclic or rhythmic manner. It is sometimes possible to represent such behavior by means of trigonometric functions, as illustrated in the next two examples.

EXAMPLE ▪ 9

The rhythmic process of breathing consists of alternating periods of inhaling and exhaling. One complete cycle normally takes place every 5 seconds. If $F(t)$ denotes the air flow rate at time t (in liters/second) and if the maximum flow rate is 0.6 liters/second, find a formula $F(t) = a\sin bt$ that fits this information.

SOLUTION If $F(t) = a\sin bt$ for some $b > 0$, then the period of F is $2\pi/b$. In this application the period is 5 seconds, and hence

$$\frac{2\pi}{b} = 5, \quad \text{or} \quad b = \frac{2\pi}{5}.$$

Since the maximum flow rate corresponds to the amplitude a of F, we let $a = 0.6$. This gives us the formula

$$F(t) = 0.6\sin\left(\frac{2\pi}{5}t\right).$$

EXAMPLE ▪ 10

The number of hours of daylight $D(t)$ at a particular time of the year can be approximated by

$$D(t) = \frac{K}{2} \sin \frac{2\pi}{365}(t - 79) + 12$$

for t in days and $t = 0$ corresponding to January 1. The constant K determines the total variation in day length and depends on the latitude of the locale.

(a) For Boston, $K \approx 6$. Sketch the graph of D for $0 \leq t \leq 365$.

(b) When is the day length the longest? the shortest?

SOLUTION

(a) If $K = 6$, then $K/2 = 3$, and we may write $D(t)$ in the form

$$D(t) = f(t) + 12$$

with

$$f(t) = 3 \sin \frac{2\pi}{365}(t - 79).$$

We shall sketch the graph of f and then apply a vertical shift through a distance 12. Let us rewrite $f(t)$ as

$$f(t) = 3 \sin \left(\frac{2\pi}{365} t - \frac{158}{365} \pi \right)$$

which has the form $f(t) = a \sin(bt + c)$ with $a = 3$, $b = 2\pi/365$, and $c = -158\pi/365$. Thus the amplitude is 3 and the period of f is

$$\frac{2\pi}{b} = \frac{2\pi}{2\pi/365} = 365 \text{ days}.$$

As in (ii) of the preceding theorem, we can obtain a t-interval containing exactly one cycle by solving the two equations

$$\frac{2\pi}{365} t - \frac{158}{365} \pi = 0 \quad \text{and} \quad \frac{2\pi}{365} t - \frac{158}{365} \pi = 2\pi.$$

The solutions are $t = 79$ and $t = 444$. Hence one sine wave occurs on the interval [79, 444]. Dividing this interval into four equal parts, we obtain the following table of values, which indicates the familiar sine wave pattern of amplitude 3.

t	79	170.25	261.5	352.75	444
$f(t)$	0	3	0	−3	0

FIGURE 59

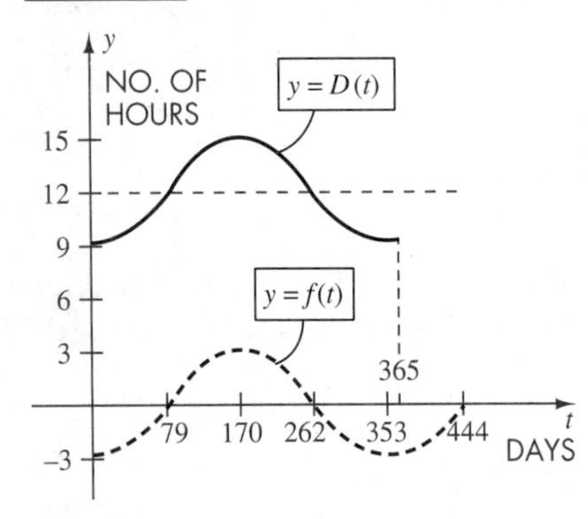

If $t = 0$,

$$f(0) = 3 \sin \frac{2\pi}{365}(-79) \approx 3 \sin(-1.36) \approx -2.9.$$

Since the period of f is 365, this implies that $f(365) \approx -2.9$.

The graph of f for the interval $[0, 444]$ is sketched with dashes in Figure 59, using different scales on the axes with t rounded off to the nearest day.

Applying a vertical shift of 12 units gives us the graph of D for $0 \le t \le 365$ shown in Figure 59.

(b) The longest day—that is, the largest value of $D(t)$—occurs 170 days after January 1. Except for leap year, this corresponds to June 20. The shortest day occurs 353 days after January 1, or December 20.

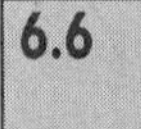

6.6 EXERCISES

1 Find the amplitude and the period, and sketch the graph of the equation:

(a) $y = 4 \sin x$ **(b)** $y = \sin 4x$

(c) $y = \frac{1}{4} \sin x$ **(d)** $y = \sin \frac{1}{4}x$

(e) $y = 2 \sin \frac{1}{4}x$ **(f)** $y = \frac{1}{2} \sin 4x$

(g) $y = -4 \sin x$ **(h)** $y = \sin(-4x)$

2 Sketch the graphs of the equations that involve the cosine and are analogous to those in (a)–(h) of Exercise 1.

3 Find the amplitude and the period, and sketch the graph of the equation:

(a) $y = 3 \cos x$ **(b)** $y = \cos 3x$

(c) $y = \frac{1}{3} \cos x$ **(d)** $y = \cos \frac{1}{3}x$

(e) $y = 2 \cos \frac{1}{3}x$ **(f)** $y = \frac{1}{2} \cos 3x$

(g) $y = -3 \cos x$ **(h)** $y = \cos(-3x)$

4 Sketch the graphs of the equations that involve the sine and are analogous to those in (a)–(h) of Exercise 3.

Exer. 5–36: Find the amplitude, the period, and the phase shift, and sketch the graph of the equation.

5 $y = \sin\left(x - \frac{\pi}{2}\right)$ **6** $y = \sin\left(x + \frac{\pi}{4}\right)$

7 $y = 3 \sin\left(x + \frac{\pi}{6}\right)$ **8** $y = 2 \sin\left(x - \frac{\pi}{3}\right)$

9 $y = \cos\left(x + \frac{\pi}{2}\right)$ **10** $y = \cos\left(x - \frac{\pi}{3}\right)$

11 $y = 4 \cos\left(x - \frac{\pi}{4}\right)$ **12** $y = 3 \cos\left(x + \frac{\pi}{6}\right)$

13 $y = \sin(2x - \pi) + 1$ **14** $y = -\sin(3x + \pi) - 1$

15 $y = -\cos(3x + \pi) - 2$ **16** $y = \cos(2x - \pi) + 2$

17 $y = -2 \sin(3x - \pi)$ **18** $y = 3 \cos(3x - \pi)$

19 $y = \sin\left(\frac{1}{2}x - \frac{\pi}{3}\right)$ **20** $y = \sin\left(\frac{1}{2}x + \frac{\pi}{4}\right)$

21 $y = 6 \sin \pi x$ **22** $y = 3 \cos \frac{\pi}{2} x$

23 $y = 2 \cos \frac{\pi}{2} x$ **24** $y = 4 \sin 3\pi x$

25 $y = \frac{1}{2} \sin 2\pi x$ **26** $y = \frac{1}{2} \cos \frac{\pi}{2} x$

27 $y = 5 \sin\left(3x - \frac{\pi}{2}\right)$ **28** $y = -4 \cos\left(2x + \frac{\pi}{3}\right)$

29 $y = 3 \cos\left(\frac{1}{2}x - \frac{\pi}{4}\right)$ **30** $y = -2 \sin\left(\frac{1}{2}x + \frac{\pi}{2}\right)$

31 $y = -5 \cos\left(\frac{1}{3}x + \frac{\pi}{6}\right)$ **32** $y = 4 \sin\left(\frac{1}{3}x - \frac{\pi}{3}\right)$

33 $y = -2 \sin(2x - \pi) + 3$ **34** $y = 3 \cos(x + 3\pi) - 2$

35 $y = 5 \cos(2x + 2\pi) + 2$ **36** $y = -4 \sin(3x - \pi) - 3$

Exer. 37–40: The graph of an equation is shown in the figure.

(a) Find the amplitude and period.

(b) Write the equation in the form $y = a \sin(bx + c)$ for the least positive real number c.

(c) Find the phase shift.

37

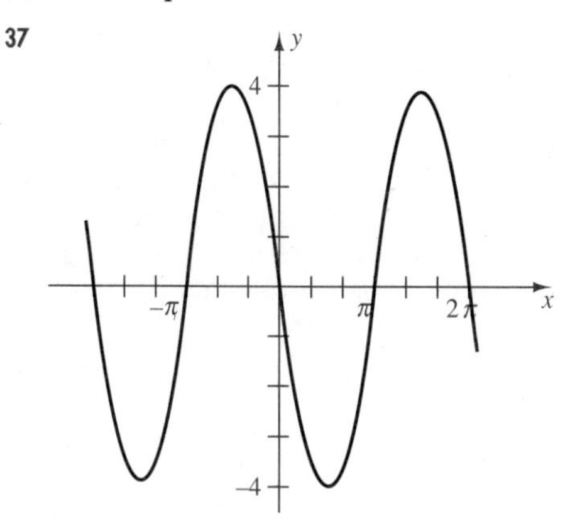

38

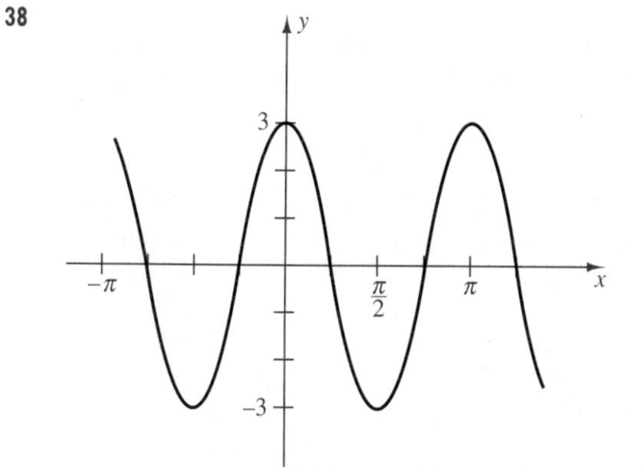

39

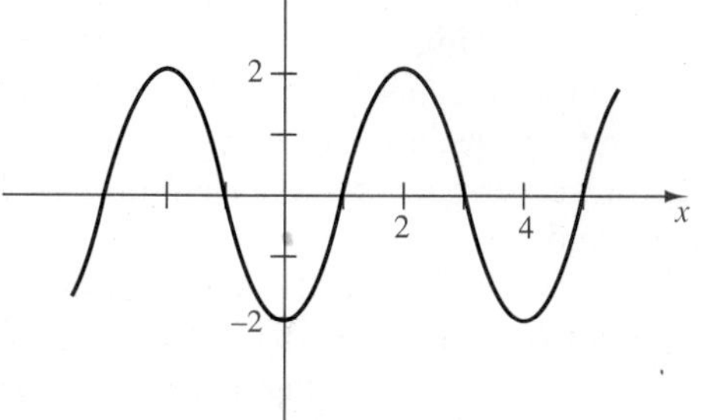

40

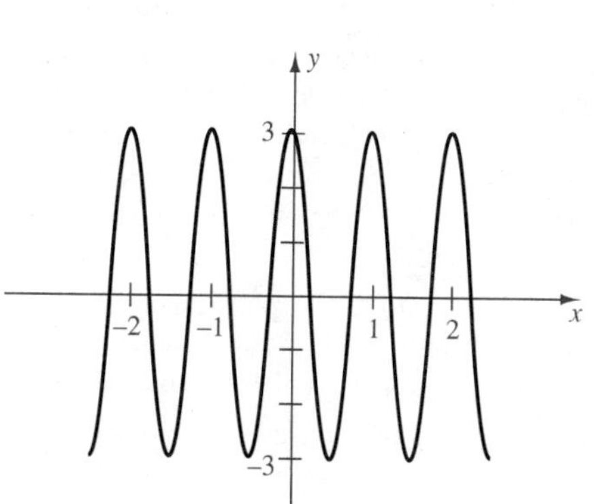

41 Shown in the figure is an electroencephalogram of human brain waves during deep sleep. If we use $W = a \sin(bt + c)$ to represent these waves, what is the value of b?

EXERCISE 41

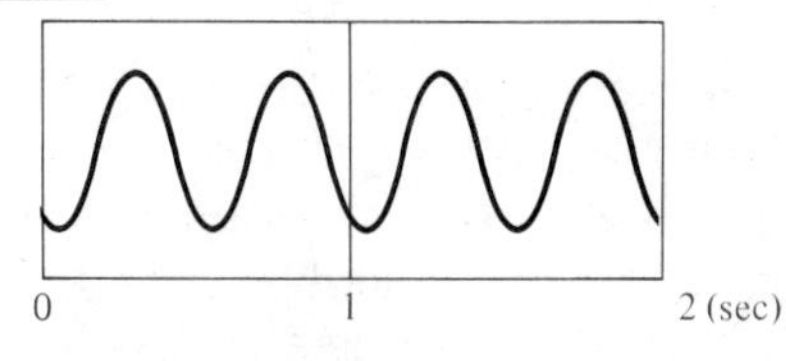

42 On a certain spring day with 12 hours of daylight, the light intensity I takes on its largest value of 510 calories per cm^2 at midday. If $t = 0$ corresponds to sunrise, find a formula $I = a \sin bt$ that fits this information.

43 The pumping action of the heart consists of a systolic phase in which blood rushes from the left ventricle into the aorta and the diastolic phase during which the heart muscle relaxes. The function whose graph is shown in the figure is sometimes used to model one complete cycle of this process. For a particular individual the systolic phase lasts $\frac{1}{4}$ second and has a maximum flow rate of 8 liters/minute. Find a and b.

EXERCISE 43

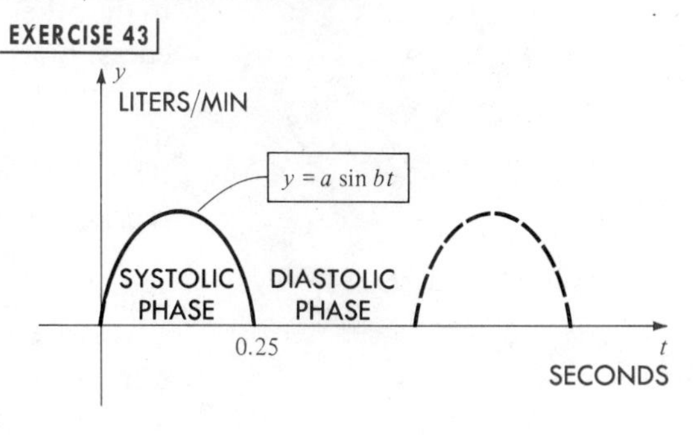

44 The popular biorhythm theory uses the graphs of three simple sine functions to make predictions about an individual's physical, emotional, and intellectual potential for a particular day. The graphs are given by $y = a \sin bt$ for t in days with $t = 0$ corresponding to birth and $a = 1$ denoting 100% potential.

(a) Find the value of b for the physical cycle having a period of 23 days; for the emotional cycle (period 28 days); and for the intellectual cycle (period 33 days).

(b) Evaluate the biorhythm cycles for a person who has just become 21 years of age and is exactly 7660 days old.

45 The height of the tide at a particular point on shore can be predicted by using seven trigonometric functions (called tidal components) of the form $f(t) = a \cos (bt + c)$. The principal lunar component may be approximated by

$$f(t) = a \cos \left(\frac{\pi}{6} t - \frac{11\pi}{12} \right)$$

with t in hours and $t = 0$ corresponding to midnight. Sketch the graph of f if $a = 0.5$ meter.

46 Refer to Exercise 45. The principal solar diurnal component may be approximated by

$$f(t) = a \cos \left(\frac{\pi}{12} t - \frac{7\pi}{12} \right).$$

Sketch the graph of f if $a = 0.2$ meter.

47 If the formula for $D(t)$ in Example 10 is used for Fairbanks, Alaska, then $K \approx 12$. Sketch the graph of D in this case for $0 \le t \le 365$.

48 Based on years of weather data, the expected low temperature T (in °F) in Fairbanks, Alaska, can be approximated by

$$T = 36 \sin \frac{2\pi}{365} (t - 101) + 14$$

for t in days and $t = 0$ corresponding to January 1.

(a) Sketch the graph of T for $0 \le t \le 365$.

(b) Predict when the coldest day of the year will occur.

Exer. 49–52: Scientists sometimes use the formula

$$f(t) = a \sin (bt + c) + d$$

to simulate temperature variations during the day with time t in hours and temperature $f(t)$ in °C with $t = 0$ corresponding to midnight. Assume that $f(t)$ is decreasing at midnight.

(a) Determine values of a, b, c, and d that fit the information.

(b) Sketch the graph of f for $0 \le t \le 24$.

49 The high temperature is 10 °C, and the low temperature of −10 °C occurs at 4 A.M.

50 The temperature at midnight is 15 °C, and the high and low temperatures are 20 °C and 10 °C.

51 The temperature varies between 10 °C and 30 °C, and the average temperature of 20 °C first occurs at 9 A.M.

52 The high temperature of 28 °C occurs at 2 P.M., and the average temperature of 20 °C occurs 6 hours later.

6.7 ADDITIONAL TRIGONOMETRIC GRAPHS

Methods we developed in Section 6.6 for the sine and cosine can be applied to the other four trigonometric functions; however, there are several differences. Since the tangent, cotangent, secant, and cosecant functions have no largest values, the notion of amplitude has no meaning. Moreover, we do not refer to cycles. For some tangent and cotangent graphs we begin by sketching the portion between successive vertical asymptotes, and then repeat that pattern to the right and to the left.

The graph of $y = a \tan x$ for $a > 0$ can be obtained by stretching or compressing the graph of $y = \tan x$ by a factor of a. If $a < 0$, then we also use a reflection. Since the tangent function has period π, it is sufficient to sketch the graph between the two successive vertical asymptotes $x = -\pi/2$ and $x = \pi/2$. The same pattern occurs to the right and to the left, as in the next example.

EXAMPLE ▪ 1

Sketch the graph of the equation:

(a) $y = 2 \tan x$ **(b)** $y = \frac{1}{2} \tan x$

SOLUTION Begin by sketching the graph of $y = \tan x$, with dashes, between the vertical asymptotes $x = -\pi/2$ and $x = \pi/2$.

(a) For $y = 2 \tan x$, we multiply the y-coordinate of each point by 2, and then extend the resulting graph to the right and left, as shown in Figure 60.

(b) For $y = \frac{1}{2} \tan x$, we multiply the y-coordinates by $\frac{1}{2}$, obtaining the sketch in Figure 61.

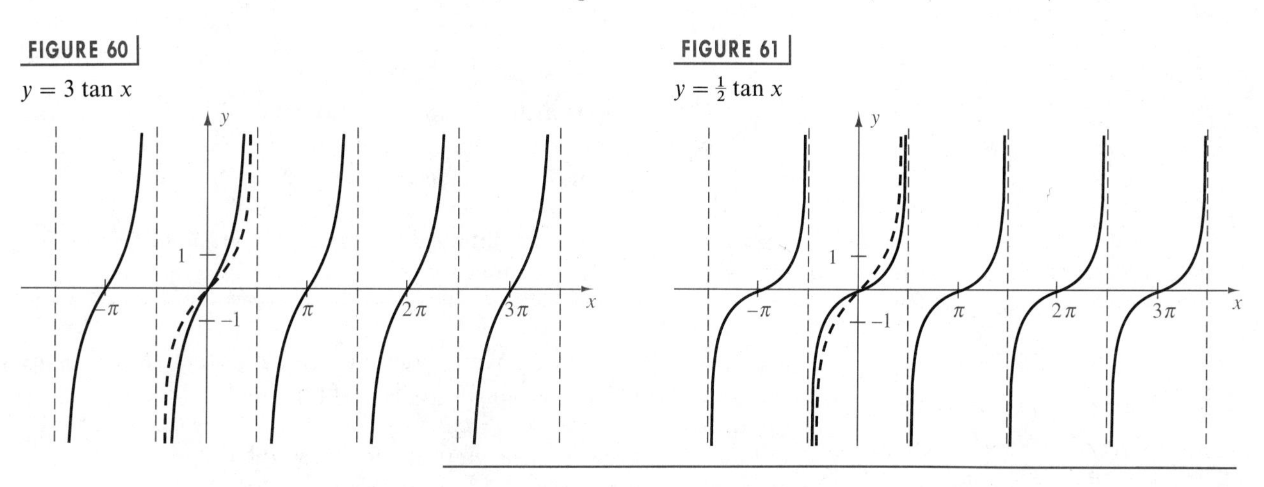

FIGURE 60 $y = 3 \tan x$

FIGURE 61 $y = \frac{1}{2} \tan x$

The method used in Example 1 can be applied to other functions. Thus, to sketch the graph of $y = 3 \sec x$ we could sketch the graph of $y = \sec x$ (with dashes), and then multiply the y-coordinate of each point by 3.

The next theorem is an analogue of the theorem stated in Section 6.6 for the sine and cosine functions.

THEOREM

If $y = a \tan (bx + c)$ for nonzero real numbers a and b, then

(i) the period is $\dfrac{\pi}{|b|}$.

(ii) successive vertical asymptotes for the graph may be found by solving the equations

$$bx + c = -\frac{\pi}{2} \quad \text{and} \quad bx + c = \frac{\pi}{2}.$$

(iii) the phase shift is $-\dfrac{c}{b}$.

FIGURE 62

$y = \frac{1}{2}\tan\left(x + \frac{\pi}{4}\right)$

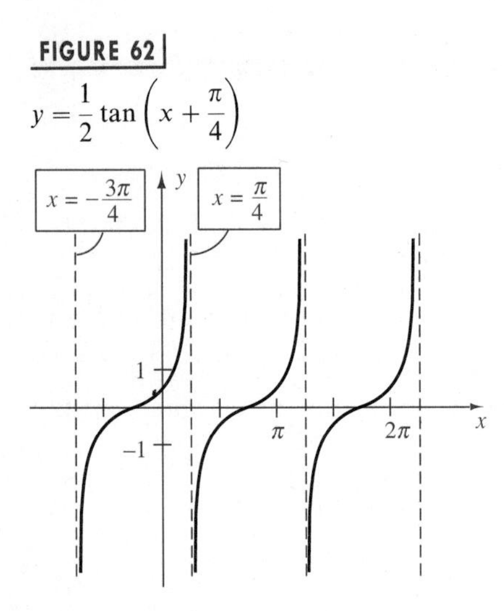

EXAMPLE ▪ 2

Find the period and sketch the graph of $y = \frac{1}{2}\tan\left(x + \frac{\pi}{4}\right)$.

SOLUTION The equation has the form given in the preceding theorem with $a = \frac{1}{2}$, $b = 1$, and $c = \pi/4$. Hence by (i), the period is $\pi/|b| = \pi/1 = \pi$.

As in (ii), to find successive vertical asymptotes we solve the two equations

$$x + \frac{\pi}{4} = -\frac{\pi}{2} \quad \text{and} \quad x + \frac{\pi}{4} = \frac{\pi}{2},$$

obtaining

$$x = -\frac{3\pi}{4} \quad \text{and} \quad x = \frac{\pi}{4}.$$

Since $a = \frac{1}{2}$, the graph of the equation on the interval $[-3\pi/4, \pi/4]$ has the shape of the graph of $y = \frac{1}{2}\tan x$ (see Figure 61). Sketching that part of the graph and extending it to the right and left gives us Figure 62.

Note that since $c = \pi/4$ and $b = 1$, the phase shift is $-c/b = -\pi/4$. Hence the graph can also be obtained by shifting the graph of $y = \frac{1}{2}\tan x$ in Figure 61 to the left a distance $\pi/4$.

If $y = a\cot(bx + c)$ we have a situation similar to that stated in the previous theorem. The only difference is part (ii). Since successive vertical asymptotes for the graph of $y = \cot x$ are $x = 0$ and $x = \pi$ (see Figure 48), we obtain successive vertical asymptotes for the graph of $y = a\cot(bx + c)$ by solving the equations

$$bx + c = 0 \quad \text{and} \quad bx + c = \pi.$$

EXAMPLE ▪ 3

Find the period and sketch the graph of $y = \cot\left(2x - \frac{\pi}{2}\right)$.

SOLUTION Using the usual notation we see that $a = 1$, $b = 2$, and $c = -\pi/2$. The period is $\pi/|b| = \pi/2$. Hence the graph repeats itself in intervals of length $\pi/2$.

As in the discussion preceding this example, to find two successive vertical asymptotes for the graph we solve the equations

$$2x - \frac{\pi}{2} = 0 \quad \text{and} \quad 2x - \frac{\pi}{2} = \pi,$$

obtaining

$$x = \frac{\pi}{4} \quad \text{and} \quad x = \frac{3\pi}{4}.$$

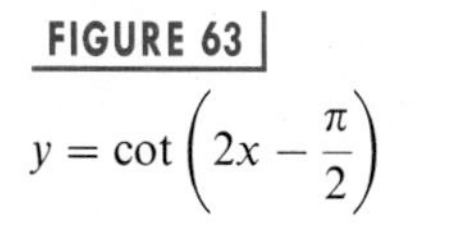
FIGURE 63 $y = \cot\left(2x - \frac{\pi}{2}\right)$

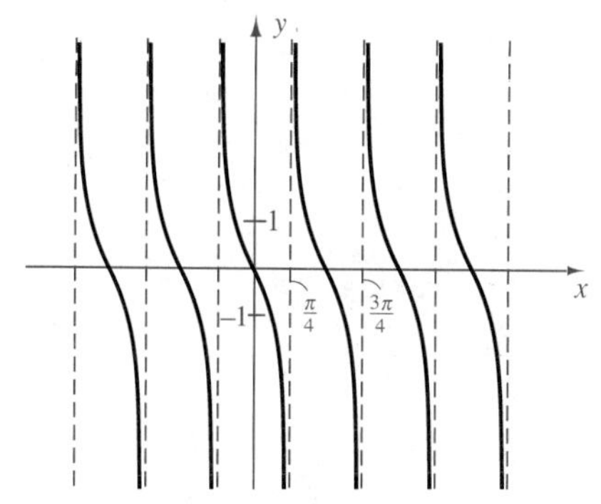

Since a is positive, we sketch a cotangent-shaped graph on the interval $[\pi/4, 3\pi/4]$, and then repeat it to the right and left in intervals of length $\pi/2$, as shown in Figure 63.

Graphs involving the secant and cosecant functions can be obtained using methods similar to those for the tangent and cotangent or by taking reciprocals of corresponding graphs of the cosine and sine functions.

EXAMPLE ▪ 4

Sketch the graph of the equation:

(a) $y = \sec\left(x - \frac{\pi}{4}\right)$ (b) $y = 2\sec\left(x - \frac{\pi}{4}\right)$

SOLUTION

(a) The graph of $y = \sec x$ is sketched with dashes in Figure 64. We can obtain the graph of $y = \sec\left(x - \frac{\pi}{4}\right)$ by shifting this graph to the right a distance $\pi/4$, as shown in Figure 64.

(b) We can sketch the graph by multiplying the y-coordinates of the graph in part (a) by 2. This gives us Figure 65.

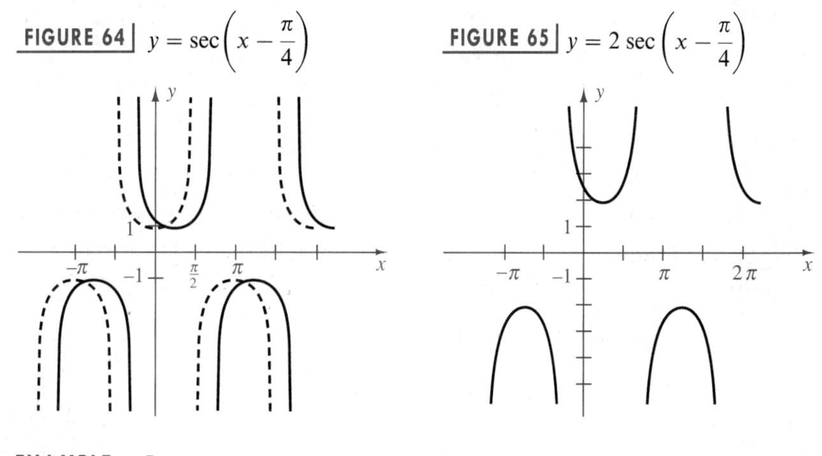

FIGURE 64 $y = \sec\left(x - \frac{\pi}{4}\right)$

FIGURE 65 $y = 2\sec\left(x - \frac{\pi}{4}\right)$

EXAMPLE ▪ 5

Sketch the graph of $y = \csc(2x + \pi)$.

SOLUTION Since $\csc\theta = 1/\sin\theta$,

$$y = \frac{1}{\sin(2x + \pi)}.$$

Thus we may obtain the graph of $y = \csc(2x + \pi)$ by finding the graph of $y = \sin(2x + \pi)$, and then taking the reciprocal of the y-coordinate of each point. Using $a = 1$, $b = 2$, and $c = \pi$, we see that the amplitude of $y = \sin(2x + \pi)$ is 1 and the period is $2\pi/|b| = 2\pi/2 = \pi$. To find an interval containing one cycle we solve the equations

$$2x + \pi = 0 \quad \text{and} \quad 2x + \pi = 2\pi,$$

obtaining

$$x = -\frac{\pi}{2} \quad \text{and} \quad x = \frac{\pi}{2}.$$

This leads to the dashed graph in Figure 66. Taking reciprocals gives us the graph of $y = \cos(2x + \pi)$ shown in the figure.

FIGURE 66 $y = \csc(2x + \pi)$

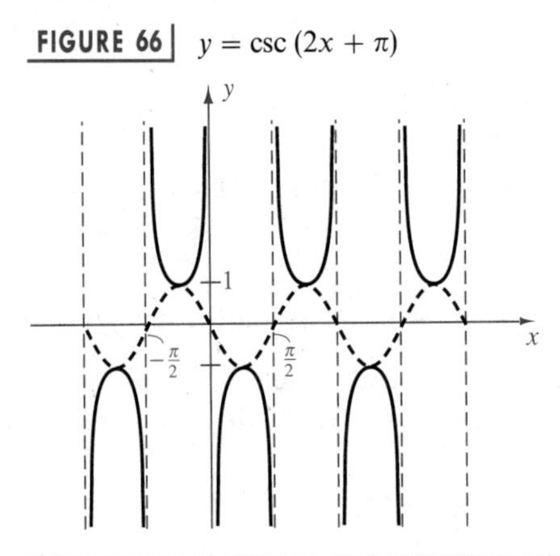

6.7 EXERCISES

Exer. 1–48: Find the period and sketch the graph of the equation. Show the asymptotes.

1 $y = 4 \tan x$

2 $y = \frac{1}{4} \tan x$

3 $y = 3 \cot x$

4 $y = \frac{1}{3} \cot x$

5 $y = 2 \csc x$

6 $y = \frac{1}{2} \csc x$

7 $y = 3 \sec x$

8 $y = \frac{1}{4} \sec x$

9 $y = \tan\left(x - \frac{\pi}{4}\right)$

10 $y = \tan\left(x + \frac{3\pi}{4}\right)$

11 $y = \tan 2x$

12 $y = \tan \frac{1}{2}x$

13 $y = \tan \frac{1}{4}x$

14 $y = \tan 4x$

15 $y = 2 \tan\left(2x + \frac{\pi}{2}\right)$

16 $y = \frac{1}{3} \tan\left(2x - \frac{\pi}{4}\right)$

17 $y = -\frac{1}{4} \tan\left(\frac{1}{2}x + \frac{\pi}{3}\right)$

18 $y = -3 \tan\left(\frac{1}{3}x - \frac{\pi}{3}\right)$

19 $y = \cot\left(x - \frac{\pi}{2}\right)$

20 $y = \cot\left(x + \frac{\pi}{4}\right)$

21 $y = \cot 2x$

22 $y = \cot \frac{1}{2}x$

23 $y = \cot \frac{1}{3}x$

24 $y = \cot 3x$

25 $y = 2\cot\left(2x + \dfrac{\pi}{2}\right)$

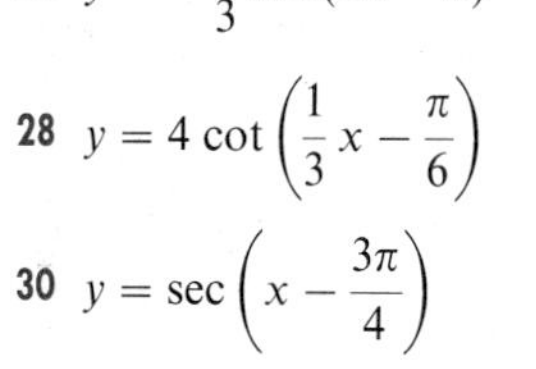

26 $y = -\dfrac{1}{3}\cot(3x - \pi)$

27 $y = -\dfrac{1}{2}\cot\left(\dfrac{1}{2}x + \dfrac{\pi}{4}\right)$

28 $y = 4\cot\left(\dfrac{1}{3}x - \dfrac{\pi}{6}\right)$

29 $y = \sec\left(x - \dfrac{\pi}{2}\right)$

30 $y = \sec\left(x - \dfrac{3\pi}{4}\right)$

31 $y = \sec 2x$

32 $y = \sec \frac{1}{2}x$

33 $y = \sec \frac{1}{3}x$

34 $y = \sec 3x$

35 $y = 2\sec\left(2x - \dfrac{\pi}{2}\right)$

36 $y = \dfrac{1}{2}\sec\left(2x - \dfrac{\pi}{2}\right)$

37 $y = -\dfrac{1}{3}\sec\left(\dfrac{1}{2}x + \dfrac{\pi}{4}\right)$

38 $y = -3\sec\left(\dfrac{1}{3}x + \dfrac{\pi}{3}\right)$

39 $y = \csc\left(x - \dfrac{\pi}{2}\right)$

40 $y = \csc\left(x + \dfrac{3\pi}{4}\right)$

41 $y = \csc 2x$

42 $y = \csc \frac{1}{2}x$

43 $y = \csc \frac{1}{3}x$

44 $y = \csc 3x$

45 $y = 2\csc\left(2x + \dfrac{\pi}{2}\right)$

46 $y = -\dfrac{1}{2}\csc(2x - \pi)$

47 $y = -\dfrac{1}{4}\csc\left(\dfrac{1}{2}x + \dfrac{\pi}{2}\right)$

48 $y = 4\csc\left(\dfrac{1}{2}x - \dfrac{\pi}{4}\right)$

49 Find an equation using the cotangent function that has the same graph as $y = \tan x$.

50 Find an equation using the cosecant function that has the same graph as $y = \sec x$.

6.8 REVIEW

Define or discuss each of the following.

Angle ■ Initial and terminal sides of an angle ■ Coterminal sides of angles ■ Standard position of an angle ■ Positive and negative angles ■ Degree measure ■ Radian measure ■ Acute or obtuse angles ■ Right angle ■ Complementary angles ■ Supplementary angles ■ The relationship between radians and degrees ■ Trigonometric functions of acute angles ■ Fundamental identities ■ Trigonometric solutions of right triangles ■ Trigonometric functions of any angle ■ Signs of the trigonometric functions ■ Trigonometric functions of real numbers ■ Reference angles ■ Domains and ranges of trigonometric functions ■ Periodic function, period, and periods of the trigonometric functions ■ Formulas for negatives ■ Graphs of the trigonometric functions ■ Sketching trigonometric graphs using amplitudes, periods, and phase shifts

6.8 EXERCISES

1 Find the radian measure that corresponds to each degree measure: 330°, 405°, −150°, 240°, 36°.

2 Find the degree measure that corresponds to each radian measure: $\dfrac{9\pi}{2}$, $-\dfrac{2\pi}{3}$, $\dfrac{7\pi}{4}$, 5π, $\dfrac{\pi}{5}$.

3 A central angle θ is subtended by an arc 20 cm long on a circle of radius 2 meters. Find the radian measure of θ.

4 If θ is an acute angle of a right triangle and if the adjacent side and hypotenuse have lengths 4 and 7, respectively, find the values of the trigonometric functions of θ.

5 **(a)** Find the reference angle for each radian measure: $\dfrac{5\pi}{4}$, $-\dfrac{5\pi}{6}$, $-\dfrac{9\pi}{8}$.

(b) Find the reference angle for each degree measure: 245°, 137°, 892°.

6 Find the quadrant containing θ if:

(a) $\sec\theta < 0$ and $\sin\theta > 0$.

(b) $\cot\theta > 0$ and $\csc\theta < 0$.

(c) $\cos\theta > 0$ and $\tan\theta < 0$.

7 Find the exact values of the remaining trigonometric functions if:

(a) $\sin\theta = -\frac{4}{5}$ and $\cos\theta = \frac{3}{5}$.

(b) $\csc\theta = \dfrac{\sqrt{13}}{2}$ and $\cot\theta = -\frac{3}{2}$.

8 Find the exact values of the trigonometric functions of θ when θ is in standard position and satisfies the stated condition.

(a) The point $(30, -40)$ is on the terminal side of θ.

(b) The terminal side of θ is in quadrant II and is parallel to the line $2x + 3y + 6 = 0$.

(c) $\theta = -90°$.

9 Without the use of tables or a calculator, find the exact values of the trigonometric functions corresponding to each value of θ:

(a) $\dfrac{9\pi}{2}$ (b) $-\dfrac{5\pi}{4}$ (c) 0 (d) $\dfrac{11\pi}{6}$

10 Find the exact value:

(a) $\cos 225°$ (b) $\tan 150°$

(c) $\sin\left(-\dfrac{\pi}{6}\right)$ (d) $\sec\dfrac{4\pi}{3}$

(e) $\cot\dfrac{7\pi}{4}$ (f) $\csc 300°$

Exer. 11–16: Find the amplitude and period, and sketch the graph of the equation.

11 $y = 5\cos x$ **12** $y = \frac{2}{3}\sin x$

13 $y = \frac{1}{3}\sin 3x$ **14** $y = -\frac{1}{2}\cos\frac{1}{3}x$

15 $y = -3\cos\frac{1}{2}x$ **16** $y = 4\sin 2x$

Exer. 17–20: The graph of an equation is shown in the figure.

(a) **Find the amplitude and period.**

(b) **Write the equation in the form $y = a\sin bx$ or $y = a\cos bx$.**

17

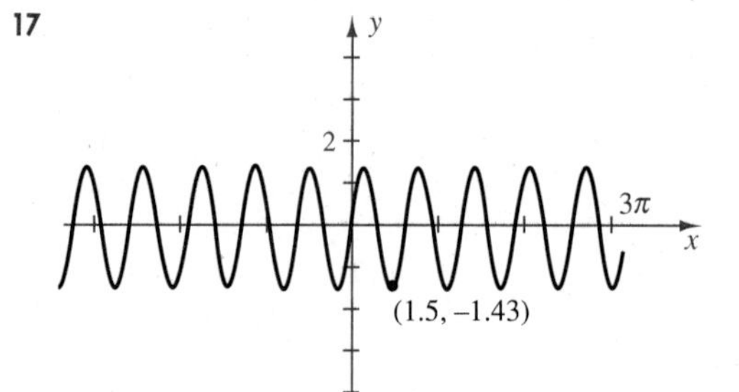

18

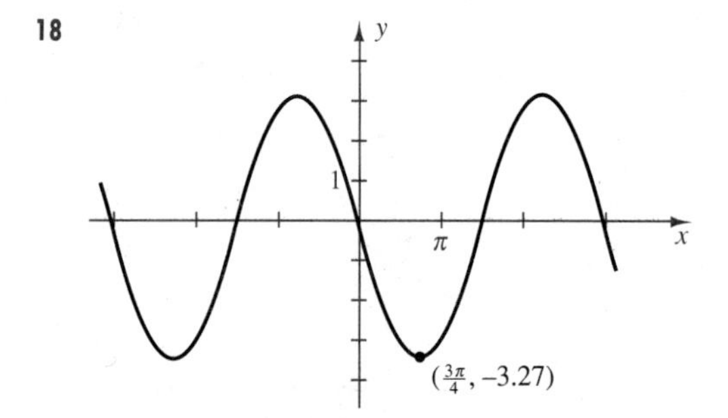

19

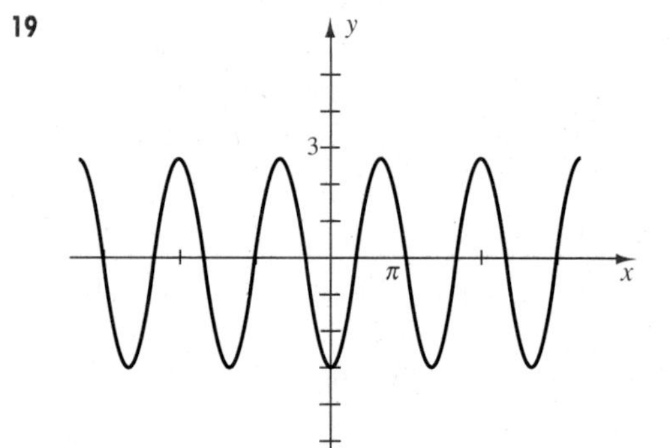

20

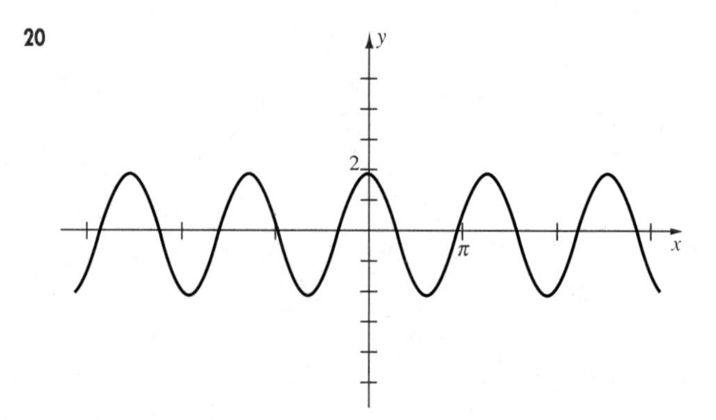

Exer. 21–32: Sketch the graph of the equation.

21 $y = 2\sin\left(x - \frac{2\pi}{3}\right)$

22 $y = -3\sin\left(\frac{1}{2}x - \frac{\pi}{4}\right)$

23 $y = -4\cos\left(x + \frac{\pi}{6}\right)$

24 $y = 5\cos\left(2x + \frac{\pi}{2}\right)$

25 $y = 2\tan\left(\frac{1}{2}x - \pi\right)$

26 $y = -3\tan\left(2x + \frac{\pi}{3}\right)$

27 $y = -4\cot\left(2x - \frac{\pi}{2}\right)$

28 $y = 2\cot\left(\frac{1}{2}x + \frac{\pi}{4}\right)$

29 $y = \sec\left(\frac{1}{2}x + \pi\right)$

30 $y = \sec\left(2x - \frac{\pi}{2}\right)$

31 $y = \csc\left(2x - \frac{\pi}{4}\right)$

32 $y = \csc\left(\frac{1}{2}x + \frac{\pi}{4}\right)$

Exer. 33–36: Given the indicated parts of triangle ABC with $\gamma = 90°$, approximate the remaining parts.

33 $\beta = 60°$, $b = 40$

34 $\alpha = 54°40'$, $b = 220$

35 $a = 62$, $b = 25$

36 $a = 9.0$, $c = 41$

37 A conical paper cup is constructed by removing a sector from a circle of radius 5 inches and attaching edge OA to OB (see figure). Find angle AOB so that the cup has a depth of 4 inches.

EXERCISE 37

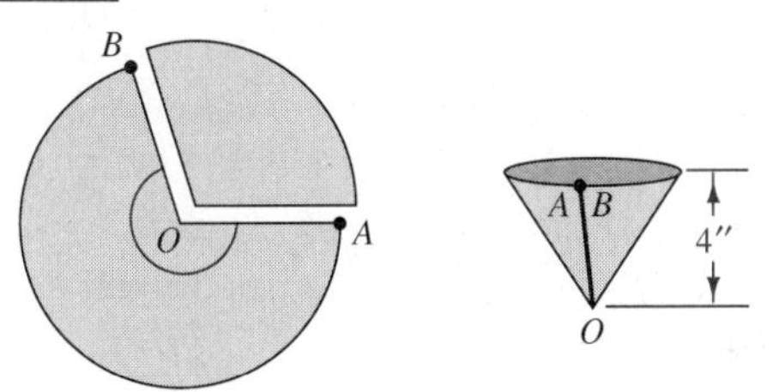

38 The length of the largest airplane propeller ever used was 22 feet 7.5 inches. The plane was powered by four engines that turned the propeller at 545 revolutions per minute.

(a) Find the angular speed of the propeller in radians per second.

(b) Approximately how fast (in mi/hr) does the tip of the propeller travel along the circle it generates?

39 To simulate the response of a structure to an earthquake, an engineer must choose a shape for the initial displacement of the beams in the building. When the beam has length L feet and the maximum displacement is a feet, the equation $y = a - a\cos\left(\frac{1}{2}\pi x/L\right)$ has been used by engineers for the displacement y (see figure). If $a = 1$ and $L = 10$, sketch the graph of the equation for $0 \le x \le 10$.

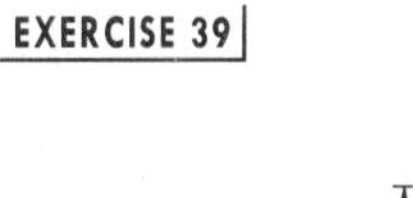
EXERCISE 39

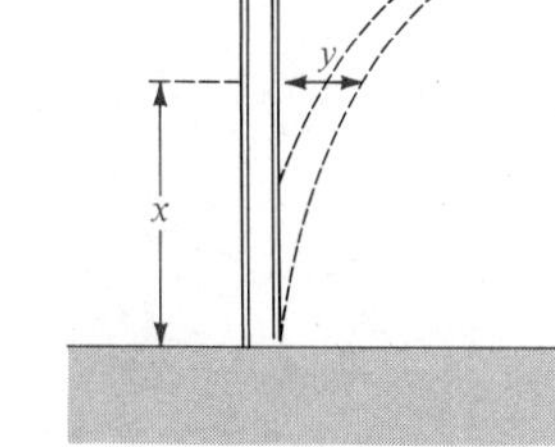

40 The variation in body temperature is an example of a circadian rhythm, a biological process that repeats itself approximately every 24 hours. Body temperature is highest around 5 P.M. and lowest at 5 A.M. Let y denote the body temperature (in °F) and let $t = 0$ correspond to midnight. If the low and high body temperatures are 98.3° and 98.9°, respectively, find an equation of the form $y = 98.6 + a\sin(bt + c)$ that fits this information.

41 The Great Pyramid of Egypt is 147 meters high with a square base of side 230 meters (see figure). Find, to the nearest degree, the angle φ formed when an observer stands at the midpoint of one of the sides and views the apex of the pyramid.

EXERCISE 41

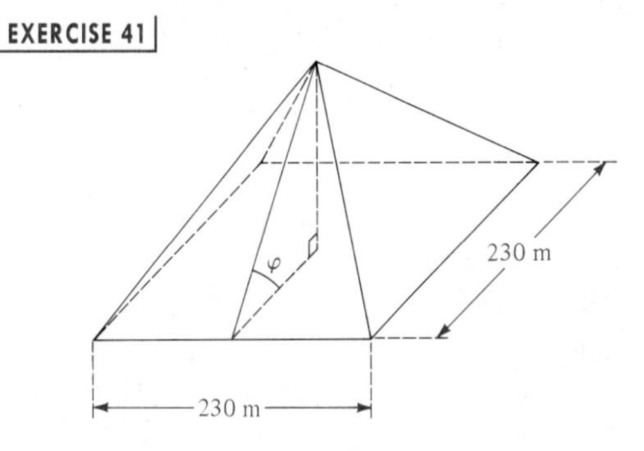

42 When the top of the Eiffel Tower is viewed at a distance of 200 feet from the base, the angle of elevation is 79.2°. Estimate the height of the tower.

43 A tunnel for a new highway is to be cut through a mountain that is 260 feet high. At a distance of 200 feet from the base of the mountain, the angle of elevation is 36° (see figure). From a distance of 150 feet on the other

side, the angle of elevation measures 47°. Find the length of the tunnel to the nearest foot.

EXERCISE 43

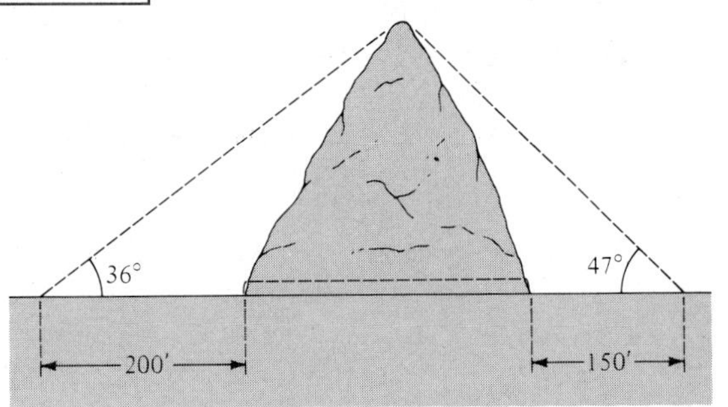

44 When a certain skyscraper is viewed from the top of a building 50 feet tall, the angle of elevation is 59° (see figure). When viewed from the street next to the shorter building, the angle of elevation is 62°.

(a) Approximately how far apart are the two structures?

(b) Approximate the height of the skyscraper to the nearest tenth of a foot.

EXERCISE 44

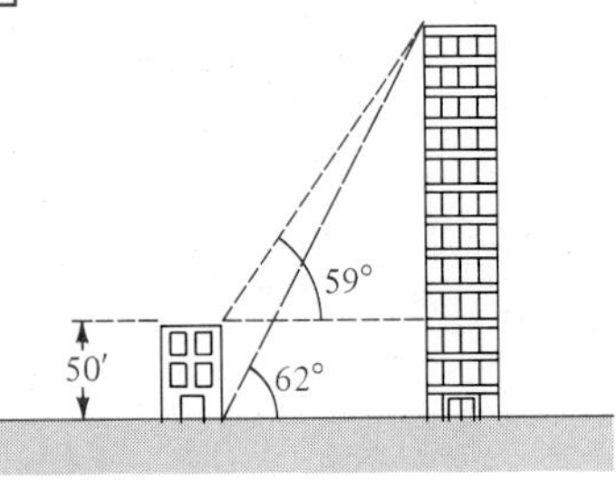

45 When viewed from the earth over a period of time, the planet Venus appears to move back and forth along a line segment with the sun at its midpoint (see figure). At the maximum apparent distance from the sun, angle SEV is approximately 47°. Using $ES = 92{,}900{,}000$ miles, estimate the distance of Venus from the sun.

EXERCISE 45

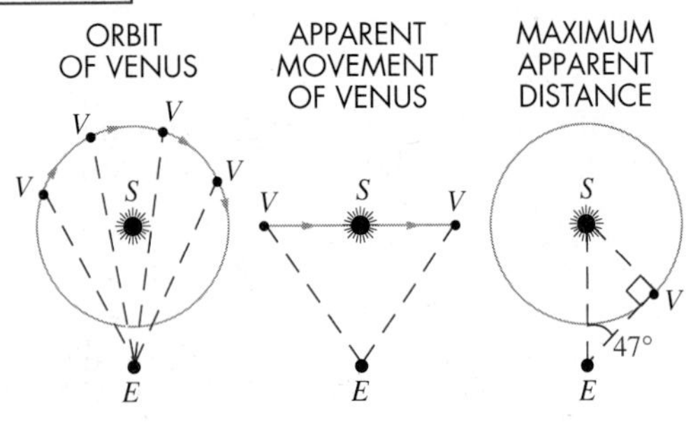

46 An observer of height h stands on an incline at a distance d from the base of a building of height T as shown in the figure. The angle of elevation from the observer to the top of the building is θ, and the incline makes an angle of α with the horizontal.

(a) Express T in terms of h, d, α, and θ.

(b) Estimate the height of the building if $h = 6$ feet, $d = 50$ feet, $\alpha = 15°$, and $\theta = 31.4°$.

EXERCISE 46

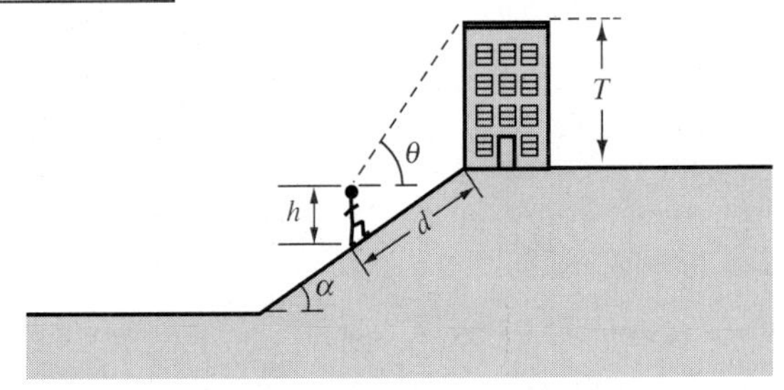

47 The annual variation in temperature T (in °C) in Ottawa, Canada, may be approximated by

$$T(t) = 5 + 15.8 \sin\left[\frac{\pi}{6}(t - 3)\right]$$

for time t in months with $t = 0$ corresponding to January 1.

(a) Sketch the graph of T for $0 \le t \le 12$.

(b) Find the highest temperature of the year and the date when it occurs.

48 A reservoir supplies water to a community. During the summer months, the demand $D(t)$ for water (in ft^3/day) is given by

$$D(t) = 4000 + 2000 \sin\left(\frac{\pi}{90} t\right),$$

for time t in days with $t = 0$ corresponding to the beginning of summer.

(a) Sketch the graph of D for $0 \le t \le 90$.

(b) When is the demand for water the greatest?

49 A spotlight with intensity 5000 candles is located 15 feet above a stage. If the spotlight is rotated through an angle

θ as shown in the figure, the illuminance E (in foot-candles) in the lighted area of the stage is given by

$$E = \frac{5000 \cos \theta}{s^2},$$

where s is the distance (in feet) that the light must travel.

(a) Find the illuminance if the spotlight is rotated through an angle of 30°.

(b) The maximum illuminance occurs when $\theta = 0°$. For what value of θ will the illuminance be one-half the maximum value?

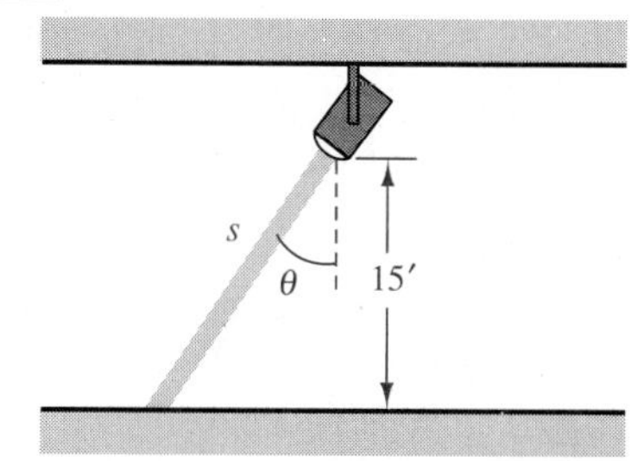

50 The manufacturer of a computerized projection system recommends that a projection unit be mounted on the ceiling as shown in the figure. The distance from the end of the mounting bracket to the center of the screen is 85.5 inches and the angle of depression is 30°.

(a) How far from the wall should the bracket be mounted? Neglect the thickness of the screen.

(b) If the bracket is 18 inches long and the screen is 6 feet high, determine the distance from the ceiling to the top edge of the screen.

EXERCISE 50

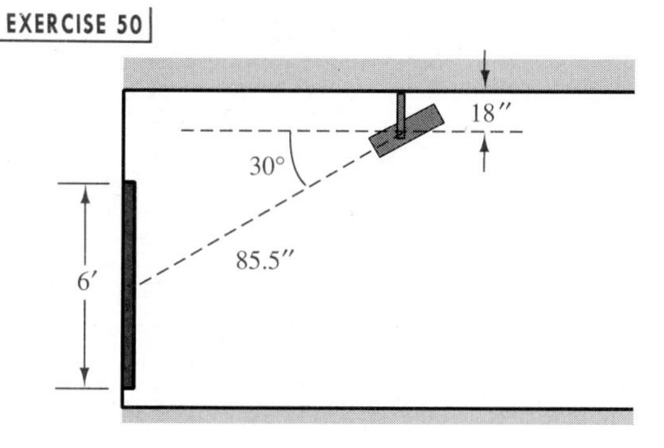

51 A pyramid has a square base and congruent triangular faces. Let θ be the angle that the altitude a of a triangular face makes with the altitude y of the pyramid, and let x be the length of a side (see figure).

(a) Express the total surface area S of the four faces in terms of a and θ.

(b) Express the volume V of the pyramid in terms of a and θ.

EXERCISE 51

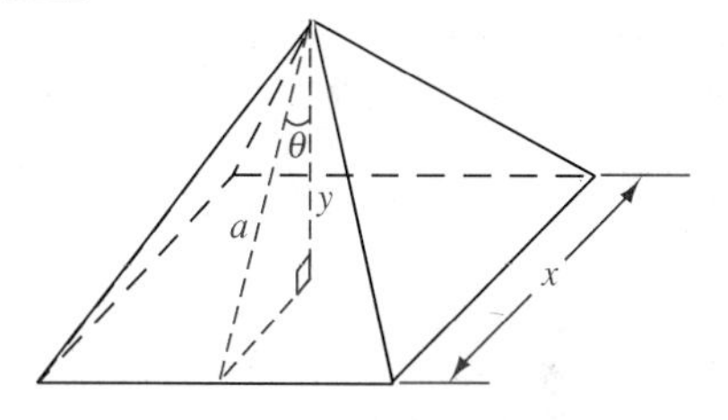

52 A surveyor, using a transit, sights the edge B of a bluff, as shown in the first figure (not drawn to scale). Due to the curvature of the earth, the true elevation h of the bluff is larger than that measured by the surveyor. A cross-sectional schematic view of the earth is shown in the second figure.

(a) If s is the length of arc $\widehat{PQ}$ and R is the distance from P to the center C of the earth, express h in terms of R and s.

(b) If $R = 4000$ miles and $s = 50$ miles, estimate the elevation of the bluff in feet.

EXERCISE 52

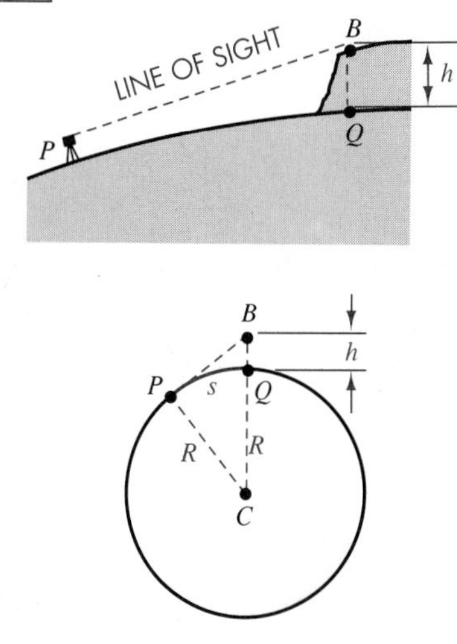

CHAPTER 7

In this chapter we shall derive many trigonometric formulas; for reference, they are listed on the inside of the back cover of the text. In addition to considering formal techniques involving trigonometric expressions, we shall also discuss numerous applications of trigonometry.

ANALYTIC TRIGONOMETRY

TRIGONOMETRIC IDENTITIES

A **trigonometric expression** contains symbols involving trigonometric functions.

ILLUSTRATION | TRIGONOMETRIC EXPRESSIONS

- $x + \sin x$
- $\dfrac{\sqrt{\theta} + 2^{\sin \theta}}{\cot \theta}$
- $\dfrac{\cos (3y + 1)}{x^2 + \tan^2 (z - y^2)}$

We assume that the domain of each variable in a trigonometric expression is the set of real numbers (or angles) for which the expression is meaningful.

We may use the fundamental identities, which were introduced in Section 6.2, to help simplify complicated trigonometric expressions. Let us begin by restating these formulas and working an example. The variable t in each fundamental identity may represent either a real number or the measure of an angle.

FUNDAMENTAL IDENTITIES

$$\csc t = \frac{1}{\sin t} \qquad \tan t = \frac{\sin t}{\cos t} \qquad \sin^2 t + \cos^2 t = 1$$

$$\sec t = \frac{1}{\cos t} \qquad \cot t = \frac{\cos t}{\sin t} \qquad 1 + \tan^2 t = \sec^2 t$$

$$\cot t = \frac{1}{\tan t} \qquad 1 + \cot^2 t = \csc^2 t$$

EXAMPLE ▪ 1

Simplify the expression $(\sec \theta + \tan \theta)(1 - \sin \theta)$.

SOLUTION We use fundamental identities:

$$\begin{aligned}(\sec \theta + \tan \theta)(1 - \sin \theta) &= \left(\frac{1}{\cos \theta} + \frac{\sin \theta}{\cos \theta}\right)(1 - \sin \theta)\\ &= \left(\frac{1 + \sin \theta}{\cos \theta}\right)(1 - \sin \theta)\\ &= \frac{(1 + \sin \theta)(1 - \sin \theta)}{\cos \theta}\end{aligned}$$

$$= \frac{1 - \sin^2 \theta}{\cos \theta}$$

$$= \frac{\cos^2 \theta}{\cos \theta}$$

$$= \cos \theta$$

There are other ways to simplify the expression in Example 1. We could first multiply the two factors, and then simplify and combine terms. The method we employed—changing all expressions to expressions that involve only sines and cosines—is often useful. However, that technique does not always lead to the shortest possible simplification.

Most of the identities we shall consider in the remainder of this section are unimportant in their own right, but we shall use them to gain manipulative practice. The ability to carry out trigonometric manipulations is essential for solving problems that occur in advanced courses in mathematics and science.

We often use the phrase *verify an identity* instead of *prove that an equation is an identity*. When verifying an identity, we usually use fundamental identities and algebraic manipulations to transform one side into the other, as illustrated in the first three examples.

Warning: *If an (assumed) identity contains fractions, do not multiply both sides by the lcd. Each side should be examined independently.*

EXAMPLE ▪ 2

Verify the identity $\dfrac{\tan t + \cos t}{\sin t} = \sec t + \cot t$.

SOLUTION By the warning preceding this example, *multiplication of both sides by* $\sin t$ *is not allowed*. We may transform the left-hand side into the right-hand side as follows:

$$\frac{\tan t + \cos t}{\sin t} = \frac{\tan t}{\sin t} + \frac{\cos t}{\sin t}$$

$$= \frac{\left(\dfrac{\sin t}{\cos t}\right)}{\sin t} + \cot t$$

$$= \frac{1}{\cos t} + \cot t$$

$$= \sec t + \cot t$$

Another way to verify the identity is to transform the right-hand side $\sec t + \cot t$ into the left-hand side.

EXAMPLE ▪ 3

Verify the identity $\sec \alpha - \cos \alpha = \sin \alpha \tan \alpha$.

SOLUTION We transform the left-hand side into the right-hand side:

$$\begin{aligned} \sec \alpha - \cos \alpha &= \frac{1}{\cos \alpha} - \cos \alpha \\ &= \frac{1 - \cos^2 \alpha}{\cos \alpha} \\ &= \frac{\sin^2 \alpha}{\cos \alpha} \\ &= \sin \alpha \left(\frac{\sin \alpha}{\cos \alpha} \right) \\ &= \sin \alpha \tan \alpha \end{aligned}$$

EXAMPLE ▪ 4

Verify the identity $\dfrac{\cos x}{1 - \sin x} = \dfrac{1 + \sin x}{\cos x}$.

SOLUTION Multiplication of both sides by the lcd $(1 - \sin x) \cos x$ is not allowed. Instead, we change the form of the fraction on the left-hand side by multiplying the numerator and denominator by $1 + \sin x$:

$$\begin{aligned} \frac{\cos x}{1 - \sin x} &= \frac{\cos x}{1 - \sin x} \cdot \frac{1 + \sin x}{1 + \sin x} \\ &= \frac{\cos x\,(1 + \sin x)}{1 - \sin^2 x} \\ &= \frac{\cos x\,(1 + \sin x)}{\cos^2 x} \\ &= \frac{1 + \sin x}{\cos x} \end{aligned}$$

Another technique for showing that an equation $p = q$ is an identity is to begin by transforming the left-hand side p into another expression s, making sure that each step is *reversible*; that is, it is possible to transform s back into p by reversing the procedure that has been used in each step.

In this case the equation $p = s$ is an identity. Next, as a *separate* exercise, we must show that the right-hand side q can also be transformed into the expression s by means of reversible steps and hence that $q = s$ is an identity. It then follows that $p = q$ is an identity. This method is illustrated in the next example.

EXAMPLE ▪ 5

Verify the identity $(\tan\theta - \sec\theta)^2 = \dfrac{1 - \sin\theta}{1 + \sin\theta}$.

SOLUTION We shall verify the identity by showing that each side of the equation can be transformed into the same expression. First we work only with the left-hand side:

$$\begin{aligned}(\tan\theta - \sec\theta)^2 &= \tan^2\theta - 2\tan\theta\sec\theta + \sec^2\theta \\ &= \left(\frac{\sin\theta}{\cos\theta}\right)^2 - 2\left(\frac{\sin\theta}{\cos\theta}\right)\left(\frac{1}{\cos\theta}\right) + \left(\frac{1}{\cos\theta}\right)^2 \\ &= \frac{\sin^2\theta}{\cos^2\theta} - \frac{2\sin\theta}{\cos^2\theta} + \frac{1}{\cos^2\theta} \\ &= \frac{\sin^2\theta - 2\sin\theta + 1}{\cos^2\theta}\end{aligned}$$

We next work only with the right-hand side, multiplying numerator and denominator by $1 - \sin\theta$:

$$\begin{aligned}\frac{1 - \sin\theta}{1 + \sin\theta} &= \frac{1 - \sin\theta}{1 + \sin\theta}\cdot\frac{1 - \sin\theta}{1 - \sin\theta} \\ &= \frac{1 - 2\sin\theta + \sin^2\theta}{1 - \sin^2\theta} \\ &= \frac{1 - 2\sin\theta + \sin^2\theta}{\cos^2\theta}\end{aligned}$$

The last expression is the same as that obtained from $(\tan\theta - \sec\theta)^2$. Since all steps are reversible, the given equation is an identity.

In calculus it is sometimes convenient to change the forms of certain algebraic expressions by making a **trigonometric substitution**, as illustrated in the following example.

EXAMPLE ▪ 6

Express $\sqrt{a^2 - x^2}$ in terms of a trigonometric function of θ without radicals by making the substitution $x = a \sin \theta$, for $-\frac{\pi}{2} \le \theta \le \frac{\pi}{2}$ and $a > 0$.

SOLUTION We let $x = a \sin \theta$:

$$\begin{aligned}\sqrt{a^2 - x^2} &= \sqrt{a^2 - (a \sin \theta)^2} \\ &= \sqrt{a^2 - a^2 \sin^2 \theta} \\ &= \sqrt{a^2(1 - \sin^2 \theta)} \\ &= \sqrt{a^2 \cos^2 \theta} \\ &= a \cos \theta.\end{aligned}$$

The last equality is true because first $\sqrt{a^2} = a$ if $a > 0$ and second if $-\pi/2 \le \theta \le \pi/2$, then $\cos \theta \ge 0$ has hence $\sqrt{\cos^2 \theta} = \cos \theta$.

EXERCISES

Exer. 1–88: Verify the identity.

1 $\cos \theta \sec \theta = 1$

2 $\tan \alpha \cot \alpha = 1$

3 $\sin \theta \sec \theta = \tan \theta$

4 $\sin \alpha \cot \alpha = \cos \alpha$

5 $\dfrac{\csc x}{\sec x} = \cot x$

6 $\cot \beta \sec \beta = \csc \beta$

7 $(1 + \cos \alpha)(1 - \cos \alpha) = \sin^2 \alpha$

8 $\cos^2 x (\sec^2 x - 1) = \sin^2 x$

9 $\cos^2 t - \sin^2 t = 2 \cos^2 t - 1$

10 $(\tan \theta + \cot \theta) \tan \theta = \sec^2 \theta$

11 $\dfrac{\sin t}{\csc t} + \dfrac{\cos t}{\sec t} = 1$

12 $1 - 2 \sin^2 x = 2 \cos^2 x - 1$

13 $(1 + \sin \alpha)(1 - \sin \alpha) = \dfrac{1}{\sec^2 \alpha}$

14 $(1 - \sin^2 t)(1 + \tan^2 t) = 1$

15 $\sec \beta - \cos \beta = \tan \beta \sin \beta$

16 $\dfrac{\sin w + \cos w}{\cos w} = 1 + \tan w$

17 $\dfrac{\csc^2 \theta}{1 + \tan^2 \theta} = \cot^2 \theta$

18 $\sin x + \cos x \cot x = \csc x$

19 $\sin t (\csc t - \sin t) = \cos^2 t$

20 $\cot t + \tan t = \csc t \sec t$

21 $\csc \theta - \sin \theta = \cot \theta \cos \theta$

22 $\cos \theta (\tan \theta + \cot \theta) = \csc \theta$

23 $\dfrac{\sec^2 u - 1}{\sec^2 u} = \sin^2 u$

24 $(\tan u + \cot u)(\cos u + \sin u) = \sec u + \csc u$

25 $(\cos^2 x - 1)(\tan^2 x + 1) = 1 - \sec^2 x$

26 $(\cot \alpha + \csc \alpha)(\tan \alpha - \sin \alpha) = \sec \alpha - \cos \alpha$

27 $\sec t \csc t + \cot t = \tan t + 2 \cos t \csc t$

28 $\dfrac{1 + \cos^2 y}{\sin^2 y} = 2 \csc^2 y - 1$

29 $\sec^2 \theta \csc^2 \theta = \sec^2 \theta + \csc^2 \theta$

30 $\dfrac{\sec x - \cos x}{\tan x} = \dfrac{\tan x}{\sec x}$

31 $\dfrac{1+\cos t}{\sin t}+\dfrac{\sin t}{1+\cos t}=2\csc t$

32 $\tan^2\alpha-\sin^2\alpha=\tan^2\alpha\sin^2\alpha$

33 $\dfrac{1+\tan^2 v}{\tan^2 v}=\csc^2 v$

34 $\dfrac{\sec\theta+\csc\theta}{\sec\theta-\csc\theta}=\dfrac{\sin\theta+\cos\theta}{\sin\theta-\cos\theta}$

35 $\dfrac{1+\sin x}{1-\sin x}-\dfrac{1-\sin x}{1+\sin x}=4\tan x\sec x$

36 $\dfrac{1}{1-\cos\gamma}+\dfrac{1}{1+\cos\gamma}=2\csc^2\gamma$

37 $\dfrac{1+\csc\beta}{\sec\beta}-\cot\beta=\cos\beta$

38 $\dfrac{\cos x\cot x}{\cot x-\cos x}=\dfrac{\cot x+\cos x}{\cos x\cot x}$

39 $(\sec u-\tan u)(\csc u+1)=\cot u$

40 $\dfrac{\cot\theta-\tan\theta}{\sin\theta+\cos\theta}=\csc\theta-\sec\theta$

41 $\dfrac{\cot\alpha-1}{1-\tan\alpha}=\cot\alpha$

42 $\dfrac{1+\sec\beta}{\tan\beta+\sin\beta}=\csc\beta$

43 $\csc^4 t-\cot^4 t=\cot^2 t+\csc^2 t$

44 $\cos^4\theta+\sin^2\theta=\sin^4\theta+\cos^2\theta$

45 $\dfrac{\cos\beta}{1-\sin\beta}=\sec\beta+\tan\beta$

46 $\dfrac{1}{\csc y-\cot y}=\csc y+\cot y$

47 $\dfrac{\tan^2 x}{\sec x+1}=\dfrac{1-\cos x}{\cos x}$

48 $\dfrac{\cot x}{\csc x+1}=\dfrac{\csc x-1}{\cot x}$

49 $\dfrac{\cot u-1}{\cot u+1}=\dfrac{1-\tan u}{1+\tan u}$

50 $\dfrac{1+\sec x}{\sin x+\tan x}=\csc x$

51 $\sin^4 r-\cos^4 r=\sin^2 r-\cos^2 r$

52 $\sin^4\theta+2\sin^2\theta\cos^2\theta+\cos^4\theta=1$

53 $\tan^4 k-\sec^4 k=1-2\sec^2 k$

54 $\sec^4 u-\sec^2 u=\tan^4 u+\tan^2 u$

55 $(\sec t+\tan t)^2=\dfrac{1+\sin t}{1-\sin t}$

56 $\sec^2\gamma+\tan^2\gamma=(1-\sin^4\gamma)\sec^4\gamma$

57 $(\sin^2\theta+\cos^2\theta)^3=1$

58 $\dfrac{\sin t}{1-\cos t}=\csc t+\cot t$

59 $\dfrac{1+\csc\beta}{\cot\beta+\cos\beta}=\sec\beta$

60 $\dfrac{\sin z\tan z}{\tan z-\sin z}=\dfrac{\tan z+\sin z}{\sin z\tan z}$

61 $\left(\dfrac{\sin^2 x}{\tan^4 x}\right)^3\left(\dfrac{\csc^3 x}{\cot^6 x}\right)^2=1$

62 $\dfrac{\cos^3 x-\sin^3 x}{\cos x-\sin x}=1+\sin x\cos x$

63 $\dfrac{\sin\theta+\cos\theta}{\tan^2\theta-1}=\dfrac{\cos^2\theta}{\sin\theta-\cos\theta}$

64 $(\csc t-\cot t)^4(\csc t+\cot t)^4=1$

65 $(a\cos t-b\sin t)^2+(a\sin t+b\cos t)^2=a^2+b^2$

66 $\sin^6 v+\cos^6 v=1-3\sin^2 v\cos^2 v$

67 $\dfrac{\sin\alpha\cos\beta+\cos\alpha\sin\beta}{\cos\alpha\cos\beta-\sin\alpha\sin\beta}=\dfrac{\tan\alpha+\tan\beta}{1-\tan\alpha\tan\beta}$

68 $\dfrac{\tan u-\tan v}{1+\tan u\tan v}=\dfrac{\cot v-\cot u}{1+\cot u\cot v}$

69 $\sqrt{\dfrac{1-\cos t}{1+\cos t}}=\dfrac{1-\cos t}{|\sin t|}$

70 $\sqrt{\dfrac{1-\sin\theta}{1+\sin\theta}}=\dfrac{|\cos\theta|}{1+\sin\theta}$

71 $\dfrac{\tan\alpha}{1+\sec\alpha}+\dfrac{1+\sec\alpha}{\tan\alpha}=2\csc\alpha$

72 $\dfrac{\csc x}{1+\csc x}-\dfrac{\csc x}{1-\csc x}=2\sec^2 x$

73 $\dfrac{1}{\tan\beta+\cot\beta}=\sin\beta\cos\beta$

74 $\dfrac{\cot y-\tan y}{\sin y\cos y}=\csc^2 y-\sec^2 y$

75 $\sec\theta+\csc\theta-\cos\theta-\sin\theta=\sin\theta\tan\theta+\cos\theta\cot\theta$

76 $\sin^3 t+\cos^3 t=(1-\sin t\cos t)(\sin t+\cos t)$

77 $(1-\tan^2\phi)^2=\sec^4\phi-4\tan^2\phi$

78 $\cos^4 w+1-\sin^4 w=2\cos^2 w$

79 $\dfrac{\tan x}{1-\cot x}+\dfrac{\cot x}{1-\tan x}=1+\sec x\csc x$

80 $\dfrac{\cos \gamma}{1 - \tan \gamma} + \dfrac{\sin \gamma}{1 - \cot \gamma} = \cos \gamma + \sin \gamma$

81 $\sin(-t)\sec(-t) = -\tan t$

82 $\dfrac{\cot(-v)}{\csc(-v)} = \cos v$

83 $\log 10^{\tan t} = \tan t$

84 $10^{\log|\sin t|} = |\sin t|$

85 $\ln \cot x = -\ln \tan x$

86 $\ln \sec \theta = -\ln \cos \theta$

87 $-\ln|\sec \theta - \tan \theta| = \ln|\sec \theta + \tan \theta|$

88 $\ln|\csc x - \cot x| = -\ln|\csc x + \cot x|$

Exer. 89–100: Show that the equation is *not* an identity. (*Hint:* Find one number in the domain of t or θ for which the equation is false.)

89 $\cos t = \sqrt{1 - \sin^2 t}$

90 $\sqrt{\sin^2 t + \cos^2 t} = \sin t + \cos t$

91 $\sqrt{\sin^2 t} = \sin t$

92 $\sec t = \sqrt{\tan^2 t + 1}$

93 $(\sin \theta + \cos \theta)^2 = \sin^2 \theta + \cos^2 \theta$

94 $\log(1/\sin t) = 1/(\log \sin t)$

95 $\cos(-t) = -\cos t$

96 $\sin(t + \pi) = \sin t$

97 $\cos(\sec t) = 1$

98 $\cot(\tan \theta) = 1$

99 $\sin^2 t - 4 \sin t - 5 = 0$

100 $3\cos^2 \theta + \cos \theta - 2 = 0$

Exer. 101–104: Refer to Example 6. Make the trigonometric substitution $x = a \sin \theta$ for $-\frac{\pi}{2} < \theta < \frac{\pi}{2}$ and $a > 0$. Use fundamental identities to simplify the resulting expression.

101 $(a^2 - x^2)^{3/2}$

102 $\dfrac{\sqrt{a^2 - x^2}}{x}$

103 $\dfrac{x^2}{\sqrt{a^2 - x^2}}$

104 $\dfrac{1}{x\sqrt{a^2 - x^2}}$

Exer. 105–108: Make the trigonometric substitution $x = a \tan \theta$ for $-\frac{\pi}{2} < \theta < \frac{\pi}{2}$ and $a > 0$. Simplify the resulting expression.

105 $\sqrt{a^2 + x^2}$

106 $\dfrac{1}{\sqrt{a^2 + x^2}}$

107 $\dfrac{1}{x^2 + a^2}$

108 $\dfrac{(x^2 + a^2)^{3/2}}{x}$

Exer. 109–112: Make the trigonometric substitution $x = a \sec \theta$ for $0 < \theta < \frac{\pi}{2}$ and $a > 0$. Simplify the resulting expression.

109 $\sqrt{x^2 - a^2}$

110 $\dfrac{1}{x^2\sqrt{x^2 - a^2}}$

111 $x^3\sqrt{x^2 - a^2}$

112 $\dfrac{\sqrt{x^2 - a^2}}{x^2}$

7.2 TRIGONOMETRIC EQUATIONS

A **trigonometric equation** is an equation that contains trigonometric expressions. Each identity considered in the preceding section is an example of a trigonometric equation with every number (or angle) in the domain of the variable a solution of the equation. If a trigonometric equation is not an identity, we often find solutions by using techniques similar to those used for algebraic equations. The main difference is that we first solve the trigonometric equation for $\sin x$, $\cos \theta$, and so on, and then find values of x or θ that satisfy the equation. Solutions may be expressed either as real numbers or angles. Throughout our work we shall use the following rule: *If degree measure is not specified, then solutions of a trigonometric* *equation should be expressed in radian measure (or as real numbers)*. If solutions in degree measure are desired, an appropriate statement will be included in an example or exercise.

EXAMPLE ▪ 1

Find the solutions of the equation $\sin\theta = \frac{1}{2}$ if

(a) θ is in the interval $[0, 2\pi)$.

(b) θ is any real number.

FIGURE 1

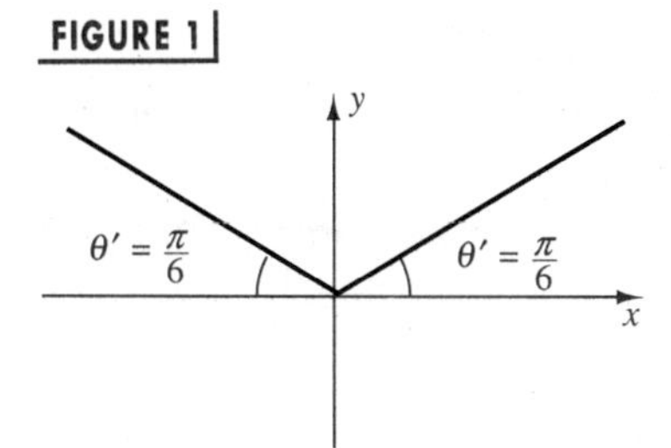

SOLUTION

(a) If $\sin\theta = \frac{1}{2}$, then the reference angle for θ is $\theta' = \pi/6$. If we regard θ as an angle in standard position, then, since $\sin\theta > 0$, the terminal side is in either quadrant I or II, as illustrated in Figure 1. Thus there are two solutions for $0 \le \theta < 2\pi$:

$$\theta = \frac{\pi}{6} \quad \text{and} \quad \theta = \pi - \frac{\pi}{6} = \frac{5\pi}{6}.$$

(b) Since the sine function has period 2π, we may obtain all solutions by adding multiples of 2π to $\pi/6$ and $5\pi/6$. This gives us

$$\theta = \frac{\pi}{6} + 2\pi n \quad \text{and} \quad \theta = \frac{5\pi}{6} + 2\pi n \quad \text{for every integer } n.$$

An alternative (graphical) solution involves determining where the graph of $y = \sin\theta$ intersects the horizontal line $y = \frac{1}{2}$, as illustrated in Figure 2.

FIGURE 2

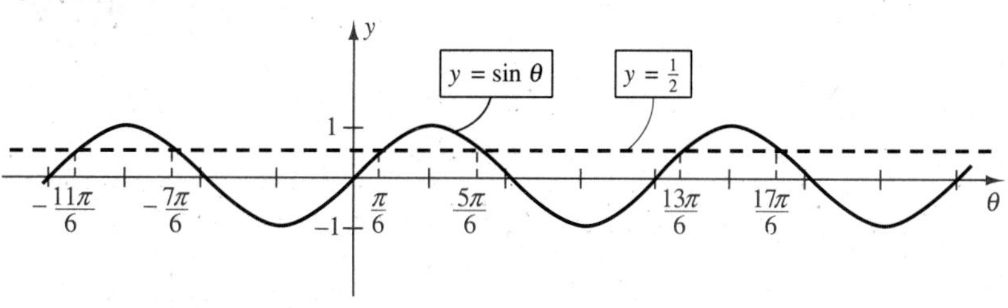

EXAMPLE ▪ 2

Find the solutions of the equation $\tan u = -1$.

SOLUTION Since the tangent function has period π, it is sufficient to find one real number u such that $\tan u = -1$ and then add multiples of π.

FIGURE 3
$y = \tan u$

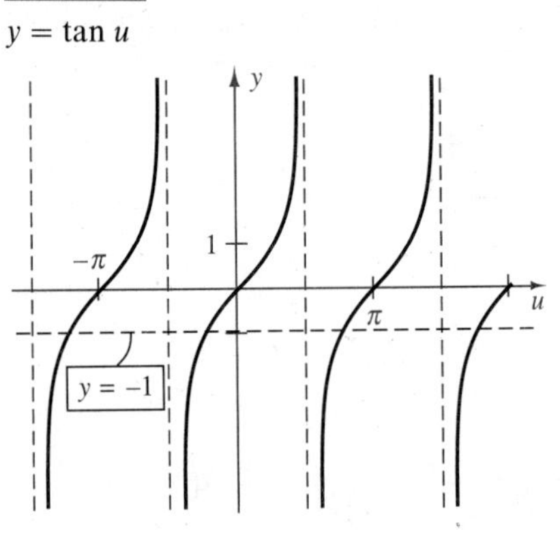

A portion of the graph of $y = \tan u$ is sketched in Figure 3. Since $\tan(3\pi/4) = -1$, one solution is $3\pi/4$, and hence

$$u = \frac{3\pi}{4} + \pi n \quad \text{for every integer } n.$$

We could also have chosen $-\pi/4$ (or some other number u such that $\tan u = -1$) for the basic solution and write

$$u = -\frac{\pi}{4} + \pi n \quad \text{for every integer } n.$$

EXAMPLE ▪ 3

(a) Solve the equation $\cos 2x = 0$, and express the solutions both in radian measure and in degree measure.

(b) Find the particular solutions that are in the intervals $[0, 2\pi)$ and $[0°, 360°)$.

SOLUTION

(a) Referring to the graph of the cosine function (see Figure 42 in Chapter 6), we see that

$$\cos\theta = 0 \quad \text{if} \quad \theta = \frac{\pi}{2} + \pi n \quad \text{for any integer } n.$$

Letting $\theta = 2x$,

$$\cos 2x = 0 \quad \text{if} \quad 2x = \frac{\pi}{2} + \pi n \quad \text{for any integer } n.$$

Dividing both sides of the last equation by 2, we obtain the solutions

$$x = \frac{\pi}{4} + \frac{\pi}{2}n \quad \text{for any integer } n.$$

Using degree measure, we may write the solutions as

$$x = 45° + 90°n \quad \text{for any integer } n.$$

(b) We may find particular solutions of the equation by substituting integers for n in either of the formulas for x obtained in part (a). Several such solutions are listed in the following table.

n	$\frac{\pi}{4}+\frac{\pi}{2}n$	$45° + 90°n$
-1	$\frac{\pi}{4}+\frac{\pi}{2}(-1)=-\frac{\pi}{4}$	$45° + 90°(-1) = -45°$
0	$\frac{\pi}{4}+\frac{\pi}{2}(0)=\frac{\pi}{4}$	$45° + 90°(0) = 45°$
1	$\frac{\pi}{4}+\frac{\pi}{2}(1)=\frac{3\pi}{4}$	$45° + 90°(1) = 135°$
2	$\frac{\pi}{4}+\frac{\pi}{2}(2)=\frac{5\pi}{4}$	$45° + 90°(2) = 225°$
3	$\frac{\pi}{4}+\frac{\pi}{2}(3)=\frac{7\pi}{4}$	$45° + 90°(3) = 315°$
4	$\frac{\pi}{4}+\frac{\pi}{2}(4)=\frac{9\pi}{4}$	$45° + 90°(4) = 405°$

Note that the solutions in the intervals $[0, 2\pi)$ and $[0°, 360°)$ are given by $n = 0$, $n = 1$, $n = 2$, and $n = 3$. These solutions are

$$\frac{\pi}{4},\quad \frac{3\pi}{4},\quad \frac{5\pi}{4},\quad \frac{7\pi}{4} \quad \text{and} \quad 45°,\quad 135°,\quad 225°,\quad 315°.$$

EXAMPLE ▪ 4

Solve the equation $\sin\theta\tan\theta = \sin\theta$.

SOLUTION The following are equivalent to the given equation:

$$\sin\theta\tan\theta - \sin\theta = 0$$

$$\sin\theta\,(\tan\theta - 1) = 0$$

To find the solutions we set each factor on the left equal to zero, obtaining

$$\sin\theta = 0 \quad \text{and} \quad \tan\theta = 1.$$

The solutions of the equation $\sin\theta = 0$ are $0, \pm\pi, \pm 2\pi, \ldots$, that is, $\theta = \pi n$ for every integer n.

The tangent function has period π and hence we find the solutions of the equation $\tan\theta = 1$ that are in the interval $(-\pi/2, \pi/2)$ and then add multiples of π. The only solution of $\tan\theta = 1$ in $(-\pi/2, \pi/2)$ is $\pi/4$, and hence every solution has the form

$$\theta = \frac{\pi}{4} + \pi n \quad \text{for some integer } n.$$

Thus the solutions of $\sin\theta\tan\theta = \sin\theta$ are given by

$$n\pi \quad \text{and} \quad \frac{\pi}{4} + \pi n \quad \text{for every integer } n.$$

Some *particular* solutions, obtained by letting $n = 0$, $n = 1$, $n = 2$, and $n = -1$, are

$$0, \quad \frac{\pi}{4}, \quad \pi, \quad \frac{5\pi}{4}, \quad 2\pi, \quad \frac{9\pi}{4}, \quad -\pi, \quad -\frac{3\pi}{4}.$$

Other values of n give other solutions.

In Example 4 it would have been incorrect to begin by dividing both sides by $\sin\theta$, since we would then lose the solutions of $\sin\theta = 0$.

EXAMPLE ▪ 5

Solve the equation $2\sin^2 t - \cos t - 1 = 0$, and express the solutions both in radian measure and in degree measure.

SOLUTION We first change the equation to an equation that involves only $\cos t$, and then factor:

$$\begin{aligned}
2\sin^2 t - \cos t - 1 &= 0\\
2(1 - \cos^2 t) - \cos t - 1 &= 0\\
2 - 2\cos^2 t - \cos t - 1 &= 0\\
-2\cos^2 t - \cos t + 1 &= 0\\
2\cos^2 t + \cos t - 1 &= 0\\
(2\cos t - 1)(\cos t + 1) &= 0
\end{aligned}$$

The solutions of the last equation are the solutions of

$$2\cos t - 1 = 0 \quad \text{and} \quad \cos t + 1 = 0$$

or, equivalently, $\cos t = \frac{1}{2}$ and $\cos t = -1$.

Since the cosine function has period 2π, we may find all solutions of these equations by adding multiples of 2π to the solutions that are in the interval $[0, 2\pi)$.

If $\cos t = \frac{1}{2}$, then the reference angle is $\pi/3$ (or $60°$). Since $\cos t$ is positive, the angle of radian measure t is in either quadrant I or IV. Hence in the interval $[0, 2\pi)$ we have the solutions

$$t = \frac{\pi}{3} \quad \text{and} \quad t = 2\pi - \frac{\pi}{3} = \frac{5\pi}{3}.$$

If $\cos t = -1$, then $t = \pi$.
Thus, the solutions of the given equation are given by

$$\frac{\pi}{3} + 2\pi n, \quad \frac{5\pi}{3} + 2\pi n, \quad \pi + 2\pi n \quad \text{for every integer } n.$$

In degree measure the solutions are

$$60^\circ + 360^\circ n, \quad 300^\circ + 360^\circ n, \quad 180^\circ + 360^\circ n.$$

EXAMPLE ▪ 6

Find the solutions of $4 \sin^2 x \tan x - \tan x = 0$ that are in the interval $[0, 2\pi)$.

SOLUTION We factor the left-hand side of the equation:

$$4 \sin^2 x \tan x - \tan x = 0$$
$$\tan x \, (4 \sin^2 x - 1) = 0$$

Setting each factor equal to zero gives us

$$\tan x = 0 \quad \text{and} \quad \sin^2 x = \tfrac{1}{4}$$

or, equivalently,

$$\tan x = 0, \qquad \sin x = \tfrac{1}{2}, \qquad \sin x = -\tfrac{1}{2}.$$

We now find the solutions of each of these three equations.

Referring to Figure 3 we see that the equation $\tan x = 0$ has solutions 0 and π in the interval $[0, 2\pi)$.

From Example 1, the equation $\sin x = \frac{1}{2}$ has solutions $\pi/6$ and $5\pi/6$ in $[0, 2\pi)$.

The equation $\sin x = -\frac{1}{2}$ leads to the angles between π and 2π that have reference angle $\pi/6$, as illustrated in Figure 4. These are

FIGURE 4

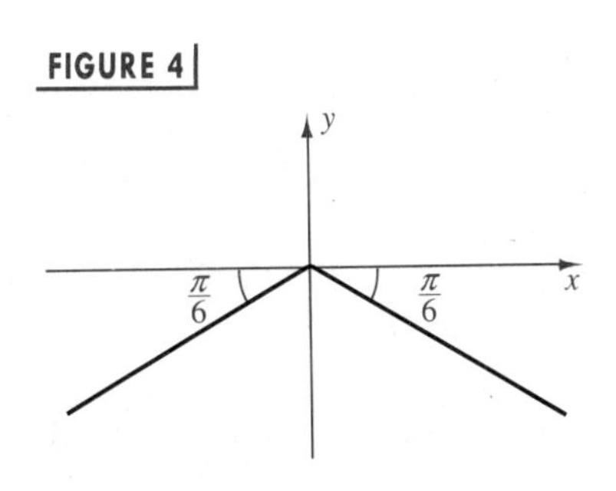

$$\pi + \frac{\pi}{6} = \frac{7\pi}{6} \quad \text{and} \quad 2\pi - \frac{\pi}{6} = \frac{11\pi}{6}.$$

We could also use the graph of the sine function in Figure 2 to find these solutions. Hence, the solutions of the given equation in the interval $[0, 2\pi)$ are

$$0, \quad \pi, \quad \frac{\pi}{6}, \quad \frac{5\pi}{6}, \quad \frac{7\pi}{6}, \quad \frac{11\pi}{6}.$$

EXAMPLE ▪ 7

Find the solutions of $\csc^4 2u - 4 = 0$.

SOLUTION Factoring the left-hand side of the equation gives us

$$(\csc^2 2u - 2)(\csc^2 2u + 2) = 0.$$

Setting each factor equal to zero, we obtain

$$\csc^2 2u = 2 \quad \text{and} \quad \csc^2 2u = -2.$$

The equation $\csc^2 2u = -2$ has no real solution. The solutions of $\csc^2 2u = 2$ are the solutions of the two equations

$$\csc 2u = \sqrt{2} \quad \text{and} \quad \csc 2u = -\sqrt{2},$$

or, equivalently,

$$\sin 2u = \frac{1}{\sqrt{2}} = \frac{\sqrt{2}}{2} \quad \text{and} \quad \sin 2u = \frac{1}{-\sqrt{2}} = -\frac{\sqrt{2}}{2}.$$

If $\sin 2u = \sqrt{2}/2$, then the reference angle for $2u$ is $\pi/4$ and hence

$$2u = \frac{\pi}{4} + 2\pi n \quad \text{or} \quad 2u = \frac{3\pi}{4} + 2\pi n$$

for some integer n. Dividing both sides of the last two equations by 2 gives us

$$u = \frac{\pi}{8} + \pi n \quad \text{or} \quad u = \frac{3\pi}{8} + \pi n.$$

Similarly, from $\sin 2u = -\sqrt{2}/2$, we obtain

$$2u = \frac{5\pi}{4} + 2\pi n \quad \text{or} \quad 2u = \frac{7\pi}{4} + 2\pi n$$

for some integer n. Dividing both sides of the last two equations by 2 gives us

$$u = \frac{5\pi}{8} + \pi n \quad \text{or} \quad u = \frac{7\pi}{8} + \pi n.$$

Analyzing the preceding information, we see that all solutions can be written in the form

$$u = \frac{\pi}{8} + \frac{\pi}{4}n \quad \text{for every integer } n.$$

The next example illustrates the use of a calculator in solving a trigonometric equation.

EXAMPLE ▪ 8

Approximate, to the nearest degree, the solutions of the equation

$$5 \sin \theta \tan \theta - 10 \tan \theta + 3 \sin \theta - 6 = 0$$

in the interval $[0°, 360°)$.

SOLUTION We first factor the equation by grouping terms:

$$(5 \sin \theta \tan \theta - 10 \tan \theta) + (3 \sin \theta - 6) = 0$$
$$5 \tan \theta\,(\sin \theta - 2) + 3(\sin \theta - 2) = 0$$
$$(5 \tan \theta + 3)(\sin \theta - 2) = 0$$

Setting each factor on the left-hand side equal to zero gives us

$$\tan \theta = -\tfrac{3}{5} \quad \text{and} \quad \sin \theta = 2.$$

The equation $\sin \theta = 2$ has no solution, since $\sin \theta \leq 1$ for every θ. Thus the solutions of the given equation are the same as those of

$$\tan \theta = -\tfrac{3}{5} = -0.6000.$$

Let us begin by approximating the reference angle, that is, the acute angle θ' such that $\tan \theta' = \frac{3}{5} = 0.6$. We use a calculator in degree mode:

Enter 0.6:	0.6	(the value of $\tan \theta'$)
Press INV TAN:	30.963757	(the degree measure of θ')

Hence $\theta' \approx 31°$. Since θ is in either quadrant II or IV, we obtain the solutions

$$\theta \approx 180° - 31°, \quad \text{or} \quad \theta \approx 149°$$

and

$$\theta \approx 360° - 31°, \quad \text{or} \quad \theta \approx 329°.$$

EXAMPLE ▪ 9

In Boston the number of hours of daylight $D(t)$ at a particular time of the year may be approximated by

$$D(t) = 3 \sin \frac{2\pi}{365}(t - 79) + 12$$

with t in days and $t = 0$ corresponding to January 1. How many days of the year have more than 10.5 hours of daylight?

FIGURE 5

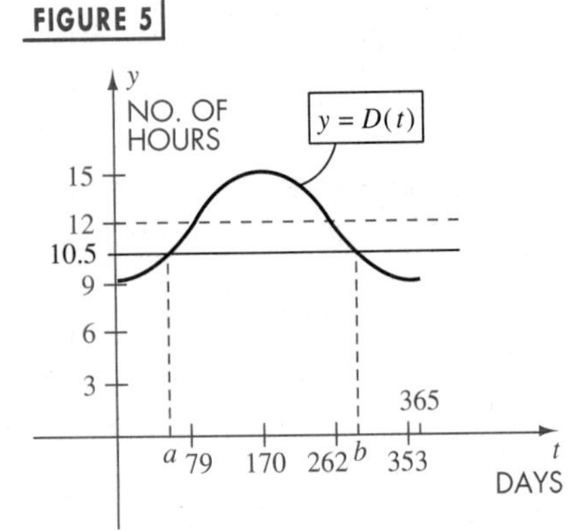

SOLUTION The graph of D was discussed in Example 10 of Section 6.6 and is resketched in Figure 5. As illustrated in the figure, if we can find two numbers a and b with $D(a) = 10.5$, $D(b) = 10.5$, and $0 < a < b < 365$, then there will be more than 10.5 hours of daylight in the tth day of the year if $a < t < b$.

Let us solve the equation $D(t) = 10.5$, that is,

$$3 \sin \frac{2\pi}{365}(t - 79) + 12 = 10.5.$$

This is equivalent to $\quad 3 \sin \dfrac{2\pi}{365}(t - 79) = -1.5$

and $\quad \sin \dfrac{2\pi}{365}(t - 79) = -0.5 = -\dfrac{1}{2}.$

If $\sin \theta = -\frac{1}{2}$, then the reference angle is $\pi/6$ and the angle θ is in either quadrant III or IV. Thus we can find the numbers a and b by solving the equations

$$\frac{2\pi}{365}(t - 79) = \frac{7\pi}{6} \quad \text{and} \quad \frac{2\pi}{365}(t - 79) = \frac{11\pi}{6}.$$

From the first of these equations we obtain

$$t - 79 = \frac{7\pi}{6} \cdot \frac{365}{2\pi} = \frac{2555}{12} \approx 213$$

and hence $\quad t \approx 213 + 79, \quad \text{or} \quad t \approx 292.$

Similarly, the second equation gives us $t \approx 414$. Since the period of the function D is 365 days (see Figure 5), we also obtain

$$t \approx 414 - 365, \quad \text{or} \quad t \approx 49.$$

Thus there will be at least 10.5 hours of daylight from $t = 49$ to $t = 292$, that is, for 243 days of the year.

7.2 EXERCISES

Exer. 1–34: Find all solutions of the equation.

1 $\sin x = -\dfrac{\sqrt{2}}{2}$

2 $\cos t = -1$

3 $\tan \theta = \sqrt{3}$

4 $\cot \alpha = -\dfrac{1}{\sqrt{3}}$

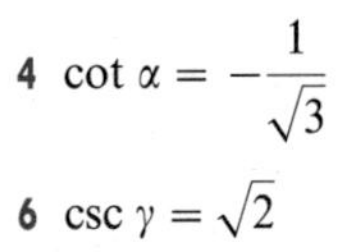

5 $\sec \beta = 2$

6 $\csc \gamma = \sqrt{2}$

7 $\sin x = \frac{\pi}{2}$

8 $\cos x = -\frac{\pi}{3}$

9 $\cos\theta = \frac{1}{\sec\theta}$

10 $\csc\theta\sin\theta = 1$

11 $2\cos 2\theta - \sqrt{3} = 0$

12 $2\sin 3\theta + \sqrt{2} = 0$

13 $\sqrt{3}\tan\frac{1}{3}t = 1$

14 $\cos\frac{1}{4}x = -\frac{\sqrt{2}}{2}$

15 $\sin\left(\theta + \frac{\pi}{4}\right) = \frac{1}{2}$

16 $\cos\left(x - \frac{\pi}{3}\right) = -1$

17 $\sin\left(2x - \frac{\pi}{3}\right) = \frac{1}{2}$

18 $\cos\left(4x - \frac{\pi}{4}\right) = \frac{\sqrt{2}}{2}$

19 $2\cos t + 1 = 0$

20 $\cot\theta + 1 = 0$

21 $\tan^2 x = 1$

22 $4\cos\theta - 2 = 0$

23 $(\cos\theta - 1)(\sin\theta + 1) = 0$

24 $2\cos x = \sqrt{3}$

25 $\sec^2\alpha - 4 = 0$

26 $3 - \tan^2\beta = 0$

27 $\sqrt{3} + 2\sin\beta = 0$

28 $4\sin^2 x - 3 = 0$

29 $\cot^2 x - 3 = 0$

30 $(\sin t - 1)\cos t = 0$

31 $(2\sin\theta + 1)(2\cos\theta + 3) = 0$

32 $(2\sin u - 1)(\cos u - \sqrt{2}) = 0$

33 $\sin 2x\,(\csc 2x - 2) = 0$

34 $\tan\alpha + \tan^2\alpha = 0$

Exer. 35–58: Find the solutions of the equation that are in the interval $[0, 2\pi)$.

35 $\cos\left(2x - \frac{\pi}{4}\right) = 0$

36 $\sin\left(3x - \frac{\pi}{4}\right) = 1$

37 $2 - 8\cos^2 t = 0$

38 $\cot^2\theta - \cot\theta = 0$

39 $2\sin^2 u = 1 - \sin u$

40 $2\cos^2 t + 3\cos t + 1 = 0$

41 $\tan^2 x \sin x = \sin x$

42 $\sec\beta\csc\beta = 2\csc\beta$

43 $2\cos^2\gamma + \cos\gamma = 0$

44 $\sin x - \cos x = 0$

45 $\sin^2\theta + \sin\theta - 6 = 0$

46 $2\sin^2 u + \sin u - 6 = 0$

47 $1 - \sin t = \sqrt{3}\cos t$

48 $\cos\theta - \sin\theta = 1$

49 $\cos\alpha + \sin\alpha = 1$

50 $\sqrt{3}\sin t + \cos t = 1$

51 $2\tan t - \sec^2 t = 0$

52 $\tan\theta + \sec\theta = 1$

53 $\cot\alpha + \tan\alpha = \csc\alpha\sec\alpha$

54 $\sin x + \cos x\cot x = \csc x$

55 $2\sin^3 x + \sin^2 x - 2\sin x - 1 = 0$

56 $\sec^5\theta = 4\sec\theta$

57 $2\tan t\csc t + 2\csc t + \tan t + 1 = 0$

58 $2\sin v\csc v - \csc v = 4\sin v - 2$

Exer. 59–64: Approximate, to the nearest 10′, the solutions of the equation in the interval [0°, 360°).

59 $\sin^2 t - 4\sin t + 1 = 0$

60 $\cos^2 t - 4\cos t + 2 = 0$

61 $\tan^2\theta + 3\tan\theta + 2 = 0$

62 $2\tan^2 x - 3\tan x - 1 = 0$

63 $12\sin^2 u - 5\sin u - 2 = 0$

64 $5\cos^2\alpha + 3\cos\alpha - 2 = 0$

65 A tidal wave of height 50 feet and period 30 minutes is approaching a sea wall that is 20 feet above sea level (see figure). From a particular point on shore, the distance y from sea level to the top of the wave is given by $y = 25\cos\frac{\pi}{15}t$ with t in minutes. For approximately how many minutes of each 30-minute period is the top of the wave above the sea wall?

EXERCISE 65

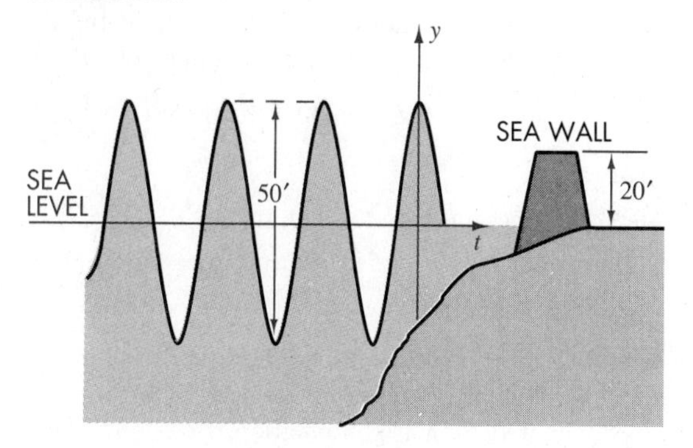

66 The expected low temperature T (in °F) in Fairbanks, Alaska, can be approximated by

$$T = 36\sin\frac{2\pi}{365}(t - 101) + 14$$

for t in days and $t = 0$ corresponding to January 1. For how many days during the year is the low temperature expected to be below −4°F?

67 On a clear day with D hours of daylight the intensity of sunlight I (in calories/cm^2) may be approximated by $I = I_M \sin^3 \frac{\pi t}{D}$ for $0 \le t \le D$ with $t = 0$ corresponding to sunrise and I_M the maximum intensity. If $D = 12$, approximately how many hours after sunrise is $I = \frac{1}{2}I_M$?

68 Refer to Exercise 67. On cloudy days, the sun intensity I is better represented by $I = I_M \sin^2 \frac{\pi t}{D}$. If $D = 12$, approximately how many hours after sunrise is $I = \frac{1}{2}I_M$?

69 Refer to Exercises 67 and 68. A skin-care specialist recommends protection from the sun when the intensity I exceeds 75% of the maximum intensity. If $D = 12$ hours, approximate the number of hours that protection is required on **(a)** a clear day; **(b)** a cloudy day.

70 In the study of frost penetration problems in highway engineering, the temperature T at time t hours and depth x feet is given by $T = T_0 e^{-\lambda x} \sin(\omega t - \lambda x)$ for constants T_0, ω, and λ. The period of T is 24 hours.

(a) Find a formula for the temperature at the surface.

(b) At what times is the surface temperature at a minimum?

(c) For $\lambda = 2.5$, find the times when the temperature is a minimum at a depth of 1 foot.

71 Many animal populations, such as rabbits, fluctuate over ten-year cycles. Suppose that the number of rabbits at time t is given by $N(t) = 4000 + 1000 \cos\left(\frac{\pi}{5}t\right)$ for t in years.

(a) Sketch the graph of N for $0 \le t \le 10$.

(b) For what values of t in part (a) does the population exceed 4500 rabbits?

72 The flow rate (or water discharge rate) at the mouth of a river frequently follows an annual cycle that can be approximated by a trigonometric expression. The flow rate for the Orinoco River in South America may be approximated by

$$F(t) = 34{,}000 + 26{,}000 \sin \frac{\pi}{6}(t - 5.5)$$

for time t in months and flow rate $F(t)$ in m^3/second. For approximately how many months each year does the flow rate exceed 55,000 m^3/second?

73 Shown in the figure is a graph of $y = \frac{1}{2}x + \sin x$ for $-2\pi \le x \le 2\pi$. Using calculus it can be shown that the x-coordinates of the turning points A, B, C and D on the graph are solutions of the equation $\frac{1}{2} + \cos x = 0$. Determine the coordinates of these points.

EXERCISE 73

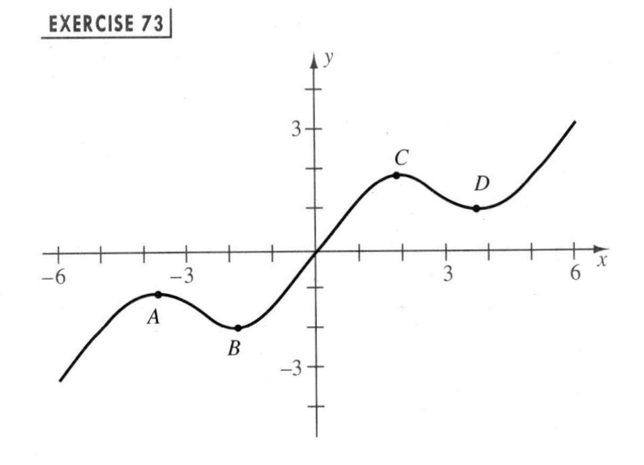

74 Damped oscillations are oscillations of decreasing magnitude that occur when frictional forces act on a vibrating object. Shown in the figure is the graph of the damped oscillation given by the equation

$$y = e^{-x/2} \sin 2x.$$

The x-coordinates of the turning points on the graph are solutions of $2 \cos 2x - \frac{1}{2} \sin 2x = 0$. Approximate the x-coordinates of these points for $x > 0$.

EXERCISE 74

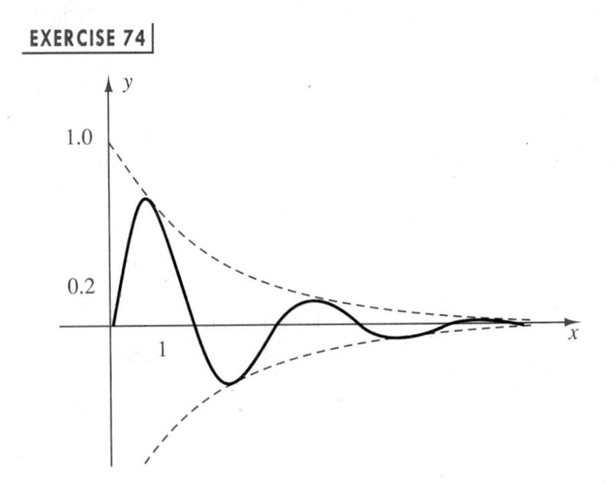

7.3 THE ADDITION AND SUBTRACTION FORMULAS

In this section we derive formulas that involve trigonometric functions of $u + v$ or $u - v$ for any real numbers or angles u and v. These formulas are known as *addition* or *subtraction formulas*, respectively. The first formula may be stated as follows.

SUBTRACTION FORMULA FOR COSINE

$$\cos(u - v) = \cos u \cos v + \sin u \sin v$$

PROOF Let u and v be any real numbers, and consider angles of radian measure u and v, respectively. Let $w = u - v$. Figure 6 illustrates one possibility with the angles in standard position on a rectangular coordinate system. For convenience we have assumed that both u and v are positive and that $0 \leq u - v < v$.

FIGURE 6 FIGURE 7

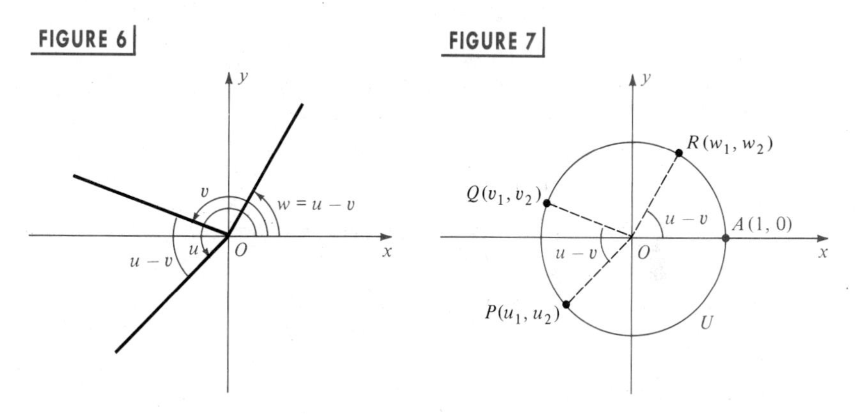

As in Figure 7, let $P(u_1, u_2)$, $Q(v_1, v_2)$, and $R(w_1, w_2)$ be the points on the terminal sides of the indicated angles that are each a distance 1 from the origin. In this case P, Q, and R are on the unit circle U with center at the origin. From the definitions of trigonometric functions of any angle (with $c = 1$),

$$(*) \qquad \begin{array}{lll} \cos u = u_1, & \cos v = v_1, & \cos(u - v) = w_1, \\ \sin u = u_2, & \sin v = v_2, & \sin(u - v) = w_2. \end{array}$$

The symbol $(*)$ has been used for later reference to these formulas.

We next observe that the distance between $A(1, 0)$ and R must equal the distance between P and Q, because angles AOR and POQ have the same measure $u - v$. Using the distance formula,

$$\sqrt{(w_1 - 1)^2 + (w_2 - 0)^2} = \sqrt{(u_1 - v_1)^2 + (u_2 - v_2)^2}.$$

Squaring both sides and expanding the terms under the radicals gives us

$$w_1^2 - 2w_1 + 1 + w_2^2 = u_1^2 - 2u_1v_1 + v_1^2 + u_2^2 - 2u_2v_2 + v_2^2.$$

Since the points (u_1, u_2), (v_1, v_2), and (w_1, w_2) are on the unit circle U, and since an equation for U is $x^2 + y^2 = 1$, we may substitute 1 for each of $u_1^2 + u_2^2$, $v_1^2 + v_2^2$ and $w_1^2 + w_2^2$. Doing this and simplifying, we obtain

$$2 - 2w_1 = 2 - 2u_1v_1 - 2u_2v_2,$$

which reduces to

$$w_1 = u_1v_1 + u_2v_2.$$

Substituting from the formulas stated in (*) gives us

$$\cos(u - v) = \cos u \cos v + \sin u \sin v,$$

which is what we wished to prove. It is possible to extend our discussion to all values of u and v. ❑

The next example demonstrates the use of the subtraction formula in finding the exact value of $\cos 15°$. Of course, if only an approximation is desired we could use a calculator or Table 4.

EXAMPLE ▪ 1

Find the exact value of $\cos 15°$ by using the fact that $15° = 60° - 45°$.

SOLUTION We use the subtraction formula with $u = 60°$ and $v = 45°$:

$$\begin{aligned}\cos 15° &= \cos(60° - 45°)\\ &= \cos 60° \cos 45° + \sin 60° \sin 45°\\ &= \frac{1}{2}\frac{\sqrt{2}}{2} + \frac{\sqrt{3}}{2}\frac{\sqrt{2}}{2}\\ &= \frac{\sqrt{2} + \sqrt{6}}{4}\end{aligned}$$

It is easy to obtain a formula for $\cos(u + v)$. We begin by writing $u + v = u - (-v)$ and then employ the subtraction formula:

$$\begin{aligned}\cos(u + v) &= \cos[u - (-v)]\\ &= \cos u \cos(-v) + \sin u \sin(-v).\end{aligned}$$

Using the formulas for negatives, $\cos(-v) = \cos v$ and $\sin(-v) = -\sin v$ gives us the following.

ADDITION FORMULA FOR COSINE

$$\cos(u + v) = \cos u \cos v - \sin u \sin v$$

EXAMPLE ▪ 2

Find $\cos \frac{7\pi}{12}$ by using $\frac{7\pi}{12} = \frac{\pi}{3} + \frac{\pi}{4}$.

SOLUTION We apply the formula for $\cos(u + v)$:

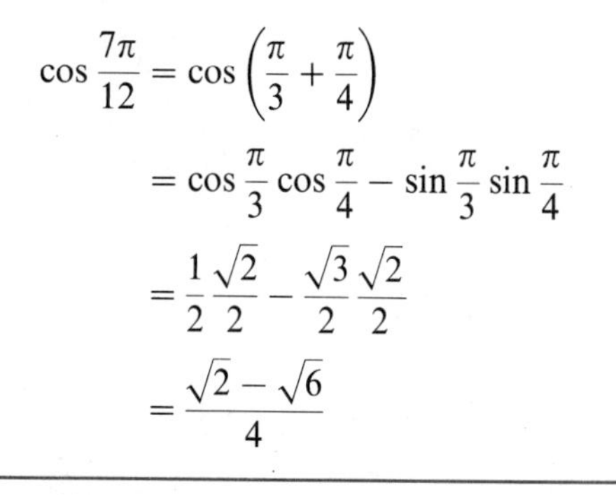

$$\begin{aligned}\cos \frac{7\pi}{12} &= \cos\left(\frac{\pi}{3} + \frac{\pi}{4}\right)\\ &= \cos \frac{\pi}{3} \cos \frac{\pi}{4} - \sin \frac{\pi}{3} \sin \frac{\pi}{4}\\ &= \frac{1}{2}\frac{\sqrt{2}}{2} - \frac{\sqrt{3}}{2}\frac{\sqrt{2}}{2}\\ &= \frac{\sqrt{2} - \sqrt{6}}{4}\end{aligned}$$

We refer to the sine and cosine functions as **cofunctions** of one another. Similarly, the tangent and cotangent functions are cofunctions, as are the secant and cosecant. If u is the radian measure of an acute angle, then the angle with radian measure $\frac{\pi}{2} - u$ is complementary to u, and we may consider the right triangle shown in Figure 8. Using ratios, we see that

FIGURE 8

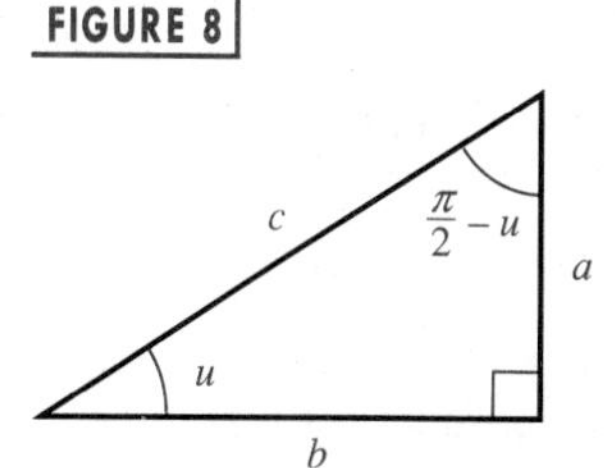

$$\sin u = \frac{a}{c} = \cos\left(\frac{\pi}{2} - u\right)$$

$$\cos u = \frac{b}{c} = \sin\left(\frac{\pi}{2} - u\right)$$

$$\tan u = \frac{a}{b} = \cot\left(\frac{\pi}{2} - u\right)$$

These three formulas and their analogues for $\sec u$, $\csc u$, and $\cot u$ state that *the function value of u equals the cofunction of the complementary angle* $\frac{\pi}{2} - u$.

In the following formulas we use subtraction formulas to extend these relationships to any real number u.

COFUNCTION FORMULAS

If u is a real number or the radian measure of an angle, then:

$$\cos\left(\frac{\pi}{2} - u\right) = \sin u \qquad \sin\left(\frac{\pi}{2} - u\right) = \cos u$$

$$\tan\left(\frac{\pi}{2} - u\right) = \cot u \qquad \cot\left(\frac{\pi}{2} - u\right) = \tan u$$

$$\sec\left(\frac{\pi}{2} - u\right) = \csc u \qquad \csc\left(\frac{\pi}{2} - u\right) = \sec u$$

PROOF Using the subtraction formula for cosine,

$$\cos\left(\frac{\pi}{2} - u\right) = \cos\frac{\pi}{2}\cos u + \sin\frac{\pi}{2}\sin u$$
$$= (0)\cos u + (1)\sin u = \sin u.$$

This gives us the first formula.

If we substitute $\frac{\pi}{2} - v$ for u in the first formula, we obtain

$$\cos\left[\frac{\pi}{2} - \left(\frac{\pi}{2} - v\right)\right] = \sin\left(\frac{\pi}{2} - v\right),$$

or

$$\cos v = \sin\left(\frac{\pi}{2} - v\right).$$

Since the symbol v is arbitrary, this equation is equivalent to

$$\cos u = \sin\left(\frac{\pi}{2} - u\right).$$

Using a fundamental identity and the first two cofunction formulas,

$$\tan\left(\frac{\pi}{2} - u\right) = \frac{\sin\left(\frac{\pi}{2} - u\right)}{\cos\left(\frac{\pi}{2} - u\right)} = \frac{\cos u}{\sin u} = \cot u.$$

The proofs of the remaining three formulas are similar. ❑

An easy way to remember the cofunction formulas is to refer to the triangle in Figure 8.

We may now prove the following identities.

ADDITION AND SUBTRACTION FORMULAS FOR SINE AND TANGENT

$$\sin(u+v) = \sin u \cos v + \cos u \sin v$$

$$\sin(u-v) = \sin u \cos v - \cos u \sin v$$

$$\tan(u+v) = \frac{\tan u + \tan v}{1 - \tan u \tan v}$$

$$\tan(u-v) = \frac{\tan u - \tan v}{1 + \tan u \tan v}$$

PROOF We shall prove the first and third formulas. Supply reasons for each of the following steps:

$$\begin{aligned}
\sin(u+v) &= \cos\left[\frac{\pi}{2} - (u+v)\right] \\
&= \cos\left[\left(\frac{\pi}{2} - u\right) - v\right] \\
&= \cos\left(\frac{\pi}{2} - u\right)\cos v + \sin\left(\frac{\pi}{2} - u\right)\sin v \\
&= \sin u \cos v + \cos u \sin v.
\end{aligned}$$

To verify the formula for $\tan(u+v)$ we begin as follows:

$$\begin{aligned}
\tan(u+v) &= \frac{\sin(u+v)}{\cos(u+v)} \\
&= \frac{\sin u \cos v + \cos u \sin v}{\cos u \cos v - \sin u \sin v}.
\end{aligned}$$

If $\cos u \cos v \neq 0$, then we may divide numerator and denominator by $\cos u \cos v$, obtaining

$$\begin{aligned}
\tan(u+v) &= \frac{\left(\dfrac{\sin u}{\cos u}\right)\left(\dfrac{\cos v}{\cos v}\right) + \left(\dfrac{\cos u}{\cos u}\right)\left(\dfrac{\sin v}{\cos v}\right)}{\left(\dfrac{\cos u}{\cos u}\right)\left(\dfrac{\cos v}{\cos v}\right) - \left(\dfrac{\sin u}{\cos u}\right)\left(\dfrac{\sin v}{\cos v}\right)} \\
&= \frac{\tan u + \tan v}{1 - \tan u \tan v}.
\end{aligned}$$

If $\cos u \cos v = 0$, then either $\cos u = 0$ or $\cos v = 0$. In this case either $\tan u$ or $\tan v$ is undefined and the formula is invalid. ❑

EXAMPLE ▪ 3

If $\sin\alpha = \frac{4}{5}$ for some angle α in quadrant I, and $\cos\beta = -\frac{12}{13}$ for some angle β in quadrant II, find **(a)** the exact values of $\sin(\alpha+\beta)$ and $\tan(\alpha+\beta)$ and **(b)** the quadrant containing $\alpha+\beta$.

FIGURE 9

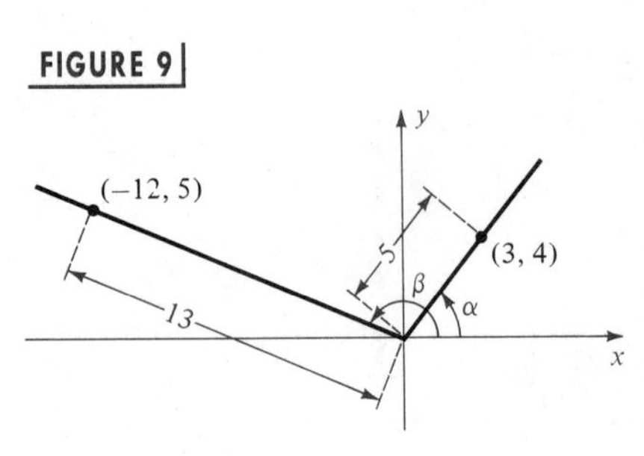

SOLUTION Angles α and β are illustrated in Figure 9. There is no loss of generality in regarding α and β as positive angles between 0 and 2π as we have done in the figure. Since $\sin\alpha = \frac{4}{5}$, we may choose the point (3, 4) on the terminal side of α. Similarly, since $\cos\beta = -\frac{12}{13}$, the point $(-12, 5)$ is on the terminal side of β. Referring to Figure 9 and using the definition of trigonometric functions of any angle,

$$\cos\alpha = \tfrac{3}{5},\qquad \tan\alpha = \tfrac{4}{3},\qquad \sin\beta = \tfrac{5}{13},\qquad \tan\beta = -\tfrac{5}{12}.$$

(a) Using addition formulas:

$$\begin{aligned}\sin(\alpha+\beta) &= \sin\alpha\cos\beta + \cos\alpha\sin\beta\\ &= \left(\tfrac{4}{5}\right)\left(-\tfrac{12}{13}\right) + \left(\tfrac{3}{5}\right)\left(\tfrac{5}{13}\right) = -\tfrac{33}{65}.\end{aligned}$$

$$\tan(\alpha+\beta) = \frac{\tan\alpha + \tan\beta}{1 - \tan\alpha\tan\beta} = \frac{\frac{4}{3} + \left(-\frac{5}{12}\right)}{1 - \left(\frac{4}{3}\right)\left(-\frac{5}{12}\right)} = \frac{33}{56}.$$

(b) Since $\sin(\alpha+\beta)$ is negative and $\tan(\alpha+\beta)$ is positive, the angle $\alpha+\beta$ is in quadrant III.

The type of simplification illustrated in the next example occurs in calculus.

EXAMPLE ▪ 4

If $f(x) = \sin x$ and $h \neq 0$, show that

$$\frac{f(x+h) - f(x)}{h} = \sin x\left(\frac{\cos h - 1}{h}\right) + \cos x\left(\frac{\sin h}{h}\right).$$

SOLUTION We use the definition of f and the addition formula for sine:

$$\begin{aligned}\frac{f(x+h) - f(x)}{h} &= \frac{\sin(x+h) - \sin x}{h}\\ &= \frac{\sin x\cos h + \cos x\sin h - \sin x}{h}\\ &= \frac{\sin x(\cos h - 1) + \cos x\sin h}{h}\\ &= \sin x\left(\frac{\cos h - 1}{h}\right) + \cos x\left(\frac{\sin h}{h}\right)\end{aligned}$$

EXAMPLE ▪ 5

Let a and b be real numbers with $a > 0$. Show that for every x

$$a \cos Bx + b \sin Bx = A \cos (Bx - C)$$

where $A = \sqrt{a^2 + b^2}$ and $\tan C = \frac{b}{a}$ with $-\frac{\pi}{2} < C < \frac{\pi}{2}$.

SOLUTION Given $a \cos Bx + b \sin Bx$, let us consider $\tan C = b/a$ with $-\pi/2 < C < \pi/2$. Thus $b = a \tan C$, and we may write

$$\begin{aligned} a \cos Bx + b \sin Bx &= a \cos Bx + (a \tan C) \sin Bx \\ &= a \cos Bx + a \frac{\sin C}{\cos C} \sin Bx \\ &= \frac{a}{\cos C} (\cos C \cos Bx + \sin C \sin Bx) \\ &= (a \sec C) \cos (Bx - C). \end{aligned}$$

We shall complete the proof by showing that $a \sec C = \sqrt{a^2 + b^2}$. Since $-\pi/2 < C < \pi/2$, it follows that $\sec C$ is positive, and hence

$$a \sec C = a\sqrt{1 + \tan^2 C}.$$

Using $\tan C = b/a$ and $a > 0$, we obtain

$$a \sec C = a\sqrt{1 + \frac{b^2}{a^2}} = \sqrt{a^2\left(1 + \frac{b^2}{a^2}\right)} = \sqrt{a^2 + b^2}.$$

EXAMPLE ▪ 6

If $f(x) = \cos x + \sin x$, use the formulas given in Example 5 to express $f(x)$ in the form $A \cos (Bx - C)$, and then sketch the graph of f.

SOLUTION Letting $a = 1$, $b = 1$, and $B = 1$ in the formulas from Example 5,

$$A = \sqrt{a^2 + b^2} = \sqrt{1 + 1} = \sqrt{2} \quad \text{and} \quad \tan C = \frac{b}{a} = \frac{1}{1} = 1.$$

Since $\tan C = 1$ and $-\pi/2 < C < \pi/2$, we obtain $C = \pi/4$. Substituting for a, b, A, B, and C in the formula

$$a \cos Bx + b \sin Bx = A \cos (Bx - C)$$

FIGURE 10

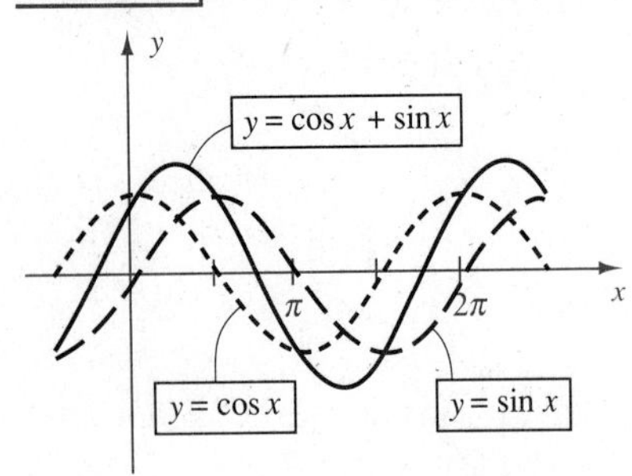

gives us

$$f(x) = \cos x + \sin x = \sqrt{2}\cos\left(x - \frac{\pi}{4}\right).$$

Comparing the last formula with the equation $y = a\cos(bx + c)$, which we discussed in Section 6.6, we see that the amplitude of the graph is $\sqrt{2}$, the period is 2π, and the phase shift is $\pi/4$. The graph of f is sketched in Figure 10, where we have also shown the graphs of $y = \sin x$ and $y = \cos x$. Note that since $f(x) = \cos x + \sin x$, we can find the y-coordinates of points on the graph of f by adding the corresponding y-coordinates of points on the graphs of $y = \cos x$ and $y = \sin x$.

7.3 EXERCISES

Exer. 1–4: Write the expression as a cofunction of a complementary angle.

1 (a) $\sin 46°37'$ (b) $\cos 73°12'$

(c) $\tan \frac{\pi}{6}$ (d) $\sec 17.28°$

2 (a) $\tan 24°12'$ (b) $\sin 89°41'$

(c) $\cos \frac{\pi}{3}$ (d) $\cot 61.87°$

3 (a) $\cos \frac{7\pi}{20}$ (b) $\sin \frac{1}{4}$

(c) $\tan 1$ (d) $\csc 0.53$

4 (a) $\sin \frac{\pi}{12}$ (b) $\cos 0.64$

(c) $\tan \sqrt{2}$ (d) $\sec 1.2$

Exer. 5–10: Find the exact values.

5 (a) $\cos \frac{\pi}{4} + \cos \frac{\pi}{6}$

(b) $\cos \frac{5\pi}{12}$ $\left(\text{use } \frac{5\pi}{12} = \frac{\pi}{4} + \frac{\pi}{6}\right)$

6 (a) $\sin \frac{2\pi}{3} + \sin \frac{\pi}{4}$

(b) $\sin \frac{11\pi}{12}$ $\left(\text{use } \frac{11\pi}{12} = \frac{2\pi}{3} + \frac{\pi}{4}\right)$

7 (a) $\tan 60° + \tan 225°$

(b) $\tan 285°$ (use $285° = 60° + 225°$)

8 (a) $\cos 135° - \cos 60°$

(b) $\cos 75°$ (use $75° = 135° - 60°$)

9 (a) $\sin \frac{3\pi}{4} - \sin \frac{\pi}{6}$

(b) $\sin \frac{7\pi}{12}$ $\left(\text{use } \frac{7\pi}{12} = \frac{3\pi}{4} - \frac{\pi}{6}\right)$

10 (a) $\tan \frac{3\pi}{4} - \tan \frac{\pi}{6}$

(b) $\tan \frac{7\pi}{12}$ $\left(\text{use } \frac{7\pi}{12} = \frac{3\pi}{4} - \frac{\pi}{6}\right)$

Exer. 11–16: Write the expression as a function of one angle.

11 $\cos 48° \cos 23° + \sin 48° \sin 23°$

12 $\cos 13° \cos 50° - \sin 13° \sin 50°$

13 $\cos 10° \sin 5° - \sin 10° \cos 5°$

14 $\sin 57° \cos 4° + \cos 57° \sin 4°$

15 $\cos 3 \sin(-2) - \cos 2 \sin 3$

16 $\sin(-5) \cos 2 + \cos 5 \sin(-2)$

17 If α and β are acute angles such that $\cos \alpha = \frac{4}{5}$ and $\tan \beta = \frac{8}{15}$, find (a) $\sin(\alpha + \beta)$, (b) $\cos(\alpha + \beta)$, and (c) the quadrant containing $\alpha + \beta$.

18 If α and β are acute angles such that $\csc \alpha = \frac{13}{12}$ and $\cot \beta = \frac{4}{3}$, find **(a)** $\sin(\alpha + \beta)$, **(b)** $\tan(\alpha + \beta)$, and **(c)** the quadrant containing $\alpha + \beta$.

19 If $\sin \alpha = -\frac{4}{5}$ and $\sec \beta = \frac{5}{3}$ for a third-quadrant angle α and a first-quadrant angle β, find **(a)** $\sin(\alpha + \beta)$, **(b)** $\tan(\alpha + \beta)$, and **(c)** the quadrant containing $\alpha + \beta$.

20 If $\tan \alpha = -\frac{7}{24}$ and $\cot \beta = \frac{3}{4}$ for α in the second quadrant and β in the third quadrant, find **(a)** $\sin(\alpha + \beta)$, **(b)** $\cos(\alpha + \beta)$, **(c)** $\tan(\alpha + \beta)$, **(d)** $\sin(\alpha - \beta)$, **(e)** $\cos(\alpha - \beta)$, and **(f)** $\tan(\alpha - \beta)$.

21 If α and β are third-quadrant angles with $\cos \alpha = -\frac{2}{5}$ and $\cos \beta = -\frac{3}{5}$, find **(a)** $\sin(\alpha - \beta)$, **(b)** $\cos(\alpha - \beta)$, and **(c)** the quadrant containing $\alpha - \beta$.

22 If α and β are second-quadrant angles such that $\sin \alpha = \frac{2}{3}$ and $\cos \beta = -\frac{1}{3}$, find **(a)** $\sin(\alpha + \beta)$, **(b)** $\tan(\alpha + \beta)$, and **(c)** the quadrant containing $\alpha + \beta$.

Exer. 23–38: Verify the identity.

23 $\cos(\theta - \pi) = -\cos \theta$

24 $\sin(\theta + \pi) = -\sin \theta$

25 $\sin\left(x + \frac{\pi}{2}\right) = \cos x$

26 $\cos\left(x + \frac{\pi}{2}\right) = -\sin x$

27 $\cos\left(\theta + \frac{3\pi}{2}\right) = \sin \theta$

28 $\sin\left(\alpha - \frac{3\pi}{2}\right) = \cos \alpha$

29 $\sin\left(\theta + \frac{\pi}{4}\right) = \frac{\sqrt{2}}{2}(\sin \theta + \cos \theta)$

30 $\cos\left(\theta + \frac{\pi}{4}\right) = \frac{\sqrt{2}}{2}(\cos \theta - \sin \theta)$

31 $\tan\left(u + \frac{\pi}{4}\right) = \frac{1 + \tan u}{1 - \tan u}$

32 $\tan\left(x - \frac{\pi}{4}\right) = \frac{\tan x - 1}{\tan x + 1}$

33 $\tan\left(u + \frac{\pi}{2}\right) = -\cot u$

34 $\tan\left(x - \frac{\pi}{2}\right) = -\cot x$

35 $\cos(u + v) + \cos(u - v) = 2 \cos u \cos v$

36 $\sin(u + v) + \sin(u - v) = 2 \sin u \cos v$

37 $\sin(u + v) \cdot \sin(u - v) = \sin^2 u - \sin^2 v$

38 $\cos(u + v) \cdot \cos(u - v) = \cos^2 u - \sin^2 v$

39 Express $\sin(u + v + w)$ in terms of functions of u, v, and w. (*Hint:* Write

$$\sin(u + v + w) = \sin[(u + v) + w]$$

and use an addition formula.)

40 Express $\tan(u + v + w)$ in terms of functions of u, v, and w.

41 Derive the formula $\cot(u + v) = \dfrac{\cot u \cot v - 1}{\cot u + \cot v}$.

42 If α and β are complementary angles, show that

$$\sin^2 \alpha + \sin^2 \beta = 1.$$

43 Derive the subtraction formula for the sine function.

44 Derive the subtraction formula for the tangent function.

45 If $f(x) = \cos x$, show that

$$\frac{f(x + h) - f(x)}{h} = \cos x\left(\frac{\cos h - 1}{h}\right) - \sin x\left(\frac{\sin h}{h}\right).$$

46 If $f(x) = \tan x$, show that

$$\frac{f(x + h) - f(x)}{h} = \sec^2 x\left(\frac{\sin h}{h}\right)\frac{1}{\cos h - \sin h \tan x}.$$

Exer. 47–52: Use an addition or a subtraction formula to find the solutions of the equation that are in the interval $[0, \pi)$.

47 $\sin 4t \cos t = \sin t \cos 4t$

48 $\cos 5t \cos 3t = \frac{1}{2} + \sin(-5t) \sin 3t$

49 $\cos 5t \cos 2t = -\sin 5t \sin 2t$

50 $\sin 3t \cos t + \cos 3t \sin t = -\frac{1}{2}$

51 $\tan 2t + \tan t = 1 - \tan 2t \tan t$

52 $\tan t - \tan 4t = 1 + \tan 4t \tan t$

Exer. 53–56:

(a) **Use the formula from Example 5 to express f in terms of the cosine function.**

(b) **Determine the amplitude, period, and phase shift of f.**

(c) **Sketch the graph of f.**

53 $f(x) = \sqrt{3} \cos 2x + \sin 2x$

54 $f(x) = \cos 4x + \sqrt{3} \sin 4x$

55 $f(x) = 2 \cos 3x - 2 \sin 3x$

56 $f(x) = 5 \cos 10x - 5 \sin 10x$

Exer. 57–58: For some applications in electrical engineering the sum of several sinusoidal signals, such as voltage signals or radio waves, of the same frequency are written in the compact form $y = A \cos(Bt - C)$. Express the signal in this form.

57 $y = 50 \sin 60\pi t + 40 \cos 60\pi t$

58 $y = 10\sin\left(120\pi t - \frac{\pi}{2}\right) + 5\sin 120\pi t$

59 If a mass that is attached to a spring is raised y_0 feet and given an initial vertical velocity of v_0 ft/sec, then the subsequent position y of the mass is given by

$$y = y_0 \cos \omega t + \frac{v_0}{\omega} \sin \omega t$$

for time t in seconds and a positive constant ω.

(a) If $\omega = 1$, $y_0 = 2$ feet, and $v_0 = 3$ ft/sec, write y in the form $A\cos(Bt - C)$ and find the amplitude and period.

(b) Determine the times when $y = 0$; that is, determine when the mass passes through the equilibrium position.

60 Refer to Exercise 59. If $y_0 = 1$ and $\omega = 2$, find the initial velocities that result in an amplitude of 4 feet.

61 If a tuning fork is struck and then held a certain distance from the eardrum, the pressure $p_1(t)$ on the outside of the eardrum at time t may be represented by $p_1(t) = A\sin\omega t$ for positive constants A and ω. If a second identical tuning fork is struck with a possibly different force and held a different distance from the eardrum (see figure), its effect may be represented by $p_2(t) = B\sin(\omega t + \tau)$ for $0 \le \tau \le 2\pi$ and a positive constant B. The total pressure $p(t)$ on the eardrum is

$$p(t) = A\sin\omega t + B\sin(\omega t + \tau).$$

(a) Show that $p(t) = a\cos\omega t + b\sin\omega t$ with $a = B\sin\tau$ and $b = A + B\cos\tau$.

(b) Show that the amplitude C of p is given by

$$C^2 = A^2 + B^2 + 2AB\cos\tau.$$

EXERCISE 61

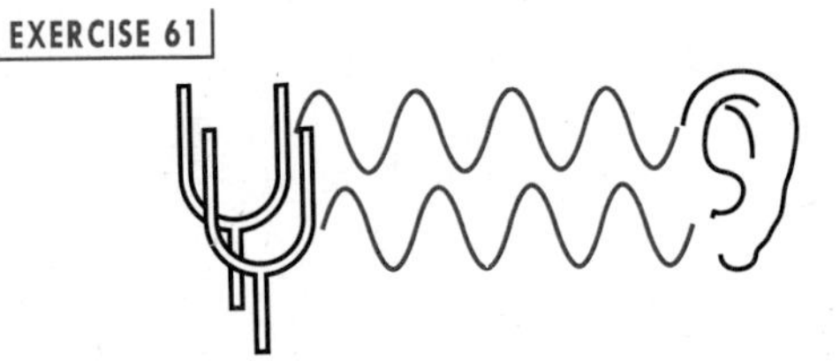

62 Refer to Exercise 61. Destructive interference occurs if the amplitude of the resulting sound wave is less than A. Suppose that two tuning forks are struck with the same force, that is, $A = B$.

(a) When total destructive interference occurs, the amplitude of p is zero and no sound is heard. Find the value of τ for which this occurs, and conclude that the low point on one wave coincides with a high point on the other wave.

(b) Determine the values of τ for which destructive interference occurs.

63 Refer to Exercise 61. If two tuning forks are struck, then constructive interference occurs if the amplitude C of the resulting sound wave is larger than either A or B (see figure).

(a) Show that $C \le A + B$.

(b) Find the values of τ such that $C = A + B$.

(c) If $A \ge B$, determine a condition for which constructive interference will occur.

EXERCISE 63

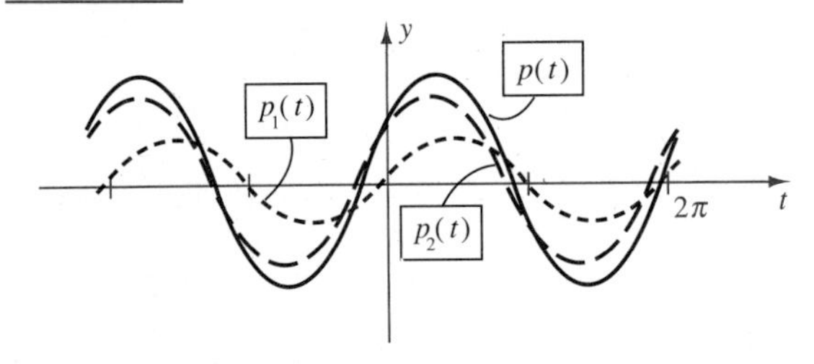

64 If a mountaintop is viewed from a point P due south of the mountain, the angle of elevation is α (see figure). If viewed from a point Q that is d miles east of P, the angle of elevation is β.

(a) Show that the height h of the mountain is given by

$$h = \frac{d\sin\alpha\sin\beta}{\sqrt{\sin^2\alpha - \sin^2\beta}}.$$

(b) If $\alpha = 30°$, $\beta = 20°$, and $d = 10$ miles, approximate h to the nearest hundredth of a mile.

EXERCISE 64

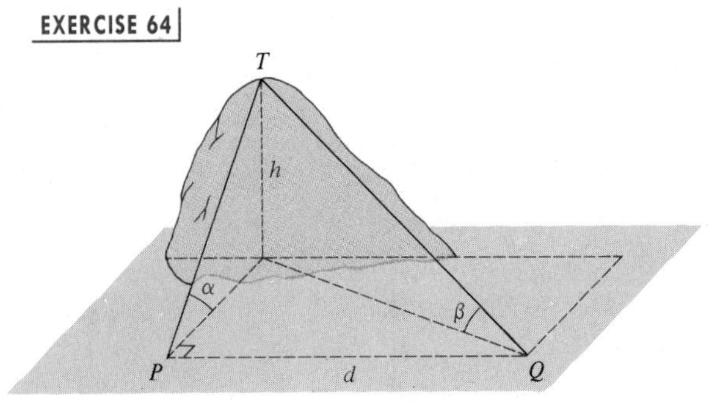

MULTIPLE-ANGLE FORMULAS

We refer to the formulas considered in this section as **multiple-angle formulas**. In particular, the following identities are **double-angle formulas**, because they contain the expression $2u$.

DOUBLE-ANGLE FORMULAS

$$\sin 2u = 2 \sin u \cos u$$

$$\cos 2u = \cos^2 u - \sin^2 u$$

$$\cos 2u = 1 - 2 \sin^2 u$$

$$\cos 2u = 2 \cos^2 u - 1$$

$$\tan 2u = \frac{2 \tan u}{1 - \tan^2 u}$$

PROOF The identities may be proved by letting $u = v$ in the appropriate addition formulas. If we use the formula for $\sin (u + v)$, then

$$\begin{aligned}\sin 2u &= \sin (u + u)\\ &= \sin u \cos u + \cos u \sin u\\ &= 2 \sin u \cos u.\end{aligned}$$

Using the formula for $\cos (u + v)$,

$$\begin{aligned}\cos 2u &= \cos (u + u)\\ &= \cos u \cos u - \sin u \sin u\\ &= \cos^2 u - \sin^2 u.\end{aligned}$$

To obtain the other two forms for $\cos 2u$ we use the fundamental identity $\sin^2 u + \cos^2 u = 1$. Thus,

$$\begin{aligned}\cos 2u &= \cos^2 u - \sin^2 u\\ &= (1 - \sin^2 u) - \sin^2 u\\ &= 1 - 2 \sin^2 u.\end{aligned}$$

Similarly, if we substitute for $\sin^2 u$ instead of $\cos^2 u$, we obtain

$$\begin{aligned}\cos 2u &= \cos^2 u - (1 - \cos^2 u)\\ &= 2 \cos^2 u - 1.\end{aligned}$$

The formula for $\tan 2u$ may be obtained by letting $u = v$ in the formula for $\tan(u + v)$. ❑

EXAMPLE ▪ 1

If $\sin\alpha = \frac{4}{5}$ and α is an acute angle, find the exact values of $\sin 2\alpha$ and $\cos 2\alpha$.

FIGURE 11

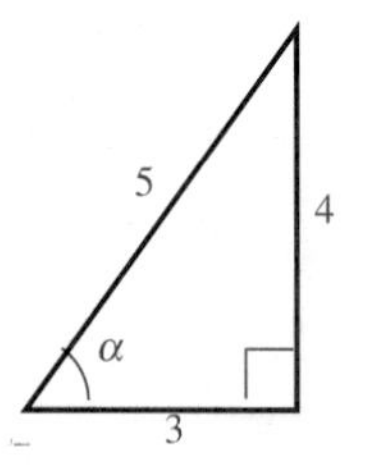

SOLUTION If we regard α as an acute angle of a right triangle, as shown in Figure 11, we obtain $\cos\alpha = \frac{3}{5}$ (see also Example 3 of Section 7.3). We now substitute in double-angle formulas:

$$\sin 2\alpha = 2\sin\alpha\cos\alpha = 2(\tfrac{4}{5})(\tfrac{3}{5}) = \tfrac{24}{25}$$

$$\cos 2\alpha = \cos^2\alpha - \sin^2\alpha = \tfrac{9}{25} - \tfrac{16}{25} = -\tfrac{7}{25}$$

EXAMPLE ▪ 2

Express $\cos 3\theta$ in terms of $\cos\theta$.

SOLUTION We proceed as follows:

$$\begin{aligned}
\cos 3\theta &= \cos(2\theta + \theta) \\
&= \cos 2\theta\cos\theta - \sin 2\theta\sin\theta \\
&= (2\cos^2\theta - 1)\cos\theta - (2\sin\theta\cos\theta)\sin\theta \\
&= 2\cos^3\theta - \cos\theta - 2\cos\theta\sin^2\theta \\
&= 2\cos^3\theta - \cos\theta - 2\cos\theta(1 - \cos^2\theta) \\
&= 2\cos^3\theta - \cos\theta - 2\cos\theta + 2\cos^3\theta \\
&= 4\cos^3\theta - 3\cos\theta
\end{aligned}$$

We call each of the next three formulas a **half-angle identity** because the number u is one-half the number $2u$.

HALF-ANGLE IDENTITIES

$$\sin^2 u = \frac{1 - \cos 2u}{2}$$

$$\cos^2 u = \frac{1 + \cos 2u}{2}$$

$$\tan^2 u = \frac{1 - \cos 2u}{1 + \cos 2u}$$

PROOF The first and second of these identities may be verified by solving the equations

$$\cos 2u = 1 - 2\sin^2 u \quad \text{and} \quad \cos 2u = 2\cos^2 u - 1$$

for $\sin^2 u$ and $\cos^2 u$, respectively. The third identity may be obtained from the first two by using the fact that

$$\tan^2 u = \left(\frac{\sin u}{\cos u}\right)^2 = \frac{\sin^2 u}{\cos^2 u}. \quad \square$$

Half-angle identities may be used to express even powers of trigonometric functions in terms of functions with exponent 1, as illustrated in the next two examples.

EXAMPLE ▪ 3

Verify the identity $\sin^2 x \cos^2 x = \frac{1}{8}(1 - \cos 4x)$.

SOLUTION Using the identities for $\sin^2 u$ and $\cos^2 u$ in the preceding theorem with $u = x$, we obtain

$$\begin{aligned}\sin^2 x \cos^2 x &= \left(\frac{1 - \cos 2x}{2}\right)\left(\frac{1 + \cos 2x}{2}\right) \\ &= \tfrac{1}{4}(1 - \cos^2 2x) \\ &= \tfrac{1}{4}\sin^2 2x.\end{aligned}$$

We next use the formula $\sin^2 u = \dfrac{1 - \cos 2u}{2}$ with $u = 2x$:

$$\begin{aligned}\sin^2 x \cos^2 x &= \frac{1}{4}\left(\frac{1 - \cos 4x}{2}\right) \\ &= \tfrac{1}{8}(1 - \cos 4x).\end{aligned}$$

Another method of proof is to use the fact that $\sin 2x = 2 \sin x \cos x$ and hence

$$\sin x \cos x = \tfrac{1}{2}\sin 2x.$$

Squaring both sides of this equation yields

$$\sin^2 x \cos^2 x = \tfrac{1}{4}\sin^2 2x.$$

The remainder of the solution is the same.

EXAMPLE ▪ 4

Express $\cos^4 t$ in terms of values of the cosine function with exponent 1.

SOLUTION We begin by writing

$$\cos^4 t = (\cos^2 t)^2 = \left(\frac{1 + \cos 2t}{2}\right)^2$$
$$= \tfrac{1}{4}(1 + 2\cos 2t + \cos^2 2t).$$

We next use the formula $\cos^2 u = \dfrac{1 + \cos 2u}{2}$ with $u = 2t$:

$$\cos^4 t = \frac{1}{4}\left(1 + 2\cos 2t + \frac{1 + \cos 4t}{2}\right).$$

This simplifies to

$$\cos^4 t = \frac{3}{8} + \frac{1}{2}\cos 2t + \frac{1}{8}\cos 4t.$$

Substituting $v/2$ for u in thc threc half-angle identities gives us

$$\sin^2 \frac{v}{2} = \frac{1 - \cos v}{2}, \qquad \cos^2 \frac{v}{2} = \frac{1 + \cos v}{2}, \qquad \tan^2 \frac{v}{2} = \frac{1 - \cos v}{1 + \cos v}.$$

Taking the square roots of both sides of each of these equations, we obtain the following **half-angle formulas**.

HALF-ANGLE FORMULAS

$$\sin \frac{v}{2} = \pm\sqrt{\frac{1 - \cos v}{2}}$$

$$\cos \frac{v}{2} = \pm\sqrt{\frac{1 + \cos v}{2}}$$

$$\tan \frac{v}{2} = \pm\sqrt{\frac{1 - \cos v}{1 + \cos v}}$$

When using half-angle formulas we choose either the + or −, depending on the quadrant containing an angle of radian measure $v/2$. Thus, for

$\sin(v/2)$ we use $+$ if $v/2$ corresponds to an angle in quadrant I or II and $-$ for quadrant III or IV. For $\cos(v/2)$ we use $+$ if $v/2$ corresponds to an angle in quadrant I or IV, and so on.

EXAMPLE ▪ 5

Find the exact values of $\sin 22.5°$ and $\cos 22.5°$.

SOLUTION We use the formula for $\sin \frac{v}{2}$ and the fact that $22.5°$ is in quadrant I:

$$\sin 22.5° = \sin \frac{45°}{2} = \sqrt{\frac{1 - \cos 45°}{2}}$$
$$= \sqrt{\frac{1 - \sqrt{2}/2}{2}} = \frac{\sqrt{2 - \sqrt{2}}}{2}$$

Similarly,

$$\cos 22.5° = \sqrt{\frac{1 + \cos 45°}{2}}$$
$$= \sqrt{\frac{1 + \sqrt{2}/2}{2}} = \frac{\sqrt{2 + \sqrt{2}}}{2}$$

We can obtain an alternative form for $\tan(v/2)$. Multiplying numerator and denominator of the radicand in the third half-angle formula by $1 - \cos v$ gives us

$$\tan \frac{v}{2} = \pm\sqrt{\frac{1 - \cos v}{1 + \cos v} \cdot \frac{1 - \cos v}{1 - \cos v}}$$
$$= \pm\sqrt{\frac{(1 - \cos v)^2}{\sin^2 v}}$$
$$= \pm\frac{1 - \cos v}{\sin v}.$$

We can eliminate the $\pm$ sign in the preceding formula. First note that the numerator $1 - \cos v$ is never negative. We can show that $\tan(v/2)$ and $\sin v$ always have the same sign. For example, if $0 < v < \pi$, then $0 < v/2 < \pi/2$, and consequently both $\sin v$ and $\tan(v/2)$ are positive. If $\pi < v < 2\pi$, then $\pi/2 < v/2 < \pi$, and hence both $\sin v$ and $\tan(v/2)$ are negative. This gives us the first of the next two identities. The second identity for $\tan(v/2)$ may be obtained by multiplying numerator and denominator of the radicand in the third half-angle formula by $1 + \cos v$.

HALF-ANGLE FORMULAS FOR TANGENT

$$\tan\frac{v}{2} = \frac{1 - \cos v}{\sin v}, \qquad \tan\frac{v}{2} = \frac{\sin v}{1 + \cos v}.$$

EXAMPLE ▪ 6

If $\tan\alpha = -\frac{4}{3}$ and α is in quadrant IV, find $\tan\dfrac{\alpha}{2}$.

FIGURE 12

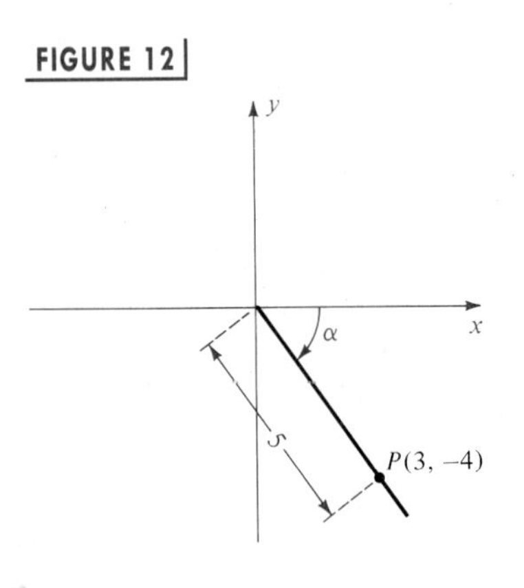

SOLUTION If we choose the point $(3, -4)$ on the terminal side of α, as illustrated in Figure 12, then $\sin\alpha = -\frac{4}{5}$ and $\cos\alpha = \frac{3}{5}$. Applying a half-angle formula,

$$\tan\frac{\alpha}{2} = \frac{1 - \cos\alpha}{\sin\alpha} = \frac{1 - \frac{3}{5}}{-\frac{4}{5}} = -\frac{1}{2}.$$

EXAMPLE ▪ 7

A graph of the equation $y = \cos 2x + \cos x$ for $0 \leq x \leq 2\pi$ is shown in Figure 13. The x-intercepts appear to be approximately 1.1, 3.1, and 5.2. Find their exact values and three-decimal-place approximations.

FIGURE 13

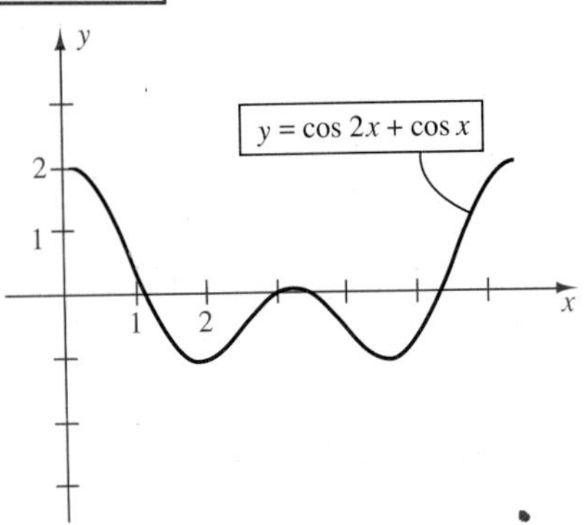

SOLUTION We may find the x-intercepts by letting $y = 0$ in the given equation. This leads to the following equivalent equations:

$$\cos 2x + \cos x = 0$$
$$(2\cos^2 x - 1) + \cos x = 0$$
$$2\cos^2 x + \cos x - 1 = 0$$
$$(2\cos x - 1)(\cos x + 1) = 0$$

Setting each factor equal to zero, we obtain

$$\cos x = \tfrac{1}{2} \quad \text{and} \quad \cos x = -1.$$

The solutions of the last two equations in the interval $[0, 2\pi]$ give us the following x-intercepts:

$$\frac{\pi}{3} \approx 1.047, \qquad \frac{5\pi}{3} \approx 5.236, \qquad \pi \approx 3.142.$$

FIGURE 14

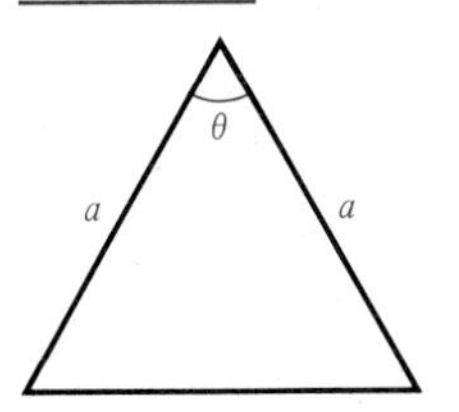

EXAMPLE ▪ 8

An isosceles triangle has two equal sides of length a, and the angle between them is θ (see Figure 14). Express the area A of the triangle in terms of a and θ.

FIGURE 15

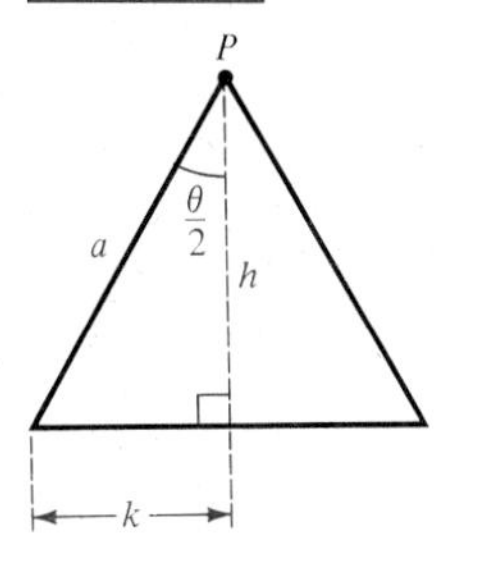

SOLUTION From Figure 15 we see that the altitude from point P bisects θ and that $A = hk$. Referring to the right triangle in this figure,

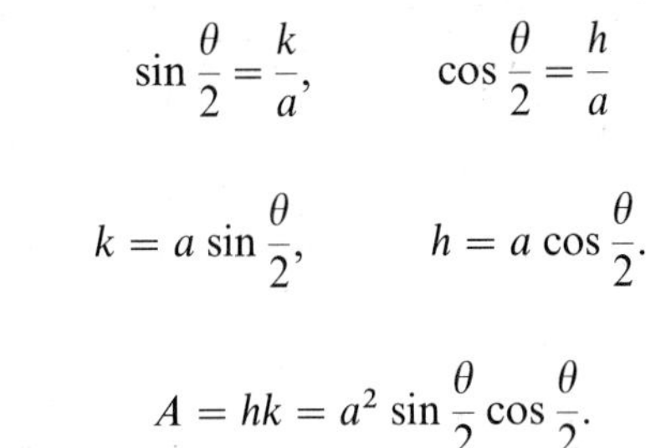

$$\sin\frac{\theta}{2} = \frac{k}{a}, \qquad \cos\frac{\theta}{2} = \frac{h}{a}$$

and

$$k = a\sin\frac{\theta}{2}, \qquad h = a\cos\frac{\theta}{2}.$$

Hence

$$A = hk = a^2 \sin\frac{\theta}{2}\cos\frac{\theta}{2}.$$

Using half-angle formulas gives us

$$A = a^2\sqrt{\frac{1-\cos\theta}{2}}\sqrt{\frac{1+\cos\theta}{2}}$$

$$A = a^2\sqrt{\frac{1-\cos^2\theta}{4}}$$

$$A = a^2\sqrt{\frac{\sin^2\theta}{4}}.$$

$$A = \tfrac{1}{2}a^2\sin\theta.$$

7.4 EXERCISES

Exer. 1–4: Find the exact values of $\sin 2\theta$, $\cos 2\theta$, and $\tan 2\theta$ for the given conditions.

1 $\cos\theta = \frac{3}{5}$ and θ acute

2 $\cot\theta = \frac{4}{3}$ and $180° < \theta < 270°$

3 $\sec\theta = -3$ and $90° < \theta < 180°$

4 $\sin\theta = -\frac{4}{5}$ and $270° < \theta < 360°$

Exer. 5–8: Find the exact values of $\sin\frac{\theta}{2}$, $\cos\frac{\theta}{2}$, and $\tan\frac{\theta}{2}$ for the given conditions.

5 $\sec\theta = \frac{5}{4}$ and θ acute

6 $\csc\theta = -\frac{5}{3}$ and $-90° < \theta < 0°$

7 $\tan \theta = 1$ and $-180° < \theta < -90°$

8 $\sec \theta = -4$ and $180° < \theta < 270°$

Exer. 9–10: Use half-angle formulas to find the exact values.

9 (a) $\cos 67°30'$ (b) $\sin 15°$ (c) $\tan \dfrac{3\pi}{8}$

10 (a) $\cos 165°$ (b) $\sin 157°30'$ (c) $\tan \dfrac{\pi}{8}$

Exer. 11–28: Verify the identity.

11 $\sin 10\theta = 2 \sin 5\theta \cos 5\theta$

12 $\cos^2 3x - \sin^2 3x = \cos 6x$

13 $4 \sin \dfrac{x}{2} \cos \dfrac{x}{2} = 2 \sin x$

14 $\dfrac{\sin^2 2\alpha}{\sin^2 \alpha} = 4 - 4 \sin^2 \alpha$

15 $(\sin t + \cos t)^2 = 1 + \sin 2t$

16 $\csc 2u = \frac{1}{2} \sec u \csc u$

17 $\sin 3u = \sin u (3 - 4 \sin^2 u)$

18 $\sin 4t = 4 \cos t \sin t (1 - 2 \sin^2 t)$

19 $\cos 4\theta = 8 \cos^4 \theta - 8 \cos^2 \theta + 1$

20 $\cos 6t = 32 \cos^6 t - 48 \cos^4 t + 18 \cos^2 t - 1$

21 $\sin^4 t = \frac{3}{8} - \frac{1}{2} \cos 2t + \frac{1}{8} \cos 4t$

22 $\cos^4 x - \sin^4 x = \cos 2x$

23 $\sec 2\theta = \dfrac{\sec^2 \theta}{2 - \sec^2 \theta}$

24 $\cot 2u = \dfrac{\cot^2 u - 1}{2 \cot u}$

25 $2 \sin^2 2t + \cos 4t = 1$

26 $\tan \theta + \cot \theta = 2 \csc 2\theta$

27 $\tan 3u = \dfrac{(3 - \tan^2 u) \tan u}{1 - 3 \tan^2 u}$

28 $\dfrac{1 + \sin 2v + \cos 2v}{1 + \sin 2v - \cos 2v} = \cot v$

Exer. 29–32: Write the expression in terms of values of the cosine function with exponent 1.

29 $\cos^4 \dfrac{\theta}{2}$

30 $\sin^4 2x$

31 $\cos^4 2x$

32 $\sin^4 \dfrac{\theta}{2}$

Exer. 33–40: Find all solutions of the equation that are in the interval $[0, 2\pi)$.

33 $\sin 2t + \sin t = 0$

34 $\cos t - \sin 2t = 0$

35 $\cos u + \cos 2u = 0$

36 $\cos 2\theta - \tan \theta = 1$

37 $\tan 2x = \tan x$

38 $\tan 2t - 2 \cos t = 0$

39 $\sin \frac{1}{2}u + \cos u = 1$

40 $2 - \cos^2 x = 4 \sin^2 \frac{1}{2}x$

41 If $a > 0$, $b > 0$, and $0 < u < \pi/2$, show that

$$a \sin u + b \cos u = \sqrt{a^2 + b^2} \sin (u + v)$$

for $0 < v < \dfrac{\pi}{2}$, $\sin v = \dfrac{b}{\sqrt{a^2 + b^2}}$, and $\cos v = \dfrac{a}{\sqrt{a^2 + b^2}}$.

42 Use Exercise 41 to express $8 \sin u + 15 \cos u$ in the form $c \sin (u + v)$ for some number c.

43 A graph of $y = \cos 2x + 2 \cos x$ for $0 \le x \le 2\pi$ is shown in the figure.

(a) Approximate the x-intercepts to two decimal places.

(b) The x-coordinates of the turning points P, Q, and R on the graph are solutions of $\sin 2x + \sin x = 0$. Find the coordinates of these points.

EXERCISE 43

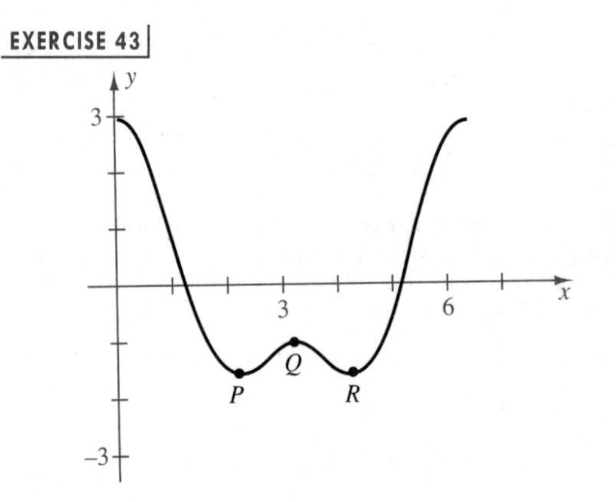

44 A graph of $y = \cos x - \sin 2x$ for $-2\pi \le x \le 2\pi$ is shown in the figure.

(a) Find the x-intercepts.

(b) The x-coordinates of the eight turning points on the graph are solutions of $\sin x + 2 \cos 2x = 0$. Approximate these x-coordinates to two decimal places.

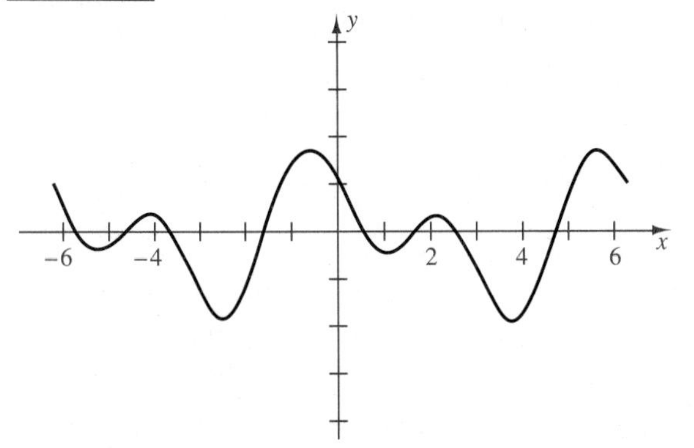

45 A graph of $y = \cos 3x - 3 \cos x$ for $-2\pi \le x \le 2\pi$ is shown in the figure.

(a) Find the x-intercepts. (*Hint:* Use the formula for $\cos 3x$ given in Example 2.)

(b) The x-coordinates of the 13 turning points on the graph are solutions of $\sin 3x - \sin x = 0$. Find these x-coordinates. (*Hint:* Use the formula for $\sin 3x$ in Exercise 17.)

EXERCISE 45

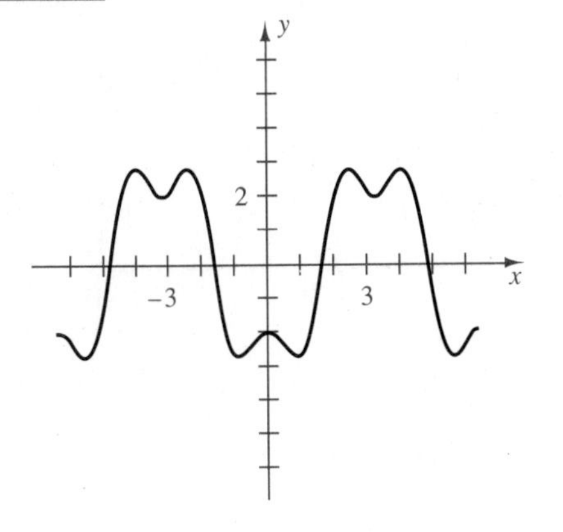

46 A graph of $y = \sin 4x - 4 \sin x$ for $-2\pi \le x \le 2\pi$ is shown in the figure. Find the x-intercepts. (*Hint:* Use the formula for $\sin 4x$ in Exercise 18.)

EXERCISE 46

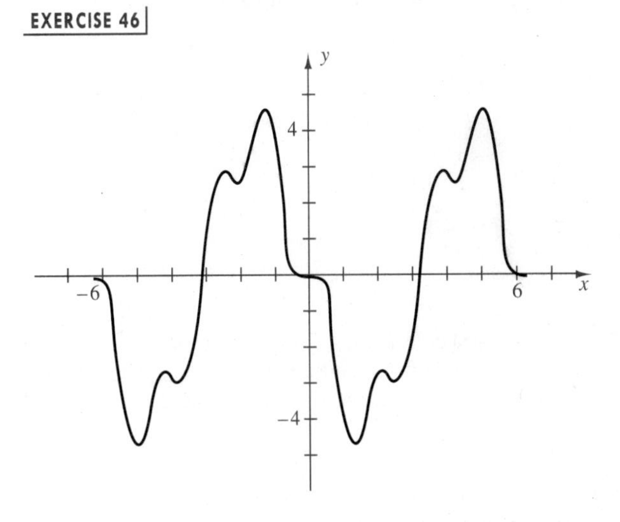

47 Shown in the figure is a proposed railroad route through three towns located at points A, B, and C. The track will branch out from B toward C at an angle θ.

(a) Show that the total distance d from A to C is given by $d = 20 \tan \dfrac{\theta}{2} + 40$.

(b) Because of mountains between A and C, the branching point B must be at least 20 miles from A. Is there a route that avoids the mountains and measures exactly 50 miles?

EXERCISE 47

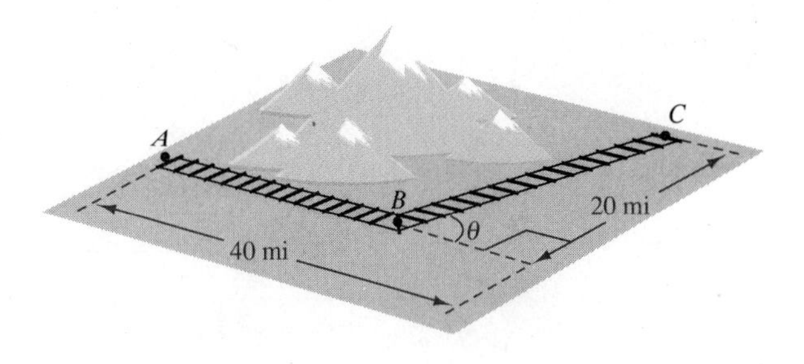

48 Shown in the figure is a design for a rain gutter.

(a) Express the volume V as a function of θ.

(b) Approximate the value of θ that results in a volume of $2\,\text{ft}^3$. (*Hint:* See Example 8.)

EXERCISE 48

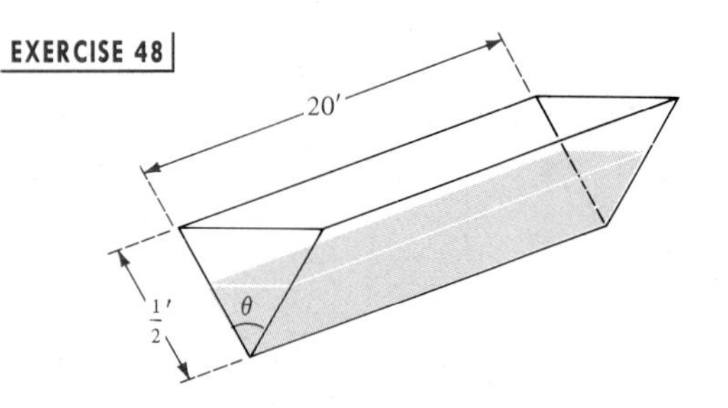

49 If a projectile is fired from ground level with an initial velocity of v ft/sec and at an angle of θ degrees with the horizontal, the range R of the projectile is given by

$$R = \frac{v^2}{16} \sin\theta \cos\theta.$$

If $v = 80$ ft/sec, approximate the angles that result in a range of 150 feet.

50 A highway engineer is designing the street curbing at an intersection where two highways meet at an angle ϕ, as shown in the figure. The curbing between points A and B is to be constructed using a circle that is tangent to the highway at these two points.

(a) Show that the relationship between the radius R of the circle and the distance d in the figure is given by

$$d = R \tan\frac{\phi}{2}.$$

EXERCISE 50

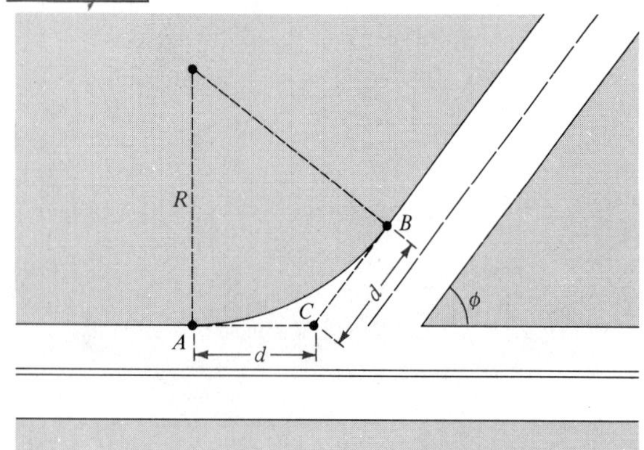

(b) If $\phi = 45°$ and $d = 20$ feet, approximate R and the length of the curbing.

51 A common form of cardiovascular branching is bifurcation, in which an artery splits into two smaller blood vessels, as shown in the figure. The bifurcation angle θ is the angle formed by the two smaller arteries.

(a) Show that the length l of the artery from A to B is given by $l = a + \frac{b}{2}\tan\frac{\theta}{4}$.

(b) Estimate l from the measurements $a = 10$ mm, $b = 6$ mm, and $\theta = 156°$.

EXERCISE 51

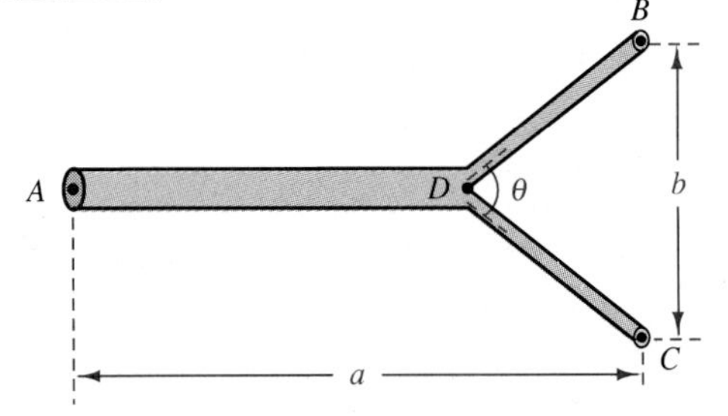

52 By definition, the *average value* of $f(t) = c + a\cos bt$ for one or more complete cycles is c (see figure).

(a) Use a double-angle formula to find the average value of $f(t) = \sin^2 \omega t$ for $0 \le t \le 2\pi/\omega$ with t in seconds.

(b) In an electrical circuit with an alternating current $I = I_0 \sin \omega t$, the rate r (in calories/sec) at which heat is produced in an R-ohm resistor is given by $r = RI^2$. Find the average rate at which heat is produced for one complete cycle.

EXERCISE 52

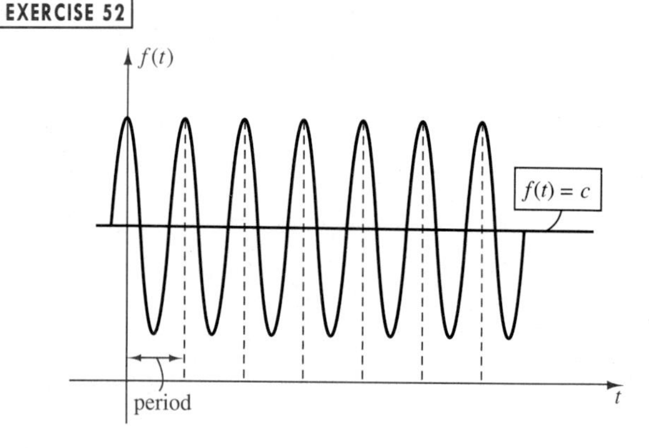

7.5 PRODUCT-TO-SUM AND SUM-TO-PRODUCT FORMULAS

The following formulas may be used to change the form of certain trigonometric expressions from products to sums. We refer to these as **product-to-sum formulas**, even though two of the formulas express a product as a difference, because any difference $x - y$ of two real numbers is also a sum $x + (-y)$.

PRODUCT-TO-SUM FORMULAS

$$\sin u \cos v = \tfrac{1}{2}[\sin (u + v) + \sin (u - v)]$$
$$\cos u \sin v = \tfrac{1}{2}[\sin (u + v) - \sin (u - v)]$$
$$\cos u \cos v = \tfrac{1}{2}[\cos (u + v) + \cos (u - v)]$$
$$\sin u \sin v = \tfrac{1}{2}[\cos (u - v) - \cos (u + v)]$$

Each of the formulas is a consequence of our work in Section 7.3. For example, to verify the first formula we merely add the left- and right-hand sides of the identities we obtained for $\sin (u + v)$ and $\sin (u - v)$ and divide by 2. The remaining product-to-sum formulas are obtained in similar fashion.

EXAMPLE ▪ 1

Express each as a sum: **(a)** $\sin 4\theta \cos 3\theta$ **(b)** $\sin 3x \sin x$

SOLUTION

(a) We use the first product-to-sum formula with $u = 4\theta$ and $v = 3\theta$:

$$\begin{aligned}\sin 4\theta \cos 3\theta &= \tfrac{1}{2}[\sin (4\theta + 3\theta) + \sin (4\theta - 3\theta)] \\ &= \tfrac{1}{2}(\sin 7\theta + \sin \theta)\end{aligned}$$

We can also obtain this relationship by using the second product-to-sum formula.

(b) We use the fourth product-to-sum formula with $u = 3x$ and $v = x$:

$$\begin{aligned}\sin 3x \sin x &= \tfrac{1}{2}[\cos (3x - x) - \cos (3x + x)] \\ &= \tfrac{1}{2}(\cos 2x - \cos 4x)\end{aligned}$$

We may use product-to-sum formulas to express a sum or difference as a product. To obtain forms that can be applied more easily, we shall change the notation as follows. If we let

$$u + v = a \quad \text{and} \quad u - v = b,$$

then $(u + v) + (u - v) = a + b$, which simplifies to

$$u = \frac{a + b}{2}.$$

Similarly, since $(u + v) - (u - v) = a - b$,

$$v = \frac{a - b}{2}.$$

We now substitute for $u + v$ and $u - v$ on the right-hand sides of the product-to-sum formulas and for u and v on the left-hand sides. If we then multiply by 2, we obtain the following.

SUM-TO-PRODUCT FORMULAS

$$\sin a + \sin b = 2 \sin \frac{a + b}{2} \cos \frac{a - b}{2}$$

$$\sin a - \sin b = 2 \cos \frac{a + b}{2} \sin \frac{a - b}{2}$$

$$\cos a + \cos b = 2 \cos \frac{a + b}{2} \cos \frac{a - b}{2}$$

$$\cos a - \cos b = -2 \sin \frac{a + b}{2} \sin \frac{a - b}{2}$$

EXAMPLE ▪ 2

Express $\sin 5x - \sin 3x$ as a product.

SOLUTION We use the second sum-to-product formula with $a = 5x$ and $b = 3x$:

$$\begin{aligned} \sin 5x - \sin 3x &= 2 \cos \frac{5x + 3x}{2} \sin \frac{5x - 3x}{2} \\ &= 2 \cos 4x \sin x \end{aligned}$$

EXAMPLE ▪ 3

Verify the identity $\dfrac{\sin 3t + \sin 5t}{\cos 3t - \cos 5t} = \cot t$.

SOLUTION We use the first and fourth sum-to-product formulas:

$$
\begin{aligned}
\frac{\sin 3t + \sin 5t}{\cos 3t - \cos 5t} &= \frac{2 \sin \dfrac{3t + 5t}{2} \cos \dfrac{3t - 5t}{2}}{-2 \sin \dfrac{3t + 5t}{2} \sin \dfrac{3t - 5t}{2}} \\
&= \frac{2 \sin 4t \cos (-t)}{-2 \sin 4t \sin (-t)} \\
&= \frac{\cos (-t)}{-\sin (-t)} \\
&= \frac{\cos t}{\sin t} = \cot t.
\end{aligned}
$$

EXAMPLE ▪ 4

Find the solutions of $\sin 5x + \sin x = 0$.

SOLUTION We use the first sum-to-product formula with $a = 5x$ and $b = x$:

$$
\begin{aligned}
\sin 5x + \sin x &= 2 \sin \frac{5x + x}{2} \sin \frac{5x - x}{2} \\
&= 2 \sin 3x \sin 2x.
\end{aligned}
$$

Hence the given equation is equivalent to

$$2 \sin 3x \sin 2x = 0.$$

Setting each factor on the left-hand side of this equation equal to zero gives us

$$\sin 3x = 0 \quad \text{and} \quad \sin 2x = 0.$$

The solutions of the last two equations are

$$3x = \pi n \quad \text{and} \quad 2x = \pi n \quad \text{for every integer } n.$$

Hence the solutions of the given equation are

$$\frac{\pi}{3} n \quad \text{and} \quad \frac{\pi}{2} n \quad \text{for every integer } n.$$

EXAMPLE ▪ 5

A graph of the equation $y = \cos x - \sin 2x - \cos 3x$ is shown in Figure 16. Find the 13 x-intercepts that are in the interval $[-2\pi, 2\pi]$.

FIGURE 16

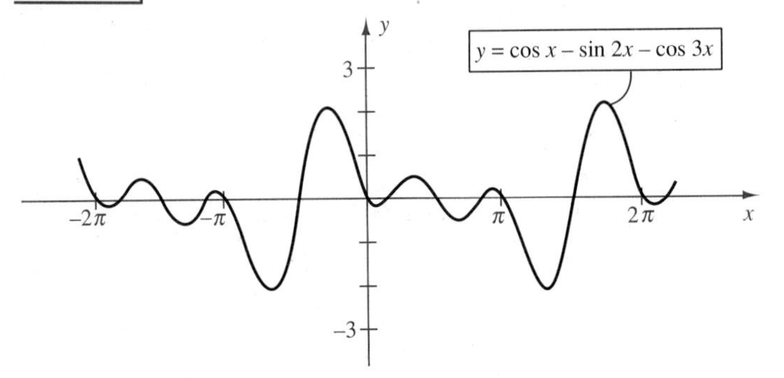

SOLUTION The x-intercepts are solutions of the equation

$$\cos x - \sin 2x - \cos 3x = 0.$$

Let us regroup the left-hand side of the equation and use the fourth sum-to-product formula as follows:

$$(\cos x - \cos 3x) - \sin 2x = 0$$

$$-2 \sin \frac{x + 3x}{2} \sin \frac{x - 3x}{2} - \sin 2x = 0$$

$$-2 \sin 2x \sin(-x) - \sin 2x = 0$$

$$2 \sin 2x \sin x - \sin 2x = 0$$

$$\sin 2x(2 \sin x - 1) = 0$$

Setting each factor on the left-hand side equal to zero gives us

$$\sin 2x = 0 \quad \text{and} \quad \sin x = \tfrac{1}{2}.$$

The equation $\sin 2x = 0$ has solutions $2x = \pi n$ or, equivalently,

$$x = \frac{\pi}{2} n \quad \text{for every integer } n.$$

If we let $n = 0, \pm 1, \pm 2, \pm 3$, and ± 4, we obtain nine x-intercepts in $[-2\pi, 2\pi]$:

$$0, \quad \pm\frac{\pi}{2}, \quad \pm\pi, \quad \pm\frac{3\pi}{2}, \quad \pm 2\pi$$

The solutions of the equation $\sin x = \frac{1}{2}$ are

$$\frac{\pi}{6} + 2\pi n \quad \text{and} \quad \frac{5\pi}{6} + 2\pi n \quad \text{for every integer } n.$$

The four solutions in $[-2\pi, 2\pi]$ are given by $n = 0$ and $n = -1$:

$$\frac{\pi}{6}, \quad \frac{5\pi}{6}, \quad -\frac{11\pi}{6}, \quad -\frac{7\pi}{6}$$

Addition formulas may also be employed to derive **reduction formulas**. Reduction formulas may be used to change expressions such as

$$\sin\left(\theta + \frac{\pi}{2}n\right) \quad \text{and} \quad \cos\left(\theta + \frac{\pi}{2}n\right) \quad \text{for any integer } n$$

to expressions involving only $\sin\theta$ or $\cos\theta$. Similar formulas are true for the other trigonometric functions. Instead of deriving general reduction formulas, we shall illustrate several special cases in the next example.

EXAMPLE ▪ 6

Express $\sin\left(\theta - \frac{3\pi}{2}\right)$ and $\cos(\theta + \pi)$ in terms of a function of θ.

SOLUTION Using addition formulas, we obtain

$$\begin{aligned}\sin\left(\theta - \frac{3\pi}{2}\right) &= \sin\theta\cos\frac{3\pi}{2} - \cos\theta\sin\frac{3\pi}{2}\\ &= \sin\theta\cdot(0) - \cos\theta\cdot(-1) = \cos\theta\\ \cos(\theta + \pi) &= \cos\theta\cos\pi - \sin\theta\sin\pi\\ &= \cos\theta\cdot(-1) - \sin\theta\cdot(0) = -\cos\theta\end{aligned}$$

We have derived many important identities in this chapter. For reference, these identities are listed on the inside of the back cover of the text.

7.5 EXERCISES

Exer. 1–8: Express the product as a sum or difference.

1 $2\sin 9\theta\cos 3\theta$

2 $2\cos 5\theta\sin 5\theta$

3 $\sin 7t\sin 3t$

4 $\sin(-4x)\cos 8x$

5 $\cos 6u\cos(-4u)$

6 $\sin 4t\sin 6t$

7 $3\cos x\sin 2x$

8 $5\cos u\sin 5u$

Exer. 9–16: Write the expression as a product.

9 $\sin 6\theta + \sin 2\theta$

10 $\sin 4\theta - \sin 8\theta$

11 $\cos 5x - \cos 3x$

12 $\cos 5t + \cos 6t$

13 $\sin 3t - \sin 7t$

14 $\cos \theta - \cos 5\theta$

15 $\cos x + \cos 2x$

16 $\sin 8t + \sin 2t$

Exer. 17–24: Verify the identity.

17 $\dfrac{\sin 4t + \sin 6t}{\cos 4t - \cos 6t} = \cot t$

18 $\dfrac{\sin \theta + \sin 3\theta}{\cos \theta + \cos 3\theta} = \tan 2\theta$

19 $\dfrac{\sin u + \sin v}{\cos u + \cos v} = \tan \frac{1}{2}(u + v)$

20 $\dfrac{\sin u - \sin v}{\cos u - \cos v} = -\cot \frac{1}{2}(u + v)$

21 $\dfrac{\sin u - \sin v}{\sin u + \sin v} = \dfrac{\tan \frac{1}{2}(u - v)}{\tan \frac{1}{2}(u + v)}$

22 $\dfrac{\cos u - \cos v}{\cos u + \cos v} = -\tan \frac{1}{2}(u + v) \tan \frac{1}{2}(u - v)$

23 $\sin 2x + \sin 4x + \sin 6x = 4 \cos x \cos 2x \sin 3x$

24 $\dfrac{\cos t + \cos 4t + \cos 7t}{\sin t + \sin 4t + \sin 7t} = \cot 4t$

Exer. 25–26: Express the product as a sum.

25 $(\sin ax)(\cos bx)$

26 $(\cos mu)(\cos nu)$

Exer. 27–34: Find the solutions of the equation by using sum-to-product formulas.

27 $\sin 5t + \sin 3t = 0$

28 $\sin t + \sin 3t = \sin 2t$

29 $\cos x = \cos 3x$

30 $\cos 4x - \cos 3x = 0$

31 $\cos 3x + \cos 5x = \cos x$

32 $\cos 3x = -\cos 6x$

33 $\sin 2x - \sin 5x = 0$

34 $\sin 5x - \sin x = 2 \cos 3x$

Exer. 35–40: Verify the reduction formula.

35 $\sin(\theta + \pi) = -\sin \theta$

36 $\sin\left(\theta + \dfrac{3\pi}{2}\right) = -\cos \theta$

37 $\cos\left(\theta - \dfrac{5\pi}{2}\right) = \sin \theta$

38 $\cos(\theta - 3\pi) = -\cos \theta$

39 $\tan(\pi - \theta) = -\tan \theta$

40 $\tan\left(\theta + \dfrac{\pi}{2}\right) = -\cot \theta$

Exer. 41–42: Shown in the figure is a graph of the function f for $0 \le x \le 2\pi$. Use a sum-to-product formula to help find the x-intercepts.

41 $f(x) = \cos x + \cos 3x$

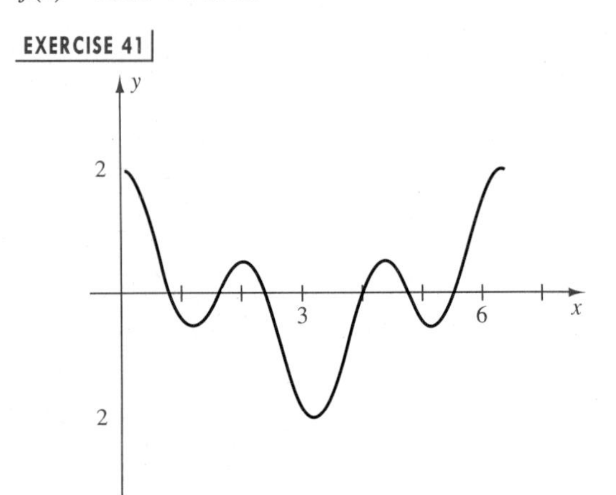

42 $f(x) = \sin 4x - \sin x$

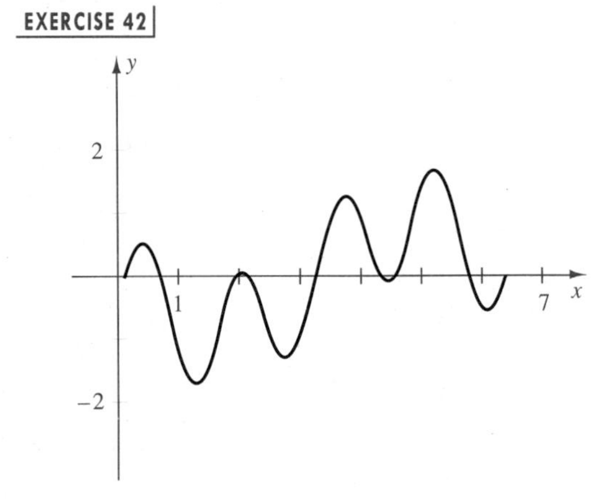

43 Refer to Exercise 45 of Section 7.4. The graph of $y = \cos 3x - 3 \cos x$ has 13 turning points for $-2\pi \le x \le 2\pi$. The x-coordinates of these points are solutions of the equation $\sin 3x - \sin x = 0$. Find these x-coordinates by using a sum-to-product formula.

44 Refer to Exercise 46 of Section 7.4. The x-coordinates of the turning points on the graph of $y = \sin 4x - 4 \sin x$ are solutions of the equation $\cos 4x - \cos x = 0$. Use a sum-to-product formula to find these x-coordinates for $-2\pi \le x \le 2\pi$.

45 Mathematical analysis of a vibrating violin string of length l involves functions such that

$$f(x) = \sin\left(\frac{\pi n}{l}x\right)\cos\left(\frac{k\pi n}{l}t\right)$$

for an integer n, a constant k, and time t. Express f as a sum of two sine functions.

46 If two tuning forks are struck simultaneously with the same force and are then held at the same distance from the eardrum, the pressure on the outside of the eardum at time t can be represented by $p(t) = a\cos\omega_1 t + a\cos\omega_2 t$ for constants a, ω_1, and ω_2. If ω_1 is close to ω_2, a tone is produced that alternates between loudness and virtual silence. This phenomenon is known as beats.

(a) Use a sum-to-product formula to express $p(t)$ as a product.

(b) Show that $p(t)$ may be considered as a cosine wave with approximate period $2\pi/\omega_1$ and variable amplitude $f(t) = 2a\cos\frac{1}{2}(\omega_1 - \omega_2)t$. Find the maximum amplitude.

(c) Shown in the figure is a graph of the equation $p(t) = \cos 4.5t + \cos 3.5t$. Near-silence occurs at points A and B, where the variable amplitude $f(t)$ in part (b) is zero. Find the coordinates of these points, and determine how frequently near-silence occurs.

(d) Use the graph to show that the function p in part (c) has period 4π. Conclude that the maximum amplitude of 2 occurs every 4π units of time.

EXERCISE 46

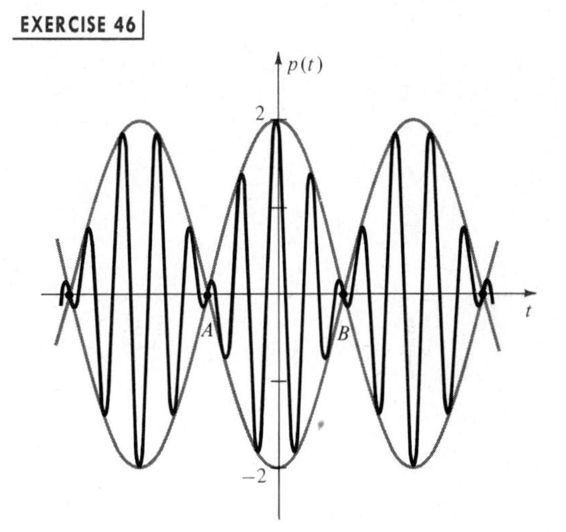

7.6 THE INVERSE TRIGONOMETRIC FUNCTIONS

1, 5, 9, 11, 13, 15, 17, 19, 21, 25, 45, 49, 53, 55, 61

The concept of *inverse function* was discussed in Section 3.7. Remember that to define the inverse function f^{-1} of a function f, it is essential that f be one-to-one; that is, if $a \neq b$ in the domain of f, then $f(a) \neq f(b)$ in the range. If f is a one-to-one function, then, as illustrated in Figure 17(i), for each number u in the range of f there is exactly one number v in the domain of f such that $u = f(v)$. We then obtain the inverse function f^{-1}, as in (ii) of the figure, by letting $v = f^{-1}(u)$. Thus

$$v = f^{-1}(u) \quad \text{if and only if} \quad u = f(v).$$

FIGURE 17

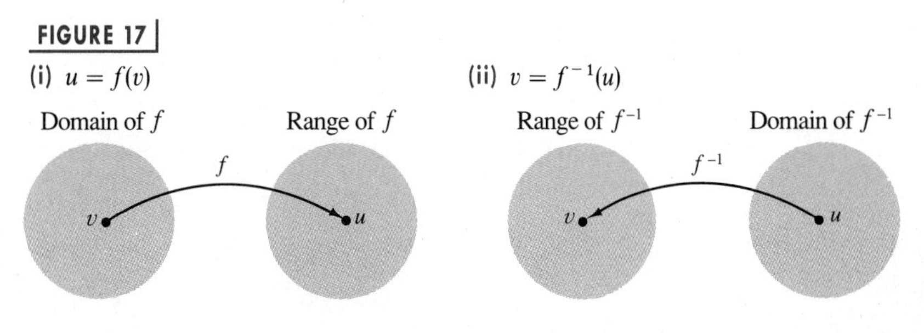

The inverse function f^{-1} *reverses* the correspondence given by f. Thus the range of f is the domain of f^{-1}, and the domain of f is the range of f^{-1}. Since domains and ranges are interchanged, it is customary to let x denote an arbitrary number from the domain of f^{-1} and y the corresponding number from the domain of f. The following general relationships are true (see Section 3.7).

RELATIONSHIPS BETWEEN f^{-1} AND f

(1) $y = f^{-1}(x)$ if and only if $x = f(y)$ for every x in the domain of f^{-1} and for every y in the domain of f.

(2) $f(f^{-1}(x)) = x$ for every x in the domain of f^{-1}.

(3) $f^{-1}(f(y)) = y$ for every y in the domain of f.

(4) The graphs of f^{-1} and f are reflections of one another through the line $y = x$.

We shall use relationship (1) to define each of the inverse trigonometric functions.

The sine function is not one-to-one since, for example, different numbers, such as $\pi/6$, $5\pi/6$, and $-7\pi/6$, yield the same function value $\frac{1}{2}$. If we restrict the domain to $[-\pi/2, \pi/2]$ then, as illustrated by the solid portion of the graph of $y = \sin x$ in Figure 18, we obtain a one-to-one (increasing) function that takes on every value of the sine function once and only once. We use this *new* function with domain $[-\pi/2, \pi/2]$ and range $[-1, 1]$ to define the *inverse sine function*.

FIGURE 18

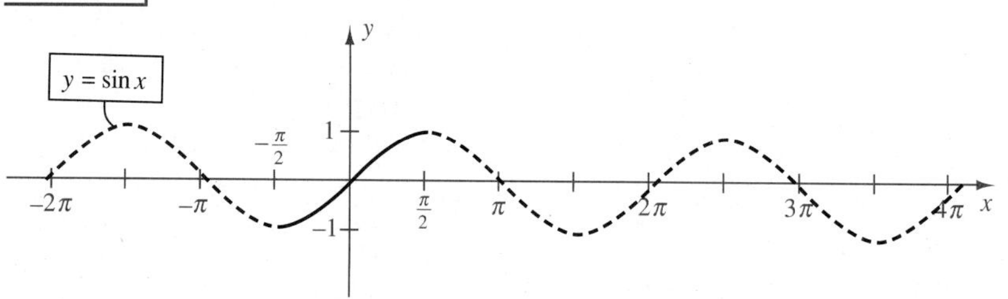

DEFINITION

The **inverse sine function**, denoted by $\sin^{-1}$, is defined by

$$y = \sin^{-1} x \quad \text{if and only if} \quad x = \sin y$$

for $-1 \leq x \leq 1$ and $-\dfrac{\pi}{2} \leq y \leq \dfrac{\pi}{2}$.

The domain of the inverse sine function is $[-1, 1]$ and the range is $[-\pi/2, \pi/2]$.

The notation $y = \sin^{-1} x$ is sometimes read *y is the inverse sine of x*. The equation $x = \sin y$ in the definition allows us to regard y as an angle, and hence $y = \sin^{-1} x$ may also be read *y is an angle whose sine is x* (with $-\pi/2 \leq y \leq \pi/2$).

ILLUSTRATION | INVERSE SINE FUNCTION

- $y = \sin^{-1} \frac{1}{2}$ means that $\sin y = \frac{1}{2}$ and $-\dfrac{\pi}{2} \leq y \leq \dfrac{\pi}{2}$. Thus, $y = \dfrac{\pi}{6}$.
- $y = \sin^{-1} (-\frac{1}{2})$ means that $\sin y = -\frac{1}{2}$ and $-\dfrac{\pi}{2} \leq y \leq \dfrac{\pi}{2}$. Thus, $y = -\dfrac{\pi}{6}$.

The inverse sine function is also called the **arcsine function**, and $\arcsin x$ may be used in place of $\sin^{-1} x$. If $t = \arcsin x$, then $\sin t = x$, and t may be interpreted as an *arc*length on the unit circle U. We will use both notations $\sin^{-1}$ and arcsin throughout our work. Additional illustrations of the definition are given in the next example.

EXAMPLE ▪ 1

Find the exact value:

(a) $\sin^{-1} \dfrac{\sqrt{2}}{2}$ (b) $\arcsin\left(-\dfrac{\sqrt{3}}{2}\right)$ (c) $\sin^{-1} 1$

SOLUTION

(a) By definition,

$$y = \sin^{-1} \frac{\sqrt{2}}{2} \quad \text{means that} \quad \sin y = \frac{\sqrt{2}}{2} \quad \text{and} \quad -\frac{\pi}{2} \leq y \leq \frac{\pi}{2}.$$

The only number y in the interval $[-\pi/2, \pi/2]$ such that $\sin y = \sqrt{2}/2$ is $y = \pi/4$. Hence

$$\sin^{-1} \frac{\sqrt{2}}{2} = \frac{\pi}{4}.$$

Warning: It is *essential* to choose y in the range $[-\pi/2, \pi/2]$ of the inverse sine function. Thus the number $y = 3\pi/4$ is not the inverse function value $\sin^{-1} (\sqrt{2}/2)$ even though $\sin (3\pi/4) = \sqrt{2}/2$.

(b) By definition,

$$y = \arcsin\left(-\frac{\sqrt{3}}{2}\right) \quad \text{means that} \quad \sin y = -\frac{\sqrt{3}}{2} \quad \text{and} \quad -\frac{\pi}{2} \le y \le \frac{\pi}{2}.$$

It follows that $y = -\pi/3$. Thus

$$\arcsin\left(-\frac{\sqrt{3}}{2}\right) = -\frac{\pi}{3}.$$

(c) $$y = \sin^{-1} 1 \quad \text{means that} \quad \sin y = 1 \quad \text{and} \quad -\frac{\pi}{2} \le y \le \frac{\pi}{2}.$$

Consequently, $y = \pi/2$, and

$$\sin^{-1} 1 = \frac{\pi}{2}.$$

In Section 6.2 we discussed how to find the value $\sin^{-1} x$ with a calculator by entering the number x and then pressing either [INV] [SIN], [SIN^{-1}], or [ARCSIN]. We usually use radian mode; however, if the degree measure of the angle $\sin^{-1} x$ is desired, then degree mode should be used. Thus, to find $\sin^{-1}(\sqrt{2}/2)$ in Example 1(a), we select radian mode and proceed as follows:

Enter $\sqrt{2} \div 2$: 0.707106781
Press [INV] [SIN]: 0.785398163

The last decimal is an approximation to $\pi/4$.

EXAMPLE ▪ 2

Find the exact value of $\sin^{-1}\left(\tan \frac{3\pi}{4}\right)$.

SOLUTION We first evaluate the inner expression $\tan(3\pi/4)$ and then find the inverse sine of that number. If we let

$$y = \sin^{-1}\left(\tan \frac{3\pi}{4}\right)$$

then $$y = \sin^{-1}(-1).$$

By definition, $$\sin y = -1 \quad \text{and} \quad -\frac{\pi}{2} \le y \le \frac{\pi}{2}.$$

FIGURE 19

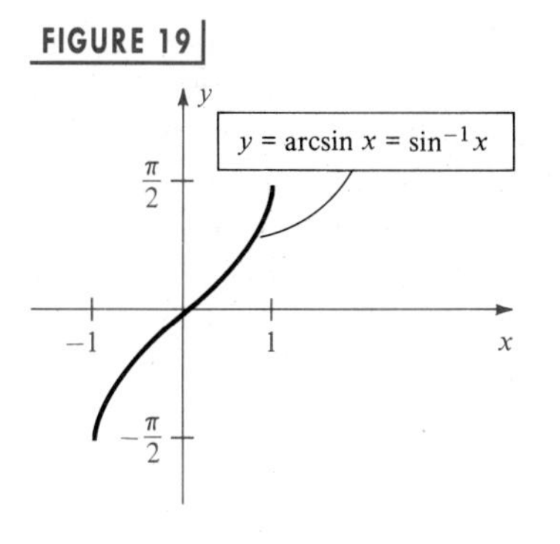

It follows that $y = -\pi/2$ and hence

$$\sin^{-1}\left(\tan\frac{3\pi}{4}\right) = -\frac{\pi}{2}.$$

By relationship (4) for the graphs of f and f^{-1}, we can sketch the graph of $y = \sin^{-1} x$ by reflecting the solid portion of Figure 18 through the line $y = x$. We could also use the equation $x = \sin y$ with the restriction $-\pi/2 \le y \le \pi/2$ to find points on the graph. This gives us Figure 19.

The relationships (2) $f(f^{-1}(x)) = x$ and (3) $f^{-1}(f(y)) = y$ that hold for any inverse function f^{-1} give us the following properties.

PROPERTIES OF $\sin^{-1}$

(i) $\sin(\sin^{-1} x) = \sin(\arcsin x) = x$ if $-1 \le x \le 1$.

(ii) $\sin^{-1}(\sin y) = \arcsin(\sin y) = y$ if $-\dfrac{\pi}{2} \le y \le \dfrac{\pi}{2}$.

EXAMPLE ▪ 3

Find the exact value:

(a) $\sin(\sin^{-1} \frac{1}{2})$ (b) $\sin^{-1}\left(\sin\dfrac{\pi}{4}\right)$ (c) $\sin^{-1}\left(\sin\dfrac{2\pi}{3}\right)$.

SOLUTION

(a) The *difficult* way to find the value is to first find the angle $\sin^{-1}\frac{1}{2}$, namely $\pi/6$, and then evaluate $\sin(\pi/6)$, obtaining $\frac{1}{2}$. The *easy* way is to use property (i) of the previous result as follows:

$$\sin(\sin^{-1}\tfrac{1}{2}) = \tfrac{1}{2}$$

(b) Since $-\pi/2 < \pi/4 < \pi/2$, we can use property (ii) of the previous result to write

$$\sin^{-1}\left(\sin\frac{\pi}{4}\right) = \frac{\pi}{4}$$

(c) Be careful! Since number $2\pi/3$ is *not* between $-\pi/2$ and $\pi/2$, we cannot use property (ii) of the previous result. Instead, we first evaluate the inner expression $\sin(2\pi/3)$ and then use the definition of $\sin^{-1}$ as follows:

$$\sin^{-1}\left(\sin\frac{2\pi}{3}\right) = \sin^{-1}\left(-\frac{\sqrt{3}}{2}\right) = -\frac{\pi}{3}$$

The other trigonometric functions may also be used to introduce inverse functions. The procedure is first to determine a convenient subset of the domain to obtain a one-to-one function. If the domain of the cosine function is restricted to the interval $[0, \pi]$, as illustrated by the solid portion of the graph of $y = \cos x$ in Figure 20, we obtain a one-to-one (decreasing) function that takes on all the values of the cosine function once and only once. We use this *new* function with domain $[0, \pi]$ and range $[-1, 1]$ to define the *inverse cosine function.*

FIGURE 20

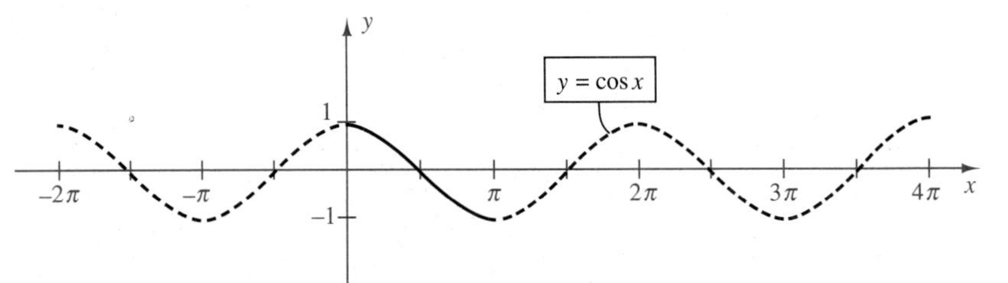

DEFINITION

The **inverse cosine function**, denoted by $\cos^{-1}$, is defined by

$$y = \cos^{-1} x \quad \text{if and only if} \quad x = \cos y$$

for $-1 \le x \le 1$ and $0 \le y \le \pi$.

The domain of the inverse cosine function is $[-1, 1]$ and the range is $[0, \pi]$.

The notation $y = \cos^{-1} x$ may be read *y is the inverse cosine of x* or *y is an angle whose cosine is x.*

The inverse cosine function is also called the **arccosine function** and the notation $\arccos x$ is used interchangeably with $\cos^{-1} x$. Using relationships (2) and (3) of general inverse functions f and f^{-1}, we obtain the following properties.

PROPERTIES OF $\cos^{-1}$

(i) $\cos(\cos^{-1} x) = \cos(\arccos x) = x \quad$ if $-1 \le x \le 1$.

(ii) $\cos^{-1}(\cos y) = \arccos(\cos y) = y \quad$ if $0 \le y \le \pi$.

We can find the graph of the inverse cosine function by reflecting the solid portion of Figure 20 through the line $y = x$. This gives us the sketch

FIGURE 21

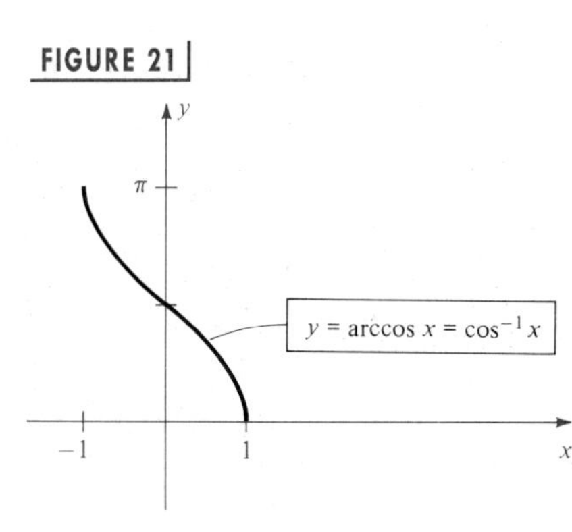

in Figure 21. We could also use the equation $x = \cos y$ with $0 \le y \le \pi$ to find points on the graph.

In the next example, we determine several values of the inverse cosine function by means of the definition.

EXAMPLE ▪ 4

Find the exact value:

(a) $\cos^{-1}\frac{1}{2}$ (b) $\cos^{-1}(-\frac{1}{2})$ (c) $\arccos 0$.

SOLUTION

(a) By definition,

$$y = \cos^{-1}\tfrac{1}{2} \quad \text{means that} \quad \cos y = \tfrac{1}{2} \quad \text{and} \quad 0 \le y \le \pi.$$

The only number y in the interval $[0, \pi]$ such that $\cos y = \frac{1}{2}$ is $y = \pi/3$. Hence

$$\cos^{-1}\tfrac{1}{2} = \frac{\pi}{3}.$$

(b) By definition,

$$y = \cos^{-1}(-\tfrac{1}{2}) \quad \text{means that} \quad \cos y = -\tfrac{1}{2} \quad \text{and} \quad 0 \le y \le \pi.$$

The reference angle for y is $\pi/3$. Since we must choose y in the interval $[0, \pi]$, we take $y = \pi - (\pi/3) = 2\pi/3$. Thus

$$\cos^{-1}(-\tfrac{1}{2}) = \frac{2\pi}{3}.$$

(c) $$y = \arccos 0 \quad \text{means that} \quad \cos y = 0 \quad \text{and} \quad 0 \le y \le \pi.$$

Consequently, $y = \pi/2$ and

$$\arccos 0 = \frac{\pi}{2}.$$

EXAMPLE ▪ 5

Find the exact value:

(a) $\cos[\cos^{-1}(-0.5)]$ (b) $\cos^{-1}(\cos 3.14)$ (c) $\cos^{-1}\left[\sin\left(-\frac{\pi}{6}\right)\right]$.

SOLUTION For parts (a) and (b) we may use the properties of $\cos^{-1}$:

(a) $$\cos[\cos^{-1}(-0.5)] = -0.5$$

(b)

$$\cos^{-1}(\cos 3.14) = 3.14$$

(c) We first find $\sin(-\pi/6)$ and then use the definition of $\cos^{-1}$ (see Example 4(a)):

$$\cos^{-1}\left[\sin\left(-\frac{\pi}{6}\right)\right] = \cos^{-1}\left(-\tfrac{1}{2}\right) = \frac{2\pi}{3}$$

If we restrict the domain of the tangent function to the open interval $(-\pi/2, \pi/2)$, we obtain a one-to-one (increasing) function (see Figure 3). We use this *new* function to define the *inverse tangent function*.

DEFINITION

> The **inverse tangent function**, or **arctangent function**, denoted by $\tan^{-1}$ or arctan, is defined by
>
> $$y = \tan^{-1} x = \arctan x \quad \text{if and only if} \quad x = \tan y$$
>
> for any real number x and $-\dfrac{\pi}{2} < y < \dfrac{\pi}{2}$.

FIGURE 22

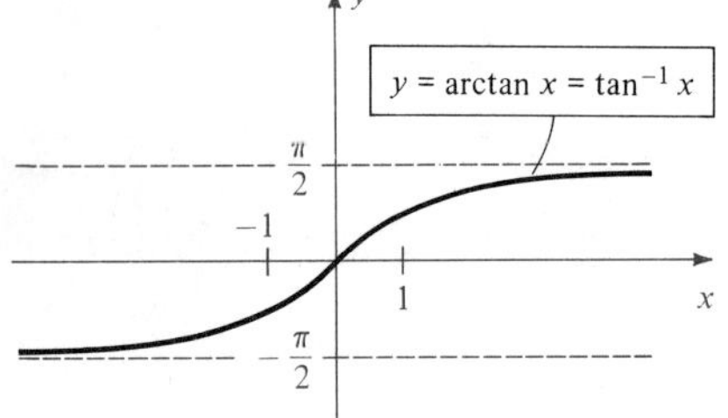

The domain of the arctangent function is $\mathbb{R}$ and the range is the open interval $(-\pi/2, \pi/2)$.

We can obtain the graph of $y = \tan^{-1} x$ in Figure 22 by sketching the graph of $x = \tan y$ for $-\pi/2 < y < \pi/2$.

EXAMPLE ■ 6

Find the exact value of $\sec(\arctan \frac{2}{3})$.

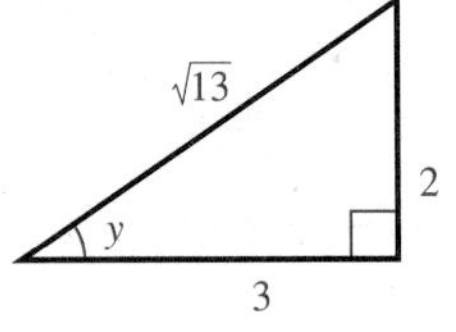

SOLUTION If we let $y = \arctan \frac{2}{3}$, then $\tan y = \frac{2}{3}$. We wish to find $\sec y$.

Since $-\pi/2 < \arctan x < \pi/2$ for every x and $\tan y > 0$, it follows that $0 < y < \pi/2$. Thus we may regard y as the radian measure of an angle of a right triangle such that $\tan y = \frac{2}{3}$, as illustrated in Figure 23. By the Pythagorean theorem, the hypotenuse is $\sqrt{3^2 + 2^2} = \sqrt{13}$. Referring to the triangle, we obtain

$$\sec(\arctan \tfrac{2}{3}) = \sec y = \frac{\sqrt{13}}{3}.$$

The following examples illustrate some of the manipulations that can be carried out with inverse trigonometric functions.

EXAMPLE ▪ 7

Find the exact value of $\sin(\arctan \frac{1}{2} - \arccos \frac{4}{5})$.

SOLUTION If we let

$$u = \arctan \tfrac{1}{2} \quad \text{and} \quad v = \arccos \tfrac{4}{5},$$

then $$\tan u = \tfrac{1}{2} \quad \text{and} \quad \cos v = \tfrac{4}{5}.$$

FIGURE 24

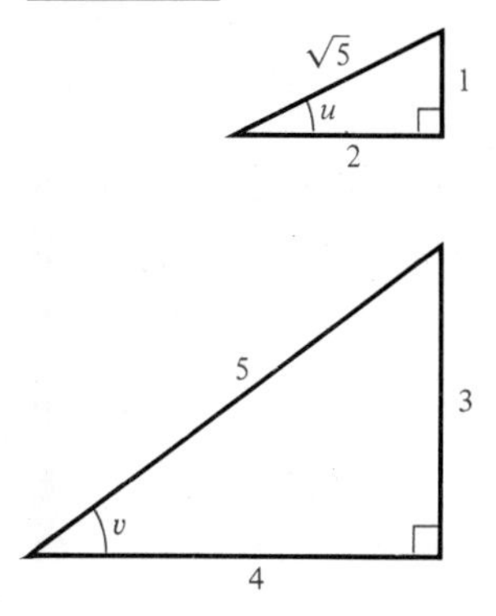

We wish to find $\sin(u - v)$. Since u and v are in the interval $(0, \pi/2)$, they can be considered as the radian measures of positive acute angles, and we may refer to the right triangles in Figure 24. This gives us

$$\sin u = \frac{1}{\sqrt{5}}, \qquad \cos u = \frac{2}{\sqrt{5}}, \qquad \sin v = \frac{3}{5}, \qquad \cos v = \frac{4}{5}.$$

Consequently,

$$\begin{aligned} \sin(u - v) &= \sin u \cos v - \cos u \sin v \\ &= \frac{1}{\sqrt{5}}\frac{4}{5} - \frac{2}{\sqrt{5}}\frac{3}{5} \\ &= \frac{-2}{5\sqrt{5}} = \frac{-2\sqrt{5}}{25}. \end{aligned}$$

EXAMPLE ▪ 8

If $-1 \leq x \leq 1$, rewrite $\cos(\sin^{-1} x)$ as an algebraic expression in x.

SOLUTION Let

$$y = \sin^{-1} x \quad \text{or, equivalently,} \quad \sin y = x.$$

We wish to express $\cos y$ in terms of x. Since $-\pi/2 \leq y \leq \pi/2$, it follows that $\cos y \geq 0$, and hence

$$\cos y = \sqrt{1 - \sin^2 y} = \sqrt{1 - x^2}.$$

Consequently, $$\cos(\sin^{-1} x) = \sqrt{1 - x^2}.$$

FIGURE 25

$y = \sin^{-1} x$

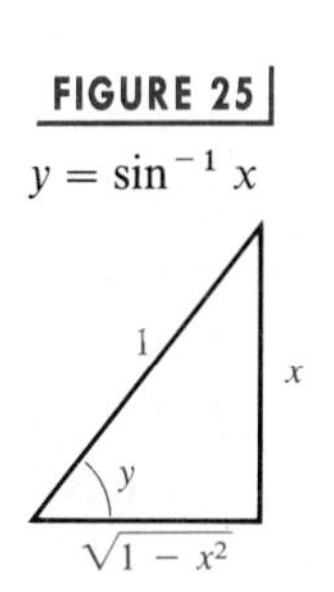

The last identity is also evident geometrically if $0 < x < 1$. In this case $0 < y < \pi/2$, and we may regard y as the radian measure of an angle of a right triangle such that $\sin y = x$, as illustrated in Figure 25. (The side of length $\sqrt{1 - x^2}$ is found by the Pythagorean theorem.) Referring to the triangle,

$$\cos(\sin^{-1} x) = \cos y = \frac{\sqrt{1 - x^2}}{1} = \sqrt{1 - x^2}.$$

Most of the trigonometric equations we considered in Section 7.2 had solutions that were rational multiples of π, such as $\pi/3$, $3\pi/4$, π, and so on. If solutions of trigonometric equations are not of that type, we can sometimes use inverse functions to express them in exact form, as illustrated in the next example.

EXAMPLE ▪ 9

Find the solutions of the equation $5\sin^2 t + 3\sin t - 1 = 0$ that are in the interval $[-\pi/2, \pi/2]$.

SOLUTION The equation may be regarded as a quadratic equation in $\sin t$. We may apply the quadratic formula:

$$\sin t = \frac{-3 \pm \sqrt{9 + 20}}{10} = \frac{-3 \pm \sqrt{29}}{10}.$$

Using the definition of the inverse sine function, we obtain the solutions

$$t = \sin^{-1} \tfrac{1}{10}(-3 + \sqrt{29})$$

and

$$t = \sin^{-1} \tfrac{1}{10}(-3 - \sqrt{29}).$$

If approximations are desired, we place a calculator in radian mode and proceed as follows:

Enter $\frac{1}{10}(-3 + \sqrt{29})$: 0.2385165

Press [INV] [SIN]: 0.240838

Hence

$$t \approx 0.2408.$$

Similarly,

$$t = \sin^{-1} \tfrac{1}{10}(-3 - \sqrt{29}) \approx \sin^{-1}(-0.8385165) \approx -0.9946.$$

The next example illustrates one of many identities that are true for inverse trigonometric functions. Other identities are given in Exercises 61–66.

EXAMPLE ▪ 10

Verify the identity $\sin^{-1} x + \cos^{-1} x = \dfrac{\pi}{2}$ for $-1 \le x \le 1$.

SOLUTION Let

$$\alpha = \sin^{-1} x \quad \text{and} \quad \beta = \cos^{-1} x.$$

We wish to show that $\alpha + \beta = \pi/2$. From the definitions of $\sin^{-1}$ and $\cos^{-1}$,

$$\sin \alpha = x \quad \text{for} \quad -\frac{\pi}{2} \le \alpha \le \frac{\pi}{2}$$

and

$$\cos \beta = x \quad \text{for} \quad 0 \le \beta \le \pi.$$

Adding the two inequalities on the right, we see that

$$-\frac{\pi}{2} \le \alpha + \beta \le \frac{3\pi}{2}.$$

Note also that

$$\cos \alpha = \sqrt{1 - \sin^2 \alpha} = \sqrt{1 - x^2}$$

and

$$\sin \beta = \sqrt{1 - \cos^2 \beta} = \sqrt{1 - x^2}.$$

Using an addition formula,

$$\begin{aligned} \sin(\alpha + \beta) &= \sin \alpha \cos \beta + \cos \alpha \sin \beta \\ &= x \cdot x + \sqrt{1 - x^2}\sqrt{1 - x^2} \\ &= x^2 + (1 - x^2) = 1. \end{aligned}$$

Since $\alpha + \beta$ is in the interval $[-\pi/2, 3\pi/2]$, the equation $\sin(\alpha + \beta) = 1$ has only one solution, $\alpha + \beta = \pi/2$, which is what we wished to show.

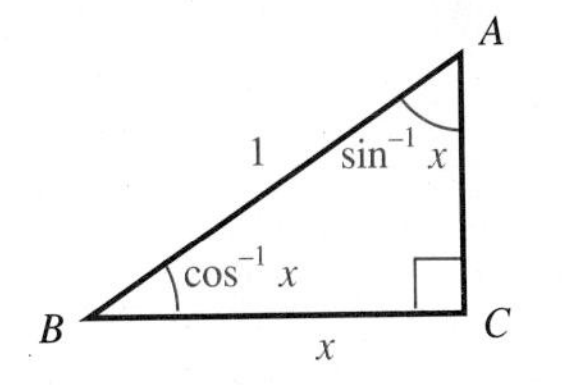

We may interpret the identity geometrically if $0 < x < 1$. If we construct a right triangle with one side of length x and hypotenuse of length 1, as illustrated in Figure 26, then angle β at B is an angle whose cosine is x; that is, $\beta = \cos^{-1} x$. Similarly, angle α at A is an angle whose sine is x; that is, $\alpha = \sin^{-1} x$. Since the acute angles of a right triangle are complementary,

$$\sin^{-1} x + \cos^{-1} x = \frac{\pi}{2}.$$

As a final remark, we can also define $\sec^{-1}$, $\csc^{-1}$, and $\cot^{-1}$; however, of these inverse trigonometric functions, only $\sec^{-1}$ is commonly used in calculus. If we consider the graph of $y = \sec x$, there are many ways to restrict x so that we obtain a one-to-one function that takes on every value of the secant function. There is no universal agreement on how this should be done. For purposes of calculus it is convenient to restrict x to the intervals $[0, \pi/2)$ and $[\pi, 3\pi/2)$, as indicated by the solid portion of the graph of $y = \sec x$ in Figure 27. This leads to the following.

DEFINITION

> The **inverse secant function**, or **arcsecant function**, denoted by $\sec^{-1}$ or arcsec, is defined by
>
> $$y = \sec^{-1} x = \operatorname{arcsec} x \quad \text{if and only if} \quad x = \sec y$$
>
> for $|x| \geq 1$ and y in $[0, \pi/2)$ or in $[\pi, 3\pi/2)$.

The graph of $y = \sec^{-1} x$ is sketched in Figure 28.

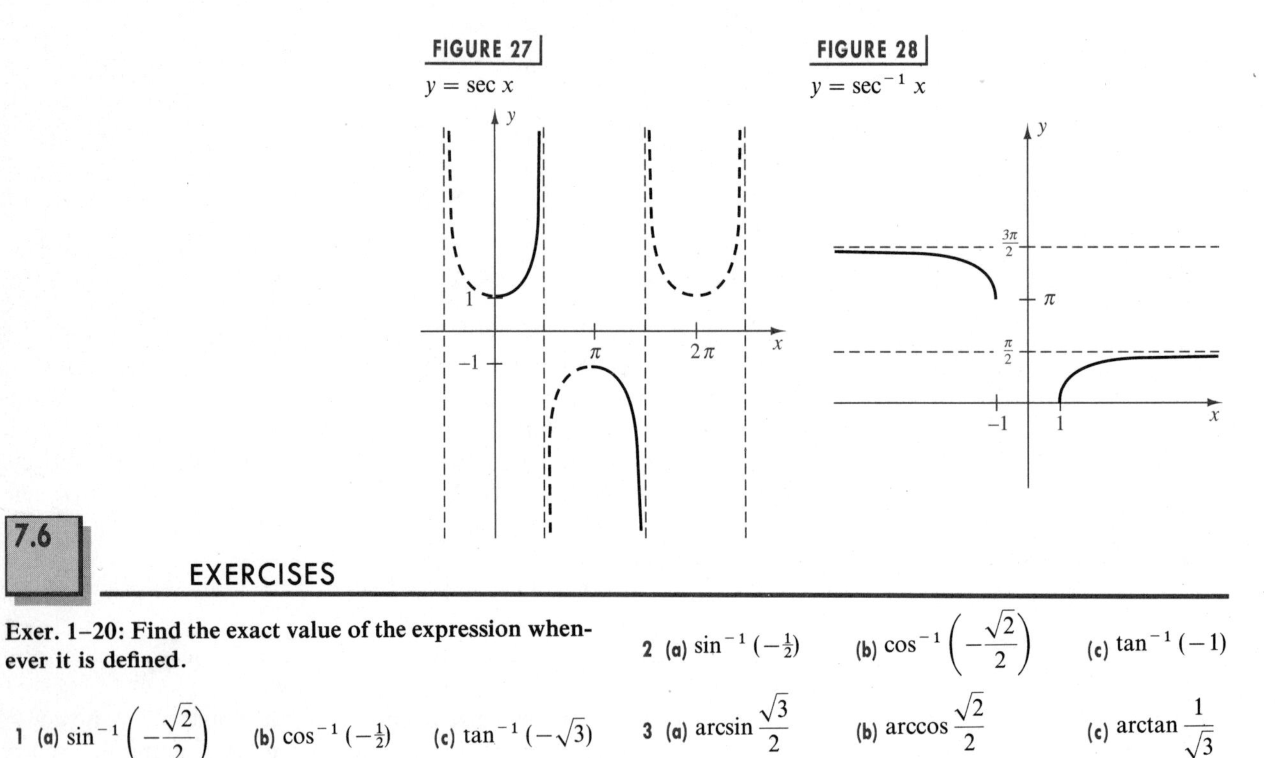

FIGURE 27 $y = \sec x$

FIGURE 28 $y = \sec^{-1} x$

7.6 EXERCISES

Exer. 1–20: Find the exact value of the expression whenever it is defined.

1 (a) $\sin^{-1}\left(-\frac{\sqrt{2}}{2}\right)$ (b) $\cos^{-1}\left(-\frac{1}{2}\right)$ (c) $\tan^{-1}(-\sqrt{3})$

2 (a) $\sin^{-1}\left(-\frac{1}{2}\right)$ (b) $\cos^{-1}\left(-\frac{\sqrt{2}}{2}\right)$ (c) $\tan^{-1}(-1)$

3 (a) $\arcsin \frac{\sqrt{3}}{2}$ (b) $\arccos \frac{\sqrt{2}}{2}$ (c) $\arctan \frac{1}{\sqrt{3}}$

4 (a) $\arcsin 0$ (b) $\arccos(-1)$ (c) $\arctan 0$

5 (a) $\sin^{-1}\frac{\pi}{3}$ (b) $\cos^{-1}\frac{\pi}{2}$ (c) $\tan^{-1} 1$

6 (a) $\arcsin\frac{\pi}{2}$ (b) $\arccos\frac{\pi}{3}$ (c) $\arctan\left(-\frac{\sqrt{3}}{3}\right)$

7 (a) $\sin[\arcsin(-\frac{3}{10})]$ (b) $\cos(\arccos\frac{1}{2})$
(c) $\tan(\arctan 14)$

8 (a) $\sin(\sin^{-1}\frac{2}{3})$ (b) $\cos[\cos^{-1}(-\frac{1}{5})]$
(c) $\tan[\tan^{-1}(-9)]$

9 (a) $\sin^{-1}\left(\sin\frac{\pi}{3}\right)$ (b) $\cos^{-1}\left[\cos\left(-\frac{\pi}{4}\right)\right]$
(c) $\tan^{-1}\left[\tan\left(-\frac{\pi}{6}\right)\right]$

10 (a) $\arcsin\left[\sin\left(-\frac{\pi}{2}\right)\right]$ (b) $\arccos(\cos 0)$
(c) $\arctan\left(\tan\frac{\pi}{4}\right)$

11 (a) $\arcsin\left(\sin\frac{5\pi}{4}\right)$ (b) $\arccos\left(\cos\frac{5\pi}{4}\right)$
(c) $\arctan\left(\tan\frac{7\pi}{4}\right)$

12 (a) $\sin^{-1}\left(\sin\frac{2\pi}{3}\right)$ (b) $\cos^{-1}\left(\cos\frac{4\pi}{3}\right)$
(c) $\tan^{-1}\left(\tan\frac{7\pi}{6}\right)$

13 (a) $\sin[\cos^{-1}(-\frac{1}{2})]$ (b) $\cos(\tan^{-1} 1)$
(c) $\tan[\sin^{-1}(-1)]$

14 (a) $\sin(\tan^{-1}\sqrt{3})$ (b) $\cos(\sin^{-1} 1)$
(c) $\tan(\cos^{-1} 0)$

15 (a) $\cot(\sin^{-1}\frac{2}{3})$ (b) $\sec[\tan^{-1}(-\frac{3}{5})]$
(c) $\csc[\cos^{-1}(-\frac{1}{4})]$

16 (a) $\cot[\sin^{-1}(-\frac{2}{5})]$ (b) $\sec(\tan^{-1}\frac{7}{4})$
(c) $\csc(\cos^{-1}\frac{1}{5})$

17 (a) $\sin(\arcsin\frac{1}{2} + \arccos 0)$
(b) $\cos[\arctan(-\frac{3}{4}) - \arcsin\frac{4}{5}]$
(c) $\tan(\arctan\frac{4}{3} + \arccos\frac{8}{17})$

18 (a) $\sin[\sin^{-1}\frac{5}{13} - \cos^{-1}(-\frac{3}{5})]$
(b) $\cos(\sin^{-1}\frac{4}{5} + \tan^{-1}\frac{3}{4})$
(c) $\tan[\cos^{-1}\frac{1}{2} - \sin^{-1}(-\frac{1}{2})]$

19 (a) $\sin[2\arccos(-\frac{3}{5})]$ (b) $\cos(\frac{1}{2}\tan^{-1}\frac{8}{15})$
(c) $\tan(2\tan^{-1}\frac{3}{4})$

20 (a) $\sin[\frac{1}{2}\sin^{-1}(-\frac{7}{25})]$ (b) $\cos(2\sin^{-1}\frac{15}{17})$
(c) $\tan(\frac{1}{2}\cos^{-1}\frac{3}{5})$

Exer. 21–28: Rewrite as an algebraic expression in x.

21 $\sin(\tan^{-1} x)$

22 $\tan(\arccos x)$

23 $\sec\left(\sin^{-1}\frac{x}{\sqrt{x^2+4}}\right)$

24 $\cot\left(\sin^{-1}\frac{\sqrt{x^2-9}}{x}\right)$

25 $\sin(2\sin^{-1} x)$

26 $\cos(2\tan^{-1} x)$

27 $\cos(\frac{1}{2}\arccos x)$

28 $\tan\left(\frac{1}{2}\cos^{-1}\frac{1}{x}\right)$

Exer. 29–30: Complete the statements.

29 (a) As $x \to -1^+$, $\sin^{-1} x \to$ __
(b) As $x \to 1^-$, $\cos^{-1} x \to$ __
(c) As $x \to -1^-$, $\sec^{-1} x \to$ __
(d) As $x \to \infty$, $\tan^{-1} x \to$ __

30 (a) As $x \to 1^-$, $\sin^{-1} x \to$ __
(b) As $x \to -1^+$, $\cos^{-1} x \to$ __
(c) As $x \to \infty$, $\sec^{-1} x \to$ __
(d) As $x \to -\infty$, $\tan^{-1} x \to$ __

Exer. 31–42: Sketch the graph of the equation.

31 $y = \sin^{-1} 2x$

32 $y = \frac{1}{2}\sin^{-1} x$

33 $y = \sin^{-1}(x+1)$

34 $y = \sin^{-1}(x-2) + \frac{\pi}{2}$

35 $y = \cos^{-1}(\frac{1}{2}x)$

36 $y = 2\cos^{-1} x$

37 $y = 2 + \tan^{-1} x$

38 $y = \tan^{-1} 2x$

39 $y = \sec^{-1}(x-1)$

40 $y = \sec^{-1} x - \frac{\pi}{2}$

41 $y = \sin(\arccos x)$

42 $y = \sin(\sin^{-1} x)$

43 (a) Define $\cot^{-1}$ by restricting the domain of the cotangent function to the interval $(0, \pi)$.
(b) Sketch the graph of $y = \cot^{-1} x$.

44 (a) Define $\csc^{-1}$ by restricting the domain of the cosecant function to $[-\pi/2, 0) \cup (0, \pi/2]$.

(b) Sketch the graph of $y = \csc^{-1} x$.

Exer. 45–52: Solve the equation for x in terms of y.

45 $y = \frac{1}{2}\sin^{-1}(x - 3)$

46 $y = 3\tan^{-1}(2x + 1)$

47 $y = 4\cos^{-1}(\frac{2}{3}x)$

48 $y = 2 + 3\sin x$

49 $y = 15 - 2\cos x$

50 $y = 6 - 3\cos x$

51 $\dfrac{\sin x}{3} = \dfrac{\sin y}{4}$

52 $\dfrac{4}{\sin x} = \dfrac{7}{\sin y}$

Exer. 53–58: (a) Use inverse trigonometric functions to find the solutions of the equation that are in the given interval, and (b) approximate the solutions to four decimal places.

53 $2\tan^2 t + 9\tan t + 3 = 0;\quad \left(-\dfrac{\pi}{2}, \dfrac{\pi}{2}\right)$

54 $3\sin^2 t + 7\sin t + 3 = 0;\quad \left[-\dfrac{\pi}{2}, \dfrac{\pi}{2}\right]$

55 $15\cos^4 x - 14\cos^2 x + 3 = 0;\quad [0, \pi]$

56 $3\tan^4\theta - 19\tan^2\theta + 2 = 0;\quad \left(-\dfrac{\pi}{2}, \dfrac{\pi}{2}\right)$

57 $6\sin^3\theta + 18\sin^2\theta - 5\sin\theta - 15 = 0;\quad \left(-\dfrac{\pi}{2}, \dfrac{\pi}{2}\right)$

58 $6\sin 2x - 8\cos x + 9\sin x - 6 = 0;\quad \left(-\dfrac{\pi}{2}, \dfrac{\pi}{2}\right)$

59 As shown in the figure, a sailboat is following a straight-line course l. The shortest distance from a tracking station T to the course is d miles. As the boat sails, the tracking station records its distance k from T and its direction θ with respect to T. Angle α specifies the direction of the sailboat.

EXERCISE 59

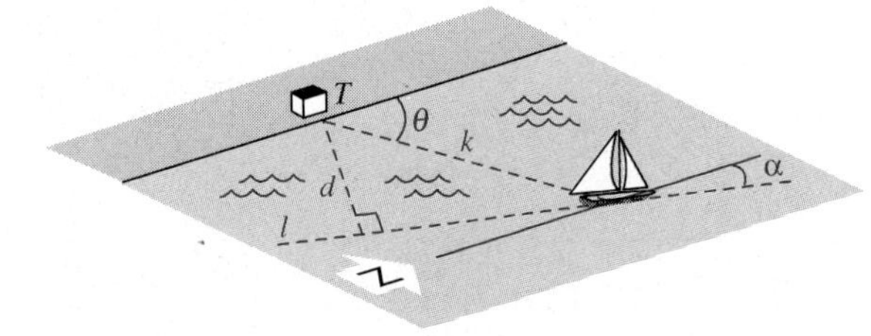

(a) Express α in terms of d, k, and θ.

(b) Estimate α to the nearest degree if $d = 50$ miles, $k = 210$ miles, and $\theta = 53.4°$.

60 An art critic views a large painting that is 10 feet in height and is mounted 4 feet above the floor.

(a) If the art critic is 6 feet tall and is standing x feet from the wall as shown in the figure, express the viewing angle θ in terms of x.

(b) Use the addition formula for the tangent to show that

$$\theta = \tan^{-1}\left(\frac{10x}{x^2 - 16}\right).$$

(c) For what value of x is $\theta = 45°$?

EXERCISE 60

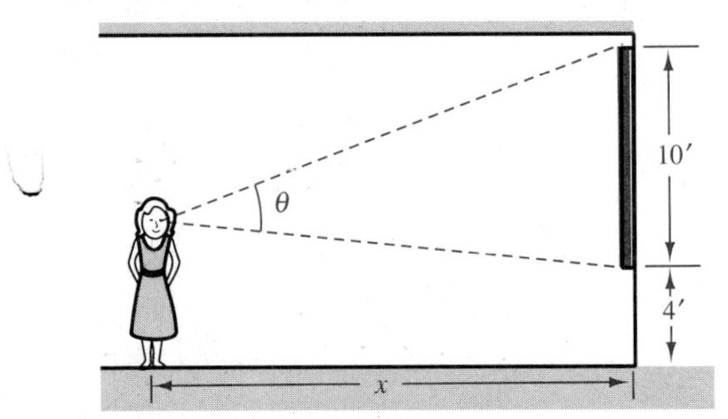

Exer. 61–66: Verify the identity.

61 $\arctan x + \arctan \dfrac{1}{x} = \dfrac{\pi}{2},\quad x > 0$

62 $\arccos x + \arccos\sqrt{1 - x^2} = \dfrac{\pi}{2},\quad |x| \le 1$

63 $\arcsin(-x) = -\arcsin x$

64 $\arccos(-x) = \pi - \arccos x$

65 $\sin^{-1} x = \tan^{-1}\dfrac{x}{\sqrt{1 - x^2}}$

66 $2\cos^{-1} x = \cos^{-1}(2x^2 - 1),\quad 0 \le x \le 1$

7.7 REVIEW

Define or discuss each of the following.

The fundamental identities ■ Verifying identities ■ Trigonometric equations ■ Addition and subtraction formulas ■ Double-angle formulas ■ Half-angle identities ■ Half-angle formulas ■ Product-to-sum formulas ■ Sum-to-product formulas ■ Reduction formulas ■ Inverse trigonometric functions

7.7 EXERCISES

Exer. 1–16: Verify the identity.

1 $(\cot^2 x + 1)(1 - \cos^2 x) = 1$

2 $\cos\theta + \sin\theta\tan\theta = \sec\theta$

3 $\dfrac{(\sec^2\theta - 1)\cot\theta}{\tan\theta\sin\theta + \cos\theta} = \sin\theta$

4 $(\tan x + \cot x)^2 = \sec^2 x \csc^2 x$

5 $\dfrac{1}{1 + \sin t} = (\sec t - \tan t)\sec t$

6 $\dfrac{\sin(\alpha - \beta)}{\cos(\alpha + \beta)} = \dfrac{\tan\alpha - \tan\beta}{1 - \tan\alpha\tan\beta}$

7 $\dfrac{2\cot u}{\csc^2 u - 2} = \tan 2u$

8 $\cos^2\dfrac{v}{2} = \dfrac{1 + \sec v}{2\sec v}$

9 $\dfrac{\tan^3\phi - \cot^3\phi}{\tan^2\phi + \csc^2\phi} = \tan\phi - \cot\phi$

10 $\dfrac{\sin u + \sin v}{\csc u + \csc v} = \dfrac{1 - \sin u \sin v}{-1 + \csc u \csc v}$

11 $\cos\left(x - \dfrac{5\pi}{2}\right) = \sin x$

12 $\tan\left(x + \dfrac{3\pi}{4}\right) = \dfrac{\tan x - 1}{\tan x + 1}$

13 $\frac{1}{4}\sin 4\beta = \sin\beta\cos^3\beta - \cos\beta\sin^3\beta$

14 $\tan\frac{1}{2}\theta = \csc\theta - \cot\theta$

15 $\sin 8\theta = 8\sin\theta\cos\theta(1 - 2\sin^2\theta)(1 - 8\sin^2\theta\cos^2\theta)$

16 $\arctan x = \frac{1}{2}\arctan\dfrac{2x}{1 - x^2},\ |x| \le 1$

Exer. 17–28: Find the solutions of the equation that are in the interval $[0, 2\pi)$.

17 $2\cos^3\theta - \cos\theta = 0$

18 $2\cos\alpha + \tan\alpha = \sec\alpha$

19 $\sin\theta = \tan\theta$

20 $\csc^5\theta - 4\csc\theta = 0$

21 $2\cos^3 t + \cos^2 t - 2\cos t - 1 = 0$

22 $\cos x \cot^2 x = \cos x$

23 $\sin\beta + 2\cos^2\beta = 1$

24 $\cos 2x + 3\cos x + 2 = 0$

25 $2\sec u \sin u + 2 = 4\sin u + \sec u$

26 $\sin 2u = \sin u$

27 $2\cos^2\frac{1}{2}\theta - 3\cos\theta = 0$

28 $\sec 2x \csc 2x = 2\csc 2x$

Exer. 29–32: Find the exact value.

29 $\cos 75^\circ$

30 $\tan 285^\circ$

31 $\sin 195^\circ$

32 $\csc\dfrac{\pi}{8}$

Exer. 33–41: For acute angles θ and ϕ such that $\csc\theta = \frac{5}{3}$ and $\cos\phi = \frac{8}{17}$, find the exact value.

33 $\sin(\theta + \phi)$ **34** $\cos(\theta + \phi)$ **35** $\tan(\theta - \phi)$

36 $\sin(\phi - \theta)$ **37** $\sin 2\phi$ **38** $\cos 2\phi$

39 $\tan 2\theta$ **40** $\sin\dfrac{\theta}{2}$ **41** $\tan\dfrac{\theta}{2}$

42 Express $\cos(\alpha + \beta + \gamma)$ in terms of functions of α, β, and γ.

43 Express the product as a sum or difference:

(a) $\sin 7t \sin 4t$

(b) $\cos\dfrac{u}{4}\cos\left(-\dfrac{u}{6}\right)$

(c) $6\cos 5x \sin 3x$

44 Write the expression as a product.

(a) $\sin 8u + \sin 2u$

(b) $\cos 3\theta - \cos 8\theta$

(c) $\sin\dfrac{t}{4} - \sin\dfrac{t}{5}$

Exer. 45–53: Find the exact value.

45 $\cos^{-1}\left(-\dfrac{\sqrt{3}}{2}\right)$ **46** $\sin^{-1}\left(-\dfrac{\sqrt{2}}{2}\right)$

47 $\arccos\left(\tan\dfrac{3\pi}{4}\right)$ **48** $\arctan\left(-\dfrac{\sqrt{3}}{3}\right)$

49 $\sin\left[\arccos\left(-\dfrac{\sqrt{3}}{2}\right)\right]$

50 $\cos(\sin^{-1}\frac{15}{17} - \sin^{-1}\frac{8}{17})$

51 $\cos(2\sin^{-1}\frac{4}{5})$ **52** $\sin(\sin^{-1}\frac{2}{3})$

53 $\cos^{-1}(\sin 0)$

Exer. 54–56: Sketch the graph of the equation.

54 $y = \cos^{-1} 3x$ **55** $y = 4\sin^{-1} x$

56 $y = 1 - \sin^{-1} x$

57 When an individual is walking, the magnitude F of the vertical force of one foot on the ground (see figure) can be described by $F = A(\cos bt - a\cos 3bt)$ for time t in seconds, $A > 0$, $b > 0$, and $0 < a < 1$.

(a) Show that $F = 0$ when $t = -\dfrac{\pi}{2b}$ and $t = \dfrac{\pi}{2b}$. (The time $t = -\dfrac{\pi}{2b}$ corresponds to the moment when the foot first touches the ground and the weight of the body is being supported by the other foot.)

(b) The maximum force occurs when $3a\sin 3bt = \sin bt$. If $a = \frac{1}{3}$, find the solutions of this equation for $-\dfrac{\pi}{2b} < t < \dfrac{\pi}{2b}$.

(c) If $a = \frac{1}{3}$, express the maximum force in terms of A.

EXERCISE 57

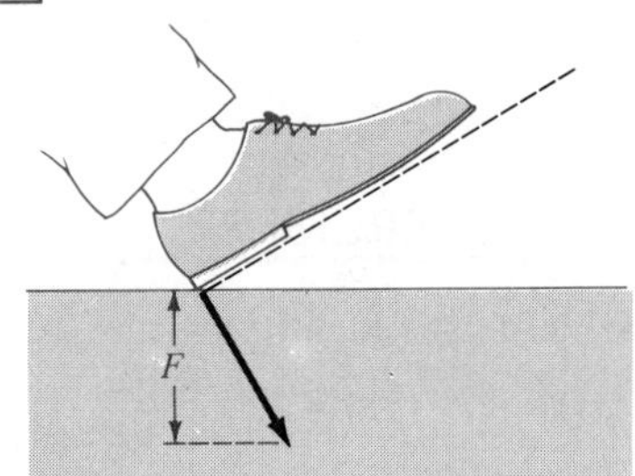

58 Shown in the figure is a graph of the equation $y = \sin x - \frac{1}{2}\sin 2x + \frac{1}{3}\sin 3x$. The x-coordinates of the turning points are solutions of the equation $\cos x - \cos 2x + \cos 3x = 0$. Use sum-to-product formulas to find these x-coordinates.

EXERCISE 58

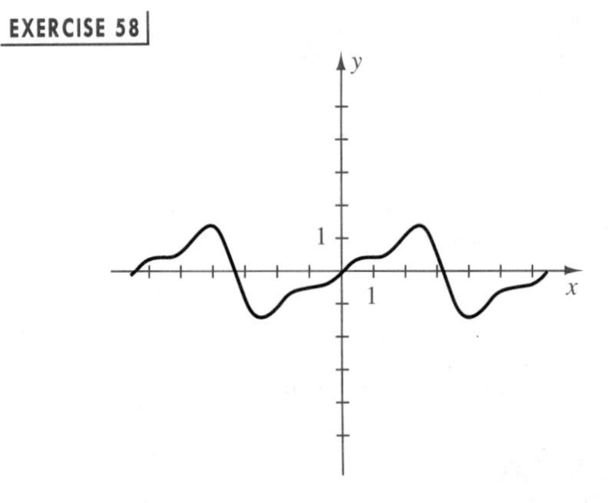

59 The human eye can distinguish between two distant points P and Q provided the angle of resolution θ is not too small. Suppose P and Q are x units apart and are d units from the eye, as illustrated in the figure.

(a) Express x in terms of d and θ.

(b) For a person with normal vision, the smallest distinguishable angle of resolution is about 0.0005 radian. If a pen 6 inches long is viewed by such an individual at a distance of d feet, for what values of d will the endpoints of the pen be distinguishable?

EXERCISE 59

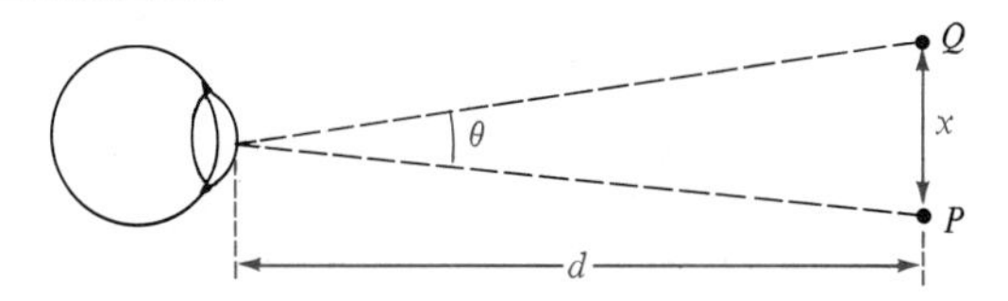

60 A satellite S circles a planet at a distance d miles from the planet's surface. The portion of the planet's surface that is visible from the satellite is determined by the angle θ shown in the figure.

(a) Assuming that the planet is spherical in shape, express d in terms of θ and the radius r of the planet.

(b) Approximate θ for a satellite 300 miles from the surface of the earth, using $r = 4000$ miles.

EXERCISE 60

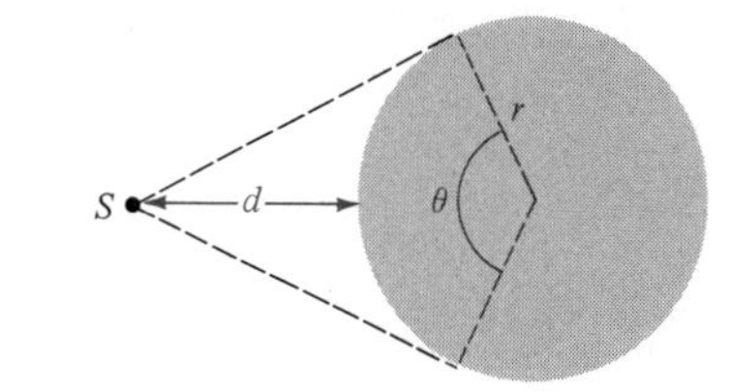

CHAPTER 8

In this chapter we discuss some applications of trigonometry to geometry and algebra. First we consider methods for solving oblique triangles. We next introduce the trigonometric form for complex numbers and obtain a result about *n*th roots. The last two sections present an introduction to vectors.

APPLICATIONS OF TRIGONOMETRY

8.1 THE LAW OF SINES

An **oblique triangle** is a triangle that does not contain a right angle. If we know two angles and a side, or two sides and an angle opposite one of them, then we may find the remaining parts of an oblique triangle by means of the formulas discussed in this section. We shall use the letters A, B, C, a, b, c, α, β, and γ for parts of triangles as we did in Chapter 6. Given triangle ABC, let us place angle α in standard position on a rectangular coordinate system so that B is on the positive x-axis. The case for α obtuse is illustrated in Figure 1; however, the following discussion is also valid if α is acute.

FIGURE 1

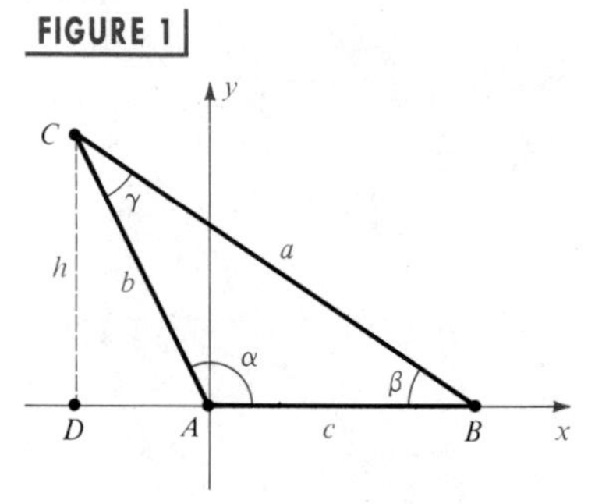

Consider the line through C parallel to the y-axis and intersecting the x-axis at point D. If we let $d(C, D) = h$, then the y-coordinate of C is h. From the definitions of trigonometric functions of any angle,

$$\sin \alpha = \frac{h}{b} \quad \text{and} \quad h = b \sin \alpha.$$

Referring to triangle BDC, we see that

$$\sin \beta = \frac{h}{a} \quad \text{and} \quad h = a \sin \beta.$$

Consequently,

$$b \sin \alpha = a \sin \beta,$$

which we may write as

$$\frac{\sin \alpha}{a} = \frac{\sin \beta}{b}.$$

If we place α in standard position with C on the positive x-axis, then by the same reasoning,

$$\frac{\sin \alpha}{a} = \frac{\sin \gamma}{c}.$$

The last two equalities give us the following result.

THE LAW OF SINES

If ABC is an oblique triangle labeled in the usual manner, then

$$\frac{\sin \alpha}{a} = \frac{\sin \beta}{b} = \frac{\sin \gamma}{c}.$$

The law of sines can also be written in the form

$$\frac{a}{\sin \alpha} = \frac{b}{\sin \beta} = \frac{c}{\sin \gamma}.$$

In examples and exercises involving triangles we shall assume that known lengths of sides and angles have been obtained by measurement, and hence are approximations to exact values. Unless directed otherwise, when finding parts of triangles we shall round off answers according to the following rule: *If known sides or angles are stated to a certain accuracy, then unknown sides or angles should be calculated to the same accuracy.* To illustrate, if known sides are stated to the nearest 0.1, then unknown sides should be calculated to the nearest 0.1. If known angles are stated to the nearest 10′, then unknown angles should be calculated to the nearest 10′. Similar remarks hold for accuracy to the nearest 0.01, 0.1°, and so on.

EXAMPLE ▪ 1

Given triangle ABC with $\alpha = 48°20'$, $\gamma = 57°30'$, and $b = 47.3$, approximate the remaining parts.

FIGURE 2

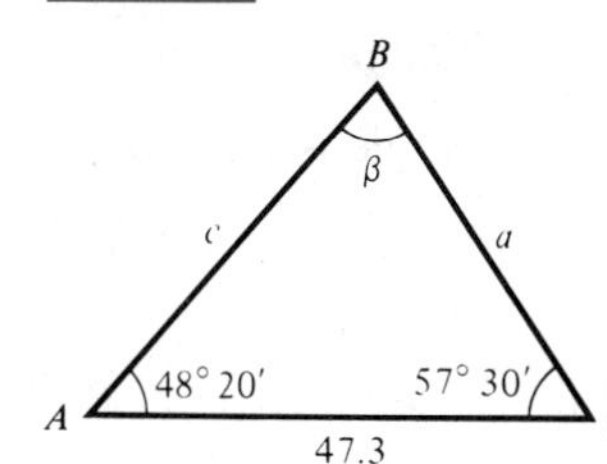

SOLUTION The triangle is sketched in Figure 2. Since the sum of the angles of a triangle is 180°,

$$\beta = 180° - (57°30' + 48°20') = 74°10'.$$

Using $$\frac{a}{\sin \alpha} = \frac{b}{\sin \beta},$$

we obtain $$a = \frac{b \sin \alpha}{\sin \beta} = \frac{47.3 \sin 48°20'}{\sin 74°10'}.$$

and $$a \approx \frac{(47.3)(0.7470)}{0.9621} \approx 36.7.$$

Similarly, $$c = \frac{b \sin \gamma}{\sin \beta} = \frac{47.3 \sin 57°30'}{\sin 74°10'}$$

and $$c \approx \frac{(47.3)(0.8434)}{0.9621} \approx 41.5.$$

Data such as that in Example 1 leads to exactly one triangle ABC. However, if two sides and an angle *opposite* one of them are given, a unique triangle is not always determined. To illustrate, suppose that a and b are to be lengths of sides of triangle ABC and that a given angle

FIGURE 3

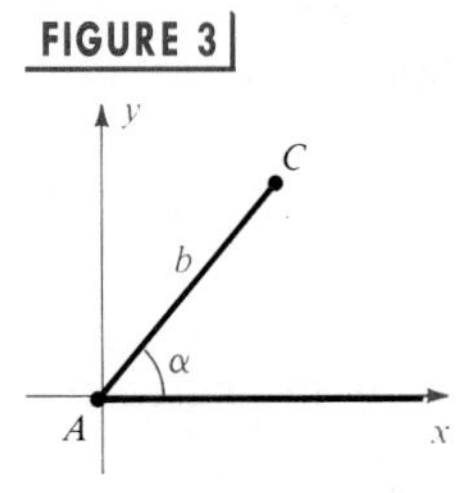

α is to be opposite the side of length a. Let us examine the case for α acute. Place α in standard position on a coordinate system and consider the line segment AC of length b on the terminal side of α, as shown in Figure 3. The third vertex, B, should be somewhere on the x-axis. Since the length a of the side opposite α is given, we may find B by striking off a circular arc of length a with center at C. The four possible outcomes are illustrated in Figure 4 (without the coordinate axes).

FIGURE 4

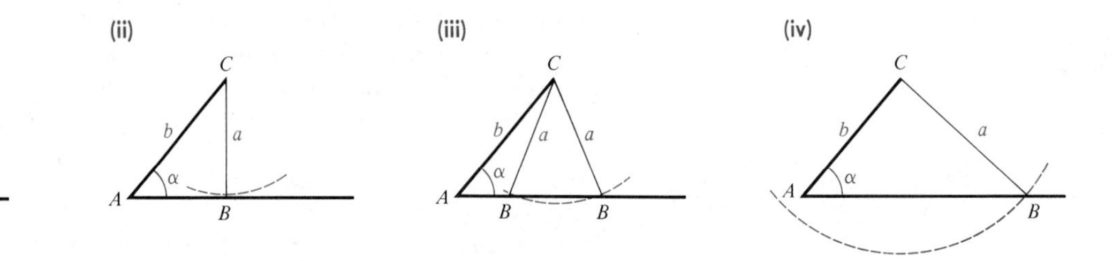

The four possibilities in the figure may be described as follows:

(i) The arc does not intersect the x-axis and no triangle is formed.

(ii) The arc is tangent to the x-axis and a right triangle is formed.

(iii) The arc intersects the positive x-axis in two distinct points and two triangles are formed.

(iv) The arc intersects both the positive and nonpositive parts of the x-axis and one triangle is formed.

The particular case that occurs in a given problem will become evident when the solution is attempted. For example, if we solve the equation

$$\frac{\sin \alpha}{a} = \frac{\sin \beta}{b}$$

and obtain $\sin \beta > 1$, then no triangle exists. If we obtain $\sin \beta = 1$, then $\beta = 90^\circ$ and hence (ii) occurs. If $\sin \beta < 1$, then we have two possible choices for the angle β. By checking both possibilities, we may determine whether (iii) or (iv) occurs.

If the measure of α is greater than 90°, then a triangle exists if and only if $a > b$ (see Figure 5). Since we may have more than one possibility when two sides and an angle opposite one of them are given, this situation is sometimes called the **ambiguous case**.

FIGURE 5

(i) $a < b$

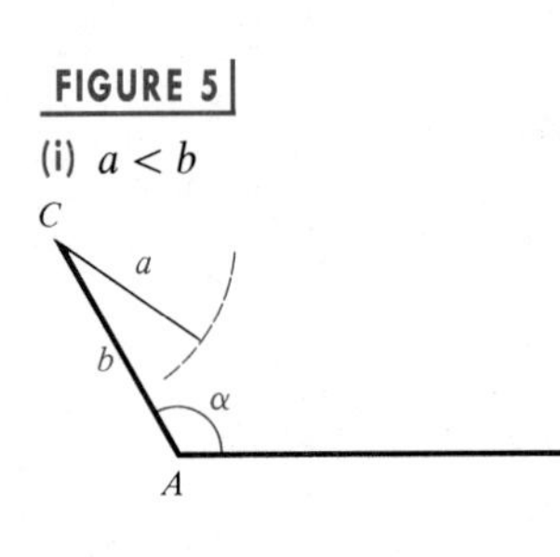

(ii) $a > b$

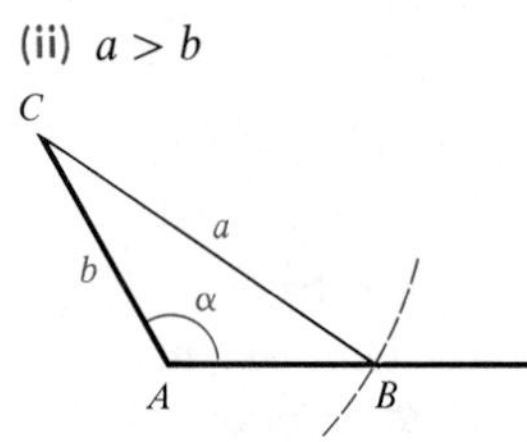

EXAMPLE ▪ 2

Approximate the remaining parts of triangle ABC if $\alpha = 67°$, $c = 125$, and $a = 100$.

SOLUTION We use $\sin \gamma = \dfrac{c \sin \alpha}{a}$:

$$\sin \gamma = \frac{125 \sin 67°}{100}$$

$$\sin \gamma \approx \frac{(125)(0.9205)}{100} \approx 1.1506$$

Since $\sin \gamma > 1$, no triangle can be constructed with the given parts.

EXAMPLE ▪ 3

Approximate the remaining parts of triangle ABC if $a = 12.4$, $b = 8.7$, and $\beta = 36°40'$.

SOLUTION We use $\sin \alpha = \dfrac{a \sin \beta}{b}$:

$$\sin \alpha = \frac{12.4 \sin 36°40'}{8.7}$$

$$\sin \alpha \approx \frac{(12.4)(0.5972)}{8.7} \approx 0.8512$$

There are two possible angles α between 0° and 180° such that $\sin \alpha \approx 0.8512$. If we let α' denote the reference angle for α, then we obtain $\alpha' \approx 58°20'$. Consequently, the two possibilities for α are

$$\alpha_1 \approx 58°20',$$

$$\alpha_2 \approx 180° - 58°20', \quad \text{or} \quad \alpha_2 \approx 121°40'.$$

FIGURE 6

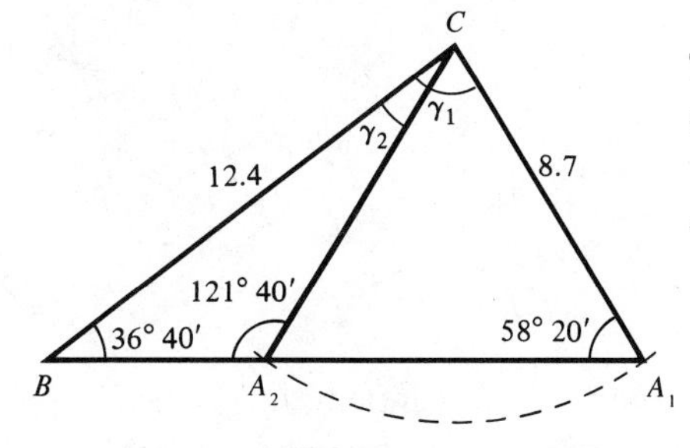

The angle $\alpha_1 \approx 58°20'$ gives us triangle A_1BC in Figure 6, and the angle $\alpha_2 \approx 121°40'$ gives us triangle A_2BC.

If we let γ_1 and γ_2 denote the third angles of the triangles A_1BC and A_2BC corresponding to the angles α_1 and α_2, respectively, then:

$$\gamma_1 \approx 180° - (36°40' + 58°20'), \quad \text{or} \quad \gamma_1 \approx 85°$$

$$\gamma_2 \approx 180° - (36°40' + 121°40'), \quad \text{or} \quad \gamma_2 \approx 21°40'$$

If c_1 is the side opposite γ_1 in triangle A_1BC, then

$$c_1 = \frac{a \sin \gamma_1}{\sin \alpha_1}$$

$$c_1 \approx \frac{12.4 \sin 85^\circ}{\sin 58^\circ 20'} \approx \frac{(12.4)(0.9962)}{0.8511} \approx 14.5.$$

Thus the remaining parts of triangle A_1BC are

$$\alpha_1 \approx 58^\circ 20', \qquad \gamma_1 \approx 85^\circ, \qquad c_1 \approx 14.5.$$

If c_2 is the side opposite angle γ_2 in triangle A_2BC,

$$c_2 = \frac{a \sin \gamma_2}{\sin \alpha_2}$$

$$c_2 \approx \frac{12.4 \sin 21^\circ 40'}{\sin 121^\circ 40'} \approx \frac{(12.4)(0.3692)}{0.8511} \approx 5.4.$$

Consequently, the remaining parts of triangle A_2BC are

$$\alpha_2 \approx 121^\circ 40', \qquad \gamma_2 \approx 21^\circ 40', \qquad c_2 \approx 5.4.$$

FIGURE 7

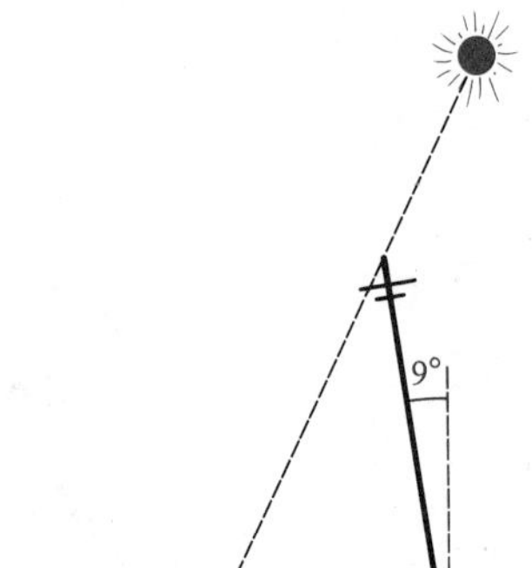

EXAMPLE ▪ 4

When the angle of elevation of the sun is 64°, a telephone pole that is tilted at an angle of 9° directly away from the sun casts a shadow 21 feet long on level ground. Approximate the length of the pole.

SOLUTION The problem is illustrated in Figure 7 (not drawn to scale). Triangle ABC in Figure 8 also displays the given facts. Either drawing is sufficient for our purpose. Note that in Figure 8 we have calculated

$$\beta = 90^\circ - 9^\circ = 81^\circ$$

and

$$\gamma = 180^\circ - (64^\circ + 81^\circ) = 35^\circ.$$

FIGURE 8

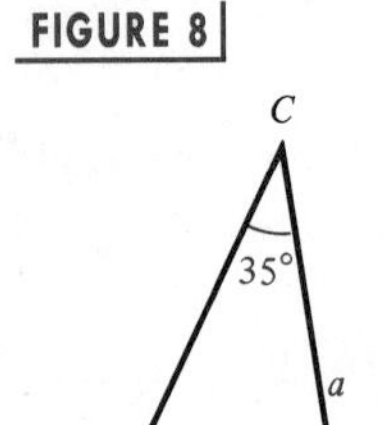

The length of the pole is side a of triangle ABC. We may apply the law of sines:

$$\frac{a}{\sin 64^\circ} = \frac{21}{\sin 35^\circ}$$

This gives us

$$a = \frac{21 \sin 64^\circ}{\sin 35^\circ} \approx \frac{(21)(0.8988)}{0.5736} \approx 33.$$

Thus the telephone pole is approximately 33 feet in length.

EXAMPLE ▪ 5

A point P on level ground is 3.0 km due north of a point Q. A runner proceeds in the direction N25°E from Q to a point R, and then from R to P in the direction S70°W. Approximate the distance run.

FIGURE 9

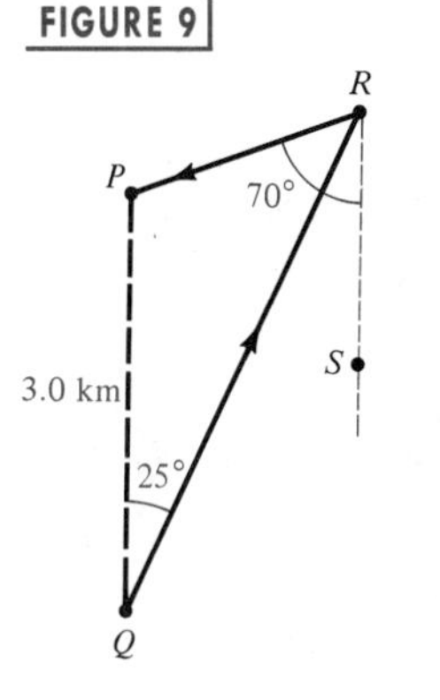

SOLUTION The notation used to specify directions was introduced in Section 6.3. The arrows in Figure 9 show the path of the runner, together with a north-south (dashed) line from R to another point S.

Since the lines through PQ and RS are parallel, it follows from geometry that the alternate interior angles $\angle PQR$ and $\angle QRS$ both have measure 25°. Hence

$$\angle PRQ = 70° - 25° = 45°.$$

These observations give us triangle PQR in Figure 10 with

$$\angle QPR = 180° - (25° + 45°) = 110°.$$

FIGURE 10

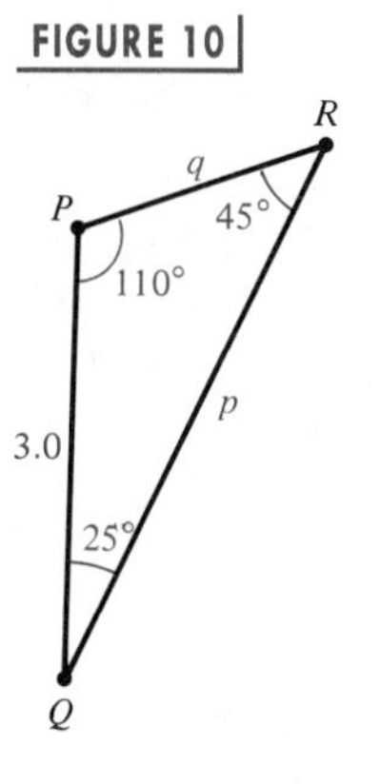

We apply the law of sines twice:

$$\frac{q}{\sin 25°} = \frac{3.0}{\sin 45°} \quad \text{and} \quad \frac{p}{\sin 110°} = \frac{3.0}{\sin 45°}.$$

Hence

$$q = \frac{3.0 \sin 25°}{\sin 45°} \approx \frac{(3.0)(0.4226)}{0.7071} \approx 1.8,$$

$$p = \frac{3.0 \sin 110°}{\sin 45°} \approx \frac{(3.0)(0.9397)}{0.7071} \approx 4.0$$

The distance run, $p + q$, is approximately $1.8 + 4.0 = 5.8$ km.

EXAMPLE ▪ 6

A commercial fishing boat uses sounding equipment to detect a school of fish 2 miles east of the boat and traveling in the direction N51°W at a rate of 8 mi/hr (see Figure 11).

FIGURE 11

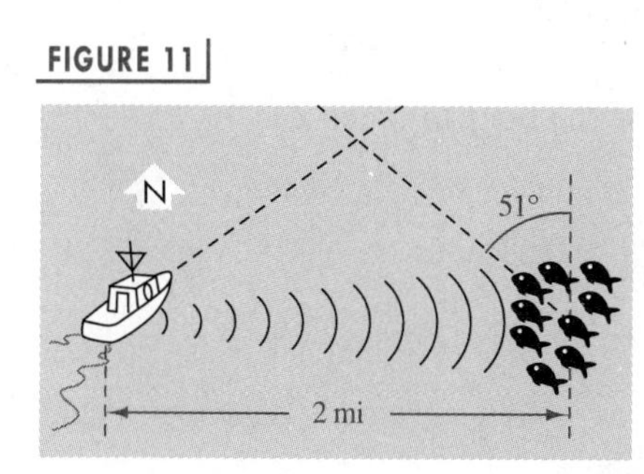

(a) If the boat travels at 20 mi/hr, approximate, to the nearest 0.1° the direction it should head to intercept the school.

(b) Find, to the nearest minute, the time it takes the boat to reach the fish.

SOLUTION

(a) The problem is illustrated by the triangle in Figure 12, with the ship at B, the school of fish at A, and the point of interception at C. Note that

FIGURE 12

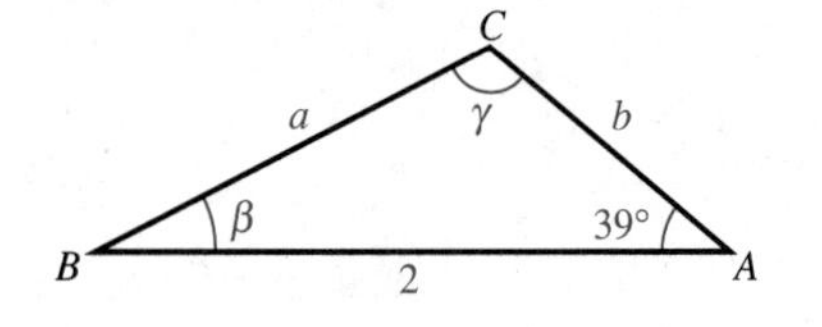

$\alpha = 90° - 51° = 39°$. We use the law of sines:

$$\frac{\sin \beta}{b} = \frac{\sin 39°}{a}$$

and

$$\sin \beta = \frac{b}{a} \sin 39°.$$

If t is the time at which the boat and fish meet at C, then, using the formula *distance* = (*rate*)(*time*), we see that

$$a = 20t \quad \text{and} \quad b = 8t.$$

Hence

$$\frac{b}{a} = \frac{8t}{20t} = \frac{2}{5}.$$

Thus

$$\sin \beta = \tfrac{2}{5} \sin 39° \approx 0.2517,$$

which gives us $\beta \approx 14.6°$. Since $90° - 14.6° = 75.4°$, the boat should travel in the (approximate) direction N75.4°E.

(b) Let us first find the distance a from B to C. We begin by noting that

$$\gamma \approx 180° - (39° + 14.6°), \quad \text{or} \quad \gamma \approx 126.4°.$$

We next apply the law of sines:

$$a = \frac{c \sin \alpha}{\sin \gamma}$$

$$a \approx \frac{2 \sin 39°}{\sin 126.4°} \approx \frac{2(0.6293)}{0.8049} \approx 1.56 \text{ miles}$$

Using $a = 20t$, we find the time t for the boat to reach C:

$$t = \frac{a}{20} \approx \frac{1.56}{20} \approx 0.08 \text{ hour}$$

The time in minutes is

$$t \approx 60(0.08) \approx 5 \text{ minutes}.$$

8.1 EXERCISES

Exer. 1–16: Approximate the remaining parts of triangle *ABC*.

1	$\alpha = 41°$,	$\gamma = 77°$,	$a = 10.5$
2	$\beta = 20°$,	$\gamma = 31°$,	$b = 210$
3	$\alpha = 27°40'$,	$\beta = 52°10'$,	$a = 32.4$
4	$\beta = 50°50'$,	$\gamma = 70°30'$,	$c = 537$
5	$\alpha = 42°10'$,	$\gamma = 61°20'$,	$b = 19.7$
6	$\alpha = 103.45°$,	$\gamma = 27.19°$,	$b = 38.84$
7	$\gamma = 81°$,	$c = 11$,	$b = 12$
8	$\alpha = 32.32°$,	$c = 574.3$,	$a = 263.6$
9	$\gamma = 53°20'$,	$a = 140$,	$c = 115$
10	$\alpha = 27°30'$,	$c = 52.8$,	$a = 28.1$
11	$\gamma = 47.74°$,	$a = 131.08$,	$c = 97.84$
12	$\alpha = 42.17°$,	$a = 5.01$,	$b = 6.12$
13	$\alpha = 65°10'$,	$a = 21.3$,	$b = 18.9$
14	$\beta = 113°10'$,	$b = 248$,	$c = 195$
15	$\beta = 121.624°$,	$b = 0.283$,	$c = 0.178$
16	$\gamma = 73.01°$,	$a = 17.31$,	$c = 20.24$

17 To find the distance between two points A and B that lie on opposite banks of a river, a surveyor lays off a line segment AC of length 240 yards along one bank, and determines that the measures of $\angle BAC$ and $\angle ACB$ are $63°20'$ and $54°10'$, respectively (see figure). Approximate the distance between A and B.

EXERCISE 17

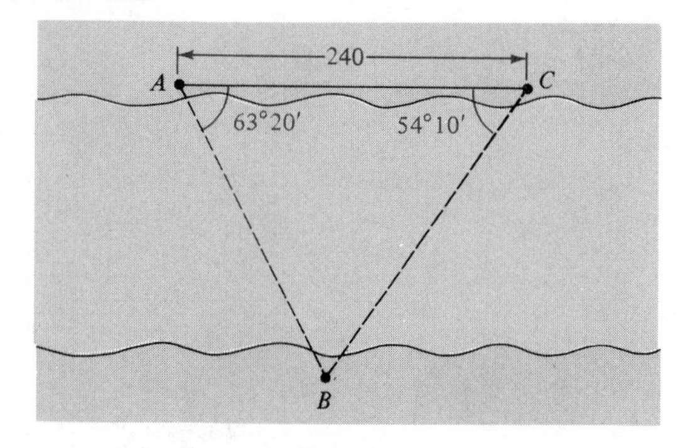

18 To determine the distance between two points A and B, a surveyor chooses a point C that is 375 yards from A and 530 yards from B. If $\angle BAC$ has measure $49°30'$, approximate the distance between A and B.

19 As shown in the figure, a cable car carries passengers from a point A, which is 1.2 miles from a point B at the base of a mountain, to a point P at the top of the mountain. The angles of elevation of P from A and B are 21° and 65°, respectively.

(a) Approximate the distance between A and P.

(b) Approximate the height of the mountain.

EXERCISE 19

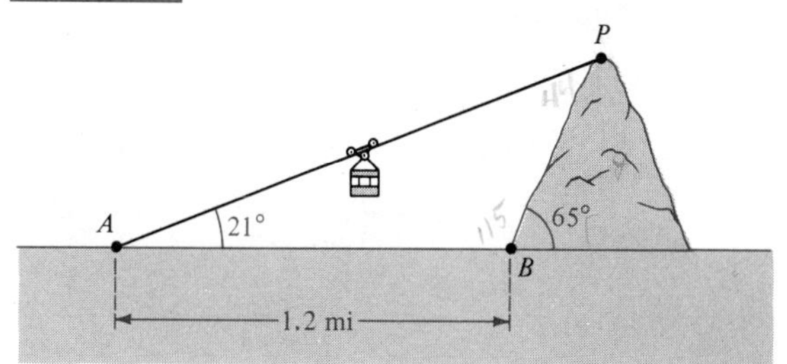

20 A straight road makes an angle of 15° with the horizontal. When the angle of elevation of the sun is 57°, a vertical pole at the side of the road casts a shadow 75 feet long directly down the road, as shown in the figure. Approximate the length of the pole.

EXERCISE 20

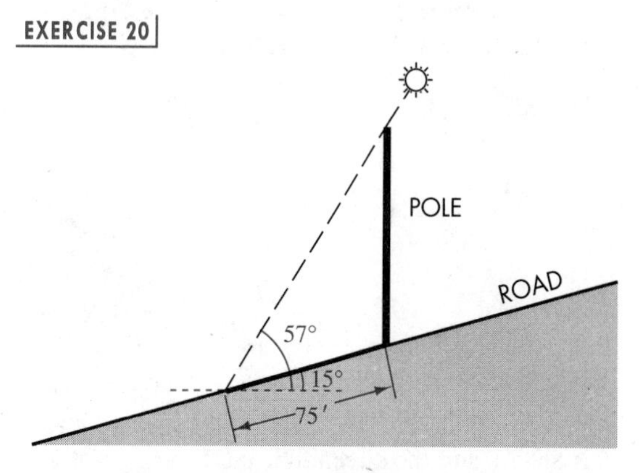

21 The angles of elevation of a balloon from two points A and B on level ground are $24°10'$ and $47°40'$, respectively.

As shown in the figure, points A and B are 8.4 miles apart and the balloon is between the points, in the same vertical plane. Approximate the height of the balloon above the ground.

EXERCISE 21

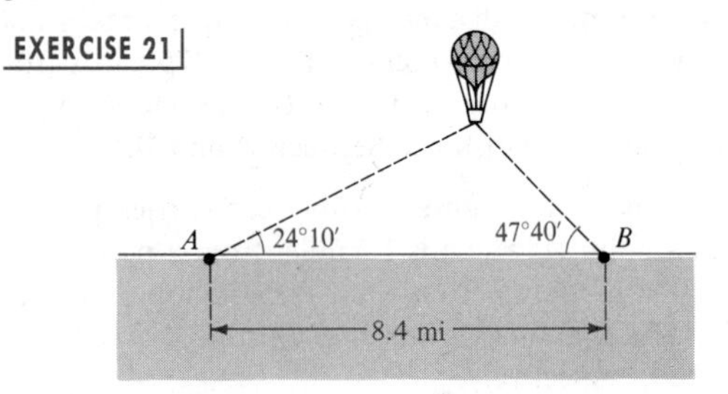

22 Shown in the figure is a solar panel that is 10 feet in width and is to be attached to a roof that makes an angle of 25° with the horizontal. Approximate the length d of the brace that is needed if the panel must make an angle of 45° with the horizontal.

EXERCISE 22

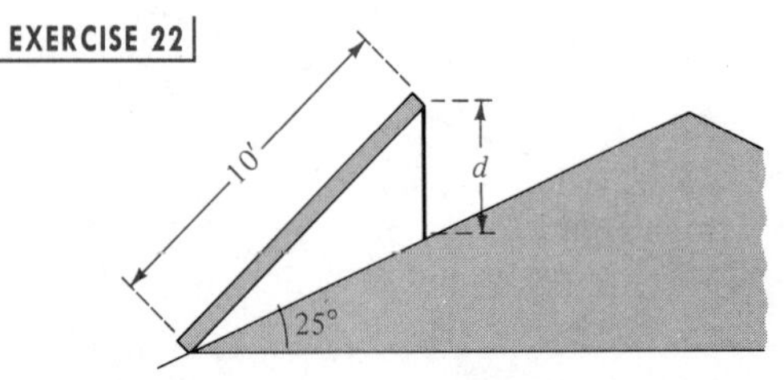

23 A straight road makes an angle of 22° with the horizontal. From a certain point P on the road the angle of elevation of an airplane at point A is 57°. At the same instant, from another point Q, 100 meters farther up the road, the angle of elevation is 63°. As indicated in the figure, the points P, Q, and A lie in the same vertical plane. Approximate the distance from P to the airplane

EXERCISE 23

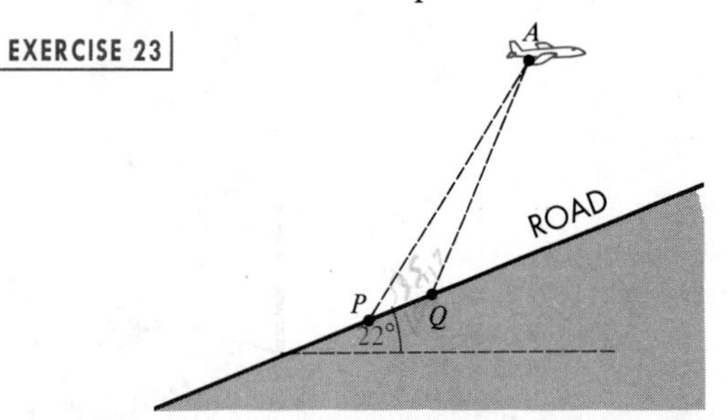

24 A surveyor notes that the direction from point A to point B is S63°W and the direction from A to C is S38°W. The distance from A to B is 239 yards and the distance from B to C is 374 yards. Approximate the distance from A to C.

25 A forest ranger at an observation point A sights a fire in the direction N27°10′E. Another ranger at an observation point B, 6.0 miles due east of A, sights the same fire at N52°40′W. Approximate the distance from each of the observation points to the fire.

26 The leaning tower of Pisa was originally 179 feet tall, but now, due to sinking into the earth, it leans at a certain angle θ from the perpendicular, as shown in the figure. When the top of the tower is viewed from a point 150 feet from the center of its base, the angle of elevation is 53.3°.

(a) Approximate the angle θ.

(b) Approximate the distance d that the center of the top of the tower has moved from the perpendicular.

EXERCISE 26

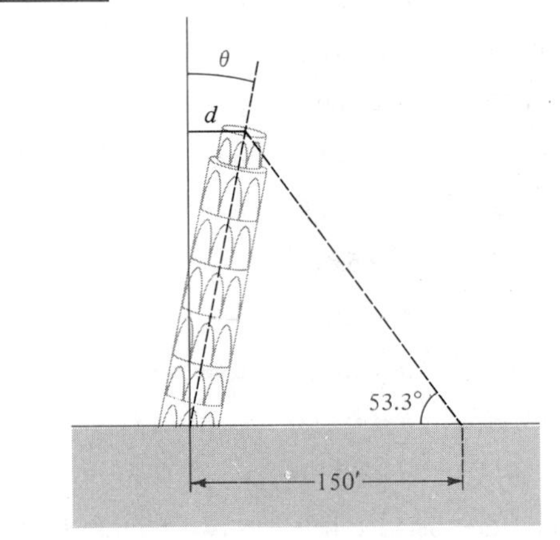

27 A cathedral is located on a hill as shown in the figure. When the top of the spire is viewed from the base of the hill, the angle of elevation is 48°. When viewed at a distance of 200 feet from the base of the hill, the angle of elevation is 41°. The hill rises at an angle of 32°. Approximate the height of the cathedral.

EXERCISE 27

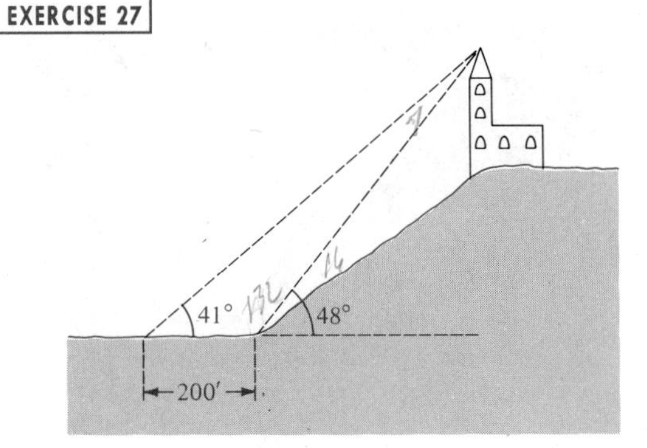

28 A helicopter hovers at an altitude that is 1000 feet above a mountain peak of altitude 5210 feet. A second, taller, peak is viewed from both the mountaintop and the helicopter. From the helicopter, the angle of depression is 43°. From the mountaintop, the angle of elevation is 18°.

(a) Approximate the distance from peak to peak.

(b) Approximate the altitude of the taller peak.

EXERCISE 28

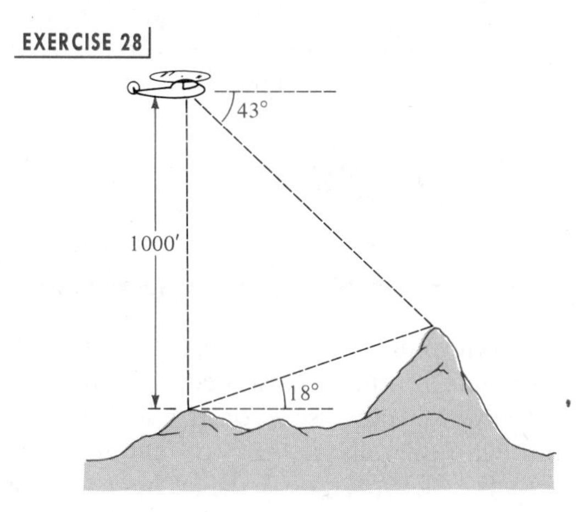

29 The volume V of the right triangular prism shown in the figure is $\frac{1}{3}Bh$, where B is the area of the base and h is the height of the prism. Approximate h and V.

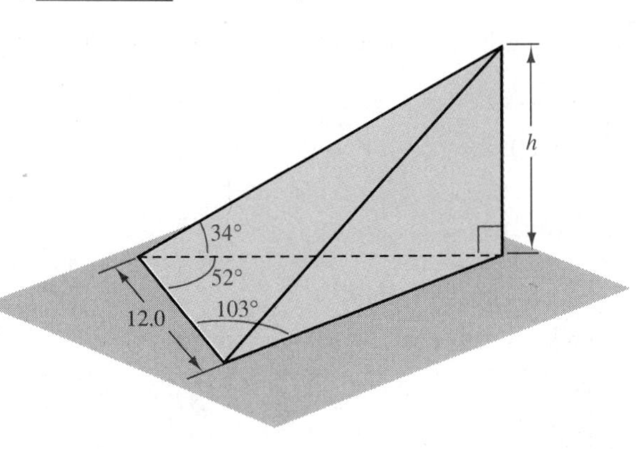

30 Shown in the figure is a top view of a Delta Dart jet with its triangular wing design.

(a) Approximate angle ϕ.

(b) If the fuselage is 4.80 feet wide, approximate the wing span CC'.

(c) Estimate the area of triangle ABC.

EXERCISE 30

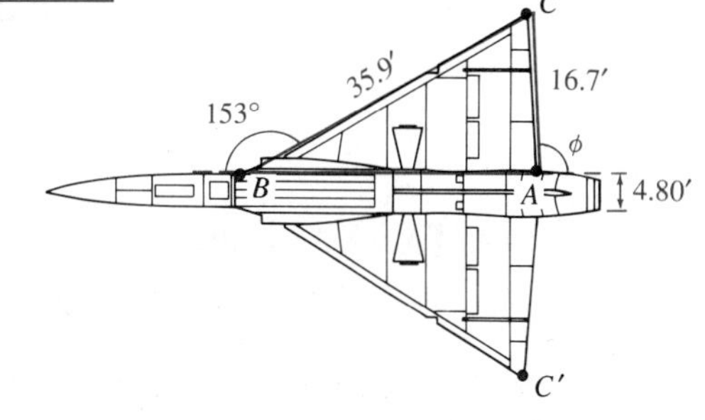

31 Computer software for surveyors makes use of coordinate systems to locate geographic positions. An offshore oil well at point R in the figure is viewed from points P and Q, and $\angle QPR$ and $\angle RQP$ are found to be 55°50′ and 65°22′, respectively. If P and Q have coordinates (1487.7, 3452.8) and (3145.8, 5127.5), respectively, approximate the coordinates of R.

EXERCISE 31

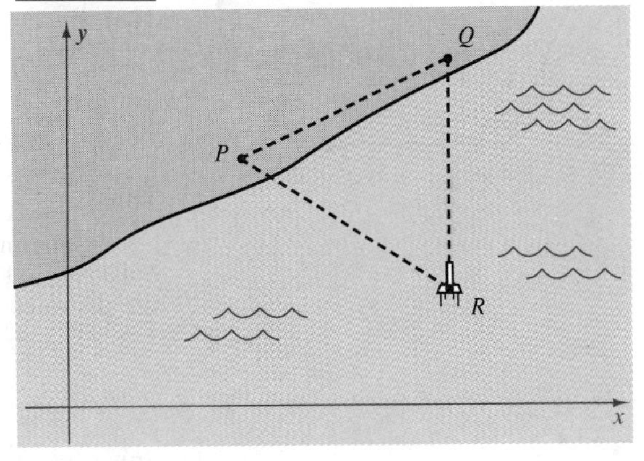

32 Mollweide's formula can be used to check solutions to triangles because it involves all angles and sides of a triangle:

$$\frac{a+b}{c} = \frac{\cos \frac{1}{2}(\alpha - \beta)}{\sin \frac{1}{2}\gamma}.$$

(a) Use the law of sines to show that $\dfrac{a+b}{c} = \dfrac{\sin \alpha + \sin \beta}{\sin \gamma}$.

(b) Use a sum-to-product formula and a double-angle formula to verify Mollweide's formula.

8.2 THE LAW OF COSINES

If two sides and the angle between them or if the three sides of a triangle are known, we cannot apply the law of sines directly to find the remaining parts. Instead, we may use the following result.

THE LAW OF COSINES

If ABC is a triangle labeled in the usual manner, then

$$a^2 = b^2 + c^2 - 2bc \cos \alpha,$$
$$b^2 = a^2 + c^2 - 2ac \cos \beta,$$
$$c^2 = a^2 + b^2 - 2ab \cos \gamma.$$

FIGURE 13

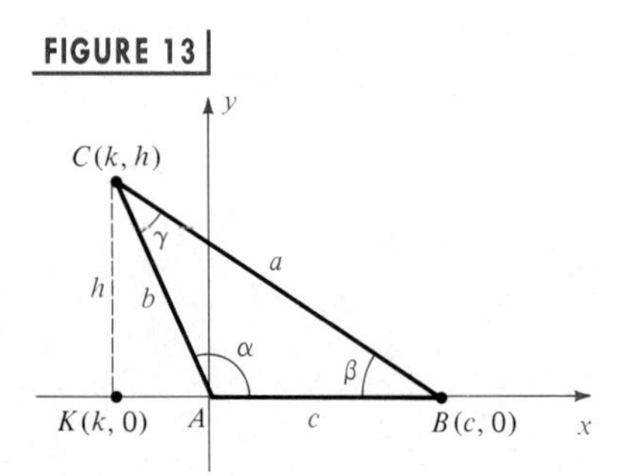

PROOF Let us prove the first formula. Given triangle ABC, place α in standard position on a coordinate system, as illustrated in Figure 13. We have pictured α as obtuse; however, our discussion is also valid if α is acute. Consider the line through C, parallel to the y-axis and intersecting the x-axis at the point $K(k, 0)$. If we let $d(C, K) = h$, then C has coordinates (k, h). By the definition of trigonometric functions of any angle,

$$\cos \alpha = \frac{k}{b} \quad \text{and} \quad \sin \alpha = \frac{h}{b}.$$

Thus $$k = b \cos \alpha \quad \text{and} \quad h = b \sin \alpha.$$

Since the segment AB has length c, the coordinates of B are $(c, 0)$. Using the distance formula,

$$a^2 = [d(B, C)]^2 = (k - c)^2 + (h - 0)^2.$$

Substituting for k and h gives us

$$\begin{aligned} a^2 &= (b \cos \alpha - c)^2 + (b \sin \alpha)^2 \\ &= b^2 \cos^2 \alpha - 2bc \cos \alpha + c^2 + b^2 \sin^2 \alpha \\ &= b^2 (\cos^2 \alpha + \sin^2 \alpha) + c^2 - 2bc \cos \alpha \\ &= b^2 + c^2 - 2bc \cos \alpha, \end{aligned}$$

which is the first formula stated in the law of cosines. The second and third formulas may be obtained by placing β and γ, respectively, in standard position on a coordinate system. ❑

Note that if $\alpha = 90°$ in Figure 13, then $\cos \alpha = 0$ and the law of cosines reduces to $a^2 = b^2 + c^2$. This shows that the Pythagorean theorem is a special case of the law of cosines.

Instead of memorizing each of the three formulas of the law of cosines, it is more convenient to remember the following statement, which takes all of them into account.

THE LAW OF COSINES (Alternative Form)

> The square of the length of any side of a triangle equals the sum of the squares of the lengths of the other two sides minus twice the product of the lengths of the other two sides and the cosine of the angle between them.

EXAMPLE ▪ 1

Approximate the remaining parts of triangle ABC if $a = 5.0$, $c = 8.0$, and $\beta = 77°10'$.

FIGURE 14

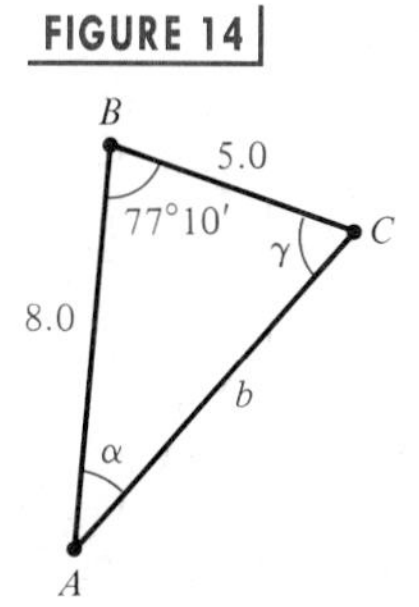

SOLUTION The triangle is sketched in Figure 14. We use the second formula of the law of cosines:

$$b^2 = (5.0)^2 + (8.0)^2 - 2(5.0)(8.0)\cos 77°10'$$

$$b^2 \approx 25 + 64 - (80)(0.2221) \approx 71.2.$$

Consequently, $$b \approx \sqrt{71.2} \approx 8.44 \approx 8.4.$$

We next use the law of sines to find α:

$$\sin \alpha = \frac{a \sin \beta}{b}$$

$$\sin \alpha \approx \frac{5.0 \sin 77°10'}{\sqrt{71.2}} \approx \frac{(5.0)(0.9750)}{8.44} \approx 0.5776.$$

This gives us

$$\alpha \approx 35°20'$$

$$\gamma \approx 180° - (77°10' + 35°20'), \quad \text{or} \quad \gamma \approx 67°30'.$$

After we found the length of the third side in Example 1, we used the law of sines to find a second angle of the triangle. Whenever this procedure is followed, it is best to find the angle opposite the shortest side (as we did), since that angle is always acute. Of course, we can also use the law of cosines to find the remaining angles.

EXAMPLE ▪ 2

If triangle ABC has sides $a = 90$, $b = 70$, and $c = 40$, approximate angles α, β, and γ to the nearest degree.

SOLUTION Use the first formula of the law of cosines:

$$\cos \alpha = \frac{b^2 + c^2 - a^2}{2bc}$$

$$\cos \alpha = \frac{4900 + 1600 - 8100}{5600} \approx -0.2857.$$

and $$\alpha \approx 107°.$$

Similarly, from the second formula of the law of cosines,

$$\cos \beta = \frac{a^2 + c^2 - b^2}{2ac}$$

$$\cos \beta = \frac{8100 + 1600 - 4900}{7200} \approx 0.6667$$

and $$\beta \approx 48°.$$

Finally,

$$\gamma \approx 180° - (107° + 48°), \quad \text{or} \quad \gamma \approx 25°.$$

After finding the first angle in the preceding solution, we could have used the law of sines to find a second angle. Whenever this procedure is followed, it is best to use the law of cosines to find the angle opposite the longer side, that is, the largest angle of the triangle (as we did). This will guarantee that the remaining two angles are acute.

EXAMPLE ▪ 3

A parallelogram has sides of lengths 30 cm and 70 cm and one angle of measure 65°. Approximate the length of each diagonal.

FIGURE 15

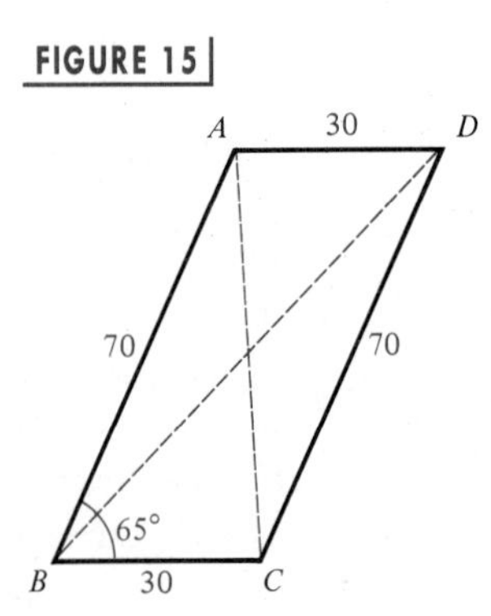

SOLUTION The parallelogram $ABCD$ and its diagonals AC and BD are shown in Figure 15. We may apply the law of cosines to triangle ABC:

$$(AC)^2 = (30)^2 + (70)^2 - 2(30)(70) \cos 65°$$

$$(AC)^2 \approx 900 + 4900 - 4200(0.4226) \approx 4025.1$$

and $$AC \approx \sqrt{4025.1} \approx 63.$$

Similarly, using triangle BAD and $\angle BAD = 180^\circ - 65^\circ = 115^\circ$,

$$(BD)^2 = (30)^2 + (70)^2 - 2(30)(70)\cos 115^\circ$$
$$(BD)^2 \approx 900 + 4900 - 4200(-0.4226) \approx 7574.9$$

and
$$BD \approx \sqrt{7574.9} \approx 87.$$

EXAMPLE ▪ 4

A vertical pole 40 feet tall stands on a hillside that makes an angle of 17° with the horizontal. Approximate the minimal length of cable that will reach from the top of the pole to a point on the hill, 72 feet downhill from the base of the pole.

FIGURE 16

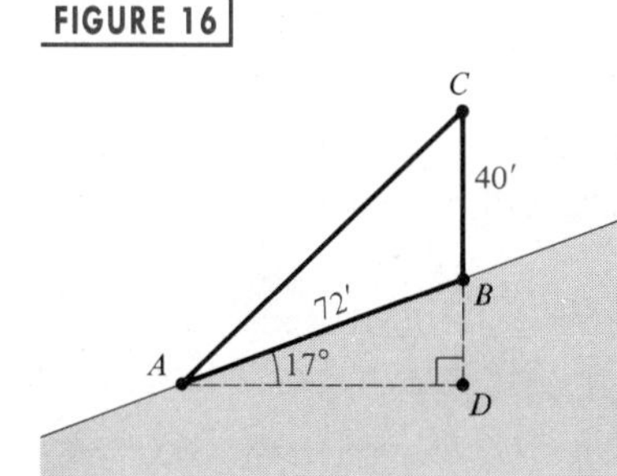

SOLUTION The sketch in Figure 16 displays the given data. We wish to find AC. Referring to the figure, we see that

$$\angle ABD = 90^\circ - 17^\circ = 73^\circ;$$
$$\angle ABC = 180^\circ - 73^\circ = 107^\circ.$$

We apply the law of cosines to triangle ABC:

$$(AC)^2 = (72)^2 + (40)^2 - 2(72)(40)\cos 107^\circ$$
$$(AC)^2 \approx 5184 + 1600 - 5760(-0.2924) \approx 8468.$$

Hence
$$AC \approx \sqrt{8468} \approx 92 \text{ feet.}$$

The law of cosines can be used to derive a formula for the area of a triangle. Let us first prove a preliminary result.

Given triangle ABC, place angle α in standard position, as in Figure 13. As shown in the proof of the law of cosines, the altitude h from vertex C is $h = b \sin \alpha$. Since the area $\mathscr{A}$ of the triangle is given by $\mathscr{A} = \frac{1}{2}ch$, we see that

$$\mathscr{A} = \tfrac{1}{2}bc \sin \alpha.$$

Our argument is independent of the specific angle that is placed in standard position. By taking β and γ in standard position, the formulas $\mathscr{A} = \frac{1}{2}ac \sin \beta$ and $\mathscr{A} = \frac{1}{2}ab \sin \gamma$ are obtained. All three formulas are included in the following result.

AREA OF A TRIANGLE

> The area of a triangle equals one-half the product of the lengths of any two sides and the sine of the angle between them.

EXAMPLE ▪ 5

Approximate the area of triangle ABC with $a = 2.20\,\text{cm}$, $b = 1.30\,\text{cm}$, and $\gamma = 43°10'$.

SOLUTION By the preceding discussion $\mathscr{A} = \frac{1}{2}ab \sin \gamma$. Hence

$$\mathscr{A} = \tfrac{1}{2}(2.20)(1.30) \sin 43°10' \approx 0.98.$$

Thus the area is approximately $0.98\,\text{cm}^2$.

We shall use the preceding result for the area of a triangle to derive *Heron's formula,* which expresses the area of a triangle in terms of the lengths of its sides.

HERON'S FORMULA

> The area $\mathscr{A}$ of a triangle with sides a, b, and c is given by
>
> $$\mathscr{A} = \sqrt{s(s-a)(s-b)(s-c)}$$
>
> with $s = \frac{1}{2}(a + b + c)$.

PROOF Using $\mathscr{A} = \frac{1}{2}bc \sin \alpha$ leads to the following equivalent equations:

$$\mathscr{A} = \sqrt{\tfrac{1}{4}b^2c^2 \sin^2 \alpha}$$
$$\mathscr{A} = \sqrt{\tfrac{1}{4}b^2c^2(1 - \cos^2 \alpha)}$$
$$\mathscr{A} = \sqrt{\tfrac{1}{2}bc(1 + \cos \alpha) \cdot \tfrac{1}{2}bc(1 - \cos \alpha)}.$$

We shall obtain Heron's formula by replacing the expressions under the radical sign by expressions involving only a, b, and c. We use the law of cosines:

$$\begin{aligned} \tfrac{1}{2}bc(1 + \cos \alpha) &= \tfrac{1}{2}bc\left(1 + \frac{b^2 + c^2 - a^2}{2bc}\right) \\ &= \tfrac{1}{2}bc\left(\frac{2bc + b^2 + c^2 - a^2}{2bc}\right) \\ &= \frac{2bc + b^2 + c^2 - a^2}{4} \\ &= \frac{(b + c)^2 - a^2}{4}, \end{aligned}$$

which may be written in the form

$$\tfrac{1}{2}bc(1 + \cos\alpha) = \frac{(b + c) + a}{2} \cdot \frac{(b + c) - a}{2}.$$

We use the same type of manipulations on the second expression under the radical sign:

$$\tfrac{1}{2}bc(1 - \cos\alpha) = \frac{a - b + c}{2} \cdot \frac{a + b - c}{2}.$$

If we now substitute for the expressions under the radical sign, we obtain

$$\mathscr{A} = \sqrt{\frac{b + c + a}{2} \cdot \frac{b + c - a}{2} \cdot \frac{a - b + c}{2} \cdot \frac{a + b - c}{2}}.$$

Letting $s = \frac{1}{2}(a + b + c)$, we see that

$$s - a = \frac{b + c - a}{2}, \qquad s - b = \frac{a - b + c}{2}, \qquad s - c = \frac{a + b - c}{2}.$$

Substitution in the last formula for $\mathscr{A}$ gives us Heron's formula. ❑

EXAMPLE ▪ 6

A farmer has a triangular field with sides of lengths 125 yards, 160 yards, and 225 yards. Approximate the number of acres in the field. (One acre is equivalent to 4840 square yards.)

SOLUTION We shall find the area of the field using Heron's formula with $a = 125$, $b = 160$, and $c = 225$:

$$s = \tfrac{1}{2}(125 + 160 + 225) = \tfrac{1}{2}(510) = 255$$

$$s - a = 255 - 125 = 130$$

$$s - b = 255 - 160 = 95$$

$$s - c = 255 - 225 = 30$$

Substituting in Heron's formula gives us

$$\mathscr{A} = \sqrt{(255)(130)(95)(30)} \approx 9720 \text{ yd}^2.$$

Since there are 4840 square yards in one acre, the number of acres is $\frac{9720}{4840}$, or approximately 2.

8.2 EXERCISES

Exer. 1–10: Approximate the remaining parts of triangle *ABC*.

1 $\alpha = 60°$, $b = 20$, $c = 30$

2 $\gamma = 45°$, $b = 10.0$, $a = 15.0$

3 $\beta = 150°$, $a = 150$, $c = 30$

4 $\beta = 73°50'$, $c = 14.0$, $a = 87.0$

5 $\gamma = 115°10'$, $a = 1.10$, $b = 2.10$

6 $\alpha = 23°40'$, $c = 4.30$, $b = 70.0$

7 $a = 2.0$, $b = 3.0$, $c = 4.0$

8 $a = 10$, $b = 15$, $c = 12$

9 $a = 25.0$, $b = 80.0$, $c = 60.0$

10 $a = 20.0$, $b = 20.0$, $c = 10.0$

11 The angle at one corner of a triangular plot of ground has measure 73°40′ and the sides that meet at this corner are 175 feet and 150 feet long. Approximate the length of the third side.

12 To find the distance between two points A and B, a surveyor chooses a point C that is 420 yards from A and 540 yards from B. Angle ACB has measure 63°10′. Approximate the distance AB.

13 Two automobiles leave from the same point and travel along straight highways that differ in direction by 84°. Their speeds are 60 mi/hr and 45 mi/hr, respectively. Approximately how far apart will the cars be at the end of 20 minutes?

14 A triangular plot of land has sides of lengths 420 feet, 350 feet, and 180 feet. Approximate the smallest angle between the sides.

15 A ship leaves port at 1:00 P.M. and travels S35°E at the rate of 24 mi/hr. Another ship leaves port at 1:30 P.M. and travels S20°W at 18 mi/hr. Approximately how far apart are the ships at 3:00 P.M.?

16 An airplane flies 165 miles from point A in the direction 130° and then travels in the direction 245° for 80 miles. Approximately how far is the airplane from A?

17 An athlete runs at a constant speed of one mile every eight minutes in the direction S40°E for 20 minutes and then in the direction N20°E for the next 16 minutes. Approximate, to the nearest tenth of a mile, the distance from the endpoint to the starting point of the athlete's course.

18 Two points P and Q on level ground are on opposite sides of a building. To find the distance d between the points, a surveyor chooses a point R that is 300 feet from P and 438 feet from Q, and then determines that angle PRQ has measure 37°40′. Approximate d.

EXERCISE 18

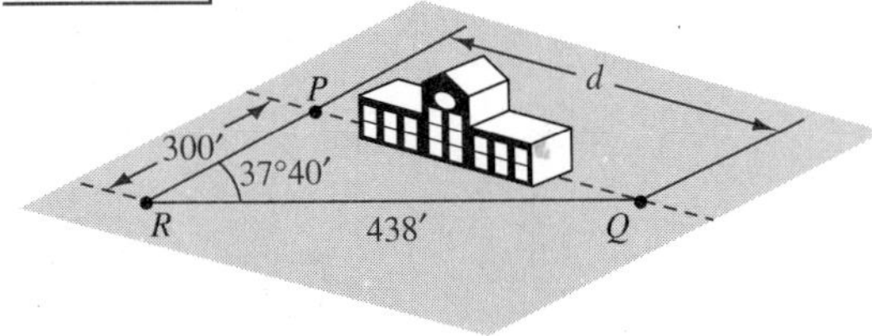

19 A motorboat traveled along a triangular course of sides 2 km, 4 km, and 3 km, respectively. The first side was traversed in the direction N20°W and the second in a direction SD°W, where D° is the degree measure of an acute angle. Approximate, to the nearest minute, the direction that the third side was traversed.

20 The rectangular box shown in the figure has dimensions $8'' \times 6'' \times 4''$. Find the angle θ formed by a diagonal of the base and a diagonal of the $6'' \times 4''$ side.

EXERCISE 20

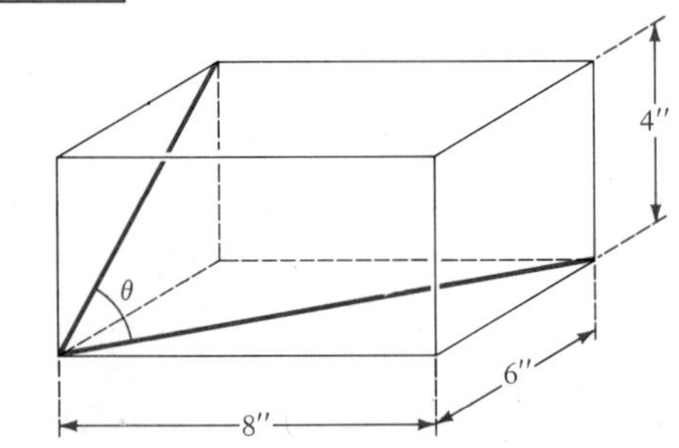

21 In major-league baseball, the four bases (forming a square) are 90 feet apart and the pitcher's mound is 60.5

feet from home plate. Approximate the distance from the pitcher's mound to each of the other three bases.

22 A rhombus has sides of length 100 cm and the angle at one of the vertices is 70°. Approximate the lengths of the diagonals to the nearest tenth of a centimeter.

23 A reconaissance jet J, flying at 10,000 feet above a point R on the surface of the water, spots a submarine S at an angle of depression of 37° and a tanker T at an angle of depression of 21°, as shown in the figure. In addition, $\angle SJT$ is found to be 130°. Approximate the distance between the submarine and the tanker.

EXERCISE 23

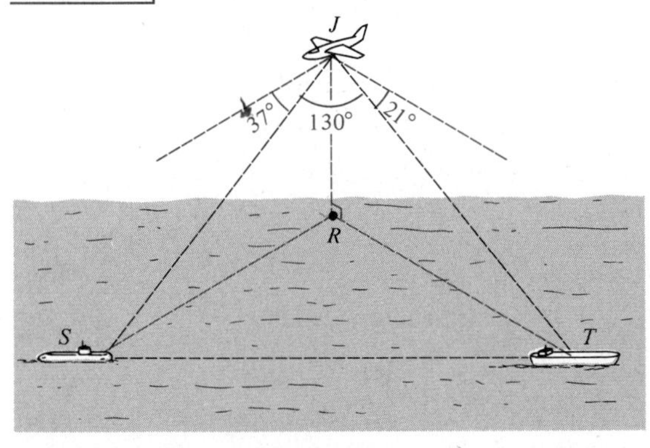

24 A cruise ship sets a course N47°E from an island to a port on the mainland, which is 150 miles away. After moving through strong currents, the ship is off course at a position P that is N33°E and 80 miles from the island, as illustrated in the figure.

(a) Approximately how far is the ship from the port?

(b) In what direction should the ship head to correct its course?

EXERCISE 24

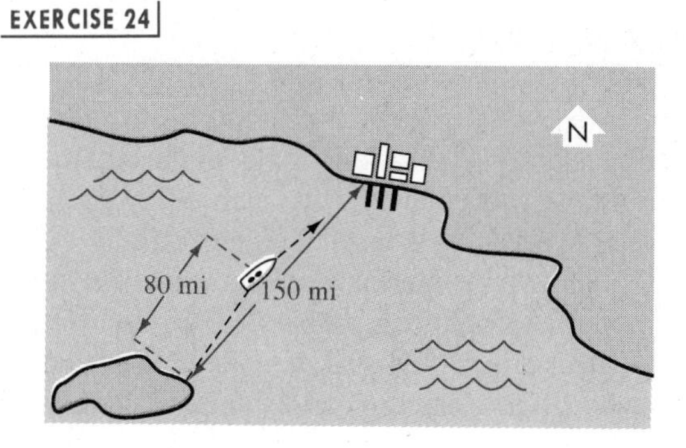

25 Shown in the figure is a top view of a jet fighter with its quadrilateral wing design $ABCD$.

(a) Approximate angle ϕ.

(b) Approximate the area of $ABCD$.

(c) If the fuselage is 5.8 feet wide, estimate the wing span CC'.

EXERCISE 25

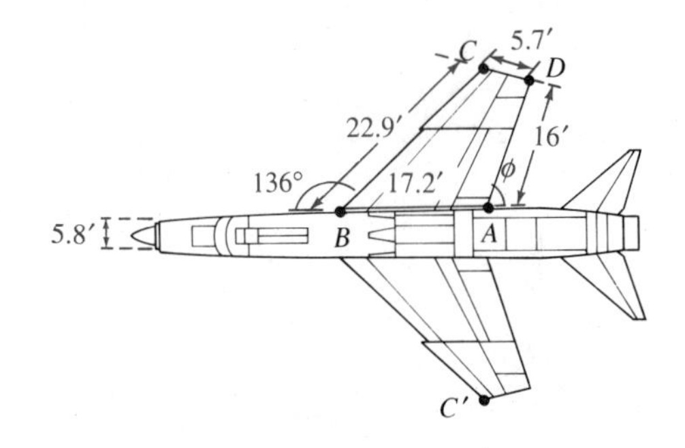

26 The distance across the river shown in the figure can be found without measuring angles. Two points B and C on the opposite shore are selected and line segments AB and AC are extended as shown. Points D and E are chosen as indicated, and the distances BC, BD, BE, CD, and CE are then measured. Suppose that $BC = 184$ feet, $BD = 102$ feet, $BE = 218$ feet, $CE = 80$ feet, and $CD = 236$ feet.

(a) Estimate the distances AB and AC.

(b) Estimate the shortest distance across the river from point A.

EXERCISE 26

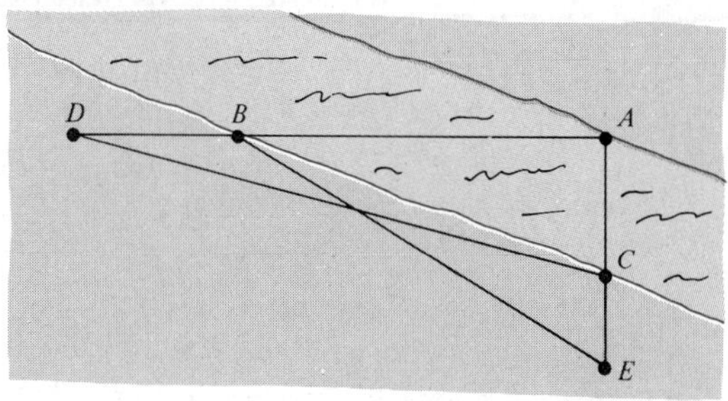

27 Penrose tiles are formed from a rhombus $ABCD$ of side 1 and an interior angle of 72° by first locating a point P that lies on the diagonal AC and is 1 unit from vertex C, and then drawing segments PB and PD to the other vertices of the diagonal, as shown in the figure. The two tiles formed are called a dart and a kite. Three-dimensional counterparts of these tiles have been applied in molecular chemistry.

(a) Find the degree measures of $\angle ABP$, $\angle APB$, and $\angle BPC$.

(b) Approximate, to the nearest 0.01, the length of segment BP.

(c) Approximate, to the nearest 0.01, the area of a kite and the area of a dart.

EXERCISE 27

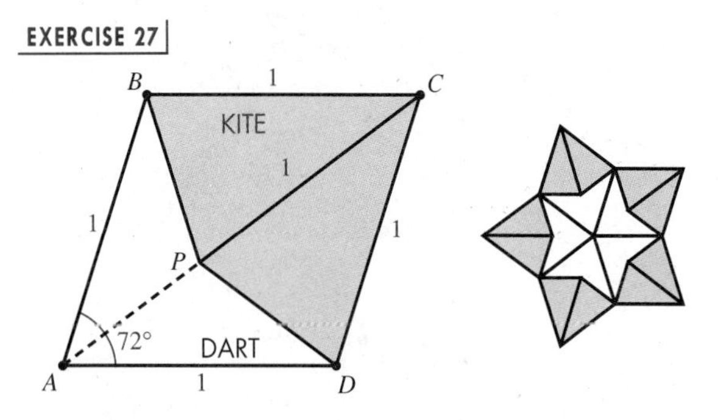

28 The rear hatchback door of an automobile is 42 inches long. A strut with fully extended length of 24 inches is to be attached to the door and the body of the car so that it is vertical and the rear clearance is 32 inches when the door is opened completely, as shown in the figure. Approximate the lengths of segments TP and TQ.

EXERCISE 28

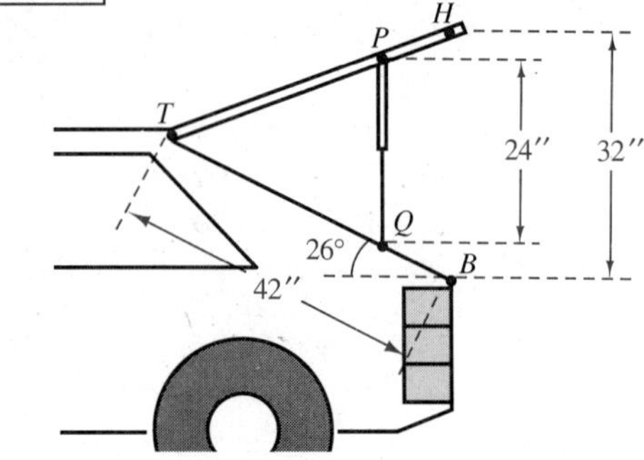

Exer. 29–36: Approximate the area of triangle ABC.

29 $\alpha = 60°$, $b = 20$, $c = 30$

30 $\gamma = 45°$, $b = 10.0$, $a = 15.0$

31 $\beta = 150°$, $a = 150$, $c = 30.0$

32 $\beta = 73°50'$, $c = 14.0$, $a = 87.0$

33 $a = 2.0$, $b = 3.0$, $c = 4.0$

34 $a = 10$, $b = 15$, $c = 12$

35 $a = 25.0$, $b = 80.0$, $c = 60.0$

36 $a = 20.0$, $b = 20.0$, $c = 10.0$

37 A triangular field has sides of lengths 115 yards, 140 yards, and 200 yards. Approximate the number of acres in the field. (One acre is equivalent to 4840 square yards.)

38 Find the area of a parallelogram that has sides of lengths 12.0 and 16.0 feet if one angle at a vertex has measure 40°.

8.3 TRIGONOMETRIC FORM FOR COMPLEX NUMBERS

In Section 1.1 we represented real numbers by points on a coordinate line. We can obtain geometric representations for complex numbers by using points in a coordinate plane. Specifically, each complex number $a + bi$ determines a unique ordered pair (a, b). The corresponding point $P(a, b)$ in a coordinate plane is the **geometric representation of** $a + bi$. To emphasize that we are assigning complex numbers to points in a plane, we may label the point $P(a, b)$ as $a + bi$. A coordinate plane with a complex number assigned to each point is referred to as a **complex plane** instead

FIGURE 17

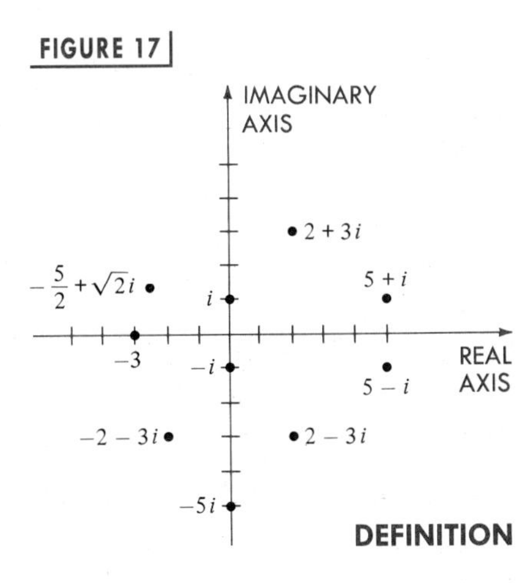

of an xy-plane. The x-axis is the **real axis** and the y-axis the **imaginary axis**. In Figure 17 we have represented several complex numbers geometrically. Note that to obtain the point corresponding to the conjugate $a - bi$ of any complex number $a + bi$, we simply reflect through the real axis.

The absolute value $|a|$ of a real number a is the distance between the origin and the point on the x-axis that corresponds to a. Thus it is natural to interpret the absolute value $|a + bi|$ of a complex number as the distance $\sqrt{a^2 + b^2}$ between the origin of a complex plane and the point (a, b) that corresponds to $a + bi$.

DEFINITION

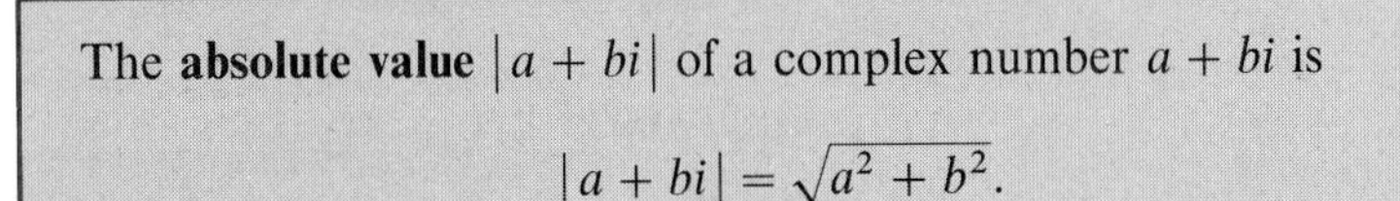
The **absolute value** $|a + bi|$ of a complex number $a + bi$ is

$$|a + bi| = \sqrt{a^2 + b^2}.$$

EXAMPLE ▪ 1

Find **(a)** $|2 - 6i|$ and **(b)** $|3i|$.

SOLUTION We use the definition of absolute value:

(a) $$|2 - 6i| = \sqrt{(2)^2 + (-6)^2} = \sqrt{4 + 36} = \sqrt{40} = 2\sqrt{10}$$

(b) $$|3i| = \sqrt{(0)^2 + (3)^2} = \sqrt{9} = 3$$

The points corresponding to all complex numbers that have a fixed absolute value k are on a circle of radius k with center at the origin in the complex plane. For example, the points corresponding to the complex numbers z with $|z| = 1$ are on a unit circle.

Let us consider a nonzero complex number $z = a + bi$ and its geometric representation $P(a, b)$ as illustrated in Figure 18. Let θ be any angle in standard position whose terminal side lies on the segment OP and let $r = |z| = \sqrt{a^2 + b^2}$. Since $\cos\theta = a/r$ and $\sin\theta = b/r$, we see that $a = r\cos\theta$ and $b = r\sin\theta$. Substituting for a and b in $z = a + bi$, we obtain

FIGURE 18

$z = a + bi = r(\cos\theta + i\sin\theta)$

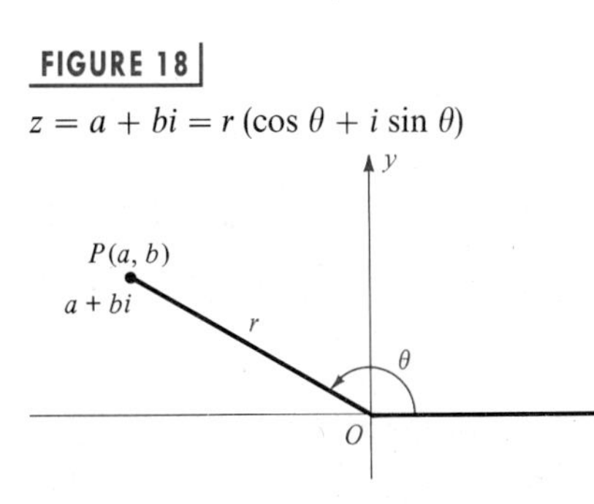

$$z = a + bi = (r\cos\theta) + (r\sin\theta)i = r(\cos\theta + i\sin\theta).$$

This is called the **trigonometric form** of the complex number $a + bi$. A common abbreviation is

$$r(\cos\theta + i\sin\theta) = r\operatorname{cis}\theta.$$

The trigonometric form for $z = a + bi$ is not unique, since there are an unlimited number of different choices for the angle θ. When the trigonometric form is used, the absolute value r of z is sometimes referred to as the **modulus** of z and an angle θ associated with z an **argument** (or **amplitude**) of z.

We may summarize our discussion as follows.

TRIGONOMETRIC FORM OF $a + bi$

Let $z = a + bi$. If $r = |z| = \sqrt{a^2 + b^2}$ and if θ is an argument of z, then

$$z = r(\cos\theta + i\sin\theta) = r \operatorname{cis}\theta.$$

EXAMPLE ▪ 2

Express the complex number in trigonometric form with $0 \le \theta < 2\pi$:

(a) $-4 + 4i$ (b) $2\sqrt{3} - 2i$ (c) $2 + 7i$ (d) $-2 + 7i$

SOLUTION We begin by representing each complex number geometrically as in Figure 19.

FIGURE 19

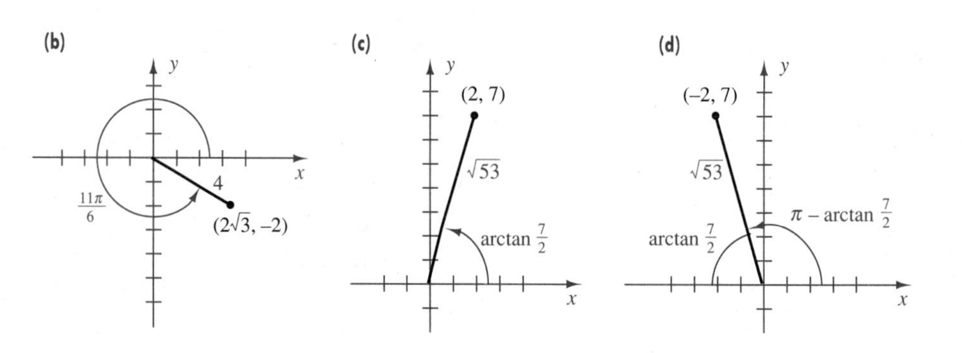

We next substitute for r and θ in the trigonometric form:

(a) $$-4 + 4i = 4\sqrt{2}\left(\cos\frac{3\pi}{4} + i\sin\frac{3\pi}{4}\right) = 4\sqrt{2}\operatorname{cis}\frac{3\pi}{4}$$

(b) $$2\sqrt{3} - 2i = 4\left(\cos\frac{11\pi}{6} + i\sin\frac{11\pi}{6}\right) = 4\operatorname{cis}\frac{11\pi}{6}$$

(c) $2 + 7i = \sqrt{53}\,[\cos(\arctan\frac{7}{2}) + i\sin(\arctan\frac{7}{2})] = \sqrt{53}\operatorname{cis}(\arctan\frac{7}{2})$

(d) $$-2 + 7i = \sqrt{53}\,[\cos(\pi - \arctan\tfrac{7}{2}) + i\sin(\pi - \arctan\tfrac{7}{2})]$$
$$= \sqrt{53}\operatorname{cis}(\pi - \arctan\tfrac{7}{2})$$

If we allow arbitrary values for θ, there are many other trigonometric forms for the complex numbers in Example 2. Thus, for $-4 + 4i$ in part (a) we could use

$$\theta = \frac{3\pi}{4} + 2\pi n \quad \text{for any integer } n.$$

If, for example, we let $n = 1$ and $n = -1$, we obtain

$$-4 + 4i = 4\sqrt{2}\left(\cos \frac{11\pi}{4} + i \sin \frac{11\pi}{4}\right) = 4\sqrt{2} \text{ cis } \frac{11\pi}{4}$$

$$-4 + 4i = 4\sqrt{2}\left[\cos\left(-\frac{5\pi}{4}\right) + i \sin\left(-\frac{5\pi}{4}\right)\right] = 4\sqrt{2} \text{ cis}\left(-\frac{5\pi}{4}\right)$$

In general, *arguments for the same complex number always differ by a multiple of 2π.*

If complex numbers are expressed in trigonometric form, then multiplication and division may be performed as indicated in the next theorem.

THEOREM

If trigonometric forms for two complex numbers z_1 and z_2 are

$$z_1 = r_1(\cos \theta_1 + i \sin \theta_1) \quad \text{and} \quad z_2 = r_2(\cos \theta_2 + i \sin \theta_2)$$

then

(i) $z_1 z_2 = r_1 r_2 [\cos (\theta_1 + \theta_2) + i \sin (\theta_1 + \theta_2)]$

(ii) $\dfrac{z_1}{z_2} = \dfrac{r_1}{r_2} [\cos (\theta_1 - \theta_2) + i \sin (\theta_1 - \theta_2)], \quad z_2 \neq 0.$

PROOF We shall prove (i) and leave (ii) as an exercise. Thus,

$$\begin{aligned} z_1 z_2 &= r_1(\cos \theta_1 + i \sin \theta_1) \cdot r_2(\cos \theta_2 + i \sin \theta_2) \\ &= r_1 r_2[(\cos \theta_1 \cos \theta_2 - \sin \theta_1 \sin \theta_2) \\ &\qquad + i(\sin \theta_1 \cos \theta_2 + \cos \theta_1 \sin \theta_2)]. \end{aligned}$$

Applying the addition formulas for $\cos (\theta_1 + \theta_2)$ and $\sin (\theta_1 + \theta_2)$ gives us (i). ❑

Part (i) of the preceding theorem states that *the modulus of a product of two complex numbers is the product of their moduli, and an argument is the sum of their arguments*. An analogous statement can be made for (ii).

EXAMPLE ▪ 3

If $z_1 = 2\sqrt{3} - 2i$ and $z_2 = -1 + \sqrt{3}\,i$, use trigonometric forms to find **(a)** $z_1 z_2$ and **(b)** z_1/z_2. Check by using algebraic methods.

SOLUTION The complex number $2\sqrt{3} - 2i$ is represented geometrically in Figure 19(b). If we use $\theta = -\pi/6$ in the trigonometric form, then

$$z_1 = 2\sqrt{3} - 2i = 4\left[\cos\left(-\frac{\pi}{6}\right) + i\sin\left(-\frac{\pi}{6}\right)\right].$$

FIGURE 20

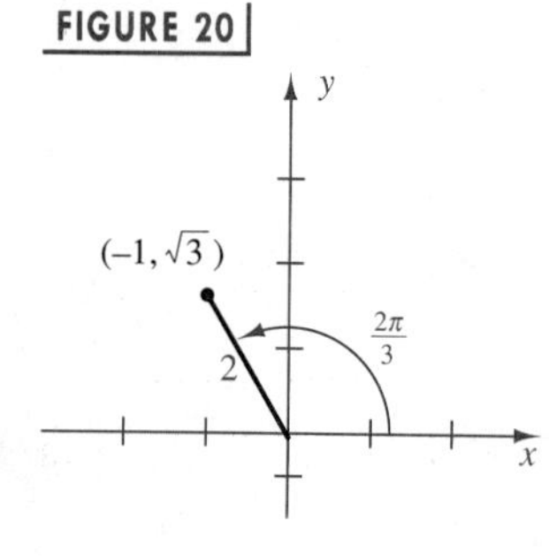

The complex number $z_2 = -1 + \sqrt{3}i$ is represented geometrically in Figure 20. A trigonometric form is

$$z_2 = -1 + \sqrt{3}\,i = 2\left(\cos\frac{2\pi}{3} + i\sin\frac{2\pi}{3}\right).$$

(a) We apply (i) of the preceding theorem:

$$\begin{aligned} z_1 z_2 &= 8\left[\cos\left(-\frac{\pi}{6} + \frac{2\pi}{3}\right) + i\sin\left(-\frac{\pi}{6} + \frac{2\pi}{3}\right)\right] \\ &= 8\left(\cos\frac{\pi}{2} + i\sin\frac{\pi}{2}\right) = 8i. \end{aligned}$$

As a check, we use algebraic methods:

$$\begin{aligned} z_1 z_2 &= (2\sqrt{3} - 2i)(-1 + \sqrt{3}i) \\ &= (-2\sqrt{3} + 2\sqrt{3}) + (2 + 6)i = 0 + 8i = 8i. \end{aligned}$$

(b) We apply (ii) of the theorem:

$$\begin{aligned} \frac{z_1}{z_2} &= 2\left[\cos\left(-\frac{\pi}{6} - \frac{2\pi}{3}\right) + i\sin\left(-\frac{\pi}{6} - \frac{2\pi}{3}\right)\right] \\ &= 2\left[\cos\left(-\frac{5\pi}{6}\right) + i\sin\left(-\frac{5\pi}{6}\right)\right] \\ &= 2\left[-\frac{\sqrt{3}}{2} + i\left(-\frac{1}{2}\right)\right] = -\sqrt{3} - i. \end{aligned}$$

Using algebraic methods,

$$\begin{aligned} \frac{z_1}{z_2} &= \frac{2\sqrt{3} - 2i}{-1 + \sqrt{3}\,i} \cdot \frac{-1 - \sqrt{3}\,i}{-1 - \sqrt{3}\,i} \\ &= \frac{(-2\sqrt{3} - 2\sqrt{3}) + (2 - 6)i}{4} = -\sqrt{3} - i. \end{aligned}$$

8.3 EXERCISES

Exer. 1–10: Find the absolute value.

1 $|3 - 4i|$ 2 $|5 + 8i|$

3 $|-6 - 7i|$ 4 $|1 - i|$

5 $|8i|$ 6 $|i^7|$

7 $|i^{500}|$ 8 $|-15i|$

9 $|0|$ 10 $|-15|$

Exer. 11–20: Represent the complex number geometrically.

11 $4 + 2i$ 12 $-5 + 3i$

13 $3 - 5i$ 14 $-2 - 6i$

15 $-(3 - 6i)$ 16 $(1 + 2i)^2$

17 $2i(2 + 3i)$ 18 $(-3i)(2 - i)$

19 $(1 + i)^2$ 20 $4(-1 + 2i)$

Exer. 21–46: Express the complex number in trigonometric form with $0 \le \theta < 2\pi$.

21 $1 - i$ 22 $\sqrt{3} + i$

23 $-4\sqrt{3} + 4i$ 24 $-2 - 2i$

25 $-20i$ 26 15

27 $-5(1 + \sqrt{3}\,i)$ 28 $-6i$

29 -7 30 $2i(1 - \sqrt{3}\,i)$

31 $-4 - 4i$ 32 $-10 + 10i$

33 $6i$ 34 -5

35 $\sqrt{3} - i$ 36 $-5 - 5\sqrt{3}\,i$

37 12 38 0

39 $2 + i$ 40 $3 + 2i$

41 $-3 + i$ 42 $-4 + 2i$

43 $-5 - 3i$ 44 $-2 - 7i$

45 $4 - 3i$ 46 $1 - 3i$

Exer. 47–56: Express in the form $a + bi$ for real numbers a and b.

47 $4\left(\cos\frac{\pi}{4} + i \sin\frac{\pi}{4}\right)$ 48 $8\left(\cos\frac{7\pi}{4} + i \sin\frac{7\pi}{4}\right)$

49 $6\left(\cos\frac{2\pi}{3} + i \sin\frac{2\pi}{3}\right)$ 50 $12\left(\cos\frac{4\pi}{3} + i \sin\frac{4\pi}{3}\right)$

51 $5(\cos \pi + i \sin \pi)$ 52 $3\left(\cos\frac{3\pi}{2} + i \sin\frac{3\pi}{2}\right)$

53 $\sqrt{34} \operatorname{cis} (\tan^{-1} \frac{3}{5})$ 54 $\sqrt{53} \operatorname{cis} [\tan^{-1} (-\frac{2}{7})]$

55 $\sqrt{5} \operatorname{cis} [\tan^{-1} (-\frac{1}{2})]$ 56 $\sqrt{10} \operatorname{cis} (\tan^{-1} 3)$

Exer. 57–64: Use trigonometric forms to find $z_1 z_2$ and z_1/z_2.

57 $z_1 = -1 + i,\quad z_2 = 1 + i$

58 $z_1 = \sqrt{3} - i,\quad z_2 = -\sqrt{3} - i$

59 $z_1 = -2 - 2\sqrt{3}\,i,\quad z_2 = 5i$

60 $z_1 = -5 + 5i,\quad z_2 = -3i$

61 $z_1 = -10,\quad z_2 = -4$

62 $z_1 = 2i,\quad z_2 = -3i$

63 $z_1 = 4,\quad z_2 = 2 - i$

64 $z_1 = -3,\quad z_2 = 5 + 2i$

65 (a) Extend (i) of the theorem in this section to the case of three complex numbers.

(b) Generalize (i) of the theorem to n complex numbers.

66 Prove (ii) of the theorem in this section.

8.4 DE MOIVRE'S THEOREM AND nTH ROOTS OF COMPLEX NUMBERS

If z is a complex number and n is a positive integer, then a complex number w is an ***n*th root** of z if $w^n = z$. We will show that every nonzero complex number has n different nth roots. Since $\mathbb{R}$ is contained in $\mathbb{C}$, it will also follow that every nonzero real number has n distinct nth (complex) roots. If a is a positive real number and $n = 2$, then we already know that the roots are $\sqrt{a}$ and $-\sqrt{a}$.

If, in the theorem on page 431, we let both z_1 and z_2 equal the complex number $z = r(\cos\theta + i\sin\theta)$, we obtain

$$z^2 = r^2(\cos 2\theta + i\sin 2\theta).$$

Applying the same theorem to z and z^2 gives us

$$z^2 \cdot z = (r^2 \cdot r)\left[\cos(2\theta + \theta) + i\sin(2\theta + \theta)\right]$$

and
$$z^3 = r^3(\cos 3\theta + i\sin 3\theta).$$

Applying the theorem to z^3 and z,

$$z^4 = r^4(\cos 4\theta + i\sin 4\theta).$$

In general, we have the following result, named after the French mathematician Abraham De Moivre (1667–1754).

DE MOIVRE'S THEOREM

$$[r(\cos\theta + i\sin\theta)]^n = r^n(\cos n\theta + i\sin n\theta) \quad \text{for every integer } n.$$

EXAMPLE ▪ 1

Find $(1 + i)^{20}$.

SOLUTION It would be tedious to find $(1 + i)^{20}$ using algebraic methods. Let us therefore introduce a trigonometric form for $1 + i$. Referring to Figure 21, we see that

FIGURE 21

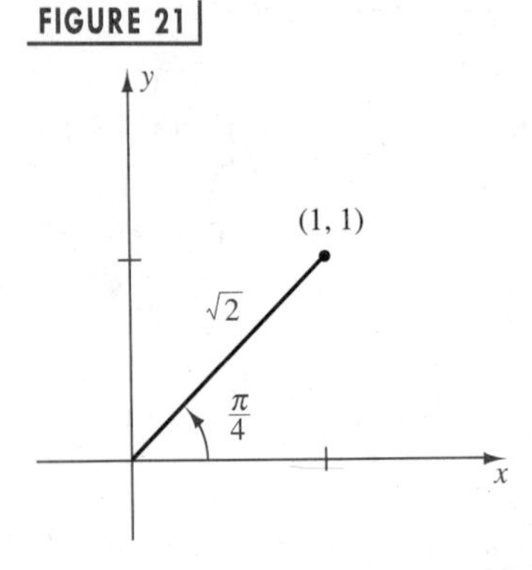

$$1 + i = \sqrt{2}\left(\cos\frac{\pi}{4} + i\sin\frac{\pi}{4}\right).$$

We now apply De Moivre's theorem:

$$\begin{aligned}(1 + i)^{20} &= (2^{1/2})^{20}\left[\cos\left(20 \cdot \frac{\pi}{4}\right) + i\sin\left(20 \cdot \frac{\pi}{4}\right)\right] \\ &= 2^{10}(\cos 5\pi + i\sin 5\pi) = -1024\end{aligned}$$

If a nonzero complex number z has an nth root w, then $w^n = z$. If trigonometric forms for w and z are

$$w = s(\cos\alpha + i\sin\alpha) \quad \text{and} \quad z = r(\cos\theta + i\sin\theta),$$

then, applying De Moivre's Theorem to $w^n = z$,

$$s^n (\cos n\alpha + i \sin n\alpha) = r (\cos \theta + i \sin \theta).$$

If two complex numbers are equal, then so are their absolute values. Consequently, $s^n = r$ and, since s and r are nonnegative, $s = \sqrt[n]{r}$. Substituting s^n for r in the last displayed equation and dividing both sides by s^n, we obtain

$$\cos n\alpha + i \sin n\alpha = \cos \theta + i \sin \theta.$$

Since the arguments of equal complex numbers differ by a multiple of 2π, there is an integer k such that $n\alpha = \theta + 2\pi k$. Dividing both sides of the last equation by n, we see that

$$\alpha = \frac{\theta + 2\pi k}{n} \quad \text{for some integer } k.$$

Substituting in the trigonometric form for w gives us the formula

$$w = \sqrt[n]{r}\left[\cos\left(\frac{\theta + 2\pi k}{n}\right) + i \sin\left(\frac{\theta + 2\pi k}{n}\right)\right].$$

If we substitute $k = 0, 1, \ldots, n - 1$ successively, we obtain n different nth roots of z. No other value of k will produce a new nth root. For example, if $k = n$, we obtain the angle $(\theta + 2\pi n)/n$, or $(\theta/n) + 2\pi$, which gives us the same nth root as $k = 0$. Similarly, $k = n + 1$ yields the same nth root as $k = 1$, and so on. The same is true for negative values of k. We have proved the following.

THEOREM ON *n*TH ROOTS

If $z = r (\cos \theta + i \sin \theta)$ is any nonzero complex number and if n is any positive integer, then z has precisely n distinct nth roots $w_0, w_1, w_2, \ldots, w_{n-1}$. These roots are

$$w_k = \sqrt[n]{r}\left[\cos\left(\frac{\theta + 2\pi k}{n}\right) + i \sin\left(\frac{\theta + 2\pi k}{n}\right)\right]$$

for $k = 0, 1, \ldots, n - 1$.

The nth roots of z all have absolute value $\sqrt[n]{r}$ and hence their geometric representations lie on a circle of radius $\sqrt[n]{r}$ with center at O. Moreover, they are equispaced on this circle, since the difference in the arguments of successive nth roots is $2\pi/n$.

The following corollary of the theorem on nth roots may be used if θ is measured in degrees.

COROLLARY

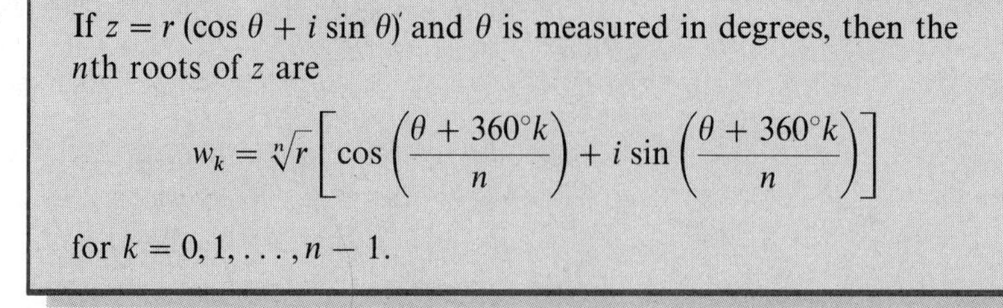

If $z = r(\cos\theta + i\sin\theta)$ and θ is measured in degrees, then the nth roots of z are

$$w_k = \sqrt[n]{r}\left[\cos\left(\frac{\theta + 360^\circ k}{n}\right) + i\sin\left(\frac{\theta + 360^\circ k}{n}\right)\right]$$

for $k = 0, 1, \ldots, n - 1$.

EXAMPLE ▪ 2

Find the four fourth roots of $-8 - 8\sqrt{3}\,i$.

FIGURE 22

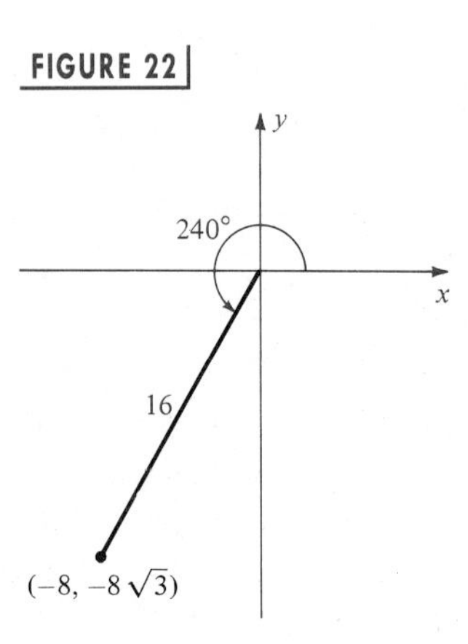

SOLUTION The geometric representation of $-8 - 8\sqrt{3}\,i$ is shown in Figure 22. Introducing trigonometric form,

$$-8 - 8\sqrt{3}\,i = 16(\cos 240^\circ + i\sin 240^\circ).$$

Using the preceding corollary with $n = 4$, and noting that $\sqrt[4]{16} = 2$, the fourth roots are

$$w_k = 2\left[\cos\left(\frac{240^\circ + 360^\circ k}{4}\right) + i\sin\left(\frac{240^\circ + 360^\circ k}{4}\right)\right]$$

for $k = 0, 1, 2, 3$. This formula may be written

$$w_k = 2[\cos(60^\circ + 90^\circ k) + i\sin(60^\circ + 90^\circ k)].$$

Substituting 0, 1, 2, and 3 for k gives us the four fourth roots:

$$w_0 = 2(\cos 60^\circ + i\sin 60^\circ) = 1 + \sqrt{3}\,i$$

$$w_1 = 2(\cos 150^\circ + i\sin 150^\circ) = -\sqrt{3} + i$$

$$w_2 = 2(\cos 240^\circ + i\sin 240^\circ) = -1 - \sqrt{3}\,i$$

$$w_3 = 2(\cos 330^\circ + i\sin 330^\circ) = \sqrt{3} - i$$

EXAMPLE ▪ 3

(a) Find the six sixth roots of -1.

(b) Represent the roots geometrically.

SOLUTION

(a) Writing $-1 = 1(\cos\pi + i\sin\pi)$ and using the theorem on nth roots with $n = 6$, we find that the sixth roots of -1 are given by

$$w_k = \cos\left(\frac{\pi + 2\pi k}{6}\right) + i \sin\left(\frac{\pi + 2\pi k}{6}\right)$$

for $k = 0, 1, 2, 3, 4, 5$. We can also write w_k in the form

$$w_k = \cos\left(\frac{\pi}{6} + \frac{\pi}{3}k\right) + i \sin\left(\frac{\pi}{6} + \frac{\pi}{3}k\right).$$

Substituting $0, 1, 2, 3, 4, 5$ for k, we obtain the six sixth roots of -1:

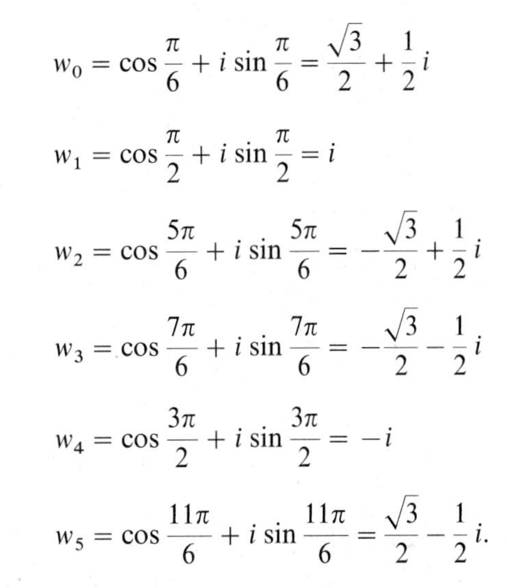

$$w_0 = \cos\frac{\pi}{6} + i \sin\frac{\pi}{6} = \frac{\sqrt{3}}{2} + \frac{1}{2}i$$

$$w_1 = \cos\frac{\pi}{2} + i \sin\frac{\pi}{2} = i$$

$$w_2 = \cos\frac{5\pi}{6} + i \sin\frac{5\pi}{6} = -\frac{\sqrt{3}}{2} + \frac{1}{2}i$$

$$w_3 = \cos\frac{7\pi}{6} + i \sin\frac{7\pi}{6} = -\frac{\sqrt{3}}{2} - \frac{1}{2}i$$

$$w_4 = \cos\frac{3\pi}{2} + i \sin\frac{3\pi}{2} = -i$$

$$w_5 = \cos\frac{11\pi}{6} + i \sin\frac{11\pi}{6} = \frac{\sqrt{3}}{2} - \frac{1}{2}i.$$

FIGURE 23

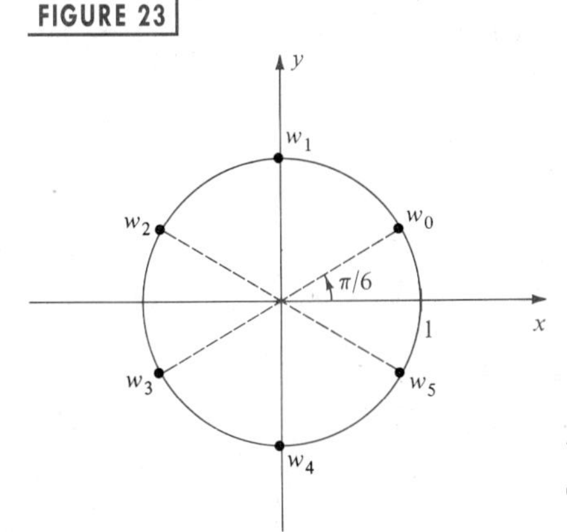

(b) Since $|-1| = 1$, the points that represent the roots of -1 all lie on the unit circle shown in Figure 23. Moreover, they are equispaced on this circle.

The special case in which $z = 1$ is of particular interest. The n distinct nth roots of 1 are called the **nth roots of unity**. The *cube roots of unity* were found algebraically in Example 6 of Section 2.4. In the following example we obtain them using the theorem on nth roots with $n = 3$.

EXAMPLE ▪ 4

Find the three cube roots of unity.

SOLUTION Writing $1 = \cos 0 + i \sin 0$ and using the theorem on nth roots with $n = 3$, we obtain the three roots:

$$w_k = \cos \frac{2\pi k}{3} + i \sin \frac{2\pi k}{3}$$

for $k = 0, 1, 2$. Substituting for k gives us

$$w_0 = \cos 0 + i \sin 0 = 1$$

$$w_1 = \cos \frac{2\pi}{3} + i \sin \frac{2\pi}{3} = -\frac{1}{2} + \frac{\sqrt{3}}{2} i$$

$$w_2 = \cos \frac{4\pi}{3} + i \sin \frac{4\pi}{3} = -\frac{1}{2} - \frac{\sqrt{3}}{2} i.$$

8.4 EXERCISES

Exer. 1–12: Use De Moivre's theorem to express the number in the form $a + bi$ for real numbers a and b.

1 $(3 + 3i)^5$

2 $(1 + i)^{12}$

3 $(1 - i)^{10}$

4 $(-1 + i)^8$

5 $(1 - \sqrt{3}\,i)^3$

6 $(1 - \sqrt{3}\,i)^5$

7 $\left(-\frac{\sqrt{2}}{2} + \frac{\sqrt{2}}{2} i\right)^{15}$

8 $\left(\frac{\sqrt{2}}{2} + \frac{\sqrt{2}}{2} i\right)^{25}$

9 $\left(-\frac{\sqrt{3}}{2} - \frac{1}{2} i\right)^{20}$

10 $\left(-\frac{\sqrt{3}}{2} - \frac{1}{2} i\right)^{50}$

11 $(\sqrt{3} + i)^7$

12 $(-2 - 2i)^{10}$

13 Find the two square roots of $1 + \sqrt{3}\,i$.

14 Find the two square roots of $-9i$.

15 Find the four fourth roots of $-1 - \sqrt{3}\,i$.

16 Find the four fourth roots of $-8 + 8\sqrt{3}\,i$.

17 Find the three cube roots of $-27i$.

18 Find the three cube roots of $64i$.

Exer. 19–22: Find the indicated roots, and represent them geometrically.

19 The six sixth roots of unity

20 The eight eighth roots of unity

21 The five fifth roots of $1 + i$

22 The five fifth roots of $-\sqrt{3} - i$

Exer. 23–30: Find the solutions of the equation.

23 $x^4 - 16 = 0$

24 $x^6 - 64 = 0$

25 $x^6 + 64 = 0$

26 $x^5 + 1 = 0$

27 $x^3 + 8i = 0$

28 $x^3 - 64i = 0$

29 $x^5 - 243 = 0$

30 $x^4 + 81 = 0$

8.5 VECTORS

Some quantities in mathematics and the sciences, such as area, volume, distance, temperature, and time, have magnitude only and can be completely characterized by a single real number (with an appropriate unit of measurement such as in.^{2}, ft^3, cm, deg, or sec). A quantity of this type is a **scalar quantity**, and the corresponding real number is a **scalar**. A concept such as velocity or force has both magnitude and direction and is often represented by a **directed line segment**, that is, a line segment to

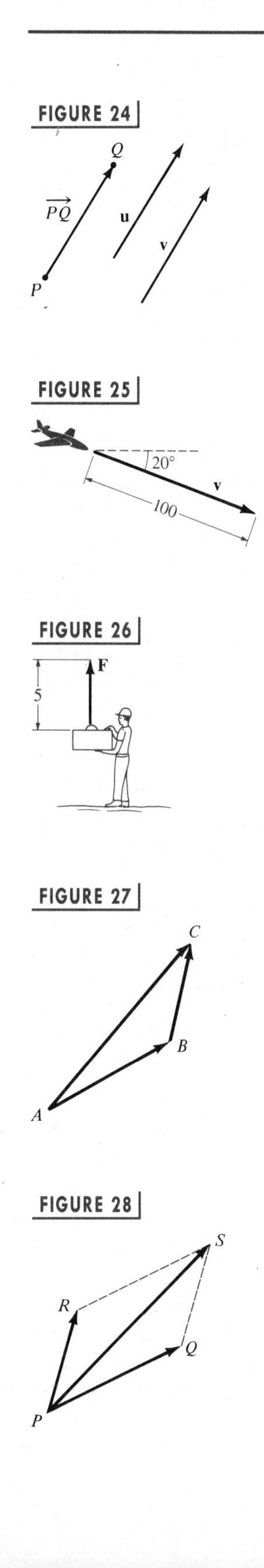

FIGURE 24
FIGURE 25
FIGURE 26
FIGURE 27
FIGURE 28

which a direction has been assigned. Another name for a directed line segment is a **vector**.

If a vector extends from a point P (the **initial point**) to a point Q (the **terminal point**), we indicate the direction by placing an arrowhead at Q on the line segment PQ, and denote the vector by $\overrightarrow{PQ}$ (see Figure 24). The **magnitude** of $\overrightarrow{PQ}$ is the length of PQ and is denoted by $\|\overrightarrow{PQ}\|$. As in Figure 24, we use boldface letters such as **u** or **v** to denote vectors whose endpoints are not specified. For written work, a notation such as $\vec{u}$ or $\vec{v}$ may be employed.

Vectors that have the same magnitude and direction are **equivalent**. Since we want a vector to be determined only by its magnitude and direction, and not by its location, we shall regard equivalent vectors, such as those in Figure 24, as **equal**, and write

$$\mathbf{u} = \overrightarrow{PQ}, \quad \mathbf{v} = \overrightarrow{PQ}, \quad \text{and} \quad \mathbf{u} = \mathbf{v}.$$

Thus, *vectors may be translated from one location to another, provided neither the magnitude nor direction is changed.*

Many physical concepts may be represented by vectors. To illustrate, suppose an airplane is descending at a constant speed of 100 mi/hr and the line of flight makes an angle of 20° with the horizontal. Both of these facts are represented by the vector **v** of magnitude 100 in Figure 25. The vector **v** is a **velocity vector**.

As a second illustration, suppose a person lifts a 5-pound weight directly upward. We may indicate this by the vector **F** of magnitude 5 in Figure 26. A vector that represents a pull or push of some type is a **force vector**.

A vector $\overrightarrow{AB}$ may be used to represent the path of a point as it moves along the line segment from A to B. We then refer to $\overrightarrow{AB}$ as a **displacement** of the point. As in Figure 27, a displacement $\overrightarrow{AB}$ followed by a displacement $\overrightarrow{BC}$ leads to the same point as the single displacement $\overrightarrow{AC}$. By definition, the vector $\overrightarrow{AC}$ is the **sum** of $\overrightarrow{AB}$ and $\overrightarrow{BC}$, and we write

$$\overrightarrow{AC} = \overrightarrow{AB} + \overrightarrow{BC}.$$

Since vectors may be translated from one location to another, *any* two vectors may be added by placing the initial point of one on the terminal point of the other and then proceeding as in Figure 27.

Another way to find the sum is to consider vectors that are equal to $\overrightarrow{AB}$ and $\overrightarrow{BC}$ and have the same initial point, as illustrated by $\overrightarrow{PQ}$ and $\overrightarrow{PR}$ in Figure 28. If we construct the parallogram $RPQS$, then since $\overrightarrow{PR} = \overrightarrow{QS}$, it follows that $\overrightarrow{PS} = \overrightarrow{PQ} + \overrightarrow{PR}$. If $\overrightarrow{PQ}$ and $\overrightarrow{PR}$ are two forces acting at P,

FIGURE 29

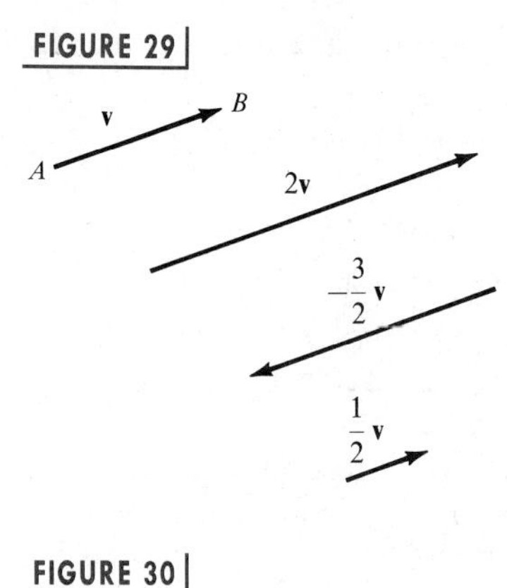

then $\overrightarrow{PS}$ is the **resultant force**, that is, the single force that produces the same effect as the two combined forces.

If c is a real number and **v** is a vector, then $c\mathbf{v}$ is defined as a vector whose magnitude is $|c|$ times the magnitude $\|\mathbf{v}\|$ of **v**, and whose direction is the same as **v** if $c > 0$, or opposite that of **v** if $c < 0$. Illustrations are given in Figure 29. We refer to $c\mathbf{v}$ as a **scalar multiple** of **v**.

FIGURE 30

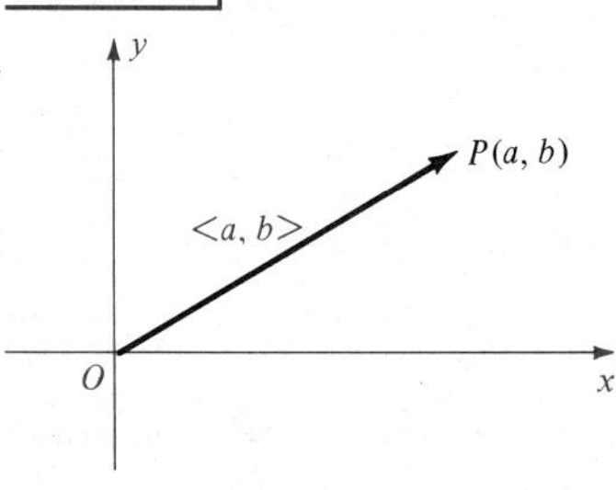

Let us next introduce a coordinate plane and assume that all vectors under discussion are in that plane. Since we may change the position of a vector, provided that the magnitude and direction are not altered, let us place the initial point at the origin. The terminal point of a typical vector $\overrightarrow{OP}$ has rectangular coordinates (a, b), as shown in Figure 30. Conversely, every ordered pair (a, b), determines the vector $\overrightarrow{OP}$, where P has rectangular coordinates (a, b). This gives us a one-to-one correspondence between vectors and ordered pairs and allows us to regard a vector as an ordered pair of real numbers instead of as a directed line segment.

To avoid confusion with the notation for open intervals or points, we use the symbol $\langle a, b\rangle$ for an ordered pair that represents a vector, and we refer to $\langle a, b\rangle$ as a vector and denote it by a boldface letter. The numbers a and b are the **components** of the vector $\langle a, b\rangle$. The magnitude of $\langle a, b\rangle$ is, by definition, the distance from the origin to the point $P(a, b)$. This may also be stated as follows.

DEFINITION

The **magnitude** $\|\mathbf{v}\|$ of the vector $\mathbf{v} = \langle a, b\rangle$ is

$$\|\mathbf{v}\| = \sqrt{a^2 + b^2}.$$

EXAMPLE ▪ 1

Sketch the vector corresponding to the ordered pair, and find the magnitude of each vector.

(a) $\mathbf{a} = \langle -3, -3\rangle$ (b) $\mathbf{b} = \langle 0, -2\rangle$ (c) $\mathbf{c} = \langle \frac{4}{5}, \frac{3}{5}\rangle$

FIGURE 31

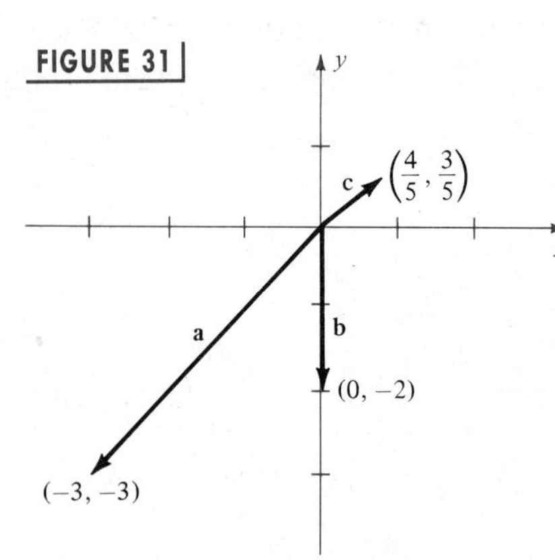

SOLUTION The vectors are sketched in Figure 31. We apply the definition of magnitude:

(a) $$\|\mathbf{a}\| = \sqrt{9 + 9} = 3\sqrt{2}$$

(b) $$\|\mathbf{b}\| = \sqrt{0 + 4} = 2$$

(c) $$\|\mathbf{c}\| = \sqrt{\tfrac{16}{25} + \tfrac{9}{25}} = 1$$

FIGURE 32

$\langle a,b\rangle + \langle c,d\rangle = \langle a+c, b+d\rangle$

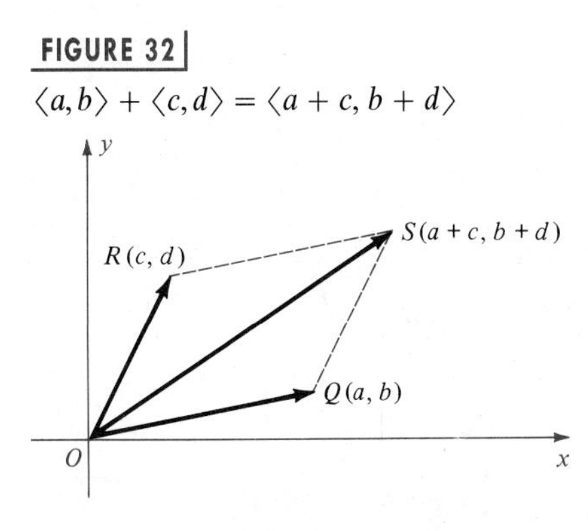

Consider vectors $\overrightarrow{OQ}$ and $\overrightarrow{OR}$ corresponding to $\langle a,b\rangle$ and $\langle c,d\rangle$, respectively, as shown in Figure 32. If we let $\overrightarrow{OS}$ be the vector corresponding to $\langle a+c, b+d\rangle$, we can show that O, Q, S, and R are vertices of a parallelogram, and hence

$$\overrightarrow{OQ} + \overrightarrow{OR} = \overrightarrow{OS}.$$

Expressing this fact in terms of ordered pairs gives us the following rule for addition of vectors.

ADDITION OF VECTORS

$$\langle a,b\rangle + \langle c,d\rangle = \langle a+c, b+d\rangle$$

The rule for a scalar multiple $k\overrightarrow{OP}$ of a vector is as follows.

SCALAR MULTIPLE OF A VECTOR

$$k\langle a,b\rangle = \langle ka, kb\rangle$$

EXAMPLE ▪ 2

If $\mathbf{a} = \langle 2,1\rangle$, find $3\mathbf{a}$ and $-2\mathbf{a}$, and represent all three vectors geometrically in a coordinate plane.

SOLUTION If $\mathbf{a} = \langle 2,1\rangle$, then, by the preceding rule for scalar multiples, $3\mathbf{a} = \langle 6,3\rangle$ and $-2\mathbf{a} = \langle -4,-2\rangle$. The geometric representations are shown in Figure 33.

FIGURE 33

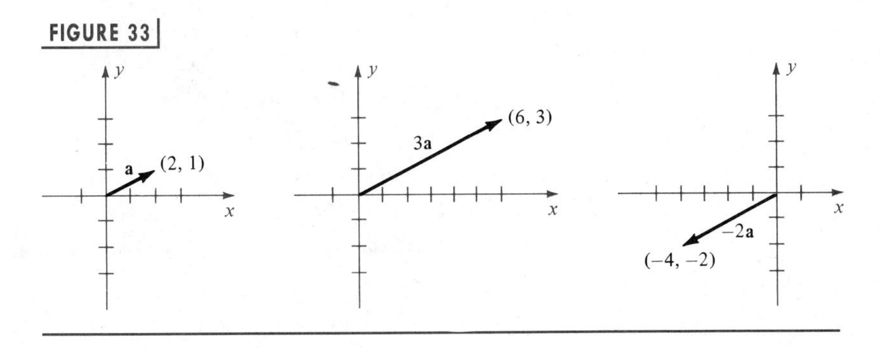

EXAMPLE ▪ 3

If $\mathbf{a} = \langle 3, -2 \rangle$ and $\mathbf{b} = \langle -6, 7 \rangle$, find $\mathbf{a} + \mathbf{b}$, $4\mathbf{a}$, and $2\mathbf{a} + 3\mathbf{b}$.

SOLUTION We use the rules for addition and scalar multiples of vectors:

$$\mathbf{a} + \mathbf{b} = \langle 3, -2 \rangle + \langle -6, 7 \rangle = \langle -3, 5 \rangle$$
$$4\mathbf{a} = 4\langle 3, -2 \rangle = \langle 12, -8 \rangle$$
$$2\mathbf{a} + 3\mathbf{b} = \langle 6, -4 \rangle + \langle -18, 21 \rangle = \langle -12, 17 \rangle$$

By definition, the **zero vector 0** corresponds to $\langle 0, 0 \rangle$. If $\mathbf{a} = \langle a, b \rangle$, we define $-\mathbf{a} = \langle -a, -b \rangle$. Using these definitions, we may establish the following properties for arbitrary vectors **a**, **b**, and **c**.

PROPERTIES OF ADDITION

$$\mathbf{a} + \mathbf{b} = \mathbf{b} + \mathbf{a}$$
$$\mathbf{a} + (\mathbf{b} + \mathbf{c}) = (\mathbf{a} + \mathbf{b}) + \mathbf{c}$$
$$\mathbf{a} + \mathbf{0} = \mathbf{a}$$
$$\mathbf{a} + (-\mathbf{a}) = \mathbf{0}$$

PROOF The proof of each property follows readily from the rule for addition of vectors and properties of real numbers. For example, if $\mathbf{a} = \langle a_1, a_2 \rangle$ and $\mathbf{b} = \langle b_1, b_2 \rangle$, then since $a_1 + b_1 = b_1 + a_1$ and $a_2 + b_2 = b_2 + a_2$,

$$\begin{aligned} \mathbf{a} + \mathbf{b} &= \langle a_1 + b_1, a_2 + b_2 \rangle \\ &= \langle b_1 + a_1, b_2 + a_2 \rangle \\ &= \mathbf{b} + \mathbf{a}. \end{aligned}$$

The remaining properties may be proved in similar fashion. ❑

Subtraction of vectors (denoted by $-$) is defined by $\mathbf{a} - \mathbf{b} = \mathbf{a} + (-\mathbf{b})$. If we use the ordered pair notation for **a** and **b**, then $-\mathbf{b} = \langle -b_1, -b_2 \rangle$, and we obtain the following.

SUBTRACTION

$$\mathbf{a} - \mathbf{b} = \langle a_1, a_2 \rangle - \langle b_1, b_2 \rangle = \langle a_1 - b_1, a_2 - b_2 \rangle$$

Thus, to find $\mathbf{a} - \mathbf{b}$, we merely subtract the components of **b** from the corresponding components of **a**.

EXAMPLE ▪ 4

If $\mathbf{a} = \langle 2, -4 \rangle$ and $\mathbf{b} = \langle 6, 7 \rangle$, find $3\mathbf{a} - 5\mathbf{b}$.

SOLUTION

$$\begin{aligned} 3\mathbf{a} - 5\mathbf{b} &= 3\langle 2, -4 \rangle - 5\langle 6, 7 \rangle \\ &= \langle 6, -12 \rangle - \langle 30, 35 \rangle = \langle -24, -47 \rangle \end{aligned}$$

PROPERTIES OF SCALAR MULTIPLES OF VECTORS

$$c(\mathbf{a} + \mathbf{b}) = c\mathbf{a} + c\mathbf{b}$$
$$(c + d)\mathbf{a} = c\mathbf{a} + d\mathbf{a}$$
$$(cd)\mathbf{a} = c(d\mathbf{a}) = d(c\mathbf{a})$$
$$1\mathbf{a} = \mathbf{a}$$
$$0\mathbf{a} = \mathbf{0} = c\mathbf{0}$$

PROOF We shall prove the first property. The proofs of the other properties are similar. Letting $\mathbf{a} = \langle a_1, a_2 \rangle$ and $\mathbf{b} = \langle b_1, b_2 \rangle$,

$$\begin{aligned} c(\mathbf{a} + \mathbf{b}) &= c\langle a_1 + b_1, a_2 + b_2 \rangle \\ &= \langle ca_1 + cb_1, ca_2 + cb_2 \rangle \\ &= \langle ca_1, ca_2 \rangle + \langle cb_1, cb_2 \rangle \\ &= c\langle a_1, a_2 \rangle + c\langle b_1, b_2 \rangle \\ &= c\mathbf{a} + c\mathbf{b}. \quad \square \end{aligned}$$

The special vectors **i** and **j** are defined as follows.

DEFINITION OF i AND j

$$\mathbf{i} = \langle 1, 0 \rangle, \qquad \mathbf{j} = \langle 0, 1 \rangle$$

A **unit vector** is a vector of magnitude 1. The vectors **i** and **j** are unit vectors, as is the vector $\mathbf{c} = \langle \frac{4}{5}, \frac{3}{5} \rangle$ in Example 1(c).

The vectors **i** and **j** can be used to obtain an alternative way of denoting vectors. Specifically, if $\mathbf{a} = \langle a_1, a_2 \rangle$, then

$$\mathbf{a} = \langle a_1, 0 \rangle + \langle 0, a_2 \rangle = a_1\langle 1, 0 \rangle + a_2\langle 0, 1 \rangle.$$

This gives us the following.

i, j FORM FOR VECTORS

$$\mathbf{a} = \langle a_1, a_2 \rangle = a_1\mathbf{i} + a_2\mathbf{j}$$

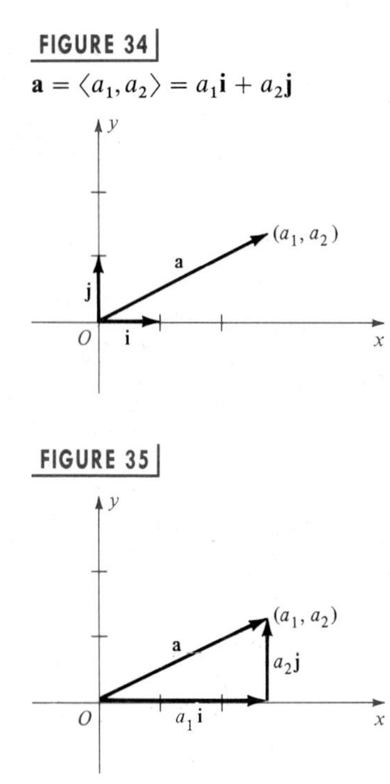

FIGURE 34 $\mathbf{a} = \langle a_1, a_2 \rangle = a_1\mathbf{i} + a_2\mathbf{j}$

FIGURE 35

Vectors corresponding to **i**, **j**, and **a** are illustrated in Figure 34. Since **i** and **j** are unit vectors, $a_1\mathbf{i}$ and $a_2\mathbf{j}$ may be represented by horizontal and vertical vectors of magnitudes $\|a_1\|$ and $\|a_2\|$, respectively, as illustrated in Figure 35. For this reason we call a_1 the **horizontal component** and a_2 the **vertical component** of the vector **a**.

The vector sum $a_1\mathbf{i} + a_2\mathbf{j}$ is a **linear combination** of **i** and **j**. Rules for addition, subtraction, and multiplication by a scalar may be written as follows, with $\mathbf{b} = \langle b_1, b_2 \rangle = b_1\mathbf{i} + b_2\mathbf{j}$:

$$(a_1\mathbf{i} + a_2\mathbf{j}) + (b_1\mathbf{i} + b_2\mathbf{j}) = (a_1 + b_1)\mathbf{i} + (a_2 + b_2)\mathbf{j}$$

$$(a_1\mathbf{i} + a_2\mathbf{j}) - (b_1\mathbf{i} + b_2\mathbf{j}) = (a_1 - b_1)\mathbf{i} + (a_2 - b_2)\mathbf{j}$$

$$c(a_1\mathbf{i} + a_2\mathbf{j}) = (ca_1)\mathbf{i} + (ca_2)\mathbf{j}.$$

These formulas show that we may regard linear combinations of **i** and **j** as algebraic sums.

EXAMPLE ▪ 5

If $\mathbf{a} = 5\mathbf{i} + \mathbf{j}$ and $\mathbf{b} = 4\mathbf{i} - 7\mathbf{j}$, express $3\mathbf{a} - 2\mathbf{b}$ as a linear combination of **i** and **j**.

SOLUTION

$$\begin{aligned} 3\mathbf{a} - 2\mathbf{b} &= 3(5\mathbf{i} + \mathbf{j}) - 2(4\mathbf{i} - 7\mathbf{j}) \\ &= (15\mathbf{i} + 3\mathbf{j}) - (8\mathbf{i} - 14\mathbf{j}) \\ &= 7\mathbf{i} + 17\mathbf{j} \end{aligned}$$

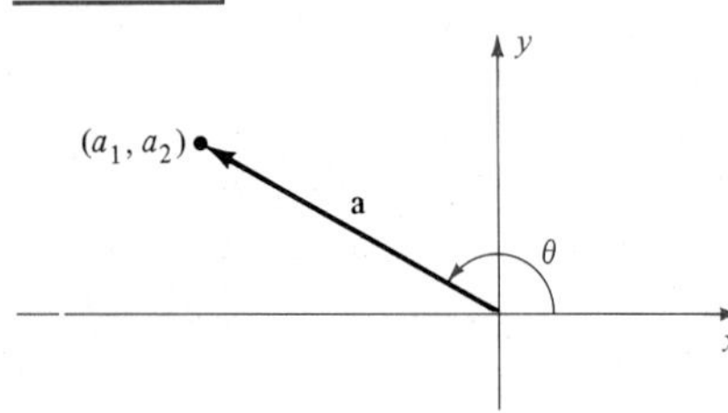

FIGURE 36

Let θ be an angle in standard position, measured from the positive x-axis to the vector $\mathbf{a} = \langle a_1, a_2 \rangle = a_1\mathbf{i} + a_2\mathbf{j}$, as illustrated in Figure 36. Since

$$\cos\theta = \frac{a_1}{\|\mathbf{a}\|} \quad \text{and} \quad \sin\theta = \frac{a_2}{\|\mathbf{a}\|},$$

we obtain the following formulas.

HORIZONTAL AND VERTICAL COMPONENTS OF $\mathbf{a} = \langle a_1, a_2 \rangle$

If $\mathbf{a} = \langle a_1, a_2 \rangle = a_1\mathbf{i} + a_2\mathbf{j}$, then

$$a_1 = \|\mathbf{a}\| \cos\theta, \qquad a_2 = \|\mathbf{a}\| \sin\theta.$$

Using the preceding formulas for a_1 and a_2,

$$\begin{aligned}\mathbf{a} = \langle a_1, a_2\rangle &= \langle \|\mathbf{a}\| \cos\theta, \|\mathbf{a}\| \sin\theta\rangle \\ &= \|\mathbf{a}\| \cos\theta\mathbf{i} + \|\mathbf{a}\| \sin\theta\mathbf{j} \\ &= \|\mathbf{a}\|(\cos\theta\mathbf{i} + \sin\theta\mathbf{j}).\end{aligned}$$

This is analogous to the trigonometric form of a complex number.

EXAMPLE ▪ 6

If the wind is blowing at 12 mi/hr in the direction N40°W, express its velocity as a vector $\mathbf{v}$.

FIGURE 37

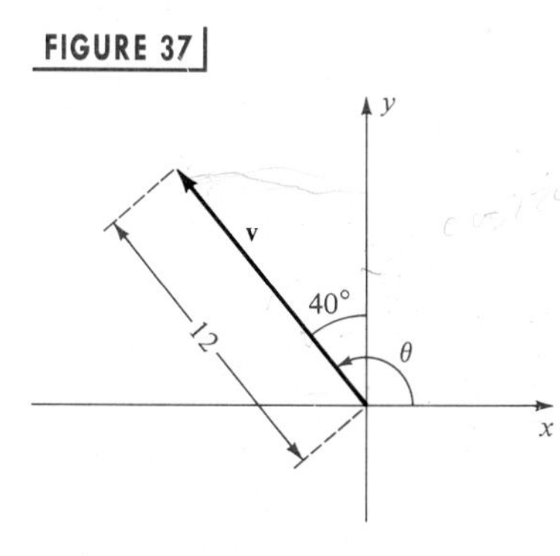

SOLUTION The vector $\mathbf{v}$ is illustrated in Figure 37, where $\theta = 90° + 40°$, or $\theta = 130°$. Using the formulas for horizontal and vertical components with $\mathbf{v} = \langle v_1, v_2\rangle$ gives us

$$v_1 = \|\mathbf{v}\| \cos\theta = 12\cos 130°, \qquad v_2 = \|\mathbf{v}\| \sin\theta = 12\sin 130°.$$

Hence $$\mathbf{v} = (12\cos 130°)\mathbf{i} + (12\sin 130°)\mathbf{j}.$$

If we desire an approximation,

$$\mathbf{v} \approx (-7.7)\mathbf{i} + (9.2)\mathbf{j}.$$

EXAMPLE ▪ 7

Two forces $\overrightarrow{PQ}$ and $\overrightarrow{PR}$ of magnitudes 5.0 kg and 8.0 kg, respectively, act at a point P. The direction of $\overrightarrow{PQ}$ is N20°E and the direction of $\overrightarrow{PR}$ is N65°E. Approximate the magnitude and direction of the resultant vector $\overrightarrow{PS}$.

FIGURE 38

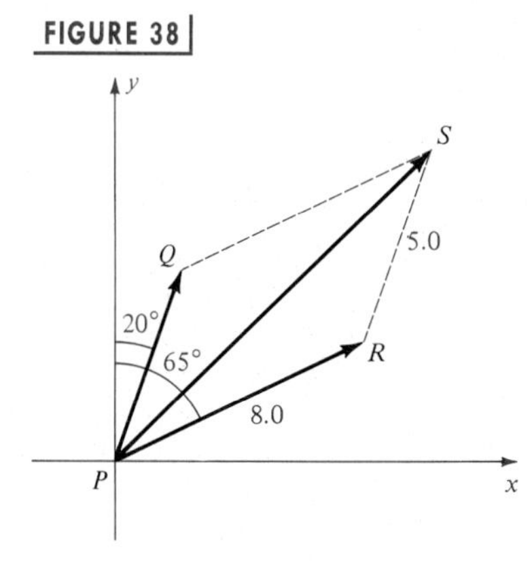

SOLUTION The forces are represented geometrically in Figure 38. Note that the angles from the x-axis to $\overrightarrow{PQ}$ and $\overrightarrow{PR}$ have measures 70° and 25°, respectively. We may use the formulas for horizontal and vertical components:

$$\begin{aligned}\overrightarrow{PQ} &= (5\cos 70°)\mathbf{i} + (5\sin 70°)\mathbf{j}, \\ \overrightarrow{PR} &= (8\cos 25°)\mathbf{i} + (8\sin 25°)\mathbf{j}.\end{aligned}$$

Since $\overrightarrow{PS} = \overrightarrow{PQ} + \overrightarrow{PR}$,

$$\overrightarrow{PS} = (5\cos 70° + 8\cos 25°)\mathbf{i} + (5\sin 70° + 8\sin 25°)\mathbf{j}.$$

$$\overrightarrow{PS} \approx (8.9606)\mathbf{i} + (8.0794)\mathbf{j} \approx (9.0)\mathbf{i} + (8.1)\mathbf{j}.$$

Consequently,

$$\|\overrightarrow{PS}\| \approx \sqrt{(9.0)^2 + (8.1)^2} \approx 12.1.$$

We can also find $\|\overrightarrow{PS}\|$ by using the law of cosines (see Example 3 of Section 8.2). Since $\angle QPR = 45°$, it follows that $\angle PRS = 135°$ and hence

$$\begin{aligned}\|\overrightarrow{PS}\|^2 &= (8.0)^2 + (5.0)^2 - 2(8.0)(5.0)\cos 135° \\ &= 64 + 25 - 80\left(-\frac{\sqrt{2}}{2}\right) \\ &= 89 + 40\sqrt{2} \approx 145.6\end{aligned}$$

Thus $$\|\overrightarrow{PS}\| \approx \sqrt{145.6} \approx 12.1.$$

If θ is the angle from the positive x-axis to the resultant $\overrightarrow{PS}$, then using the (approximate) coordinates (8.9606, 8.0794) of S, we obtain

$$\tan\theta \approx \frac{8.0794}{8.9606} \approx 0.9017$$

and $\theta \approx 42°$. Hence the direction of $\overrightarrow{PS}$ is approximately N48°E.

8.5 EXERCISES

Exer. 1–10: Find a + b, a − b, 4a + 5b, and 4a − 5b.

1 $\mathbf{a} = \langle 2, -3\rangle$, $\mathbf{b} = \langle 1, 4\rangle$

2 $\mathbf{a} = \langle -2, 6\rangle$, $\mathbf{b} = \langle 2, 3\rangle$

3 $\mathbf{a} = -\langle 7, -2\rangle$, $\mathbf{b} = 4\langle -2, 1\rangle$

4 $\mathbf{a} = 2\langle 5, -4\rangle$, $\mathbf{b} = -\langle 6, 0\rangle$

5 $\mathbf{a} = \mathbf{i} + 2\mathbf{j}$, $\mathbf{b} = 3\mathbf{i} - 5\mathbf{j}$

6 $\mathbf{a} = -3\mathbf{i} + \mathbf{j}$, $\mathbf{b} = -3\mathbf{i} + \mathbf{j}$

7 $\mathbf{a} = -(4\mathbf{i} - \mathbf{j})$, $\mathbf{b} = 2(\mathbf{i} - 3\mathbf{j})$

8 $\mathbf{a} = 8\mathbf{j}$, $\mathbf{b} = (-3)(-2\mathbf{i} + \mathbf{j})$

9 $\mathbf{a} = 2\mathbf{j}$, $\mathbf{b} = -3\mathbf{i}$

10 $\mathbf{a} = \mathbf{0}$, $\mathbf{b} = \mathbf{i} + \mathbf{j}$

Exer. 11–14: Sketch vectors corresponding to a, b, a + b, 2a, and −3b.

11 $\mathbf{a} = 3\mathbf{i} + 2\mathbf{j}$, $\mathbf{b} = -\mathbf{i} + 5\mathbf{j}$

12 $\mathbf{a} = -5\mathbf{i} + 2\mathbf{j}$, $\mathbf{b} = \mathbf{i} - 3\mathbf{j}$

13 $\mathbf{a} = \langle -4, 6\rangle$, $\mathbf{b} = \langle -2, 3\rangle$

14 $\mathbf{a} = \langle 2, 0\rangle$ $\mathbf{b} = \langle -2, 0\rangle$

Exer. 15–20: Use components to express the sum or difference as a scalar multiple of one of the vectors a, b, c, d, e, or f shown in the figure.

15 $\mathbf{a} + \mathbf{b}$

16 $\mathbf{c} - \mathbf{d}$

17 $\mathbf{b} + \mathbf{e}$

18 $\mathbf{f} - \mathbf{b}$

19 $\mathbf{b} + \mathbf{d}$

20 $\mathbf{e} + \mathbf{c}$

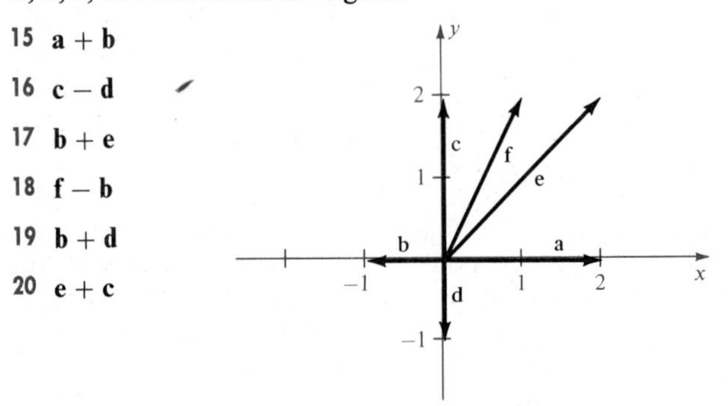

Exer. 21–30: If $\mathbf{a} = \langle a_1, a_2 \rangle$, $\mathbf{b} = \langle b_1, b_2 \rangle$, $\mathbf{c} = \langle c_1, c_2 \rangle$, and m and n are real numbers, prove the property.

21 $\mathbf{a} + (\mathbf{b} + \mathbf{c}) = (\mathbf{a} + \mathbf{b}) + \mathbf{c}$

22 $\mathbf{a} + \mathbf{0} = \mathbf{a}$

23 $\mathbf{a} + (-\mathbf{a}) = \mathbf{0}$

24 $(m + n)\mathbf{a} = m\mathbf{a} + n\mathbf{a}$

25 $(mn)\mathbf{a} = m(n\mathbf{a}) = n(m\mathbf{a})$

26 $1\mathbf{a} = \mathbf{a}$

27 $0\mathbf{a} = \mathbf{0} = m\mathbf{0}$

28 $(-m)\mathbf{a} = -m\mathbf{a}$

29 $-(\mathbf{a} + \mathbf{b}) = -\mathbf{a} - \mathbf{b}$

30 $m(\mathbf{a} - \mathbf{b}) = m\mathbf{a} - m\mathbf{b}$

31 If $\mathbf{v} = \langle a, b \rangle$, prove each of the following:

(a) The magnitude of $2\mathbf{v}$ is twice the magnitude of $\mathbf{v}$.

(b) The magnitude of $\frac{1}{2}\mathbf{v}$ is one-half the magnitude of $\mathbf{v}$.

(c) The magnitude of $-2\mathbf{v}$ is twice the magnitude of $\mathbf{v}$.

(d) If k is any real number, then the magnitude of $k\mathbf{v}$ is $|k|$ times the magnitude of $\mathbf{v}$.

32 If $\mathbf{v} = \langle a, b \rangle$ and $\mathbf{w} = \langle c, d \rangle$, give a geometric interpretation for $\mathbf{v} - \mathbf{w}$.

Exer. 33–40: Find the magnitude of a and the smallest positive angle θ from the positive x-axis to the vector $\overrightarrow{OP}$ that corresponds to a.

33 $\mathbf{a} = \langle 3, -3 \rangle$

34 $\mathbf{a} = \langle -2, -2\sqrt{3} \rangle$

35 $\mathbf{a} = \langle -5, 0 \rangle$

36 $\mathbf{a} = \langle 0, 10 \rangle$

37 $\mathbf{a} = -4\mathbf{i} + 5\mathbf{j}$

38 $\mathbf{a} = 10\mathbf{i} - 10\mathbf{j}$

39 $\mathbf{a} = -18\mathbf{j}$

40 $\mathbf{a} = 2\mathbf{i} - 3\mathbf{j}$

Exer. 41–44: The vectors a and b represent two forces acting at the same point, and θ is the smallest positive angle between a and b. Approximate the magnitude of the resultant force.

41 $\mathbf{a} = 40$ lb, $\mathbf{b} = 70$ lb, $\theta = 45^\circ$

42 $\mathbf{a} = 2.0$ lb, $\mathbf{b} = 8.0$ lb, $\theta = 120^\circ$

43 $\mathbf{a} = 5.5$ kg, $\mathbf{b} = 6.2$ kg, $\theta = 60^\circ$

44 $\mathbf{a} = 30$ kg, $\mathbf{b} = 50$ kg, $\theta = 150^\circ$

Exer. 45–48: The magnitudes and directions of two forces acting at a point P are given in (a) and (b). Approximate the magnitude and direction of the resultant.

45 (a) 90 kg, N75°W (b) 60 kg, S5°E

46 (a) 20 kg, S17°W (b) 50 kg, N82°W

47 (a) 6.0 lb, 110° (b) 2.0 lb, 215°

48 (a) 70 lb, 320° (b) 40 lb, 30°

Exer. 49–52: Approximate the horizontal and vertical components of the vector that is described.

49 A quarterback releases the football with a velocity of 50 ft/sec at an angle of 35° with the horizontal.

50 A girl pulls a sled through the snow by exerting a force of 20 pounds at an angle of 40° with the horizontal.

51 The biceps muscle, in supporting the forearm and a weight held in the hand, exerts a force of 200 pounds. As shown in the figure, the muscle makes an angle of 108° with the forearm.

EXERCISE 51

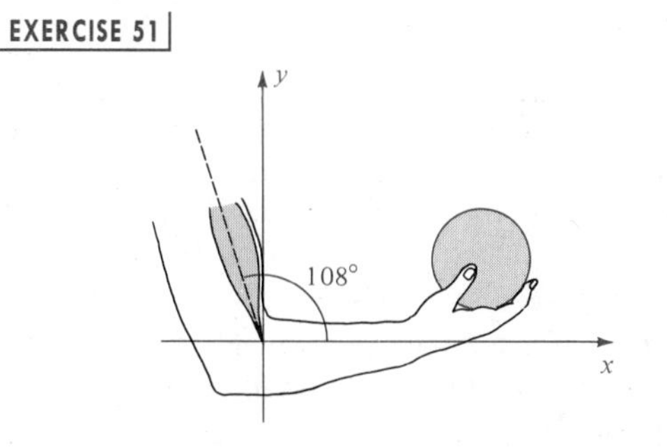

52 A jet airplane approaches a runway at an angle of 7.5° with the horizontal, traveling at a velocity of 160 mi/hr.

Exer. 53–56: When forces $\mathbf{F}_1, \mathbf{F}_2, \ldots, \mathbf{F}_n$ act at a point P, the net (or resultant) force F is the sum $\mathbf{F}_1 + \mathbf{F}_2 + \cdots + \mathbf{F}_n$. If $\mathbf{F} = \mathbf{0}$, the system of forces is *in equilibrium*. The given forces act at the origin O of an xy-plane.

(a) Find the net force F.

(b) Find an additional force G so that equilibrium occurs.

53 $\mathbf{F}_1 = \langle 4, 3 \rangle$, $\mathbf{F}_2 = \langle -2, -3 \rangle$, $\mathbf{F}_3 = \langle 5, 2 \rangle$

54 $\mathbf{F}_1 = \langle -3, -1 \rangle$, $\mathbf{F}_2 = \langle 0, -3 \rangle$, $\mathbf{F}_3 = \langle 3, 4 \rangle$

55 56

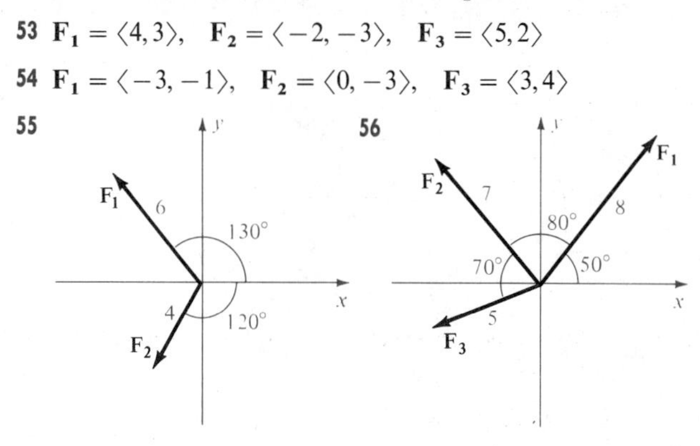

57 Two tugboats are towing a large ship into port, as shown in the figure. The larger tug exerts a force of 4000 pounds on its cable and the smaller tug exerts a force of 3200 pounds on its cable. If the ship is to travel along the line l, approximate the angle θ.

EXERCISE 57

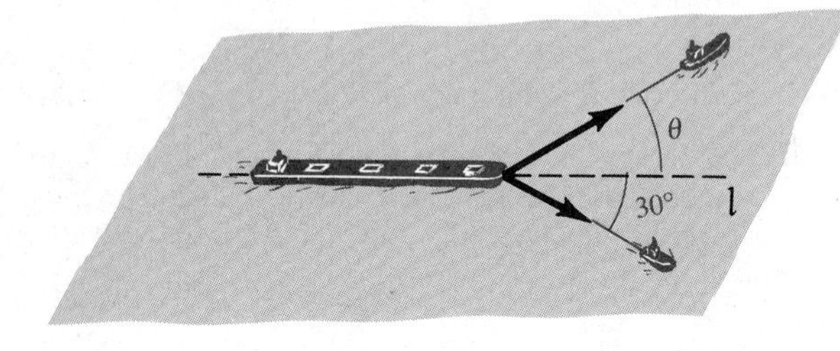

58 Shown in the figure is an apparatus used to simulate gravity conditions on other planets. A rope is attached to an astronaut who maneuvers on an inclined plane that makes an angle of θ degrees with the horizontal.

(a) If the astronaut weighs 160 pounds, find the x- and y-components of this downward force with respect to the axes in the figure.

(b) The y-component in part (a) is the weight of the astronaut relative to the inclined plane. The astronaut would weigh 27 pounds on the moon and 60 pounds on Mars. Approximate the angles θ, to the nearest 0.01°, so that the inclined-plane apparatus will simulate walking on these surfaces.

EXERCISE 58

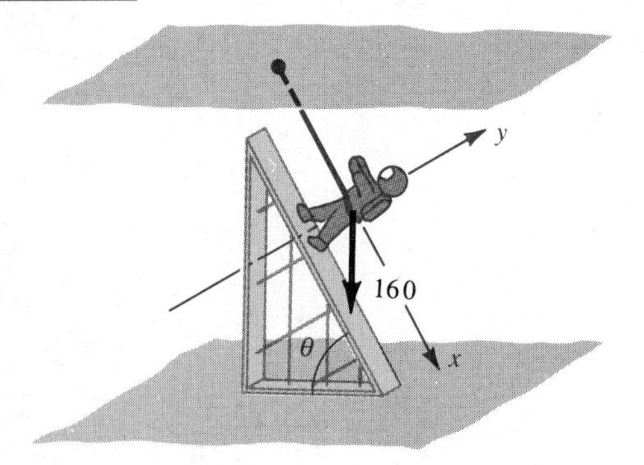

59 An airplane with an airspeed of 200 mi/hr is flying in the direction 50°, and a 40 mi/hr wind is blowing directly from the west. As shown in the figure, these facts may be represented by vectors **p** and **w** of magnitudes 200 and 40, respectively. The direction of the resultant $\mathbf{p} + \mathbf{w}$ gives the *true course* of the airplane relative to the ground, and the magnitude $\|\mathbf{p} + \mathbf{w}\|$ is the *ground speed* of the airplane. Approximate the true course and ground speed.

EXERCISE 59

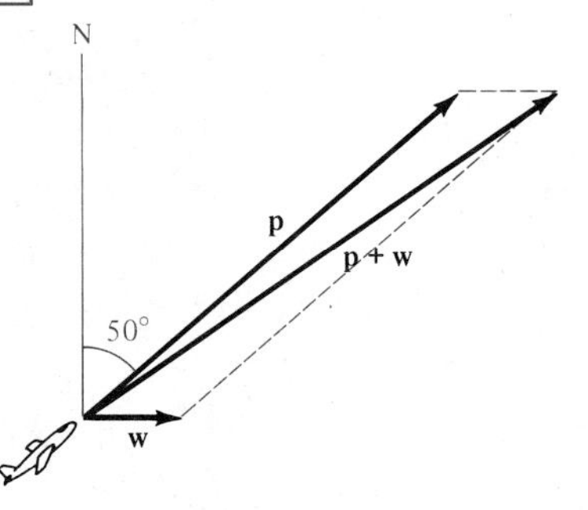

60 Refer to Exercise 59. An airplane is flying in the direction 140° with an airspeed of 500 mi/hr, and a 30 mi/hr wind is blowing in the direction 65°. Approximate the true course and ground speed of the airplane.

61 An airplane pilot wishes to maintain a true course in the direction 250° with a ground speed of 400 mi/hr when the wind is blowing directly north at 50 mi/hr. Approximate the required airspeed and compass heading.

62 An airplane is flying in the direction 20° with an airspeed of 300 mi/hr. Its ground speed and true course are 350 mi/hr and 30°, respectively. Approximate the direction and speed of the wind.

63 The current in a river flows directly from the west at a rate of 1.5 ft/sec. A person who rows a boat at a rate of 4 ft/sec in still water wishes to row directly north across the river. Approximate, to the nearest 10′, the direction in which the person should row.

64 For a motorboat moving at a speed of 30 mi/hr to travel directly north across a river, it must aim at a point that has a bearing of N15°E. If the current is flowing directly west, approximate the rate at which it flows.

8.6 THE DOT PRODUCT

The *dot product* of two vectors has many applications. We begin with an algebraic definition.

DEFINITION

Let $\mathbf{a} = \langle a_1, a_2 \rangle = a_1\mathbf{i} + a_2\mathbf{j}$ and $\mathbf{b} = \langle b_1, b_2 \rangle = b_1\mathbf{i} + b_2\mathbf{j}$. The **dot product** $\mathbf{a} \cdot \mathbf{b}$ of $\mathbf{a}$ and $\mathbf{b}$ is

$$\mathbf{a} \cdot \mathbf{b} = a_1 b_1 + a_2 b_2.$$

The symbol $\mathbf{a} \cdot \mathbf{b}$ is read **a** *dot* **b**. We also refer to the dot product as the **scalar product**, or **inner product**. Note that $\mathbf{a} \cdot \mathbf{b}$ is a real number, and not a vector, as illustrated in the following example.

EXAMPLE ▪ 1

Find $\mathbf{a} \cdot \mathbf{b}$:

(a) $\mathbf{a} = \langle -5, 3 \rangle, \quad \mathbf{b} = \langle 2, 6 \rangle$

(b) $\mathbf{a} = 4\mathbf{i} + 6\mathbf{j}, \quad \mathbf{b} = 3\mathbf{i} - 7\mathbf{j}$

SOLUTION

(a) $\langle -5, 3 \rangle \cdot \langle 2, 6 \rangle = (-5)(2) + (3)(6) = -10 + 18 = 8$

(b) $(4\mathbf{i} + 6\mathbf{j}) \cdot (3\mathbf{i} - 7\mathbf{j}) = (4)(3) + (6)(-7) = 12 - 42 = -30$

PROPERTIES OF THE DOT PRODUCT

If $\mathbf{a}$, $\mathbf{b}$, $\mathbf{c}$ are vectors and c is a real number, then

(i) $\mathbf{a} \cdot \mathbf{a} = \|\mathbf{a}\|^2$

(ii) $\mathbf{a} \cdot \mathbf{b} = \mathbf{b} \cdot \mathbf{a}$

(iii) $\mathbf{a} \cdot (\mathbf{b} + \mathbf{c}) = \mathbf{a} \cdot \mathbf{b} + \mathbf{a} \cdot \mathbf{c}$

(iv) $(c\mathbf{a}) \cdot \mathbf{b} = c(\mathbf{a} \cdot \mathbf{b}) = \mathbf{a} \cdot (c\mathbf{b})$

(v) $\mathbf{0} \cdot \mathbf{a} = 0.$

PROOF The proof of each property follows from the definition of dot product and properties of real numbers. Thus, if $\mathbf{a} = \langle a_1, a_2 \rangle$, $\mathbf{b} = \langle b_1, b_2 \rangle$, and $\mathbf{c} = \langle c_1, c_2 \rangle$, then

$$\begin{aligned}\mathbf{a} \cdot (\mathbf{b} + \mathbf{c}) &= \langle a_1, a_2 \rangle \cdot \langle b_1 + c_1, b_2 + c_2 \rangle \\ &= a_1(b_1 + c_1) + a_2(b_2 + c_2) \\ &= (a_1 b_1 + a_2 b_2) + (a_1 c_1 + a_2 c_2) \\ &= \mathbf{a} \cdot \mathbf{b} + \mathbf{a} \cdot \mathbf{c},\end{aligned}$$

which proves property (iii). The proofs of the remaining properties are left as exercises. ❑

FIGURE 39

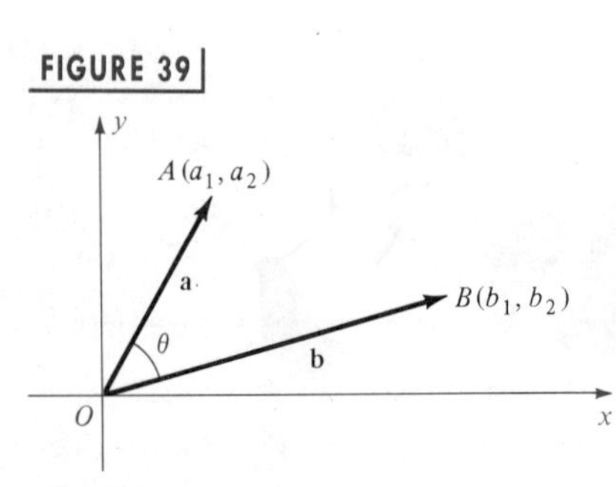

Any two nonzero vectors $\mathbf{a} = \langle a_1, a_2 \rangle$ and $\mathbf{b} = \langle b_1, b_2 \rangle$ may be represented in a coordinate system by directed line segments from the origin O to the points $A(a_1, a_2)$ and $B(b_1, b_2)$, respectively. The **angle θ between a and b** is, by definition, $\angle AOB$ (see Figure 39). Note that $0 \leq \theta \leq \pi$, and that $\theta = 0$ if **a** and **b** have the same direction or $\theta = \pi$ if **a** and **b** have the opposite direction.

DEFINITION

Let θ be the angle between two nonzero vectors **a** and **b**.

(a) **a** and **b** are **parallel** if $\theta = 0$ or $\theta = \pi$.

(b) **a** and **b** are **orthogonal** if $\theta = \dfrac{\pi}{2}$.

The vectors **a** and **b** in Figure 39 are parallel if and only if they lie on the same line that passes through the origin. In this case, $\mathbf{b} = c\mathbf{a}$ for some real number c. The vectors are orthogonal if and only if they lie on mutually perpendicular lines that pass through the origin. We assume that the zero vector **0** is parallel and orthogonal to *every* vector **a**.

The next theorem shows the close relationship between the angle between two vectors and their dot product.

THEOREM

If θ is the angle between two nonzero vectors **a** and **b**, then

$$\mathbf{a} \cdot \mathbf{b} = \|\mathbf{a}\| \, \|\mathbf{b}\| \cos \theta.$$

PROOF If **a** and **b** are not parallel, we have a situation similar to that illustrated in Figure 39. We may then apply the law of cosines to triangle AOB. Since the lengths of the three sides of the triangle are $\|\mathbf{a}\|$, $\|\mathbf{b}\|$, and $d(A, B)$,

$$[d(A, B)]^2 = \|\mathbf{a}\|^2 + \|\mathbf{b}\|^2 - 2\|\mathbf{a}\| \, \|\mathbf{b}\| \cos \theta.$$

Using the distance formula and the definition of magnitude of a vector, we obtain

$$(b_1 - a_1)^2 + (b_2 - a_2)^2 = (a_1^2 + a_2^2) + (b_1^2 + b_2^2) - 2\|\mathbf{a}\|\,\|\mathbf{b}\| \cos\theta,$$

which reduces to

$$-2a_1b_1 - 2a_2b_2 = -2\|\mathbf{a}\|\,\|\mathbf{b}\| \cos\theta.$$

Dividing both sides of the last equation by -2 give us

$$a_1b_1 + a_2b_2 = \|\mathbf{a}\|\,\|\mathbf{b}\| \cos\theta,$$

which is what we wished to prove.

If **a** and **b** are parallel, then either $\theta = 0$ or $\theta = \pi$, and therefore, $\mathbf{b} = c\mathbf{a}$ for some real number c with $c > 0$ if $\theta = 0$ and $c < 0$ if $\theta = \pi$. We can show, using properties of the dot product, that $\mathbf{a} \cdot (c\mathbf{a}) = \|\mathbf{a}\|\,\|c\mathbf{a}\| \cos\theta$, and hence the theorem is true for all nonzero vectors **a** and **b**. ❑

COROLLARY

If θ is the angle between two nonzero vectors **a** and **b**, then

$$\cos\theta = \frac{\mathbf{a} \cdot \mathbf{b}}{\|\mathbf{a}\|\,\|\mathbf{b}\|}.$$

EXAMPLE ▪ 2

Find the angle between $\mathbf{a} = \langle 4, -3 \rangle$ and $\mathbf{b} = \langle 1, 2 \rangle$.

FIGURE 40

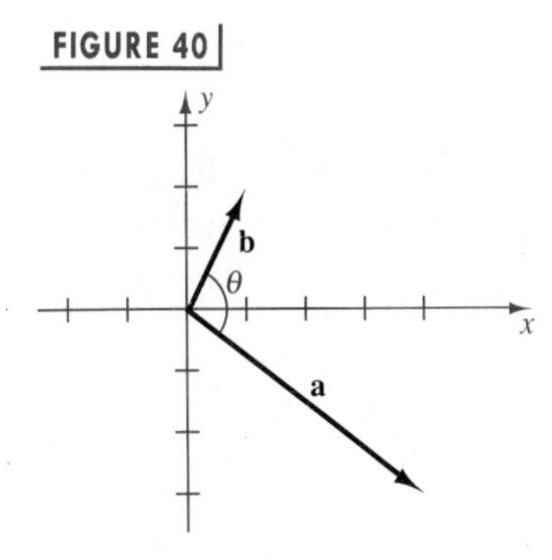

SOLUTION The vectors are sketched in Figure 40. We apply the preceding corollary:

$$\cos\theta = \frac{\mathbf{a} \cdot \mathbf{b}}{\|\mathbf{a}\|\,\|\mathbf{b}\|} = \frac{(4)(1) + (-3)(2)}{\sqrt{16+9}\sqrt{1+4}} = \frac{-2}{5\sqrt{5}} = \frac{-2\sqrt{5}}{25}$$

Hence,
$$\theta = \arccos\left(\frac{-2\sqrt{5}}{25}\right) \approx 100°18'.$$

Using the formula $\mathbf{a} \cdot \mathbf{b} = \|\mathbf{a}\|\,\|\mathbf{b}\| \cos\theta$, together with the fact that two vectors are orthogonal if and only if the angle between them is $\pi/2$ (or one of the vectors is **0**), gives us the following result.

THEOREM

Two vectors $\mathbf{a}$ and $\mathbf{b}$ are orthogonal if and only if $\mathbf{a} \cdot \mathbf{b} = 0$.

EXAMPLE ▪ 3

Prove that the pair of vectors is orthogonal:

(a) $\mathbf{i}, \mathbf{j}$ (b) $2\mathbf{i} + 3\mathbf{j},\ 6\mathbf{i} - 4\mathbf{j}$

SOLUTION We may use the preceding theorem to prove orthogonality by showing that the dot product of each pair is zero:

(a) $\mathbf{i} \cdot \mathbf{j} = \langle 1, 0\rangle \cdot \langle 0, 1\rangle = (1)(0) + (0)(1) = 0 + 0 = 0$

(b) $(2\mathbf{i} + 3\mathbf{j}) \cdot (6\mathbf{i} - 4\mathbf{j}) = (2)(6) + (3)(-4) = 12 - 12 = 0$

DEFINITION

Let θ be the angle between two nonzero vectors $\mathbf{a}$ and $\mathbf{b}$. The **component of a along b**, denoted by $\text{comp}_{\mathbf{b}}\, \mathbf{a}$, is given by

$$\text{comp}_{\mathbf{b}}\, \mathbf{a} = \|\mathbf{a}\| \cos \theta.$$

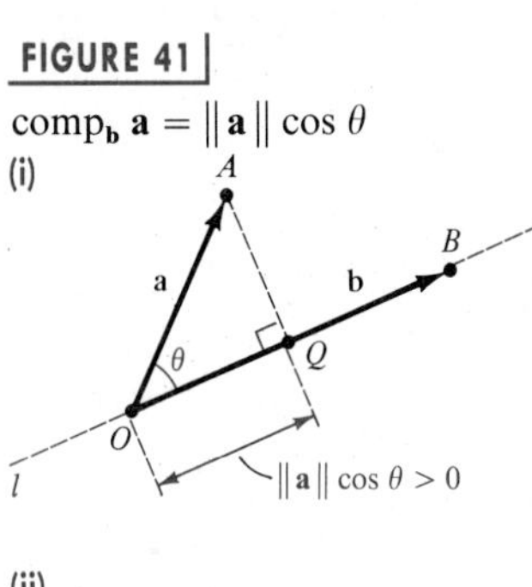

FIGURE 41 $\text{comp}_{\mathbf{b}}\, \mathbf{a} = \|\mathbf{a}\| \cos \theta$ (i) (ii)

The geometric significance of the preceding definition with θ acute or obtuse is illustrated in Figure 41, where the x- and y-axes are not shown.

If angle θ is acute, then, as in Figure 41(i), we may form a right triangle by constructing a line segment AQ perpendicular to the line l through O and B. Note that the vector $\overrightarrow{OQ}$ has the same direction as $\overrightarrow{OB}$. Referring to (i) of the figure, we see that

$$\cos \theta = \frac{d(O, Q)}{\|\mathbf{a}\|} \quad \text{and} \quad \|\mathbf{a}\| \cos \theta = d(O, Q).$$

If θ is obtuse, then, as in Figure 41(ii), we again construct AQ perpendicular to l. In this case, the direction of the vector $\overrightarrow{OQ}$ is opposite that of $\overrightarrow{OB}$, and since $\cos \theta$ is negative,

$$\cos \theta = \frac{-d(O, Q)}{\|\mathbf{a}\|} \quad \text{and} \quad \|\mathbf{a}\| \cos \theta = -d(O, Q).$$

If $\theta = \pi/2$, then $\mathbf{a}$ is orthogonal to $\mathbf{b}$ and $\text{comp}_{\mathbf{b}}\, \mathbf{a} = 0$.
If $\theta = 0$, then $\mathbf{a}$ has the same direction as $\mathbf{b}$ and $\text{comp}_{\mathbf{b}}\, \mathbf{a} = \|\mathbf{a}\|$.
If $\theta = \pi$, then $\mathbf{a}$ and $\mathbf{b}$ have opposite direction and $\text{comp}_{\mathbf{b}}\, \mathbf{a} = -\|\mathbf{a}\|$.

The preceding discussion shows that the component of **a** along **b** may be found by *projecting* the endpoint of **a** onto the line l containing **b**. For this reason, $\|\mathbf{a}\| \cos \theta$ is sometimes called the **projection of a on b**, and is denoted by $\text{proj}_{\mathbf{b}}\, \mathbf{a}$.

THEOREM

If **a** and **b** are nonzero vectors, then

$$\text{comp}_{\mathbf{b}}\, \mathbf{a} = \frac{\mathbf{a} \cdot \mathbf{b}}{\|\mathbf{b}\|}.$$

PROOF If θ is the angle between **a** and **b**, then, from the first theorem of this section,

$$\mathbf{a} \cdot \mathbf{b} = \|\mathbf{a}\| \|\mathbf{b}\| \cos \theta.$$

Dividing both sides of this equation by $\|\mathbf{b}\|$ gives us

$$\frac{\mathbf{a} \cdot \mathbf{b}}{\|\mathbf{b}\|} = \|\mathbf{a}\| \cos \theta = \text{comp}_{\mathbf{b}}\, \mathbf{a}. \quad \square$$

EXAMPLE ▪ 4

If $\mathbf{c} = 10\mathbf{i} + 4\mathbf{j}$ and $\mathbf{d} = 3\mathbf{i} - 2\mathbf{j}$, find $\text{comp}_{\mathbf{d}}\, \mathbf{c}$ and $\text{comp}_{\mathbf{c}}\, \mathbf{d}$, and illustrate these numbers graphically.

FIGURE 42

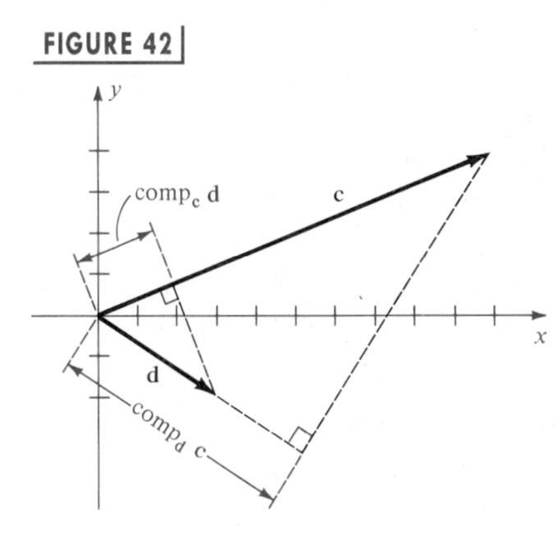

SOLUTION The vectors **c**, **d** and the desired components are illustrated in Figure 42. We may use the preceding theorem:

$$\text{comp}_{\mathbf{d}}\, \mathbf{c} = \frac{\mathbf{c} \cdot \mathbf{d}}{\|\mathbf{d}\|} = \frac{(10)(3) + (4)(-2)}{\sqrt{9+4}} = \frac{22}{\sqrt{13}} \approx 6.10$$

$$\text{comp}_{\mathbf{c}}\, \mathbf{d} = \frac{\mathbf{d} \cdot \mathbf{c}}{\|\mathbf{c}\|} = \frac{(3)(10) + (-2)(4)}{\sqrt{100+16}} = \frac{22}{\sqrt{116}} \approx 2.04$$

We shall conclude this section with a physical application of the dot product. First let us briefly discuss the scientific concept of *work*.

A **force** may be thought of as the physical entity that is used to describe a push or pull on an object. For example, a force is needed to push or pull an object along a horizontal plane, to lift an object off the ground, or to move a charged particle through an electromagnetic field. Forces are often measured in pounds. If an object weighs ten pounds, then by definition the force required to lift it (or hold it off the ground) is ten pounds. A force of this type is a **constant force**, since its magnitude does not change while it is applied to the given object.

If a constant force F is applied to an object, moving it a distance d in the direction of the force, then by definition the **work** W done is

$$W = Fd.$$

If F is measured in pounds and d in feet, then the units for W are foot-pounds (ft-lb). In the cgs system a **dyne** is used as the unit of force. If F is expressed in dynes and d in centimeters, then the unit for W is the dyne-centimeter, or **erg**. In the mks system, the **Newton** is used as the unit of force. If F is in Newtons and d is in meters, then the unit for W is the Newton-meter, or **joule**.

EXAMPLE ▪ 5

Find the work done in pushing an autombile along a level road from a point A to another point B, 40 feet from A, while exerting a constant force of 90 pounds.

FIGURE 43

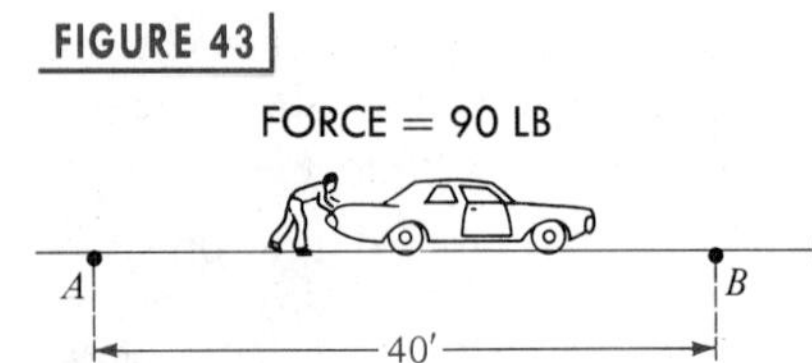

SOLUTION The problem is illustrated in Figure 43, where we have pictured the road as part of a line l. Since the constant force is $F = 90$ pounds and the distance the automobile moves is $d = 40$ feet, the work done is

$$W = (90)(40) = 3600 \text{ ft-lb}.$$

FIGURE 44

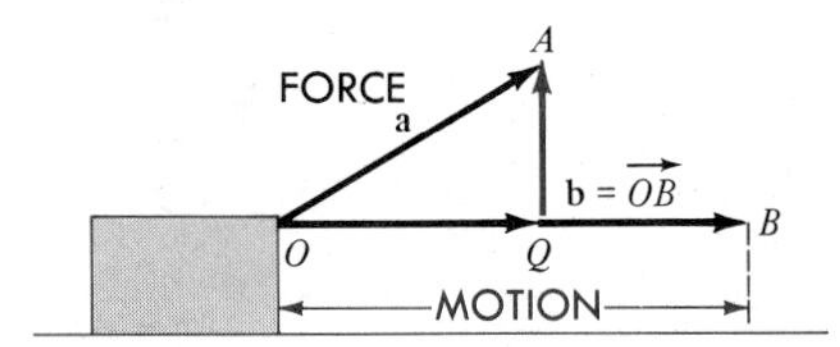

The formula $W = Fd$ is very restrictive, since it can only be used if the force is applied along the line of motion. More generally, suppose that a vector **a** represents a force and that its point of application moves along a vector **b**. This is illustrated in Figure 44, where the force **a** is used to pull an object along a level path from O to B, and $\mathbf{b} = \overrightarrow{OB}$.

The vector **a** is the sum of the vectors $\overrightarrow{OQ}$ and $\overrightarrow{QA}$, where $\overrightarrow{QA}$ is orthogonal to **b**. Since $\overrightarrow{QA}$ does not contribute to the horizontal movement, we may assume that the motion from O to B is caused by $\overrightarrow{OQ}$ alone. Applying $W = Fd$, the work is the product of $\|\overrightarrow{OQ}\|$ and $\|\mathbf{b}\|$. Since $\|\overrightarrow{OQ}\| = \text{comp}_{\mathbf{b}}\, \mathbf{a}$, we obtain

$$W = (\text{comp}_{\mathbf{b}}\, \mathbf{a})\|\mathbf{b}\| = (\|\mathbf{a}\| \cos \theta)\|\mathbf{b}\| = \mathbf{a} \cdot \mathbf{b},$$

where θ represents $\angle AOQ$. This leads to the following definition.

DEFINITION

> The **work** W done by a constant force $\mathbf{a}$ as its point of application moves along a vector $\mathbf{b}$ is $W = \mathbf{a} \cdot \mathbf{b}$.

EXAMPLE ▪ 6

The magnitude and direction of a constant force are given by $\mathbf{a} = 2\mathbf{i} + 5\mathbf{j}$. Find the work done if the point of application of the force moves from the origin to the point $P(4, 1)$.

FIGURE 45

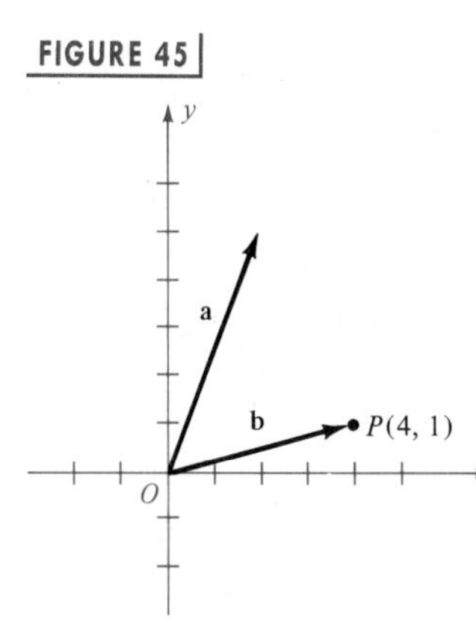

SOLUTION The force $\mathbf{a}$ and the vector $\mathbf{b} = \overrightarrow{OP}$ are sketched in Figure 45. Since $\mathbf{b} = \langle 4, 1\rangle = 4\mathbf{i} + \mathbf{j}$, we have, from the preceding definition,

$$\begin{aligned} W = \mathbf{a} \cdot \mathbf{b} &= (2\mathbf{i} + 5\mathbf{j}) \cdot (4\mathbf{i} + \mathbf{j}) \\ &= (2)(4) + (5)(1) = 13. \end{aligned}$$

If, for example, the unit of length is feet and the magnitude of the force is measured in pounds, then the work done is 13 ft-lb.

EXAMPLE ▪ 7

A small cart weighing 100 pounds is pushed up an incline that makes an angle of 30° with the horizontal, as shown in Figure 46. Find the work done against gravity in pushing the cart a distance of 80 feet.

FIGURE 46

SOLUTION Let us introduce an xy-coordinate system as shown in Figure 47. The vector $\overrightarrow{PQ}$ represents the force of gravity acting vertically downward with a magnitude of 100 pounds. The corresponding vector $\mathbf{F}$ is $0\mathbf{i} - 100\mathbf{j}$. The point of application of this force moves along the vector $\overrightarrow{PR}$ of magnitude 80. If $\overrightarrow{PR}$ corresponds to $\mathbf{a} = a_1\mathbf{i} + a_2\mathbf{j}$, then referring to triangle PTR, we see that

FIGURE 47

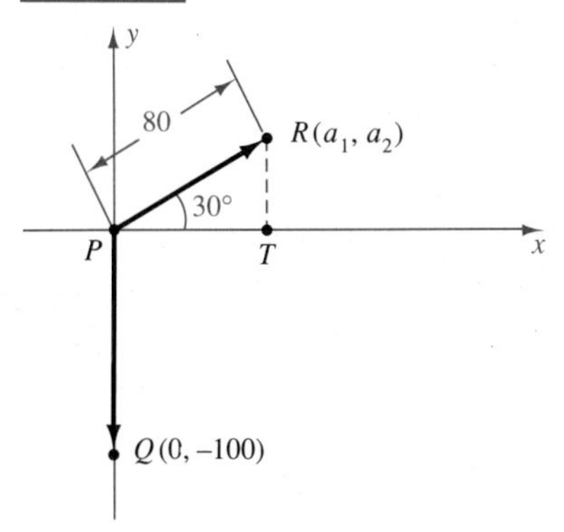

$$a_1 = 80 \cos 30° = 40\sqrt{3},$$

$$a_2 = 80 \sin 30° = 40$$

and hence

$$\mathbf{a} = 40\sqrt{3}\,\mathbf{i} + 40\mathbf{j}.$$

Applying the definition, the work done *by* gravity is

$$\mathbf{F} \cdot \mathbf{a} = (0\mathbf{i} - 100\mathbf{j}) \cdot (40\sqrt{3}\,\mathbf{i} + 40\mathbf{j}) = 0 - 4000 = -4000 \text{ ft-lb}.$$

The work done *against* gravity is

$$-\mathbf{F} \cdot \mathbf{a} = 4000 \text{ ft-lb}.$$

8.6 EXERCISES

Exer. 1–6: Find (a) the dot product of the two vectors, and (b) the angle between the two vectors.

1 $\langle -2, 5\rangle,\ \ \langle 3, 6\rangle$
2 $\langle 4, -7\rangle,\ \ \langle -2, 3\rangle$
3 $4\mathbf{i} - \mathbf{j},\ \ -3\mathbf{i} + 2\mathbf{j}$
4 $8\mathbf{i} - 3\mathbf{j},\ \ 2\mathbf{i} - 7\mathbf{j}$
5 $9\mathbf{i},\ \ 5\mathbf{i} + 4\mathbf{j}$
6 $6\mathbf{j},\ \ -4\mathbf{i}$

Exer. 7–10: Show that the vectors are orthogonal.

7 $\langle 4, -1\rangle,\ \ \langle 2, 8\rangle$
8 $\langle 3, 6\rangle,\ \ \langle 4, -2\rangle$
9 $-4\mathbf{j},\ \ -7\mathbf{i}$
10 $8\mathbf{i} - 4\mathbf{j},\ \ -6\mathbf{i} - 12\mathbf{j}$

Exer. 11–14: Determine c such that the two vectors are orthogonal.

11 $3\mathbf{i} - 2\mathbf{j},\ \ 4\mathbf{i} + 5c\mathbf{j}$
12 $4c\mathbf{i} + \mathbf{j},\ \ 9c\mathbf{i} - 25\mathbf{j}$
13 $9\mathbf{i} - 16c\mathbf{j},\ \ \mathbf{i} + 4c\mathbf{j}$
14 $5c\mathbf{i} + 3\mathbf{j},\ \ 2\mathbf{i} + 7\mathbf{j}$

Exer. 15–22: If $\mathbf{a} = \langle 2, -3\rangle$, $\mathbf{b} = \langle 3, 4\rangle$, and $\mathbf{c} = \langle -1, 5\rangle$, find the number.

15 (a) $\mathbf{a} \cdot (\mathbf{b} + \mathbf{c})$ (b) $\mathbf{a} \cdot \mathbf{b} + \mathbf{a} \cdot \mathbf{c}$
16 (a) $\mathbf{b} \cdot (\mathbf{a} - \mathbf{c})$ (b) $\mathbf{b} \cdot \mathbf{a} - \mathbf{b} \cdot \mathbf{c}$
17 $(2\mathbf{a} + \mathbf{b}) \cdot (3\mathbf{c})$
18 $(\mathbf{a} - \mathbf{b}) \cdot (\mathbf{b} + \mathbf{c})$
19 $\text{comp}_{\mathbf{c}}\, \mathbf{b}$
20 $\text{comp}_{\mathbf{b}}\, \mathbf{c}$
21 $\text{comp}_{\mathbf{b}}\, (\mathbf{a} + \mathbf{c})$
22 $\text{comp}_{\mathbf{c}}\, \mathbf{c}$

Exer. 23–26: If a represents a constant force, find the work done when the point of application of a moves along the line segment from P to Q.

23 $\mathbf{a} = 3\mathbf{i} + 4\mathbf{j};\ \ P(0, 0),\ \ Q(5, -2)$
24 $\mathbf{a} = -10\mathbf{i} + 12\mathbf{j};\ \ P(0, 0),\ \ Q(4, 7)$
25 $\mathbf{a} = 6\mathbf{i} + 4\mathbf{j};\ \ P(2, -1),\ \ Q(4, 3)$
(*Hint:* Find a vector $\mathbf{b} = \langle b_1, b_2\rangle$ such that $\mathbf{b} = \overrightarrow{PQ}$.)
26 $\mathbf{a} = -\mathbf{i} + 7\mathbf{j};\ \ P(-2, 5),\ \ Q(6, 1)$

27 A constant force of magnitude 4 has the same direction as $\mathbf{j}$. Find the work done when its point of application moves from $P(0, 0)$ to $Q(8, 3)$.

28 A constant force of magnitude 10 has the same direction as $-\mathbf{i}$. Find the work done when its point of application moves from $P(0, 1)$ to $Q(1, 0)$.

Exer. 29–34: Prove the property if a and b are vectors and c is a real number.

29 $\mathbf{a} \cdot \mathbf{a} = \|\mathbf{a}\|^2$
30 $\mathbf{a} \cdot \mathbf{b} = \mathbf{b} \cdot \mathbf{a}$
31 $(c\mathbf{a}) \cdot \mathbf{b} = c(\mathbf{a} \cdot \mathbf{b})$
32 $c(\mathbf{a} \cdot \mathbf{b}) = \mathbf{a} \cdot (c\mathbf{b})$
33 $\mathbf{0} \cdot \mathbf{a} = 0$
34 $(\mathbf{a} + \mathbf{b}) \cdot (\mathbf{a} - \mathbf{b}) = \mathbf{a} \cdot \mathbf{a} - \mathbf{b} \cdot \mathbf{b}$

35 A child pulls a wagon along level ground by exerting a force of 20 pounds on a handle that makes an angle of 30° with the horizontal, as shown in the figure. Find the work done in pulling the cart 100 feet.

EXERCISE 35

36 Refer to Exercise 35. Find the work done if the cart is pulled, with the same force, 100 feet up an incline that makes an angle of 30° with the horizontal, as shown in the figure.

EXERCISE 36

8.7 REVIEW

Define or discuss each of the following.

The law of sines ■ The law of cosines ■ Heron's formula ■ Geometric representation of a complex number ■ Complex plane ■ Absolute value of a complex number ■ De Moivre's theorem ■

nth roots of a complex number ■ Vector ■ Magnitude of a vector ■ Addition of vectors ■ Scalar multiple of a vector ■ Vectors as ordered pairs ■ Components of a vector ■ Subtraction of vectors ■ Unit vector ■ The vectors **i** and **j** ■ Linear combination of **i** and **j** ■ Dot product ■ Parallel vectors ■ Orthogonal vectors ■ Component of **a** along **b** ■ Work

8.7 EXERCISES

Exer. 1–4: Find the remaining parts of triangle ABC.

1 $\alpha = 60°, \quad \beta = 45°, \quad b = 100$

2 $\gamma = 30°, \quad a = 2\sqrt{3}, \quad c = 2$

3 $\alpha = 60°, \quad b = 6, \quad c = 7$

4 $a = 2, \quad b = 3, \quad c = 4$

Exer. 5–8: Approximate the remaining parts of triangle ABC.

5 $\beta = 67°, \quad \gamma = 75°, \quad b = 12$

6 $\alpha = 23°30', \quad c = 125, \quad a = 152$

7 $\beta = 115°, \quad a = 4.6, \quad c = 7.3$

8 $a = 37, \quad b = 55, \quad c = 43$

Exer. 9–10: Approximate the area of triangle ABC to the nearest 0.1 square unit.

9 $\alpha = 75°, \quad b = 20, \quad c = 30$

10 $a = 4, \quad b = 7, \quad c = 10$

Exer. 11–16: Express the complex number in trigonometric form with $0 \le \theta < 2\pi$.

11 $-10 + 10i$

12 $2 - 2\sqrt{3}\,i$

13 -17

14 $-12i$

15 $-5\sqrt{3} - 5i$

16 $4 + 5i$

Exer. 17–20: Use De Moivre's theorem to express the number in the form $a + bi$ for real numbers a and b.

17 $(-\sqrt{3} + i)^9$

18 $\left(\frac{\sqrt{2}}{2} - \frac{\sqrt{2}}{2}i\right)^{30}$

19 $(3 - 3i)^5$

20 $(2 + 2\sqrt{3}\,i)^{10}$

21 Find the three cube roots of -27.

22 Find the solutions of the equation $x^5 - 32 = 0$.

23 If $\mathbf{a} = \langle -4, 5\rangle$ and $\mathbf{b} = \langle 2, -8\rangle$, sketch vectors corresponding to:

(a) $\mathbf{a} + \mathbf{b}$ (b) $\mathbf{a} - \mathbf{b}$

(c) $2\mathbf{a}$ (d) $-\frac{1}{2}\mathbf{b}$

24 If $\mathbf{a} = 2\mathbf{i} + 5\mathbf{j}$ and $\mathbf{b} = 4\mathbf{i} - \mathbf{j}$, find the vector or number corresponding to:

(a) $4\mathbf{a} + \mathbf{b}$ (b) $2\mathbf{a} - 3\mathbf{b}$

(c) $\|\mathbf{a} - \mathbf{b}\|$ (d) $\|\mathbf{a}\| - \|\mathbf{b}\|$

25 If $\mathbf{a} = 6\mathbf{i} - 2\mathbf{j}$ and $\mathbf{b} = \mathbf{i} + 3\mathbf{j}$, find each of the following:

(a) $(2\mathbf{a} - 3\mathbf{b}) \cdot \mathbf{a}$

(b) the angle between $\mathbf{a}$ and $\mathbf{a} + \mathbf{b}$

(c) $\text{comp}_{\mathbf{a}}\,(\mathbf{a} + \mathbf{b})$

26 A constant force has the magnitude and direction of $\mathbf{a} = 7\mathbf{i} + 4\mathbf{j}$. Find the work done when the point of application of $\mathbf{a}$ moves along the x-axis from $P(-5, 0)$ to $Q(3, 0)$.

27 If $\mathbf{a} = \langle a_1, a_2\rangle$, $\mathbf{r} = \langle x, y\rangle$, and $c > 0$, describe the set of all points $P(x, y)$ such that $\|\mathbf{r} - \mathbf{a}\| = c$.

28 If $\mathbf{a}$ and $\mathbf{b}$ are vectors with the same initial point and θ is the angle between them, prove that

$$\|\mathbf{a} - \mathbf{b}\|^2 = \|\mathbf{a}\|^2 + \|\mathbf{b}\|^2 - 2\|\mathbf{a}\|\|\mathbf{b}\|\cos\theta.$$

29 A ship is sailing at a speed of 14 mi/hr in the direction S50°E. Express its velocity $\mathbf{v}$ as a vector.

30 The magnitudes and directions of two forces are 72 kg, S60°E and 46 kg, N74°E, respectively. Approximate the magnitude and direction of the resultant force.

31 When a mountaintop is viewed from the point P shown in the figure, the angle of elevation is α. From a point Q which is d units closer to the mountain, the angle of elevation increases to β.

EXERCISE 31

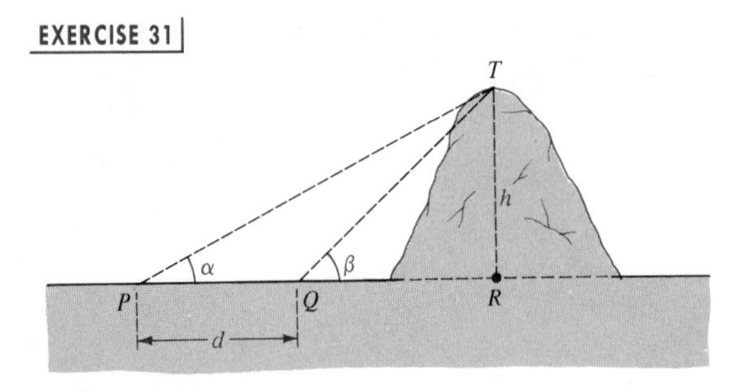

(a) Show that the height h of the mountain is given by

$$h = \frac{d}{\cot \alpha - \cot \beta}.$$

(b) Approximate the height of the mountain if $d = 2$ miles, $\alpha = 15^\circ$, and $\beta = 20^\circ$.

32 A course for a skateboard race consists of a 200-meter downhill run and a 150-meter level portion. When the starting point of the race is spotted from the finishing line, the angle of elevation is 27.4°. What angle does the hill make with the horizontal?

33 If a skyscraper is viewed from the top of a 50-foot building, the angle of elevation is 59°. If viewed from street level, the angle of elevation measures 62° (see Exercise 44 of Section 6.8).

(a) Use the law of sines to approximate the distance between the tops of the two buildings.

(b) Estimate the height of the skyscraper.

34 The beach communities of San Clemente and Long Beach are 41 miles apart, along a fairly straight stretch of coastline. Shown in the figure is the triangle formed by the two cities and the town of Avalon on the southeast corner of Santa Catalina Island. Angles ALS and ASL are found to be 66.4° and 47.2°, respectively. Approximate

(a) the distance from Avalon to each of the two cities.

(b) the shortest distance from Avalon to the coast.

EXERCISE 34

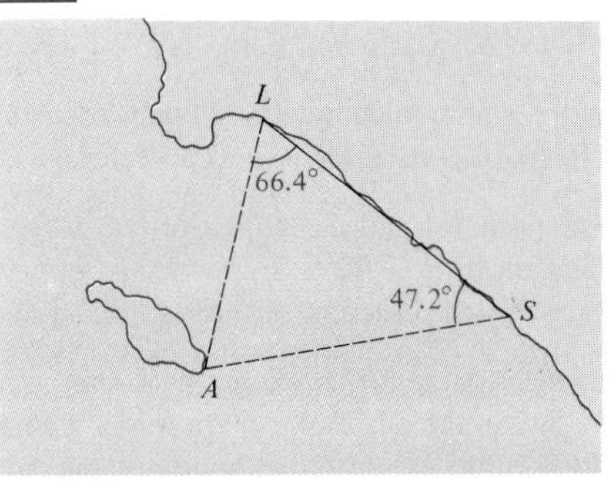

35 A surveyor wishes to find the distance between two inaccessible points A and B. As shown in the figure, two points C and D are selected from which it is possible to view both A and B. The distance CD together with angles ACD, ACB, BDC, and BDA are then measured. If $CD = 120$ ft, $\angle ACD = 115^\circ$, $\angle ACB = 92^\circ$, $\angle BDC = 125^\circ$, and $\angle BDA = 100^\circ$, estimate the distance AB.

EXERCISE 35

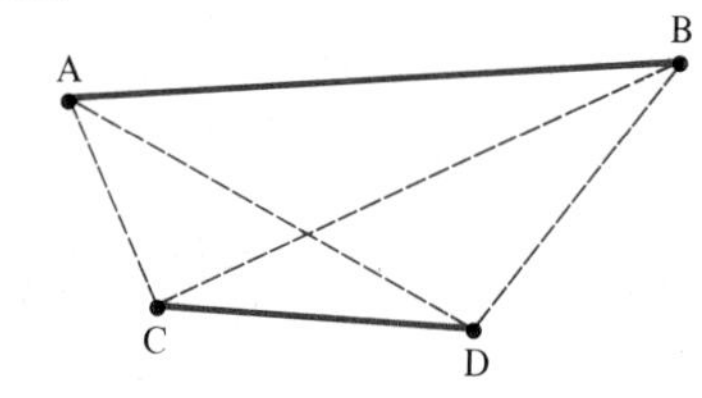

36 Two girls with two-way radios are at the intersection of two country roads that meet at a 105° angle (see figure). One girl begins walking in a northerly direction along one road at the rate of 5 mi/hr, and at the same time the other girl walks east along the other road at the same rate. If each radio has a range of 10 miles, how long will the girls maintain contact?

EXERCISE 36

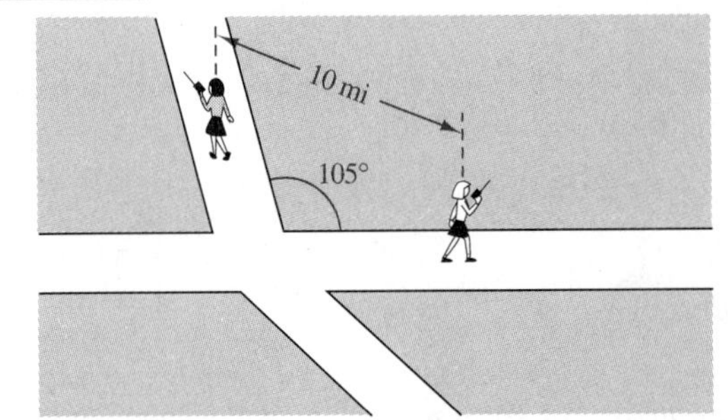

37 Shown in the figure is a design for a robotic arm with two moving parts. The dimensions are chosen to emulate a human arm. The upper arm AC and lower arm CP rotate through angles θ_1 and θ_2, respectively, to hold an object at point $P(x, y)$.

(a) Show that $\angle ACP = 180^\circ - (\theta_2 - \theta_1)$.

(b) Find $d(A, P)$, and then use part (a) and the law of cosines to show that

$$1 + \cos(\theta_2 - \theta_1) = \frac{x^2 + (26 - y)^2}{578}.$$

(c) If $x = 25$, $y = 4$, and $\theta_1 = 135°$, approximate θ_2.

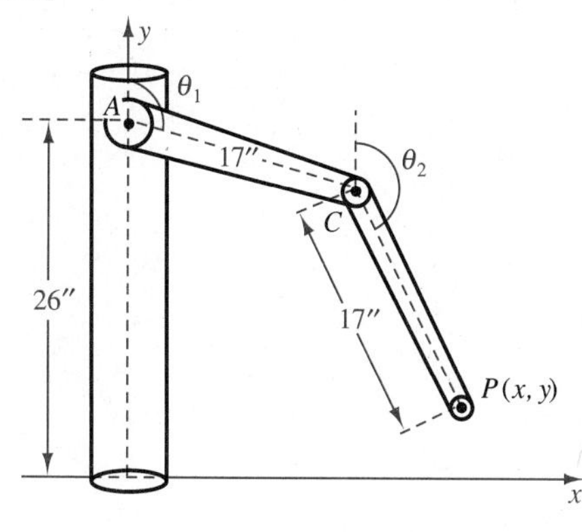

38 A child is trapped 45 feet down an abandoned mine shaft that slants at an angle of 78° from the horizontal. A rescue tunnel is to be dug 50 feet from the shaft opening (see figure).

(a) At what angle should the tunnel be dug?

(b) If the tunnel can be dug at a rate of 3 ft/hr, how many hours will it take to reach the child?

EXERCISE 38

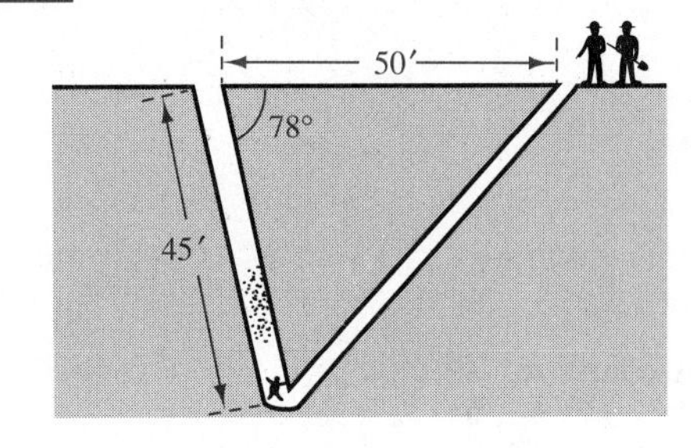

CHAPTER 9

It is sometimes necessary to work simultaneously with more than one equation in several variables, that is, with a *system* of equations. In this chapter we shall develop methods for finding solutions that are common to all equations in a system. Of particular importance are the matrix techniques introduced for systems of linear equations. We shall also briefly discuss systems of inequalities and linear programming.

SYSTEMS OF EQUATIONS AND INEQUALITIES

9.1 SYSTEMS OF EQUATIONS

A **system** of two equations in x and y is any two equations in those variables. An ordered pair (a, b) is a **solution of the system** if (a, b) is a solution of both equations. The points that correspond to the solutions are the points at which the graphs of the two equations intersect.

As an example, consider the system

$$\begin{cases} x^2 - y = 0 \\ y - 2x - 3 = 0. \end{cases}$$

The brace indicates that the two equations are to be treated simultaneously.

The following table lists some solutions of the equation $x^2 - y = 0$, or equivalently, $y = x^2$.

x	-2	-1	0	1	2	3	4
y	4	1	0	1	4	9	16

FIGURE 1

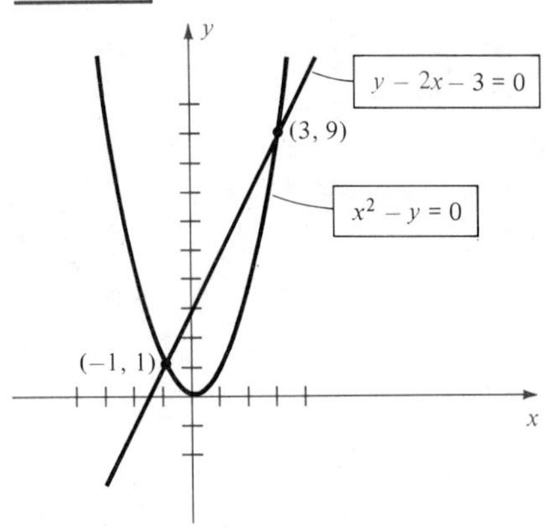

The graph is the parabola sketched in Figure 1.

Several solutions for $y - 2x - 3 = 0$, or equivalently, $y = 2x + 3$, are listed in the following table.

x	-2	-1	0	1	2	3	4
y	-1	1	3	5	7	9	11

The graph is the line in Figure 1.

The pairs $(3, 9)$ and $(-1, 1)$ are solutions of both of the preceding equations and hence are solutions of the system. Note that the solutions of the system correspond to the points of intersection of the two graphs.

We can also find the solutions of the system algebraically, that is, without reference to graphs. Let us begin by solving the equation $x^2 - y = 0$ for y in terms of x, obtaining

$$y = x^2.$$

Thus, if (x, y) is a solution of the system, then the variable y in the equation $y - 2x - 3 = 0$ must satisfy the condition $y = x^2$. Hence we *substitute* x^2

for y in $y - 2x - 3 = 0$:

$$x^2 - 2x - 3 = 0$$

$$(x - 3)(x + 1) = 0$$

Setting each factor equal to zero gives us

$$x = 3, \qquad x = -1.$$

The corresponding values for y (obtained from $y = x^2$) are 9 and 1, respectively. Thus, the only possible solutions of the system are the ordered pairs $(3, 9)$ and $(-1, 1)$. That these are actually solutions can be seen by substitution.

To find the solutions algebraically we first solved one equation for y in terms of x, and then we substituted for y in the other equation, obtaining an equation in one variable, x. The solutions of the equation in x are the only possible x-values for the solutions of the system. The corresponding y-values are found by means of the equation that expressed y in terms of x. This technique is the **method of substitution**.

We could have started by solving the *second* equation of the system for y, obtaining

$$y = 2x + 3.$$

We then substitute for y in the *first* equation:

$$x^2 - (2x + 3) = 0$$

$$x^2 - 2x - 3 = 0$$

The remaining steps are the same.

Sometimes we may first solve one of the equations for x in terms of y, and then substitute for x in the other equation, as illustrated in the next example.

EXAMPLE ▪ 1

Find the solutions of the system

$$\begin{cases} x^2 - y = 0 \\ y - 2x - 3 = 0. \end{cases}$$

SOLUTION We first solve the second equation for x in terms of y:

$$2x = y - 3$$

$$x = \frac{y - 3}{2}$$

We then substitute for x in the first equation, $x^2 - y = 0$:

$$\left(\frac{y-3}{2}\right)^2 - y = 0$$

$$\frac{y^2 - 6y + 9}{4} - y = 0$$

$$y^2 - 6y + 9 - 4y = 0$$

$$y^2 - 10y + 9 = 0$$

$$(y - 9)(y - 1) = 0$$

The last equation has solutions $y = 9$ and $y = 1$. If we substitute 9 and 1 for y in the equation $x = (y - 3)/2$, we obtain the values $x = 3$ and $x = -1$, respectively. Hence, as in the discussion preceding this example, the solutions of the system are $(3, 9)$ and $(-1, 1)$.

EXAMPLE ▪ 2

Find the solutions of the following system and then sketch the graph of each equation, showing the points of intersection:

$$\begin{cases} x^2 + y^2 = 25 \\ x^2 + y = 19 \end{cases}$$

SOLUTION Solving the second equation for y, we obtain

$$y = 19 - x^2.$$

Substituting for y in the first equation leads to the following equivalent equations:

$$x^2 + (19 - x^2)^2 = 25$$

$$x^2 + (361 - 38x^2 + x^4) = 25$$

$$x^4 - 37x^2 + 336 = 0$$

$$(x^2 - 16)(x^2 - 21) = 0$$

Setting each factor equal to zero, we obtain

$$x^2 = 16, \qquad x^2 = 21.$$

The solutions of these equations are $x = \pm 4$ and $y = \pm\sqrt{21}$. We find the corresponding values of y by substituting for x in $y = 19 - x^2$. Substitution of 4 or -4 for x gives us $y = 3$. Substitution of $\sqrt{21}$ or $-\sqrt{21}$ gives us $y = -2$. Hence, the only possible solutions of the system are

$$(4, 3), \qquad (-4, 3), \qquad (\sqrt{21}, -2), \qquad (-\sqrt{21}, -2).$$

FIGURE 2

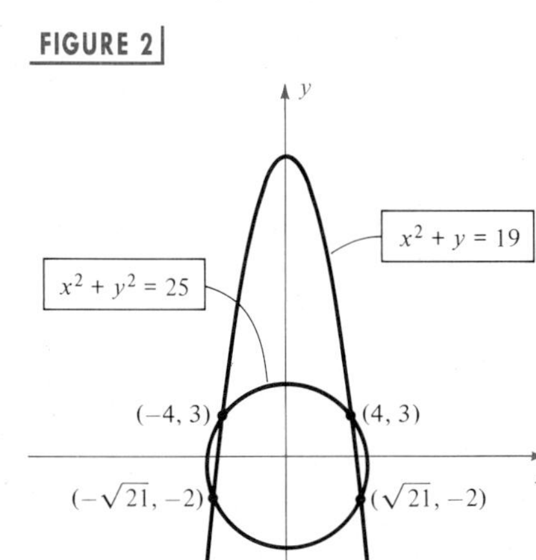

We can check by substitution in the given equations that all four pairs are solutions.

The graph of $x^2 + y^2 = 25$ is a circle of radius 5 with center at the origin, and the graph of $y = 19 - x^2$ is a parabola with a vertical axis. The graphs are sketched in Figure 2. The points of intersection correspond to the solutions of the system.

We can also consider equations in three variables x, y, and z, such as

$$x^2y + xz + 3^y = 4z^3.$$

Such an equation has a **solution** (a, b, c) if substitution of a, b, and c, for x, y, and z, respectively, produces a true statement. We refer to (a, b, c) as an **ordered triple** of real numbers. Equivalent equations have the same solutions. A system of equations in three variables and the solutions of such a system are defined as in the two-variable case. Similarly, we can consider systems of *any* number of equations in *any* number of variables.

The method of substitution can be extended to these more complicated systems. For example, given three equations in three variables, suppose that it is possible to solve one of the equations for one variable in terms of the remaining two variables. By substituting that expression in each of the other equations, we obtain a system of two equations in two variables. The solutions of the two-variable system can then be used to find the solutions of the original system.

EXAMPLE ▪ 3

Find the solutions of the system

$$\begin{cases} x - y + z = 2 \\ xyz = 0 \\ 2y + z = 1. \end{cases}$$

SOLUTION We first solve the third equation for z:

$$z = 1 - 2y$$

Substituting for z in the first two equations of the system, we obtain the following system of two equations in two variables:

$$\begin{cases} x - y + (1 - 2y) = 2 \\ xy(1 - 2y) = 0 \end{cases}$$

This system is equivalent to

$$\begin{cases} x - 3y - 1 = 0 \\ xy(1 - 2y) = 0. \end{cases}$$

We now find the solutions of the last system. Solving the first equation for x in terms of y gives us

$$x = 3y + 1.$$

Substituting $3y + 1$ for x in the second equation $xy(1 - 2y) = 0$, we obtain

$$(3y + 1)y(1 - 2y) = 0,$$

which has solutions

$$y = -\tfrac{1}{3}, \qquad y = 0, \qquad y = \tfrac{1}{2}.$$

These are the only possible y-values for the solutions of the system.

To obtain the corresponding x-values, we substitute for y in the equation $x = 3y + 1$, obtaining

$$x = 0, \qquad x = 1, \qquad x = \tfrac{5}{2}.$$

Using $z = 1 - 2y$ gives us the corresponding z values

$$z = \tfrac{5}{3}, \qquad z = 1, \qquad z = 0.$$

Thus the solutions of the original system consist of the ordered triples

$$(0, -\tfrac{1}{3}, \tfrac{5}{3}), \qquad (1, 0, 1), \qquad (\tfrac{5}{2}, \tfrac{1}{2}, 0).$$

EXAMPLE ▪ 4

Is it possible to construct an aquarium with a glass top and two square ends that holds $16\,\text{ft}^3$ of water and requires $40\,\text{ft}^2$ of glass? (Neglect the thickness of the glass.)

FIGURE 3

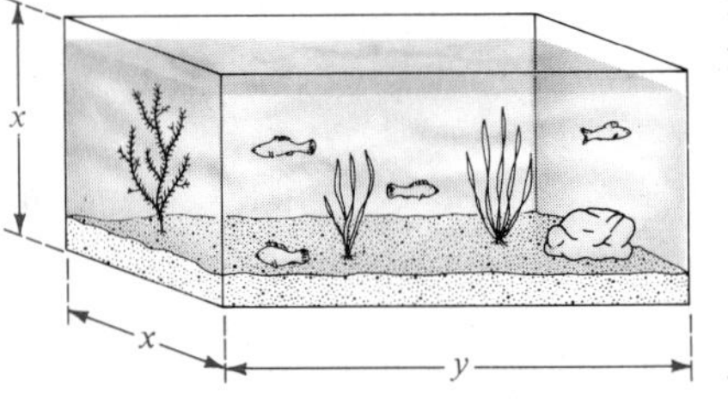

SOLUTION We begin by sketching a typical aquarium and labeling it as in Figure 3, with x and y in feet. Referring to the figure and using formulas for volume and area, we see that

$$\text{volume of the aquarium} = x^2y$$

$$\text{square feet of glass required} = 2x^2 + 4xy.$$

Since the volume and the glass required are 16 and 40, respectively, we obtain the following system of equations:

$$\begin{cases} x^2y = 16 \\ 2x^2 + 4xy = 40 \end{cases}$$

Solving the first equation for y gives us $y = 16/x^2$. Substituting for y in the second equation leads to

$$2x^2 + 4x\left(\frac{16}{x^2}\right) = 40$$

$$2x^2 + \frac{64}{x} = 40$$

$$2x^3 + 64 = 40x$$

$$x^3 - 20x + 32 = 0$$

We next look for rational solutions of this equation. Dividing the polynomial $x^3 - 20x + 32$ synthetically by $x - 2$ gives us

$$\begin{array}{r|rrrr} 2 & 1 & 0 & -20 & 32 \\ & & 2 & 4 & -32 \\ \hline & 1 & 2 & -16 & 0 \end{array}$$

Thus, one solution of $x^3 - 20x + 32 = 0$ is 2, and the remaining two solutions are zeros of the quotient $x^2 + 2x - 16$; that is, roots of the equation

$$x^2 + 2x - 16 = 0.$$

Using the quadratic formula,

$$x = \frac{-2 \pm \sqrt{4 + 64}}{2} = \frac{-2 \pm 2\sqrt{17}}{2} = -1 \pm \sqrt{17}.$$

Since x is positive we may discard $-1 - \sqrt{17}$. Hence, the only possible values of x are

$$x = 2 \quad \text{and} \quad x = -1 + \sqrt{17} \approx 3.12.$$

The corresponding y-values can found by substituting for x in the equation $y = 16/x^2$. Letting $x = 2$ gives us $y = \frac{16}{4} = 4$. Using these values, we obtain the dimensions 2 feet by 2 feet by 4 feet for an aquarium.

Letting $x = -1 + \sqrt{17}$ in $y = 16/x^2$, we obtain $y = 16/(-1 + \sqrt{17})^2$, which simplifies to $y = \frac{1}{8}(9 + \sqrt{17}) \approx 1.64$. Thus (approximate) dimensions for another aquarium are 3.12 feet by 3.12 feet by 1.64 feet.

9.1 EXERCISES

Exer. 1–30: Use the method of substitution to find the solutions of the system of equations.

1 $\begin{cases} y = x^2 - 4 \\ y = 2x - 1 \end{cases}$

2 $\begin{cases} y = x^2 + 1 \\ x + y = 3 \end{cases}$

3 $\begin{cases} y^2 = 1 - x \\ x + 2y = 1 \end{cases}$

4 $\begin{cases} y^2 = x \\ x + 2y + 3 = 0 \end{cases}$

5 $\begin{cases} 2y = x^2 \\ y = 4x^3 \end{cases}$

6 $\begin{cases} x - y^3 = 1 \\ 2x = 9y^2 + 2 \end{cases}$

7 $\begin{cases} x + 2y = -1 \\ 2x - 3y = 12 \end{cases}$

8 $\begin{cases} 3x - 4y + 20 = 0 \\ 3x + 2y + 8 = 0 \end{cases}$

9 $\begin{cases} 2x - 3y = 1 \\ -6x + 9y = 4 \end{cases}$

10 $\begin{cases} 4x - 5y = 2 \\ 8x - 10y = -5 \end{cases}$

11 $\begin{cases} x + 3y = 5 \\ x^2 + y^2 = 25 \end{cases}$

12 $\begin{cases} 3x - 4y = 25 \\ x^2 + y^2 = 25 \end{cases}$

13 $\begin{cases} x^2 + y^2 = 8 \\ y - x = 4 \end{cases}$

14 $\begin{cases} x^2 + y^2 = 25 \\ 3x + 4y = -25 \end{cases}$

15 $\begin{cases} x^2 + y^2 = 9 \\ y - 3x = 2 \end{cases}$

16 $\begin{cases} x^2 + y^2 = 16 \\ y + 2x = -1 \end{cases}$

17 $\begin{cases} x^2 + y^2 = 16 \\ 2y - x = 4 \end{cases}$

18 $\begin{cases} x^2 + y^2 = 1 \\ y + 2x = -3 \end{cases}$

19 $\begin{cases} (x-1)^2 + (y+2)^2 = 10 \\ x + y = 1 \end{cases}$

20 $\begin{cases} xy = 2 \\ 3x - y + 5 = 0 \end{cases}$

21 $\begin{cases} y = 20/x^2 \\ y = 9 - x^2 \end{cases}$

22 $\begin{cases} x = y^2 - 4y + 5 \\ x - y = 1 \end{cases}$

23 $\begin{cases} y^2 - 4x^2 = 4 \\ 9y^2 + 16x^2 = 140 \end{cases}$

24 $\begin{cases} 25y^2 - 16x^2 = 400 \\ 9y^2 - 4x^2 = 36 \end{cases}$

25 $\begin{cases} x^2 - y^2 = 4 \\ x^2 + y^2 = 12 \end{cases}$

26 $\begin{cases} 6x^3 - y^3 = 1 \\ 3x^3 + 4y^3 = 5 \end{cases}$

27 $\begin{cases} x + 2y - z = -1 \\ 2x - y + z = 9 \\ x + 3y + 3z = 6 \end{cases}$

28 $\begin{cases} 2x - 3y - z^2 = 0 \\ x - y - z^2 = -1 \\ x^2 - xy = 0 \end{cases}$

29 $\begin{cases} x^2 + z^2 = 5 \\ 2x + y = 1 \\ y + z = 1 \end{cases}$

30 $\begin{cases} x + 2z = 1 \\ 2y - z = 4 \\ xyz = 0 \end{cases}$

31 The perimeter of a rectangle is 40 inches and its area is 96 in.^{2}. Find its length and width.

32 Find the values of b such that the system

$$\begin{cases} x^2 + y^2 = 4 \\ y = x + b \end{cases}$$

has **(a)** one solution, **(b)** two solutions, **(c)** no solution. Interpret (a)–(c) graphically.

33 Is there a real number x such that $x = 2^{-x}$? Decide by displaying graphically the system

$$\begin{cases} y - x = 0 \\ y - 2^{-x} = 0. \end{cases}$$

34 Is there a real number x such that $x = \log x$? Decide by displaying graphically the system

$$\begin{cases} y - \log x = 0 \\ y - x = 0. \end{cases}$$

35 Sections of cylindrical tubing are to be made from thin rectangular sheets that have an area of 200 in.^{2} (see figure). Is it possible to construct a tube that has a volume of 200 in.^{3}? If so, find the dimensions of the rectangular sheet.

EXERCISE 35

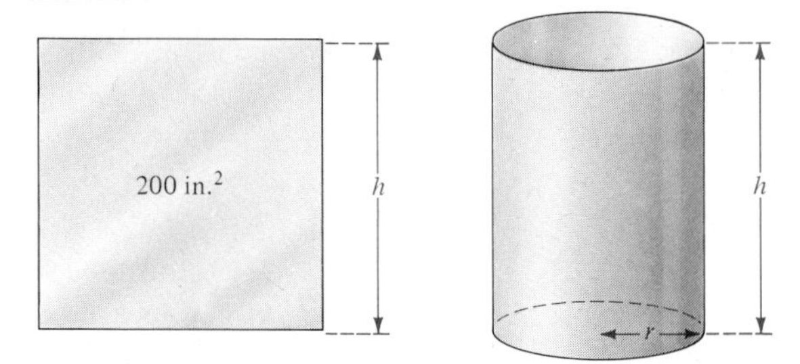

36 Shown in the figure is the graph of $y = x^2$ and a line of slope m that passes through the point (1, 1). Find a value of m such that the line intersects the graph only at (1, 1).

EXERCISE 36

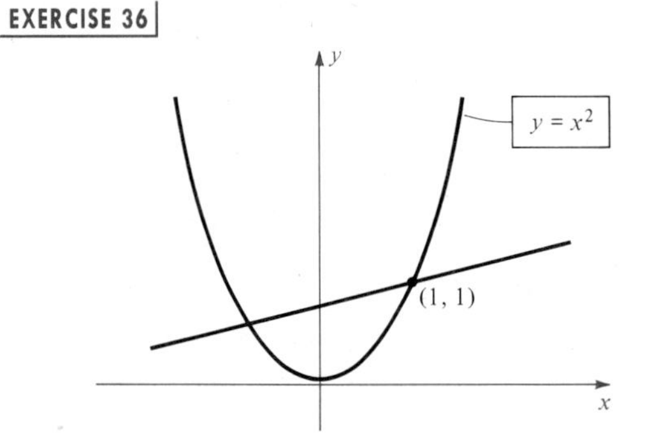

37 In fishery science, spawner-recruit functions are used to predict the number of adult fish R in next year's breeding population from an estimate S of the number of fish presently spawning.

(a) For a certain species of fish, $R = aS/(S + b)$. Estimate a and b from the data in the following table.

Year	1987	1988	1989
Number spawning	40,000	60,000	72,000

(b) Predict the breeding population for the year 1990.

38 Refer to Exercise 37. Ricker's spawner-recruit function is given by

$$R = aSe^{-bS}$$

for positive constants a and b. This relationship predicts low recruitment from very high stocks and has been found to be appropriate for many species, such as arctic cod. Rework Exercise 37 using Ricker's spawner-recruit function.

39 A competition model is a collection of equations that specifies how two or more species interact in competition for the food resources of an ecosystem. Let x and y denote the numbers (in hundreds) of two competing species, and suppose that the respective rates of growth R_1 and R_2 are given by

$$R_1 = 0.01x(50 - x - y),$$
$$R_2 = 0.02y(100 - y - 0.5x).$$

Determine the population levels (x, y) at which both rates of growth are zero. (Such population levels are called stationary points.)

40 A rancher has 2420 feet of fence to enclose a rectangular region that lies along a straight river. If no fence is used along the river (see figure), is it possible to enclose 10 acres of land? Recall that 1 acre = 43,560 ft^2.

EXERCISE 40

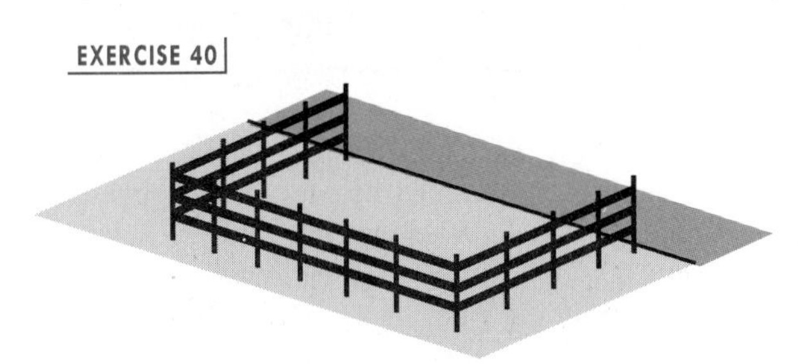

41 Refer to Example 4. Is it possible to construct a small aquarium with *open* top and two square ends that holds 2 ft^3 of water and requires 8 ft^2 of glass? If so, approximate the dimensions. (Neglect the thickness of the glass.)

42 The isoperimetric problem is to prove that of all plane geometric figures with the same perimeter (isoperimetric figures), the circle has the greatest area. Show that no rectangle has both the same area and perimeter as any circle.

43 A moiré pattern is formed when two geometrically regular patterns are superimposed. Shown in the figure is a pattern obtained from the family of circles $x^2 + y^2 = n^2$ and the family of horizontal lines $y = m$ for integers m and n.

(a) Show that the points of intersection of the circle $x^2 + y^2 = n^2$ and the line $y = n - 1$ lie on a parabola.

(b) Work part (a) using the line $y = n - 2$.

EXERCISE 43

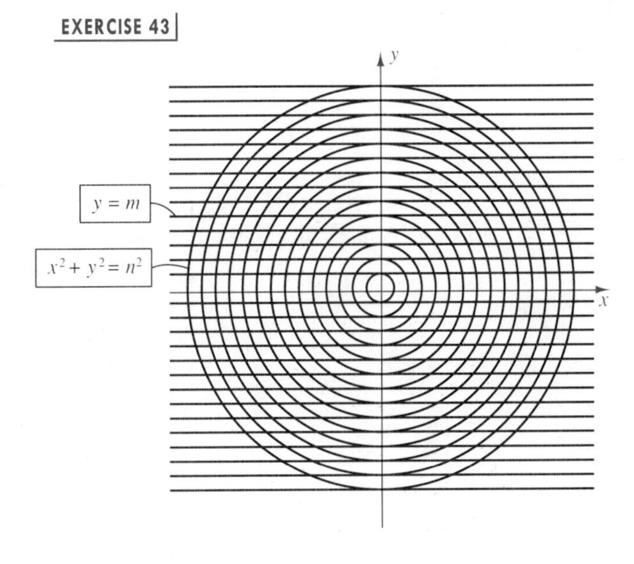

44 A spherical pill has diameter 1 cm. A second pill in the shape of a right circular cylinder is to be manufactured with the same volume and twice the surface area of the spherical pill.

(a) If r is the radius and h is the height of the cylindrical pill, show that $6r^2h = 1$ and $r^2 + rh = 1$. Conclude that $6r^3 - 6r + 1 = 0$.

(b) The positive solutions of $6r^3 - 6r + 1 = 0$ are approximately 0.172 and 0.903. Find the corresponding heights, and interpret these results.

9.2 SYSTEMS OF LINEAR EQUATIONS IN TWO VARIABLES

An equation $ax + by + c = 0$ (or, equivalently, $ax + by = -c$) with a and b not both zero is a linear equation in the variables x and y. Similarly, $ax + by + cz = d$ is a linear equation in three variables x, y, and z. We may also consider linear equations in four, five, or *any* number of variables. The most common systems of equations are those in which every equation is linear. In this section we shall consider only systems of two linear equations in two variables. Systems involving more than two variables are discussed in the next section.

Two systems of equations are equivalent if they have the same solutions. To find the solutions of a system, we may manipulate the equations until we obtain an equivalent system of simple equations for which the solutions can be found readily. Some manipulations (or *transformations*) that lead to equivalent systems are stated without proof in the next theorem.

THEOREM ON EQUIVALENT SYSTEMS

Given a system of equations, an equivalent system results if:
(i) two equations are interchanged.
(ii) an equation is multiplied by a nonzero constant.
(iii) a constant multiple of one equation is added to another equation.

The word *constant* in (ii) and (iii) means a real (or complex) number. A *constant multiple* of an equation is obtained by multiplying *each* term of the equation by the same constant k. When applying (iii) we often use the phrase *add to one equation, k times any other equation.* To *add* two equations means to add corresponding sides of the equations.

EXAMPLE ▪ 1

Find the solutions of the system

$$\begin{cases} x + 3y = -1 \\ 2x - y = 5. \end{cases}$$

SOLUTION By (ii) of the preceding theorem we may multiply the second equation by 3. This gives us the equivalent system

$$\begin{cases} x + 3y = -1 \\ 6x - 3y = 15. \end{cases}$$

Next, by (iii) of the theorem, we may add to the second equation 1 times the first equation; that is, *we may add the first equation to the second*, obtaining the equivalent system

$$\begin{cases} x + 3y = -1 \\ 7x \qquad = 14. \end{cases}$$

We see from the last equation that $x = 2$. To find the corresponding value for y, we substitute 2 for x in the equation $x + 3y = -1$:

$$2 + 3y = -1$$
$$3y = -3$$
$$y = -1$$

Thus, the system has one solution, $(2, -1)$.

There are other ways to use the theorem on equivalent systems to find the solution $(2, -1)$. For example, we could begin by multiplying the first equation by -2, obtaining

$$\begin{cases} -2x - 6y = 2 \\ 2x - y = 5. \end{cases}$$

We next add the first equation to the second:

$$\begin{cases} -2x - 6y = 2 \\ \qquad - 7y = 7 \end{cases}$$

FIGURE 4

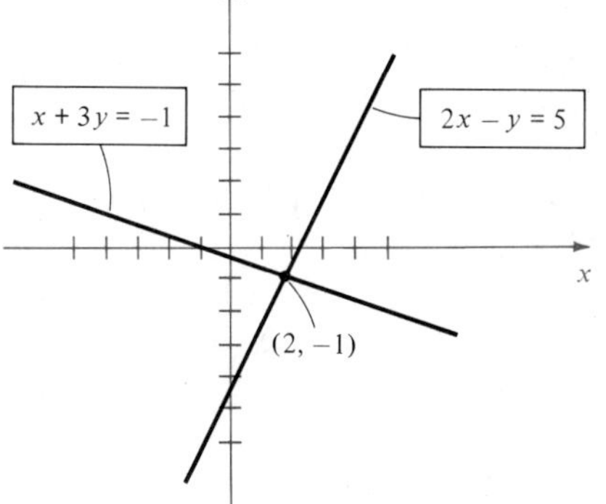

The last equation implies that $y = -1$. Hence, we substitute -1 for y in the equation $-2x - 6y = 2$:

$$-2x - 6(-1) = 2$$
$$-2x = -4$$
$$x = 2$$

Again we see that the solution is $(2, -1)$.

The graphs of the two equations are lines that intersect at the point $(2, -1)$, as shown in Figure 4.

The technique used in Example 1 is called the **method of elimination**, since it involves the elimination of a variable from one of the equations. The method of elimination usually leads to solutions in fewer steps than the method of substitution discussed in the preceding section.

EXAMPLE ▪ 2

Find the solutions of the system

$$\begin{cases} 3x + y = 6 \\ 6x + 2y = 12. \end{cases}$$

FIGURE 5

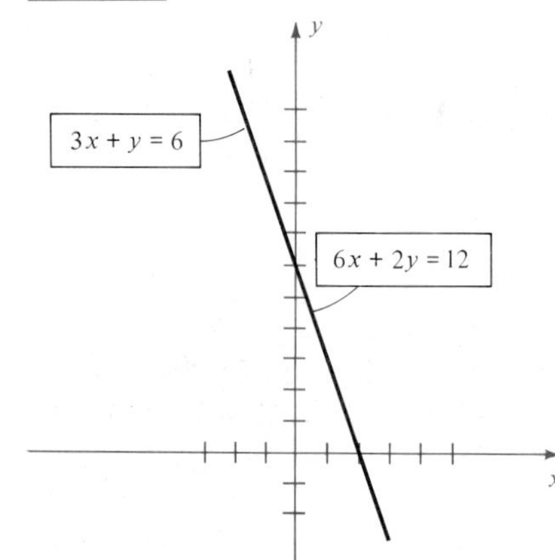

SOLUTION Multiplying the second equation by $\frac{1}{2}$ gives us

$$\begin{cases} 3x + y = 6 \\ 3x + y = 6. \end{cases}$$

Thus, (a, b) is a solution if and only if $3a + b = 6$; that is, $b = 6 - 3a$. It follows that the solutions consist of all ordered pairs of the form $(a, 6 - 3a)$ for a real number a. If we wish to find particular solutions we may substitute various values for a. A few solutions are $(0, 6)$, $(1, 3)$, $(3, -3)$, $(-2, 12)$ and $(\sqrt{2}, 6 - 3\sqrt{2})$.

The graph of each equation is the same line, as shown in Figure 5.

EXAMPLE ▪ 3

Find the solutions of the system

$$\begin{cases} 3x + y = 6 \\ 6x + 2y = 20. \end{cases}$$

FIGURE 6

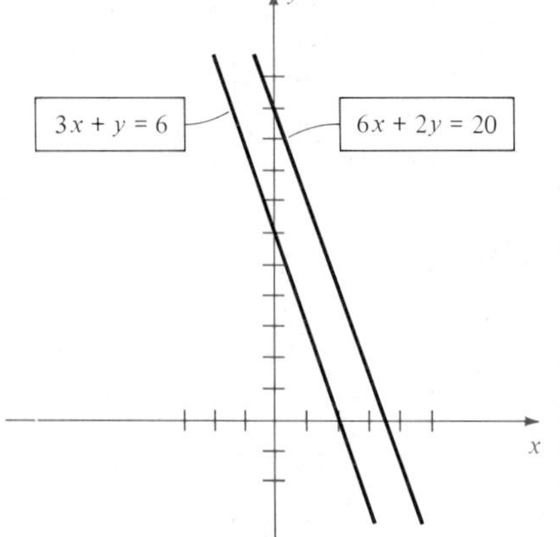

SOLUTION If we add to the second equation -2 times the first equation, we obtain the equivalent system

$$\begin{cases} 3x + y = 6 \\ 0 = 8. \end{cases}$$

The last equation can be written $0x + 0y = 8$, which is false for every ordered pair (x, y). Thus, the system has no solution.

The graphs of the two equations in the given system are lines that have the same slope, and hence are parallel (see Figure 6). The conclusion that the system has no solution corresponds to the fact that these lines do not intersect.

The preceding three examples illustrate typical outcomes of solving a system of two linear equations in two variables: there is either exactly one solution, an infinite number of solutions, or no solution. A system is **consistent** if it has at least one solution. A system with an infinite number of solutions is **dependent and consistent**. A system is **inconsistent** if it has no solution.

Since the graph of any linear equation $ax + by = c$ is a line, *exactly one* of the three cases listed in the following table holds for any system of two such equations.

CHARACTERISTICS OF A SYSTEM OF TWO LINEAR EQUATIONS IN TWO VARIABLES

Graphs	Number of solutions	Classification
Nonparallel lines	One solution	Consistent system
Identical lines	Infinite number of solutions	Dependent and consistent system
Parallel lines	No solution	Inconsistent system

In practice, we should have little difficulty determining which of the three cases occurs. The case of the unique solution will become apparent when suitable transformations are applied to the system, as illustrated in Example 1. The case of an infinite number of solutions is similar to that for Example 2, where one of the equations can be transformed into the other. The case of no solution is indicated by a contradiction, such as the statement $0 = 8$, which appeared in Example 3.

Certain applied problems can be solved by introducing systems of two linear equations, as illustrated in the next two examples.

EXAMPLE ▪ 4

A produce company has a 100-acre farm on which it grows lettuce and cabbage. Each acre of cabbage requires 600 hours of labor, and each acre of lettuce needs 400 hours of labor. If 45,000 hours are available and if all land and labor resources are to be used, find the number of acres of each crop that should be planted.

SOLUTION Let us introduce variables to denote the unknown quantities as follows:

$$x = \text{acres of cabbage},$$

$$y = \text{acres of lettuce}.$$

Thus the hours of labor required for each crop are:

$$600x = \text{hours required for cabbage}$$

$$400y = \text{hours required for lettuce}$$

Using the facts that the total number of acres is 100 and the total number of hours available is 45,000 leads to the following system:

$$\begin{cases} x + y = 100 \\ 600x + 400y = 45{,}000 \end{cases}$$

Multiplying both sides of the second equation by $\frac{1}{100}$, that is, dividing by 100, gives us

$$\begin{cases} x + y = 100 \\ 6x + 4y = 450 \end{cases}$$

Multiplying the first equation by -6, we obtain

$$\begin{cases} -6x - 6y = -600 \\ 6x + 4y = 450 \end{cases}$$

Adding the two equations, we see that $-2y = -150$, or $y = 75$. Finally, using $x + y = 100$, we obtain $x = 100 - y = 25$. Hence the company should plant 25 acres of cabbage and 75 acres of lettuce.

EXAMPLE ▪ 5

A motor boat, operating at full throttle, made a trip 4 miles upstream (against a constant current) in 15 minutes. The return trip (with the same current and at full throttle) took 12 minutes. Find the speed of the current and the equivalent speed of the boat in still water.

SOLUTION We begin by introducing variables to denote the unknown quantities. Thus, let

$$x = \text{speed of boat (in mi/hr)},$$
$$y = \text{speed of current (in mi/hr)}.$$

We plan to use the formula $d = rt$, where d denotes the distance traveled, r the rate, and t the time. Since the current slows the boat as it travels upstream, but adds to its speed as it travels downstream, we obtain

$$\text{upstream rate} = x - y \quad \text{(in mi/hr)}$$
$$\text{downstream rate} = x + y \quad \text{(in mi/hr)}.$$

The time (in hours) traveled in each direction is

$$\text{upstream time} = \tfrac{15}{60} = \tfrac{1}{4}\,\text{hr},$$
$$\text{downstream time} = \tfrac{12}{60} = \tfrac{1}{5}\,\text{hr}.$$

The distance is 4 miles for each trip. Substituting in $d = rt$ gives us the system

$$\begin{cases} 4 = (x - y)(\frac{1}{4}) \\ 4 = (x + y)(\frac{1}{5}) \end{cases}$$

or, equivalently,

$$\begin{cases} x - y = 16 \\ x + y = 20. \end{cases}$$

Adding the last two equations, we see that $2x = 36$, or $x = 18$. Consequently, $y = 20 - x = 20 - 18 = 2$. Hence, the speed of the boat in still water is 18 mi/hr and the speed of the current is 2 mi/hr.

EXERCISES

Exer. 1–20: Find the solutions of the system.

1 $\begin{cases} 2x + 3y = 2 \\ x - 2y = 8 \end{cases}$

2 $\begin{cases} 4x + 5y = 13 \\ 3x + y = -4 \end{cases}$

3 $\begin{cases} 2x + 5y = 16 \\ 3x - 7y = 24 \end{cases}$

4 $\begin{cases} 7x - 8y = 9 \\ 4x + 3y = -10 \end{cases}$

5 $\begin{cases} 3r + 4s = 3 \\ r - 2s = -4 \end{cases}$

6 $\begin{cases} 9u + 2v = 0 \\ 3u - 5v = 17 \end{cases}$

7 $\begin{cases} 5x - 6y = 4 \\ 3x + 7y = 8 \end{cases}$

8 $\begin{cases} 2x + 8y = 7 \\ 3x - 5y = 4 \end{cases}$

9 $\begin{cases} \frac{1}{3}c + \frac{1}{2}d = 5 \\ c - \frac{2}{3}d = -1 \end{cases}$

10 $\begin{cases} \frac{1}{2}t - \frac{1}{5}v = \frac{3}{2} \\ \frac{2}{3}t + \frac{1}{4}v = \frac{5}{12} \end{cases}$

11 $\begin{cases} \sqrt{3}x - \sqrt{2}y = 2\sqrt{3} \\ 2\sqrt{2}x + \sqrt{3}y = \sqrt{2} \end{cases}$

12 $\begin{cases} 0.11x - 0.03y = 0.25 \\ 0.12x + 0.05y = 0.70 \end{cases}$

13 $\begin{cases} 2x - 3y = 5 \\ -6x + 9y = 12 \end{cases}$

14 $\begin{cases} 3p - q = 7 \\ -12p + 4q = 3 \end{cases}$

15 $\begin{cases} 3m - 4n = 2 \\ -6m + 8n = -4 \end{cases}$

16 $\begin{cases} x - 5y = 2 \\ 3x - 15y = 6 \end{cases}$

17 $\begin{cases} 2y - 5x = 0 \\ 3y + 4x = 0 \end{cases}$

18 $\begin{cases} 3x + 7y = 9 \\ y = 5 \end{cases}$

19 $\begin{cases} \dfrac{2}{x} + \dfrac{3}{y} = -2 \\ \dfrac{4}{x} - \dfrac{5}{y} = 1 \end{cases}$ $\left(\textit{Hint: } \text{Let } u = \dfrac{1}{x} \text{ and } v = \dfrac{1}{y}.\right)$

20 $\begin{cases} \dfrac{3}{x-1} + \dfrac{4}{y+2} = 2 \\ \dfrac{6}{x-1} - \dfrac{7}{y+2} = -3 \end{cases}$

21 The price of admission for a high school play was \$1.50 for students and \$2.25 for nonstudents. If 450 tickets were sold for a total of \$777.75, how many of each kind were purchased?

22 An airline that flies from Los Angeles to Albuquerque with a stopover in Phoenix charges a fare of \$45 to Phoenix and a fare of \$60 from Los Angeles to Albuquerque. A total of 185 passengers boarded the plane in Los Angeles and fares totaled \$10,500. How many passengers got off the plane in Phoenix?

23 A crayon is to be 8 cm in length, 1 cm in diameter, and will be made from 5 cm^3 of colored wax. The crayon is to have the shape of a cylinder surmounted by a small conical tip (see figure). Find the length x of the cylinder and the height y of the cone.

EXERCISE 23

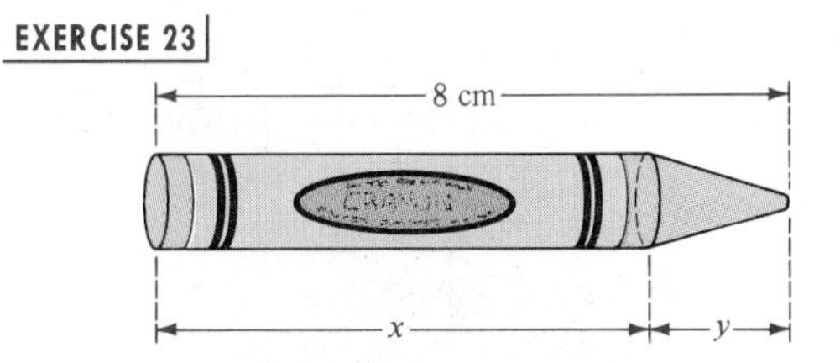

24 A man rows a boat 500 feet upstream against a constant current in 10 minutes. He then rows 300 feet downstream (with the same current) in 5 minutes. Find the speed of the current and the equivalent rate at which he can row in still water.

25 A large table for a conference room is to be constructed in the shape of a rectangle with two semicircles at the ends (see figure). The table is to have a perimeter of 40 feet and the area of the rectangular portion is to be twice the sum of the areas of the two ends. Find the length l and width w of the rectangular portion.

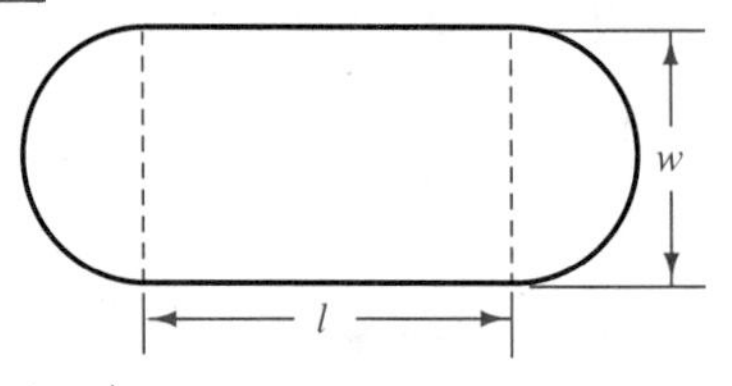

26 A woman has $15,000 to invest in two funds that pay simple interest at the rates of 6% and 8%. Interest on the 6% fund is tax-exempt; however, income tax must be paid on the 8% fund, and being in a high tax bracket, she does not wish to invest the entire sum in this account. Is there a way of investing the money so that she will receive $1000 in interest at the end of one year?

27 A bobcat population is classified by age as kittens (less than one year old) and adults (at least one year old). All adult females, including those born the prior year, have a litter each June, with an average litter size of three kittens. The springtime population of bobcats in a certain area is estimated to be 6000 and the male-female ratio is one. Estimate the number of adults and kittens in the population.

28 A 300-gallon water storage tank is filled by a single inlet pipe, and two identical outlet pipes can be used to supply water to the surrounding fields (see figure). It takes 5 hours to fill an empty tank when both outlet pipes are open. When one outlet pipe is closed, it takes 3 hours to fill the tank. Find the flow rates (in gallons per hour) in and out of the pipes.

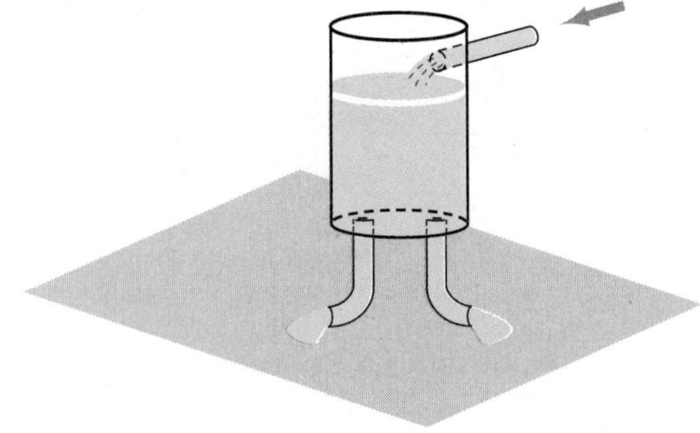

29 A silversmith has two alloys, one containing 35% silver, and the other 60% silver. How much of each should be melted and combined to obtain 100 grams of an alloy containing 50% silver?

30 A merchant wishes to mix peanuts costing $3 per pound with cashews costing $8 per pound to obtain 60 pounds of a mixture costing $5 per pound. How many pounds of each variety should be mixed?

31 An airplane, flying with a tail wind, travels 1200 miles in 2 hours. The return trip, against the wind, takes $2\frac{1}{2}$ hours. Find the cruising speed of the plane and the speed of the wind (assume that both rates are constant).

32 A stationery company sells two types of notepads to college bookstores, the first wholesaling for 50¢ and the second for 70¢. The company receives an order for 500 notepads together with a check for $286. If the order fails to specify the number of each type, how should the company fill the order?

33 As a ball rolls down an inclined plane, its velocity $v(t)$ (in cm/sec) at time t (in seconds) is given by $v(t) = v_0 + at$ for initial velocity v_0 and acceleration a (in cm/sec^2). If $v(2) = 16$ and $v(5) = 25$, find v_0 and a.

34 If an object is projected vertically upward from an altitude of s_0 feet with an initial velocity of v_0 ft/sec, then its distance $s(t)$ above the ground after t seconds is

$$s(t) = -16t^2 + v_0 t + s_0.$$

If $s(1) = 84$ and $s(2) = 116$, what are v_0 and s_0?

35 A small furniture company manufactures sofas and recliners. Each sofa requires 8 hours of labor and $60 in materials, while a recliner can be built for $35 in 6 hours. The company has 340 hours of labor available each week and can afford to buy $2250 in materials. How many recliners and sofas can be produced if all labor hours and all materials will be used?

36 A rancher is preparing an oat-cornmeal mixture for livestock. Each ounce of oats provides 4 grams of protein and 18 grams of carbohydrates, and an ounce of corn meal provides 3 grams of protein and 24 grams of carbohydrates. How many ounces of each can be used to meet the nutritional goals of 200 grams of protein and 1320 grams of carbohydrates per feeding?

37 A plumber and an electrician are each doing repairs on their offices and agree to swap services. The number of hours spent on each of the projects is shown in the following table.

	Plumber's office	Electrician's office
Plumber's hours	6	4
Electrician's hours	5	6

They would prefer to call the matter even, but due to tax laws, they must charge for all work performed. They agree to select hourly wage rates so that the bill on each project will match the income that each person would ordinarily receive for a comparable job.

(a) If x and y denote the hourly wages of the plumber and electrician, respectively, show that $6x + 5y = 10x$ and $4x + 6y = 11y$. Describe the solutions to this system.

(b) If the plumber ordinarily makes $20 per hour, what should the electrician charge?

38 Find equations for the altitudes of the triangle with vertices $A(-3, 2)$, $B(5, 4)$, $C(3, -8)$, and find the point at which the altitudes intersect.

9.3 SYSTEMS OF LINEAR EQUATIONS IN MORE THAN TWO VARIABLES

For systems of linear equations containing more than two variables, we can use either the method of substitution explained in Section 9.1 or the method of elimination developed in Section 9.2. The method of elimination is the shorter and more straightforward technique for finding solutions. In addition, it leads to the matrix technique, discussed in this section.

EXAMPLE ▪ 1

Find the solutions of the system

$$\begin{cases} x - 2y + 3z = 4 \\ 2x + y - 4z = 3 \\ -3x + 4y - z = -2. \end{cases}$$

SOLUTION We begin by eliminating x from the second and third equations. If we add -2 times the first equation to the second equation, we get the equivalent system

$$\begin{cases} x - 2y + 3z = 4 \\ 5y - 10z = -5 \\ -3x + 4y - z = -2. \end{cases}$$

Next we add 3 times the first equation to the third equation:

$$\begin{cases} x - 2y + 3z = 4 \\ 5y - 10z = -5 \\ -2y + 8z = 10. \end{cases}$$

To simplify computations we multiply the second equation by $\frac{1}{5}$:

$$\begin{cases} x - 2y + 3z = 4 \\ y - 2z = -1 \\ - 2y + 8z = 10. \end{cases}$$

We now eliminate y from the third equation by adding to it 2 times the second equation. This gives us

$$\begin{cases} x - 2y + 3z = 4 \\ y - 2z = -1 \\ 4z = 8. \end{cases}$$

Finally, multiplying the third equation by $\frac{1}{4}$, we obtain

$$\begin{cases} x - 2y + 3z = 4 \\ y - 2z = -1 \\ z = 2. \end{cases}$$

The solutions of the last system are easy to find by **back substitution**. From the third equation we see that $z = 2$. Substituting 2 for z in the second equation, $y - 2z = -1$, we get

$$y - 2(2) = -1$$
$$y = 3.$$

Finally, we find the x-value by substituting for y and z in the first equation:

$$x - 2(3) + 3(2) = 4$$
$$x = 4.$$

Thus, there is one solution, $(4, 3, 2)$.

Any system of three linear equations in three variables has either a *unique solution*, an *infinite number of solutions*, or *no solution*. As for two equations in two variables, the terminology used to describe these is *consistent*, *dependent and consistent*, or *inconsistent*, respectively.

If we analyze the method of solution in Example 1, we see that the symbols used for the variables are immaterial. The *coefficients* of the variables are what we must consider. Thus, if different symbols such as r, s, and t are used for the variables, we obtain the system

$$\begin{cases} r - 2s + 3t = 4 \\ 2r + s - 4t = 3 \\ -3r + 4s - t = -2. \end{cases}$$

The method of elimination could then proceed exactly as in the example. Since this is true, it is possible to simplify the process. Specifically, we introduce a scheme for keeping track of the coefficients in such a way that we do not have to write down the variables. Referring to the preceding system, we first check that variables appear in the same order in each equation and that terms not involving variables are to the right of the equal signs. We then list the numbers that are involved in the equations as follows:

$$\begin{bmatrix} 1 & -2 & 3 & 4 \\ 2 & 1 & -4 & 3 \\ -3 & 4 & -1 & -2 \end{bmatrix}$$

An array of numbers of this type is a **matrix**. The **rows** of the matrix are the numbers that appear next to one another *horizontally*. Thus, the first row R_1 is $1 \ -2 \ 3 \ 4$, the second row R_2 is $2 \ 1 \ -4 \ 3$, and the third row R_3 is $-3 \ 4 \ -1 \ -2$. The **columns** of the matrix are the numbers that appear *vertically*. For example, the first column consists of the numbers $1, 2, -3$ (in that order); the second column consist of $-2, 1, 4$; and so on.

The matrix obtained from a system of linear equations in the preceding manner is the **matrix of the system**. If we delete the last column of this matrix, the remaining array of numbers is the **coefficient matrix**. Since the matrix of the system can be obtained from the coefficient matrix by adjoining one column, we call it the **augmented coefficient matrix**, or simply the **augmented matrix**. Later, when we use matrices to find the solutions of a system of linear equations, we shall introduce a vertical rule in the augmented matrix to indicate where the equal signs would appear in the corresponding system of equations, as in the next illustration.

ILLUSTRATION | COEFFICIENT MATRIX AND AUGMENTED MATRIX

System	Coefficient matrix	Augmented matrix
$\begin{cases} x - 2y + 3z = 4 \\ 2x + y - 4z = 3 \\ -3x + 4y - z = -2 \end{cases}$	$\begin{bmatrix} 1 & -2 & 3 \\ 2 & 1 & -4 \\ -3 & 4 & -1 \end{bmatrix}$	$\left[\begin{array}{ccc\|c} 1 & -2 & 3 & 4 \\ 2 & 1 & -4 & 3 \\ -3 & 4 & -1 & -2 \end{array}\right]$

Before discussing a matrix method of solving a system of linear equations, let us state a general definition of matrix. We shall use a **double subscript notation**, denoting the number that appears in row i and column j by a_{ij}. The **row subscript** of a_{ij} is i and the **column subscript** is j.

DEFINITION

Let m and n be positive integers. An $\boldsymbol{m \times n}$ **matrix** is an array of the form

$$\begin{bmatrix} a_{11} & a_{12} & a_{13} & \cdots & a_{1n} \\ a_{21} & a_{22} & a_{23} & \cdots & a_{2n} \\ a_{31} & a_{32} & a_{33} & \cdots & a_{3n} \\ \vdots & \vdots & \vdots & & \vdots \\ a_{m1} & a_{m2} & a_{m3} & \cdots & a_{mn} \end{bmatrix}$$

where each a_{ij} is a real number.

The notation $m \times n$ in the definition is read *m by n*. It is possible to consider matrices in which the symbols a_{ij} represent complex numbers, polynomials, or other mathematical objects; however, we shall not do so in this text. The rows and columns of a matrix are defined as before. Thus, the matrix in the definition has m rows and n columns. Note that a_{23} is in row 2 and column 3 and a_{32} is in row 3 and column 2. Each a_{ij} is an **element of the matrix**. The elements $a_{11}, a_{22}, a_{33}, \ldots$ are the **main diagonal elements**. If $m = n$, the matrix is a **square matrix of order** $\boldsymbol{n}$.

To find the solutions of a system of linear equations we begin with the augmented matrix. If a variable does not appear in an equation, we assume that the coefficient is zero. We then work with the rows of the matrix *just as though they were equations*. The only items missing are the symbols for the variables, the addition signs used between terms, and the equal signs. We simply keep in mind that the numbers in the first column are the coefficients of the first variable, the numbers in the second column are the coefficients of the second variable, and so on. The rules for transforming a matrix are formulated so that they always produce a matrix of an equivalent system of equations.

The next theorem is a restatement, in terms of matrices, of the theorem on equivalent systems in Section 9.2. In (ii) of the theorem, the terminology *a row is multiplied by a nonzero constant* means that each element in the row is multiplied by the constant. To *add* two rows of a matrix, as in (iii), we add corresponding elements in each row.

MATRIX ROW TRANSFORMATION THEOREM

Given a matrix of a system of linear equations, a matrix of an equivalent system results if
(i) two rows are interchanged.
(ii) a row is multiplied by a nonzero constant.
(iii) a constant multiple of one row is added to another row.

We refer to (i)–(iii) as the **elementary row transformations** of a matrix. If a matrix is obtained from another matrix by one or more elementary row transformations, the two matrices are said to be **equivalent**, or more precisely, **row equivalent**. We shall use the following symbols to denote elementary row transformations of a matrix.

ELEMENTARY ROW TRANSFORMATIONS

Symbol	Meaning
$R_i \leftrightarrow R_j$	Interchange rows R_i and R_j
$kR_i \to R_i$	Multiply row R_i by k
$kR_i + R_j \to R_j$	Add kR_i to row R_j

The arrow $\to$ may be read *replaces*. Thus, for the transformation $kR_i \to R_i$, the constant multiple kR_i *replaces* R_i. Similarly, for $kR_i + R_j \to R_j$ the sum $kR_i + R_j$ *replaces* R_j. For convenience, we shall write $(-1)R_i$ as $-R_i$.

We shall next rework Example 1 using matrices. You should compare the two solutions, since analogous steps are employed in each case.

EXAMPLE ▪ 2

Find the solutions of the system

$$\begin{cases} x - 2y + 3z = 4 \\ 2x + y - 4z = 3 \\ -3x + 4y - z = -2. \end{cases}$$

SOLUTION We begin with the matrix of the system, that is, with the augmented matrix:

$$\left[\begin{array}{ccc|c} 1 & -2 & 3 & 4 \\ 2 & 1 & -4 & 3 \\ -3 & 4 & -1 & -2 \end{array}\right]$$

We next apply elementary row transformations to obtain another (simpler) matrix of an equivalent system of equations. These transformations correspond to the manipulations used for equations in Example 1. We will place appropriate symbols between equivalent matrices.

$$\left[\begin{array}{rrr|r} 1 & -2 & 3 & 4 \\ 2 & 1 & -4 & 3 \\ -3 & 4 & -1 & -2 \end{array}\right] \begin{array}{r} \\ -2R_1 + R_2 \to R_2 \\ 3R_1 + R_3 \to R_3 \end{array} \left[\begin{array}{rrr|r} 1 & -2 & 3 & 4 \\ 0 & 5 & -10 & -5 \\ 0 & -2 & 8 & 10 \end{array}\right] \quad \begin{array}{l} \\ \text{Add } -2R_1 \text{ to } R_2 \\ \text{Add } 3R_1 \text{ to } R_3 \end{array}$$

$$\tfrac{1}{5}R_2 \to R_2 \left[\begin{array}{rrr|r} 1 & -2 & 3 & 4 \\ 0 & 1 & -2 & -1 \\ 0 & -2 & 8 & 10 \end{array}\right] \quad \text{Multiply } R_2 \text{ by } \tfrac{1}{5}$$

$$2R_2 + R_3 \to R_3 \left[\begin{array}{rrr|r} 1 & -2 & 3 & 4 \\ 0 & 1 & -2 & -1 \\ 0 & 0 & 4 & 8 \end{array}\right] \quad \text{Add } 2R_2 \text{ to } R_3$$

$$\tfrac{1}{4}R_3 \to R_3 \left[\begin{array}{rrr|r} 1 & -2 & 3 & 4 \\ 0 & 1 & -2 & -1 \\ 0 & 0 & 1 & 2 \end{array}\right] \quad \text{Multiply } R_3 \text{ by } \tfrac{1}{4}$$

We use the final matrix to return to the system of equations

$$\begin{cases} x - 2y + 3z = 4 \\ y - 2z = -1 \\ z = 2, \end{cases}$$

which is equivalent to the original system. The solution $x = 4$, $y = 3$, $z = 2$ may now be found by back substitution as in Example 1.

The final matrix in the solution of Example 2 is in **echelon form**. In general, a matrix is in echelon form if it satisfies the following conditions.

ECHELON FORM OF A MATRIX

(i) The first nonzero number in each row, reading from left to right, is 1.

(ii) The column containing the first nonzero number in any row is to the left of the column containing the first nonzero number in the row below.

(iii) Rows consisting entirely of zeros appear at the bottom of the matrix.

The following is an illustration of a 6×7 matrix in echelon form. The symbols a_{ij} represent real numbers.

ILLUSTRATION | ECHELON FORM

$$\begin{bmatrix} 1 & a_{12} & a_{13} & a_{14} & a_{15} & a_{16} & a_{17} \\ 0 & 1 & a_{23} & a_{24} & a_{25} & a_{26} & a_{27} \\ 0 & 0 & 0 & 1 & a_{35} & a_{36} & a_{37} \\ 0 & 0 & 0 & 0 & 0 & 1 & a_{47} \\ 0 & 0 & 0 & 0 & 0 & 0 & 0 \\ 0 & 0 & 0 & 0 & 0 & 0 & 0 \end{bmatrix}$$

To find an echelon form for a matrix, we may apply elementary row transformations until we get the number 1 into the first position of the first row. Next we apply elementary row transformations of the type $kR_1 + R_j \to R_j$ for $j > 1$ to obtain the number 0 *underneath* that 1 in each of the remaining rows.

We then repeat this procedure for the second row as follows: *disregarding the first row completely*, we use elementary row transformations to get 1 into the first nonzero position of the second row, and then apply elementary row transformations of the type $kR_2 + R_j \to R_j$ for $j > 2$ to obtain the number 0 *underneath* that 1 in the third and remaining rows.

We then *disregard the first and second rows*, and repeat the procedure for the third row. Continuing this process, we eventually arrive at an echelon form. Not all echelon forms contain rows consisting only of zeros (see Example 2).

We can use elementary row operations to transform the matrix of any system of linear equations to echelon form. The echelon form can then be used to produce a system of equations that is equivalent to the original system. The solutions of the given system may be found by back substitution. The next example illustrates this technique for a system of four linear equations.

EXAMPLE ▪ 3

Find the solutions of the system

$$\begin{cases} -2x + 3y + 4z \phantom{{}+2w} = -1 \\ x \phantom{{}+3y} - 2z + 2w = 1 \\ y + z - w = 0 \\ 3x + y - 2z - w = 3. \end{cases}$$

SOLUTION We have arranged the equations so that the same variables appear in vertical columns. We begin with the augmented matrix and then obtain an echelon form as described in the preceding discussion.

$$\left[\begin{array}{rrrr|r} -2 & 3 & 4 & 0 & -1 \\ 1 & 0 & -2 & 2 & 1 \\ 0 & 1 & 1 & -1 & 0 \\ 3 & 1 & -2 & -1 & 3 \end{array}\right] \xrightarrow{R_1 \leftrightarrow R_2} \left[\begin{array}{rrrr|r} 1 & 0 & -2 & 2 & 1 \\ -2 & 3 & 4 & 0 & -1 \\ 0 & 1 & 1 & -1 & 0 \\ 3 & 1 & -2 & -1 & 3 \end{array}\right]$$

$$\begin{array}{r} \\ 2R_1 + R_2 \rightarrow R_2 \\ \\ -3R_1 + R_4 \rightarrow R_4 \end{array} \left[\begin{array}{rrrr|r} 1 & 0 & -2 & 2 & 1 \\ 0 & 3 & 0 & 4 & 1 \\ 0 & 1 & 1 & -1 & 0 \\ 0 & 1 & 4 & -7 & 0 \end{array}\right]$$

$$R_2 \leftrightarrow R_3 \left[\begin{array}{rrrr|r} 1 & 0 & -2 & 2 & 1 \\ 0 & 1 & 1 & -1 & 0 \\ 0 & 3 & 0 & 4 & 1 \\ 0 & 1 & 4 & -7 & 0 \end{array}\right]$$

$$\begin{array}{r} \\ \\ -3R_2 + R_3 \rightarrow R_3 \\ -R_2 + R_4 \rightarrow R_4 \end{array} \left[\begin{array}{rrrr|r} 1 & 0 & -2 & 2 & 1 \\ 0 & 1 & 1 & -1 & 0 \\ 0 & 0 & -3 & 7 & 1 \\ 0 & 0 & 3 & -6 & 0 \end{array}\right]$$

$$\begin{array}{r} \\ \\ \\ R_3 + R_4 \rightarrow R_4 \end{array} \left[\begin{array}{rrrr|r} 1 & 0 & -2 & 2 & 1 \\ 0 & 1 & 1 & -1 & 0 \\ 0 & 0 & -3 & 7 & 1 \\ 0 & 0 & 0 & 1 & 1 \end{array}\right]$$

$$\begin{array}{r} \\ \\ -\tfrac{1}{3}R_3 \rightarrow R_3 \\ \\ \end{array} \left[\begin{array}{rrrr|r} 1 & 0 & -2 & 2 & 1 \\ 0 & 1 & 1 & -1 & 0 \\ 0 & 0 & 1 & -\tfrac{7}{3} & -\tfrac{1}{3} \\ 0 & 0 & 0 & 1 & 1 \end{array}\right]$$

The final matrix is in echelon form and corresponds to the following system of equations:

$$\begin{cases} x \quad\;\; - 2z + 2w = 1 \\ \quad y + z - w = 0 \\ \qquad\quad z - \tfrac{7}{3}w = -\tfrac{1}{3} \\ \qquad\qquad\quad w = 1 \end{cases}$$

We now use back substitution to find the solution. From the last equation we see that $w = 1$. Substituting in the third equation $z - \frac{7}{3}w = -\frac{1}{3}$, we get

$$z - \tfrac{7}{3}(1) = -\tfrac{1}{3}, \quad \text{or} \quad z = \tfrac{6}{3} = 2.$$

Substituting $w = 1$ and $z = 2$ in the second equation $y + z - w = 0$, we obtain

$$y + 2 - 1 = 0, \quad \text{or} \quad y = -1.$$

Finally, from the first equation $x - 2z + 2w = 1$, we have

$$x - 2(2) + 2(1) = 1, \quad \text{or} \quad x = 3.$$

Hence, the system has one solution, $x = 3$, $y = -1$, $z = 2$, and $w = 1$.

After obtaining an echelon form it is often convenient to apply additional elementary row operations of the type $kR_i + R_j \to R_j$ so that 0 also appears *above* the first 1 in each row. We refer to the resulting matrix as being in **reduced echelon form**. The following is an illustration of a 6×7 matrix in reduced echelon form. (Compare with the echelon form on page 505.)

ILLUSTRATION | REDUCED ECHELON FORM

$$\blacksquare \quad \begin{bmatrix} 1 & 0 & a_{13} & 0 & a_{15} & 0 & a_{17} \\ 0 & 1 & a_{23} & 0 & a_{25} & 0 & a_{27} \\ 0 & 0 & 0 & 1 & a_{35} & 0 & a_{37} \\ 0 & 0 & 0 & 0 & 0 & 1 & a_{47} \\ 0 & 0 & 0 & 0 & 0 & 0 & 0 \\ 0 & 0 & 0 & 0 & 0 & 0 & 0 \end{bmatrix}$$

EXAMPLE ▪ 4

Find the solutions of the system in Example 3 using reduced echelon form.

SOLUTION We begin with the echelon form obtained in Example 3 and apply additional row operations as follows:

$$\left[\begin{array}{cccc|c} 1 & 0 & -2 & 2 & 1 \\ 0 & 1 & 1 & -1 & 0 \\ 0 & 0 & 1 & -\frac{7}{3} & -\frac{1}{3} \\ 0 & 0 & 0 & 1 & 1 \end{array}\right] \begin{array}{r} -2R_4 + R_1 \to R_1 \\ R_4 + R_2 \to R_2 \\ \frac{7}{3}R_4 + R_3 \to R_3 \\ \end{array} \left[\begin{array}{cccc|c} 1 & 0 & -2 & 0 & -1 \\ 0 & 1 & 1 & 0 & 1 \\ 0 & 0 & 1 & 0 & 2 \\ 0 & 0 & 0 & 1 & 1 \end{array}\right]$$

$$\begin{array}{r} 2R_3 + R_1 \to R_1 \\ -R_3 + R_2 \to R_2 \\ \\ \end{array} \left[\begin{array}{cccc|c} 1 & 0 & 0 & 0 & 3 \\ 0 & 1 & 0 & 0 & -1 \\ 0 & 0 & 1 & 0 & 2 \\ 0 & 0 & 0 & 1 & 1 \end{array}\right]$$

The system of equations corresponding to the reduced echelon form gives us the solution *without* using back substitution:

$$x = 3, \qquad y = -1, \qquad z = 2, \qquad w = 1$$

Sometimes it is necessary to consider systems in which the number of equations is not the same as the number of variables. The same matrix techniques are applicable, as illustrated in the next example.

EXAMPLE ▪ 5

Find the solutions of the system

$$\begin{cases} 2x + 3y + 4z = 1 \\ 3x + 4y + 5z = 3. \end{cases}$$

SOLUTION We shall begin with the augmented matrix and then find a reduced echelon form. There are many different ways of getting the number 1 into the first position of the first row. For example, the elementary row transformations $\frac{1}{2}R_1 \to R_1$ or $-\frac{1}{3}R_2 + R_1 \to R_1$ would accomplish this in one step. Another way, which does not involve fractions, is the following:

$$\left[\begin{array}{ccc|c} 2 & 3 & 4 & 1 \\ 3 & 4 & 5 & 3 \end{array}\right] \xrightarrow{R_1 \leftrightarrow R_2} \left[\begin{array}{ccc|c} 3 & 4 & 5 & 3 \\ 2 & 3 & 4 & 1 \end{array}\right]$$

$$\xrightarrow{-R_2 + R_1 \to R_1} \left[\begin{array}{ccc|c} 1 & 1 & 1 & 2 \\ 2 & 3 & 4 & 1 \end{array}\right]$$

$$\xrightarrow{-2R_1 + R_2 \to R_2} \left[\begin{array}{ccc|c} 1 & 1 & 1 & 2 \\ 0 & 1 & 2 & -3 \end{array}\right]$$

$$\xrightarrow{-R_2 + R_1 \to R_1} \left[\begin{array}{ccc|c} 1 & 0 & -1 & 5 \\ 0 & 1 & 2 & -3 \end{array}\right]$$

The reduced echelon form is the matrix of the system

$$\begin{cases} x \qquad - z = 5 \\ \qquad y + 2z = -3 \end{cases}$$

or, equivalently,

$$\begin{cases} x = z + 5 \\ y = -2z - 3. \end{cases}$$

There are an infinite number of solutions to this system; they can be found by assigning z any value c and then using the last two equations to express x and y in terms of c. This gives us

$$x = c + 5, \qquad y = -2c - 3, \qquad z = c.$$

Thus, the solutions of the system consist of all ordered triples of the form

$$(c + 5, -2c - 3, c)$$

for any real number c. The solutions may be checked by substituting $c + 5$ for x, $-2c - 3$ for y, and c for z in the two given equations.

We can obtain any number of solutions for the system by substituting specific real numbers for c. For example, if $c = 0$, we obtain $(5, -3, 0)$; if $c = 2$, we have $(7, -7, 2)$; and so on.

There are other ways to specify the general solution. For example, starting with $x = z + 5$ and $y = -2z - 3$, we could let $z = d - 5$ for any real number d. In this case,

$$x = z + 5 = (d - 5) + 5 = d,$$

$$y = -2z - 3 = -2(d - 5) - 3 = -2d + 7,$$

and the solutions of the system have the form

$$(d, -2d + 7, d - 5).$$

These triples produce the same solutions as $(c + 5, -2c - 3, c)$. For example, if $d = 5$, we get $(5, -3, 0)$; if $d = 7$, we obtain $(7, -7, 2)$; and so on.

A system of linear equations is **homogeneous** if all the terms that do not contain variables, that is, the *constant terms*, are zero. A system of homogeneous equations always has the **trivial solution** obtained by substituting zero for each variable. Nontrivial solutions sometimes exist. The procedure for finding solutions is the same as that used for nonhomogeneous systems.

EXAMPLE ▪ 6

Find the solutions of the homogeneous system

$$\begin{cases} x - y + 4z = 0 \\ 2x + y - z = 0 \\ -x - y + 2z = 0. \end{cases}$$

SOLUTION We begin with the augmented matrix and find a reduced echelon form:

$$\left[\begin{array}{rrr|r} 1 & -1 & 4 & 0 \\ 2 & 1 & -1 & 0 \\ -1 & -1 & 2 & 0 \end{array}\right] \begin{array}{r} \\ -2R_1 + R_2 \to R_2 \\ R_1 + R_3 \to R_3 \end{array} \left[\begin{array}{rrr|r} 1 & -1 & 4 & 0 \\ 0 & 3 & -9 & 0 \\ 0 & -2 & 6 & 0 \end{array}\right]$$

$$\begin{array}{r} \\ \tfrac{1}{3}R_2 \to R_2 \\ \\ \end{array} \left[\begin{array}{rrr|r} 1 & -1 & 4 & 0 \\ 0 & 1 & -3 & 0 \\ 0 & -2 & 6 & 0 \end{array}\right]$$

$$\begin{array}{r} R_2 + R_1 \to R_1 \\ \\ 2R_2 + R_3 \to R_3 \end{array} \left[\begin{array}{rrr|r} 1 & 0 & 1 & 0 \\ 0 & 1 & -3 & 0 \\ 0 & 0 & 0 & 0 \end{array}\right]$$

The reduced echelon form corresponds to the system

$$\begin{cases} x + z = 0 \\ y - 3z = 0 \end{cases}$$

or, equivalently,

$$\begin{cases} x = -z \\ y = 3z \end{cases}$$

Assigning any value c to z, we obtain $x = -c$ and $y = 3c$. The solutions consist of all ordered triples of the form $(-c, 3c, c)$ for any real number c.

EXAMPLE ■ 7

Find the solutions of the system

$$\begin{cases} x + y + z = 0 \\ x - y + z = 0 \\ x - y - z = 0. \end{cases}$$

SOLUTION We begin with the augmented matrix and find a reduced echelon form:

$$\left[\begin{array}{ccc|c} 1 & 1 & 1 & 0 \\ 1 & -1 & 1 & 0 \\ 1 & -1 & -1 & 0 \end{array}\right] \begin{array}{l} \\ -R_1 + R_2 \to R_2 \\ -R_1 + R_3 \to R_3 \end{array} \left[\begin{array}{ccc|c} 1 & 1 & 1 & 0 \\ 0 & -2 & 0 & 0 \\ 0 & -2 & -2 & 0 \end{array}\right]$$

$$-\tfrac{1}{2}R_2 \to R_2 \left[\begin{array}{ccc|c} 1 & 1 & 1 & 0 \\ 0 & 1 & 0 & 0 \\ 0 & -2 & -2 & 0 \end{array}\right]$$

$$\begin{array}{l} -R_2 + R_1 \to R_1 \\ \\ 2R_2 + R_3 \to R_3 \end{array} \left[\begin{array}{ccc|c} 1 & 0 & 1 & 0 \\ 0 & 1 & 0 & 0 \\ 0 & 0 & -2 & 0 \end{array}\right]$$

$$-\tfrac{1}{2}R_3 \to R_3 \left[\begin{array}{ccc|c} 1 & 0 & 1 & 0 \\ 0 & 1 & 0 & 0 \\ 0 & 0 & 1 & 0 \end{array}\right]$$

$$-R_3 + R_1 \to R_1 \left[\begin{array}{ccc|c} 1 & 0 & 0 & 0 \\ 0 & 1 & 0 & 0 \\ 0 & 0 & 1 & 0 \end{array}\right]$$

The reduced echelon form is the matrix of the system

$$x = 0, \qquad y = 0, \qquad z = 0.$$

Thus the only solution for the given system is the trivial one, $(0, 0, 0)$.

The next two examples illustrate applied problems.

EXAMPLE ▪ 8

A manufacturer of electrical equipment has the following information about the weekly profit from the production and sale of a type of electric motor.

Production level x	25	50	100
Profit $P(x)$ (dollars)	5250	7500	4500

(a) Determine a, b, and c so that the graph of $P(x) = ax^2 + bx + c$ fits this information.

(b) According to the quadratic function P in part (a), how many motors should be produced each week for maximum profit? What is the maximum weekly profit?

SOLUTION

(a) We see from the table that the graph of $P(x) = ax^2 + bx + c$ contains the points (25, 5250), (50, 7500), and (100, 4500). This gives us the system of equations

$$\begin{cases} 5250 = 625a + 25b + c \\ 7500 = 2500a + 50b + c \\ 4500 = 10000a + 100b + c \end{cases}$$

We can verify that the solution is $a = -2$, $b = 240$, $c = 500$.

(b) From part (a),

$$P(x) = -2x^2 + 240x + 500.$$

Since $a = -2 < 0$, the graph of the quadratic function P is a parabola that opens downward. Using the formula on page 187, the x-coordinate of the vertex is

$$x = \frac{-b}{2a} = \frac{-240}{2(-2)} = \frac{-240}{-4} = 60.$$

Hence the manufacturer should produce and sell 60 motors per week. The maximum weekly profit is

$$P(60) = -2(60)^2 + 240(60) + 500 = \$7700.$$

EXAMPLE ▪ 9

A merchant wishes to mix two grades of peanuts costing \$3 and \$4 per pound, respectively, with cashews costing \$8 per pound, to obtain 140 pounds of a mixture costing \$6 per pound. If the merchant also wants the amount of cheaper-grade peanuts to be twice that of the better grade, how many pounds of each variety should be mixed?

SOLUTION Let us introduce three variables as follows:

x = pounds of peanuts at \$3 per pound,

y = pounds of peanuts at \$4 per pound,

z = pounds of cashews at \$8 per pound.

We refer to the statement of the problem and obtain the following system:

$$\begin{cases} x + y + z = 140 \\ 3x + 4y + 8z = 6(140) \\ x = 2y. \end{cases}$$

You may verify that the solution of this system is $x = 40$, $y = 20$, $z = 80$. Thus, the merchant should use 40 pounds of the \$3 peanuts, 20 pounds of the \$4 peanuts, and 80 pounds of cashews.

9.3 EXERCISES

Exer. 1–26: Use matrices to solve the system.

1 $\begin{cases} x - 2y - 3z = -1 \\ 2x + y + z = 6 \\ x + 3y - 2z = 13 \end{cases}$

2 $\begin{cases} x + 3y - z = -3 \\ 3x - y + 2z = 1 \\ 2x - y + z = -1 \end{cases}$

3 $\begin{cases} 5x + 2y - z = -7 \\ x - 2y + 2z = 0 \\ 3y + z = 17 \end{cases}$

4 $\begin{cases} 4x - y + 3z = 6 \\ -8x + 3y - 5z = -6 \\ 5x - 4y = -9 \end{cases}$

5 $\begin{cases} 2x + 6y - 4z = 1 \\ x + 3y - 2z = 4 \\ 2x + y - 3z = -7 \end{cases}$

6 $\begin{cases} x + 3y - 3z = -5 \\ 2x - y + z = -3 \\ -6x + 3y - 3z = 4 \end{cases}$

7 $\begin{cases} 2x - 3y + 2z = -3 \\ -3x + 2y + z = 1 \\ 4x + y - 3z = 4 \end{cases}$

8 $\begin{cases} 2x - 3y + z = 2 \\ 3x + 2y - z = -5 \\ 5x - 2y + z = 0 \end{cases}$

9 $\begin{cases} x + 3y + z = 0 \\ x + y - z = 0 \\ x - 2y - 4z = 0 \end{cases}$

10 $\begin{cases} 2x - y + z = 0 \\ x - y - 2z = 0 \\ 2x - 3y - z = 0 \end{cases}$

11 $\begin{cases} 2x + y + z = 0 \\ x - 2y - 2z = 0 \\ x + y + z = 0 \end{cases}$

12 $\begin{cases} x + y - 2z = 0 \\ x - y - 4z = 0 \\ y + z = 0 \end{cases}$

13 $\begin{cases} 3x - 2y + 5z = 7 \\ x + 4y - z = -2 \end{cases}$

14 $\begin{cases} 2x - y + 4z = 8 \\ -3x + y - 2z = 5 \end{cases}$

15 $\begin{cases} 4x - 2y + z = 5 \\ 3x + y - 4z = 0 \end{cases}$

16 $\begin{cases} 5x + 2y - z = 10 \\ y + z = -3 \end{cases}$

17 $\begin{cases} x + 2y - z - 3w = 2 \\ 3x + y - 2z - w = 6 \\ x + y + 3z - 2w = -3 \\ -2x - 2y + 3z + w = -9 \end{cases}$

18 $\begin{cases} x - 2y - 5z + w = -1 \\ 2x - y + z + w = 1 \\ 3x - 2y - 4z - 2w = 1 \\ x + y + 3z - 2w = 2 \end{cases}$

19 $\begin{cases} 2x - y - 2z + 2s - 5t = 2 \\ x + 3y - 2z + s - 2t = -5 \\ -x + 4y + 2z - 3s + 8t = -4 \\ 3x - 2y - 4z + s - 3t = -3 \\ 4x - 6y + z - 2s + t = 10 \end{cases}$

20 $\begin{cases} 3x + 2y + z + 3u + v + w = 1 \\ 2x + y - 2z + 3u - v + 4w = 6 \\ 6x + 3y + 4z - u + 2v + w = -6 \\ x + y + z + u - v - w = 8 \\ -2x - 2y + z - 3u + 2v - 3w = -10 \\ x - 3y + 2z + u + 3v + w = -1 \end{cases}$

21 $\begin{cases} 5x + 2z = 1 \\ y - 3z = 2 \\ 2x + y = 3 \end{cases}$

22 $\begin{cases} 2x - 3y = 12 \\ 3y + z = -2 \\ 5x - 3z = 3 \end{cases}$

23 $\begin{cases} 4x - 3y = 1 \\ 2x + y = -7 \\ -x + y = -1 \end{cases}$

24 $\begin{cases} 2x + 3y = -2 \\ x + y = 1 \\ x - 2y = 13 \end{cases}$

25 $\begin{cases} 2x + 3y = 5 \\ x - 3y = 4 \\ x + y = -2 \end{cases}$

26 $\begin{cases} 4x - y = 2 \\ 2x + 2y = 1 \\ 4x - 5y = 3 \end{cases}$

27 Three solutions contain a certain acid. The first contains 10% acid, the second 30%, and the third 50%. A chemist wishes to use all three solutions to obtain a mixture of 50 liters containing 32% acid, using twice as much of the 50% solution as the 30% solution. How many liters of each solution should be used?

28 A swimming pool can be filled by three pipes A, B, and C. Pipe A alone can fill the pool in 8 hours. If pipes A and C are used together, the pool can be filled in 6 hours. If B and C are used together, it takes 10 hours. How long does it take to fill the pool if all three pipes are used?

29 A company has three machines A, B, and C that are each capable of producing a certain item. However, because of a lack of skilled operators, only two of the machines can be used simultaneously. The following table indicates production over a three-day period using various combinations of the machines.

Machines used	Hours used	Items produced
A and B	6	4500
A and C	8	3600
B and C	7	4900

How long would it take each machine, if used alone, to produce 1000 items?

30 In electrical circuits, the formula $1/R = (1/R_1) + (1/R_2)$ is used to find the total resistance R if two resistors R_1 and R_2 are connected in parallel. Given three resistors A, B, and C, suppose that the total resistance is 48 ohms if A and B are connected in parallel, 80 ohms if B and C are connected in parallel, and 60 ohms if A and C are connected in parallel. Find the resistances of A, B, and C.

31 A supplier of lawn products has three types of grass fertilizer, G_1, G_2, and G_3, having nitrogen contents of 30%, 20%, and 15%, respectively. The supplier plans to mix them, obtaining 600 pounds of fertilizer with a 25% nitrogen content. In addition, the mixture is to contain 100 pounds more of type G_3 than of type G_2. How much of each type should be used?

32 If a particle moves along a coordinate line with a constant acceleration a (in cm/sec^2), then at time t (in seconds) its distance $s(t)$ (in cm) from the origin is

$$s(t) = \tfrac{1}{2}at^2 + v_0 t + s_0$$

for the velocity v_0 and distance s_0 from the origin at $t = 0$. If the distances of the particle from the origin at $t = \frac{1}{2}$, $t = 1$, and $t = \frac{3}{2}$ are 7, 11, and 17, respectively, find a, v_0, and s_0.

33 Shown in the figure is an electrical circuit containing three resistors, a 6-volt battery, and a 12-volt battery. It can be shown, using Kirchhoff's laws, that the three

EXERCISE 33

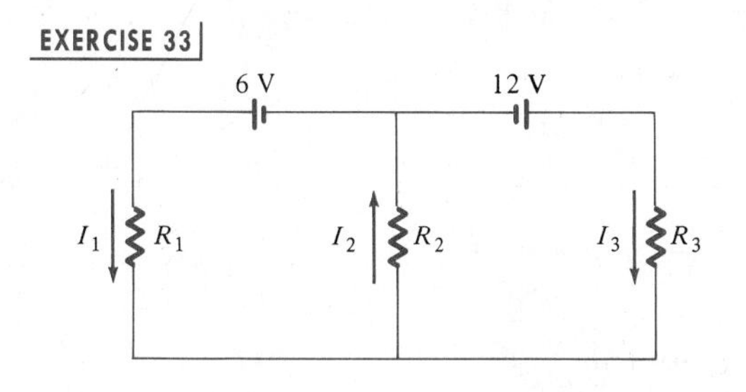

currents I_1, I_2, and I_3 are solutions of the following system of equations:

$$\begin{cases} I_1 - I_2 + I_3 = 0 \\ R_1 I_1 + R_2 I_2 = 6 \\ R_2 I_2 + R_3 I_3 = 12. \end{cases}$$

Find the three currents if

(a) $R_1 = R_2 = R_3 = 3$ ohms.

(b) $R_1 = 4$ ohms, $R_2 = 1$ ohm, and $R_3 = 4$ ohms.

34 A stable population of 35,000 birds lives on three islands. Each year 10% of the population on island A migrates to island B, 20% of the population on island B migrates to island C, and 5% of the population on island C migrates to A. Find the number of birds on each island if the population count on each island does not vary from year to year.

35 A shop specializes in preparing blends of gourmet coffees. The owner wishes to prepare 1-pound bags that will sell for $8.50 from Columbian, Brazilian, and Kenyan coffees. The cost per pound of these coffees is $10, $6, and $8, respectively. The amount of Columbian is to be three times the amount of Brazilian. Find the amount of each type of coffee in the blend.

36 A rancher has 750 head of cattle consisting of 400 adults (aged 2 or more years), 150 yearlings, and 200 calves. The following information is known about this particular species. Each spring an adult female gives birth to a single calf and 75% of these calves will survive the first year. The yearly survival percentages for yearlings and adults are 80% and 90%, respectively. The male-female ratio is one in all age classes. Estimate the population of each age class

(a) next spring. **(b)** last spring.

37 Shown in the figure is a system of four one-way streets leading into the center of a city. The numbers in the figure denote the average number of vehicles per hour that travel in the directions shown. A total of 300 vehicles enter the area and 300 vehicles leave the area every hour. Signals at intersections A, B, C, and D are to be timed in order to avoid congestion, and this timing will determine traffic flow rates x_1, x_2, x_3 and x_4.

(a) If the number of vehicles entering an intersection per hour must equal the number leaving the intersection per hour, describe the traffic flow rates at each intersection with a system of equations.

(b) If the signal at intersection C is timed so that $x_3 = 100$, find x_1, x_2, and x_4.

(c) Use the system in part (a) to explain why $75 \le x_3 \le 150$.

EXERCISE 37

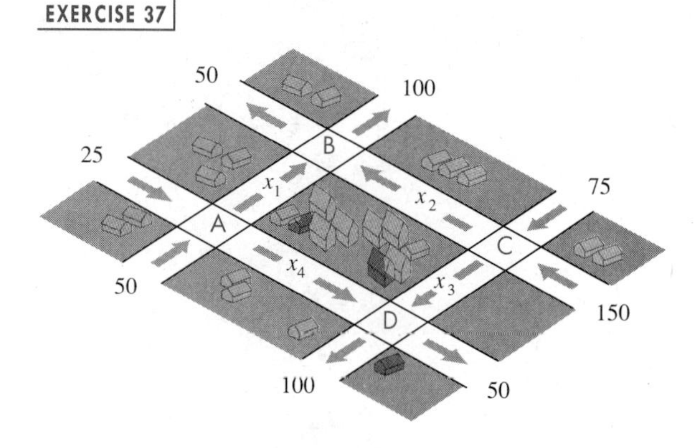

38 If $f(x) = ax^3 + bx + c$, determine a, b, and c such that the graph of f passes through the points $(-3, -12)$, $(-1, 22)$, and $(2, 13)$.

39 Find an equation of the circle that passes through the three points $P_1(2, 1)$, $P_2(-1, -4)$, and $P_3(3, 0)$. (*Hint:* An equation of the circle is $x^2 + y^2 + ax + by + c = 0$.)

40 Determine a, b, and c such that the graph of the equation $y = ax^2 + bx + c$ passes through the points $P_1(3, -1)$, $P_2(1, -7)$, and $P_3(-2, 14)$.

9.4 PARTIAL FRACTIONS

In this section we show how systems of equations can be used to help decompose rational expressions into sums of simpler expressions. This technique is useful in advanced mathematics courses.

It is easy to verify that

$$\frac{2}{x^2 - 1} = \frac{1}{x - 1} + \frac{-1}{x + 1}.$$

The expression on the right-hand side of this equation is the *partial fraction decomposition* of $2/(x^2 - 1)$.

It is theoretically possible to write *any* rational expression as a sum of rational expressions whose denominators involve powers of polynomials of degree not greater than two. Specifically, if $f(x)$ and $g(x)$ are polynomials *and the degree of* $f(x)$ *is less than the degree of* $g(x)$, it can be proved that

$$\frac{f(x)}{g(x)} = F_1 + F_2 + \cdots + F_r$$

such that each F_k has one of the forms

$$\frac{A}{(px + q)^m} \quad \text{or} \quad \frac{Ax + B}{(ax^2 + bx + c)^n}$$

for some real numbers A and B and nonnegative integers m and n, and where the quadratic polynomial $ax^2 + bx + c$ is **irreducible**, that is, it has no real zero. The sum $F_1 + F_2 + \cdots + F_r$ is the **partial fraction decomposition** of $f(x)/g(x)$ and each F_k is a **partial fraction**. We shall not prove this result but will, instead, give rules for obtaining the decomposition.

To find the partial fraction decomposition of $f(x)/g(x)$ *it is essential that* $f(x)$ *have lower degree than* $g(x)$. If this is not the case, use long division to obtain such an expression. For example, given

$$\frac{x^3 - 6x^2 + 5x - 3}{x^2 - 1},$$

we obtain

$$\frac{x^3 - 6x^2 + 5x - 3}{x^2 - 1} = x - 6 + \frac{6x - 9}{x^2 - 1}.$$

We then find the partial fraction decomposition of $(6x - 9)/(x^2 - 1)$.

GUIDELINES FINDING PARTIAL FRACTION DECOMPOSITIONS OF $\dfrac{f(x)}{g(x)}$

1 If the degree of $f(x)$ is not lower than the degree of $g(x)$, use long division to obtain the proper form.

2 Express $g(x)$ as a product of linear factors $px + q$ or irreducible quadratic factors $ax^2 + bx + c$, and collect repeated factors so that $g(x)$ is a product of *different* factors of the form $(px + q)^m$ or $(ax^2 + bx + c)^n$ for a nonnegative integer m or n.

3 Apply the following rules.

RULE A For each factor of the form $(px + q)^m$ with $m \geq 1$, the partial fraction decomposition contains a sum of m partial fractions of the form

$$\frac{A_1}{px + q} + \frac{A_2}{(px + q)^2} + \cdots + \frac{A_m}{(px + q)^m}$$

where each numerator A_k is a real number.

RULE B For each factor of the form $(ax^2 + bx + c)^n$ with $n \geq 1$ and $ax^2 + bx + c$ irreducible, the partial fraction decomposition contains a sum of n partial fractions of the form

$$\frac{A_1x + B_1}{ax^2 + bx + c} + \frac{A_2x + B_2}{(ax^2 + bx + c)^2} + \cdots + \frac{A_nx + B_n}{(ax^2 + bx + c)^n}$$

where each A_k and B_k is a real number. ❑

EXAMPLE ▪ 1

Find the partial fraction decomposition of

$$\frac{4x^2 + 13x - 9}{x^3 + 2x^2 - 3x}.$$

SOLUTION The denominator has the factored form $x(x + 3)(x - 1)$. Each factor has the form stated in Rule A, with $m = 1$. Thus, for the factor x there corresponds a partial fraction of the form A/x. Similarly, for the factors $x + 3$ and $x - 1$ there correspond partial fractions $B/(x + 3)$ and $C/(x - 1)$, respectively. The partial fraction decomposition has the form

$$\frac{4x^2 + 13x - 9}{x(x + 3)(x - 1)} = \frac{A}{x} + \frac{B}{x + 3} + \frac{C}{x - 1}.$$

We multiply both sides by the lcd, $x(x+3)(x-1)$:

$$\begin{aligned} 4x^2 + 13x - 9 &= A(x+3)(x-1) + Bx(x-1) + Cx(x+3) \\ &= A(x^2 + 2x - 3) + B(x^2 - x) + C(x^2 + 3x) \\ &= (A + B + C)x^2 + (2A - B + 3C)x - 3A \end{aligned}$$

If we equate the coefficients of like powers of x on each side of the last equation, we obtain the system of equations

$$\begin{cases} A + B + C = 4 \\ 2A - B + 3C = 13 \\ -3A = -9. \end{cases}$$

The solution is $A = 3$, $B = -1$, $C = 2$, and the partial fraction decomposition is

$$\frac{4x^2 + 13x - 9}{x(x+3)(x-1)} = \frac{3}{x} + \frac{-1}{x+3} + \frac{2}{x-1}.$$

There is an alternative way to find A, B, and C if all factors of the denominator are linear and nonrepeated, as in this example. Instead of equating coefficients and using a system of equations, we begin with the equation

$$4x^2 + 13x - 9 = A(x+3)(x-1) + Bx(x-1) + Cx(x+3).$$

We next substitute values for x that make the factors, $x, x - 1$, and $x - 3$, zero. If we let $x = 0$ and simplify, we obtain

$$-9 = -3A, \quad \text{or} \quad A = 3.$$

Letting $x = 1$ in the equation leads to $8 = 4C$, or $C = 2$. Finally, letting $x = -3$, we obtain $-12 = 12B$, or $B = -1$.

EXAMPLE ▪ 2

Find the partial fraction decomposition of

$$\frac{x^2 + 10x - 36}{x(x-3)^2}.$$

SOLUTION By Rule A with $m = 1$, there is a partial fraction A/x corresponding to the factor x. Next, applying Rule A with $m = 2$, the factor $(x-3)^2$ determines a sum of two partial fractions of the form $B/(x-3) + C/(x-3)^2$.

Thus, the partial fraction decomposition has the form

$$\frac{x^2 + 10x - 36}{x(x-3)^2} = \frac{A}{x} + \frac{B}{x-3} + \frac{C}{(x-3)^2}.$$

We multiply both sides by the lcd, $x(x-3)^2$:

$$\begin{aligned} x^2 + 10x - 36 &= A(x-3)^2 + Bx(x-3) + Cx \\ &= A(x^2 - 6x + 9) + B(x^2 - 3x) + Cx \\ &= (A + B)x^2 + (-6A - 3B + C)x + 9A \end{aligned}$$

We next equate the coefficients of like powers of x, obtaining the system

$$\begin{cases} A + B = 1 \\ -6A - 3B + C = 10 \\ 9A = -36. \end{cases}$$

This system of equations has the solution $A = -4$, $B = 5$, $C = 1$. The partial fraction decomposition is therefore

$$\frac{x^2 + 10x - 36}{x(x-3)^2} = \frac{-4}{x} + \frac{5}{x-3} + \frac{1}{(x-3)^2}.$$

As in Example 1, we could also obtain A and C by beginning with the equation

$$x^2 + 10x - 36 = A(x-3)^2 + Bx(x-3) + Cx$$

and then substituting values for x that make the factors, $x - 3$ and x, zero. Thus, letting $x = 3$, we obtain $3 = 3C$, or $C = 1$. Letting $x = 0$ gives us $-36 = 9A$, or $A = -4$. The value of B may then be found by using one of the equations in the system.

EXAMPLE ▪ 3

Find the partial fraction decomposition of

$$\frac{x^2 - x - 21}{2x^3 - x^2 + 8x - 4}.$$

SOLUTION The denominator may be factored by grouping, as follows:

$$2x^3 - x^2 + 8x - 4 = x^2(2x-1) + 4(2x-1) = (x^2 + 4)(2x - 1)$$

Applying Rule B to the irreducible quadratic factor $x^2 + 4$, we see that one of the partial fractions has the form $(Ax + B)/(x^2 + 4)$. By Rule A, there is also a partial fraction $C/(2x - 1)$ corresponding to $2x - 1$. Consequently,

$$\frac{x^2 - x - 21}{2x^3 - x^2 + 8x - 4} = \frac{Ax + B}{x^2 + 4} + \frac{C}{2x - 1}.$$

As in previous examples, this leads to

$$\begin{aligned} x^2 - x - 21 &= (Ax + B)(2x - 1) + C(x^2 + 4) \\ &= 2Ax^2 - Ax + 2Bx - B + Cx^2 + 4C \\ &= (2A + C)x^2 + (-A + 2B)x - B + 4C. \end{aligned}$$

This gives us the system

$$\begin{cases} 2A \qquad + C = 1 \\ -A + 2B \qquad = -1 \\ \qquad - B + 4C = -21, \end{cases}$$

which has the solution $A = 3$, $B = 1$, $C = -5$. Thus, the partial fraction decomposition is

$$\frac{x^2 - x - 21}{2x^3 - x^2 + 8x - 4} = \frac{3x + 1}{x^2 + 4} + \frac{-5}{2x - 1}.$$

EXAMPLE ▪ 4

Find the partial fraction decomposition of

$$\frac{5x^3 - 3x^2 + 7x - 3}{(x^2 + 1)^2}.$$

SOLUTION We apply Rule B with $n = 2$:

$$\frac{5x^3 - 3x^2 + 7x - 3}{(x^2 + 1)^2} = \frac{Ax + B}{x^2 + 1} + \frac{Cx + D}{(x^2 + 1)^2}.$$

Multiplying both sides by $(x^2 + 1)^2$ gives us

$$\begin{aligned} 5x^3 - 3x^2 + 7x - 3 &= (Ax + B)(x^2 + 1) + Cx + D \\ &= Ax^3 + Bx^2 + (A + C)x + (B + D). \end{aligned}$$

Comparing the coefficients of x^3 and x^2, we obtain $A = 5$ and $B = -3$.

From the coefficients of x we see that $A + C = 7$. Thus $C = 7 - A = 7 - 5 = 2$. Finally, the constant terms give us $B + D = -3$ and $D = -3 - B = -3 - (-3) = 0$. Therefore,

$$\frac{5x^3 - 3x^2 + 7x - 3}{(x^2 + 1)^2} = \frac{5x - 3}{x^2 + 1} + \frac{2x}{(x^2 + 1)^2}.$$

9.4 EXERCISES

Exer. 1–26: Find the partial fraction decomposition.

1 $\dfrac{8x - 1}{(x - 2)(x + 3)}$

2 $\dfrac{x - 29}{(x - 4)(x + 1)}$

3 $\dfrac{x + 34}{x^2 - 4x - 12}$

4 $\dfrac{5x - 12}{x^2 - 4x}$

5 $\dfrac{4x^2 - 15x - 1}{(x - 1)(x + 2)(x - 3)}$

6 $\dfrac{x^2 + 19x + 20}{x(x + 2)(x - 5)}$

7 $\dfrac{4x^2 - 5x - 15}{x^3 - 4x^2 - 5x}$

8 $\dfrac{37 - 11x}{(x + 1)(x^2 - 5x + 6)}$

9 $\dfrac{2x + 3}{(x - 1)^2}$

10 $\dfrac{5x^2 - 4}{x^2(x + 2)}$

11 $\dfrac{19x^2 + 50x - 25}{3x^3 - 5x^2}$

12 $\dfrac{10 - x}{x^2 + 10x + 25}$

13 $\dfrac{x^2 - 6}{(x + 2)^2(2x - 1)}$

14 $\dfrac{2x^2 + x}{(x - 1)^2(x + 1)^2}$

15 $\dfrac{3x^3 + 11x^2 + 16x + 5}{x(x + 1)^3}$

16 $\dfrac{4x^3 + 3x^2 + 5x - 2}{x^3(x + 2)}$

17 $\dfrac{x^2 + x - 6}{(x^2 + 1)(x - 1)}$

18 $\dfrac{x^2 - x - 21}{(x^2 + 4)(2x - 1)}$

19 $\dfrac{9x^2 - 3x + 8}{x^3 + 2x}$

20 $\dfrac{2x^3 + 2x^2 + 4x - 3}{x^4 + x^2}$

21 $\dfrac{4x^3 - x^2 + 4x + 2}{(x^2 + 1)^2}$

22 $\dfrac{3x^3 + 13x - 1}{(x^2 + 4)^2}$

23 $\dfrac{2x^4 - 2x^3 + 6x^2 - 5x + 1}{x^3 - x^2 + x - 1}$

24 $\dfrac{x^3}{x^3 - 3x^2 + 9x - 27}$

25 $\dfrac{4x^3 + 4x^2 - 4x + 2}{2x^2 - x - 1}$

26 $\dfrac{x^5 - 5x^4 + 7x^3 - x^2 - 4x + 12}{x^3 - 3x^2}$

9.5 THE ALGEBRA OF MATRICES

In this section we discuss algebraic properties of matrices. In the following definition the symbol (a_{ij}) denotes an $m \times n$ matrix A of the type displayed in the definition on page 502. We use similar notations for the matrices B and C.

EQUALITY AND ADDITION OF MATRICES

> Let $A = (a_{ij})$, $B = (b_{ij})$, and $C = (c_{ij})$ be $m \times n$ matrices.
> (i) $A = B$ means that $a_{ij} = b_{ij}$ for every i and j.
> (ii) $C = A + B$ means that $c_{ij} = a_{ij} + b_{ij}$ for every i and j.

By definition, two matrices are equal if and only if corresponding elements are identical.

ILLUSTRATION | EQUALITY

■ $$\begin{bmatrix} 1 & 0 & 5 \\ \sqrt[3]{8} & 3^2 & -2 \end{bmatrix} = \begin{bmatrix} (-1)^2 & 0 & \sqrt{25} \\ 2 & 9 & -2 \end{bmatrix}$$

To add two matrices we add the elements in corresponding positions in each matrix. *Two matrices can be added only if they have the same number of rows and the same number of columns.*

ILLUSTRATION | ADDITION

■ $$\begin{bmatrix} 4 & -5 \\ 0 & 4 \\ -6 & 1 \end{bmatrix} + \begin{bmatrix} 3 & 2 \\ 7 & -4 \\ -2 & 1 \end{bmatrix} = \begin{bmatrix} 7 & -3 \\ 7 & 0 \\ -8 & 2 \end{bmatrix}$$

The $\boldsymbol{m \times n}$ **zero matrix**, denoted by O, is the matrix with m rows and n columns in which every element is 0.

ILLUSTRATION | ZERO MATRICES

■ $$\begin{bmatrix} 0 & 0 \\ 0 & 0 \end{bmatrix}$$ ■ $$\begin{bmatrix} 0 & 0 \\ 0 & 0 \\ 0 & 0 \end{bmatrix}$$ ■ $$\begin{bmatrix} 0 & 0 & 0 & 0 \\ 0 & 0 & 0 & 0 \end{bmatrix}$$

The **additive inverse** $\boldsymbol{-A}$ of the matrix $A = (a_{ij})$ is the matrix $(-a_{ij})$ obtained by changing the sign of each element of A.

ILLUSTRATION | ADDITIVE INVERSE

■ $$-\begin{bmatrix} 2 & -3 & 4 \\ -1 & 0 & 5 \end{bmatrix} = \begin{bmatrix} -2 & 3 & -4 \\ 1 & 0 & -5 \end{bmatrix}$$

The proof of the next theorem follows directly from the definition of addition of matrices.

THEOREM

If A, B, and C are $m \times n$ matrices and if O is the $m \times n$ zero matrix, then:

(i) $A + B = B + A$

(ii) $A + (B + C) = (A + B) + C$

(iii) $A + O = A$

(iv) $A + (-A) = O$

Subtraction of two $m \times n$ matrices is defined by

$$A - B = A + (-B).$$

Using the parentheses notation for matrices,

$$(a_{ij}) - (b_{ij}) = (a_{ij}) + (-b_{ij}) = (a_{ij} - b_{ij}).$$

Thus, to subtract two matrices, we subtract the elements in corresponding positions.

ILLUSTRATION | SUBTRACTION

■ $$\begin{bmatrix} 4 & -5 \\ 0 & 4 \\ -6 & 1 \end{bmatrix} - \begin{bmatrix} 3 & 2 \\ 7 & -4 \\ -2 & 1 \end{bmatrix} = \begin{bmatrix} 1 & -7 \\ -7 & 8 \\ -4 & 0 \end{bmatrix}$$

DEFINITION

> The **product** of a real number c and an $m \times n$ matrix $A = (a_{ij})$ is
>
> $$cA = (ca_{ij}).$$

To find cA, we multiply each element of A by c.

ILLUSTRATION | PRODUCT OF A NUMBER AND A MATRIX

■ $$3\begin{bmatrix} 4 & -1 \\ 2 & 3 \end{bmatrix} = \begin{bmatrix} 12 & -3 \\ 6 & 9 \end{bmatrix}$$

We can prove the following.

THEOREM

> If A and B are $m \times n$ matrices and if c and d are real numbers,
>
> (i) $c(A + B) = cA + cB$
>
> (ii) $(c + d)A = cA + dA$
>
> (iii) $(cd)A = c(dA)$

To define the product AB of two matrices A and B, *the number of columns of A must be the same as the number of rows of B*; that is, if A is

$m \times n$ for integers m and n, then B must be $n \times p$ for an integer p. If $C = AB$, then the element c_{ij} in row i and column j of the product may be found using the following procedure:

(i) Single out row i of A and column j of B:

$$\begin{bmatrix} a_{11} & a_{12} & \cdots & a_{1n} \\ \vdots & \vdots & & \vdots \\ a_{i1} & a_{i2} & \cdots & a_{in} \\ \vdots & \vdots & & \vdots \\ a_{m1} & a_{m2} & \cdots & a_{mn} \end{bmatrix} \begin{bmatrix} b_{11} & \cdots & b_{1j} & \cdots & b_{1p} \\ b_{21} & \cdots & b_{2j} & \cdots & b_{2p} \\ \vdots & & \vdots & & \vdots \\ b_{n1} & \cdots & b_{nj} & \cdots & b_{np} \end{bmatrix}$$

(ii) *Simultaneously* move to the right along row i of A and down column j of B, multiplying pairs of elements, obtaining

$$a_{i1}b_{1j}, \quad a_{i2}b_{2j}, \quad a_{i3}b_{3j}, \quad \ldots, \quad a_{in}b_{nj}.$$

(iii) Add the products of the pairs in (ii) to obtain c_{ij}:

$$c_{ij} = a_{i1}b_{1j} + a_{i2}b_{2j} + \cdots + a_{in}b_{nj}$$

Using this procedure we see that the element c_{11} in the first row and the first column of AB is

$$c_{11} = a_{11}b_{11} + a_{12}b_{21} + \cdots + a_{1n}b_{n1}.$$

The element c_{12} in the first row and second column of AB is

$$c_{12} = a_{11}b_{12} + a_{12}b_{22} + \cdots + a_{1n}b_{n2}.$$

We summarize the preceding discussion in the next definition.

PRODUCT OF TWO MATRICES

Let $A = (a_{ij})$ be an $m \times n$ matrix and let $B = (b_{ij})$ be an $n \times p$ matrix. The **product** $\mathbf{AB}$ is the $m \times p$ matrix $C = (c_{ij})$ such that

$$c_{ij} = a_{i1}b_{1j} + a_{i2}b_{2j} + \cdots + a_{in}b_{nj}$$

for $i = 1, 2, \ldots, m$ and $j = 1, 2, \ldots, p$.

A special case of this definition for the product of a 2×3 matrix and a 3×4 matrix is given in the next illustration. Note that the product is a 2×4 matrix.

ILLUSTRATION | PRODUCT OF MATRICES

■ $$\begin{bmatrix} 1 & 2 & -3 \\ 4 & 0 & -2 \end{bmatrix}\begin{bmatrix} 5 & -4 & 2 & 0 \\ -1 & 6 & 3 & 1 \\ 7 & 0 & 4 & 8 \end{bmatrix} = \begin{bmatrix} -18 & 8 & -4 & -22 \\ 6 & -16 & 0 & -16 \end{bmatrix}$$

Some typical computations of the elements c_{ij} in the preceding illustration are:

$$c_{11} = (1)(5) + (2)(-1) + (-3)(7) = 5 - 2 - 21 = -18$$

$$c_{13} = (1)(2) + (2)(3) + (-3)(4) = 2 + 6 - 12 = -4$$

$$c_{23} = (4)(2) + (0)(3) + (-2)(4) = 8 + 0 - 8 = 0$$

$$c_{24} = (4)(0) + (0)(1) + (-2)(8) = 0 + 0 - 16 = -16$$

You should calculate the remaining elements.

The product operation for matrices is not commutative. For example, if A is 2×3 and B is 3×4, then AB may be found, but BA is undefined, since the number of columns of B is different from the number of rows of A. Even if AB and BA are both defined, it is often true that these products are different. This is illustrated in the next example, along with the fact that the product of two nonzero matrices may equal a zero matrix.

EXAMPLE ■ 1

If $A = \begin{bmatrix} 2 & 2 \\ -1 & -1 \end{bmatrix}$ and $B = \begin{bmatrix} 1 & 2 \\ 1 & 2 \end{bmatrix}$, show that $AB \neq BA$.

SOLUTION Using the definition of product we obtain

$$AB = \begin{bmatrix} 4 & 8 \\ -2 & -4 \end{bmatrix} \quad \text{and} \quad BA = \begin{bmatrix} 0 & 0 \\ 0 & 0 \end{bmatrix}.$$

Hence, $AB \neq BA$. Note that BA is a zero matrix.

Although matrix multiplication is not commutative, it is associative. Thus, if A is $m \times n$, B is $n \times p$, and C is $p \times q$, then

$$A(BC) = (AB)C.$$

The distributive properties also hold if the matrices involved have the proper number of rows and columns. If A_1 and A_2 are $m \times n$ matrices, and if B_1 and B_2 are $n \times p$ matrices, then:

$$A_1(B_1 + B_2) = A_1B_1 + A_1B_2$$

$$(A_1 + A_2)B_1 = A_1B_1 + A_2B_1.$$

As a special case, if all matrices are square, of order n, then both the associative and distributive properties are true.

Throughout the remainder of this section we shall concentrate on square matrices. The symbol I_n will denote the square matrix of order n that has 1 in each position on the main diagonal and 0 elsewhere. We call I_n the **identity matrix of order *n*.**

ILLUSTRATION | IDENTITY MATRICES

- $I_2 = \begin{bmatrix} 1 & 0 \\ 0 & 1 \end{bmatrix}$ ■ $I_3 = \begin{bmatrix} 1 & 0 & 0 \\ 0 & 1 & 0 \\ 0 & 0 & 1 \end{bmatrix}$

We can show that if A is any square matrix of order n, then

$$AI_n = A = I_nA.$$

ILLUSTRATION | $AI_n = A = I_nA$

- $\begin{bmatrix} a_{11} & a_{12} \\ a_{21} & a_{22} \end{bmatrix}\begin{bmatrix} 1 & 0 \\ 0 & 1 \end{bmatrix} = \begin{bmatrix} a_{11} & a_{12} \\ a_{21} & a_{22} \end{bmatrix} = \begin{bmatrix} 1 & 0 \\ 0 & 1 \end{bmatrix}\begin{bmatrix} a_{11} & a_{12} \\ a_{21} & a_{22} \end{bmatrix}$

Sometimes an $n \times n$ matrix A has an **inverse**—a matrix B such that $AB = I_n = BA$. If A has an inverse, we denote it by A^{-1} and write

$$AA^{-1} = I_n = A^{-1}A.$$

The symbol A^{-1} is read A *inverse*. For matrices the symbol $1/A$ does not represent the inverse A^{-1}.

If a square matrix A has an inverse, we can calculate A^{-1} using elementary row operations. If $A = (a_{ij})$ is $n \times n$, we begin with the $n \times 2n$ matrix

$$\left[\begin{array}{cccc|cccc} a_{11} & a_{12} & \cdots & a_{1n} & 1 & 0 & \cdots & 0 \\ a_{21} & a_{22} & \cdots & a_{2n} & 0 & 1 & \cdots & 0 \\ \vdots & \vdots & & \vdots & \vdots & \vdots & & \vdots \\ a_{n1} & a_{n2} & \cdots & a_{nn} & 0 & 0 & \cdots & 1 \end{array}\right]$$

in which the $n \times n$ identity matrix I_n appears to the right of the vertical rule. We next apply a succession of elementary row transformations, as we did in Section 9.3 to find reduced echelon forms, until we arrive at a matrix of the form

$$\left[\begin{array}{cccc|cccc} 1 & 0 & \cdots & 0 & b_{11} & b_{12} & \cdots & b_{1n} \\ 0 & 1 & \cdots & 0 & b_{21} & b_{22} & \cdots & b_{2n} \\ \vdots & \vdots & & \vdots & \vdots & \vdots & & \vdots \\ 0 & 0 & \cdots & 1 & b_{n1} & b_{n2} & \cdots & b_{nn} \end{array}\right]$$

in which the identity matrix I_n appears to the left of the vertical rule. The $n \times n$ matrix (b_{ij}) is the inverse A^{-1}.

EXAMPLE ▪ 2

Find A^{-1} if $A = \begin{bmatrix} 3 & 5 \\ 1 & 4 \end{bmatrix}$.

SOLUTION We begin with the matrix

$$\left[\begin{array}{cc|cc} 3 & 5 & 1 & 0 \\ 1 & 4 & 0 & 1 \end{array}\right].$$

Next we perform elementary row transformations until the identity matrix I_2 appears on the left of the vertical rule, as follows:

$$\left[\begin{array}{cc|cc} 3 & 5 & 1 & 0 \\ 1 & 4 & 0 & 1 \end{array}\right] \xrightarrow{\mathbf{R_1 \leftrightarrow R_2}} \left[\begin{array}{cc|cc} 1 & 4 & 0 & 1 \\ 3 & 5 & 1 & 0 \end{array}\right]$$

$$\mathbf{-3R_1 + R_2 \rightarrow R_2} \left[\begin{array}{cc|cc} 1 & 4 & 0 & 1 \\ 0 & -7 & 1 & -3 \end{array}\right]$$

$$\mathbf{-\tfrac{1}{7}R_2 \rightarrow R_2} \left[\begin{array}{cc|cc} 1 & 4 & 0 & 1 \\ 0 & 1 & -\frac{1}{7} & \frac{3}{7} \end{array}\right]$$

$$\mathbf{-4R_2 + R_1 \rightarrow R_1} \left[\begin{array}{cc|cc} 1 & 0 & \frac{4}{7} & -\frac{5}{7} \\ 0 & 1 & -\frac{1}{7} & \frac{3}{7} \end{array}\right]$$

By the previous discussion,

$$A^{-1} = \begin{bmatrix} \frac{4}{7} & -\frac{5}{7} \\ -\frac{1}{7} & \frac{3}{7} \end{bmatrix}.$$

Let us verify that $AA^{-1} = I_2 = A^{-1}A$:

$$\begin{bmatrix} 3 & 5 \\ 1 & 4 \end{bmatrix}\begin{bmatrix} \frac{4}{7} & -\frac{5}{7} \\ -\frac{1}{7} & \frac{3}{7} \end{bmatrix} = \begin{bmatrix} 1 & 0 \\ 0 & 1 \end{bmatrix} = \begin{bmatrix} \frac{4}{7} & -\frac{5}{7} \\ -\frac{1}{7} & \frac{3}{7} \end{bmatrix}\begin{bmatrix} 3 & 5 \\ 1 & 4 \end{bmatrix}$$

EXAMPLE ▪ 3

Find A^{-1} if $A = \begin{bmatrix} -1 & 3 & 1 \\ 2 & 5 & 0 \\ 3 & 1 & -2 \end{bmatrix}$.

SOLUTION

$$\left[\begin{array}{rrr|rrr} -1 & 3 & 1 & 1 & 0 & 0 \\ 2 & 5 & 0 & 0 & 1 & 0 \\ 3 & 1 & -2 & 0 & 0 & 1 \end{array}\right] \begin{array}{l} -R_1 \to R_1 \\ \\ \\ \end{array} \left[\begin{array}{rrr|rrr} 1 & -3 & -1 & -1 & 0 & 0 \\ 2 & 5 & 0 & 0 & 1 & 0 \\ 3 & 1 & -2 & 0 & 0 & 1 \end{array}\right]$$

$$\begin{array}{r} \\ -2R_1 + R_2 \to R_2 \\ -3R_1 + R_3 \to R_3 \end{array} \left[\begin{array}{rrr|rrr} 1 & -3 & -1 & -1 & 0 & 0 \\ 0 & 11 & 2 & 2 & 1 & 0 \\ 0 & 10 & 1 & 3 & 0 & 1 \end{array}\right]$$

$$\begin{array}{r} \\ -R_3 + R_2 \to R_2 \\ \\ \end{array} \left[\begin{array}{rrr|rrr} 1 & -3 & -1 & -1 & 0 & 0 \\ 0 & 1 & 1 & -1 & 1 & -1 \\ 0 & 10 & 1 & 3 & 0 & 1 \end{array}\right]$$

$$\begin{array}{r} 3R_2 + R_1 \to R_1 \\ \\ -10R_2 + R_3 \to R_3 \end{array} \left[\begin{array}{rrr|rrr} 1 & 0 & 2 & -4 & 3 & -3 \\ 0 & 1 & 1 & -1 & 1 & -1 \\ 0 & 0 & -9 & 13 & -10 & 11 \end{array}\right]$$

$$\begin{array}{r} \\ \\ -\frac{1}{9}R_3 \to R_3 \end{array} \left[\begin{array}{rrr|rrr} 1 & 0 & 2 & -4 & 3 & -3 \\ 0 & 1 & 1 & -1 & 1 & -1 \\ 0 & 0 & 1 & -\frac{13}{9} & \frac{10}{9} & -\frac{11}{9} \end{array}\right]$$

$$\begin{array}{r} -2R_3 + R_1 \to R_1 \\ -R_3 + R_2 \to R_2 \\ \\ \end{array} \left[\begin{array}{rrr|rrr} 1 & 0 & 0 & -\frac{10}{9} & \frac{7}{9} & -\frac{5}{9} \\ 0 & 1 & 0 & \frac{4}{9} & -\frac{1}{9} & \frac{2}{9} \\ 0 & 0 & 1 & -\frac{13}{9} & \frac{10}{9} & -\frac{11}{9} \end{array}\right]$$

Consequently,

$$A^{-1} = \begin{bmatrix} -\frac{10}{9} & \frac{7}{9} & -\frac{5}{9} \\ \frac{4}{9} & -\frac{1}{9} & \frac{2}{9} \\ -\frac{13}{9} & \frac{10}{9} & -\frac{11}{9} \end{bmatrix} = \frac{1}{9}\begin{bmatrix} -10 & 7 & -5 \\ 4 & -1 & 2 \\ -13 & 10 & -11 \end{bmatrix}.$$

You should verify that $AA^{-1} = I_3 = A^{-1}A$.

If the procedure used in Examples 2 and 3 does not lead to an identity matrix to the left of the vertical rule, then the matrix A has no inverse.

We may apply inverses of matrices to solutions of systems of linear equations. Consider the case of two linear equations in two unknowns:

$$\begin{cases} a_{11}x + a_{12}y = k_1 \\ a_{21}x + a_{22}y = k_2 \end{cases}$$

which we can write in terms of matrices as

$$\begin{bmatrix} a_{11}x + a_{12}y \\ a_{21}x + a_{22}y \end{bmatrix} = \begin{bmatrix} k_1 \\ k_2 \end{bmatrix}.$$

If we let

$$A = \begin{bmatrix} a_{11} & a_{12} \\ a_{21} & a_{22} \end{bmatrix}, \quad X = \begin{bmatrix} x \\ y \end{bmatrix}, \quad \text{and} \quad B = \begin{bmatrix} k_1 \\ k_2 \end{bmatrix},$$

then

$$AX = B.$$

If A^{-1} exists, then multiplying both sides of the last equation by A^{-1} gives us $A^{-1}AX = A^{-1}B$. Since $A^{-1}A = I_2$ and $I_2X = X$, this leads to

$$X = A^{-1}B,$$

from which the solution (x, y) may be found. This technique may be extended to systems of n linear equations in n unknowns.

EXAMPLE ▪ 4

Solve the system of equations:

$$\begin{cases} -x + 3y + z = 1 \\ 2x + 5y = 3 \\ 3x + y - 2z = -2 \end{cases}$$

SOLUTION If we let

$$A = \begin{bmatrix} -1 & 3 & 1 \\ 2 & 5 & 0 \\ 3 & 1 & -2 \end{bmatrix}, \quad X = \begin{bmatrix} x \\ y \\ z \end{bmatrix}, \quad \text{and} \quad B = \begin{bmatrix} 1 \\ 3 \\ -2 \end{bmatrix},$$

then $AX = B$. This implies that $X = A^{-1}B$. The matrix A^{-1} was found in Example 3. Hence

$$\begin{bmatrix} x \\ y \\ z \end{bmatrix} = \frac{1}{9}\begin{bmatrix} -10 & 7 & -5 \\ 4 & -1 & 2 \\ -13 & 10 & -11 \end{bmatrix}\begin{bmatrix} 1 \\ 3 \\ -2 \end{bmatrix} = \frac{1}{9}\begin{bmatrix} 21 \\ -3 \\ 39 \end{bmatrix} = \begin{bmatrix} \frac{7}{3} \\ -\frac{1}{3} \\ \frac{13}{3} \end{bmatrix}.$$

Thus $x = \frac{7}{3}$, $y = -\frac{1}{3}$, $z = \frac{13}{3}$, and the ordered triple $(\frac{7}{3}, -\frac{1}{3}, \frac{13}{3})$ is the solution of the given system.

The method of solution in Example 4 is beneficial only if A^{-1} is known or if many systems with the same coefficient matrix are to be considered. The preferred technique for solving an arbitrary system of linear equations is the matrix method discussed in Section 9.3.

9.5 EXERCISES

Exer. 1–8: Find, if possible, $A + B$, $A - B$, $2A$, and $-3B$.

1 $A = \begin{bmatrix} 5 & -2 \\ 1 & 3 \end{bmatrix}$, $B = \begin{bmatrix} 4 & 1 \\ -3 & 2 \end{bmatrix}$

2 $A = \begin{bmatrix} 3 & 0 \\ -1 & 2 \end{bmatrix}$, $B = \begin{bmatrix} 3 & -4 \\ 1 & 1 \end{bmatrix}$

3 $A = \begin{bmatrix} 6 & -1 \\ 2 & 0 \\ -3 & 4 \end{bmatrix}$, $B = \begin{bmatrix} 3 & 1 \\ -1 & 5 \\ 6 & 0 \end{bmatrix}$

4 $A = \begin{bmatrix} 0 & -2 & 7 \\ 5 & 4 & -3 \end{bmatrix}$, $B = \begin{bmatrix} 8 & 4 & 0 \\ 0 & 1 & 4 \end{bmatrix}$

5 $A = [4 \quad -3 \quad 2]$, $B = [7 \quad 0 \quad -5]$

6 $A = \begin{bmatrix} 7 \\ -16 \end{bmatrix}$, $B = \begin{bmatrix} -11 \\ 9 \end{bmatrix}$

7 $A = \begin{bmatrix} 3 & -2 & 2 \\ 0 & 1 & -4 \\ -3 & 2 & -1 \end{bmatrix}$, $B = \begin{bmatrix} 4 & 0 \\ 2 & -1 \\ -1 & 3 \end{bmatrix}$

8 $A = [2 \quad 1]$, $B = [3 \quad -1 \quad 5]$

Exer. 9–20: Find, if possible, AB and BA.

9 $A = \begin{bmatrix} 2 & 6 \\ 3 & -4 \end{bmatrix}$, $B = \begin{bmatrix} 5 & -2 \\ 1 & 7 \end{bmatrix}$

10 $A = \begin{bmatrix} 4 & -2 \\ -2 & 1 \end{bmatrix}$, $B = \begin{bmatrix} 2 & 1 \\ 4 & 2 \end{bmatrix}$

11 $A = \begin{bmatrix} 3 & 0 & -1 \\ 0 & 4 & 2 \\ 5 & -3 & 1 \end{bmatrix}$, $B = \begin{bmatrix} 1 & -5 & 0 \\ 4 & 1 & -2 \\ 0 & -1 & 3 \end{bmatrix}$

12 $A = \begin{bmatrix} 5 & 0 & 0 \\ 0 & -3 & 0 \\ 0 & 0 & 2 \end{bmatrix}$, $B = \begin{bmatrix} 3 & 0 & 0 \\ 0 & 4 & 0 \\ 0 & 0 & -2 \end{bmatrix}$

13 $A = \begin{bmatrix} 4 & -3 & 1 \\ -5 & 2 & 2 \end{bmatrix}$, $B = \begin{bmatrix} 2 & 1 \\ 0 & 1 \\ -4 & 7 \end{bmatrix}$

14 $A = \begin{bmatrix} 2 & 1 & -1 & 0 \\ 3 & -2 & 0 & 5 \\ -2 & 1 & 4 & 2 \end{bmatrix}$, $B = \begin{bmatrix} 5 & -3 & 1 \\ 1 & 2 & 0 \\ -1 & 0 & 4 \\ 0 & -2 & 3 \end{bmatrix}$

15 $A = \begin{bmatrix} 1 & 2 & 3 \\ 4 & 5 & 6 \\ 7 & 8 & 9 \end{bmatrix}$, $B = \begin{bmatrix} 1 & 0 & 0 \\ 0 & 1 & 0 \\ 0 & 0 & 1 \end{bmatrix}$

16 $A = \begin{bmatrix} 1 & 2 & 3 \\ 2 & 3 & 1 \\ 3 & 1 & 2 \end{bmatrix}$, $B = \begin{bmatrix} 2 & 0 & 0 \\ 0 & 2 & 0 \\ 0 & 0 & 2 \end{bmatrix}$

17 $A = [-3 \quad 7 \quad 2]$, $B = \begin{bmatrix} 1 \\ 4 \\ -5 \end{bmatrix}$

18 $A = [4 \quad 8]$, $B = \begin{bmatrix} -3 \\ 2 \end{bmatrix}$

19 $A = \begin{bmatrix} 2 & 0 & 1 \\ -1 & 2 & 0 \end{bmatrix}$, $B = \begin{bmatrix} 1 & -1 & 2 \\ 3 & 1 & 0 \\ 0 & 2 & 1 \end{bmatrix}$

20 $A = [3 \quad -1 \quad 4]$, $B = \begin{bmatrix} -2 \\ 5 \end{bmatrix}$

Exer. 21–24: Find AB.

21 $A = \begin{bmatrix} 4 & -2 \\ 0 & 3 \\ -7 & 5 \end{bmatrix}, \quad B = \begin{bmatrix} 3 \\ 4 \end{bmatrix}$

22 $A = \begin{bmatrix} 4 \\ -3 \\ 2 \end{bmatrix}, \quad B = \begin{bmatrix} 5 & 1 \end{bmatrix}$

23 $A = \begin{bmatrix} 2 & 1 & 0 & -3 \\ -7 & 0 & -2 & 4 \end{bmatrix}, \quad B = \begin{bmatrix} 4 & -2 & 0 \\ 1 & 1 & -2 \\ 0 & 0 & 5 \\ -3 & -1 & 0 \end{bmatrix}$

24 $A = \begin{bmatrix} 1 & 2 & -3 \\ 4 & -5 & 6 \end{bmatrix}, \quad B = \begin{bmatrix} 1 & -1 & 0 & 2 \\ -2 & 3 & 1 & 0 \\ 0 & 4 & 0 & -3 \end{bmatrix}$

Exer. 25–28: Let

$$A = \begin{bmatrix} 1 & 2 \\ 0 & -3 \end{bmatrix}, \quad B = \begin{bmatrix} 2 & -1 \\ 3 & 1 \end{bmatrix}, \quad C = \begin{bmatrix} 3 & 1 \\ -2 & 0 \end{bmatrix}.$$

Verify the statement.

25 $(A + B)(A - B) \neq A^2 - B^2$, where $A^2 = AA$ and $B^2 = BB$.

26 $(A + B)(A + B) \neq A^2 + 2AB + B^2$

27 $A(B + C) = AB + AC$

28 $A(BC) = (AB)C$

Exer. 29–32: Verify the identity for

$$A = \begin{bmatrix} a & b \\ c & d \end{bmatrix}, \quad B = \begin{bmatrix} p & q \\ r & s \end{bmatrix}, \quad C = \begin{bmatrix} w & x \\ y & z \end{bmatrix}$$

and real numbers m and n.

29 $m(A + B) = mA + mB$

30 $(m + n)A = mA + nA$

31 $A(B + C) = AB + AC$

32 $A(BC) = (AB)C$

Exer. 33–42: Find the inverse of the matrix, if it exists.

33 $\begin{bmatrix} 2 & -4 \\ 1 & 3 \end{bmatrix}$

34 $\begin{bmatrix} 3 & 2 \\ 4 & 5 \end{bmatrix}$

35 $\begin{bmatrix} 2 & 4 \\ 4 & 8 \end{bmatrix}$

36 $\begin{bmatrix} 3 & -1 \\ 6 & -2 \end{bmatrix}$

37 $\begin{bmatrix} 3 & -1 & 0 \\ 2 & 2 & 0 \\ 0 & 0 & 4 \end{bmatrix}$

38 $\begin{bmatrix} 3 & 0 & 2 \\ 0 & 1 & 0 \\ -4 & 0 & 2 \end{bmatrix}$

39 $\begin{bmatrix} -2 & 2 & 3 \\ 1 & -1 & 0 \\ 0 & 1 & 4 \end{bmatrix}$

40 $\begin{bmatrix} 1 & 2 & 3 \\ -2 & 1 & 0 \\ 3 & -1 & 1 \end{bmatrix}$

41 $\begin{bmatrix} 2 & 0 & 0 \\ 0 & 4 & 0 \\ 0 & 0 & 6 \end{bmatrix}$

42 $\begin{bmatrix} 1 & 1 & 1 \\ 2 & 2 & 2 \\ 3 & 3 & 3 \end{bmatrix}$

43 State conditions on a and b that guarantee that the matrix $\begin{bmatrix} a & 0 \\ 0 & b \end{bmatrix}$ has an inverse, and find a formula for the inverse, if it exists.

44 If $abc \neq 0$, find the inverse of $\begin{bmatrix} a & 0 & 0 \\ 0 & b & 0 \\ 0 & 0 & c \end{bmatrix}$.

45 If $A = \begin{bmatrix} a_{11} & a_{12} & a_{13} \\ a_{21} & a_{22} & a_{23} \\ a_{31} & a_{32} & a_{33} \end{bmatrix}$, show that $AI_3 = A = I_3A$.

46 Show that $AI_4 = A = I_4A$ for every square matrix A of order 4.

Exer. 47–50: Solve the system using the method in Example 4. (Refer to Exercises 33–34 and 39–40.)

47 $\begin{cases} 2x - 4y = c \\ x + 3y = d \end{cases}$

(a) $\begin{bmatrix} c \\ d \end{bmatrix} = \begin{bmatrix} 3 \\ 1 \end{bmatrix}$ (b) $\begin{bmatrix} c \\ d \end{bmatrix} = \begin{bmatrix} -2 \\ 5 \end{bmatrix}$

48 $\begin{cases} 3x + 2y = c \\ 4x + 5y = d \end{cases}$

(a) $\begin{bmatrix} c \\ d \end{bmatrix} = \begin{bmatrix} -1 \\ 1 \end{bmatrix}$ (b) $\begin{bmatrix} c \\ d \end{bmatrix} = \begin{bmatrix} 4 \\ 3 \end{bmatrix}$

49 $\begin{cases} -2x + 2y + 3z = c \\ x - y = d \\ y + 4z = e \end{cases}$

(a) $\begin{bmatrix} c \\ d \\ e \end{bmatrix} = \begin{bmatrix} 1 \\ 3 \\ -2 \end{bmatrix}$ (b) $\begin{bmatrix} c \\ d \\ e \end{bmatrix} = \begin{bmatrix} -1 \\ 0 \\ 4 \end{bmatrix}$

50 $\begin{cases} x + 2y + 3z = c \\ -2x + y = d \\ 3x - y + z = e \end{cases}$

(a) $\begin{bmatrix} c \\ d \\ e \end{bmatrix} = \begin{bmatrix} -1 \\ 4 \\ 2 \end{bmatrix}$ (b) $\begin{bmatrix} c \\ d \\ e \end{bmatrix} = \begin{bmatrix} -3 \\ -2 \\ 1 \end{bmatrix}$

9.6 DETERMINANTS

Throughout this section and the next we will assume that all matrices under discussion are *square* matrices. Associated with each square matrix A is a number called the **determinant of A**, denoted by $|A|$. This notation should not be confused with the symbol for the absolute value of a real number. To avoid any misunderstanding, the expression $\det A$ is sometimes used in place of $|A|$. We shall define $|A|$ by beginning with the case in which A has order 1, and then by increasing the order a step at a time.

If A is a square matrix of order 1, then A has only one element. Thus, $A = [a_{11}]$ and we define $|A| = a_{11}$. If A is a square matrix of order 2, then

$$A = \begin{bmatrix} a_{11} & a_{12} \\ a_{21} & a_{22} \end{bmatrix}$$

and the determinant of A is defined by

$$|A| = a_{11}a_{22} - a_{21}a_{12}.$$

Another notation for $|A|$ is obtained by replacing the brackets used for A with vertical bars as follows.

DETERMINANT OF A 2 × 2 MATRIX A

$$|A| = \begin{vmatrix} a_{11} & a_{12} \\ a_{21} & a_{22} \end{vmatrix} = a_{11}a_{22} - a_{21}a_{12}$$

EXAMPLE ▪ 1

Find $|A|$ if $A = \begin{bmatrix} 2 & -1 \\ 4 & -3 \end{bmatrix}$.

SOLUTION By definition,

$$|A| = \begin{vmatrix} 2 & -1 \\ 4 & -3 \end{vmatrix} = (2)(-3) - (4)(-1) = -6 + 4 = -2.$$

For square matrices of order $n > 1$ it is convenient to introduce the following terminology.

DEFINITION

Let $A = (a_{ij})$ be a square matrix of order $n > 1$.
(i) The **minor** M_{ij} of the element a_{ij} is the determinant of the matrix of order $n - 1$ obtained by deleting row i and column j.
(ii) The **cofactor** A_{ij} of the element a_{ij} is $A_{ij} = (-1)^{i+j}M_{ij}$.

To determine the minor of an element, we delete the row and column in which the element appears and then find the determinant of the resulting square matrix. To illustrate, for the 3×3 matrix

$$A = \begin{bmatrix} a_{11} & a_{12} & a_{13} \\ a_{21} & a_{22} & a_{23} \\ a_{31} & a_{32} & a_{33} \end{bmatrix}$$

we obtain

$$M_{11} = \begin{vmatrix} a_{22} & a_{23} \\ a_{32} & a_{33} \end{vmatrix} = a_{22}a_{33} - a_{32}a_{23}$$

$$M_{12} = \begin{vmatrix} a_{21} & a_{23} \\ a_{31} & a_{33} \end{vmatrix} = a_{21}a_{33} - a_{31}a_{23}$$

$$M_{13} = \begin{vmatrix} a_{21} & a_{22} \\ a_{31} & a_{32} \end{vmatrix} = a_{21}a_{32} - a_{31}a_{22}$$

$$M_{23} = \begin{vmatrix} a_{11} & a_{12} \\ a_{31} & a_{32} \end{vmatrix} = a_{11}a_{32} - a_{31}a_{12}.$$

The other minors M_{21}, M_{22}, M_{31}, M_{32}, and M_{33} are obtained in similar fashion.

To obtain the cofactor of a_{ij} of a square matrix $A = (a_{ij})$, we find the minor and multiply it by 1 or -1, depending on whether the sum of i and j is even or odd, respectively. An easy way to remember the sign $(-1)^{i+j}$ associated with the cofactor A_{ij} is to consider the following checkerboard of plus and minus signs:

$$\begin{bmatrix} + & - & + & - & \cdots \\ - & + & - & + & \cdots \\ + & - & + & - & \cdots \\ - & + & - & + & \cdots \\ \vdots & \vdots & \vdots & \vdots & \end{bmatrix}$$

EXAMPLE ▪ 2

If $A = \begin{bmatrix} 1 & -3 & 3 \\ 4 & 2 & 0 \\ -2 & -7 & 5 \end{bmatrix}$, find M_{11}, M_{21}, M_{22}, A_{11}, A_{21}, and A_{22}.

SOLUTION

$$M_{11} = \begin{vmatrix} 2 & 0 \\ -7 & 5 \end{vmatrix} = (2)(5) - (-7)(0) = 10$$

$$M_{21} = \begin{vmatrix} -3 & 3 \\ -7 & 5 \end{vmatrix} = (-3)(5) - (-7)(3) = 6$$

$$M_{22} = \begin{vmatrix} 1 & 3 \\ -2 & 5 \end{vmatrix} = (1)(5) - (-2)(3) = 11.$$

To obtain the cofactors, we prefix the corresponding minors with the proper signs. Thus, using the definition of cofactor,

$$A_{11} = (-1)^{1+1}M_{11} = (1)(10) = 10$$
$$A_{21} = (-1)^{2+1}M_{21} = (-1)(6) = -6$$
$$A_{22} = (-1)^{2+2}M_{22} = (1)(11) = 11.$$

We can also use the checkerboard of plus and minus signs to determine the proper signs.

The determinant $|A|$ of a square matrix of order 3 is defined as follows.

DETERMINANT OF A 3 × 3 MATRIX A

$$|A| = \begin{vmatrix} a_{11} & a_{12} & a_{13} \\ a_{21} & a_{22} & a_{23} \\ a_{31} & a_{32} & a_{33} \end{vmatrix} = a_{11}A_{11} + a_{12}A_{12} + a_{13}A_{13}$$

Since $A_{11} = (-1)^{1+1}M_{11} = M_{11}$, $A_{12} = (-1)^{1+2}M_{12} = -M_{12}$, and $A_{13} = (-1)^{1+3}M_{13} = M_{13}$, the preceding definition may also be written

$$|A| = a_{11}M_{11} - a_{12}M_{12} + a_{13}M_{13}.$$

If we express M_{11}, M_{12}, and M_{13} in terms of elements of A, we obtain the following formula for $|A|$:

$$|A| = a_{11}a_{22}a_{33} - a_{11}a_{23}a_{32} - a_{12}a_{21}a_{33} + a_{12}a_{23}a_{31} + a_{13}a_{21}a_{32} - a_{13}a_{22}a_{31}$$

The definition of $|A|$ for a square matrix A of order 3 displays a pattern of multiplying each element in row 1 by its cofactor, and then adding to find $|A|$. This is referred to as *expanding* $|A|$ *by the first row*. By actually carrying out the computations, we can show that $|A|$ *can be expanded in similar fashion by using any row or column.* As an illustration, the expansion by the second column is

$$\begin{aligned}|A| &= a_{12}A_{12} + a_{22}A_{22} + a_{32}A_{32}\\ &= a_{12}\left(-\begin{vmatrix} a_{21} & a_{23}\\ a_{31} & a_{33}\end{vmatrix}\right) + a_{22}\left(+\begin{vmatrix} a_{11} & a_{13}\\ a_{31} & a_{33}\end{vmatrix}\right) + a_{32}\left(-\begin{vmatrix} a_{11} & a_{13}\\ a_{21} & a_{23}\end{vmatrix}\right).\end{aligned}$$

Applying the definition to the determinants in parentheses, multiplying as indicated, and rearranging the terms in the sum, we could arrive at the formula for $|A|$ in terms of the elements of A. Similarly, the expansion by the third row is

$$|A| = a_{31}A_{31} + a_{32}A_{32} + a_{33}A_{33}.$$

Once again we can show that this agrees with previous expansions.

EXAMPLE ▪ 3

Find $|A|$ if $A = \begin{bmatrix} -1 & 3 & 1\\ 2 & 5 & 0\\ 3 & 1 & -2\end{bmatrix}$.

SOLUTION Since the second row contains a zero, we shall expand $|A|$ by that row, because then we need to evaluate only two cofactors. Thus,

$$|A| = (2)A_{21} + (5)A_{22} + (0)A_{23}.$$

Using the definition of cofactor,

$$A_{21} = (-1)^3M_{21} = -\begin{vmatrix} 3 & 1\\ 1 & -2\end{vmatrix} = -[(3)(-2) - (1)(1)] = 7$$

$$A_{22} = (-1)^4M_{22} = \begin{vmatrix} -1 & 1\\ 3 & -2\end{vmatrix} = [(-1)(-2) - (3)(1)] = -1.$$

Consequently,

$$|A| = (2)(7) + (5)(-1) + (0)A_{23} = 14 - 5 + 0 = 9.$$

The following definition of the determinant of a matrix of arbitrary order n is patterned after that used for order 3.

DETERMINANT OF AN $n \times n$ MATRIX A

The **determinant $|A|$ of a matrix A of order n** is the cofactor expansion by the first row:

$$|A| = a_{11}A_{11} + a_{12}A_{12} + \cdots + a_{1n}A_{1n}.$$

In terms of minors:

$$|A| = a_{11}M_{11} - a_{12}M_{12} + \cdots + a_{1n}(-1)^{1+n}M_{1n}.$$

The number $|A|$ may be found by using *any* row or column, as stated in the following theorem.

EXPANSION THEOREM FOR DETERMINANTS

If A is a square matrix of order $n > 1$, then the determinant $|A|$ may be found by multiplying the elements of any row (or column) by their respective cofactors, and adding the resulting products.

The proof of this theorem may be found in texts on matrix theory. The theorem is quite useful if many zeros appear in a row or column, as illustrated in the following example.

EXAMPLE ▪ 4

Find $|A|$ if $A = \begin{bmatrix} 1 & 0 & 2 & 5 \\ -2 & 1 & 5 & 0 \\ 0 & 0 & -3 & 0 \\ 0 & -1 & 0 & 3 \end{bmatrix}$.

SOLUTION Note that all but one of the elements in the third row is zero. Hence if we expand $|A|$ by the third row, there will be at most one nonzero term. Specifically,

$$|A| = (0)A_{31} + (0)A_{32} + (-3)A_{33} + (0)A_{34} = -3A_{33}$$

with

$$A_{33} = \begin{vmatrix} 1 & 0 & 5 \\ -2 & 1 & 0 \\ 0 & -1 & 3 \end{vmatrix}.$$

We expand A_{33} by column 1:

$$A_{33} = (1)\begin{vmatrix} 1 & 0 \\ -1 & 3 \end{vmatrix} + (-2)\left(-\begin{vmatrix} 0 & 5 \\ -1 & 3 \end{vmatrix}\right) + (0)\begin{vmatrix} 0 & 5 \\ 1 & 0 \end{vmatrix}$$
$$= 3 + 10 + 0 = 13.$$

Thus, $$|A| = -3A_{33} = (-3)(13) = -39.$$

In general, if all but one element in some row (or column) of A is zero, and if the determinant $|A|$ is expanded by that row (or column), then all terms drop out except the product of that element with its cofactor. If *all* elements in a row (or column) are zero we have the following.

THEOREM If every element of a row (or column) of a square matrix A is zero, then $|A| = 0$.

PROOF If every element in a row (or column) of a matrix A is zero, the expansion by that row (or column) is zero. ❑

9.6 EXERCISES

Exer. 1–4: Find all the minors and cofactors of the elements in the matrix.

1 $\begin{bmatrix} 2 & 4 & -1 \\ 0 & 3 & 2 \\ -5 & 7 & 0 \end{bmatrix}$

2 $\begin{bmatrix} 5 & -2 & 1 \\ 4 & 7 & 0 \\ -3 & 4 & -1 \end{bmatrix}$

3 $\begin{bmatrix} 7 & -1 \\ 5 & 0 \end{bmatrix}$

4 $\begin{bmatrix} -6 & 4 \\ 3 & 2 \end{bmatrix}$

5–8 Find the determinants of the matrices in Exercises 1–4.

Exer. 9–20: Find the determinant of the matrix.

9 $\begin{bmatrix} -5 & 4 \\ -3 & 2 \end{bmatrix}$

10 $\begin{bmatrix} 6 & 4 \\ -3 & 2 \end{bmatrix}$

11 $\begin{bmatrix} a & -a \\ b & -b \end{bmatrix}$

12 $\begin{bmatrix} c & d \\ -d & c \end{bmatrix}$

13 $\begin{bmatrix} 3 & 1 & -2 \\ 4 & 2 & 5 \\ -6 & 3 & -1 \end{bmatrix}$

14 $\begin{bmatrix} 2 & -5 & 1 \\ -3 & 1 & 6 \\ 4 & -2 & 3 \end{bmatrix}$

15 $\begin{bmatrix} -5 & 4 & 1 \\ 3 & -2 & 7 \\ 2 & 0 & 6 \end{bmatrix}$

16 $\begin{bmatrix} 2 & 7 & -3 \\ 1 & 0 & 4 \\ 4 & -1 & -2 \end{bmatrix}$

17 $\begin{bmatrix} 3 & -1 & 2 & 0 \\ 4 & 0 & -3 & 5 \\ 0 & 6 & 0 & 0 \\ 1 & 3 & -4 & 2 \end{bmatrix}$

18 $\begin{bmatrix} 2 & 5 & 1 & 0 \\ -4 & 0 & -3 & 0 \\ 3 & -2 & 1 & 6 \\ -1 & 4 & 2 & 0 \end{bmatrix}$

19 $\begin{bmatrix} 0 & b & 0 & 0 \\ 0 & 0 & c & 0 \\ a & 0 & 0 & 0 \\ 0 & 0 & 0 & d \end{bmatrix}$

20 $\begin{bmatrix} a & u & v & w \\ 0 & b & x & y \\ 0 & 0 & c & z \\ 0 & 0 & 0 & d \end{bmatrix}$

Exer. 21–28: Verify the identity by expanding each determinant.

21 $\begin{vmatrix} a & b \\ c & d \end{vmatrix} = -\begin{vmatrix} c & d \\ a & b \end{vmatrix}$

22 $\begin{vmatrix} a & b \\ c & d \end{vmatrix} = -\begin{vmatrix} b & a \\ d & c \end{vmatrix}$

23 $\begin{vmatrix} a & kb \\ c & kd \end{vmatrix} = k\begin{vmatrix} a & b \\ c & d \end{vmatrix}$

24 $\begin{vmatrix} a & b \\ kc & kd \end{vmatrix} = k\begin{vmatrix} a & b \\ c & d \end{vmatrix}$

25 $\begin{vmatrix} a & b \\ c & d \end{vmatrix} = \begin{vmatrix} a & b \\ ka + c & kb + d \end{vmatrix}$

26 $\begin{vmatrix} a & b \\ c & d \end{vmatrix} = \begin{vmatrix} a & ka + b \\ c & kc + d \end{vmatrix}$

27 $\begin{vmatrix} a & b \\ c & d \end{vmatrix} + \begin{vmatrix} a & e \\ c & f \end{vmatrix} = \begin{vmatrix} a & b + e \\ c & d + f \end{vmatrix}$

28 $\begin{vmatrix} a & b \\ c & d \end{vmatrix} + \begin{vmatrix} a & b \\ e & f \end{vmatrix} = \begin{vmatrix} a & b \\ c + e & d + f \end{vmatrix}$

29 Let $A = (a_{ij})$ be a square matrix of order n such that $a_{ij} = 0$ if $i < j$. Show that $|A| = a_{11}a_{22} \cdots a_{nn}$.

30 If $A = (a_{ij})$ is any 2×2 matrix such that $|A| \neq 0$, show that A has an inverse, and find a general formula for A^{-1}.

Exer. 31–34: Let $I = I_2$ be the identity matrix of order 2, and let $f(x) = |A - xI|$. Find (a) the polynomial $f(x)$ and (b) the zeros of $f(x)$. (In the study of matrices, $f(x)$ is the *characteristic polynomial* of A, and the zeros of $f(x)$ are the *characteristic values* (*eigenvalues*) of A.)

31 $A = \begin{bmatrix} 1 & 2 \\ 3 & 2 \end{bmatrix}$

32 $A = \begin{bmatrix} 3 & 1 \\ 2 & 2 \end{bmatrix}$

33 $A = \begin{bmatrix} -3 & -2 \\ 2 & 2 \end{bmatrix}$

34 $A = \begin{bmatrix} 2 & -4 \\ -3 & 5 \end{bmatrix}$

Exer. 35–38: Let $I = I_3$ and let $f(x) = |A - xI|$. Find (a) the polynomial $f(x)$ and (b) the zeros of $f(x)$.

35 $A = \begin{bmatrix} 1 & 0 & 0 \\ 1 & 0 & -2 \\ -1 & 1 & -3 \end{bmatrix}$

36 $A = \begin{bmatrix} 2 & 1 & 0 \\ -1 & 0 & 0 \\ 1 & 3 & 2 \end{bmatrix}$

37 $A = \begin{bmatrix} 0 & 2 & -2 \\ -1 & 3 & 1 \\ -3 & 3 & 1 \end{bmatrix}$

38 $A = \begin{bmatrix} 3 & 2 & 2 \\ 1 & 0 & 2 \\ -1 & -1 & 0 \end{bmatrix}$

Exer. 39–42: Express the determinant in the form $ai + bj + ck$ for real numbers a, b, and c.

39 $\begin{vmatrix} i & j & k \\ 2 & -1 & 6 \\ -3 & 5 & 1 \end{vmatrix}$

40 $\begin{vmatrix} i & j & k \\ 1 & -2 & 3 \\ 2 & 1 & -4 \end{vmatrix}$

41 $\begin{vmatrix} i & j & k \\ 5 & -6 & -1 \\ 3 & 0 & 1 \end{vmatrix}$

42 $\begin{vmatrix} i & j & k \\ 4 & -6 & 2 \\ -2 & 3 & -1 \end{vmatrix}$

9.7 PROPERTIES OF DETERMINANTS

Evaluating a determinant by using the expansion theorem stated in Section 9.6 is inefficient for matrices of high order. For example, if a deter minant of a matrix of order 10 is expanded by any row, a sum of 10 terms is obtained, and each term contains the determinant of a matrix of order 9, which is a cofactor of the original matrix. If any of the latter determinants are expanded by a row (or column), a sum of 9 terms is obtained, each containing the determinant of a matrix of order 8. Hence, at this stage there are 90 determinants of matrices of order 8 to evaluate. The process could be continued until only determinants of matrices of order 2 remain. Unless many elements of the original matrix are zero, it is an enormous task to carry out all of the computations.

In this section we discuss rules that simplify the process of evaluating determinants. The main use for these rules is to introduce zeros into the determinant. They may also be used to change the determinant to **echelon form**, that is, a form in which the elements below the main diagonal elements are all zero (see Section 9.3). The transformations on rows stated in the next theorem are the same as the elementary row transformations of a matrix introduced in Section 9.3. However, for determinants we may also employ similar transformations on columns.

THEOREM ON ROW AND COLUMN TRANSFORMATIONS OF A DETERMINANT

Let A be a square matrix of order n.

(i) If a matrix B is obtained from A by interchanging two rows (or columns), then $|B| = -|A|$.

(ii) If B is obtained from A by multiplying every element of one row (or column) of A by a real number k, then $|B| = k|A|$.

(iii) If B is obtained from A by adding k times any row (or column) of A to another row (or column) for a real number k, then $|B| = |A|$.

When using the theorem, we refer to the rows (or columns) of the *determinant* in the obvious way. For example, property (iii) may be phrased: *Adding k times any row (or column) to another row (or column) of a determinant does not affect the value of the determinant.*

Row transformations of determinants will be specified by means of the symbols $R_i \leftrightarrow R_j$, $kR_i \rightarrow R_i$, and $kR_i + R_j \rightarrow R_j$ which were introduced in Section 9.3. Analogous symbols are used for column transformations. For example, $kC_i + C_j \rightarrow C_j$ means: Add k times the ith column to the jth column. The following are illustrations of the preceding theorem, with the reason for each equality stated at the right.

ILLUSTRATION | TRANSFORMATIONS OF DETERMINANTS

■ $\begin{vmatrix} 2 & 0 & 1 \\ 6 & 4 & 3 \\ 0 & 3 & 5 \end{vmatrix} = -\begin{vmatrix} 6 & 4 & 3 \\ 2 & 0 & 1 \\ 0 & 3 & 5 \end{vmatrix}$ $\quad \mathbf{R_1 \leftrightarrow R_2}$

■ $\begin{vmatrix} 2 & 0 & 1 \\ 6 & 4 & 3 \\ 0 & 3 & 5 \end{vmatrix} = -\begin{vmatrix} 1 & 0 & 2 \\ 3 & 4 & 6 \\ 5 & 3 & 0 \end{vmatrix}$ $\quad \mathbf{C_1 \leftrightarrow C_3}$

■ $\begin{vmatrix} 1 & -3 & 4 \\ 2 & -1 & 0 \\ 3 & 1 & 6 \end{vmatrix} = \begin{vmatrix} 1 & -3 & 4 \\ 0 & 5 & -8 \\ 3 & 1 & 6 \end{vmatrix}$ $\quad \mathbf{-2R_1 + R_2 \rightarrow R_2}$

■ $\begin{vmatrix} 1 & -3 & 4 \\ 2 & -1 & 0 \\ 3 & 1 & 6 \end{vmatrix} = \begin{vmatrix} -5 & -3 & 4 \\ 0 & -1 & 0 \\ 5 & 1 & 6 \end{vmatrix}$ $\quad \mathbf{2C_2 + C_1 \rightarrow C_1}$

THEOREM

> If two rows (or columns) of a square matrix A are identical, then $|A| = 0$.

PROOF If B is the matrix obtained from A by interchanging the two identical rows (or columns), then B and A are the same and, consequently, $|B| = |A|$. However, by (i) of the theorem on row and column transformations of a determinant, $|B| = -|A|$ and hence $-|A| = |A|$. Thus, $2|A| = 0$, and therefore $|A| = 0$. ❑

EXAMPLE ▪ 1

Find $|A|$ if $A = \begin{bmatrix} 2 & 3 & 0 & 4 \\ 0 & 5 & -1 & 6 \\ 1 & 0 & -2 & 3 \\ -3 & 2 & 0 & -5 \end{bmatrix}$.

SOLUTION We plan to use (iii) of the theorem on row and column transformations to introduce many zeros in some row or column. To do this, it is convenient to work with an element of the matrix that equals 1, since this enables us to avoid the use of fractions. If 1 is not an element of the original matrix, it is always possible to introduce the number 1 by using (ii) or (iii) of the theorem. In this example 1 appears in row 3, and we proceed as follows, with the reason for each equality stated at the right.

$$\begin{vmatrix} 2 & 3 & 0 & 4 \\ 0 & 5 & -1 & 6 \\ 1 & 0 & -2 & 3 \\ -3 & 2 & 0 & -5 \end{vmatrix} = \begin{vmatrix} 0 & 3 & 4 & -2 \\ 0 & 5 & -1 & 6 \\ 1 & 0 & -2 & 3 \\ 0 & 2 & -6 & 4 \end{vmatrix} \qquad \begin{matrix} \mathbf{-2R_3 + R_1 \to R_1} \\ \\ \\ \mathbf{3R_3 + R_4 \to R_4} \end{matrix}$$

$$= (1)\begin{vmatrix} 3 & 4 & -2 \\ 5 & -1 & 6 \\ 2 & -6 & 4 \end{vmatrix} \qquad \textbf{Expand by the first column.}$$

$$= \begin{vmatrix} 23 & 4 & 22 \\ 0 & -1 & 0 \\ -28 & -6 & -32 \end{vmatrix} \qquad \begin{matrix} \mathbf{5C_2 + C_1 \to C_1} \\ \mathbf{6C_2 + C_3 \to C_3} \end{matrix}$$

$$= (-1)\begin{vmatrix} 23 & 22 \\ -28 & -32 \end{vmatrix} \qquad \textbf{Expand by the second row.}$$

$$= (-1)[(23)(-32) - (-28)(22)] \qquad \textbf{Definition of determinant}$$

$$= 120$$

Part (ii) of the theorem on row and column transformations is useful for finding factors of determinants. To illustrate, for a determinant of a matrix of order 3 we have the following:

$$\begin{vmatrix} a_{11} & a_{12} & a_{13} \\ ka_{21} & ka_{22} & ka_{23} \\ a_{31} & a_{32} & a_{33} \end{vmatrix} = k \begin{vmatrix} a_{11} & a_{12} & a_{13} \\ a_{21} & a_{22} & a_{23} \\ a_{31} & a_{32} & a_{33} \end{vmatrix}.$$

Similar formulas hold if k is a common factor of the elements of any other row or column. When referring to this manipulation, we often use the phrase *k is a common factor in the row* (*or column*).

EXAMPLE ▪ 2

Find $|A|$ if $A = \begin{bmatrix} 14 & -6 & 4 \\ 4 & -5 & 12 \\ -21 & 9 & -6 \end{bmatrix}$.

SOLUTION

$$|A| - 2 \begin{vmatrix} 7 & -3 & 2 \\ 4 & -5 & 12 \\ -21 & 9 & -6 \end{vmatrix}$$ **2 is a common factor in row 1.**

$$= (2)(-3) \begin{vmatrix} 7 & -3 & 2 \\ 4 & -5 & 12 \\ 7 & -3 & 2 \end{vmatrix}$$ **−3 is a common factor in row 3.**

$$= 0$$ **Two rows are identical.**

EXAMPLE ▪ 3

Without expanding, show that $a - b$ is a factor of $\begin{vmatrix} 1 & 1 & 1 \\ a & b & c \\ a^2 & b^2 & c^2 \end{vmatrix}$.

SOLUTION

$$\begin{vmatrix} 1 & 1 & 1 \\ a & b & c \\ a^2 & b^2 & c^2 \end{vmatrix} = \begin{vmatrix} 0 & 1 & 1 \\ a - b & b & c \\ a^2 - b^2 & b^2 & c^2 \end{vmatrix}$$ $-C_2 + C_1 \rightarrow C_1$

$$= (a - b) \begin{vmatrix} 0 & 1 & 1 \\ 1 & b & c \\ a + b & b^2 & c^2 \end{vmatrix}$$ **$a - b$ is a common factor of column 1.**

9.7 EXERCISES

Exer. 1–14: Without expanding, explain why the statement is true.

1 $\begin{vmatrix} 1 & 0 & 1 \\ 0 & 1 & 1 \\ 1 & 1 & 0 \end{vmatrix} = -\begin{vmatrix} 1 & 0 & 1 \\ 1 & 1 & 0 \\ 0 & 1 & 1 \end{vmatrix}$

2 $\begin{vmatrix} 1 & 0 & 1 \\ 0 & 1 & 1 \\ 1 & 1 & 0 \end{vmatrix} = -\begin{vmatrix} 1 & 1 & 0 \\ 0 & 1 & 1 \\ 1 & 0 & 1 \end{vmatrix}$

3 $\begin{vmatrix} 1 & 0 & 1 \\ 2 & 1 & 0 \\ 1 & 1 & 2 \end{vmatrix} = \begin{vmatrix} 1 & 0 & 1 \\ 2 & 1 & 0 \\ 0 & 1 & 1 \end{vmatrix}$

4 $\begin{vmatrix} 1 & 1 & 2 \\ 1 & 0 & 1 \\ 2 & 1 & 1 \end{vmatrix} = \begin{vmatrix} 0 & 1 & 1 \\ 1 & 0 & 1 \\ 2 & 1 & 1 \end{vmatrix}$

5 $\begin{vmatrix} 2 & 4 & 2 \\ 1 & 2 & 4 \\ 2 & 6 & 4 \end{vmatrix} = 4\begin{vmatrix} 1 & 2 & 1 \\ 1 & 2 & 4 \\ 1 & 3 & 2 \end{vmatrix}$

6 $\begin{vmatrix} 2 & 1 & 6 \\ 4 & 3 & 3 \\ 2 & 1 & 3 \end{vmatrix} = 6\begin{vmatrix} 1 & 1 & 2 \\ 2 & 3 & 1 \\ 1 & 1 & 1 \end{vmatrix}$

7 $\begin{vmatrix} 1 & -1 & 2 \\ 1 & 2 & -1 \\ 1 & -1 & 2 \end{vmatrix} = 0$

8 $\begin{vmatrix} 1 & -1 & 1 \\ 0 & 1 & 0 \\ -1 & 1 & -1 \end{vmatrix} = 0$

9 $\begin{vmatrix} 1 & 5 \\ -3 & 2 \end{vmatrix} = -\begin{vmatrix} 1 & 5 \\ 3 & -2 \end{vmatrix}$

10 $\begin{vmatrix} 2 & -2 \\ 1 & 1 \end{vmatrix} = -\begin{vmatrix} -2 & 2 \\ 1 & 1 \end{vmatrix}$

11 $\begin{vmatrix} 0 & 0 & 1 \\ 1 & 0 & 0 \\ 0 & 0 & 2 \end{vmatrix} = 0$

12 $\begin{vmatrix} 1 & 0 & 1 \\ 0 & 0 & 0 \\ 1 & 1 & 0 \end{vmatrix} = 0$

13 $\begin{vmatrix} 1 & -1 & -2 \\ -1 & 2 & 1 \\ 0 & 1 & 1 \end{vmatrix} = \begin{vmatrix} 1 & -1 & 0 \\ -1 & 2 & -1 \\ 0 & 1 & 1 \end{vmatrix}$

14 $\begin{vmatrix} a & 0 & 0 \\ 0 & b & 0 \\ 0 & 0 & c \end{vmatrix} = -\begin{vmatrix} 0 & 0 & a \\ 0 & b & 0 \\ c & 0 & 0 \end{vmatrix}$

Exer. 15–24: Find the determinant of the matrix after introducing zeros, as in Example 1.

15 $\begin{bmatrix} 3 & 1 & 0 \\ -2 & 0 & 1 \\ 1 & 3 & -1 \end{bmatrix}$

16 $\begin{bmatrix} -3 & 0 & 4 \\ 1 & 2 & 0 \\ 4 & 1 & -1 \end{bmatrix}$

17 $\begin{bmatrix} 5 & 4 & 3 \\ -3 & 2 & 1 \\ 0 & 7 & -2 \end{bmatrix}$

18 $\begin{bmatrix} 0 & 2 & -6 \\ 5 & 1 & -3 \\ 6 & -2 & 5 \end{bmatrix}$

19 $\begin{bmatrix} 2 & 2 & -3 \\ 3 & 6 & 9 \\ -2 & 5 & 4 \end{bmatrix}$

20 $\begin{bmatrix} 3 & 8 & 5 \\ 5 & 3 & -6 \\ 2 & 4 & -2 \end{bmatrix}$

21 $\begin{bmatrix} 3 & 1 & -2 & 2 \\ 2 & 0 & 1 & 4 \\ 0 & 1 & 3 & 5 \\ -1 & 2 & 0 & -3 \end{bmatrix}$

22 $\begin{bmatrix} 3 & 2 & 0 & 4 \\ -2 & 0 & 5 & 0 \\ 4 & -3 & 1 & 6 \\ 2 & -1 & 2 & 0 \end{bmatrix}$

23 $\begin{bmatrix} 2 & -2 & 0 & 0 & -3 \\ 3 & 0 & 3 & 2 & -1 \\ 0 & 1 & -2 & 0 & 2 \\ -1 & 2 & 0 & 3 & 0 \\ 0 & 4 & 1 & 0 & 0 \end{bmatrix}$

24 $\begin{bmatrix} 2 & 0 & -1 & 0 & 2 \\ 1 & 3 & 0 & 0 & 1 \\ 0 & 4 & 3 & 0 & -1 \\ -1 & 2 & 0 & -2 & 0 \\ 0 & 1 & 5 & 0 & -4 \end{bmatrix}$

25 Show that

$$\begin{vmatrix} 1 & 1 & 1 \\ a & b & c \\ a^2 & b^2 & c^2 \end{vmatrix} = (a-b)(b-c)(c-a).$$

(*Hint:* See Example 3.)

26 Show that

$$\begin{vmatrix} 1 & 1 & 1 \\ a & b & c \\ a^3 & b^3 & c^3 \end{vmatrix} = (a-b)(b-c)(c-a)(a+b+c).$$

27 If

$$A = \begin{bmatrix} a_{11} & a_{12} & a_{13} & a_{14} \\ 0 & a_{22} & a_{23} & a_{24} \\ 0 & 0 & a_{33} & a_{34} \\ 0 & 0 & 0 & a_{44} \end{bmatrix}$$

show that $|A| = a_{11}a_{22}a_{33}a_{44}$.

28 If

$$A = \begin{bmatrix} a & b & 0 & 0 \\ c & d & 0 & 0 \\ 0 & 0 & e & f \\ 0 & 0 & g & h \end{bmatrix}$$

show that

$$|A| = \begin{vmatrix} a & b \\ c & d \end{vmatrix} \begin{vmatrix} e & f \\ g & h \end{vmatrix}.$$

29 If $A = (a_{ij})$ and $B = (b_{ij})$ are arbitrary square matrices of order 2, show that $|AB| = |A||B|$.

30 If $A = (a_{ij})$ is a square matrix of order n and k is any real number, show that $|kA| = k^n|A|$. (*Hint:* Use (ii) of the theorem on row and column transformations of a determinant.)

31 Use properties of determinants to show that the following is an equation of a line through the points (x_1, y_1) and (x_2, y_2):

$$\begin{vmatrix} x & y & 1 \\ x_1 & y_1 & 1 \\ x_2 & y_2 & 1 \end{vmatrix} = 0$$

32 Use properties of determinants to show that the following is an equation of a circle through three noncollinear points (x_1, y_1), (x_2, y_2), and (x_3, y_3):

$$\begin{vmatrix} x^2 + y^2 & x & y & 1 \\ x_1^2 + y_1^2 & x_1 & y_1 & 1 \\ x_2^2 + y_2^2 & x_2 & y_2 & 1 \\ x_3^2 + y_3^2 & x_3 & y_3 & 1 \end{vmatrix} = 0$$

9.8 CRAMER'S RULE

Determinants occur in the study of solutions of systems of linear equations. To illustrate, let us consider two linear equations in two variables x and y:

$$\begin{cases} a_{11}x + a_{12}y = k_1 \\ a_{21}x + a_{22}y = k_2 \end{cases}$$

such that at least one nonzero coefficient appears in each equation. We may assume that $a_{11} \neq 0$, for otherwise $a_{12} \neq 0$, and we could then regard y as the first variable instead of x. We shall use elementary row transformations to obtain the matrix of an equivalent system as follows:

$$\left[\begin{array}{cc|c} a_{11} & a_{12} & k_1 \\ a_{21} & a_{22} & k_2 \end{array}\right] \xrightarrow{-\frac{a_{21}}{a_{11}}\mathbf{R}_1 + \mathbf{R}_2 \to \mathbf{R}_2} \left[\begin{array}{cc|c} a_{11} & a_{12} & k_1 \\ 0 & a_{22} - \left(\dfrac{a_{12}a_{21}}{a_{11}}\right) & k_2 - \left(\dfrac{a_{21}k_1}{a_{11}}\right) \end{array}\right]$$

$$\xrightarrow{a_{11}\mathbf{R}_2 \to \mathbf{R}_2} \left[\begin{array}{cc|c} a_{11} & a_{12} & k_1 \\ 0 & (a_{11}a_{22} - a_{12}a_{21}) & (a_{11}k_2 - a_{21}k_1) \end{array}\right]$$

Thus, the system is equivalent to

$$\begin{cases} a_{11}x + a_{12}y = k_1 \\ (a_{11}a_{22} - a_{12}a_{21})y = a_{11}k_2 - a_{21}k_1 \end{cases}$$

which may also be written

$$\begin{cases} a_{11}x + a_{12}y = k_1 \\ \begin{vmatrix} a_{11} & a_{12} \\ a_{21} & a_{22} \end{vmatrix} y = \begin{vmatrix} a_{11} & k_1 \\ a_{21} & k_2 \end{vmatrix}. \end{cases}$$

If $\begin{vmatrix} a_{11} & a_{12} \\ a_{21} & a_{22} \end{vmatrix} \neq 0$, we can solve the second equation for y, obtaining

$$y = \frac{\begin{vmatrix} a_{11} & k_1 \\ a_{21} & k_2 \end{vmatrix}}{\begin{vmatrix} a_{11} & a_{12} \\ a_{21} & a_{22} \end{vmatrix}}$$

The corresponding value for x may be found by substituting for y in the first equation. This leads to

$$x = \frac{\begin{vmatrix} k_1 & a_{12} \\ k_2 & a_{22} \end{vmatrix}}{\begin{vmatrix} a_{11} & a_{12} \\ a_{21} & a_{22} \end{vmatrix}}$$

This proves that *if the determinant of the coefficient matrix of a system of two linear equations in two variables is not zero, then the system has a unique solution.* The last two formulas for x and y as quotients of determinants constitute **Cramer's rule** for two variables.

There is an easy way to remember Cramer's rule. Let

$$D = \begin{bmatrix} a_{11} & a_{12} \\ a_{21} & a_{22} \end{bmatrix}$$

be the coefficient matrix of the system and let D_x denote the matrix obtained from D by replacing the coefficients a_{11}, a_{21} of x by the numbers k_1, k_2, respectively. Similarly, let D_y denote the matrix obtained from D by replacing the coefficients a_{12}, a_{22} of y by the numbers k_1, k_2, respectively. Thus,

$$D_x = \begin{bmatrix} k_1 & a_{12} \\ k_2 & a_{22} \end{bmatrix}, \qquad D_y = \begin{bmatrix} a_{11} & k_1 \\ a_{21} & k_2 \end{bmatrix}.$$

If $|D| \neq 0$, the solution (x, y) is given by the following formulas.

CRAMER'S RULE FOR TWO VARIABLES

$$x = \frac{|D_x|}{|D|}, \qquad y = \frac{|D_y|}{|D|}$$

EXAMPLE ▪ 1

Use Cramer's rule to solve the system

$$\begin{cases} 2x - 3y = -4 \\ 5x + 7y = 1. \end{cases}$$

SOLUTION The determinant of the coefficient matrix is

$$|D| = \begin{vmatrix} 2 & -3 \\ 5 & 7 \end{vmatrix} = 29.$$

Using the notation introduced previously,

$$|D_x| = \begin{vmatrix} -4 & -3 \\ 1 & 7 \end{vmatrix} = -25, \qquad |D_y| = \begin{vmatrix} 2 & -4 \\ 5 & 1 \end{vmatrix} = 22.$$

Hence,
$$x = \frac{|D_x|}{|D|} = \frac{-25}{29}, \qquad y = \frac{|D_y|}{|D|} = \frac{22}{29}.$$

Thus, the system has the unique solution $(-\frac{25}{29}, \frac{22}{29})$.

Cramer's rule can be extended to systems of n linear equations in n variables $x_1, x_2, \ldots, x_n$, where each equation has the form

$$a_1x_1 + a_2x_2 + \cdots + a_nx_n = k.$$

To solve such a system, let D denote the coefficient matrix and let D_{x_i} denote the matrix obtained by replacing the coefficients of x_i in D by the numbers $k_1, \ldots, k_n$ that appear in the column to the right of the equal signs in the system. If $|D| \neq 0$, then the system has the following unique solution.

CRAMER'S RULE (GENERAL FORM)

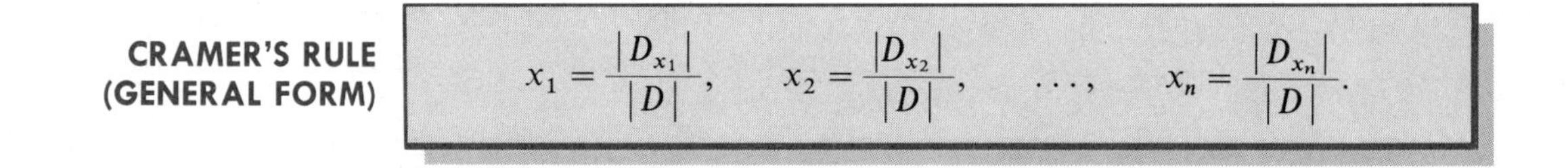

$$x_1 = \frac{|D_{x_1}|}{|D|}, \qquad x_2 = \frac{|D_{x_2}|}{|D|}, \qquad \ldots, \qquad x_n = \frac{|D_{x_n}|}{|D|}.$$

EXAMPLE ▪ 2

Use Cramer's rule to solve the system

$$\begin{cases} x \phantom{{}-y} - 2z = 3 \\ -y + 3z = 1 \\ 2x \phantom{{}-y} + 5z = 0. \end{cases}$$

SOLUTION We shall merely list the various determinants. You should check the results.

$$|D| = \begin{vmatrix} 1 & 0 & -2 \\ 0 & -1 & 3 \\ 2 & 0 & 5 \end{vmatrix} = -9, \qquad |D_x| = \begin{vmatrix} 3 & 0 & -2 \\ 1 & -1 & 3 \\ 0 & 0 & 5 \end{vmatrix} = -15,$$

$$|D_y| = \begin{vmatrix} 1 & 3 & -2 \\ 0 & 1 & 3 \\ 2 & 0 & 5 \end{vmatrix} = 27, \qquad |D_z| = \begin{vmatrix} 1 & 0 & 3 \\ 0 & -1 & 1 \\ 2 & 0 & 0 \end{vmatrix} = 6.$$

By Cramer's rule, the solution is

$$x = \frac{|D_x|}{|D|} = \frac{-15}{-9} = \frac{5}{3},$$

$$y = \frac{|D_y|}{|D|} = \frac{27}{-9} = -3,$$

$$z = \frac{|D_z|}{|D|} = \frac{6}{-9} = -\frac{2}{3}.$$

Cramer's rule is an inefficient method to apply if the system has a large number of equations, since many determinants of matrices of high order must be evaluated. Note also that Cramer's rule cannot be used directly if $|D| = 0$ or if the number of equations is not the same as the number of variables. For numerical calculations, the matrix method is superior to Cramer's rule; however, the Cramer's rule formulation is theoretically useful.

9.8 EXERCISES

Exer. 1–30: Use Cramer's rule, whenever applicable, to find the solution of the system.

1 $\begin{cases} 2x + 3y = 2 \\ x - 2y = 8 \end{cases}$

2 $\begin{cases} 4x + 5y = 13 \\ 3x + y = -4 \end{cases}$

3 $\begin{cases} 2x + 5y = 16 \\ 3x - 7y = 24 \end{cases}$

4 $\begin{cases} 7x - 8y = 9 \\ 4x + 3y = -10 \end{cases}$

5 $\begin{cases} 3r + 4s = 3 \\ r - 2s = -4 \end{cases}$

6 $\begin{cases} 9u + 2v = 0 \\ 3u - 5v = 17 \end{cases}$

7 $\begin{cases} 5x - 6y = 4 \\ 3x + 7y = 8 \end{cases}$

8 $\begin{cases} 2x + 8y = 7 \\ 3x - 5y = 4 \end{cases}$

9 $\begin{cases} \frac{1}{3}c + \frac{1}{2}d = 5 \\ c - \frac{2}{3}d = -1 \end{cases}$

10 $\begin{cases} \frac{1}{2}t - \frac{1}{5}v = \frac{3}{2} \\ \frac{2}{3}t + \frac{1}{4}v = \frac{5}{12} \end{cases}$

11 $\begin{cases} \sqrt{3}x - \sqrt{2}y = 2\sqrt{3} \\ 2\sqrt{2}x + \sqrt{3}y = \sqrt{2} \end{cases}$

12 $\begin{cases} 0.11x - 0.03y = 0.25 \\ 0.12x + 0.05y = 0.70 \end{cases}$

13 $\begin{cases} 2x - 3y = 5 \\ -6x + 9y = 12 \end{cases}$

14 $\begin{cases} 3p - q = 7 \\ -12p + 4q = 3 \end{cases}$

15 $\begin{cases} 3m - 4n = 2 \\ -6m + 8n = -4 \end{cases}$

16 $\begin{cases} x - 5y = 2 \\ 3x - 15y = 6 \end{cases}$

17 $\begin{cases} 2y - 5x = 0 \\ 3y + 4x = 0 \end{cases}$

18 $\begin{cases} 3x + 7y = 9 \\ y = 5 \end{cases}$

19 $\begin{cases} x - 2y - 3z = -1 \\ 2x + y + z = 6 \\ x + 3y - 2z = 13 \end{cases}$

20 $\begin{cases} x + 3y - z = -3 \\ 3x - y + 2z = 1 \\ 2x - y + z = -1 \end{cases}$

21 $\begin{cases} 5x + 2y - z = -7 \\ x - 2y + 2z = 0 \\ 3y + z = 17 \end{cases}$

22 $\begin{cases} 4x - y + 3z = 6 \\ -8x + 3y - 5z = -6 \\ 5x - 4y = -9 \end{cases}$

23 $\begin{cases} 2x + 6y - 4z = 1 \\ x + 3y - 2z = 4 \\ 2x + y - 3z = -7 \end{cases}$

24 $\begin{cases} x + 3y - 3z = -5 \\ 2x - y + z = -3 \\ -6x + 3y - 3z = 4 \end{cases}$

25 $\begin{cases} 2x - 3y + 2z = -3 \\ -3x + 2y + z = 1 \\ 4x + y - 3z = 4 \end{cases}$

26 $\begin{cases} 2x - 3y + z = 2 \\ 3x + 2y - z = -5 \\ 5x - 2y + z = 0 \end{cases}$

27 $\begin{cases} x + 2y - z - 3w = 2 \\ 3x + y - 2z - w = 6 \\ x + y + 3z - 2w = -3 \\ -2x - 2y + 3z + w = -9 \end{cases}$

28 $\begin{cases} x - 2y - 5z + w = -1 \\ 2x - y + z + w = 1 \\ 3x - 2y - 4z - 2w = 1 \\ x + y + 3z - 2w = 2 \end{cases}$

29 $\begin{cases} 5x + 2z = 1 \\ y - 3z = 2 \\ 2x + y = 3 \end{cases}$

30 $\begin{cases} 2x - 3y = 12 \\ 3y + z = -2 \\ 5x - 3z = 3 \end{cases}$

9.9 SYSTEMS OF INEQUALITIES

In Chapter 2 we restricted our discussion to inequalities in one variable. We shall now consider inequalities in *several* variables:

ILLUSTRATION | INEQUALITIES IN x AND y

- $3x + y < 5y^2 + 1$
- $2x^2 \geq 4 - 3y$
- $2y - x < 8$

A **solution** of an inequality in x and y is an ordered pair (a, b) that results in a true statement if a and b are substituted for x and y, respectively. The **graph of an inequality** is the set of all points (a, b) in an xy-plane that correspond to the solutions. Two inequalities are **equivalent** if they have the same solutions. An inequality in x and y can often be simplified by adding an expression in x and y to both sides or by multiplying both sides by some expression (provided we are careful about signs). Similar remarks apply to inequalities in more than two variables.

EXAMPLE ▪ 1

Find the solutions and sketch the graph of the inequality $3x - 3 < 5x - y$.

SOLUTION The following inequalities are equivalent:

$$3x - 3 < 5x - y$$
$$y + 3x - 3 < 5x$$
$$y < 5x - (3x - 3)$$
$$y < 2x + 3$$

Hence, the solutions are all ordered pairs (x, y) such that $y < 2x + 3$. It is convenient to write the solutions as

$$\{(x, y): y < 2x + 3\}.$$

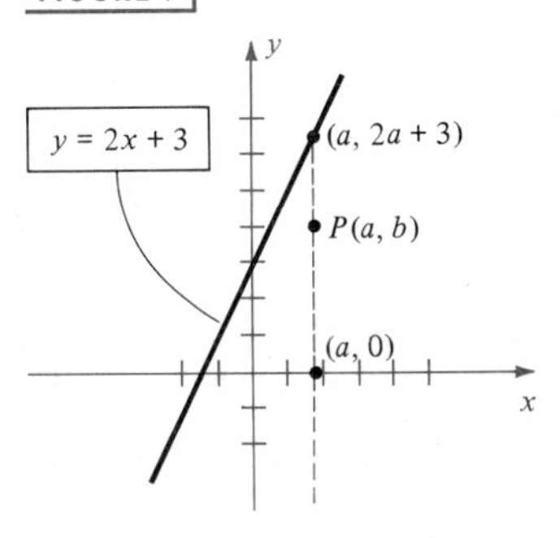

The graph of the inequality $y < 2x + 3$ is closely related to the graph of the equation $y = 2x + 3$. The graph of the equation is the line sketched in Figure 7. For each real number a, the point on the line with x-coordinate a has coordinates $(a, 2a + 3)$. A point $P(a, b)$ belongs to the graph of the *inequality* if and only if $b < 2a + 3$, that is, if and only if the point $P(a, b)$ lies directly below the point with coordinates $(a, 2a + 3)$, as shown in Figure 7. Thus the graph of the inequality $y < 2x + 3$ consists of all points that lie below the line $y = 2x + 3$. In Figure 8 we have shaded a portion of the graph of the inequality. Dashes used for the line indicate that it is not part of the graph.

FIGURE 8

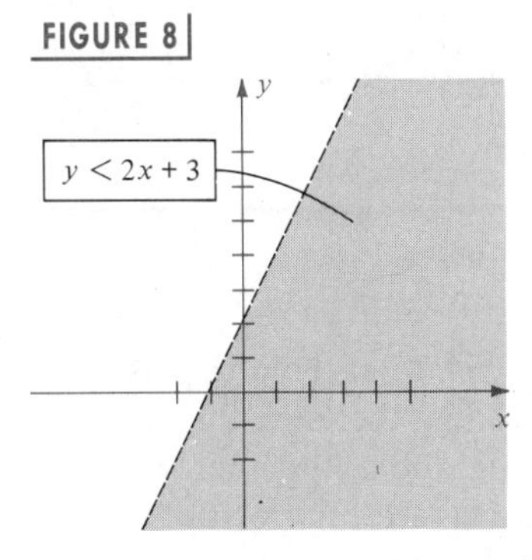

The region sketched in Figure 8 is a **half-plane**. More precisely, if the line is *not* included, the region is an **open half-plane**. If the line *is* included, as for the graph of the inequality $y \leq 2x + 3$, then the region is a **closed half-plane**.

An inequality that involves only polynomials of the first degree in x and y, as in Example 1, is a **linear inequality**.

The procedure used in Example 1 can be generalized to inequalities of the form $y < f(x)$ for any function f. Specifically, we have the following theorem.

THEOREM

Let f be a function.

(i) The graph of the inequality $y < f(x)$ is the set of points that lie *below* the graph of the equation $y = f(x)$.

(ii) The graph of $y > f(x)$ is the set of points that lie *above* the graph of $y = f(x)$.

EXAMPLE ▪ 2

Find the solutions and sketch the graph of $x(x + 1) - 2y > 3(x - y)$.

SOLUTION The inequality is equivalent to

$$x^2 + x - 2y > 3x - 3y.$$

Adding $3y - x^2 - x$ to both sides, we obtain

$$y > 2x - x^2.$$

Hence, the solutions are given by $\{(x, y): y > 2x - x^2\}$.

FIGURE 9

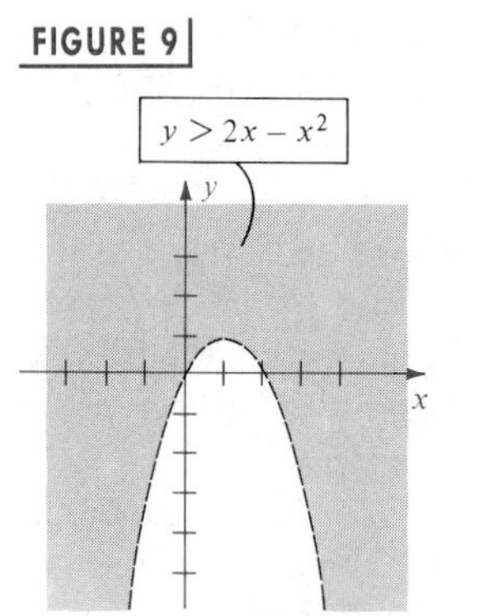

To find the graph of $y > 2x - x^2$, we begin by sketching the graph of $y = 2x - x^2$ (a parabola) with dashes as illustrated in Figure 9. By (ii) of the preceding theorem, the graph is the region above the parabola, as indicated by the shaded portion of the figure.

We can also prove the following.

THEOREM

Let g be a function.

(i) The graph of the inequality $x < g(y)$ is the set of points to the *left* of the graph of the equation $x = g(y)$.

(ii) The graph of $x > g(y)$ is the set of points to the *right* of the graph of $x = g(y)$.

EXAMPLE ▪ 3

Sketch the graph of $x \geq y^2$.

FIGURE 10

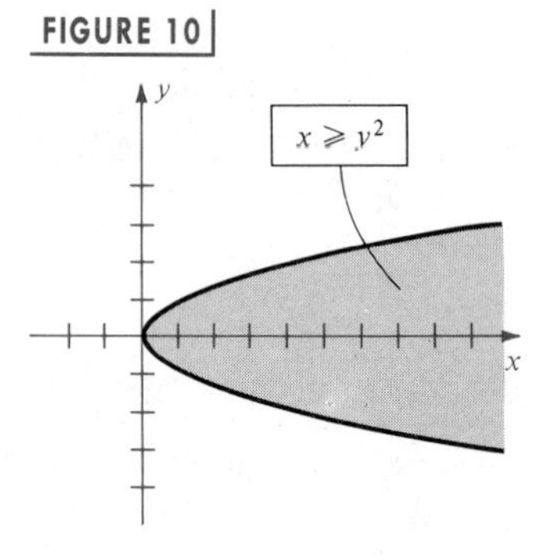

SOLUTION The graph of the equation $x = y^2$ is a parabola that is symmetric with respect to the x-axis. By the preceding theorem, the graph of the inequality consists of all points on the parabola, together with the points in the region to the right of the parabola (see Figure 10).

We sometimes work simultaneously with several inequalities in two variables, that is, with a **system of inequalities**. The **solutions of a system** of inequalities are the solutions common to all inequalities in the system. It should be clear how to define **equivalent systems** and the **graph of a system** of inequalities. The following examples illustrate a method for solving systems of inequalities.

EXAMPLE ▪ 4

Sketch the graph of the system

$$\begin{cases} x + y \leq 4 \\ 2x - y \leq 4. \end{cases}$$

SOLUTION The system is equivalent to

$$\begin{cases} y \leq 4 - x \\ y \geq 2x - 4. \end{cases}$$

FIGURE 11

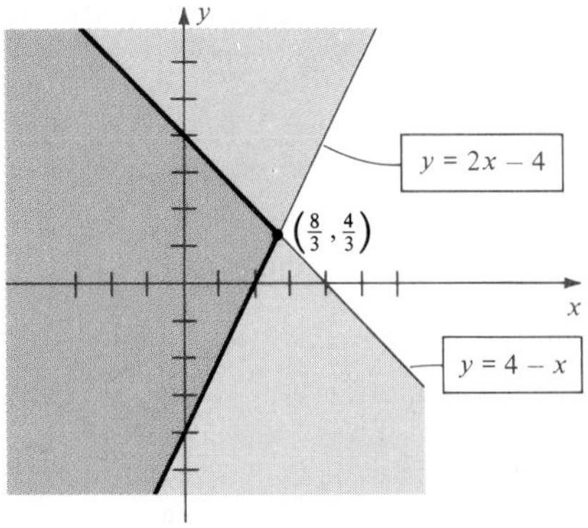

We begin by sketching the graphs of the lines $y = 4 - x$ and $y = 2x - 4$. The lines intersect at the point $(\frac{8}{3}, \frac{4}{3})$ shown in Figure 11. The graph of $y \leq 4 - x$ includes the points on the graph of $y = 4 - x$ together with the points that lie below this line. The graph of $y \geq 2x - 4$ includes the points on the graph of $y = 2x - 4$ together with the points that lie above this line. A portion of each of these regions is shown in Figure 11. The graph of the system consists of the points that are in *both* regions, as indicated by the double-shaded portion of the figure.

EXAMPLE ▪ 5

Sketch the graph of the system

$$\begin{cases} x + y \leq 4 \\ 2x - y \leq 4 \\ x \geq 0 \\ y \geq 0. \end{cases}$$

FIGURE 12

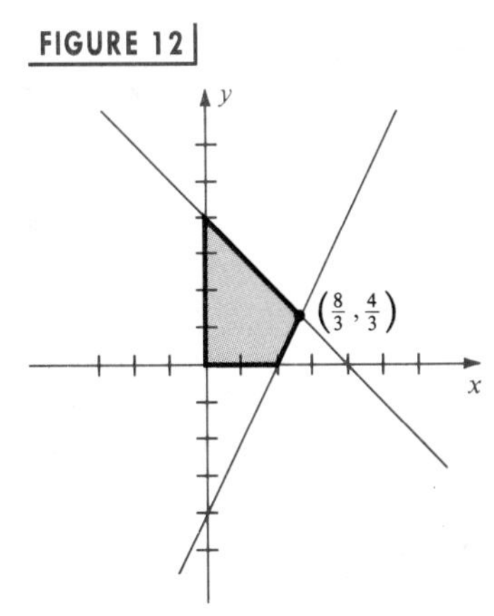

SOLUTION The first two inequalities are the same as those considered in Example 4, and hence the points on the graph of the system must lie within the double-shaded region shown in Figure 11. In addition, the third and fourth inequalities in the system tell us that the points must lie in the first quadrant or on its boundaries. This gives us the region shown in Figure 12.

EXAMPLE ▪ 6

Sketch the graph of the system

$$\begin{cases} x^2 + y^2 \leq 1 \\ (x - 1)^2 + y^2 \leq 1. \end{cases}$$

FIGURE 13

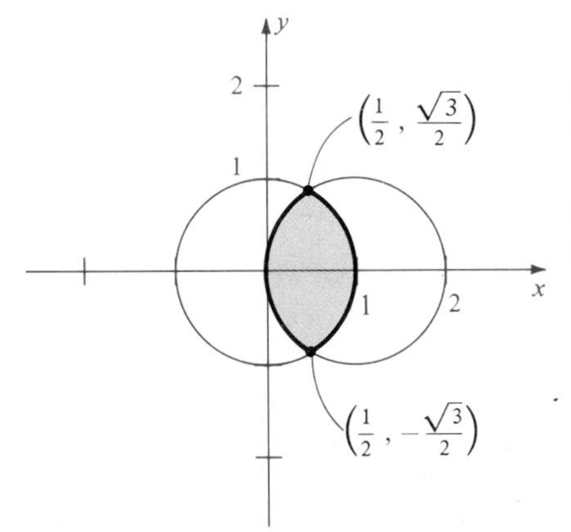

SOLUTION The graph of the equation $x^2 + y^2 = 1$ is a unit circle with center at the origin, and the graph of $(x - 1)^2 + y^2 = 1$ is a unit circle with center at the point $C(1, 0)$ (see Figure 13).

To find the points of intersection of the two circles, let us solve the equation $x^2 + y^2 = 1$ for y^2, obtaining $y^2 = 1 - x^2$. Substituting for y^2 in $(x - 1)^2 + y^2 = 1$ leads to the following equations:

$$\begin{aligned} (x - 1)^2 + (1 - x)^2 &= 1 \\ x^2 - 2x + 1 + 1 - x^2 &= 1 \\ -2x &= -1 \\ x &= \tfrac{1}{2}. \end{aligned}$$

The corresponding values for y are given by

$$y^2 = 1 - x^2 = 1 - (\tfrac{1}{2})^2 = \tfrac{3}{4}$$

and hence $y = \pm\sqrt{3}/2$. Thus, the points of intersection are $(\frac{1}{2}, \sqrt{3}/2)$ and $(\frac{1}{2}, -\sqrt{3}/2)$ as shown in Figure 13.

By the distance formula, the graphs of the inequalities are the regions within and on the two circles. The graph of the system consists of the points common to both regions, as indicated by the shaded portion of the figure.

EXAMPLE ▪ 7

The manager of a baseball team wishes to buy bats and balls costing \$12 and \$3 each, respectively. At least five bats and ten balls are required, and the total cost is not to exceed \$180. Find a system of inequalities that describes all possibilities, and sketch the graph.

SOLUTION We begin by introducing the following variables:

$$x = \text{number of bats}$$

$$y = \text{number of balls}$$

Since the cost of a bat is \$12 and the cost of a ball is \$3, we see that

$$12x = \text{cost of } x \text{ bats}$$

$$3y = \text{cost of } y \text{ balls.}$$

Since the total cost is not to exceed \$180, we must have

$$12x + 3y \leq 180$$

$$4x + y \leq 60.$$

The other restrictions are

$$x \geq 5 \quad \text{and} \quad y \geq 10.$$

The graph of $4x + y \leq 60$ is the closed half-plane that lies *below* the line with y-intercept 60 and x-intercept 15 shown in Figure 14.

The graph of $x \geq 5$ is the closed half-plane that lies to the *right* of the vertical line $x = 5$. The graph of $y \geq 10$ is the closed half-plane that lies *above* the horizontal line $y = 10$.

The graph of the system, that is, the points common to the three half-planes, is the triangular region sketched in Figure 14.

FIGURE 14

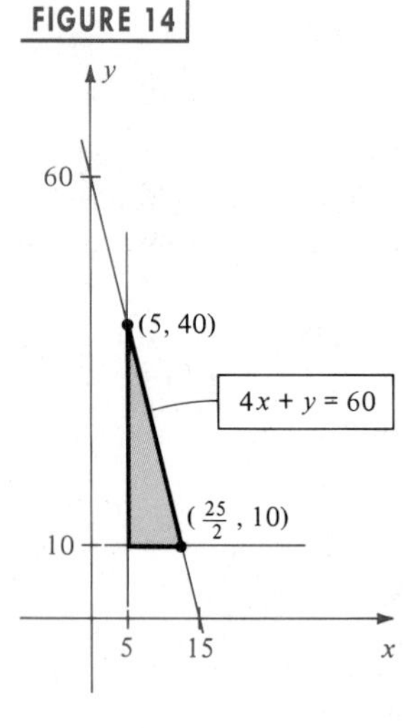

9.9 EXERCISES

Exer. 1–10: Sketch the graph of the inequality.

1 $3x - 2y < 6$

2 $4x + 3y < 12$

3 $2x + 3y \geq 2y + 1$

4 $2x - y > 3$

5 $y + 2 < x^2$

6 $y^2 - x \leq 0$

7 $x^2 + 1 \leq y$

8 $y - x^3 < 1$

9 $yx^2 \geq 1$

10 $x^2 + 4 \geq y$

Exer. 11–20: Sketch the graph of the system of inequalities.

11 $\begin{cases} 3x + y < 3 \\ 4 - y < 2x \end{cases}$

12 $\begin{cases} y + 2 < 2x \\ y - x > 4 \end{cases}$

13 $\begin{cases} y - x < 0 \\ 2x + 5y < 10 \end{cases}$

14 $\begin{cases} 2y - x \leq 4 \\ 3y + 2x < 6 \end{cases}$

15 $\begin{cases} 3x + y \leq 6 \\ y - 2x \geq 1 \\ x \geq -2 \\ y \leq 4 \end{cases}$

16 $\begin{cases} 3x - 4y \geq 12 \\ x - 2y \leq 2 \\ x \geq 9 \\ y \leq 5 \end{cases}$

17 $\begin{cases} x^2 + y^2 \leq 4 \\ x + y \geq 1 \end{cases}$

18 $\begin{cases} x^2 + y^2 > 1 \\ x^2 + y^2 < 4 \end{cases}$

19 $\begin{cases} x^2 \leq 1 - y \\ x \geq 1 + y \end{cases}$

20 $\begin{cases} x - y^2 < 0 \\ x + y^2 > 0 \end{cases}$

Exer. 21–28: Find a system of inequalities whose graph is shown.

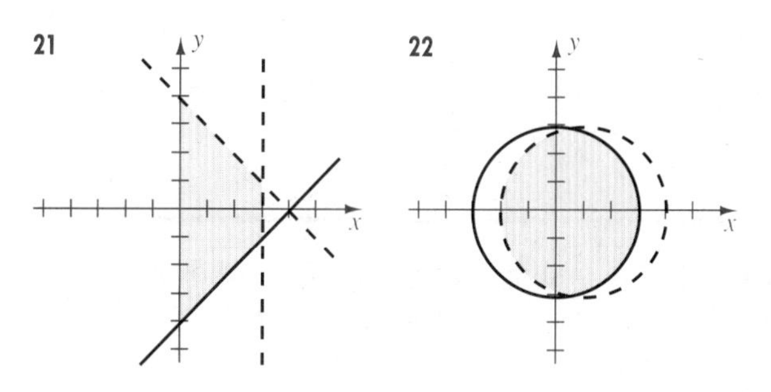

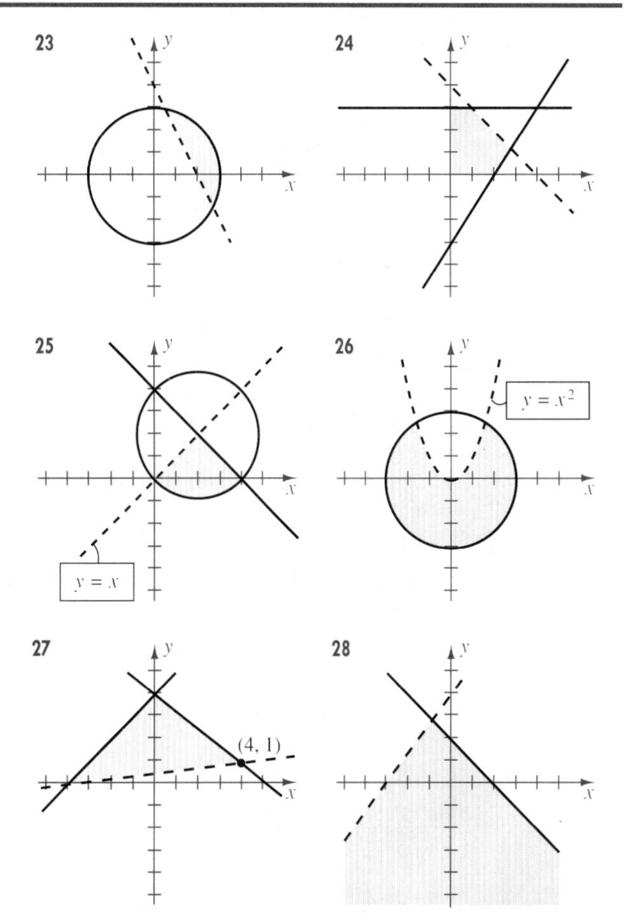

29 A store sells two brands of television sets. Customer demand indicates that it is necessary to stock at least twice as many sets of brand A as of brand B. It is also necessary to have on hand at least 20 of brand A and 10 of brand B. There is room for not more than 100 sets in the store. Find and graph a system of inequalities that describes all possibilities for stocking the two brands.

30 An auditorium contains 600 seats. For an upcoming event tickets will be priced at \$8.00 for some seats and \$5.00 for others. At least 225 tickets are to be priced at \$5.00, and total sales of more than \$3000 are desired. Find and graph a system of inequalities that describes all possibilities for pricing the two types of tickets.

31 A woman with $15,000 to invest decides to place at least $2000 in a high-risk, high-yield investment and at least three times that amount in a low-risk, low-yield investment. Find and graph a system of inequalities that describes all possibilities for placing the money in the two investments.

32 The manager of a college bookstore stocks two types of notebooks, the first wholesaling for 55 cents and the second for 85 cents. The maximum amount to be spent is $600 and an inventory of at least 300 of the 85-cent variety and 400 of the 55-cent variety is desired. Find and graph a system of inequalities that describes all possibilities for stocking the two types of notebooks.

33 An aerosol can is to be constructed in the shape of a circular cylinder with a small cone on the top. The total height of the can is to be no more than 9 inches and the cylinder must contain at least 75% of the total volume. In addition, the height of the conical top must be at least 1 inch. Find and graph a system of inequalities that describes all possibilities for the relationship between the height y of the cylinder and the height x of the cone.

34 A stained-glass window is to be constructed in the form of a rectangle surmounted by a semicircle (see figure). The total height h of the window can be no more than 6 feet, and the area of the rectangular part must be at least twice the area of the semicircle. In addition, the diameter d of the semicircle must be at least 2 feet. Find and graph a system of inequalities that describes the possibilities for the length and width of the rectangle.

EXERCISE 34

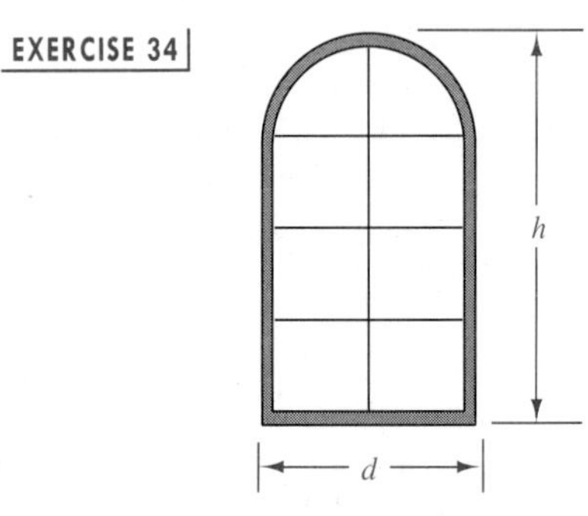

35 A nuclear power plant will be constructed to serve the power needs of cities A and B. City B is 100 miles due east of A. The state has promised that the plant will be at least 60 miles from each city. It is not possible, however, to locate the plant south of either city because of rough terrain, and the plant must be within 100 miles of both A and B. Assuming A is at the origin, find and graph a system of inequalities that describes all possible locations for the plant.

36 A man has a rectangular back yard that is 50 feet wide and 60 feet deep. He plans to construct a pool area and a patio area, as shown in the figure, and can spend at most $12,000 on the project. The patio area must be at least as large as the pool area. The pool area will cost $5 per square foot and the patio will cost $3 per square foot. Find and graph a system of inequalities that describes the possibilities for the width of the patio and pool areas.

EXERCISE 36

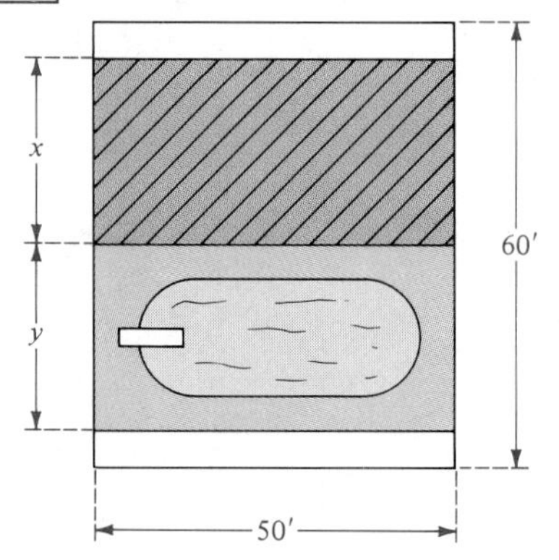

9.10 LINEAR PROGRAMMING

Some applied problems require finding particular solutions of systems of inequalities. A typical problem consists of finding maximum or minimum values of expressions involving variables that are subject to various constraints. If all the expressions and inequalities are linear in the variables, then **linear programming** may be used to help solve such a problem. This technique has become very important in business decisions concerning the best use of stock, parts, or manufacturing processes. Usually the goal of management is to maximize profit or to minimize cost. Since there are often many variables, it may be difficult to make correct decisions to achieve such goals. Linear programming can simplify the task. The logical development of the theorems and methods that are needed would take us beyond the objectives of this text, so we shall limit this discussion to several examples.

EXAMPLE ▪ 1

A distributor of compact disk players has two warehouses, W_1 and W_2, with 80 units stored at W_1 and 70 at W_2. Two customers, A and B, order 35 units and 60 units, respectively. The shipping cost from each warehouse to A and B is determined according to the following table. How should the order be filled to minimize the total shipping cost?

Warehouse	Customer	Shipping cost per unit
W_1	A	\$ 8
W_1	B	12
W_2	A	10
W_2	B	13

SOLUTION Let

$$x = \text{number of units sent to A from } W_1,$$

$$y = \text{number of units sent to B from } W_1.$$

Thus, to fill the orders we must have

$$35 - x = \text{number sent to A from } W_2,$$

$$60 - y = \text{number sent to B from } W_2.$$

FIGURE 15

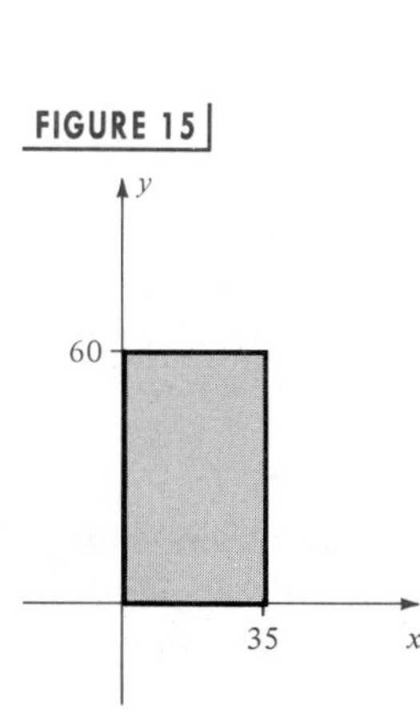

We wish to determine values for x and y that make the total shipping cost minimal. Since x and y are between 35 and 60, respectively, the pair (x, y) must be a solution of the following system of inequalities:

$$\begin{cases} 0 \le x \le 35 \\ 0 \le y \le 60 \end{cases}$$

The graph of this system is the rectangular region shown in Figure 15.

Additional constraints on x and y make it possible to reduce the size of the region in Figure 15. Since the total number of units shipped from W_1 cannot exceed 80 and the total shipped from W_2 cannot exceed 70, the

FIGURE 16

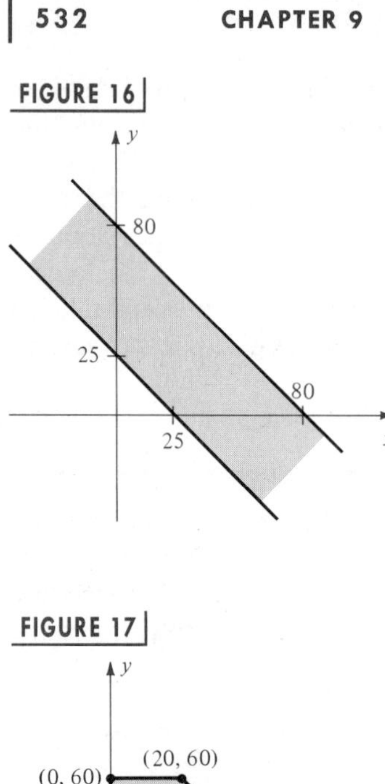

FIGURE 17

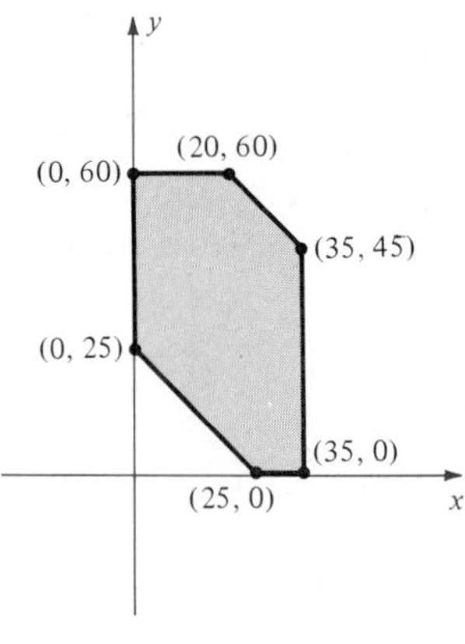

pair (x, y) must also be a solution of the system

$$\begin{cases} x + y \leq 80 \\ (35 - x) + (60 - y) \leq 70. \end{cases}$$

This system is equivalent to

$$\begin{cases} x + y \leq 80 \\ x + y \geq 25. \end{cases}$$

The graph of this system is the region between the parallel lines $x + y = 80$ and $x + y = 25$ (see Figure 16). Since the pair (x, y) that we seek must be a solution of this system, and also of the system $0 \leq x \leq 35$, $0 \leq y \leq 60$, the corresponding point must lie in the region shown in Figure 17.

Let C denote the total cost (in dollars) of shipping the disk players to customers A and B. We see from the table of shipping costs that the following are true:

$$\text{cost of shipping 35 units to A} = 8x + 10(35 - x)$$

$$\text{cost of shipping 60 units to B} = 12y + 13(60 - y)$$

The total cost is:

$$C = 8x + 10(35 - x) + 12y + 13(60 - y)$$

$$C = 8x + 350 - 10x + 12y + 780 - 13y$$

$$C = 1130 - 2x - y.$$

For each point (x, y) of the region shown in Figure 17 there corresponds a value for C. For example,

$$\text{at } (20, 40)\text{:} \quad C = 1130 - 40 - 40 = 1050,$$

$$\text{at } (10, 50)\text{:} \quad C = 1130 - 20 - 50 = 1060.$$

Since x and y are integers, we have a finite number of possible values for C. By checking each possibility, we could find the pair (x, y) that produces the smallest cost. However, since the number of pairs is very large, the task of checking each one would be very tedious. This is where linear programming is helpful. It can be shown that if we are interested in the value C of a linear expression $ax + by + c$, and if each pair (x, y) is a solution of a system of linear inequalities, and hence corresponds to a point that is common to several half-planes, then C takes on its maximum or minimum value at a point of intersection of the lines that determine the half-planes. Thus to determine the minimum (or maximum) value of C we need check only the corner points $(0, 25)$, $(0, 60)$, $(20, 60)$, $(35, 45)$, $(35, 0)$, and $(25, 0)$ shown in Figure 17. The values are displayed in the following table.

Point	$1130 - 2x - y = C$
(0, 25)	$1130 - 2(0) - 25 = 1105$
(0, 60)	$1130 - 2(0) - 60 = 1070$
(20, 60)	$1130 - 2(20) - 60 = 1030$
(35, 45)	$1130 - 2(35) - 45 = 1015$
(35, 0)	$1130 - 2(35) - 0 = 1060$
(25, 0)	$1130 - 2(25) - 0 = 1080$

According to our remarks, the minimal shipping cost \$1015 occurs if $x = 35$ and $y = 45$. This means that the distributor should ship all of the disk players to A from W_1 and none from W_2. In addition, the distributor should ship 45 units to B from W_1 and 15 units to B from W_2. (Note that the *maximum* shipping cost will occur if $x = 0$ and $y = 25$, that is, if all 35 units are shipped to A from W_2 and if B receives 25 units from W_1 and 35 units from W_2.)

Example 1 demonstrates how linear programming can be used to minimize the cost in a certain situation. The next example illustrates maximization of profit.

EXAMPLE ▪ 2

A firm manufactures two products X and Y. For each product it is necessary to use three different machines, A, B, and C. To manufacture one unit of product X, machine A must be used for 3 hours, machine B for 1 hour, and machine C for 1 hour. To manufacture one unit of product Y requires 2 hours on A, 2 hours on B, and 1 hour on C. The profit on product X is \$500 per unit and the profit on product Y is \$350 per unit. Machine A is available for a total of 24 hours per day; however, B can only be used for 16 hours and C for 9 hours. Assuming the machines are available when needed (subject to the noted total hour restrictions), determine the number of units of each product that should be manufactured each day in order to maximize the profit.

SOLUTION The following table summarizes the data given in the statement of the problem.

Machine	Hours required for one unit of X	Hours required for one unit of Y	Hours available
A	3	2	24
B	1	2	16
C	1	1	9

Let x and y denote the number of units of products X and Y, respectively, to be produced per day. Since each unit of product X requires 3 hours on machine A, x units require $3x$ hours. Similarly, since each unit of product Y requires 2 hours on A, y units require $2y$ hours. Hence, the total number of hours per day that machine A must be used is $3x + 2y$. Since A can be used for at most 24 hours per day,

$$3x + 2y \leq 24.$$

Using the same type of reasoning on rows two and three of the table, we see that

$$x + 2y \leq 16$$

$$x + y \leq 9.$$

This system of three linear inequalities together with the obvious inequalities

$$x \geq 0, \qquad y \geq 0$$

states, in mathematical form, the constraints that occur in the manufacturing process. The graph of the system of all five linear inequalities is sketched in Figure 18.

FIGURE 18

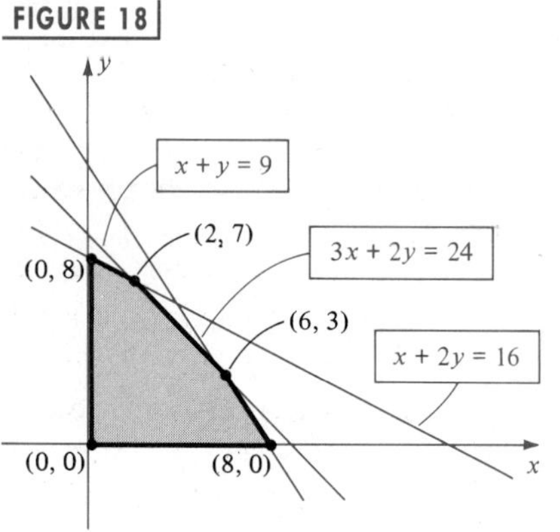

The points shown in the figure are found by solving systems of linear equations. Specifically, (6, 3) is the solution of the system $3x + 2y = 24$, $x + y = 9$, and (2, 7) is the solution of the system $x + 2y = 16$, $x + y = 9$.

Since the production of each unit of product X yields a profit of \$500, and each unit of product Y yields a profit of \$350, the profit P obtained by producing x units of X together with y units of Y is

$$P = 500x + 350y.$$

The maximum value of P must occur at one of the corner points shown in Figure 18. The values of P at all the points are shown in the following table.

(x, y)	$500x + 350y = P$
(0, 8)	$500(0) + 350(8) = 2800$
(2, 7)	$500(2) + 350(7) = 3450$
(6, 3)	$500(6) + 350(3) = 4050$
(8, 0)	$500(8) + 350(0) = 4000$
(0, 0)	$500(0) + 350(0) = 0$

We see from the table that a maximum profit of \$4050 occurs for a daily production of 6 units of product X and 3 units of product Y.

The two illustrations given in this section are elementary linear programming problems that can be solved by basic methods. The much more complicated problems that occur in practice involving, for example, more than two items, are usually solved by employing matrix techniques that are adapted for solution by computers.

9.10 EXERCISES

1 A manufacturer of tennis rackets makes a profit of \$15 on each oversized racket and \$8 on each standard racket. To meet dealer demand, daily production of standard rackets should be between 30 and 80, and production of oversized rackets should be between 10 and 30. To maintain high quality, the total number of rackets produced should not exceed 80 per day. How many of each type should be manufactured daily to maximize the profit?

2 A manufacturer of CB radios makes a profit of \$25 on a deluxe model and \$30 on a standard model. The company wishes to produce at least 80 deluxe models and at least 100 standard models per day. To maintain high quality, the daily production should not exceed 200 radios. How many of each type should be produced daily in order to maximize the profit?

3 Two substances S and T each contain two types of ingredients I and G. One pound of S contains 2 ounces of I and 4 ounces of G. One pound of T contains 2 ounces of I and 6 ounces of G. A manufacturer plans to combine quantities of the two substances to obtain a mixture that contains at least 9 ounces of I and 20 ounces of G. If the cost of S is \$3.00 per pound and the cost of T is \$4.00 per pound, how much of each substance should be used to keep the cost to a minimum?

4 A stationery company makes two types of notebooks: a deluxe notebook, with subject dividers, that sells for \$1.25, and a regular notebook that sells for \$0.90. The production cost is \$1.00 for each deluxe notebook and \$0.75 for each regular notebook. The company has the facilities to manufacture between 2000 and 3000 deluxe and between 3000 and 6000 regular, but not more than 7000 altogether. How many notebooks of each type should be manufactured to maximize the difference between the selling prices and the production cost?

5 Refer to Example 1 of this section. If the shipping costs are \$12 per unit from W_1 to A, \$10 per unit from W_2 to A, \$16 per unit from W_1 to B, and \$12 per unit from W_2 to B, determine how the order should be filled to minimize shipping cost.

6 A coffee company purchases mixed lots of coffee beans and then grades them into premium, regular, and unusable beans. The company needs at least 280 tons of premium-grade and 200 tons of regular-grade coffee beans. The company can purchase ungraded coffee from two suppliers in any amount desired. Samples from the two suppliers contain the following percentages of premium, regular, and unusable beans:

Supplier	Premium	Regular	Unusable
A	20%	50%	30%
B	40%	20%	40%

If supplier A charges \$125 per ton and B charges \$200 per ton, how much should the company purchase from each supplier to fulfill its needs at minimum cost?

7 A farmer, in the business of growing fodder for livestock, has 100 acres available for planting alfalfa and corn. The cost of seed per acre is \$4 for alfalfa and \$6 for corn. The total cost of labor will amount to \$20 per acre for alfalfa and \$10 per acre for corn. The expected income from alfalfa is \$110 per acre, and from corn, \$150 per acre. If the farmer does not wish to spend more than \$480 for seed and \$1400 for labor, how many acres of each crop should be planted to obtain the maximum profit?

8 A small firm manufactures bookshelves and desks for microcomputers. For each product it is necessary to use a table saw and a power router. To manufacture each bookshelf, the saw must be used for $\frac{1}{2}$ hour and the router for 1 hour. A desk requires the use of each machine for 2 hours. The profits are \$20 per bookshelf and \$50 per desk. If the saw can be used 8 hours per day, and the router for 12 hours per day, how many bookshelves and desks should be manufactured each day to maximize the profit?

9 Three substances X, Y, and Z each contain four ingredients A, B, C, and D. The percentage of each ingredient and the cost in cents per ounce of each substance are given in the following table.

	Ingredients				
Substance	A	B	C	D	Cost per ounce
X	20%	10%	25%	45%	25¢
Y	20%	40%	15%	25%	35¢
Z	10%	20%	25%	45%	50¢

If the cost is to be minimal, how many ounces of each substance should be combined to obtain a mixture of 20 ounces containing at least 14% A, 16% B, and 20% C? What combination would make the cost greatest?

10 A man plans to operate a stand at a one-day fair at which he will sell bags of peanuts and bags of candy. He has \$400 available to purchase his stock, which will cost 40¢ per bag of peanuts and 80¢ per bag of candy. He intends to sell the peanuts at \$1.00 and the candy at \$1.60 per bag. His stand can accommodate up to 500 bags of peanuts and 400 bags of candy. From past experience he knows that he will sell no more than a total of 700 bags. Find the number of bags of each that he should have available in order to maximize his profit. What is the maximum profit?

11 A small community wishes to purchase used vans and small buses for its public transportation system. The community can spend no more than \$100,000 for the vehicles and no more than \$500 per month for maintenance. The vans sell for \$10,000 each and average \$100 per month in maintenance costs. The corresponding cost estimates for each bus are \$20,000 and \$75 per month. If each van can carry 15 passengers and each bus can accommodate 25 riders, determine the number of vans and buses that should be purchased to maximize the passenger capacity of the system.

12 Refer to Exercise 11. The monthly fuel cost (based on 5000 miles of service) is \$550 for each van and \$850 for each bus. Find the number of vans and buses that should be purchased to minimize the monthly fuel costs if the passenger capacity of the system must be at least 75.

13 A fish farmer will purchase no more than 5000 young trout and bass from the hatchery and will feed them a special diet for the next year. The cost of food per fish will be \$0.50 for trout and \$0.75 for bass, and the total cost is not to exceed \$3000. At the end of the year, a typical trout will weigh 3 pounds and a bass will weigh 4 pounds. How many fish of each type should be stocked in the pond in order to maximize the total number of pounds of fish at the end of the year?

14 A hospital dietician wishes to prepare a corn-squash vegetable dish that will provide at least 3 grams of protein and cost no more than 36 cents per serving. An ounce of creamed corn provides $\frac{1}{2}$ gram of protein and costs 4 cents. An ounce of squash supplies $\frac{1}{4}$ gram of protein and costs 3 cents. For taste, there must be at least 2 ounces of corn and at least as much squash as corn. It is important to keep the total number of ounces in a serving as small as possible. Find the combination of corn and squash that will minimize the amount of ingredients used per serving.

15 A contractor has a large building that he wishes to convert into a series of rental storage spaces. He will construct basic 8 ft × 10 ft units and deluxe 12 ft × 10 ft units that contain extra shelves and a clothes closet. Market considerations dictate that there be at least twice as many smaller units as larger units and that the smaller units rent for \$40 per month and the deluxe units for \$75 per month. At most 7200 ft^2 is available for the storage spaces and no more than \$30,000 can be spent on construction. If each small unit will cost \$300 to make and each deluxe unit will cost \$600, how many units of each type should be constructed to maximize monthly revenue?

16 A moose feeding primarily on tree leaves and aquatic plants is capable of digesting no more than 33 kg of these foods daily. Although the aquatic plants are lower in energy content, the animal must eat at least 17 kg to satisfy its sodium requirement. A kilogram of leaves provides four times the energy as a kilogram of aquatic plants. Find the combination of foods that maximizes the daily energy intake.

9.11 REVIEW

Define or discuss each of the following.

System of equations ■ Solution of a system of equations ■ Equivalent systems of equations ■ System of linear equations ■ Partial fraction decomposition ■ An $m \times n$ matrix ■ Square matrix of order n ■ Coefficient matrix and augmented matrix of a system of linear equations ■ Elementary row transformations ■ Homogeneous system of linear equations ■ Sum and product of two matrices ■ Zero matrix ■ Identity matrix ■ Inverse of a matrix ■ Determinant ■ Minor ■ Cofactor ■ Properties of determinants ■ Cramer's rule ■ System of inequalities ■ Linear programming

9.11 EXERCISES

Exer. 1–16: Find the solutions of the system.

1 $\begin{cases} 2x - 3y = 4 \\ 5x + 4y = 1 \end{cases}$

2 $\begin{cases} x - 3y = 4 \\ -2x + 6y = 2 \end{cases}$

3 $\begin{cases} y + 4 = x^2 \\ 2x + y = -1 \end{cases}$

4 $\begin{cases} x^2 + y^2 = 25 \\ x - y = 7 \end{cases}$

5 $\begin{cases} 9x^2 + 16y^2 = 140 \\ x^2 - 4y^2 = 4 \end{cases}$

6 $\begin{cases} 2x = y^2 + 3z \\ x = y^2 + z - 1 \\ x^2 = xz \end{cases}$

7 $\begin{cases} \dfrac{1}{x} + \dfrac{3}{y} = 7 \\ \dfrac{4}{x} - \dfrac{2}{y} = 1 \end{cases}$

8 $\begin{cases} 2^x + 3^{y+1} = 10 \\ 2^{x+1} - 3^y = 5 \end{cases}$

9 $\begin{cases} 3x + y - 2z = -1 \\ 2x - 3y + z = 4 \\ 4x + 5y - z = -2 \end{cases}$

10 $\begin{cases} x + 3y = 0 \\ y - 5z = 3 \\ 2x + z = -1 \end{cases}$

11 $\begin{cases} 4x - 3y - z = 0 \\ x - y - z = 0 \\ 3x - y + 3z = 0 \end{cases}$

12 $\begin{cases} 2x + y - z = 0 \\ x - 2y + z = 0 \\ 3x + 3y + 2z = 0 \end{cases}$

13 $\begin{cases} 4x + 2y - z = 1 \\ 3x + 2y + 4z = 2 \end{cases}$

14 $\begin{cases} 2x + y = 6 \\ x - 3y = 17 \\ 3x + 2y = 7 \end{cases}$

15 $\begin{cases} \dfrac{4}{x} + \dfrac{1}{y} + \dfrac{2}{z} = 4 \\ \dfrac{2}{x} + \dfrac{3}{y} - \dfrac{1}{z} = 1 \\ \dfrac{1}{x} + \dfrac{1}{y} + \dfrac{1}{z} = 4 \end{cases}$

16 $\begin{cases} 2x - y + 3z - w = -3 \\ 3x + 2y - z + w = 13 \\ x - 3y + z - 2w = -4 \\ -x + y + 4z + 3w = 0 \end{cases}$

Exer. 17–20: Sketch the graph of the system.

17 $\begin{cases} x^2 + y^2 < 16 \\ y - x^2 > 0 \end{cases}$

18 $\begin{cases} y - x \leq 0 \\ y + x \geq 2 \\ x \leq 5 \end{cases}$

19 $\begin{cases} x - 2y \leq 2 \\ y - 3x \leq 4 \\ 2x + y \leq 4 \end{cases}$

20 $\begin{cases} x^2 - y < 0 \\ y - 2x < 5 \\ xy < 0 \end{cases}$

Exer. 21–30: Find the determinant of the matrix.

21 $[-6]$

22 $\begin{bmatrix} 3 & 4 \\ -6 & -5 \end{bmatrix}$

23 $\begin{bmatrix} 3 & -4 \\ 6 & 8 \end{bmatrix}$

24 $\begin{bmatrix} 0 & 4 & -3 \\ 2 & 0 & 4 \\ -5 & 1 & 0 \end{bmatrix}$

25 $\begin{bmatrix} 2 & -3 & 5 \\ -4 & 1 & 3 \\ 3 & 2 & -1 \end{bmatrix}$

26 $\begin{bmatrix} 3 & 1 & -2 \\ -5 & 2 & -4 \\ 7 & 3 & -6 \end{bmatrix}$

27 $\begin{bmatrix} 5 & 0 & 0 & 0 \\ 6 & -3 & 0 & 0 \\ 1 & 4 & -4 & 0 \\ 7 & 2 & 3 & 2 \end{bmatrix}$

28 $\begin{bmatrix} 1 & 2 & 0 & 3 & 1 \\ -2 & -1 & 4 & 1 & 2 \\ 3 & 0 & -1 & 0 & -1 \\ 2 & -3 & 2 & -4 & 2 \\ -1 & 1 & 0 & 1 & 3 \end{bmatrix}$

29 $\begin{bmatrix} 2 & 0 & 1 & 0 & -1 \\ 0 & 1 & 0 & 1 & 2 \\ 2 & -2 & 1 & -2 & 0 \\ 0 & 0 & -2 & 0 & 1 \\ 1 & -1 & 0 & -1 & 0 \end{bmatrix}$

30 $\begin{bmatrix} 1 & 2 & 0 & 0 & 0 \\ 3 & 4 & 0 & 0 & 0 \\ 0 & 0 & 1 & 2 & 3 \\ 0 & 0 & 2 & -1 & 1 \\ 0 & 0 & 1 & 3 & -1 \end{bmatrix}$

31 Find the determinant of the $n \times n$ matrix (a_{ij}) with $a_{ij} = 0$ for $i \neq j$.

32 Without expanding, show that

$$\begin{vmatrix} 1 & a & b+c \\ 1 & b & a+c \\ 1 & c & a+b \end{vmatrix} = 0.$$

Exer. 33–36: Find the inverse of the matrix.

33 $\begin{bmatrix} 5 & -4 \\ -3 & 2 \end{bmatrix}$

34 $\begin{bmatrix} 2 & -1 & 0 \\ 1 & 4 & 2 \\ 3 & -2 & 1 \end{bmatrix}$

35 $\begin{bmatrix} 3 & -1 & 0 & 0 \\ 1 & 2 & 0 & 0 \\ 0 & 0 & -1 & -2 \\ 0 & 0 & 5 & 3 \end{bmatrix}$

36 $\begin{bmatrix} 2 & 0 & 0 & 0 \\ 0 & 3 & 0 & 0 \\ 0 & 0 & 4 & 0 \\ 0 & 0 & 0 & 5 \end{bmatrix}$

Exer. 37–46: Express as a single matrix.

37 $\begin{bmatrix} 2 & -1 & 0 \\ 3 & 0 & -2 \end{bmatrix} \begin{bmatrix} 2 & -1 & 3 \\ 0 & 3 & 0 \\ 1 & 4 & 2 \end{bmatrix}$

38 $\begin{bmatrix} 4 & 2 \\ 5 & -3 \end{bmatrix} \begin{bmatrix} 3 \\ 7 \end{bmatrix}$

39 $\begin{bmatrix} 2 & 0 \\ 1 & 4 \\ -2 & 3 \end{bmatrix} \begin{bmatrix} 0 & 2 & -3 \\ 4 & 5 & 1 \end{bmatrix}$

40 $\begin{bmatrix} 0 & -2 & 3 \\ 4 & 1 & 2 \end{bmatrix} \begin{bmatrix} 2 & 0 \\ 3 & 8 \\ 2 & -7 \end{bmatrix}$

41 $2\begin{bmatrix} 0 & -1 & -4 \\ 3 & 2 & 1 \end{bmatrix} - 3\begin{bmatrix} 4 & -2 & 1 \\ 0 & 5 & -1 \end{bmatrix}$

42 $\begin{bmatrix} 1 & 3 \\ 2 & 4 \end{bmatrix} \begin{bmatrix} a & 0 \\ 0 & a \end{bmatrix}$

43 $\begin{bmatrix} a & 0 \\ 0 & b \end{bmatrix} \begin{bmatrix} 1 & 3 \\ 2 & 4 \end{bmatrix}$

44 $\begin{bmatrix} 3 & 2 \\ 0 & 0 \end{bmatrix} \begin{bmatrix} -2 & 0 \\ 3 & 0 \end{bmatrix}$

45 $\begin{bmatrix} 1 & 2 \\ 3 & 4 \end{bmatrix} \left\{ \begin{bmatrix} 2 & -4 \\ 3 & 7 \end{bmatrix} + \begin{bmatrix} 1 & 5 \\ -2 & -3 \end{bmatrix} \right\}$

46 $\begin{bmatrix} 3 & 2 & 5 \\ -3 & 4 & 7 \\ 6 & 5 & 1 \end{bmatrix} \begin{bmatrix} 3 & 2 & 5 \\ -3 & 4 & 7 \\ 6 & 5 & 1 \end{bmatrix}^{-1}$

Exer. 47–48: Verify without expanding the determinants.

47 $\begin{vmatrix} 2 & 4 & 6 \\ 1 & 4 & 3 \\ 2 & 2 & 0 \end{vmatrix} = 12 \begin{vmatrix} 1 & 1 & -1 \\ 1 & 2 & 1 \\ 2 & 1 & 0 \end{vmatrix}$

48 $\begin{vmatrix} a & b & c \\ d & e & f \\ g & h & k \end{vmatrix} = \begin{vmatrix} d & e & f \\ g & h & k \\ a & b & c \end{vmatrix}$

Exer. 49–50: Find the solutions of the equation $|A - xI| = 0$.

49 $A = \begin{bmatrix} 2 & 3 \\ 1 & -4 \end{bmatrix}, \quad I = I_2$

50 $A = \begin{bmatrix} 2 & -1 & 3 \\ 0 & 4 & 0 \\ 1 & 0 & -2 \end{bmatrix}, \quad I = I_3$

Exer. 51–52: Find the partial fraction decomposition.

51 $\dfrac{4x^2 + 54x + 134}{(x+3)(x^2 + 4x - 5)}$

52 $\dfrac{x^2 + 14x - 13}{x^3 + 5x^2 + 4x + 20}$

53 A rotating sprinkler head with a range of 50 feet is to be placed in the center of a rectangular field (see figure). If the area of the field is $4000\ \text{ft}^2$ and the water is to just reach the corners, find the dimensions of the field.

EXERCISE 53

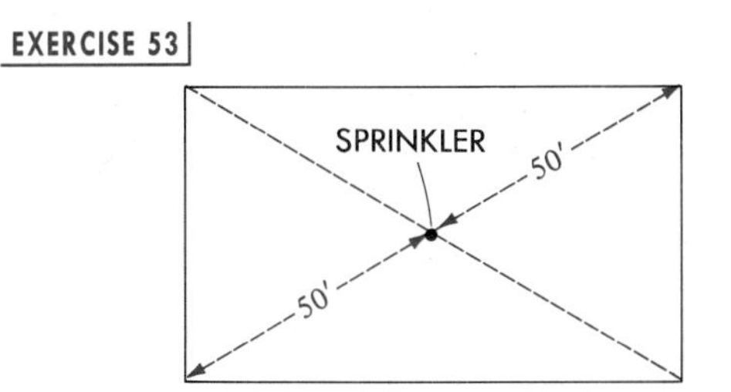

54 Find equations of the two lines that are tangent to the circle $x^2 + y^2 = 1$ and pass through the point $(0, 3)$. (*Hint:* Let $y = mx + 3$ and determine conditions on m that will ensure that the system has only one solution.)

55 An accountant must pay taxes and payroll bonuses to employees from the company's profits of \$50,000. The total tax is 40% of the amount left after bonuses are paid, and the total paid in bonuses is 10% of the amount left after taxes. Find the total tax and the total bonus amount.

56 A circular track is to have a 10-foot wide running lane around the outside (see figure). The inside distance around the track is to be 90% of the outside distance. Find the dimensions of the track.

EXERCISE 56

57 Three inlet pipes A, B, and C can be used to fill a 1000-ft^3 water storage tank. When all three pipes are in operation, the tank can be filled in 10 hours. When only A and B are used, the time increases to 20 hours. Using pipes A and C the tank can be filled in 12.5 hours. Find the individual flow rates (in ft^3/hr) for each of the three pipes.

58 To fill an order for 150 office desks, a furniture distributor must ship the desks from two warehouses. The shipping cost per desk is \$24 from the western warehouse and \$35 from the eastern warehouse. If the total shipping charge is \$4205, how many desks are shipped from each location?

59 An express-mail company charges \$15 for overnight delivery of a letter provided the dimensions of the standard envelope satisfy the following three conditions: (a) the length, the larger of the two dimensions, must be less than 12 inches; (b) the width must be less than 8 inches; (c) the width must be greater than one-half the length. Find and graph a system of inequalities that describes all the possibilities for dimensions of a standard envelope.

60 A deer spends the day in three basic activities: rest, searching for food, and grazing. At least 6 hours each day must be spent resting, and the number of hours spent searching for food will be at least two times the number of hours spent grazing. Using x as the number of hours spent searching for food and y as the number of hours spent grazing, find and graph the system of inequalities that describes the possible divisions of the day.

61 A company manufactures a power lawn mower and a power edger. These two products are of such high quality that the company can sell all the products it makes, but production capability is limited in the areas of machining, welding, and assembly. Each week, the company has 600 hours available for machining, 300 hours for welding, and 550 hours for assembly. The number of hours required for the production of a single item is shown in the following table:

Product	Machining	Welding	Assembly
Lawn mower	6	2	5
Edger	4	3	5

The profits from the sale of a mower and an edger are \$100 and \$80, respectively. How many mowers and edgers should be produced each week in order to maximize the profit?

62 A retired couple wishes to invest \$100,000, diversifying the investment in three areas: a high-risk stock that has an expected annual rate of return (or interest) of 15%; a low-risk stock that has an expected annual return of 10%; and government-issued bonds that pay annual interest of 8% and involve no risk. To protect the value of the investment, the couple wishes to place at least twice as much in the low-risk stock as in the high-risk stock, and use the remainder to buy bonds. How should the money be invested to maximize the expected annual return?

CHAPTER 10

We begin by considering the method of *mathematical induction*, which is needed to prove that certain statements are true for every positive integer. This is followed by a discussion of *sequences* and *summation notation*. Of special interest are *arithmetic* and *geometric sequences*. The last part of the chapter deals with counting processes that occur frequently in mathematics and everyday life. These include the concepts of *permutations*, *combinations*, and *probability*.

SEQUENCES, SERIES, AND PROBABILITY

10.1 MATHEMATICAL INDUCTION

If n is a positive integer, let P_n denote the mathematical statement $(xy)^n = x^n y^n$. Thus we have

$$\begin{aligned}
&\text{Statement } P_1\text{:} && (xy)^1 = x^1 y^1 \\
&\text{Statement } P_2\text{:} && (xy)^2 = x^2 y^2 \\
&\text{Statement } P_3\text{:} && (xy)^3 = x^3 y^3 \\
&\quad\vdots && \quad\vdots \\
&\text{Statement } P_n\text{:} && (xy)^n = x^n y^n \\
&\quad\vdots && \quad\vdots
\end{aligned}$$

It is easy to show that P_1, P_2, and P_3 are *true* statements. However, since the set of positive integers is infinite, it is impossible to check the validity of P_n for *every* positive integer n. To show that P_n is true requires the following principle.

PRINCIPLE OF MATHEMATICAL INDUCTION

If with each positive integer n there is associated a statement P_n, then all the statements P_n are true, provided the following two conditions are satisfied:

(i) P_1 is true.

(ii) Whenever k is a positive integer such that P_k is true, then P_{k+1} is also true.

To help understand this principle, consider a collection of statements labeled

$$P_1, \; P_2, \; P_3, \; \ldots, \; P_n, \; \ldots$$

that satisfy conditions (i) and (ii). Thus by (i), statement P_1 is true. Since condition (ii) holds, then whenever a statement P_k is true the *next* statement P_{k+1} is also true. Hence, since P_1 is true, then by (ii), P_2 is also true. However, if P_2 is true, then by (ii) we see that the next statement P_3 is true. Once again, if P_3 is true, then by (ii), P_4 is also true. If we continue in this manner, we can argue that if n is any *particular* integer, then

P_n is true, since we can use condition (ii) one step at a time, eventually reaching P_n. Although this type of reasoning does not actually *prove* the principle of mathematical induction, it certainly makes it plausible. The principle is proved in advanced algebra using postulates for the positive integers.

When applying the principle of mathematical induction, we always follow these two steps:

STEP (i) Show that P_1 is true.

STEP (ii) *Assume* that P_k is true and then prove that P_{k+1} is true.

Step (ii) often causes confusion. Note that we do not *prove* that P_k is true (except for $k = 1$). Instead, we show that *if* P_k happens to be true, then the statement P_{k+1} is also true. We refer to the assumption that P_k is true as the **induction hypothesis**.

EXAMPLE ▪ 1

Prove that for every positive integer n, the sum of the first n positive integers is

$$\frac{n(n+1)}{2}.$$

SOLUTION If n is any positive integer, let P_n denote the statement

$$1 + 2 + 3 + \cdots + n = \frac{n(n+1)}{2}.$$

The following are some special cases of P_n:

If $n = 1$, then P_1 is

$$1 = \frac{1(1+1)}{2}, \quad \text{that is,} \quad 1 = 1.$$

If $n = 2$, then P_2 is

$$1 + 2 = \frac{2(2+1)}{2}, \quad \text{that is,} \quad 3 = 3.$$

If $n = 3$, then P_3 is

$$1 + 2 + 3 = \frac{3(3+1)}{2}, \quad \text{that is,} \quad 6 = 6.$$

If $n = 4$, then P_4 is

$$1 + 2 + 3 + 4 = \frac{4(4+1)}{2}, \quad \text{that is,} \quad 10 = 10.$$

Although it is instructive to check the validity of P_n for several values of n as we have done, it is unnecessary to do so. We need only apply the two-step process outlined prior to this example. Thus, we proceed as follows:

STEP (i) If we substitute $n = 1$ in P_n, then the left-hand side contains only the number 1 and the right-hand side is $\frac{1(1+1)}{2}$, which also equals 1. Hence, P_1 is true.

STEP (ii) Assume that P_k is true. Thus, the induction hypothesis is

$$1 + 2 + 3 + \cdots + k = \frac{k(k+1)}{2}.$$

Our goal is to show that P_{k+1} is true, that is,

$$1 + 2 + 3 + \cdots + (k+1) = \frac{(k+1)[(k+1)+1]}{2}.$$

By the induction hypothesis we already have a formula for the sum of the first k positive integers. Hence, to find a formula for the sum of the first $k + 1$ positive integers we may simply add $(k + 1)$ to both sides of the induction hypothesis P_k. Doing so, we obtain

$$\begin{aligned} 1 + 2 + 3 + \cdots + k + (k+1) &= \frac{k(k+1)}{2} + (k+1) \\ &= \frac{k(k+1) + 2(k+1)}{2} \end{aligned}$$

We factor the numerator by grouping terms:

$$\begin{aligned} 1 + 2 + 3 + \cdots + k + (k+1) &= \frac{(k+1)(k+2)}{2} \\ &= \frac{(k+1)[(k+1)+1]}{2}. \end{aligned}$$

We have shown that P_{k+1} is true and, therefore, the proof by mathematical induction is complete.

EXAMPLE ▪ 2

Prove that for each positive integer n,

$$1^2 + 3^2 + \cdots + (2n - 1)^2 = \frac{n(2n - 1)(2n + 1)}{3}.$$

SOLUTION For each positive integer n, let P_n denote the given statement. Note that this is a formula for the sum of the squares of the first n odd positive integers. We again follow the two-step procedure.

STEP (i) Substituting 1 for n in P_n, we obtain

$$1^2 = \frac{(1)(2 - 1)(2 + 1)}{3} = \frac{3}{3} = 1.$$

This shows that P_1 is true.

STEP (ii) Assume that P_k is true. Thus, the induction hypothesis is

$$1^2 + 3^2 + \cdots + (2k - 1)^2 = \frac{k(2k - 1)(2k + 1)}{3}.$$

We wish to show that P_{k+1} is true, that is,

$$1^2 + 3^2 + \cdots + [2(k + 1) - 1]^2 = \frac{(k + 1)[2(k + 1) - 1][2(k + 1) + 1]}{3}.$$

This equation for P_{k+1} simplifies to

$$1^2 + 3^2 + \cdots + (2k + 1)^2 = \frac{(k + 1)(2k + 1)(2k + 3)}{3}.$$

Observe that the second from the last term on the left-hand side is $(2k - 1)^2$. In a manner similar to the solution of Example 1, we may obtain the left-hand side of P_{k+1} by adding $(2k + 1)^2$ to both sides of the induction hypothesis P_k. This gives us

$$1^2 + 3^2 + \cdots + (2k - 1)^2 + (2k + 1)^2 = \frac{k(2k - 1)(2k + 1)}{3} + (2k + 1)^2.$$

The right-hand side of the preceding equation simplifies to $\frac{1}{3}(k + 1)(2k + 1)(2k + 3)$, which gives us the same form as P_{k+1}. This shows that P_{k+1} is true, and hence P_n is true for every n.

Let j be a positive integer and suppose that with each integer $n \geq j$ there is associated a statement P_n. For example, if $j = 6$, then the statements are numbered $P_6, P_7, P_8, \ldots$. The principle of mathematical induction may be extended to cover this situation. To prove that the statements

P_n are true for $n \geq j$, we use the following two steps, in the same manner as we did for $n \geq 1$.

EXTENDED PRINCIPLE OF MATHEMATICAL INDUCTION FOR P_k, $k \geq j$

> (i) Show that P_j is true.
>
> (ii) Assume that P_k is true with $k \geq j$ and prove that P_{k+1} is true.

EXAMPLE ▪ 3

Let a be a nonzero real number such that $a > -1$. Prove that

$$(1 + a)^n > 1 + na$$

for every integer $n \geq 2$.

SOLUTION For each positive integer n, let P_n denote the inequality $(1 + a)^n > 1 + na$. Note that P_1 is *false*, since $(1 + a)^1 = 1 + (1)(a)$. However, we can show that P_n is true for $n \geq 2$ by using the extended principle with $j = 2$.

STEP (i) We first note that $(1 + a)^2 = 1 + 2a + a^2$. Since $a \neq 0$, we have $a^2 > 0$ and therefore $1 + 2a + a^2 > 1 + 2a$, or $(1 + a)^2 > 1 + 2a$. Hence P_2 is true.

STEP (ii) Assume that P_k is true. Thus, the induction hypothesis is

$$(1 + a)^k > 1 + ka.$$

We wish to show that P_{k+1} is true, that is,

$$(1 + a)^{k+1} > 1 + (k + 1)a.$$

Since $a > -1$, we see that $1 + a > 0$. Hence, multiplying both sides of the induction hypothesis by $1 + a$ does not change the inequality sign. This multiplication leads to the following equivalent inequalities:

$$(1 + a)^k(1 + a) > (1 + ka)(1 + a)$$

$$(1 + a)^{k+1} > 1 + ka + a + ka^2$$

$$(1 + a)^{k+1} > 1 + (k + 1)a + ka^2$$

Since $ka^2 > 0$,

$$1 + (k + 1)a + ka^2 > 1 + (k + 1)a,$$

and therefore

$$(1 + a)^{k+1} > 1 + (k + 1)a.$$

Thus, P_{k+1} is true and the proof is complete.

The binomial theorem was stated without proof in Section 1.6. Let us conclude this section by restating this important result and giving a proof using mathematical induction.

BINOMIAL THEOREM

$$(a+b)^n = a^n + na^{n-1}b + \frac{n(n-1)}{2!}a^{n-2}b^2 + \cdots + \frac{n(n-1)(n-2)\cdots(n-r+1)}{r!}a^{n-r}b^r + \cdots + nab^{n-1} + b^n$$

PROOF For each positive integer n, let P_n denote the statement given in the binomial theorem.

STEP (i) If $n = 1$, the statement reduces to $(a+b)^1 = a^1 + b^1$. Consequently, P_1 is true.

STEP (ii) Assume that P_k is true. Thus, the induction hypothesis is

$$(a+b)^k = a^k + ka^{k-1}b + \frac{k(k-1)}{2!}a^{k-2}b^2 + \cdots + \frac{k(k-1)(k-2)\cdots(k-r+2)}{(r-1)!}a^{k-r+1}b^{r-1} + \frac{k(k-1)(k-2)\cdots(k-r+1)}{r!}a^{k-r}b^r + \cdots + kab^{k-1} + b^k.$$

We have shown both the rth and the $(r+1)$st terms in the expansion.

We multiply both sides of the induction hypothesis by $(a+b)$:

$$(a+b)^{k+1} = \left[a^{k+1} + ka^kb + \frac{k(k-1)}{2!}a^{k-1}b^2 + \cdots + \frac{k(k-1)\cdots(k-r+1)}{r!}a^{k-r+1}b^r + \cdots + ab^k\right] + \left[a^kb + ka^{k-1}b^2 + \cdots + \frac{k(k-1)\cdots(k-r+2)}{(r-1)!}a^{k-r+1}b^r + \cdots + kab^k + b^{k+1}\right]$$

where the terms in the first pair of brackets result from multiplying the right side of the induction hypothesis by a, and the terms in the second pair of brackets result from multiplying by b. We next rearrange and combine terms:

$$(a+b)^{k+1} = a^{k+1} + (k+1)a^k b + \left[\frac{k(k-1)}{2!} + k\right]a^{k-1}b^2 + \cdots$$
$$+ \left[\frac{k(k-1)\cdots(k-r+1)}{r!} + \frac{k(k-1)\cdots(k-r+2)}{(r-1)!}\right]a^{k-r+1}b^r$$
$$+ \cdots + (1+k)ab^k + b^{k+1}.$$

If the coefficients are simplified, we obtain statement P_n with $k+1$ substituted for n. Thus, P_{k+1} is true and therefore P_n holds for every positive integer n, which completes the proof. ❑

10.1 EXERCISES

Exer. 1–22: Prove that the statement is true for every positive integer *n*.

1 $2 + 4 + 6 + \cdots + 2n = n(n+1)$

2 $1 + 4 + 7 + \cdots + (3n-2) = \dfrac{n(3n-1)}{2}$

3 $1 + 3 + 5 + \cdots + (2n-1) = n^2$

4 $3 + 9 + 15 + \cdots + (6n-3) = 3n^2$

5 $2 + 7 + 12 + \cdots + (5n-3) = \frac{1}{2}n(5n-1)$

6 $2 + 6 + 18 + \cdots + 2\cdot 3^{n-1} = 3^n - 1$

7 $1 + 2\cdot 2 + 3\cdot 2^2 + \cdots + n\cdot 2^{n-1} = 1 + (n-1)\cdot 2^n$

8 $(-1)^1 + (-1)^2 + (-1)^3 + \cdots + (-1)^n = \dfrac{(-1)^n - 1}{2}$

9 $1^2 + 2^2 + 3^2 + \cdots + n^2 = \dfrac{n(n+1)(2n+1)}{6}$

10 $1^3 + 2^3 + 3^3 + \cdots + n^3 = \left[\dfrac{n(n+1)}{2}\right]^2$

11 $\dfrac{1}{1\cdot 2} + \dfrac{1}{2\cdot 3} + \dfrac{1}{3\cdot 4} + \cdots + \dfrac{1}{n(n+1)} = \dfrac{n}{n+1}$

12 $\dfrac{1}{1\cdot 2\cdot 3} + \dfrac{1}{2\cdot 3\cdot 4} + \dfrac{1}{3\cdot 4\cdot 5} + \cdots + \dfrac{1}{n(n+1)(n+2)} = \dfrac{n(n+3)}{4(n+1)(n+2)}$

13 $3 + 3^2 + 3^3 + \cdots + 3^n = \frac{3}{2}(3^n - 1)$

14 $1^3 + 3^3 + 5^3 + \cdots + (2n-1)^3 = n^2(2n^2 - 1)$

15 $n < 2^n$

16 $1 + 2n \le 3^n$

17 $1 + 2 + 3 + \cdots + n < \frac{1}{8}(2n+1)^2$

18 If $0 < a < b$, then $\left(\dfrac{a}{b}\right)^{n+1} < \left(\dfrac{a}{b}\right)^n$.

19 3 is a factor of $n^3 - n + 3$.

20 2 is a factor of $n^2 + n$.

21 4 is a factor of $5^n - 1$.

22 9 is a factor of $10^{n+1} + 3\cdot 10^n + 5$.

Exer. 23–30: Find the smallest positive integer *j* for which the statement is true. Use the extended principle of mathematical induction to prove that the formula is true for every integer greater than *j*.

23 $n + 12 \le n^2$

24 $n^2 + 18 \le n^3$

25 $5 + \log_2 n \le n$

26 $n^2 \le 2^n$

27 $2^n \le n!$

28 $10^n \le n^n$

29 $2n + 2 \le 2^n$

30 $n \log_2 n + 20 \le n^2$

31 Use mathematical induction to prove that if a is any real number greater than 1, then $a^n > 1$ for every positive integer n.

32 Prove that

$$a + ar + ar^2 + \cdots + ar^{n-1} = \frac{a(1 - r^n)}{1 - r}$$

for every positive integer n and real numbers a and r with $r \neq 1$.

33 Use mathematical induction to prove that $a - b$ is a factor of $a^n - b^n$ for every positive integer n. (*Hint:* $a^{k+1} - b^{k+1} = a^k(a - b) + (a^k - b^k)b$.)

34 Prove that $a + b$ is a factor of $a^{2n-1} + b^{2n-1}$ for every positive integer n.

35 Use mathematical induction to prove De Moivre's theorem:

$$[r(\cos\theta + i\sin\theta)]^n = r^n(\cos n\theta + i\sin n\theta)$$

for every positive integer n.

36 Prove that for every positive integer $n \geq 3$, the sum of the interior angles of a polygon of n sides is $(n - 2) \cdot 180°$.

Exer. 37–38: Prove that the formula is true for every positive integer *n*.

37 $\sin(\theta + n\pi) = (-1)^n \sin\theta$

38 $\cos(\theta + n\pi) = (-1)^n \cos\theta$

10.2 INFINITE SEQUENCES AND SUMMATION NOTATION

An arbitrary *infinite sequence* is often denoted by

$$a_1,\ a_2,\ a_3,\ \ldots,\ a_n,\ \ldots$$

and may be regarded as a collection of real numbers that is in one-to-one correspondence with the positive integers. For convenience we sometimes refer to infinite sequences merely as *sequences*. Each real number a_k is a **term** of the sequence. The sequence is *ordered*: it has a **first term** a_1, a **second term** a_2, a **forty-fifth term** a_{45}, and, if n denotes an arbitrary positive integer, an ***n*th term** a_n.

Infinite sequences occur frequently in mathematics. For example, the sequence

$$0.6,\ 0.66,\ 0.666,\ 0.6666,\ 0.66666,\ \ldots$$

may be used to represent the rational number $\frac{2}{3}$. In this case the nth term gets closer and closer to $\frac{2}{3}$ as n increases.

FIGURE 1

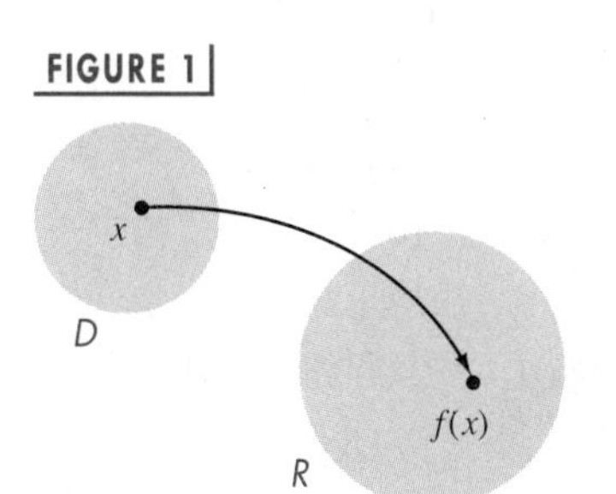

We may regard an infinite sequence as a function. Recall from Section 3.4 that a function f is a correspondence that associates with each number x in the domain D, a unique number $f(x)$ in the range R as illustrated in Figure 1. If we restrict the domain to the positive integers $1, 2, 3, \ldots$, then we obtain an infinite sequence as in the following definition.

DEFINITION

> An **infinite sequence** is a function whose domain is the set of positive integers.

In our work the range of an infinite sequence will be a set of real numbers.

If f is an infinite sequence, then to each positive integer n there corresponds a real number $f(n)$. These numbers in the range of f may be represented by writing

$$f(1),\ f(2),\ f(3),\ \ldots,\ f(n),\ \ldots$$

To obtain the subscript form of a sequence, as shown at the beginning of this section, we let $a_n = f(n)$ for every positive integer n.

From the definition of equality of functions we see that a sequence

$$a_1,\ a_2,\ a_3,\ \ldots,\ a_n,\ \ldots$$

is **equal** to a sequence $b_1,\ b_2,\ b_3,\ \ldots,\ b_n,\ \ldots$

if and only if $a_k = b_k$ for every positive integer k. Infinite sequences are often defined by stating a formula for the nth term, as in the following example.

EXAMPLE ▪ 1

List the first four terms and the tenth term of the sequence whose nth term is given:

(a) $a_n = \dfrac{n}{n+1}$ (b) $a_n = 2 + (0.1)^n$

(c) $a_n = (-1)^{n+1}\dfrac{n^2}{3n-1}$ (d) $a_n = 4$

SOLUTION To find the first four terms we substitute, successively, $n = 1, 2, 3$, and 4 in the formula for a_n. The tenth term is found by substituting 10 for n. Doing this and simplifying gives us the following results:

	nth term	first four terms	tenth term
(a)	$\dfrac{n}{n+1}$	$\dfrac{1}{2}, \dfrac{2}{3}, \dfrac{3}{4}, \dfrac{4}{5}$	$\dfrac{10}{11}$
(b)	$2 + (0.1)^n$	2.1, 2.01, 2.001, 2.0001	2.0000000001
(c)	$(-1)^{n+1}\dfrac{n^2}{3n-1}$	$\dfrac{1}{2}, -\dfrac{4}{5}, \dfrac{9}{8}, -\dfrac{16}{11}$	$-\dfrac{100}{29}$
(d)	4	4, 4, 4, 4	4

Some infinite sequences are described by stating the first term a_1, together with a rule that shows how to obtain any term a_{k+1} from the preceding term a_k whenever $k \geq 1$. A description of this type is a **recursive definition**, and the sequence is defined **recursively**.

EXAMPLE ▪ 2

Find the first four terms and the nth term of the infinite sequence defined recursively as follows:

$$a_1 = 3, \quad a_{k+1} = 2a_k \text{ for } k \geq 1.$$

SOLUTION The first four terms are:

$$a_1 = 3$$

$$a_2 = 2a_1 = 2 \cdot 3 = 6$$

$$a_3 = 2a_2 = 2 \cdot 2 \cdot 3 = 2^2 \cdot 3 = 12$$

$$a_4 = 2a_3 = 2 \cdot 2 \cdot 2 \cdot 3 = 2^3 \cdot 3 = 24$$

We have written the terms as products to gain some insight into the nature of the nth term. Continuing, we obtain $a_5 = 2^4 \cdot 3$ and $a_6 = 2^5 \cdot 3$, and it appears that

$$a_n = 2^{n-1} \cdot 3$$

for every positive integer n. We can prove that this guess is correct by mathematical induction.

If only the first few terms of an infinite sequence are known, then it is impossible to predict additional terms. For example, if we were given $3, 6, 9, \ldots$ and asked to find the fourth term, we could not proceed without further information. The infinite sequence with nth term

$$a_n = 3n + (1 - n)^3(2 - n)^2(3 - n)$$

has for its first four terms 3, 6, 9, and 120. It is possible to describe sequences in which the first three terms are 3, 6, and 9, and the fourth term is *any* given number. This shows that when we work with an infinite sequence it is essential to have specific information about the nth term or a general scheme for obtaining each term from the preceding one.

We sometimes need to find the sum of many terms of an infinite sequence. To express such sums easily we use **summation notation**. Given an infinite sequence

$$a_1, a_2, a_3, \ldots, a_n, \ldots$$

the symbol $\sum_{k=1}^{m} a_k$ represents the sum of the first m terms.

SUMMATION NOTATION

$$\sum_{k=1}^{m} a_k = a_1 + a_2 + a_3 + \cdots + a_m$$

The Greek capital letter $\sum$ (sigma) indicates a sum and the symbol a_k represents the kth term. The letter k is the **index of summation**, or the **summation variable**, and the numbers 1 and m indicate the smallest and largest values of the summation variable.

EXAMPLE ▪ 3

Find the sum $\sum_{k=1}^{4} k^2(k-3)$.

SOLUTION In this case, $a_k = k^2(k-3)$. To find the sum we merely substitute, in succession, the integers 1, 2, 3, and 4 for k and add the resulting terms:

$$\begin{aligned}\sum_{k=1}^{4} k^2(k-3) &= 1^2(1-3) + 2^2(2-3) + 3^2(3-3) + 4^2(4-3) \\ &= (-2) + (-4) + 0 + 16 = 10\end{aligned}$$

The letter we use for the summation variable is immaterial. To illustrate, if j is the summation variable, then

$$\sum_{j=1}^{m} a_j = a_1 + a_2 + a_3 + \cdots + a_m,$$

which is the same sum as $\sum_{k=1}^{m} a_k$. We may also use other symbols. For example, the sum in Example 3 can be written

$$\sum_{j=1}^{4} j^2(j-3).$$

If n is a positive integer, then the sum of the first n terms of an infinite sequence will be denoted by S_n. For example, given $a_1, a_2, a_3, \ldots, a_n, \ldots,$

$$\begin{aligned}S_1 &= a_1 \\ S_2 &= a_1 + a_2 \\ S_3 &= a_1 + a_2 + a_3 \\ S_4 &= a_1 + a_2 + a_3 + a_4\end{aligned}$$

and, in general,

$$S_n = \sum_{k=1}^{n} a_k = a_1 + a_2 + \cdots + a_n.$$

The real number S_n is the **nth partial sum** of the infinite sequence $a_1, a_2, a_3, \ldots, a_n, \ldots$, and the sequence

$$S_1,\ S_2,\ S_3,\ \ldots,\ S_n,\ \ldots$$

is a **sequence of partial sums**. Sequences of partial sums are important in the study of *infinite series*, a topic in calculus. We shall discuss some special types of infinite series in Section 10.4.

EXAMPLE ▪ 4

Find the first four terms and the nth term of the sequence of partial sums associated with the sequence $1, 2, 3, \ldots, n, \ldots$ of positive integers.

SOLUTION The first four terms of the sequence of partial sums are:

$$S_1 = 1$$
$$S_2 = 1 + 2 = 3$$
$$S_3 = 1 + 2 + 3 = 6$$
$$S_4 = 1 + 2 + 3 + 4 = 10.$$

From Example 1 of Section 10.1,

$$S_n = 1 + 2 + 3 + \cdots + n = \frac{n(n+1)}{2}.$$

If a_k is the same for every positive integer k, say $a_k = c$ for a real number c, then

$$\sum_{k=1}^{n} a_k = a_1 + a_2 + a_3 + \cdots + a_n$$
$$= c + c + c + \cdots + c = nc.$$

We have proved the following result.

THEOREM

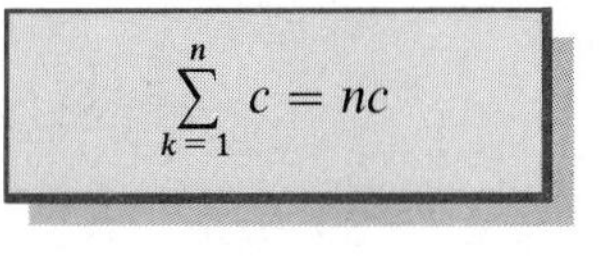

$$\sum_{k=1}^{n} c = nc$$

ILLUSTRATION $\sum_{k=1}^{n} c$

- $\sum_{k=1}^{4} 7 = 7 + 7 + 7 + 7 = 4 \cdot 7 = 28$
- $\sum_{k=1}^{10} \pi = 10\pi$

The domain of the summation variable does not have to begin at 1. For example, the following is self-explanatory:

$$\sum_{k=4}^{8} a_k = a_4 + a_5 + a_6 + a_7 + a_8.$$

As another variation, if the first term of an infinite sequence is a_0, as in

$$a_0,\ a_1,\ a_2,\ \ldots,\ a_n,\ \ldots,$$

then we may consider sums of the form

$$\sum_{k=0}^{n} a_k = a_0 + a_1 + a_2 + \cdots + a_n,$$

which is the sum of the first $n + 1$ terms of the sequence.

EXAMPLE ▪ 5

Find the sum $\sum_{k=0}^{3} \frac{2^k}{k+1}$.

SOLUTION

$$\sum_{k=0}^{3} \frac{2^k}{k+1} = \frac{2^0}{0+1} + \frac{2^1}{1+1} + \frac{2^2}{2+1} + \frac{2^3}{3+1}$$

$$= 1 + 1 + \frac{4}{3} + 2 = \frac{16}{3}$$

Summation notation can be used to denote polynomials. Thus if

$$f(x) = a_0 + a_1x + a_2x^2 + \cdots + a_nx^n,$$

then

$$f(x) = \sum_{k=0}^{n} a_kx^k.$$

As another illustration, we can write the formula for the binomial theorem in Section 1.6 as

$$(a + b)^n = \sum_{k=0}^{n} \binom{n}{k} a^{n-k}b^k.$$

The following theorem concerning sums has many uses.

THEOREM ON SUMS

If $a_1, a_2, \ldots, a_n, \ldots$ and $b_1, b_2, \ldots, b_n, \ldots$ are infinite sequences, then for every positive integer n,

(i) $\sum_{k=1}^{n} (a_k + b_k) = \sum_{k=1}^{n} a_k + \sum_{k=1}^{n} b_k$

(ii) $\sum_{k=1}^{n} (a_k - b_k) = \sum_{k=1}^{n} a_k - \sum_{k=1}^{n} b_k$

(iii) $\sum_{k=1}^{n} ca_k = c\left(\sum_{k=1}^{n} a_k\right)$ for every real number c.

PROOF The theorem can be proved by mathematical induction. We can also use the following argument. Begin with

$$\sum_{k=1}^{n} (a_k + b_k) = (a_1 + b_1) + (a_2 + b_2) + (a_3 + b_3) + \cdots + (a_n + b_n).$$

Using commutative and associative properties many times, we may rearrange the terms on the right-hand side to produce

$$\sum_{k=1}^{n} (a_k + b_k) = (a_1 + a_2 + a_3 + \cdots + a_n) + (b_1 + b_2 + b_3 + \cdots + b_n).$$

Expressing the right-hand side in summation notation gives us formula (i).

For formula (iii) we may write

$$\begin{aligned} \sum_{k=1}^{n} (ca_k) &= ca_1 + ca_2 + ca_3 + \cdots + ca_n \\ &= c(a_1 + a_2 + a_3 + \cdots + a_n) \\ &= c\left(\sum_{k=1}^{n} a_k\right). \end{aligned}$$

The proof of (ii) is left as an exercise. ❑

10.2 EXERCISES

Exer. 1–16: Find the first five terms and the eighth term of the sequence whose *n*th term is given.

1 $a_n = 12 - 3n$

2 $a_n = \dfrac{3}{5n - 2}$

3 $a_n = \dfrac{3n - 2}{n^2 + 1}$

4 $a_n = 10 + \dfrac{1}{n}$

5 $a_n = 9$

6 $a_n = (n - 1)(n - 2)(n - 3)$

7 $a_n = 2 + (-0.1)^n$

8 $a_n = 4 + (0.1)^n$

9 $a_n = (-1)^{n-1}\dfrac{n + 7}{2n}$

10 $a_n = (-1)^n \dfrac{6 - 2n}{\sqrt{n + 1}}$

11 $a_n = 1 + (-1)^{n+1}$

12 $a_n = (-1)^{n+1} + (0.1)^{n-1}$

13 $a_n = \dfrac{2^n}{n^2 + 2}$ **14** $a_n = \sqrt{2}$

15 a_n is the number of decimal places in $(0.1)^n$.

16 a_n is the number of positive integers less than n^3.

Exer. 17–24: Find the first five terms of the recursively defined infinite sequence.

17 $a_1 = 2, \quad a_{k+1} = 3a_k - 5$ **18** $a_1 = 5, \quad a_{k+1} = 7 - 2a_k$

19 $a_1 = -3, \quad a_{k+1} = a_k^2$ **20** $a_1 = 128, \quad a_{k+1} = \frac{1}{4}a_k$

21 $a_1 = 5, \quad a_{k+1} = ka_k$ **22** $a_1 = 3, \quad a_{k+1} = 1/a_k$

23 $a_1 = 2, \quad a_{k+1} = (a_k)^k$ **24** $a_1 = 2, \quad a_{k+1} = (a_k)^{1/k}$

25 A test question lists the first four terms of a sequence as 2, 4, 6, and 8, and asks for the fifth term. Show that the fifth term can be any real number a by finding the nth term of a sequence that has for its first five terms, 2, 4, 6, 8, and a.

26 The number of bacteria in a certain culture is initially 500, and the culture doubles in size every day.

(a) Find the number of bacteria present after one day; two days; three days.

(b) Find a formula for the number of bacteria present after n days.

Exer. 27–42: Find the sum.

27 $\sum_{k=1}^{5} (2k - 7)$ **28** $\sum_{k=1}^{6} (10 - 3k)$

29 $\sum_{k=1}^{4} (k^2 - 5)$ **30** $\sum_{k=1}^{10} [1 + (-1)^k]$

31 $\sum_{k=0}^{5} k(k - 2)$ **32** $\sum_{k=0}^{4} (k - 1)(k - 3)$

33 $\sum_{k=3}^{6} \dfrac{k - 5}{k - 1}$ **34** $\sum_{k=1}^{6} \dfrac{3}{k + 1}$

35 $\sum_{k=1}^{5} (-3)^{k-1}$ **36** $\sum_{k=0}^{4} 3(2^k)$

37 $\sum_{k=1}^{100} 100$ **38** $\sum_{k=1}^{1000} 5$

39 $\sum_{k=1}^{n} (k^2 + 3k + 5)$ (*Hint:* Use the theorem on sums to write the sum as $\sum_{k=1}^{n} k^2 + 3\sum_{k=1}^{n} k + \sum_{k=1}^{n} 5$. Next employ Exercise 9 of Section 10.1, Example 1 of Section 10.1, and the formula for $\sum_{k=1}^{n} c$.)

40 $\sum_{k=1}^{n} (3k^2 - 2k + 1)$ **41** $\sum_{k=1}^{n} (2k - 3)^2$

42 $\sum_{k=1}^{n} (k^3 + 2k^2 - k + 4)$

(*Hint:* See Exercise 10 of Section 10.1.)

Exer. 43–56: Express the sum in terms of summation notation. (Answers are not unique.)

43 $1 + 3 + 5 + 7$ **44** $2 + 4 + 6 + 8 + 10$

45 $1 + 5 + 9 + 13 + 17$ **46** $2 + 5 + 8 + 11 + 14$

47 $\frac{1}{2} + \frac{2}{5} + \frac{3}{8} + \frac{4}{11}$ **48** $\frac{1}{4} + \frac{2}{9} + \frac{3}{14} + \frac{4}{19}$

49 $1 - \dfrac{x^2}{2} + \dfrac{x^4}{4} - \dfrac{x^6}{6} + \cdots + (-1)^n \dfrac{x^{2n}}{2n}$

50 $2 - 4 + 8 - 16 + 32 - 64$

51 $1 - \frac{1}{2} + \frac{1}{3} - \frac{1}{4} + \frac{1}{5} - \frac{1}{6} + \frac{1}{7}$

52 $1 + 3 + 5 + \cdots + 73$

53 $2 + 4 + 6 + \cdots + 150$

54 $1 + x + \dfrac{x^2}{2} + \dfrac{x^3}{3} + \cdots + \dfrac{x^n}{n}$

55 $\dfrac{1}{1 \cdot 2} + \dfrac{1}{2 \cdot 3} + \dfrac{1}{3 \cdot 4} + \cdots + \dfrac{1}{99 \cdot 100}$

56 $\dfrac{1}{1 \cdot 2 \cdot 3} + \dfrac{1}{2 \cdot 3 \cdot 4} + \dfrac{1}{3 \cdot 4 \cdot 5} + \cdots + \dfrac{1}{98 \cdot 99 \cdot 100}$

57 Prove (ii) of the theorem on sums.

58 Extend (i) of the theorem on sums to $\sum_{k=1}^{n} (a_k + b_k + c_k)$.

Exer. 59–65: Use a calculator.

59 Terms of the sequence defined recursively by $a_1 = 5$, $a_{k+1} = \sqrt{a_k}$ may be obtained by entering 5 and pressing the $\boxed{\sqrt{x}}$ key repeatedly. Describe what happens to the terms of the sequence as k increases.

60 Approximations to $\sqrt{N}$ may be obtained from the recursively defined sequence

$$x_1 = \frac{N}{2}, \quad x_{k+1} = \frac{1}{2}\left(x_k + \frac{N}{x_k}\right).$$

Approximate x_2, x_3, x_4, x_5, x_6 if $N = 10$.

61 Bode's sequence, defined by

$$a_1 = 0.4 \quad a_k = 0.1(3 \cdot 2^{k-2} + 4) \text{ for } k \geq 2,$$

can be used to approximate distances of planets from the sun. These distances are measured in astronomical units, with 1 AU = 92,900,000 miles. For example, the third term corresponds to earth and the fifth term to the minor planet Ceres. Approximate the first five terms of the sequence.

62 The famous Fibonacci sequence is defined recursively by

$$a_1 = 1, \quad a_2 = 1, \quad a_{k+1} = a_k + a_{k-1} \text{ for } k \geq 2.$$

(a) Find the first ten terms of the sequence.

(b) The terms of the sequence $r_k = a_{k+1}/a_k$ give progressively better approximations to τ, the golden ratio. Approximate the first ten terms of this sequence.

63 The discrete logistic sequence is defined recursively by

$$y_{k+1} = y_k + \frac{r}{K} y_k(K - y_k)$$

for positive constants r and K. This sequence is used to predict the number y_n in a seasonally breeding animal population after n years.

(a) Find the terms in the sequence if $y_1 = K$.

(b) Approximate the first ten terms in the sequence if $y_1 = 400$, $r = 2$, and $K = 500$. Describe the behavior of the terms of this sequence as k increases.

64 Terms of the sequence defined recursively by $a_1 = 1$, $a_{k+1} = \cos a_k$ may be obtained by entering 1 and pressing the COS key (in radian mode) repeatedly. Describe what happens to the terms of the sequence as k increases.

65 Approximations to π may be obtained from the sequence

$$x_1 = 3, \quad x_{k+1} = x_k - \tan x_k.$$

(a) Find the first five terms of this sequence.

(b) What happens to the terms of the sequence when $x_1 = 6$?

10.3 ARITHMETIC SEQUENCES

In this section and the next we consider two special types of sequences. The first may be defined as follows.

DEFINITION

A sequence $a_1, a_2, \ldots, a_n, \ldots$ is an **arithmetic sequence** if there is a real number d such that

$$a_{k+1} = a_k + d$$

for every positive integer k.

If we rewrite the equation in the definition as

$$a_{k+1} - a_k = d,$$

we see that the difference of any two successive terms of the sequence is d. We refer to d as the **common difference** associated with the arithmetic sequence.

EXAMPLE ▪ 1

Show that the sequence

$$1, 4, 7, 10, \ldots, 3n - 2, \ldots$$

is arithmetic, and find the common difference.

SOLUTION If $a_n = 3n - 2$ then, for every positive integer k,

$$\begin{aligned} a_{k+1} - a_k &= [3(k+1) - 2] - (3k - 2) \\ &= 3k + 3 - 2 - 3k + 2 = 3. \end{aligned}$$

Hence, the given sequence is arithmetic with common difference 3.

Given an arithmetic sequence, we know that

$$a_{k+1} = a_k + d$$

for every positive integer k. This gives us a recursive formula for obtaining successive terms. Beginning with any real number a_1, we can obtain an arithmetic sequence with common difference d simply by adding d to a_1, then to $a_1 + d$, and so, obtaining

$$a_1, \quad a_1 + d, \quad a_1 + 2d, \quad a_1 + 3d, \quad a_1 + 4d, \quad \ldots$$

The nth term a_n of this sequence is given by the next formula.

***n*th TERM OF AN ARITHMETIC SEQUENCE**

$$a_n = a_1 + (n - 1)d$$

EXAMPLE ▪ 2

The first three terms of an arithmetic sequence are 20, 16.5, and 13. Find the fifteenth term.

SOLUTION The common difference is

$$16.5 - 20 = -3.5.$$

Substituting $a_1 = 20$, $d = -3.5$, and $n = 15$ in the formula $a_n = a_1 + (n - 1)d$,

$$a_{15} = 20 + (15 - 1)(-3.5) = 20 - 49 = -29.$$

EXAMPLE ▪ 3

If the fourth term of an arithmetic sequence is 5 and the ninth term is 20, find the sixth term.

SOLUTION Substituting $n = 4$ and $n = 9$ in $a_n = a_1 + (n - 1)d$, and using the fact that $a_4 = 5$ and $a_9 = 20$, we obtain the following equivalent systems of linear equations in the variables a_1 and d:

$$\begin{cases} 5 = a_1 + (4 - 1)d \\ 20 = a_1 + (9 - 1)d \end{cases}$$

$$\begin{cases} 5 = a_1 + 3d \\ 20 = a_1 + 8d \end{cases}$$

This system has the solution $d = 3$ and $a_1 = -4$. Substitution in the formula $a_n = a_1 + (n - 1)d$ gives us

$$a_6 = (-4) + (6 - 1)(3) = 11.$$

THEOREM

If $a_1, a_2, \ldots, a_n, \ldots$ is an arithmetic sequence with common difference d, then the nth partial sum S_n is given by both

$$S_n = \frac{n}{2}[2a_1 + (n - 1)d] \quad \text{and} \quad S_n = \frac{n}{2}(a_1 + a_n).$$

PROOF We may write

$$\begin{aligned} S_n &= a_1 + a_2 + a_3 + \cdots + a_n \\ &= a_1 + (a_1 + d) + (a_1 + 2d) + \cdots + [a_1 + (n - 1)d]. \end{aligned}$$

Employing commutative and associative properties many times, we obtain

$$S_n = (a_1 + a_1 + a_1 + \cdots + a_1) + [d + 2d + \cdots + (n - 1)d]$$

with a_1 appearing n times within the first parentheses. Thus

$$S_n = na_1 + d[1 + 2 + \cdots + (n - 1)].$$

The expression within brackets is the sum of the first $n - 1$ positive integers. From Example 1 of Section 10.1 (with $n - 1$ in place of n),

$$1 + 2 + \cdots + (n - 1) = \frac{(n - 1)n}{2}.$$

Substituting in the last equation for S_n and factoring,

$$\begin{aligned} S_n &= na_1 + d\frac{n(n - 1)}{2} \\ &= \frac{n}{2}[2a_1 + (n - 1)d]. \end{aligned}$$

Since $a_n = a_1 + (n - 1)d$, the last equation is equivalent to

$$S_n = \frac{n}{2}(a_1 + a_n). \quad \square$$

EXAMPLE ▪ 4

Find the sum of all the even integers from 2 through 100.

SOLUTION This problem is equivalent to finding the sum of the first 50 terms of the arithmetic sequence $2, 4, 6, \ldots, 2n, \ldots$. Substituting $n = 50$, $a_1 = 2$, and $a_{50} = 100$ in the formula $S_n = (n/2)(a_1 + a_n)$,

$$S_{50} = \tfrac{50}{2}(2 + 100) = 2550.$$

As a check we use $S_n = \dfrac{n}{2}[2a_1 + (n - 1)d]$ with $d = 2$:

$$S_{50} = \tfrac{50}{2}[2(2) + (50 - 1)2] = 25[4 + 98] = 2550$$

The **arithmetic mean** of two numbers a and b is defined as $(a + b)/2$. This is the **average** of a and b. Note that

$$a, \quad \frac{a + b}{2}, \quad b$$

is an arithmetic sequence. This concept may be generalized as follows: If $c_1, c_2, \ldots, c_k$ are real numbers such that

$$a, \ c_1, \ c_2, \ \ldots, \ c_k, \ b$$

is an arithmetic sequence, then $c_1, c_2, \ldots, c_k$ are **k arithmetic means** between the numbers a and b. The process of determining these numbers is referred to as *inserting* k arithmetic means between a and b.

EXAMPLE ▪ 5

Insert three arithmetic means between 2 and 9.

SOLUTION We wish to find three real numbers c_1, c_2, and c_3 such that 2, c_1, c_2, c_3, 9 is an arithmetic sequence. We may find the common difference d by using the formula $a_n = a_1 + (n - 1)d$ with $n = 5$, $a_1 = 2$, and $a_5 = 9$. This gives us

$$9 = 2 + (5 - 1)d, \quad \text{or} \quad d = \tfrac{7}{4}.$$

We then find the three arithmetic means:

$$c_1 = a_1 + d = 2 + \tfrac{7}{4} = \tfrac{15}{4}$$
$$c_2 = c_1 + d = \tfrac{15}{4} + \tfrac{7}{4} = \tfrac{22}{4} = \tfrac{11}{2}$$
$$c_3 = c_2 + d = \tfrac{11}{2} + \tfrac{7}{4} = \tfrac{29}{4}$$

FIGURE 2

$a_1 = 18$
a_2
a_3
a_4
a_5
a_6
a_7
a_8
$a_9 = 24$

EXAMPLE ▪ 6

A carpenter wishes to construct a ladder with nine rungs whose lengths decrease uniformly from 24 inches at the base to 18 inches at the top. Determine the lengths of the seven intermediate rungs.

SOLUTION The ladder is sketched in Figure 2. The lengths of the rungs are to form an arithmetic sequence $a_1, a_2, \ldots, a_9$ with $a_1 = 18$ and $a_9 = 24$. Hence we need to insert seven arithmetic means between 18 and 24. Using $a_n = a_1 + (n-1)d$ with $n = 9$, $a_1 = 18$, and $a_9 = 24$ gives us

$$24 = 18 + 8d, \quad \text{or} \quad 8d = 6.$$

Hence $d = \frac{6}{8} = 0.75$ and the intermediate rungs have lengths (in inches)

$$18.75, \quad 19.5, \quad 20.25, \quad 21, \quad 21.75, \quad 22.5, \quad 23.25.$$

10.3 EXERCISES

Exer. 1–8: Find the fifth term, the tenth term, and the *n*th term of the arithmetic sequence.

1 $2, 6, 10, 14, \ldots$

2 $16, 13, 10, 7, \ldots$

3 $3, 2.7, 2.4, 2.1, \ldots$

4 $-6, -4.5, -3, -1.5, \ldots$

5 $-7, -3.9, -0.8, 2.3, \ldots$

6 $x - 8, x - 3, x + 2, x + 7, \ldots$

7 $\ln 3, \ln 9, \ln 27, \ln 81, \ldots$

8 $\log 1000, \log 100, \log 10, \log 1, \ldots$

9 Find the twelfth term of the arithmetic sequence whose first two terms are 9.1 and 7.5.

10 Find the eleventh term of the arithmetic sequence whose first two terms are $2 + \sqrt{2}$ and 3.

11 The sixth and seventh terms of an arithmetic sequence are 2.7 and 5.2. Find the first term.

12 Given an arithmetic sequence with $a_3 = 7$ and $a_{20} = 43$, find a_{15}.

Exer. 13–16: Find the sum S_n of the arithmetic sequence that satisfies the stated conditions.

13 $a_1 = 40, \quad d = -3, \quad n = 30$

14 $a_1 = 5, \quad d = 0.1, \quad n = 40$

15 $a_1 = -9, \quad a_{10} = 15, \quad n = 10$

16 $a_7 = \frac{7}{3}, \quad d = -\frac{2}{3}, \quad n = 15$

Exer. 17–20: Find the sum.

17 $\sum_{k=1}^{20} (3k - 5)$

18 $\sum_{k=1}^{12} (7 - 4k)$

19 $\sum_{k=1}^{18} (\frac{1}{2}k + 7)$

20 $\sum_{k=1}^{10} (\frac{1}{4}k + 3)$

21 **(a)** Find the number of integers between 32 and 395 that are divisible by 6, and **(b)** find their sum.

22 **(a)** Find the number of negative integers greater than -500 that are divisible by 33, and **(b)** find their sum.

Exer. 23–24: Find the number of terms in the arithmetic sequence with the given conditions.

23 $a_1 = -2, \quad d = \frac{1}{4}, \quad S = 21$

24 $a_6 = -3, \quad d = 0.2, \quad S = -33$

25 Insert five arithmetic means between 2 and 10.

26 Insert three arithmetic means between 3 and -5.

27 A pile of logs has 24 logs in the bottom layer, 23 in the next, 22 in the third, and so on. The top layer contains 10 logs. Find the total number of logs in the pile.

28 The first ten rows of seating in a certain section of a stadium have 30 seats, 32 seats, 34 seats, and so on. The eleventh through the twentieth rows each contain 50 seats. Find the total number of seats in the section.

29 A grain bin is to be constructed in the shape of a frustum of a cone (see figure). The bin is to be 10 feet tall with 11 metal rings around the bin positioned uniformly from the 4-foot opening at the bottom to the 24-foot opening at the top. Find the total length of the metal rings.

EXERCISE 29

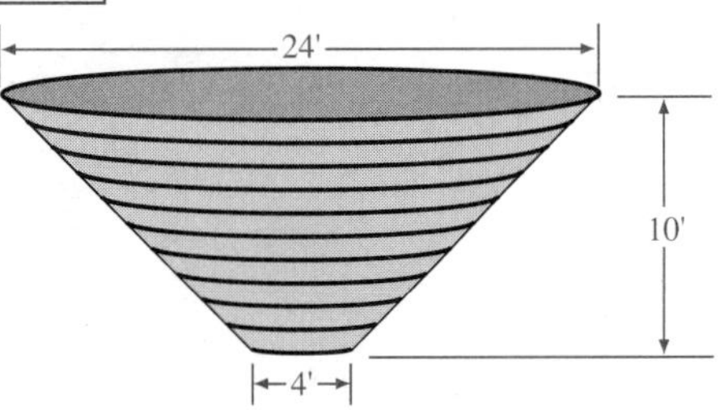

30 A bicycle rider coasts downhill, traveling 4 feet the first second and, in each succeeding second, 5 feet more than in the preceding second. If he reaches the bottom of the hill in 11 seconds, find the total distance traveled.

31 A contest will have five cash prizes totaling \$5000, and there will be a \$100 difference between successive prizes. Find the first prize.

32 A company is to distribute \$46,000 in bonuses to its top ten salespeople. The tenth salesperson on the list will receive \$1000, and the difference in bonus money between successively ranked salespeople is to be constant. Find the bonus for each salesperson.

33 The sum of the interior angles of a polygon with n sides is $(n - 2) \cdot 180°$. For certain polygons with n sides, the measures of the interior angles can be arranged in an arithmetic sequence. If, in a polygon of this type, the smallest angle is 20° and the largest angle is 160°, determine the number of sides of the polygon.

34 If f is a linear function, show that the sequence with nth term $a_n = f(n)$ is an arithmetic sequence.

35 The sequence defined recursively by $x_{k+1} = x_k/(1 + x_k)$ occurs in genetics in the study of the elimination of a deficient gene from a population. Show that the sequence whose nth term is $1/x_n$ is arithmetic.

36 Find the total length of the broken-line curve in the figure if the width of the maze formed by the curve is 16 inches and all halls in the maze have width 1 inch. What is the length if the width of the maze is 32 inches?

EXERCISE 36

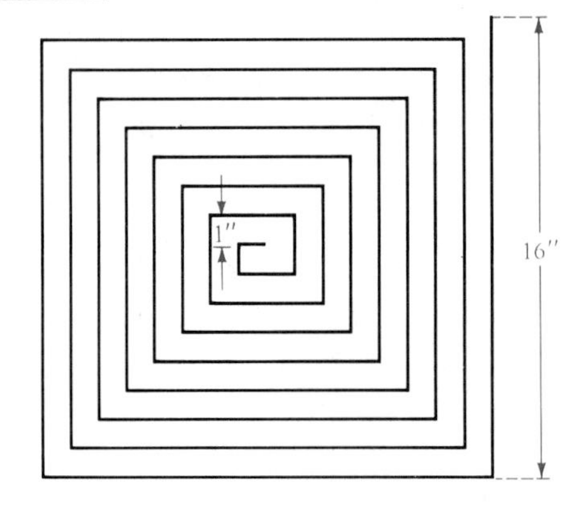

10.4 GEOMETRIC SEQUENCES

The following type of sequence occurs frequently in applied problems.

DEFINITION

A sequence $a_1, a_2, \ldots, a_n, \ldots$ is a **geometric sequence** if $a_1 \neq 0$ and if there is a real number $r \neq 0$ such that

$$a_{k+1} = a_k r$$

for every positive integer k.

If we rewrite the equation in the definition as

$$\frac{a_{k+1}}{a_k} = r,$$

we see that the ratio of any two successive terms of the sequence is r. We refer to r as the **common ratio** associated with the geometric sequence.

The formula $a_{k+1} = a_k r$ provides a recursive method for obtaining terms of a geometric sequence. Beginning with any nonzero real number a_1, we multiply by the number r successively, obtaining

$$a_1,\ a_1r,\ a_1r^2,\ a_1r^3,\ \ldots$$

The nth term a_n of this sequence is given by the next formula.

nth TERM OF A GEOMETRIC SEQUENCE

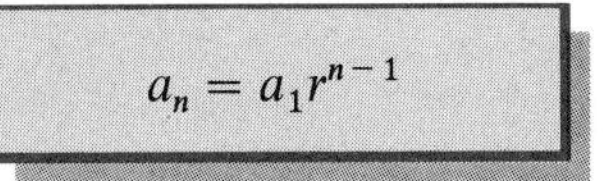

$$a_n = a_1r^{n-1}$$

EXAMPLE ▪ 1

A geometric sequence has first term 3 and common ratio $-\frac{1}{2}$. Find the first five terms and the tenth term.

SOLUTION If we let $a_1 = 3$ and $r = -\frac{1}{2}$, then the first five terms are

$$3,\ -\tfrac{3}{2},\ \tfrac{3}{4},\ -\tfrac{3}{8},\ \tfrac{3}{16}.$$

Using the formula $a_n = a_1r^{n-1}$ with $n = 10$, the tenth term is

$$a_{10} = 3(-\tfrac{1}{2})^9 = -\tfrac{3}{512}.$$

EXAMPLE ▪ 2

The third term of a geometric sequence is 5 and the sixth term is -40. Find the eighth term.

SOLUTION We are given $a_3 = 5$ and $a_6 = -40$. Substituting $n = 3$ and $n = 6$ in the formula $a_n = a_1r^{n-1}$ leads to the following system of equations:

$$\begin{cases} 5 = a_1r^2 \\ -40 = a_1r^5. \end{cases}$$

Since $r \neq 0$, the first equation is equivalent to $a_1 = 5/r^2$. Substituting for a_1 in the second equation,

$$-40 = \left(\frac{5}{r^2}\right) \cdot r^5 = 5r^3.$$

Hence, $r^3 = -8$ and $r = -2$. If we now substitute -2 for r in the equation $5 = a_1r^2$, we obtain $a_1 = \frac{5}{4}$. Finally, using $a_n = a_1r^{n-1}$ with $n = 8$,

$$a_8 = (\tfrac{5}{4})(-2)^7 = -160.$$

The next theorem contains a formula for the nth partial sum S_n of a geometric sequence.

THEOREM

> The nth partial sum S_n of a geometric sequence with first term a_1 and common ratio $r \neq 1$ is
>
> $$S_n = a_1 \frac{1 - r^n}{1 - r}.$$

PROOF By definition, the nth partial sum S_n is

$$S_n = a_1 + a_1r + a_1r^2 + \cdots + a_1r^{n-2} + a_1r^{n-1}.$$

Multiplying both sides of this equation by r, we obtain

$$rS_n = a_1r + a_1r^2 + a_1r^3 + \cdots + a_1r^{n-1} + a_1r^n.$$

If we subtract this equation from the equation for S_n, many terms on the right-hand side cancel:

$$S_n - rS_n = a_1 - a_1r^n$$
$$(1 - r)S_n = a_1(1 - r^n)$$

Since $r \neq 1$, we have $1 - r \neq 0$, and dividing both sides by $1 - r$ gives us

$$S_n = \frac{a_1(1 - r^n)}{1 - r}. \quad \square$$

EXAMPLE ▪ 3

If $1, 0.3, 0.09, 0.027, \ldots$ is a geometric sequence, find the sum of the first five terms.

SOLUTION If we let $a_1 = 1$, $r = 0.3$, and $n = 5$ in the formula for S_n stated in the preceding theorem, we obtain

$$S_5 = (1)\frac{1 - (0.3)^5}{1 - 0.3}.$$

This simplifies to $S_5 = 1.4251$.

EXAMPLE ■ 4

A man wishes to save money by setting aside 1 cent the first day, 2 cents the second day, 4 cents the third day and so on.

(a) If he continues to double the amount set aside each day, how much must he set aside on the fifteenth day?

(b) Assuming he does not run out of money, what is the total amount the man has saved at the end of 30 days?

SOLUTION

(a) The amount (in cents) set aside on successive days forms a geometric sequence

$$1, 2, 4, 8, \ldots$$

with first term 1 and common ratio 2. We find the amount to be set aside on the fifteenth day by using $a_n = a_1 r^{n-1}$ with $a_1 = 1$ and $n = 15$. This gives us $1 \cdot 2^{14}$, or \$163.84.

(b) To find the total amount saved after 30 days, we use the formula for S_n with $n = 30$. Thus,

$$S_{30} = (1)\frac{1 - 2^{30}}{1 - 2},$$

which simplifies to \$10,737,418.23.

Given the geometric series with first term a_1 and common ratio $r \neq 1$, we may write the formula for S_n of the preceding theorem in the form

$$S_n = \frac{a_1}{1 - r} - \frac{a_1}{1 - r} r^n.$$

If $|r| < 1$, then r^n *approaches* 0 *as n increases* without bound. Thus, S_n approaches $a_1/(1 - r)$ as n increases without bound. Using the notation we developed for rational functions in Section 4.6,

$$S_n \to \frac{a_1}{1 - r} \quad \text{as} \quad n \to \infty.$$

The number $a_1/(1 - r)$ is called the *sum* of the **infinite geometric series**

$$a_1 + a_1 r + a_1 r^2 + \cdots + a_1 r^{n-1} + \cdots$$

This gives us the next result.

THEOREM

If $|r| < 1$, then the infinite geometric series

$$a_1 + a_1 r + a_1 r^2 + \cdots + a_1 r^{n-1} + \cdots$$

has the sum $\dfrac{a_1}{1-r}$.

The preceding theorem implies that if we add more and more terms of the indicated infinite geometric series, the sums get closer and closer to $a_1/(1-r)$. The next example illustrates how the theorem can be used to show that every real number represented by a repeating decimal is rational.

EXAMPLE ▪ 5

Find the rational number that corresponds to the infinite repeating decimal $5.4\overline{27}$, where the block of digits underneath the bar is repeated indefinitely.

SOLUTION From the decimal expression 5.4272727 . . . , we obtain the infinite series

$$5.4 + 0.027 + 0.00027 + 0.0000027 + \cdots$$

The part of the expression after the first term is

$$0.027 + 0.00027 + 0.0000027 + \cdots,$$

which has the form given in the previous theorem with $a_1 = 0.027$ and $r = 0.01$. Hence, the sum of this infinite geometric series is

$$\frac{0.027}{1-0.01} = \frac{0.027}{0.99} = \frac{27}{990} = \frac{3}{110}.$$

Thus, it appears that the desired number is $5.4 + \frac{3}{110}$, or $\frac{597}{110}$. A check by division shows that $\frac{597}{110}$ does lead to the given repeating decimal.

In general, given any infinite sequence $a_1, a_2, \ldots, a_n, \ldots$, the expression

$$a_1 + a_2 + \cdots + a_n + \cdots$$

is called an **infinite series**, or simply a **series**. We denote this series by

$$\sum_{n=1}^{\infty} a_n.$$

Each number a_k is a **term** of the series and a_n is the ***n*th term**. Since only *finite* sums may be added algebraically, it is necessary to define what is meant by an *infinite sum*. Consider the sequence of partial sums

$$S_1,\ S_2,\ \ldots,\ S_n,\ \ldots$$

If there is a number S such that $S_n \to S$ as $n \to \infty$, then, as in our discussion of infinite geometric series, S is the **sum** of the infinite series and we write

$$S = a_1 + a_2 + \cdots + a_n + \cdots$$

In Example 5 we found that the infinite repeating decimal 5.4272727 . . . corresponds to the rational number $\frac{597}{110}$. Since $\frac{597}{110}$ is the sum of an infinite series determined by the decimal, we may write

$$\tfrac{597}{110} = 5.4 + 0.027 + 0.00027 + 0.0000027 + \cdots$$

If the terms of an infinite sequence are alternately positive and negative, and if we consider the expression

$$a_1 + (-a_2) + a_3 + (-a_4) + \cdots + [(-1)^{n+1}a_n] + \cdots$$

for positive real numbers a_k, then this expression is an **alternating infinite series**, and we write it in the form

$$a_1 - a_2 + a_3 - a_4 + \cdots + (-1)^{n+1}a_n + \cdots$$

The most common types of alternating infinite series are infinite geometric series in which the common ratio r is negative.

EXAMPLE ▪ 6

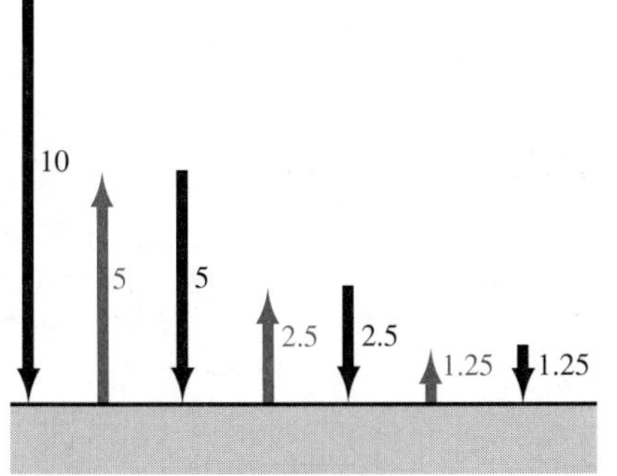

A rubber ball is dropped from a height of 10 meters. Suppose it rebounds one-half the distance after each fall, as illustrated by the arrows in Figure 3. Find the total distance the ball travels before coming to rest.

SOLUTION If the ball always rebounds half the distance it falls, then, theoretically, it *never* comes to rest. However, the sum of the distances it travels downward plus the sum of the distances it travels on the rebounds form two infinite geometric series:

$$\begin{aligned}&\text{downward series:} && 10 + 5 + 2.5 + 1.25 + 0.625 + \cdots\\ &\text{upward series:} && 5 + 2.5 + 1.25 + 0.625 + \cdots\end{aligned}$$

We assume that the total distance S the ball travels can be found by adding the sums of these infinite series. This gives us

$$\begin{aligned}S &= 10 + 2[5 + 2.5 + 1.25 + 0.625 + \cdots]\\ &= 10 + 2[5 + 5(\tfrac{1}{2}) + 5(\tfrac{1}{2})^2 + 5(\tfrac{1}{2})^3 + \cdots].\end{aligned}$$

Using the formula $S = a_1/(1 - r)$ with $a_1 = 5$ and $r = \frac{1}{2}$, we obtain

$$S = 10 + 2\left[\frac{5}{1 - \frac{1}{2}}\right] = 10 + 2(10) = 30 \text{ meters.}$$

10.4 EXERCISES

Exer. 1–12: Find the fifth term, the eighth term, and the *n*th term of the geometric sequence.

1 $8, 4, 2, 1, \ldots$

2 $4, 1.2, 0.36, 0.108, \ldots$

3 $300, -30, 3, -0.3, \ldots$

4 $1, -\sqrt{3}, 3, -\sqrt{27}, \ldots$

5 $5, 25, 125, 625, \ldots$

6 $2, 6, 18, 54, \ldots$

7 $4, -6, 9, -13.5, \ldots$

8 $162, -54, 18, -6, \ldots$

9 $1, -x^2, x^4, -x^6, \ldots$

10 $1, -\frac{x}{3}, \frac{x^2}{9}, -\frac{x^3}{27}, \ldots$

11 $2, 2^{x+1}, 2^{2x+1}, 2^{3x+1}, \ldots$

12 $10, 10^{2x-1}, 10^{4x-3}, 10^{6x-5}, \ldots$

13 Find the sixth term of the geometric sequence whose first two terms are 4 and 6.

14 Find the seventh term of the geometric sequence whose second and third terms are 2 and $-\sqrt{2}$.

15 Given a geometric sequence such that $a_5 = \frac{1}{16}$ and $r = \frac{3}{2}$, find a_1 and S_5.

16 Given a geometric sequence such that $a_4 = 4$ and $a_7 = 12$, find r and a_{10}.

Exer. 17–20: Find the sum.

17 $\sum_{k=1}^{10} 3^k$

18 $\sum_{k=1}^{9} (-\sqrt{5})^k$

19 $\sum_{k=0}^{9} (-\frac{1}{2})^{k+1}$

20 $\sum_{k=1}^{7} (3^{-k})$

21 A vacuum pump removes one-half of the air in a container with each stroke. After 10 strokes, what percentage of the original amount of air remains in the container?

22 The yearly depreciation of a certain machine is 25% of its value at the beginning of the year. If the original cost of the machine is $20,000, what is its value after 6 years?

23 A certain culture initially contains 10,000 bacteria and increases by 20% every hour.

(a) Find a formula for the number $N(t)$ of bacteria present after t hours.

(b) How many bacteria are in the culture at the end of 10 hours?

24 An amount of money P is deposited in a savings account that pays interest at a rate of r percent per year compounded quarterly, and the principal and accumulated interest are left in the account. Find a formula for the total amount in the account after n years.

Exer. 25–30: Find the sum of the infinite geometric series, if it exists.

25 $1 - \frac{1}{2} + \frac{1}{4} - \frac{1}{8} + \cdots$

26 $2 + \frac{2}{3} + \frac{2}{9} + \frac{2}{27} + \cdots$

27 $1.5 + 0.015 + 0.00015 + \cdots$

28 $1 - 0.1 + 0.01 - 0.001 + \cdots$

29 $\sqrt{2} - 2 + \sqrt{8} - 4 + \cdots$

30 $250 - 100 + 40 - 16 + \cdots$

Exer. 31–38: Find the rational number represented by the repeating decimal.

31 $0.\overline{23}$

32 $0.0\overline{71}$

33 $2.4\overline{17}$

34 $10.\overline{55}$

35 $5.\overline{146}$

36 $3.2\overline{394}$

37 $1.\overline{6124}$

38 $123.61\overline{83}$

39 A rubber ball is dropped from a height of 60 feet. If it rebounds approximately two-thirds the distance after each fall, use an infinite geometric series to approximate the total distance the ball travels before coming to rest.

40 The bob of a pendulum swings through an arc 24 cm long on its first swing. If each successive swing is approximately five-sixths the length of the preceding swing, use an infinite geometric series to approximate the total distance it travels before coming to rest.

41 A manufacturing company that has just located in a small community will pay two million dollars per year in salaries. It has been estimated that 60% of these salaries will be spent in the local area, and 60% of this money spent will again change hands within the community. This process will be repeated ad infinitum. This is called the multiplier effect. Find the total amount of spending that will be generated by company salaries.

42 In a pest eradication program, N sterilized male flies are released into the general population each day, and 90% of these flies will survive a given day.

(a) Show that the number of sterilized flies in the population n days after the program has begun is

$$N + (0.9)N + (0.9)^2N + \cdots + (0.9)^{n-1}N.$$

(b) If the *long-range* goal of the program is to keep 20,000 sterilized males in the population, how many flies should be released each day?

43 A certain drug has a half-life of about 2 hours in the bloodstream. The drug is formulated to be administered in doses of D mg every 4 hours, but D is yet to be determined.

(a) Show that the number of milligrams of drug in the bloodstream after the nth dose has been administered is

$$D + \tfrac{1}{4}D + \cdots + (\tfrac{1}{4})^{n-1}\,D,$$

and that this sum is approximately $\frac{4}{3}D$ for large values of n.

(b) A level of more than 500 mg of the drug in the bloodstream is considered to be dangerous. Find the largest possible dose that can be given repeatedly over a long period of time.

44 Shown in the figure is a family tree displaying the current generation (you) and 3 prior generations and a total of 12 grandparents. If you were to trace your family history back 10 generations, how many grandparents would you find?

EXERCISE 44

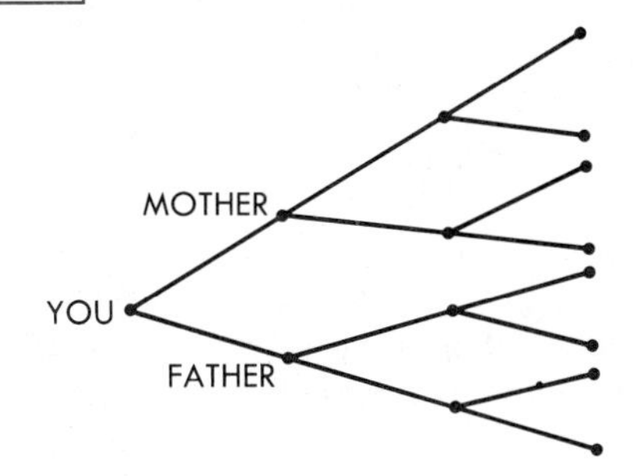

45 The first figure shows some terms of a sequence of squares $S_1, S_2, \ldots, S_k, \ldots$. Let a_k, A_k, and P_k denote the side, area, and perimeter, respectively, of the square S_k. The square S_{k+1} is constructed from S_k by connecting four points on S_k, with each point a distance of $\frac{1}{4}a_k$ from a vertex, as shown in the second figure.

(a) Find the relationship between a_{k+1} and a_k.

(b) Find a_n, A_n, and P_n.

(c) Calculate $\sum_{n=1}^{\infty} P_n$.

EXERCISE 45

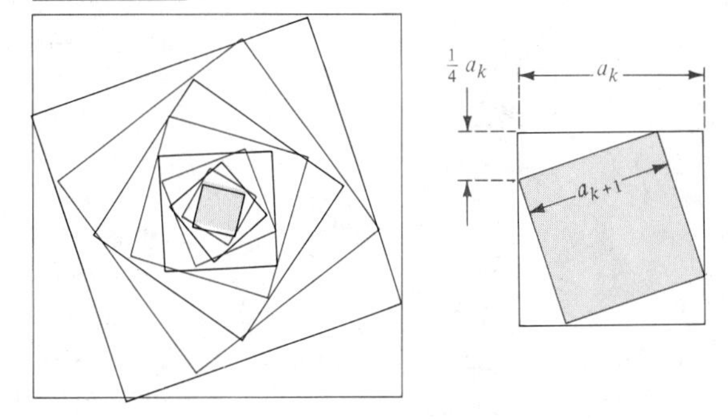

46 The figure shows several terms of a sequence consisting of alternating circles and squares. Each circle is inscribed in a square and each square (excluding the largest) is inscribed in a circle. Let S_n denote the area of the nth square and C_n the area of the nth circle.

(a) Find the relationships between S_n and C_n and between C_n and S_{n+1}.

(b) What portion of the largest square is shaded in the figure?

EXERCISE 46

10.5 PERMUTATIONS

Suppose that four teams are involved in a tournament in which first, second, third, and fourth places will be determined. For identification purposes, we label the teams a, b, c, and d. Let us find the number of different ways that first and second place can be decided. It is convenient to use a **tree diagram** as in Figure 4. Beginning at the word *Start*, the four possibilities for first place are listed. From each of these an arrow points to a possible second-place finisher. The final standings list the possible outcomes, from left to right. They are found by following the different paths (*branches* of the tree) that lead from the word *Start* to the second-place team. The total number of outcomes is 12, which is the product of the number of choices (4) for first place and the number of choices (3) for second place (after the first has been determined).

FIGURE 4

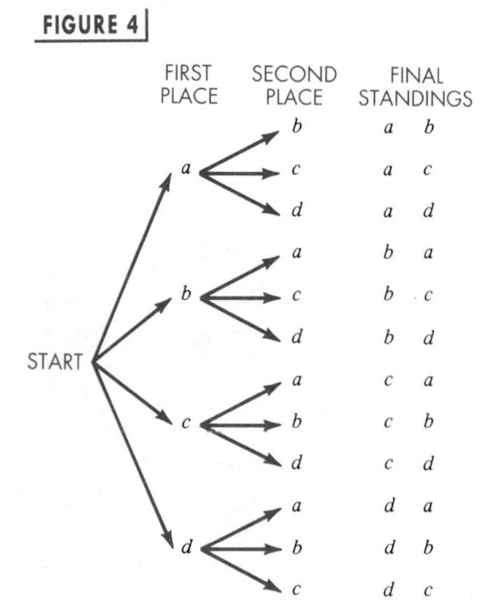

Let us now find the total number of ways that first, second, third, and fourth positions can be filled. To sketch a tree diagram we may begin by drawing arrows from the word *Start* to each possible first-place finisher a, b, c, or d. Next we draw arrows from those to possible second-place finishers, as was done in Figure 4. From each second-place position we then draw arrows indicating the possible third-place positions. Finally, we draw arrows to the fourth-place team. If we consider only the case in which team a finishes in first place, we have the diagram shown in Figure 5.

FIGURE 5

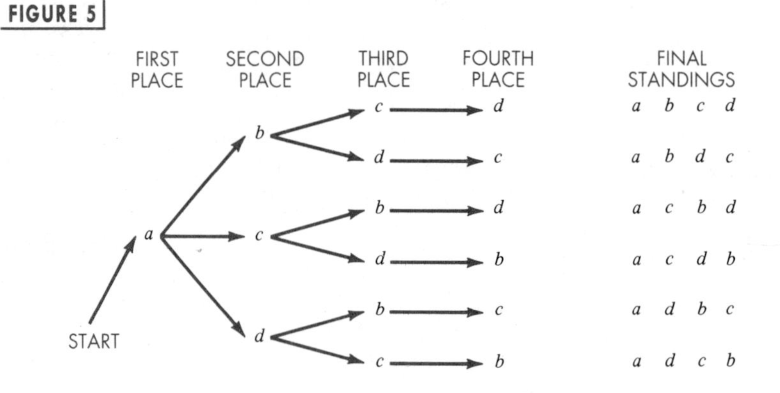

Note that there are six possible final standings in which team a occupies first place. In a complete tree diagram there would also be three other branches of this type corresponding to first place for b, c, and d. A complete diagram would display the following 24 possibilities for the final standings:

$$\begin{array}{cccccc} abcd, & abdc, & acbd, & acdb, & adbc, & adcb, \\ bacd, & badc, & bcad, & bcda, & bdac, & bdca, \\ cabd, & cadb, & cbad, & cbda, & cdab, & cdba, \\ dabc, & dacb, & dbac, & dbca, & dcab, & dcba. \end{array}$$

Note that the number of possibilities (24) is the product of the number of ways (4) that first place may occur, the number of ways (3) that second place may occur (after first place has been determined), the number of possible outcomes (2) for third place (after the first two places have been decided), and the number of ways (1) that fourth place can occur (after the first three places have been taken).

The preceding discussion illustrates the following general rule, which we accept as a basic axiom of counting.

FUNDAMENTAL COUNTING PRINCIPLE

> Let $E_1, E_2, \ldots, E_k$ be a sequence of k events. If for each i, the event E_i can occur in m_i ways, then the total number of ways all the events may take place is the product $m_1 m_2 \cdots m_k$.

Returning to our first illustration, we let E_1 represent the determination of the first-place team, so that $m_1 = 4$. If E_2 denotes the determination of the second place team, then $m_2 = 3$. Hence, the number of outcomes for the sequence E_1, E_2 is $4 \cdot 3 = 12$, which is the same as that found by means of the tree diagram. If we proceed to E_3, the determination of the third-place team, then $m_3 = 2$, and hence, $m_1 m_2 m_3 = 24$. Finally, if E_1, E_2, and E_3 have occurred, there is only one possible outcome for E_4. Thus, $m_4 = 1$ and $m_1 m_2 m_3 m_4 = 24$.

Instead of teams, let us now regard a, b, c, and d merely as symbols and consider the various *orderings*, or *arrangements*, that may be assigned to these symbols, taking them either two at a time, three at a time, or four at a time. By abstracting in this way we may apply our methods to other similar situations. The arrangements we have discussed are **arrangements without repetitions**, since a symbol may not be used more than once in an arrangement. In Example 1 we shall consider arrangements in which repetitions *are* allowed.

Previously we defined ordered pairs and ordered triples. Similarly, an **ordered 4-tuple** is a set containing four elements x_1, x_2, x_3, x_4 in which an ordering has been specified, so that one of the elements may be referred to as the *first element*, another as the *second element*, and so on. The symbol (x_1, x_2, x_3, x_4) is used for the ordered 4-tuple having first element x_1, second element x_2, third element x_3, and fourth element x_4. In general, for any positive integer r, we speak of the **ordered r-tuple**

$$(x_1, x_2, \ldots, x_r)$$

as a set of elements in which x_1 is designated as the first element, x_2 as the second element, and so on.

EXAMPLE ▪ 1

Using only the letters a, b, c, and d, determine how many of the following can be obtained:

(a) ordered triples (b) ordered 4-tuples (c) ordered r-tuples

SOLUTION

(a) We must determine the number of symbols of the form (x_1, x_2, x_3) that can be obtained using only the letters a, b, c, and d. This is not the same as listing first, second, and third place as in our previous illustration, since we have not ruled out the possibility of repetitions. For example, (a, b, a), (a, a, b), and (b, a, a) are different ordered triples. If for $i = 1, 2, 3$, we let E_i represent the determination of x_i in the ordered triple (x_1, x_2, x_3), then, since repetitions are allowed, there are four possibilities—a, b, c and d—for each of E_1, E_2, and E_3. Hence, by the fundamental counting principle, the total number of ordered triples is $4 \cdot 4 \cdot 4$, or 64.

(b) The number of possible ordered 4-tuples (x_1, x_2, x_3, x_4) is $4 \cdot 4 \cdot 4 \cdot 4$, or 256.

(c) The number of ordered r-tuples is the product $4 \cdot 4 \cdot 4 \cdots 4$, with 4 appearing as a factor r times. That product equals 4^r.

EXAMPLE ▪ 2

A class consists of 60 girls and 40 boys. In how many ways can a president, vice-president, treasurer, and secretary be chosen if the treasurer must be a girl, the secretary must be a boy, and a student may not hold more than one office?

SOLUTION If an event is specialized in some way (for example, the treasurer *must* be a girl), then that event should be performed before any nonspecialized events. Thus, we let E_1 represent the choice of treasurer and E_2 the choice of secretary. Next we let E_3 and E_4 denote the choices for president and vice-president, respectively. As in the fundamental counting principle, we let m_i denote the number of different ways E_i can occur for $i = 1, 2, 3$, and 4. It follows that $m_1 = 60$, $m_2 = 40$, $m_3 = 98$, and $m_4 = 97$. By the fundamental counting principle, the total number of possibilities is

$$60 \cdot 40 \cdot 98 \cdot 97 = 22{,}814{,}400.$$

When working with sets, we are usually not concerned about the order or arrangement of the elements. In the remainder of this section, however, the arrangement of the elements will be our main concern.

DEFINITION

Let S be a set of n elements and let $1 \leq r \leq n$. A **permutation** of r elements of S is an arrangement, without repetitions, of r elements.

We also use the phrase **permutation of n elements taken r at a time**. The symbol $P(n, r)$ will denote the number of different permutations of r elements that can be obtained from a set containing n elements. As a special case, $P(n, n)$ denotes the number of arrangements of n elements of S, that is, the number of ways of arranging *all* the elements of S.

In our first illustration involving the four teams a, b, c, and d, we had $P(4, 2) = 12$, since there were 12 different ways of arranging the four teams in groups of two. We also showed that the number of ways to arrange all the elements a, b, c, and d is 24. In permutation notation we would write this as $P(4, 4) = 24$.

The next theorem gives us a general formula for $P(n, r)$.

THEOREM ON PERMUTATIONS

Let S be a set of n elements and let $1 \leq r \leq n$. The number of different permutations of r elements of S is

$$P(n, r) = n(n-1)(n-2) \cdots (n-r+1)$$

PROOF The problem of determining $P(n, r)$ is equivalent to determining the number of different r-tuples $(x_1, x_2, \ldots, x_r)$, such that each x_i is an element of S and no element of S appears twice in the same r-tuple. We may find this number by means of the fundamental counting principle. For each $i = 1, 2, \ldots, r$, let E_i represent the determination of the element x_i and let m_i be the number of different ways of choosing x_i. We wish to apply the sequence $E_1, E_2, \ldots, E_r$. We have n possible choices for x_1 and, consequently, $m_1 = n$. Since repetitions are not allowed, we have $n - 1$ choices for x_2, so $m_2 = n - 1$. Continuing in this manner, we successively obtain $m_3 = n - 2$, $m_4 = n - 3$, and ultimately $m_r = n - (r - 1)$, or $m_r = n - r + 1$. Hence, using the fundamental counting principle, we obtain the formula for $P(n, r)$. ❑

Note that *the formula for $P(n, r)$ in the previous theorem contains exactly r factors on the right-hand side.*

ILLUSTRATION $P(n, r)$

- $P(n, 1) = n$
- $P(n, 2) = n(n-1)$
- $P(n, 3) = n(n-1)(n-2)$
- $P(n, 4) = n(n-1)(n-2)(n-3)$

EXAMPLE ▪ 3

Find $P(5, 2)$, $P(6, 4)$, and $P(5, 5)$.

SOLUTION We use the formula for $P(n, r)$ in the preceding theorem:

$$P(5, 2) = 5 \cdot 4 = 20$$
$$P(6, 4) = 6 \cdot 5 \cdot 4 \cdot 3 = 360$$
$$P(5, 5) = 5 \cdot 4 \cdot 3 \cdot 2 \cdot 1 = 120$$

EXAMPLE ▪ 4

A baseball team consists of nine players. Find the number of ways of arranging the first four positions in the batting order if the pitcher is excluded.

SOLUTION We wish to find the number of permutations of 8 objects taken 4 at a time. Using the formula for $P(n, r)$ with $n = 8$ and $r = 4$,

$$P(8, 4) = 8 \cdot 7 \cdot 6 \cdot 5 = 1680.$$

The next result gives us a form for $P(n, r)$ that involves the factorial symbol.

FACTORIAL FORM FOR $P(n, r)$

> If n is a positive integer, and $1 \le r \le n$, then
>
> $$P(n, r) = \frac{n!}{(n-r)!}.$$

PROOF If we let $r = n$ in the formula for $P(n, r)$ in the theorem on permutations, we obtain the number of different arrangements of *all* the elements of a set consisting of n elements. In this case

$$n - r + 1 = n - n + 1 = 1,$$

and
$$P(n, n) = n(n-1)(n-2) \cdots 3 \cdot 2 \cdot 1 = n!$$

Consequently, $P(n, n)$ is the product of the first n positive integers. This is also given by the factorial form, for if $r = n$, then

$$P(n, n) = \frac{n!}{(n-n)!} = \frac{n!}{0!} = \frac{n!}{1} = n!$$

If $1 \leq r < n$, then

$$\frac{n!}{(n-r)!} = \frac{n(n-1)(n-2)\cdots(n-r+1)\cdot[(n-r)!]}{(n-r)!}$$
$$= n(n-1)(n-2)\cdots(n-r+1).$$

This agrees with the formula for $P(n, r)$ in the theorem on permutations. ❑

EXAMPLE ▪ 4

Use the factorial form for $P(n, r)$ to find $P(5, 2)$, $P(6, 4)$, and $P(5, 5)$.

SOLUTION

$$P(5, 2) = \frac{5!}{(5-2)!} = \frac{5!}{3!} = \frac{5 \cdot 4 \cdot 3 \cdot 2 \cdot 1}{3 \cdot 2 \cdot 1} = 5 \cdot 4 = 20$$

$$P(6, 4) = \frac{6!}{(6-4)!} = \frac{6!}{2!} = 6 \cdot 5 \cdot 4 \cdot 3 = 360$$

$$P(5, 5) = \frac{5!}{(5-5)!} = \frac{5!}{0!} = \frac{5!}{1} = 120$$

10.5 EXERCISES

Exer. 1–8: Find the number.

1 $P(7, 3)$ 2 $P(8, 5)$

3 $P(9, 6)$ 4 $P(5, 3)$

5 $P(5, 5)$ 6 $P(4, 4)$

7 $P(6, 1)$ 8 $P(5, 1)$

Exer. 9–22: Solve by using a tree diagram, the fundamental counting principle, or a permutation.

9 How many three-digit numbers can be formed from the digits 1, 2, 3, 4, and 5 if repetitions (a) are not allowed? (b) are allowed?

10 Work Exercise 9 for four-digit numbers.

11 How many numbers can be formed from the digits 1, 2, 3, and 4 if repetitions are not allowed? (*Note:* 42 and 231 are examples of such numbers.)

12 Work Exercise 11 assuming repetitions are allowed.

13 If eight basketball teams are in a tournament, find the number of different ways that first, second, and third place can be decided, assuming ties are not allowed.

14 Work Exercise 13 for 12 teams.

15 A girl has four skirts and six blouses. How many different skirt-blouse combinations can she wear?

16 Refer to Exercise 15. If the girl also has three sweaters, how many different skirt-blouse-sweater combinations can she wear?

17 In a certain state, automobile license plates start with one letter of the alphabet followed by five numerals, using the digits $0, 1, 2, \ldots, 9$. Find how many different license plates are possible if:

(a) the first digit following the letter cannot be 0.

(b) the first letter cannot be *O* or *I* and the first digit cannot be 0.

18 Two dice are tossed, one after the other. In how many different ways can they fall? List the number of different ways the sum of the dots can equal:

(a) 3 (b) 5 (c) 7 (d) 9 (e) 11

19 A row of six seats in a classroom is to be filled by selecting individuals from a group of ten students.

(a) In how many different ways can the seats be occupied?

(b) If there are six boys and four girls in the group, and if boys and girls are to be alternated, find the number of different seating arrangements.

20 A student in a certain college may take mathematics at 8, 10, 11, or 2 o'clock; English at 9, 10, 1, or 2; and History at 8, 11, 2, or 3. Find the number of different ways in which the student can schedule the three courses.

21 In how many different ways can a test consisting of ten true-or-false questions be completed?

22 A test consists of six multiple-choice questions, and the number of choices for each question is five. In how many different ways can the test be completed?

23 In how many different ways can eight people be seated in a row?

24 In how many different ways can ten books be arranged on a shelf?

25 Using six different flags, how many different signals can be sent by placing three flags, one above the other, on a flag pole?

26 In how many different ways can five books be selected from a 12-volume set of books?

27 How many three-digit numbers can be formed from the digits 2, 4, 6, 8, and 9 if repetitions (a) are not allowed? (b) are allowed?

28 There are 24 letters in the Greek alphabet. How many fraternities may be specified by choosing three Greek letters if repetitions (a) are not allowed? (b) are allowed?

29 How many seven-digit phone numbers can be formed from the digits $0, 1, 2, 3, \ldots, 9$ if the first digit may not be 0?

30 After selecting nine players for a baseball game, the manager of the team arranges the batting order so that the pitcher bats last and the best hitter bats third. In how many different ways can the remainder of the batting order be arranged?

31 A customer remembers that 2, 4, 7, and 9 are the digits of his four-digit access code for an automatic bank-teller machine. Unfortunately, the customer has forgotten the order of the digits. Find the largest possible number of trials necessary to obtain the correct code.

32 Work Exercise 31 if the digits are 2, 4, and 7, and one of these digits is repeated in the four-digit code.

33 Three married couples have purchased tickets for a play. Spouses are to be seated next to each other and the six seats are in a row. In how many ways can the six people be seated?

34 Ten horses are entered in a race. If the possibility of a tie for any place is ignored, in how many ways can the first-, second-, and third-place winners be determined?

35 The commutative and associative laws of addition guarantee that the sum of integers 1 through 10 is independent of the order in which the numbers are added. In how many different ways can these integers be summed?

36 In how many ways can a standard deck of 52 cards be shuffled? In how many ways can the cards be shuffled so that the four aces appear on the top of the deck?

37 A palindrome is an integer, such as 45654, that reads the same backward or forward.

(a) How many five-digit palindromes are there?

(b) How many n-digit palindromes are there?

38 Each of the six squares shown in the figure is to be filled with any one of ten possible colors. How many ways are there of coloring the strip shown in the figure so that no two adjacent squares have the same color?

EXERCISE 38

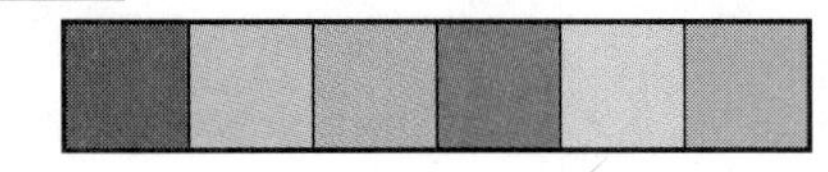

10.6 DISTINGUISHABLE PERMUTATIONS AND COMBINATIONS

Certain problems involve finding different arrangements of objects, some of which are indistinguishable. For example, suppose we are given five disks of the same size, of which three are black, one is white, and one is red. Let us find the number of ways they can be arranged in a row so that different color arrangements are obtained. If the disks were all different colors, then the number of arrangements would be 5!, or 120. However, since some of the disks have the same appearance, we cannot obtain 120 different arrangements. To clarify this point, let us write

B B B W R

for the arrangement having black disks in the first three positions in the row, the white disk in the fourth position, and the red disk in the fifth position. The first three disks can be arranged in 3!, or 6, different ways, but these arrangements cannot be distinguished from one another because the first three disks look alike. We say that those 3! permutations are **nondistinguishable**. Similarly, given any other arrangement, say

B R B W B,

there are 3! different ways of arranging the three black disks, but again each such arrangement is nondistinguishable from the others. Let us call two arrangements of objects **distinguishable permutations** if one arrangement cannot be obtained from the other by rearranging like objects. Thus, B B B W R and B R B W B are distinguishable permutations of the five disks. Let k denote the number of distinguishable permutations. Since with each such arrangement there correspond 3! *nondistinguishable* permutations, we must have $3!k = 5!$, the number of permutations of five *different* objects. Hence, $k = 5!/3! = 5 \cdot 4 = 20$. By the same type of reasoning we can obtain the following extension of this discussion.

THEOREM If r objects in a collection of n objects are alike, and if the remaining objects are different from each other and from the r objects, then the number of distinguishable permutations of the n objects is

$$\frac{n!}{r!}.$$

We can generalize this theorem to the case in which there are several subcollections of indistinguishable objects. For example, consider eight disks, of which four are black, three are white, and one is red. In this case, with each arrangement, such as

B W B W B W B R,

there are 4! arrangements of the black disks and 3! arrangements of the white disks that have no effect on the color arrangement. Hence 4!3! possible arrangements of the disks will not produce distinguishable permutations. If we let k denote the number of *distinguishable* permutations, then $4!3!k = 8!$, since 8! is the number of permutations we would obtain if the disks were all different. Thus, the number of distinguishable permutations is

$$k = \frac{8!}{4!3!} = \frac{8 \cdot 7 \cdot 6 \cdot 5}{3!} \cdot \frac{4!}{4!} = 280.$$

The following general result can be proved.

THEOREM ON DISTINGUISHABLE PERMUTATIONS

If, in a collection of n objects, n_1 are alike, n_2 are alike of another kind, . . . , n_k are alike of a further kind, and

$$n = n_1 + n_2 + \cdots + n_k,$$

then the number of distinguishable permutations of the n objects is

$$\frac{n!}{n_1!n_2! \cdots n_k!}.$$

EXAMPLE ▪ 1

Find the number of distinguishable permutations of the letters in the word *Mississippi*.

SOLUTION In this example we are given a collection of eleven objects in which four are of one kind (the letter s), four are of another kind (i), two are of a third kind (p), and one is of a fourth kind (M). Hence, by the preceding theorem, the number of distinguishable permutations is

$$\frac{11!}{4!4!2!1!} = 34{,}650.$$

When we work with permutations our concern is with the orderings or arrangements of elements. Let us now ignore the order or arrangement of elements and consider the following question: Given a set containing n distinct elements, in how many ways can a subset of r elements be chosen with $r \leq n$? Before answering, let us state a definition.

DEFINITION

Let S be a set of n elements and let $1 \leq r \leq n$. A **combination** of r elements of S is a subset of S that contains r distinct elements.

If S contains n elements, we also use the phrase **combination of n elements taken r at a time**. The symbol $C(n, r)$ will denote the number of combinations of r elements that can be obtained from a set of n elements.

THEOREM ON COMBINATIONS

The number of combinations of r elements that can be obtained from a set of n elements is

$$C(n, r) = \frac{n!}{(n-r)!r!}, \qquad 1 \leq r \leq n.$$

PROOF If S contains n elements, then, to find $C(n, r)$, we must find the total number of subsets of the form

$$\{x_1, x_2, \ldots, x_r\}$$

such that the x_k are *different* elements of S. Since the elements $x_1, x_2, \ldots, x_r$ can be arranged in $r!$ different ways, each such subset produces $r!$ different r-tuples. Thus the total number of different r-tuples is $r!C(n, r)$. However, in the previous section we found that the total number of r-tuples is

$$P(n, r) = \frac{n!}{(n-r)!}.$$

Hence,

$$r!C(n, r) = \frac{n!}{(n-r)!}$$

Dividing both sides of the last equation by $r!$ gives us the formula for $C(n, r)$. ❑

The formula for $C(n, r)$ *is identical to the formula for the binomial coefficient* $\binom{n}{r}$ in Section 1.6.

EXAMPLE ▪ 2

A little league baseball squad has six outfielders, seven infielders, five pitchers, and two catchers. Each outfielder can play any of the three outfield positions and each infielder can play any of the four infield positions. In how many ways can a team of nine players be chosen?

SOLUTION The number of ways of choosing three outfielders from the six candidates is

$$C(6,3) = \frac{6!}{(6-3)!3!} = \frac{6!}{3!3!} = \frac{(6 \cdot 5 \cdot 4)3!}{(3 \cdot 2 \cdot 1)3!} = 20.$$

The number of ways of choosing the four infielders is

$$C(7,4) = \frac{7!}{(7-4)!4!} = \frac{(7 \cdot 6 \cdot 5)4!}{3!4!} = \frac{7 \cdot 6 \cdot 5}{3 \cdot 2 \cdot 1} = 35.$$

There are five ways of choosing a pitcher and two choices for the catcher. It follows from the fundamental counting principle that the total number of ways to choose a team is

$$20 \cdot 35 \cdot 5 \cdot 2 = 7000.$$

Note that if $r = n$, then the formula for $C(n, r)$ becomes

$$C(n,n) = \frac{n!}{(n-n)!n!} = \frac{n!}{0!n!} = 1.$$

It is convenient to assign a meaning to $C(n, r)$ if $r = 0$. If the formula is to be true in this case, then we must have

$$C(n,0) = \frac{n!}{n!0!} = 1.$$

Hence, we *define* $C(n, 0) = 1$, which is the same as $C(n, n)$. Finally, for consistency we also *define* $C(0, 0) = 1$. Thus, $C(n, r)$ has meaning for all nonnegative integers n and r with $r \leq n$.

EXAMPLE ▪ 3

Let S be a set of n elements. Find the number of distinct subsets of S.

SOLUTION Let r be any nonnegative integer such that $r \leq n$. From our previous work, the number of subsets of S that consist of r elements is

$C(n, r)$, or $\binom{n}{r}$. Hence, to find the total number of subsets, we find the sum

$$\binom{n}{0} + \binom{n}{1} + \binom{n}{2} + \binom{n}{3} + \cdots + \binom{n}{n}.$$

This is precisely the binomial expansion of $(1 + 1)^n$. Thus, there are 2^n subsets of a set of n elements. In particular, a set of 3 elements has 2^3, or 8, different subsets. A set of 4 elements has 2^4, or 16, subsets. A set of 10 elements has 2^{10}, or 1024, subsets.

10.6 EXERCISES

Exer. 1–8: Find the number.

1 $C(7, 3)$

2 $C(8, 4)$

3 $C(9, 8)$

4 $C(6, 2)$

5 $C(n, n - 1)$

6 $C(n, 1)$

7 $C(7, 0)$

8 $C(5, 5)$

Exer. 9–10: Find the number of possible color arrangements for the 12 given disks, arranged in a row.

9 5 black, 3 red, 2 white, 2 green

10 3 black, 3 red, 3 white, 3 green

11 Find the number of distinguishable permutations of the letters in the word *bookkeeper*.

12 Find the number of distinguishable permutations of the letters in the word *moon*. List all the permutations.

13 Ten boys wish to play a basketball game. In how many different ways can two teams consisting of five players each be formed?

14 A student may answer any six of ten questions on an examination.

(a) In how many ways can six questions be selected?

(b) How many selections are possible if the first two questions must be answered?

Exer. 15–16: Consider any eight points such that no three are collinear.

15 How many lines are determined?

16 How many triangles are determined?

17 A student has five mathematics books, four history books, and eight fiction books. In how many different ways can they be arranged on a shelf if books in the same category are kept next to one another?

18 A basketball squad consists of twelve players.

(a) Disregarding positions, in how many ways can a team of five be selected?

(b) If the center of a team must be selected from two specific individuals on the squad and the other four members of the team from the remaining ten players, find the number of different teams possible.

19 A football squad consists of three centers, ten linemen who can play either guard or tackle, three quarterbacks, six halfbacks, four ends, and four fullbacks. A team must have one center, two guards, two tackles, two ends, two halfbacks, a quarterback, and a fullback. In how many different ways can a team be selected from the squad?

20 In how many different ways can seven keys be arranged on a key ring if the keys can slide completely around the ring?

21 A committee of 3 men and 2 women is to be chosen from a group of 12 men and 8 women. Determine the number of different ways of selecting the committee.

22 For sibling birth order, the letters G and B denote a girl birth and boy birth, respectively. For a family of three boys and three girls, one possible birth order is G G G B B B. How many other birth orders are possible for these six children?

Exer. 23–24: Shown in the figure is a street map and a possible path from point A to point B. How many other possible paths are there from A to B if moves are restricted to the right and up? (*Hint:* If R denotes a move one unit right and U denotes a move one unit up, then the path in Exercise 23 can be specified by R U U R R R U R.)

23

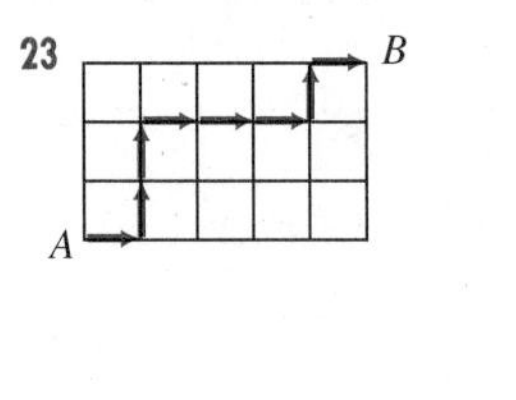

24

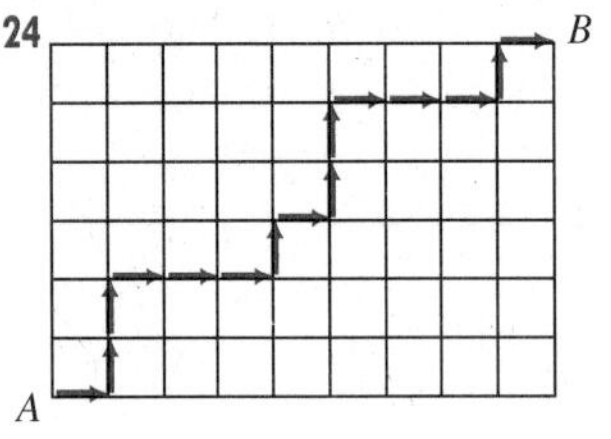

25 To win a state lottery game, Lotto, a player must correctly select six numbers from the numbers 1 through 49.

(a) Find the total number of selections possible.

(b) Work part (a) if a player selects only even numbers.

26 A mathematics department has ten faculty members but only nine offices, so one office must be shared by two individuals. In how many different ways can the offices be assigned?

27 In a round-robin tennis tournament, every player meets every other player exactly once. How many players can participate in a tournament of 45 matches?

28 A true-false test has 20 questions.

(a) In how many different ways can the test be completed?

(b) In how many different ways can a student answer 10 questions correctly?

29 The winner of the N.B.A. championship series is the first team to win four games. In how many different ways can the series be extended to seven games?

30 A geometric design is determined by joining every pair of vertices of an octagon (see figure).

(a) How many triangles in the design have their three vertices on the octagon?

(b) How many quadrilaterals in the design have their four vertices on the octagon?

EXERCISE 30

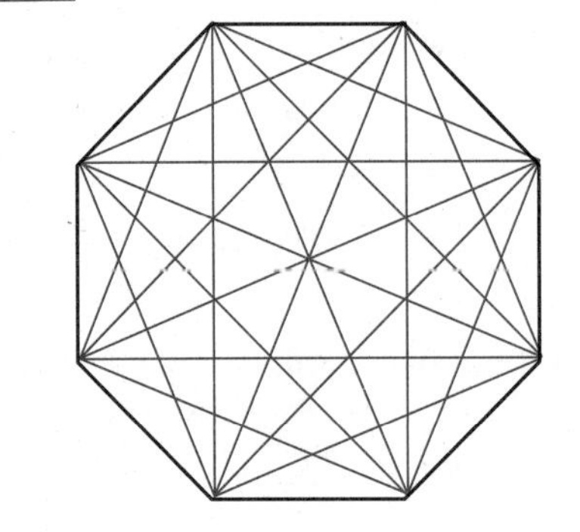

10.7 PROBABILITY

If two dice are tossed, what are the chances of rolling a 7? If a person is dealt five cards from a standard deck of 52 playing cards, what is the likelihood of obtaining three aces? In the 17th century, similar questions about games of chance led to the study of *probability*. Since that time, the theory of probability has grown extensively. It is now used to predict outcomes of a large variety of situations that arise in the natural and social sciences.

Any chance process, such as flipping a coin, rolling a die, being dealt a card from a deck, determining if a manufactured item is defective, or finding the blood pressure of an individual, is an **experiment**. A result of an experiment is an **outcome**. We shall restrict our discussion to experiments for which outcomes are **equally likely**. This means, for example, that if a coin is flipped, we assume that the possibility of obtaining a head is the same as that of obtaining a tail. Similarly, if a die is tossed, we assume that the die is *fair*, that is, there is an equal chance of obtaining either a 1, 2, 3, 4, 5, or 6. The set S of all possible outcomes of an experiment is the **sample space** of the experiment. Thus, if the experiment consists of flipping a coin, and we let H or T denote the outcome of obtaining a head or tail, respectively, then the sample space S may be denoted by

$$S = \{\mathrm{H}, \mathrm{T}\}.$$

If a fair die is tossed as an experiment, then the set S of all possible outcomes (the sample space) is

$$S = \{1, 2, 3, 4, 5, 6\}.$$

The following definition expresses, in mathematical terms, the notion of obtaining *particular* outcomes of an experiment.

DEFINITION

> Let S be the sample space of an experiment. An **event** associated with the experiment is any subset E of S.

Let us consider the experiment of tossing a single die, so that the sample space is $S = \{1, 2, 3, 4, 5, 6\}$. If $E = \{4\}$, then the event E associated with the experiment consists of the outcome of obtaining a 4 on the toss. Different events may be associated with the same experiment. For example, if we let $E = \{1, 3, 5\}$, then this event consists of obtaining an odd number on a toss of the die.

As another illustration, suppose the experiment consists of flipping two coins, one after the other. If we let HH denote the outcome of two heads appearing, HT that of a head appearing on the first coin and a tail on the second, and so on, then the sample space S of the experiment may be denoted by

$$S = \{\mathrm{HH}, \mathrm{HT}, \mathrm{TH}, \mathrm{TT}\}.$$

If we let

$$E = \{\mathrm{HT}, \mathrm{TH}\}$$

then the event E consists of a head appearing on one of the coins and a tail on the other.

Next we shall define what is meant by the *probability* of an event. Throughout our discussion we will assume that the sample space S of an experiment contains only a finite number of elements. If E is an event, *the symbols $n(E)$ and $n(S)$ will denote the number of elements in E and S, respectively.*

DEFINITION

Let S be the sample space of an experiment and E an event. The **probability** $P(E)$ of E is given by

$$P(E) = \frac{n(E)}{n(S)}.$$

Since E is a subset of S, we see that

$$0 \le n(E) \le n(S).$$

Dividing by $n(S)$, we obtain

$$0 \le P(E) \le 1.$$

Note that $P(E) = 1$ if $E = S$, and $P(E) = 0$ if E contains no element.

The next example provides several illustrations of the preceding definition if E contains exactly one element.

EXAMPLE ▪ 1

(a) If a coin is flipped, find the probability that a head will turn up.

(b) If a fair die is tossed, find the probability of obtaining a 4.

(c) If two coins are flipped, find the probability that both coins turn up heads.

SOLUTION For each experiment we shall list sets S and E, and then use the definition of probability to find $P(E)$.

(a) $S = \{\text{H}, \text{T}\}, \quad E = \{\text{H}\}, \quad P(E) = \dfrac{n(E)}{n(S)} = \dfrac{1}{2}$

(b) $S = \{1, 2, 3, 4, 5, 6\}, \quad E = \{4\}, \quad P(E) = \dfrac{n(E)}{n(S)} = \dfrac{1}{6}$

(c) $S = \{\text{HH}, \text{HT}, \text{TH}, \text{TT}\}, \quad E = \{\text{HH}\}, \quad P(E) = \dfrac{n(E)}{n(S)} = \dfrac{1}{4}$

In (a) of Example 1 we found that the probability of obtaining a head on a flip of a coin is $\frac{1}{2}$. We take this to mean that if a coin is flipped many times, the number of times that a head turns up should be approximately one-half the total number of flips. Thus, for 100 flips, a head should turn up approximately 50 times. It is unlikely that this number will be *exactly* 50. A probability of $\frac{1}{2}$ implies that if we let the number of flips increase, then the number of times a head turns up *approaches* $\frac{1}{2}$ the total number of flips. Similar remarks can be made for (b) and (c) of Example 1.

In the next two examples we consider experiments in which an event contains more than one element.

EXAMPLE ▪ 2

If two dice are tossed, what is the probability of rolling a sum of

(a) 7? **(b)** 9?

SOLUTION Let us refer to one die as *the first die* and the other as *the second die*. We shall use ordered pairs to represent outcomes as follows: (2, 4) will denote the outcome of obtaining a 2 on the first die and a 4 on the second; (5, 3) represents a 5 on the first die and a 3 on the second, and so on. Since there are six different possibilities for the first number of the ordered pair and, with each of these, six possibilities for the second number, the total number of ordered pairs is 36. Hence, if S is the sample space, then $n(S) = 36$.

(a) The event E corresponding to rolling a sum of 7 is given by

$$E = \{(1, 6), (2, 5), (3, 4), (4, 3), (5, 2), (6, 1)\}$$

and, consequently,
$$P(E) = \frac{n(E)}{n(S)} = \frac{6}{36} = \frac{1}{6}.$$

(b) If E is the event corresponding to rolling a sum of 9, then

$$E = \{(3, 6), (4, 5), (5, 4), (6, 3)\}$$

and
$$P(E) = \frac{n(E)}{n(S)} = \frac{4}{36} = \frac{1}{9}.$$

In the next example (and in the exercises), when it is stated that one or more cards are drawn from a deck, we mean that each card is removed from a standard 52-card deck and is *not* replaced before the next card is drawn.

EXAMPLE ▪ 3

Suppose five cards are drawn from a standard deck of 52 playing cards. Find the probability that all five cards are hearts.

SOLUTION The sample space S of the experiment is the set of all possible five-card hands that can be formed from the 52 cards in the deck. It follows from our work in the preceding section that $n(S) = C(52, 5)$.

Since there are 13 cards in the heart suit, the number of different ways of obtaining a hand that contains five hearts is $C(13, 5)$. Hence, if E

represents this event, then

$$P(E) = \frac{n(E)}{n(S)} = \frac{C(13, 5)}{C(52, 5)}$$

$$= \frac{\dfrac{13!}{5!8!}}{\dfrac{52!}{5!47!}}$$

We may show that $$P(E) = \frac{1287}{2{,}598{,}960} \approx 0.0005.$$

Thus $$P(E) \approx \frac{1}{2000}.$$

This implies that if the experiment is performed many times, then a five-card heart hand should be drawn approximately once every 2000 times.

Suppose S is the sample space of an experiment, and E_1 and E_2 are two events associated with the experiment. Suppose further that E_1 and E_2 have no elements in common, that is, E_1 and E_2 are *disjoint* sets. If $E = E_1 \cup E_2$, then

$$n(E) = n(E_1 \cup E_2) = n(E_1) + n(E_2)$$

Hence $$P(E) = \frac{n(E_1) + n(E_2)}{n(S)}$$

$$P(E) = \frac{n(E_1)}{n(S)} + \frac{n(E_2)}{n(S)}$$

$$P(E) = P(E_1) + P(E_2).$$

Thus, the probability of E is the sum of the probabilities of E_1 and E_2.

The next theorem states that this result can be extended to any number of events $E_1, E_2, \ldots, E_k$ that are *mutually disjoint*; that is, if $i \neq j$, then E_i and E_j have no element in common.

THEOREM

Let $E_1, E_2, \ldots, E_k$ be mutually disjoint events associated with the same experiment. If

$$E = E_1 \cup E_2 \cup \cdots \cup E_k,$$

then $$P(E) = P(E_1) + P(E_2) + \cdots + P(E_k).$$

The theorem may be proved by mathematical induction.

EXAMPLE ▪ 4

If two dice are tossed, find the probability of rolling a sum of either 7 or 9.

SOLUTION Let E_1 denote the event of rolling 7, and E_2 that of rolling 9. We wish to find the probability of the event $E = E_1 \cup E_2$. From Example 3, we know that $P(E_1) = \frac{1}{6}$ and $P(E_2) = \frac{1}{9}$. Hence, by the last theorem,

$$P(E) = P(E_1) + P(E_2) = \frac{1}{6} + \frac{1}{9} = \frac{3+2}{18} = \frac{5}{18}.$$

The results of this section provide merely an introduction to the theory of probability. Many other types of problems may be considered. For example, given two events E_1 and E_2, we have not discussed how to find $P(E_1 \cup E_2)$ if E_1 and E_2 have elements in common, or how to find $P(E_2)$ *after* E_1 has occurred. You may refer to a textbook or course on probability for methods that can be used to investigate these and other problems.

10.7 EXERCISES

Exer. 1–2: A single card is drawn from a 52-card deck. Find the probability that the card is as specified.

1 (a) a king

(b) a king or a queen

(c) a king, a queen, or a jack

2 (a) a heart

(b) a heart or a diamond

(c) a heart, a diamond, or a club

3 If a single die is tossed, find the probability of obtaining (a) a 4, (b) a 6, (c) a 4 or a 6.

4 An urn contains five red balls, six green balls, and four white balls. If a single ball is drawn, find the probability that it is (a) red, (b) green, (c) red or white.

5 If two dice are tossed, find the probability of rolling a sum of (a) 11, (b) 8, (c) 11 or 8.

6 If two dice are tossed, find the probability that the sum of the dots is greater than 5.

7 If three dice are tossed, find the probability that the sum of the dots is 5.

8 If three dice are tossed, find the probability that a 6 turns up on exactly one die.

9 If three coins are flipped, find the probability that exactly two heads turn up.

10 If four coins are flipped, find the probability of obtaining two heads and two tails.

Exer. 11–16: Suppose five cards are drawn from a 52-card deck. Find the probability of obtaining the indicated cards.

11 Four of a kind (such as four aces or four kings)

12 Three aces and two kings

13 Four diamonds and one spade

14 Five face cards

15 A flush (five cards, all of the same suit)

16 A royal flush (an ace, king, queen, jack, and 10 of the same suit)

17 A true-or-false test contains eight questions. If a student guesses the answer for each question, find the probability that:

(a) eight answers are correct.

(b) seven answers are correct and one is incorrect.

(c) six answers are correct and two are incorrect.

(d) at least six answers are correct.

18 A six-member committee is to be chosen by drawing names of individuals from a hat. If the hat contains the names of eight men and fourteen women, find the probability that the committee will consist of three men and three women.

19 **(a)** If S is the sample space of an experiment and E is an event, let E' denote the elements of S that are not in E. Show that $P(E') = 1 - P(E)$.

(b) If five cards are drawn from a 52-card deck, use part (a) to find the probability of obtaining at least one ace.

20 If five cards are drawn from a 52-card deck, use Exercise 19 to find the probability of obtaining at least one heart.

21 In the popular dice game of craps, the shooter rolls two die and wins on the first roll if a sum of 7 or 11 is obtained. The shooter loses on the first roll if the sum is 2, 3, or 12. Find the probability of

(a) winning on the first roll.

(b) losing on the first roll.

22 A standard slot machine contains three reels and each reel contains 20 symbols. If the first reel has five bells, the middle reel four bells, and the last reel two bells, find the probability of obtaining three bells in a row.

23 Refer to Exercise 19. Assuming that girl-boy births are equiprobable, find the probability that a family with five children has **(a)** all boys; **(b)** at least one girl.

24 Three cards are placed in a hat. One card is red on both sides, the second card is black on both sides, and the final card is red on one side and black on the other. A card is selected from the hat and placed on a table. If the card shows red, what is the probability that the other side is also red?

25 In a simple experiment designed to test ESP, four cards (jack, queen, king, and ace) are shuffled and then placed face down on a table. The subject then attempts to identify each of the four cards, giving a different name to each of the cards. If the individual is guessing, find the probability of correctly identifying **(a)** all four cards; **(b)** exactly two of the four cards.

26 Three dice are tossed.

(a) Find the probability that all dice show the same number of dots.

(b) Find the probability that the number of dots on the dice are all different.

(c) Work parts (a) and (b) for n dice.

27 For a normal die, the sum of the dots on opposite faces is 7. Shown in the figure is a pair of trick dice in which the *same* number of dots appear on opposite faces. Find the probability of rolling a sum of **(a)** 7 and **(b)** 8.

EXERCISE 27

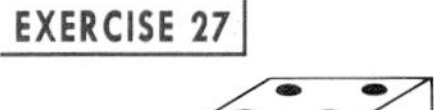

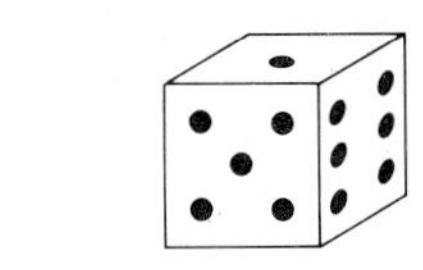

28 In a common carnival game, three balls are rolled down an incline into slots numbered 1 through 9 (see figure). Because the slots are so narrow, players have no control over where the balls will collect. A prize is given if the sum of the resulting three numbers is less than 7. Find the probability of winning a prize.

EXERCISE 28

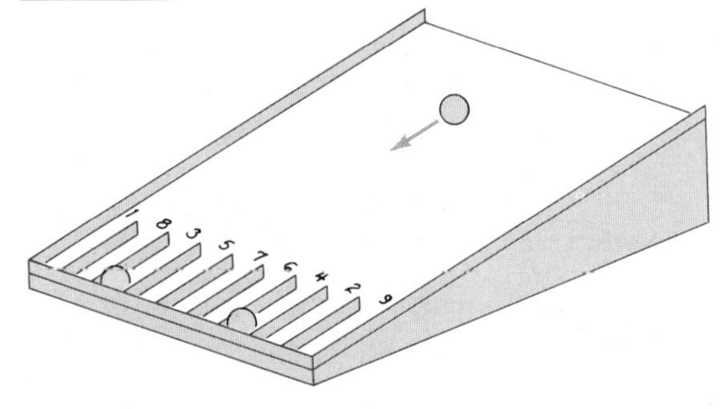

29 Shown in the figure is a small version of a probability demonstration device. A small ball is dropped into the top of the maze and tumbles to the bottom. Each time the ball strikes an obstacle, there is a 50% chance that the ball will move to the left. Find the probability that the ball ends up in the slot **(a)** on the far left; **(b)** in the middle.

EXERCISE 29

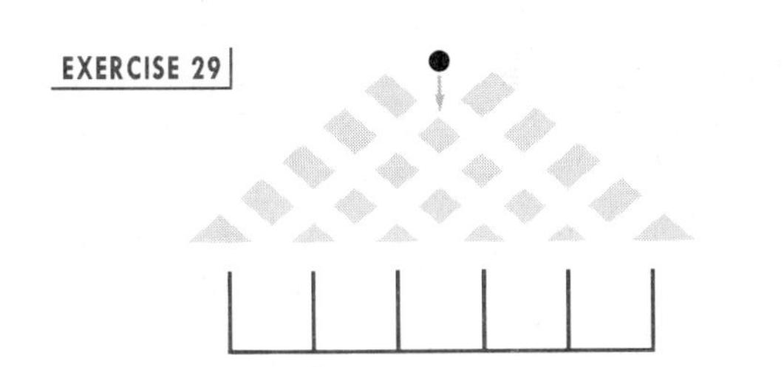

EXERCISE 30

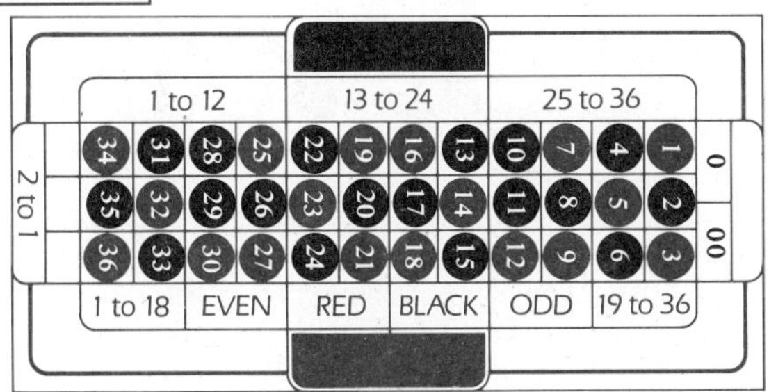

30 In the American version of roulette, a ball is spun around a wheel and has an equal chance of landing in any one of 38 slots numbered $0, 00, 1, 2, \ldots, 36$. Shown in the figure is a standard betting layout for roulette. Find the probability that the ball lands **(a)** in a black slot; **(b)** in a black slot two times in succession.

10.8 REVIEW

Define or discuss each of the following.

Principle of mathematical induction ■ Infinite sequence ■ Summation notation ■ The nth partial sum of an infinite sequence ■ Arithmetic sequence ■ Arithmetic mean of two numbers ■ Geometric sequence ■ Infinite geometric series ■ Infinite series ■ Tree diagram ■ Fundamental counting principle ■ Ordered r-tuple ■ Permutation ■ $P(n, r)$ ■ Nondistinguishable permutations ■ Distinguishable permutations ■ Combination ■ $C(n, r)$ ■ Sample space ■ Event ■ Probability of an event

10.8 EXERCISES

Exer. 1–5: Prove that the statement is true for every positive integer n.

1 $2 + 5 + 8 + \cdots + (3n - 1) = \dfrac{n(3n + 1)}{2}$

2 $2^2 + 4^2 + 6^2 + \cdots + (2n)^2 = \dfrac{2n(2n + 1)(n + 1)}{3}$

3 $\dfrac{1}{1 \cdot 3} + \dfrac{1}{3 \cdot 5} + \dfrac{1}{5 \cdot 7} + \cdots + \dfrac{1}{(2n - 1)(2n + 1)} = \dfrac{n}{2n + 1}$

4 $1 \cdot 2 + 2 \cdot 3 + 3 \cdot 4 + \cdots + n(n + 1) = \dfrac{n(n + 1)(n + 2)}{3}$

5 3 is a factor of $n^3 + 2n$.

6 Prove that $n^2 + 3 < 2^n$ for every positive integer $n \geq 5$.

Exer. 7–10: Find the first four terms and the seventh term of the sequence that has the given nth term.

7 $a_n = \dfrac{5n}{3 - 2n^2}$

8 $a_n = (-1)^{n+1} - (0.1)^n$

9 $a_n = 1 + (-\frac{1}{2})^{n-1}$

10 $a_n = \dfrac{2^n}{(n + 1)(n + 2)(n + 3)}$

Exer. 11–14: Find the first five terms of the recursively defined infinite sequence.

11 $a_1 = 10, \quad a_{k+1} = 1 + (1/a_k)$

12 $a_1 = 2, \quad a_{k+1} = a_k!$

13 $a_1 = 9, \quad a_{k+1} = \sqrt{a_k}$

14 $a_1 = 1, \quad a_{k+1} = (1 + a_k)^{-1}$

Exer. 15–18: Find the sum.

15 $\sum_{k=1}^{5} (k^2 + 4)$ **16** $\sum_{k=2}^{6} \dfrac{2k - 8}{k - 1}$

17 $\sum_{k=1}^{100} 10$ **18** $\sum_{k=1}^{4} (2^k - 10)$

Exer. 19–22: Use summation notation to represent the sum.

19 $3 + 6 + 9 + 12 + 15$

20 $2 + 4 + 8 + 16 + 32 + 64 + 128$

21 $100 - 95 + 90 - 85 + 80$

22 $a_0 + a_4x^4 + a_8x^8 + \cdots + a_{100}x^{100}$

23 Find the tenth term and the sum of the first ten terms of the arithmetic sequence whose first two terms are $4 + \sqrt{3}$ and 3.

24 Find the sum of the first eight terms of an arithmetic sequence in which the fourth term is 9 and the common difference is -5.

25 The fifth and thirteenth terms of an arithmetic sequence are 5 and 77, respectively. Find the first term and the tenth term.

26 Insert four arithmetic means between 20 and -10.

27 Find the tenth term of the geometric sequence whose first two terms are $\frac{1}{8}$ and $\frac{1}{4}$.

28 If a geometric sequence has 3 and -0.3 as its third and fourth terms, find the eighth term.

29 Find a positive number c such that 4, c, 8 are successive terms of a geometric sequence.

30 In a certain geometric sequence the eighth term is 100 and the common ratio is $-\frac{3}{2}$. Find the first term.

Exer. 31–34: Find the sum.

31 $\sum_{k=1}^{15} (5k - 2)$ **32** $\sum_{k=1}^{10} (6 - \frac{1}{2}k)$

33 $\sum_{k=1}^{10} (2^k - \frac{1}{2})$ **34** $\sum_{k=1}^{8} (\frac{1}{2} - 2^k)$

35 Find the sum of the infinite geometric series

$$1 - \frac{2}{5} + \frac{4}{25} - \frac{8}{125} + \cdots.$$

36 Find the rational number whose decimal representation is $6.\overline{274}$.

37 Ten-foot lengths of 2×2 lumber are to be cut into five pieces to form children's building blocks, and the lengths of the five blocks are to form an arithmetic sequence.

(a) Show that the difference d in lengths must be less than 1 foot.

(b) If the smallest block is to have a length of 6 inches, find the lengths of the other four pieces.

38 When a ball is dropped from a height of h feet, it reaches the ground in $\sqrt{h}/4$ seconds. The ball rebounds to a height of d feet in $\sqrt{d}/4$ seconds. If a rubber ball is dropped from a height of 10 feet and rebounds to three-fourths of its height after each fall, how many seconds elapse before the ball comes to rest?

39 Shown in the first figure is a broken-line curve obtained by taking two adjacent sides of a square, each of length s_n, decreasing the length of the side by a factor f with $0 < f < 1$, and forming two sides of a smaller square of length $s_{n+1} = f \cdot s_n$. The process is then repeated ad infinitum. If $s_1 = 1$ in the second figure, express the length of the resulting (infinite) broken-line curve in terms of f.

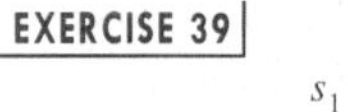
EXERCISE 39

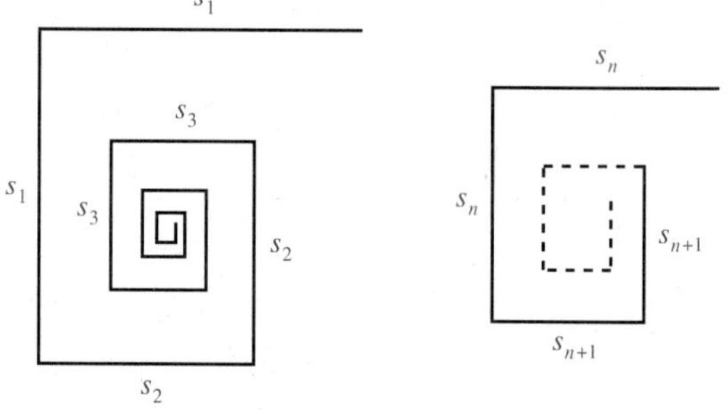

40 In the game of backgammon, a player is allowed to move his counters the same number of spaces as the sum of the dots on two dice. However, if a double is rolled (that is, both dice show the same number of dots), then the player may move his counters twice the sum of the dots. What is the probability that a player will be able to move his counters at least 10 spaces on a given roll?

41 Find the probability that the coins will match if

(a) two boys each toss a coin.

(b) three boys each toss a coin.

42 If four cards are dealt from a 52-card deck, find the probability that

(a) all four cards will be the same color.

(b) the cards dealt will alternate red-black-red-black.

43 **(a)** In how many ways can 13 cards be selected from a deck of 52 cards?

(b) In how many ways can 13 cards be selected to obtain five spades, three hearts, three clubs, and two diamonds?

44 How many four-digit numbers can be formed from the digits 1, 2, 3, 4, 5, and 6 if repetitions **(a)** are not allowed? **(b)** are allowed?

45 **(a)** If a student must answer eight of twelve questions on an examination, how many different selections of questions are possible?

(b) How many selections are possible if the first three questions must be answered?

46 If six black, five red, four white, and two green disks are to be arranged in a row, what is the number of possible color arrangements?

47 If 1000 tickets are sold for a raffle, find the probability of winning if an individual purchases **(a)** 1 ticket, **(b)** 10 tickets, **(c)** 50 tickets.

48 If four coins are flipped, find the probability of obtaining one head and three tails.

49 A quiz consists of six true-or-false questions, and at least four correct answers are required for a passing grade. If a student guesses at each answer, what is the probability of **(a)** passing? **(b)** failing?

50 If a single die is tossed and then a card is drawn from a 52-card deck, what is the probability of obtaining

(a) a 6 on the die and the king of hearts?

(b) *either* a 6 on the die or the king of hearts?

CHAPTER 11

Plane geometry includes the study of figures, such as lines, circles, and triangles, that lie in a plane. Theorems are proved by reasoning deductively from certain postulates. In *analytic* geometry, plane geometric figures are investigated by introducing coordinate systems and then using equations and formulas. If the study of analytic geometry were to be summarized by means of one statement, perhaps the following would be appropriate: *Given an equation, find its graph and, conversely, given a graph, find its equation*. In this chapter we shall apply coordinate methods to several basic plane figures.

TOPICS FROM ANALYTIC GEOMETRY

11.1 CONIC SECTIONS

Each of the geometric figures discussed in Sections 11.2–11.5 can be obtained by intersecting a double-napped right circular cone with a plane. For this reason we call these figures **conic sections** or, simply, **conics**. If, as in Figure 1(i), the plane cuts entirely across one nappe of the cone and is perpendicular to the axis l of the cone, then the curve of intersection is a circle. If the plane is not perpendicular to the axis of the cone, as in (ii) of the figure, an **ellipse** results. If the plane does not cut across one entire nappe and does not intersect both nappes, as illustrated in Figure 1(iii), then the curve of intersection is a **parabola**. If the plane cuts through both nappes of the cone, as in (iv), we obtain a **hyperbola**.

FIGURE 1

(i) Circle

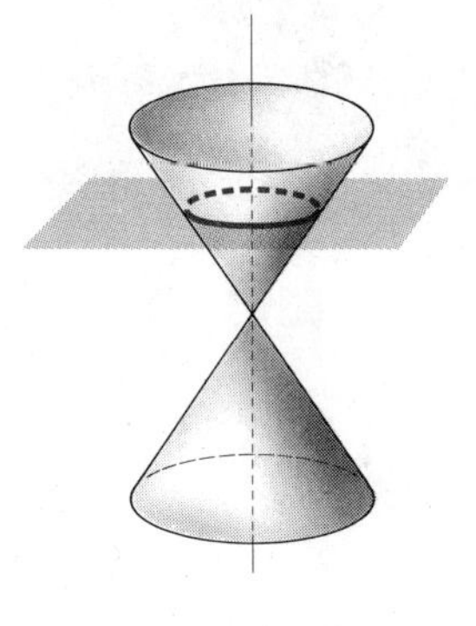

(ii) Ellipse (iii) Parabola (iv) Hyperbola

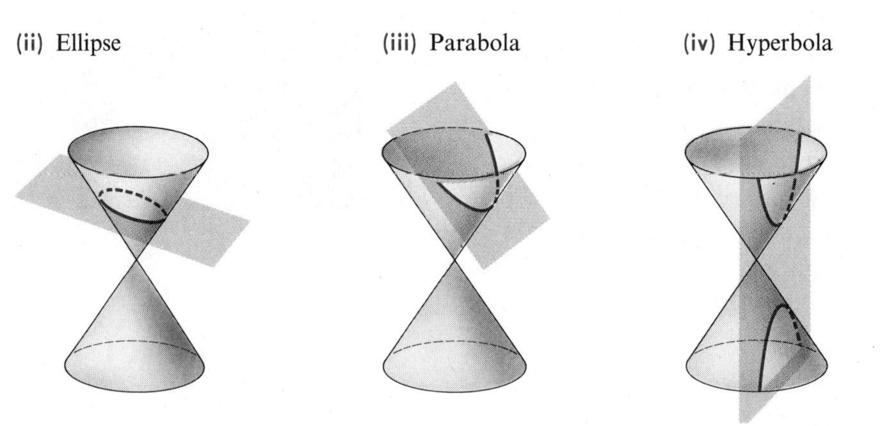

By changing the shape of the cone and the position of the plane, conics can be made to vary considerably. Certain positions of the plane result in **degenerate conics**. For example, if the plane intersects the cone only at the vertex, then the conic consists of one point. If the axis of the cone lies on the plane, then we obtain a pair of intersecting lines. Finally, we can begin with the case of a parabola, as in Figure 1(iii), and move the plane parallel to its initial position until it has only one line in common with the cone.

Conic sections were studied extensively by the ancient Greeks, who discovered properties that enable us to define conics in terms of points and lines. A remarkable fact about conic sections is that although they were studied thousands of years ago, they are far from obsolete. They are important tools for present-day investigations in outer space and for the study of the behavior of atomic particles. From physics we know that if a particle moves under the influence of an *inverse square force field*, then its path may be described by means of a conic section. Examples of inverse square fields are gravitational and electromagnetic fields. Planetary orbits are elliptical. If the ellipse is very flat, the curve resembles the path of a comet. Parabolic mirrors are sometimes used to collect solar energy. The hyperbola is useful for describing the path of an alpha particle in the electric field of the nucleus of an atom.

11.2 PARABOLAS

Parabolas were discussed in Section 4.1; however, the definition was not stated at that time. Moreover, we concentrated on parabolas with vertical axes. We shall now define *parabola* and derive equations for parabolas that have either vertical or horizontal axes.

DEFINITION

A **parabola** is the set of all points in a plane equidistant from a fixed point F (the **focus**) and a fixed line l (the **directrix**) in the plane.

FIGURE 2

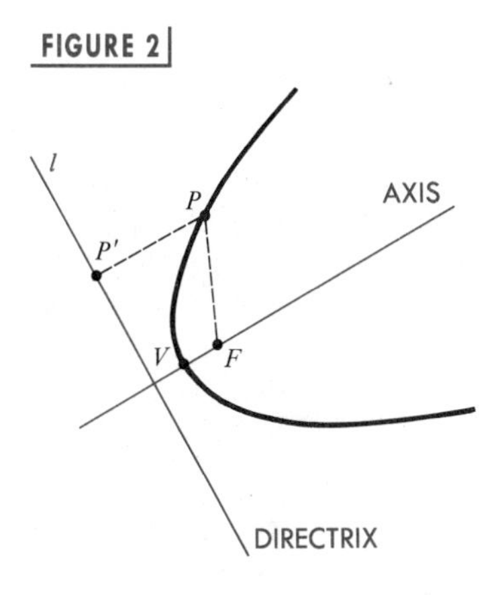

We shall assume that F is not on l, for this would result in a line. If P is a point in the plane and P' is the point on l determined by a line through P that is perpendicular to l (see Figure 2), then by definition, P is on the parabola if and only if $d(P, F) = d(P, P')$. The line through F, perpendicular to the directrix, is the **axis** of the parabola. The point V on the axis, half-way from F to l, is the **vertex** of the parabola.

By introducing a coordinate system, we can obtain a simple equation for a parabola. Place the y-axis along the axis of the parabola, with the origin at the vertex V, as shown in Figure 3. In this case, the focus F has coordinates $(0, p)$ for some real number $p \neq 0$, and the equation of the directrix is $y = -p$. By the distance formula, a point $P(x, y)$ is on the parabola if and only if

$$\sqrt{(x-0)^2 + (y-p)^2} = \sqrt{(x-x)^2 + (y+p)^2}.$$

We square both sides and simplify:

$$\begin{aligned}(x-0)^2 + (y-p)^2 &= (y+p)^2\\ x^2 + y^2 - 2py + p^2 &= y^2 + 2py + p^2\\ x^2 &= 4py\end{aligned}$$

FIGURE 3

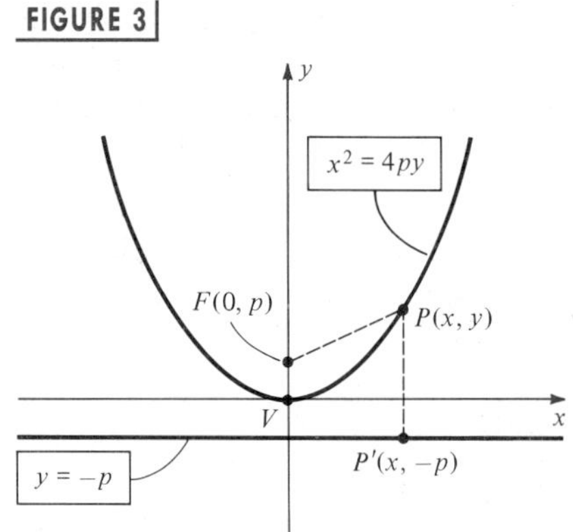

We have shown that the coordinates of every point (x, y) on the parabola satisfy $x^2 = 4py$. Conversely, if (x, y) is a solution of $x^2 = 4py$, then by reversing the previous steps we see that the point (x, y) is on the parabola. If $p > 0$ the parabola opens upward, as in Figure 3, and if $p < 0$ the parabola opens downward. The graph is symmetric with respect to the y-axis, since substitution of $-x$ for x does not change the equation $x^2 = 4py$.

We sometimes place the axis of the parabola along the x-axis. If the vertex is $V(0,0)$, the focus is $F(p,0)$, and the directrix has equation $x = -p$ (see Figure 4), then we obtain the equation $y^2 = 4px$. If $p > 0$ the parabola opens to the right, and if $p < 0$ it opens to the left. In this case the graph is symmetric with respect to the x-axis.

FIGURE 4

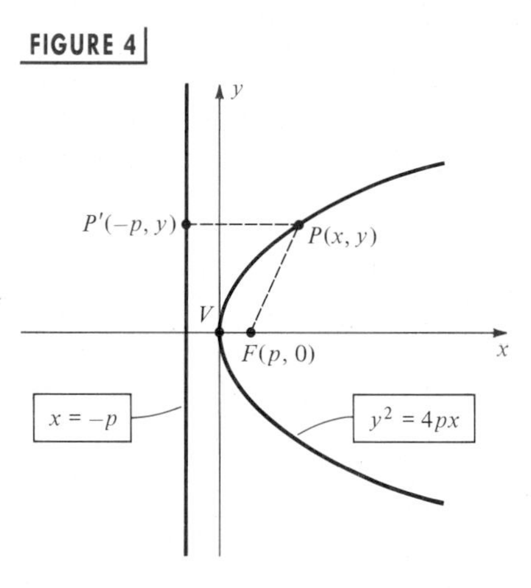

We have proved the following.

PARABOLAS WITH VERTEX (0, 0)

> The graph of each of the following equations is a parabola that has its vertex at the origin and has the indicated focus and directrix.
> (i) $x^2 = 4py$: focus $F(0, p)$, directrix $y = -p$.
> (ii) $y^2 = 4px$: focus $F(p, 0)$, directrix $x = -p$.

For convenience we often refer to *the parabola* $x^2 = 4py$ (or $y^2 = 4px$) instead of *the parabola with equation* $x^2 = 4py$ (or $y^2 = 4px$).

EXAMPLE ▪ 1

Find the focus and directrix of the parabola $y^2 = -6x$, and sketch its graph.

FIGURE 5

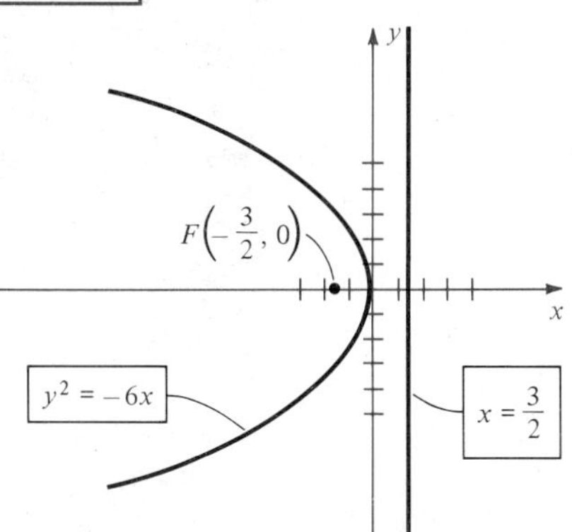

SOLUTION The equation $y^2 = -6x$ is of type (ii) in the preceding box with $4p = -6$ or, equivalently, $p = -\frac{3}{2}$. Hence the graph is a parabola that has its vertex at the origin and opens to the left. The focus and directrix are as follows:

$$\text{focus } F(p, 0): \quad F(-\tfrac{3}{2}, 0)$$

$$\text{directrix } x = -p: \quad x = \tfrac{3}{2}$$

The graph is sketched in Figure 5.

EXAMPLE ▪ 2

Find an equation of the parabola that has its vertex at the origin, opens upward, and passes through the point $P(-3, 7)$.

SOLUTION The general form of the equation is $x^2 = 4py$. If $P(-3, 7)$ is on the parabola, then $(-3, 7)$ is a solution of the equation. Hence we must have

$$(-3)^2 = 4p(7).$$

Solving for p gives us

$$p = \tfrac{9}{28}.$$

We now substitute for p in $x^2 = 4py$, obtaining

$$x^2 = \tfrac{9}{7}y$$

Multiplying by 7 gives us the equivalent equation $7x^2 = 9y$.

FIGURE 6

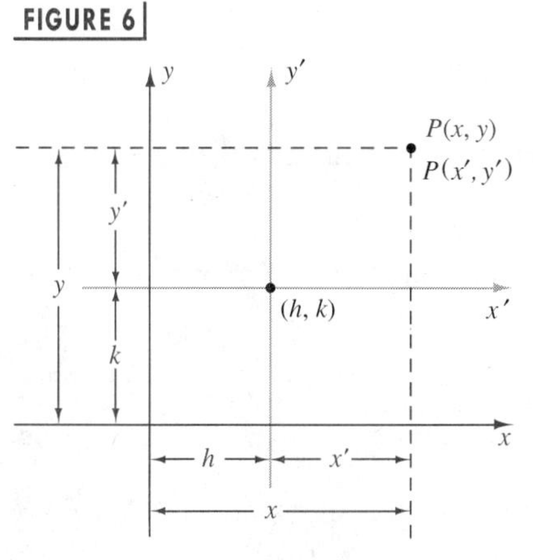

To extend our discussion to the case in which the vertex of the parabola is not at the origin, we use a **translation of axes**, as illustrated in Figure 6, where the x- and y-axes are shifted to positions, denoted by x' and y', that are parallel to their original positions. Every point P in the plane then has two different ordered pair representations: $P(x, y)$ in the xy-system and $P(x', y')$ in the $x'y'$-system. If the origin of the new $x'y'$-system has coordinates (h, k) in the xy-plane, as illustrated in Figure 6, we see that

$$x = x' + h \quad \text{and} \quad y = y' + k.$$

These formulas are true for all values of h and k. Equivalent formulas are

$$x' = x - h \quad \text{and} \quad y' = y - k.$$

This gives us the following.

TRANSLATION OF AXES FORMULAS

If (x, y) are the coordinates of a point P in an xy-plane and if (x', y') are the coordinates of P in an $x'y'$-plane with origin at the point (h, k) of the xy-plane, then:

$$\text{(i)}\quad x = x' + h, \quad y = y' + k$$

$$\text{(ii)}\quad x' = x - h, \quad y' = y - k$$

If, in the xy-plane, a certain collection of points is the graph of an equation in x and y, then to find an equation in x' and y' that has the same graph in the $x'y'$-plane, we substitute $x' + h$ for x and $y' + k$ for y. Conversely, if a set of points in the $x'y'$-plane is the graph of an equation in x' and y', then to find the corresponding equation in x and y, we substitute $x - h$ for x' and $y - k$ for y'.

As a simple illustration, the equation

$$(x')^2 + (y')^2 = r^2$$

has for its graph in the $x'y'$-plane a circle of radius r with center at the origin. Using translation of axes formulas (ii), an equation for this circle in the xy-plane is

$$(x - h)^2 + (y - k)^2 = r^2,$$

which is in agreement with the formula for a circle of radius r with center at $C(h, k)$ in the xy-plane.

As another illustration, we know that

$$(x')^2 = 4py'$$

is an equation of a parabola with vertex at the origin O' of the $x'y'$-plane. Using translation of axes formulas (ii), we see that

$$(x - h)^2 = 4p(y - k)$$

is an equation of the same parabola in the xy-plane with vertex $V(h, k)$. The focus is $F(h, k + p)$ and the directrix is $y = k - p$. Similarly, starting with $(y')^2 = 4px'$ gives us $(y - k)^2 = 4p(x - h)$.

We may summarize our discussion as follows.

PARABOLAS WITH VERTEX (h, k)

The graph of each of the following equations is a parabola that has vertex $V(h, k)$ and has the indicated focus and directrix.

(i) $(x - h)^2 = 4p(y - k)$: focus $F(h, k + p)$, directrix $y = k - p$.

(ii) $(y - k)^2 = 4p(x - h)$: focus $F(h + p, k)$, directrix $x = h - p$.

In each case the axis of the parabola is parallel to a coordinate axis. The parabola in (i) opens upward or downward, and the parabola in (ii) opens to the right or left. Typical graphs are sketched in Figure 7.

FIGURE 7

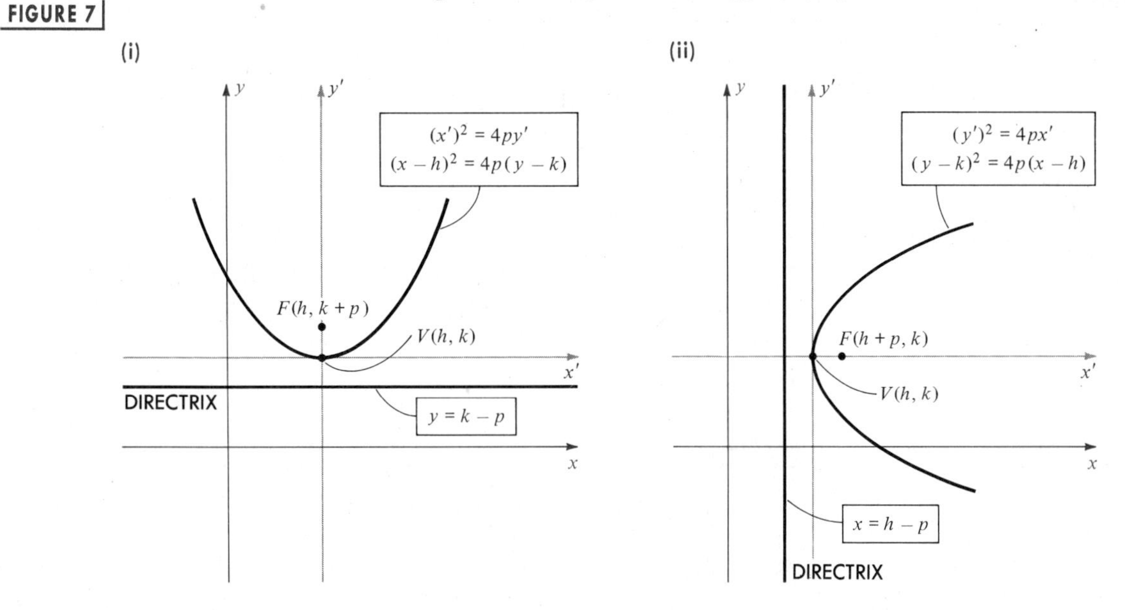

Squaring the left side of $(x - h)^2 = 4p(y - k)$ and simplifying leads to an equation of the form

$$y = ax^2 + bx + c$$

for real numbers a, b, and c. Conversely, the graph of $y = ax^2 + bx + c$ is a parabola with a vertical axis. As in Section 4.1, we may complete the square in x to find the vertex (see Examples 3 and 4 of Section 4.1).

Similarly, the equation $(y - k)^2 = 4p(x - h)$ may be written

$$x = ay^2 + by + c$$

for real numbers a, b, and c. Conversely, the last equation can be expressed in form (ii) by completing the square in y as illustrated in the following example. Hence, if $a \neq 0$, the graph of $x = ay^2 + by + c$ is a parabola with a horizontal axis.

EXAMPLE ▪ 3

Discuss and sketch the graph of $2x = y^2 + 8y + 22$.

SOLUTION The graph is a parabola with a horizontal axis. Writing

$$y^2 + 8y = 2x - 22,$$

we complete the square on the left by adding 16 to both sides:

$$y^2 + 8y + 16 = 2x - 6.$$
$$(y + 4)^2 = 2(x - 3).$$

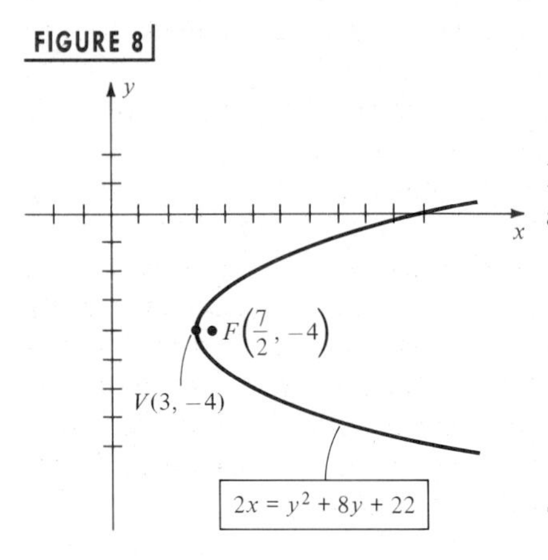

FIGURE 8

Referring to form (ii) in the last box, we see that $h = 3$, $k = -4$, $4p = 2$, and $p = \frac{1}{2}$. Thus the following statements are true:

The parabola has a horizontal axis.
The vertex $V(h, k)$ is $V(3, -4)$.
The focus $F(h + p, k)$ is $F(\frac{7}{2}, -4)$.
The directrix $x = h - p$ is $x = \frac{5}{2}$.

The parabola is sketched in Figure 8.

EXAMPLE ▪ 4

Find an equation of the parabola with vertex $V(-4, 2)$ and directrix $y = 5$.

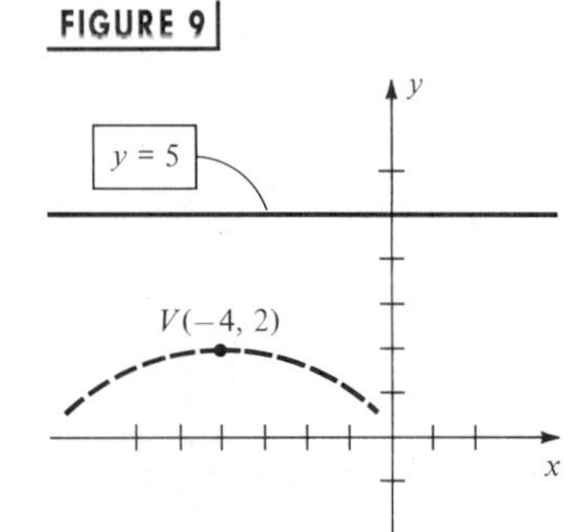

FIGURE 9

SOLUTION The vertex and directrix are shown in Figure 9. The dashes indicate a possible position for the parabola. An equation of the parabola is

$$(x - h)^2 = 4p(y - k)$$

with $h = -4$, $k = 2$, and $p = -3$. This gives us

$$(x + 4)^2 = -12(y - 2).$$

The last equation can be expressed in the form $y = ax^2 + bx + c$ as follows:

$$x^2 + 8x + 16 = -12y + 24$$
$$12y = -x^2 - 8x + 8$$
$$y = -\tfrac{1}{12}x^2 - \tfrac{2}{3}x + \tfrac{2}{3}$$

FIGURE 10
Path of a baseball

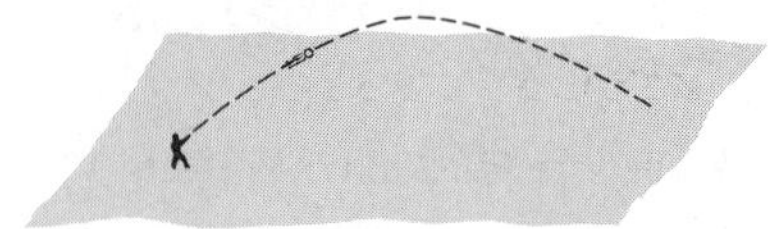

If, as shown in Figure 10, a baseball player throws a ball in a nonvertical direction, and if gravity is the only force acting on the ball (that is, air resistance and other outside factors are negligible), then the path of the ball is parabolic. We can use this fact to determine where the ball will land, to find its maximum height, and so on. Problems of this type were considered in Section 4.1.

*A *tangent line* to a parabola is a line that has exactly one point in common with the parabola.

FIGURE 11

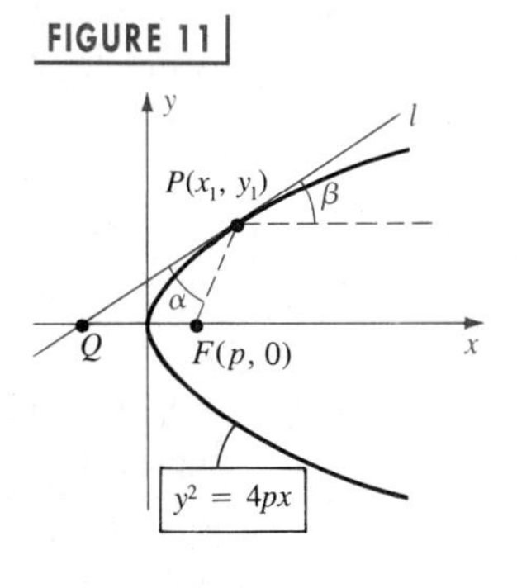

An important property is associated with a tangent line to a parabola*. Suppose l is the tangent line at a point $P(x_1, y_1)$ on the graph of $y^2 = 4px$, and let F be the focus. As in Figure 11, let α denote the angle between l and the line segment FP, and let β denote the angle between l and the horizontal half-line with endpoint P. In Exercise 40, you are asked to prove that $\alpha = \beta$. This *reflective property* has many applications. For example, the shape of the mirror in a searchlight is obtained by revolving a parabola about its axis. If a light source is placed at F, then by a law of physics (*the angle of reflection equals the angle of incidence*), a beam of light will be reflected along a line parallel to the axis (see Figure 12(i)). The same principle is employed in the construction of mirrors for telescopes or solar ovens—a beam of light coming toward the parabolic mirror, and parallel to the axis, will be reflected into the focus (see Figure 12(ii)). Antennas for radar systems, radio telescopes, and field microphones used at football games also make use of this property.

FIGURE 12

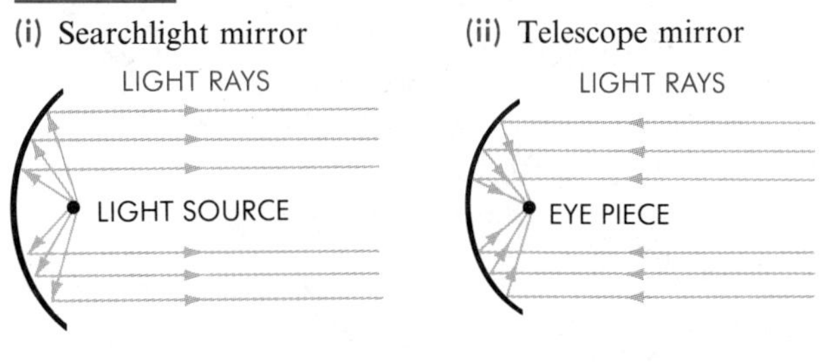

11.2 EXERCISES

Exer. 1–20: Find the vertex, focus, and directrix of the parabola, and sketch its graph.

1 $x^2 = -12y$

2 $y^2 = \frac{1}{2}x$

3 $2y^2 = -3x$

4 $x^2 = -3y$

5 $8x^2 = y$

6 $y^2 = -100x$

7 $(x + 2)^2 = -8(y - 1)$

8 $(x - 3)^2 = \frac{1}{2}(y + 1)$

9 $(y - 2)^2 = \frac{1}{4}(x - 3)$

10 $(y + 1)^2 = -12(x + 2)$

11 $y^2 - 12 = 12x$

12 $y = 40x - 97 - 4x^2$

13 $y = x^2 - 4x + 2$

14 $y = 8x^2 + 16x + 10$

15 $y^2 - 4y - 2x - 4 = 0$

16 $y^2 + 14y + 4x + 45 = 0$

17 $4x^2 + 40x + y + 106 = 0$

18 $y^2 - 20y + 100 = 6x$

19 $x^2 + 20y = 10$

20 $4x^2 + 4x + 4y + 1 = 0$

Exer. 21–30: Find an equation of the parabola that satisfies the given conditions.

21 Focus $F(2, 0)$, directrix $x = -2$

22 Focus $F(0, -4)$, directrix $y = 4$

23 Focus $F(6, 4)$, directrix $y = -2$

24 Focus $F(-3, -2)$, directrix $y = 1$

25 Vertex $V(3, -5)$, directrix $x = 2$

26 Vertex $V(-2, 3)$, directrix $y = 5$

27 Vertex $V(-1, 0)$, Focus $F(-4, 0)$

28 Vertex $V(1, -2)$, Focus $F(1, 0)$

29 Vertex at the origin, symmetric to the y-axis, and passing through the point $(2, -3)$

30 Vertex $V(-3,5)$, axis parallel to the x-axis, and passing through the point $(5,9)$

31 A searchlight reflector is designed so that a cross section through its axis is a parabola and the light source is at the focus. Find the focus if the reflector is 3 feet across at the opening and 1 foot deep.

Exer. 32–33: Find an equation of the parabola that has a horizontal axis and passes through the given points.

32 $A(2,1)$, $B(6,2)$, $C(12,-1)$

33 $A(-1,1)$, $B(11,-2)$, $C(5,-1)$

34 Find an equation of a parabola that has a vertical axis and passes through $A(2,5)$, $B(-2,-3)$, and $C(1,6)$.

35 A paraboloid of revolution, which is formed by revolving a parabola about its axis, is the basic shape of a wide variety of collectors and reflectors. Shown in the figure is a (finite) paraboloid of height h and radius r.

(a) The focal length of the paraboloid is the distance p between the vertex and the focus of the parabola. Express p in terms of r and h.

(b) A reflector is to be constructed with a focal length of 10 feet and a height of 5 feet. Find the radius of the reflector.

EXERCISE 35

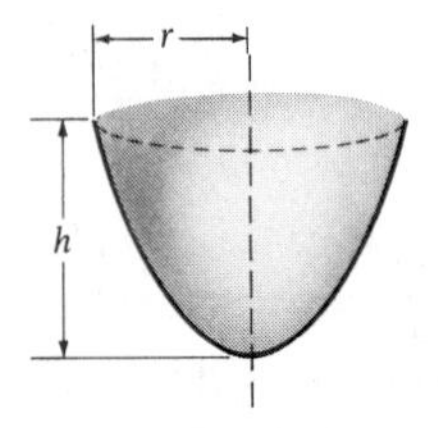

36 The parabola $y^2 = 4p(x + p)$ has its focus at the origin and the x-axis as its axis. By assigning different values to p, a family of confocal parabolas is obtained (see figure). Families of this type occur in the theories of electricity and magnetism. Show that the family includes exactly two parabolas through any point $P(x_1, y_1)$ for $y_1 \neq 0$.

EXERCISE 36

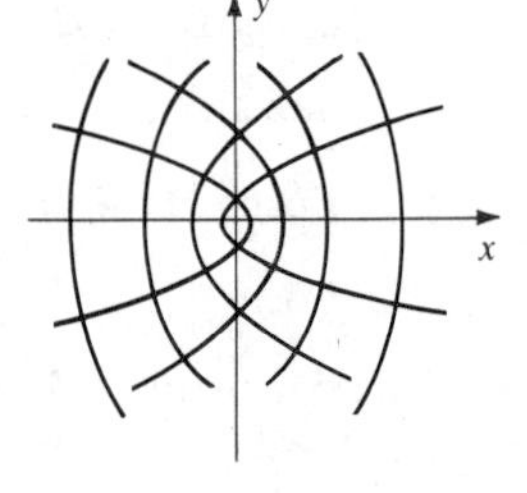

37 Refer to Exercise 35. A radio telescope has the shape of a paraboloid of revolution with diameter $2a$ and focal length p. From calculus, the surface area S available for collecting radio waves is

$$S = \frac{8\pi p^2}{3}\left[\left(1 + \frac{a^2}{4p^2}\right)^{3/2} - 1\right].$$

One of the largest radio telescopes, located in Jodrell Bank, Cheshire, England, has diameter 250 feet and focal length 50 feet. Approximate S to the nearest square foot.

38 A cylindrical container is partially filled with a liquid such as mercury. The container is then rotated about its axis so that the angular speed of each cross section is ω radians/second (see Example 7 of Section 6.1). From physics, the function f whose graph generates the inside surface of the liquid (see figure) is given by

$$f(x) = \tfrac{1}{64}\omega^2 x^2 + k$$

for some constant k. Determine the angular speed ω that will result in a focal length of 2 feet.

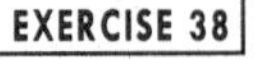
EXERCISE 38

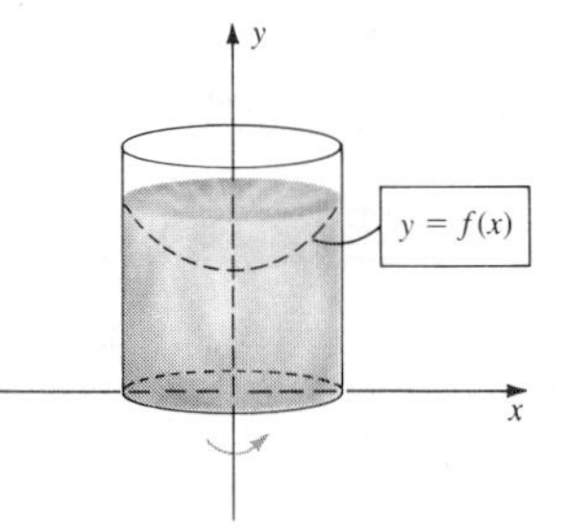

39 Let m and b be nonzero real numbers. Show that if the line $y = mx + b$ intersects the parabola $y^2 = 4px$ in only one point, then $p = mb$. Conclude that the line tangent to $y^2 = 4px$ at the point $P(x_1, y_1)$ on the parabola has slope $y_1/(2x_1)$.

40 Establish the reflective property of the parabola. (*Hint:* Show that $d(Q, F) = d(F, P)$ in Figure 11 and use the result in Exercise 39.)

11.3 ELLIPSES

An ellipse may be defined as follows. (The word *foci* is the plural of *focus*.)

DEFINITION

> An **ellipse** is the set of all points in a plane, the sum of whose distances from two fixed points (the **foci**) in the plane is constant.

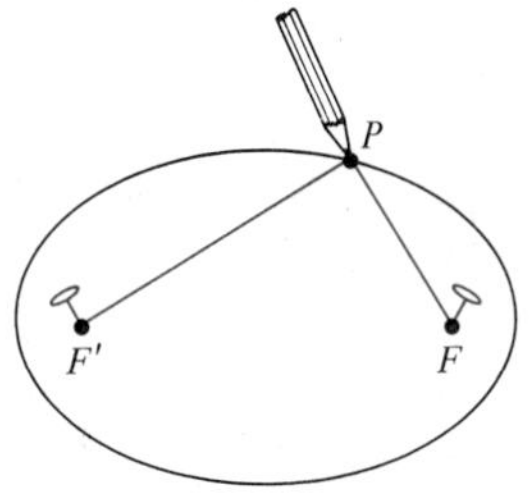

An ellipse can easily be constructed on paper. Begin by inserting two pushpins in the paper at arbitrary points F and F', and fasten the ends of a piece of string to the pins. If the string is now looped around a pencil and drawn taut at point P, as in Figure 13, then moving the pencil and at the same time keeping the string taut, the sum of the distances $d(F, P)$ and $d(F', P)$ is the length of the string and hence is constant. The pencil will trace out an ellipse with foci at F and F'. By changing the positions of F and F', but keeping the length of string fixed, we can vary the shape of the ellipse considerably. If F and F' are far apart, so that $d(F, F')$ is almost the same as the length of the string, then the ellipse is flat. If $d(F, F')$ is close to zero, the ellipse is almost circular. If $F = F'$, we obtain a circle with center F.

FIGURE 13

By introducing suitable coordinate systems, we may derive simple equations for ellipses. Let us choose the x-axis as the line through the two foci F and F', with the origin at the midpoint of the segment $F'F$. This point is called the **center** of the ellipse. If F has coordinates $(c, 0)$ with $c > 0$, then, as shown in Figure 14, F' has coordinates $(-c, 0)$. Hence the distance between F and F' is $2c$. The constant sum of the distances of P from F and F' will be denoted by $2a$. To obtain points that are not on the x-axis, we must have $2a > 2c$, that is, $a > c$. By definition, $P(x, y)$ is on the ellipse if and only if

FIGURE 14

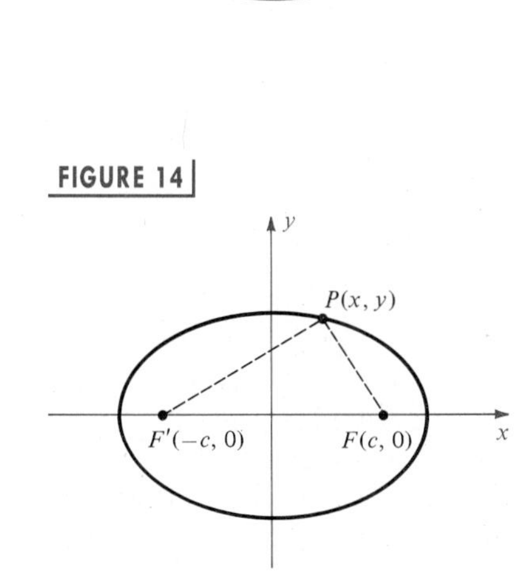

$$d(P, F) + d(P, F') = 2a$$

or, by the distance formula,

$$\sqrt{(x - c)^2 + (y - 0)^2} + \sqrt{(x + c)^2 + (y - 0)^2} = 2a.$$

Writing the preceding equation as

$$\sqrt{(x - c)^2 + y^2} = 2a - \sqrt{(x + c)^2 + y^2}$$

and squaring both sides, we obtain

$$x^2 - 2cx + c^2 + y^2 = 4a^2 - 4a\sqrt{(x+c)^2 + y^2} + x^2 + 2cx + c^2 + y^2,$$

which simplifies to

$$a\sqrt{(x+c)^2 + y^2} = a^2 + cx.$$

Squaring both sides gives us

$$a^2(x^2 + 2cx + c^2 + y^2) = a^4 + 2a^2cx + c^2x^2,$$

which may be written

$$x^2(a^2 - c^2) + a^2y^2 = a^2(a^2 - c^2).$$

Dividing both sides by $a^2(a^2 - c^2)$ leads to

$$\frac{x^2}{a^2} + \frac{y^2}{a^2 - c^2} = 1.$$

Recalling that $a > c$ and therefore $a^2 - c^2 > 0$, we let

$$b = \sqrt{a^2 - c^2}, \quad \text{or} \quad b^2 = a^2 - c^2.$$

This gives us the equation

$$\frac{x^2}{a^2} + \frac{y^2}{b^2} = 1.$$

Since $c > 0$ and $b^2 = a^2 - c^2$, it follows that $a^2 > b^2$ and hence $a > b$.

We have shown that the coordinates of every point (x, y) on the ellipse in Figure 15 satisfy the equation $(x^2/a^2) + (y^2/b^2) = 1$. Conversely, if (x, y) is a solution of this equation, then by reversing the preceding steps we see that the point (x, y) is on the ellipse.

FIGURE 15

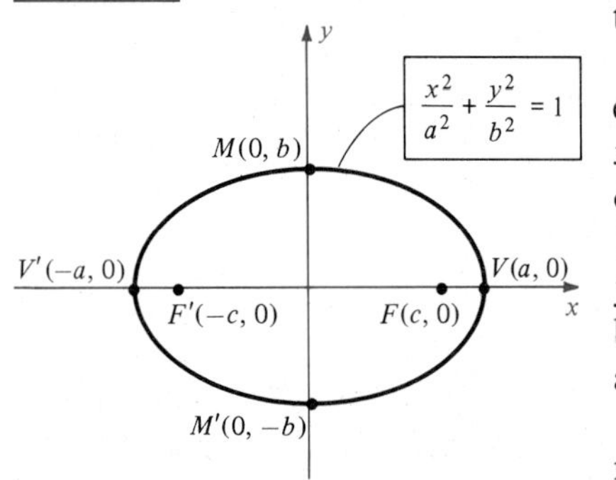

We may find the x-intercepts of the ellipse by letting $y = 0$ in the equation. Doing so gives us $x^2/a^2 = 1$, or $x^2 = a^2$, and consequently the x-intercepts are a and $-a$. The corresponding points $V(a, 0)$ and $V'(-a, 0)$ on the graph are the **vertices** of the ellipse (see Figure 15). The line segment $V'V$ is the **major axis**. Similarly, letting $x = 0$ in the equation, we obtain $y^2/b^2 = 1$, or $y^2 = b^2$. Hence the y-intercepts are b and $-b$. The segment between $M'(0, -b)$ and $M(0, b)$ is the **minor axis** of the ellipse. The major axis is always longer than the minor axis, since $a > b$.

Applying tests for symmetry, we see that the ellipse is symmetric with respect to the x-axis, the y-axis, and the origin.

The preceding discussion may be summarized as follows.

THEOREM

> The graph of the equation
>
> $$\frac{x^2}{a^2} + \frac{y^2}{b^2} = 1,$$
>
> for $a^2 > b^2$, is an ellipse with vertices $(\pm a, 0)$. The endpoints of the minor axis are $(0, \pm b)$. The foci are $(\pm c, 0)$ with $c^2 = a^2 - b^2$.

EXAMPLE ▪ 1

Discuss and sketch the graph of $4x^2 + 18y^2 = 36$.

SOLUTION To obtain the form in the theorem we divide both sides of the equation by 36 and simplify. This leads to

$$\frac{x^2}{9} + \frac{y^2}{2} = 1,$$

FIGURE 16

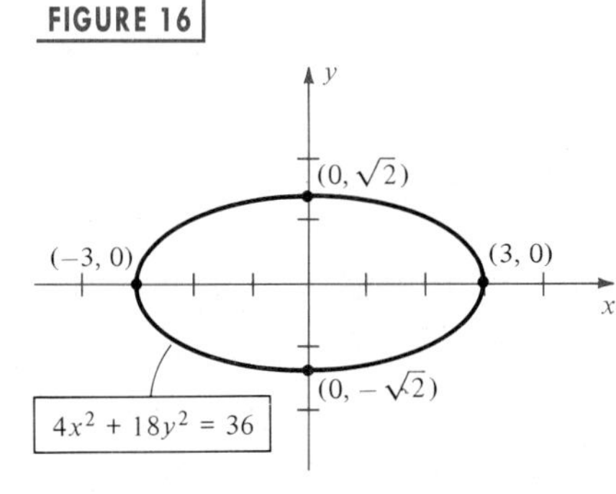

which is in the proper form with $a^2 = 9$ and $b^2 = 2$. Thus $a = 3$, $b = \sqrt{2}$, and hence the endpoints of the major axis are $(\pm 3, 0)$ and the endpoints of the minor axis are $(0, \pm\sqrt{2})$. Since

$$c^2 = a^2 - b^2 = 9 - 2 = 7, \quad \text{or} \quad c = \sqrt{7},$$

the foci are $(\pm\sqrt{7}, 0)$. The graph is sketched in Figure 16.

EXAMPLE ▪ 2

Find an equation of the ellipse with vertices $(\pm 4, 0)$ and foci $(\pm 2, 0)$.

SOLUTION Using the notation of the theorem, $a = 4$ and $c = 2$. Since $c^2 = a^2 - b^2$, we see that $b^2 = a^2 - c^2 = 16 - 4 = 12$. This gives us

$$\frac{x^2}{16} + \frac{y^2}{12} = 1.$$

We sometimes choose the major axis of the ellipse along the y-axis. If the foci are $(0, \pm c)$, then by the same type or argument used previously, we obtain the following.

THEOREM

> The graph of the equation
>
> $$\frac{x^2}{b^2} + \frac{y^2}{a^2} = 1,$$
>
> for $a^2 > b^2$, is an ellipse with vertices $(0, \pm a)$. The endpoints of the minor axis are $(\pm b, 0)$. The foci are $(0, \pm c)$ with $c^2 = a^2 - b^2$.

FIGURE 17

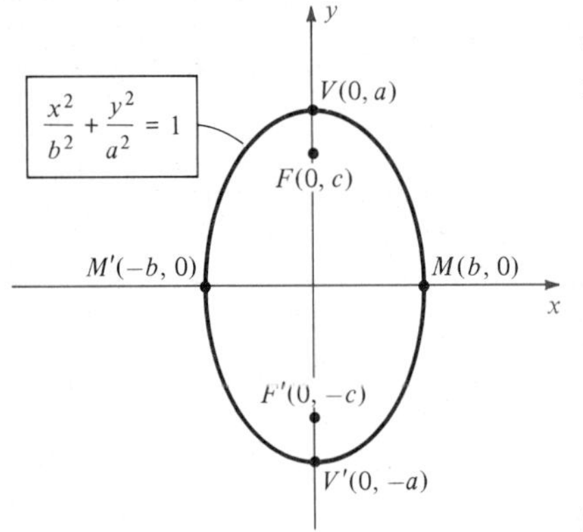

A typical graph is sketched in Figure 17.

The preceding discussion shows that an equation of an ellipse with center at the origin and foci on a coordinate axis can always be written in the form

$$\frac{x^2}{p} + \frac{y^2}{q} = 1, \quad \text{or} \quad qx^2 + py^2 = pq$$

with p and q positive and $p \neq q$. If $p > q$ the major axis is on the x-axis, and if $q > p$ the major axis is on the y-axis. It is unnecessary to memorize these facts, since in any given problem the major axis can be determined by examining the x- and y-intercepts.

EXAMPLE ▪ 3

Sketch the graph of $9x^2 + 4y^2 = 25$.

FIGURE 18

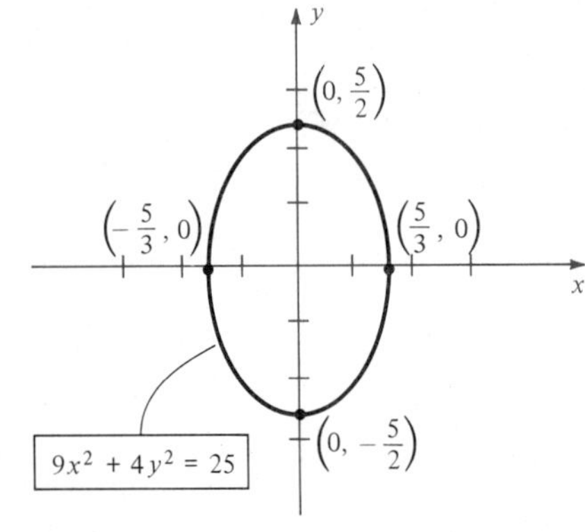

SOLUTION The graph is an ellipse with center at the origin and foci on one of the coordinate axes. To find x-intercepts, we let $y = 0$, obtaining $9x^2 = 25$, or $x = \pm\frac{5}{3}$. Similarly, to find the y-intercepts, we let $x = 0$, obtaining $4y^2 = 25$, or $y = \pm\frac{5}{2}$. This enables us to sketch the ellipse (see Figure 18). Since $\frac{5}{3} < \frac{5}{2}$, the major axis is on the y-axis.

To find the foci we first calculate

$$c^2 = a^2 - b^2 = (\tfrac{5}{2})^2 - (\tfrac{5}{3})^2 = \tfrac{125}{36}.$$

Thus $c = 5\sqrt{5}/6$ and the foci are $(0, \pm 5\sqrt{5}/6)$.

We can use the translation of axes formulas to extend our work to an ellipse with center at any point $C(h, k)$ in the xy-plane. For example, since the graph of

$$\frac{(x')^2}{a^2} + \frac{(y')^2}{b^2} = 1$$

FIGURE 19

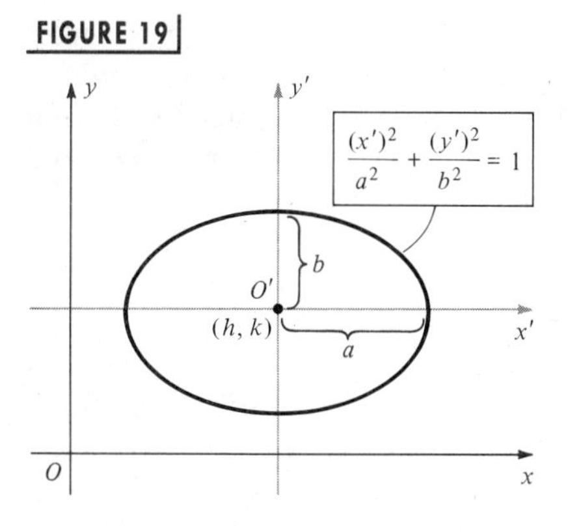

is an ellipse with center at O' in an $x'y'$-plane (see Figure 19), its equation relative to the xy-coordinate system is

$$\frac{(x-h)^2}{a^2} + \frac{(y-k)^2}{b^2} = 1.$$

Squaring the indicated terms in this equation and simplifying gives us an equation of the form

$$Ax^2 + Cy^2 + Dx + Ey + F = 0$$

such that the coefficients are real numbers and A and C are positive. Conversely, if we start with such an equation, then by completing squares we can obtain a form that displays the center of the ellipse and the lengths of the major and minor axes. This technique is illustrated in the next example.

EXAMPLE ▪ 4

Discuss and sketch the graph of $16x^2 + 9y^2 + 64x - 18y - 71 = 0$.

SOLUTION We begin by writing the equation in the form

$$16(x^2 + 4x \qquad) + 9(y^2 - 2y \qquad) = 71.$$

Next, we complete the squares for the expressions within parentheses:

$$16(x^2 + 4x + 4) + 9(y^2 - 2y + 1) = 71 + 64 + 9.$$

By adding 4 to the expression within the first parentheses we have added 64 to the left-hand side of the equation, and hence must compensate by adding 64 to the right-hand side. Similarly, by adding 1 to the expression within the second parentheses, we have added 9 to the left side and, consequently, we must also add 9 to the right side. The last equation may be written

$$16(x + 2)^2 + 9(y - 1)^2 = 144.$$

Dividing by 144, we obtain

$$\frac{(x+2)^2}{9} + \frac{(y-1)^2}{16} = 1,$$

which is of the form $$\frac{(x')^2}{9} + \frac{(y')^2}{16} = 1$$

with $x' = x + 2$ and $y' = y - 1$. This corresponds to letting $h = -2$ and $k = 1$ in the translation of axes formulas.

FIGURE 20

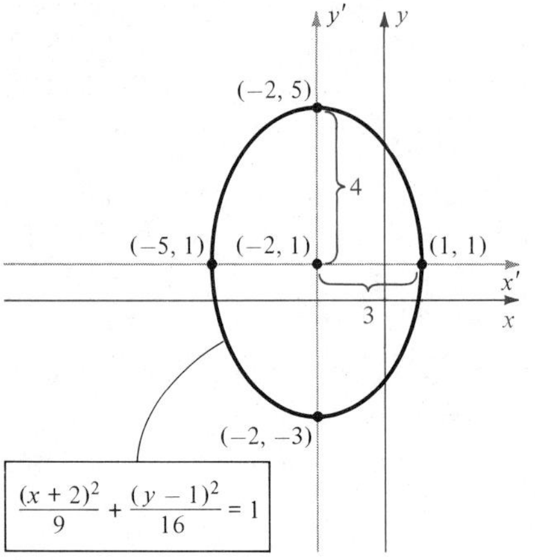

The graph of the equation $(x')^2/9 + (y')^2/16 = 1$ is an ellipse with center at the origin O' in the $x'y'$-plane and major axis on the y'-axis. It follows that the graph of the given equation is an ellipse with center $C(-2, 1)$ in the xy-plane and major axis on the vertical line $x = -2$. Using $a = 4$ and $b = 3$ gives us the ellipse in Figure 20.

To find the foci, we first calculate

$$c^2 = a^2 - b^2 = 16 - 9 = 7.$$

The distance from the center of the ellipse to the foci is $c = \sqrt{7}$. Since the center is $(-2, 1)$, the foci are $(-2, 1 \pm \sqrt{7})$.

Ellipses can be very flat or almost circular. To obtain information about the "roundness" of an ellipse we sometimes use the *eccentricity e* (not to be confused with the base of the natural logarithms). Eccentricity is defined as follows, with a, b, and c having the same meanings as before.

DEFINITION

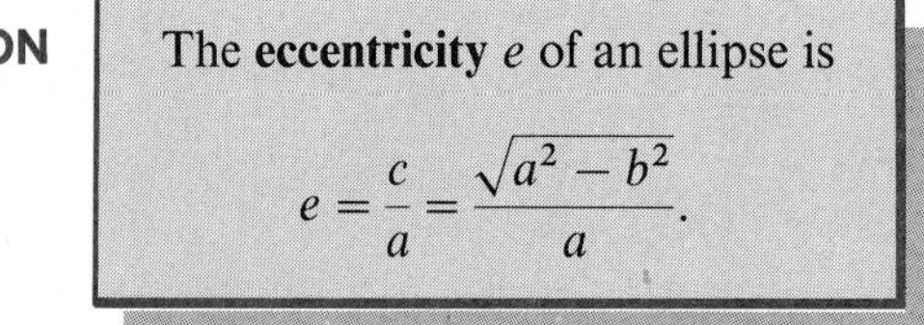
The **eccentricity** e of an ellipse is

$$e = \frac{c}{a} = \frac{\sqrt{a^2 - b^2}}{a}.$$

Consider the ellipse $(x^2/a^2) + (y^2/b^2) = 1$, and suppose that the length $2a$ of the major axis is fixed, and the length $2b$ of the minor axis is variable. Since $\sqrt{a^2 - b^2} < a$, we see that $0 < e < 1$. If $e \approx 1$, then $\sqrt{a^2 - b^2} \approx a$, and hence $b \approx 0$. In this case the ellipse is very flat. If $e \approx 0$, then $\sqrt{a^2 - b^2} \approx 0$ and $a \approx b$. In this case the ellipse is almost circular.

After many years of analyzing an enormous amount of empirical data, the German astronomer Johannes Kepler (1571–1630) formulated three laws that describe the motion of planets about the sun. Kepler's first law states that the orbit of each planet in the solar system is an ellipse with the sun at one focus. Most of these orbits are almost circular, and hence their corresponding eccentricities are close to 0. To illustrate, for Earth, $e \approx 0.017$; for Mars, $e \approx 0.093$; and for Uranus, $e \approx 0.046$. The orbits of Mercury and Pluto are less circular, with eccentricities of 0.206 and 0.249, respectively.

Many comets have elliptical orbits with the sun at a focus. In this case the eccentricity e is close to 1 and the ellipse is very flat. In the next example we use the **astronomical unit** (1 AU = 92,000,000 miles) to specify large distances.

EXAMPLE ▪ 5

Halley's comet has an elliptical orbit with eccentricity $e = 0.967$. Measuring distance in astronomical units, the closest that Halley's comet comes to the sun is 0.587 AU. Approximate the maximum distance of the comet from the sun, to the nearest 0.1 AU.

FIGURE 21
HALLEY'S COMET

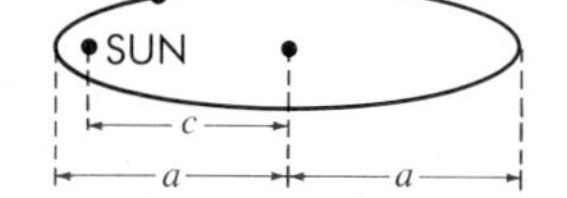

SOLUTION Figure 21 illustrates the orbit of the comet, where c is the distance from the center of the ellipse to a focus (the sun) and $2a$ is the length of the major axis.

Since $a - c$ is the minimum distance between the sun and the comet, we have (in AU)

$$a - c = 0.587, \quad \text{or} \quad a = c + 0.587.$$

Since $e = c/a = 0.967$,

$$c = 0.967a = 0.967(c + 0.587)$$
$$c \approx 0.967c + 0.568.$$

Thus

$$0.033c \approx 0.568 \quad \text{and} \quad c \approx \frac{0.568}{0.033} \approx 17.2.$$

Consequently,

$$a = c + 0.587$$
$$a \approx 17.2 + 0.587 \approx 17.8$$

and the maximum distance between the sun and the comet is

$$a + c \approx 17.8 + 17.2, \quad \text{or} \quad a + c \approx 35.0 \text{ AU}.$$

An ellipse has a reflective property analogous to that of the parabola discussed at the end of Section 11.2. To illustrate, let l denote the tangent line at a point P on an ellipse with foci F and F', as shown in Figure 22. If α is the angle between $F'P$ and l, and if β is the angle between FP and l, it can be shown that $\alpha = \beta$. Thus, if a ray of light or sound emanates from one focus, it is reflected to the other focus. This property is used in the design of certain types of optical equipment. It is also evident in *whispering galleries*; that is, rooms with elliptically shaped ceilings, in which a person who whispers at one focus can be heard at the other focus. Some examples of whispering galleries may be found in the Rotunda of the Capitol Building in Washington, D.C., and in the Mormon Tabernacle in Salt Lake City.

FIGURE 22

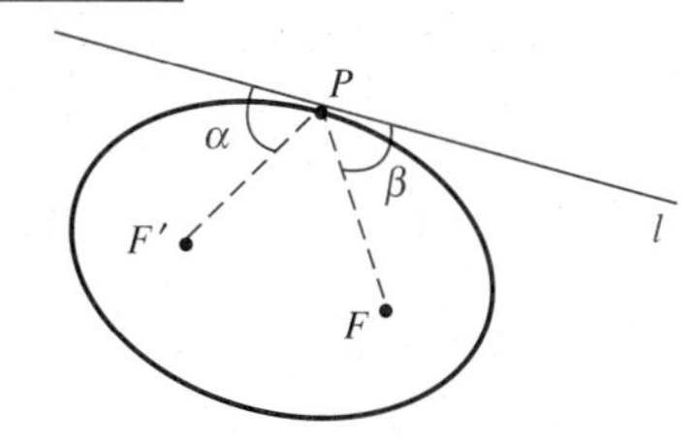

11.3 EXERCISES

Exer. 1–16: Find the vertices and foci of the ellipse, and sketch its graph.

1 $\dfrac{x^2}{9} + \dfrac{y^2}{4} = 1$

2 $\dfrac{x^2}{25} + \dfrac{y^2}{16} = 1$

3 $4x^2 + y^2 = 16$

4 $y^2 + 9x^2 = 9$

5 $5x^2 + 2y^2 = 10$

6 $\frac{1}{2}x^2 + 2y^2 = 8$

7 $4x^2 + 25y^2 = 1$

8 $10y^2 + x^2 = 5$

9 $\dfrac{(x-3)^2}{16} + \dfrac{(y+4)^2}{9} = 1$

10 $\dfrac{(x+2)^2}{25} + \dfrac{(y-3)^2}{4} = 1$

11 $4x^2 + 9y^2 - 32x - 36y + 64 = 0$

12 $x^2 + 2y^2 + 2x - 20y + 43 = 0$

13 $9x^2 + 16y^2 + 54x - 32y - 47 = 0$

14 $4x^2 + 9y^2 + 24x + 18y + 9 = 0$

15 $25x^2 + 4y^2 - 250x - 16y + 541 = 0$

16 $4x^2 + y^2 = 2y$

Exer. 17–28: Find an equation for the ellipse that has its center at the origin and satisfies the given conditions.

17 Vertices $V(\pm 8, 0)$, foci $F(\pm 5, 0)$

18 Vertices $V(0, \pm 7)$, foci $F(0, \pm 2)$

19 Vertices $V(0, \pm 5)$, minor axis of length 3

20 Foci $F(\pm 3, 0)$, minor axis of length 2

21 Vertices $V(0, \pm 6)$, passing through $(3, 2)$

22 Center at the origin, passing through $(2, 3)$ and $(6, 1)$

23 Eccentricity $\frac{3}{4}$, vertices $V(0, \pm 4)$

24 Eccentricity $\frac{1}{2}$, center at the origin, vertices on the x-axis, passing through $(1, 3)$

25 x-intercepts ± 2, y-intercepts $\pm \frac{1}{3}$

26 x-intercepts $\pm \frac{1}{2}$, y-intercepts ± 4

27 Horizontal major axis of length 8, minor axis of length 5

28 Vertical major axis of length 7, minor axis of length 6

Exer. 29–30: Find the points of intersection of the graphs of the equations. Sketch both graphs on the same coordinate plane, and show the points of intersection.

29 $\begin{cases} x^2 + 4y^2 = 20 \\ x + 2y = 6 \end{cases}$

30 $\begin{cases} x^2 + 4y^2 = 36 \\ x^2 + y^2 = 12 \end{cases}$

31 An arch of a bridge is semielliptical with major axis horizontal. The base of the arch is 30 feet across and the highest part of the arch is 10 feet above the horizontal roadway, as shown in the figure. Find the height of the arch 6 feet from the center of the base.

EXERCISE 31

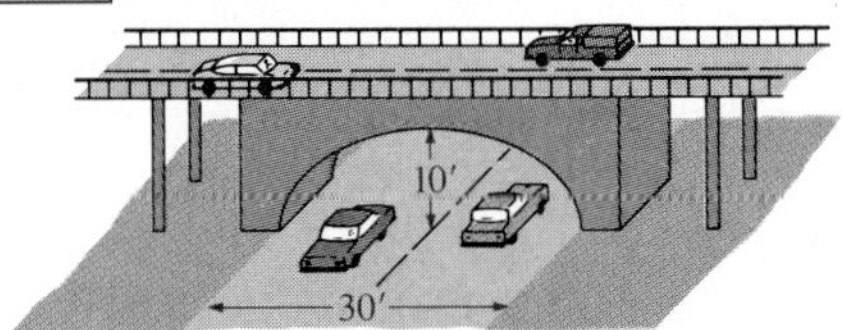

32 The planet Mercury travels in an elliptical orbit that has eccentricity 0.206 and semimajor axis 0.387 AU. Find the maximum and minimum distances between Mercury and the sun.

33 Assume that the length of the major axis of the earth's orbit is 186,000,000 miles and the eccentricity is 0.017. Find, to the nearest 1000 miles, the maximum and minimum distances between the earth and the sun.

34 An ellipse has a vertex at the origin and foci $F_1(p, 0)$ and $F_2(p + 2c, 0)$, as shown in the figure. If the focus at F_1 is fixed and (x, y) is on the ellipse, show that y^2 approaches $4px$ as $c \to \infty$. Thus, a parabola may be considered as an ellipse with "one focus at infinity."

EXERCISE 34

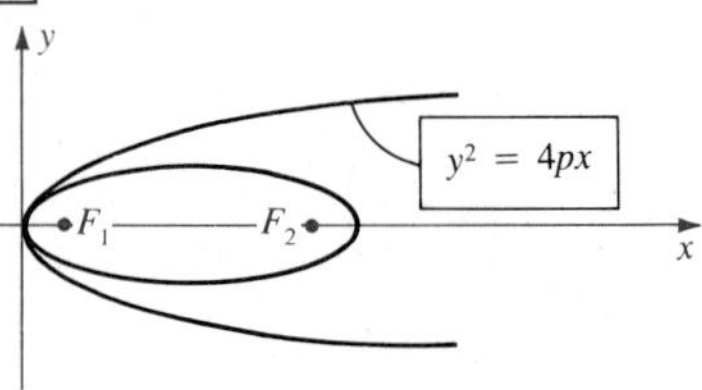

11.4 HYPERBOLAS

The definition of a hyperbola is similar to that of an ellipse. The only change is that instead of using the *sum* of distances from two fixed points, we use the *difference*.

DEFINITION

A **hyperbola** is the set of all points in a plane, the difference of whose distances from two fixed points in the plane (the **foci**) is a positive constant.

FIGURE 23

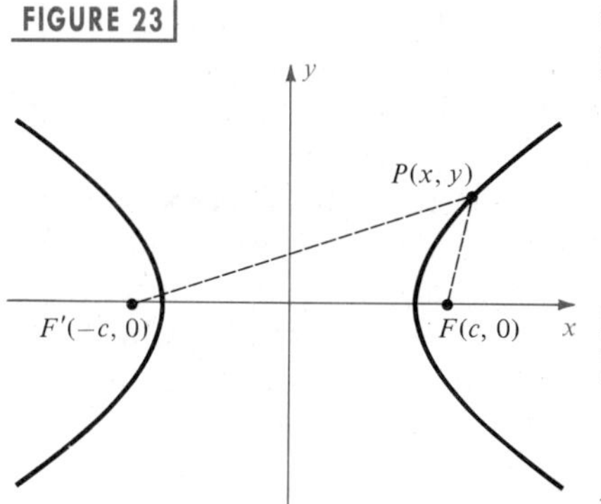

To find a simple equation for a hyperbola, we choose a coordinate system with foci at $F(c, 0)$ and $F'(-c, 0)$, and denote the (constant) distance by $2a$. Referring to Figure 23, we see that a point $P(x, y)$ is on the hyperbola if and only if either one of the following is true:

$$d(P, F') - d(P, F) = 2a$$

$$d(P, F) - d(P, F') = 2a.$$

For hyperbolas (unlike ellipses), we need $a < c$ to obtain points on the hyperbola that are not on the x-axis, for if P is such a point, then from Figure 23, we see that

$$d(P, F) < d(F', F) + d(P, F')$$

because the length of one side of a triangle is always less than the sum of the lengths of the other two sides. Similarly,

$$d(P, F') < d(F', F) + d(P, F).$$

Equivalent forms for the previous two inequalities are

$$d(P, F) - d(P, F') < d(F', F),$$

$$d(P, F') - d(P, F) < d(F', F).$$

Since the differences on the left-hand sides of these inequalities both equal $2a$, and since $d(F', F) = 2c$, the last two inequalities imply that $2a < 2c$, or $a < c$.

The equations $d(P, F') - d(P, F) = 2a$ and $d(P, F) - d(P, F') = 2a$ may be replaced by the single equation

$$|d(P, F) - d(P, F')| = 2a.$$

Using the distance formula to find $d(P, F)$ and $d(P, F')$, we obtain an equation of the hyperbola:

$$\left|\sqrt{(x-c)^2+(y-0)^2}-\sqrt{(x+c)^2+(y-0)^2}\right| = 2a.$$

Employing the type of simplification procedure that we used to derive an equation for an ellipse, we can rewrite the preceding equation as

$$\frac{x^2}{a^2}-\frac{y^2}{c^2-a^2}=1.$$

If, in this equation, we let

$$b^2 = c^2 - a^2 \quad \text{with } b > 0,$$

we obtain

$$\frac{x^2}{a^2}-\frac{y^2}{b^2}=1.$$

We have shown that the coordinates of every point (x, y) on the hyperbola in Figure 23 satisfy the last equation. Conversely, if (x, y) is a solution of that equation, then by reversing the preceding steps we see that the point (x, y) is on the hyperbola.

By tests for symmetry, the hyperbola is symmetric with respect to both axes and the origin. The x-intercepts are $\pm a$. The corresponding points $V(a, 0)$ and $V'(-a, 0)$ are the **vertices**, and the line segment $V'V$ is the **transverse axis** of the hyperbola. The origin is the **center** of the hyperbola. The graph has no y-intercept, since the equation $-y^2/b^2 = 1$ has no solution.

The preceding discussion may be summarized as follows.

THEOREM

The graph of the equation

$$\frac{x^2}{a^2}-\frac{y^2}{b^2}=1$$

is a hyperbola with vertices $(\pm a, 0)$. The foci are $(\pm c, 0)$, with $c^2 = a^2 + b^2$.

If we solve the equation $(x^2/a^2) - (y^2/b^2) = 1$ for y, we obtain

$$y = \pm\frac{b}{a}\sqrt{x^2-a^2}.$$

There are no points (x, y) on the graph if $x^2 - a^2 < 0$, or, equivalently, $-a < x < a$. There *are* points $P(x, y)$ on the graph if $x \geq a$ or $x \leq -a$.

If $x \geq a$, we may write the equation $y = \pm(b/a)\sqrt{x^2 - a^2}$ in the form

$$y = \pm\frac{b}{a}\sqrt{x^2\left(1 - \frac{a^2}{x^2}\right)} = \pm\frac{b}{a}x\sqrt{1 - \frac{a^2}{x^2}}.$$

If x is large (in comparison to a), then $1 - (a^2/x^2) \approx 1$, and hence the y-coordinate of the point $P(x, y)$ on the hyperbola is close to either $(b/a)x$ or $-(b/a)x$. Thus the point $P(x, y)$ is close to the line $y = (b/a)x$ when y is positive, or close to the line $y = -(b/a)x$ when y is negative. As x increases, we say that the point $P(x, y)$ *approaches* one of these lines. A corresponding situation exists if $x \leq -a$. The lines with equations

$$y = \pm\frac{b}{a}x$$

are the **asymptotes** of the hyperbola $(x^2/a^2) - (y^2/b^2) = 1$.

The asymptotes serve as excellent guides for sketching the graph. A convenient way to sketch the asymptotes is to first plot the vertices $V(a, 0)$, $V'(-a, 0)$ and the points $W(0, b)$, $W'(0, -b)$ (see Figure 24). The line segment $W'W$ of length $2b$ is the **conjugate axis** of the hyperbola. If horizontal and vertical lines are drawn through the endpoints of the conjugate and transverse axes, respectively, then the diagonals of the resulting rectangle have slopes b/a and $-b/a$. Hence, by extending these diagonals we obtain lines with equations $y = (\pm b/a)x$. The hyperbola is then sketched as in Figure 24, using the asymptotes as guides. The two curves that make up the hyperbola are the **branches** of the hyperbola.

FIGURE 24 $\frac{x^2}{a^2} - \frac{y^2}{b^2} = 1$

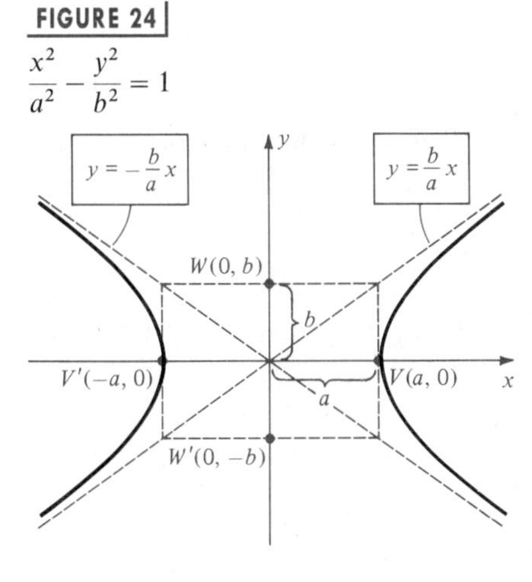

EXAMPLE ▪ 1

Discuss and sketch the graph of $9x^2 - 4y^2 = 36$.

SOLUTION Dividing both sides by 36 gives us

$$\frac{x^2}{4} - \frac{y^2}{9} = 1,$$

which is of the form stated in the theorem with $a^2 = 4$ and $b^2 = 9$. Hence, $a = 2$ and $b = 3$. The vertices $(\pm 2, 0)$ and the endpoints $(0, \pm 3)$ of the conjugate axis determine a rectangle whose diagonals (extended) give us the asymptotes. The graph of the equation is sketched in Figure 25.

The equations of the asymptotes, $y = \pm\frac{3}{2}x$, can be found by referring to the graph or to the equations $y = \pm(b/a)x$.

To find the foci we calculate

$$c^2 = a^2 + b^2 = 4 + 9 = 13.$$

Thus $c = \sqrt{13}$ and the foci are $(\pm\sqrt{13}, 0)$.

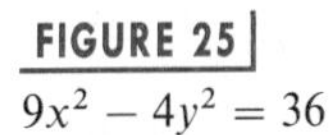
FIGURE 25 $9x^2 - 4y^2 = 36$

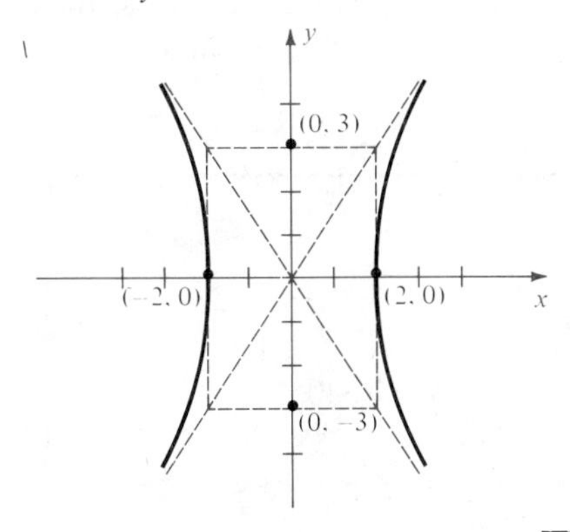

The preceding example indicates that for hyperbolas it is not always true that $a > b$, as is the case for ellipses. Indeed, we may have $a < b$, $a > b$, or $a = b$.

EXAMPLE ▪ 2

A hyperbola has vertices $(\pm 3, 0)$ and passes through the point $P(5, 2)$. Find its equation, foci, and asymptotes.

FIGURE 26

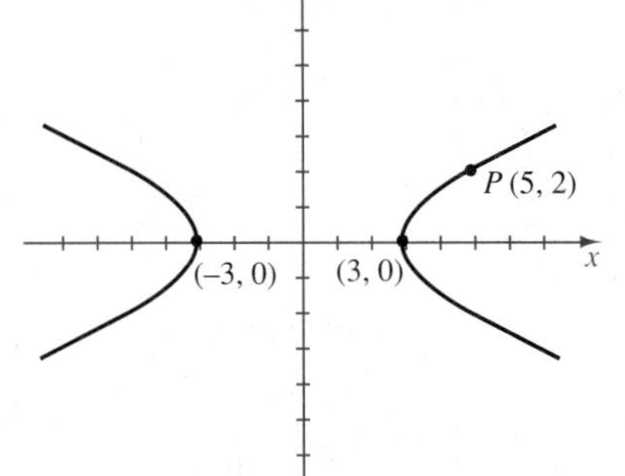

SOLUTION We begin by sketching a hyperbola with vertices $(\pm 3, 0)$ that passes through the point $P(5, 2)$ as in Figure 26.

An equation of the hyperbola has the form

$$\frac{x^2}{9} - \frac{y^2}{b^2} = 1.$$

Since $P(5, 2)$ is on the hyperbola, the x- and y-coordinates satisfy the last equation, that is,

$$\frac{25}{9} - \frac{4}{b^2} = 1.$$

Solving for b^2 gives us $b^2 = \frac{9}{4}$, and hence the desired equation is

$$\frac{x^2}{9} - \frac{4y^2}{9} = 1$$

or, equivalently,

$$x^2 - 4y^2 = 9.$$

To find the foci we first calculate

$$c^2 = a^2 + b^2 = 9 + \tfrac{9}{4} = \tfrac{45}{4}.$$

Hence $c = \sqrt{\frac{45}{4}} = \frac{3}{2}\sqrt{5}$, and the foci are $(\pm\frac{3}{2}\sqrt{5}, 0)$.

The general equations of the asymptotes are $y = \pm(b/a)x$. Substituting $a = 3$ and $b = \frac{3}{2}$ gives us $y = \pm\frac{1}{2}x$.

If the foci of a hyperbola are the points $(0, \pm c)$ on the y-axis, then by the same type of argument used previously, we obtain the following theorem.

THEOREM

> The graph of the equation
>
> $$\frac{y^2}{a^2} - \frac{x^2}{b^2} = 1$$
>
> is a hyperbola with vertices $(0, \pm a)$. The foci are $(0, \pm c)$, with $c^2 = a^2 + b^2$.

For the hyperbola in the preceding theorem the endpoints of the conjugate axis are $W(b, 0)$ and $W'(-b, 0)$. We find the asymptotes as before, by using the diagonals of the rectangle determined by these points, the vertices, and lines parallel to the coordinate axes. The graph is sketched in Figure 27. The equations of the asymptotes are $y = \pm(a/b)x$. Note the difference between these equations and the equations $y = \pm(b/a)x$ for the asymptotes of the hyperbola considered first in this section.

FIGURE 27

$$\frac{y^2}{a^2} - \frac{x^2}{b^2} = 1$$

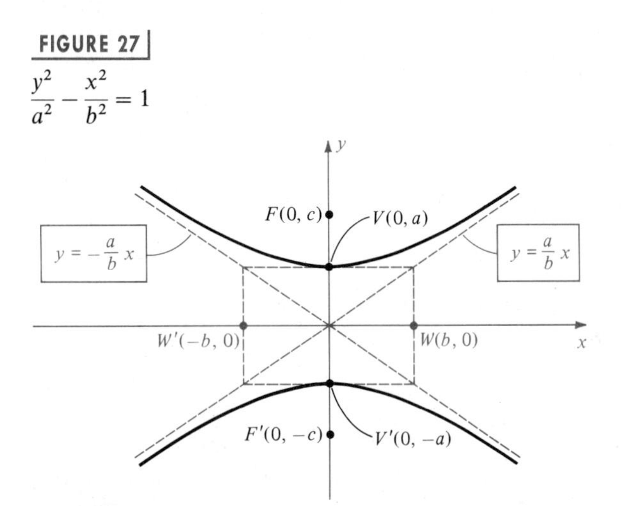

EXAMPLE ▪ 3

Discuss and sketch the graph of $4y^2 - 2x^2 = 1$.

SOLUTION We may obtain the form in the last theorem by writing the equation as

$$\frac{y^2}{\frac{1}{4}} - \frac{x^2}{\frac{1}{2}} = 1.$$

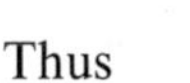

Thus $\qquad a^2 = \frac{1}{4}, \qquad b^2 = \frac{1}{2}, \qquad c^2 = a^2 + b^2 = \frac{3}{4}$

and, consequently,

$$a = \tfrac{1}{2}, \qquad b = \frac{\sqrt{2}}{2}, \qquad c = \frac{\sqrt{3}}{2}.$$

FIGURE 28

$$\frac{y^2}{\frac{1}{4}} - \frac{x^2}{\frac{1}{2}} = 1$$

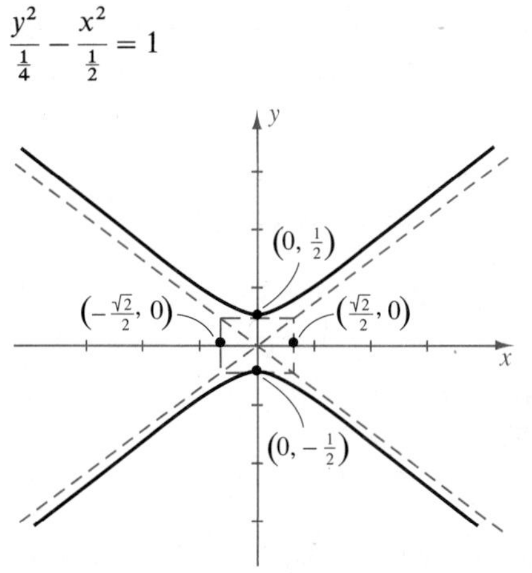

The vertices are $(0, \pm\frac{1}{2})$, the foci are $(0, \pm\sqrt{3}/2)$, and the endpoints of the conjugate axes are $(\pm\sqrt{2}/2, 0)$. The graph is sketched in Figure 28.

To find equations of the asymptotes we can use $y = \pm(a/b)x$, obtaining $y = \pm(\sqrt{2}/2)x$.

As was the case for ellipses, we may use translations of axes to generalize our work. The following example illustrates this technique.

EXAMPLE ▪ 4

Discuss and sketch the graph of $9x^2 - 4y^2 - 54x - 16y + 29 = 0$.

SOLUTION We arrange our work as follows:

$$9(x^2 - 6x \qquad) - 4(y^2 + 4y \qquad) = -29$$
$$9(x^2 - 6x + 9) - 4(y^2 + 4y + 4) = -29 + 81 - 16$$
$$9(x - 3)^2 - 4(y + 2)^2 = 36$$
$$\frac{(x-3)^2}{4} - \frac{(y+2)^2}{9} = 1,$$

FIGURE 29 $\dfrac{(x-3)^2}{4} - \dfrac{(y+2)^2}{9} = 1$

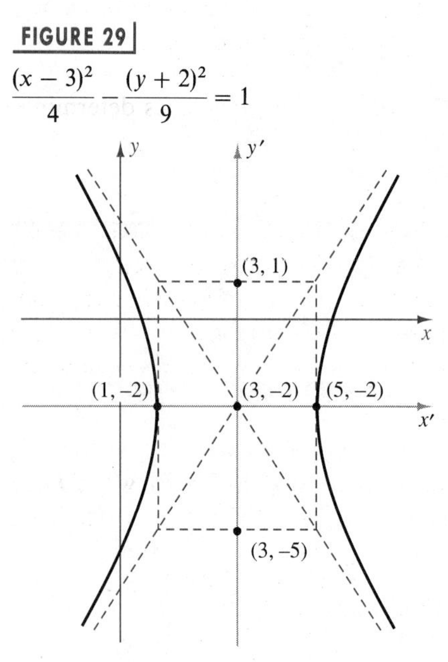

which is of the form

$$\frac{(x')^2}{4} - \frac{(y')^2}{9} = 1$$

with $x' = x - 3$ and $y' = y + 2$. Thus we translate the x- and y-axes to the new origin $C(3, -2)$. The graph is a hyperbola with vertices on the x'-axis (the line $y = -2$) and

$$a^2 = 4, \qquad b^2 = 9, \qquad c^2 = a^2 + b^2 = 13.$$

Thus $$a = 2, \qquad b = 3, \qquad c = \sqrt{13}.$$

As illustrated in Figure 29, the vertices are $(3 \pm 2, -2)$, that is, $(5, -2)$ and $(1, -2)$. The endpoints of the conjugate axis are $(3, -2 \pm 3)$, that is, $(3, 1)$ and $(3, -5)$. The foci are $(3 \pm \sqrt{13}, -2)$, and equations of the asymptotes are

$$y + 2 = \pm\tfrac{3}{2}(x - 3).$$

The results of the last three sections indicate that the graph of every equation of the form

$$Ax^2 + Cy^2 + Dx + Ey + F = 0$$

is a conic, except for certain degenerate cases in which a point, one or two lines, or no graph is obtained. Although we have only considered special examples, our methods are perfectly general. If A and C are equal and not zero, then the graph, when it exists, is a circle or, in exceptional cases, a point. If A and C are unequal but have the same sign, then by completing squares and properly translating axes, we obtain an equation whose graph, when it exists, is an ellipse (or a point). If A and C have opposite signs, an equation of a hyperbola is obtained, or possibly, in the degenerate case, two intersecting straight lines. If either A or C (but not both) is zero, the graph is a parabola or, in certain cases, a pair of parallel lines.

The definition of hyperbola is based on the difference of the distances from two fixed points (the foci). This property is used in the navigational system LORAN (for *Long Range Navigation*). This system involves two pairs of radio transmitters, such as those located at T, T' and S, S' in Figure 30. Suppose that signals sent out by the transmitters at T and T' reach a radio receiver in a ship located at some point P. The difference in time of arrival of the signals can be used to determine the difference of the distances of P from T and from T'. Thus P lies on one branch of a hyperbola with foci at T and T'. Repeating this process for the other pair of transmitters, we see that P also lies on one branch of a hyperbola with foci at S and S'. The intersection of these two branches determines the position of P.

FIGURE 30

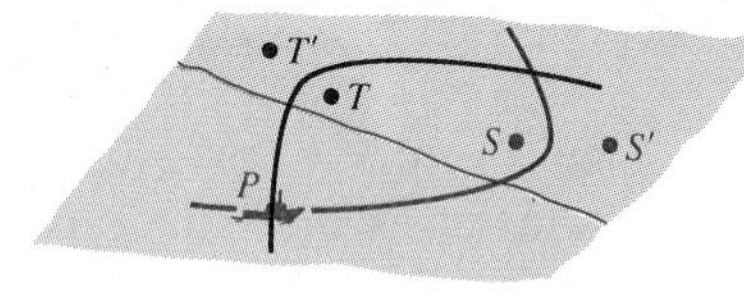

EXAMPLE ▪ 5

A LORAN station A is 200 miles directly east of another station B. A ship is sailing on a line parallel to, and 50 miles north of, the line through stations A and B. Signals are sent out from A and B at the rate of 980 ft/μsec (microsecond). If the signal from B reaches the ship 400 μsec after the signal from A, locate the position of the ship.

FIGURE 31

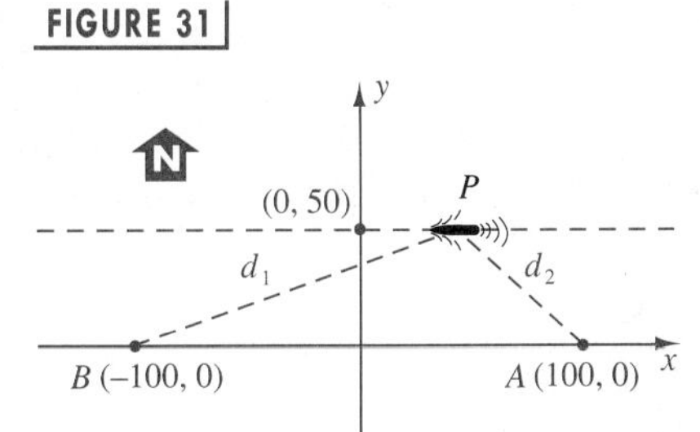

SOLUTION Let us introduce a coordinate system as shown in Figure 31, with the stations at points A and B on the x-axis, and the ship at P on the line $y = 50$. Since it takes 400 μsec longer for the signal to arrive from B than from A, the difference $d_1 - d_2$ in the indicated distances is

$$d_1 - d_2 = (980)(400) = 392{,}000 \text{ feet.}$$

Dividing by 5280 (ft/mi),

$$d_1 - d_2 = \frac{392{,}000}{5280} = 74.2424\ldots \text{ miles.}$$

Recalling that in our derivation of the equation $(x^2/a^2) - (y^2/b^2) = 1$, we let $d_1 - d_2 = 2a$, it follows that in the present situation,

$$a = \frac{74.2424\ldots}{2} = 37.1212\ldots \quad \text{and} \quad a^2 \approx 1378.$$

Since $c = 100$,

$$b^2 = c^2 - a^2 \approx 10{,}000 - 1378, \quad \text{or} \quad b^2 \approx 8622.$$

Hence an (approximate) equation for the hyperbola that has foci A and B and passes through P is

$$\frac{x^2}{1378} - \frac{y^2}{8622} = 1.$$

If we let $y = 50$ (the y-coordinate of P), we obtain

$$\frac{x^2}{1378} - \frac{2500}{8622} = 1.$$

Solving for x gives us $x \approx 42.16$. Rounding off to the nearest mile, the coordinates of P are approximately $(42, 50)$.

FIGURE 32

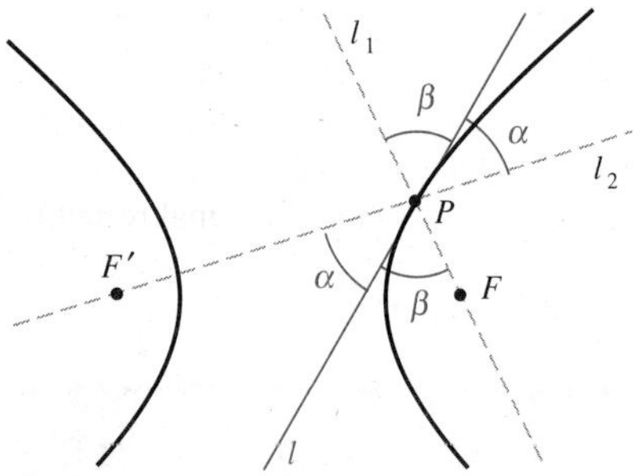

A hyperbola has a reflective property analogous to that of the ellipse discussed at the end of Section 11.3. To illustrate, let l denote the tangent line at a point P on a hyperbola with foci F and F', as shown in Figure 32. If α is the acute angle between $F'P$ and l, and if β is the angle between FP and l, then $\alpha = \beta$. If a ray of light is directed along the line l_1 through F and P, it will be reflected back from P along the line l_2 through F' and P. This property is used in the design of telescopes of the Cassegrain type (see Exercise 40).

11.4 EXERCISES

Exer. 1–20: Find the vertices, the foci, and equations of the asymptotes of the hyperbola, and sketch its graph.

1 $\dfrac{x^2}{9} - \dfrac{y^2}{4} = 1$

2 $\dfrac{y^2}{49} - \dfrac{x^2}{16} = 1$

3 $\dfrac{y^2}{9} - \dfrac{x^2}{4} = 1$

4 $\dfrac{x^2}{49} - \dfrac{y^2}{16} = 1$

5 $y^2 - 4x^2 = 16$

6 $x^2 - 2y^2 = 8$

7 $x^2 - y^2 = 1$

8 $y^2 - 16x^2 = 1$

9 $x^2 - 5y^2 = 25$

10 $4y^2 - 4x^2 = 1$

11 $3x^2 - y^2 = -3$

12 $16x^2 - 36y^2 = 1$

13 $\dfrac{(y+2)^2}{9} - \dfrac{(x+2)^2}{4} = 1$

14 $\dfrac{(x-3)^2}{25} - \dfrac{(y-1)^2}{4} = 1$

15 $25x^2 - 16y^2 + 250x + 32y + 109 = 0$

16 $y^2 - 4x^2 - 12y - 16x + 16 = 0$

17 $4y^2 - x^2 + 40y - 4x + 60 = 0$

18 $25x^2 - 9y^2 + 100x - 54y + 10 = 0$

19 $9y^2 - x^2 - 36y + 12x - 36 = 0$

20 $4x^2 - y^2 + 32x - 8y + 49 = 0$

Exer. 21–32: Find an equation for the hyperbola that has its center at the origin and satisfies the given conditions.

21 Foci $F(0, \pm 4)$, vertices $V(0, \pm 1)$

22 Foci $F(\pm 8, 0)$, vertices $V(\pm 5, 0)$

23 Foci $F(\pm 5, 0)$, vertices $V(\pm 3, 0)$

24 Foci $F(0, \pm 3)$, vertices $V(0, \pm 2)$

25 Foci $F(0, \pm 5)$, conjugate axis of length 4

26 Vertices $V(\pm 4, 0)$, passing through $(8, 2)$

27 Vertices $V(\pm 3, 0)$, asymptotes $y = \pm 2x$

28 Foci $F(0, \pm 10)$, asymptotes $y = \pm \frac{1}{3}x$

29 x-intercepts ± 5, asymptotes $y = \pm 2x$

30 y-intercepts ± 2, asymptotes $y = \pm \frac{1}{4}x$

31 Vertical transverse axis of length 10, conjugate axis of length 14

32 Horizontal transverse axis of length 6, conjugate axis of length 2

Exer. 33–34: Find the points of intersection of the graphs of the equations. Sketch both graphs on the same coordinate plane, and show points of intersection.

33 $\begin{cases} y^2 - 4x^2 = 16 \\ y - x = 4 \end{cases}$ 34 $\begin{cases} x^2 - y^2 = 4 \\ y^2 - 3x = 0 \end{cases}$

35 The graphs of the following equations are *conjugate hyperbolas*:

$$\frac{x^2}{a^2} - \frac{y^2}{b^2} = 1; \qquad \frac{x^2}{a^2} - \frac{y^2}{b^2} = -1$$

(a) Sketch the graphs of both equations on the same coordinate plane if $a = 5$ and $b = 3$.

(b) Describe the relationship between the two graphs.

36 Find an equation of the hyperbola with foci $(h \pm c, k)$ and vertices $(h \pm a, k)$ with $0 < a < c$ and $c^2 = a^2 + b^2$.

37 The physicist Ernest Rutherford discovered that when alpha particles are shot toward the nucleus of an atom, they are eventually repulsed away from the nucleus along hyperbolic paths. The figure illustrates the path of a particle that starts toward the origin along the line $y = \frac{1}{2}x$ and comes within 3 units of the nucleus. Find an equation of the path.

EXERCISE 37

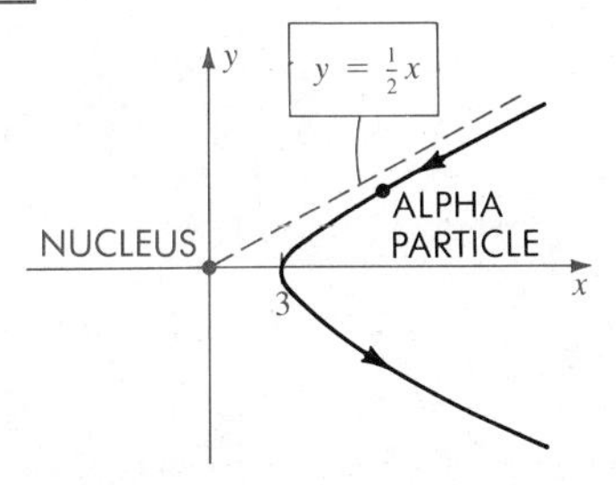

38 A jet airplane is executing a high-speed maneuver along the hyperbolic path illustrated in the figure. If an equation of the path is $2y^2 - x^2 = 8$, determine how close it comes to a town located at $(3, 0)$. (*Hint:* Let S denote the square of the distance from a point (x, y) on the path to $(3, 0)$, and find the minimum value of S.)

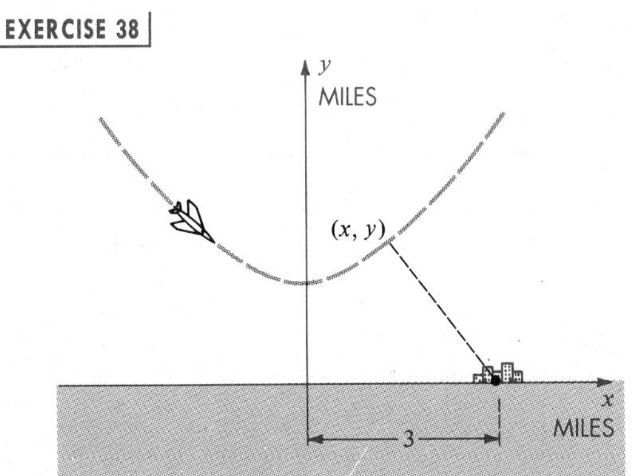

39 A ship is traveling a course that is 100 miles from, and parallel to, a straight shoreline. The ship sends out a distress signal, which is received by two Coast Guard stations A and B, located 200 miles apart, as shown in the figure. By measuring the difference in signal reception times, officials determine that the ship is 160 miles closer to station B than to A. Where is the ship?

EXERCISE 39

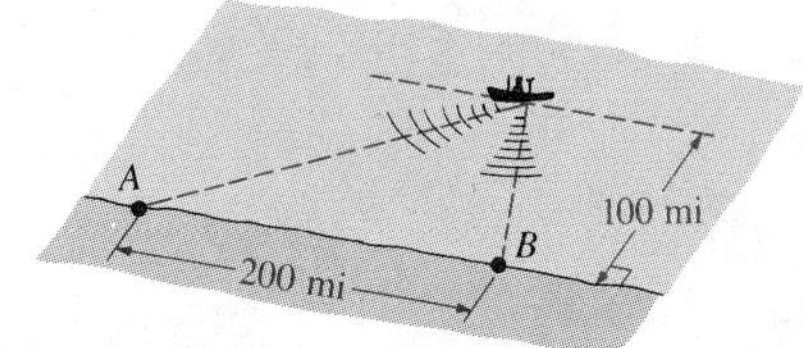

40 The Cassegrain telescope design (c. 1672) makes use of the reflective properties of both the parabola and hyperbola (see Figures 11 and 32). Shown in the figure is a (split) parabolic mirror with focus at F_1 and axis the line l, and a second, hyperbolic mirror with one focus also at F_1 and transverse axis along the axis of the parabola. Determine where incoming light waves parallel to the common axis will finally collect.

EXERCISE 40

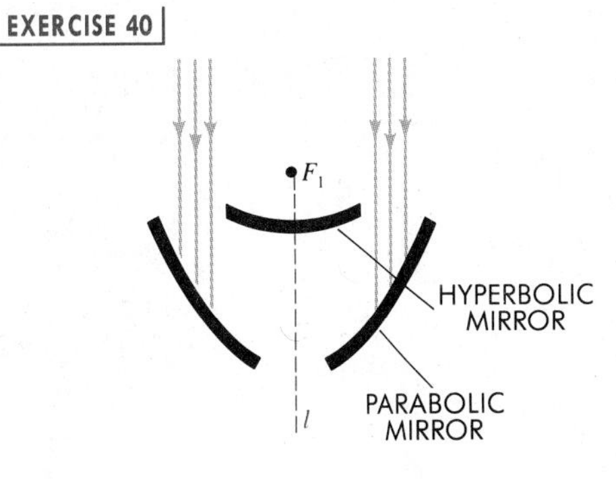

11.5 ROTATION OF AXES

We may obtain the $x'y'$-plane used in a translation of axes by moving the origin O of the xy-plane to a new position $C(h, k)$ while, at the same time, not changing the positive directions of the axes or the units of length. We shall next introduce a new coordinate plane by keeping the origin O fixed and rotating the x- and y-axes about O to another position denoted by x' and y'. A transformation of this type is a **rotation of axes**.

FIGURE 33

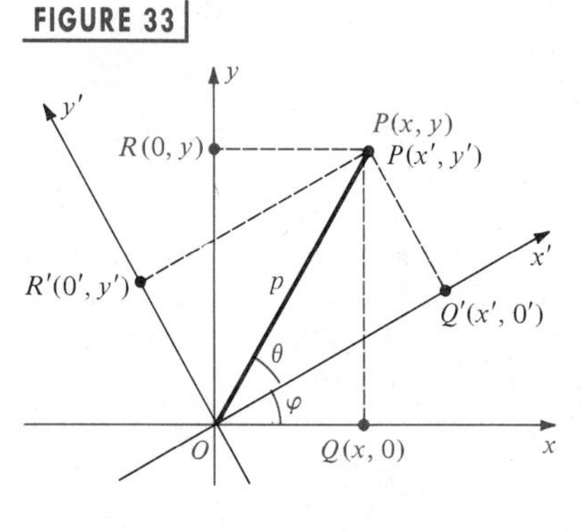

Consider the rotation of axes in Figure 33, and let φ denote the acute angle through which the positive x-axis must be rotated in order to coincide with the positive x'-axis. If (x, y) are the coordinates of a point P relative to the xy-plane, then (x', y') will denote its coordinates relative to the new $x'y'$-plane.

Let the projections of P on the various axes be denoted as in Figure 33, and let θ denote angle POQ'. If $p = d(O, P)$, then

$$x' = p \cos \theta, \qquad y' = p \sin \theta,$$
$$x = p \cos (\theta + \varphi), \qquad y = p \sin (\theta + \varphi).$$

Applying the addition formulas for the sine and cosine, we see that

$$x = p \cos \theta \cos \varphi - p \sin \theta \sin \varphi,$$
$$y = p \sin \theta \cos \varphi + p \cos \theta \sin \varphi.$$

Using the fact that $x' = p \cos \theta$ and $y' = p \sin \theta$ gives us (i) of the next theorem. The formulas in (ii) may be obtained from (i) by solving for x' and y'.

ROTATION OF AXES FORMULAS

If the x- and y-axes are rotated about the origin O, through an acute angle φ, then the coordinates (x, y) and (x', y') of a point P in the xy- and $x'y'$-planes are related as follows:

(i) $x = x' \cos \varphi - y' \sin \varphi, \qquad y = x' \sin \varphi + y' \cos \varphi$

(ii) $x' = x \cos \varphi + y \sin \varphi, \qquad y' = -x \sin \varphi + y \cos \varphi.$

EXAMPLE ▪ 1

The graph of $xy = 1$ or, equivalently, $y = 1/x$, is sketched in Figure 34. If the coordinate axes are rotated through an angle of 45°, find an equation of the graph relative to the new $x'y'$-plane.

FIGURE 34
$y = 1/x$

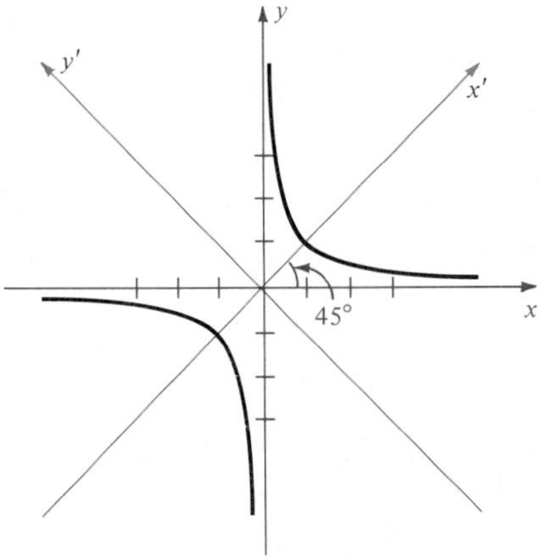

SOLUTION We let $\varphi = 45°$ in rotation of axes formulas (i):

$$x = x'\left(\frac{\sqrt{2}}{2}\right) - y'\left(\frac{\sqrt{2}}{2}\right) = \frac{\sqrt{2}}{2}(x' - y')$$
$$y = x'\left(\frac{\sqrt{2}}{2}\right) + y'\left(\frac{\sqrt{2}}{2}\right) = \frac{\sqrt{2}}{2}(x' + y').$$

Substituting for x and y in the equation $xy = 1$ gives us

$$\frac{\sqrt{2}}{2}(x' - y') \cdot \frac{\sqrt{2}}{2}(x' + y') = 1.$$

This reduces to $$\frac{(x')^2}{2} - \frac{(y')^2}{2} = 1,$$

which is an equation of a hyperbola with vertices $(\pm\sqrt{2}, 0)$ on the x'-axis. Note that the asymptotes for the hyperbola have equations $y' = \pm x'$ in the new system. These correspond to the original x- and y-axes.

Example 1 illustrates a method for eliminating a term of an equation that contains the product xy. This method can be used to transform any equation of the form

$$Ax^2 + Bxy + Cy^2 + Dx + Ey + F = 0$$

with $B \neq 0$ into an equation in x' and y' that contains no $x'y'$ term. Let us prove that this may always be done. If we rotate the axes through an angle φ, then using rotation of axes formulas (i) to substitute for x and y gives us

$$\begin{aligned} &A(x' \cos \varphi - y' \sin \varphi)^2 \\ &\qquad + B(x' \cos \varphi - y' \sin \varphi)(x' \sin \varphi + y' \cos \varphi) \\ &\qquad + C(x' \sin \varphi + y' \cos \varphi)^2 + D(x' \cos \varphi - y' \sin \varphi) \\ &\qquad + E(x' \sin \varphi + y' \cos \varphi) + F = 0. \end{aligned}$$

By performing the multiplications and rearranging terms, we may write this equation in the form

$$A'(x')^2 + B'x'y' + C'(y')^2 + D'x' + E'y' + F' = 0$$

with

$$\begin{aligned}
A' &= A\cos^2\varphi + B\cos\varphi\sin\varphi + C\sin^2\varphi\\
B' &= 2(C - A)\sin\varphi\cos\varphi + B(\cos^2\varphi - \sin^2\varphi)\\
C' &= A\sin^2\varphi - B\sin\varphi\cos\varphi + C\cos^2\varphi\\
D' &= D\cos\varphi + E\sin\varphi\\
E' &= -D\sin\varphi + E\cos\varphi\\
F' &= F.
\end{aligned}$$

To eliminate the $x'y'$ term, we must select φ such that $B' = 0$, that is,

$$2(C - A)\sin\varphi\cos\varphi + B(\cos^2\varphi - \sin^2\varphi) = 0.$$

Using double-angle formulas, we may write this equation as

$$(C - A)\sin 2\varphi + B\cos 2\varphi = 0,$$

which is equivalent to $\cot 2\varphi = \dfrac{A - C}{B}$.

This proves the next result.

THEOREM

To eliminate the xy-term from the equation

$$Ax^2 + Bxy + Cy^2 + Dx + Ey + F = 0,$$

where $B \neq 0$, choose an angle φ such that

$$\cot 2\varphi = \frac{A - C}{B} \quad \text{with } 0^\circ < 2\varphi < 180^\circ,$$

and use the rotation of axes formulas.

The graph of any equation in x and y of the type displayed in the preceding theorem is a conic, except for certain degenerate cases.

When using the preceding theorem, note that $\sin 2\varphi > 0$, since $0^\circ < 2\varphi < 180^\circ$. Moreover, since $\cot 2\varphi = \cos 2\varphi/\sin 2\varphi$, the signs of $\cot 2\varphi$ and $\cos 2\varphi$ are always the same.

EXAMPLE ▪ 2

Discuss and sketch the graph of $41x^2 - 24xy + 34y^2 - 25 = 0$.

SOLUTION Using the notation of the preceding theorem,

$$A = 41, \qquad B = -24, \qquad C = 34,$$

and

$$\cot 2\varphi = \frac{41 - 34}{-24} = -\frac{7}{24}.$$

Since $\cot 2\varphi$ is negative, we choose 2φ such that $90^\circ < 2\varphi < 180^\circ$, and, consequently, $\cos 2\varphi = -\frac{7}{25}$. We now use the half-angle formulas to obtain

$$\sin \varphi = \sqrt{\frac{1 - \cos 2\varphi}{2}} = \sqrt{\frac{1 - (-\frac{7}{25})}{2}} = \frac{4}{5}$$

$$\cos \varphi = \sqrt{\frac{1 + \cos 2\varphi}{2}} = \sqrt{\frac{1 + (-\frac{7}{25})}{2}} = \frac{3}{5}.$$

FIGURE 35

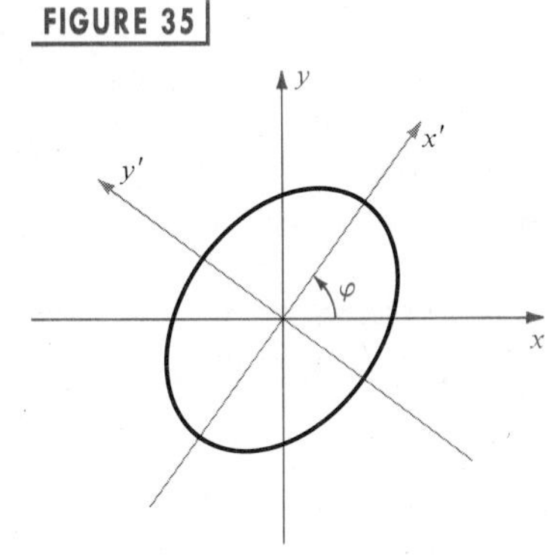

Thus, the desired rotation of axes formulas are

$$x = \tfrac{3}{5}x' - \tfrac{4}{5}y', \qquad y = \tfrac{4}{5}x' + \tfrac{3}{5}y'.$$

After substituting for x and y in the given equation and simplifying, we obtain the equation

$$(x')^2 + 2(y')^2 = 1.$$

Thus the graph is an ellipse with vertices at $(\pm 1, 0)$ on the x'-axis. Since $\tan \varphi = \sin \varphi / \cos \varphi = (\frac{4}{5})/(\frac{3}{5}) = \frac{4}{3}$, we obtain $\varphi = \tan^{-1}(\frac{4}{3})$. To the nearest minute, $\varphi \approx 53^\circ 8'$. The graph is sketched in Figure 35.

The next theorem states rules that we can apply to identify the type of conic *before* rotating the axes.

IDENTIFICATION THEOREM

The graph of the equation

$$Ax^2 + Bxy + Cy^2 + Dx + Ey + F = 0$$

is either a conic or a degenerate conic. If the graph is a conic, then it is

(i) a parabola if $B^2 - 4AC = 0$.
(ii) an ellipse if $B^2 - 4AC < 0$.
(iii) a hyperbola if $B^2 - 4AC > 0$.

PROOF If the x and y axes are rotated through an angle φ, then using the rotation of axes formulas gives us

$$A'(x')^2 + B'x'y' + C'(y')^2 + D'x' + E'y' + F' = 0.$$

Using the formulas for A', B', and C' on page 622, we can show that

$$(B')^2 - 4A'C' = B^2 - 4AC.$$

For a suitable roation of axes, we obtain $B' = 0$ and

$$A'(x')^2 + C'(y')^2 + D'x' + E'y' + F' = 0.$$

Except for degenerate cases, the graph of this equation is an ellipse if $A'C' > 0$ (A' and C' have the same sign), a hyperbola if $A'C' < 0$ (A' and C' have opposite signs), or a parabola if $A'C' = 0$ (either $A' = 0$ or $C' = 0$). However, if $B' = 0$, then $B^2 - 4AC = -4A'C'$, and hence the graph is an ellipse if $B^2 - 4AC < 0$, a hyperbola if $B^2 - 4AC > 0$, or a parabola if $B^2 - 4AC = 0$. ❑

The expression $B^2 - 4AC$ is called the **discriminant** of the equation in the identification theorem. We say that this discriminant is **invariant** under a rotation of axes, because it is unchanged by any such rotation.

EXAMPLE ▪ 3

Use the identification theorem to determine if the graph of the equation

$$41x^2 - 24xy + 34y^2 - 25 = 0$$

is a parabola, an ellipse, or a hyperbola.

SOLUTION The equation was considered in Example 2, where we performed a rotation of axes. Since $A = 41$, $B = -24$, and $C = 34$, the discriminant is

$$B^2 - 4AC = 576 - 4(41)(34) = -5000 < 0.$$

Hence, by the identification theorem, the graph is an ellipse.

In some cases, after eliminating the xy term, it may be necessary to translate the axes of the $x'y'$-coordinate system to obtain the graph, as illustrated in the next example.

EXAMPLE ▪ 4

Discuss and sketch the graph of the equation

$$x^2 + 2\sqrt{3}\,xy + 3y^2 + 8\sqrt{3}\,x - 8y + 32 = 0.$$

SOLUTION Using $A = 1$, $B = 2\sqrt{3}$, and $C = 3$, we see that

$$B^2 - 4AC = 12 - 12 = 0.$$

By the identification theorem, the graph is a parabola.

To apply a rotation of axes we calculate

$$\cot 2\varphi = \frac{A - C}{B} = \frac{1 - 3}{2\sqrt{3}} = -\frac{1}{\sqrt{3}}.$$

Hence $2\varphi = 120°$, $\varphi = 60°$, and

$$\sin\varphi = \frac{\sqrt{3}}{2}, \qquad \cos\varphi = \frac{1}{2}.$$

The rotation of axes formulas are

$$x = \frac{1}{2}x' - \frac{\sqrt{3}}{2}y' = \frac{1}{2}(x' - \sqrt{3}y'),$$

$$y = \frac{\sqrt{3}}{2}x' + \frac{1}{2}y' = \frac{1}{2}(\sqrt{3}x' + y').$$

FIGURE 36

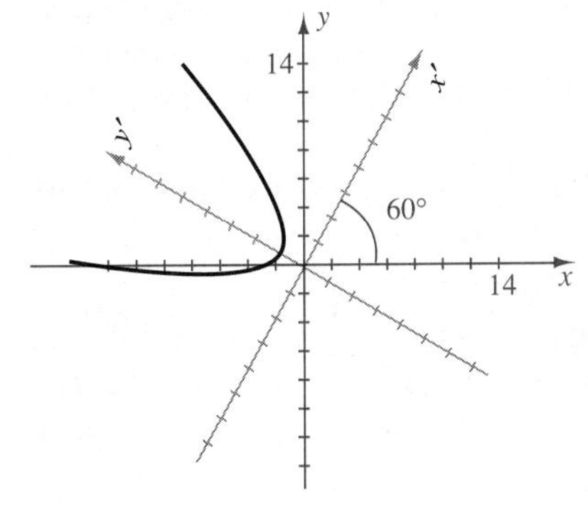

Substituting for x and y in the given equation and simplifying leads to

$$4(x')^2 - 16y' + 32 = 0$$

or, equivalently,

$$(x')^2 = 4(y' - 2).$$

The parabola is sketched in Figure 36. Note that the vertex is at the point $(0, 2)$ in the $x'y'$-plane, and the graph is symmetric with respect to the y'-axis.

11.5 EXERCISES

Exer. 1–13:

(a) Use the identification theorem to determine if the graph of the equation is a parabola, ellipse, or hyperbola.

(b) Use a suitable rotation of axes to find an equation for the graph in an $x'y'$-plane, and then sketch the graph, labeling vertices.

1 $x^2 - 2xy + y^2 - 2\sqrt{2}x - 2\sqrt{2}y = 0$

2 $x^2 - 2xy + y^2 + 4x + 4y = 0$

3 $5x^2 - 8xy + 5y^2 = 9$

4 $x^2 - xy + y^2 = 3$

5 $11x^2 + 10\sqrt{3}xy + y^2 = 4$

6 $7x^2 - 48xy - 7y^2 = 225$

7 $16x^2 - 24xy + 9y^2 - 60x - 80y + 100 = 0$

8 $x^2 + 4xy + 4y^2 + 6\sqrt{5}\,x - 18\sqrt{5}\,y + 45 = 0$

9 $40x^2 - 36xy + 25y^2 - 8\sqrt{13}\,x - 12\sqrt{13}\,y = 0$

10 $18x^2 - 48xy + 82y^2 + 6\sqrt{10}\,x + 2\sqrt{10}\,y - 80 = 0$

11 $5x^2 + 6\sqrt{3}\,xy - y^2 + 8x - 8\sqrt{3}\,y - 12 = 0$

12 $15x^2 + 20xy - 4\sqrt{5}\,x + 8\sqrt{5}\,y - 100 = 0$

13 $32x^2 - 72xy + 53y^2 = 80$

11.6 POLAR COORDINATES

In a rectangular coordinate system the ordered pair (a, b) denotes the point whose directed distances from the x- and y-axes are b and a, respectively. Another method for representing points is to use *polar coordinates.* We begin with a fixed point O (the **origin**, or **pole**) and a directed half-line (the **polar axis**) with endpoint O. Next we consider any point P in the plane different from O. If, as illustrated in Figure 37, $r = d(O, P)$ and θ denotes the measure of any angle determined by the polar axis and OP, then r and θ are **polar coordinates** of P, and the symbols (r, θ) or $P(r, \theta)$ are used to denote P. As usual, θ is considered positive if the angle is generated by a counterclockwise rotation of the polar axis and negative if the rotation is clockwise. Either radian or degree measure may be used for θ.

FIGURE 37

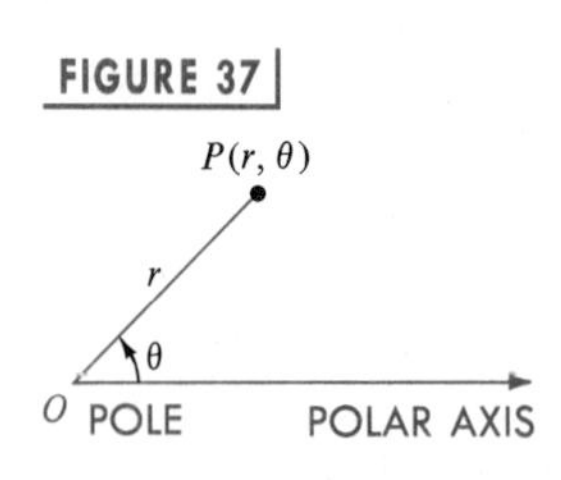

The polar coordinates of a point are not unique. For example $(3, \pi/4)$, $(3, 9\pi/4)$, and $(3, -7\pi/4)$ all represent the same point (see Figure 38). We shall also allow r to be negative. In this case, instead of measuring $|r|$ units along the terminal side of the angle θ, we measure along the half-line with endpoint O that has direction *opposite* that of the terminal side. The points corresponding to the pairs $(-3, 5\pi/4)$ and $(-3, -3\pi/4)$ are also plotted in Figure 38.

FIGURE 38

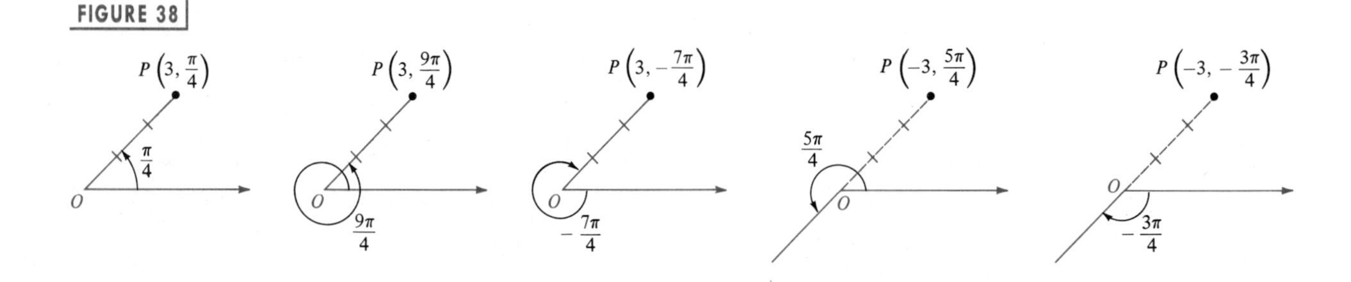

We agree that the pole O has polar coordinates $(0, \theta)$ for *any* θ. An assignment of ordered pairs of the form (r, θ) to points in a plane is a **polar coordinate system** and the plane is an ***rθ*-plane**.

A **polar equation** is an equation in r and θ. A **solution** of a polar equation is an ordered pair (a, b) that leads to equality if a is substituted for r and b for θ. The **graph** of a polar equation is the set of all points (in an $r\theta$-plane) that correspond to the solutions.

EXAMPLE ▪ 1

Sketch the graph of the polar equation $r = 4 \sin \theta$.

SOLUTION The following table displays some solutions of the equation. We have included a third row in the table that contains one-decimal-place approximations to r.

θ	0	$\frac{\pi}{6}$	$\frac{\pi}{4}$	$\frac{\pi}{3}$	$\frac{\pi}{2}$	$\frac{2\pi}{3}$	$\frac{3\pi}{4}$	$\frac{5\pi}{6}$	π
r	0	2	$2\sqrt{2}$	$2\sqrt{3}$	4	$2\sqrt{3}$	$2\sqrt{2}$	2	0
r (approx.)	0	2	2.8	3.4	4	3.4	2.8	2	0

FIGURE 39

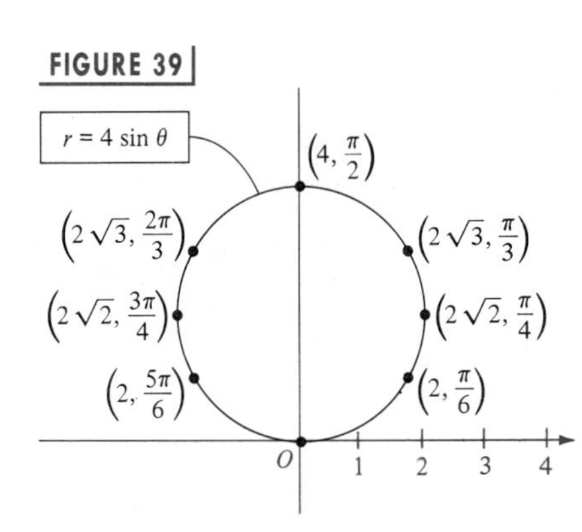

In rectangular coordinates, the graph of the equation consists of sine waves of amplitude 4 and period 2π. However, if polar coordinates are used, then the points that correspond to the pairs in the table appear to lie on a circle of radius 2, and we draw the graph accordingly (see Figure 39). As an aid to plotting points, we have extended the polar axis in the negative direction and introduced a vertical line through the pole.

The proof that the graph of $r = 4 \sin \theta$ is a circle is given in Example 6. Additional points obtained by letting θ vary from π to 2π lie on the same circle. For example, the solution $(-2, 7\pi/6)$ gives us the same point as $(2, \pi/6)$; the point corresponding to $(-2\sqrt{2}, 5\pi/4)$ is the same as that obtained from $(2\sqrt{2}, \pi/4)$; and so on. If we let θ increase through all real numbers, we obtain the same points again and again because of the periodicity of the sine function.

EXAMPLE ▪ 2

Sketch the graph of the polar equation $r = 2 + 2 \cos \theta$.

SOLUTION Since the cosine function decreases from 1 to -1 as θ varies from 0 to π, it follows that r decreases from 4 to 0 in this θ-interval. The following table exhibits some solutions of the equation, together with approximations to r.

FIGURE 40
$r = 2 + 2\cos\theta$

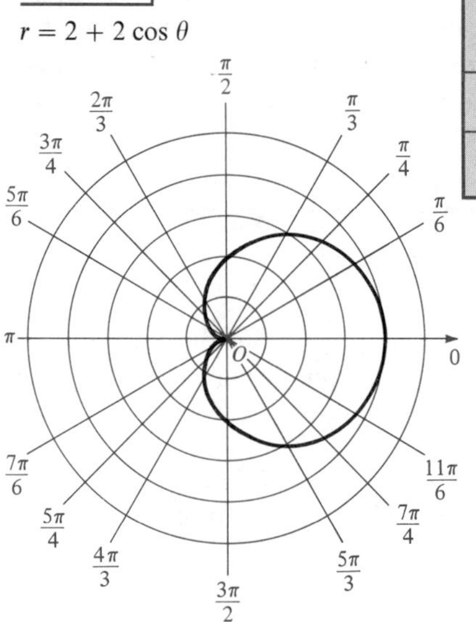

θ	0	$\frac{\pi}{6}$	$\frac{\pi}{4}$	$\frac{\pi}{3}$	$\frac{\pi}{2}$	$\frac{2\pi}{3}$	$\frac{3\pi}{4}$	$\frac{5\pi}{6}$	π
r	4	$2+\sqrt{3}$	$2+\sqrt{2}$	3	2	1	$2-\sqrt{2}$	$2-\sqrt{3}$	0
r (approx.)	4	3.7	3.4	3	2	1	0.6	0.3	0

Plotting points gives us the upper half of the graph sketched in Figure 40, where we have used polar coordinate graph paper, which displays lines through O at various angles and concentric circles with centers at the pole.

If θ increases from π to 2π, then $\cos\theta$ increases from -1 to 1 and, consequently, r increases from 0 to 4. Plotting points for $\pi \leq \theta \leq 2\pi$ gives us the lower half of the graph.

The same graph may be obtained by taking other intervals of length 2π for θ.

The heart-shaped graph in Example 2 is a **cardioid**. In general, the graph of any polar equation of the form

$$r = a(1 + \cos\theta), \qquad r = a(1 + \sin\theta),$$
$$r = a(1 - \cos\theta), \qquad r = a(1 - \sin\theta),$$

for a nonzero real number a, is a cardioid.

The graph of a polar equation of the form

$$r = a + b\cos\theta \quad \text{or} \quad r = a + b\sin\theta,$$

for $|a| \neq |b|$, is a **limaçon**. The graph is similar in shape to a cardioid; however, there may be an added "loop," as shown in the next example.

EXAMPLE ▪ 3

Sketch the graph of the polar equation $r = 2 + 4\cos\theta$.

SOLUTION Coordinates of some points corresponding to $0 \leq \theta \leq \pi$ are listed in the following table.

θ	0	$\frac{\pi}{6}$	$\frac{\pi}{4}$	$\frac{\pi}{3}$	$\frac{\pi}{2}$	$\frac{2\pi}{3}$	$\frac{3\pi}{4}$	$\frac{5\pi}{6}$	π
r	6	$2+2\sqrt{3}$	$2+2\sqrt{2}$	4	2	0	$2-2\sqrt{2}$	$2-2\sqrt{3}$	-2
r (approx.)	6	5.4	4.8	4	2	0	-0.8	-1.4	-2

FIGURE 41

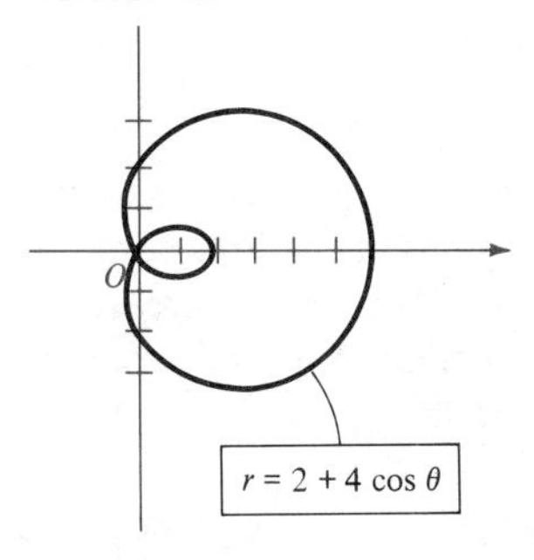

Note that $r = 0$ at $\theta = 2\pi/3$. The values of r are negative if $2\pi/3 < \theta \le \pi$, and this leads to the lower half of the small loop in Figure 41. Letting θ range from π to 2π gives us the upper half of the small loop and the lower half of the large loop.

EXAMPLE ▪ 4

Sketch the graph of the polar equation $r = a \sin 2\theta$ for $a > 0$.

FIGURE 42

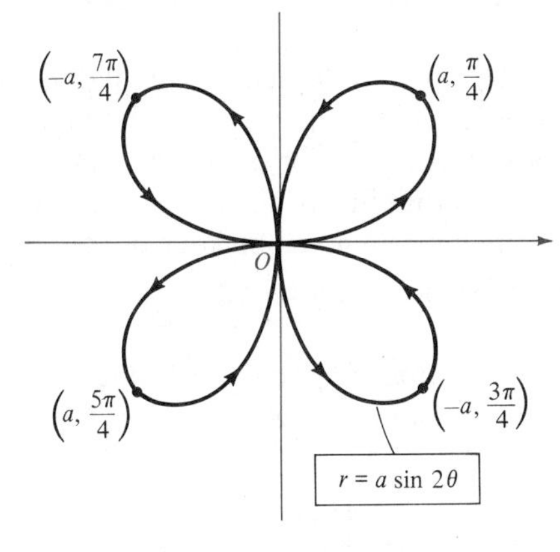

SOLUTION Instead of tabulating solutions, let us reason as follows: If θ increases from 0 to $\pi/4$, then 2θ varies from 0 to $\pi/2$ and hence $\sin 2\theta$ increases from 0 to 1. It follows that r increases from 0 to a in the θ-interval $[0, \pi/4]$. If we next let θ increase from $\pi/4$ to $\pi/2$, then 2θ changes from $\pi/2$ to π and hence $\sin 2\theta$ decreases from 1 to 0. Thus r decreases from a to 0 in the θ-interval $[\pi/4, \pi/2]$. The corresponding points on the graph constitute the first-quadrant loop illustrated in Figure 42. Note that the point $P(r, \theta)$ traces the loop in a *counterclockwise* direction (indicated by the arrows) as θ increases from 0 to $\pi/2$.

If $\pi/2 \le \theta \le \pi$, then $\pi \le 2\theta \le 2\pi$ and, therefore, $r = a \sin 2\theta \le 0$. Thus, *if* $\pi/2 < \theta < \pi$, *then* r *is negative and the points* $P(r, \theta)$ *are in the fourth quadrant*. If θ increases from $\pi/2$ to π, then we can show, by plotting points, that $P(r, \theta)$ traces (in a counterclockwise direction) the loop shown in the fourth quadrant.

Similarly, for $\pi \le \theta \le 3\pi/2$ we get the loop in the third quadrant, and for $3\pi/2 \le \theta \le 2\pi$ we get the loop in the second quadrant. Both loops are traced in a counterclockwise direction as θ increases. You should verify these facts by plotting some points with, say, $a = 1$. In Figure 42 we have plotted only those points on the graph that correspond to the largest numerical values of r.

The graph in Example 4 is a **four-leafed rose**. In general, a polar equation of the form

$$r = a \sin n\theta \quad \text{or} \quad r = a \cos n\theta,$$

for any positive integer n greater than 1 and any nonzero real number a, has a graph that consists of a number of loops attached to the origin. If n is even there are $2n$ loops, and if n is odd there are n loops (see, for example, Exercises 5, 6, and 21–24).

The graph of the polar equation $r = a\theta$ for any real number a is a **spiral of Archimedes**. The case $a = 1$ is considered in the next example.

EXAMPLE ■ 5

Sketch the graph of the polar equation $r = \theta$ for $\theta \geq 0$.

FIGURE 43

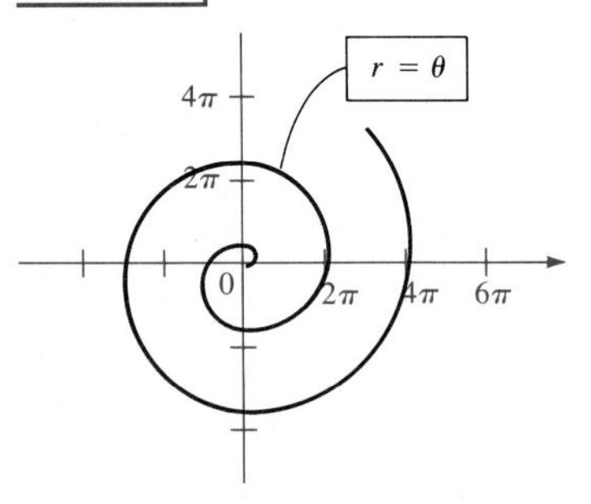

SOLUTION The graph consists of all points that have polar coordinates of the form (c, c) for every real number $c \geq 0$. Thus the graph contains the points $(0, 0)$, $(\pi/2, \pi/2)$, (π, π), and so on. As θ increases, r increases at the same rate, and the spiral winds around the origin in a counterclockwise direction, intersecting the polar axis at $0, 2\pi, 4\pi, \ldots$, as illustrated in Figure 43.

If θ is allowed to be negative, then as θ decreases through negative values, the resulting spiral winds around the origin and is the symmetric image, with respect to the vertical axis, of the curve sketched in Figure 43.

Polar coordinates are useful in applied problems that involve circles with centers at the origin or lines that pass through the origin, since equations that have these graphs may be written in the simple forms $r = k$ or $\theta = k$ for some constant k.

Let us next superimpose an xy-plane on an $r\theta$-plane such that the positive x-axis coincides with the polar axis. Any point P in the plane may then be assigned rectangular coordinates (x, y) or polar coordinates (r, θ). If $r > 0$ we have a situation similar to that illustrated in Figure 44(i). If $r < 0$ we have that shown in (ii) of the figure where, for later purposes, we have also plotted the point P' having polar coordinates $(|r|, \theta)$ and rectangular coordinates $(-x, -y)$.

The following result specifies relationships between (x, y) and (r, θ), where it is assumed that the positive x-axis coincides with the polar axis.

RELATIONSHIPS BETWEEN RECTANGULAR AND POLAR COORDINATES

The rectangular coordinates (x, y) and polar coordinates (r, θ) of a point P are related as follows:

(i) $x = r \cos \theta, \qquad y = r \sin \theta$

(ii) $r^2 = x^2 + y^2, \qquad \tan \theta = \dfrac{y}{x} \quad \text{if } x \neq 0.$

PROOF Although we have pictured θ as an acute angle in Figure 44, the discussion that follows is valid for all angles. If $r > 0$ as in Figure 44(i), then $\cos \theta = x/r$, $\sin \theta = y/r$, and hence

$$x = r \cos \theta, \qquad y = r \sin \theta.$$

FIGURE 44

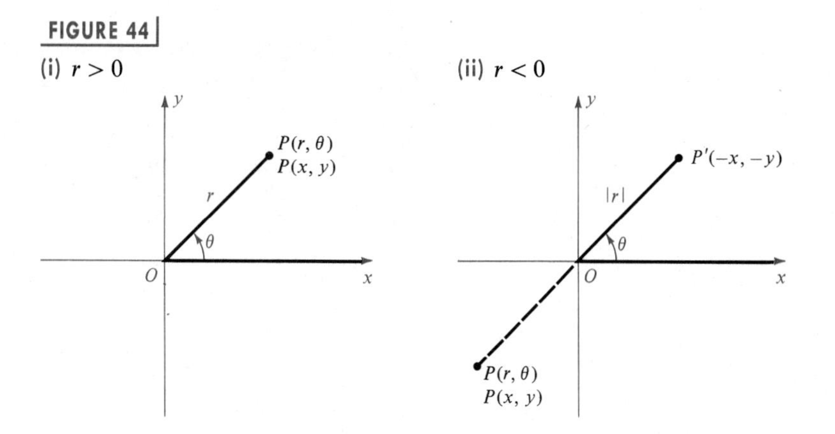

If $r < 0$ then $|r| = -r$ and from Figure 44(ii) we see that

$$\cos\theta = \frac{-x}{|r|} = \frac{-x}{-r} = \frac{x}{r},$$

$$\sin\theta = \frac{-y}{|r|} = \frac{-y}{-r} = \frac{y}{r}.$$

Multiplication by r gives us relationship (i) and, therefore, these formulas hold if r is either positive or negative. If $r = 0$, then the point is the pole and we again see that the formulas in (i) are true.

The formulas in (ii) follow readily from Figure 44. ❑

We may use the preceding result to change from one system of coordinates to the other. A more important use is for transforming a polar equation to an equation in x and y, and vice versa. This is illustrated in the next three examples.

EXAMPLE ▪ 6

Find an equation in x and y that has the same graph as the polar equation $r = 4\sin\theta$.

SOLUTION The equation $r = 4\sin\theta$ was considered in Example 1. If we multiply both sides by r, we obtain

$$r^2 = 4r\sin\theta.$$

Using $r^2 = x^2 + y^2$ and $y = r\sin\theta$ gives us

$$x^2 + y^2 = 4y$$

or, equivalently, $$x^2 + (y-2)^2 = 4.$$

Thus, the graph is a circle of radius 2 with center at $(0, 2)$ in the xy-plane.

EXAMPLE ▪ 7

Find a polar equation for the hyperbola $x^2 - y^2 = 16$.

SOLUTION Using the formulas $x = r\cos\theta$ and $y = r\sin\theta$, we obtain the following polar equations:

$$(r\cos\theta)^2 - (r\sin\theta)^2 = 16$$
$$r^2\cos^2\theta - r^2\sin^2\theta = 16$$
$$r^2(\cos^2\theta - \sin^2\theta) = 16$$
$$r^2\cos 2\theta = 16$$
$$r^2 = \frac{16}{\cos 2\theta}$$
$$r^2 = 16\sec 2\theta$$

The division by $\cos 2\theta$ is allowable, because $\cos 2\theta \neq 0$. (Note that if $\cos 2\theta = 0$, then $r^2\cos 2\theta \neq 16$.)

EXAMPLE ▪ 8

Find a polar equation of an arbitrary line.

SOLUTION Every line in an xy-coordinate plane is the graph of a linear equation $ax + by + c = 0$. Using the formulas $x = r\cos\theta$ and $y = r\sin\theta$ gives us the following equivalent polar equations:

$$ar\cos\theta + br\sin\theta + c = 0$$
$$r(a\cos\theta + b\sin\theta) + c = 0$$

If we superimpose an xy-plane on an $r\theta$-plane, then the graph of a polar equation may be symmetric with respect to the x-axis, the y-axis, or the origin. Some typical symmetries are illustrated in Figure 45. This leads to the next result.

FIGURE 45

Symmetries of graphs of polar equations
(i) x-axis (ii) y-axis (iii) Origin

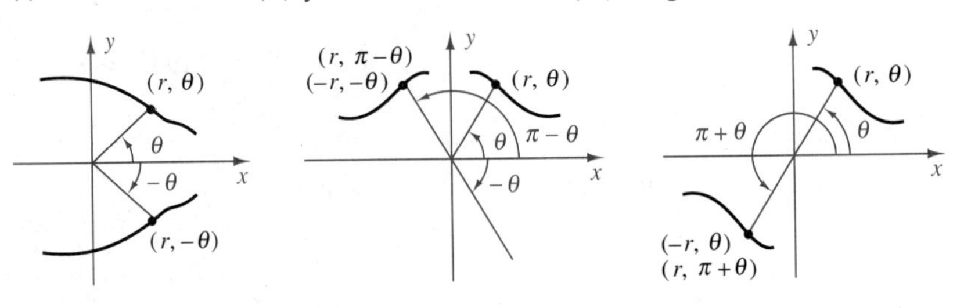

TESTS FOR SYMMETRY

(i) The graph of $r = f(\theta)$ is symmetric with respect to the x-axis if substitution of $-\theta$ for θ leads to an equivalent equation.

(ii) The graph of $r = f(\theta)$ is symmetric with respect to the y-axis if substitution of either (a) $\pi - \theta$ for θ or (b) $-r$ for r and $-\theta$ for θ leads to an equivalent equation.

(iii) The graph of $r = f(\theta)$ is symmetric with respect to the origin if substitution of either (a) $-r$ for r or (b) $\pi + \theta$ for θ leads to an equivalent equation.

To illustrate, since $\cos(-\theta) = \cos\theta$, the graph of the polar equation $r = 2 + 4\cos\theta$ in Example 3 is symmetric with respect to the x-axis by test (i). Since $\sin(\pi - \theta) = \sin\theta$, the graph in Example 1 is symmetric with respect to the y-axis by test (ii). The graph in Example 4 is symmetric to both axes and to the origin. Other tests for symmetry may be stated; however, those we have listed are among the easiest to apply.

Unlike the graph of an equation in x and y, it is possible for the graph of a polar equation $r = f(\theta)$ to be symmetric with respect to an axis or the origin *without* satisfying one of the preceding tests for symmetry. This is true because of the many different ways of specifying a point in polar coordinates.

Another difference between rectangular and polar coordinate systems is that the points of intersection of two graphs cannot always be found by solving the polar equations simultaneously. To illustrate, from Example 1, the graph of $r = 4\sin\theta$ is a circle of diameter 4 with center at $(2, \pi/2)$ (see Figure 39). Similarly, the graph of $r = 4\cos\theta$ is a circle of diameter 4 with center at $(2, 0)$ on the polar axis. Referring to Figure 46, the coordinates of the point of intersection $P(2\sqrt{2}, \pi/4)$ in quadrant I satisfy both equations; however, the origin O, which is on each circle, *cannot* be found by solving the equations simultaneously. Thus, when searching for points of intersection of polar graphs, it is sometimes necessary to refer to the graphs themselves, *in addition* to solving the two equations simultaneously. An alternative method is to use different (equivalent) equations for the graphs.

FIGURE 46

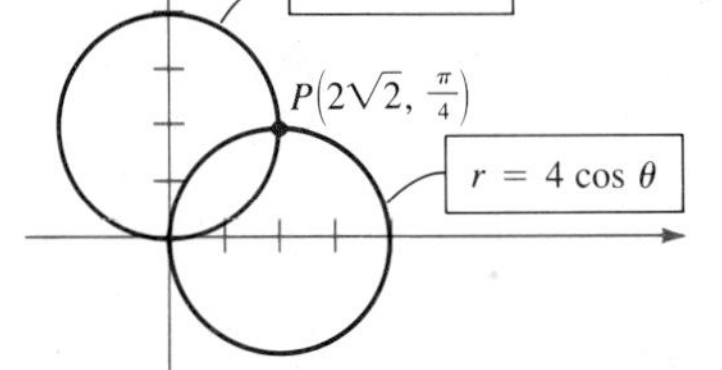

11.6 EXERCISES

Exer. 1–34: Sketch the graph of the polar equation.

1 $r = 5$

2 $r = -2$

3 $\theta = -\dfrac{\pi}{6}$

4 $\theta = \dfrac{\pi}{4}$

5 $r = 3\cos\theta$

6 $r = -2\sin\theta$

7 $r = 4\cos\theta + 2\sin\theta$

8 $r = 6\cos\theta - 2\sin\theta$

9 $r = 4(1 - \sin\theta)$

10 $r = 3(1 + \cos\theta)$

11 $r = -6(1 + \cos\theta)$

12 $r = 2(1 + \sin\theta)$

13 $r = 2 + 4 \sin \theta$

14 $r = 1 + 2 \cos \theta$

15 $r = \sqrt{3} - 2 \sin \theta$

16 $r = 2\sqrt{3} - 4 \cos \theta$

17 $r = 2 - \cos \theta$

18 $r = 5 + 3 \sin \theta$

19 $r = 4 \csc \theta$

20 $r = -3 \sec \theta$

21 $r = 8 \cos 3\theta$

22 $r = 2 \sin 4\theta$

23 $r = 3 \sin 2\theta$

24 $r = 8 \cos 5\theta$

25 $r^2 = 4 \cos 2\theta$ (lemniscate)

26 $r^2 = -16 \sin 2\theta$

27 $r = 2^\theta, \quad \theta \geq 0$ (spiral)

28 $r = e^{2\theta}, \quad \theta \geq 0$ (logarithmic spiral)

29 $r = 2\theta, \theta \geq 0$

30 $r\theta = 1, \quad \theta > 0$ (spiral)

31 $r = 6 \sin^2 \left(\frac{\theta}{2}\right)$

32 $r = -4 \cos^2 \left(\frac{\theta}{2}\right)$

33 $r = 2 + 2 \sec \theta$ (conchoid)

34 $r = 1 - \csc \theta$

Exer. 35–44: Find a polar equation that has the same graph as the equation in x and y.

35 $x = -3$

36 $y - 2$

37 $x^2 + y^2 = 16$

38 $x^2 = 8y$

39 $2y = -x$

40 $y = 6x$

41 $y^2 - x^2 = 4$

42 $xy = 8$

43 $(x - 1)^2 + y^2 = 1$

44 $(x + 2)^2 + (y - 3)^2 = 13$

Exer. 45–62: Find an equation in x and y that has the same graph as the polar equation. Use it to help sketch the graph in an $r\theta$-plane.

45 $r \cos \theta = 5$

46 $r \sin \theta = -2$

47 $r - 6 \sin \theta = 0$

48 $r = 2$

49 $\theta = \frac{\pi}{4}$

50 $r = 4 \sec \theta$

51 $r^2(4 \sin^2 \theta - 9 \cos^2 \theta) = 36$

52 $r^2 (\cos^2 \theta + 4 \sin^2 \theta) = 16$

53 $r^2 \cos 2\theta = 1$

54 $r^2 \sin 2\theta = 4$

55 $r (\sin \theta - 2 \cos \theta) = 16$

56 $r(3 \cos \theta - 4 \sin \theta) = 12$

57 $r (\sin \theta + r \cos^2 \theta) = 1$

58 $r(r \sin^2 \theta - \cos \theta) = 3$

59 $r = 8 \sin \theta - 2 \cos \theta$

60 $r = 2 \cos \theta - 4 \sin \theta$

61 $r = \tan \theta$

62 $r = 6 \cot \theta$

63 If $P_1(r_1, \theta_1)$ and $P_2(r_2, \theta_2)$ are points in an $r\theta$-plane, use the law of cosines to prove that

$$[d(P_1, P_2)]^2 = r_1^2 + r_2^2 - 2r_1r_2 \cos (\theta_2 - \theta_1).$$

64 Prove that the graph of each polar equation is a circle, and find its center and radius:

(a) $r = a \sin \theta, \quad a \neq 0$

(b) $r = b \cos \theta, \quad b \neq 0$

(c) $r = a \sin \theta + b \cos \theta, \quad ab \neq 0$

11.7 POLAR EQUATIONS OF CONICS

The following theorem combines the definitions of parabola, ellipse, and hyperbola into a unified description of the conic sections.

THEOREM

Let F be a fixed point and l a fixed line in a plane. The set of all points P in the plane, such that the ratio $d(P, F)/d(P, Q)$ is a positive constant e with $d(P, Q)$ the distance from P to l, is a conic section. The conic is a parabola if $e = 1$, an ellipse if $0 < e < 1$, and a hyperbola if $e > 1$.

FIGURE 47

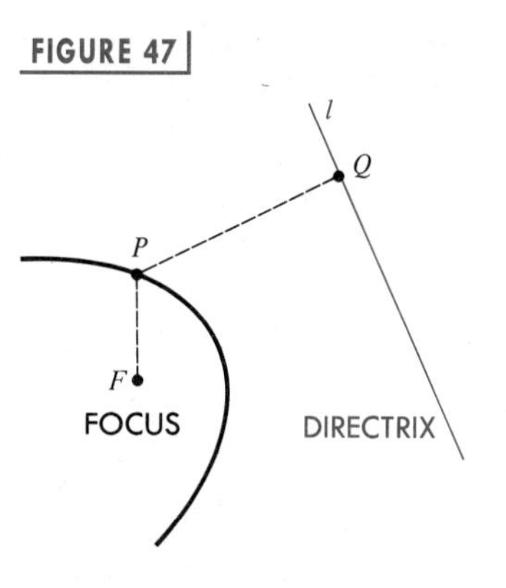

FIGURE 48

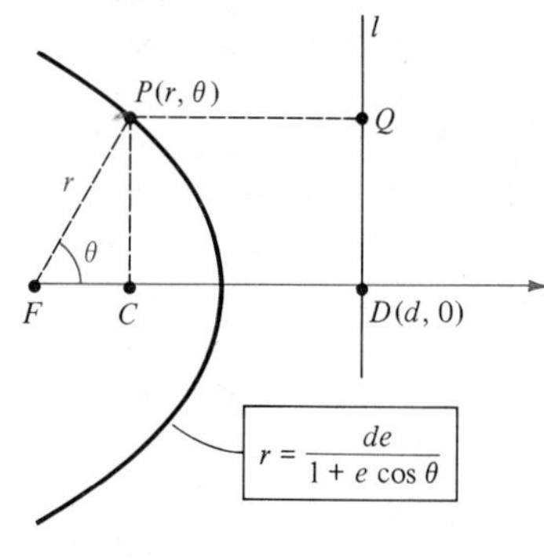

The constant e is the **eccentricity** of the conic. We will see that *the point* F *is a* **focus** *of the conic*. The line l is a **directrix**. A typical situation is illustrated in Figure 47.

We shall prove the theorem if $e \leq 1$ and leave the case $e > 1$ as an exercise.

If $e = 1$, then $d(P, F) = d(P, Q)$ and, by definition, the resulting conic is a parabola with focus F and directrix l.

Suppose next that $0 < e < 1$. It is convenient to introduce a polar coordinate system in the plane with F as the pole and with l perpendicular to the polar axis at the point $D(d, 0)$ with $d > 0$. If $P(r, \theta)$ is a point in the plane such that $d(P, F)/d(P, Q) = e < 1$, then from Figure 48 we see that P lies to the left of l. Let C be the projection of P on the polar axis. Since

$$d(P, F) = r \quad \text{and} \quad d(P, Q) = d - r\cos\theta,$$

it follows that P satisfies the condition in the theorem if and only if any of the following are identities:

$$\frac{r}{d - r\cos\theta} = e$$

$$r = de - er\cos\theta$$

$$r(1 + e\cos\theta) = de$$

$$r = \frac{de}{1 + e\cos\theta}$$

The same equations are obtained if $e = 1$; however, there is no point (r, θ) on the graph if $1 + \cos\theta = 0$.

An equation in x and y corresponding to $r = de - er\cos\theta$ is

$$\sqrt{x^2 + y^2} = de - ex.$$

Squaring both sides and rearranging terms leads to

$$(1 - e^2)x^2 + 2de^2x + y^2 = d^2e^2.$$

Completing the square and simplifying, we obtain

$$\left(x + \frac{de^2}{1 - e^2}\right)^2 + \frac{y^2}{1 - e^2} = \frac{d^2e^2}{(1 - e^2)^2}.$$

Finally, dividing both sides by $d^2e^2/(1 - e^2)^2$ gives us an equation of the form

$$\frac{(x - h)^2}{a^2} + \frac{y^2}{b^2} = 1$$

with $h = -de^2/(1 - e^2)$. Consequently, the graph is an ellipse with center at the point $(h, 0)$ on the x-axis and with

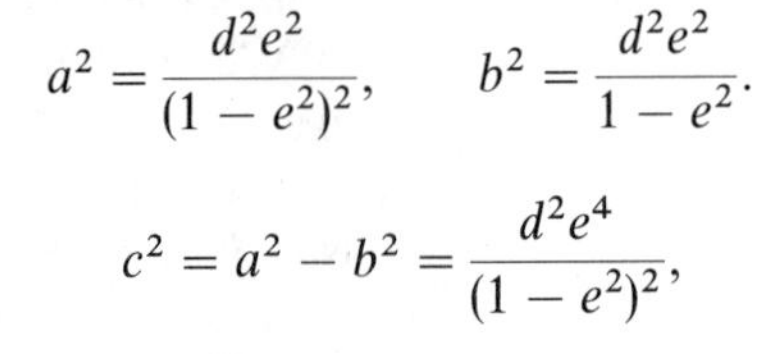

$$a^2 = \frac{d^2e^2}{(1 - e^2)^2}, \qquad b^2 = \frac{d^2e^2}{1 - e^2}.$$

Since

$$c^2 = a^2 - b^2 = \frac{d^2e^4}{(1 - e^2)^2},$$

we obtain $c = de^2/(1 - e^2)$ and hence $|h| = c$. This proves that F is a focus of the ellipse. It also follows that $e = c/a$. A similar proof may be given for the case $e > 1$.

FIGURE 49

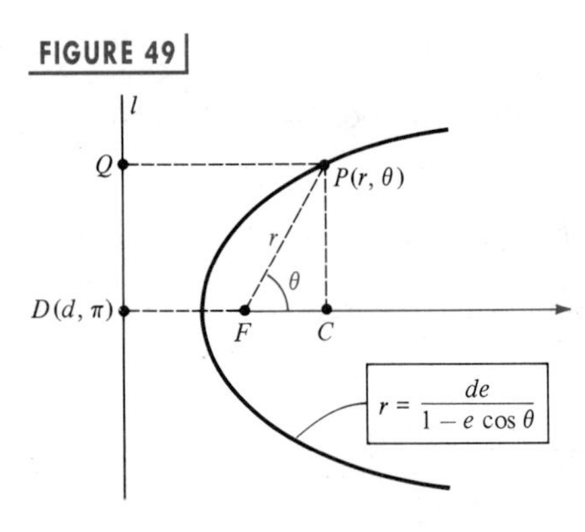

We can show, conversely, that every conic that is not degenerate may be described by means of the statement in the theorem. This gives us a formulation of conic sections that is equivalent to the approach used previously. Since the theorem includes all three types of conics, it is sometimes regarded as a definition for the conic sections.

If we had chosen the focus F to the *right* of the directrix, as illustrated in Figure 49 (with $d > 0$), then the equation $r = de/(1 - e\cos\theta)$ would have resulted. (Note the minus sign in place of the plus sign.) Other sign changes occur if d is allowed to be negative.

If we had taken l *parallel* to the polar axis through one of the points $(d, \pi/2)$ or $(d, 3\pi/2)$, as illustrated in Figure 50, then the corresponding equations would contain $\sin\theta$ instead of $\cos\theta$.

FIGURE 50

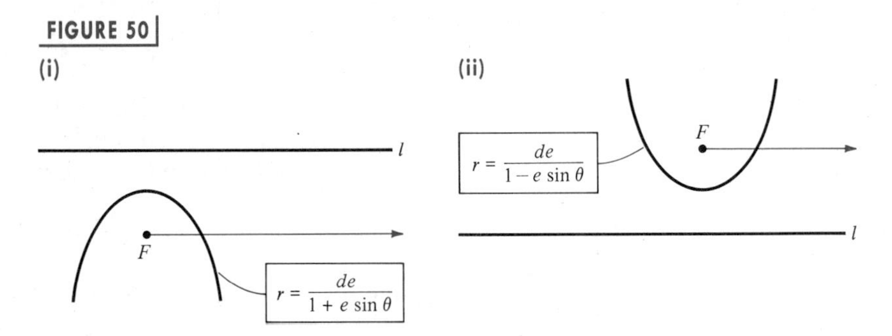

The following theorem summarizes our discussion.

THEOREM

> A polar equation that has one of the four forms
>
> $$r = \frac{de}{1 \pm e\cos\theta}, \qquad r = \frac{de}{1 \pm e\sin\theta}$$
>
> is a conic section. The conic is a parabola if $e = 1$, an ellipse if $0 < e < 1$, or a hyperbola if $e > 1$.

EXAMPLE ▪ 1

Describe and sketch the graph of the polar equation $r = \dfrac{10}{3 + 2\cos\theta}$.

SOLUTION We first divide numerator and denominator of the fraction by 3:

$$r = \frac{\frac{10}{3}}{1 + \frac{2}{3}\cos\theta}$$

This equation has one of the forms in the theorem with $e = \frac{2}{3}$. Thus the graph is an ellipse with focus F at the pole and major axis along the polar axis. We find the endpoints of the major axis by letting $\theta = 0$ and $\theta = \pi$. This gives us $V(2, 0)$ and $V'(10, \pi)$. Hence

$$2a = d(V', V) = 12, \quad \text{or} \quad a = 6.$$

FIGURE 51

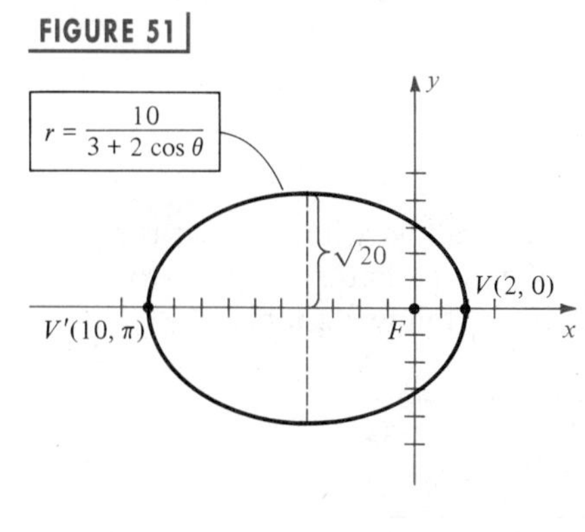

The center of the ellipse is the midpoint $(4, \pi)$ of the segment $V'V$. Using the fact that $e = c/a$, we obtain

$$c = ae = 6(\tfrac{2}{3}) = 4.$$

Hence,
$$b^2 = a^2 - c^2 = 36 - 16 = 20.$$

Thus, $b = \sqrt{20}$. The graph is sketched in Figure 51. For reference, we have superimposed an xy-coordinate system on the polar system.

EXAMPLE ▪ 2

Describe and sketch the graph of the polar equation $r = \dfrac{10}{2 + 3\sin\theta}$.

SOLUTION To express the equation in the proper form we divide numerator and denominator of the fraction by 2:

$$r = \frac{5}{1 + \frac{3}{2}\sin\theta}$$

Thus $e = \frac{3}{2}$ and the graph is a hyperbola with a focus at the pole. The expression $\sin\theta$ tells us that the transverse axis of the hyperbola is perpendicular to the polar axis. To find the vertices we let $\theta = \pi/2$ and $\theta = 3\pi/2$ in the given equation. This gives us the points $V(2, \pi/2)$, $V'(-10, 3\pi/2)$, and hence

$$2a = d(V, V') = 8, \quad \text{or} \quad a = 4.$$

The points $(5, 0)$ and $(5, \pi)$ on the graph can be used to sketch the lower branch of the hyperbola. The upper branch is obtained by symmetry, as

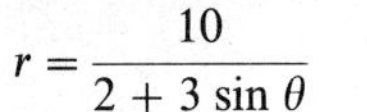

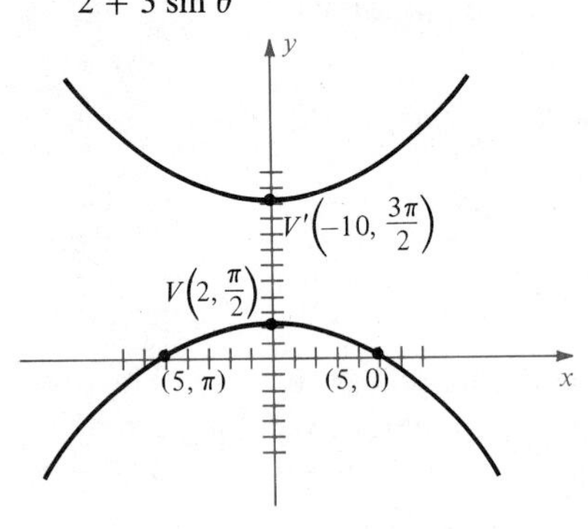

illustrated in Figure 52. If we desire more accuracy or additional information, we may calculate

$$c = ae = 4(\tfrac{3}{2}) = 6$$

and

$$b^2 = c^2 - a^2 = 36 - 16 = 20.$$

Asymptotes may then be constructed in the usual way.

EXAMPLE ▪ 3

Sketch the graph of the polar equation $r = \dfrac{15}{4 - 4 \cos \theta}$.

SOLUTION To obtain the proper form we divide numerator and denominator by 4:

$$r = \frac{\frac{15}{4}}{1 - \cos \theta}$$

Consequently, $e = 1$ and the graph is a parabola with focus at the pole. We may obtain a sketch by plotting the points that correspond to the x- and y-intercepts. These are indicated in the following table.

θ	0	$\frac{\pi}{2}$	π	$\frac{3\pi}{2}$
r	undefined	$\frac{15}{4}$	$\frac{15}{8}$	$\frac{15}{4}$

FIGURE 53

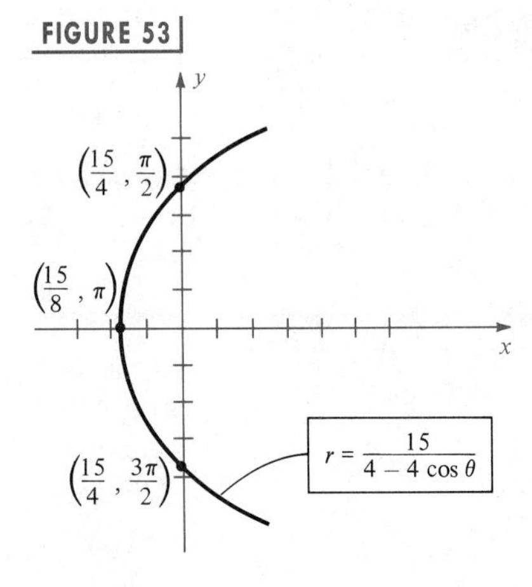

Note that there is no point on the graph corresponding to $\theta = 0$, since the denominator $1 - \cos \theta$ is 0 for that value. Plotting the three points and using the fact that the graph is a parabola with focus at the pole gives us the sketch in Figure 53.

If we desire only a rough sketch of a conic, then the technique employed in Example 3 is recommended. To use this method we plot (if possible) points corresponding to $\theta = 0, \pi/2, \pi$, and $3\pi/2$. These points, together with the type of conic (obtained from the value of e), readily lead to the sketch.

EXAMPLE ▪ 4

Find an equation in x and y that has the same graph as the polar equation

$$r = \frac{15}{4 - 4 \sin \theta}.$$

SOLUTION We multiply both sides of the polar equation by the lcd, $4 - 4 \sin \theta$, and then use relationships between rectangular and polar coordinates:

$$r(4 - 4 \sin \theta) = 15$$

$$4r = 15 + 4r \sin \theta$$

$$4(\pm\sqrt{x^2 + y^2}) = 15 + 4y$$

We next square both sides and simplify:

$$16(x^2 + y^2) = 225 + 120y + 16y^2$$

$$16x^2 = 120y + 225$$

EXAMPLE ▪ 5

Find a polar equation of the conic with a focus at the pole, eccentricity $e = \frac{1}{2}$, and directrix $r = -3 \sec \theta$.

SOLUTION The equation $r = -3 \sec \theta$ of the directrix may be written $r \cos \theta = -3$, which is equivalent to $x = -3$ in a rectangular coordinate system. This gives us the situation illustrated in Figure 49 with $d = 3$. Hence a polar equation has the form

$$r = \frac{de}{1 - e \cos \theta}.$$

We now substitute $d = 3$ and $e = \frac{1}{2}$:

$$r = \frac{3(\frac{1}{2})}{1 - \frac{1}{2} \cos \theta}$$

$$r = \frac{3}{2 - \cos \theta}$$

11.7 EXERCISES

Exer. 1–12: Find the eccentricity and classify the conic. Sketch the graph and label the vertices.

1 $r = \dfrac{12}{6 + 2 \sin \theta}$

2 $r = \dfrac{12}{6 - 2 \sin \theta}$

3 $r = \dfrac{12}{2 - 6 \cos \theta}$

4 $r = \dfrac{12}{2 + 6 \cos \theta}$

5 $r = \dfrac{3}{2 + 2 \cos \theta}$

6 $r = \dfrac{3}{2 - 2 \sin \theta}$

7 $r = \dfrac{4}{\cos \theta - 2}$

8 $r = \dfrac{4 \sec \theta}{2 \sec \theta - 1}$

9 $r = \dfrac{6 \csc \theta}{2 \csc \theta + 3}$

10 $r = \dfrac{8 \csc \theta}{2 \csc \theta - 5}$

11 $r = \dfrac{4 \csc \theta}{1 + \csc \theta}$

12 $r = \csc \theta (\csc \theta - \cot \theta)$

13–24 Find equations in x and y for the polar equations in Exercises 1–12.

Exer. 25–32: Find a polar equation of the conic with focus at the pole and the given eccentricity and directrix.

25 $e = \frac{1}{3}, \quad r = 2 \sec \theta$

26 $e = 1, \quad r \cos \theta = 5$

27 $e = \frac{4}{3}, \quad r \cos \theta = -3$

28 $e = 3, \quad r = -4 \sec \theta$

29 $e = 1, \quad r \sin \theta = -2$

30 $e = 4, \quad r = -3 \csc \theta$

31 $e = \frac{2}{5}, \quad r = 4 \csc \theta$

32 $e = \frac{3}{4}, \quad r \sin \theta = 5$

Exer. 33–34: Find a polar equation of the parabola with focus at the pole and the given vertex.

33 $V\left(4, \dfrac{\pi}{2}\right)$

34 $V(5, 0)$

Exer. 35–36: An ellipse has a focus at the pole with the given center C and vertex V. Find (a) the eccentricity and (b) a polar equation for the ellipse.

35 $C\left(3, \dfrac{\pi}{2}\right), \quad V\left(1, \dfrac{3\pi}{2}\right)$

36 $C(2, \pi), \quad V(1, 0)$

37 Kepler's first law asserts that planets travel in elliptical orbits with the sun at one focus. To find an equation of an orbit, place the pole O at the center of the sun and the polar axis along the major axis of the ellipse (see figure).

(a) Show that an equation of the orbit is $r = \dfrac{(1 - e^2)a}{1 - e \cos \theta}$ for eccentricity e and major axis of length $2a$.

(b) The perihelion distance r_{per} and aphelion distance r_{aph} are defined as the minimum and maximum distances, respectively, of a planet from the sun. Show that

$$r_{per} = a(1 - e) \text{ and } r_{aph} = a(1 + e).$$

EXERCISE 37

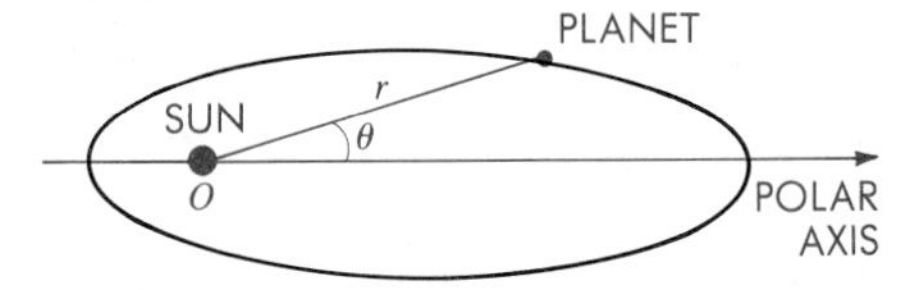

38 Refer to Exercise 37. The planet Pluto travels in an elliptical orbit of eccentricity 0.249. If the perihelion distance is 29.62 AU, find a polar equation for the orbit, and estimate the aphelion distance.

11.8 PLANE CURVES AND PARAMETRIC EQUATIONS

If f is a function, the graph of the equation $y = f(x)$ is often called a *plane curve*. However, this definition is restrictive, since it rules out many useful graphs. The following definition is more general.

DEFINITION

A **plane curve** is a set C of ordered pairs of the form $(f(t), g(t))$ for functions f and g that are defined on an interval I.

For simplicity, we often refer to a plane curve as a **curve**. The **graph** of the curve C in the definition consists of all points $P(t) = (f(t), g(t))$ in an xy-plane for t in I. Each $P(t)$ is a *point* on the curve. We shall use the term *curve* interchangeably with *graph of a curve*. We sometimes regard the point $P(t)$ as tracing the curve C as t varies through the interval

I. This is especially true in applications where t represents time and $P(t)$ is the position of a moving particle at time t.

The graphs of several curves are sketched in Figure 54 for the case where I is a closed interval $[a, b]$. If, as in (i) of the figure, $P(a) \neq P(b)$, then $P(a)$ and $P(b)$ are the **endpoints** of C. The curve illustrated in (i) intersects itself, that is, two different values of t produce the same point. If $P(a) = P(b)$, as illustrated in Figure 54(ii), then C is a **closed curve**. If $P(a) = P(b)$ and C does not intersect itself at any other point, as illustrated in (iii), then C is a **simple closed curve**.

FIGURE 54

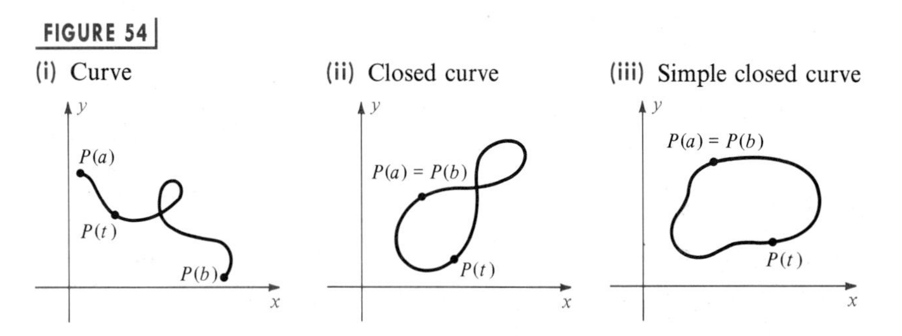

A convenient way to represent curves is given in the next definition.

DEFINITION

Let C be the curve consisting of all ordered pairs $(f(t), g(t))$ with f and g defined on an interval I. The equations

$$x = f(t), \qquad y = g(t),$$

for t in I, are **parametric equations** for C with **parameter** t.

The curve C in this definition is referred to as a **parametrized curve**, and the parametric equations are a **parametrization** for C. We often use the notation

$$x = f(t), \quad y = g(t); \qquad t \text{ in } I$$

to indicate the domain I of f and g. We sometimes obtain a sketch of the graph by plotting points and connecting them in the order of increasing t, as illustrated in Example 1. The example also shows that it may be possible to eliminate the parameter and obtain an equation for C that involves only the variables x and y.

EXAMPLE ▪ 1

Sketch the graph of the curve C with parametrization

$$x = 2t, \quad y = t^2 - 1; \qquad -1 \leq t \leq 2.$$

SOLUTION We may use the parametric equations to tabulate coordinates of points $P(x, y)$ on C as in the following table.

t	-1	$-\frac{1}{2}$	0	$\frac{1}{2}$	1	$\frac{3}{2}$	2
x	-2	-1	0	1	2	3	4
y	0	$-\frac{3}{4}$	-1	$-\frac{3}{4}$	0	$\frac{5}{4}$	3

FIGURE 55

$x = 2t;\ y = t^2 - 1;\ -1 \leq t \leq 2$

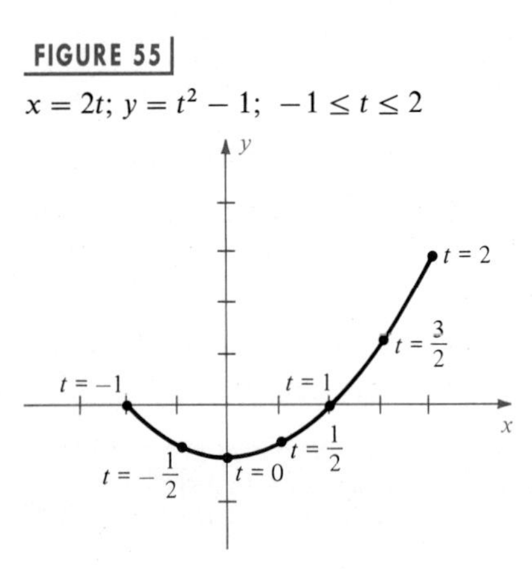

Plotting points gives us the sketch in Figure 55.

We may obtain a clearer description of the graph by eliminating the parameter. Solving the first parametric equation for t, we obtain $t = \frac{1}{2}x$. Substituting for t in the second equation gives us

$$y = (\tfrac{1}{2}x)^2 - 1, \quad \text{or} \quad y + 1 = \tfrac{1}{4}x^2.$$

The graph of this equation is a parabola with vertical axis and vertex at the point $(0, -1)$; however, since $x = 2t$ and $-1 \leq t \leq 2$, we see that $-2 \leq x \leq 4$. Hence the curve C is that part of the parabola between the points $(-2, 0)$ and $(4, 3)$ shown in Figure 55.

A curve has many different parametrizations. To illustrate, the curve C in Example 1 is given parametrically by any of the following:

$$x = 2t, \quad y = t^2 - 1; \qquad -1 \leq t \leq 2$$

$$x = t, \quad y = \tfrac{1}{4}t^2 - 1; \qquad -2 \leq t \leq 4$$

$$x = t^3, \quad y = \tfrac{1}{4}t^6 - 1; \qquad \sqrt[3]{-2} \leq t \leq \sqrt[3]{4}$$

The next example illustrates the fact that it is often useful to eliminate the parameter *before* plotting points.

EXAMPLE ▪ 2

Sketch the graph of the curve C that has the parametrization

$$x = -2 + t^2, \quad y = 1 + 2t^2; \qquad t \text{ in } \mathbb{R}.$$

FIGURE 56

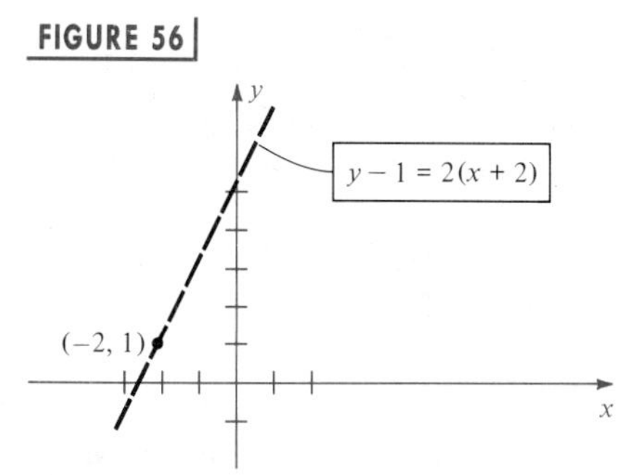

SOLUTION To eliminate the parameter, we note from the first equation that $t^2 = x + 2$. We substitute for t^2 in the second equation:

$$y = 1 + 2(x + 2)$$
$$y - 1 = 2(x + 2).$$

This is an equation of the line of slope 2 through the point $(-2, 1)$, as indicated by the dashes in Figure 56. However, since $t^2 \geq 0$,

$$x = -2 + t^2 \geq -2 \quad \text{and} \quad y = 1 + 2t^2 \geq 1.$$

FIGURE 57

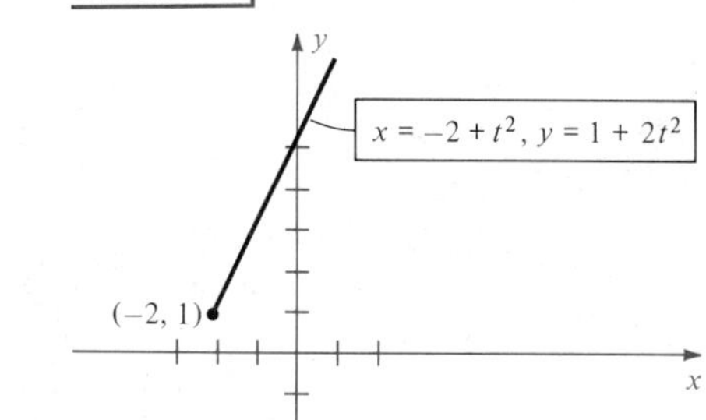

Thus the graph of C is that part of the line to the right of the point $(-2, 1)$, corresponding to $t = 0$, as shown in Figure 57. This fact may also be verified by plotting several points.

EXAMPLE ▪ 3

A particle moves in a plane such that its position $P(x, y)$ at time t is given by

$$x = a \cos t, \quad y = a \sin t; \qquad t \text{ in } \mathbb{R}$$

for a positive constant a. Describe the path of the particle.

SOLUTION If we rewrite the parametric equations as

$$\frac{x}{a} = \cos t, \qquad \frac{y}{a} = \sin t$$

and use the identity $\cos^2 t + \sin^2 t = 1$, we obtain

$$\left(\frac{x}{a}\right)^2 + \left(\frac{y}{a}\right)^2 = 1$$

and

$$x^2 + y^2 = a^2.$$

FIGURE 58

$x = a \cos t,\ y = a \sin t;\ t \text{ in } \mathbb{R}$

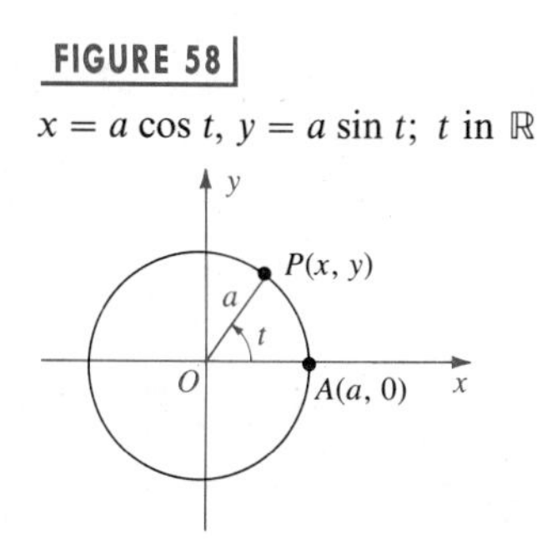

Thus the particle moves on the circle of radius a with center at the origin (see Figure 58). In particular, the particle is at the point $A(a, 0)$ when $t = 0$; at $(0, a)$ when $t = \pi/2$; at $(-a, 0)$ when $t = 3\pi/2$; and so on. In general, as t increases, the particle moves around the circle in a counterclockwise direction, making one revolution every 2π units of time.

Note that in this application, we may interpret the parameter t geometrically as the radian measure of the angle generated by the line segment OP.

If a curve C is described by an equation $y = f(x)$ for a function f, then an easy way to obtain parametric equations for C is to let

$$x = t, \quad y = f(t)$$

for t in the domain of f. For example, if $y = x^3$, then parametric equations are

$$x = t, \quad y = t^3; \qquad t \text{ in } \mathbb{R}.$$

We can use many different substitutions for x, provided that as t varies through some interval, x takes on all values in the domain of f. Thus the graph of $y = x^3$ is also given by

$$x = t^{1/3}, \quad y = t; \qquad t \text{ in } \mathbb{R}.$$

Note, however, that the parametric equations

$$x = \sin t, \quad y = \sin^3 t; \qquad t \text{ in } \mathbb{R}$$

give only that part of the graph of $y = x^3$ that lies between the points $(-1, -1)$ and $(1, 1)$.

EXAMPLE ▪ 4

Find three parametrizations for the line of slope m through the point (x_1, y_1).

SOLUTION By the point-slope form, an equation for the line is

$$y - y_1 = m(x - x_1).$$

If we let $x = t$, then $y - y_1 = m(t - x_1)$ and we obtain the parametric equations

$$x = t, \quad y = y_1 + m(t - x_1); \qquad t \text{ in } \mathbb{R}.$$

We obtain another pair of parametric equations for the line if we let $x - x_1 = t$. In this case $y - y_1 = mt$, and we have

$$x = x_1 + t, \quad y = y_1 + mt; \qquad t \text{ in } \mathbb{R}.$$

As a third illustration, if we let $x - x_1 = \tan t$, then

$$x = x_1 + \tan t, \quad y = y_1 + m \tan t; \qquad -\frac{\pi}{2} < t < \frac{\pi}{2}.$$

There are many other parametrizations for the line.

Parametric equations of the form

$$x = a \sin \omega_1 t, \quad y = b \cos \omega_2 t; \qquad t \geq 0,$$

for constants a, b, ω_1, and ω_2, occur in electrical theory. The variables x and y usually represent voltages or currents at time t. The resulting curve is often difficult to graph by plotting points or eliminating the parameter; however, using an oscilloscope and imposing voltages (or currents) on the horizontal and vertical input terminals, respectively, we can represent the graph by a figure on the screen of the oscilloscope. A figure of this type is called a **Lissajous figure**. Computers are also useful in obtaining these complicated graphs.

EXAMPLE ▪ 5

(a) Sketch the Lissajous figure corresponding to

$$x = \sin 2t, \quad y = \cos t; \qquad 0 \leq t \leq 2\pi.$$

(b) Find an equation in x and y for the curve.

SOLUTION

(a) The graph in Figure 59 was generated by a computer. To sketch the graph by plotting points, we could make a table as in Example 1. The following table gives some values of x and y for $0 \leq t \leq \pi$.

FIGURE 59

$x = \sin 2t$; $y = \cos t$; $0 \leq t \leq 2t$

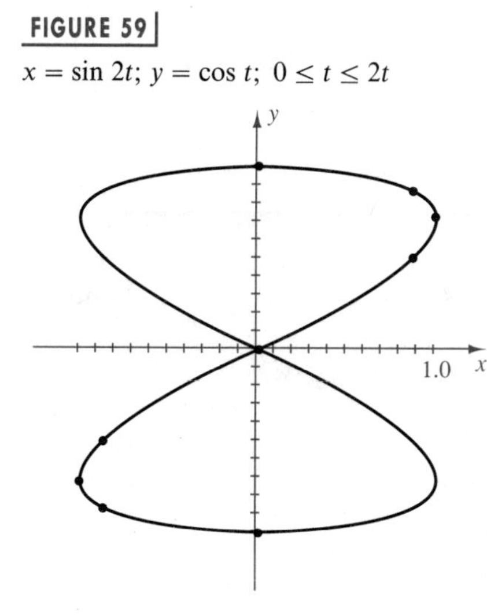

t	0	$\frac{\pi}{6}$	$\frac{\pi}{4}$	$\frac{\pi}{3}$	$\frac{\pi}{2}$	$\frac{2\pi}{3}$	$\frac{3\pi}{4}$	$\frac{5\pi}{6}$	π
x	0	$\frac{\sqrt{3}}{2}$	1	$\frac{\sqrt{3}}{2}$	0	$-\frac{\sqrt{3}}{2}$	-1	$-\frac{\sqrt{3}}{2}$	0
y	1	$\frac{\sqrt{3}}{2}$	$\frac{\sqrt{2}}{2}$	$\frac{1}{2}$	0	$-\frac{1}{2}$	$-\frac{\sqrt{2}}{2}$	$-\frac{\sqrt{3}}{2}$	-1

The points $P(x, y)$ with the indicated t-values are plotted in Figure 59. Note that as t increases from 0 to $\pi/2$, the point $P(x, y)$ traces the part of the curve in quadrant I (in a generally clockwise direction); as t increases from $\pi/2$ to π, $P(x, y)$ traces the part in quadrant III (in a counterclockwise direction); for $\pi \leq t \leq 3\pi/2$ we obtain the part in quadrant IV; and $3\pi/2 \leq t \leq 2\pi$ gives us the part in quadrant II.

(b) We may find an equation in x and y for the curve by employing trigonometric identities and algebraic manipulations. Writing $x = 2 \sin t \cos t$

and squaring gives us

$$x^2 = 4\sin^2 t \cos^2 t$$

and

$$x^2 = 4(1 - \cos^2 t)\cos^2 t.$$

Since $y = \cos t$, this leads to

$$x^2 = 4(1 - y^2)y^2,$$

which is an equation for the curve.

If we wish to express y in terms of x, we could rewrite the last equation as

$$4y^4 - 4y^2 + x^2 = 0$$

and use the quadratic formula to solve for y^2 as follows:

$$y^2 = \frac{4 \pm \sqrt{16 - 16x^2}}{8} = \frac{1 \pm \sqrt{1 - x^2}}{2}$$

Taking square roots,

$$y = \pm\sqrt{\frac{1 \pm \sqrt{1 - x^2}}{2}}.$$

These complicated equations clearly indicate the advantage of expressing the curve in parametric form.

EXAMPLE ▪ 6

The curve traced by a fixed point P on the circumference of a circle as the circle rolls along a line in a plane is called a **cycloid**. Find parametric equations for a cycloid.

SOLUTION Suppose the circle has radius a and that it rolls along (and above) the x-axis in the positive direction. If one position of P is the origin, then Figure 60 displays part of the curve and a possible position of the circle.

Let K denote the center of the circle and T the point of tangency with the x-axis. We introduce a parameter t as the radian measure of angle TKP. The distance the circle has rolled is $d(O, T) = at$. Consequently, the coordinates of K are (at, a). If we consider an $x'y'$-coordinate system with origin at $K(at, a)$ and if $P(x', y')$ denotes the point P relative to this system, then by the translation of axes formulas with $h = at$ and $k = a$,

$$x = at + x', \qquad y = a + y'.$$

FIGURE 60

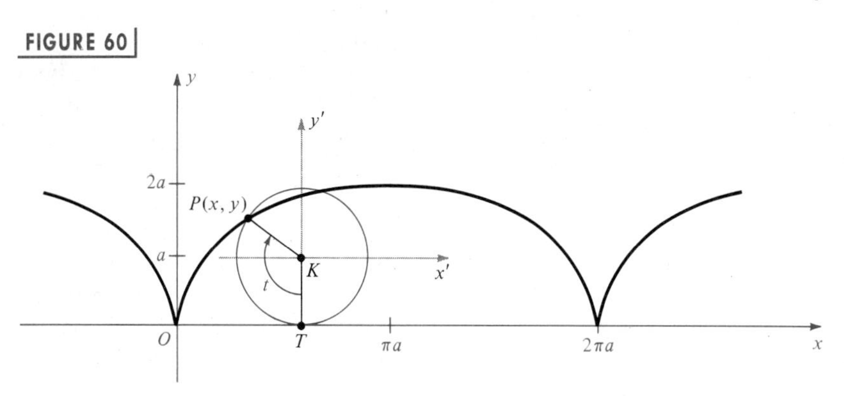

FIGURE 61

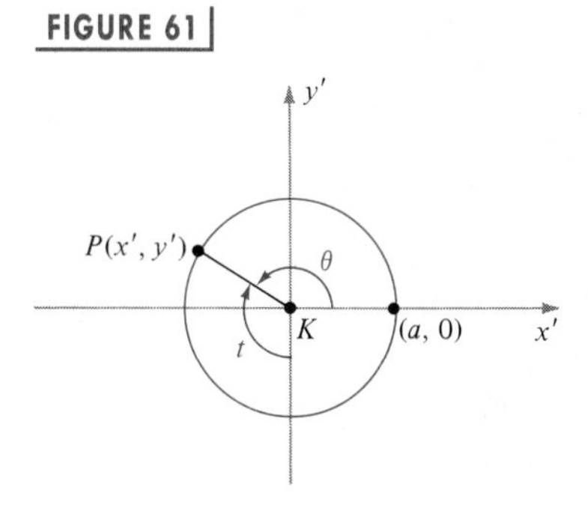

If, as in Figure 61, θ denotes an angle in standard position on the $x'y'$-plane, then $\theta = (3\pi/2) - t$. Hence

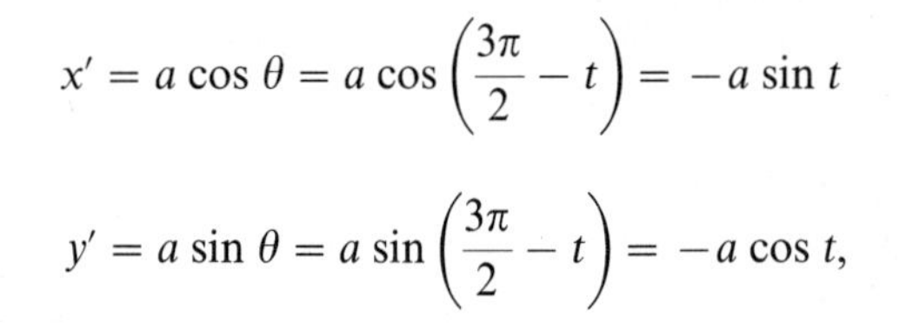

$$x' = a\cos\theta = a\cos\left(\frac{3\pi}{2} - t\right) = -a\sin t$$

$$y' = a\sin\theta = a\sin\left(\frac{3\pi}{2} - t\right) = -a\cos t,$$

and substitution in $x = at + x'$, $y = a + y'$ gives us parametric equations for the cycloid:

$$x = a(t - \sin t), \quad y = a(1 - \cos t); \qquad t \text{ in } \mathbb{R}.$$

FIGURE 62

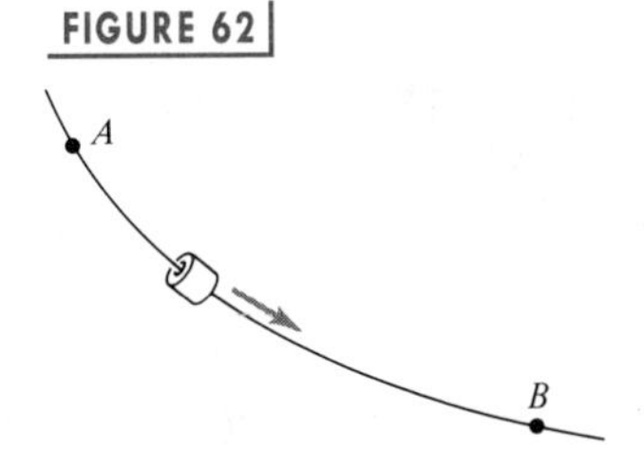

If $a < 0$, then the graph of $x = a(t - \sin t)$, $y = a(1 - \cos t)$ is the inverted cycloid that results if the circle of Example 6 rolls *below* the x-axis. This curve has a number of important physical properties. In particular, suppose a thin wire passes through two fixed points A and B as illustrated in Figure 62, and that the shape of the wire can be changed by bending it in any manner. Suppose further that a bead is allowed to slide along the wire and the only force acting on the bead is gravity. We now ask which of all the possible paths will allow the bead to slide from A to B in the least amount of time. It is natural to believe that the desired path is the straight line segment from A to B; however, this is not the correct answer. The path that requires the least time coincides with the graph of the inverted cycloid with A at the origin. Because the velocity of the bead increases more rapidly along the cycloid than along the line through A and B, the bead reaches B more rapidly, even though the distance is greater.

There is another interesting property of this **curve of least descent**. Suppose that A is the origin and B is the point with x-coordinate $\pi|a|$, that is, the lowest point on the cycloid in the first arc to the right of A. If the bead is released at *any* point between A and B, the time required for it to reach B is always the same.

Variations of the cycloid occur in the investigation of applications. For example, if a motorcycle wheel rolls along a straight road, then the curve traced by a fixed point on one of the spokes is a cycloidlike curve. In this case, the curve does not have sharp corners, nor does it intersect the road (the x-axis) as does the graph of a cycloid. If the wheel of a train rolls along a railroad track, then the curve traced by a fixed point on the circumference of the wheel (which extends below the track) contains loops at regular intervals (see Exercise 38). Other cycloids are defined in Exercises 43 and 44.

11.8 EXERCISES

Exer. 1–18:

(a) Sketch the graph of the curve C.

(b) Find an equation in x and y whose graph contains the points on C.

1 $x = t - 2, \quad y = 2t + 3; \qquad 0 \le t \le 5$

2 $x = 1 - 2t, \quad y = 1 + t; \qquad -1 \le t \le 4$

3 $x = t^2 + 1, \quad y = t^2 - 1; \qquad -2 \le t \le 2$

4 $x = t^3 + 1, \quad y = t^3 - 1; \qquad -2 \le t \le 2$

5 $x = 4t^2 - 5, \quad y = 2t + 3; \qquad t \text{ in } \mathbb{R}$

6 $x = t^3, \quad y = t^2; \quad t \text{ in } \mathbb{R}$

7 $x = e^t, \quad y = e^{-2t}; \qquad t \text{ in } \mathbb{R}$

8 $x = \sqrt{t}, \quad y = 3t + 4; \qquad t \ge 0$

9 $x = 2 \sin t, \quad y = 3 \cos t; \qquad 0 \le t \le 2\pi$

10 $x = \cos t - 2, \quad y = \sin t + 3; \qquad 0 \le t \le 2\pi$

11 $x = \sec t, \quad y = \tan t; \qquad -\frac{\pi}{2} < t < \frac{\pi}{2}$

12 $x = \cos 2t, \quad y = \sin t; \qquad -\pi \le t \le \pi$

13 $x = t^2, \quad y = 2 \ln t; \qquad t > 0$

14 $x = \cos^3 t, \quad y = \sin^3 t; \qquad 0 \le t \le 2\pi$

15 $x = \sin t, \quad y = \csc t; \qquad 0 < t \le \frac{\pi}{2}$

16 $x = e^t, \quad y = e^{-t}; \qquad t \text{ in } \mathbb{R}$

17 $x = t, \quad y = \sqrt{t^2 - 1}; \qquad |t| \ge 1$

18 $x = -2\sqrt{1 - t^2}, \quad y = t; \qquad |t| \le 1$

Exer. 19–22: Sketch the graph of the curve C.

19 $x = t, \quad y = \sqrt{t^2 - 2t + 1}; \qquad 0 \le t \le 4$

20 $x = 2t, \quad y = 8t^3; \qquad -1 \le t \le 1$

21 $x = (t + 1)^3, \quad y = (t + 2)^2; \qquad 0 \le t \le 2$

22 $x = \tan t, \quad y = 1; \qquad -\frac{\pi}{2} < t < \frac{\pi}{2}$

Exer. 23–24: Curves C_1, C_2, C_3, and C_4 are given parametrically for t in $\mathbb{R}$. Sketch their graphs, and discuss their similarities and differences.

23 $C_1: x = t^2, \quad y = t$
$C_2: x = t^4, \quad y = t^2$
$C_3: x = \sin^2 t, \quad y = \sin t$
$C_4: x = e^{2t}, \quad y = -e^t$

24 $C_1: x = t, \quad y = 1 - t$
$C_2: x = 1 - t^2, \quad y = t^2$
$C_3: x = \cos^2 t, \quad y = \sin^2 t$
$C_4: x = \ln t - t, \quad y = 1 + t - \ln t$

25 Show that

$$x = a \cos t + h, \quad y = b \sin t + k; \qquad 0 \le t \le 2\pi$$

are parametric equations of an ellipse with center at the point (h, k) and axes of lengths a and b.

26 Show that

$$x = a \sec t + h, \quad y = b \tan t + k;$$

$$-\frac{\pi}{2} < t < \frac{3\pi}{2} \text{ and } t \neq \frac{\pi}{2}$$

are parametric equations of a hyperbola with center (h, k), transverse axis of length $2a$, and conjugate axis of length $2b$. Determine the values of t for each branch.

27 A circle C of radius b rolls on the outside of a second circle having equation $x^2 + y^2 = a^2$ with $b < a$. Let P be a fixed point on C and let the initial position of P be $A(a, 0)$ (see figure). If the parameter t is the angle from the positive x-axis to the line segment from O to the center of C, show that parametric equations for the curve traced by P (an *epicycloid*) are

$$x = (a + b) \cos t - b \cos\left(\frac{a + b}{b} t\right),$$

$$y = (a + b) \sin t - b \sin\left(\frac{a + b}{b} t\right);$$

$$0 \leq t \leq 2\pi.$$

EXERCISE 27

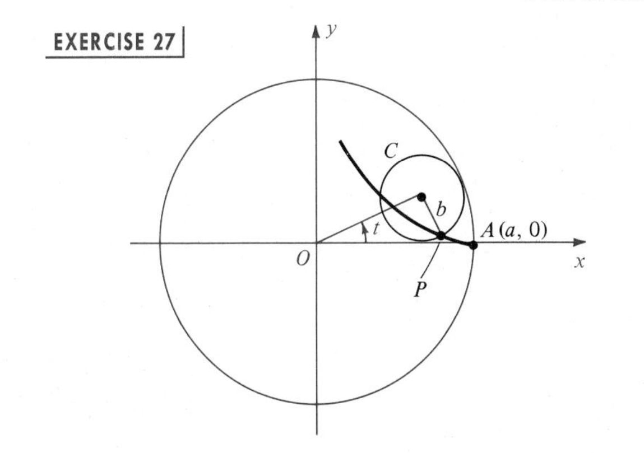

28 If the circle C of Exercise 27 rolls on the inside of the second circle (see figure), then the curve traced by P is a *hypocycloid*.

(a) Show that parametric equations for this curve are

$$x = (a - b) \cos t + b \cos\left(\frac{a - b}{b} t\right),$$

$$y = (a - b) \sin t - b \sin\left(\frac{a - b}{b} t\right);$$

$$0 \leq t \leq 2\pi.$$

(b) If $b = \frac{1}{4}a$, show that

$$x = a \cos^3 t, \quad y = a \sin^3 t,$$

and sketch the graph.

EXERCISE 28

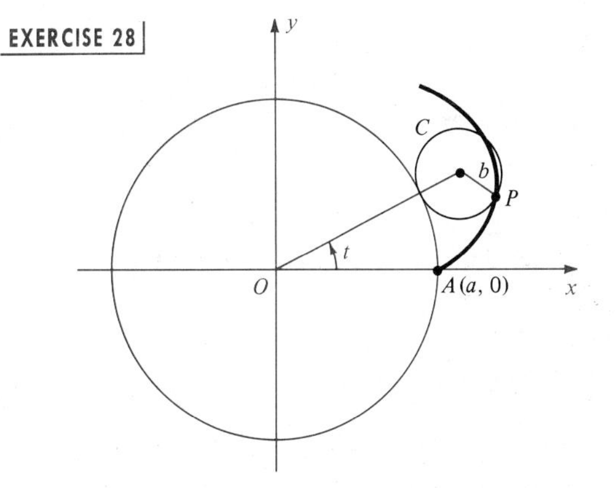

29 If $b = \frac{1}{3}a$ in Exercise 27, find parametric equations for the epicycloid, and sketch the graph.

30 The radius of circle B is one-third that of circle A. How many revolutions will circle B make as it rolls around circle A until it reaches its starting point? (*Hint:* Use Exercise 29.)

11.9 REVIEW

Define or discuss each of the following.

Conic sections ■ Parabola: focus directrix, vertex, axis ■ Ellipse: foci, vertices, major and minor axes ■ Hyperbola: foci, vertices, transverse and conjugate axes, asymptotes, branches ■ Translation of axes ■ Rotation of axes ■ Polar coordinates of a point ■ The relationship between polar coordinates and rectangular coordinates ■ Graph of a polar equation ■ Polar equations of conics ■ Plane curve ■ Parametric equations for a curve

11.9 EXERCISES

Exer. 1–16: Find the vertices and foci of the conic, and sketch its graph.

1 $y^2 = 64x$

2 $y - 1 = 8(x + 2)^2$

3 $9y^2 = 144 - 16x^2$

4 $9y^2 = 144 + 16x^2$

5 $x^2 - y^2 - 4 = 0$

6 $25x^2 + 36y^2 = 1$

7 $25y = 100 - x^2$

8 $3x^2 + 4y^2 - 18x + 8y + 19 = 0$

9 $x^2 - 9y^2 + 8x + 90y - 210 = 0$

10 $x = 2y^2 + 8y + 3$

11 $4x^2 + 9y^2 + 24x - 36y + 36 = 0$

12 $4x^2 - y^2 - 40x - 8y + 88 = 0$

13 $y^2 - 8x + 8y + 32 = 0$

14 $4x^2 + y^2 - 24x + 4y + 36 = 0$

15 $x^2 - 9y^2 + 8x + 7 = 0$

16 $y^2 - 2x^2 + 6y + 8x - 3 = 0$

Exer. 17–26: Find an equation for the conic that satisfies the given conditions.

17 The hyperbola with vertices $V(0, \pm 7)$ and endpoints of conjugate axes $(\pm 3, 0)$

18 The parabola with focus $F(-4, 0)$ and directrix $x = 4$

19 The parabola with focus $F(0, -10)$ and directrix $y = 10$

20 The parabola with vertex at the origin, symmetric to the x-axis, and passing through the point $(5, -1)$

21 The ellipse with vertices $V(0, \pm 10)$ and foci $F(0, \pm 5)$

22 The hyperbola with foci $F(\pm 10, 0)$ and vertices $V(\pm 5, 0)$

23 The hyperbola with vertices $V(0, \pm 6)$ and asymptotes $y = \pm 9x$

24 The ellipse with foci $F(\pm 2, 0)$ and passing through the point $(2, \sqrt{2})$

25 The ellipse with eccentricity $\frac{2}{3}$ and endpoints of minor axis $(\pm 5, 0)$

26 The ellipse with eccentricity $\frac{3}{4}$ and foci $F(\pm 12, 0)$

Exer. 27–34: Sketch the graph of the polar equation.

27 $r = -4 \sin \theta$

28 $r = 3 \sin 5\theta$

29 $r = 6 - 3 \cos \theta$

30 $r^2 = 9 \sin 2\theta$

31 $2r = \theta$

32 $r = \dfrac{8}{1 - 3 \sin \theta}$

33 $r = 6 - r \cos \theta$

34 $r = 8 \sec \theta$

Exer. 35–38: Find a polar equation that has the same graph as the equation in x and y.

35 $y^2 = 4x$

36 $x^2 + y^2 - 3x + 4y = 0$

37 $2x - 3y = 8$

38 $x^2 + y^2 = 2xy$

Exer. 39–42: Find an equation in x and y that has the same graph as the polar equation.

39 $r^2 = \tan \theta$

40 $r = 2 \cos \theta + 3 \sin \theta$

41 $r^2 = 4 \sin 2\theta$

42 $\theta = \sqrt{3}$

Exer. 43–45:

(a) Sketch the graph of the curve C.

(b) Find an equation in x and y whose graph contains the points on C.

43 $x = \dfrac{1}{t} + 1, \quad y = \dfrac{2}{t} - t; \qquad 0 < t \le 4$

44 $x = \cos^2 t - 2, \quad y = \sin t + 1; \qquad 0 \le t \le 2\pi$

45 $x = \sqrt{t}, \quad y = 2^{-t}; \qquad t \ge 0$

46 Curves C_1, C_2, C_3, and C_4 are given parametrically for t in $\mathbb{R}$. Sketch their graphs, and discuss their similarities and differences.

$C_1: x = t, \quad y = \sqrt{t}$

$C_2: x = t^2, \quad y = t$

$C_3: x = 1 - \sin^2 t, \quad y = \cos t$

$C_4: x = e^{2t}, \quad y = -e^t$

Exer. 47–49: Use a suitable rotation of axes to find an equation for the graph in an $x'y'$-plane, and then sketch the graph, labeling vertices.

47 $x^2 - 8xy + 16y^2 - 12\sqrt{17}\,x - 3\sqrt{17}\,y = 0$

48 $8x^2 + 12xy + 17y^2 - 16\sqrt{5}\,x - 12\sqrt{5}\,y = 0$

49 $11x^2 - 10\sqrt{3}\,xy + y^2 = 20$

50 Use the discriminant to identify the graph of each equation. (Do not sketch the graph.)

(a) $2x^2 - 3xy + 4y^2 + 6x - 2y - 6 = 0$

(b) $3x^2 + 2xy - y^2 - 2x + y + 4 = 0$

(c) $x^2 - 6xy + 9y^2 + x - 3y + 5 = 0$

51 A bridge is to be constructed across a river that is 200 feet wide. The arch of the bridge is to be semielliptical and must be constructed so that a ship less than 50 feet wide and 30 feet high can pass safely through the arch, as shown in the figure.

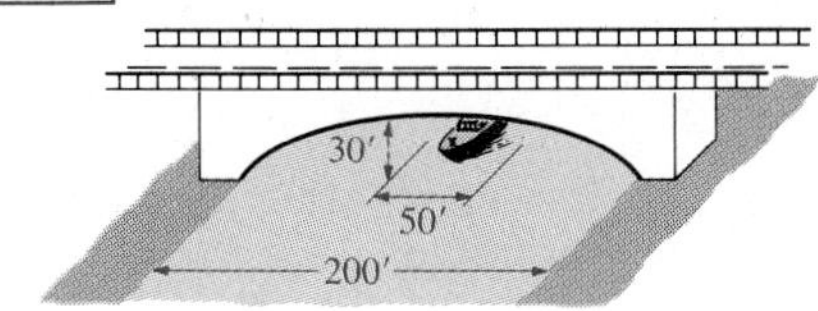

(a) Find an equation for the arch.

(b) Approximate the height of the arch in the middle of the bridge.

52 As shown in the figure, point $P(x, y)$ is the same distance from the point $(4, 0)$ as it is from the circle $x^2 + y^2 = 4$. Show that the collection of all such points forms a branch of a hyperbola, and sketch its graph.

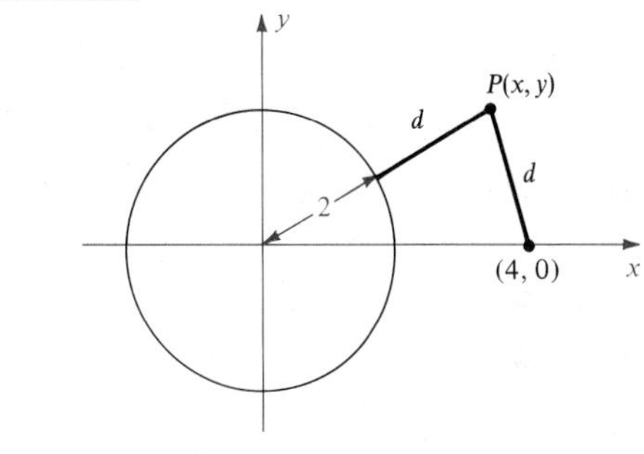

APPENDICES

I USING LOGARITHMIC AND TRIGONOMETRIC TABLES

II TABLES

1 Common Logarithms

2 Natural Exponential Function

3 Natural Logarithms

4 Values of the Trigonometric Functions

5 Trigonometric Functions of Radians and Real Numbers

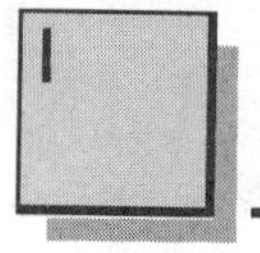

USING LOGARITHMIC AND TRIGONOMETRIC TABLES

In this appendix we discuss how to use logarithmic and trigonometric tables.

TABLE 1 ▪ COMMON LOGARITHMS

If x is any positive real number and we write

$$x = c \cdot 10^k$$

for $1 \leq c < 10$ and an integer k, then we may apply laws of logarithms, obtaining

$$\log x = \log c + \log 10^k.$$

$$\log x = \log c + k.$$

From the last equation we see that to find $\log x$ for any positive real number x it is sufficient to know the logarithms of numbers between 1 and 10. The number $\log c$, for $1 \leq c < 10$, is the **mantissa** and the integer k is the **characteristic** of $\log x$.

If $1 \leq c < 10$, then, since $\log x$ increases as x increases,

$$\log 1 \leq \log c < \log 10,$$

or, equivalently,

$$0 \leq \log c < 1.$$

Hence, the mantissa of a logarithm is a number between 0 and 1.

In numerical problems it is usually necessary to approximate logarithms. For example,

$$\log 2 = 0.3010299957\ldots$$

where the decimal is nonrepeating and nonterminating. We often round off such logarithms to four decimal places and write

$$\log 2 \approx 0.3010.$$

If a number between 0 and 1 is written as a finite decimal, it is sometimes referred to as a **decimal fraction**. Thus, the equation $\log x = \log c + k$ implies that if x is any positive real number, then $\log x$ *may be approximated by the sum of a positive decimal fraction (the mantissa) and an integer k (the characteristic)*. We shall refer to this representation as the **standard form** for $\log x$.

Common logarithms of many numbers between 1 and 10 have been calculated. Table 1 of Appendix II contains four-decimal-place approximations for logarithms of numbers between 1.00 and 9.99 at intervals of 0.01. This table can be used to find the common logarithm of any three-digit number to four-decimal-place accuracy. The use of Table 1 is illustrated in the following examples.

EXAMPLE ▪ 1

Approximate the logarithm:

(a) $\log 43.6$ (b) $\log 43{,}600$ (c) $\log 0.0436$

SOLUTION

(a) Since $43.6 = (4.36)10^1$, the characteristic of $\log 43.6$ is 1. Referring to Table 1, we find that the mantissa of $\log 4.36$ may be approximated by the decimal fraction 0.6395. Hence, as in the preceding discussion,

$$\log 43.6 \approx 0.6395 + 1$$

$$\log 43.6 \approx 1.6395.$$

(b) Since $43{,}600 = (4.36)10^4$, the mantissa is the same as in part (a); however, the characteristic is 4. Consequently,

$$\log 43{,}600 \approx 0.6395 + 4$$

$$\log 43{,}600 \approx 4.6395.$$

(c) If we write $0.0436 = (4.36)10^{-2}$, then

$$\log 0.0436 = \log 4.36 + (-2).$$

Hence, the standard form is

$$\log 0.0436 \approx 0.6395 + (-2).$$

If we subtract 2 from 0.6395, we obtain

$$\log 0.0436 \approx -1.3605.$$

Note that this is not standard form, since $-1.3605 = -0.3605 + (-1)$, a number in which the decimal part is *negative.*

In Example 1, after obtaining $\log 0.0436 \approx 0.6395 + (-2)$, a common error is to write this as -2.6395. This is incorrect, since

$$-2.6395 = -0.6395 + (-2),$$

which is not the same as $0.6395 + (-2)$.

If a logarithm has a negative characteristic, we usually either leave it in standard form or rewrite the logarithm, keeping the decimal part positive. To illustrate the second technique, let us add and subtract 8 on the right side of the equation as follows:

$$\log 0.0436 \approx 0.6395 + (-2)$$
$$\log 0.0436 \approx 0.6395 + (8 - 8) + (-2)$$
$$\log 0.0436 \approx 8.6395 - 10$$

We could also write

$$\log 0.0436 \approx 18.6395 - 20$$
$$\log 0.0436 \approx 28.6395 - 30$$
$$\log 0.0436 \approx 43.6395 - 45$$

and so on, as long as the sum of the positive integer to the left of the decimal and the negative integer to the right of the decimal equals the characteristic of the logarithm.

EXAMPLE ▪ 2

Approximate the logarithm:

(a) $\log (0.00652)^2$ (b) $\log (0.00652)^{-2}$ (c) $\log (0.00652)^{1/2}$

SOLUTION

(a) By (iii) of the laws of logarithms (see Section 5.3),

$$\log (0.00652)^2 = 2 \log 0.00652.$$

Since $0.00652 = (6.52)10^{-3}$,

$$\log 0.00652 = \log 6.52 + (-3).$$

Referring to Table 1, we see that $\log 6.52 \approx 0.8142$ and, therefore,

$$\log 0.00652 \approx 0.8142 + (-3).$$

Hence,

$$\log (0.00652)^2 = 2 \log 0.00652$$
$$\log (0.00652)^2 \approx 2[0.8142 + (-3)]$$
$$\log (0.00652)^2 \approx 1.6284 + (-6)$$
$$\log (0.00652)^2 \approx 0.6284 + (-5).$$

The last number is the standard form for the logarithm.

(b) Again we use law (iii) and the value for log 0.00652 found in part (a):

$$\log (0.00652)^{-2} = -2 \log 0.00652$$
$$\log (0.00652)^{-2} \approx -2[0.8142 + (-3)]$$
$$\log (0.00652)^{-2} \approx -1.6284 + 6$$

It is important to note that -1.6284 means $-0.6284 + (-1)$ and, consequently, the decimal part is negative. To obtain the standard form, we may write

$$\begin{aligned} -1.6284 + 6 &= 6.0000 - 1.6284 \\ &= 4.3716. \end{aligned}$$

This shows that the mantissa is 0.3716 and the characteristic is 4.

By law (iii),

$$\log (0.00652)^{1/2} = \tfrac{1}{2} \log 0.00652$$
$$\log (0.00652)^{1/2} \approx \tfrac{1}{2}[0.8142 + (-3)].$$

If we multiply by $\frac{1}{2}$, the standard form is not obtained, since neither number in the resulting sum is the characteristic. To avoid this, we may adjust the expression within brackets by adding and subtracting a suitable number. If we use 1 in this way, we obtain

$$\log (0.00652)^{1/2} \approx \tfrac{1}{2}[1.8142 + (-4)]$$
$$\log (0.00652)^{1/2} \approx 0.9071 + (-2)$$

We could also have added and subtracted any number other than 1. For example,

$$\begin{aligned} \tfrac{1}{2}[0.8142 + (-3)] &= \tfrac{1}{2}[17.8142 + (-20)] \\ &= 8.9071 + (-10). \end{aligned}$$

If $\log x$ is given, we can use Table 1 to find an approximation to x, as illustrated in the following example.

EXAMPLE ▪ 3

Find a decimal approximation to x:

(a) $\log x = 1.7959$ **(b)** $\log x = -3.5918$

SOLUTION

(a) The mantissa 0.7959 determines the sequence of digits in x and the characteristic determines the position of the decimal point. Referring to the *body* of Table 1, we see that the mantissa 0.7959 is the logarithm of

6.25. Since the characteristic is 1, we know that x lies between 10 and 100. Consequently, $x \approx 62.5$.

(b) To find x from Table 1, we must express $\log x$ in standard form. To change $\log x = -3.5918$ to standard form, we may add and subtract 4, obtaining

$$\begin{aligned}\log x &= (4 - 3.5918) - 4 \\ &= 0.4082 - 4.\end{aligned}$$

Referring to Table 1, we see that the mantissa 0.4082 is the logarithm of 2.56. Since the characteristic of $\log x$ is -4, it follows that $x \approx 0.000256$.

If we use a calculator with a [LOG] key to determine common logarithms, then the standard form for $\log x$ is obtained only if $x \geq 1$. For example, to find log 43.6 on a typical calculator, we enter 43.6 and press [LOG], obtaining the standard form

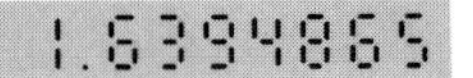

If we find log 0.0436 in similar fashion, then the following number appears on the display panel:

-1.3605135

This is not the standard form for the logarithm, since the decimal part is negative (compare with Example 1(c)). To find the standard form we could add 2 to the logarithm (using a calculator) and then subtract 2 as follows:

$$\begin{aligned}\log 0.0436 &\approx -1.3605135 \\ \log 0.0436 &\approx (-1.3605135 + 2) - 2 \\ \log 0.0436 &\approx 0.6394865 - 2 \\ \log 0.0436 &\approx 0.6394865 + (-2)\end{aligned}$$

The only common logarithms that can be found *directly* from Table 1 are logarithms of numbers that contain at most three nonzero digits. If *four* nonzero digits are involved, then it is possible to obtain an approximation by using the method of linear interpolation described next. The term **linear interpolation** is used because, as we shall see, the method is based upon approximating portions of the graph of $y = \log x$ by line segments.

To illustrate the process of linear interpolation, and at the same time give some justification for it, let us consider the specific example log 12.64.

FIGURE 1

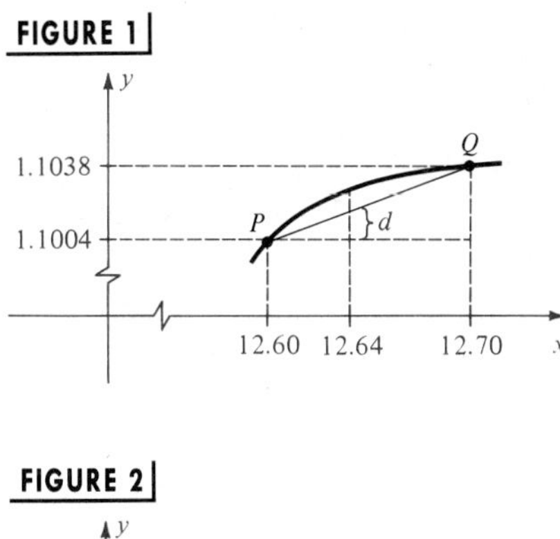

FIGURE 2

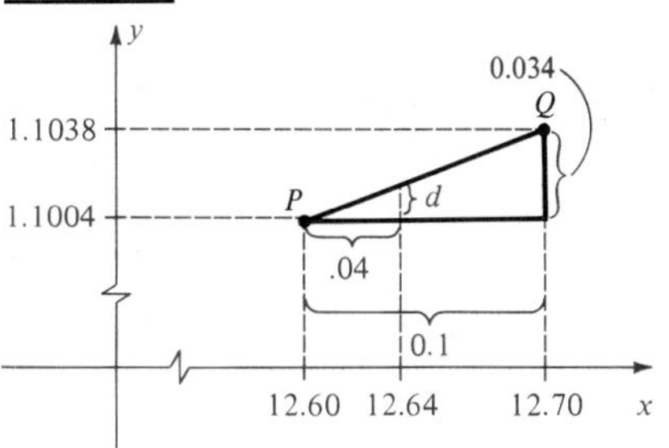

Since the logarithmic function with base 10 is increasing, this number lies between log 12.60 ≈ 1.1004 and log 12.70 ≈ 1.1038. Examining the graph of $y = \log x$, we have the situation shown in Figure 1, where we have distorted the units on the x- and y-axes and also the portion of the graph shown. (A more accurate drawing would indicate that the graph of $y = \log x$ is much closer to the line segment joining $P(12.60, 1.1004)$ to $Q(12.70, 1.1038)$ than is shown in the figure.) Since log 12.64 is the y-coordinate of the point on the graph having x-coordinate 12.64, it can be approximated by the y-coordinate of the point with x-coordinate 12.64 on the *line segment PQ*. Referring to Figure 1, we see that the latter y-coordinate is $1.1004 + d$. The number d can be approximated by using similar triangles. Referring to Figure 2, where the graph of $y = \log x$ has been deleted, we may form the following proportion:

$$\frac{d}{0.0034} = \frac{0.04}{0.1}$$

Hence $$d = \frac{(0.04)(0.0034)}{0.1} = 0.00136.$$

When using this technique, we always round off decimals to the same number of places as appear in the body of the table. Consequently, $d \approx 0.0014$ and

$$\log 12.64 \approx 1.1004 + 0.0014$$
$$\log 12.64 \approx 1.1018.$$

Hereafter we shall not sketch a graph when interpolating. Instead we shall use the scheme illustrated in the next example.

EXAMPLE ▪ 4

Approximate log 572.6.

SOLUTION It is convenient to arrange our work as follows:

$$1.0\left\{\begin{array}{l} 0.6\left\{\begin{array}{l}\log 572.0 \approx 2.7574 \\ \log 572.6 = \,?\end{array}\right\}d \\ \qquad \log 573.0 \approx 2.7582\end{array}\right\}0.0008$$

We have indicated differences next to the braces. This leads to the following:

$$\frac{d}{0.0008} = \frac{0.6}{1.0} = \frac{6}{10}$$

$$d = \tfrac{6}{10}(0.0008) = 0.00048 \approx 0.0005$$

Hence,
$$\log 572.6 \approx 2.7574 + 0.0005$$
$$\log 572.6 \approx 2.7579.$$

Another way of working this type of problem is to reason that since 572.6 is $\frac{6}{10}$ of the way from 572.0 to 573.0, then log 572.6 is (approximately) $\frac{6}{10}$ of the way from 2.7574 to 2.7582. Hence,

$$\log 572.6 \approx 2.7574 + \tfrac{6}{10}(0.0008)$$
$$\log 572.6 \approx 2.7574 + 0.0005$$
$$\log 572.6 \approx 2.7579.$$

EXAMPLE ■ 5

Approximate log 0.003678.

SOLUTION We begin by arranging our work as in the solution of Example 1. Thus,

$$10\left\{\begin{array}{l} 8\left\{\begin{array}{l}\log 0.003670 \approx 0.5647 + (-3) \\ \log 0.003678 = ?\end{array}\right\}d \\ \log 0.003680 \approx 0.5658 + (-3)\end{array}\right\}0.0011$$

Since we are only interested in ratios, we have used the numbers 8 and 10 on the left side because their ratio is the same as the ratio of 0.000008 to 0.000010. This leads to the following:

$$\frac{d}{0.0011} = \frac{8}{10} = 0.8$$
$$d = (0.0011)(0.8) = 0.00088 \approx 0.0009.$$

Hence,
$$\log 0.003678 \approx [0.5647 + (-3)] + 0.0009$$
$$\log 0.003678 \approx 0.5656 + (-3).$$

If a number x is written in the form $x = c \cdot 10^k$ with $1 \leq c < 10$, then before using Table 1 to find $\log x$ by interpolation, c should be rounded off to three decimal places. Another way of saying this is that x should be rounded off to four **significant figures**. Some examples will help to clarify the procedure. If $x = 36.4635$, we round off to 36.46 before approximating $\log x$. The number 684,279 should be rounded off to 684,300. For a decimal such as 0.096202 we use 0.09620. The reason for doing this is that Table 1 does not guarantee more than four-digit accuracy, since the mantissas that appear in it are approximations. This means that if *more* than four-digit accuracy is required in a problem, then Table 1 cannot be used. If, in more extensive tables, the logarithm of a number containing n digits can be found directly, then interpolation is allowed for numbers involving $n + 1$ digits, and numbers should be rounded off accordingly.

The method of interpolation can also be used to find x when we are given $\log x$. If we use Table 1, then x may be found to four significant figures. In this case we are given the *y-coordinate* of a point on the graph of $y = \log x$ and are asked to find the *x-coordinate*. A geometric argument similar to the one given earlier can be used to justify the procedure illustrated in the next example.

EXAMPLE ▪ 6

Find x to four significant figures if $\log x = 1.7949$.

SOLUTION The mantissa 0.7949 does not appear in Table 1, but it can be isolated between adjacent entries for the mantissas corresponding to 6.230 and 6.240. We shall arrange our work as follows:

$$0.1\left\{\begin{array}{l} r\left\{\begin{array}{ll} \log 62.30 & \approx 1.7945 \\ \log x & \approx 1.7949 \end{array}\right\} 0.0004 \\ \quad \log 62.40 \approx 1.7952 \end{array}\right\} 0.0007.$$

This leads to the proportion

$$\frac{r}{0.1} = \frac{0.0004}{0.0007} = \frac{4}{7}$$

$$r = (0.1)\left(\frac{4}{7}\right) \approx 0.06.$$

Hence,

$$x \approx 62.30 + 0.06$$

$$x \approx 62.36.$$

TABLE 4 ▪ VALUES OF THE TRIGONOMETRIC FUNCTIONS

Methods of calculus can be employed to approximate, to any degree of accuracy, all values of the trigonometric functions in the t-interval $[0, \pi/2]$ or, equivalently, in the degree interval $[0°, 90°]$. Table 4, part of which is reproduced here, gives four-decimal-place approximations to such values.

In Table 4, $0 \le t \le 1.5708$. The number 1.5708 is a four-decimal-place approximation to $\pi/2$. The table is arranged so that function values corresponding to angles in degree measure may be found directly. Angular measures are given in $10'$ intervals from $0°$ to $90°$. The reason that t varies at intervals of approximately 0.0029 is because $10' \approx 0.0029$ radians.

t	t degrees	sin t	cos t	tan t	cot t	sec t	csc t		
.4887	**28°00′**	.4695	.8829	.5317	1.881	1.133	2.130	**62°00′**	1.0821
.4916	10	.4720	.8816	.5354	1.868	1.134	2.118	50	1.0792
.4945	20	.4746	.8802	.5392	1.855	1.136	2.107	40	1.0763
.4974	30	.4772	.8788	.5430	1.842	1.138	2.096	30	1.0734
.5003	40	.4797	.8774	.5467	1.829	1.140	2.085	20	1.0705
.5032	50	.4823	.8760	.5505	1.816	1.142	2.074	10	1.0676
.5061	**29°00′**	.4848	.8746	.5543	1.804	1.143	2.063	**61°00′**	1.0647
.5091	10	.4874	.8732	.5581	1.792	1.145	2.052	50	1.0617
.5120	20	.4899	.8718	.5619	1.780	1.147	2.041	40	1.0588
.5149	30	.4924	.8704	.5658	1.767	1.149	2.031	30	1.0559
.5178	40	.4950	.8689	.5696	1.756	1.151	2.020	20	1.0530
.5207	50	.4975	.8675	.5735	1.744	1.153	2.010	10	1.0501
.5236	**30°00′**	.5000	.8660	.5774	1.732	1.155	2.000	**60°00′**	1.0472
.5265	10	.5025	.8646	.5812	1.720	1.157	1.990	50	1.0443
.5294	20	.5050	.8631	.5851	1.709	1.159	1.980	40	1.0414
.5323	30	.5075	.8616	.5890	1.698	1.161	1.970	30	1.0385
.5352	40	.5100	.8601	.5930	1.686	1.163	1.961	20	1.0356
.5381	50	.5125	.8587	.5969	1.675	1.165	1.951	10	1.0327
		cos t	sin t	cot t	tan t	csc t	sec t	t degrees	t

To find values of trigonometric functions if $0 \leq t \leq 0.7854 \approx \pi/4$ or $0° \leq \theta \leq 45°$, we use the labels at the *top* of the columns. For example, from the displayed portion of the table,

$$\sin 0.5003 \approx 0.4797 \qquad \tan 28°30' \approx 0.5430$$

$$\cos 0.4945 \approx 0.8802 \qquad \sec 29°00' \approx 1.143$$

$$\cot 0.5120 \approx 1.780 \qquad \csc 29°40' \approx 2.020.$$

If $0.7854 \leq t \leq 1.5708$, or if $45° \leq \theta \leq 90°$, then we use the labels at the *bottom* of the columns. For example,

$$\sin 1.0705 \approx 0.8774 \qquad \csc 62°00' \approx 1.133$$

$$\cos 1.0530 \approx 0.4950 \qquad \cot 61°10' \approx 0.5505$$

$$\tan 1.0821 \approx 1.881 \qquad \sec 60°30' \approx 2.031.$$

The reason that the table can be so arranged follows from the fact that

$$\sin t = \cos\left(\frac{\pi}{2} - t\right), \qquad \cot t = \tan\left(\frac{\pi}{2} - t\right), \qquad \csc t = \sec\left(\frac{\pi}{2} - t\right)$$

or, equivalently,

$$\sin\theta = \cos(90° - \theta), \qquad \cot\theta = \tan(90° - \theta), \qquad \csc\theta = \sec(90° - \theta).$$

For example, as shown in the table,

$$\sin 29^\circ = \cos(90^\circ - 29^\circ) = \cos 61^\circ$$

$$\cot 28^\circ 20' = \tan(90^\circ - 28^\circ 20') = \tan 61^\circ 40'.$$

To find function values when t lies *between* numbers given in the table, we may use the method of linear interpolation. Similarly, given a value such as $\sin t = 0.6371$, we may refer to the body of Table 4 and use linear interpolation, if necessary, to obtain an approximation to t. If t is measured in degrees we round off to the nearest minute.

The following examples illustrate the use of interpolation in Table 4.

EXAMPLE ▪ 7

Approximate $\tan 24^\circ 16'$.

SOLUTION Consult Table 4 and interpolate:

$$10' \left\{ \begin{array}{l} 6' \left\{ \begin{array}{l} \tan 24^\circ 10' \approx 0.4487 \\ \tan 24^\circ 16' = ? \end{array} \right\} d \\ \tan 24^\circ 20' \approx 0.4522 \end{array} \right\} 0.0035$$

$$\frac{d}{0.0035} = \frac{6}{10}$$

$$d = \tfrac{6}{10}(0.0035) \approx 0.0021$$

Thus

$$\tan 24^\circ 16' \approx 0.4487 + 0.0021$$

$$\tan 24^\circ 16' \approx 0.4508.$$

EXAMPLE ▪ 8

Approximate $\cos(-117^\circ 47')$.

SOLUTION The angle is in quadrant III. You should check that the reference angle is $62^\circ 13'$. Consequently, $\cos(-117^\circ 47') = -\cos 62^\circ 13'$. We then interpolate in Table 4:

$$10' \left\{ \begin{array}{l} 3' \left\{ \begin{array}{l} \cos 62^\circ 10' \approx 0.4669 \\ \cos 62^\circ 13' = ? \end{array} \right\} d \\ \cos 62^\circ 20' \approx 0.4643 \end{array} \right\} 0.0026$$

$$\frac{d}{0.0026} = \frac{3}{10}$$

$$d = \tfrac{3}{10}(0.0026) \approx 0.0008$$

Since the cosine function is decreasing,

$$\cos 62°13' \approx 0.4669 - 0.0008$$

$$\cos 62°13' \approx 0.4661$$

$$\cos(-117°47') \approx -0.4661.$$

EXAMPLE ▪ 9

Approximate the smallest positive real number t such that $\sin t = 0.6635$.

SOLUTION We locate 0.6635 between successive entries in the sine column of Table 4 and interpolate:

$$0.0029\left\{\begin{matrix} d\left\{\begin{matrix}\sin 0.7243 \approx 0.6626 \\ \sin t \qquad = 0.6635\end{matrix}\right\}0.0009 \\ \sin 0.7272 \approx 0.6648\end{matrix}\right\}0.0022$$

$$\frac{d}{0.0029} = \frac{0.0009}{0.0022}$$

$$d = \tfrac{9}{22}(0.0029) \approx 0.0012$$

Hence,

$$t \approx 0.7243 + 0.0012$$

$$t \approx 0.7255.$$

EXAMPLE ▪ 10

If $\sin\theta = -0.7963$, approximate the degree measure of all angles θ that are in the interval $[0°, 360°)$.

SOLUTION Let θ' be the reference angle so that $\sin\theta' = 0.7963$. We interpolate in Table 4:

$$10'\left\{\begin{matrix} d\left\{\begin{matrix}\sin 52°40' \approx 0.7951 \\ \sin\theta' \qquad = 0.7963\end{matrix}\right\}0.0012 \\ \sin 52°50' \approx 0.7969\end{matrix}\right\}0.0018$$

$$\frac{d}{10} = \frac{0.0012}{0.0018}$$

$$d = 10\left(\tfrac{12}{18}\right) \approx 7'$$

Thus

$$\theta' = 52°47'$$

Since $\sin\theta$ is negative, θ lies in quadrant III or IV. Using the reference angle $52°47'$:

$$\theta \approx 180° + 52°47', \quad \text{or} \quad \theta \approx 232°47'$$

$$\theta \approx 360° - 52°47', \quad \text{or} \quad \theta \approx 307°13'.$$

A.1 EXERCISES

Exer. 1–16: Use Table 1 and laws of logarithms to approximate the common logarithms of the numbers.

1 347; 0.00347; 3.47

2 86.2; 8620; 0.862

3 0.54; 540; 540,000

4 208; 2.08; 20,800

5 60.2; 0.0000602; 602

6 5; 0.5; 0.0005

7 $(44.9)^2$; $(44.9)^{1/2}$; $(44.9)^{-2}$

8 $(1810)^4$; $(1810)^{40}$; $(1810)^{1/4}$

9 $(0.943)^3$; $(0.943)^{-3}$; $(0.943)^{1/3}$

10 $(0.017)^{10}$; $10^{0.017}$; $10^{1.43}$

11 $(638)(17.3)$

12 $\dfrac{(2.73)(78.5)}{621}$

13 $\dfrac{(47.4)^3}{(29.5)^2}$

14 $\dfrac{(897)^4}{\sqrt{17.8}}$

15 $\sqrt[3]{20.6}(371)^3$

16 $\dfrac{(0.0048)^{10}}{\sqrt{0.29}}$

Exer. 17–30: Use Table 1 to find a decimal approximation to x.

17 $\log x = 3.6274$

18 $\log x = 1.8965$

19 $\log x = 0.9469$

20 $\log x = 0.5729$

21 $\log x = 5.2095$

22 $\log x = 6.7300 - 10$

23 $\log x = 9.7348 - 10$

24 $\log x = 7.6739 - 10$

25 $\log x = 8.8306 - 10$

26 $\log x = 4.9680$

27 $\log x = 2.2765$

28 $\log x = 3.0043$

29 $\log x = -1.6253$

30 $\log x = -2.2118$

Exer. 31–50: Use interpolation in Table 1 to approximate the common logarithm of the number.

31 25.48

32 421.6

33 5363

34 0.3817

35 0.001259

36 69,450

37 123,400

38 0.0212

39 0.7786

40 1.203

41 384.7

42 54.44

43 0.9462

44 7259

45 66,590

46 0.001428

47 0.04321

48 300,100

49 3.003

50 1.236

Exer. 51–70: Use interpolation in Table 1 to approximate x.

51 $\log x = 1.4437$

52 $\log x = 3.7455$

53 $\log x = 4.6931$

54 $\log x = 0.5883$

55 $\log x = 9.1664 - 10$

56 $\log x = 8.3902 - 10$

57 $\log x = 3.8153 - 6$

58 $\log x = 5.9306 - 9$

59 $\log x = 2.3705$

60 $\log x = 4.2867$

61 $\log x = 0.1358$

62 $\log x = 0.0194$

63 $\log x = 8.9752 - 10$

64 $\log x = 2.4979 - 5$

65 $\log x = 5.0409$

66 $\log x = 1.3796$

67 $\log x = -2.8712$

68 $\log x = -1.8164$

69 $\log x = -0.6123$

70 $\log x = -3.1426$

Exer. 71–82: Use interpolation in Table 4 to approximate the number.

71 $\sin 0.46$

72 $\cos 0.82$

73 $\tan 3$

74 $\cot 6$

75 $\sec \frac{1}{4}$

76 $\csc 1.54$

77 $\cos 37°43'$

78 $\sin 22°34'$

79 $\cot 62°27'$

80 $\tan 57°16'$

81 $\csc 16°55'$

82 $\sec 9°12'$

Exer. 83–88: Use interpolation in Table 4 to approximate the smallest positive number t for which the equality is true.

83 $\cos t = 0.8620$

84 $\sin t = 0.6612$

85 $\tan t = 4.501$

86 $\sec t = 3.641$

87 $\csc t = 1.436$

88 $\cot t = 1.165$

Exer. 89–96: Use interpolation in Table 4 to approximate, to the nearest minute, the degree measure of all angles θ that lie in the interval $[0°, 360°)$.

89 $\sin \theta = 0.3672$

90 $\cos \theta = 0.8426$

91 $\tan \theta = 0.5042$

92 $\cot \theta = 1.348$

93 $\cos \theta = 0.3465$

94 $\csc \theta = 1.219$

95 $\sec \theta = 1.385$

96 $\sin \theta = 0.7534$

TABLE 1 | COMMON LOGARITHMS

N	0	1	2	3	4	5	6	7	8	9
1.0	.0000	.0043	.0086	.0128	.0170	.0212	.0253	.0294	.0334	.0374
1.1	.0414	.0453	.0492	.0531	.0569	.0607	.0645	.0682	.0719	.0755
1.2	.0792	.0828	.0864	.0899	.0934	.0969	.1004	.1038	.1072	.1106
1.3	.1139	.1173	.1206	.1239	.1271	.1303	.1335	.1367	.1399	.1430
1.4	.1461	.1492	.1523	.1553	.1584	.1614	.1644	.1673	.1703	.1732
1.5	.1761	.1790	.1818	.1847	.1875	.1903	.1931	.1959	.1987	.2014
1.6	.2041	.2068	.2095	.2122	.2148	.2175	.2201	.2227	.2253	.2279
1.7	.2304	.2330	.2355	.2380	.2405	.2430	.2455	.2480	.2504	.2529
1.8	.2553	.2577	.2601	.2625	.2648	.2672	.2695	.2718	.2742	.2765
1.9	.2788	.2810	.2833	.2856	.2878	.2900	.2923	.2945	.2967	.2989
2.0	.3010	.3032	.3054	.3075	.3096	.3118	.3139	.3160	.3181	.3201
2.1	.3222	.3243	.3263	.3284	.3304	.3324	.3345	.3365	.3385	.3404
2.2	.3424	.3444	.3464	.3483	.3502	.3522	.3541	.3560	.3579	.3598
2.3	.3617	.3636	.3655	.3674	.3692	.3711	.3729	.3747	.3766	.3784
2.4	.3802	.3820	.3838	.3856	.3874	.3892	.3909	.3927	.3945	.3962
2.5	.3979	.3997	.4014	.4031	.4048	.4065	.4082	.4099	.4116	.4133
2.6	.4150	.4166	.4183	.4200	.4216	.4232	.4249	.4265	.4281	.4298
2.7	.4314	.4330	.4346	.4362	.4378	.4393	.4409	.4425	.4440	.4456
2.8	.4472	.4487	.4502	.4518	.4533	.4548	.4564	.4579	.4594	.4609
2.9	.4624	.4639	.4654	.4669	.4683	.4698	.4713	.4728	.4742	.4757
3.0	.4771	.4786	.4800	.4814	.4829	.4843	.4857	.4871	.4886	.4900
3.1	.4914	.4928	.4942	.4955	.4969	.4983	.4997	.5011	.5024	.5038
3.2	.5051	.5065	.5079	.5092	.5105	.5119	.5132	.5145	.5159	.5172
3.3	.5185	.5198	.5211	.5224	.5237	.5250	.5263	.5276	.5289	.5302
3.4	.5315	.5328	.5340	.5353	.5366	.5378	.5391	.5403	.5416	.5428
3.5	.5441	.5453	.5465	.5478	.5490	.5502	.5514	.5527	.5539	.5551
3.6	.5563	.5575	.5587	.5599	.5611	.5623	.5635	.5647	.5658	.5670
3.7	.5682	.5694	.5705	.5717	.5729	.5740	.5752	.5763	.5775	.5786
3.8	.5798	.5809	.5821	.5832	.5843	.5855	.5866	.5877	.5888	.5899
3.9	.5911	.5922	.5933	.5944	.5955	.5966	.5977	.5988	.5999	.6010
4.0	.6021	.6031	.6042	.6053	.6064	.6075	.6085	.6096	.6107	.6117
4.1	.6128	.6138	.6149	.6160	.6170	.6180	.6191	.6201	.6212	.6222
4.2	.6232	.6243	.6253	6263	.6274	.6284	.6294	.6304	.6314	.6325
4.3	.6335	.6345	.6355	.6365	.6375	.6385	.6395	.6405	.6415	.6425
4.4	.6435	.6444	.6454	.6464	.6474	.6484	.6493	.6503	.6513	.6522
4.5	.6532	.6542	.6551	.6561	.6571	.6580	.6590	.6599	.6609	.6618
4.6	.6628	.6637	.6646	.6656	.6665	.6675	.6684	.6693	.6702	.6712
4.7	.6721	.6730	.6739	.6749	.6758	.6767	.6776	.6785	.6794	.6803
4.8	.6812	.6821	.6830	.6839	.6848	.6857	.6866	.6875	.6884	.6893
4.9	.6902	.6911	.6920	.6928	.6937	.6946	.6955	.6964	.6972	.6981
5.0	.6990	.6998	.7007	.7016	.7024	.7033	.7042	.7050	.7059	.7067
5.1	.7076	.7084	.7093	.7101	.7110	.7118	.7126	.7135	.7143	.7152
5.2	.7160	.7168	.7177	.7185	.7193	.7202	.7210	.7218	.7226	.7235
5.3	.7243	.7251	.7259	.7267	.7275	.7284	.7292	.7300	.7308	.7316
5.4	.7324	.7332	.7340	.7348	.7356	.7364	.7372	.7380	.7388	.7396
5.5	.7404	.7412	.7419	.7427	.7435	.7443	.7451	.7459	.7466	.7474
5.6	.7482	.7490	.7497	.7505	.7513	.7520	.7528	.7536	.7543	.7551
5.7	.7559	.7566	.7574	.7582	.7589	.7597	.7604	.7612	.7619	.7627
5.8	.7634	.7642	.7649	.7657	.7664	.7672	.7679	.7686	.7694	.7701
5.9	.7709	.7716	.7723	.7731	.7738	.7745	.7752	.7760	.7767	.7774
6.0	.7782	.7789	.7796	.7803	.7810	.7818	.7825	.7832	.7839	.7846
6.1	.7853	.7860	.7868	.7875	.7882	.7889	.7896	.7903	.7910	.7917
6.2	.7924	.7931	.7938	.7945	.7952	.7959	.7966	.7973	.7980	.7987
6.3	.7993	.8000	.8007	.8014	.8021	.8028	.8035	.8041	.8048	.8055
6.4	.8062	.8069	.8075	.8082	.8089	.8096	.8102	.8109	.8116	.8122
6.5	.8129	.8136	.8142	.8149	.8156	.8162	.8169	.8176	.8182	.8189
6.6	.8195	.8202	.8209	.8215	.8222	.8228	.8235	.8241	.8248	.8254
6.7	.8261	.8267	.8274	.8280	.8287	.8293	.8299	.8306	.8312	.8319
6.8	.8325	.8331	.8338	.8344	.8351	.8357	.8363	.8370	.8376	.8382
6.9	.8388	.8395	.8401	.8407	.8414	.8420	.8426	.8432	.8439	.8445
7.0	.8451	.8457	.8463	.8470	.8476	.8482	.8488	.8494	.8500	.8506
7.1	.8513	.8519	.8525	.8531	.8537	.8543	.8549	.8555	.8561	.8567
7.2	.8573	.8579	.8585	.8591	.8597	.8603	.8609	.8615	.8621	.8627
7.3	.8633	.8639	.8645	.8651	.8657	.8663	.8669	.8675	.8681	.8686
7.4	.8692	.8698	.8704	.8710	.8716	.8722	.8727	.8733	.8739	.8745
7.5	.8751	.8756	.8762	.8768	.8774	.8779	.8785	.8791	.8797	.8802
7.6	.8808	.8814	.8820	.8825	.8831	.8837	.8842	.8848	.8854	.8859
7.7	.8865	.8871	.8876	.8882	.8887	.8893	.8899	.8904	.8910	.8915
7.8	.8921	.8927	.8932	.8938	.8943	.8949	.8954	.8960	.8965	.8971
7.9	.8976	.8982	.8987	.8993	.8998	.9004	.9009	.9015	.9020	.9025
8.0	.9031	.9036	.9042	.9047	.9053	.9058	.9063	.9069	.9074	.9079
8.1	.9085	.9090	.9096	.9101	.9106	.9112	.9117	.9122	.9128	.9133
8.2	.9138	.9143	.9149	.9154	.9159	.9165	.9170	.9175	.9180	.9186
8.3	.9191	.9196	.9201	.9206	.9212	.9217	.9222	.9227	.9232	.9238
8.4	.9243	.9248	.9253	.9258	.9263	.9269	.9274	.9279	.9284	.9289
8.5	.9294	.9299	.9304	.9309	.9315	.9320	.9325	.9330	.9335	.9340
8.6	.9345	.9350	.9355	.9360	.9365	.9370	.9375	.9380	.9385	.9390
8.7	.9395	.9400	.9405	.9410	.9415	.9420	.9425	.9430	.9435	.9440
8.8	.9445	.9450	.9455	.9460	.9465	.9469	.9474	.9479	.9484	.9489
8.9	.9494	.9499	.9504	.9509	.9513	.9518	.9523	.9528	.9533	.9538
9.0	.9542	.9547	.9552	.9557	.9562	.9566	.9571	.9576	.9581	.9586
9.1	.9590	.9595	.9600	.9605	.9609	.9614	.9619	.9624	.9628	.9633
9.2	.9638	.9643	.9647	.9652	.9657	.9661	.9666	.9671	.9675	.9680
9.3	.9685	.9689	.9694	.9699	.9703	.9708	.9713	.9717	.9722	.9727
9.4	.9731	.9736	.9741	.9745	.9750	.9754	.9759	.9763	.9768	.9773
9.5	.9777	.9782	.9786	.9791	.9795	.9800	.9805	.9809	.9814	.9818
9.6	.9823	.9827	.9832	.9836	.9841	.9845	.9850	.9854	.9859	.9863
9.7	.9868	.9872	.9877	.9881	.9886	.9890	.9894	.9899	.9903	.9908
9.8	.9912	.9917	.9921	.9926	.9930	.9934	.9939	.9943	.9948	.9952
9.9	.9956	.9961	.9965	.9969	.9974	.9978	.9983	.9987	.9991	.9996

TABLE 2 NATURAL EXPONENTIAL FUNCTION

x	e^x	e^{-x}	x	e^x	e^{-x}
0.00	1.0000	1.0000	2.50	12.182	0.0821
0.05	1.0513	0.9512	2.60	13.464	0.0743
0.10	1.1052	0.9048	2.70	14.880	0.0672
0.15	1.1618	0.8607	2.80	16.445	0.0608
0.20	1.2214	0.8187	2.90	18.174	0.0550
0.25	1.2840	0.7788	3.00	20.086	0.0498
0.30	1.3499	0.7408	3.10	22.198	0.0450
0.35	1.4191	0.7047	3.20	24.533	0.0408
0.40	1.4918	0.6703	3.30	27.113	0.0369
0.45	1.5683	0.6376	3.40	29.964	0.0334
0.50	1.6487	0.6065	3.50	33.115	0.0302
0.55	1.7333	0.5769	3.60	36.598	0.0273
0.60	1.8221	0.5488	3.70	40.447	0.0247
0.65	1.9155	0.5220	3.80	44.701	0.0224
0.70	2.0138	0.4966	3.90	49.402	0.0202
0.75	2.1170	0.4724	4.00	54.598	0.0183
0.80	2.2255	0.4493	4.10	60.340	0.0166
0.85	2.3396	0.4274	4.20	66.686	0.0150
0.90	2.4596	0.4066	4.30	73.700	0.0136
0.95	2.5857	0.3867	4.40	81.451	0.0123
1.00	2.7183	0.3679	4.50	99.017	0.0111
1.10	3.0042	0.3329	4.60	99.484	0.0101
1.20	3.3201	0.3012	4.70	109.95	0.0091
1.30	3.6693	0.2725	4.80	121.51	0.0082
1.40	4.0552	0.2466	4.90	134.29	0.0074
1.50	4.4817	0.2231	5.00	148.41	0.0067
1.60	4.9530	0.2019	6.00	403.43	0.0025
1.70	5.4739	0.1827	7.00	1096.6	0.0009
1.80	6.0496	0.1653	8.00	2981.0	0.0003
1.90	6.6859	0.1496	9.00	8103.1	0.0001
2.00	7.3891	0.1353	10.00	22026.0	0.00005
2.10	8.1662	0.1225			
2.20	9.0250	0.1108			
2.30	9.9742	0.1003			
2.40	11.0232	0.0907			

TABLE 3 NATURAL LOGARITHMS

n	0.0	0.1	0.2	0.3	0.4	0.5	0.6	0.7	0.8	0.9
0*		7.697	8.391	8.796	9.084	9.307	9.489	9.643	9.777	9.895
1	0.000	0.095	0.182	0.262	0.336	0.405	0.470	0.531	0.588	0.642
2	0.693	0.742	0.788	0.833	0.875	0.916	0.956	0.993	1.030	1.065
3	1.099	1.131	1.163	1.194	1.224	1.253	1.281	1.308	1.335	1.361
4	1.386	1.411	1.435	1.459	1.482	1.504	1.526	1.548	1.569	1.589
5	1.609	1.629	1.649	1.668	1.686	1.705	1.723	1.740	1.758	1.775
6	1.792	1.808	1.825	1.841	1.856	1.872	1.887	1.902	1.917	1.932
7	1.946	1.960	1.974	1.988	2.001	2.015	2.028	2.041	2.054	2.067
8	2.079	2.092	2.104	2.116	2.128	2.140	2.152	2.163	2.175	2.186
9	2.197	2.208	2.219	2.230	2.241	2.251	2.262	2.272	2.282	2.293
10	2.303	2.313	2.322	2.332	2.342	2.351	2.361	2.370	2.380	2.389

* Subtract 10 if $n < 1$; for example, $\ln 0.3 \approx 8.796 - 10 = -1.204$.

TABLE 4 | VALUES OF THE TRIGONOMETRIC FUNCTIONS

t	t degrees	$\sin t$	$\cos t$	$\tan t$	$\cot t$	$\sec t$	$\csc t$		
.0000	**0°00′**	.0000	1.0000	.0000	—	1.000	—	**90°00′**	1.5708
.0029	10	.0029	1.0000	.0029	343.8	1.000	343.8	50	1.5679
.0058	20	.0058	1.0000	.0058	171.9	1.000	171.9	40	1.5650
.0087	30	.0087	1.0000	.0087	114.6	1.000	114.6	30	1.5621
.0116	40	.0116	.9999	.0116	85.94	1.000	85.95	20	1.5592
.0145	50	.0145	.9999	.0145	68.75	1.000	68.76	10	1.5563
.0175	**1°00′**	.0175	.9998	.0175	57.29	1.000	57.30	**89°00′**	1.5533
.0204	10	.0204	.9998	.0204	49.10	1.000	49.11	50	1.5504
.0233	20	.0233	.9997	.0233	42.96	1.000	42.98	40	1.5475
.0262	30	.0262	.9997	.0262	38.19	1.000	38.20	30	1.5446
.0291	40	.0291	.9996	.0291	34.37	1.000	34.38	20	1.5417
.0320	50	.0320	.9995	.0320	31.24	1.001	31.26	10	1.5388
.0349	**2°00′**	.0349	.9994	.0349	28.64	1.001	28.65	**88°00′**	1.5359
.0378	10	.0378	.9993	.0378	26.43	1.001	26.45	50	1.5330
.0407	20	.0407	.9992	.0407	24.54	1.001	24.56	40	1.5301
.0436	30	.0436	.9990	.0437	22.90	1.001	22.93	30	1.5272
.0465	40	.0465	.9989	.0466	21.47	1.001	21.49	20	1.5243
.0495	50	.0494	.9988	.0495	20.21	1.001	20.23	10	1.5213
.0524	**3°00′**	.0523	.9986	.0524	19.08	1.001	19.11	**87°00′**	1.5184
.0553	10	.0552	.9985	.0553	18.07	1.002	18.10	50	1.5155
.0582	20	.0581	.9983	.0582	17.17	1.002	17.20	40	1.5126
.0611	30	.0610	.9981	.0612	16.35	1.002	16.38	30	1.5097
.0640	40	.0640	.9980	.0641	15.60	1.002	15.64	20	1.5068
.0669	50	.0669	.9978	.0670	14.92	1.002	14.96	10	1.5039
.0698	**4°00′**	.0698	.9976	.0699	14.30	1.002	14.34	**86°00′**	1.5010
.0727	10	.0727	.9974	.0729	13.73	1.003	13.76	50	1.4981
.0756	20	.0756	.9971	.0758	13.20	1.003	13.23	40	1.4952
.0785	30	.0785	.9969	.0787	12.71	1.003	12.75	30	1.4923
.0814	40	.0814	.9967	.0816	12.25	1.003	12.29	20	1.4893
.0844	50	.0843	.9964	.0846	11.83	1.004	11.87	10	1.4864
.0873	**5°00′**	.0872	.9962	.0875	11.43	1.004	11.47	**85°00′**	1.4835
.0902	10	.0901	.9959	.0904	11.06	1.004	11.10	50	1.4806
.0931	20	.0929	.9957	.0934	10.71	1.004	10.76	40	1.4777
.0960	30	.0958	.9954	.0963	10.39	1.005	10.43	30	1.4748
.0989	40	.0987	.9951	.0992	10.08	1.005	10.13	20	1.4719
.1018	50	.1016	.9948	.1022	9.788	1.005	9.839	10	1.4690
.1047	**6°00′**	.1045	.9945	.1051	9.514	1.006	9.567	**84°00′**	1.4661
.1076	10	.1074	.9942	.1080	9.255	1.006	9.309	50	1.4632
.1105	20	.1103	.9939	.1110	9.010	1.006	9.065	40	1.4603
.1134	30	.1132	.9936	.1139	8.777	1.006	8.834	30	1.4573
.1164	40	.1161	.9932	.1169	8.556	1.007	8.614	20	1.4544
.1193	50	.1190	.9929	.1198	8.345	1.007	8.405	10	1.4515
.1222	**7°00′**	.1219	.9925	.1228	8.144	1.008	8.206	**83°00′**	1.4486
		$\cos t$	$\sin t$	$\cot t$	$\tan t$	$\csc t$	$\sec t$	t degrees	t

t	t degrees	$\sin t$	$\cos t$	$\tan t$	$\cot t$	$\sec t$	$\csc t$		
.1222	**7°00′**	.1219	.9925	.1228	8.144	1.008	8.206	**83°00′**	1.4486
.1251	10	.1248	.9922	.1257	7.953	1.008	8.016	50	1.4457
.1280	20	.1276	.9918	.1287	7.770	1.008	7.834	40	1.4428
.1309	30	.1305	.9914	.1317	7.596	1.009	7.661	30	1.4399
.1338	40	.1334	.9911	.1346	7.429	1.009	7.496	20	1.4370
.1367	50	.1363	.9907	.1376	7.269	1.009	7.337	10	1.4341
.1396	**8°00′**	.1392	.9903	.1405	7.115	1.010	7.185	**82°00′**	1.4312
.1425	10	.1421	.9899	.1435	6.968	1.010	7.040	50	1.4283
.1454	20	.1449	.9894	.1465	6.827	1.011	6.900	40	1.4254
.1484	30	.1478	.9890	.1495	6.691	1.011	6.765	30	1.4224
.1513	40	.1507	.9886	.1524	6.561	1.012	6.636	20	1.4195
.1542	50	.1536	.9881	.1554	6.435	1.012	6.512	10	1.4166
.1571	**9°00′**	.1564	.9877	.1584	6.314	1.012	6.392	**81°00′**	1.4137
.1600	10	.1593	.9872	.1614	6.197	1.013	6.277	50	1.4108
.1629	20	.1622	.9868	.1644	6.084	1.013	6.166	40	1.4079
.1658	30	.1650	.9863	.1673	5.976	1.014	6.059	30	1.4050
.1687	40	.1679	.9858	.1703	5.871	1.014	5.955	20	1.4021
.1716	50	.1708	.9853	.1733	5.769	1.015	5.855	10	1.3992
.1745	**10°00′**	.1736	.9848	.1763	5.671	1.015	5.759	**80°00′**	1.3963
.1774	10	.1765	.9843	.1793	5.576	1.016	5.665	50	1.3934
.1804	20	.1794	.9838	.1823	5.485	1.016	5.575	40	1.3904
.1833	30	.1822	.9833	.1853	5.396	1.017	5.487	50	1.3875
.1862	40	.1851	.9827	.1883	5.309	1.018	5.403	20	1.3846
.1891	50	.1880	.9822	.1914	5.226	1.018	5.320	10	1.3817
.1920	**11°00′**	.1908	.9816	.1944	5.145	1.019	5.241	**79°00′**	1.3788
.1949	10	.1937	.9811	.1974	5.066	1.019	5.164	50	1.3759
.1978	20	.1965	.9805	.2004	4.989	1.020	5.089	40	1.3730
.2007	30	.1994	.9799	.2035	4.915	1.020	5.016	30	1.3701
.2036	40	.2022	.9793	.2065	4.843	1.021	4.945	20	1.3672
.2065	50	.2051	.9787	.2095	4.773	1.022	4.876	10	1.3643
.2094	**12°00′**	.2079	.9781	.2126	4.705	1.022	4.810	**78°00′**	1.3614
.2123	10	.2108	.9775	.2156	4.638	1.023	4.745	50	1.3584
.2153	20	.2136	.9769	.2186	4.574	1.024	4.682	40	1.3555
.2182	30	.2164	.9763	.2217	4.511	1.024	4.620	30	1.3526
.2211	40	.2193	.9757	.2247	4.449	1.025	4.560	20	1.3497
.2240	50	.2221	.9750	.2278	4.390	1.026	4.502	10	1.3468
.2269	**13°00′**	.2250	.9744	.2309	4.331	1.026	4.445	**77°00′**	1.3439
.2298	10	.2278	.9737	.2339	4.275	1.027	4.390	50	1.3410
.2327	20	.2306	.9730	.2370	4.219	1.028	4.336	40	1.3381
.2356	30	.2334	.9724	.2401	4.165	1.028	4.284	30	1.3352
.2385	40	.2363	.9717	.2432	4.113	1.029	4.232	20	1.3323
.2414	50	.2391	.9710	.2462	4.061	1.030	4.182	10	1.3294
.2443	**14°00′**	.2419	.9703	.2493	4.011	1.031	4.134	**76°00′**	1.3265
		$\cos t$	$\sin t$	$\cot t$	$\tan t$	$\csc t$	$\sec t$	t degrees	t

TABLE 4 VALUES OF THE TRIGONOMETRIC FUNCTIONS (Cont'd.)

t	t degrees	sin t	cos t	tan t	cot t	sec t	csc t		
.2443	**14°00′**	.2419	.9703	.2493	4.011	1.031	4.134	**76°00′**	1.3265
.2473	10	.2447	.9696	.2524	3.962	1.031	4.086	50	1.3235
.2502	20	.2476	.9689	.2555	3.914	1.032	4.039	40	1.3206
.2531	30	.2504	.9681	.2586	3.867	1.033	3.994	30	1.3177
.2560	40	.2532	.9674	.2617	3.821	1.034	3.950	20	1.3148
.2589	50	.2560	.9667	.2648	3.776	1.034	3.906	10	1.3119
.2618	**15°00′**	.2588	.9659	.2679	3.732	1.035	3.864	**75°00′**	1.3090
.2647	10	.2616	.9652	.2711	3.689	1.036	3.822	50	1.3061
.2676	20	.2644	.9644	.2742	3.647	1.037	3.782	40	1.3032
.2705	30	.2672	.9636	.2773	3.606	1.038	3.742	30	1.3003
.2734	40	.2700	.9628	.2805	3.566	1.039	3.703	20	1.2974
.2763	50	.2728	.9621	.2836	3.526	1.039	3.665	10	1.2945
.2793	**16°00′**	.2756	.9613	.2867	3.487	1.040	3.628	**74°00′**	1.2915
.2822	10	.2784	.9605	.2899	3.450	1.041	3.592	50	1.2886
.2851	20	.2812	.9596	.2931	3.412	1.042	3.556	40	1.2857
.2880	30	.2840	.9588	.2962	3.376	1.043	3.521	30	1.2828
.2909	40	.2868	.9580	.2994	3.340	1.044	3.487	20	1.2799
.2938	50	.2896	.9572	.3026	3.305	1.045	3.453	10	1.2770
.2967	**17°00′**	.2924	.9563	.3057	3.271	1.046	3.420	**73°00′**	1.2741
.2996	10	.2952	.9555	.3089	3.237	1.047	3.388	50	1.2712
.3025	20	.2979	.9546	.3121	3.204	1.048	3.356	40	1.2683
.3054	30	.3007	.9537	.3153	3.172	1.049	3.326	30	1.2654
.3083	40	.3035	.9528	.3185	3.140	1.049	3.295	20	1.2625
.3113	50	.3062	.9520	.3217	3.108	1.050	3.265	10	1.2595
.3142	**18°00′**	.3090	.9511	.3249	3.078	1.051	3.236	**72°00′**	1.2566
.3171	10	.3118	.9502	.3281	3.047	1.052	3.207	50	1.2537
.3200	20	.3145	.9492	.3314	3.018	1.053	3.179	40	1.2508
.3229	30	.3173	.9483	.3346	2.989	1.054	3.152	30	1.2479
.3258	40	.3201	.9474	.3378	2.960	1.056	3.124	20	1.2450
.3287	50	.3228	.9465	.3411	2.932	1.057	3.098	10	1.2421
.3316	**19°00′**	.3256	.9455	.3443	2.904	1.058	3.072	**71°00′**	1.2392
.3345	10	.3283	.9446	.3476	2.877	1.059	3.046	50	1.2363
.3374	20	.3311	.9436	.3508	2.850	1.060	3.021	40	1.2334
.3403	30	.3338	.9426	.3541	2.824	1.061	2.996	30	1.2305
.3432	40	.3365	.9417	.3574	2.798	1.062	2.971	20	1.2275
.3462	50	.3393	.9407	.3607	2.773	1.063	2.947	10	1.2246
.3491	**20°00′**	.3420	.9397	.3640	2.747	1.064	2.924	**70°00′**	1.2217
.3520	10	.3448	.9387	.3673	2.723	1.065	2.901	50	1.2188
.3549	20	.3475	.9377	.3706	2.699	1.066	2.878	40	1.2159
.3578	30	.3502	.9367	.3739	2.675	1.068	2.855	30	1.2130
.3607	40	.3529	.9356	.3772	2.651	1.069	2.833	20	1.2101
.3636	50	.3557	.9346	.3805	2.628	1.070	2.812	10	1.2072
.3665	**21°00′**	.3584	.9336	.3839	2.605	1.071	2.790	**69°00′**	1.2043
		cos t	sin t	cot t	tan t	csc t	sec t	t degrees	t

t	t degrees	sin t	cos t	tan t	cot t	sec t	csc t		
.3665	**21°00′**	.3584	.9336	.3839	2.605	1.071	2.790	**69°00′**	1.2043
.3694	10	.3611	.9325	.3872	2.583	1.072	2.769	50	1.2014
.3723	20	.3638	.9315	.3906	2.560	1.074	2.749	40	1.1985
.3752	30	.3665	.9304	.3939	2.539	1.075	2.729	30	1.1956
.3782	40	.3692	.9293	.3973	2.517	1.076	2.709	20	1.1926
.3811	50	.3719	.9283	.4006	2.496	1.077	2.689	10	1.1897
.3840	**22°00′**	.3746	.9272	.4040	2.475	1.079	2.669	**68°00′**	1.1868
.3869	10	.3773	.9261	.4074	2.455	1.080	2.650	50	1.1839
.3898	20	.3800	.9250	.4108	2.434	1.081	2.632	40	1.1810
.3927	30	.3827	.9239	.4142	2.414	1.082	2.613	30	1.1781
.3956	40	.3854	.9228	.4176	2.394	1.084	2.595	20	1.1752
.3985	50	.3881	.9216	.4210	2.375	1.085	2.577	10	1.1723
.4014	**23°00′**	.3907	.9205	.4245	2.356	1.086	2.559	**67°00′**	1.1694
.4043	10	.3934	.9194	.4279	2.337	1.088	2.542	50	1.1665
.4072	20	.3961	.9182	.4314	2.318	1.089	2.525	40	1.1636
.4102	30	.3987	.9771	.4348	2.300	1.090	2.508	30	1.1606
.4131	40	.4014	.9159	.4383	2.282	1.092	2.491	20	1.1577
.4160	50	.4041	.9147	.4417	2.264	1.093	2.475	10	1.1548
.4189	**24°00′**	.4067	.9135	.4452	2.246	1.095	2.459	**66°00′**	1.1519
.4218	10	.4094	.9124	.4487	2.229	1.096	2.443	50	1.1490
.4247	20	.4120	.9112	.4522	2.211	1.097	2.427	40	1.1461
.4276	30	.4147	.9100	.4557	2.194	1.099	2.411	30	1.1432
.4305	40	.4173	.9088	.4592	2.177	1.100	2.396	20	1.1403
.4334	50	.4200	.9075	.4628	2.161	1.102	2.381	10	1.1374
.4363	**25°00′**	.4226	.9063	.4663	2.145	1.103	2.366	**65°00′**	1.1345
.4392	10	.4253	.9051	.4699	2.128	1.105	2.352	50	1.1316
.4422	20	.4279	.9038	.4734	2.112	1.106	2.337	40	1.1286
.4451	30	.4305	.9026	.4770	2.097	1.108	2.323	30	1.1257
.4480	40	.4331	.9013	.4806	2.081	1.109	2.309	20	1.1228
.4509	50	.4358	.9001	.4841	2.066	1.111	2.295	10	1.1199
.4538	**26°00′**	.4384	.8988	.4877	2.050	1.113	2.281	**64°00′**	1.1170
.4567	10	.4410	.8975	.4913	2.035	1.114	2.268	50	1.1141
.4596	20	.4436	.8962	.4950	2.020	1.116	2.254	40	1.1112
.4625	30	.4462	.8949	.4986	2.006	1.117	2.241	30	1.1083
.4654	40	.4488	.8936	.5022	1.991	1.119	2.228	20	1.1054
.4683	50	.4514	.8923	.5059	1.977	1.121	2.215	10	1.1025
.4712	**27°00′**	.4540	.8910	.5095	1.963	1.122	2.203	**63°00′**	1.0996
.4741	10	.4566	.8897	.5132	1.949	1.124	2.190	50	1.0966
.4771	20	.4592	.8884	.5169	1.935	1.126	2.178	40	1.0937
.4800	30	.4617	.8870	.5206	1.921	1.127	2.166	30	1.0908
.4829	40	.4643	.8857	.5243	1.907	1.129	2.154	20	1.0879
.4858	50	.4669	.8843	.5280	1.894	1.131	2.142	10	1.0850
.4887	**28°00′**	.4695	.8829	.5317	1.881	1.133	2.130	**62°00′**	1.0821
		cos t	sin t	cot t	tan t	csc t	sec t	t degrees	t

TABLE 4 VALUES OF THE TRIGONOMETRIC FUNCTIONS (Cont'd.)

t	t degrees	sin t	cos t	tan t	cot t	sec t	csc t		
.4887	**28°00′**	.4695	.8829	.5317	1.881	1.133	2.130	**62°00′**	1.0821
.4916	10	.4720	.8816	.5354	1.868	1.134	2.118	50	1.0792
.4945	20	.4746	.8802	.5392	1.855	1.136	2.107	40	1.0763
.4974	30	.4772	.8788	.5430	1.842	1.138	2.096	30	1.0734
.5003	40	.4797	.8774	.5467	1.829	1.140	2.085	20	1.0705
.5032	50	.4823	.8760	.5505	1.816	1.142	2.074	10	1.0676
.5061	**29°00′**	.4848	.8746	.5543	1.804	1.143	2.063	**61°00′**	1.0647
.5091	10	.4874	.8732	.5581	1.792	1.145	2.052	50	1.0617
.5120	20	.4899	.8718	.5619	1.780	1.147	2.041	40	1.0588
.5149	30	.4924	.8704	.5658	1.767	1.149	2.031	30	1.0559
.5178	40	.4950	.8689	.5696	1.756	1.151	2.020	20	1.0530
.5207	50	.4975	.8675	.5735	1.744	1.153	2.010	10	1.0501
.5236	**30°00′**	.5000	.8660	.5774	1.732	1.155	2.000	**60°00′**	1.0472
.5265	10	.5025	.8646	.5812	1.720	1.157	1.990	50	1.0443
.5294	20	.5050	.8631	.5851	1.709	1.159	1.980	40	1.0414
.5323	30	.5075	.8616	.5890	1.698	1.161	1.970	30	1.0385
.5352	40	.5100	.8601	.5930	1.686	1.163	1.961	20	1.0356
.5381	50	.5125	.8587	.5969	1.675	1.165	1.951	10	1.0327
.5411	**31°00′**	.5150	.8572	.6009	1.664	1.167	1.942	**59°00′**	1.0297
.5440	10	.5175	.8557	.6048	1.653	1.169	1.932	50	1.0268
.5469	20	.5200	.8542	.6088	1.643	1.171	1.923	40	1.0239
.5498	30	.5225	.8526	.6128	1.632	1.173	1.914	30	1.0210
.5527	40	.5250	.8511	.6168	1.621	1.175	1.905	20	1.0181
.5556	50	.5275	.8496	.6208	1.611	1.177	1.896	10	1.0152
.5585	**32°00′**	.5299	.8480	.6249	1.600	1.179	1.887	**58°00′**	1.0123
.5614	10	.5324	.8465	.6289	1.590	1.181	1.878	50	1.0094
.5643	20	.5348	.8450	.6330	1.580	1.184	1.870	40	1.0065
.5672	30	.5373	.8434	.6371	1.570	1.186	1.861	30	1.0036
.5701	40	.5398	.8418	.6412	1.560	1.188	1.853	20	1.0007
.5730	50	.5422	.8403	.6453	1.550	1.190	1.844	10	.9977
.5760	**33°00′**	.5446	.8387	.6494	1.540	1.192	1.836	**57°00′**	.9948
.5789	10	.5471	.8371	.6536	1.530	1.195	1.828	50	.9919
.5818	20	.5495	.8355	.6577	1.520	1.197	1.820	40	.9890
.5847	30	.5519	.8339	.6619	1.511	1.199	1.812	30	.9861
.5876	40	.5544	.8323	.6661	1.501	1.202	1.804	20	.9832
.5905	50	.5568	.8307	.6703	1.492	1.204	1.796	10	.9803
.5934	**34°00′**	.5592	.8290	.6745	1.483	1.206	1.788	**56°00′**	.9774
.5963	10	.5616	.8274	.6787	1.473	1.209	1.781	50	.9745
.5992	20	.5640	.8258	.6830	1.464	1.211	1.773	40	.9716
.6021	30	.5664	.8241	.6873	1.455	1.213	1.766	30	.9687
.6050	40	.5688	.8225	.6916	1.446	1.216	1.758	20	.9657
.6080	50	.5712	.8208	.6959	1.437	1.218	1.751	10	.9628
.6109	**35°00′**	.5736	.8192	.7002	1.428	1.221	1.743	**55°00′**	.9599
		cos t	sin t	cot t	tan t	csc t	sec t	t degrees	t

t	t degrees	sin t	cos t	tan t	cot t	sec t	csc t		
.6109	**35°00′**	.5736	.8192	.7002	1.428	1.221	1.743	**55°00′**	.9599
.6138	10	.5760	.8175	.7046	1.419	1.223	1.736	50	.9570
.6167	20	.5783	.8158	.7089	1.411	1.226	1.729	40	.9541
.6196	30	.5807	.8141	.7133	1.402	1.228	1.722	30	.9512
.6225	40	.5831	.8124	.7177	1.393	1.231	1.715	20	.9483
.6254	50	.5854	.8107	.7221	1.385	1.233	1.708	10	.9454
.6283	**36°00′**	.5878	.8090	.7265	1.376	1.236	1.701	**54°00′**	.9425
.6312	10	.5901	.8073	.7310	1.368	1.239	1.695	50	.9396
.6341	20	.5925	.8056	.7355	1.360	1.241	1.688	40	.9367
.6370	30	.5948	.8039	.7400	1.351	1.244	1.681	30	.9338
.6400	40	.5972	.8021	.7445	1.343	1.247	1.675	20	.9308
.6429	50	.5995	.8004	.7490	1.335	1.249	1.668	10	.9279
.6458	**37°00′**	.6018	.7986	.7536	1.327	1.252	1.662	**53°00′**	.9250
.6487	10	.6041	.7969	.7581	1.319	1.255	1.655	50	.9221
.6516	20	.6065	.7951	.7627	1.311	1.258	1.649	40	.9192
.6545	30	.6088	.7934	.7673	1.303	1.260	1.643	30	.9163
.6574	40	.6111	.7916	.7720	1.295	1.263	1.636	20	.9134
.6603	50	.6134	.7898	.7766	1.288	1.266	1.630	10	.9105
.6632	**38°00′**	.6157	.7880	.7813	1.280	1.269	1.624	**52°00′**	.9076
.6661	10	.6180	.7862	.7860	1.272	1.272	1.618	50	.9047
.6690	20	.6202	.7844	.7907	1.265	1.275	1.612	40	.0918
.6720	30	.6225	.7826	.7954	1.257	1.278	1.606	30	.8988
.6749	40	.6248	.7808	.8002	1.250	1.281	1.601	20	.8959
.6778	50	.6271	.7790	.8050	1.242	1.284	1.595	10	.8930
.6807	**39°00′**	.6293	.7771	.8098	1.235	1.287	1.589	**51°00′**	.8901
.6836	10	.6316	.7753	.8146	1.228	1.290	1.583	o0	.8872
.6865	20	.6338	.7735	.8195	1.220	1.293	1.578	40	.8843
.6894	30	.6361	.7716	.8243	1.213	1.296	1.572	30	.8814
.6923	40	.6383	.7698	.8292	1.206	1.299	1.567	20	.8785
.6952	50	.6406	.7679	.8342	1.199	1.302	1.561	10	.8756
.6981	**40°00′**	.6428	.7660	.8391	1.192	1.305	1.556	**50°00′**	.8727
.7010	10	.6450	.7642	.8441	1.185	1.309	1.550	50	.8698
.7039	20	.6472	.7623	.8491	1.178	1.312	1.545	40	.8668
.7069	30	.6494	.7604	.8541	1.171	1.315	1.540	30	.8639
.7098	40	.6417	.7585	.8591	1.164	1.318	1.535	20	.8610
.7127	50	.6539	.7566	.8642	1.157	1.322	1.529	10	.8581
.7156	**41°00′**	.6561	.7547	.8693	1.150	1.325	1.524	**49°00′**	.8552
.7185	10	.6583	.7528	.8744	1.144	1.328	1.519	50	.8523
.7214	20	.6604	.7509	.8796	1.137	1.332	1.514	40	.8494
.7243	30	.6626	.7490	.8847	1.130	1.335	1.509	30	.8465
.7272	40	.6648	.7470	.8899	1.124	1.339	1.504	20	.8436
.7301	50	.6670	.7451	.8952	1.117	1.342	1.499	10	.8407
.7330	**42°00′**	.6691	.7431	.9004	1.111	1.346	1.494	**48°00′**	.8378
		cos t	sin t	cot t	tan t	csc t	sec t	t degrees	t

TABLE 4 VALUES OF THE TRIGONOMETRIC FUNCTIONS (Cont'd.)

t	t degrees	$\sin t$	$\cos t$	$\tan t$	$\cot t$	$\sec t$	$\csc t$		
.7330	**42°00′**	.6691	.7431	.9004	1.111	1.346	1.494	**48°00′**	.8378
.7359	10	.6713	.7412	.9057	1.104	1.349	1.490	50	.8348
.7389	20	.6734	.7392	.9110	1.098	1.353	1.485	40	.8319
.7418	30	.6756	.7373	.9163	1.091	1.356	1.480	30	.8290
.7447	40	.6777	.7353	.9217	1.085	1.360	1.476	20	.8261
.7476	50	.6799	.7333	.9271	1.079	1.364	1.471	10	.8232
.7505	**43°00′**	.6820	.7314	.9325	1.072	1.367	1.466	**47°00′**	.8203
.7534	10	.6841	.7294	.9380	1.066	1.371	1.462	50	.8174
.7563	20	.6862	.7274	.9435	1.060	1.375	1.457	40	.8145
.7592	30	.6884	.7254	.9490	1.054	1.379	1.453	30	.8116
.7621	40	.6905	.7234	.9545	1.048	1.382	1.448	20	.8087
.7650	50	.6926	.7214	.9601	1.042	1.386	1.444	10	.8058
.7679	**44°00′**	.6947	.7193	.9657	1.036	1.390	1.440	**46°00′**	.8029
.7709	10	.6967	.7173	.9713	1.030	1.394	1.435	50	.7999
.7738	20	.6988	.7153	.9770	1.024	1.398	1.431	40	.7970
.7767	30	.7009	.7133	.9827	1.018	1.402	1.427	30	.7941
.7796	40	.7030	.7112	.9884	1.012	1.406	1.423	20	.7912
.7825	50	.7050	.7092	.9942	1.006	1.410	1.418	10	.7883
.7854	**45°00′**	.7071	.7071	1.0000	1.0000	1.414	1.414	**45°00′**	.7854
		$\cos t$	$\sin t$	$\cot t$	$\tan t$	$\csc t$	$\sec t$	t degrees	

TABLE 5 TRIGONOMETRIC FUNCTIONS OF RADIANS AND REAL NUMBERS

t	$\sin t$	$\cos t$	$\tan t$	$\cot t$	$\sec t$	$\csc t$
.00	.0000	1.0000	.0000	—	1.000	—
.01	.0100	1.0000	.0100	99.997	1.000	100.00
.02	.0200	.9998	.0200	49.993	1.000	50.00
.03	.0300	.9996	.0300	33.323	1.000	33.34
.04	.0400	.9992	.0400	24.987	1.001	25.01
.05	.0500	.9988	.0500	19.983	1.001	20.01
.06	.0600	.9982	.0601	16.647	1.002	16.68
.07	.0699	.9976	.0701	14.262	1.002	14.30
.08	.0799	.9968	.0802	12.473	1.003	12.51
.09	.0899	.9960	.0902	11.081	1.004	11.13
.10	.0998	.9950	.1003	9.967	1.005	10.02
.11	.1098	.9940	.1104	9.054	1.006	9.109
.12	.1197	.9928	.1206	8.293	1.007	8.353
.13	.1296	.9916	.1307	7.649	1.009	7.714
.14	.1395	.9902	.1409	7.096	1.010	7.166
.15	.1494	.9888	.1511	6.617	1.011	6.692
.16	.1593	.9872	.1614	6.197	1.013	6.277
.17	.1692	.9856	.1717	5.826	1.015	5.911
.18	.1790	.9838	.1820	5.495	1.016	5.586
.19	.1889	.9820	.1923	5.200	1.018	5.295
.20	.1987	.9801	.2027	4.933	1.020	5.033
.21	.2085	.9780	.2131	4.692	1.022	4.797
.22	.2182	.9759	.2236	4.472	1.025	4.582
.23	.2280	.9737	.2341	4.271	1.027	4.386
.24	.2377	.9713	.2447	4.086	1.030	4.207
.25	.2474	.9689	.2553	3.916	1.032	4.042
.26	.2571	.9664	.2660	3.759	1.035	3.890
.27	.2667	.9638	.2768	3.613	1.038	3.749
.28	.2764	.9611	.2876	3.478	1.041	3.619
.29	.2860	.9582	.2984	3.351	1.044	3.497
.30	.2955	.9553	.3093	3.233	1.047	3.384
.31	.3051	.9523	.3203	3.122	1.050	3.278
.32	.3146	.9492	.3314	3.018	1.053	3.179
.33	.3240	.9460	.3425	2.920	1.057	3.086
.34	.3335	.9428	.3537	2.827	1.061	2.999
.35	.3429	.9394	.3650	2.740	1.065	2.916
.36	.3523	.9359	.3764	2.657	1.068	2.839
.37	.3616	.9323	.3879	2.578	4.073	2.765
.38	.3709	.9287	.3994	2.504	1.077	2.696
.39	.3802	.9249	.4111	2.433	1.081	2.630

t	$\sin t$	$\cos t$	$\tan t$	$\cot t$	$\sec t$	$\csc t$
.40	.3894	.9211	.4228	2.365	1.086	2.568
.41	.3986	.9171	.4346	2.301	1.090	2.509
.42	.4078	.9131	.4466	2.239	1.095	2.452
.43	.4169	.9090	.4586	2.180	1.100	2.399
.44	.4259	.9048	.4708	2.124	1.105	2.348
.45	.4350	.9004	.4831	2.070	1.111	2.299
.46	.4439	.8961	.4954	2.018	1.116	2.253
.47	.4529	.8916	.5080	1.969	1.122	2.208
.48	.4618	.8870	.5206	1.921	1.127	2.166
.49	.4706	.8823	.5334	1.875	1.133	2.125
.50	.4794	.8776	.5463	1.830	1.139	2.086
.51	.4882	.8727	.5594	1.788	1.146	2.048
.52	.4969	.8678	.5726	1.747	1.152	2.013
.53	.5055	.8628	.5859	1.707	1.159	1.978
.54	.5141	.8577	.5994	1.668	1.166	1.945
.55	.5227	.8525	.6131	1.631	1.173	1.913
.56	.5312	.8473	.6269	1.595	1.180	1.883
.57	.5396	.8419	.6410	1.560	1.188	1.853
.58	.5480	.8365	.6552	1.526	1.196	1.825
.59	.5564	.8309	.6696	1.494	1.203	1.797
.60	.5646	.8253	.6841	1.462	1.212	1.771
.61	.5729	.8196	.6989	1.431	1.220	1.746
.62	.5810	.8139	.7139	1.401	1.229	1.721
.63	.5891	.8080	.7291	1.372	1.238	1.697
.64	.5972	.8021	.7445	1.343	1.247	1.674
.65	.6052	.7961	.7602	1.315	1.256	1.652
.66	.6131	.7900	.7761	1.288	1.266	1.631
.67	.6210	.7838	.7923	1.262	1.276	1.610
.68	.6288	.7776	.8087	1.237	1.286	1.590
.69	.6365	.7712	.8253	1.212	1.297	1.571
.70	.6442	.7648	.8423	1.187	1.307	1.552
.71	.6518	.7584	.8595	1.163	1.319	1.534
.72	.6594	.7518	.8771	1.140	1.330	1.517
.73	.6669	.7452	.8949	1.117	1.342	1.500
.74	.6743	.7385	.9131	1.095	1.354	1.483
.75	.6816	.7317	.9316	1.073	1.367	1.467
.76	.6889	.7248	.9505	1.052	1.380	1.452
.77	.6961	.7179	.9697	1.031	1.393	1.437
.78	.7033	.7109	.9893	1.011	1.407	1.422
.79	.7104	.7038	1.009	.9908	1.421	1.408

TABLE 5 TRIGONOMETRIC FUNCTIONS OF RADIANS AND REAL NUMBERS (Cont'd.)

t	$\sin t$	$\cos t$	$\tan t$	$\cot t$	$\sec t$	$\csc t$
.80	.7174	.6967	1.030	.9712	1.435	1.394
.81	.7243	.6895	1.050	.9520	1.450	1.381
.82	.7311	.6822	1.072	.9331	1.466	1.368
.83	.7379	.6749	1.093	.9146	1.482	1.355
.84	.7446	.6675	1.116	.8964	1.498	1.343
.85	.7513	.6600	1.138	.8785	1.515	1.331
.86	.7578	.6524	1.162	.8609	1.533	1.320
.87	.7643	.6448	1.185	.8437	1.551	1.308
.88	.7707	.6372	1.210	.8267	1.569	1.297
.89	.7771	.6294	1.235	.8100	1.589	1.287
.90	.7833	.6216	1.260	.7936	1.609	1.277
.91	.7895	.6137	1.286	.7774	1.629	1.267
.92	.7956	.6058	1.313	.7615	1.651	1.257
.93	.8016	.5978	1.341	.7458	1.673	1.247
.94	.8076	.5898	1.369	.7303	1.696	1.238
.95	.8134	.5817	1.398	.7151	1.719	1.229
.96	.8192	.5735	1.428	.7001	1.744	1.221
.97	.8249	.5653	1.459	.6853	1.769	1.212
.98	.8305	.5570	1.491	.6707	1.795	1.204
.99	.8360	.5487	1.524	.6563	1.823	1.196
1.00	.8415	.5403	1.557	.6421	1.851	1.188
1.01	.8468	.5319	1.592	.6281	1.880	1.181
1.02	.8521	.5234	1.628	.6142	1.911	1.174
1.03	.8573	.5148	1.665	.6005	1.942	1.166
1.04	.8624	.5062	1.704	.5870	1.975	1.160
1.05	.8674	.4976	1.743	.5736	2.010	1.153
1.06	.8724	.4889	1.784	.5604	2.046	1.146
1.07	.8772	.4801	1.827	.5473	2.083	1.140
1.08	.8820	.4713	1.871	.5344	2.122	1.134
1.09	.8866	.4625	1.917	.5216	2.162	1.128
1.10	.8912	.4536	1.965	.5090	2.205	1.122
1.11	.8957	.4447	2.014	.4964	2.249	1.116
1.12	.9001	.4357	2.066	.4840	2.295	1.111
1.13	.9044	.4267	2.120	.4718	2.344	1.106
1.14	.9086	.4176	2.176	.4596	2.395	1.101
1.15	.9128	.4085	2.234	.4475	2.448	1.096
1.16	.9168	.3993	2.296	.4356	2.504	1.091
1.17	.9208	.3902	2.360	.4237	2.563	1.086
1.18	.9246	.3809	2.427	.4120	2.625	1.082
1.19	.9284	.3717	2.498	.4003	2.691	1.077
1.20	.9320	.3624	2.572	.3888	2.760	1.073
1.21	.9356	.3530	2.650	.3773	2.833	1.069
1.22	.9391	.3436	2.733	.3659	2.910	1.065
1.23	.9425	.3342	2.820	.3546	2.992	1.061
1.24	.9458	.3248	2.912	.3434	3.079	1.057
1.25	.9490	.3153	3.010	.3323	3.171	1.054
1.26	.9521	.3058	3.113	.3212	3.270	1.050
1.27	.9551	.2963	3.224	.3102	3.375	1.047
1.28	.9580	.2867	3.341	.2993	3.488	1.044
1.29	.9608	.2771	3.467	.2884	3.609	1.041
1.30	.9636	.2675	3.602	.2776	3.738	1.038
1.31	.9662	.2579	3.747	.2669	3.878	1.035
1.32	.9687	.2482	3.903	.2562	4.029	1.032
1.33	.9711	.2385	4.072	.2456	4.193	1.030
1.34	.9735	.2288	4.256	.2350	4.372	1.027
1.35	.9757	.2190	4.455	.2245	4.566	1.025
1.36	.9779	.2092	4.673	.2140	4.779	1.023
1.37	.9799	.1994	4.913	.2035	5.014	1.021
1.38	.9819	.1896	5.177	.1931	5.273	1.018
1.39	.9837	.1798	5.471	.1828	5.561	1.017
1.40	.9854	.1700	5.798	.1725	5.883	1.015
1.41	.9871	.1601	6.165	.1622	6.246	1.013
1.42	.9887	.1502	6.581	.1519	6.657	1.011
1.43	.9901	.1403	7.055	.1417	7.126	1.010
1.44	.9915	.1304	7.602	.1315	7.667	1.009
1.45	.9927	.1205	8.238	.1214	8.299	1.007
1.46	.9939	.1106	8.989	.1113	9.044	1.006
1.47	.9949	.1006	9.887	.1011	9.938	1.005
1.48	.9959	.0907	10.983	.0910	11.029	1.004
1.49	.9967	.0807	12.350	.0810	12.390	1.003
1.50	.9975	.0707	14.101	.0709	14.137	1.003
1.51	.9982	.0608	16.428	.0609	16.458	1.002
1.52	.9987	.0508	19.670	.0508	19.695	1.001
1.53	.9992	.0408	24.498	.0408	24.519	1.001
1.54	.9995	.0308	32.461	.0308	32.476	1.000
1.55	.9998	.0208	48.078	.0208	48.089	1.000
1.56	.9999	.0108	92.620	.0108	92.626	1.000
1.57	1.0000	.0008	1255.8	.0008	1255.8	1.000

ANSWERS TO ODD-NUMBERED EXERCISES

CHAPTER ▪ 1

EXERCISES 1.2 ▪ PAGE 12

1 (a) positive (b) negative (c) negative (d) positive (e) positive (f) positive
3 (a) $<$ (b) $>$ (c) $=$ **5** (a) $>$ (b) $>$ (c) $>$
7 $x < 0$ **9** $1/z \leq 0$ **11** $q \leq \pi$ **13** $|x| > 7$
15 $t \geq 5$ **17** $1/f \leq 14$ **19** (a) 5 (b) 3 (c) 11
21 (a) -15 (b) -3 (c) 11
23 (a) $4 - \pi$ (b) $4 - \pi$ (c) $1.5 - \sqrt{2}$
25 (a) 4 (b) 12 (c) 12 (d) 8
27 (a) 10 (b) 9 (c) 9 (d) 19 **29** $-x - 3$
31 $2 - x$ **33** $b - a$ **35** $x^2 + 4$ **37** $\neq$ **39** $=$
41 $\neq$ **43** $=$
45 Construct a right triangle with sides of lengths $\sqrt{2}$ and 1. The hypotenuse will have length $\sqrt{3}$. Next construct a right triangle with sides of lengths $\sqrt{3}$ and $\sqrt{2}$. The hypotenuse will have length $\sqrt{5}$.
47 The large rectangle has area $a(b + c)$. The sum of the areas of the two small rectangles is $ab + ac$.

EXERCISES 1.3 ▪ PAGE 23

1 $\frac{16}{81}$ **3** $\frac{9}{8}$ **5** $-71/9$ **7** $\frac{1}{8}$ **9** $\frac{1}{25}$ **11** $8x^9$
13 $\dfrac{6}{x}$ **15** $-2a^{14}$ **17** $\frac{9}{2}$ **19** $\dfrac{12u^{11}}{v^2}$ **21** $\dfrac{4}{xy}$
23 $\dfrac{9y^6}{x^8}$ **25** $\dfrac{81y^6}{64}$ **27** $\dfrac{s^6}{4r^8}$ **29** $\dfrac{20y}{x^3}$ **31** $9x^{10}y^{14}$
33 $8a^2$ **35** $24x^{3/2}$ **37** $\dfrac{1}{9a^4}$ **39** $\dfrac{8}{x^{1/2}}$ **41** $4x^2y^4$
43 $\dfrac{3}{x^3y^2}$ **45** 1 **47** $x^{3/4}$ **49** $(a + b)^{2/3}$
51 $(x^2 + y^2)^{1/2}$ **53** (a) $4x\sqrt{x}$ (b) $8x\sqrt{x}$
55 (a) $8 - \sqrt[3]{y}$ (b) $\sqrt[3]{8 - y}$ **57** 9 **59** $-2\sqrt[5]{2}$
61 $\frac{1}{2}\sqrt[3]{4}$ **63** $\dfrac{3y^3}{x^2}$ **65** $\dfrac{2a^2}{b}$ **67** $\dfrac{1}{2y^2}\sqrt{6xy}$
69 $\dfrac{xy}{3}\sqrt[3]{6y}$ **71** $\dfrac{x}{3}\sqrt[4]{15x^2y^3}$ **73** $\frac{1}{2}\sqrt[5]{20x^4y^2}$
75 $\dfrac{3x^5}{y^2}$ **77** $\dfrac{2x}{y^2}\sqrt[5]{x^2y^4}$ **79** $-3tv^2$
81 $\neq$; $(a^r)^2 = a^{2r} \neq a^{(r^2)}$ **83** $\neq$; $(ab)^{xy} = a^{xy}b^{xy} \neq a^x b^y$
85 $=$; $\sqrt[n]{\dfrac{1}{c}} = \left(\dfrac{1}{c}\right)^{1/n} = \dfrac{(1)^{1/n}}{(c)^{1/n}} = \dfrac{1}{\sqrt[n]{c}}$ **87** 1.7×10^{-24}
89 (a) 4.27×10^5 (b) 9.8×10^{-8} (c) 8.1×10^8
91 (a) 830,000 (b) 0.0000000000029 (c) 563,000,000
93 5.87×10^{12} mi **95** 1.678×10^{-24} g
97 4.1472×10^6 frames **99** \$232,825.78 **101** 2.82 m
103 The 120-kg lifter

EXERCISES 1.4 ▪ PAGE 34

1 $12x^3 - 13x + 1$ **3** $x^3 - 2x^2 + 4$
5 $6u^2 - 13u - 12$ **7** $6x^3 + 37x^2 + 30x - 25$
9 $3t^4 + 5t^3 - 15t^2 + 9t - 10$
11 $2x^6 + 2x^5 - 2x^4 + 8x^3 + 10x^2 - 10x - 10$
13 $4y^2 - 5x$ **15** $3v^2 - 2u^2 + uv^2$ **17** $6x^2 + x - 35$
19 $15x^2 + 31xy + 14y^2$ **21** $4x^2 - 9y^2$
23 $x^4 - 8x^2 + 16$ **25** $x - y$ **27** $x^4 + 5x^2 - 36$
29 $9x^2 + 12xy + 4y^2$ **31** $x^4 - 6x^2y^2 + 9y^4$
33 $x - y$ **35** $x^3 - 6x^2y + 12xy^2 - 8y^3$
37 $8x^3 + 36x^2y + 54xy^2 + 27y^3$
39 $a^2 + b^2 + c^2 + 2ab - 2ac - 2bc$
41 $4x^2 + y^2 + 9z^2 + 4xy - 12xz - 6yz$ **43** $s(r + 4t)$
45 $3a^2b(b - 2)$ **47** $3x^2y^2(y - 3x)$

49 $5x^3y^2(3y^3 - 5x + 2x^3y^2)$ **51** $(8x + 3)(x - 7)$
53 Irreducible **55** $(3x - 4)(2x + 5)$
57 $(3x - 5)(4x - 3)$ **59** $(2x - 5)^2$ **61** $(5z + 3)^2$
63 $(5x + 2y)(9x + 4y)$ **65** $(6r + 5t)(6r - 5t)$
67 $(z^2 + 8w)(z^2 - 8w)$ **69** $x^2(x + 2)(x - 2)$
71 Irreducible **73** $3(5x + 4y)(5x - 4y)$
75 $(4x + 3)(16x^2 - 12x + 9)$
77 $(4x - y^2)(16x^2 + 4xy^2 + y^4)$
79 $(7x + y^3)(49x^2 - 7xy^3 + y^6)$ **81** $(2x + y)(a - 3b)$
83 $3(x + 3)(x - 3)(x + 1)$
85 $(x - 1)(x + 2)(x^2 + x + 1)$ **87** $(a^2 + b^2)(a - b)$
89 $(a + b)(a - b)(a^2 - ab + b^2)(a^2 + ab + b^2)$
91 $(x + 2 + 3y)(x + 2 - 3y)$
93 $(y + 2)(y^2 - 2y + 4)(y - 1)(y^2 + y + 1)$
95 $(x^8 + 1)(x^4 + 1)(x^2 + 1)(x + 1)(x - 1)$
97 Area of I is $(x - y)x$, area of II is $(x - y)y$, and $A = x^2 - y^2 = (x - y)x + (x - y)y = (x - y)(x + y)$.

EXERCISES 1.5 ▪ PAGE 42

1 $\frac{22}{75}$ **3** $\frac{7}{120}$ **5** $\dfrac{x + 3}{x - 4}$ **7** $\dfrac{y + 5}{y^2 + 5y + 25}$

9 $\dfrac{4 - r}{r^2}$ **11** $\dfrac{x}{x - 1}$ **13** $\dfrac{a}{(a^2 + 4)(5a + 2)}$ **15** $\dfrac{-3}{x + 2}$

17 $\dfrac{6s - 7}{(3s + 1)^2}$ **19** $\dfrac{5x^2 + 2}{x^3}$ **21** $\dfrac{4(2t + 5)}{t + 2}$

23 $\dfrac{2(2x + 3)}{3x - 4}$ **25** $\dfrac{2x - 1}{x}$ **27** $\dfrac{p^2 + 2p + 4}{p - 3}$

29 $\dfrac{11u^2 + 18u + 5}{u(3u + 1)}$ **31** $\dfrac{-x - 5}{(x + 2)^2}$ **33** $a + b$

35 $\dfrac{x^2 + xy + y^2}{x + y}$ **37** $\dfrac{2x^2 + 7x + 15}{x^2 + 10x + 7}$ **39** $2x + h - 3$

41 $-\dfrac{3x^2 + 3xh + h^2}{x^3(x + h)^3}$ **43** $\dfrac{-12}{(3x + 3h - 1)(3x - 1)}$

45 $(3x + 2)^3(36x^2 - 37x + 6)$

47 $\dfrac{(2x + 1)^2(8x^2 + x - 24)}{(x^2 - 4)^{1/2}}$ **49** $\dfrac{(3x + 1)^5(39x - 89)}{(2x - 5)^{1/2}}$

51 $\dfrac{27x^2 - 24x + 2}{(6x + 1)^4}$ **53** $\dfrac{x^2 + 12}{(x^2 + 4)^{4/3}}$ **55** $\dfrac{6(3 - 2x)}{(4x^2 + 9)^{3/2}}$

57 $\dfrac{t + 10\sqrt{t} + 25}{t - 25}$ **59** $(9x + 4y)(3\sqrt{x} + 2\sqrt{y})$

61 $\dfrac{\sqrt[3]{a^2} + \sqrt[3]{ab} + \sqrt[3]{b^2}}{a - b}$ **63** $\dfrac{1}{(a + b)(\sqrt{a} + \sqrt{b})}$

65 $\dfrac{2}{\sqrt{2(x + h) + 1} + \sqrt{2x + 1}}$ **67** $\dfrac{-1}{\sqrt{1 - x - h} + \sqrt{1 - x}}$

EXERCISES 1.6 ▪ PAGE 50

1 5040 **3** 144 **5** $\frac{1}{12}$ **7** 1 **9** 21 **11** 715
13 $a^6 + 6a^5b + 15a^4b^2 + 20a^3b^3 + 15a^2b^4 + 6ab^5 + b^6$
15 $a^4 + 4a^3b + 6a^2b^2 + 4ab^3 + b^4$
17 $81x^4 - 540x^3y + 1350x^2y^2 - 1500xy^3 + 625y^4$
19 $\frac{1}{243}x^5 + \frac{5}{81}x^4y^2 + \frac{10}{27}x^3y^4 + \frac{10}{9}x^2y^6 + \frac{5}{3}xy^8 + y^{10}$
21 $x^{-12} + 18x^{-9} + 135x^{-6} + 540x^{-3} + 1215 + 1458x^3 + 729x^6$
23 $x^{5/2} + 5x^{3/2} + 10x^{1/2} + 10x^{-1/2} + 5x^{-3/2} + x^{-5/2}$
25 $3^{25}c^{10} + 25 \cdot 3^{24}c^{52/5} + 300 \cdot 3^{23}c^{54/5}$
27 $-1680 \cdot 3^{13}b^{11} + 60 \cdot 3^{14}b^{13} - 3^{15}b^{15}$ **29** $\frac{189}{1024}c^8$
31 $\frac{114,688}{9}u^2v^6$ **33** $70x^2y^2$ **35** $-\frac{135}{16}$
37 $448y^3x^{10}$ **39** $-216b^9a^2$ **41** 4.8, 6.19
43 $4x^3 + 6x^2h + 4xh^2 + h^3$

EXERCISES 1.7 ▪ PAGE 51

1 (a) $-\frac{5}{12}$ (b) $\frac{39}{20}$ (c) $-\frac{13}{56}$ (d) $\frac{5}{8}$
3 (a) $x < 0$ (b) $\frac{1}{3} < a < \frac{1}{2}$ (c) $|x| \leq 4$

5 (a) 5 (b) 5 (c) 7 **7** $18a^5b^5$ **9** $\dfrac{xy^5}{9}$

11 $-\dfrac{p^8}{2q}$ **13** $\dfrac{x^3z}{y^{10}}$ **15** $\dfrac{b^6}{a^2}$ **17** $s + r$ **19** $\dfrac{x^8}{y^2}$

21 $\frac{1}{2}\sqrt[3]{2}$ **23** $2x^2y\sqrt[3]{x}$ **25** $\dfrac{1 - \sqrt{t}}{t}$ **27** $\dfrac{2x}{y^2}$

29 $\dfrac{1 - 2\sqrt{x} + x}{1 - x}$ **31** $(9x + y)(3\sqrt{x} - \sqrt{y})$

33 (a) 9.37×10^{10} (b) 4.02×10^{-6} **35** 2.75×10^{13}
37 $x^4 + x^3 - x^2 + x - 2$ **39** $-x^2 + 18x + 7$
41 $3y^5 - 2y^4 - 8y^3 + 10y^2 - 3y - 12$ **43** $a^4 - b^4$
45 $6a^2 + 11ab - 35b^2$ **47** $169a^4 - 16b^2$
49 $8a^3 + 12a^2b + 6ab^2 + b^3$ **51** $81x^4 - 72x^2y^2 + 16y^4$
53 $10w(6x + 7)$ **55** $(14x + 9)(2x - 1)$
57 $(2w + 3x)(y - 4z)$ **59** $8(x + 2y)(x^2 - 2xy + 4y^2)$
61 $(p^4 + q^4)(p^2 + q^2)(p + q)(p - q)$

63 $(w^2 + 1)(w^4 - w^2 + 1)$ **65** $\dfrac{3x - 5}{2x + 1}$ **67** $\dfrac{3x + 2}{x(x - 2)}$

69 $\dfrac{5x^2 - 6x - 20}{x(x + 2)^2}$ **71** $\dfrac{-(2x^2 + x + 3)}{x(x + 1)(x + 3)}$ **73** $x + 5$

75 $(x^2 + 1)^{1/2}(x + 5)^3(7x^2 + 15x + 4)$
77 $x^{12} - 18x^{10}y + 135x^8y^2 - 540x^6y^3 + 1215x^4y^4 - 1458x^2y^5 + 729y^6$
79 $-\frac{63}{16}b^{12}c^{10}$

CHAPTER ▪ 2

EXERCISES 2.1 ▪ PAGE 60

1 $\frac{5}{3}$ **3** $\frac{26}{7}$ **5** $\frac{35}{17}$ **7** $\frac{23}{18}$ **9** $-\frac{1}{40}$ **11** $\frac{49}{4}$
13 $\frac{4}{3}$ **15** $-\frac{24}{29}$ **17** $\frac{7}{31}$ **19** $-\frac{3}{61}$ **21** $\frac{29}{4}$
23 $\frac{31}{18}$ **25** No solution **27** All real numbers except $\frac{1}{2}$
29 $\frac{5}{9}$ **31** $-\frac{2}{3}$ **33** No solution **35** 0
37 All real numbers except ± 2 **39** No solution
41 No solution
43 $(4x-3)^2 - 16x^2 = (16x^2 - 24x + 9) - 16x^2 = 9 - 24x$
45 $\frac{x^2-9}{x+3} = \frac{(x+3)(x-3)}{x+3} = x - 3$
47 $\frac{3x^2+8}{x} = \frac{3x^2}{x} + \frac{8}{x} = \frac{8}{x} + 3x$ **49** $-\frac{19}{3}$
51 Choose any a and b such that $b = -\frac{5}{3}a$. For example, let $a = 3$ and $b = -5$.
53 (a) No, -5 is not a solution of $x = 5$. (b) Yes
55 (a) Yes (b) No, 7 is not a solution of the first equation.
57 $x + 2 = x + 1$

EXERCISES 2.2 ▪ PAGE 71

1 $P = \frac{I}{rt}$ **3** $h = \frac{2A}{b}$ **5** $w = \frac{P - 2l}{2}$ **7** $I = \frac{V}{R}$
9 $m = \frac{Fd^2}{gM}$ **11** $h = \frac{1}{\pi r}\sqrt{S^2 - \pi^2 r^4}$
13 $b_1 = \frac{2A - hb_2}{h}$ **15** 88 **17** -40
19 Approximately 23 weeks **21** $\frac{19}{2} - \frac{3\pi}{8} \approx 8.32$ ft
23 (a) After 64 sec (b) 96 m and 128 m, respectively
25 60.3 g
27 $\frac{14}{3}$ oz of 30% glucose solution and $\frac{7}{3}$ oz of water
29 194.6 g of British sterling silver and 5.4 g of copper
31 Not possible **33** 55 ft **35** 6 mi/hr **37** 36 min
39 400 mi **41** (a) 4050 ft^2 (b) 2592 ft^2 (c) 3600 ft^2
43 36 min **45** (a) $\frac{5}{9}$ mi/hr (b) $2\frac{2}{9}$ mi
47 (a) 125 (b) 21 **49** 180 months (or 15 yr)
51 After an additional 50 games

EXERCISES 2.3 ▪ PAGE 82

1 ± 13 **3** $\pm\frac{3}{5}$ **5** (a) $\frac{81}{4}$ (b) 16 (c) ± 12 (d) ± 7
7 $-\frac{3}{2}, \frac{4}{3}$ **9** $-\frac{6}{5}, \frac{2}{3}$ **11** $-\frac{9}{2}, \frac{3}{4}$ **13** $-\frac{2}{3}, \frac{1}{5}$
15 $-\frac{5}{2}$ **17** $-\frac{1}{2}$ **19** $-\frac{34}{5}$ **21** $-\frac{1}{2}, \frac{2}{3}$
23 $-2 \pm \sqrt{2}$ **25** $\frac{3}{4} \pm \frac{1}{4}\sqrt{41}$ **27** $\frac{4}{3} \pm \frac{1}{3}\sqrt{22}$
29 $\frac{5}{2} \pm \frac{1}{2}\sqrt{15}$ **31** $3 + \frac{1}{2}\sqrt{14} \approx 4.9$ mi or $3 - \frac{1}{2}\sqrt{14} \approx 1.1$ mi
33 2 ft **35** (a) $d = 100\sqrt{20t^2 + 4t + 1}$ (b) 3:30 P.M.
37 Until 9:24 A.M.
39 (a) After 1 sec and after 3 sec (b) After 4 sec (c) 64 ft
41 (a) 206.25 ft (b) 40 mi/hr
43 (a) 4320 m (b) 96.86 °C
45 1.5 in. at sides and top, 3 in. at bottom
47 14 in. by 27 in. **49** 2 ft **51** 7 mi/hr
53 8 teams

EXERCISES 2.4 ▪ PAGE 91

1 $2 + 4i$ **3** $18 - 3i$ **5** $41 - 11i$ **7** $17 - i$
9 $21 - 20i$ **11** $-24 - 7i$ **13** 25 **15** $-i$ **17** i
19 $\frac{3}{10} - \frac{3}{5}i$ **21** $\frac{1}{2} - i$ **23** $\frac{34}{53} + \frac{40}{53}i$ **25** $\frac{2}{5} + \frac{4}{5}i$
27 $-142 - 65i$ **29** $-2 - 14i$ **31** $-\frac{44}{113} + \frac{95}{113}i$
33 $\frac{21}{2}i$ **35** $x = 4, y = -16$ **37** $x = -2, y = -1$
39 $3 \pm 2i$ **41** $-2 \pm 3i$ **43** $\frac{5}{2} \pm \frac{1}{2}\sqrt{55}\,i$
45 $-\frac{1}{8} \pm \frac{1}{8}\sqrt{47}\,i$ **47** $-5, \frac{5}{2} \pm \frac{5}{2}\sqrt{3}\,i$ **49** $\pm 4, \pm 4i$
51 $\pm 2i, \pm\frac{3}{2}i$ **53** $0, -\frac{3}{2} \pm \frac{1}{2}\sqrt{7}\,i$
55 If $w = c + di$, then $\overline{z + w} = \overline{(a + bi) + (c + di)} = \overline{(a + c) + (b + d)i} = (a + c) - (b + d)i = (a - bi) + (c - di) = \bar{z} + \bar{w}$
57 $\overline{z^2} = \overline{(a + bi)^2} = \overline{a^2 + 2abi - b^2} = \overline{(a^2 - b^2) + 2abi} = (a^2 - b^2) - 2abi = a^2 - 2abi - b^2 = (a - bi)^2 = (\bar{z})^2$
59 $\bar{\bar{z}} = \overline{\overline{a + bi}} = \overline{a - bi} = a + bi = z$

EXERCISES 2.5 ▪ PAGE 98

1 $\pm\frac{2}{3}, 2$ **3** $\pm\frac{1}{2}\sqrt{6}, -\frac{5}{2}, 0$ **5** 0, 25 **7** $-\frac{57}{5}$
9 $\frac{9}{5}$ **11** $\pm\frac{1}{2}\sqrt{62}$ **13** 6 **15** 6 **17** 5, 7
19 -3 **21** -1 **23** $-\frac{5}{4}$ **25** 3 **27** 0, 4
29 $\pm 3, \pm 4$ **31** $\pm\frac{1}{10}\sqrt{70 \pm 10\sqrt{29}}$ **33** $\frac{8}{27}, -8$
35 $\frac{25}{4}, \frac{16}{9}$ **37** $-\frac{4}{3}, -\frac{2}{3}$ **39** $2\sqrt[3]{432/\pi} \approx 10.3$ cm
41 53.4%
43 There are two possible routes corresponding to $x \approx 0.6743$ mi and $x \approx 2.2887$ mi.

EXERCISES 2.6 ▪ PAGE 106

1 (a) $-2 < 2$ (b) $-11 < -7$ (c) $-\frac{7}{3} < -1$ (d) $1 < \frac{7}{3}$
3 $(-\infty, -2)$ **5** $[4, \infty)$

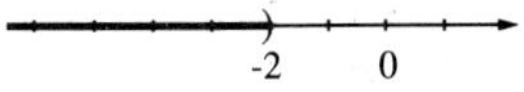

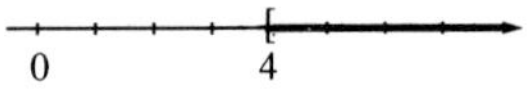

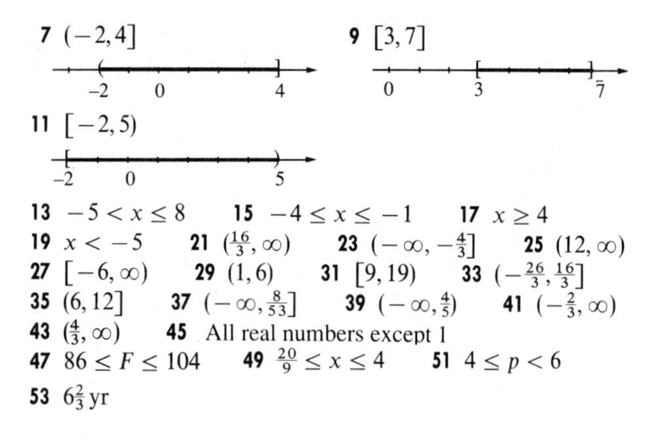

7 $(-2, 4]$ **9** $[3, 7]$

11 $[-2, 5)$

13 $-5 < x \leq 8$ **15** $-4 \leq x \leq -1$ **17** $x \geq 4$
19 $x < -5$ **21** $(\frac{16}{3}, \infty)$ **23** $(-\infty, -\frac{4}{3}]$ **25** $(12, \infty)$
27 $[-6, \infty)$ **29** $(1, 6)$ **31** $[9, 19)$ **33** $(-\frac{26}{3}, \frac{16}{3}]$
35 $(6, 12]$ **37** $(-\infty, \frac{8}{53}]$ **39** $(-\infty, \frac{4}{5})$ **41** $(-\frac{2}{3}, \infty)$
43 $(\frac{4}{3}, \infty)$ **45** All real numbers except 1
47 $86 \leq F \leq 104$ **49** $\frac{20}{9} \leq x \leq 4$ **51** $4 \leq p < 6$
53 $6\frac{2}{3}$ yr

EXERCISES 2.7 ■ PAGE 114

1 $(-3, 3)$ **3** $(-\infty, -5] \cup [5, \infty)$ **5** $(-3.01, -2.99)$
7 $(-\infty, -2.001] \cup [-1.999, \infty)$ **9** $(-\frac{9}{2}, -\frac{1}{2})$
11 $[\frac{3}{5}, \frac{9}{5}]$ **13** $(-\infty, \infty)$ **15** $(-\infty, 3) \cup (3, \infty)$
17 $(-\infty, -\frac{8}{3}] \cup [4, \infty)$ **19** $(-\infty, \frac{7}{4}) \cup (\frac{13}{4}, \infty)$
21 $(-\frac{1}{3}, \frac{1}{2})$ **23** $[-2, 1] \cup [4, \infty)$ **25** $(-2, 3)$
27 $(-\infty, -2) \cup (4, \infty)$ **29** $(-\infty, -\frac{5}{2}] \cup [1, \infty)$
31 $(2, 4)$ **33** $(-4, 4)$ **35** $(-\frac{3}{5}, \frac{3}{5})$
37 $(-\infty, 0] \cup [\frac{9}{16}, \infty)$ **39** $(-\infty, -2] \cup [2, \infty)$
41 $\{-2\} \cup [2, \infty)$ **43** $(-2, 2] \cup (5, \infty)$
45 $(-\infty, -3) \cup (0, 3)$ **47** $(\frac{3}{2}, \frac{7}{3})$
49 $(-\infty, -1) \cup (2, \frac{7}{2}]$ **51** $(-1, \frac{2}{3}) \cup [4, \infty)$
53 $(1, \frac{5}{3}) \cup [2, 5]$ **55** $(-1, 0) \cup (1, \infty)$ **57** $\frac{1}{2}$ sec
59 $0 \leq v < 30$ **61** $0 < S < 4000$ **63** $t > \frac{3}{4}$ hr

EXERCISES 2.8 ■ PAGE 116

1 $-\frac{5}{6}$ **3** -32 **5** Every $x > 0$ **7** $-\frac{2}{3} \pm \frac{1}{3}\sqrt{19}$
9 $\frac{1}{2} \pm \frac{1}{2}\sqrt{21}$ **11** $-27, 125$ **13** $\frac{1}{5} \pm \frac{1}{5}\sqrt{14}\,i$
15 $\pm\frac{1}{2}\sqrt{14}\,i, \pm\frac{2}{3}\sqrt{3}\,i$ **17** $\frac{1}{4}, \frac{1}{9}$ **19** 2 **21** 5
23 $(-\frac{11}{4}, \frac{9}{4})$ **25** $(-\infty, -\frac{3}{10})$ **27** $(-\infty, \frac{11}{3}] \cup [7, \infty)$
29 $(-\infty, -\frac{3}{2}) \cup (\frac{2}{5}, \infty)$ **31** $(-\infty, -\frac{3}{2}) \cup (2, 9)$

33 $(1, \infty)$ **35** $r = \sqrt[3]{\dfrac{3V}{4\pi}}$

37 $r = \dfrac{-\pi h + \sqrt{\pi^2 h^2 + 2\pi A}}{2\pi}$ **39** $t = \dfrac{-v_0 \pm \sqrt{v_0^2 + 2gs}}{g}$

41 $-1 + 8i$ **43** $-55 + 48i$ **45** $-\frac{9}{53} - \frac{48}{53}i$
47 $10 - 5\sqrt{3} \approx 1.34$ mi **49** 64 mi/hr
51 315.8 g of ethyl alcohol and 84.2 g of water
53 $R_2 = \frac{10}{3}$ ohms **55** 75 mi **57** 10 ft by 4 ft

59 (a) $d = \sqrt{4 - 200t + 2900t^2}$
(b) $t = \dfrac{5 + 2\sqrt{19{,}603}}{145} \approx 1.97$, or approximately 11:58 A.M.

61 (a) $2\sqrt{2}$ ft (b) 2 ft **63** After $7\frac{2}{3}$ yr **65** $4 \leq p \leq 8$

CHAPTER ■ 3

EXERCISES 3.1 ■ PAGE 123

1

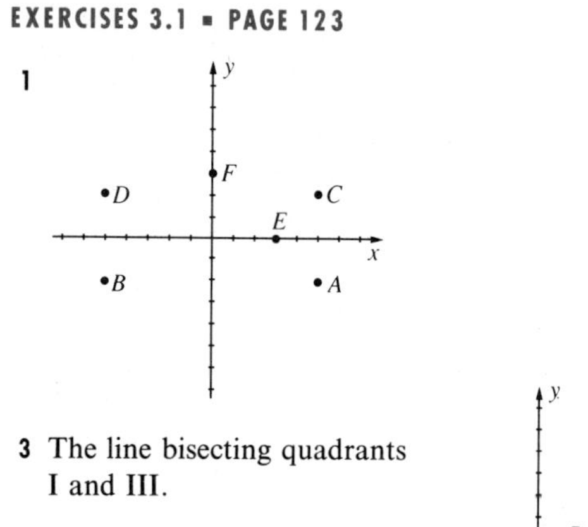

3 The line bisecting quadrants I and III.

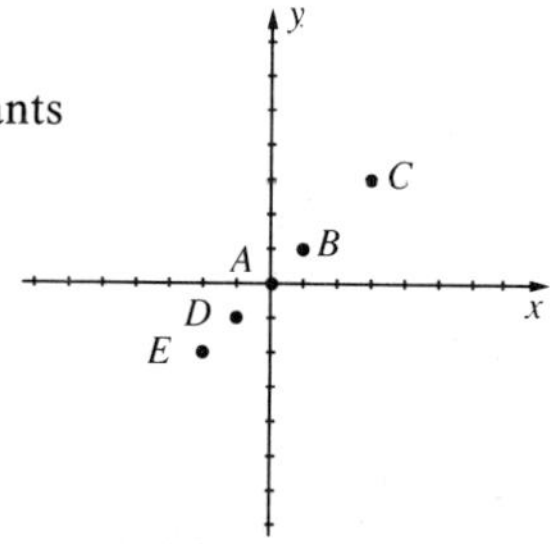

5 (a) The line parallel to the y-axis that intersects the x-axis at $(-2, 0)$
(b) The line parallel to the x-axis that intersects the y-axis at $(0, 3)$
(c) All points to the right of, and on, the y-axis
(d) All points in quadrants I and III
(e) All points below the x-axis
(f) All points on the y-axis

7 (a) $\sqrt{29}$ (b) $(5, -\frac{1}{2})$ **9** (a) $\sqrt{13}$ (b) $(-\frac{7}{2}, -1)$
11 (a) 4 (b) $(5, -3)$
13 $d(A, C)^2 = d(A, B)^2 + d(B, C)^2$; area = 28
15 $d(A, B) = d(B, C) = d(C, D) = d(D, A)$ and $d(A, C)^2 = d(A, B)^2 + d(B, C)^2$
17 $(13, -28)$ **19** $d(A, P) = d(B, P)$
21 $\sqrt{x^2 + y^2} = 5$; a circle of radius 5 with center at the origin
23 $(0, 3 + \sqrt{11}), (0, 3 - \sqrt{11})$ **25** $(-2, -1)$
27 $a < \frac{2}{5}$ or $a > 4$
29 Let M be the midpoint of the hypotenuse. Show that $d(A, M) = d(B, M) = d(O, M) = \frac{1}{2}\sqrt{a^2 + b^2}$.

EXERCISES 3.2 ▪ PAGE 130

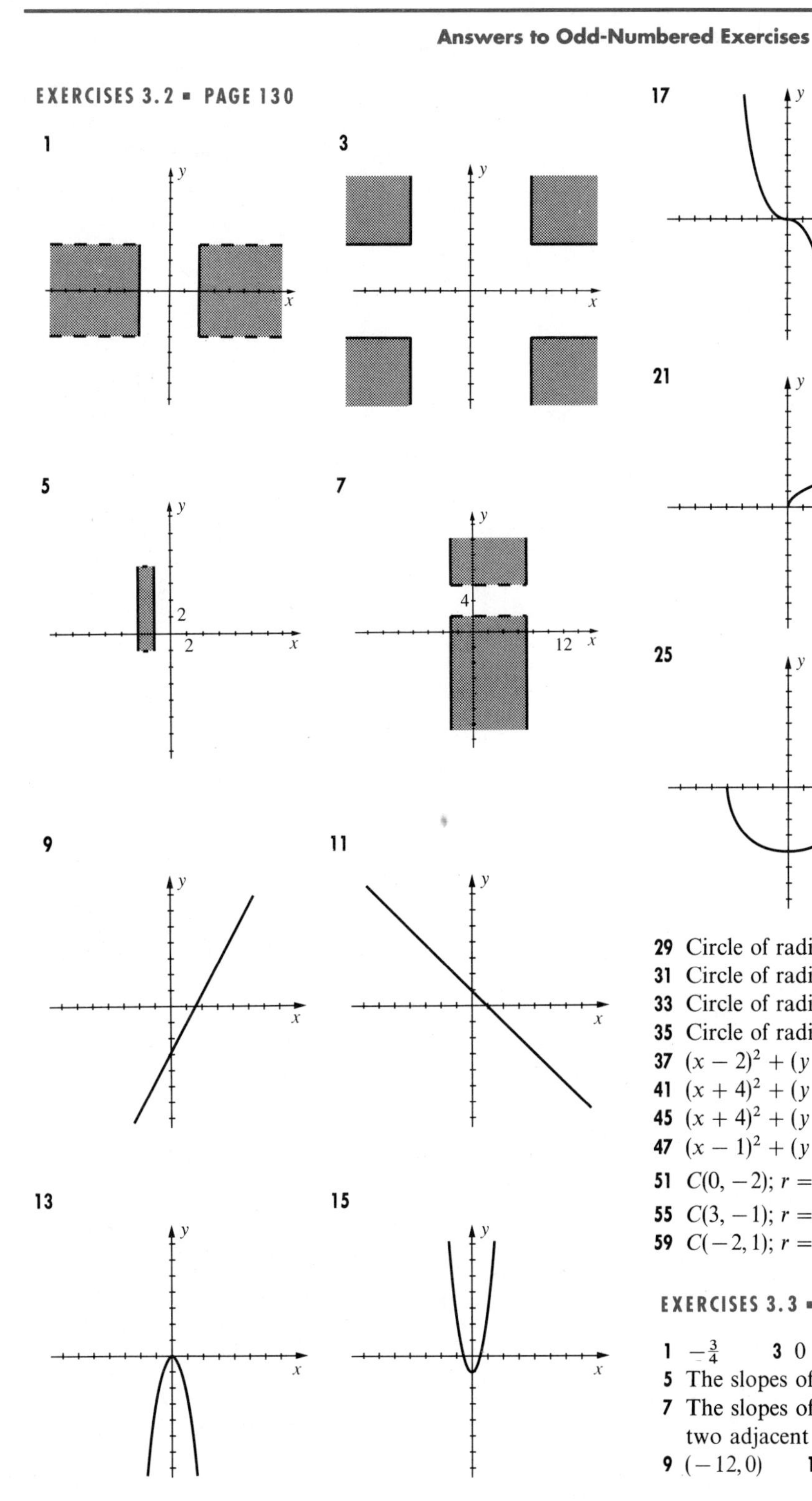

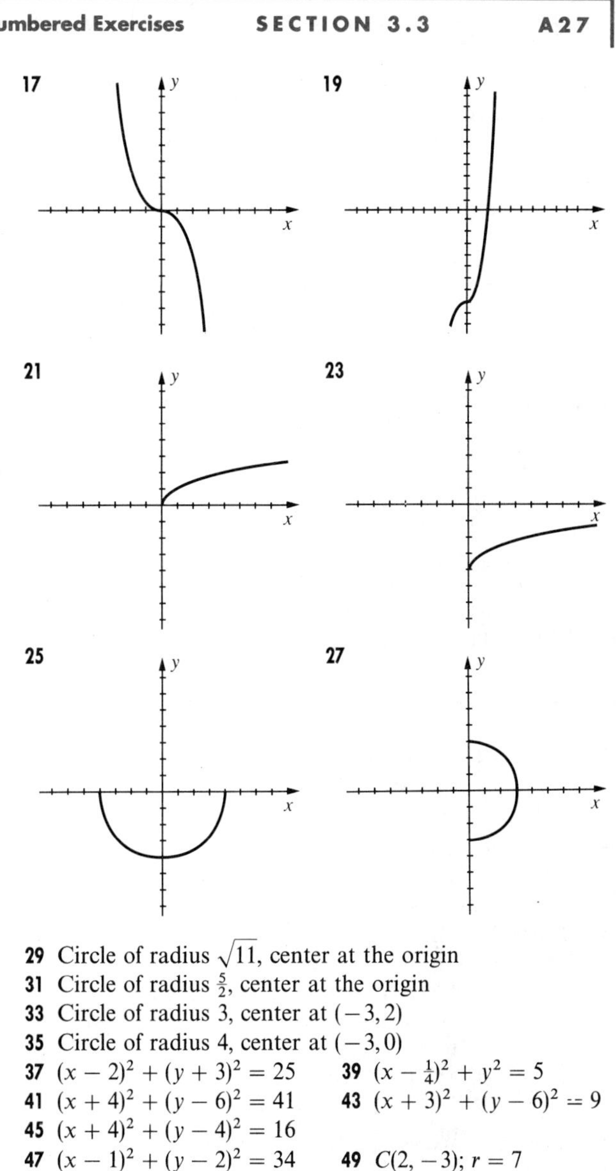

29 Circle of radius $\sqrt{11}$, center at the origin
31 Circle of radius $\frac{5}{2}$, center at the origin
33 Circle of radius 3, center at $(-3, 2)$
35 Circle of radius 4, center at $(-3, 0)$
37 $(x-2)^2 + (y+3)^2 = 25$ **39** $(x-\frac{1}{4})^2 + y^2 = 5$
41 $(x+4)^2 + (y-6)^2 = 41$ **43** $(x+3)^2 + (y-6)^2 = 9$
45 $(x+4)^2 + (y-4)^2 = 16$
47 $(x-1)^2 + (y-2)^2 = 34$ **49** $C(2, -3)$; $r = 7$
51 $C(0, -2)$; $r = 11$ **53** $C(5, 0)$; $r = \sqrt{7}$
55 $C(3, -1)$; $r = \frac{1}{2}\sqrt{70}$ **57** $C(-\frac{2}{3}, \frac{1}{3})$; $r = \frac{1}{3}$
59 $C(-2, 1)$; $r = 0$ (a point) **61** Not a circle

EXERCISES 3.3 ▪ PAGE 140

1 $-\frac{3}{4}$ **3** 0
5 The slopes of opposite sides are equal.
7 The slopes of opposite sides are equal and the slopes of two adjacent sides are negative reciprocals.
9 $(-12, 0)$ **11** (a) $x = 5$ (b) $y = -2$

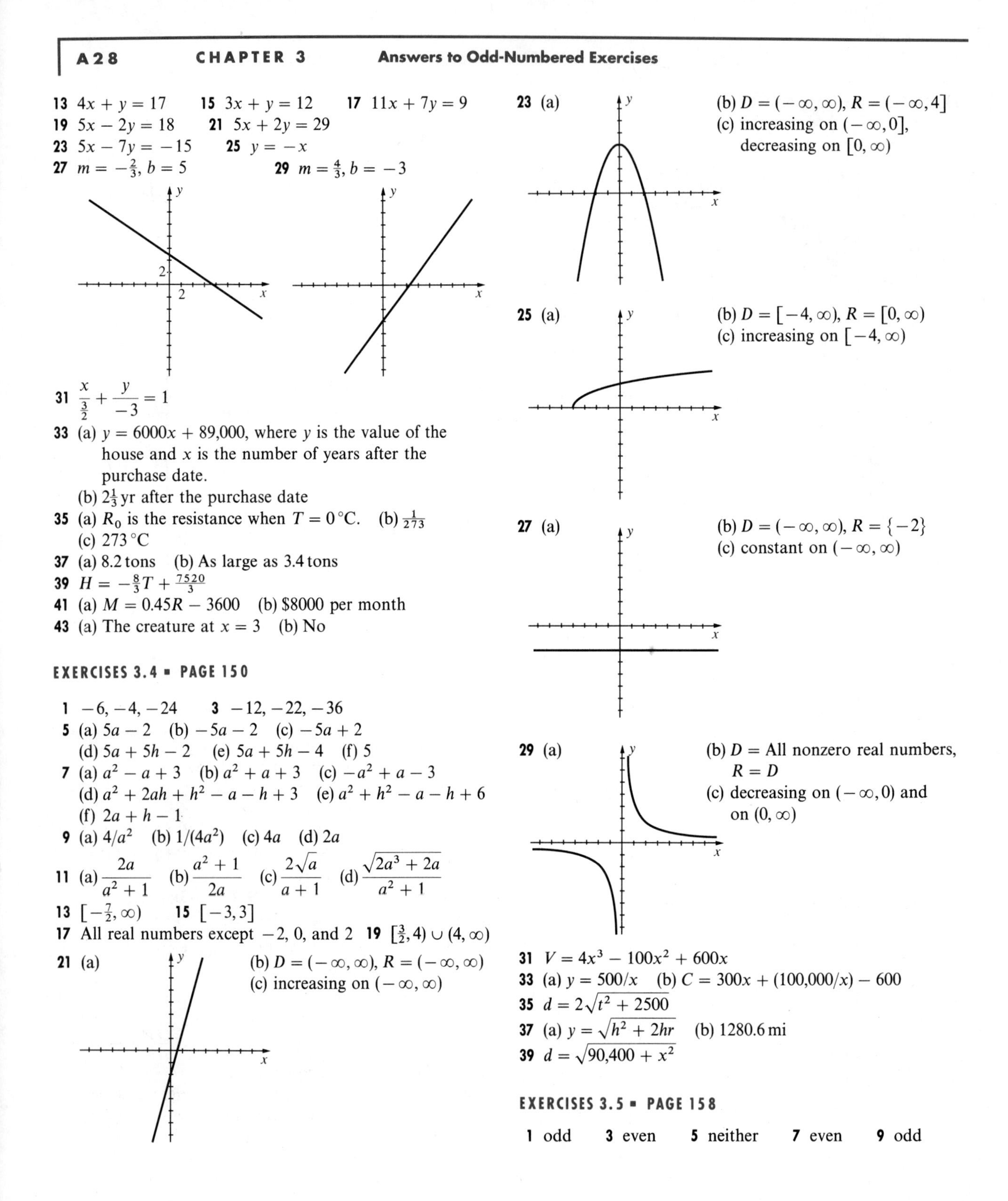

13 $4x + y = 17$ **15** $3x + y = 12$ **17** $11x + 7y = 9$
19 $5x - 2y = 18$ **21** $5x + 2y = 29$
23 $5x - 7y = -15$ **25** $y = -x$
27 $m = -\frac{2}{3}, b = 5$ **29** $m = \frac{4}{3}, b = -3$

31 $\frac{x}{\frac{3}{2}} + \frac{y}{-3} = 1$
33 (a) $y = 6000x + 89{,}000$, where y is the value of the house and x is the number of years after the purchase date.
(b) $2\frac{1}{3}$ yr after the purchase date
35 (a) R_0 is the resistance when $T = 0\,°\text{C}$. (b) $\frac{1}{273}$
(c) $273\,°\text{C}$
37 (a) 8.2 tons (b) As large as 3.4 tons
39 $H = -\frac{8}{3}T + \frac{7520}{3}$
41 (a) $M = 0.45R - 3600$ (b) \$8000 per month
43 (a) The creature at $x = 3$ (b) No

EXERCISES 3.4 ▪ PAGE 150

1 $-6, -4, -24$ **3** $-12, -22, -36$
5 (a) $5a - 2$ (b) $-5a - 2$ (c) $-5a + 2$
(d) $5a + 5h - 2$ (e) $5a + 5h - 4$ (f) 5
7 (a) $a^2 - a + 3$ (b) $a^2 + a + 3$ (c) $-a^2 + a - 3$
(d) $a^2 + 2ah + h^2 - a - h + 3$ (e) $a^2 + h^2 - a - h + 6$
(f) $2a + h - 1$
9 (a) $4/a^2$ (b) $1/(4a^2)$ (c) $4a$ (d) $2a$
11 (a) $\frac{2a}{a^2 + 1}$ (b) $\frac{a^2 + 1}{2a}$ (c) $\frac{2\sqrt{a}}{a + 1}$ (d) $\frac{\sqrt{2a^3 + 2a}}{a^2 + 1}$
13 $[-\frac{7}{2}, \infty)$ **15** $[-3, 3]$
17 All real numbers except -2, 0, and 2 **19** $[\frac{3}{2}, 4) \cup (4, \infty)$
21 (a) (b) $D = (-\infty, \infty)$, $R = (-\infty, \infty)$
(c) increasing on $(-\infty, \infty)$

23 (a) (b) $D = (-\infty, \infty)$, $R = (-\infty, 4]$
(c) increasing on $(-\infty, 0]$, decreasing on $[0, \infty)$

25 (a) (b) $D = [-4, \infty)$, $R = [0, \infty)$
(c) increasing on $[-4, \infty)$

27 (a) (b) $D = (-\infty, \infty)$, $R = \{-2\}$
(c) constant on $(-\infty, \infty)$

29 (a) (b) $D =$ All nonzero real numbers, $R = D$
(c) decreasing on $(-\infty, 0)$ and on $(0, \infty)$

31 $V = 4x^3 - 100x^2 + 600x$
33 (a) $y = 500/x$ (b) $C = 300x + (100{,}000/x) - 600$
35 $d = 2\sqrt{t^2 + 2500}$
37 (a) $y = \sqrt{h^2 + 2hr}$ (b) 1280.6 mi
39 $d = \sqrt{90{,}400 + x^2}$

EXERCISES 3.5 ▪ PAGE 158

1 odd **3** even **5** neither **7** even **9** odd

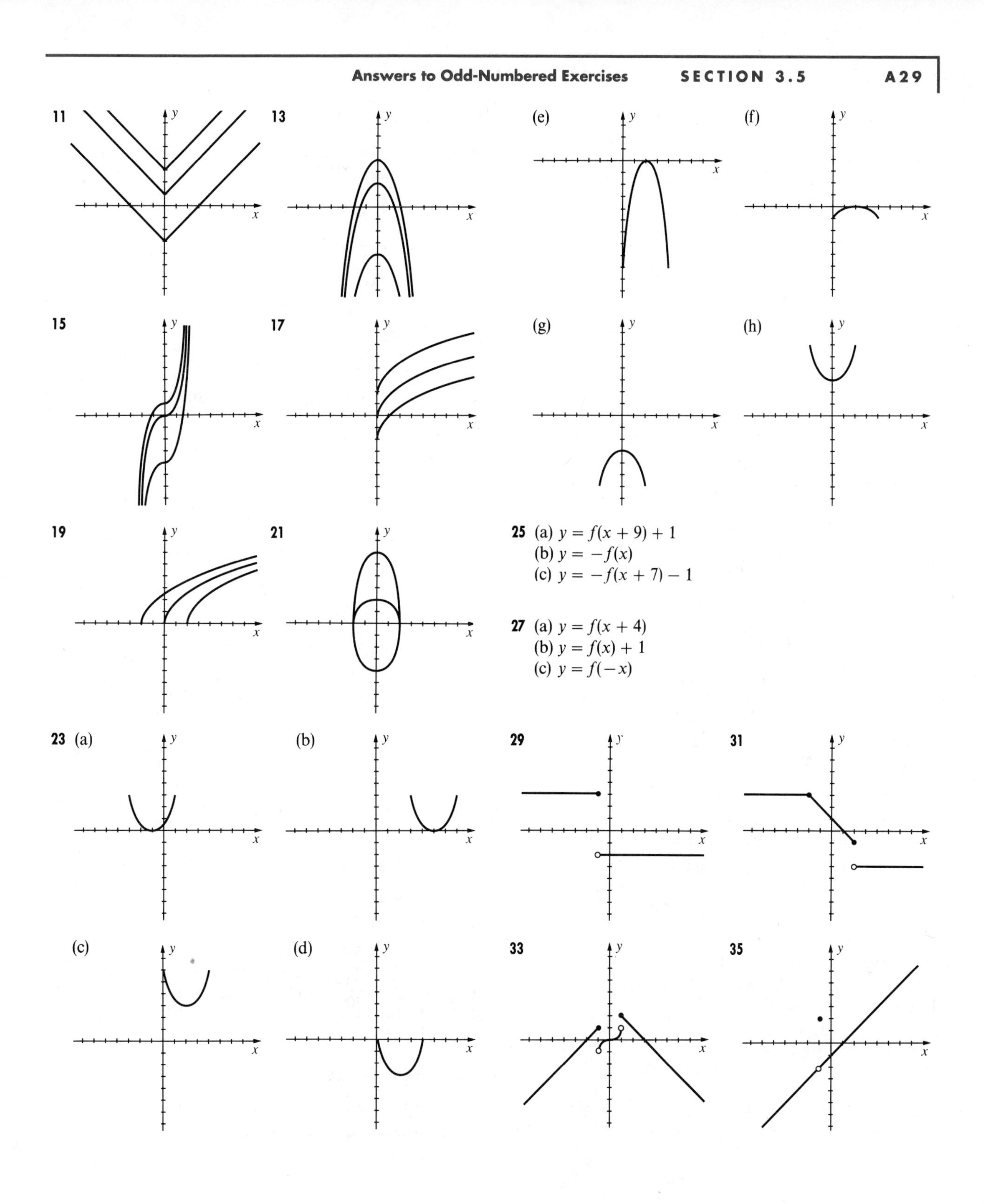

25 (a) $y = f(x + 9) + 1$
(b) $y = -f(x)$
(c) $y = -f(x + 7) - 1$

27 (a) $y = f(x + 4)$
(b) $y = f(x) + 1$
(c) $y = f(-x)$

37 (a) (b) (c) (d)

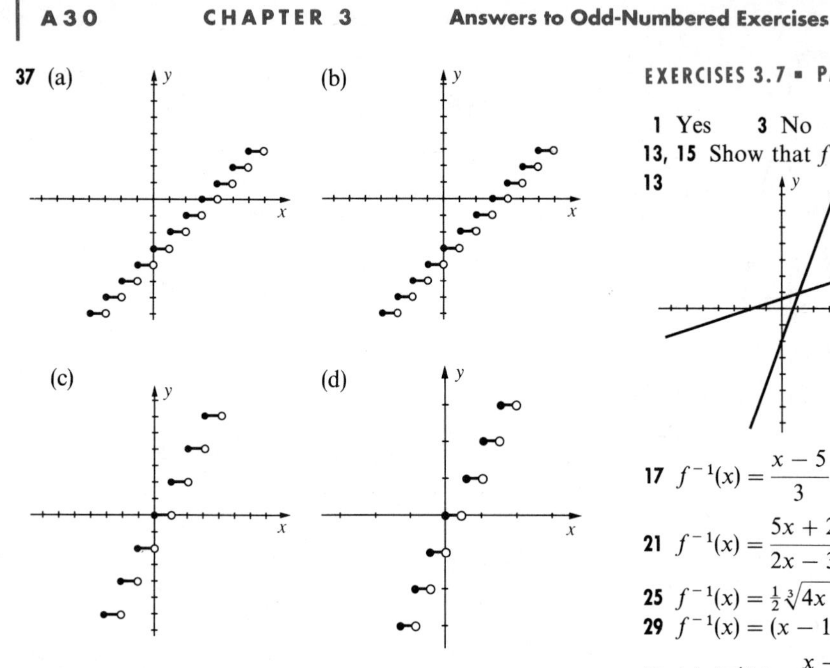

39 If $x > 0$, two different points on the graph have x-coordinate x.

EXERCISES 3.6 ▪ PAGE 165

Exer. 1–5: Answers are listed in the order $(f+g)(x)$; $(f-g)(x)$; $(fg)(x)$; the domain of the first three functions; $(f/g)(x)$, domain.

1 $3x^2+1$; $3-x^2$; $2x^4+3x^2-2$; $(-\infty,\infty)$; $(x^2+2)/(2x^2-1)$, all real numbers except $\pm\frac{1}{2}\sqrt{2}$

3 $2\sqrt{x+5}$; 0; $x+5$; $[-5,\infty)$; 1, $(-5,\infty)$

5 $\dfrac{3x^2+6x}{(x-4)(x+5)}$; $\dfrac{x^2+14x}{(x-4)(x+5)}$; $\dfrac{2x^2}{(x-4)(x+5)}$; all real numbers except -5 and 4; $\dfrac{2x+10}{x-4}$, all real numbers except -5, 0, and 4

7 (a) $6x+9$ (b) $6x-8$

9 (a) $75x^2+4$ (b) $15x^2+20$

11 (a) $8x^2-2x-5$ (b) $4x^2+6x-9$

13 (a) $8x^3-20x$ (b) $128x^3-20x$

15 (a) $x+2-3\sqrt{x+2}$ (b) $\sqrt{x^2-3x+2}$

17 (a) $\dfrac{1}{x+3}$ (b) $\dfrac{6x+4}{x}$ **19** (a) 7 (b) -7

21 (a) $1/x^6$ (b) $1/x^6$ **23** (a) x (b) x **25** $36\pi t^2$

27 $h=5\sqrt{t^2+8t}$ **29** $h=(14/\sqrt{821})t+2$

EXERCISES 3.7 ▪ PAGE 173

1 Yes **3** No **5** Yes **7** No **9** No **11** Yes

13, 15 Show that $f(g(x)) = x = g(f(x))$

13 **15**

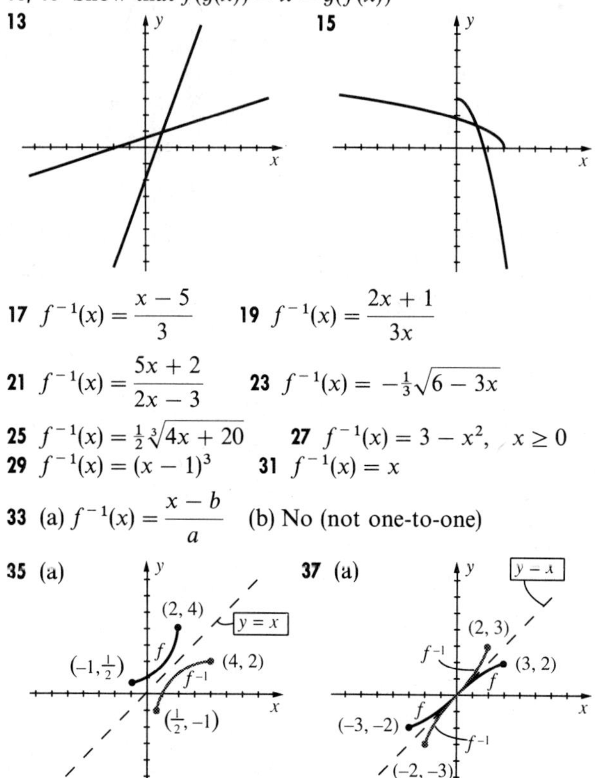

17 $f^{-1}(x)=\dfrac{x-5}{3}$ **19** $f^{-1}(x)=\dfrac{2x+1}{3x}$

21 $f^{-1}(x)=\dfrac{5x+2}{2x-3}$ **23** $f^{-1}(x)=-\frac{1}{3}\sqrt{6-3x}$

25 $f^{-1}(x)=\frac{1}{2}\sqrt[3]{4x+20}$ **27** $f^{-1}(x)=3-x^2,\ x\geq 0$

29 $f^{-1}(x)=(x-1)^3$ **31** $f^{-1}(x)=x$

33 (a) $f^{-1}(x)=\dfrac{x-b}{a}$ (b) No (not one-to-one)

35 (a) **37** (a)

35 (b) $D=[-1,2]$; $R=[\frac{1}{2},4]$ (c) $R=[\frac{1}{2},4]$; $D=[-1,2]$

37 (b) $D=[-3,3]$; $R=[-2,2]$ (c) $R=[-2,2]$; $D=[-3,3]$

EXERCISES 3.8 ▪ PAGE 177

1 $a=kv$, $k=\frac{2}{5}$ **3** $r=ks/t$, $k=-14$

5 $y=kx^2/z^3$, $k=27$ **7** $c=ka^2b^3$, $k=-\frac{2}{49}$

9 $y=kx/z^2$, $k=36$ **11** $y=k\sqrt{x}/z^3$, $k=\frac{40}{3}$

13 295 lb/ft² **15** $\frac{50}{9}$ ohms **17** $\frac{3}{2}\sqrt{3}$ sec

19 223 days **21** 60.6 mi/hr **23** 154 lb

25 2.05 times as hard

EXERCISES 3.9 ▪ PAGE 179

1 $d(A,B)^2+d(A,C)^2=d(B,C)^2$; area = 10

3 The points in quadrants II and IV

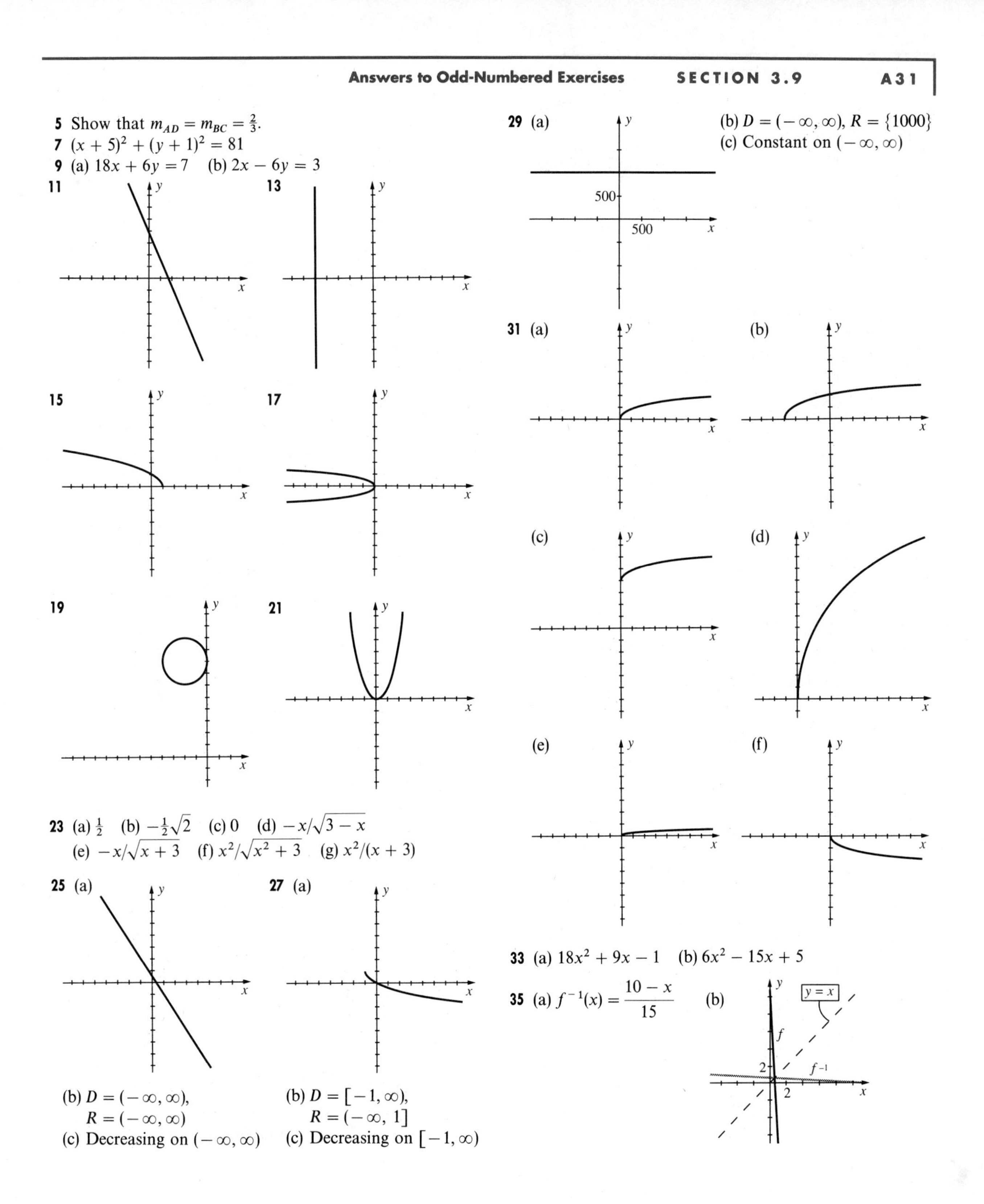

5 Show that $m_{AD} = m_{BC} = \frac{2}{3}$.

7 $(x + 5)^2 + (y + 1)^2 = 81$

9 (a) $18x + 6y = 7$ (b) $2x - 6y = 3$

11

13

15

17

19

21

23 (a) $\frac{1}{2}$ (b) $-\frac{1}{2}\sqrt{2}$ (c) 0 (d) $-x/\sqrt{3 - x}$
(e) $-x/\sqrt{x + 3}$ (f) $x^2/\sqrt{x^2 + 3}$ (g) $x^2/(x + 3)$

25 (a)

(b) $D = (-\infty, \infty)$,
$R = (-\infty, \infty)$
(c) Decreasing on $(-\infty, \infty)$

27 (a)

(b) $D = [-1, \infty)$,
$R = (-\infty, 1]$
(c) Decreasing on $[-1, \infty)$

29 (a)

(b) $D = (-\infty, \infty)$, $R = \{1000\}$
(c) Constant on $(-\infty, \infty)$

31 (a) (b) (c) (d) (e) (f)

33 (a) $18x^2 + 9x - 1$ (b) $6x^2 - 15x + 5$

35 (a) $f^{-1}(x) = \dfrac{10 - x}{15}$ (b)

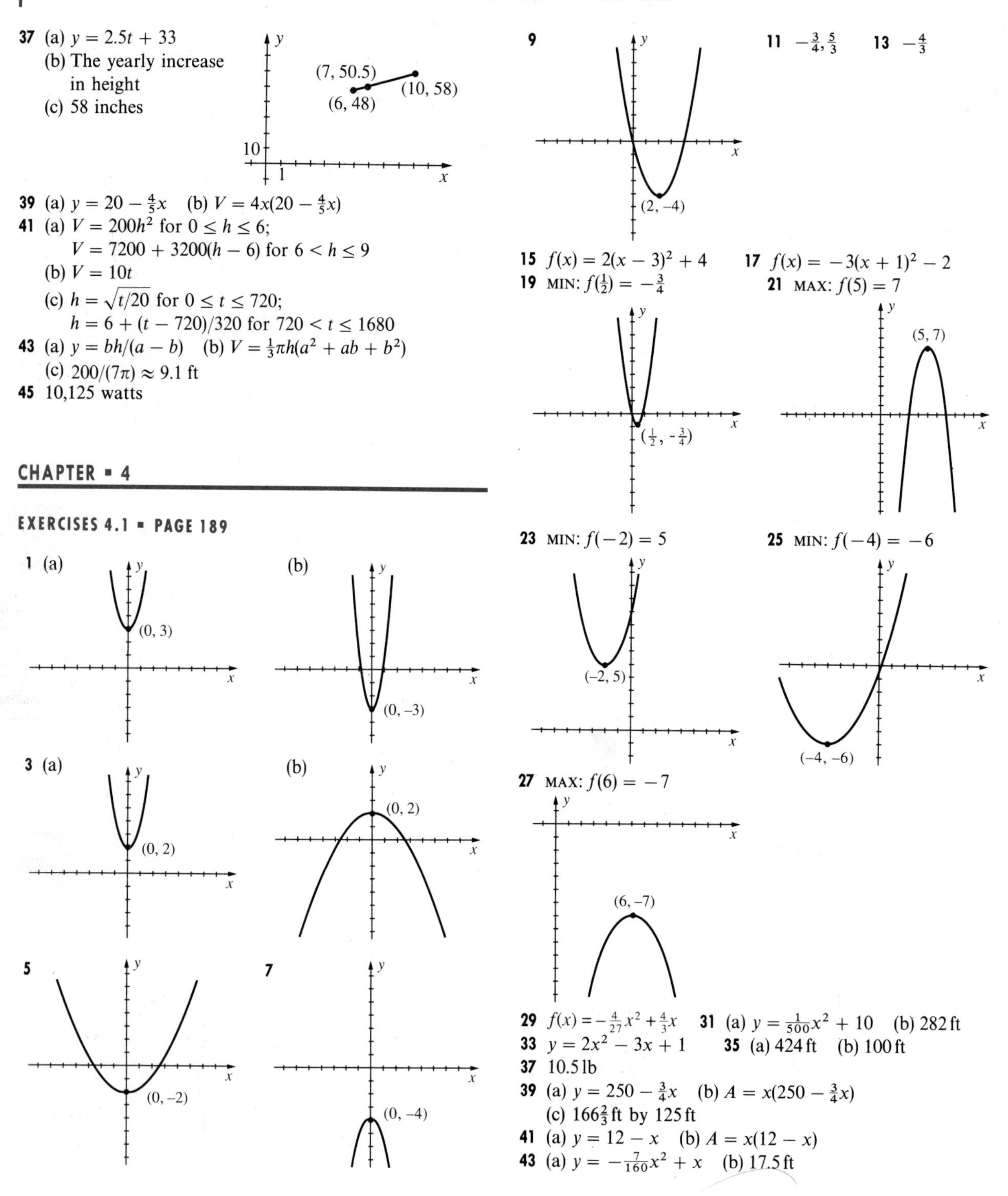

37 (a) $y = 2.5t + 33$
(b) The yearly increase in height
(c) 58 inches

39 (a) $y = 20 - \frac{4}{5}x$ (b) $V = 4x(20 - \frac{4}{5}x)$

41 (a) $V = 200h^2$ for $0 \le h \le 6$;
$V = 7200 + 3200(h - 6)$ for $6 < h \le 9$
(b) $V = 10t$
(c) $h = \sqrt{t/20}$ for $0 \le t \le 720$;
$h = 6 + (t - 720)/320$ for $720 < t \le 1680$

43 (a) $y = bh/(a - b)$ (b) $V = \frac{1}{3}\pi h(a^2 + ab + b^2)$
(c) $200/(7\pi) \approx 9.1$ ft

45 10,125 watts

CHAPTER ▪ 4

EXERCISES 4.1 ▪ PAGE 189

1 (a) (b)

3 (a) (b)

5

7

9

11 $-\frac{3}{4}, \frac{5}{3}$ **13** $-\frac{4}{3}$

15 $f(x) = 2(x - 3)^2 + 4$ **17** $f(x) = -3(x + 1)^2 - 2$

19 MIN: $f(\frac{1}{2}) = -\frac{3}{4}$ **21** MAX: $f(5) = 7$

23 MIN: $f(-2) = 5$ **25** MIN: $f(-4) = -6$

27 MAX: $f(6) = -7$

29 $f(x) = -\frac{4}{27}x^2 + \frac{4}{3}x$ **31** (a) $y = \frac{1}{500}x^2 + 10$ (b) 282 ft

33 $y = 2x^2 - 3x + 1$ **35** (a) 424 ft (b) 100 ft

37 10.5 lb

39 (a) $y = 250 - \frac{3}{4}x$ (b) $A = x(250 - \frac{3}{4}x)$
(c) $166\frac{2}{3}$ ft by 125 ft

41 (a) $y = 12 - x$ (b) $A = x(12 - x)$

43 (a) $y = -\frac{7}{160}x^2 + x$ (b) 17.5 ft

EXERCISES 4.2 ■ PAGE 197

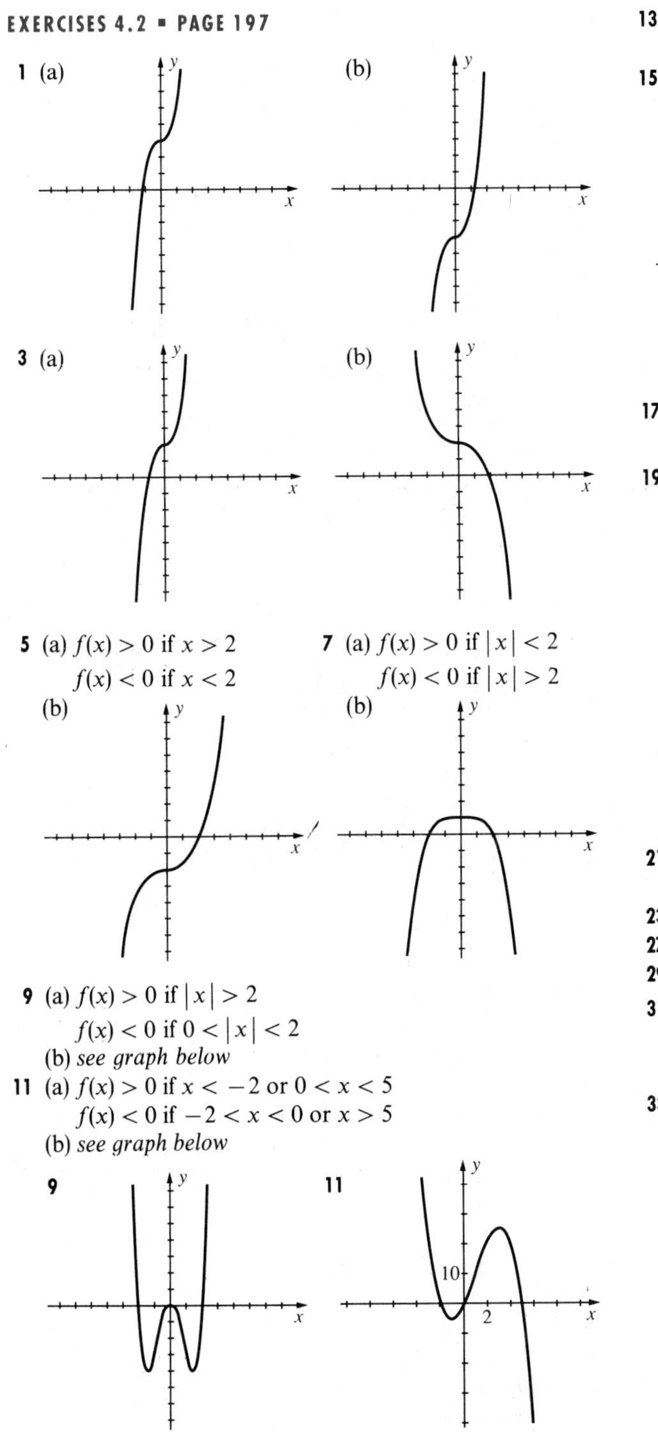

1 (a) (b)

3 (a) (b)

5 (a) $f(x) > 0$ if $x > 2$
$f(x) < 0$ if $x < 2$
(b)

7 (a) $f(x) > 0$ if $|x| < 2$
$f(x) < 0$ if $|x| > 2$
(b)

9 (a) $f(x) > 0$ if $|x| > 2$
$f(x) < 0$ if $0 < |x| < 2$
(b) *see graph below*

11 (a) $f(x) > 0$ if $x < -2$ or $0 < x < 5$
$f(x) < 0$ if $-2 < x < 0$ or $x > 5$
(b) *see graph below*

9 **11**

13 (a) $f(x) > 0$ if $-2 < x < 3$ or $x > 4$ (b) *see graph below*
$f(x) < 0$ if $x < -2$ or $3 < x < 4$

15 (a) $f(x) > 0$ if $x > 2$ (b) *see graph below*
$f(x) < 0$ if $x < -2$ or $|x| < 2$

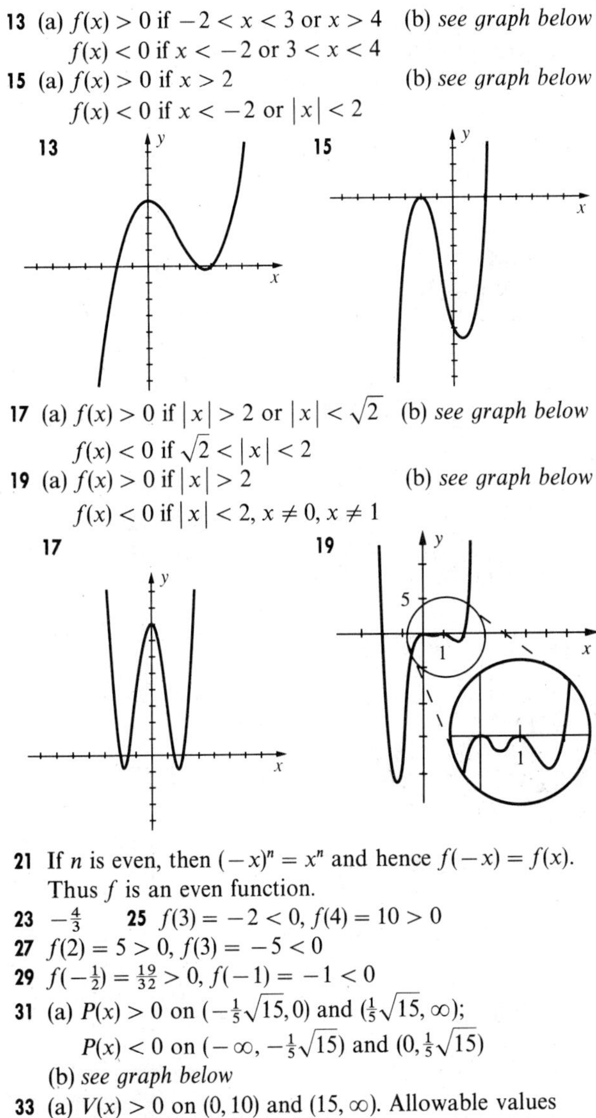

13 **15**

17 (a) $f(x) > 0$ if $|x| > 2$ or $|x| < \sqrt{2}$ (b) *see graph below*
$f(x) < 0$ if $\sqrt{2} < |x| < 2$

19 (a) $f(x) > 0$ if $|x| > 2$ (b) *see graph below*
$f(x) < 0$ if $|x| < 2$, $x \neq 0$, $x \neq 1$

17 **19**

21 If n is even, then $(-x)^n = x^n$ and hence $f(-x) = f(x)$. Thus f is an even function.

23 $-\frac{4}{3}$ **25** $f(3) = -2 < 0$, $f(4) = 10 > 0$

27 $f(2) = 5 > 0$, $f(3) = -5 < 0$

29 $f(-\frac{1}{2}) = \frac{19}{32} > 0$, $f(-1) = -1 < 0$

31 (a) $P(x) > 0$ on $(-\frac{1}{5}\sqrt{15}, 0)$ and $(\frac{1}{5}\sqrt{15}, \infty)$;
$P(x) < 0$ on $(-\infty, -\frac{1}{5}\sqrt{15})$ and $(0, \frac{1}{5}\sqrt{15})$
(b) *see graph below*

33 (a) $V(x) > 0$ on $(0, 10)$ and $(15, \infty)$. Allowable values for x are in $(0, 10)$. (b) *see graph below*

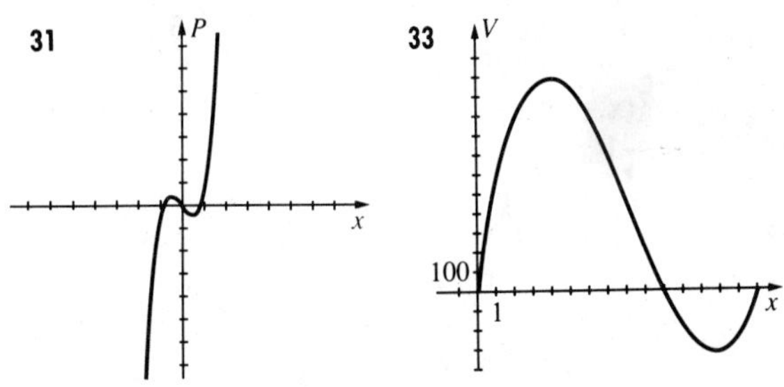

35 (a) $T > 0$ for $0 < t < 12$; $T < 0$ for $12 < t < 24$
(b) *see graph below*
(c) $T(6) = 32.4 > 32$, $T(7) = 29.75 < 32$

37 (a) $N > 0$ for $0 < t < 5$. The population becomes extinct after 5 years.
(b) *see graph below*

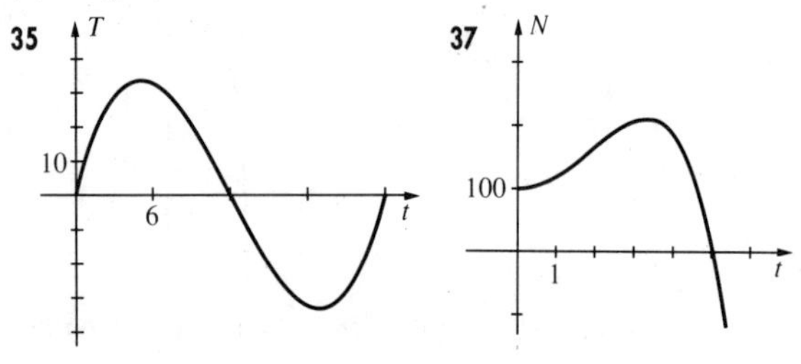

39 (a) $f(1) = -1 < 0$, $f(2) = 3 > 0$
(b) $f(1.5) = -0.125 < 0$, $f(1.6) = 0.296 > 0$
(c) $f(1.53) = -0.008423 < 0$, $f(1.54) = 0.032264 > 0$

41 $f(2.51) = -0.103898 < 0$, $f(2.52) = 0.044416 > 0$

43 $f(-1.10) = -0.03051 < 0$, $f(-1.09) = 0.08517605 > 0$

EXERCISES 4.3 ▪ PAGE 204

1 $2x^2 - x + 3; 4x - 3$ **3** $\frac{3}{2}x; \frac{1}{2}x - 4$
5 $0; 7x + 2$ **7** 26 **9** 7 **11** $f(-3) = 0$
13 $f(-2) = 0$ **15** $2x^2 + x + 6; 7$
17 $x^2 - 3x + 1; -8$
19 $3x^4 - 6x^3 + 12x^2 - 18x + 36; -65$
21 $4x^3 + 2x^2 - 4x - 2; 0$ **23** 73 **25** -0.0444
27 $8 + 7\sqrt{3}$ **29** $f(-2) = 0$ **31** $f(\frac{1}{2}) = 0$
33 3, 5 **35** $f(c) > 0$ **37** -14
39 If $f(x) = x^n - y^n$ and n is even, then $f(-y) = 0$.
41 (a) $V = \pi x^2(6 - x)$ (b) $(\frac{1}{2}(5 + \sqrt{45}), \frac{1}{2}(7 - \sqrt{45}))$
43 (a) $A = 8x - 2x^3$ (b) $\sqrt{13} - 1 \approx 2.6$

EXERCISES 4.4 ▪ PAGE 213

1 $-4x^3 + 16x^2 - 4x - 24$ **3** $3x^3 + 3x^2 - 36x$
5 $-2x^3 + 6x^2 - 8x + 24$
7 $a(x^4 + 2x^3 - 23x^2 - 24x + 144)$ for $a \neq 0$
9 $3x^6 - 27x^5 + 81x^4 - 81x^3$
11 $f(x) = \frac{7}{9}(x + 1)(x - \frac{3}{2})(x - 3)$
13 $f(x) = -1(x - 1)^2(x - 3)$
15 $f(t) = \frac{5}{3528}t(t - 5)(t - 19)(t - 24)$
17 $-\frac{2}{3}$ (multiplicity 1); 0 (multiplicity 2); $\frac{5}{2}$ (multiplicity 3)
19 $-\frac{3}{2}$ (multiplicity 2); 0 (multiplicity 3)
21 -4 (multiplicity 3); -3 (multiplicity 2); 3 (multiplicity 5)
23 $\pm 4i$, ± 3 (each of multiplicity 1)
25 $f(x) = (x + 3)^2(x + 2)(x - 1)$
27 $f(x) = (x - 1)^5(x + 1)$

Exer. 29–35: The types of possible solutions are listed in the order positive, negative, nonreal complex.

29 3, 0, 0 or 1, 0, 2 **31** 0, 1, 2
33 2, 2, 0; 2, 0, 2; 0, 2, 2; 0, 0, 4
35 2, 3, 0; 2, 1, 2; 0, 3, 2; 0, 1, 4
37 Upper 5, lower -2 **39** Upper 2, lower -2
41 Upper 3, lower -3
43 Let $h(x) = f(x) - g(x) = (a_n - b_n)x^n + \cdots + (a_0 - b_0)$. *Suppose* $h(x)$ is not the zero polynomial. Thus $h(x)$ is a polynomial of degree m with $0 < m \leq n$. If $f(x)$ and $g(x)$ are equal for more than n distinct values of x, then each of these values is a zero of $h(x)$, and hence $h(x)$ has more than m distinct zeros. This is a contradiction, since a polynomial of degree m can have at most m different zeros. Hence our supposition is *false*, and therefore $h(x)$ is the zero polynomial. Thus each coefficient $a_k - b_k$ of $h(x)$ is 0; that is, $a_k = b_k$ for every k.

EXERCISES 4.5 ▪ PAGE 221

1 $a(x^2 - 6x + 13)$ for $a \neq 0$
3 $a(x^3 + 2x^2 + 21x - 58)$ for $a \neq 0$
5 $a(x^4 - 5x^3 + 4x^2 + 10x)$ for $a \neq 0$
7 $a(x^4 - 4x^3 - 2x^2 + 60x + 125)$ for $a \neq 0$
9 $a(x^5 - 2x^4 + 6x^3 - 8x^2 + 8x)$ for $a \neq 0$
11 No. If i is a root, then $-i$ is also a root. Hence, the polynomial would have factors $x - 1$, $x + 1$, $x - i$, $x + i$ and, therefore, would be of degree greater than 3.
13 $-1, -2, 4$ **15** $2, -3, \frac{5}{2}$ **17** $4, -7, \pm\sqrt{2}$
19 $4, -2, \frac{3}{2}$ **21** $\frac{1}{2}, -\frac{2}{3}, -3$ **23** $-\frac{3}{4}, -\frac{3}{4} \pm \frac{3}{4}\sqrt{7}\,i$
25 $4, -3, \pm\sqrt{3}\,i$ **27** $1, -2, -\frac{4}{3}, \frac{2}{3}$
29 $3, -4, -2, \frac{1}{2}, -\frac{1}{2}$

Exer. 31–37: Show that none of the possible rational roots listed satisfy the equation.

31 $\pm 1, \pm 2, \pm 3, \pm 6$ **33** $\pm 1, \pm 2$
35 $\pm 1, \pm\frac{1}{3}, \pm 2, \pm\frac{2}{3}, \pm 5, \pm\frac{5}{3}, \pm 10, \pm\frac{10}{3}$
37 $\pm 1, \pm\frac{1}{2}, \pm\frac{1}{4}, \pm\frac{1}{8}, \pm 3, \pm\frac{3}{2}, \pm\frac{3}{4}, \pm\frac{3}{8}, \pm 5, \pm\frac{5}{2}, \pm\frac{5}{4}, \pm\frac{5}{8}, \pm 15, \pm\frac{15}{2}, \pm\frac{15}{4}, \pm\frac{15}{8}$
39 Since n is odd and nonreal complex zeros occur in conjugate pairs for polynomials with real coefficients, there must be at least one real zero.
43 (a) The two boxes correspond to $x = 5$ and $x = 5(2 - \sqrt{2})$.
(b) The box corresponding to $x = 5$
45 $t = 4$ (10:00 A.M.) and $t \approx 6.2020$ (12:12 P.M.)
47 (c) In feet: 5, 12, and 13 **49** (b) 4 ft

EXERCISES 4.6 ■ PAGE 232

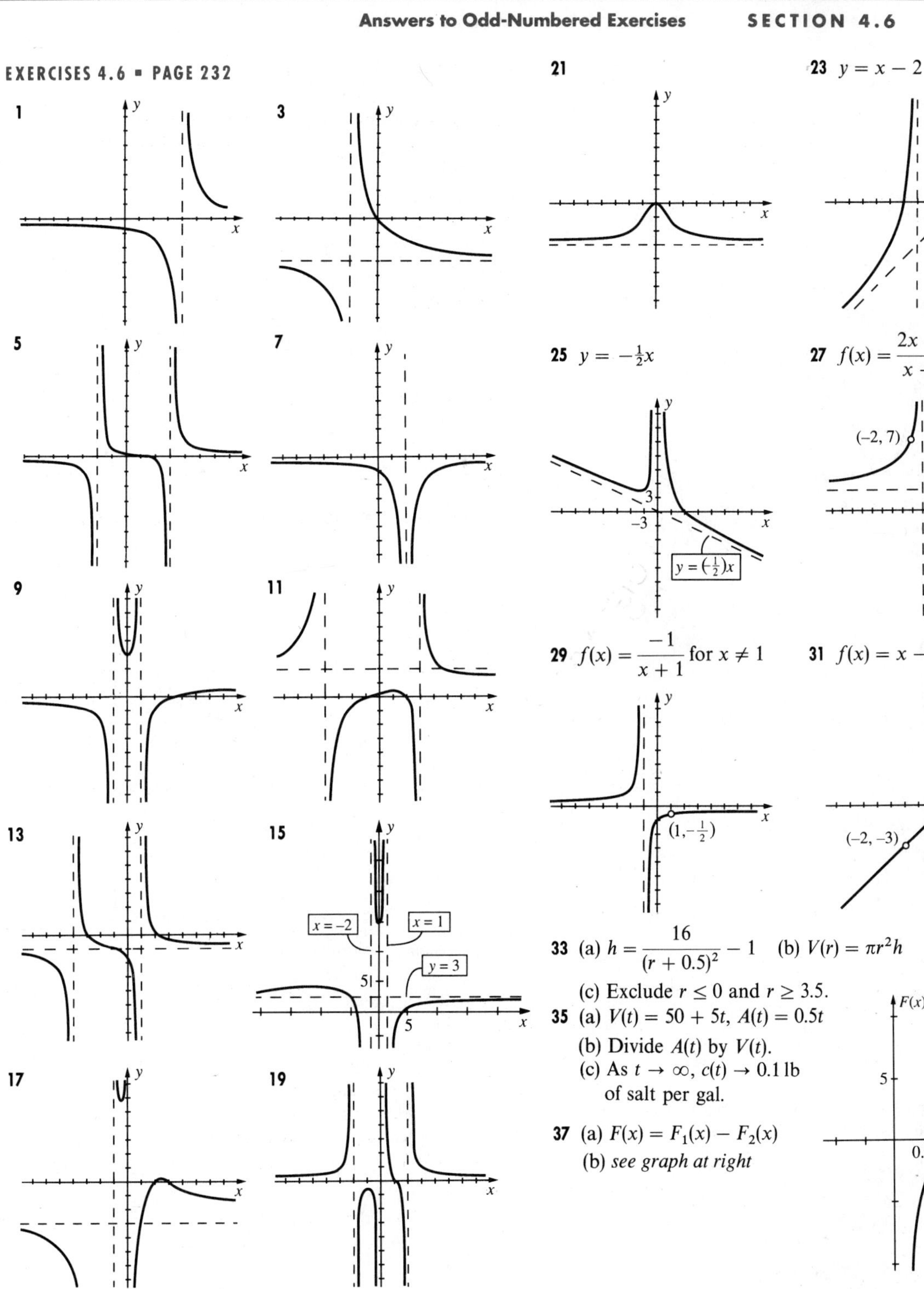

21

23 $y = x - 2$

25 $y = -\frac{1}{2}x$

27 $f(x) = \dfrac{2x - 3}{x + 1}$ for $x \neq -2$

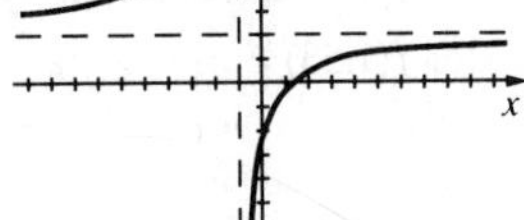

29 $f(x) = \dfrac{-1}{x + 1}$ for $x \neq 1$

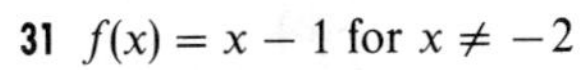

31 $f(x) = x - 1$ for $x \neq -2$

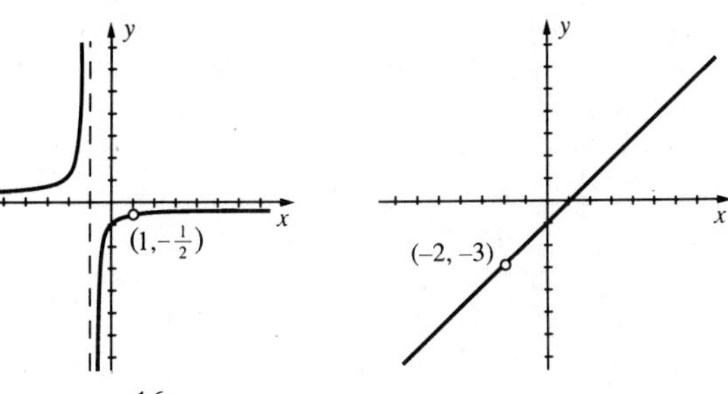

33 (a) $h = \dfrac{16}{(r + 0.5)^2} - 1$ (b) $V(r) = \pi r^2 h$

(c) Exclude $r \leq 0$ and $r \geq 3.5$.

35 (a) $V(t) = 50 + 5t$, $A(t) = 0.5t$

(b) Divide $A(t)$ by $V(t)$.

(c) As $t \to \infty$, $c(t) \to 0.1$ lb of salt per gal.

37 (a) $F(x) = F_1(x) - F_2(x)$

(b) *see graph at right*

EXERCISES 4.7 ▪ PAGE 234

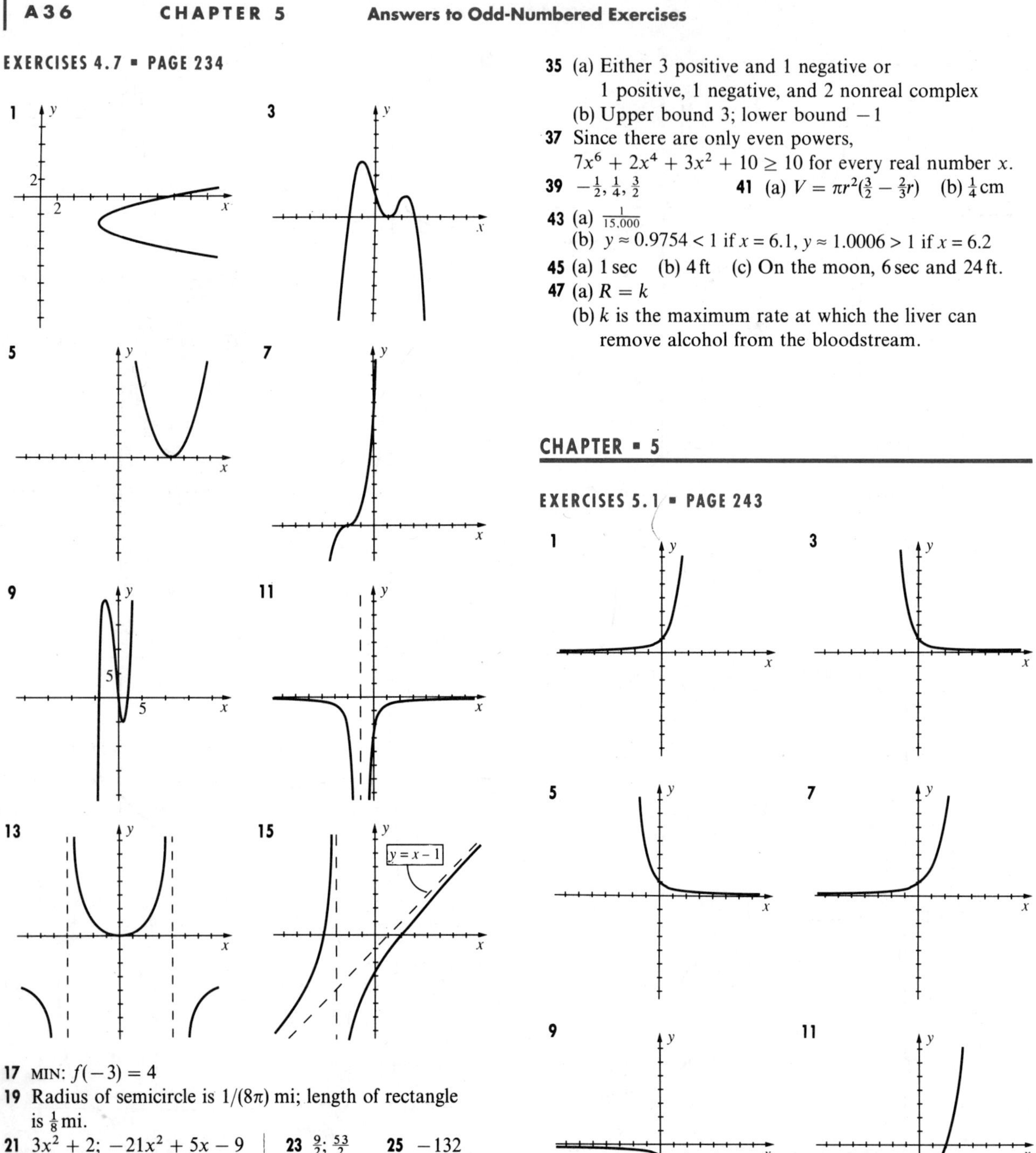

17 MIN: $f(-3) = 4$

19 Radius of semicircle is $1/(8\pi)$ mi; length of rectangle is $\frac{1}{8}$ mi.

21 $3x^2 + 2$; $-21x^2 + 5x - 9$ | **23** $\frac{9}{2}$; $\frac{53}{2}$ **25** -132

27 $6x^4 - 12x^3 + 24x^2 - 52x + 104$; -200

29 $\frac{2}{41}x^3 + \frac{14}{41}x^2 + \frac{80}{41}x + \frac{68}{41}$

31 $x^7 + 6x^6 + 9x^5$

33 1 (multiplicity 5); -3 (multiplicity 1)

35 (a) Either 3 positive and 1 negative or 1 positive, 1 negative, and 2 nonreal complex
(b) Upper bound 3; lower bound -1

37 Since there are only even powers, $7x^6 + 2x^4 + 3x^2 + 10 \geq 10$ for every real number x.

39 $-\frac{1}{2}, \frac{1}{4}, \frac{3}{2}$ **41** (a) $V = \pi r^2(\frac{3}{2} - \frac{2}{3}r)$ (b) $\frac{1}{4}$ cm

43 (a) $\frac{1}{15{,}000}$
(b) $y \approx 0.9754 < 1$ if $x = 6.1$, $y \approx 1.0006 > 1$ if $x = 6.2$

45 (a) 1 sec (b) 4 ft (c) On the moon, 6 sec and 24 ft.

47 (a) $R = k$
(b) k is the maximum rate at which the liver can remove alcohol from the bloodstream.

CHAPTER ▪ 5

EXERCISES 5.1 ▪ PAGE 243

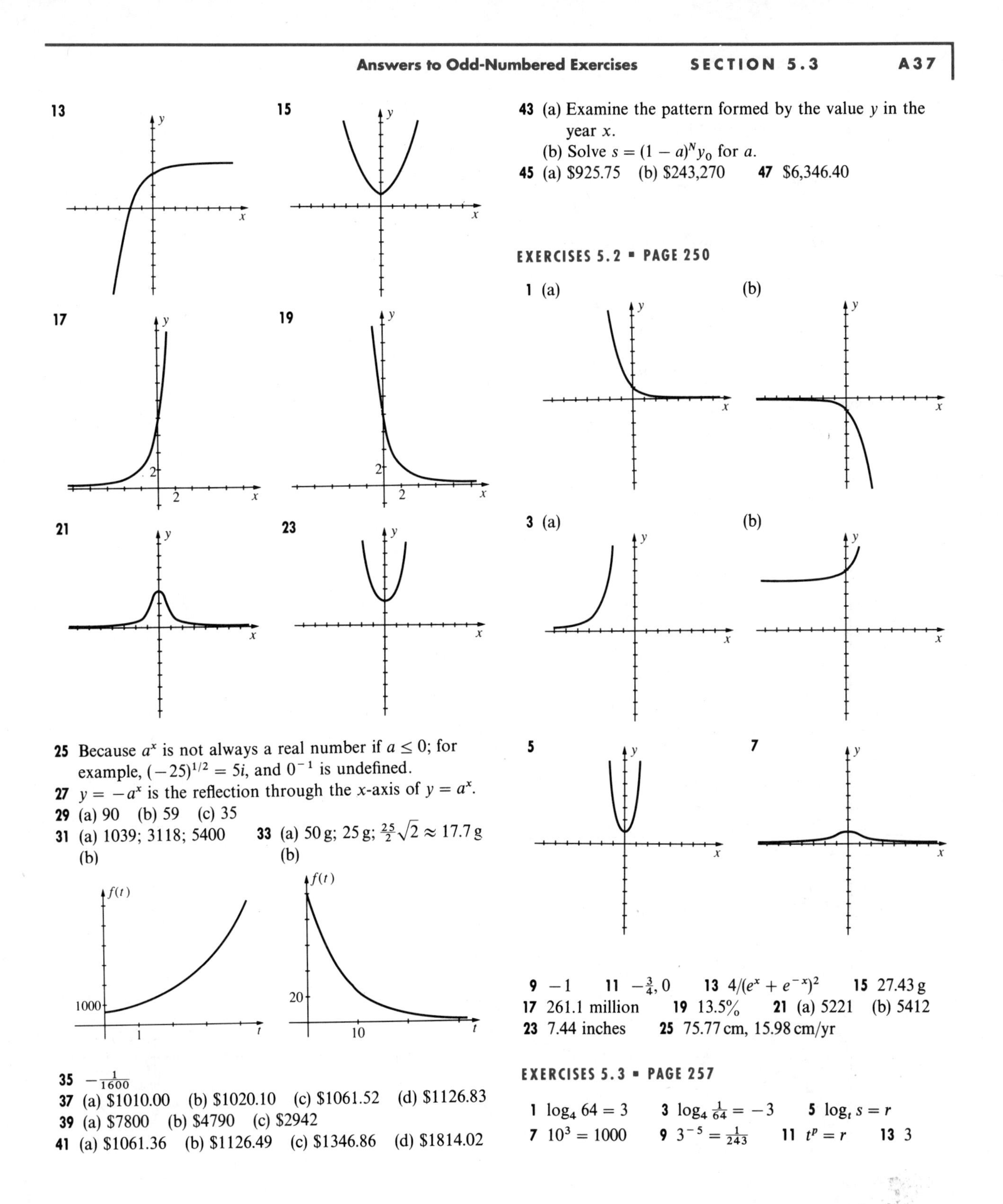

13 **15** **17** **19** **21** **23**

25 Because a^x is not always a real number if $a \leq 0$; for example, $(-25)^{1/2} = 5i$, and 0^{-1} is undefined.

27 $y = -a^x$ is the reflection through the x-axis of $y = a^x$.

29 (a) 90 (b) 59 (c) 35

31 (a) 1039; 3118; 5400 (b)

33 (a) 50 g; 25 g; $\frac{25}{2}\sqrt{2} \approx 17.7$ g (b)

35 $-\frac{1}{1600}$

37 (a) \$1010.00 (b) \$1020.10 (c) \$1061.52 (d) \$1126.83

39 (a) \$7800 (b) \$4790 (c) \$2942

41 (a) \$1061.36 (b) \$1126.49 (c) \$1346.86 (d) \$1814.02

43 (a) Examine the pattern formed by the value y in the year x.
(b) Solve $s = (1 - a)^N y_0$ for a.

45 (a) \$925.75 (b) \$243,270 **47** \$6,346.40

EXERCISES 5.2 ▪ PAGE 250

1 (a) (b)

3 (a) (b)

5 **7**

9 -1 **11** $-\frac{3}{4}, 0$ **13** $4/(e^x + e^{-x})^2$ **15** 27.43 g

17 261.1 million **19** 13.5% **21** (a) 5221 (b) 5412

23 7.44 inches **25** 75.77 cm, 15.98 cm/yr

EXERCISES 5.3 ▪ PAGE 257

1 $\log_4 64 = 3$ **3** $\log_4 \frac{1}{64} = -3$ **5** $\log_t s = r$

7 $10^3 = 1000$ **9** $3^{-5} = \frac{1}{243}$ **11** $t^p = r$ **13** 3

15 -2 **17** 3

19 $3 \log_a x + \log_a w - 2 \log_a y - 4 \log_a z$

21 $\frac{1}{3} \log_a z - \log_a x - \frac{1}{2} \log_a y$

23 $\frac{7}{4} \log_a x - \frac{5}{4} \log_a y - \frac{1}{4} \log_a z$ **25** $\log_a \dfrac{x^2 \sqrt[3]{x-2}}{(2x+3)^5}$

27 $\log_a \dfrac{y^{13/3}}{x^2}$ **29** $\neq$, $\log_a x + \log_a y = \log_a (xy)$

31 $=$, logarithm law (ii)

33 $\neq$, $\log_{ab} xy = \log_{ab} x + \log_{ab} y$ **35** 13 **37** 27

39 $\pm\frac{1}{5}$ **41** $\frac{7}{2}$ **43** $5\sqrt{5}$ **45** No solution

47 -7 **49** 1 **51** -2 **53** $t = \dfrac{1}{C} \log_2 \left(\dfrac{A-D}{B} \right)$

55 $t = -1600 \log_2 (q/q_0)$ **57** $D = c/p^k$

59 (a) $t = \log_{10} (F/F_0)/\log_{10} (1-m)$ (b) 13,863 generations

EXERCISES 5.4 ■ PAGE 261

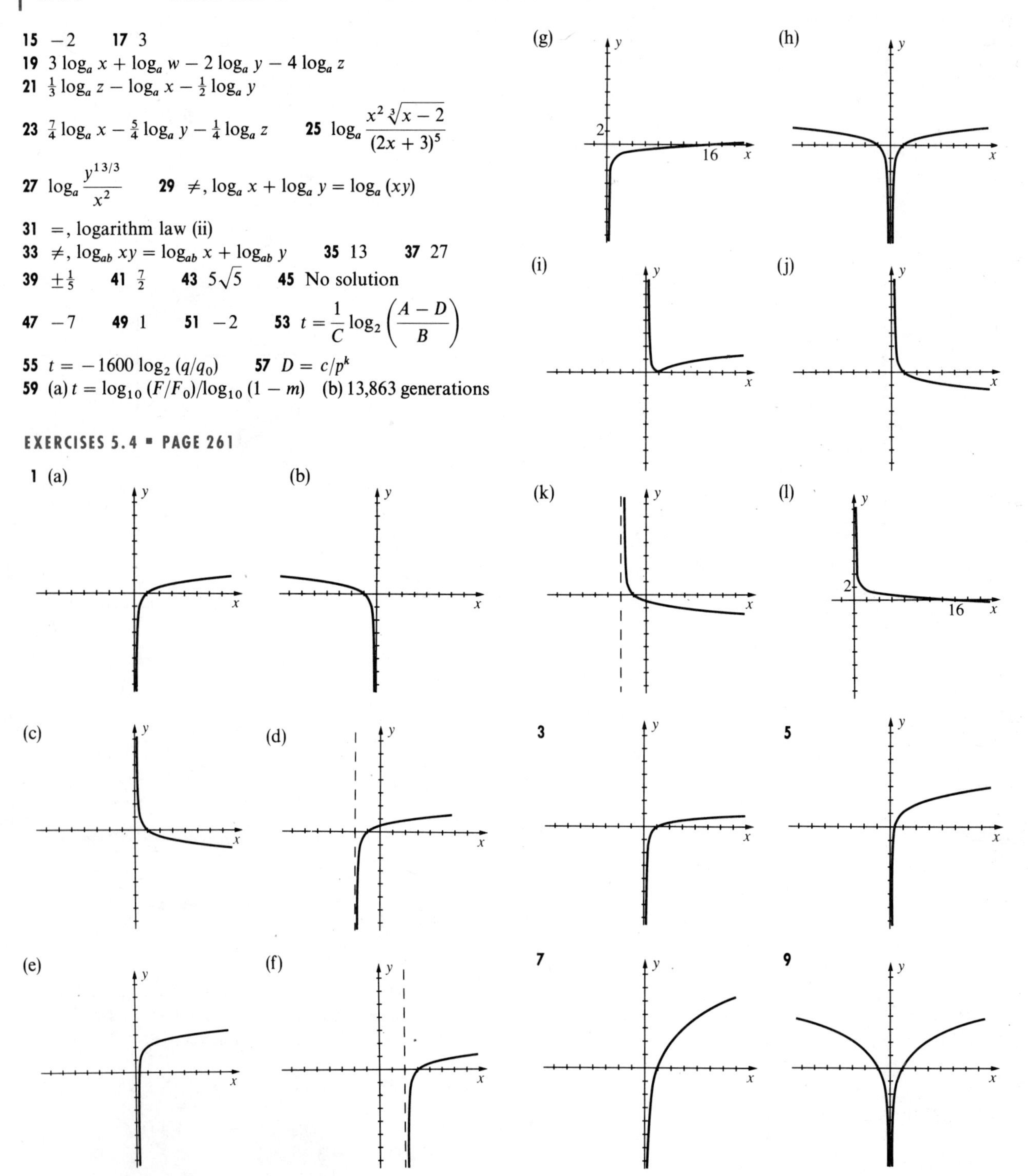

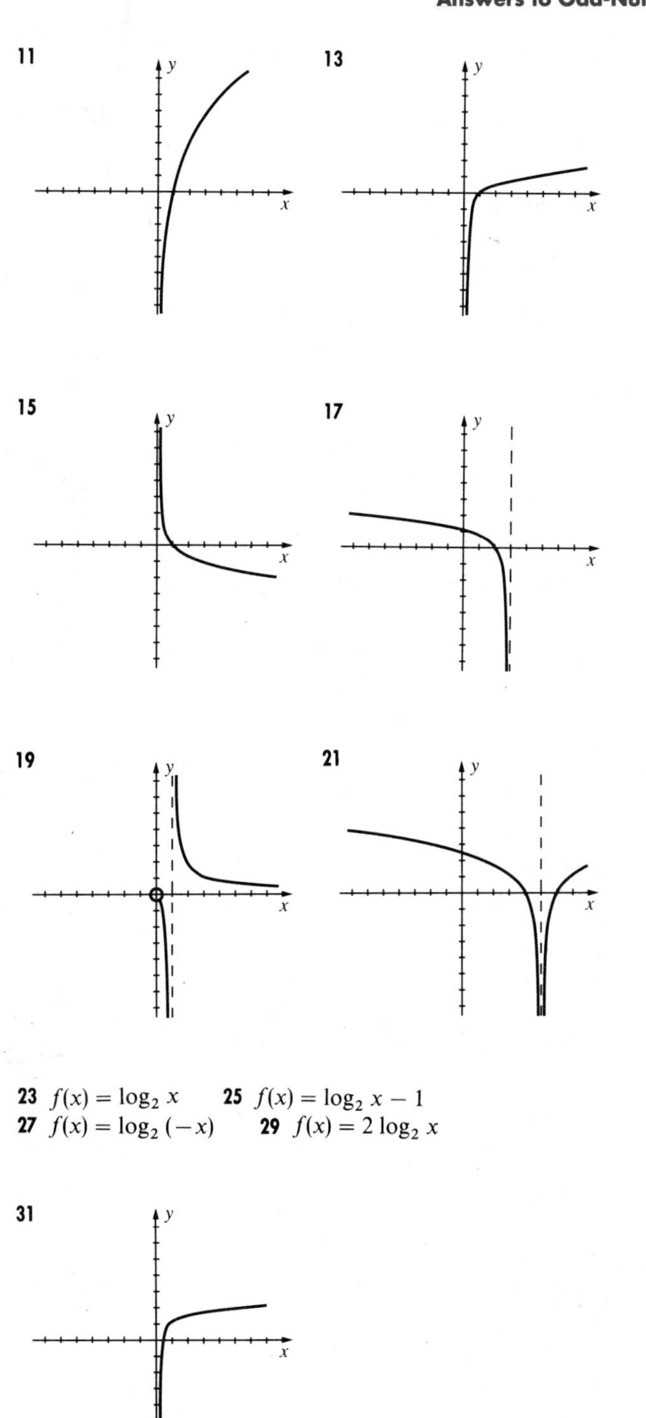

23 $f(x) = \log_2 x$ **25** $f(x) = \log_2 x - 1$
27 $f(x) = \log_2(-x)$ **29** $f(x) = 2\log_2 x$

EXERCISES 5.5 ▪ PAGE 268

1 (a) 2 (b) 4 (c) 5 **3** $A = 10^{(R+5.1)/2.3} - 3{,}000$
5 (a) 10 (b) 30 (c) 40 **7** (a) 2.2 (b) 5 (c) 8.3
9 Acidic if pH < 7, basic if pH > 7
11 $t = -(L/R)\ln(I/20)$ **13** $h = \ln(29/p)/0.000034$
15 $W = 2.4e^{1.84h}$
17 (a) $n = 10^{7.7-0.9R}$ (b) 12,589; 1585; 200
19 110 days **21** 11.58 yr ≈ 11 yr 7 mo
23 In the year 2079 **25** 86.8 cm, 9.715 cm/yr
27 (a) $R = 0$ (b) $R(2x) = R(x) + a\log 2$ **29** 4240
31 8.85 **33** 0.0237 **35** 9.97 **37** 1.05
39 0.202 **41** 1.1133 **43** 2

45 y-intercept $= \log_2 3 \approx 1.5850$ **47** x-intercept $= \log_4 3 \approx 0.7925$

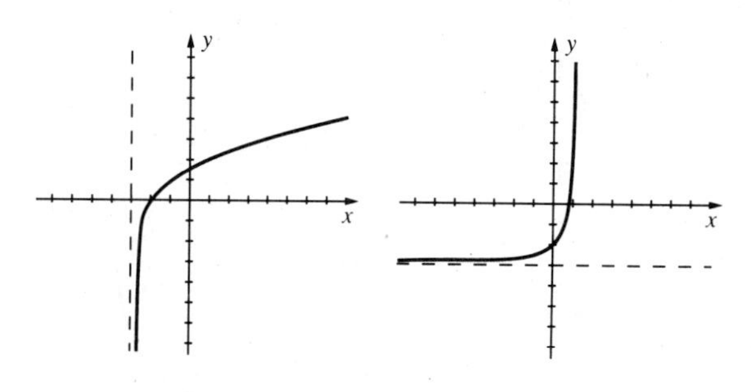

EXERCISES 5.6 ▪ PAGE 274

1 $\log 8/\log 5 \approx 1.29$ **3** $4 - \log 5/\log 3 \approx 2.54$
5 $\log \frac{2}{81}/\log 24 \approx -1.16$ **7** $\log \frac{8}{25}/\log \frac{4}{5} \approx 5.11$
9 -3 **11** 5 **13** $\frac{2}{3}\sqrt{\frac{1001}{111}} \approx 2.00$ **15** 1, 2
17 $\dfrac{\log(4+\sqrt{19})}{\log 4} \approx 1.53$ **19** 1 or 100 **21** 10^{100}
23 10,000 **25** $x = \log(y \pm \sqrt{y^2 - 1})$
27 $x = \frac{1}{2}\log\left(\dfrac{1+y}{1-y}\right)$ **29** $x = \ln(y + \sqrt{y^2 + 1})$
31 $x = \frac{1}{2}\ln\left(\dfrac{y+1}{y-1}\right)$ **33** $t = -\dfrac{L}{R}\ln\left(\dfrac{V - RI}{V}\right)$
35 86.4 m **37** (a) 7.21 hr (b) 3.11 hr
39 (a) 26,749 yr (b) 30% **41** 16.91 yr **43** $x > 10.1$ mi

EXERCISES 5.7 ▪ PAGE 276

1 -4 **3** 4 **5** 6 **7** $\frac{1}{2}$ **9** 5

11 **13** **15** **17** **19** **21** **23**

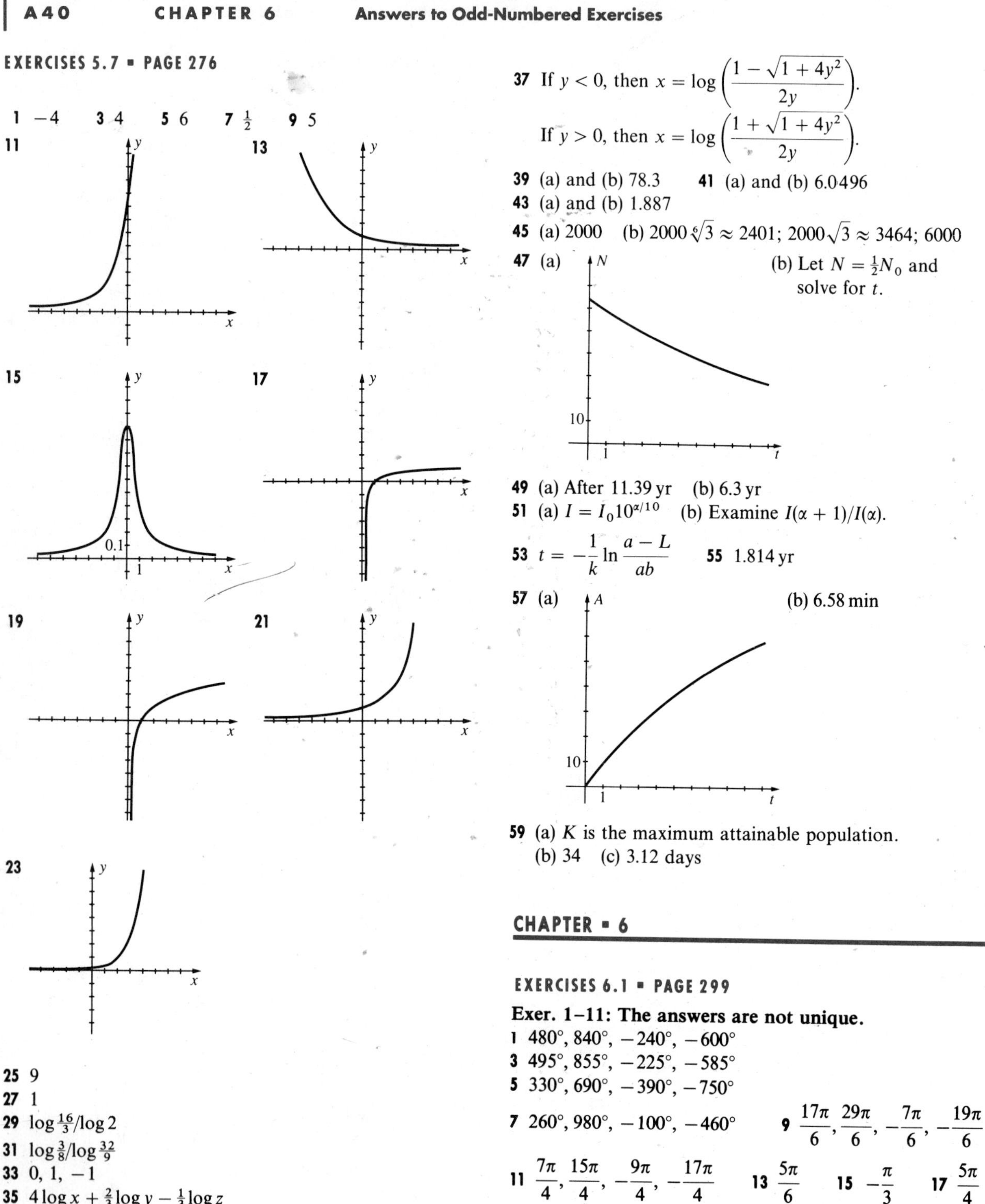

25 9
27 1
29 $\log\frac{16}{3}/\log 2$
31 $\log\frac{3}{8}/\log\frac{32}{9}$
33 0, 1, -1
35 $4\log x + \frac{2}{3}\log y - \frac{1}{3}\log z$

37 If $y < 0$, then $x = \log\left(\dfrac{1 - \sqrt{1 + 4y^2}}{2y}\right)$.
If $y > 0$, then $x = \log\left(\dfrac{1 + \sqrt{1 + 4y^2}}{2y}\right)$.

39 (a) and (b) 78.3 **41** (a) and (b) 6.0496
43 (a) and (b) 1.887
45 (a) 2000 (b) $2000\sqrt[6]{3} \approx 2401$; $2000\sqrt{3} \approx 3464$; 6000
47 (a) (b) Let $N = \frac{1}{2}N_0$ and solve for t.

49 (a) After 11.39 yr (b) 6.3 yr
51 (a) $I = I_0 10^{\alpha/10}$ (b) Examine $I(\alpha + 1)/I(\alpha)$.
53 $t = -\dfrac{1}{k}\ln\dfrac{a - L}{ab}$ **55** 1.814 yr
57 (a) (b) 6.58 min

59 (a) K is the maximum attainable population.
(b) 34 (c) 3.12 days

CHAPTER ▪ 6

EXERCISES 6.1 ▪ PAGE 299

Exer. 1–11: The answers are not unique.
1 480°, 840°, $-240°$, $-600°$
3 495°, 855°, $-225°$, $-585°$
5 330°, 690°, $-390°$, $-750°$
7 260°, 980°, $-100°$, $-460°$ **9** $\dfrac{17\pi}{6}, \dfrac{29\pi}{6}, -\dfrac{7\pi}{6}, -\dfrac{19\pi}{6}$
11 $\dfrac{7\pi}{4}, \dfrac{15\pi}{4}, -\dfrac{9\pi}{4}, -\dfrac{17\pi}{4}$ **13** $\dfrac{5\pi}{6}$ **15** $-\dfrac{\pi}{3}$ **17** $\dfrac{5\pi}{4}$

19 $\frac{5\pi}{2}$ **21** $\frac{2\pi}{5}$ **23** $\frac{5\pi}{9}$ **25** 120° **27** 330°
29 135° **31** −630° **33** 1260° **35** 20°
37 114°35′30″ **39** 286°28′44″ **41** 37.6833°
43 115.4408° **45** 63°10′8″ **47** 310°37′17″
49 1.75, 100.27° **51** 6.98 m **53** 2.5 cm
55 (In mi): (a) 4189 (b) 3142 (c) 2094 (d) 698 (e) 70
57 $\frac{1}{8}$ radian $\approx 7°10'$ **59** (a) 80π (b) $\frac{100\pi}{3} \approx 104.72$
61 (a) $\frac{200\pi}{3}$, 90π (b) $\frac{100\pi}{3}$, $\frac{105\pi}{4}$
63 (a) $\frac{21\pi}{8} \approx 8.25$ ft (b) $\frac{2}{3}d$ **65** 192.1

EXERCISES 6.2 ■ PAGE 298

	sin θ	cos θ	tan θ	cot θ	sec θ	csc θ
1	$\frac{4}{5}$	$\frac{3}{5}$	$\frac{4}{3}$	$\frac{3}{4}$	$\frac{5}{3}$	$\frac{5}{4}$
3	$\frac{2\sqrt{13}}{13}$	$\frac{3\sqrt{13}}{13}$	$\frac{2}{3}$	$\frac{3}{2}$	$\frac{\sqrt{13}}{3}$	$\frac{\sqrt{13}}{2}$
5	$\frac{2}{5}$	$\frac{\sqrt{21}}{5}$	$\frac{2\sqrt{21}}{21}$	$\frac{\sqrt{21}}{2}$	$\frac{5\sqrt{21}}{21}$	$\frac{5}{2}$
7	$\frac{a}{\sqrt{a^2+b^2}}$	$\frac{b}{\sqrt{a^2+b^2}}$	$\frac{a}{b}$	$\frac{b}{a}$	$\frac{\sqrt{a^2+b^2}}{b}$	$\frac{\sqrt{a^2+b^2}}{a}$
9	$\frac{b}{c}$	$\frac{\sqrt{c^2-b^2}}{c}$	$\frac{b}{\sqrt{c^2-b^2}}$	$\frac{\sqrt{c^2-b^2}}{b}$	$\frac{c}{\sqrt{c^2-b^2}}$	$\frac{c}{b}$
11	$\frac{3}{5}$	$\frac{4}{5}$	$\frac{3}{4}$	$\frac{4}{3}$	$\frac{5}{4}$	$\frac{5}{3}$
13	$\frac{\sqrt{11}}{6}$	$\frac{5}{6}$	$\frac{\sqrt{11}}{5}$	$\frac{5\sqrt{11}}{11}$	$\frac{6}{5}$	$\frac{6\sqrt{11}}{11}$
15	$\frac{\sqrt{2}}{2}$	$\frac{\sqrt{2}}{2}$	1	1	$\sqrt{2}$	$\sqrt{2}$

17 0.783 **19** 6.197 **21** 2.650 **23** 0.778
25 0.472 **27** 1.035 **29** (a) 30.46° (b) 30°27′
31 (a) 74.88° (b) 74°53′ **33** (a) 24.94° (b) 24°57′
35 (a) 76.38° (b) 76°23′ **37** $\cot\theta = \frac{\sqrt{1-\sin^2\theta}}{\sin\theta}$
39 $\sec\theta = \frac{1}{\sqrt{1-\sin^2\theta}}$ **41** $\tan\theta = \sqrt{\sec^2\theta - 1}$
43 $\sin\theta = \frac{\sqrt{\sec^2\theta-1}}{\sec\theta}$
45 (a) $\sin\theta\csc\theta = \frac{\text{opp}}{\text{hyp}}\cdot\frac{\text{hyp}}{\text{opp}} = 1$
(b) $\cos\theta\sec\theta = \frac{\text{adj}}{\text{hyp}}\cdot\frac{\text{hyp}}{\text{adj}} = 1$
(c) $\tan\theta\cot\theta = \frac{\text{opp}}{\text{adj}}\cdot\frac{\text{adj}}{\text{opp}} = 1$
47 $\log\csc\theta = \log\left(\frac{1}{\sin\theta}\right) = \log 1 - \log\sin\theta$
$= 0 - \log\sin\theta = -\log\sin\theta$
49 Press, in succession: (a) cos, x^2 (b) sin, $1/x$
(c) tan, $1/x$, x^2, x^2

EXERCISES 6.3 ■ PAGE 304

1 $\beta = 60°$, $a = \frac{20\sqrt{3}}{3}$, $c = \frac{40\sqrt{3}}{3}$
3 $\alpha = 45°$, $a = b = 15\sqrt{2}$ **5** $\alpha = \beta = 45°$, $c = 5\sqrt{2}$
7 $\alpha = 60°$, $\beta = 30°$, $a = 15$ **9** $\beta = 53°$, $a \approx 18$, $c \approx 30$
11 $\alpha = 18°9'$, $a \approx 78.7$, $c \approx 252.6$
13 $\alpha \approx 29°$, $\beta \approx 61°$, $c \approx 51$ **15** $\alpha \approx 69°$, $\beta \approx 21°$, $a \approx 5.4$
17 $b = c\sin\alpha$ **19** $a = b\tan\beta$ **21** $c = a\sec\alpha$
23 $b = \sqrt{c^2 - a^2}$ **25** 51°20′
27 $250\sqrt{3} + 4 \approx 437$ ft **29** 20.2 m **31** 29.7 km
33 1,459,379 ft² **35** 160 m **37** 9659 ft
39 126.2 mi/hr **41** 28,800 ft **43** (a) 55 mi (b) S63°E
45 324 mi **47** (a) 58 ft (b) 27 ft **49** 30°
51 21.8° **53** 3944 mi
55 (a) 45%
(b) Each satellite has a signal range of more than 120°.
57 $x = d\cot\alpha$ **59** $P = 2nr\sin(180/n)°$
61 $h = d(\tan\beta - \tan\alpha)$

EXERCISES 6.4 ■ PAGE 317

	sin θ	cos θ	tan θ	cot θ	sec θ	csc θ
1	$-\frac{3}{5}$	$\frac{4}{5}$	$-\frac{3}{4}$	$-\frac{4}{3}$	$\frac{5}{4}$	$-\frac{5}{3}$
3	$-\frac{5\sqrt{29}}{29}$	$-\frac{2\sqrt{29}}{29}$	$\frac{5}{2}$	$\frac{2}{5}$	$-\frac{\sqrt{29}}{2}$	$-\frac{\sqrt{29}}{5}$
5	$\frac{4\sqrt{17}}{17}$	$-\frac{\sqrt{17}}{17}$	−4	$-\frac{1}{4}$	$-\sqrt{17}$	$\frac{\sqrt{17}}{4}$
7	$-\frac{7\sqrt{53}}{53}$	$-\frac{2\sqrt{53}}{53}$	$\frac{7}{2}$	$\frac{2}{7}$	$-\frac{\sqrt{53}}{2}$	$-\frac{\sqrt{53}}{7}$
9	$-\frac{8}{17}$	$\frac{15}{17}$	$-\frac{8}{15}$	$-\frac{15}{8}$	$\frac{17}{15}$	$-\frac{17}{8}$
11	$\frac{4}{5}$	$\frac{3}{5}$	$\frac{4}{3}$	$\frac{3}{4}$	$\frac{5}{3}$	$\frac{5}{4}$
13	1	0	undefined	0	undefined	1
15	0	−1	0	undefined	−1	undefined
17	0	1	0	undefined	1	undefined
19	−1	0	undefined	0	undefined	−1

21 IV **23** III **25** II **27** III **29** I

31 (a) 60° (b) 20° (c) 70°

33 (a) 49°20′ (b) 45° (c) 80°35′

35 (a) $\frac{\pi}{4}$ (b) $\frac{\pi}{3}$ (c) $\frac{\pi}{6}$ **37** (a) $\frac{\pi}{4}$ (b) $\frac{\pi}{6}$ (c) $\frac{\pi}{3}$

39 (a) $\frac{\sqrt{3}}{2}$ (b) $-\frac{\sqrt{3}}{2}$ **41** (a) -1 (b) -1

43 (a) $-\frac{2\sqrt{3}}{3}$ (b) -2 **45** -0.836 **47** -3.566

49 2.443 **51** 0.362 **53** 0.111 **55** -0.143

57 214°20′, 325°40′ **59** 70°20′, 250°20′

61 153°40′, 206°20′ **63** 48°40′, 311°20′

EXERCISES 6.5 ▪ PAGE 325

1 $-\frac{1}{2}$ **3** $-\frac{\sqrt{2}}{2}$ **5** -1 **7** 1 **9** $\frac{2\sqrt{3}}{3}$ **11** 2

13 (a) 0 (b) -1 **15** (a) $\frac{\sqrt{2}}{2}$ (b) -1

17 (a) 1 (b) $-\infty$ **19** (a) -1 (b) ∞

21 (a) ∞ (b) $\sqrt{2}$ **23** (a) $-\infty$ (b) 1 **25** $\frac{3\pi}{2}, \frac{7\pi}{2}$

27 $0, 2\pi, 4\pi$ **29** $\frac{\pi}{4}, \frac{7\pi}{4}, \frac{9\pi}{4}, \frac{15\pi}{4}$ **31** $\frac{\pi}{6}, \frac{5\pi}{6}, \frac{13\pi}{6}, \frac{17\pi}{6}$

33 $\frac{\pi}{4}, \frac{5\pi}{4}$ **35** $-\frac{\pi}{4}, \frac{3\pi}{4}$ **37** $-\frac{\pi}{3}, \frac{2\pi}{3}$ **39** $0, \pi$

41 0 **43** $\frac{\pi}{4}, \frac{3\pi}{4}$ **45** No value **47** $\frac{3\pi}{4}, \frac{5\pi}{4}$

49 (a) $-\frac{11\pi}{6}, -\frac{7\pi}{6}, \frac{\pi}{6}, \frac{5\pi}{6}$

(b) $-\frac{11\pi}{6} < \theta < -\frac{7\pi}{6}$ and $\frac{\pi}{6} < \theta < \frac{5\pi}{6}$

(c) $-2\pi \le \theta < -\frac{11\pi}{6}$, $-\frac{7\pi}{6} < \theta < \frac{\pi}{6}$, and $\frac{5\pi}{6} < \theta \le 2\pi$

51 (a) $-\frac{4\pi}{3}, -\frac{2\pi}{3}, \frac{2\pi}{3}, \frac{4\pi}{3}$

(b) $-2\pi \le \theta < -\frac{4\pi}{3}$, $-\frac{2\pi}{3} < \theta < \frac{2\pi}{3}$, and $\frac{4\pi}{3} < \theta \le 2\pi$

(c) $-\frac{4\pi}{3} < \theta < -\frac{2\pi}{3}$ and $\frac{2\pi}{3} < \theta < \frac{4\pi}{3}$

53 $(-\infty, -1] \cup [1, \infty)$

55 (a) $\left[-2\pi, -\frac{3\pi}{2}\right), \left(-\frac{3\pi}{2}, -\pi\right], \left[0, \frac{\pi}{2}\right), \left(\frac{\pi}{2}, \pi\right]$

(b) $\left[-\pi, -\frac{\pi}{2}\right), \left(-\frac{\pi}{2}, 0\right], \left[\pi, \frac{3\pi}{2}\right), \left(\frac{3\pi}{2}, 2\pi\right]$

57 (a) The tangent function increases on *all* intervals on which it is defined. Between -2π and 2π, these intervals are

$$\left[-2\pi, -\frac{3\pi}{2}\right), \left(-\frac{3\pi}{2}, -\frac{\pi}{2}\right), \left(-\frac{\pi}{2}, \frac{\pi}{2}\right), \left(\frac{\pi}{2}, \frac{3\pi}{2}\right), \left(\frac{3\pi}{2}, 2\pi\right].$$

(b) The tangent function is *never* decreasing on any interval for which it is defined.

59 **61**

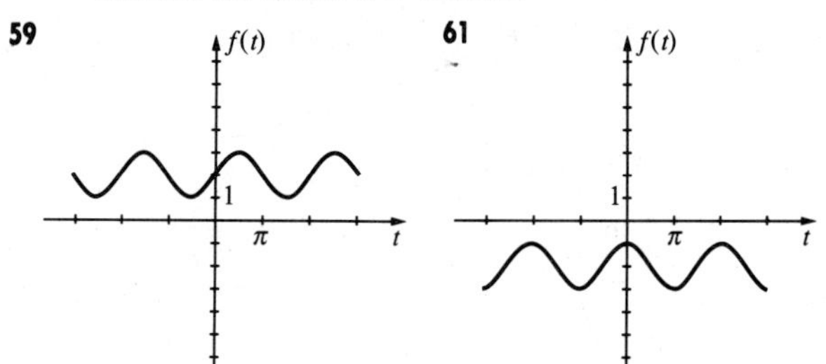

63

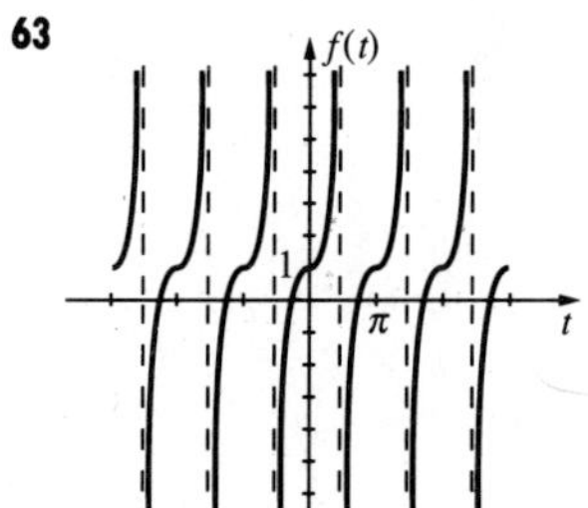

65

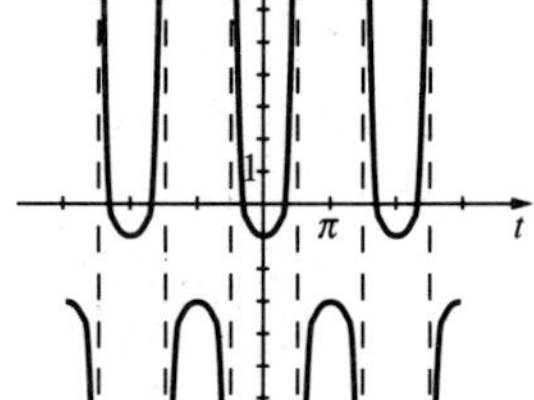

67

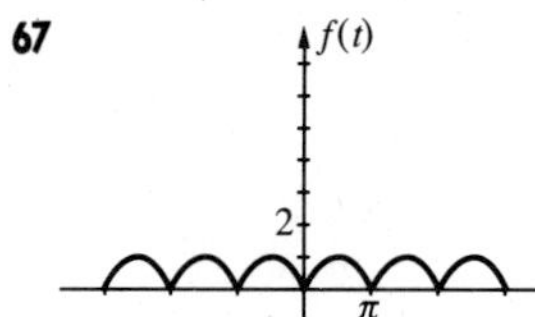

69

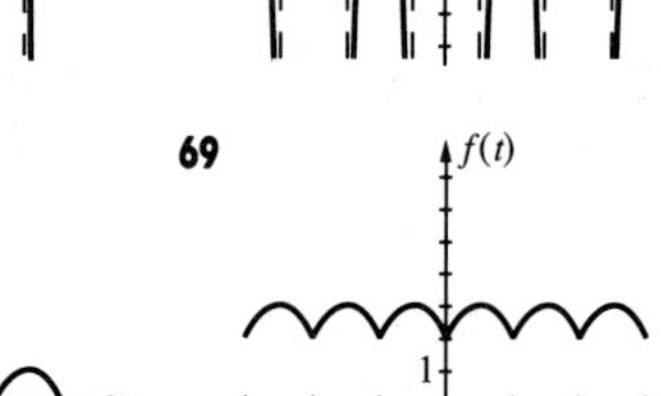

71

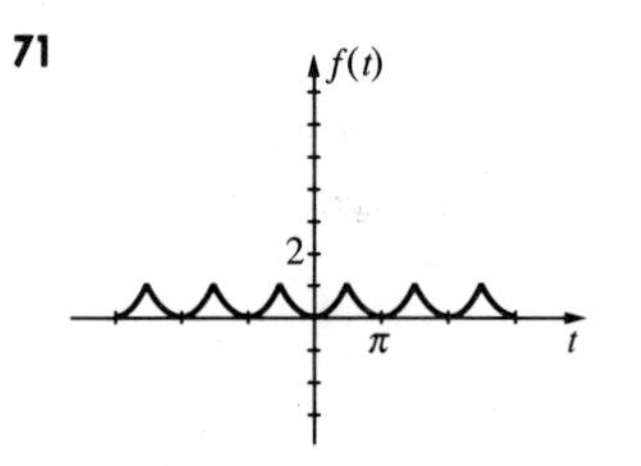

75 If $\tan\theta = y/x = c$, with $x^2 + y^2 = 1$, then $c = \pm\sqrt{1-x^2}/x$. Solving for x, we obtain $xc = \pm\sqrt{1-x^2}$, or $x^2c^2 = 1 - x^2$. Consequently, $x^2(c^2+1) = 1$, or $x = \pm 1/\sqrt{c^2+1}$. If $c > 0$, choose a point $P(x, y)$ on U such that $x = 1/\sqrt{c^2+1}$. If $c < 0$, choose $P(x, y)$ such that $x = -1/\sqrt{c^2+1}$. If $c = 0$, choose $P(1, 0)$. Thus, there is always a point $P(x, y)$ on U such that $y/x = c$.

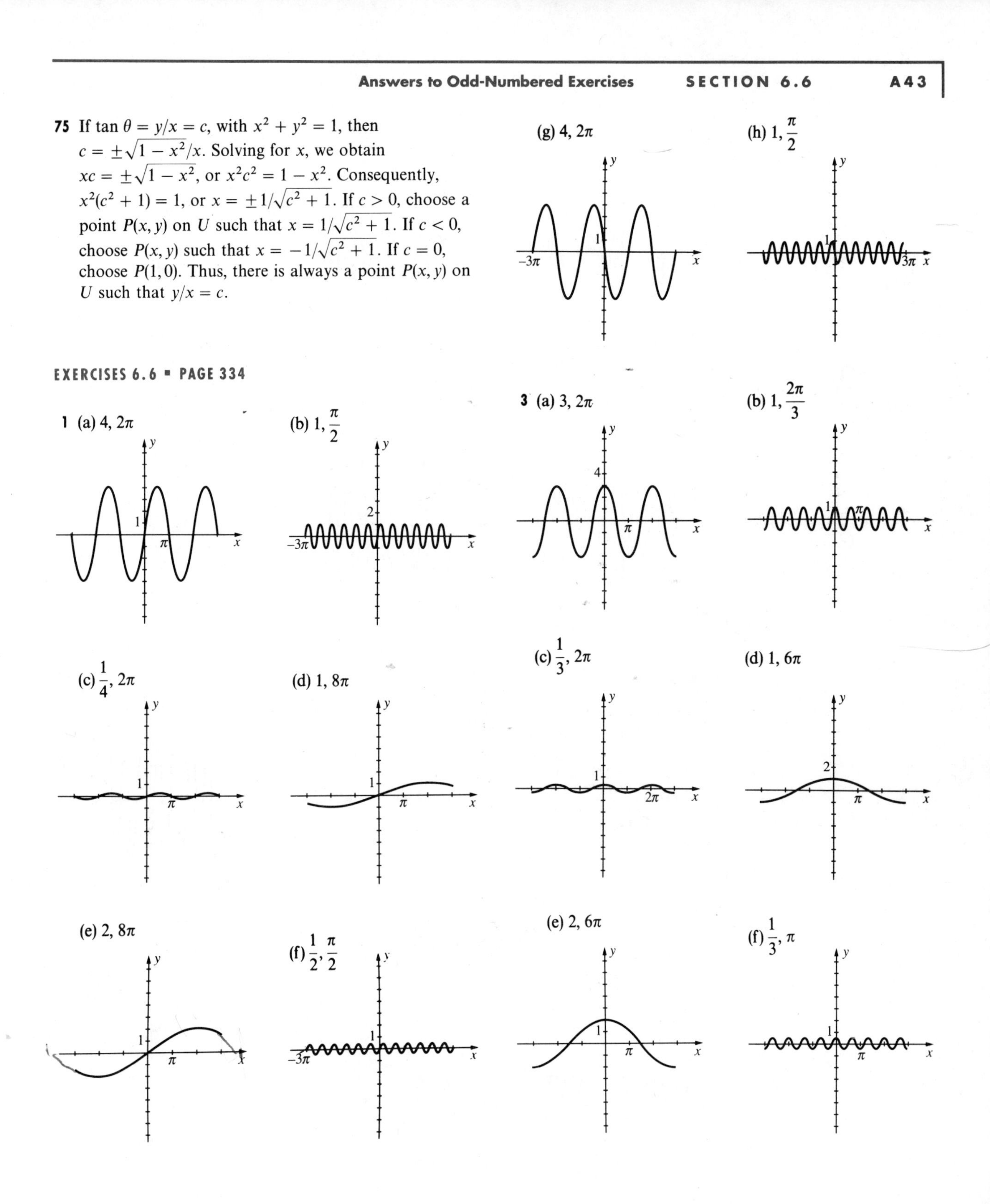

(g) $4, 2\pi$ (h) $1, \frac{\pi}{2}$

EXERCISES 6.6 ▪ PAGE 334

1 (a) $4, 2\pi$ (b) $1, \frac{\pi}{2}$ (c) $\frac{1}{4}, 2\pi$ (d) $1, 8\pi$ (e) $2, 8\pi$ (f) $\frac{1}{2}, \frac{\pi}{2}$

3 (a) $3, 2\pi$ (b) $1, \frac{2\pi}{3}$ (c) $\frac{1}{3}, 2\pi$ (d) $1, 6\pi$ (e) $2, 6\pi$ (f) $\frac{1}{3}, \pi$

(g) $3, 2\pi$

(h) $1, \frac{2\pi}{3}$

17 $2, \frac{2\pi}{3}, \frac{\pi}{3}$

19 $1, 4\pi, \frac{2\pi}{3}$

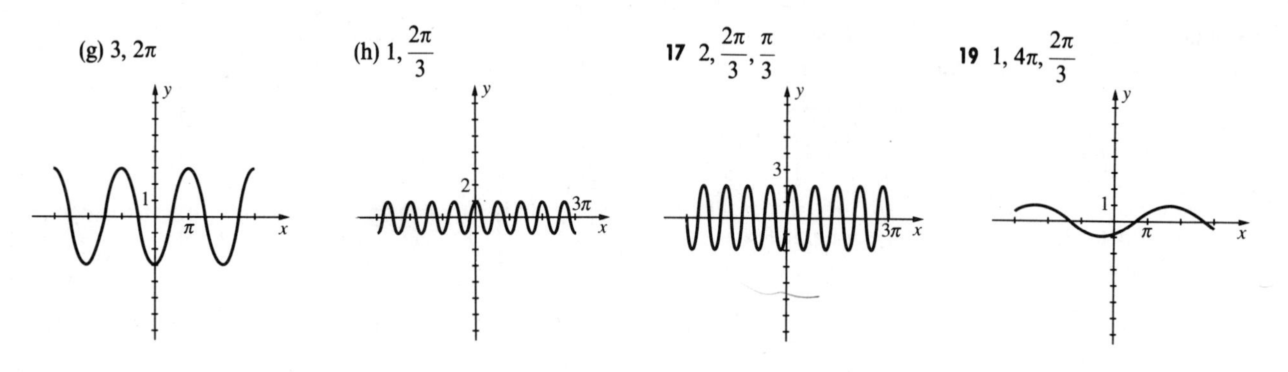

5 $1, 2\pi, \frac{\pi}{2}$

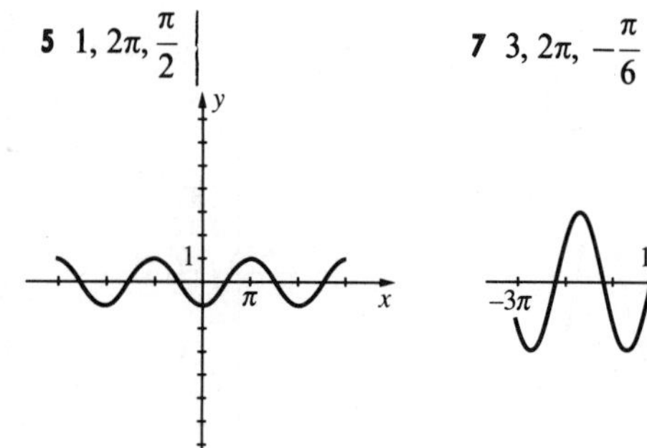

7 $3, 2\pi, -\frac{\pi}{6}$

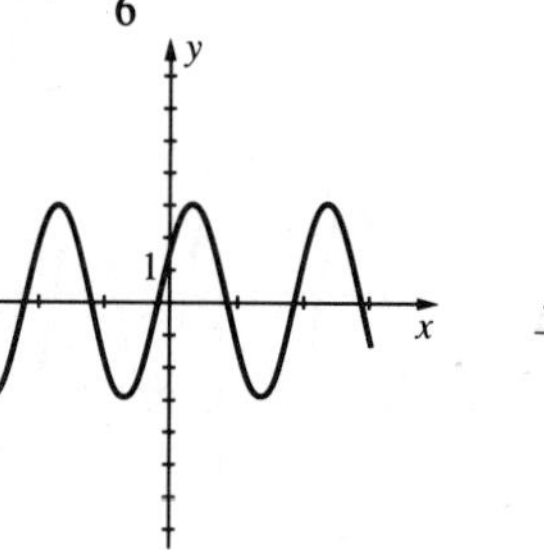

21 $6, 2, 0$

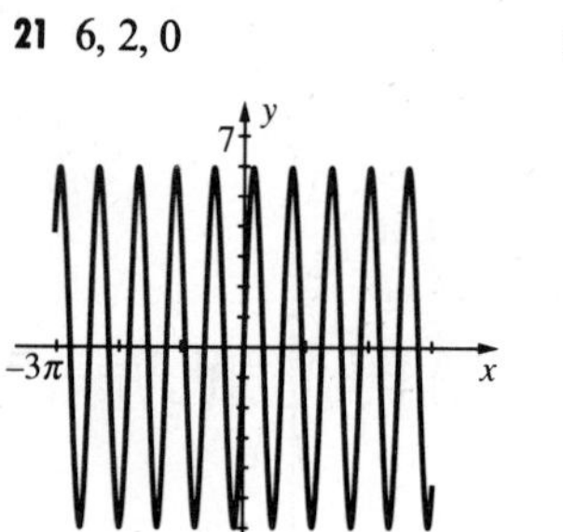

23 $2, 4, 0$

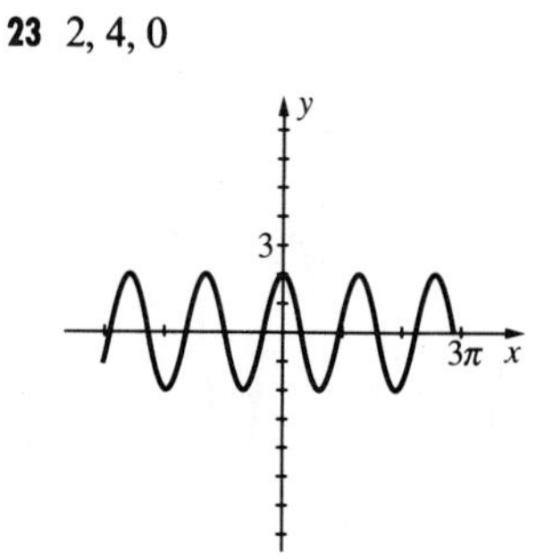

9 $1, 2\pi, -\frac{\pi}{2}$

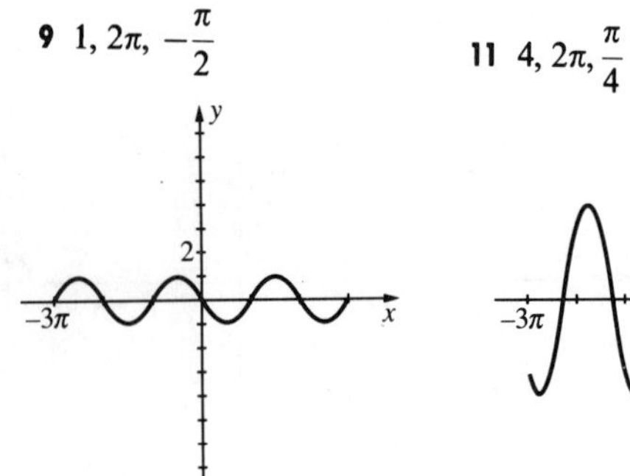

11 $4, 2\pi, \frac{\pi}{4}$

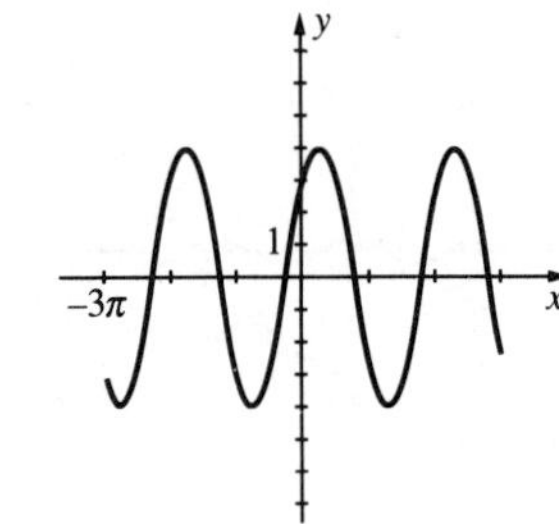

25 $\frac{1}{2}, 1, 0$

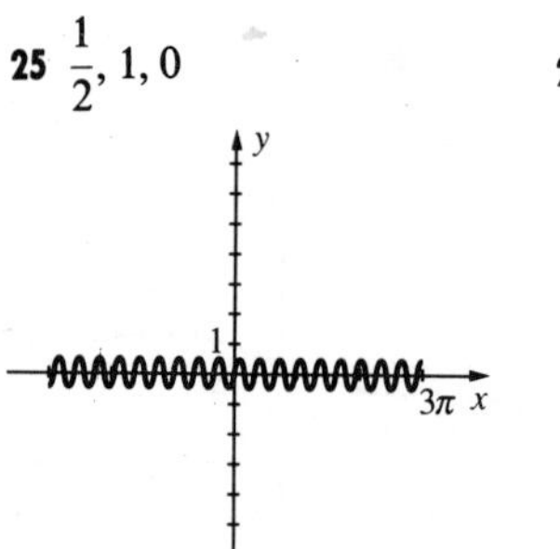

27 $5, \frac{2\pi}{3}, \frac{\pi}{6}$

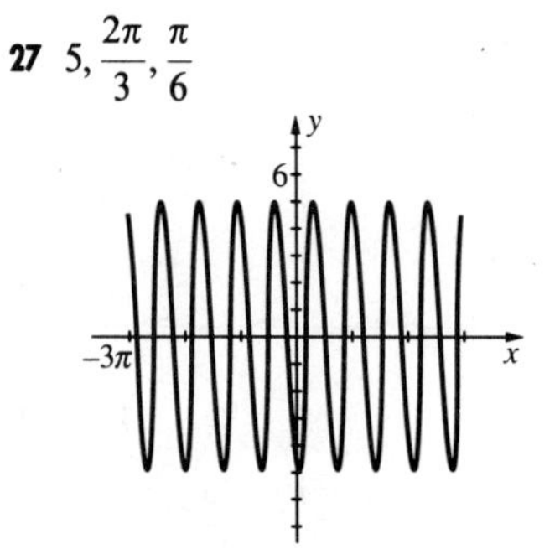

13 $1, \pi, \frac{\pi}{2}$

15 $1, \frac{2\pi}{3}, -\frac{\pi}{3}$

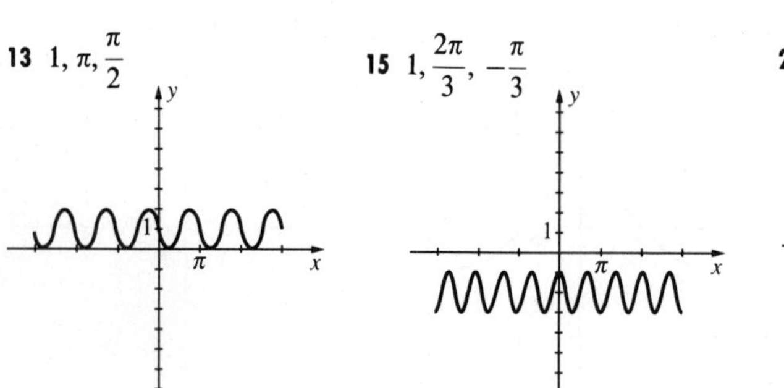

29 $3, 4\pi, \frac{\pi}{2}$

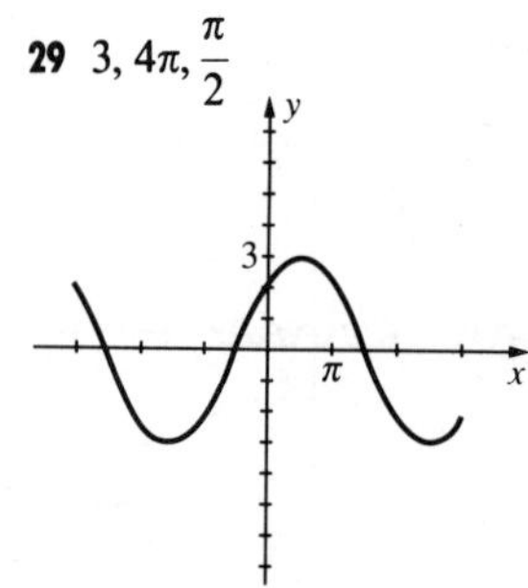

31 $5, 6\pi, -\frac{\pi}{2}$

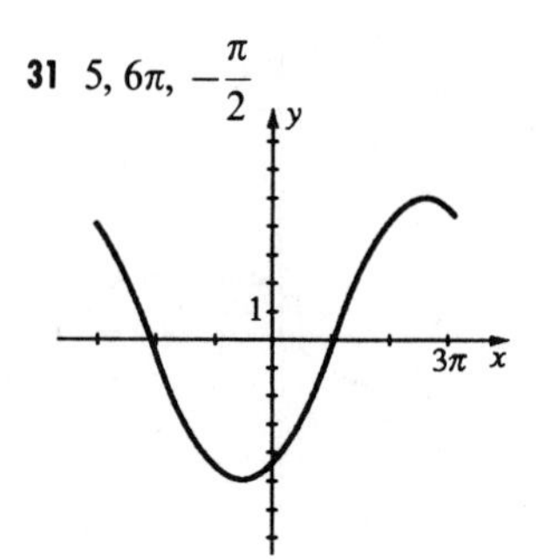

33 $2, \pi, \frac{\pi}{2}$

35 $5, \pi, -\pi$

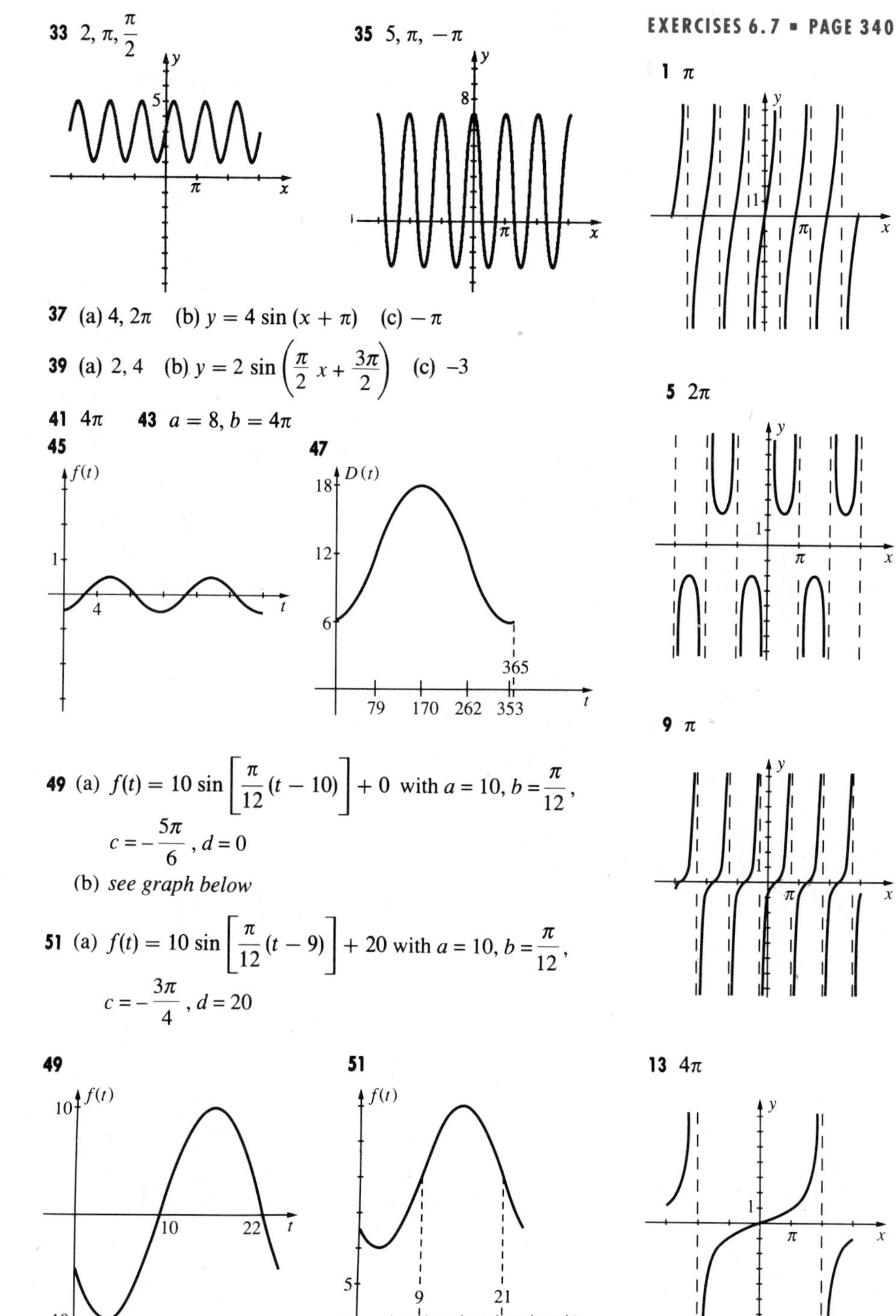

37 (a) $4, 2\pi$ (b) $y = 4 \sin (x + \pi)$ (c) $-\pi$

39 (a) 2, 4 (b) $y = 2 \sin \left(\frac{\pi}{2} x + \frac{3\pi}{2}\right)$ (c) -3

41 4π **43** $a = 8, b = 4\pi$

45

47

49 (a) $f(t) = 10 \sin \left[\frac{\pi}{12}(t - 10)\right] + 0$ with $a = 10, b = \frac{\pi}{12}$, $c = -\frac{5\pi}{6}, d = 0$

(b) *see graph below*

51 (a) $f(t) = 10 \sin \left[\frac{\pi}{12}(t - 9)\right] + 20$ with $a = 10, b = \frac{\pi}{12}$, $c = -\frac{3\pi}{4}, d = 20$

49

51

EXERCISES 6.7 ▪ PAGE 340

1 π

3 π

5 2π

7 2π

9 π

11 $\frac{\pi}{2}$

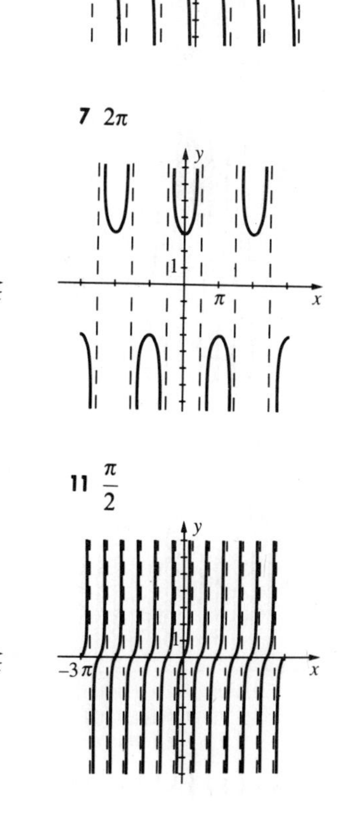

13 4π

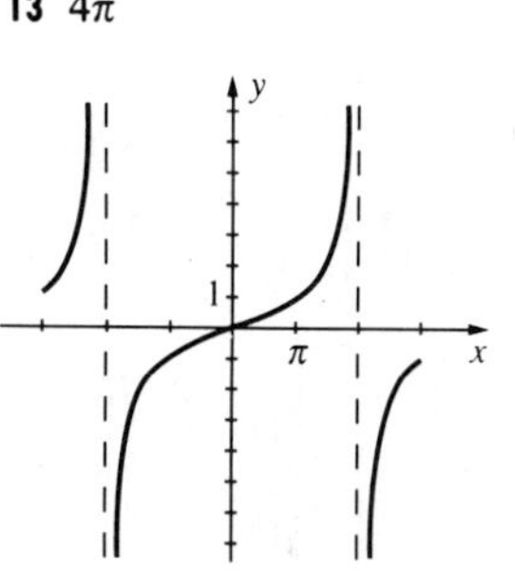

15 $\frac{\pi}{2}$

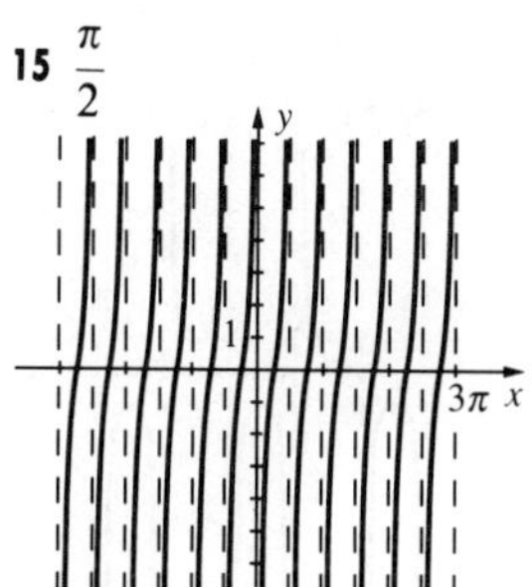

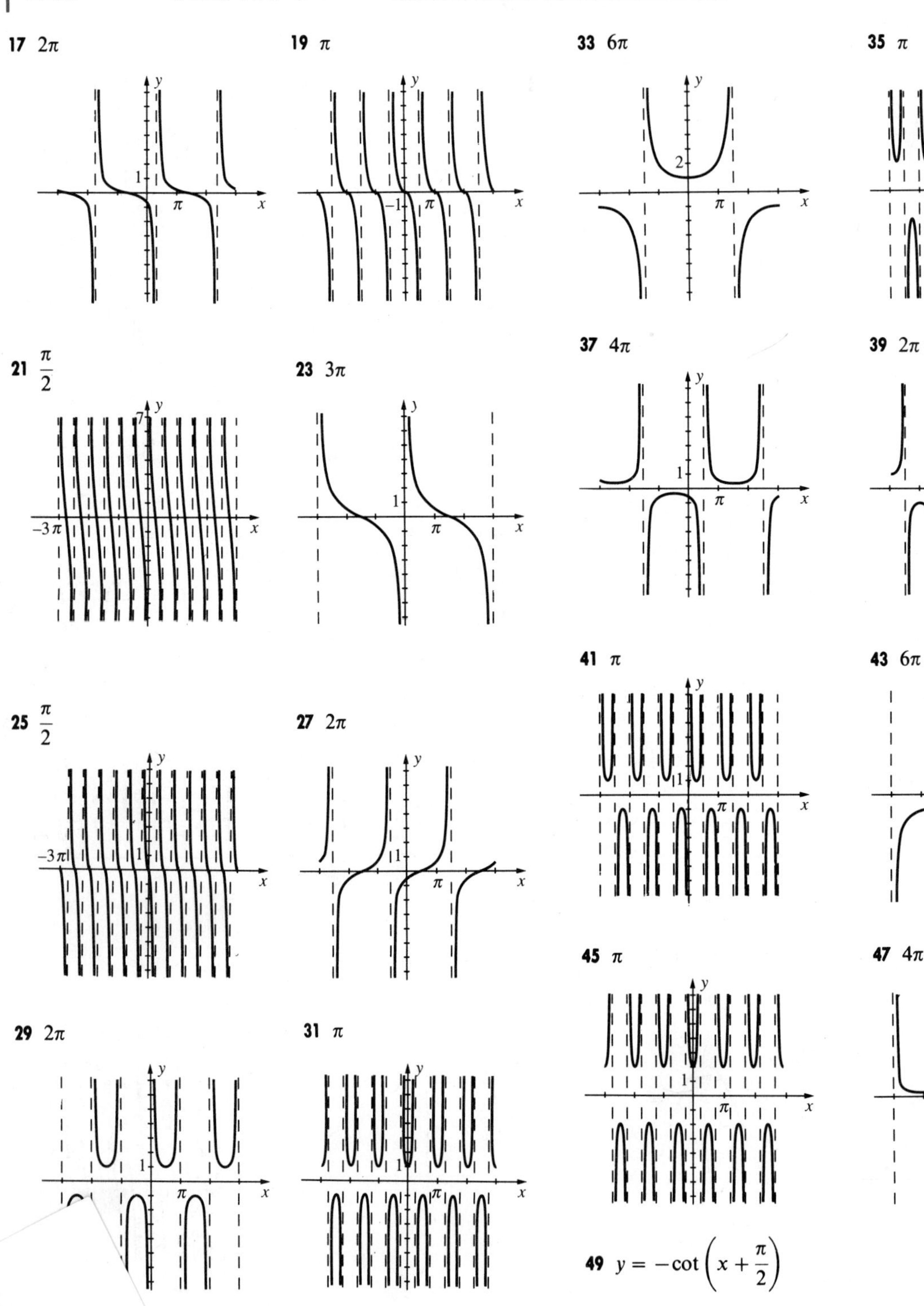

17 2π

19 π

21 $\frac{\pi}{2}$

23 3π

25 $\frac{\pi}{2}$

27 2π

29 2π

31 π

33 6π

35 π

37 4π

39 2π

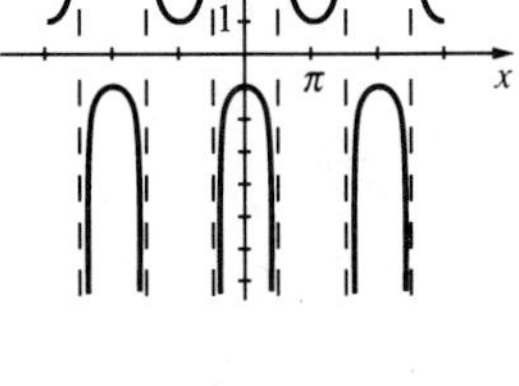

41 π

43 6π

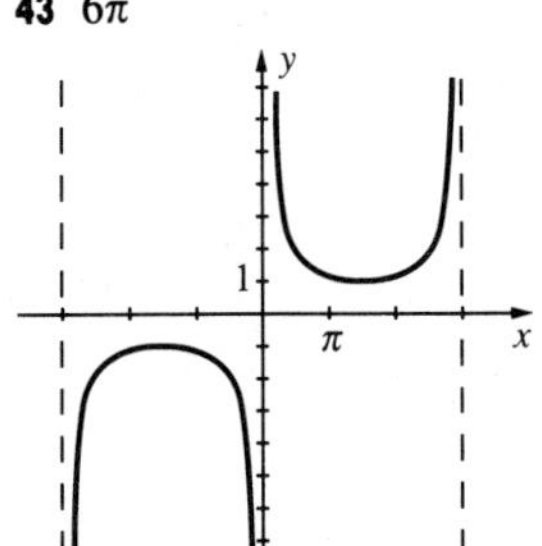

45 π

47 4π

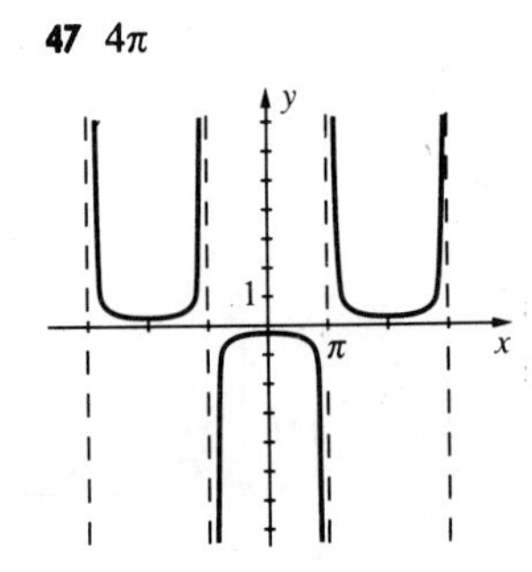

49 $y = -\cot\left(x + \frac{\pi}{2}\right)$

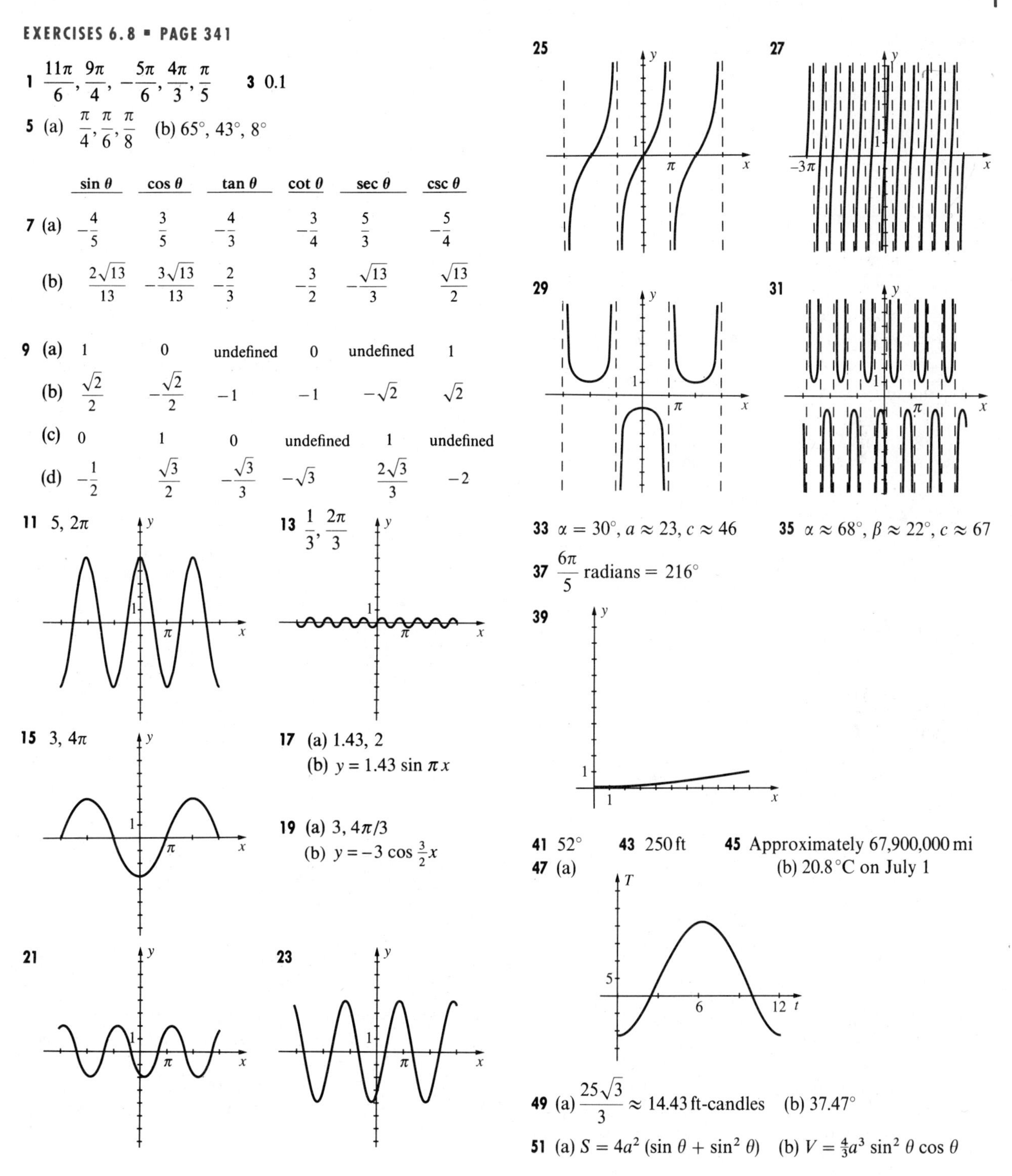

EXERCISES 6.8 ▪ PAGE 341

1 $\frac{11\pi}{6}, \frac{9\pi}{4}, -\frac{5\pi}{6}, \frac{4\pi}{3}, \frac{\pi}{5}$ **3** 0.1

5 (a) $\frac{\pi}{4}, \frac{\pi}{6}, \frac{\pi}{8}$ (b) 65°, 43°, 8°

	$\sin\theta$	$\cos\theta$	$\tan\theta$	$\cot\theta$	$\sec\theta$	$\csc\theta$
7 (a)	$-\frac{4}{5}$	$\frac{3}{5}$	$-\frac{4}{3}$	$-\frac{3}{4}$	$\frac{5}{3}$	$-\frac{5}{4}$
(b)	$\frac{2\sqrt{13}}{13}$	$-\frac{3\sqrt{13}}{13}$	$-\frac{2}{3}$	$-\frac{3}{2}$	$-\frac{\sqrt{13}}{3}$	$\frac{\sqrt{13}}{2}$
9 (a)	1	0	undefined	0	undefined	1
(b)	$\frac{\sqrt{2}}{2}$	$-\frac{\sqrt{2}}{2}$	-1	-1	$-\sqrt{2}$	$\sqrt{2}$
(c)	0	1	0	undefined	1	undefined
(d)	$-\frac{1}{2}$	$\frac{\sqrt{3}}{2}$	$-\frac{\sqrt{3}}{3}$	$-\sqrt{3}$	$\frac{2\sqrt{3}}{3}$	-2

11 5, 2π

13 $\frac{1}{3}, \frac{2\pi}{3}$

15 3, 4π

17 (a) 1.43, 2
(b) $y = 1.43 \sin \pi x$

19 (a) 3, $4\pi/3$
(b) $y = -3 \cos \frac{3}{2}x$

21

23

25

27

29

31

33 $\alpha = 30°$, $a \approx 23$, $c \approx 46$ **35** $\alpha \approx 68°$, $\beta \approx 22°$, $c \approx 67$

37 $\frac{6\pi}{5}$ radians = 216°

39

41 52° **43** 250 ft **45** Approximately 67,900,000 mi

47 (a) (b) 20.8 °C on July 1

49 (a) $\frac{25\sqrt{3}}{3} \approx 14.43$ ft-candles (b) 37.47°

51 (a) $S = 4a^2 (\sin\theta + \sin^2\theta)$ (b) $V = \frac{4}{3}a^3 \sin^2\theta \cos\theta$

CHAPTER ■ 7

EXERCISES 7.1 ■ PAGE 352

Exer. 1–87: Typical verifications are given for Exercises 1, 5, 9, . . . , 85.

1 $\cos\theta\sec\theta = \cos\theta\left(\dfrac{1}{\cos\theta}\right) = 1$

5 $\dfrac{\csc x}{\sec x} = \dfrac{1/\sin x}{1/\cos x} = \dfrac{\cos x}{\sin x} = \cot x$

9 $\cos^2 t - \sin^2 t = \cos^2 t - (1 - \cos^2 t) = 2\cos^2 t - 1$

13 $(1 + \sin\alpha)(1 - \sin\alpha) = 1 - \sin^2\alpha = \cos^2\alpha = 1/\sec^2\alpha$

17 $\dfrac{\csc^2\theta}{1 + \tan^2\theta} = \dfrac{\csc^2\theta}{\sec^2\theta} = \dfrac{1/\sin^2\theta}{1/\cos^2\theta} = \dfrac{\cos^2\theta}{\sin^2\theta} = \cot^2\theta$

21 $\csc\theta - \sin\theta = \dfrac{1}{\sin\theta} - \sin\theta = \dfrac{1 - \sin^2\theta}{\sin\theta} = \dfrac{\cos^2\theta}{\sin\theta}$
$= \dfrac{\cos\theta}{\sin\theta}\cdot\cos\theta = \cot\theta\cos\theta$

25 $(\cos^2 x - 1)(\tan^2 x + 1) = (\cos^2 x - 1)(\sec^2 x)$
$= \cos^2 x\sec^2 x - \sec^2 x$
$= 1 - \sec^2 x$

29 $\sec^2\theta\csc^2\theta = (1 + \tan^2\theta)(1 + \cot^2\theta)$
$= 1 + \tan^2\theta + \cot^2\theta + 1$
$= \sec^2\theta + \csc^2\theta$

33 $\dfrac{1 + \tan^2 v}{\tan^2 v} = \dfrac{1}{\tan^2 v} + 1 = \cot^2 v + 1 = \csc^2 v$

37 $\dfrac{1 + \csc\beta}{\sec\beta} - \cot\beta = \dfrac{1 + \csc\beta - \cot\beta\sec\beta}{\sec\beta}$
$= \dfrac{1 + \csc\beta - \dfrac{\cos\beta}{\sin\beta}\dfrac{1}{\cos\beta}}{\sec\beta}$
$= \dfrac{1 + \csc\beta - \csc\beta}{\sec\beta} = \dfrac{1}{\sec\beta} = \cos\beta$

41 $\dfrac{\cot\alpha - 1}{1 - \tan\alpha} = \dfrac{\dfrac{\cos\alpha}{\sin\alpha} - 1}{1 - \dfrac{\sin\alpha}{\cos\alpha}} = \dfrac{\dfrac{\cos\alpha - \sin\alpha}{\sin\alpha}}{\dfrac{\cos\alpha - \sin\alpha}{\cos\alpha}}$
$= \dfrac{(\cos\alpha - \sin\alpha)\cos\alpha}{(\cos\alpha - \sin\alpha)\sin\alpha} = \dfrac{\cos\alpha}{\sin\alpha} = \cot\alpha$

45 $\dfrac{\cos\beta}{1 - \sin\beta}\cdot\dfrac{1 + \sin\beta}{1 + \sin\beta} = \dfrac{\cos\beta\,(1 + \sin\beta)}{1 - \sin^2\beta}$
$= \dfrac{\cos\beta\,(1 + \sin\beta)}{\cos^2\beta} = \dfrac{1 + \sin\beta}{\cos\beta}$
$= \dfrac{1}{\cos\beta} + \dfrac{\sin\beta}{\cos\beta} = \sec\beta + \tan\beta$

49 $\dfrac{\cot u - 1}{\cot u + 1} = \dfrac{(1/\tan u) - 1}{(1/\tan u) + 1} = \dfrac{(1 - \tan u)/\tan u}{(1 + \tan u)/\tan u}$
$= \dfrac{1 - \tan u}{1 + \tan u}$

53 $\tan^4 k - \sec^4 k = (\tan^2 k - \sec^2 k)(\tan^2 k + \sec^2 k)$
$= (-1)(\sec^2 k - 1 + \sec^2 k)$
$= 1 - 2\sec^2 k$

57 $(\sin^2\theta + \cos^2\theta)^3 = (1)^3 = 1$

61 $\left(\dfrac{\sin^2 x}{\tan^4 x}\right)^3\left(\dfrac{\csc^3 x}{\cot^6 x}\right)^2 = \left(\dfrac{\sin^6 x}{\tan^{12} x}\right)\left(\dfrac{\csc^6 x}{\cot^{12} x}\right)$
$= \dfrac{(\sin x\csc x)^6}{(\tan x\cot x)^{12}} = \dfrac{(1)^6}{(1)^{12}} = 1$

65 $(a\cos t - b\sin t)^2 + (a\sin t + b\cos t)^2$
$= a^2\cos^2 t - 2ab\cos t\sin t + b^2\sin^2 t$
$+ a^2\sin^2 t + 2ab\sin t\cos t + b^2\cos^2 t$
$= a^2(\cos^2 t + \sin^2 t) + b^2(\sin^2 t + \cos^2 t)$
$= a^2 + b^2$

69 $\sqrt{\dfrac{(1 - \cos t)}{(1 + \cos t)}\cdot\dfrac{(1 - \cos t)}{(1 - \cos t)}} = \sqrt{\dfrac{(1 - \cos t)^2}{1 - \cos^2 t}}$
$= \sqrt{\dfrac{(1 - \cos t)^2}{\sin^2 t}}$
$= \dfrac{|1 - \cos t|}{|\sin t|} = \dfrac{1 - \cos t}{|\sin t|}$,
since $(1 - \cos t) \geq 0$.

73 $\dfrac{1}{\tan\beta + \cot\beta} = \dfrac{1}{\dfrac{\sin\beta}{\cos\beta} + \dfrac{\cos\beta}{\sin\beta}}$
$= \dfrac{1}{\dfrac{\sin^2\beta + \cos^2\beta}{\cos\beta\sin\beta}} = \sin\beta\cos\beta$

77 $\sec^4\phi - 4\tan^2\phi = (1 + \tan^2\phi)^2 - 4\tan^2\phi$
$= 1 + 2\tan^2\phi + \tan^4\phi - 4\tan^2\phi$
$= 1 - 2\tan^2\phi + \tan^4\phi$
$= (1 - \tan^2\phi)^2$

81 $\sin(-t)\sec(-t) = (-\sin t)\sec t = (-\sin t)(1/\cos t)$
$= -\tan t$

85 $\ln(\cot x) = \ln(\tan x)^{-1} = -\ln(\tan x)$

Exer. 89–99: A typical value of t and the resulting nonequality are given.

89 $\pi,\ -1 \neq 1$ **91** $\frac{3\pi}{2},\ 1 \neq -1$ **93** $\frac{\pi}{4},\ 2 \neq 1$

95 $\pi,\ -1 \neq 1$ **97** $\frac{\pi}{4},\ \cos\sqrt{2} \neq 1$ **99** $\pi,\ -5 \neq 0$

101 $a^3 \cos^3\theta$ **103** $a \tan\theta \sin\theta$ **105** $a \sec\theta$
107 $(1/a^2)\cos^2\theta$ **109** $a \tan\theta$ **111** $a^4 \sec^3\theta \tan\theta$

EXERCISES 7.2 ▪ PAGE 362

Exer. 1–33: n denotes any integer.

1 $\frac{5\pi}{4} + 2\pi n,\ \frac{7\pi}{4} + 2\pi n$ **3** $\frac{\pi}{3} + \pi n$

5 $\frac{\pi}{3} + 2\pi n,\ \frac{5\pi}{3} + 2\pi n$ **7** No solution, since $\frac{\pi}{2} > 1$.

9 All θ except $\theta = \frac{\pi}{2} + \pi n$ **11** $\frac{\pi}{12} + \pi n,\ \frac{11\pi}{12} + \pi n$

13 $\frac{\pi}{2} + 3\pi n$ **15** $-\frac{\pi}{12} + 2\pi n,\ \frac{19\pi}{12} + 2\pi n$

17 $\frac{\pi}{4} + \pi n,\ \frac{7\pi}{12} + \pi n$ **19** $\frac{2\pi}{3} + 2\pi n,\ \frac{4\pi}{3} + 2\pi n$

21 $\frac{\pi}{4} + \frac{\pi}{2}n$ **23** $2\pi n,\ \frac{3\pi}{2} + 2\pi n$ **25** $\frac{\pi}{3} + \pi n,\ \frac{2\pi}{3} + \pi n$

27 $\frac{4\pi}{3} + 2\pi n,\ \frac{5\pi}{3} + 2\pi n$ **29** $\frac{\pi}{6} + \pi n,\ \frac{5\pi}{6} + \pi n$

31 $\frac{7\pi}{6} + 2\pi n,\ \frac{11\pi}{6} + 2\pi n$ **33** $\frac{\pi}{12} + \pi n,\ \frac{5\pi}{12} + \pi n$

35 $\frac{3\pi}{8}, \frac{7\pi}{8}, \frac{11\pi}{8}, \frac{15\pi}{8}$ **37** $\frac{\pi}{3}, \frac{2\pi}{3}, \frac{4\pi}{3}, \frac{5\pi}{3}$ **39** $\frac{\pi}{6}, \frac{5\pi}{6}, \frac{3\pi}{2}$

41 $0, \pi, \frac{\pi}{4}, \frac{3\pi}{4}, \frac{5\pi}{4}, \frac{7\pi}{4}$ **43** $\frac{\pi}{2}, \frac{3\pi}{2}, \frac{2\pi}{3}, \frac{4\pi}{3}$

45 No solution **47** $\frac{11\pi}{6}, \frac{\pi}{2}$ **49** $0, \frac{\pi}{2}$ **51** $\frac{\pi}{4}, \frac{5\pi}{4}$

53 All α in $[0, 2\pi)$ except $0, \frac{\pi}{2}, \pi$, and $\frac{3\pi}{2}$.

55 $\frac{\pi}{2}, \frac{3\pi}{2}, \frac{7\pi}{6}, \frac{11\pi}{6}$ **57** $\frac{3\pi}{4}, \frac{7\pi}{4}$ **59** 15°30′, 164°30′

61 135°, 315°, 116°30′, 296°30′

63 41°50′, 138°10′, 194°30′, 345°30′ **65** 6.145 min

67 $t \approx 3.50$ and $t \approx 8.50$ **69** (a) 3.29 hr (b) 4 hr

71 (a)

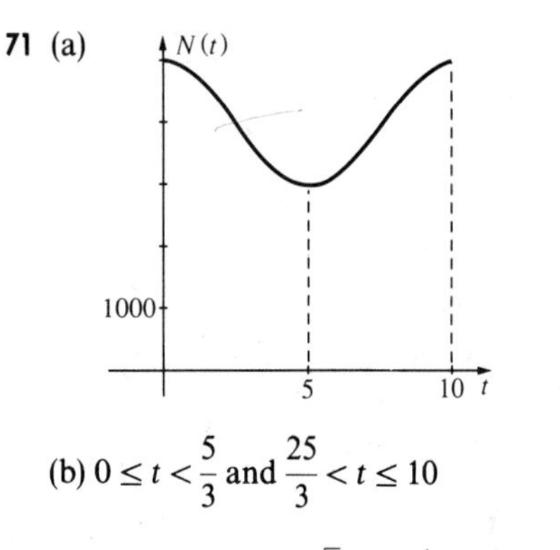

(b) $0 \le t < \frac{5}{3}$ and $\frac{25}{3} < t \le 10$

73 $A\left(-\frac{4\pi}{3}, -\frac{2\pi}{3} + \frac{\sqrt{3}}{2}\right), B\left(-\frac{2\pi}{3}, -\frac{\pi}{3} - \frac{\sqrt{3}}{2}\right),$
$C\left(\frac{2\pi}{3}, \frac{\pi}{3} + \frac{\sqrt{3}}{2}\right), D\left(\frac{4\pi}{3}, \frac{2\pi}{3} - \frac{\sqrt{3}}{2}\right)$

EXERCISES 7.3 ▪ PAGE 372

1 (a) cos 43°23′ (b) sin 16°48′ (c) $\cot\frac{\pi}{3}$ (d) csc 72.72°

3 (a) $\sin\frac{3\pi}{20}$ (b) $\cos\left(\frac{2\pi - 1}{4}\right)$ (c) $\cot\left(\frac{\pi - 2}{2}\right)$

(d) $\sec\left(\frac{\pi}{2} - 0.53\right)$

5 (a) $\frac{\sqrt{2} + \sqrt{3}}{2}$ (b) $\frac{\sqrt{6} - \sqrt{2}}{4}$

7 (a) $\sqrt{3} + 1$ (b) $-2 - \sqrt{3}$

9 (a) $\frac{\sqrt{2} - 1}{2}$ (b) $\frac{\sqrt{6} + \sqrt{2}}{4}$ **11** cos 25°

13 $\sin(-5°)$ **15** $\sin(-5)$ **17** (a) $\frac{36}{85}$ (b) $\frac{77}{85}$ (c) I

19 (a) $-\frac{24}{25}$ (b) $-\frac{24}{7}$ (c) IV

21 (a) $\frac{3\sqrt{21} - 8}{25} \approx 0.23$ (b) $\frac{4\sqrt{21} + 6}{25} \approx 0.97$ (c) I

Exer. 23–37: Typical verifications are given.

23 $\cos(\theta - \pi) = \cos\theta\cos\pi + \sin\theta\sin\pi$
$= \cos\theta(-1) + \sin\theta(0) = -\cos\theta$

25 $\sin\left(x + \frac{\pi}{2}\right) = \sin x \cos\frac{\pi}{2} + \cos x \sin\frac{\pi}{2} = \cos x$

27 $\cos\left(\theta + \frac{3\pi}{2}\right) = \cos\theta\cos\frac{3\pi}{2} - \sin\theta\sin\frac{3\pi}{2} = \sin\theta$

29 $\sin\left(\theta + \frac{\pi}{4}\right) = \sin\theta\cos\frac{\pi}{4} + \cos\theta\sin\frac{\pi}{4}$
$= \frac{\sqrt{2}}{2}\sin\theta + \frac{\sqrt{2}}{2}\cos\theta = \frac{\sqrt{2}}{2}(\sin\theta + \cos\theta)$

31 $\tan\left(u + \frac{\pi}{4}\right) = \dfrac{\tan u + \tan\frac{\pi}{4}}{1 - \tan u \tan\frac{\pi}{4}} = \dfrac{1 + \tan u}{1 - \tan u}$

33 $\tan\left(u + \frac{\pi}{2}\right) = \cot\left[\frac{\pi}{2} - \left(u + \frac{\pi}{2}\right)\right] = \cot(-u) = -\cot u$

35 $\cos(u + v) + \cos(u - v)$
$= (\cos u\cos v - \sin u\sin v) + (\cos u\cos v + \sin u\sin v)$
$= 2\cos u\cos v$

37 $\sin(u + v) \cdot \sin(u - v)$
$= (\sin u\cos v + \cos u\sin v)(\sin u\cos v - \cos u\sin v)$
$= \sin^2 u\cos^2 v - \cos^2 u\sin^2 v$
$= \sin^2 u(1 - \sin^2 v) - (1 - \sin^2 u)\sin^2 v$
$= \sin^2 u - \sin^2 u\sin^2 v - \sin^2 v + \sin^2 u\sin^2 v$
$= \sin^2 u - \sin^2 v$

39 $\sin(u + v + w) = \sin[(u + v) + w]$
$= \sin(u + v)\cos w + \cos(u + v)\sin w$
$= (\sin u\cos v + \cos u\sin v)\cos w$
$+ (\cos u\cos v - \sin u\sin v)\sin w$
$= \sin u\cos v\cos w + \cos u\sin v\cos w$
$+ \cos u\cos v\sin w - \sin u\sin v\sin w$

41 $\cot(u + v) = \dfrac{\cos(u + v)}{\sin(u + v)}$
$= \dfrac{(\cos u\cos v - \sin u\sin v)/(1/\sin u\sin v)}{(\sin u\cos v + \cos u\sin v)(1/\sin u\sin v)}$
$= \dfrac{\cot u\cot v - 1}{\cot v + \cot u}$

43 $\sin(u - v) = \sin[u + (-v)]$
$= \sin u\cos(-v) + \cos u\sin(-v)$
$= \sin u\cos v - \cos u\sin v$

45 $\dfrac{f(x + h) - f(x)}{h} = \dfrac{\cos(x + h) - \cos x}{h}$
$= \dfrac{\cos x\cos h - \sin x\sin h - \cos x}{h}$
$= \dfrac{\cos x\cos h - \cos x}{h} - \dfrac{\sin x\sin h}{h}$
$= \cos x\left(\dfrac{\cos h - 1}{h}\right) - \sin x\left(\dfrac{\sin h}{h}\right)$

47 $0, \frac{\pi}{3}, \frac{2\pi}{3}$ **49** $\frac{\pi}{6}, \frac{\pi}{2}, \frac{5\pi}{6}$ **51** $\frac{\pi}{12}, \frac{5\pi}{12}$; $\frac{3\pi}{4}$ is extraneous

53 (a) $f(x) = 2\cos\left(2x - \frac{\pi}{6}\right)$
(b) $2, \pi, \frac{\pi}{12}$
(c) *see graph at right*

55 (a) $f(x) = 2\sqrt{2}\cos\left(3x + \frac{\pi}{4}\right)$
(b) $2\sqrt{2}, \frac{2\pi}{3}, -\frac{\pi}{12}$
(c) *see graph at right*

57 $y = 10\sqrt{41}\cos(60\pi t - C)$ with $\tan C = \frac{5}{4}$, or $y \approx 10\sqrt{41}\cos(60\pi t - 0.8961)$

59 (a) $y = \sqrt{13}\cos(t - C)$ with $\tan C = \frac{3}{2}$; $\sqrt{13}, 2\pi$
(b) $t = C + (\pi/2) + \pi n \approx 2.55 + \pi n$ for every nonnegative integer n.

61 (a) $p(t) = A\sin\omega t + B\sin(\omega t + \tau)$
$= A\sin\omega t + B(\sin\omega t\cos\tau + \cos\omega t\sin\tau)$
$= (B\sin\tau)\cos\omega t + (A + B\cos\tau)\sin\omega t$
$= a\cos\omega t + b\sin\omega t$
with $a = B\sin\tau$ and $b = A + B\cos\tau$
(b) $C^2 = (B\sin\tau)^2 + (A + B\cos\tau)^2$
$= B^2\sin^2\tau + A^2 + 2AB\cos\tau + B^2\cos^2\tau$
$= A^2 + B^2(\sin^2\tau + \cos^2\tau) + 2AB\cos\tau$
$= A^2 + B^2 + 2AB\cos\tau$

63 (a) $C^2 = A^2 + B^2 + 2AB\cos\tau \le A^2 + B^2 + 2AB$, since $\cos\tau \le 1$ and $A > 0$, $B > 0$. Thus, $C^2 \le (A + B)^2$ and hence $C \le A + B$.
(b) $0, 2\pi$ (c) $\cos\tau > -B/2A$

EXERCISES 7.4 ▪ PAGE 381

1 $\frac{24}{25}, -\frac{7}{25}, -\frac{24}{7}$ **3** $-\frac{4\sqrt{2}}{9}, -\frac{7}{9}, \frac{4\sqrt{2}}{7}$

5 $\frac{\sqrt{10}}{10}, \frac{3\sqrt{10}}{10}, \frac{1}{3}$ **7** $-\frac{\sqrt{2 + \sqrt{2}}}{2}, \frac{\sqrt{2 - \sqrt{2}}}{2}, -\sqrt{2} - 1$

9 (a) $\dfrac{\sqrt{2-\sqrt{2}}}{2}$ (b) $\dfrac{\sqrt{2-\sqrt{3}}}{2}$ (c) $\sqrt{2}+1$

11 $\sin 10\theta = \sin(2 \cdot 5\theta) = 2 \sin 5\theta \cos 5\theta$

13 $4 \sin\dfrac{x}{2} \cos\dfrac{x}{2} = 2 \cdot 2 \sin\dfrac{x}{2} \cos\dfrac{x}{2} = 2 \sin\left(2 \cdot \dfrac{x}{2}\right)$
$= 2 \sin x$

15 $(\sin t + \cos t)^2 = \sin^2 t + 2 \sin t \cos t + \cos^2 t$
$= 1 + \sin 2t$

17 $\sin 3u = \sin(2u + u) = \sin 2u \cos u + \cos 2u \sin u$
$= (2 \sin u \cos u) \cos u + (1 - 2 \sin^2 u) \sin u$
$= 2 \sin u \cos^2 u + \sin u - 2 \sin^3 u$
$= 2 \sin u(1 - \sin^2 u) + \sin u - 2 \sin^3 u$
$= 2 \sin u - 2 \sin^3 u + \sin u - 2 \sin^3 u$
$= 3 \sin u - 4 \sin^3 u$
$= \sin u(3 - 4 \sin^2 u)$

19 $\cos 4\theta = \cos(2 \cdot 2\theta) = 2 \cos^2 2\theta - 1$
$= 2(2 \cos^2 \theta - 1)^2 - 1$
$= 2(4 \cos^4 \theta - 4 \cos \theta + 1) - 1$
$= 8 \cos^4 \theta - 8 \cos^2 \theta + 1$

21 $\sin^4 t = (\sin^2 t)^2 = \left(\dfrac{1 - \cos 2t}{2}\right)^2$
$= \frac{1}{4}(1 - 2 \cos 2t + \cos^2 2t)$
$= \frac{1}{4} - \frac{1}{2} \cos 2t + \dfrac{1}{4}\left(\dfrac{1 + \cos 4t}{2}\right)$
$= \frac{3}{8} - \frac{1}{2} \cos 2t + \frac{1}{8} \cos 4t$

23 $\sec 2\theta = \dfrac{1}{\cos 2\theta} = \dfrac{1}{2 \cos^2 \theta - 1} = \dfrac{1}{2(1/\sec^2 \theta) - 1}$
$= \dfrac{1}{(2 - \sec^2 \theta)/\sec^2 \theta} = \dfrac{\sec^2 \theta}{2 - \sec^2 \theta}$

25 $2 \sin^2 2t + \cos 4t = 2\left(\dfrac{1 - \cos 4t}{2}\right) + \cos 4t$
$= 1 - \cos 4t + \cos 4t = 1$

27 $\tan 3u = \tan(2u + u) = \dfrac{\tan 2u + \tan u}{1 - \tan 2u \tan u}$
$= \dfrac{\dfrac{2 \tan u}{1 - \tan^2 u} + \tan u}{1 - \dfrac{2 \tan u}{1 - \tan^2 u} \cdot \tan u}$
$= \dfrac{\dfrac{2 \tan u + \tan u - \tan^3 u}{1 - \tan^2 u}}{\dfrac{1 - \tan^2 u - 2 \tan^2 u}{1 - \tan^2 u}}$
$= \dfrac{3 \tan u - \tan^3 u}{1 - 3 \tan^2 u} = \dfrac{(3 - \tan^2 u) \tan u}{1 - 3 \tan^2 u}$

29 $\frac{3}{8} + \frac{1}{2} \cos \theta + \frac{1}{8} \cos 2\theta$ **31** $\frac{3}{8} + \frac{1}{2} \cos 4x + \frac{1}{8} \cos 8x$

33 $0, \pi, \dfrac{2\pi}{3}, \dfrac{4\pi}{3}$ **35** $\dfrac{\pi}{3}, \dfrac{5\pi}{3}, \pi$ **37** $0, \pi$ **39** $0, \dfrac{\pi}{3}, \dfrac{5\pi}{3}$

43 (a) 1.20, 5.09
(b) $P\left(\dfrac{2\pi}{3}, -1.5\right), Q(\pi, -1), R\left(\dfrac{4\pi}{3}, -1.5\right)$

45 (a) $-\dfrac{3\pi}{2}, -\dfrac{\pi}{2}, \dfrac{\pi}{2}, \dfrac{3\pi}{2}$
(b) $0, \pm\pi, \pm 2\pi, \pm\dfrac{\pi}{4}, \pm\dfrac{3\pi}{4}, \pm\dfrac{5\pi}{4}, \pm\dfrac{7\pi}{4}$

47 (b) Yes, point B is 25 miles from A.
49 24.30°, 65.70° **51** (b) 12.43 mm

EXERCISES 7.5 ▪ PAGE 389

1 $\sin 12\theta + \sin 6\theta$ **3** $\frac{1}{2} \cos 4t - \frac{1}{2} \cos 10t$
5 $\frac{1}{2} \cos 10u + \frac{1}{2} \cos 2u$ **7** $\frac{3}{2} \sin 3x + \frac{3}{2} \sin x$
9 $2 \sin 4\theta \cos 2\theta$ **11** $-2 \sin 4x \sin x$
13 $-2 \cos 5t \sin 2t$ **15** $2 \cos \frac{3}{2}x \cos \frac{1}{2}x$

17 $\dfrac{\sin 4t + \sin 6t}{\cos 4t - \cos 6t} = \dfrac{2 \sin 5t \cos t}{2 \sin 5t \sin t} = \cot t$

19 $\dfrac{\sin u + \sin v}{\cos u + \cos v} = \dfrac{2 \sin \frac{1}{2}(u + v) \cos \frac{1}{2}(u - v)}{2 \cos \frac{1}{2}(u + v) \cos \frac{1}{2}(u - v)} = \tan \frac{1}{2}(u + v)$

21 $\dfrac{\sin u - \sin v}{\sin u + \sin v} = \dfrac{2 \cos \frac{1}{2}(u + v) \sin \frac{1}{2}(u - v)}{2 \sin \frac{1}{2}(u + v) \cos \frac{1}{2}(u - v)}$
$= \cot \frac{1}{2}(u + v) \tan \frac{1}{2}(u - v)$
$= \dfrac{\tan \frac{1}{2}(u - v)}{\tan \frac{1}{2}(u + v)}$

23 $4 \cos x \cos 2x \sin 3x = 2 \cos 2x(2 \cos x \sin 3x)$
$= 2 \cos 2x(\sin 4x - \sin 2x)$
$= 2 \cos 2x \sin 4x - 2 \cos 2x \sin 2x$
$= \sin 6x + \sin 2x + \sin 4x - \sin 0$
$= \sin 2x + \sin 4x + \sin 6x$

25 $\frac{1}{2} \sin(a + b)x + \frac{1}{2} \sin(a - b)x$ **27** $\dfrac{\pi}{4} n$ **29** $\dfrac{\pi}{2} n$

31 $\dfrac{\pi}{2} + \pi n, \dfrac{\pi}{12} + \dfrac{\pi}{2} n, \dfrac{5\pi}{12} + \dfrac{\pi}{2} n$ **33** $\dfrac{\pi}{7} + \dfrac{2\pi}{7} n, \dfrac{2\pi}{3} n$

35 $\sin(\theta + \pi) = \sin \theta \cos \pi + \cos \theta \sin \pi = -\sin \theta$

37 $\cos\left(\theta - \dfrac{5\pi}{2}\right) = \cos \theta \cos \dfrac{5\pi}{2} + \sin \theta \sin \dfrac{5\pi}{2} = \sin \theta$

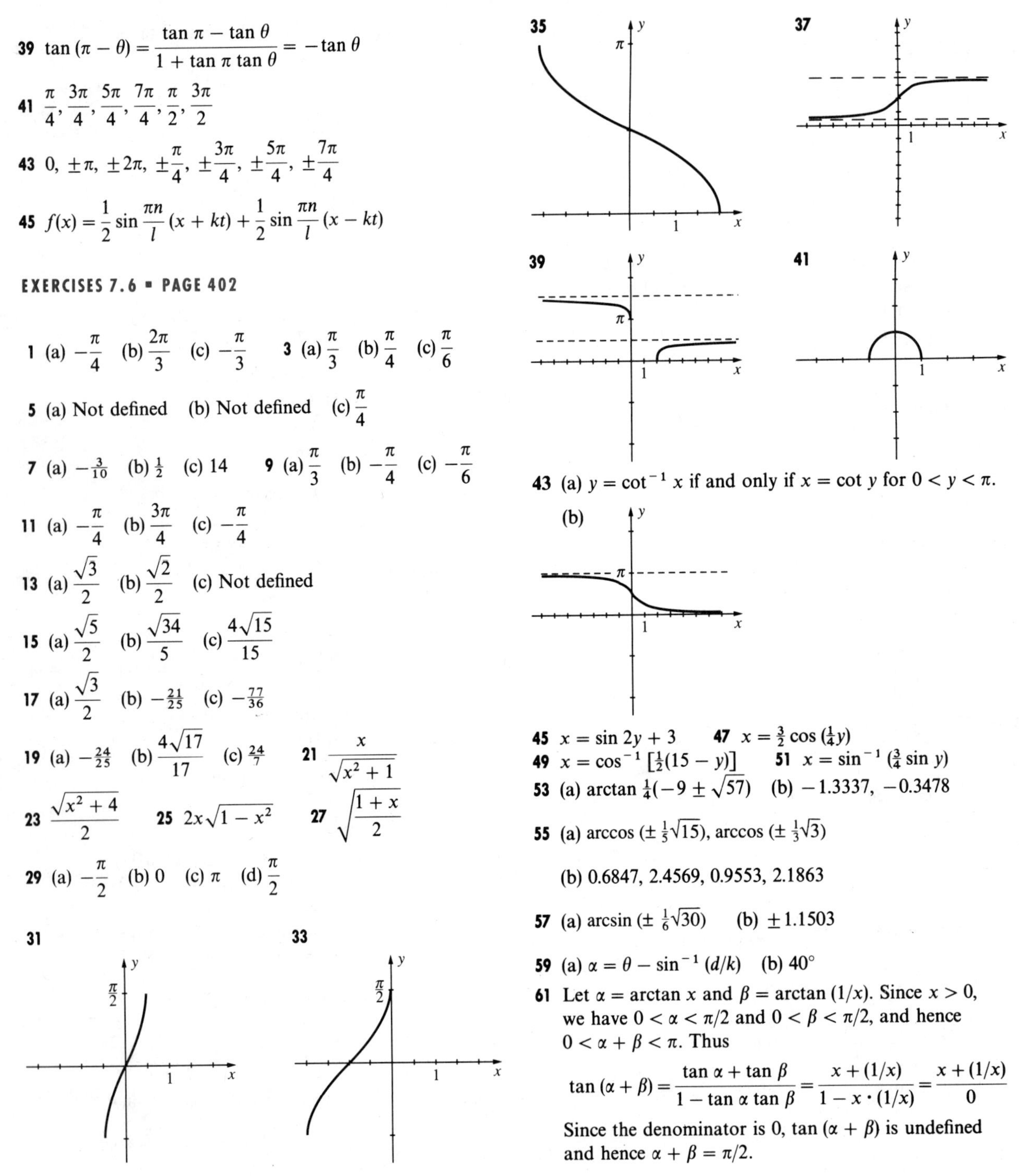

39 $\tan(\pi - \theta) = \dfrac{\tan \pi - \tan \theta}{1 + \tan \pi \tan \theta} = -\tan \theta$

41 $\dfrac{\pi}{4}, \dfrac{3\pi}{4}, \dfrac{5\pi}{4}, \dfrac{7\pi}{4}, \dfrac{\pi}{2}, \dfrac{3\pi}{2}$

43 $0, \pm\pi, \pm 2\pi, \pm\dfrac{\pi}{4}, \pm\dfrac{3\pi}{4}, \pm\dfrac{5\pi}{4}, \pm\dfrac{7\pi}{4}$

45 $f(x) = \dfrac{1}{2} \sin \dfrac{\pi n}{l}(x + kt) + \dfrac{1}{2} \sin \dfrac{\pi n}{l}(x - kt)$

EXERCISES 7.6 ▪ PAGE 402

1 (a) $-\dfrac{\pi}{4}$ (b) $\dfrac{2\pi}{3}$ (c) $-\dfrac{\pi}{3}$ **3** (a) $\dfrac{\pi}{3}$ (b) $\dfrac{\pi}{4}$ (c) $\dfrac{\pi}{6}$

5 (a) Not defined (b) Not defined (c) $\dfrac{\pi}{4}$

7 (a) $-\frac{3}{10}$ (b) $\frac{1}{2}$ (c) 14 **9** (a) $\dfrac{\pi}{3}$ (b) $-\dfrac{\pi}{4}$ (c) $-\dfrac{\pi}{6}$

11 (a) $-\dfrac{\pi}{4}$ (b) $\dfrac{3\pi}{4}$ (c) $-\dfrac{\pi}{4}$

13 (a) $\dfrac{\sqrt{3}}{2}$ (b) $\dfrac{\sqrt{2}}{2}$ (c) Not defined

15 (a) $\dfrac{\sqrt{5}}{2}$ (b) $\dfrac{\sqrt{34}}{5}$ (c) $\dfrac{4\sqrt{15}}{15}$

17 (a) $\dfrac{\sqrt{3}}{2}$ (b) $-\frac{21}{25}$ (c) $-\frac{77}{36}$

19 (a) $-\frac{24}{25}$ (b) $\dfrac{4\sqrt{17}}{17}$ (c) $\frac{24}{7}$ **21** $\dfrac{x}{\sqrt{x^2+1}}$

23 $\dfrac{\sqrt{x^2+4}}{2}$ **25** $2x\sqrt{1-x^2}$ **27** $\sqrt{\dfrac{1+x}{2}}$

29 (a) $-\dfrac{\pi}{2}$ (b) 0 (c) π (d) $\dfrac{\pi}{2}$

31

33

35

37

39

41

43 (a) $y = \cot^{-1} x$ if and only if $x = \cot y$ for $0 < y < \pi$.

(b)

45 $x = \sin 2y + 3$ **47** $x = \frac{3}{2}\cos(\frac{1}{4}y)$

49 $x = \cos^{-1}[\frac{1}{2}(15 - y)]$ **51** $x = \sin^{-1}(\frac{3}{4}\sin y)$

53 (a) $\arctan \frac{1}{4}(-9 \pm \sqrt{57})$ (b) $-1.3337, -0.3478$

55 (a) $\arccos(\pm\frac{1}{5}\sqrt{15}), \arccos(\pm\frac{1}{3}\sqrt{3})$

(b) 0.6847, 2.4569, 0.9553, 2.1863

57 (a) $\arcsin(\pm\frac{1}{6}\sqrt{30})$ (b) ± 1.1503

59 (a) $\alpha = \theta - \sin^{-1}(d/k)$ (b) 40°

61 Let $\alpha = \arctan x$ and $\beta = \arctan(1/x)$. Since $x > 0$, we have $0 < \alpha < \pi/2$ and $0 < \beta < \pi/2$, and hence $0 < \alpha + \beta < \pi$. Thus

$$\tan(\alpha + \beta) = \frac{\tan \alpha + \tan \beta}{1 - \tan \alpha \tan \beta} = \frac{x + (1/x)}{1 - x \cdot (1/x)} = \frac{x + (1/x)}{0}$$

Since the denominator is 0, $\tan(\alpha + \beta)$ is undefined and hence $\alpha + \beta = \pi/2$.

63 Let $\alpha = \arcsin(-x)$ and $\beta = \arcsin x$ with $-\pi/2 \le \alpha \le \pi/2$ and $-\pi/2 \le \beta \le \pi/2$. Thus $\sin \alpha = -x$ and $\sin \beta = x$. Consequently, $\sin \alpha = -\sin \beta = \sin(-\beta)$. Since the sine function is one-to-one on $[-\pi/2, \pi/2]$, we have $\alpha = -\beta$.

65 Let $\alpha = \sin^{-1} x$ and $\beta = \tan^{-1}(x/\sqrt{1-x^2})$ with $-\pi/2 < \alpha < \pi/2$ and $-\pi/2 < \beta < \pi/2$. Thus $\sin \alpha = x$ and $\sin \beta = x$. Since the sine function is one-to-one on $[-\pi/2, \pi/2]$, we have $\alpha = \beta$.

EXERCISES 7.7 ▪ PAGE 405

1 $(\cot^2 x + 1)(1 - \cos^2 x) = (\csc^2 x)(\sin^2 x) = 1$

3 $$\frac{(\sec^2\theta - 1)\cot\theta}{\tan\theta\sin\theta + \cos\theta} = \frac{(\tan^2\theta)\cot\theta}{(\sin\theta/\cos\theta)\cdot\sin\theta + \cos\theta} = \frac{\tan\theta}{(\sin^2\theta + \cos^2\theta)/\cos\theta} = \frac{\sin\theta/\cos\theta}{1/\cos\theta} = \sin\theta$$

5 $$\frac{1}{1+\sin t}\cdot\frac{1-\sin t}{1-\sin t} = \frac{1-\sin t}{1-\sin^2 t} = \frac{1-\sin t}{\cos^2 t} = \frac{1-\sin t}{\cos t}\cdot\frac{1}{\cos t} = \left(\frac{1}{\cos t} - \frac{\sin t}{\cos t}\right)\cdot\sec t = (\sec t - \tan t)\sec t$$

7 $$\tan 2u = \frac{2\tan u}{1-\tan^2 u} = \frac{2(1/\cot u)}{1-(1/\cot^2 u)} = \frac{2/\cot u}{(\cot^2 u - 1)/\cot^2 u} = \frac{2\cot u}{\cot^2 u - 1} = \frac{2\cot u}{(\csc^2 u - 1) - 1} = \frac{2\cot u}{\csc^2 u - 2}$$

9 $$\frac{\tan^3\phi - \cot^3\phi}{\tan^2\phi + \csc^2\phi} = \frac{(\tan\phi - \cot\phi)[(\tan^2\phi + \tan\phi\cot\phi + \cot^2\phi)]}{[\tan^2\phi + (1+\cot^2\phi)]} = \tan\phi - \cot\phi$$

11 $$\cos\left(x - \frac{5\pi}{2}\right) = \cos x\cos\frac{5\pi}{2} + \sin x\sin\frac{5\pi}{2} = \sin x$$

13 $$\tfrac{1}{4}\sin 4\beta = \tfrac{1}{4}\sin(2\cdot 2\beta) = \tfrac{1}{4}(2\sin 2\beta\cos 2\beta) = \tfrac{1}{2}(2\sin\beta\cos\beta)(\cos^2\beta - \sin^2\beta) = \sin\beta\cos^3\beta - \sin^3\beta\cos\beta$$

15 $$\sin 8\theta = 2\sin 4\theta\cos 4\theta = 2(2\sin 2\theta\cos 2\theta)(1 - 2\sin^2 2\theta) = 8\sin\theta\cos\theta(1-2\sin^2\theta)[1 - 2(2\sin\theta\cos\theta)^2] = 8\sin\theta\cos\theta(1-2\sin^2\theta)(1 - 8\sin^2\theta\cos^2\theta)$$

17 $\frac{\pi}{2}, \frac{3\pi}{2}, \frac{\pi}{4}, \frac{7\pi}{4}, \frac{3\pi}{4}, \frac{5\pi}{4}$ **19** $0, \pi$ **21** $0, \pi, \frac{2\pi}{3}, \frac{4\pi}{3}$

23 $\frac{7\pi}{6}, \frac{11\pi}{6}, \frac{\pi}{2}$ **25** $\frac{\pi}{6}, \frac{5\pi}{6}, \frac{\pi}{3}, \frac{5\pi}{3}$ **27** $\frac{\pi}{3}, \frac{5\pi}{3}$

29 $\frac{\sqrt{6}-\sqrt{2}}{4}$ **31** $\frac{\sqrt{2}-\sqrt{6}}{4}$ **33** $\frac{84}{85}$ **35** $-\frac{36}{77}$

37 $\frac{240}{289}$ **39** $\frac{24}{7}$ **41** $\frac{1}{3}$

43 (a) $$\sin 7t\sin 4t = \tfrac{1}{2}(2\sin 7t\sin 4t) = \tfrac{1}{2}[\cos(7t-4t) - \cos(7t+4t)] = \tfrac{1}{2}\cos 3t - \tfrac{1}{2}\cos 11t$$

(b) $$\cos\frac{u}{4}\cos\left(-\frac{u}{6}\right) = \tfrac{1}{2}\left(2\cos\frac{u}{4}\cos\frac{u}{6}\right) = \tfrac{1}{2}\left[\cos\left(\frac{u}{4}+\frac{u}{6}\right) + \cos\left(\frac{u}{4}-\frac{u}{6}\right)\right] = \tfrac{1}{2}\cos\frac{5u}{12} + \tfrac{1}{2}\cos\frac{u}{12}$$

(c) $$6\cos 5x\sin 3x = 3(2\cos 5x\sin 3x) = 3[\sin(5x+3x) - \sin(5x-3x)] = 3\sin 8x - 3\sin 2x$$

45 $\frac{5\pi}{6}$ **47** π **49** $\frac{1}{2}$ **51** $-\frac{7}{25}$ **53** $\frac{\pi}{2}$

55

57 (b) $0, \pm\frac{\pi}{4b}$ (c) $\frac{2\sqrt{2}}{3}A$

59 (a) $x = 2d\tan\frac{\theta}{2}$ (b) $d \le 1000$ ft

CHAPTER ▪ 8

EXERCISES 8.1 ▪ PAGE 417

1 $\beta = 62°$, $b \approx 14.1$, $c \approx 15.6$
3 $\gamma = 100°10'$, $b \approx 55.1$, $c \approx 68.7$
5 $\beta = 76°30'$, $a \approx 13.6$, $c \approx 17.8$ **7** No triangle exists
9 $\alpha \approx 77°30'$, $\beta \approx 49°10'$, $b \approx 108$;
$\alpha \approx 102°30'$, $\beta \approx 24°10$, $b \approx 59$
11 $\alpha \approx 82.54°$, $\beta \approx 49.72°$, $b \approx 100.85°$;
$\alpha \approx 97.46°$, $\beta \approx 34.80°$, $b \approx 75.45$
13 $\beta \approx 53°40'$, $\gamma \approx 61°10'$, $c \approx 20.6$
15 $\alpha \approx 25.993°$, $\gamma \approx 32.383°$, $a \approx 0.146$ **17** 219 yd
19 (a) 1.6 mi (b) 0.6 mi **21** 2.7 mi
23 628 m **25** 3.7 mi from A and 5.4 mi from B
27 350 ft **29** (a) 18.7 (b) 814 **31** (3949.8, 2994.3)

EXERCISES 8.2 ▪ PAGE 426

1 $a \approx 26$, $\beta \approx 41°$, $\gamma \approx 79°$
3 $b \approx 177$, $\alpha \approx 25°10'$, $\gamma \approx 4°50'$
5 $c \approx 2.75$, $\alpha \approx 21°10'$, $\beta \approx 43°40'$
7 $\alpha \approx 29°$, $\beta \approx 47°$, $\gamma \approx 104°$
9 $\alpha \approx 12°30'$, $\beta \approx 136°30'$, $\gamma \approx 31°$ **11** 196 ft
13 24 mi **15** 39 mi **17** 2.3 mi **19** N55°31′E
21 63.7 ft from first and third base; 66.8 ft from second base
23 40,630 ft or 7.7 mi
25 (a) 72° (b) 181.6 ft^2 (c) 37.6 ft
27 (a) 36°, 108°, 72° (b) 0.62 (c) 0.59, 0.36

Exer. 29–35: Answer is in square units.
29 260 **31** 1125 **33** 2.9 **35** 517 **37** 1.62 acres

EXERCISES 8.3 ▪ PAGE 433

1 5 **3** $\sqrt{85}$ **5** 8 **7** 1 **9** 0

Exer. 11–19: Point P is the point corresponding to the geometric representation.
11 $P(4, 2)$ **13** $P(3, -5)$ **15** $P(-3, 6)$ **17** $P(-6, 4)$
19 $P(0, 2)$ **21** $\sqrt{2}\,\text{cis}\,\frac{7\pi}{4}$ **23** $8\,\text{cis}\,\frac{5\pi}{6}$ **25** $20\,\text{cis}\,\frac{3\pi}{2}$
27 $10\,\text{cis}\,\frac{4\pi}{3}$ **29** $7\,\text{cis}\,\pi$ **31** $4\sqrt{2}\,\text{cis}\,\frac{5\pi}{4}$ **33** $6\,\text{cis}\,\frac{\pi}{2}$
35 $2\,\text{cis}\,\frac{11\pi}{6}$ **37** $12\,\text{cis}\,0$ **39** $\sqrt{5}\,\text{cis}\,(\tan^{-1}\frac{1}{2})$
41 $\sqrt{10}\,\text{cis}\,[\tan^{-1}(-\frac{1}{3}) + \pi]$ **43** $\sqrt{34}\,\text{cis}\,(\tan^{-1}\frac{3}{5} + \pi)$
45 $5\,\text{cis}\,[\tan^{-1}(-\frac{3}{4}) + 2\pi]$ **47** $2\sqrt{2} + 2\sqrt{2}\,i$
49 $-3 + 3\sqrt{3}\,i$ **51** -5 **53** $5 + 3i$ **55** $2 - i$
57 $-2; i$ **59** $10\sqrt{3} - 10i;\ -\frac{2\sqrt{3}}{5} + \frac{2}{5}i$ **61** $40; \frac{5}{2}$
63 $8 - 4i; \frac{8}{5} + \frac{4}{5}i$
65 (a) $z_1z_2z_3 = (r_1r_2r_3)\,\text{cis}\,(\theta_1 + \theta_2 + \theta_3)$
(b) $z_1z_2\cdots z_n = (r_1r_2\cdots r_n)\,\text{cis}\,(\theta_1 + \theta_2 + \cdots + \theta_n)$

EXERCISES 8.4 ▪ PAGE 438

1 $-972 - 972i$ **3** $-32i$ **5** -8
7 $-\frac{\sqrt{2}}{2} - \frac{\sqrt{2}}{2}i$ **9** $-\frac{1}{2} - \frac{\sqrt{3}}{2}i$
11 $-64\sqrt{3} - 64i$ **13** $\pm\left(\frac{\sqrt{6}}{2} + \frac{\sqrt{2}}{2}i\right)$
15 $\pm\left(\frac{\sqrt[4]{2}}{2} + \frac{\sqrt[4]{18}}{2}i\right), \pm\left(\frac{\sqrt[4]{18}}{2} - \frac{\sqrt[4]{2}}{2}i\right)$
17 $3i, \pm\frac{3\sqrt{3}}{2} - \frac{3}{2}i$
19 $\pm 1, \frac{1}{2} \pm \frac{\sqrt{3}}{2}i, -\frac{1}{2} \pm \frac{\sqrt{3}}{2}i$ *(see graph below)*
21 $\sqrt[10]{2}\,\text{cis}\,\theta$ with $\theta = 9°, 81°, 153°, 225°, 297°$

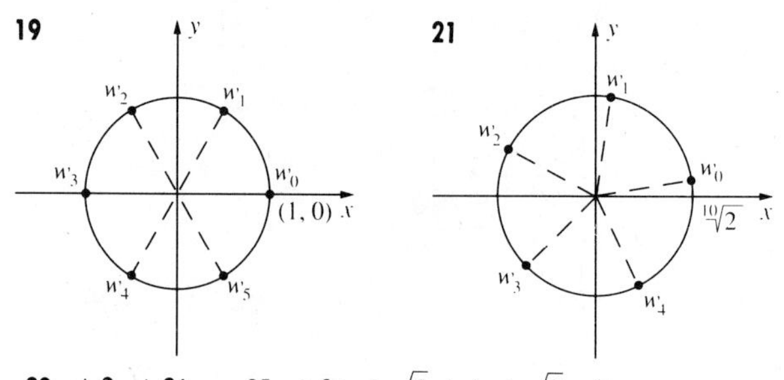

23 $\pm 2, \pm 2i$ **25** $\pm 2i, \pm\sqrt{3} + i, \pm\sqrt{3} - i$
27 $2i, \pm\sqrt{3} - i$
29 $3\,\text{cis}\,\theta$ with $\theta = 0°, 72°, 144°, 216°, 288°$

EXERCISES 8.5 ▪ PAGE 446

1 $\langle 3, 1\rangle, \langle 1, -7\rangle, \langle 13, 8\rangle, \langle 3, -32\rangle$
3 $\langle -15, 6\rangle, \langle 1, -2\rangle, \langle -68, 28\rangle, \langle 12, -12\rangle$
5 $4\mathbf{i} - 3\mathbf{j}, -2\mathbf{i} + 7\mathbf{j}, 19\mathbf{i} - 17\mathbf{j}, -11\mathbf{i} + 33\mathbf{j}$
7 $-2\mathbf{i} - 5\mathbf{j}, -6\mathbf{i} + 7\mathbf{j}, -6\mathbf{i} - 26\mathbf{j}, -26\mathbf{i} + 34\mathbf{j}$
9 $-3\mathbf{i} + 2\mathbf{j}, 3\mathbf{i} + 2\mathbf{j}, -15\mathbf{i} + 8\mathbf{j}, 15\mathbf{i} + 8\mathbf{j}$
11 Terminal points are (3, 2), (−1, 5), (2, 7), (6, 4), (3, −15)
13 Terminal points are (−4, 6), (−2, 3), (−6, 9), (−8, 12), (6, −9).
15 $-\mathbf{b}$ **17** $\mathbf{f}$ **19** $-\frac{1}{2}\mathbf{e}$

21 $\mathbf{a} + (\mathbf{b} + \mathbf{c}) = \langle a_1, a_2\rangle + (\langle b_1, b_2\rangle + \langle c_1, c_2\rangle)$
$= \langle a_1, a_2\rangle + \langle b_1 + c_1, b_2 + c_2\rangle$
$= \langle a_1 + b_1 + c_1, a_2 + b_2 + c_2\rangle$
$= \langle a_1 + b_1, a_2 + b_2\rangle + \langle c_1, c_2\rangle$
$= (\langle a_1, a_2\rangle + \langle b_1, b_2\rangle) + \langle c_1, c_2\rangle$
$= (\mathbf{a} + \mathbf{b}) + \mathbf{c}$

23 $\mathbf{a} + (-\mathbf{a}) = \langle a_1, a_2\rangle + (-\langle a_1, a_2\rangle)$
$= \langle a_1, a_2\rangle + \langle -a_1, -a_2\rangle$
$= \langle a_1 - a_1, a_2 - a_2\rangle$
$= \langle 0, 0\rangle = \mathbf{0}$

25 $(mn)\mathbf{a} = (mn)\langle a_1, a_2\rangle = \langle (mn)a_1, (mn)a_2\rangle$
$= \langle mna_1, mna_2\rangle$
$= m\langle na_1, na_2\rangle$ or $n\langle ma_1, ma_2\rangle$
$= m(n\langle a_1, a_2\rangle)$ or $n(m\langle a_1, a_2\rangle)$
$= m(n\mathbf{a})$ or $n(m\mathbf{a})$

27 $0\mathbf{a} = 0\langle a_1, a_2\rangle = \langle 0a_1, 0a_2\rangle = \langle 0, 0\rangle = \mathbf{0}$
Also, $c\mathbf{0} = c\langle 0, 0\rangle = \langle c0, c0\rangle = \langle 0, 0\rangle = \mathbf{0}$

29 $-(\mathbf{a} + \mathbf{b}) = -(\langle a_1, a_2\rangle + \langle b_1, b_2\rangle)$
$= -(\langle a_1 + b_1, a_2 + b_2\rangle)$
$= \langle -(a_1 + b_1), -(a_2 + b_2)\rangle$
$= \langle -a_1 - b_1, -a_2 - b_2\rangle$
$= \langle -a_1, -a_2\rangle + \langle -b_1, -b_2\rangle$
$= -\mathbf{a} + (-\mathbf{b}) = -\mathbf{a} - \mathbf{b}$

31 (a) $\|2\mathbf{v}\| = \|2\langle a, b\rangle\| = \|\langle 2a, 2b\rangle\| = \sqrt{(2a)^2 + (2b)^2}$
$= \sqrt{4a^2 + 4b^2} = 2\sqrt{a^2 + b^2} = 2\|\langle a, b\rangle\|$
$= 2\|\mathbf{v}\|$
(b) and (c) Proofs are similar to (a).
(d) $\|k\mathbf{v}\| = \|k\langle a, b\rangle\| = \|\langle ka, kb\rangle\| = \sqrt{(ka)^2 + (kb)^2}$
$= \sqrt{k^2a^2 + k^2b^2} = \sqrt{k^2}\sqrt{a^2 + b^2}$
$= |k|\|\langle a, b\rangle\| = |k|\|\mathbf{v}\|$

33 $3\sqrt{2}, \dfrac{7\pi}{4}$ **35** $5, \pi$ **37** $\sqrt{41}, \arctan(-\frac{5}{4}) + \pi$

39 $18, \dfrac{3\pi}{2}$ **41** 102 lb **43** 10.1 kg **45** 89 kg, S66°W

47 5.8 lb, 129° **49** 40.96, 28.68 **51** −61.80, 190.21

53 (a) $\mathbf{F} = \langle 7, 2\rangle$ (b) $\mathbf{G} = -\mathbf{F} = \langle -7, -2\rangle$

55 (a) $\mathbf{F} \approx \langle -5.86, 1.13\rangle$ (b) $\mathbf{G} = -\mathbf{F} \approx \langle 5.86, -1.13\rangle$

57 23.6° **59** 59°, 232 mi/hr **61** 420 mi/hr, 244°

63 N22°W

EXERCISES 8.6 ▪ PAGE 456

1 (a) 24 (b) $\arccos\left(\dfrac{24}{\sqrt{29}\sqrt{45}}\right) \approx 48°22'$

3 (a) −14 (b) $\arccos\left(\dfrac{-14}{\sqrt{17}\sqrt{13}}\right) \approx 160°21'$

5 (a) 45 (b) $\arccos\left(\dfrac{45}{\sqrt{81}\sqrt{41}}\right) \approx 38°40'$

7 $\langle 4, -1\rangle \cdot \langle 2, 8\rangle = 0$ **9** $(-4\mathbf{j}) \cdot (-7\mathbf{i}) = 0$ **11** $\frac{6}{5}$

13 $\pm\frac{3}{8}$ **15** (a) −23 (b) −23 **17** −51

19 $\dfrac{17\sqrt{26}}{26} \approx 3.34$ **21** $\frac{11}{5}$ **23** 7 **25** 28 **27** 12

29 $\mathbf{a} \cdot \mathbf{a} = \langle a_1, a_2\rangle \cdot \langle a_1, a_2\rangle = a_1^2 + a_2^2 = (\sqrt{a_1^2 + a_2^2})^2 = \|\mathbf{a}\|^2$

31 $(c\mathbf{a}) \cdot \mathbf{b} = (c\langle a_1, a_2\rangle) \cdot \langle b_1, b_2\rangle = \langle ca_1, ca_2\rangle \cdot \langle b_1, b_2\rangle$
$= ca_1b_1 + ca_2b_2 = c(a_1b_1 + a_2b_2) = c(\mathbf{a} \cdot \mathbf{b})$

33 $\mathbf{0} \cdot \mathbf{a} = \langle 0, 0\rangle \cdot \langle a_1, a_2\rangle = 0(a_1) + 0(a_2) = 0 + 0 = 0.$

35 $1000\sqrt{3} \approx 1732$ ft-lb

EXERCISES 8.7 ▪ PAGE 457

1 $\gamma = 75°, a = 50\sqrt{6}, c = 50(1 + \sqrt{3})$

3 $\beta = \arccos\left(\dfrac{4\sqrt{43}}{43}\right), \gamma = \arccos\left(\dfrac{5\sqrt{43}}{86}\right), a = \sqrt{43}$

5 $\alpha = 38°, a \approx 8, c \approx 13$ **7** $\alpha \approx 24°, \gamma \approx 41°, b \approx 10.1$

9 289.8 **11** $10\sqrt{2}\,\text{cis}\,\dfrac{3\pi}{4}$ **13** $17\,\text{cis}\,\pi$

15 $10\,\text{cis}\,\dfrac{7\pi}{6}$ **17** $-512i$ **19** $-972 + 972i$

21 $-3, \dfrac{3}{2} \pm \dfrac{3\sqrt{3}}{2}i$

23 Terminal points are $(-2, -3), (-6, 13), (-8, 10), (-1, 4)$.

25 (a) 80 (b) $\arccos\left(\dfrac{40}{\sqrt{40}\sqrt{50}}\right) \approx 26°34'$ (c) $2\sqrt{10}$

27 Circle with center (a_1, a_2) and radius c

29 $\langle 14\cos 40°, -14\sin 40°\rangle$ **31** (b) 2 mi

33 (a) 449 ft (b) 434 ft **35** 204 **37** (c) 158°

CHAPTER ▪ 9

EXERCISES 9.1 ▪ PAGE 467

1 $(3, 5), (-1, -3)$ **3** $(1, 0), (-3, 2)$ **5** $(0, 0), (\frac{1}{8}, \frac{1}{128})$

7 $(3, -2)$ **9** No solution **11** $(-4, 3), (5, 0)$

13 $(-2, 2)$

15 $(-\frac{3}{5} + \frac{1}{10}\sqrt{86}, \frac{1}{5} + \frac{3}{10}\sqrt{86}), (-\frac{3}{5} - \frac{1}{10}\sqrt{86}, \frac{1}{5} - \frac{3}{10}\sqrt{86})$

17 $(-4, 0), (\frac{12}{5}, \frac{16}{5})$ **19** $(0, 1), (4, -3)$

21 $(\pm 2, 5), (\pm\sqrt{5}, 4)$ **23** $(\sqrt{2}, \pm 2\sqrt{3}), (-\sqrt{2}, \pm 2\sqrt{3})$

25 $(2\sqrt{2}, \pm 2), (-2\sqrt{2}, \pm 2)$ **27** $(3, -1, 2)$
29 $(1, -1, 2), (-1, 3, -2)$ **31** 12 inches × 8 inches
33 Yes; a solution occurs between 0 and 1.

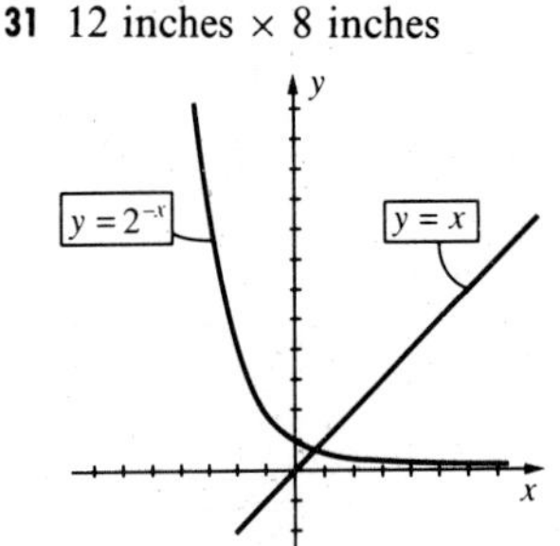

1 $(4, -2)$ **3** $(8, 0)$ **5** $(-1, \frac{3}{2})$ **7** $(\frac{76}{53}, \frac{28}{53})$
9 $(\frac{51}{13}, \frac{96}{13})$ **11** $(\frac{8}{7}, -\frac{3}{7}\sqrt{6})$ **13** No solution
15 All ordered pairs (m, n) such that $3m - 4n = 2$
17 $(0, 0)$ **19** $(-\frac{22}{7}, -\frac{11}{5})$
21 313 students, 137 nonstudents
23 $x = (30/\pi) - 4 \approx 5.55$ cm, $y = 12 - (30/\pi) \approx 2.45$ cm
25 $l = 10$ ft, $w = 20/\pi$ ft **27** 2400 adults, 3600 kittens
29 40 g of 35% alloy, 60 g of 60% alloy
31 540 mi/hr, 60 mi/hr **33** $v_0 = 10$, $a = 3$
35 20 sofas, 30 recliners
37 (a) $(c, \frac{4}{5}c)$ for an arbitrary $c > 0$ (b) \$16 per hour

EXERCISES 9.2 ▪ PAGE 475

35 Yes; $r = 2$ inches, $h = 50/\pi$ inches
37 (a) $a = 120{,}000$, $b = 40{,}000$ (b) 77,143
39 $(0, 0), (0, 100), (50, 0)$; the fourth solution $(-100, 150)$ is not meaningful.
41 Yes; 1 ft × 1 ft × 2 ft or
$$\frac{\sqrt{13}-1}{2}\text{ ft} \times \frac{\sqrt{13}-1}{2}\text{ ft} \times \frac{8}{(\sqrt{13}-1)^2}\text{ ft} \approx$$
1.30 ft × 1.30 ft × 1.18 ft
43 The points are on the parabola (a) $y = \frac{1}{2}x^2 - \frac{1}{2}$ and (b) $y = \frac{1}{4}x^2 - 1$.

EXERCISES 9.3 ▪ PAGE 490

1 $(2, 3, -1)$ **3** $(-2, 4, 5)$ **5** No solution **7** $(\frac{2}{3}, \frac{31}{21}, \frac{1}{21})$

Exer. 9–15: There are other forms for the answers; c is any real number.

9 $(2c, -c, c)$ **11** $(0, -c, c)$ **13** $(\frac{12}{7} - \frac{9}{7}c, \frac{4}{7}c - \frac{13}{14}, c)$
15 $(\frac{7}{10}c + \frac{1}{2}, \frac{19}{10}c - \frac{3}{2}, c)$ **17** $(1, 3, -1, 2)$
19 $(2, -1, 3, 4, 1)$ **21** $(\frac{1}{11}, \frac{31}{11}, \frac{3}{11})$ **23** $(-2, -3)$
25 No solution **27** 17 of 10%, 11 of 30%, 22 of 50%
29 4 hr for A, 2 hr for B, 5 hr for C
31 380 lb of G_1, 60 lb of G_2, 160 lb of G_3
33 (a) $I_1 = 0, I_2 = 2, I_3 = 2$ (b) $I_1 = \frac{3}{4}, I_2 = 3, I_3 = \frac{9}{4}$
35 $\frac{3}{8}$ lb Columbian, $\frac{1}{8}$ lb Brazilian, $\frac{1}{2}$ lb Kenyan
37 (a) A: $x_1 + x_4 = 75$, B: $x_1 + x_2 = 150$, C: $x_2 + x_3 = 225$, D: $x_3 + x_4 = 150$
(b) $x_1 = 25, x_2 = 125, x_4 = 50$
(c) $x_3 = 150 - x_4 \leq 150$;
$x_3 = 225 - x_2 = 225 - (150 - x_1) = 75 + x_1 \geq 75$
39 $x^2 + y^2 - x + 3y - 6 = 0$

EXERCISES 9.4 ▪ PAGE 498

1 $\dfrac{3}{x-2} + \dfrac{5}{x+3}$ **3** $\dfrac{5}{x-6} - \dfrac{4}{x+2}$

5 $\dfrac{2}{x-1} + \dfrac{3}{x+2} - \dfrac{1}{x-3}$ **7** $\dfrac{3}{x} + \dfrac{2}{x-5} - \dfrac{1}{x+1}$

9 $\dfrac{2}{x-1} + \dfrac{5}{(x-1)^2}$ **11** $-\dfrac{7}{x} + \dfrac{5}{x^2} + \dfrac{40}{3x-5}$

13 $\dfrac{\frac{24}{25}}{x+2} + \dfrac{\frac{2}{5}}{(x+2)^2} - \dfrac{\frac{23}{25}}{2x-1}$ **15** $\dfrac{5}{x} - \dfrac{2}{x+1} + \dfrac{3}{(x+1)^3}$

17 $-\dfrac{2}{x-1} + \dfrac{3x+4}{x^2+1}$ **19** $\dfrac{4}{x} + \dfrac{5x-3}{x^2+2}$

21 $\dfrac{4x-1}{x^2+1} + \dfrac{3}{(x^2+1)^2}$ **23** $2x + \dfrac{1}{x-1} + \dfrac{3x}{x^2+1}$

25 $2x + 3 + \dfrac{2}{x-1} - \dfrac{3}{2x+1}$

EXERCISES 9.5 ▪ PAGE 507

1 $\begin{bmatrix} 9 & -1 \\ -2 & 5 \end{bmatrix}, \begin{bmatrix} 1 & -3 \\ 4 & 1 \end{bmatrix}, \begin{bmatrix} 10 & -4 \\ 2 & 6 \end{bmatrix}, \begin{bmatrix} -12 & -3 \\ 9 & -6 \end{bmatrix}$

3 $\begin{bmatrix} 9 & 0 \\ 1 & 5 \\ 3 & 4 \end{bmatrix}, \begin{bmatrix} 3 & -2 \\ 3 & -5 \\ -9 & 4 \end{bmatrix}, \begin{bmatrix} 12 & -2 \\ 4 & 0 \\ -6 & 8 \end{bmatrix}, \begin{bmatrix} -9 & -3 \\ 3 & -15 \\ -18 & 0 \end{bmatrix}$

5 $[11 \quad -3 \quad -3], [-3 \quad -3 \quad 7], [8 \quad -6 \quad 4], [-21 \quad 0 \quad 15]$

7 Not possible, not possible,
$\begin{bmatrix} 6 & -4 & 4 \\ 0 & 2 & -8 \\ -6 & 4 & -2 \end{bmatrix}, \begin{bmatrix} -12 & 0 \\ -6 & 3 \\ 3 & -9 \end{bmatrix}$

9 $\begin{bmatrix} 16 & 38 \\ 11 & -34 \end{bmatrix}, \begin{bmatrix} 4 & 38 \\ 23 & -22 \end{bmatrix}$

11 $\begin{bmatrix} 3 & -14 & -3 \\ 16 & 2 & -2 \\ -7 & -29 & 9 \end{bmatrix}, \begin{bmatrix} 3 & -20 & -11 \\ 2 & 10 & -4 \\ 15 & -13 & 1 \end{bmatrix}$

13 $\begin{bmatrix} 4 & 8 \\ -18 & 11 \end{bmatrix}$, $\begin{bmatrix} 3 & -4 & 4 \\ -5 & 2 & 2 \\ -51 & 26 & 10 \end{bmatrix}$

15 $\begin{bmatrix} 1 & 2 & 3 \\ 4 & 5 & 6 \\ 7 & 8 & 9 \end{bmatrix}$, $\begin{bmatrix} 1 & 2 & 3 \\ 4 & 5 & 6 \\ 7 & 8 & 9 \end{bmatrix}$

17 $[15]$, $\begin{bmatrix} -3 & 7 & 2 \\ -12 & 28 & 8 \\ 15 & -35 & -10 \end{bmatrix}$

19 $\begin{bmatrix} 2 & 0 & 5 \\ 5 & 3 & -2 \end{bmatrix}$, not possible **21** $\begin{bmatrix} 4 \\ 12 \\ -1 \end{bmatrix}$

23 $\begin{bmatrix} 18 & 0 & 2 \\ -40 & 10 & -10 \end{bmatrix}$ **33** $\frac{1}{10}\begin{bmatrix} 3 & 4 \\ -1 & 2 \end{bmatrix}$

35 Does not exist **37** $\frac{1}{8}\begin{bmatrix} 2 & 1 & 0 \\ -2 & 3 & 0 \\ 0 & 0 & 2 \end{bmatrix}$

39 $\frac{1}{3}\begin{bmatrix} -4 & -5 & 3 \\ -4 & -8 & 3 \\ 1 & 2 & 0 \end{bmatrix}$ **41** $\begin{bmatrix} \frac{1}{2} & 0 & 0 \\ 0 & \frac{1}{4} & 0 \\ 0 & 0 & \frac{1}{6} \end{bmatrix}$

43 $ab \neq 0$; $\begin{bmatrix} 1/a & 0 \\ 0 & 1/b \end{bmatrix}$ **47** (a) $(\frac{13}{10}, -\frac{1}{10})$ (b) $(\frac{7}{5}, \frac{6}{5})$

49 (a) $(-\frac{25}{3}, -\frac{34}{3}, \frac{7}{3})$ (b) $(\frac{16}{3}, \frac{16}{3}, -\frac{1}{3})$

EXERCISES 9.6 ■ PAGE 514

1 $M_{11} = -14 = A_{11}$; $M_{12} = 10$; $A_{12} = -10$; $M_{13} = 15 = A_{13}$; $M_{21} = 7$; $A_{21} = -7$; $M_{22} = -5 = A_{22}$; $M_{23} = 34$; $A_{23} = -34$; $M_{31} = 11 = A_{31}$; $M_{32} = 4$; $A_{32} = -4$; $M_{33} = 6 = A_{33}$
3 $M_{11} = 0 = A_{11}$; $M_{12} = 5$; $A_{12} = -5$; $M_{21} = -1$; $A_{21} = 1$; $M_{22} = 7 = A_{22}$
5 -83 **7** 5 **9** 2 **11** 0 **13** -125 **15** 48
17 -216 **19** $abcd$ **31** (a) $x^2 - 3x - 4$ (b) $-1, 4$
33 (a) $x^2 + x - 2$ (b) $-2, 1$
35 (a) $-x^3 - 2x^2 + x + 2$ (b) $-2, -1, 1$
37 (a) $-x^3 + 4x^2 + 4x - 16$ (b) $-2, 2, 4$
39 $-31i - 20j + 7k$ **41** $-6i - 8j + 18k$

EXERCISES 9.7 ■ PAGE 519

1 $R_2 \leftrightarrow R_3$ **3** $-R_1 + R_3 \to R_3$
5 2 is a common factor of R_1 and R_3
7 R_1 and R_3 are identical **9** -1 is a factor of R_2
11 Every number in C_2 is 0 **13** $2C_1 + C_3 \to C_3$
15 -10 **17** -142 **19** -183 **21** 44 **23** 359

EXERCISES 9.8 ■ PAGE 523

1 $(4, -2)$ **3** $(8, 0)$ **5** $(-1, \frac{3}{2})$ **7** $(\frac{76}{53}, \frac{28}{53})$
9 $(\frac{51}{13}, \frac{96}{13})$ **11** $(\frac{8}{7}, -\frac{3}{7}\sqrt{6})$
13, 15 $|D| = 0$ so Cramer's rule cannot be used.
17 $(0, 0)$ **19** $(2, 3, -1)$ **21** $(-2, 4, 5)$
23 $|D| = 0$ so Cramer's rule cannot be used.
25 $(\frac{2}{3}, \frac{31}{21}, \frac{1}{21})$ **27** $(1, 3, -1, 2)$ **29** $(\frac{1}{11}, \frac{31}{11}, \frac{3}{11})$

EXERCISES 9.9 ■ PAGE 529

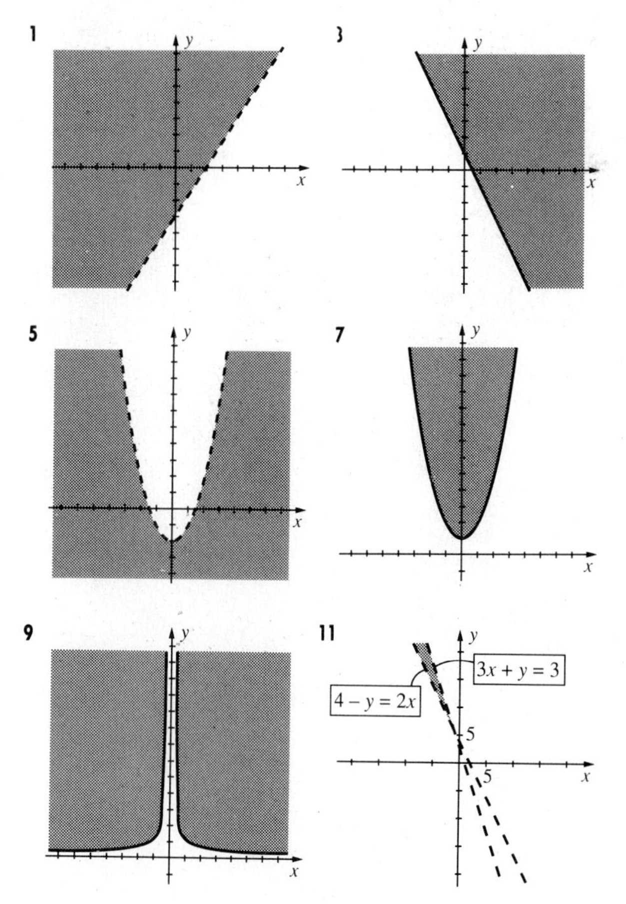

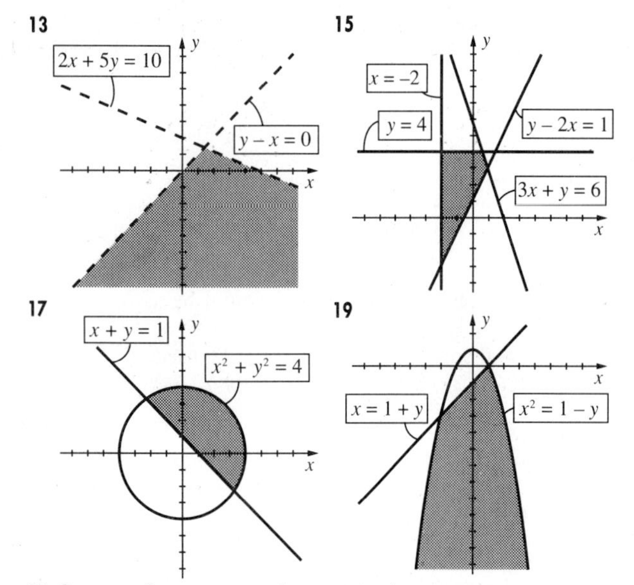

21 $0 \le x < 3,\ y < -x + 4,\ y \ge x - 4$

23 $x^2 + y^2 \le 9,\ y > -2x + 4$

25 $y < x,\ y \le -x + 4,\ (x - 2)^2 + (y - 2)^2 \le 8$

27 $y > \frac{1}{8}x + \frac{1}{2},\ y \le x + 4,\ y \le -\frac{3}{4}x + 4$

29 If x and y denote the number of brand A and brand B, respectively, then a system is $x \ge 20,\ y \ge 10,\ x \ge 2y,\ x + y \le 100$. *(see graph below)*

31 If x and y denote the amount placed in the high-risk and low-risk investment, respectively, then a system is $x \ge 2000,\ y \ge 3x,\ x + y \le 15{,}000$. *(see graph below)*

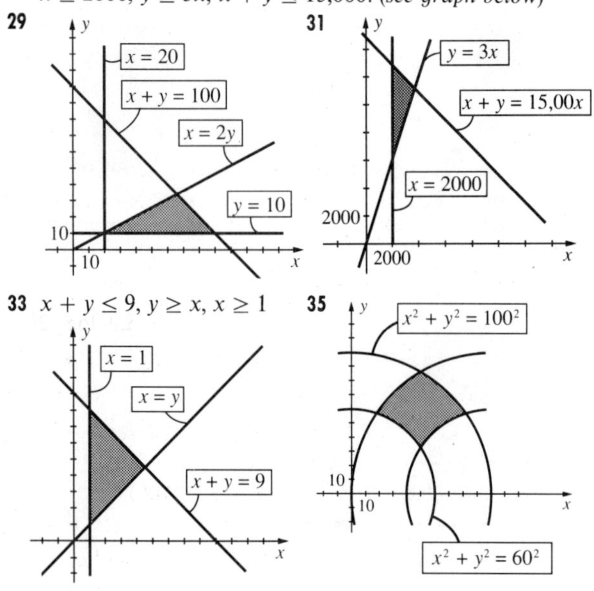

33 $x + y \le 9,\ y \ge x,\ x \ge 1$

35 If the plant is located at (x, y), then $(60)^2 \le x^2 + y^2 \le (100)^2$ and $3600 \le (x - 100)^2 + y^2 \le 10{,}000$ and $y \ge 0$. *(see graph at bottom of preceding column)*

EXERCISES 9.10 ▪ PAGE 535

1 50 standard and 30 oversized

3 3.5 lb of S and 1 lb of T

5 Send 25 from W_1 to A and 0 from W_1 to B. Send 10 from W_2 to A and 60 from W_2 to B.

7 None of alfalfa and 80 acres of corn

9 Minimum cost: 16 oz X, 4 oz Y, 0 oz Z; maximum cost: 0 oz X, 8 oz Y, 12 oz Z

11 2 vans and 4 buses **13** 3000 trout and 2000 bass

15 60 small units and 20 deluxe units

EXERCISES 9.11 ▪ PAGE 537

1 $(\frac{19}{23}, -\frac{18}{23})$ **3** $(-3, 5), (1, -3)$

5 $(2\sqrt{3}, \pm\sqrt{2}), (-2\sqrt{3}, \pm\sqrt{2})$ **7** $(\frac{14}{17}, \frac{14}{27})$

9 $(\frac{6}{11}, -\frac{7}{11}, 1)$ **11** $(-2c, -3c, c)$ for any real number c

13 $(5c - 1, -\frac{19}{2}c + \frac{5}{2}, c)$ for any real number c

15 $(-1, \frac{1}{2}, \frac{1}{3})$

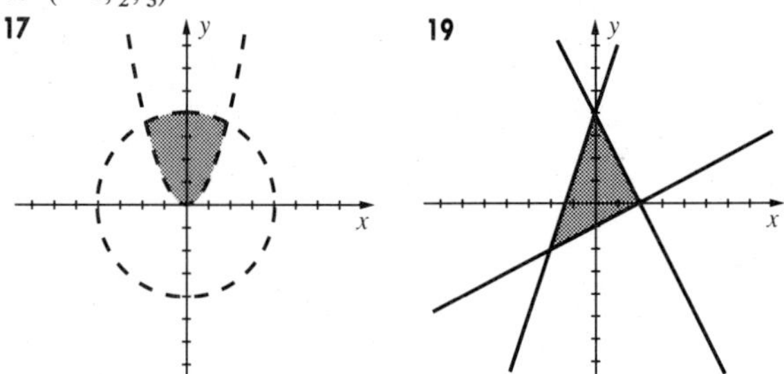

21 -6 **23** 48 **25** -84 **27** 120 **29** 0

31 $a_{11}a_{22}a_{33} \cdots a_{nn}$ **33** $-\frac{1}{2}\begin{bmatrix} 2 & 4 \\ 3 & 5 \end{bmatrix}$

35 $\frac{1}{7}\begin{bmatrix} 2 & 1 & 0 & 0 \\ -1 & 3 & 0 & 0 \\ 0 & 0 & 3 & 2 \\ 0 & 0 & -5 & -1 \end{bmatrix}$ **37** $\begin{bmatrix} 4 & -5 & 6 \\ 4 & -11 & 5 \end{bmatrix}$

39 $\begin{bmatrix} 0 & 4 & -6 \\ 16 & 22 & 1 \\ 12 & 11 & 9 \end{bmatrix}$ **41** $\begin{bmatrix} -12 & 4 & -11 \\ 6 & -11 & 5 \end{bmatrix}$

43 $\begin{bmatrix} a & 3a \\ 2b & 4b \end{bmatrix}$ **45** $\begin{bmatrix} 5 & 9 \\ 13 & 19 \end{bmatrix}$ **49** $-1 \pm 2\sqrt{3}$

51 $\dfrac{8}{x-1} - \dfrac{3}{x+5} - \dfrac{1}{x+3}$

53 $40\sqrt{5}$ ft $\times$ $20\sqrt{5}$ ft

55 Tax = \$18,750; bonus = \$3,125

57 In ft^3/hr: A, 30; B, 20; C, 50

59

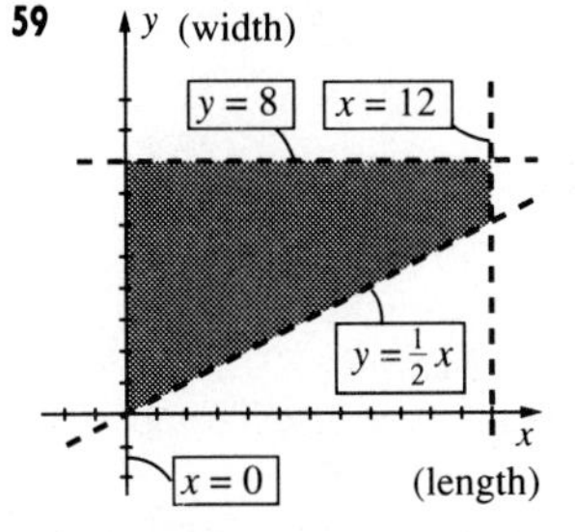

61 80 mowers and 30 edgers

CHAPTER ▪ 10

EXERCISES 10.1 ▪ PAGE 548

Exer. 1–33: A typical proof is given for Exercise 1, 5, 9, . . . , 33.

1 (i) P_1 is true, since $2(1) = 1(1+1) = 2$.

(ii) Assume P_k is true:

$2 + 4 + 6 + \cdots + 2k = k(k+1)$. Hence

$$\begin{aligned} 2 + 4 + 6 + \cdots + 2k + 2(k+1) &= k(k+1) + 2(k+1) \\ &= (k+1)(k+2) \\ &= (k+1)(k+1+1) \end{aligned}$$

Thus, P_{k+1} is true and the proof is complete.

5 (i) P_1 is true, since $5(1) - 3 = \frac{1}{2}(1)[5(1) - 1] = 2$.

(ii) Assume P_k is true:

$2 + 7 + 12 + \cdots + (5k-3) = \frac{1}{2}k(5k-1)$. Hence

$$\begin{aligned} 2 + 7 + 12 + \cdots + (5k-3) &+ 5(k+1) - 3 \\ &= \tfrac{1}{2}k(5k-1) + 5(k+1) - 3 \\ &= \tfrac{5}{2}k^2 + \tfrac{9}{2}k + 2 \\ &= \tfrac{1}{2}(5k^2 + 9k + 4) \\ &= \tfrac{1}{2}(k+1)(5k+4) \\ &= \tfrac{1}{2}(k+1)[5(k+1)+1] \end{aligned}$$

Thus, P_{k+1} is true and the proof is complete.

9 (i) P_1 is true, since $(1)^1 = \dfrac{1(1+1)[2(1)+1]}{6} = 1$.

(ii) Assume P_k is true:

$1^2 + 2^2 + 3^2 + \cdots + k^2 = \dfrac{k(k+1)(2k+1)}{6}$. Hence

$$\begin{aligned} 1^2 + 2^2 + 3^2 + \cdots + k^2 &+ (k+1)^2 \\ &= \frac{k(k+1)(2k+1)}{6} + (k+1)^2 \\ &= (k+1)\left[\frac{k(2k+1)}{6} + \frac{6(k+1)}{6}\right] \\ &= \frac{(k+1)(2k^2+7k+6)}{6} \\ &= \frac{(k+1)(k+2)(2k+3)}{6} \end{aligned}$$

Thus, P_{k+1} is true and the proof is complete.

13 (i) P_1 is true, since $3^1 = \frac{3}{2}(3^1 - 1) = 3$.

(ii) Assume P_k is true:

$3 + 3^2 + 3^3 + \cdots + 3^k = \frac{3}{2}(3^k - 1)$. Hence

$$\begin{aligned} 3 + 3^2 + 3^3 + \cdots + 3^k + 3^{k+1} &= \tfrac{3}{2}(3^k - 1) + 3^{k+1} \\ &= \tfrac{3}{2}\cdot 3^k - \tfrac{3}{2} + 3\cdot 3^k \\ &= \tfrac{9}{2}\cdot 3^k - \tfrac{3}{2} \\ &= \tfrac{3}{2}(3\cdot 3^k - 1) \\ &= \tfrac{3}{2}(3^{k+1} - 1) \end{aligned}$$

Thus, P_{k+1} is true and the proof is complete.

17 (i) P_1 is true, since $1 < \frac{1}{8}[2(1)+1]^2 = \frac{9}{8}$.

(ii) Assume P_k is true:

$1 + 2 + 3 + \cdots + k < \frac{1}{8}(2k+1)^2$. Hence

$$\begin{aligned} 1 + 2 + 3 + \cdots + k + (k+1) &< \\ \tfrac{1}{8}(2k+1)^2 + (k+1) &= \tfrac{1}{2}k^2 + \tfrac{3}{2}k + \tfrac{9}{8} \\ &= \tfrac{1}{8}(4k^2 + 12k + 9) \\ &= \tfrac{1}{8}(2k+3)^2 \\ &= \tfrac{1}{8}[2(k+1)+1]^2 \end{aligned}$$

Thus, P_{k+1} is true and the proof is complete.

21 (i) For $n = 1$, $5^n - 1 = 4$ and 4 is a factor of 4.

(ii) Assume 4 is a factor of $5^k - 1$. The $(k+1)$st term is

$$\begin{aligned} 5^{k+1} - 1 &= 5\cdot 5^k - 1 \\ &= 5\cdot 5^k - 5 + 4 = 5(5^k - 1) + 4 \end{aligned}$$

By the induction hypothesis, 4 is a factor of 5^{k-1} and 4 is a factor of 4, so 4 is a factor of the $(k+1)$st term. Thus, P_{k+1} is true and the proof is complete.

25 (i) P_8 is true, since $5 + \log_2 8 \le 8$.

(ii) Assume P_k is true: $5 + \log_2 k \le k$. Hence

$$\begin{aligned} 5 + \log_2(k+1) < 5 + \log_2(k+k) &= 5 + \log_2 2k \\ &= 5 + \log_2 2 + \log_2 k \\ &= (5 + \log_2 k) + 1 \le k + 1 \end{aligned}$$

Thus, P_{k+1} is true and the proof is complete.

29 (i) P_3 is true, since $2(3) + 2 \le 2^3$.

(ii) Assume P_k is true: $2k + 2 \le 2^k$. Hence

$2(k+1) + 2 = (2k+2) + 2 \le 2^k + 2^k = 2\cdot 2^k = 2^{k+1}$

Thus, P_{k+1} is true and the proof is complete.

33 (i) For $n = 1$, $a - b$ is a factor of $a^1 - b^1$.
(ii) Assume $a - b$ is a factor of $a^k - b^k$. Following the hint for the $(k+1)$st term, $a^{k+1} - b^{k+1} = a^k \cdot a - b \cdot a^k + b \cdot a^k - b^k \cdot b = a^k(a - b) + (a^k - b^k)b$. Since $(a - b)$ is a factor of $a^k(a - b)$ and since by the induction hypothesis, $a - b$ is a factor of $(a^k - b^k)$, it follows that $a - b$ is a factor of the $(k+1)$st term. Thus, P_{k+1} is true and the proof is complete.

37 (i) For $n = 1$, $\sin(\theta + 1\pi) = \sin\theta\cos\pi + \cos\theta\sin\pi = -\sin\theta = (-1)^1 \sin\theta$.
(ii) Assume P_k is true: $\sin(\theta + k\pi) = (-1)^k \sin\theta$. Hence
$$\begin{aligned}\sin[\theta + (k+1)\pi] &= \sin[(\theta + k\pi) + \pi] \\ &= \sin(\theta + k\pi)\cos\pi + \cos(\theta + k\pi)\sin\pi \\ &= [(-1)^k \sin\theta] \cdot (-1) + \cos(\theta + k\pi) \cdot (0) \\ &= (-1)^{k+1} \sin\theta.\end{aligned}$$
Thus, P_{k+1} is true and the proof is complete.

EXERCISES 10.2 ▪ PAGE 555

1 9, 6, 3, 0, −3; −12 **3** $\frac{1}{2}, \frac{4}{5}, \frac{7}{10}, \frac{10}{17}, \frac{1}{2}; \frac{22}{65}$
5 9, 9, 9, 9, 9; 9
7 1.9, 2.01, 1.999, 2.0001, 1.99999; 2.00000001
9 $4, -\frac{9}{4}, \frac{5}{3}, -\frac{11}{8}, \frac{6}{5}; -\frac{15}{16}$ **11** 2, 0, 2, 0, 2; 0
13 $\frac{2}{3}, \frac{2}{3}, \frac{8}{11}, \frac{8}{9}, \frac{32}{27}; \frac{128}{33}$ **15** 1, 2, 3, 4, 5; 8
17 2, 1, −2, −11, −38 **19** $-3, 3^2, 3^4, 3^8, 3^{16}$
21 5, 5, 10, 30, 120 **23** $2, 2, 4, 4^3, 4^{12}$
23 $a_n = 2n + \frac{1}{24}(n-1)(n-2)(n-3)(n-4)(a-10)$ (The answer is not unique.)
27 −5 **29** 10 **31** 25 **33** $-\frac{17}{15}$ **35** 61
37 10,000 **39** $\frac{1}{3}(n^3 + 6n^2 + 20n)$
40 $\frac{1}{3}(4n^3 - 12n^2 + 11n)$ **43** $\sum_{k=0}^{3} (2k+1)$
45 $\sum_{k=1}^{5} (4k-3)$ **47** $\sum_{k=1}^{4} \frac{k}{3k-1}$
49 $1 + \sum_{k=1}^{n} (-1)^k \frac{x^{2k}}{2k}$ **51** $\sum_{k=1}^{7} (-1)^{k-1} \frac{1}{k}$
53 $\sum_{k=1}^{75} 2k$ **55** $\sum_{k=1}^{99} \frac{1}{k(k+1)}$

57
$$\begin{aligned}\sum_{k=1}^{n} (a_k - b_k) &= (a_1 - b_1) + (a_2 - b_2) + \cdots + (a_n - b_n) \\ &= (a_1 + a_2 + \cdots + a_n) + (-b_1 - b_2 - \cdots - b_n) \\ &= (a_1 + a_2 + \cdots + a_n) - (b_1 + b_2 + \cdots + b_n) \\ &= \sum_{k=1}^{n} a_k - \sum_{k=1}^{n} b_k\end{aligned}$$
59 As k increases, the terms approach 1.
61 0.4, 0.7, 1, 1.6, 2.8
63 (a) $y_n = K$ for every n
(b) 400, 560, 425.6, 552.3, 436.8, 547.2, 443.9, 543.5, 448.9, 540.7; the terms appear to be oscillating about 500.
65 3, 3.142546543, 3.141592653, 3.141592654, 3.141592654; the terms approach 2π.

EXERCISES 10.3 ▪ PAGE 561

1 18; 38; $4n - 2$ **3** 1.8; 0.3; $3.3 - 0.3n$
5 5.4; 20.9; $(3.1)n - 10.1$ **7** $\ln 3^5$; $\ln 3^{10}$; $\ln 3^n$
9 −8.5 **11** −9.8 **13** −105 **15** 30 **17** 530
19 $\frac{423}{2}$ **21** (a) 60 (b) 12,780 **23** 24
25 $\frac{10}{3}, \frac{14}{3}, 6, \frac{22}{3}, \frac{26}{3}$ **27** 255 logs **29** 154π ft
31 \$1,200 **33** 4

EXERCISES 10.4 ▪ PAGE 568

1 $\frac{1}{2}; \frac{1}{6}; 8(\frac{1}{2})^{n-1} = 2^{4-n}$ **3** 0.03; −0.00003; $300(-0.1)^{n-1}$
5 3125; 5^8; 5^n **7** $\frac{81}{4}$; $-3^7/2^5$; $4(-\frac{3}{2})^{n-1}$
9 x^8; $-x^{14}$; $(-1)^{n-1}x^{2n-2}$ **11** 2^{4x+1}; 2^{7x+1}; 2^{nx-x+1}
13 $\frac{243}{8}$ **15** $a_1 = \frac{1}{81}$, $S_5 = \frac{211}{1296}$ **17** 88,572
19 $-\frac{1023}{3072}$ **21** $\frac{25}{256}\% \approx 0.1\%$
23 (a) $N(t) = 10{,}000(1.2)^t$ (b) 61,917 **25** $\frac{2}{3}$ **27** $\frac{50}{33}$
29 Since $|r| = \sqrt{2} > 1$, the sum does not exist. **31** $\frac{23}{99}$
33 $\frac{2393}{990}$ **35** $\frac{5141}{999}$ **37** $\frac{16{,}123}{9999}$ **39** 300 ft
41 \$3,000,000 **43** (b) 375 mg
45 (a) $a_{k+1} = \frac{1}{4}\sqrt{10}\,a_k$
(b) $a_n = (\frac{1}{4}\sqrt{10})^{n-1}a_1$, $A_n = (\frac{5}{8})^{n-1}A_1$, $P_n = (\frac{1}{4}\sqrt{10})^{n-1}P_1$
(c) $16a_1/(4 - \sqrt{10})$

EXERCISES 10.5 ▪ PAGE 575

1 210 **3** 60,480 **5** 120 **7** 6
9 (a) 60 (b) 125 **11** 64 **13** $P(8, 3) = 336$
15 24 **17** (a) 2,340,000 (b) 2,160,000
19 (a) 151,200 (b) 5760 **21** 1024
23 $P(8, 8) = 40{,}320$ **25** $P(6, 3) = 120$

27 (a) $P(5, 3) = 60$ (b) 125 **29** 9,000,000
31 $P(4, 4) = 24$ **33** 48 **35** 3,628,800
37 (a) 900
(b) If n is even, $9 \cdot 10^{(n/2)-1}$; if n is odd, $9 \cdot 10^{n/2}$.

EXERCISES 10.6 ■ PAGE 581

1 35 **3** 9 **5** n **7** 1 **9** $\dfrac{12!}{5!3!2!2!} = 166{,}320$

11 $\dfrac{10!}{3!2!2!1!1!1!} = 151{,}200$ **13** $C(10, 5) = 252$

15 $C(8, 2) = 28$ **17** $(5! \cdot 4! \cdot 8!) \cdot 3! = 696{,}729{,}600$
19 $3 \cdot C(10, 2) \cdot C(8, 2) \cdot C(4, 2) \cdot C(6, 2) \cdot 3 \cdot 4 = 4{,}082{,}400$
21 $C(12, 3) \cdot C(8, 2) = 6160$ **23** $C(8, 3) = 56$
25 (a) $C(49, 6) = 13{,}983{,}816$ (b) $C(24, 6) = 134{,}596$
27 $C(n, 2) = 45$ and hence $n = 10$ **29** $C(6, 3) = 20$

EXERCISES 10.7 ■ PAGE 587

1 (a) $\frac{4}{52} = \frac{1}{13}$ (b) $\frac{8}{52} = \frac{2}{13}$ (c) $\frac{12}{52} = \frac{3}{13}$
3 (a) $\frac{1}{6}$ (b) $\frac{1}{6}$ (c) $\frac{2}{6} = \frac{1}{3}$ **5** (a) $\frac{2}{36} = \frac{1}{18}$ (b) $\frac{5}{36}$ (c) $\frac{7}{36}$

7 $\frac{6}{216} = \frac{1}{36}$ **9** $\frac{3}{8}$ **11** $\dfrac{48 \cdot 13}{C(52, 5)} = \dfrac{1}{4165}$

13 $\dfrac{C(13, 4) \cdot C(13, 1)}{C(52, 5)} = \dfrac{143}{39{,}984}$ **15** $\dfrac{C(13, 5) \cdot 4}{C(52, 5)} = \dfrac{33}{16{,}660}$

17 (a) $\dfrac{C(8, 8)}{2^8} = \dfrac{1}{256}$ (b) $\dfrac{C(8, 7)}{2^8} = \dfrac{1}{32}$ (c) $\dfrac{C(8, 6)}{2^8} = \dfrac{7}{64}$

(d) $\dfrac{C(8, 6) + C(8, 7) + C(8, 8)}{2^8} = \dfrac{37}{256}$

19 (a) Since $n(E') + n(E) = n(S)$, we obtain $\dfrac{n(E')}{n(S)} + \dfrac{n(E)}{n(S)} = 1$

and hence $P(E') = 1 - P(E)$.

(b) $1 - P(\text{no aces}) = 1 - \dfrac{C(48, 5)}{C(52, 5)} = 1 - \dfrac{35{,}673}{54{,}145} = \dfrac{18{,}472}{54{,}145}$

21 (a) $\dfrac{6 + 2}{36} = \dfrac{2}{9}$ (b) $\dfrac{1 + 2 + 1}{36} = \dfrac{1}{9}$

23 (a) $\dfrac{1}{2^5} = \dfrac{1}{32}$ (b) $1 - \frac{1}{32} = \frac{31}{32}$

25 (a) $\dfrac{C(4, 4)}{4!} = \dfrac{1}{24}$ (b) $\dfrac{C(4, 2)}{4!} = \dfrac{1}{4}$

27 (a) 0 (b) $\dfrac{1}{3^2} = \dfrac{1}{9}$ **29** (a) $\dfrac{1}{2^4} = \dfrac{1}{16}$ (b) $\dfrac{C(4, 2)}{2^4} = \dfrac{3}{8}$

EXERCISES 10.8 ■ PAGE 589

1 (i) P_1 is true, since $3(1) - 1 = \dfrac{1[3(1) + 1]}{2} = 2$.

(ii) Assume P_k is true:

$2 + 5 + 8 + \cdots + (3k - 1) = \dfrac{k(3k + 1)}{2}$. Hence

$$\begin{aligned} 2 + 5 + 8 + \cdots &+ (3k - 1) + 3(k + 1) - 1 \\ &= \frac{k(3k + 1)}{2} + 3(k + 1) - 1 \\ &= \frac{3k^2 + k + 6k + 4}{2} \\ &= \frac{3k^2 + 7k + 4}{2} \\ &= \frac{(k + 1)(3k + 4)}{2} \\ &= \frac{(k + 1)[3(k + 1) + 1]}{2} \end{aligned}$$

Thus, P_{k+1} is true and the proof is complete.

3 (i) P_1 is true, since $\dfrac{1}{[2(1) - 1][2(1) + 1]} = \dfrac{1}{2(1) + 1} = \dfrac{1}{3}$.

(ii) Assume P_k is true:

$\dfrac{1}{1 \cdot 3} + \dfrac{1}{3 \cdot 5} + \dfrac{1}{5 \cdot 7} + \cdots + \dfrac{1}{(2k - 1)(2k + 1)} = \dfrac{k}{2k + 1}$.

Hence

$$\begin{aligned} \frac{1}{1 \cdot 3} + \frac{1}{3 \cdot 5} + \frac{1}{5 \cdot 7} + \cdots &+ \frac{1}{(2k - 1)(2k + 1)} \\ + \frac{1}{(2k + 1)(2k + 3)} &= \frac{k}{2k + 1} + \frac{1}{(2k + 1)(2k + 3)} \\ &= \frac{k(2k + 3) + 1}{(2k + 1)(2k + 3)} \\ &= \frac{2k^2 + 3k + 1}{(2k + 1)(2k + 3)} \\ &= \frac{(2k + 1)(k + 1)}{(2k + 1)(2k + 3)} \\ &= \frac{k + 1}{2(k + 1) + 1} \end{aligned}$$

Thus, P_{k+1} is true and the proof is complete.

5 (i) For $n = 1$, $n^3 + 2n = 3$ and 3 is a factor of 3.
(ii) Assume 3 is a factor of $k^3 + 2k$. The $(k + 1)$st term is

$$\begin{aligned} (k + 1)^3 + 2(k + 1) &= k^3 + 3k^2 + 5k + 3 \\ &= (k^3 + 2k) + (3k^2 + 3k + 3) \\ &= (k^3 + 2k) + 3(k^2 + k + 1) \end{aligned}$$

By the induction hypothesis, 3 is a factor of $k^3 + 2k$. Since 3 is also a factor of $3(k^2 + k + 1)$, it is a factor of the $(k + 1)$st term. Thus, P_{k+1} is true and the proof is complete.

7 $5, -2, -1, -\frac{20}{29}; -\frac{7}{19}$ **9** $2, \frac{1}{2}, \frac{5}{4}, \frac{7}{8}; \frac{65}{64}$

11 $10, \frac{11}{10}, \frac{21}{11}, \frac{32}{11}, \frac{53}{32}$ **13** $9, 3, \sqrt{3}, \sqrt[4]{3}, \sqrt[8]{3}$ **15** 75

17 1000 **19** $\sum_{k=1}^{5} 3k$ **21** $\sum_{k=1}^{5} (-1)^{k+1}(105 - 5k)$

23 $-5 - 8\sqrt{3}; -5 - 35\sqrt{3}$ **25** $-31; 50$ **27** 64

29 $4\sqrt{2}$ **31** 570 **33** 2041 **35** $\frac{5}{7}$

37 (a) $d = 1 - \frac{1}{2}a_1$ (b) In ft: $1\frac{1}{4}, 2, 2\frac{3}{4}, 3\frac{1}{2}$

39 $\dfrac{2}{1 - f}$ **41** (a) $\frac{2}{4} = \frac{1}{2}$ (b) $\frac{4}{8} = \frac{1}{2}$

43 (a) $P(52, 13) \approx 3.954 \times 10^{21}$
(b) $P(13, 5) \cdot P(13, 3) \cdot P(13, 3) \cdot P(13, 2) \approx 7.094 \times 10^{13}$

45 (a) $C(12, 8) = 495$ (b) $C(9, 5) = 126$

47 (a) $\frac{1}{1000}$ (b) $\frac{10}{1000} = \frac{1}{100}$ (c) $\frac{50}{1000} = \frac{1}{20}$

49 (a) $\dfrac{C(6,4) + C(6,5) + C(6,6)}{2^6} = \dfrac{11}{32}$ (b) $1 - \frac{11}{32} = \frac{21}{32}$

CHAPTER ▪ 11

EXERCISES 11.2 ▪ PAGE 601

1 $V(0, 0); F(0, -3); y = 3$ **3** $V(0, 0); F(-\frac{3}{8}, 0); x = \frac{3}{8}$

5 $V(0, 0); F(0, \frac{1}{32}); y = -\frac{1}{32}$ **7** $V(-2, 1); F(-2, -1); y = 3$

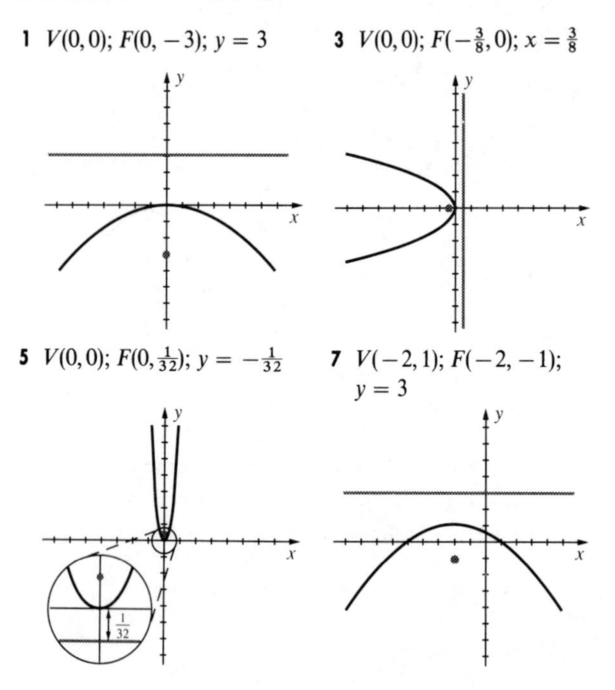

9 $V(3, 2); F(\frac{49}{16}, 2); x = \frac{47}{16}$ **11** $V(-1, 0); F(2, 0); x = -4$

13 $V(2, -2); F(2, -\frac{7}{4}); y = -\frac{9}{4}$ **15** $V(-4, 2); F(-\frac{7}{2}, 2); x = -\frac{9}{2}$

17 $V(-5, -6); F(-5, -\frac{97}{16}); y = -\frac{95}{16}$ **19** $V(0, \frac{1}{2}); F(0, -\frac{9}{2}); y = \frac{11}{2}$

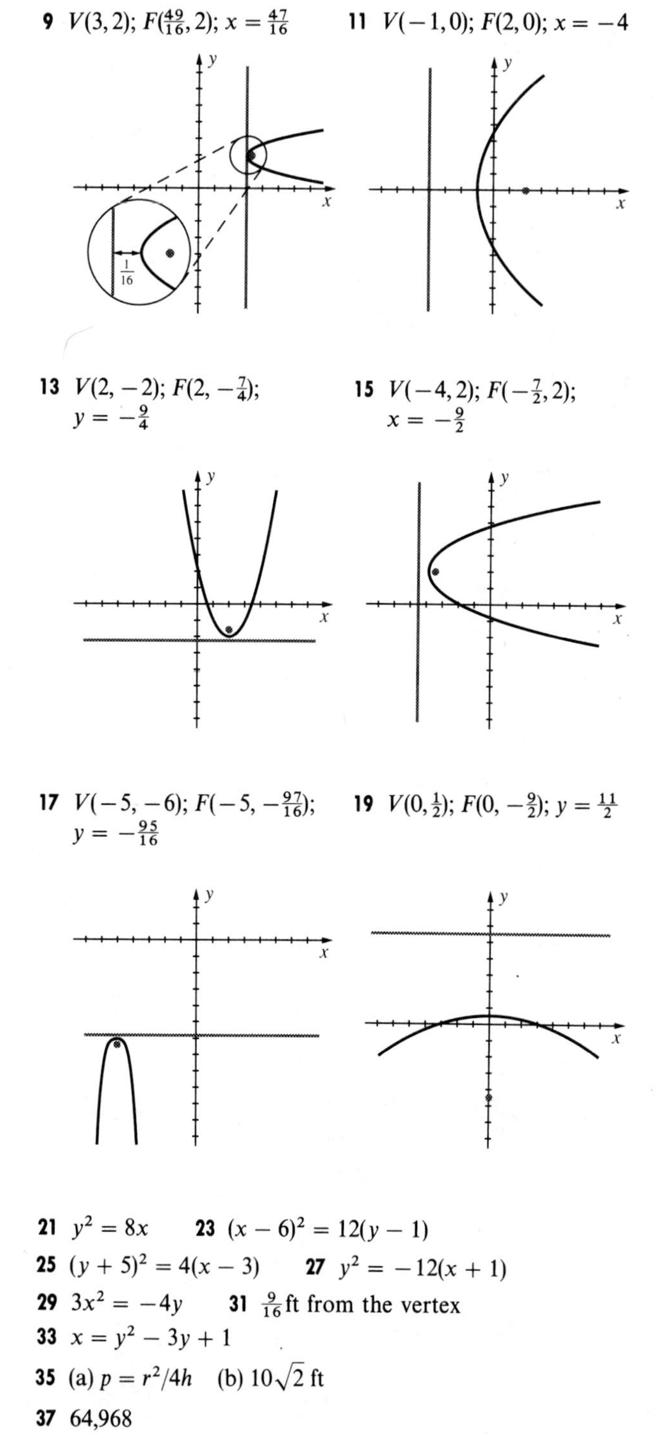

21 $y^2 = 8x$ **23** $(x - 6)^2 = 12(y - 1)$

25 $(y + 5)^2 = 4(x - 3)$ **27** $y^2 = -12(x + 1)$

29 $3x^2 = -4y$ **31** $\frac{9}{16}$ ft from the vertex

33 $x = y^2 - 3y + 1$

35 (a) $p = r^2/4h$ (b) $10\sqrt{2}$ ft

37 64,968

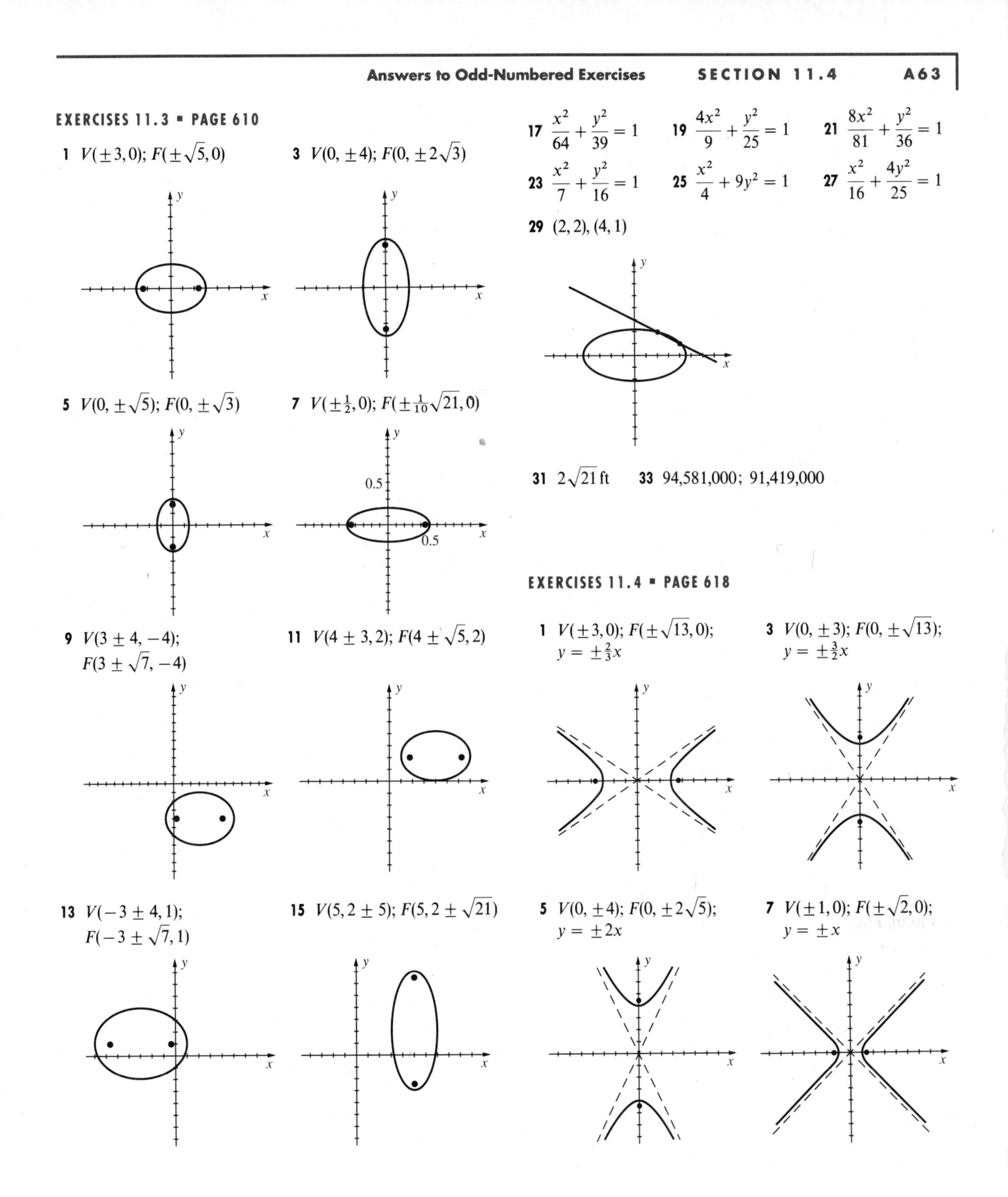

EXERCISES 11.3 ▪ PAGE 610

1 $V(\pm 3, 0)$; $F(\pm\sqrt{5}, 0)$

3 $V(0, \pm 4)$; $F(0, \pm 2\sqrt{3})$

5 $V(0, \pm\sqrt{5})$; $F(0, \pm\sqrt{3})$

7 $V(\pm\frac{1}{2}, 0)$; $F(\pm\frac{1}{10}\sqrt{21}, 0)$

9 $V(3 \pm 4, -4)$; $F(3 \pm \sqrt{7}, -4)$

11 $V(4 \pm 3, 2)$; $F(4 \pm \sqrt{5}, 2)$

13 $V(-3 \pm 4, 1)$; $F(-3 \pm \sqrt{7}, 1)$

15 $V(5, 2 \pm 5)$; $F(5, 2 \pm \sqrt{21})$

17 $\dfrac{x^2}{64} + \dfrac{y^2}{39} = 1$ **19** $\dfrac{4x^2}{9} + \dfrac{y^2}{25} = 1$ **21** $\dfrac{8x^2}{81} + \dfrac{y^2}{36} = 1$

23 $\dfrac{x^2}{7} + \dfrac{y^2}{16} = 1$ **25** $\dfrac{x^2}{4} + 9y^2 = 1$ **27** $\dfrac{x^2}{16} + \dfrac{4y^2}{25} = 1$

29 $(2, 2), (4, 1)$

31 $2\sqrt{21}$ ft **33** 94,581,000; 91,419,000

EXERCISES 11.4 ▪ PAGE 618

1 $V(\pm 3, 0)$; $F(\pm\sqrt{13}, 0)$; $y = \pm\frac{2}{3}x$

3 $V(0, \pm 3)$; $F(0, \pm\sqrt{13})$; $y = \pm\frac{3}{2}x$

5 $V(0, \pm 4)$; $F(0, \pm 2\sqrt{5})$; $y = \pm 2x$

7 $V(\pm 1, 0)$; $F(\pm\sqrt{2}, 0)$; $y = \pm x$

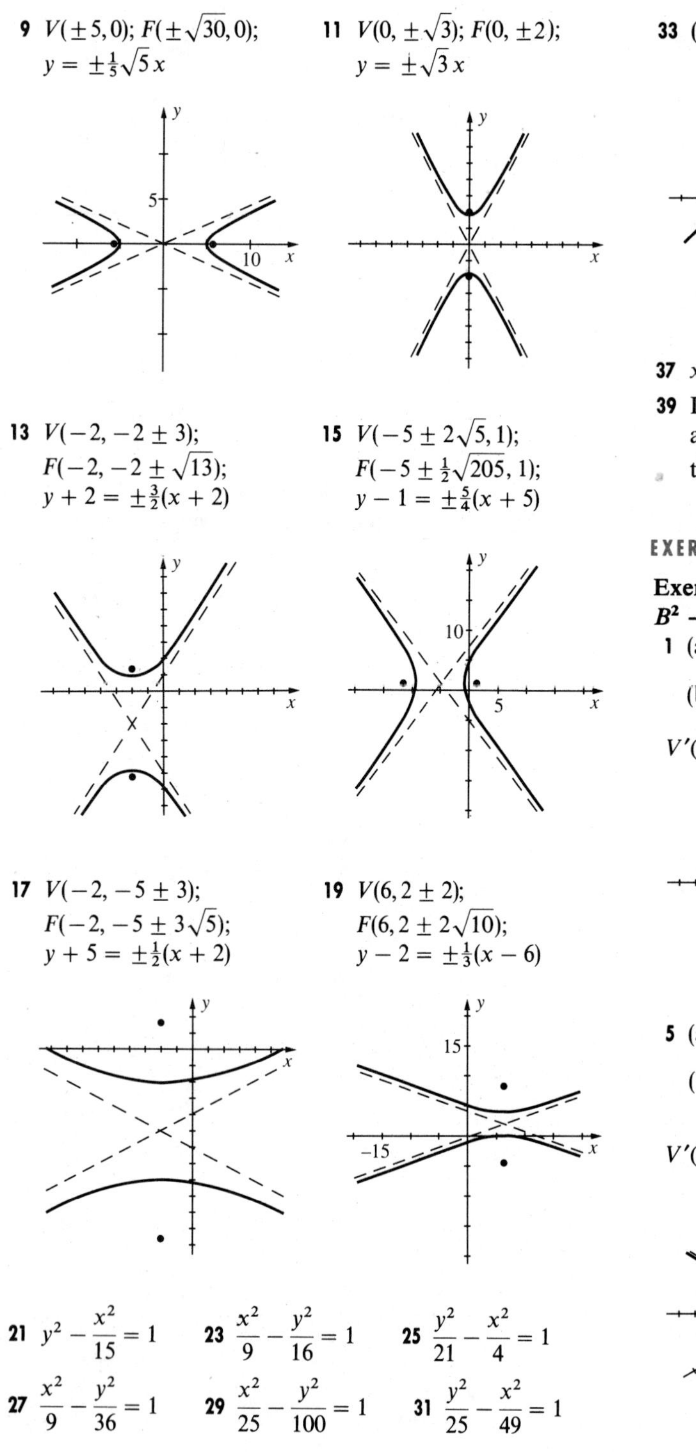

9 $V(\pm 5, 0)$; $F(\pm\sqrt{30}, 0)$; $y = \pm\frac{1}{5}\sqrt{5}\,x$

11 $V(0, \pm\sqrt{3})$; $F(0, \pm 2)$; $y = \pm\sqrt{3}\,x$

13 $V(-2, -2 \pm 3)$; $F(-2, -2 \pm \sqrt{13})$; $y + 2 = \pm\frac{3}{2}(x + 2)$

15 $V(-5 \pm 2\sqrt{5}, 1)$; $F(-5 \pm \frac{1}{2}\sqrt{205}, 1)$; $y - 1 = \pm\frac{5}{4}(x + 5)$

17 $V(-2, -5 \pm 3)$; $F(-2, -5 \pm 3\sqrt{5})$; $y + 5 = \pm\frac{1}{2}(x + 2)$

19 $V(6, 2 \pm 2)$; $F(6, 2 \pm 2\sqrt{10})$; $y - 2 = \pm\frac{1}{3}(x - 6)$

21 $y^2 - \dfrac{x^2}{15} = 1$ **23** $\dfrac{x^2}{9} - \dfrac{y^2}{16} = 1$ **25** $\dfrac{y^2}{21} - \dfrac{x^2}{4} = 1$

27 $\dfrac{x^2}{9} - \dfrac{y^2}{36} = 1$ **29** $\dfrac{x^2}{25} - \dfrac{y^2}{100} = 1$ **31** $\dfrac{y^2}{25} - \dfrac{x^2}{49} = 1$

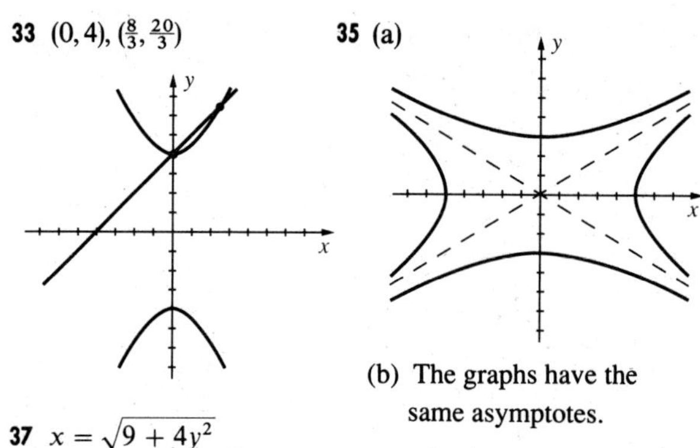

33 $(0, 4)$, $(\frac{8}{3}, \frac{20}{3})$

35 (a) (graph)

(b) The graphs have the same asymptotes.

37 $x = \sqrt{9 + 4y^2}$

39 If a coordinate system is introduced with the x-axis along AB and the y-axis through the midpoint of AB, then the ship is at the point $(\frac{80}{3}\sqrt{34}, 100) \approx (155, 100)$.

EXERCISES 11.5 ▪ PAGE 625

Exer. 1–13: The answer in part (a) gives the value of $B^2 - 4AC$ in the identification theorem.

1 (a) 0, parabola (b) $2(x') = (y')^2$ $V'(0, 0)$

3 (a) -36, ellipse (b) $\dfrac{(x')^2}{9} + (y')^2 = 1$ $V'(\pm 3, 0)$

5 (a) 256, hyperbola (b) $\dfrac{(x')^2}{\frac{1}{4}} - (y')^2 = 1$ $V'(\pm\frac{1}{2}, 0)$

7 (a) 0, parabola (b) $(y')^2 = 4(x' - 1)$ $V'(1, 0)$

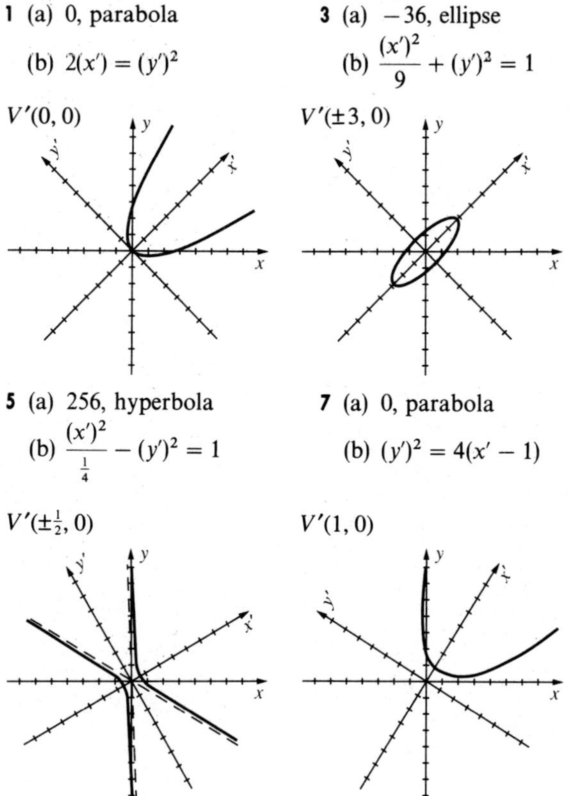

9 (a) -2704, ellipse

(b) $\dfrac{(x'-2)^2}{4} + (y')^2 = 1$

$V'(2 \pm 2, 0)$

11 (a) 128, hyperbola

(b) $(y'+2)^2 - \dfrac{(x')^2}{\frac{1}{2}} = 1$

$V'(0, -2 \pm 1)$

13 (a) -1600, ellipse (b) $\dfrac{(x')^2}{16} + (y')^2 = 1$

$V'(\pm 4, 0)$

EXERCISES 11.6 ▪ PAGE 633

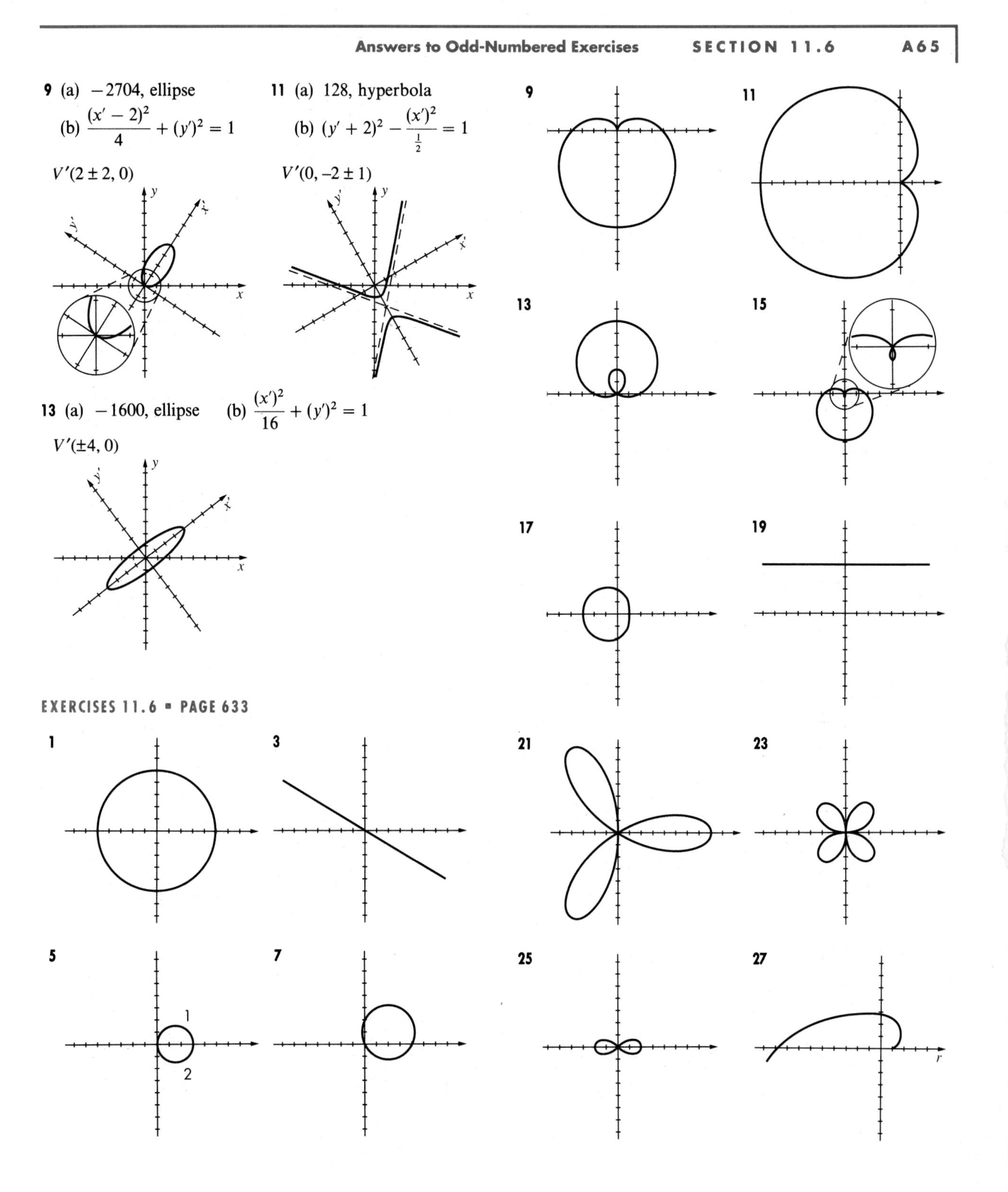

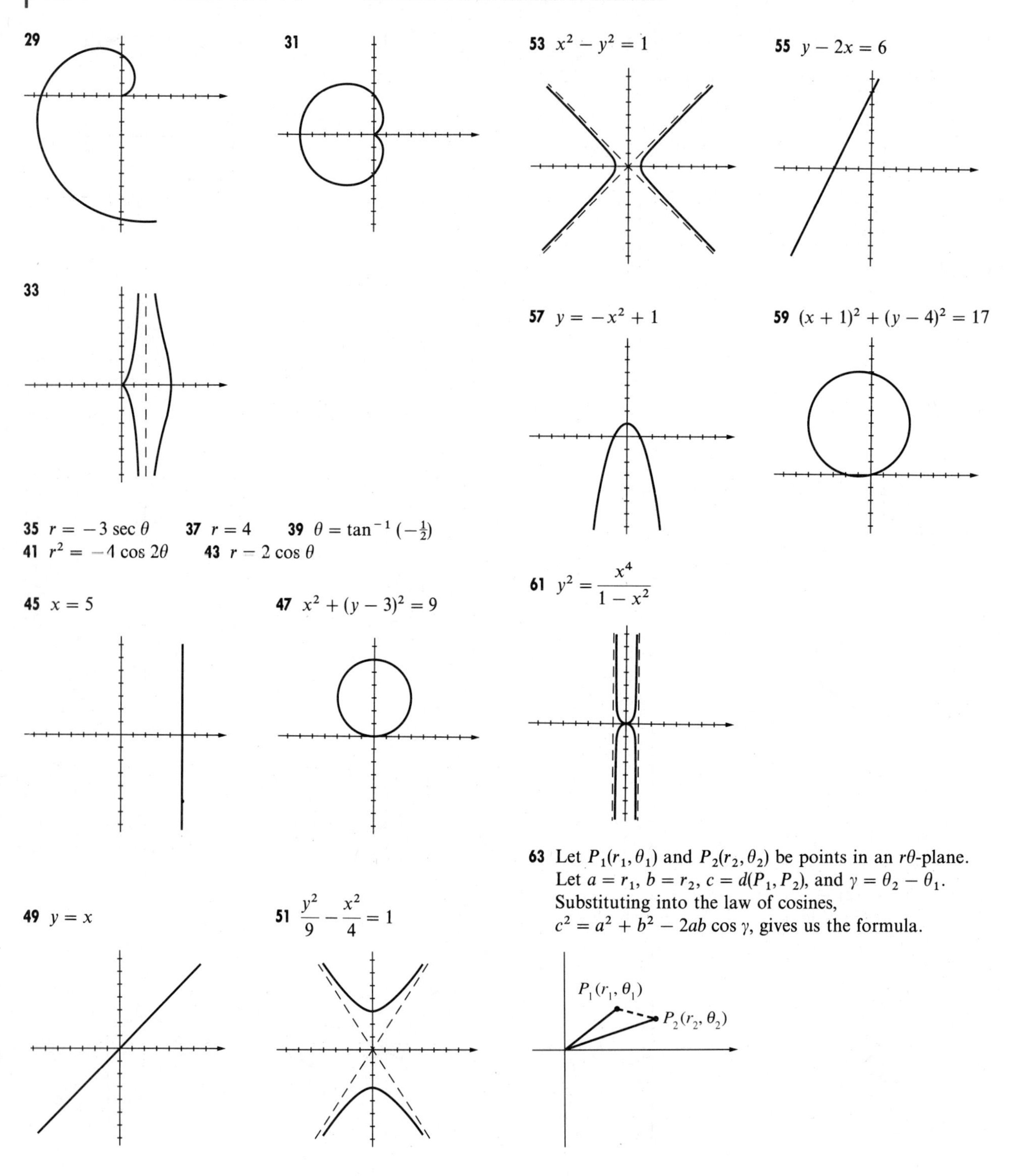

29

31

53 $x^2 - y^2 = 1$

55 $y - 2x = 6$

33

57 $y = -x^2 + 1$

59 $(x + 1)^2 + (y - 4)^2 = 17$

35 $r = -3 \sec \theta$ 37 $r = 4$ 39 $\theta = \tan^{-1}(-\frac{1}{2})$
41 $r^2 = -4 \cos 2\theta$ 43 $r = 2 \cos \theta$

61 $y^2 = \dfrac{x^4}{1 - x^2}$

45 $x = 5$

47 $x^2 + (y - 3)^2 = 9$

63 Let $P_1(r_1, \theta_1)$ and $P_2(r_2, \theta_2)$ be points in an $r\theta$-plane. Let $a = r_1$, $b = r_2$, $c = d(P_1, P_2)$, and $\gamma = \theta_2 - \theta_1$. Substituting into the law of cosines, $c^2 = a^2 + b^2 - 2ab \cos \gamma$, gives us the formula.

49 $y = x$

51 $\dfrac{y^2}{9} - \dfrac{x^2}{4} = 1$

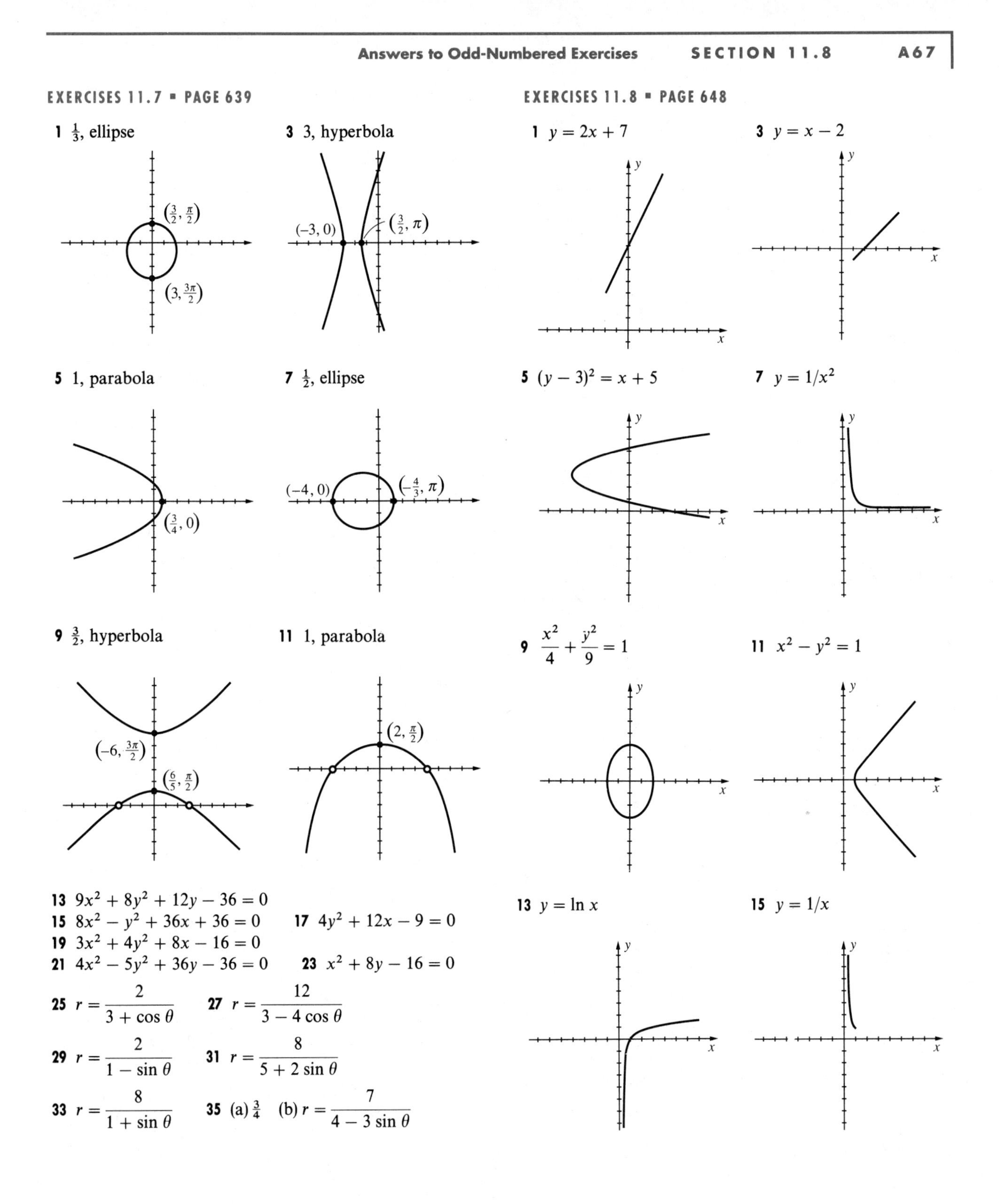

EXERCISES 11.7 ■ PAGE 639

1 $\frac{1}{3}$, ellipse

3 3, hyperbola

5 1, parabola

7 $\frac{1}{2}$, ellipse

9 $\frac{3}{2}$, hyperbola

11 1, parabola

13 $9x^2 + 8y^2 + 12y - 36 = 0$

15 $8x^2 - y^2 + 36x + 36 = 0$ **17** $4y^2 + 12x - 9 = 0$

19 $3x^2 + 4y^2 + 8x - 16 = 0$

21 $4x^2 - 5y^2 + 36y - 36 = 0$ **23** $x^2 + 8y - 16 = 0$

25 $r = \dfrac{2}{3 + \cos\theta}$ **27** $r = \dfrac{12}{3 - 4\cos\theta}$

29 $r = \dfrac{2}{1 - \sin\theta}$ **31** $r = \dfrac{8}{5 + 2\sin\theta}$

33 $r = \dfrac{8}{1 + \sin\theta}$ **35** (a) $\frac{3}{4}$ (b) $r = \dfrac{7}{4 - 3\sin\theta}$

EXERCISES 11.8 ■ PAGE 648

1 $y = 2x + 7$

3 $y = x - 2$

5 $(y - 3)^2 = x + 5$

7 $y = 1/x^2$

9 $\dfrac{x^2}{4} + \dfrac{y^2}{9} = 1$

11 $x^2 - y^2 = 1$

13 $y = \ln x$

15 $y = 1/x$

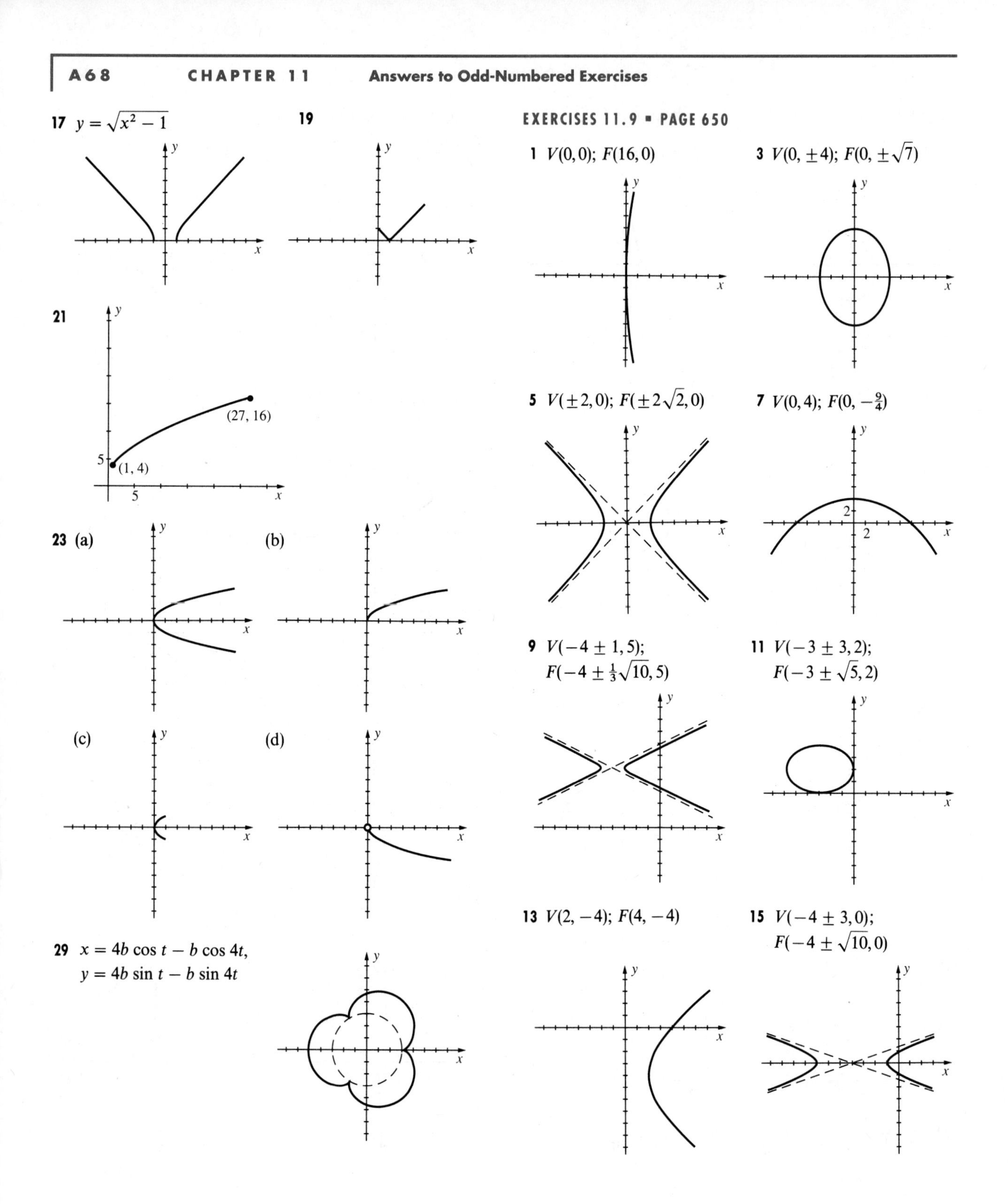
A68
CHAPTER 11
Answers to Odd-Numbered Exercises
17 $y = \sqrt{x^2 - 1}$
19
21
(27, 16)
(1, 4)
5
5
23 (a)
(b)
(c)
(d)
29 $x = 4b \cos t - b \cos 4t$,
$y = 4b \sin t - b \sin 4t$
EXERCISES 11.9 ■ PAGE 650
1 $V(0,0)$; $F(16,0)$
3 $V(0, \pm 4)$; $F(0, \pm\sqrt{7})$
5 $V(\pm 2, 0)$; $F(\pm 2\sqrt{2}, 0)$
7 $V(0,4)$; $F(0, -\frac{9}{4})$
2
2
9 $V(-4 \pm 1, 5)$;
$F(-4 \pm \frac{1}{3}\sqrt{10}, 5)$
11 $V(-3 \pm 3, 2)$;
$F(-3 \pm \sqrt{5}, 2)$
13 $V(2, -4)$; $F(4, -4)$
15 $V(-4 \pm 3, 0)$;
$F(-4 \pm \sqrt{10}, 0)$

17 $\dfrac{y^2}{49} - \dfrac{x^2}{9} = 1$ **19** $x^2 = -40y$ **21** $\dfrac{x^2}{75} + \dfrac{y^2}{100} = 1$

23 $\dfrac{y^2}{36} - \dfrac{x^2}{\frac{4}{9}} = 1$ **25** $\dfrac{x^2}{25} + \dfrac{y^2}{45} = 1$

27 **29** **31** **33**

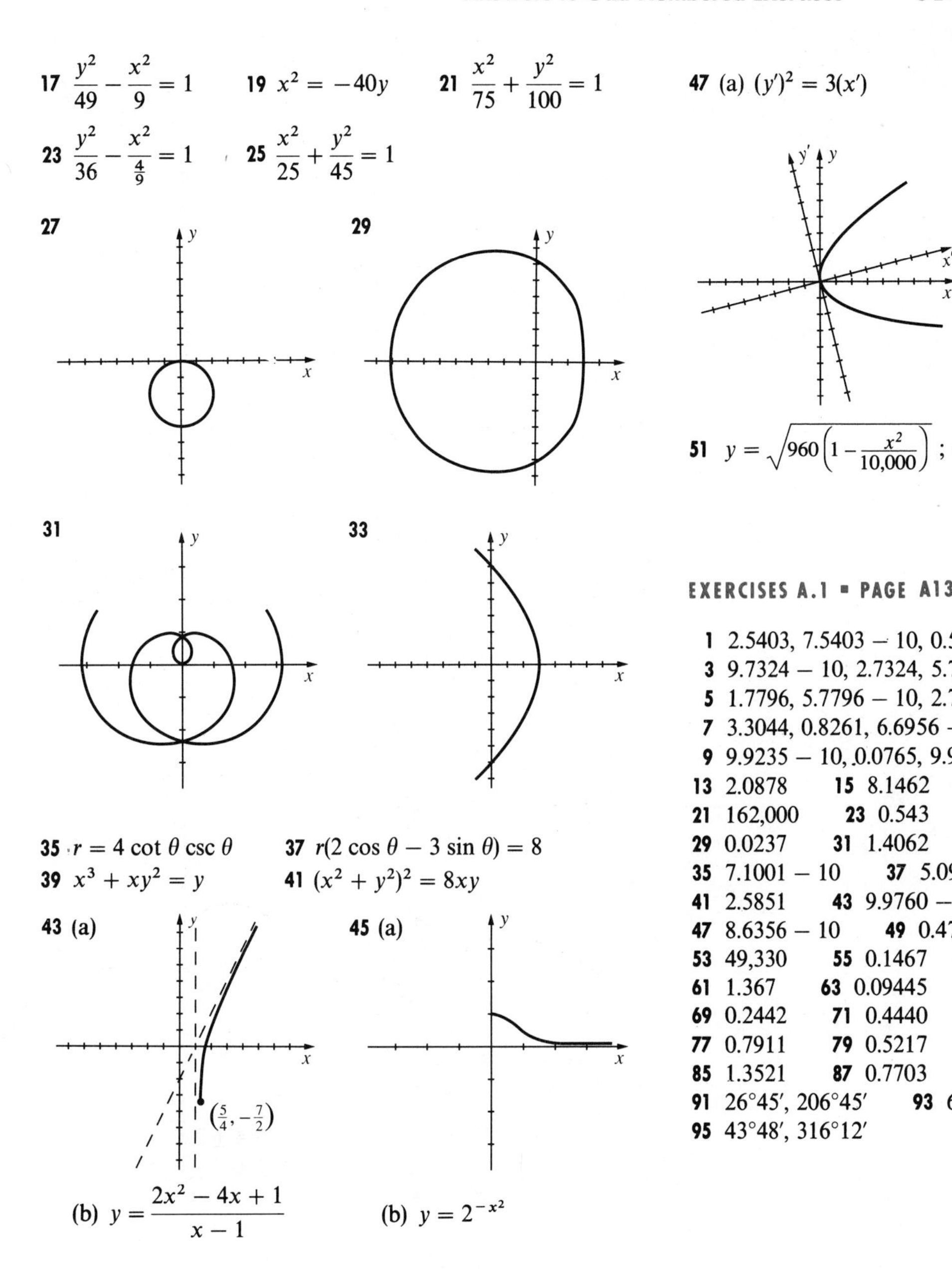

35 $r = 4 \cot \theta \csc \theta$ **37** $r(2 \cos \theta - 3 \sin \theta) = 8$

39 $x^3 + xy^2 = y$ **41** $(x^2 + y^2)^2 = 8xy$

43 (a) (b) $y = \dfrac{2x^2 - 4x + 1}{x - 1}$

45 (a) (b) $y = 2^{-x^2}$

47 (a) $(y')^2 = 3(x')$ **49** (a) $\dfrac{(y')^2}{\frac{5}{4}} - \dfrac{(x')^2}{5} = 1$

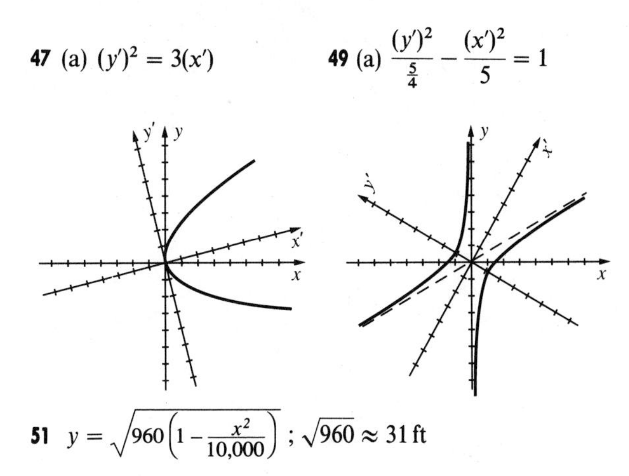

51 $y = \sqrt{960\left(1 - \frac{x^2}{10{,}000}\right)}$; $\sqrt{960} \approx 31$ ft

EXERCISES A.1 ▪ PAGE A13

1 2.5403, 7.5403 − 10, 0.5403
3 9.7324 − 10, 2.7324, 5.7324
5 1.7796, 5.7796 − 10, 2.7796
7 3.3044, 0.8261, 6.6956 − 10
9 9.9235 − 10, 0.0765, 9.9915 − 10 **11** 4.0428
13 2.0878 **15** 8.1462 **17** 4240 **19** 8.85
21 162,000 **23** 0.543 **25** 0.0677 **27** 189
29 0.0237 **31** 1.4062 **33** 3.7294
35 7.1001 − 10 **37** 5.0913 **39** 9.8913 − 10
41 2.5851 **43** 9.9760 − 10 **45** 4.8234
47 8.6356 − 10 **49** 0.4776 **51** 27.78
53 49,330 **55** 0.1467 **57** 0.006536 **59** 234.7
61 1.367 **63** 0.09445 **65** 109,900 **67** 0.001345
69 0.2442 **71** 0.4440 **73** −0.1426 **75** 1.032
77 0.7911 **79** 0.5217 **81** 3.437 **83** 0.5315
85 1.3521 **87** 0.7703 **89** 21°33′, 158°27′
91 26°45′, 206°45′ **93** 69°44′, 290°16′
95 43°48′, 316°12′

INDEX